HISTOCHEMISTRY

Theoretical and Applied

Volume 2

HISTOCHEMISTRY

Theoretical and Applied

by

A. G. EVERSON PEARSE

M.A. (Cambridge), M.D. (Cambridge & Basel),
F.R.C.P., D.C.P. (London), F.R.C.Path.

Professor of Histochemistry in the University of London
(Royal Postgraduate Medical School)
Hon. Consultant Pathologist, Hammersmith Hospital
Fulbright Fellow and Visiting Professor of Pathology
in the University of Alabama
Guest Instructor in Histochemistry, University of Kansas
Visiting Lecturer in Histochemistry, Vanderbilt University

THIRD EDITION

VOLUME 2

With 15 Coloured Plates and
119 Black and White Illustrations

CHURCHILL LIVINGSTONE

EDINBURGH AND LONDON

1972

First Edition	1953
Reprinted	1954
Translated into Polish	
Translated into Spanish	
Translated into Russian	
Second Edition	1960
Reprinted	1961
Translated into Russian	1962
Third Edition (vol. 1)	1968
Third Edition (vol. 2)	1972
Italian Edition in preparation	

International Standard Book Number
01 443 00865 5

Any correspondence relating to this volume should be directed to the publishers at 104 Gloucester Place, London, W1H 4AE.

PRINTED IN ENGLAND BY
THE WHITEFRIARS PRESS LTD
LONDON & TONBRIDGE

ACKNOWLEDGEMENTS

As in the case of Volume 1 of this edition, I should like to record the great amount of help provided by colleagues in my own, and other, Departments of the Royal Postgraduate Medical School, and by the various authorities quoted in the appropriate section of that volume.

I would like to mention, especially, the invaluable assistance I have received from my Lecturer, Dr. F. W. D. Rost. He has written Chapters 29 and 31 and the improvement of the text of these two chapters, conferred by his special knowledge in the field of microspectrofluorimetry, is clearly noticeable.

My sincere thanks go to my secretary, Miss Shirley Green, who has made all the new formulae, equations and diagrams and who is the creator, also, of the Source List (with the help of Mr. L. Maryon) and the Author Index.

For the following illustrations I am indebted to the persons (and sources) mentioned hereunder:

Plate XXIIa, Mr. J. D. Bancroft; Plate XVId, Dr. Imelda Campo-Aasen and the Editor, *Medicina Cutanea;* Plate XXIIIb and Fig. 230, Mrs. S. A. M. Cross; Fig. 213, Dr. Roger Daoust; Fig. 259, Dr. L. E. Ericson and the Editor and Publishers, *J. Ultrastructure Research;* Fig. 255, Dr. H. Flodh; Fig. 271, Dr. J. S. Hanker; Plates XVc, XVIc and XXVIIIc, Dr. R. Hess and Ciba-Geigy; Plate XXIIIc, Dr. Margaret Johnson; Plate XX, Dr. Robert Jones; Fig. 272, Dr. P. Kása; Fig. 273, Dr. G. B. Koelle; Fig. 267, E. Leitz Instruments Ltd., London; Plate XVIId, Dr. Z. Lojda; Fig. 205, Mr. Edwin Ngaha; Fig. 274, Dr. Milton Okun; Plate XXVd, Dr. J. S. Ploem; Plate XXIVb, Dr. S. Raia; Plates XVIIIa and XXVIIIa, Dr. Ernst Riecken; Fig. 298, Dr. R. Rigler and the Editor and Publishers, *Acta Physiologica Scandinavica;* Fig. 257, Dr. R. Speirs; Figs. 256, 280 and 281, Dr. W. E. Stumpf and the Editors, Oak Ridge Symposium, "Radioisotopes in Medicine", 1968; Figs. 277 and 278, Dr. C. J. Tandler and the Editor, *J. Cell Biology;* Fig. 276, Dr. J.-P. Tranzer; Fig. 262, Vickers Ltd.; Plates XVIIIb and XXIId, Dr. Klaus Wächtler; and Fig. 261, Carl Zeiss, Oberkochen and Degenhardt Ltd., London.

To the Editors and Publishers of the following Journals I am grateful for permission to reproduce the following illustrations from the work of my own Department. Figs. 198 and 199, *Histochemie;* Figs. 203 and 204, *Experimental Cell Research;* Fig. 218, *Virchows Archiv. Abteilung A;* Fig. 244, Gut; Fig. 275, *Journal of Histochemistry and Cytochemistry.*

On account of increasing professional commitments my wife, Dr. Elizabeth Pearse, has been unable to provide all the practical assistance which she gave me in the case of earlier volumes. For her continued patience and forbearance I am, as ever, eternally grateful.

Addenda et Corrigenda to Volume 1

p. vi line 14 for "Wellcome Foundation" read "Wellcome Trust".

p. vii line 23 delete second "had".

p. 18 line 9 for "atained" read "attained".

p. 87 11th line from foot should read "Chapter 27".

p. 306 line 4 from foot—for "done" read "due".

Plate V (p. 320) transpose Vc and Vd legends.

p. 379 delete Sylvén, B. (1959).

Plate IX (p. 387) transpose IXa and IXc legends.

p. 586 second paragraph—line 2—for "nucleic acid" read "maleic acid".

p. 628 Appendix 6. Fluorescent Methyl Ketone Method (5) at asterisk read "at least 18 hours old".

p. 629 line 2 delete "2 minutes" and substitute "10 seconds".

p. 637 line 2 should read "1·00mg/ml".

p. 644 Appendix 8—Bargmann's Chrome Haematoxylin for NSS. After "3 per cent aqueous chrome-alum" add "-ammonium".

p. 659 The first "14" in this method should read "13".

p. 709 Footnote, last name should read "Frehden".

CONTENTS

Chapter		Page
17	Carboxylic Ester Hydrolases	761
18	Glycoside Hydrolases and Glycosyltransferases	808
19	Oxidoreductases I (Oxidases and Peroxidases)	841
20	Principles of Oxidoreductase Histochemistry	880
21	Oxidoreductases II (Diaphorases and Dehydrogenases)	921
22	Peptidases	962
23	Other Enzymes and Co-enzymes	981
24	Substrate Film Techniques	1009
25	Enzymes as Histochemical Reagents	1016
26	Pigments and Pigment Precursors	1050
27	Biogenic Amines	1101
28	Inorganic Constituents and Foreign Substances	1128
29	Fluorescence Microscopy	1171
30	Autoradiography and its Applications	1207
31	Quantitative Histochemistry	1225
32	Electron Histochemistry (Ultrahistochemistry)	1260
	Appendices	1303
	List of Suppliers	1454
	Molecular Weight and Source List	1458
	Authors Index	1471
	Subject Index	1485

CHAPTER 17

CARBOXYLIC ESTER HYDROLASES

IN Chapters 15 and 16 those particular esterases which hydrolyse phosphate esters have been dealt with. There remain to be described in this chapter a group of esterases, of probably diverse function *in vivo*, whose common link is that their substrates are esters of carboxylic acids. Out of the 20 enzymes in this group recorded in the I.U.B. Enzyme Commission Report (1961) histochemical techniques exist for six but it needs little imagination to see extensions to many others. The present list of six includes the so-called carboxylesterases, the arylesterases, the lipases, acetylcholinesterase, pseudocholinesterase and phospholipase B.

Carboxylesterases, Arylesterases and Lipases

In dealing with these three classes of enzyme we are dealing with a large and widespread group acting at optima between pH 5 and 9 on a variety of substrates. All these enzymes catalyse reactions of the type

$$RCOOR' + H_2O \rightleftharpoons RCOOH + R'OH$$

and the majority are both hydrolytic and synthetic in action. If a simple ester such as methylbutyrate or naphthyl acetate is the substrate concerned, the enzyme is classed as a *non-specific esterase*. The non-specific esterases can be broken down into several different types, of which the *carboxyl esterases* and the *arylesterases* are the most important histochemically. Recently a group of *acetylesterases* has been separated from the non-specific esterases whose members hydrolyse esters of acetic acid with particular efficiency. They hydrolyse aromatic esters and are resistant to most of the usual esterase inhibitors.

If the substrate is the ester of a higher fatty acid with glycerol, or some other alcohol, the enzyme concerned is a *lipase*. The esterases, however, also act on fatty acid esters so that there is some degree of overlap between the two.

Distinction between Lipases and Esterases (Historical Methods)

As early as 1906 Loevenhart distinguished between pancreatic "lipase" and liver "esterase," but despite this the terms lipase and esterase have continued to be used interchangeably. Broadly speaking, the esters of short-chained fatty acids (C_2—C_4) are acted upon by esterases, and the long-chained esters (C_8 and upwards) by lipases. A very large number of studies have been made, using various fatty acid esters and various chemical methods of estimation, on the differences between these enzymes. The differences recorded include substrate-specificity, stereochemical specificity, behaviour towards activators and inhibitors, and pH optima. Gomori (1948) concluded that owing to multiple overlapping of differences and similarities it could not be decided whether the various effects were due to several well-defined individual enzymes or to the presence of unidentified accompanying substances. Seligman and Nachlas (1950), however, surveying the methods used for measuring the esterolytic activity of serum, declared that at least two distinct esterolytic enzymes existed; these were esterase and lipase. The former, abundant in liver, kidney, serum and pancreas, acted preferentially on short-chained fatty acids and was partially inhibited by sodium taurocholate. Lipase, which was abundant almost exclusively in pancreas, acted preferentially on esters of long-chained fatty acids and was activated by taurocholate. This activation of the true lipases by salts of the bile acids is characteristic, and they have also been shown to be activated by amino-acids and by peptides such as leucylglycylglycine, for instance (Dawson, 1927). Esterases are also activated by amino-acids (particularly arginine, lysine and histidine) and peptides. Although a large number of inhibitors of esterases and lipases have been described (e.g. strychnine, quinine, atoxyl), the differential inhibition caused by these compounds is not sufficient to allow their use for distinguishing between the two enzymes histochemically.

Biochemistry of Non-specific Esterases and Lipases

Until the turn of the century biochemical information about the substrate specificity of the esterases was scanty and inadequate. Fodor (1950) suggested that the hydrolysis of methyl esters such as methyl butyrate was due largely to cholesterol esterase, an enzyme which was distinguished from lipase by Kelsey (1939). The methods used by Fodor were not adequate, however, to permit the definition of relative substrate specificities. A great advance was made by Webb (1948) with the suggestion that the so-called organophosphorus compounds could be used to distinguish lipases from esterases. This work was continued and extended by Aldridge (1953a and b, 1954), by Mounter and Whittaker (1953) and by Myers and his associates (1955a and b), with the result that we can now distinguish at least three separate non-specific esterases as well as a lipase of pancreatic type. The organophosphorus inhibitors were fully considered in an admirable monograph by Heath (1963), and their pharmacology in an excellent review by Holmstedt (1959).

The older methods of distinguishing esterase from lipase activity were found by Myers *et al.* (1955b) to be completely without value, although high concentrations of atoxyl were found to inhibit esterase activity (phenyl butyrate) without affecting lipase activity (tributyrin). Quinine, on the other hand, was found to inhibit both enzymes equally. The work of Aldridge (1954) made it clear that the use of a variety of substrates (as in most of the older work) was far less useful in distinguishing types of esterase than the employment of activators and inhibitors.

It is still useful, however, to have some idea of the substrate specificities of esterases and lipases. Investigations by Barrowman and Borgström (1968) on the enzymes of the pancreatic juice showed that lipase hydrolysed most actively triolein, β-naphthyl laurate and β-naphthyl oleate and, less actively mono-olein. The latter was not hydrolysed by esterase unless a conjugated bile salt was added. Substrates actively hydrolysed by esterase (but not by lipase) included β-naphthyl acetate, and cholesterol oleate in dispersed form, with added bile salt.

At the present time the two most important inhibitors in the esterase field are DFP (diisopropyl fluorophosphate) and E600 (diethyl *p*-nitrophenyl phosphate). With the aid of the latter compound Aldridge (1953a and b, 1954) distinguished two basic types of carboxyl ester hydrolases. These he called A-esterases and B-esterases. The former, which are not inhibited by concentrations of E600 up to 1mM, hydrolyse acetates at a faster rate than butyrates. The B-esterases, on the other hand, are inhibited by concentrations of E600 as low as 10 μM and they hydrolyse butyrates at an equal rate, or faster than acetates. Myers *et al.*, (1955b) showed that 0·3mM E600, preincubated with suspensions of rat pancreas, inhibited 95 per cent of the activity of lipase on tributyrin. In his second paper (1953b) Aldridge showed that his A-esterase was identical with the previously recorded E600-esterase which was capable of hydrolysing a variety of organophosphorus compounds.

A third type of enzyme was described by Bergmann, Segal and Rimon (1957) which they called C-esterase. This enzyme does not hydrolyse DFP, which nevertheless fails to inhibit it. In contrast to A and B-esterases, C-esterase is activated rather than inhibited by low concentrations of *p*-mercuribenzoate and other sulphydryl blocking agents. The C-esterases were recognized and described on a histochemical basis (Mietkiewski and Malendowicz, 1967).

Terminology. Some authors (e.g. Hobbiger, 1957) preferred to use the terms organophosphate-resistant and organophosphate-sensitive while others continued to use the term ali-esterase (aliphatic esterase) in conjunction with Aldridge's term arom-esterase (aromatic esterase). Others, yet again, retained the use of specific substrate nomenclature as in the DFP-ase of Mounter *et al.*, (1957). The following terms are, however, approximately synonymous:

A-esterase = Arom-esterase = Arylesterase = Organophosphate-resistant esterase = E600-esterase = DFP-ase

B-esterase = Ali-esterase = Carboxylesterase = Organophosphate-sensitive esterase

C-esterase = Acetylesterase = Organophosphate-resistant, sulphydryl inhibitor-resistant esterase

These esterases are distinguished from the cholinesterases (see below) by their insensitivity to esterine.

Histochemists should use the convenient terms A-, B- and C-esterase and, provided it is recognized that these terms may include a variety of similar but not necessarily identical enzymes, there can be little objection to such usage. If one includes the two cholinesterases, the full list of esterase types which are histochemically demonstrable is as follows:

A-esterase, Carboxylesterase	E.C.3.1.1.1
B-esterase, Arylesterase	E.C.3.1.1.2
C-esterase, Acetylesterase	E.C.3.1.1.6
Acetylcholinesterase	E.C.3.1.1.7
Cholinesterase	E.C.3.1.1.8

Activity of the A-esterases is believed to be bound up with the presence of an SH group in the active centre since these enzymes are sensitive to heavy metals, particularly Hg^{2+}, Ni^{2+}, Cu^{2+}, and to rare earth metals such as La^{3+} (Augustinsson, 1961, 1964). They are also inhibited to some extent by iodoacetate (Aldridge, 1953b). It was assumed by Bergmann, Segal and Rimon (1957) that the active centre of A-esterases also contained an imidazole ring since the pH activity curves for the hydrolysis of E600 by rabbit serum A-esterase and of DFP by hog kidney showed a maximum at pH 7·6. The B-esterases were also considered to contain an imidazole ring in their active centre (Wilson and Bergmann, 1950) and probably also serine (Schaffer *et al.*, 1953, 1954).

The C-esterase, referred to above, was obtained by Bergmann *et al.*, (1957) from hog kidney extracts and further details were given by Bergmann and Rimon (1958). Their enzyme was inhibited by Hg^{2+} ions over a wide range of substrate concentrations whereas DFP-ase exhibited a narrow zone of activation. This is shown in Fig. 162 below.

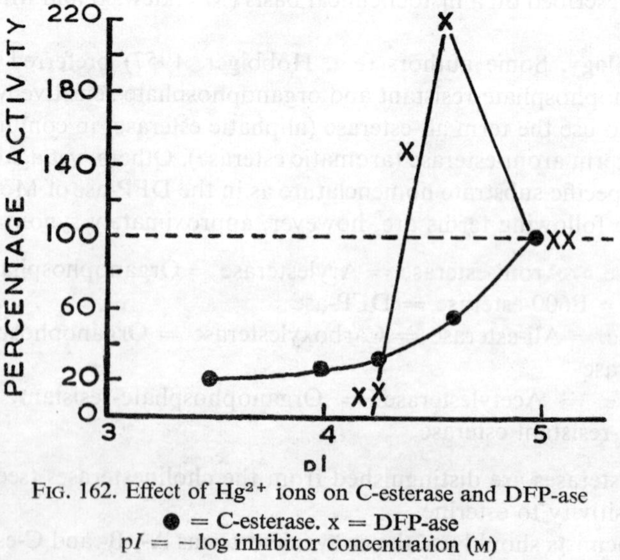

FIG. 162. Effect of Hg^{2+} ions on C-esterase and DFP-ase

● = C-esterase. x = DFP-ase

p*I* = −log inhibitor concentration (M)

Distinction between carboxylic esterases and certain proteolytic enzymes is not always easy. Considerable overlap has been observed between the properties of esterases on the one hand, and trypsin, chymotrypsin and thrombin, for example, on the other (Myers, 1960).

Isoenzyme Groups. Within each of the 4 main esterase groups (carboxylesterases, arylesterases, acetylesterases, cholinesterases) there are multiple forms of the enzymes. These are the so-called isoenzymes, described originally by Hunter and Markert (1957) and Markert and Hunter (1959), and revealed by the now well-established zymogram technique which they devised. The number of different enzyme forms varies greatly between different species, even when the latter are phylogenetically closely related (Coutinho *et al.*, 1965; Holmes and Masters, 1967a and b, 1968). In rat liver, for instance, these last authors found 14 forms of carboxylesterase and 4 forms of acetylesterase were found in the adult male genital tissues of sheep and ox and also in guinea-pig testis and epididymis (Holmes and Masters, 1967a).

Although the greater proportion of esterase histochemistry is still carried out without parallel isoenzyme studies it may be convenient to list here the isoenzyme status of the various esterases and their performance with the substrates and inhibitors used in histochemical practice. Table 56, below, is derived substantially from Holmes and Masters (1967a). It forms the essential basis for all biochemically correlated esterase studies.

TABLE 56

Classification of Types of Esterase Activity (Isoenzymes)
(After Holmes and Masters, 1967a)

Substrate or Inhibitor	Aryl- A	Carboxyl- 'Slow' B	Carboxyl- 'Fast' B	Acetyl- C	Cholin- ChE	Acetylcholin- AChE
α-Napth. Acet.	+++	+++	+++	+++	+++	+++
β-Napth. Acet.	+++	+++	+++	+++	+++	+++
α-Napth. Butyr.	+	++	++	(+)	+++	++
β-Napth. Laur.	−	−	−	−	−	−
Indoxyl. Acet.	++	−	++	++	++	++
Acetylthioch.	−	−	−	−	(+)	++
DFP 10^{-4} M	−	+++	+++	−	+++	+++
E 600 10^{-5} M	−	+++	+++	−	+++	+++
Eserine 10^{-5} M	−	−	−	−	+++	+++
P.C.M.B 10^{-3} M	++	−	−	−−	−	−

Activity + to +++
Inhibition + to +++
P.C.M.B. = parachloromercuribenzoate.

It is unfortunately true, as Myers (1960) has pointed out, that the metabolic functions of the simple esterases are not known. Pancreatic esterase I, for instance, has been identified on the basis of inhibitor studies with pancreatic cholesterol esterase (Fodor, 1950) and pancreatic esterase II closely resembles carboxypeptidase (Myers *et al.*, 1955). Liver aliesterase has been stated to be identical with liver amidase. Recently Patrick and Lake (1969) have identified the lysosomal E600-resistant esterase as "acid lipase". Histochemists working on the distribution of esterases in particular cells can contribute by identifying as clearly as possible the number and type of esterases present, as well as their precise localization. They are, at present, free to speculate within wide limits on the metabolic function of the esterases which they have identified.

Biochemistry of the Cholinesterases

These enzymes are separated from the other two groups of esterases on account of their ability to hydrolyse esters of choline. Many classifications exist for this group; one of the earliest and best was that of Augustinsson and Nachmansohn (1949). These authors distinguished the true, or acetyl cholinesterase (AChE) from non-specific, or pseudocholinesterase (ChE). The former has an affinity for its physiological substrate, acetylcholine, higher than for any other ester (except acetylthiocholine); it is capable of splitting acetyl-β-methylcholine but not benzoylcholine, and is inhibited by high concentrations of acetylcholine. Non-specific cholinesterase, on the other hand, hydrolyses acetylcholine more rapidly as the concentration of the latter rises and it splits long-chained choline esters at a greater velocity than acetylcholine. It can hydrolyse benzoylcholine but not acetyl-β-methylcholine (Mendel, Mundell and Rudney, 1943; Nachmansohn and Rothenberg, 1945).

It is now assumed, therefore, that various types of choline ester-splitting enzymes exist and various methods have been worked out for their distinction *in vitro*. One method is based on the observation that acetyl-β-methylcholine is split only by AChE and another on the fact that butyrylcholine is hydrolysed only by ChE. The selective inhibition of either AChE or ChE by various compounds has also been used for distinguishing the two types. The use of both selective substrates and of selective inhibitors has been applied to the histochemical distinction between AChE and ChE but the position, as will be seen, is complicated by the fact that unnatural substrates must be employed and in many cases these have been inadequately tested *in vitro*. The pH optima of the two enzymes scarcely differ, that of AChE is 7·5–8·0, while ChE is most active between 8·0 and 8·5. A more modern comprehensive classification is given by Augustinsson (1963).

Inhibitors of the Cholinesterases. According to Wilson and Bergmann (1950a and b) the behaviour of acetylcholinesterase can be explained by postulating that the active site of the enzyme contains two subsites (Fig. 163, below). The first of these is an anionic site which binds and thus orientates the basic groups of certain types of reversible inhibitor molecules (Wilson and Alexander, 1962) or, for that matter, of acetylcholine itself. The second is an esteratic site containing a basic group, which is considered to be acetylated and deacetylated as part of the hydrolytic process, and an acidic group. These are represented in the diagram as H—G.

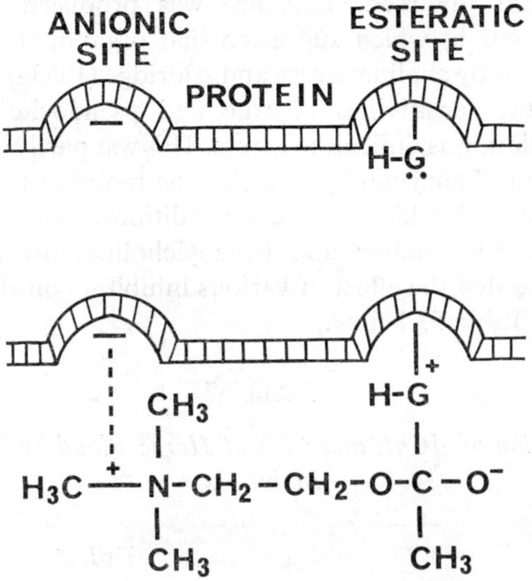

FIG. 163. Anionic and Esteratic Sites.

The formation of the enzyme-substrate complex is the first stage in the proposed reaction which proceeds as follows:

Acetylcholine + enzyme ⇌ acetyl enzyme + choline

Acetyl enzyme + water ⇌ acetic acid + enzyme

Organophosphorus inhibitors (such as DFP) react only at the esteratic site (Wilson and Bergmann, 1950a). Eserine and the choline esters react at both sites (Goldstein and Hamlisch, 1952).

The anti-cholinesterases which are exceedingly numerous, can be divided into three groups: (*a*) quaternary ammonium bases; (*b*) physostigmine and urethans; (*c*) organic phosphorus compounds. Comprehensive lists of substances in the first two groups may be found in papers by Ammon (1943) and Augustinsson (1948), and of those in the last group in papers by Whittaker (1951) and Holmstedt (1951). Burgen and Hobbiger (1951) discussed particularly the organic phosphate inhibitors and Adams and Thompson (1948) gave a list of inhibitors in all three groups. Classical examples in the first group are Nu-1250 (*N-p*-chlorophenyl-*N*-methylcarbamate of *m*-hydroxyphenyl-trimethylammonium bromide) and Nu-683 (dimethylcarbamate of 2-hydroxy-phenyl-benzyl-trimethylammonium bromide). In the second group is eserine (physostigmine) itself, and in the third group DFP (diisopropyl fluoro-phosphate). This compound was at one time extensively used to distinguish between true and pseudocholinesterases since it was found that, in some species at least, the latter were inhibited at a much lower concentration (Mazur and Bodansky, 1946; Mendel and Hawkins, 1947; Adams and Thompson, 1948). In other species, however (Ord and Thompson, 1950), this difference was not observed. Evidence was produced by Aldridge and Davison (1952a and b) which suggested that organophosphorus inhibitors could be hydrolysed by cholinesterase and Aldridge (1953a) pointed out that a comparison of the sensitivities of true and pseudocholinesterase, in an unpurified condition, was difficult to assess. This was partly due to the presence in tissues or serum of enzymes hydrolysing the inhibitor and this effect is, of course, inevitable under histochemical conditions. Using the specific substrates (acetyl-β-methylcholine and benzoylcholine) for the two enzymes Aldridge (1953c) tested the effect of various inhibitors on their activities. His results appear in Table 57 below.

TABLE 57

Inhibition of AChE and ChE of Horse Blood by Various Inhibitors

Inhibitor	pI_{50}*		Inhibitor Ratio
	ChE	AChE	(AChE/ChE
Dimethyl *p*-nitrophenyl phosphate	6·10	6·27	0·67
Diethyl *p*-nitrophenyl phosphate (E600)	6·85	6·38	2·9
Diisopropyl *p*-nitrophenyl phosphate (DINP)	6·50	5·50	10
Tetraethyl pyrophosphate (TEPP)	8·39	6·52	73
Tetraisopropyl pyrophosphate	8·68	5·96	520
Diisopropyl phosphorofluoridate	8·18	5·75	270
Diethyl fluorophosphonate	7·80	6·09	52
Bisdimethylamino fluorophosphine oxide	3·31	1·96	22
Bismonoisopropylamino fluorophosphine oxide (Mipafox)	7·42	3·82	3950
Tetraisopropylpyrophosphoramide (iso-OMPA)	6·48	2·51	9400

* $pI_{50} = -\log_{10}M$ concentration of inhibitor to produce 50 per cent inhibition.

Table 57 contains the concept of the *Inhibitor Safety factor* which can be expressed more clearly in graphical form, as in Fig. 164, below.
For many inhibitors there is thus a clear gap between the concentration required for complete inhibition of one type of cholinesterase and the concentration at which inhibition of the other type begins.

The highest inhibitor ratio was obtained by Aldridge with the compound

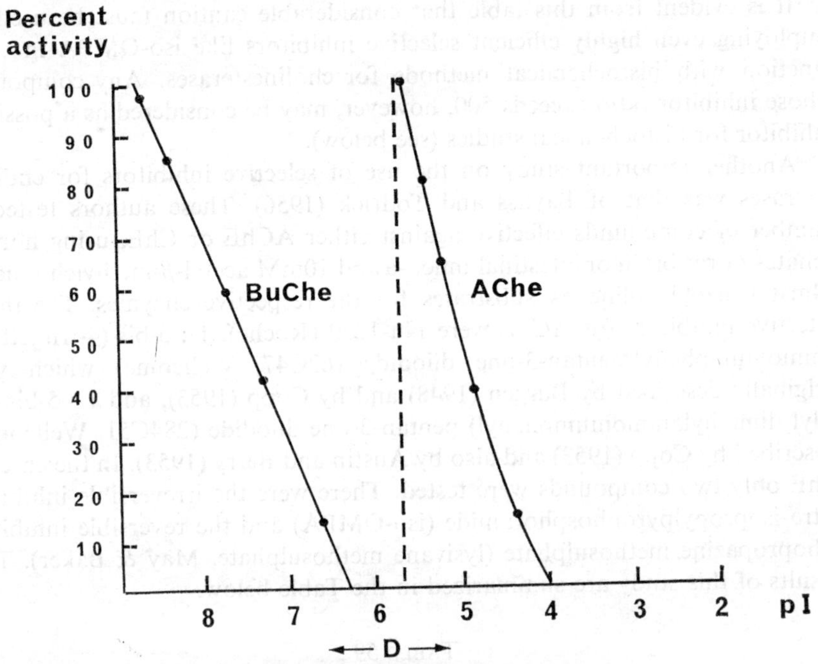

FIG. 164. Inhibitor Safety Factor (*Mipafox*).
Vertical line = rec. p*I* value (4×10^{-6} M).
Safety Factor = D.

iso-OMPA and this compound was therefore further tested against AChE (erythrocyte) and ChE (plasma or serum) in a number of different animals. In the case of the rat, brain and heart muscle homogenates were used instead of red cells and serum. The results of this study are given in Table 58.

TABLE 58

*Sensitivity of AChE and ChE of Various Species to iso*OMPA

Species	pI_{50} AChE	pI_{50} ChE	Inhibitor Ratio (AChE/ChE)
Horse	2·47	6·52	11,300
Man	3·52	5·27	56
Guinea-pig	2·34	5·89	3600
Dog	2·42	6·54	13,200
Rat	3·60	6·32	530

It is evident from this table that considerable caution must be used in employing even highly efficient selective inhibitors like iso-OMPA in conjunction with histochemical methods for cholinesterases. Any compound whose inhibitor ratio exceeds 500, however, may be considered as a possible inhibitor for histochemical studies (see below).

Another important study on the use of selective inhibitors for cholinesterases was that of Bayliss and Todrick (1956). These authors tested a number of compounds effective against either AChE or ChE, using homogenates of rat brain or intestinal mucosa and 10mM acetyl-β-methylcholine or 14mM benzoylcholine as substrates for the respective enzymes. The three selective inhibitors for AChE were Nu-1250 (Roche), 1 : 5-bis-(4-trimethylammoniumphenyl)pentan-3-one diiodide (62C47, Wellcome) which was originally described by Burgen (1948) and by Copp (1953), and 1 : 5-bis-(4-allyl dimethylammoniumphenyl) pentan-3-one diiodide (284C51, Wellcome) described by Copp (1953) and also by Austin and Berry (1953). In the case of ChE only two compounds were tested. There were the irreversible inhibitor tetra-isopropylpyrophosphoramide (iso-OMPA) and the reversible inhibitor ethopropazine methosulphate (lysivane methosulphate, May & Baker). The results of this study are summarized in the Table below.

TABLE 59

Inhibition of AChE and ChE of Rat Brain and Intestine by Selective Inhibitors

Inhibitor	Concn. (M)	Percentage Inhibition AChE	Percentage Inhibition ChE
Nu-1250 . .	5×10^{-7}	81	1
	2×10^{-6}	98	8
62C47 . . .	5×10^{-6}	94	1
	2×10^{-5}	101	3
284C51 . . .	3×10^{-5}	100	2
	1×10^{-4}	102	5
Lysivane . .	3×10^{-5}	−1	90
	8×10^{-5}	9	95
iso-OMPA . .	1×10^{-5}	0	90
	3×10^{-5}	6	92
	1×10^{-4}	24	—

Bayliss and Todrick suggested that Nu-1250 was insufficiently selective and that although 284C51 appeared to be slightly more efficient than 62C47 the difference might well be more apparent than real. Little difference was observed between iso-OMPA and lysivane but the authors preferred the latter on account of the fact that with irreversible inhibitors the degree of inhibition varies markedly with the time of contact. Histochemical users have expressed varied preferences in their choice of inhibitors, which has sometimes had to be based on availability rather than on theoretically derived preferences. Holmstedt and Sjöqvist (1961) suggested that BW 284C51 and Mipafox could not and should not be substituted for each other in histochemical studies unless satisfactory inhibition curves were presented. Main and Iverson (1966) found 150-fold stronger inhibition of ChE than AChE by DFP, and Cimasoni (1966) indicated that while ChE was first inhibited by 15—25 μM fluoride, ten times this level was required to initiate inhibition of AChE. This inhibition is reversible.

Two other selective inhibitors used in contemporary studies (Hobbiger, 1954, 1957) were 3-(diethoxyphosphinyloxy)-N-methylquinolinium methyl sulphate (3–0422, Roche) and 3-oxo-1 : 5-diphenylpentane-*pp*′-bis(propyl-dimethylammonium) dibromide (285C51, Wellcome). The former inhibits both cholinesterases at very low concentrations (10^{-8}M) while the latter is selective for AChE at a concentration of about 10^{-4}M. Heilbron (1954) drew attention to the pH dependence of ChE activity, at various substrate and inhibitor concentrations. She showed that the optimum substrate concentration was inversely related to pH and that the degree of inhibition obtained was effectively reduced by lowering the pH. A number of different types of compound have subsequently been investigated by different workers. Walsh and Nielsen (1960), for instance, found a mixed inhibition of AChE by cetyltrimethyl ammonium bromide and esters of methane sulphonic acid were found to be irreversible inhibitors of AChE by Kitz and Wilson (1962). The whole question of AChE inhibition by quaternary ammonium ions was considered by Coleman and Leley (1962). Using the red cell enzyme, these authors evaluated the inhibitor dissociation constants (Ki) for a wide range of monoquaternary ammonium ion inhibitors of the type $CH_3(CH_2)_{n-1}$ $N^+(CH_3)_3$ Br^- and diquaternary ammonium ion inhibitors, $Br^-(CH_3)_3$ N^+ $(CH_2)_n$ N^+ $(CH_3)_3$ Br^-.

They found that pKi versus ‘n’ was a straight line for both series, the slopes of the lines corresponding to changes in the free energy of dissociation per CH_2 group of 310 cal/mole for the monoquaternaries and 730 cal/mole for the diquaternaries. It was suggested that the higher dissociation energy of the latter was due to their lying flat along the enzyme surface.

Competitive inhibition of AChE by a number of thiazolines and oxazolines, substituted in the 2-position, was observed by Gawron and Keil (1960). Inhibition constants calculated on the basis of cation concentration were all of the same order of magnitude (10^{-4}M). The similarity between these observations and those found for alkyl ammonium ions suggested that thiazolines and oxazolines combine reversibly with the anionic site of the enzyme.

A warning against the use of Tris as a buffer for cholinesterase studies was given by Pavlic (1967) who found appreciable (competitive) inhibition of both

AChE and ChE by concentrations of Tris such as those normally employed in buffers. The observed inhibition constants were the same (13—14mM) for both enzymes.

Distribution of the Cholinesterases. Cholinesterases have been found in all vertebrates so far examined and in most of the invertebrates. AChE occurs as the predominant enzyme in nervous and muscular tissues and in erythrocytes. ChE is found in the blood serum, and in pancreas and salivary glands especially. There is, however, some degree of overlap between the two. Sympathetic ganglia have been reported to contain a mixture of AChE and ChE (Mendel and Rudney, 1944; Sawyer and Hollinshead, 1945), and cholinergic nerves have a higher activity than sensory ones (Umbrath and Hellauer, 1948). In muscles AChE is present in high concentrations at the motor end plates, where it was reported to be localized in the postsynaptic membrane (Couteaux, 1942, 1947). Erythrocyte AChE varies considerably in different species, being highest in man and in the guinea-pig. ChE has been reported in pancreas, ovary, Harderian glands, brown fat, liver, testis and parotid glands, *inter alia.* The esterases present in rat and guinea-pig liver are capable of splitting benzoylcholine although they will not hydrolyse acetylcholine (Blaschko, Chou and Wajda, 1947). They need not be characterized as ChE, however, since they split non-choline esters at a higher rate than benzoylcholine.

Methods for Non-specific Esterases (E.C. 3.1.1.1, 3.1.1.2, and 3.1.1.6)

Most of the substrates for the histochemical demonstration of esterases are hydrolysed by a number of different enzymes. For this reason it is difficult accurately to classify the methods which employ them. These are therefore set out in this chapter in sections which correspond to the enzyme or group of enzymes for which the method is *predominantly* employed. After consideration of the individual methods the histochemical use of inhibitors and activators will be discussed in succeeding sections. I noted in the 2nd Edition of this book (1960) that in applied esterase histochemistry insufficient attention had been paid to separation of the various types of enzyme by the use of inhibitors. There has clearly been considerable improvement since this time and histochemists have been enabled to catch up with biochemistry by the relatively slow progress made by that science in recent years in the esterase field.

Simultaneous Coupling Azo Dye Methods

Naphthyl Acetates. The first simultaneous coupling azo dye method for non-specific esterase was devised by Nachlas and Seligman (1949a) using β-naphthyl acetate as substrate. This ester and the α-naphthyl ester were equally rapidly hydrolysed by the tissues, but the authors expressed a preference for the former on account of the brilliance of the colours obtainable on coupling with diazonium salts. In this case they used a stabilized α-naphthylamine diazotate. Using cold acetone-fixed sections and avoiding contact with alcohol, incubation was carried out at pH 7·8 and at room temperature. A small amount of acetone (1 per cent in the final solution) was needed in order to dissolve the substrate in the buffer solution. Incubation was continued for 20 minutes and, after washing in water, permanent mounts were made in an aqueous medium since the red azo dye was soluble in organic solvents.

Distribution of Enzyme. The distribution of esterase in the tissues of mammals, demonstrated by this method, was described by Nachlas and Seligman (1949b). In normal rat tissues esterase activity was found by these authors to be highest in lung (bronchial epithelium), liver, pancreas and kidney. These results agreed with those of Huggins and Moulton (1948) derived from tissue assay, using *p*-nitrophenyl acetate as substrate. Liver cells were uniformly stained with some accentuation in the region of the central vein.

Modifications of the Method. The original method of Nachlas and Seligman was found to be somewhat unsatisfactory. Even with the substitution of other diazonium salts (such as Fast blue B salt) for the diazotate of α-naphthylamine, the final azo dye reaction product was not only soluble in alcohol and xylene but also in water. Gomori (1950a) first drew attention to this point, observing that part of the azo dyes produced by β-naphthol remained soluble for many minutes, *in vitro*, especially in the presence of acetone. The azo dyes produced by coupling with α-naphthol, on the contrary, precipitated instantly at pH 7. Gomori, therefore, suggested a modification of the technique using α-naphthyl acetate and the minimum of acetone necessary for its solution. He found by this method that several sites of esterase activity recorded by the β-naphthyl acetate method were uniformly negative.

When the β-naphthyl and α-naphthyl acetate methods were tested for diffusion by the method of face to face incubation of sections, one active, the other inactivated by treatment with heat and Lugol's iodine, diffusion of reaction product in the first case exceeded 150 μ. With the α-naphthyl acetate method, using frozen sections, no diffusion of reaction product occurred at 75 μ, the lowest practicable space between the sections, and from examination of such sections it seemed clear that any diffusion which occurred must be considerably less than this. This method became one of the standard methods in applied histochemistry although it was always doubtful whether the localization obtained could be considered accurate at the cytological level.

Distribution of Enzyme by the Modified Method. Structures which were positive by the original method, but which were reported by Gomori to be negative with α-naphthyl acetate, were the parietal cells of the stomach (man, dog, rabbit), leucocytes in the dog and the elastic fibres of large vessels in many animals. Gomori found that these sites could be stained by brief incubation with either α- or β-naphthol, followed by washing and coupling with a diazonium salt, suggesting that they were produced as artifacts and did not represent sites of esterase activity. The standard α-naphthyl acetate modification of the Nachlas-Seligman technique is given in Appendix 17, p. 1303. This method has now been superseded by a further modification using HPR (hexazotized pararosanilin) as diazonium salt. Details are given in Appendix 17. The α-naphthyl acetate/HPR technique is to be regarded as the base-line esterase technique for all studies in this field.

Naphthol AS Acetates. Naphthol AS (2-hydroxy-3-naphthoic acid anilide) is used for the production of fast azo dyes in place of β-naphthol. Its acetate,

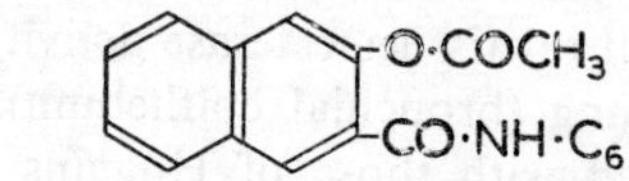

which is easily prepared, was first used by Gomori (1952a and b) in place of α- and β-naphthyl acetates in the coupling azo dye method for esterases.

When coupled with a variety of stable diazotates Naphthol AS produces colours resembling those given by β-naphthol. It is far less soluble in water than α- or β-naphthol, however, so that the Naphthol AS acetate methods possess the advantage that the initial reaction product (Naphthol AS) is less likely than the other two naphthols to diffuse from its site of production before coupling occurs.

Using Naphthol AS acetate, with a variety of stable diazotates, in a standard method for esterase the final product was particulate in every case. Plate XVA shows the results obtained, with guinea-pig kidney fixed in cold formalin and cut on the freezing microtome, using Naphthol AS acetate and the stable diazotate of 4-chloro-*o*-anisidine. Considerable improvement in the precision of esterase localization was produced by the use of substituted AS-naphthol acetates (Pearse, 1954). Three substrates of this type were prepared from Brenthols* AS-BG, AS-OT and AS-FR. These were, respectively, 2-acetoxy-3-naphthoic-2,5′-dimethoxyanilide, 2-acetoxy-3-naphthoic-*o*-toluidide and 2-acetoxy-3-naphthoic-*o*-anisidide. In Germany and the United States the equivalents of these last two substrates are AS-D acetate and AS-OL acetate. With the first, the final reaction product (FRP) was invariably macrocrystalline, but with the other two the results, especially in free-floating frozen sections, were much better than those obtained with Naphthol AS-acetate.

Further tests with acetates of substituted AS-naphthols were carried out by Burstone (1957a and b) and by Goessner (1958). Both these authors obtained excellent results with AS-D acetate. Goessner used his substrates in 0·1M-phosphate buffer at pH 6·8 with Fast blue BB (Plate XVB) or Fast Garnet GBC salt (18). Burstone (1957a) used AS-D acetate in 0·2M-tris buffer at pH 7·1 with Fast red RC (7) or Fast Garnet salts. Later (1957b) he employed

* I.C.I. Ltd. equivalents of AS-Naphthols.

AS-LC acetate at the same pH with Fast red-violet LB salt. Burstone observed with this last substrate that the distribution of activity was essentially the same as with other naphtholic substrates but that sharper localizations were obtained. In the case of freeze-dried sections of developing bone, however, esterolytic activity was observed in the bone matrix which could not be demonstrated with the usual naphtholic substrates. The activity of this enzyme was resistant to E-600 and to DFP but it was inhibited to some extent by Cu^{2+}, Cd^{2+} and Zn^{2+} ions and also by sodium fluoride. It thus fell into the category of an A-esterase, but further studies may well show that it belongs to the C-esterases and functions as a cathepsin (Bergmann *et al.*, 1957) (see Chapter 22, p. 963).

Hexazotized Pararosanilin and AS-Naphthols. The greater coupling energy of freshly prepared diazonium salts, and the greater insolubility of the resulting azo dye products, made it certain that the HPR variant of the standard esterase technique would be employed likewise with AS-Naphthols. The procedure was strongly recommended by Thybusch *et al.*, (1966) but the majority of users of Naphthol AS carboxylesters as substrates apparently still prefer to use stable diazotates.

Interpretation of Results. The interpretation of results obtained when using Naphthol AS substrates was clarified by the work of Shnitka and Seligman (1961). These authors used a double-staining technique in conjunction with reversible esterase inhibitors, washed out between the two main stages of the reaction. The inhibitor-resistant "droplet" esterase was demonstrated with a blue product (Fast blue BB salt) and the inhibitor-sensitive 'diffuse' esterase in red (Fast red violet LB salt). The first of these enzymes is equivalent to type A esterase and the second to type B. It is now clear that it is possible to show both types of enzyme with a variety of esterase techniques. If may be necessary, on occasion, to use a double-staining method like that of Shnitka and Seligman but in most cases a single-stage method can be employed. Very clear results are obtained by using an indoxyl substrate (see below) and HPR.

Other Substituted Naphthol Acetates. An alternative substrate in this class which has been employed in a simultaneous coupling method for esterases is 6-bromo-2-naphthyl acetate. This was used by Burstone (1957a), who found that it was not hydrolysed by the bone matrix enzyme. In frozen sections the results with this substrate are comparable with those obtained with simple naphthol esters. It has no real advantage over the latter although theoretical considerations (Chapter 14, p. 490) would lead one to expect that it might have

Another substituted naphthol acetate substrate originally described by Gomori (1953) still finds application, particularly in haematological studies. This is the chloroacetate of Naphthol AS-D which, in the original author's hands, gave a distinct reaction in developing blood cells of the myeloid series. Using essentially the same method as that given in Appendix 17, p. 1305, Moloney *et al.* (1960) found strong reactivity in practically all neutrophil leucocytes in peripheral blood smears and also in promyelocytes, leukaemic myeloid cells, mast cells and histiocytes.

Further observations on these lines were made by Leder (1964) who found tissue mast cells, in paraffin sections, were also strongly stained. Meister (1965) described the reaction of histiocytes in histiocytomas of the skin but subsequent studies by Schaefer and Fischer (1966) indicated that in these tumours the only cells which stained, except for neutrophil leucocytes, were mast cells. A Naphthol AS-D chloroacetate-hydrolysing activity was observed in Auer bodies in immature neutrophils by Fischer *et al.*, (1966). This suggested a close relationship between Auer bodies and the normal cytoplasmic granules of the neutrophils.

The true identity of the enzyme concerned with the above reactions is unknown. Because of its restricted distribution it is possibly a peptidase, rather than an esterase. In view of the demonstrably high affinity of mast cells for diazonium salts (Geyer, 1962; and see Vol 1, p. 340), recently emphasized by Enerbäck and Hansson (1968), it is necessary to consider the possibility that such staining is artifactual.

Naphthol AS and AS-D esters of ε-aminocaproic acid were used by Glenner *et al.*, (1962) for the demonstration of trypsin-like enzymes but Hopsu and Glenner (1963) found that they were hydrolysed by a number of esterases, in various tissues, including the motor end plates. The enzyme responsible for the reaction, in this last situation, was shown by Hopsu and Pontinen (1964) to be non-specific cholinesterase. It is necessary, therefore, carefully to exclude a number of possible hydrolytic activities before concluding that a specific peptidase is responsible for splitting aminocaproates of AS-Naphthols.

Metal Chelation Method for A-Esterases

The histochemical localization of A-esterases in mouse kidney, using a principle formerly employed for the demonstration of β-glucuronidase, was described by Deimling (1965). This author's substrate, 8-propionoxy-5-nitroquinoline, was observed to be hydrolysed rapidly by A-esterases between pH 7 and 8. Under histochemical conditions the liberated 8-hydroxy-5-nitroquinoline then takes part in a simultaneous chelation reaction with Bi^{3+} ions, present in the medium:

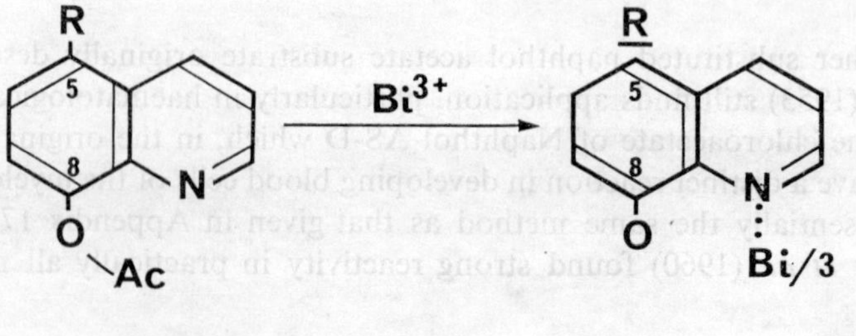

Subsequently, treatment with yellow ammonium sulphide converts the golden-yellow quinoline chelate into brown bismuth sulphide (Bi_2S_3). This method is described in Appendix 17, together with technical details for the preparation of the substrate.

As far as can be judged, a number of objections must be overcome before the method can be recommended, for light microscopy, as an alternative to the azo dye or indoxylazo techniques. Metal chelation methods are always interesting, however, since they can usually be adapted for use at E.M. level. Alternative conversion techniques for the final stage must also be considered. Feigl (1954) gives two possible routes. In the first, treatment with potassium thiocyanatochromate, $K_3Cr(SCN)_6$, in mineral acid solution, gives an insoluble red precipitate. In the second, reduction with sodium stannite (Na_2SnO_2) gives black metallic bismuth.

Methods Depending on the Production of Indigo

Histochemical methods based on the use of an entirely new principle were described, independently, by Barrnett and Seligman (1951) and by Holt (1952). In both cases the methods evolved were initially applied to the demonstration and localization of esterases and for this reason the whole principle of enzyme demonstration by means of indoxyl substrates is considered at this point. With the synthesis of new indoxyl substrates, satisfactory methods have been produced not only for esterases but also for phosphatases, glucosidases and glucuronidases. Extension to the peptidases has also been achieved. The acetates of indoxyl esters were employed by Holt in a long series of studies culminating in the production of an esterase substrate (5-bromo-4-chloro-indoxyl acetate) capable of very precise localization of enzyme activity. Reviews of these studies were given by Holt (1956, 1958), and these should be consulted by those whose interests lie in this field.

Principles of Indigogenic Indoxyl Methods. In all these methods the principle involved is the hydrolysis of a soluble ester of indoxyl, with the liberation of the free indoxyl. This normally exists as a tautomeric mixture of substances I and II.

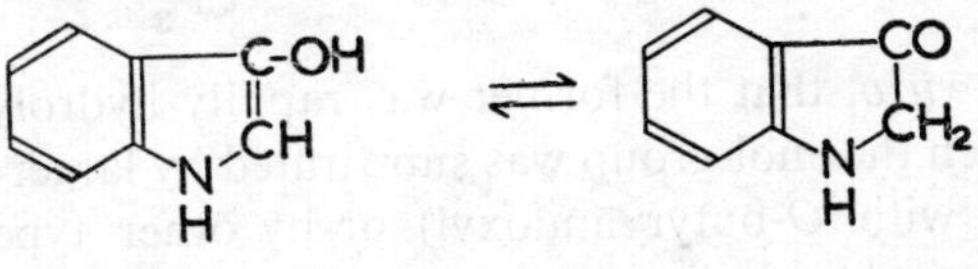

I II

The equilibrium normally favours the production of II, which is oxidized by atmospheric oxygen to the completely insoluble bright blue indigo III.

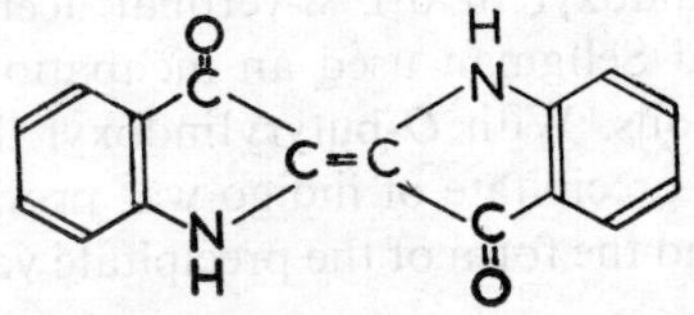

The demonstrating agent is thus the simplest possible, molecular oxygen, and the formation of indigo in the presence of air will be induced by hydrolytic enzymes catalysing either of the two processes, A and B, for which the equations are given below.

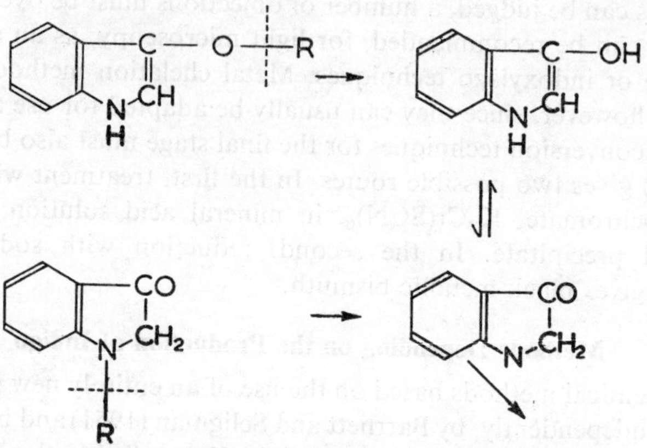

As will be observed from the above equations, indoxyl has two main reactive centres, one in the enol-hydroxy group (A) and the other in the imino-group (B). From the point of view of histochemistry the most important indoxyls are those derivatives which are substituted on the enol-group.

Application of Indigogenic Indoxyl Methods. Two substrates were used by Barrnett and Seligman (1951) for histochemical application, (1) indoxyl acetate (*O*-acetylindoxyl) and (2) indoxyl butyrate (*O*-butyrylindoxyl). Holt (1952) used these two compounds and later (Holt and Sadler, 1958a) some forty other mono-, di- and tri- substituted indoxyls. He drew particular attention to the structural resemblance of *O*-acetylindoxyl to acetylcholine,

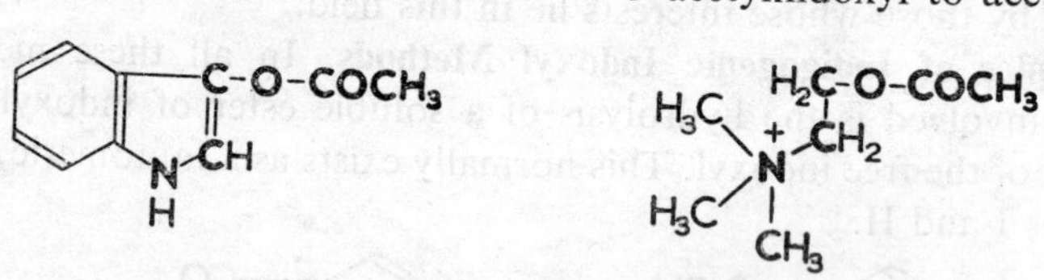

and he showed, *in vitro*, that the former was rapidly hydrolysed by specific cholinesterase. When the enol-group was substituted by larger acyl groups (to some extent even with *O*-butyrylindoxyl) or by other types of grouping altogether, hydrolysis by cholinesterases became negligible. Esterases and lipases, however, continued to hydrolyse the molecule with the consequent production of indigo.

With mM-*O*-acetylindoxyl in 0·1 M-veronal acetate buffer (pH 8·5) as substrate, Barrnett and Seligman used an incubation period of 10 minutes with fresh frozen sections. With *O*-butyrylindoxyl the incubation time was 15–20 minutes. A blue precipitate of indigo was produced in the majority of tissues in these times and the form of the precipitate varied between numerous small particles, in kidney for instance, to a few large star-shaped crystals in single fat cells. Seligman, who apparently hoped that the indigo would be deposited in submicroscopic particles, observed that the cause of the large size of the granules was not altogether clear. Holt (1952) also remarked that the reasons for the precipitation of the dye in large granules or crystals were not clear. Certainly localization of esterases with the two substrates employed by Barrnett and Seligman was extremely poor (see Fig. 165).

Improvement was to be expected from three sources; first, by the addition to the substrate solution of substances altering the character of the precipitate; secondly by the provision of an efficient capture reaction (see Chapter 14, p. 481), and thirdly, by modification of the substrate molecule at points other than those which are the sites of enzyme attack.

Physical Modifications. The growth of crystals is favoured by an acid pH and by a high lipid content of the tissues. Excessive solvent (ethanol) for the indoxyl substrate also favours this type of crystallization. The process is diminished by using an alkaline pH, and only above pH 9·0 are the precipitates from unsubstituted indoxyls of reasonable size.

Improvements of an entirely different order were brought about by the use of mono- or dihalogen-substituted indoxyl acetates as substrates. Holt and Withers (1952) used the first of a long series of halogen-substituted indoxyl acetates, 5-bromoindoxyl acetate, which gave rise to extremely small crystals of 5-5′-dibromoindigo. This is one of the more insoluble indigoids, but it has considerable solubility in lipids and, where these are present in large amounts, crystallization can still occur. Elsewhere the microcrystals are small enough not to be distinguished except under the higher powers of the microscope (see Plates XVc and XVd and Fig. 165). Further studies on the desirable properties of indigogenic substrates (Holt and Sadler, 1958b) led to the selection of 5-bromo-4-chloroindoxyl acetate as the optimal indigogenic reagent for the histochemical demonstration of esterases by oxidation techniques. This substrate gives rise to a highly substantive indigoid dye whose deposits in the tissues are not crystalline. They are, however, less strongly blue than those of some of the other halogen-substituted indigos. Although 5,5′-dibrom-4,4′-dichloroindigo is lipid soluble it does not form large crystals in fat cells as does 5,5′-dibromindigo. This effect is considered to be due to its substantivity for protein.

Oxidation Catalysts. It was observed by Holt and Withers (1952) that indoxyl itself, and halogen-substituted indoxyls, liberated at sites of high enzyme activity, could diffuse widely before complete oxidation to insoluble indigoid dyes could take place. They found that this process was greatly reduced by traces of Cu^{2+} ions in the incubating medium and that this effect was due to the catalytic action of the metal on oxidation of the liberated indoxyl. A series of oxidation catalysts was tested and potassium ferricyanide was found to be the most satisfactory in practice. A tendency for further oxidation to colourless dehydroindigo was prevented by addition of an equimolar solution of potassium ferrocyanide. The kinetics of aerial oxidation of indoxyl and some of its halogen derivatives were the subject of a special study by Cotson and Holt (1958) who found that, in the presence of excess O_2, oxidation was first order with respect to indoxyl under alkaline conditions. Under acid conditions, however, the reaction was much slower. In the absence of a catalyst the oxidation rates were found in all cases to be inadequate for accurate histochemical localization of enzyme. Holt and Withers (1958) reported the effect of changes in oxidation rate on the staining pattern and showed, for instance, that diffuse staining in many types of tissue was reduced to discrete intracellular localization of indigo deposits when the concentration of oxidation catalyst was sufficiently high.

The obvious alternative explanation was the "differential inhibition" of diffuse (microsomal) esterase, as pointed out in the 2nd Edition of this book (p. 470). This view was put forward much more forcibly by Shnitka and Seligman (1961). These authors presented evidence to show that the droplet (lysosomal) localization of indoxyl esterases, observed when using high concentrations of the ferro-ferricyanide oxidation catalyst, was due to selective inhibition of sensitive esterases. They were able to produce identical appearances with the unrelated naphthol AS-esterase system (p. 774), in combination with organophosphorus and other inhibitors. Shnitka and Seligman suggested that the correct localization of esterases could be obtained with the indoxyl methods only by reducing the concentration of the ferro-ferricyanide redox buffer from 5 to 0·5 mM. This procedure was recommended.

Further observations on the effects of the ferro-ferricyanide reagent were presented by Ahlqvist (1963), working in my laboratory, and also by Tsou and Su (1963). Ahlqvist correctly observed that the theory of Shnitka and Seligman could be accepted only if increased formation of colourless by-products could be excluded. Tsou and Su provided chemical evidence that ferro-ferricyanide inhibited the oxidation of 5-bromoindoxyl to 5,5′-dibromoindigo. They concluded that this system should not be used in histochemical methodology.

It is a fact that when azo-coupling is substituted for oxidation (see below) the droplet type of localization predominates although the incubation period remains very short. This is in contradistinction to the very prolonged time required to obtain the same droplet localization when using the ferro-ferricyanide reagent. There is thus no indication that inhibition of a "sensitive" esterase is responsible for the type of localization observed.

Substrate Modifications. A number of mono and dihalogen-substituted indoxyl acetates have been used as histochemical substrates and most of these give rise to microcrystalline rather than grossly crystalline indigos. Technical details of their synthesis have been given by Holt (1958), by Su and Tsou (1960) and, most recently, by Kambli (1964). The first of these authors gave details for the synthesis of six substrates (indoxyl acetate, propionate and butyrate, 5-bromoacetylindoxyl, 5-bromo-4-chloroacetylindoxyl and 5-bromo-

PLATE XV

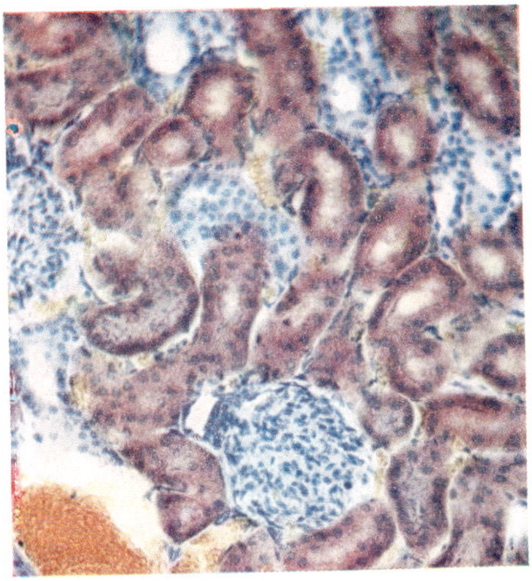

XVa. Rat kidney (10 µ frozen section). AS-esterase in the convoluted tubules. Incubation 30 minutes. Coupling azo dye method. Mayer's haemalum. × 130.

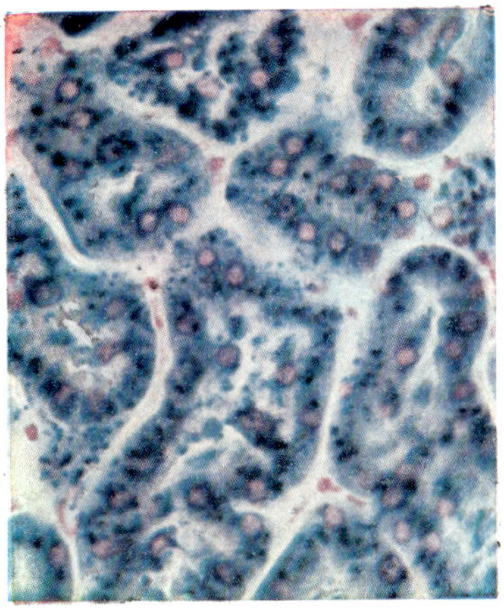

XVb. Rat kidney. Intracellular localization of non-specific esterase. Naphthol AS-D acetate—Fast blue BB, Feulgen. × 100.

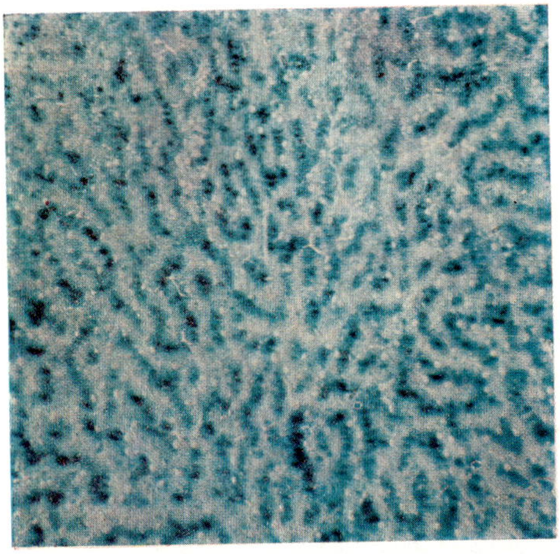

XVc. Rat liver. Localization of non-specific esterase using 5-bromo-4-chloroindoxyl acetate. Both the lysosomal and the diffuse esterase are shown. × 90.

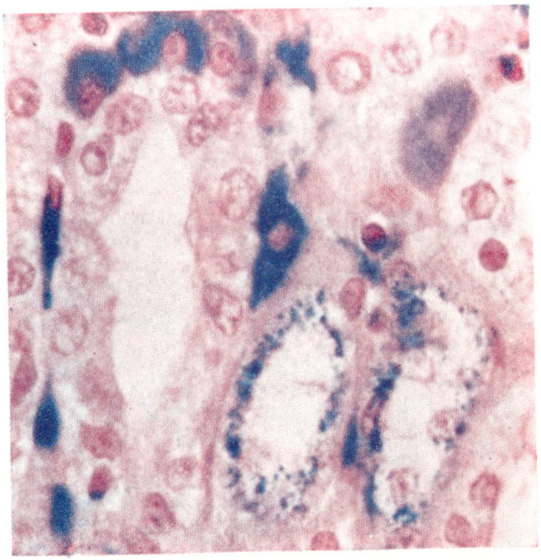

XVd. Rat kidney. Twenty days' potassium depletion. Large interstitial cells containing a strong esterase appear between the tubules. 5-bromoindoxyl acetate-Carmalum. × 335.

PLATE XVI

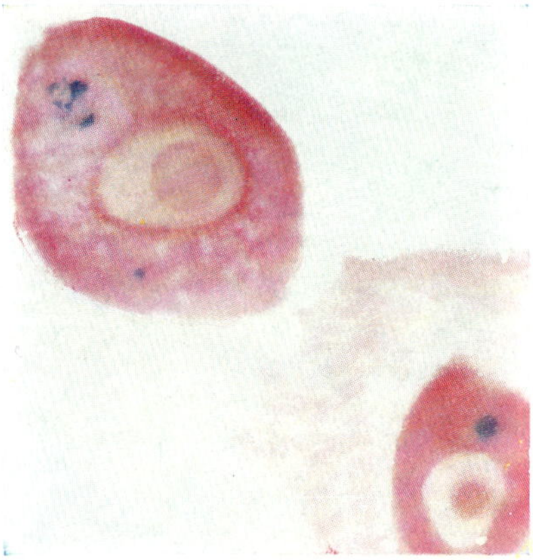

XVIa. *Buccinum undatum* (whelk). Oocyte. Non-specific esterase restricted to a few circumscribed zones, possibly centrosomes. 5-bromoindoxyl acetate—Neutral red. × 460.

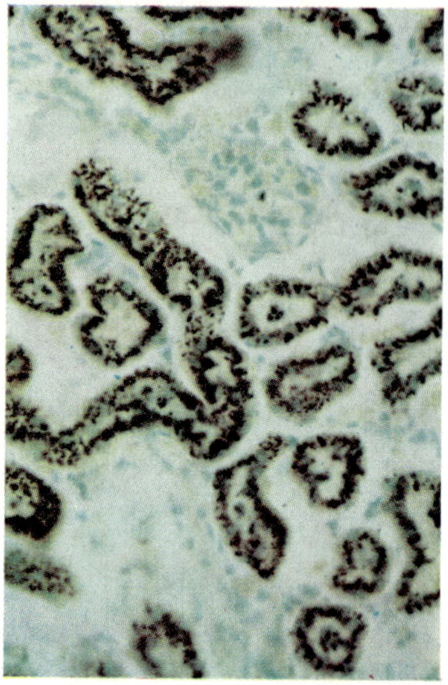

XVIb. Rat kidney. Fixed in cold formalin. Cryostat sections (5 μ). The indoxylazo reaction, using 4-chloro-5-bromoacetyl indoxyl as substrate and HPR as diazonium salt, shows exclusively the lysosomal non-specific esterase. × 254.

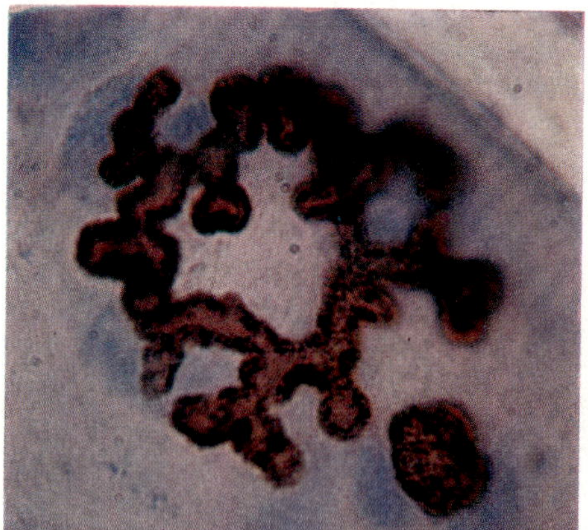

XVIc. Rat muscle. Localization of specific AChE in the motor end plate using the copper-thiocholine technique. × 2000.

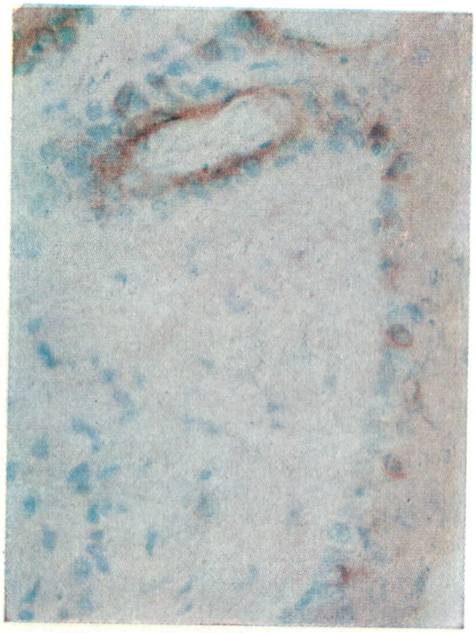

XVId. Rat skin. A strong lipase reaction in the Langerhans cells and in the epidermal cells surrounding a hair follicle. Naphthol AS nonanoate—Methyl Green. × 150.

6-chloroacetylindoxyl. Su and Tsou gave details for the preparation of 5-bromoindoxyl acetate, acrylate, methacrylate and crotonate.

Improvements on Holt and Sadler's (1958a) method for synthesizing 4-chloro-5-bromoacetylindoxyl were described by Kambli (1964). These methods avoided difficulties arising from the formation of isomeric intermediate products by the selection of different starting compounds. Holt and Sadler produced their isatins by means of the Sandmeyer method, starting from arylamines *via* isonitrosoacetarylides. By this method, however, the production of 4-chloro-5-bromoisatin is very difficult because there are many possibilities for the formation of isomeric compounds.

The methods developed by Kambli overcame these difficulties in two different ways. In the first method an entirely new principle was adopted for the production of the dihalogen-substituted isatin and in the second advantage was taken of the fact that the Sandmeyer synthesis for isatins leads to the production of uniform compounds if in the initial amine the *para* position to the chlorine atom is substituted. Technical details of Kambli's two methods are given in Appendix 17, p. 1307.

Some indication of the scale of improvement produced by the substrate modifications listed above can be obtained by comparing Figs 165 and 166, and Plates XVc, XVd, and XVIa.

Staining of Red Cells. In the first edition of this book the view was expressed that staining of red blood cells, which takes place when indoxyl acetates are used in the absence of oxidation catalysts, was due to the latter preventing the substrate from gaining access to the contained red cell esterase. This view was derived from Holt and Withers (1952). The subsequent studies of Cotson and Holt (1958) showed that aerial oxidation of indoxyls results in the production of hydrogen peroxide and the current view of the mechanism of red cell staining is that a peroxidase-like compound within the red cell (? haemoglobin) oxidizes the diffusing indoxyl to indigo. Considerable reduction of the red cell reaction can be obtained by addition of 10^{-2} M-azide, cyanide, sulphide or hydroxylamine, and also by excess H_2O_2. The effect of these peroxidase inhibitors suggests that the above explanation is correct. The action of ferricyanide on the red cell reaction may be by inhibition of the peroxidase effect or, possibly, by prevention of diffusion of the indoxyl into the red cell. A somewhat similar view of the mechanism of erythrocyte staining was recorded by Tsou and Su (1963). Since it is not observed when the azo-coupling alternative to oxidation is used with halogen-substituted indoxyls the matter may well be considered of academic interest only. Davis and Ornstein (1959), in their original publication, noted that with indoxyl acetate and hexazotized pararosanilin there was a clear demonstration of erythrocyte esterase.

Results of Indoxyl Methods. Indoxyl acetates are hydrolysed by carboxylic esterases of many different types (Underhay *et al.*, 1956; Hobbiger, 1957). They can therefore be used for the histochemical demonstration of any or all of these esterases in a variety of types of preparation. The best results are obtained with formol-calcium-fixed frozen sections which can be processed by the special procedures of Holt and Withers (1958). These involve impregnation of the block, after fixation, in a gum-sucrose solution at 0–2° for 24 hours before cutting sections. The latter are preferably used as free-floating sections rather than mounted on slides. Besides the substrate and sufficient ethanol to carry it into solution, the incubating medium (Holt, 1958) contains the oxidation catalyst, 1·0 M-NaCl, and 0·01 M-$CaCl_2$. The sodium chloride is presumed to act by reducing diffusion of the lyoesterase into the incubating medium and it might well be replaced by PVP or methyl cellulose. The precise function of the calcium ions is not clear. Their omission causes blurring of the sharp localization usually obtained and Kambli (1958) suggested that they might act by forming an insoluble calcium-indoxyl complex which is subsequently oxidized to indigo.

As well as the blue indigo reaction product most workers with indoxyl substrates have observed a red reaction product more marked in some tissues than in others. This product is especially noticeable in brain tissues where it stains the white matter a delicate pink colour, increasing in intensity on storage in glycerine jelly. Cotson and Holt (1958) showed that during the formation of indigoid dyes from indoxyls small amounts of isatins are formed. It is these which are apparently responsible for the pink and red colours observed in sections incubated with indoxyl substrates.

Formation of Indigo as a Function of pH. An important point was made by Holt (1958) which re-emphasized his earlier views on the mechanism and rate of production of indigo. When ferricyanide is present in excess the oxidation of indoxyls proceeds in two directions, one towards indigo and the other towards colourless dehydroindigos. As the reaction medium becomes more alkaline the amount of indigo decreases and the amount of dehydroindigo increases. At acid pH levels the converse is true and the amount of indigo deposited is greater than at alkaline levels. According to Holt the effect of changes of pH on indigo production is much greater than their effect on esterase activity and he stated that Pearson and Defendi (1957) were thus led erroneously to conclude that the pH optimum for indoxyl esterase was 4·7. In rat brain, Pepler and Pearse (1957a) found that indigo production after hydrolysis of 5-bromoacetylindoxyl was maximal at pH 6·2, irrespective of the type of esterase responsible. However, they also found that a sharp fall in indigo production with four of the six esterases, was produced by fractional lowering of the pH values. These observations supported Holt's views to some extent but suggested that he underrated the importance of enzyme activity.

Principles of Indoxylazo Methods. As an alternative capture reaction to aerial or catalysed oxidation it is preferable to use diazonium coupling. This was first suggested by Davis and Ornstein (1959) and used in combination with a wide range of halogen-substituted indoxyl substrates by Hess and Pearse

(1961) who employed principally either Fast dark blue R salt or hexazotized pararosanilin (HPR). The reaction product from combination of the former with 4-chloro-5-bromindoxyl is shown below:

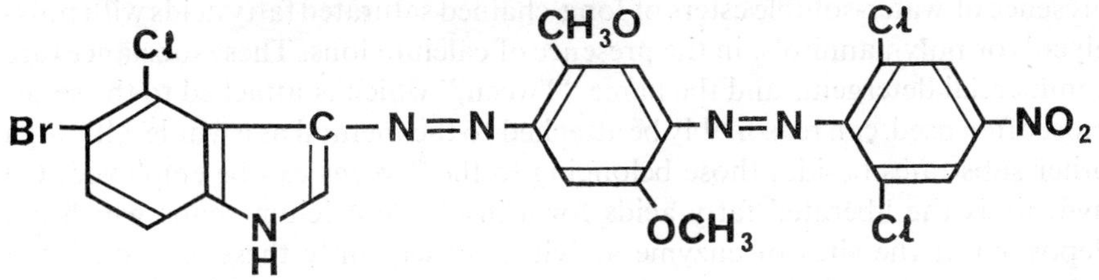

While the coupling rates of indoxyls with diazonium salts have not been measured it is evident that with both the above-mentioned salts azo-coupling provides a very much more efficient capture reaction than metal-catalysed oxidation. In view of the large and growing number of objections to the latter procedure it is certain that the future of indoxyl ester hydrolase histochemistry lies with the azo-coupling procedure.

Application of Indoxylazo Methods. In addition to those listed by Hess and Pearse (1961) a number of combinations of indoxyl and thioindoxyl esters with HPR and hexazotized new fuchsin (HNF) have been tested. For the majority of purposes I have found the most reliable pair to be 5-bromoindoxyl acetate and HPR, at pH levels between 5·8 and 8·6. The results obtained with this system are illustrated in Plate XVIb, p. 780. At acid pH levels the main esterase demonstrated is the particle-bound variety. At alkaline pH levels a diffuse esterase component is increasingly visible. Since the effect, if any, of raising the coupling rate of the diazonium salt would be to prevent or reduce diffusion of a primary reaction product, it would appear that the observed pH effect distinguishes the two types of esterase by their different pH optima.

Using 5-bromoindoxyl acetate and HPR Delellis and Fishman (1965) observed that at pH 6·1 both droplet *and* diffuse esterase were demonstrated and they recommended this pH as part of a standard procedure. While agreeing absolutely with their findings, and substantially with their conclusions, I think that the final choice of pH must be left to the experimenter who must choose the optimal pH, having regard to the species and tissue being examined and to available biochemical data.

Methods for Lipases (E.C. 3.1.1.3 etc)

The " Tween " Methods

Gomori (1945) observed that lipase was not destroyed by acetone fixation. Since it appeared to be a small and moderately diffusible molecule it was essential to fix tissues as soon as possible after removal if accurate histochemical localization was desired. After fixation and dehydration in acetone lipase was found to be moderately resistant to the action of heat, and a paraffin embedding technique was therefore employed. The method evolved by Gomori, which underwent many modifications in the hands of its original author, depended on the incubation of acetone-fixed paraffin sections in the presence of water-soluble esters of long-chained saturated fatty acids with polyglycols or polymannitols, in the presence of calcium ions. These substances are commercial detergents and the name "Tween," which is attached to the series most often used, can reasonably be attached to the method as a whole, although other substrates besides those belonging to the Tweens can be employed. On hydrolysis the liberated fatty acids form insoluble calcium soaps which are deposited at the sites of enzyme activity; subsequently these are converted into lead soaps and thence to the brown sulphide of lead. The duration of incubation varied from 12–48 hours, depending on the tissue employed, and control, sections were provided by slides inactivated either by short heating (boiling water, 10 minutes), Lugol's iodine (1 minute), or 5 per cent phenol (10 minutes). In the original method glycerol was used as an activator and maleate buffer at pH 7·4. Later, glycerol was found to have no activating effect under the conditions of incubation and, since the buffering power of maleate was found to be inadequate, a tris(hydroxymethyl)aminomethane, $(CH_2OH)_3CNH_2$, buffer (pH 7·2 to 7·4) was used in its place (Gomori, 1946, 1949, 1950a). The final method (Gomori, 1952b) is given in Appendix 17.

Many other authors published modifications of the original lipase technique including Wachstein (1946), Stowell and Lee (1950), Mark (1950), and Martin (1953a). Mark observed that acetone fixation destroyed a large amount of lipase and he therefore used formalin-fixed frozen sections instead of the more usual acetone-fixed ones. His modification of Gomori's method is also given in Appendix 17. The effects of fixation (alcohol-acetic, formol-calcium, formalin, Zenker) on lipase activity were investigated by Eapen (1960). Comparison was made with control tissue (100 per cent) and with unfixed gelatin-embedded material (106 per cent). The highest percentage survival of lipase was 14 (alcohol-acetic) compared with 7 (formol-calcium) 5 (formalin) and 0 (Zenker).

Sneath (1950) made the comment that the lipase method of Gomori had been little used in morbid histology. As the result of a critical study of the original method, and Wachstein's modification, he concluded that both methods required further improvement before they would be reliable for diagnostic work. Richterich (1951), however, claimed satisfactory results in various tissues. The same technique was used by Buño and Marino (1952), with Tween 60 as substrate, in a study of the "lipases" of the developing chick embryo. These authors stated that they were unable to distinguish lipase from non-specific esterase activity. Later studies by Martin (1953b), using Tween 40, revealed strong "lipase" activity in many organs in a variety of mammalian species, in addition to pancreas and liver. Especially strong activity was found in the ducts of the salivary glands in all species except the rat. No inhibitors or activators were used in these studies and therefore no distinction between true lipases and non-specific esterases could be made. A later study by Siurala (1956) of the "lipase" activity of human and monkey gastric mucosa employed Tweens 20, 40, 60 and 80. Again no attempt was made to distinguish lipase from esterase activity.

The variability of the results obtained by the lipase method appeared to be due to several factors. Postmortem autolysis certainly caused diffusion of enzyme in the tissues just as prolonged fixation diminished the amount of enzyme available. A similar result was effected by excessive or prolonged heating during embedding or mounting on the slides (no more than 58° for 3 hours should be permitted for the former or than 37° for the latter). Floating sections on water instead of acetone caused some loss of enzyme or disturbance of distribution, and storage in the paraffin block also causes a gradual diminution in activity. The maintenance of the pH in the region of 7·4 during incubation appeared to be one of the main difficulties of the method, although Tris buffer is very efficient in the physiological range. Another factor influencing the result was the presence of varying amounts of free fatty acid in different batches of the same substrate. If large amounts of these were present the incubating mixture gradually became turbid, due to the formation of calcium soaps which settled on the slides and were indistinguishable from the products of enzymic activity in the tissues. Gomori recommended removal of free fatty acids by incubating the turbid mixture at 37° for a few days and filtering through glass wool. Bacterial decomposition of the substrate was avoided by the use of a preservative such as thymol or chloretone. Merthiolate, which was originally used, was found to be a weak inhibitor of lipase.

In the course of his long-term studies of skeletal muscle lipase George (George and Iype, 1960; Bokdawala and George, 1964; George, 1964) introduced a number of modifications of the original lipase technique. In the earliest of these a 4-hour fixation in cold buffered formalin was followed by coating of the sections, mounted on slides, with 5–10 per cent gelatin. A further hour's fixation in formalin preceded incubation in a medium containing Tween 80. In this case the standard (PbS) method of demonstrating the calcium soap product of the reaction was adopted.

Later, however, Bokdawala and George (1964) employed Alizarin red S to stain the reaction product after removing any calcium originally present in the sections by treatment with EDTA. These authors claimed that precise (mitochondrial) localization of lipase was possible. While there must be considerable doubt as to the accuracy of this (rather unlikely) localization, the method offers some advantages and it is therefore given in full in Appendix 17, p. 1310.

In spite of the studies of Eapen, and of Bokdawala and George, Piantelli and Rebollo (1967) used acetone-fixed frozen sections (and Tween 60) in their studies of the lipases in developing skeletal muscle of the chick embryo.

Specificity of the Tween Methods. As pointed out by Desnuelle (1951) the only true substrates for lipases are higher triglycerides. Biochemical investigations carried out by Evans and Stansfield (1961) indicated that true lipase (pig pancreas), hydrolysing triolein, was inhibited by 0·2 mM-*p*-chloromercuribenzoate but not by the same concentration of *NN'*-diisopropyl phosphodiamidic fluoride (Mipafox). The enzyme hydrolysing Tween 20 (Tweenase), on the other hand, was inhibited by Mipafox and unaffected by the mercurial. It seems that these findings have not been applied at the histochemical level.

Gomori (1950b) remarked that the topography of enzymic activity by the Tween method was not identical with that shown by the Nachlas-Seligman method for esterase. Certain cells of the cardiac portion of the mouse stomach and cellular debris in the tubules of the testis, were intensely Tween-positive but esterase-negative. On the other hand, the superficial cells of the gastric mucosa and the interstitial cells of the testis (man, rat rabbit) stained much more intensely by the azo dye method. As the result of some studies on the histochemical specificity of his method, in which thirty-five different water-soluble long-chained (C_{12} to C_{18}) fatty acid esters were employed, Gomori (1949) reported that the true lipases could hydrolyse esters of unsaturated fatty acids rapidly, a property not possessed by the non-specific esterases. This difference, reinforced by the difference in response to small concentrations of taurocholate, could serve to distinguish the two enzymes and this second principle was indeed used by Abe, Kramer and Seligman (1964) in their azo dye technique for pancreatic lipase. This method is described below.

Specific Method for Pancreatic Lipase (E.C.3.1.1.3)

Observing that the Tween techniques produced variable results in a variety of tissues, Darnton and Barrowman (1969) used an emulsified long-chain triglyceride as substrate, in the presence of sodium taurocholate as activator and emulsifier, in designing a method specific for pancreatic glycerol ester hydrolase (lipase). Only the true pancreatic enzyme was demonstrated by their method, which produced a brownish-black precipitate at the sites of enzyme activity.

Application of the Tween Method. For routine application to pathological problems, or even for normal tissues, the Tween method cannot be recommended. Except in the pancreas, where uniform results are consistently obtainable, I found that the final precipitates of lead sulphide are always patchily distributed. Fig. 167 shows the kind of result obtained by application of the method to formalin-fixed frozen sections, in this case a section of rat liver, along the lines of the modification introduced by Mark (1950). In the first and second editions of this book (1953, 1960) I stated that as a starting point for further research on the Tween methods a standard 16-hour fixation in 6 per cent neutral formalin at 4° should be used, and that much further work would be necessary before the theoretical advantages of using substrates resembling the natural substrates for the lipases could be translated success-

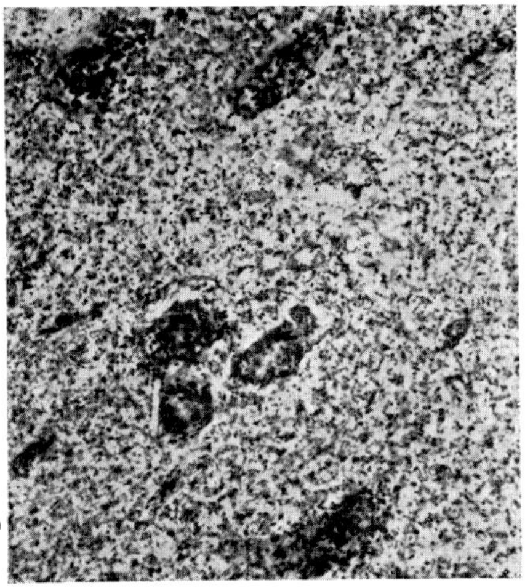

FIG. 165. Frozen section (10 μ) of rat spinal cord. Showing crystalline precipitates of indigo in the anterior horn cells and elsewhere. Indoxyl method for cholinesterase (substrate *O*-acetyl-indoxyl). × 270.

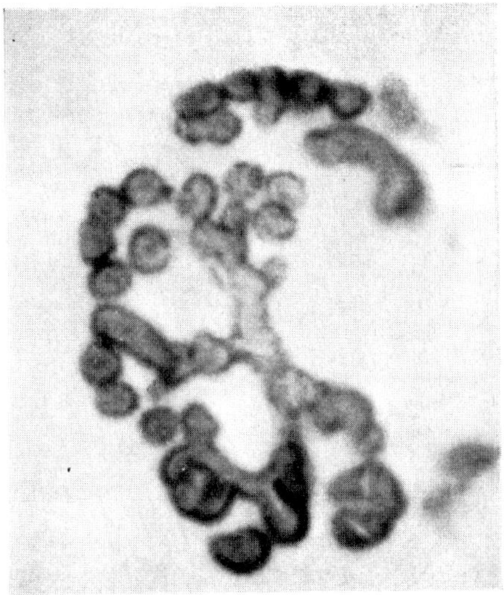

FIG. 166. Rat diaphragm. Motor end plate. Shows concentration of specific acetylcholinesterase in the subneural apparatus. 4-Chloro-5-bromoindoxyl acetate. × 2000.

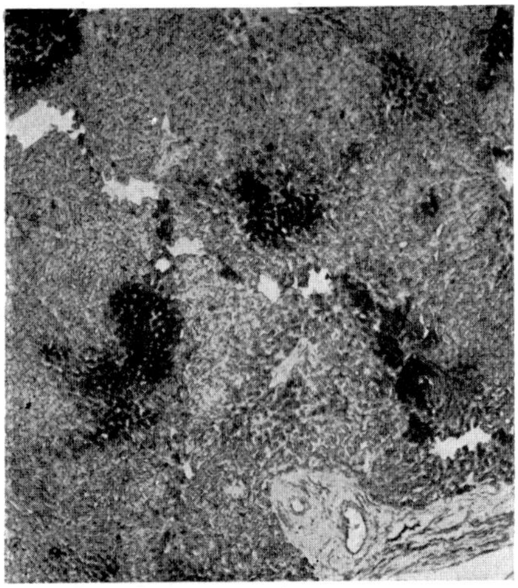

FIG. 167. Formalin-fixed frozen section (10 μ) of rat liver. Tween method for lipase (substrate Tween 60), showing patchy localization. × 24.

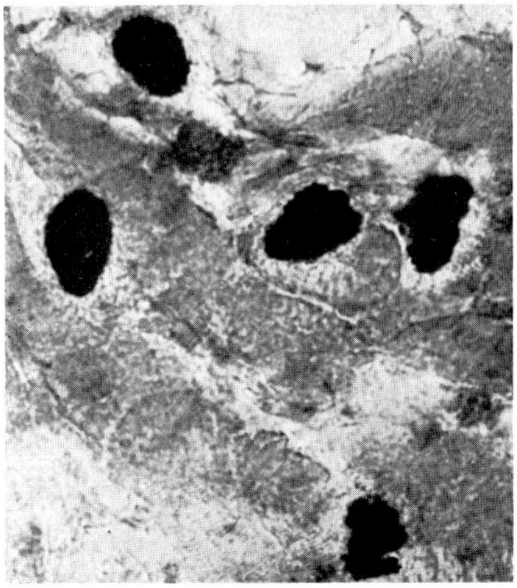

FIG. 168. Rat diaphragm, unfixed frozen section (20 μ). Showing a group of motor end plates. (Incubation 60 minutes.) Myristoyl choline method. × 430.

[To face p. 786

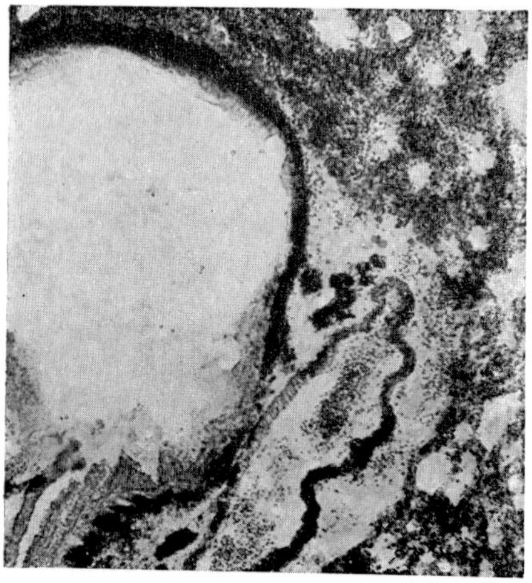

FIG. 169. Mouse lung, acetone-fixed paraffin section (5·5 μ). Showing, at upper left, the medial coat of an artery and, below, bronchial epithelium. (Incubation 16 hours.) Myristoyl choline method. × 20.

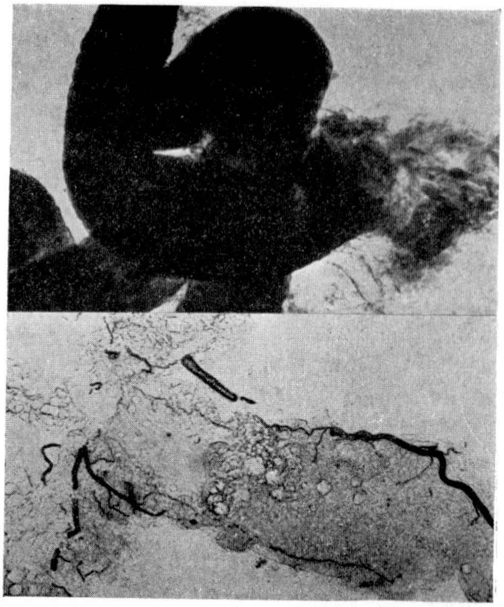

FIG. 171. Cockroach (*Periplaneta americana*). Right and left colleterial glands. Strong β-glucosidase activity in the right gland (top) with only traces of activity in the left gland. Post-coupling method. × 75.

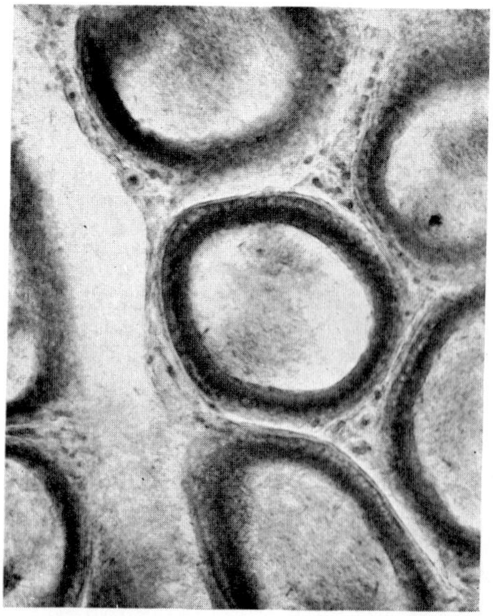

FIG. 172. Rat epididymis. Strong activity of acetyl-β-glucosaminidase. Coupling azo dye method. × 136.

fully to routine histochemistry. I cannot see any reason, at the present time, to modify this view, except to substitute alcohol-acetic fixation.

The Naphthol AS-Nonanoate Method.

A new series of lipase substrates was prepared by Abe *et al.* (1964) and of these the best for histochemical purposes proved to be the AS nonanoate, used in a simultaneous coupling technique with Fast blue BB salt to give a blue azo dye product. Details of this method are given in Appendix 17, p. 1310.

In order to produce a product clearly related to "pancreatic-like" lipase activity the original authors observed that it was necessary to include sodium taurocholate (5 mM) in the medium although members of the same group (Kramer *et al.*, 1963) had reported cholate to be more reliable than taurocholate as an activator of human pancreatic lipase. It was considered probable that some, at least, of the activity observed in the absence of taurocholate was due to lipase rather than to non-specific esterases.

The pH recommended by Abe *et al.* (7·4) is lower than the optimum for purified pancreatic lipase which, according to Baskys *et al.* (1963), is between 8·0 and 8·1. It appears, however, that bile salt activators such as taurocholate shift the pH optimum from 8 to 9 down to 6·0 to 6·5 (Desnuelle, 1961).

In a study of the enzymology of Langerhans cells in the epidermis Campo-Aasen & Pearse (1966) used Naphthol AS nonanoate, with hexazotized pararosanilin (Chapter 15, p. 518). The enzyme hydrolysing nonanoate was inhibited rather than activated by taurocholate (Plate XVId, p. 781). It could not, therefore, be regarded as a true lipase and other inhibitor studies indicated that it was an unusual type of non-specific esterase with high affinity for Naphthol AS esters.

Cobalt Technique for Phospholipase B. (E.C. 3.1.1.5)

Practically all tissues contain an active phospholipase B which can deacylate lysolecithin with the liberation of glycerylphosphorylcholine and free fatty acid (Dawson, 1956; Marples and Thompson, 1960). Since lysolecithin has a destructive effect on cell membranes it has been suggested that the function of phospholipase B is to protect the tissue from damage by this mechanism.

The phosphatidyl residue in lecithin, and in many other phospholipids can be hydrolysed at all ester bonds by phospholipases from various sources. The following diagram, derived from Dawson (1966), indicates the bonds which are attacked by phospholipases A, B, C & D.

It can be observed that phospholipase A (E.C. 3.1.1.4) attacks only the acyl ester group on the *β*-carbon of the glycerol residue, giving a free fatty acid and lysolecithin. Phospholipase B can hydrolyse both acyl ester bonds, yielding free fatty acids and glycerylphosphorylcholine. The bacterial phospholipase C (E.C. 3.1.4.3) produces a diglyceride and phosphorylcholine while phospholipase D, found in the plant kingdom, removes the base (choline or ethanolamine) leaving phosphatidic acid.

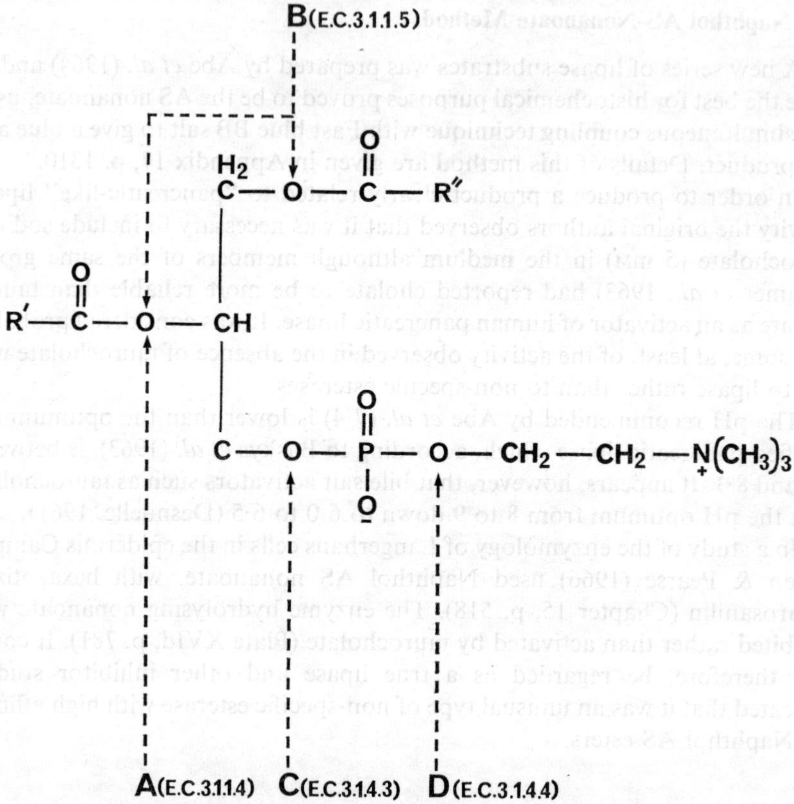

Hydrolysis of Lecithin by Phospholipases.

A cytochemical technique for phospholipase B, based on the original lipase technique of Gomori (1945) (See Chapter 16), was described by Ottolenghi *et al.* (1967). Cold formol-calcium fixed frozen sections are incubated at pH 6·6 with lysolecithin and 1 per cent cobalt acetate. The free fatty acid is captured as its cobalt soap and the latter is rendered visible by conversion to the sulphide. Details of the method are given in Appendix 17.

Methods for Cholinesterases (E.C. 3.1.1.7 & 3.1.1.8)

Methods using Higher Fatty Acid Esters of Choline

Gomori (1948) produced a method for demonstrating cholinesterase in the tissues, resembling his method for lipases, in which a variety of higher fatty acid esters of choline were used as substrates. The four alternative substrates which he evolved were lauroyl, myristoyl, palmitoyl and stearoyl choline, and the second was found to be the most satisfactory for histochemical purposes. Thin slices of tissue were fixed in cold acetone and double embedded in celloidin and paraffin. Sections from these blocks were incubated for 2–16 hours at 37° in 0·1 M-tris(hydroxymethyl)aminomethane buffer, at pH 7·5 to 7·8, in the presence of the substrate (0·4 mM-myristoyl choline), cobalt ions, and traces of $CaCl_2$, $MgCl_2$ and $MnCl_2$. On hydrolysis the free fatty acid combined with the cobalt to give a fine granular white precipitate, easily rendered visible by subsequent treatment with yellow ammonium sulphide. Calcium, strontium, barium and magnesium were originally tried in place of cobalt but all gave a coarse crystalline precipitate unsuitable for precise localization. Ca^{2+}, Mg^{2+} and Mn^{2+} ions were found to activate cholinesterases *in vitro* (Nachmansohn, 1940) and they were employed by Gomori for this reason. Higher concentrations of substrate than 1 mM inhibited the enzymes and the particular buffer was chosen because it was found to cause minimal inhibition.

Theoretical Considerations. Gomori showed that crude extracts of various animal tissues would hydrolyse the higher choline esters at varying rates, but notably poor hydrolysis was produced by rat brain homogenates. Brain tissue was stated by Mendel and Rudney (1943) to contain only AChE, and Gomori's finding, therefore, suggested that the higher choline esters were not attacked with any ease by AChE. In no case did the rate of hydrolysis of the artificial substrates by nervous tissue approach that for acetylcholine. Purified bovine erythrocyte AChE, and a purified AChE from the electric organ of *Torpedo*, failed to hydrolyse the myristoyl, palmitoyl or stearoyl esters although the first two did hydrolyse the lauroyl ester. *Torpedo* AChE, probably the purest of the three, would not hydrolyse lauroyl choline. In view of these *in vitro* results Koelle and Friedenwald (1949) concluded that the myristoyl choline method localized "only non-specific esterases," meaning non-specific cholinesterases, but in spite of Gomori's *in vitro* results, Denz (1954) found that AChE (red cell) hydrolysed myristoyl choline at a slow rate. This was apparently sufficient to produce a positive histochemical result, even with short incubation, since in Denz' fresh frozen sections the enzyme was fully active and the sensitivity of the method was therefore high.

Distribution of Enzyme. The distribution of enzyme shown by the myristoyl choline method differs to some extent from that shown by a non-specific esterase method such as that of Nachlas and Seligman, or its α-naphthyl acetate modification. In particular, tissues such as brain and spinal cord, in which various components react by the Gomori method, have been reported as negative by the Nachlas-Seligman technique. Some interesting locations are demonstrated by the former, including ganglion cells, structures in striated muscle identified as muscle spindles, and the smooth muscle of the bladder, uterus and gastro-intestinal tract in several species; Gomori, however, did not give a full list of localizations in his paper. Dog and mouse tissues gave most intense pictures with the lauroyl and myristoyl esters and human tissues with palmitoyl choline. Rat tissues seem to be unable to hydrolyse myristoyl choline at a sufficiently rapid rate to give a positive histochemical result.

Hard and Peterson (1950) used the myristoyl choline method to study the distribution of cholinesterase in nerve tissues of the dog. They obtained largely consistent results and attributed occasional erratic or incomplete reactions to variations in concentration and pH of the substrate solution. In the central nervous system the reaction was confined almost entirely to nerve fibres, in nerve cells the reaction was slight but the nuclei and nucleoli often stained. In the peripheral nervous system a strong reaction was localized at the margins of both spinal and sympathetic ganglia. Since the enzyme concerned was inhibited both by 10^{-5} M-prostigmine and by excess of substrate, the authors were inclined to regard it as a specific cholinesterase. In a series of further studies with myristoyl choline as substrate, Hard, Peterson and Fox (1951) obtained evidence that the abolition of the histochemical reaction in sympathetic ganglia after transection of the cord, was closely related to the fall in AChE measured in the same tissues by biochemical means. These findings were regarded as further evidence that the enzyme demonstrated was AChE. Nevertheless, none of the evidence offered precluded ChE being responsible for their reactions.

Application of the Method. Denz (1953) found that the myristoyl choline method was exceedingly capricious with acetone-fixed tissues, sections from the majority of blocks giving negative results. He used instead fresh frozen sections and with these the method was absolutely reliable and reproducible. With fresh tissues the incubation time is reduced from an average of 8 hours to less than 30 minutes; localization is sharp and diffusion artifacts are not in evidence. Fig. 168 shows motor end plates in rat diaphragm stained by the myristoyl choline technique, and Fig. 169 illustrates the application of the method to an acetone-fixed section of mouse lung. Staining is most intense in the arterial walls (upper left) and much less in the bronchial mucosa (lower right).

The Gomori technique for cholinesterases, if used with fresh or suitably fixed material, is thus seen to be capable of demonstrating both AChE and ChE. With the former, the rate of hydrolysis is far slower than that obtainable with other histochemical substrates, but with ChE the rate of hydrolysis is excellent. If a positive result is obtained with the myristoyl choline method in formalin-fixed frozen sections, using short incubation (Appendix 17), the participation of AChE is largely excluded. It is still necessary to distinguish between ChE and non-specific esterases and this can be done by using specific inhibitors for ChE in suitable concentrations.

The Gomori method has not been found useful for routine investigations but it may still be found valuable as a control method for other methods for cholinesterases described below, especially where ChE is concerned. The only extensive histochemical study using higher fatty acid esters of choline as substrates is that of Carbonell (1956). This author demonstrated the conducting system of the heart in several mammalian species, including man, with palmitoyl-, myristoyl-, and pentadecanoyl-choline. The extent and distribution of the conducting system of the heart is difficult to study by any method other than one capable of showing cholinesterase activity. Carbonell found that the best results were obtained with pentadecanoyl-choline. Activators ($MgCl_2$, $MnCl_2$ and $CaCl_2$) employed in the original method were not found useful and the incubation period was 5–6 hours.

Methods Using Acetyl- or Butyrylthiocholine

The Original Method. Koelle and Friedenwald (1949) employed a substrate (acetylthiocholine) which is hydrolysed by AChE and ChE at a more rapid rate than acetylcholine itself. It was presumed by these authors that the increased rate of hydrolysis was due to the $-\overset{\overset{O}{\|}}{C}-S-$ linkage being weaker than the $-\overset{\overset{O}{\|}}{C}-O-$ linkage of acetylcholine. The original histochemical procedure consisted of incubating fresh frozen sections (or teased preparations) in a medium containing 4 mM-acetylthiocholine or butyrylthiocholine, 2 mM-copper sulphate and 8 mM-glycine at pH 8·06 for 10–60 minutes. On hydrolysis the liberated thiocholine was presumed to react with the copper salt to give a relatively insoluble product, copper thiocholine.

In order to facilitate precipitation, and to prevent diffusion, the incubating medium was saturated with copper thiocholine before the sections were placed in it. This was a critical step in the performance of the original method and it remained of importance in subsequent modifications of the method introduced by Koelle (1950, 1951). Unless complete saturation with copper thiocholine was achieved subsequent localization of the enzyme was poor and diffusion evident. Much sharper localization of the precipitate was obtained by controlling the pH. It was considered that no false reactions were caused by initial saturation with the reaction product since it was observed that inactivated control sections, incubated in the same solution, showed no staining. These controls were prepared by allowing freshly-cut sections to stand in 1 mM-di-isopropylfluorophosphate (DFP)

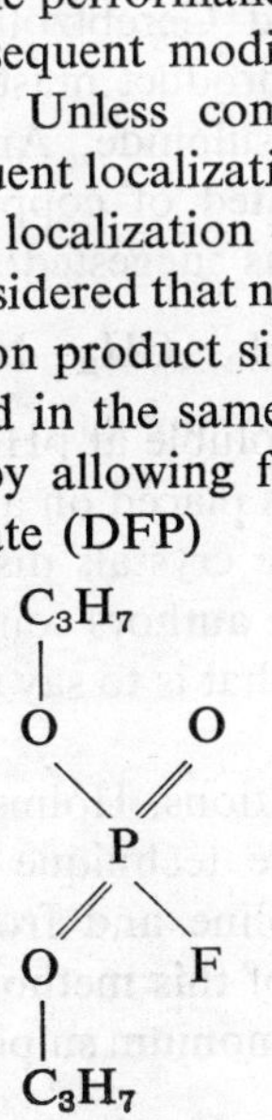

in 0·85 per cent saline, for 30 minutes at room temperature. (At this concentration DFP irreversibly inhibits both AChE and ChE.) After thorough washing, the control sections were transferred to the incubating medium and, after incubation, both test and control slides were washed and treated with yellow ammonium sulphide to convert the white precipitate of copper thiocholine into a brown amorphous deposit of copper sulphide.

Modifications of the Thiocholine Techniques. Very many modifications have been made to the original technique. Space allows consideration only of some of the most important of these, and of the theoretical considerations which prompted some of them.

In order to reduce diffusion of the primary reaction product (PRP) Koelle (1950) lowered the pH of incubation to 6·4, using phosphate buffer, and Couteaux (1951) employed pH levels as low as 4·0, finding the optimum to lie between 5·3 and 5·6 (acetate buffer).

Fixation was found to be necessary at an early stage in the development of the technique and cold buffered formalin became the most popular medium. Considerably better results can be obtained with Pearson's (1963) cold formalin-sucrose-ammonia (Chapter 5, p. 98).

High concentrations of sodium sulphate in the incubating medium were used by Koelle (1951) and at the same time the phosphate buffer was replaced by maleate, since the latter did not precipitate copper in the concentrations employed.

The mechanism of the thiocholine method was critically examined by Malmgren and Sylvén (1955) who were prompted to make their investigations when Zajicek, Sylvén and Datta (1954) found that the reaction product was deposited in the form of crystals. These had also been noted by Couteaux and Taxi (1952) and by Mohr and Gerebtzoff (1954). It appeared that these crystals of the initial reaction product must be dissolved and redeposited as the final product of copper sulphide. Analyses of the primary reaction product showed that it consisted of copper thiocholine sulphate and the following empirical formula was suggested:

$$[Cu^{+}—S—CH_2—CH_2—N^{+}(CH_3)_3]SO_4^{-}$$

It was observed to be slightly soluble at pH 8 but less soluble in acid media, and when a purified sample was placed on a coverslip and treated with dilute yellow ammonium sulphide the crystals dissolved and were replaced by an amorphous deposit of CuS. The authors considered that this was not topically related to the original crystals, that is to say there was no direct conversion of one to the other, *in situ.*

Acting on the above suggestions, Holmstedt (1957b) introduced a much modified version of the Koelle technique in which the medium was not saturated with copper thiocholine and from which conversion to copper sulphide was omitted. Details of this method are given in Appendix 17.

Further criticisms of the ammonium sulphide stage were made by Palkama

(1961) but few workers have chosen to employ direct observation of the PRP by phase contrast microscopy. The dangers of redistribution have been over-estimated. Henderson's (1967) silver intensification stage (see Appendix 17, p. 1316) allows the incubation period to be cut by a third, reducing the degree of artifact caused by crystallization of the PRP. A modified $(NH_4)_2S$ solution in acid alcohol was employed by Bull *et al.* (1957) and when this was substituted for the usual aqueous reagent no alteration was detectable in the distribution of the PRP before and after conversion.

Alternative thiocholine esters have been considered but only one, propionylthiocholine, has been employed in histochemical practice. This was used by Atherton (1963) particularly for the demonstration of avian cholinesterases. These hydrolyse propionylcholine more rapidly than the acetyl or butyryl esters (Myers, 1953).

A number of attempts have been made to replace the cupric ion in the incubating medium by other metals. Silver sulphate was used in place of copper sulphate by Birks and Brown (1960) and Pb^{2+} by Joó *et al.* (1965). The latter authors clearly distinguished their product from the "lead-reactive material" described in peripheral synapses by Sávay and Csillik (1959). Another alternative metal, gold in its aurous form, was employed by Koelle and Gromadzki (1966) in the gold-thiocholine technique. The main advantages of this technique, which is given in Appendix 17, p. 1316, lies in its application to the electron microscope.

The "Direct-Colouring" Thiocholine Method. This modification of the original thiocholine technique (Karnovsky and Roots, 1964) makes use of a new and original principle. Ferricyanide is added to the medium, together with copper sulphate, and the Cu^{2+} ions are complexed with citrate to prevent the formation of copper ferricyanide. Enzymatically released thiocholine reduces the ferricyanide ion, *in situ*, to ferrocyanide and the latter combines with Cu^{2+} to form insoluble copper ferrocyanide (Hatchett's Brown).

The true importance of this innovation lies neither in the fine, precise, localization of ChE and AChE which it affords (Martinez Rodriguez *et al.*, 1964; Bell, 1966) at optical microscope level, nor in the fact that it demonstrates them clearly also at the E.M. level. It lies rather in the fact that the system can be used as an electron acceptor in a variety of cytochemical systems to produce a fine granular, insoluble, visible product which, at the same time, is electron opaque. Details of the method are given in Appendix 17.

A modification of this modification was introduced by Eränkö *et al.* (1967). These authors employed Pb^{2+}, complexed with Tris-acetate buffer, to trap ferrocyanide ion produced by the reduction of ferricyanide by enzymatically released thiocholine. For light microscopical investigations the Pb^{2+} modification has no real advantages over the Cu^{2+} method. At the E.M. level however (Chapter 32, p. 1282) it produces a finer precipitate with a higher degree of electron density. Details are given in Appendix 17.

The Dimethylthioacetate Method. The quaternary carbon analogue of acetylcholine (3,3′-dimethylthioacetate) was synthesized by Aarseth *et al.* (1969) and employed as a histochemical substrate. This compound

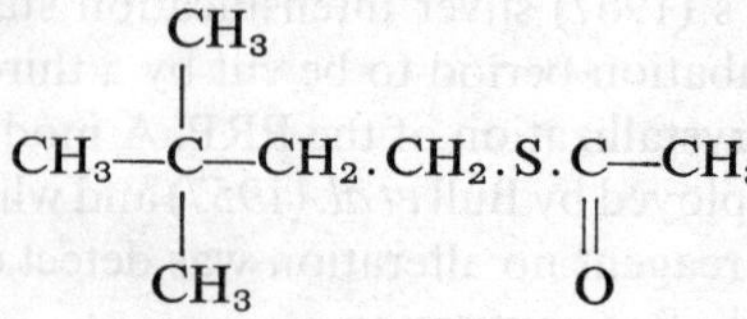

is hydrolysed by AChE at 25 per cent of the rate of AThCh at a concentration of 0·6 mM. Comparisons between this substrate (DMBTA) and AThCh have yet to be made but, because of its lack of electrical charge and the ability to pass rapidly through biological membranes, the former may well be found to have advantages in histochemical practice.

False Localizations due to Metal Affinities. The non-specific binding of lead ions to the subneural apparatus of motor-end-plates was first described by Sávay and Csillik (1959a) and later Zenker (1964) showed that treatment of sections with copper sulphate produced non-specific metal-binding in the region of Ranvier's nodes. This second observation gave rise to the view that the apparent localization of AChE in tissue components having metal-binding capacity was a non-specific artifact (Mladenov, 1965; Herbst, 1965).

As far as the subneural apparatus is concerned, the work of Nakamura *et al.* (1967) makes it clear that metal-binding (Pb^{2+}, Sn^{2+}, Cd^{2+}, Zn^{2+}, Cu^{2+}) in this structure is post-synaptic, as is the AChE activity. The two activities are quite distinct, however, since metal-binding is reversed by formalin fixation, or by freezing and thawing, whereas the enzyme activity is not.

With regard to Ranvier's nodes Gerebtzoff and Mladenov (1967) concluded that, with the histochemical techniques in use, it was not possible to ascertain the exact nodal localization of AChE. On the other hand Adams *et al.* (1968) showed very clearly that at low pH levels copper ions were bound by acidic mucosubstances in the region of the node but that at pH 6·0, as in ChE methods, they were not.

While it is important, especially at the E.M. level, to be aware of the possibility of metal-binding artifacts, the various metal precipitation techniques for AChE and ChE can be used with complete confidence in optical microscopical studies.

Other Metal Precipitation Techniques

The Thiolacetic Acid Methods. It was shown by Wilson (1951) that AChE could catalyse the hydrolysis of thiolacetic acid (CH_3COSH). This fact was made use of by Crevier and Bélanger (1955) who evolved a histochemical method for AChE using thiolacetic acid as substrate. The liberated H_2S was trapped by Pb^{2+} ions added as lead nitrate or acetate. The authors used either fresh frozen or cold formalin-fixed frozen sections mounted on slides and thoroughly dried to ensure adherence. They obtained clear localization of AChE in motor end plates, in neurones, and along certain fibre tracts in the C.N.S.

The prime advantage of the thiolacetic (ThAc) method would appear to be its reliance on a single stage capture reaction. Its specificity, however, is very wide indeed (Wachstein *et al.*, 1961). In the case of the cholinesterases, under histochemical conditions, it is the esterase site (see p. 767) which is involved and hence blocking of the anionic site with specific inhibitors has no effect on subsequent hydrolysis. It is thus impossible with the ThAc method to determine whether a component reacts because it possesses both ChE and non-specific esterase activities.

Between the first (1953) and second (1960) editions of this book there were practically no reports of the application of the method at the optical microscope level. Since then a few papers have appeared, incorporating some worthwhile modifications. The first of these (Sávay and Csillik, 1959b) used barbiturate buffer and a final pH of 7·2–7·8, in place of the phosphate-citrate buffer of the original method. (Phosphate buffers should not be employed in media containing Pb^{2+}, or other metal cations, having insoluble phosphates).

Wachstein *et al.* (1961) found that the Sávay and Csillik buffering system failed to hold the pH with any degree of constancy. They therefore brought their ThAc solution to pH 5·5 with 0·1 N-NaOH and maintained this pH with acetate buffer. Zacks and Blumberg (1961) used sodium cacodylate to buffer their ThAc to between pH 5·6 and 6·0.

In a number of laboratories the ThAc method was found to be capricious, its performance apparently varying with the particular batch of substrate employed. The mystery was solved by Koelle and Horn (1968). These authors found that triple-distilled ThAc was much less effective as a substrate for AChE than either the original impure preparation or the residue from the first distillation. They identified the active contaminant as acetyl disulphide, $(CH_3COS)_2$. In a subsequent paper (Koelle, Davis and Devlin, 1968) its properties as a substrate for AChE were described and suggestions as to the mechanisms involved were put forward. Technical details of the preparation of acetyl disulphide, and of the alternative substrate bis-(thioacetoxy)aurate, $Au(CH_3COS)_2$, together with the appropriate histochemical techniques, are not given here since it is unlikely that either method will be employed other than for localization of cholinesterases at the E.M. level.

Inhibitors and Activators in Esterase Histochemistry

Before discussing the application of inhibitors and activators to the histochemical distinction of the various carboxylic esterases it is necessary to emphasize the very great differences which are observed between one species and another. Unless this is understood there may be disappointment when a given inhibitor fails to act as expected. The closest possible adherence to the results of biochemical studies is essential.

Distinction of Esterases from Lipases. Distinction between true lipases and non-specific esterases was at one time made by observing the activation of the former by low concentrations of the salts of the bile acids, especially sodium taurocholate. Although an enzyme which is activated by this substance may be a true lipase, there is evidence that some types of non-specific esterase are similarly activated. As pointed out by Darnton and Barrowman (1969), the Tween-type substrates are hydrolysed by a number of enzymes in addition to the true lipases.Distinction can be made, according to these authors, only by the use of a specific lipase substrate (see p. 786).

Distinction of Types of Non-specific Esterase. In an extensive series of histochemical studies Gomori and Chessick (1953a and b) found that the esterases of mammalian tissues showed marked organ and species specificities, the latter being very great for some tissues (e.g. brain) and much smaller for others (e.g. placenta). Gomori (1952a) divided the esterases into lipase, esterase, α-esterase, AS-esterase and cholinesterase. He continued (1952b) to use this classification but later in the case of human esterases, proposed a division into hepatic, pancreatic and general types on the basis of histochemical substrate preference.

I do not consider that classifications of these types should be used by histochemists and I prefer to divide the esterases, following established

TABLE 60

Non-specific Esterases

(Esterases Insensitive to 2 μM-Eserine)

Enzyme	Substrates	Complete Inhibition by	Activation by
A-esterase	α-naphthyl acetate	E 600, 10 mM	Cysteine, 1 mM
Arom-esterase	Naphthol AS acetates	$CuSO_4$, 1 mM $Pb(NO_3)_2$, 1 mM	
Organophosphate-resistant esterase I.	Indoxyl acetates	$AgNO_3$, 10 mM PCMB†, 100 μM	
B-esterase	α-naphthyl acetate	E 600, 10 μM Mipafox, 1 mM	Cysteine, 1 mM
Ali-esterase	Naphthol AS acetates	Eserine, 100 μM Ro3-0422, 0·1 μM	
Organophosphate-sensitive esterase	Indoxyl acetates	$AgNO_3$, 10 mM	
C-esterase	Indoxyl acetates	β-phenylpropionic acid, 10 mM	Phenylmercuric chloride, 100 μM
Organophosphate-resistant esterase II	Naphthol AS acetates	$AgNO_3$, 10 mM $CuSO_4$, 1 mM	Trypsin, 1/2 sat. sol. PCMB†, 100 μM

* Approximately 50 per cent inhibition.
† *p*-chloromercuribenzoate.

biochemical practice, into A-, B-, and C-esterases. This can be done on the basis of inhibitor and activator studies such as those used by Pepler and Pearse (1957a) and by Hess and Pearse (1958). Table 60, above, indicates the lines on which histochemical differentiation between the three types of esterase can be attempted.

In practice sections (usually cold formalin-fixed frozen sections) are incubated with 10 μM-E 600 in buffer at pH 5·3 for one hour at 37°. B-type esterases are irreversibly inhibited by this treatment and, after washing, the sections are incubated by the chosen method to demonstrate the two organophosphate-resistant esterases.

Distinction between A- and C-esterases is made by following up the treatment with E 600 (or Mipafox) with one hour's incubation in buffer containing inhibitors or activators as indicated in Table 60. Thereafter the sections are incubated in substrate medium containing the chosen inhibitor or activator. The best compound for distinguishing A- and C- esterases seems to be *p*-chloromercuribenzoate though phenylmercuric chloride can be used as an alternative. At a concentration of 100 μM these substances inhibit A-esterase but activate C-esterase to the extent of about 25 per cent. This effect is easily observable under histochemical conditions and Hess and Pearse (1958) showed that in rat kidney a moderate amount of the total indoxyl-esterase activity was due to a C-esterase.

Distinction of Cholinesterases. All the esterase substrates mentioned in this chapter can be hydrolysed by specific and non-specific cholinesterases and three of the methods described have mainly been used for the demonstration of these enzymes. They are the choline ester methods (usually myristoyl choline), the thiocholine methods and the thiolacetic acid methods. Most popular have been the thiocholine methods, on account of their high specificity. Although a number of workers have used the specific substrates (acetylthiocholine, butyrylthiocholine) for distinguishing between the two enzymes, others, myself included, have preferred to use a single substrate in conjunction with the specific inhibitors. As mentioned above, the provision of a selective method of inhibiting the non-specific esterases would allow the use of alternatives to the Koelle method for routine cholinesterase studies. In Table 61, below, are recorded the various inhibitors which have been, and are being, used in cholinesterase histochemistry.

As with the various non-specific esterase inhibitors, those listed in Table 61 are usually employed by pre-incubating sections for 1 hour in buffer before transferring them to the substrate medium. Reversible inhibitors are included in the latter as well as being used for pre-incubation but irreversible inhibitors are omitted. I believe that at the present time acetylcholinesterase is best identified by its ability to hydrolyse acetylthiocholine in the presence of the selective ChE inhibitor isoOMPA. In regions where significant concentrations of non-specific esterase are absent, as in the caudate nucleus for instance, the indoxyl acetate methods can equally well be employed. In

TABLE 61
Cholinesterases
(Esterases Sensitive to 10 μM Eserine)

Enzyme	Substrates	Complete Inhibition by
All cholinesterases	α-naphthyl acetate Acetylthiocholine *O*-acetylindoxyls Myristoylcholine	Eserine, 10 μM Ro3-0422, 0·01 μM E 600, 0·1 μM DFP, 10 μM
Acetylcholinesterase Acetocholinesterase Specific cholinesterase True cholinesterase	α-naphthyl acetate Acetylthiocholine O-acetylindoxyls	DFP, 10 μM Nu-683, 100 μM Nu-1250, 0·01μM 62C47, 10 μM 285C51, 100 μM 284C51, 50 μM Mipafox, 5 μM Ambenonium, ·001 μM
	Myristoyl choline	Poorly hydrolysed
	Butyrylthiocholine	Very slowly hydrolysed
Pseudocholinesterase Propionocholinesterase Butyrocholinesterase	α-naphthyl acetate Acetylthiocholine Butyrylthiocholine Myristoylcholine *O*-acetylindoxyls	DFP, 0·1 μM Nu-683, 0·01 μM Nu-1250, 10 μM isoOMPA, 1 μM NaF, 15–25 μM Ethopropazine HCl, 10 mM

the supraoptic and paraventricular nuclei of the hypothalamus, on the other hand, non-specific esterases are highly active and only the acetylthiocholine method gives a true appreciation of changes in AChE.

Holmstedt (1957a) especially recommended Mipafox as the selective inhibitor of choice for ChE and 284C51 for AChE. As recorded earlier in this chapter, there is little to choose between the latter and 62C47. Both are more than 500 times as effective against AChE as against ChE. Mipafox, which was used as a selective inhibitor for ChE by Davison (1953) and by Denz (1954), is only 10 times as effective against this enzyme as against AChE. This is scarcely adequate. Judging by their I_{50} ratios lysivane and isoOMPA (see p. 770) are very much superior to Mipafox as ChE inhibitors. Furthermore, the concentration of Mipafox with which Holmstedt observed complete inhibition of human serum ChE (4 μM) was found by Pepler and Pearse (1957a) to have no effect on rat brain ChE. Little effect was observed, in fact, with concentrations as high as 100 μM. (See paper by Diegenbach, 1965.)

Discussing the principles of thiocholine methods Holmstedt and Sjöqvist (1961) maintained that Mipafox and BW 284C51 could not be substituted unless satisfactory inhibition curves were presented. Cavanagh and Holland (1961), however, used Mipafox (1–5 μM) as an AChE inhibitor in their studies of cholinesterases in the C.N.S. of the chicken.

Low concentrations of DFP are known to inactivate ChE, leaving AChE relatively unimpaired in most species (Adrian, Feldberg and Kilby, 1947; Hawkins and Mendel, 1947; Adams and Thompson, 1948) and after some years of unpopularity this compound has begun to find regular employment in histochemical studies.

Main and Iverson (1966), using human serum cholinesterase and bovine erythrocyte AChE, found a 150-fold difference in the inhibitory power of DFP for ChE and AChE and this was attributed entirely to differences in affinity (affinity constant Ka at 5° for AChE was $1{\cdot}58 \times 10^{-3}$ M and for ChE it was $9{\cdot}95 \times 10^{-6}$ M)

The reversible inactivation of ChE (human serum) by NaF (15–25 μM) was studied by Cimasoni (1966) who found that the effect was independent of the presence of metal ions. The pI_{50} (mM fluoride) was 0·4 for serum ChE and 2·9 for red cell AChE. Siegel *et al.* (1966) recommended ambenonium chloride (0·001 μM) or edrophonium chloride (1·0 μM) as specific total inhibitors of AChE. Ambenonium is *N,N′-bis* (2-diethylaminoethyl)oxamide-*bis*-2-chlorobenzyl chloride.

Hawkins and Gunter (1946) tested a neostigmine analogue (Nu-683) and found that it gave marked selective inhibition for ChE while Hawkins and Mendel (1949) showed that a similar selective inhibition of AChE could be obtained with Nu-1250, a substituted trimethyl ammonium salt like Nu-683. Both these compounds are effective in histochemical use, but the margin of safety is less than with isoOMPA and 62C47 and I recommended (1960) the use of these last two inhibitors. In view of present difficulties in obtaining 62C47 it is reasonable to substitute 284C51 (50 μM). When isoOMPA is not available ethopropazine HCl (10 μM) is an effective substitute.

Applied Esterase Histochemistry

The majority of applied studies in the esterase field have been concerned with the cholinesterases and the Koelle technique has been almost exclusively employed for this purpose. The list of works which appears below contains some of the more important applications of cholinesterase histochemistry, but it is not intended to be in any sense complete.

General Studies. Gerebtzoff *et al.* (1954), survey of cholinesterases; Röhlich (1956), AChE after embedding in polyethylene glycols, Koelle (1962), Arvy (1964).

Electrophorus (*Electric Eel*). Couceiro *et al.* (1953); Couceiro, d'Almeida and Miranda (1955).

Salivary Glands. Snell and Garrett (1965), Strömblad (1959). *Sweat Glands.* Hurley and Mescon (1956a). *Harderian Gland.* Fourman and Ballantyne (1967).

Motor End Plates and Muscle. Couteaux (1951); Hellmann (1952), Harris (1954), with good list of references, including Russian ones; Mohr and

Gerebtzoff (1954); Dumont and Drouin (1954); Gerebtzoff (1954a); Coers (1954), amyotonia congenita; Snell and McIntyre (1956); Durante (1956, 1957); Zelená and Szentágothai (1956–57); Lundin (1958); Hines (1960); Maynard and Maynard (1960); Hirano (1967); Eränkö and Teräväinen (1967); Teräväinen (1968); Khera and Latham (1965).

Synapses, Ganglia and Nerve Endings. Kupfer (1951); Szentágothai *et al.* (1954–55); Hebb and Hill (1955); Hurley and Mescon (1956a, 1956b); Koelle (1955b); Bergner (1957); Tewari and Bourne (1960); Cauna *et al.*, (1961); Lubinska *et al.* (1961); Cauna (1960); Fredricsson and Sjöqvist (1962); Kokko (1965); Giacobini *et al.* (1967); Lubinska and Zelena (1967); Machida *et al.* (1967). *Retina*. Francis (1953); Hebb *et al.* (1953); Leplat and Gerebtzoff (1956); Esilä (1963); Ehinger (1966). *Liver*. Gerebtzoff (1954a); Gürtner *et al.* (1963). *Congenital Megacolon*. Niemi *et al.* (1961).

Blood and Bone Marrow Cells. Zajicek *et al.* (1954); Davies and Rutland (1956); Rogister (1956).

Developing Embryo. Rossi *et al.* (1964); Kussäther *et al.* (1967); Turbow and Burkhalter (1968).

Central and Peripheral Nervous System. Hull (1954); Koelle (1954, 1955a); Csillik and Sávay (1954); Abrahams *et al.* (1957), biochemical study; Giacobini and Zajicek (1956); Hardwick and Hebb (1956); de Giacomo (1962); Shute and Lewis (1963); Barry and Leonardelli (1967); Adams *et al.* (1967); Rodriguez (1967). *Parasites*. Pepler (1958), miracidium of *Schistosoma mansoni*; Durante (1956), *Ciona intestinalis*; Panitz and Knapp (1967), *Fasciola hepatica*; Douglas (1966), *Schistoma mansoni* and *Hymenolepis diminuta*. *Placenta*. Gerebtzoff (1957). *Thyroid*. Weber (1954); Dejardin (1955); Dumont (1956); Pepler and Pearse (1957b); Sandritter *et al.* (1956); Carvalheira and Pearse (1967). *Skin*. Magnus and Thompson (1954), biochemical study; Bourlond *et al.* (1967), melanocytes. *Adrenal*. Hagen (1955); Allen *et al.* (1957); Eränkö (1958).

Studies of the non-specific esterases and lipases have been less frequently reported in the literature, though the α-naphthyl acetate methods have, in fact, been widely employed in general studies of enzyme histochemistry in various physiological and pathological states. The list of references given below, again incomplete, refers to papers largely or wholly concerned with the non-specific esterases and lipases.

General Studies. Nachlas and Seligman (1949b); Bernsohn *et al.* (1954); Curri (1954); Malaty and Bourne (1954); Verne (1954).

Central Nervous System. Gomori and Chessick (1953); Sávay *et al.* (1953). *Embryos*. Zacks (1954); Reale (1955); Sakae (1955); Rossi, Pescetto and Reale (1957). *Skin*. Wells (1957); Winkelmann and Schmit (1959); Raekallio and Levonen (1963); Campo-Aasen and Pearse (1966), Langerhans cells. *Spleen*. Argyris (1956). *Liver*. Underhay *et al.* (1956); Svanborg (1954); Mietkiewski and Malendowicz (1967), beryllium-treated. *Kidney*. Holt and Withers (1952); Pearse (1953); Holt (1956); Hess and Pearse (1958); Pedersen and Dalgaard (1960). *Hypophysis*. Fand (1955); Pearse (1956). *Thyroid*. Weber (1954). *Connective Tissues*. Steigleder and Löffler (1956a and b). *Salivary Glands*. Hill and Bourne (1954); Burstone (1956). *Lipofuscins*. Gedigk and Bontke (1956). *Teeth*. Yoshioka *et al.* (1960); Itoiz *et al.* (1967). *Blood Cells*. Wachstein and Wolf (1958); Hosoda and Takase (1961); Moloney *et al.* (1960); Monis and Weinberg (1961). *Protozoa and Parasites*. Fennell and Pastor (1958), *Tetrahymena*; Halton (1967), *Fasciola hepatica*. *Testis*. Baust *et al.* (1966). *Prostate*. Frost and Brandes (1967). *Root tips*. Beneš (1962).

REFERENCES

Aarseth, P., Barstad, J. A. B., Rogne, O., and Øksne, S. (1969). *Histochemie*, **15**, 229.
Abe, M., Kramer, S. P., and Seligman, A. M. (1964) *J. Histochem. Cytochem.*, **12**, 364.
Abrahams, V. C., Koelle, G. B., and Smart, P. (1957). *J. Physiol. Lond.*, **139**, 137.
Adams, C. W. M., Grant, R. T., and Bayliss, O. B. (1967). *Brain Res.*, **5**, 366.
Adams, C. W. M., Bayliss, O. B., and Grant, R. T. (1968). *Histochem. J.*, **1**, 1968.
Adams, D. H., and Thompson, R. H. S. (1948). *Biochem. J.*, **42**, 170.
Adrian, E. D., Feldberg, E., and Kilby, B. A. (1947). *Br. J. Pharmac.*, **2**, 56.
Ahlqvist, J. (1963). *Acta path. scand.*, **57**, 353.
Aldridge, W. N. (1953a). *Biochem. J.*, **53**, 110.
Aldridge, W. N. (1953b). *Biochem. J.*, **53**, 117.
Aldridge, W. N. (1953c). *Biochem. J.*, **53**, 62.
Aldridge, W. N. (1954). *Biochem. J.*, **57**, 692.
Aldridge, W. N., and Davison, A. N. (1952a). *Biochem. J.*, **51**, 62.
Aldridge, W. N., and Davison, A. N. (1952b). *Biochem. J.*, **52**, 663.
Allen, J. M., Eränkö, O., and Hunter, R. L. (1957). *Anat. Rec.*, **128**, 514.
Ammon, R. (1943). *Ergebn. Enzymforsch.*, **9**, 35.
Argyris, T. S. (1956). *Anat. Rec.*, **126**, 1.
Arvy, L. (1964). *Revue fr. etudes clin. biol.*, **9**, 234.
Atherton, G. W. (1963). *Histochemie*, **3**, 214.
Augustinsson, K.-B. (1948). *Acta physiol. scand.*, **15**, Suppl. 52.
Augustinsson, K.-B. (1949). *Archs. Biochem.*, **23**, 111.
Augustinsson, K.-B. (1961). *Ann. N.Y. Acad. Sci.*, **94**, 844.
Augustinsson, K.-B. (1963). In "Cholinesterases and Anticholinesterase Agents." Ed. Koelle, G. B., *Handb. exp. Pharmakol.*, Suppl. 15. Springer, Berlin, pp. 101–102.
Augustinsson, K.-B. (1964). *J. Histochem. Cytochem.*, **12**, 744.
Augustinsson, K.-B. and Nachmansohn, D. (1949). *Science, N.Y.*, **110**, 98.
Austin, L., and Berry, W. K. (1953). *Biochem. J.*, **54**, 695.
Barrnett, R. J., and Seligman, A. M. (1951). *Science, N.Y.*, **114**, 576.
Barrowman, J. A., and Borgström, B. (1968). *Gastroenterology*, **55**, 601.
Barry, J., and Leonardelli, J. (1967). *C. r. Soc. Séane. Biol.*, **161**, 561.
Baskys, S. B., Klein, E., and Lever, W. F. (1963). *Archs. Biochem. Biophys.*, **102**, 207.
Baust, P., Goslar, H. G., and Tonutti, E. (1966). *Zellforsch, Mikrosk. Anat.*, **69**, 686.
Bayliss, B. J., and Todrick, A. (1956). *Biochem. J.* **62**, 62.
Bell, C. (1966). *J. Histochem. Cytochem.*, **14**, 567.
Beneš, K. (1962). *Biol. plantarum. (Praha).*, **4**, 211.
Bergmann, F., and Rimon, S. (1958). *Biochem. J.*, **70**, 339.
Bergmann, F., Segal, R., and Rimon, S. (1957). *Biochem. J.*, **67**, 481.
Bergner, A. D. (1957). *Br. J. exp. Path.*, **38**, 160.

BERNSOHN, J., ZELLER, E. A., and COCHRANE, L. S. G. (1954). *Fedn. Proc. Fedn. Am. Socs. exp.*, **13**, 183.
BIRKS, R. I., and BROWN, L. M. (1960). *J. Physiol., Lond.*, **152**, 145.
BLASCHKO, K., CHOU, T. C., and WAJDA, I. (1947). *Br. J. Pharmac.*, **2**, 108, 116.
BOKDAWALA, F. D., and GEORGE, J. C. (1964). *J. Histochem. Cytochem.*, **12**, 768.
BOURLOND, A., WOLFF, K., and WINKELMANN, R. K. (1967). *Nature, Lond.*, **214**, 846
BULL, G., LAWES, M., and LEONARD, M. (1957). *Stain Technol.*, **32**, 59.
BUÑO, W., and MARIÑO, R. G. (1952). *Acta anat.*, **16**, Separatum, p. 85.
BURGEN, A. S. V. (1948). *Br. J. Pharmac.*, **4**, 219.
BURGEN, A. S. V., and HOBBIGER, F. (1951). *Br. J. Pharmac.*, **6**, 593.
BURSTONE, M. S. (1956). *J. Histochem. Cytochem.*, **4**, 130.
BURSTONE, M. S. (1957a). *Archs. Path.*, **63**, 164.
BURSTONE, M. S. (1957b). *J. natn. Cancer Inst.*, **18**, 167.
BURSTONE, M. S. (1962). "Enzyme Histochemistry," Academic Press, New York.
CAMPO-AASEN, I., and PEARSE, A. G. E. (1966). *Med. Cután.*, **1**, 35.
CARBONELL, L. M. (1956). *J. Histochem. Cytochem.*, **4**, 87.
CARVALHEIRA, A. F., and PEARSE, A. G. E. (1967). *Histochemie*, **8**, 175.
CAUNA, N. (1960). *J. Histochem. Cytochem.*, **8**, 367.
CAUNA, N., NAIK, N. T., LEAMING, D. B., and ALBERTI, P. (1961). *Biblphie. anat.* **2**, 90.
CAVANAGH, J. B., and HOLLAND, P. (1961). *Br. J. Pharmac. Chemother.*, **16**, 218.
CHRISTIE, G. A. (1968). *Histochemie*, **12**, 189.
CIMASONI, G. (1966). *Biochem. J.*, **99**, 133.
COERS, C. (1954). *Acta neurol. psychiat. belg.*, **54**, 69.
COLEMAN, M. H., and LELEY, D. D. (1962). *Biochim. biophys. Acta*, **58**, 231.
COPP, F. C. (1953). *J. chem. Soc.*, p. 3116.
COTSON, S., and HOLT, S. J. (1958). *Proc. Roy. Soc. Biol.*, **148**, 507.
COUCEIRO, A., d'ALMEIDA, D. F., and FREIRE, J. R. C. (1953). *Anais. Acad. brasil. Ciencias*, **25**, 205.
COUCEIRO, A., d'ALMEIDA, D. F., and MIRANDA, M. (1955). *Anais. Acad. brasil. Ciencias*, **27**, 49.
COUTEAUX, R. (1942). *Bull. biol. Fr. Belg.*, **76**, 14.
COUTEAUX, R. (1947). *Rev. canad. Biol.*, **6**, 563.
COUTEAUX, R. (1951). *Archs. int. Physiol.*, **59**, 526.
COUTEAUX, R., and TAXI, J. (1952). *Archs. Anat. microsc. Morph. exp.*, **41**, 352.
COUTINHO, H. B., PADILHA, M. C. S., GOMES, J. M., and ALMEIDA ALVES, J. J. (1965). *J. Histochem. Cytochem.*, **13**, 3990.
CREVIER, M., and BÉLANGER, L. F. (1955). *Science, N.Y.*, **122**, 556.
CSILLIK, B., and SÁVAY, G. (1954). *Acta morphol. Hung.*, **4**, 103.
CURRI, S. B. (1954). *Riv. Anat. patol. oncol.*, **8**, 1.
DARNTON, S. J., and BARROWMAN, J. A. (1969). *Histochem. J.*, **1**, 551.
DAVIES, D. R., and RUTLAND, J. P. (1956). *Nature, Lond.*, **178**, 697.
DAVIS, B. J., and ORNSTEIN, L. (1959). *J. Histochem. Cytochem.*, **7**, 297.
DAVISON, A. N. (1953). *Biochem. J.*, **54**, 583.
DAWSON, E. R. (1927). *Biochem. J.*, **21**, 398.
DAWSON, R. M. C. (1956). *Biochem. J.*, **64**, 192.
DAWSON, R. M. C. (1966). In "Essays in Biochemistry." Eds. Campbell, P. N., and Greville, G. D. Acad. Press, London, Vol. 2, p. 69.
DE GIACOMO, P. (1962). *Lav. Neuropsichiat.*, **33**, 1.
DEIMLING, O. (1965). *Histochemie*, **5**, 141.
DEJARDIN, M. (1955). *C. r. Séanc. Soc. Biol.*, **149**, 621.
DELELLIS, R., and FISHMAN, W. H. (1965). *J. Histochem. Cytochem.*, **13**, 297.
DENZ, F. A. (1953). *Br. J. exp. Path.*, **34**, 329.
DENZ, F. A. (1954). *Br. J. exp. Path.*, **35**, 459.
DESNUELLE, P. (1951). *Bull. Soc. Chim. biol.*, **33**, 909.
DESNUELLE, P. (1961). *Adv. Enzymol.*, **23**, 129.
DIEGENBACH, P. C. (1965). *Nature, Lond.*, **207**, 308.
DIXON, M., and WEBB, E. C. (1958). "Enzymes," Longmans, London.
DOUGLAS, L. T. (1966). *Acta histochem.*, **24**, 301.

DREWS, U., KUSSÄTHER, E., and USADEL, K. H. (1967). *Histochemie*, **8**, 65.
DUMONT, L. (1956) *Annls. Endocr.*, **17**, 700.
DUMONT, L., and DROUIN, M. (1954). *C. r. hebd. Séanc. Acad. Sci.*, **238**, 274.
DURANTE, M. (1956). *Experientia*, **12**, 307.
DURANTE, M. (1957). *Acta Embryol. Morphol. exp.*, **1**, 131.
EAPEN, J. (1960). *Stain Technol.*, **35**, 227.
EHINGER, B. (1966). *Acta Univ. lund*, Sect. 2, No. 2.
ENERBÄCK, L., and HANSSON, C. G. (1968). *J. Histochem. Cytochem.*, **16**, 441.
ERÄNKÖ, O. (1958). *Nature, Lond.*, **182**, 183.
ERÄNKÖ, O., KOELLE, G. B., and RÄISÄNEN, L. (1967). *J. Histochem. Cytochem.* **15**, 674.
ERÄNKÖ, O., and TERÄVÄINEN, H. (1967). *J. Histochem. Cytochem.*, **15**, 399.
ESILA, R. (1963). *Acta Ophthalmol.*, Suppl. 77.
EVANS, R. A. and STANSFIELD, D. A. (1961). *Biochem. J.*, **78**, 6P.
FAND, S. B. (1955). *J. clin. Endocr. Metab.*, **15**, 685.
FEIGL, F. (1954). "Spot Tests," Vol. I, p. 76. Elsevier, Amsterdam.
FENNELL, R. A., and PASTOR, E. P. (1958). *J. Morph.*, **103**, 187.
FISCHER, R., HENNEKEUSER, H. H., and KÄUFER, C. (1966). *Klin. Wschr.*, **24**, 1401.
FODOR, P. J. (1950). *Archs. biochem. Biophys.*, **26**, 331.
FOURMAN, J., and BALLANTYNE, B. (1967). *Anat. Rec.*, **159**, 17.
FRANCIS, C. M. (1953). *J. Physiol.*, **120**, 435.
FREDRICSSON, B., and SJÖQVIST, F. (1962). *Acta morph. neerl.-scand.*, **5**, 1.
FROST, J. L., and BRANDES, D. (1967). *J. Histochem. Cytochem.*, **15**, 589.
FULLER, E. G., and OWCZARZAK, A. (1967). *Biol. Bull.*, **133**, 539.
GAWRON, O., and KEIL, J. (1960). *Archs. biochem. Biophys.*, **89**, 293.
GEDIGK, P., and BONTKE, E. (1956). *Z. Zellforsch. mikrosk. Anat.*, **44**, 495.
GEORGE, J. C. (1964). *J. Anim. Morph. Physiol.*, **11**, 233.
GEORGE, J. C., and IYPE, P. T. (1960). *Stain Technol.*, **35**, 151.
GEREBTZOFF, M. A. (1954a). *C. R. Soc. biol., Paris*, **148**, 397.
GEREBTZOFF, M. A. (1954b). *C. r. Séanc. Soc. biol.*, **148**, 632.
GEREBTZOFF, M. A. (1957). *Ann. Histochim.*, **2**, 3.
GEREBTZOFF, M. A., PHILIPPOT, E., and DALLEMAGNE, M. J. (1954). *Acta anat.*, **20**, 234.
GEREBTZOFF, M. A., and MLADENOV, S. (1967). *Acta histochem.*, **26**, 318.
GEYER, G. (1962). *Acta Histochemi*, **14**, 26.
GIACOBINI, E., and ZAJICEK, J. (1956). *Nature, Lond.*, **177**, 185.
GIACOBINI, E., PALMBORG, B., and SJÖQVIST, F. (1967). *Acta physiol. scand.*, **69**, 355.
GLENNER, G. G., HOPSU, V. K., and COHEN, L. A. (1962). *J. Histochem. Cytochem.* **10**, 109.
GOESSNER, W. (1958). *Histochemie*, **1**, 48.
GOLDSTEIN, A., and HAMLISCH, R. E. (1952). *Archs. biochem. Biophys.*, **35**, 12.
GOMORI, G. (1945). *Proc. Soc. exp. Biol. Med.*, **58**, 362.
GOMORI, G. (1946). *Archs. Path.*, **41**, 121.
GOMORI, G. (1948). *Proc. Soc. exp. Biol. Med.*, **68**, 354.
GOMORI, G. (1949). *Ibid.*, **72**, 697.
GOMORI, G. (1950a). In "Menstruation and its Disorders," (E. T. Engle, Edit.).
GOMORI, G. (1950b). *J. Lab. clin. Med.*, **35**, 802.
GOMORI, G. (1951). In "Methods in Medical Research," **4**, 1.
GOMORI, G. (1952a). *Int. Rev. Cytol.*, **1**, 323.
GOMORI, G. (1952b). "Microscopic Histochemistry," Univ. Press, Chicago.
GOMORI, G. (1953). *J. Histochem. Cytochem.*, **1**, 469.
GOMORI, G., and CHESSICK, R. D. (1953a). *J. cell. comp. Physiol.*, **41**, 51.
GOMORI, G., and CHESSICK, R. D. (1953b). *J. Neuropath. exp. Neurol.*, **12**, 387.
GÜRTNER, Th., KREUTZBERG, G., and DOENICKE, A. (1963). *Acta Anaesthesiol. Scand.*, **7**, 69.
HAGEN, P. (1955). *J. Physiol., Lond.*, **29**, 50.
HALTON, S. W. (1967). *J. Parasit.*, **53**, 1210.
HARD, W. L., and PETERSON, A. C. (1950). *Anat. Rec.*, **108**, 57.

HARD, W. L., PETERSON, A. C., and FOX, M. D. (1951). *J. Neuropath. exp. Neurol.*, **10**, 48.
HARDWICK, D. C., and HEBB, C. (1956). *Nature, Lond.*, **177**, 667.
HARRIS, C. (1954). *Am. J. Path.*, **30**, 501.
HAWKINS, R. D., and GUNTER, J. M. (1946). *Biochem. J.*, **40**, 192.
HAWKINS, R. D., and MENDEL, B. (1947). *Br. J. Pharmac.*, **2**, 173.
HAWKINS, R. D., and MENDEL, B. (1949). *Biochem. J.*, **44**, 260.
HEATH, D. F. (1963). Int. Monographs. *Pure and Appl. Biol.* No. 13. "Organophosphorus Poisons". Pergamon, Oxford.
HEBB, C., and HILL, K. J. (1955). *Nature, Lond.*, **175**, 597.
HEBB, C., SILVER, A., SWAN, A. A. B., and WALSH, E. G. (1953). *Q. J. exp. Physiol.*, **38**, 185.
HEILBRON, E. (1954). *Acta chem. scand.*, **8**, 1368.
HELLMANN, K. (1952). *J. cell. comp. Physiol.*, **40**, 421.
HELLMANN, K. (1952). *Nature, Lond.*, **169**, 113.
HENDERSON, J. R. (1967). *Stain Technol.*, **42**, 101.
HERBST, F. (1965). *Acta histochem.*, **22**, 223.
HESS, R., and PEARSE, A. G. E. (1958). *Br. J. exp. Path.*, **39**, 292.
HESS, R., and PEARSE, A. G. E. (1961). *Enzymol. biol. clin.*, **1**, 87.
HILL, C. R., and BOURNE, G. H. (1954). *Acta anat.*, **20**, 116.
HINES, M. (1960). *Anat. Rec.*, **138**, 283.
HIRANO, H. (1967). *Archhistol. jap.*, **28**, 89.
HOBBIGER, E. E. (1957). *Biochem. J.*, **67**, 600.
HOBBIGER, F. (1954). *Br. J. Pharmac.*, **9**, 159.
HOLMES, R. S., and MASTERS, C. J. (1967a). *Biochim. biophys. Acta*, **132**, 379.
HOLMES, R. S., and MASTERS, C. J. (1967b). *Biochim. biophys. Acta*, **146**, 138.
HOLMES, R. S., and MASTERS, C. J. (1968). *Biochim. biophys. Acta*, **151**, 147.
HOLMSTEDT, B. (1951). *Acta physiol. scand.*, **25**, Suppl., 90.
HOLMSTEDT, B. (1957a). *Ibid.*, **40**, 331.
HOLMSTEDT, B. (1957b). *Ibid.*, **40**, 322.
HOLMSTEDT, B. (1959). *Pharmac. Rev.*, **11**, 567.
HOLMSTEDT, B., and SJÖQVIST, F. (1961). *Bibliotheca Anat.*, *Fasc.*, **2**, p. 1.
HOLT, S. J. (1952). *Nature, Lond.*, **169**, 271.
HOLT, S. J. (1956). *J. Histochem. Cytochem.*, **4**, 541.
HOLT, S. J. (1958). In "General Cytochemical Methods," Ed. J. F. Danielli, Vol. I, p. 375.
HOLT, S. J. and SADLER, P. W. (1958a). *Proc. Roy. Soc. B.*, **148**, 481.
HOLT, S. J., and SADLER, P. W. (1958b). *Ibid.*, **148**, 495.
HOLT, S. J., and WITHERS, R. F. J. (1952). *Nature, Lond.*, **170**, 1012.
HOLT, S. J., and WITHERS, R. F. J. (1958). *Proc. Roy. Soc. B.*, **148**, 520.
HOPSU, V. K., and GLENNER, G. G. (1963). *J. Histochem. Cytochem.*, **11**, 520.
HOPSU, V. K., and PONTINEN, J. (1964). *J. Histochem. Cytochem.*, **12**, 853.
HOSODA, S., and TAKASE, S. (1961). *Nature, Lond.*, **190**, 927.
HUGGINS, C., and MOULTON, S. H. (1948). *J. exp. Med.*, **88**, 169.
HULL, S. (1954). *Stain Technol.*, **29**, 219.
HUNTER, R. L., and MARKERT, C. L. (1957). *Science, N.Y.*, **125**, 1294.
HURLEY, H. J., and MESCON, H. (1956a). *Proc. Soc. exp. Biol.*, **91**, 103.
HURLEY, H. J., and MESCON, H. (1956b). *Br. J. Derm. Syph.*, **68**, 290.
ITOIZ, M. E., CARRANZA, F. A., jr., and CABRINI, R. L. (1967). *J. Periodontol.*, **38**, 130.
JOÓ, F., SÁVAY, G., and CSILLIK, B. (1965). *Acta histochem.*, **22**, 40.
KAMBLI, E. (1958). Personal Communication.
KAMBLI, E. (1964). *Helv. chim. Acta*, **47**, 2155.
KARNOVSKY, M. J., and ROOTS, L. (1964). *J. Histochem. Cytochem.*, **12**, 219.
KELSEY, F. E. (1939). *J. biol. Chem.*, **130**, 199.
KHERA, K. S., and LAHAM, Q. N. (1965). *J. Histochem. Cytochem.*, **13**, 559.
KITZ, R., and WILSON, I. B. (1962). *J. biol. Chem.*, **237**, 3245.
KOELLE, G. B. (1950). *J. Pharmac. exp. Ther.*, **100**, 158.
KOELLE, G. B. (1951). *J. Pharmac. exp. Ther.*, **103**, 153.

KOELLE, G. B. (1954). *J. comp. Neurol.*, **100**, 211.
KOELLE, G. B. (1955a). *J. Neuropath. exp. Neurol.*, **14**, 23.
KOELLE, G. B. (1955b). *J. Pharmac.*, **114**, 167.
KOELLE, G. B. (1962). In "Handbuch der experimentellen Pharmakologie Ergänzungswerk". Vol. XV, Springer, Berlin, p. 187.
KOELLE, G. B., DAVIS, R., and DEVLIN, M. (1968). *J. Histochem. Cytochem.*, **16**, 754.
KOELLE, G. B., and FRIEDENWALD, J. S. (1949). *Proc. Soc. exp. Biol. Med.*, **70**, 617.
KOELLE, G. B., and GROMADZKI, C. G. (1966). *J. Histochem. Cytochem.*, **14**, 443.
KOELLE, G. B., and HORN, R. S. (1968). *J. Histochem. Cytochem.*, **16**, 743.
KOELLE, G. B., and KOELLE, F. S., and FRIEDENWALD, J. S. (1950). *J. Pharmac. exp. Ther.*, **100**, 180.
KOKKO, A. (1965). *Acta physiol. scand.*, **66**, Suppl. 261.
KRAMER, S. P., ARONSON, L. D., ROSENFELD, M. G., SULKIN, M. D., CHANG, A., and SELIGMAN, A. M. (1963). *Archs. Biochem.*, **102**, 1.
KUPFER, C. (1951). *J. cell. comp. Physiol.*, **38**, 469.
KUSSÄTHER, E., USADEL, K. H., and DREWS, U. (1967). *Histochemie*, **8**, 237.
LAKE, B. D. (1970). *J. Pediatrics*, **76**, 262.
LEDER, L. D. (1964). *Klin. Wschr.*, **42**, 553.
LEPLAT, G., and GEREBTZOFF, M. A. (1956). *Ann. Oculist.*, **189**, 121.
LOEVENHART, A. S. (1906-7). *J. biol. Chem.*, **2**, 427.
LUBIŃSKA, L., NIEMIERKO, S., and ODERFELD, B. (1961). *Nature, Lond.*, **189**, 122.
LUBIŃSKA, L., and ZELENA, J. (1967). *J. Anat.*, **101**, 295.
LUNDIN, S. J. (1958). *Experientia*, **14**, 131.
MACHIDA, H., GIACOMETTI, L., and ALLEGRA, F. (1967). *J. invest. Dermatol.*, **48**, 304.
MAGNUS, I. A., and THOMPSON, R. H. S. (1954). *Br. J. Derm. Syph.*, **66**, 163.
MAIN, A. R., and IVERSON, F. (1966). *Biochem. J.*, **100**, 525.
MALATY, H. A., and BOURNE, G. H. (1954). *Acta Anat.*, **21**, 249.
MALMGREN, H., and SYLVÉN, B. (1955). *J. Histochem. Cytochem.*, **3**, 441.
MARK, D. D. (1950). *Archs Path.*, **49**, 548.
MARKERT, C. L., and HUNTER, R. L. (1959). *J. Histochem. Cytochem.*, **7**, 42.
MARPLES, E. A. and THOMPSON, R. H. S. (1960). *Biochem. J.*, **74**, 123.
MARTIN, B. F. (1953a). *J. Physiol.*, **119**, 24p.
MARTIN, B. F. (1953b). *Nature, Lond.*, **172**, 1048.
MARTINEZ RODRÍGUEZ, R., RIBA SOTO, A., and MOYA MANGAS, J. (1964). *Trab. Inst. Cajal Madrid*, **56**, 27.
MAYNARD, E. A., and MAYNARD, D. M. (1960). *J. Histochem. Cytochem.*, **8**, 376.
MAZUR, A., and BODANSKY, O. (1946). *J. biol. Chem.*, **163**, 261.
MEISTER, H. (1965). *Zbl. allg. Path. Path. Anat.*, **108**, 105.
MENDEL, B., and HAWKINS, R. D. (1947). *Biochem. J.*, **41**, xxii.
MENDEL, B., and RUDNEY, H. (1943). *Science, N.Y.*, **98**, 201.
MENDEL, B., and RUDNEY, H. (1944). *Ibid.*, **100**, 499.
MENDEL, B., MUNDELL, D. B., and RUDNEY, H. (1943). *Biochem. J.*, **37**, 473.
MIETKIEWSKI, K., and MALENDOWICZ, L. (1967). *Histochemie*, **11**, 223.
MLADENOV, S. (1965). *C. R. Acad. bulg. Sci.*, **18**, 1047.
MOHR, E., and GEREBTZOFF, M. A. (1954). *Acta anat.*, **22**, 143.
MOLONEY, W. C., MCPHERSON, K., and FLIEGELMAN, L. (1960). *J. Histochem. Cytochem.*, **8**, 200.
MONIS, B., and WEINBERG, T. (1961). *Cancer*, **14**, 369.
MOUNTER, L. A., ALEXANDER, H. C., 3rd., TUCK, K. D., and DIEN, L. T. H. (1957). *J. biol. Chem.*, **226**, 867.
MOUNTER, L. A., and WHITTAKER, V. P. (1953). *Biochem. J.*, **54**, 551.
MUMENTHALER, M., and ENGEL, W. K. (1961). *Acta anat.*, **47**, 274.
MYERS, D. K. (1953). *Biochem. J.*, **55**, 67.
MYERS, D. K. (1960). "Carboxyl Ester Cleavage." In "The Enzymes". Vol. **4**, p. 475. Eds. Boyer, P. D., Lardy, H., and Myrbäck, K. Academic Press, New York.
MYERS, D. K., SCHOTTE, A., and MENDEL, B. (1955a). *Biochem. J.*, **60**, 481.
MYERS, D. K., SCHOTTE, A., BOER, H., and BORSJE-BAKKER, H. (1955b). *Ibid.*, **61**, 521.

Nachlas, M. M., and Seligman, A. M. (1949a). *J. natn. Cancer Inst.*, **9**, 415.
Nachlas, M. M., and Seligman, A. M. (1949b). *Anat. Rec.*, **105**, 677.
Nachmansohn, D. (1940). *Nature, Lond.*, **145**, 513.
Nachmansohn, D., and Rothenberg, M. A. (1945). *J. biol. Chem.*, **158**, 653.
Nakamura, T., Namba, T., and Grob, D. (1967). *J. Histochem. Cytochem.*, **15**, 276.
Niemi, M., Kouvalainen, K., and Hjelt, L. (1961). *J. Path. Bact.*, **82**, 363.
Ord, M. G., and Thompson, R. H. S. (1950). *Biochem. J.*, **46**, 346.
Ottolenghi, A., Pickett, J. P., and Greene, W. B. (1966). *J. Histochem. Cytochem.*, **14**, 907.
Palkama, A. (1961). *Ann. Med. exp. Biol. Fenn.*, **39**, 273.
Panitz, E., and Knapp, S. E. (1967). *J. Parasit.*, **53**, 354.
Patrick, A. D., and Lake, B. D. (1969). *Nature, Lond.*, **222**, 1067.
Patrick, A. D., and Lake, B. D. (1969). *Biochem. J.* **112**, 29p.
Pavlic, M. (1967). *Biochim. biophys. Acta*, **139**, 133.
Pearse, A. G. E. (1953). *J. Path. Bact.*, **66**, 331.
Pearse, A. G. E. (1954). *Int. Rev. Cytol.*, **3**, 329.
Pearse, A. G. E. (1956)., *J. Path. Bact.*, **72**, 471.
Pearson, B., Defendi, V. (1957). *J. Histochem. Cytochem.*, **5**, 72.
Pearson, C. K. (1963). *J. Histochem. Cytochem.*, **11**, 664.
Pedersen, K. J., and Dalgaard, O. Z. (1960). *Acta. path. scand.*, **50**, 228.
Pepler, W. J. (1958). *J. Histochem. Cytochem.*, **6**, 139.
Pepler, W. J., and Pearse, A. G. E. (1957a). *J. Neurochem.*, **1**, 193.
Pepler, W. J., and Pearse, A. G. E. (1957b). *Br. J. exp. Path.*, **38**, 221.
Piantelli, A., Rebollo, M. A. (1967) *Acta histochem.*, **26**, 1.
Raekallio, J., and Levonen, E. (1963). *Ann. Med. exp. Biol. Fenn.*, **41**, 305.
Ravin, H. A., Zacks, S. I., and Seligman, A. M. (1953). *J. Pharmac. exp. Ther.* **107**, 37.
Reale, E. (1955). *Riv. Istochim.*, **1**, 551.
Richterich, R. (1951). *Enzymologia*, **15**, 40.
Rodriguez, R. M. (1967). *Acta histochem.*, **27**, 172.
Rogister, G. (1956). *Revue fr. Étud. clin. biol.*, **1**, 1078.
Röhlich, P. (1956). *Nature, Lond.*, **178**, 1398.
Rossi, F., Pescetto, G., and Reale, E. (1957). *Ann Histochim.*, **2**, 113.
Rossi, F., Schippers, M., and Grossi, C. E. (1964). *Histochemie*, **4**, 222.
Sakae, Y. (1955). *Riv. Istochim.*, **2**, 5.
Sandritter, W., Kummer, E., Pillat, G., and Rowe, L. (1956). *Klin. Wschr.*, **34** 871.
Saunders, L., and Thomas, I. L. (1958). *J. chem. Soc.*, 483.
Sávay, G., Csillik, B., and Bondray, O. (1953). *Acta morphol.*, **3**, 207.
Sávay, G., and Csillik, B. (1959a). *Experientia*, **15**, 396.
Sávay, G., and Csillik, B. (1959b). *Acta histochem.*, **6**, 307.
Sawyer, C. H., and Hollinshead, W. H. (1945). *J. Neurophysiol.*, **8**, 137.
Schaefer, H. E., and Fischer, R. (1966). *Zbl. allg. Path. Path. Anat.*, **109**, 397.
Schaffer, N. K., May, S. C. jr., and Summerson, W. H. (1953). *J. biol. Chem.*, **202** 69.
Schaffer, N. K., May, S. C. jr., and Summerson, W. H. (1954). *Ibid.*, **206**, 201.
Schiffer, D., Fabiani, A., and Monticone, G. F. (1967). *Acta neuropath.*, **9**, 316.
Seligman, A. M., and Nachlas, M. M. (1950). *J. clin. Invest.*, **29**, 31.
Shnitka, T. K., and Seligman, A. M. (1961). *J. Histochem. Cytochem.*, **9**, 504.
Shute, C. C. D., and Lewis, P. R. (1963). *Nature, Lond.*, **199**, 1160.
Siegel, G. J., Lehrer, G. M., and Silides, D. (1966). *J. Histochem. Cytochem.*, **14**, 473.
Siurala, M. (1956). *Acta Path. scand.*, **39**, 268.
Sneath, P. H. A. (1950). *Nature, Lond.*, **166**, 699.
Snell, R. S., and Garrett, J. R. (1956). *Nature, Lond.*, **178**, 1177.
Snell, R. S., and McIntyre, N. (1956). *Br. J. exp. Path.*, **37**, 44.
Steigleder, G. K., and Löffler, H. (1956a). *Archs klin. exp. Derm.*, **203**, 41.
Steigleder, G. K., and Löffler, H. (1956b). *Ibid.*, **203**, 448.
Stowell, R. E., and Lee, C. S. (1950). *Archs Path.*, **50**, 519.

STRÖMBLAD, B. C. R. (1959). *Experientia*, **15**, 426.
SU, H. C. F., and TSOU, K. C. (1960). *J. Amer. chem. Soc.*, **82**, 1187.
SVANBORG, A. (1954). *Acta med. scand.*, **149**, 349.
SZENTÁGOTHAI, J., DONHOFFER, Á., and RAJKOVITS, K. (1954–55). *Acta Histochem.* **1**, 272.
TERÄVÄINEN, H. (1968). *Histochemie*, **12**, 307.
TEWARI, H. B., and BOURNE, G. H. (1960). *Expl. Cell Res.*, **21**, 245.
THYBUSCH, D., BROSOWSKI, K. H., and WOOHSMANN, H. (1966). *Acta histochem.*, **23**, 127.
TSOU, K. C., and SU, H. C. F. (1963). *J. Histochem. Cytochem.*, **11**, 561.
TURBOW, M. M., and BURKHALTER, A. (1968). *Devel. Biol.*, **17**, 233.
UMBRATH, G., and HELLAUER, H. F. (1948). *Pflügers. Arch. ges. Physiol.*, **250**, 737.
UNDERHAY, E., HOLT, S. J., BEAUFAY, H., and DE DUVE, C. (1956). *J. biophys. biochem. Cytol.*, **2**, 635.
VANHA-PERTTULA, T. (1966). *Acta physiol. scand.*, **69**, Suppl. 283.
VERNE, J. (1954). *Ann. Biol.*, **30**, 129.
VILLIGER, V. (1909). *Ber. dt. chem. Ges.*, **42**, 3529.
WACHSTEIN, M. (1946). *J. exp. Med.*, **84**, 25.
WACHSTEIN, M., and WOLF, G. (1958). *J. Histochem. Cytochem.*, **6**, 457.
WACHSTEIN, M., MEISEL, E., and FALCON, C. (1961). *J. Histochem. Cytochem.*, **9**, 325.
WALSH, R. R., and NIELSEN, T. W. (1960). *Biochim. biophys. Acta*, **44**, 280.
WEBB, E. C. (1948). *Symp. biochem. Soc.*, **2**, 50.
WEBER, G. (1954). *Arch. de Vecchi*, **22**, 929.
WELLS, G. C. (1957). *Br. J. Derm. Syph.*, **69**, 415.
WHITTAKER, V. P. (1951). *Physiol. Rev.*, **31**, 312.
WILSON, I. B. (1951). *Biochim. biophys. Acta*, **7**, 520.
WILSON, I. B., and BERGMANN, F. (1950a). *J. biol. Chem.*, **135**, 479.
WILSON, I. B., and BERGMANN, F. (1950b). *J. biol. Chem.*, **136**, 683.
WILSON, I. B., and ALEXANDER, J. (1962). *J. biol. Chem.*, **237**, 1323.
WINKELMANN, R. K., and SCHMIT, R. W. (1959). *Archs Dermat.*, **80**, 543.
YOSHIOKA, W., MORI, M., MIZUSHIMA, T., and AMATSU, T. (1960). *Arch. histol. jap.*, **20**, 523.
ZACKS, S. I. (1954). *Anat. Rec.*, **118**, 509.
ZACKS, S. I., and BLUMBERG, J. M. (1961). *J. Histochem. Cytochem.*, **9**, 317.
ZAJICEK, J., SYLVÉN, B., and DATTA, N. (1954). *J. Histochem. Cytochem.*, **2**, 115.
ZFLENÁ, J., and SZENTÁGOTHAI, J. (1956–7). *Acta histochem.*, **3**, 84.
ZENKER, W. (1964). *Acta histochem.*, **19**, 67.

CHAPTER 18

GLYCOSIDE HYDROLASES AND GLYCOSYLTRANSFERASES

Introduction

Two main groups of enzymes are dealt with in this chapter. The first, the glycoside hydrolases, contains such enzymes as the α- and β-amylases, amylo-1, 6-glucosidases, muramidase, neuraminidase, β-acetylamino-glucosidase, α- and β-glucosidases and galactosidases, and the β-glucuronidases. From the histochemical point of view the most important is the last of these but methods exist for several others, including α- and β-glucosidase, β-galactosidase, acetyl-β-glucosaminidase and disaccharidases.

The second group contains a large number of enzymes capable of transferring glycosyl groups and some of these can build up long chain polymers of the type used in vertebrate and invertebrate tissues for carbohydrate storage and for structural purposes (starch, dextran, chitin). Out of the 19 enzymes in this group histochemical methods are available for 3. These are α-1, 4-glucan: orthophosphate glucosyltransferase (α-glucan phosphorylase), commonly referred to as phosphorylase, UDP glucose: α-1, 4-glucan α-4-glucosyltransferase (α-glucan-UDP glucosyltransferase), known histochemically as UDPG transglycosylase, and α-1, 4-glucan: α-1, 4-glucan 6-glycosyltransferase (α-glucan-branching glycosyltransferase), commonly called branching enzyme.

A third group of transferring enzymes, the phosphotransferases, is also considered here because the sole histochemical example, phosphoglucomutase, is closely associated with the synthesis and breakdown of glycogen.

The Glucoside Hydrolases. I Glucuronidases

Biochemistry of the β-D-glucuronide glucuronohydrolases.

These enzymes are commonly known as β-glucuronidases. They form a group of enzymes with specificity for the β-glycoside linkage of a variety of naturally occurring and synthetic glucuronides. The earliest studies of any importance were made by Oshima (1934, 1936), using ox spleen and menthyl glucuronide as substrate. In the next few years the β-glucuronidases came to occupy an important position in the biochemical literature and this early work was adequately summarized by Fishman (1950). It was originally suggested that the β-glucuronidases might have three functions in the body: (1) the conjugation of steroid hormones, (2) the hydrolysis of conjugated glucuronides, and (3) participation in cellular proliferation. Fishman (1940) was of the opinion that the β-glucuronidases played a synthetic role in the body and later (1947) he proposed that these enzymes catalysed the conjugation of oestrogens with glucuronic acid, a process regarded as one of "metabolic conjugation", essential for the function of such steroids, and not simply a detoxication. Levvy (1947, 1948) took the opposite view, that the β-glucuronidases were concerned solely with the hydrolysis of glucuronides, and strong evidence against Fishman's theory was given by Karunairatnam and Levvy (1949), who showed that saccharic acid, which inhibits β-glucuronidases, failed to influence the formation of glucuronides by mouse liver slices *in vitro*. Levvy, Kerr and Campbell (1948) related the β-glucuronidase activity of mouse liver and kidney to the degree of cellular proliferation which was taking place, and Kerr, Campbell and Levvy (1949) similarly showed that uterine β-glucuronidase, in the mouse, was an index of the growth of that organ. Hollinger and Rossiter (1952), in their studies on regenerating nerve, found a peak of cellular proliferation (DNA) at 4 days and a peak enzyme activity at 16 days. At the same time Fishman and his co-workers (Fishman and Anlyan, 1947; Fishman, Anlyan and Gordon, 1947; Fishman and Bigelow, 1950) reported increased amounts of β-glucuronidase in cancerous tissues which they regarded as characteristic of neoplasia. Mills and Smith (1951), on the other hand, believed that the raised β-glucuronidase activity of tumours was related to a change in cell type rather than to the neoplastic change itself. They showed that rat hepatomas produced by feeding dimethylaminoazobenzene had a lower β-glucuronidase activity than normal rat liver.

These early differences of opinion were slow to be resolved. Pellegrino and Villani (1956) observed a rise in β-glucuronidase in lymphatic tissues during fasting, a process accompanied by marked atrophy of the tissues concerned. A later study by these authors (1957) showed a similar atrophy and a similar rise in β-glucuronidase in tissues following irradiation. Pellegrino and Villani considered that their results indicated a correlation between atrophy and β-glucuronidase activity. Further work designed to test the cellular proliferation hypothesis of Levvy and his associates was undertaken by Mills, Paul and Smith (1953b). These authors measured enzyme activity and DNA concentration in rat liver from the 16th day of embryonic life until the 220th day of adult life. From a very low level at birth activity rose to a maximum between the 25th and 30th days, falling slightly thereafter to reach a constant adult level. Using the allometric method of analysis (Huxley, 1924) they compared β-glucuronidase levels with DNA content and found two phases of activity. In the first phase the enzyme increased faster than the number of cells but in the second phase the two rates were identical. These results suggest a species difference in the behaviour of liver β-glucuronidase as between the mouse (Levvy *et al.*, 1948) and the rat (Mills *et al.*, 1953b). A study of β-glucuronidase activity during the early development of *Xenopus laevis* embryos was made by Billett (1957). He found that a slight fall in enzyme levels occurred at the time of formation of the neural plates but could find no evidence to connect β-glucuronidase with cell proliferation.

The hypothesis of Fishman (1947), that β-glucuronidase activity was related to steroid metabolism, depended on the transferase activity of the enzyme which was later demonstrated *in vitro* by Fishman and Green (1957). The subsequent demonstration that glucuronide synthesis was unaffected by inhibition of β-glucuronidase, using the specific inhibitor saccharo-1, 4-lactone (see below), apparently excludes the participation of the enzyme in transferase activities. There is no doubt, however, that the natural substrates of the enzyme include steroid glucuronides and also the acyl glucuronide formed from bilirubin (conjugated bilirubin; Chapter 26, p. 1071), first described by Billing, Cole and Lathe (1957).

The observation by Levvy, McAllan and Marsh (1958) that the preputial gland of the female rat is the richest known source of β-glucuronidase is of considerable interest and it provides a convenient histochemical control tissue (see Fishman and Baker, 1956). The connection between atrophic changes and high β-glucuronidase levels is no more surprising than the rise of acid phosphatase which occurs under similar circumstances and which is apparently due to the release of enzyme from intracellular organelles. During pregnancy and growth of the rat uterus Woessner (1965) noted a fall in β-glucuronidase activity. With post-partum involution he found a 3–4 fold rise in the level of enzyme. The majority of experimental work at present clearly favours a catabolic role for β-glucuronidase.

pH. *Optima.* Enzymes hydrolysing β-glucuronides have their optimum pH in the acid range. Mills (1948) described two glucuronidases in ox spleen, one having its optimum at pH 4·5, the other at pH 5·2, and Kerr *et al.* (1949) found both enzymes in mouse spleen but only the first in mouse uterus.

In contrast to the mammalian enzymes the β-glucuronidase of the limpet (*Patella vulgata*), has a broad pH optimum in the region of pH 4·0 (Dodgson *et al.*, 1953) and the position of the peak can be altered, without marked change in optimum activity, by varying the concentration of buffer or substrate (Smith and Mills, 1953; Levvy, 1954; Levvy, Hay and Marsh, 1957). Histochemical tests are usually carried out in the region 5·0–5·2 and no reaction is obtainable at pH 4·5, with any of the available methods. Since the original work of Mills it has become apparent that the overall activity of ox spleen extracts cannot be explained on the basis of two glucuronidases. Further work by Mills, Paul and Smith (1953a) showed that there were three fractions, with optima at pH 3·4, 4·5 and 5·2 and examination of the effects of variation in substrate concentration on reaction velocity, and other considerations, allowed the conclusion that the three fractions could be considered as separate enzymes. A full list of known sources of β-glucuronidases, together with their relative activities with different substrates, was provided by Levvy and Marsh (1959). There still remains, obviously, a wide gulf between biochemical and histochemical studies in the field of mammalian β-glucuronidases.

Activators and Inhibitors. The earliest known inhibitors of the β-glucuronidases were organic acids such as citric acid (Oshima, 1936). Later Karunairatnam and Levvy (1949) showed that mucic and saccharic acids could inhibit the enzyme and that D-saccharic acid, the stronger of the two, acted competitively at pH 5·2. Becker and Friedenwald (1949) found that ascorbic acid and heparin were glucuronidase inhibitors and Spencer and Williams (1951) demonstrated inhibition of the enzyme by glucuronate at pH 5·2.

In their comprehensive study (1953a) Mills and his associates found that citric acid would, in fact, inhibit only the two enzymes active at pH 3·4 and 4·5. This is shown in Fig, 170, below.

The most effective organic acids were saccharate (10^{-4}M) mucate (10^{-3}M) and oxalate (10^{-2}M), in that order but Karunairatnam and Levvy (1949) observed that saccharo-1:4-lactone was more effective than saccharate itself as a competitive inhibitor for *β*-glucuronidase. Later Levvy (1952) showed that the high activity of saccharate was due to traces of the lactone present in

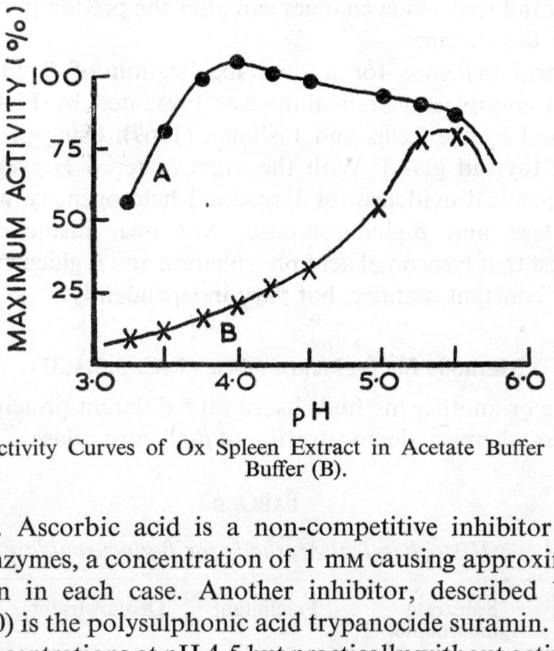

FIG. 170. pH Activity Curves of Ox Spleen Extract in Acetate Buffer (A) and Citrate Buffer (B).

most samples. Ascorbic acid is a non-competitive inhibitor for all three mammalian enzymes, a concentration of 1 mM causing approximately 50 per cent inhibition in each case. Another inhibitor, described by Wills and Wormall (1950) is the polysulphonic acid trypanocide suramin. This is inhibitory in low concentrations at pH 4·5 but practically without activity at pH 5·2. Above this level activation is observed. Inhibition by suramin is noncompetitive. Mills *et al.* (1953a) observed that 15 per cent formaldehyde caused only 15 per cent inhibition of the fractions with optima at pH 4·5 and 5·2. Activation of the enzyme by phthalate was noted by Smith and Mills (1953), and a specific activation of purified spleen *β*-glucuronidase by DNA was reported by Bernfeld, Guarino and Fishman (1950). Later work, summarized by Levvy, McAllan and Marsh (1958), showed that *β*-glucuronidase at high dilutions required the presence of proteins for full activity. They did not regard DNA as a satisfactory substitute for protein since its activity was influenced by other tissue constituents and also by the pH.

Intracellular Localization. According to de Duve, Wattiaux and Baudhuin (1962) the principal localization of β-glucuronidase is lysosomal and this fact is generally accepted. There are several reports, however, which indicate a bimodal distribution of the enzyme (Van Lancker, 1964; Shibko and Tappel, 1965). At an early stage in the development of the lysosome concept de Duve and Beaufay (1959) showed that in the left lobe of the (rat) liver, after cutting off its blood supply, there was a rise in unsedimentable (soluble) forms of the acid hydrolases. Weber (1957) had earlier demonstrated lysosomal rupture during the resorption phase in the tadpole tail and Schrieb (1963) showed similar changes in the regressing Mullerian ducts of the chick embryo. Rapid increases in the amounts of both free and lysosomal β-glucuronidase were found by Greenbaum *et al.* (1965) during the early phases of involution in the rat mammary gland. It is thus reasonable to conclude that many physiological or experimentally induced tissue changes can alter the predominantly lysosomal localization of the enzyme.

Histochemical evidence for a dual localization of β-glucuronidase in lysosomes and endoplasmic reticulum was presented by Fishman and De Lellis (1967) and by De Lellis and Fishman (1967), using propylthiouracil-stimulated rat thyroid gland. With the same material Herveg *et al.* (1966) provided biochemical evidence for lysosomal heterogeneity with respect to acid phosphatase and β-glucuronidase. My own histochemical studies certainly suggest that lysosomal acid phosphatase and β-glucuronidase are not related in any constant manner, but vary independently.

Methods for β-glucuronidase (E.C. 3.2.1.31)

At one time or another methods based on 5 different principles have been used for the histochemical demonstration of β-glucuronidases. These methods

TABLE 62

Histochemical Methods for β-glucuronidase

Type of Method	Substrate (glucuronide)	Precipitant	Demonstrator	Author
Non-coupling azo	1-*o*-hydroxyphenyl-azo-2-naphthyl	Nil	Nil	Friedenwald & Becker (1948)
Non-azo dye	8-hydroxyquinolyl	Fe^{3+}	Perls' reagent	Friedenwald & Becker (1948)
Simultaneous coupling azo, A	1- or 2-naphthyl	Fast blue B	Nil	Seligman *et al.* (1949)
Simultaneous coupling azo B	8-hydroxyquinolyl	Fast blue RR	Nil	Pearse (unpublished)
Simultaneous coupling azo C	AS-LC naphthyl	Fast Garnet GBC	Nil	Pugh & Walker (1961)
Indigogenic A	5-bromoindoxyl	Nil (O_2)	Nil	Holt (unpublished)
Indigogenic B	5-bromo-4-chloro indoxyl	Nil (O_2)	Nil	Pearson *et al.* (1967)

fall into the five main divisions of Table 62, where they are classified as: (1) *non-coupling azo dye* methods; (2) *non-azo dye* methods; (3) *simultaneous coupling azo dye* methods; (4) *post-coupling azo dye* methods; and *indigogenic* methods. These terms have already been used in Chapters 13 to 16 (Vol I) with reference to the azo dye methods.

Non-coupling Azo Dye Method

This method, the first of two described in the same paper by Friedenwald and Becker (1948), comes under the heading of a non-coupling azo dye technique. It employed as substrate the glucuronide of a water-insoluble azo dye, 1-*o*-hydroxyphenylazo-2-naphthol, and the sites of glucuronidase activity were supposedly indicated by the precipitation of a bright red dye. Campbell (1949), however, showed that there was no appreciable difference in staining intensity between active control sections and sections previously incubated with, or incubated in the presence of, the specific inhibitor saccharic acid. Burton and Pearse (1952) confirmed this finding and showed additionally (1) that with heat-inactivated sections incubation with the substrate produced results comparable with those in active controls, and (2) that in active or inactivated sections a dilute alcoholic solution of the reaction product (1-*o*-hydroxyphenolazo-2-naphthol) produced a similar result. This method does not localize, or even necessarily demonstrate, β-glucuronidase in the tissues.

The Ferric Hydroxyquinoline Method

The second method produced by Friedenwald and Becker (1948) for the demonstration of β-glucuronidase depended on the hydrolysis of 8-hydroxyquinoline glucuronide, by unfixed tissue sections, in the presence of a ferric salt; it was postulated that the latter combined with the liberated 8-hydroxyquinoline to form an insoluble greenish-black precipitate of ferric hydroxyquinoline. This was subsequently converted to Prussian blue by means of Perls' acid-ferrocyanide reagent, partly to enable permanent preparations to be made by converting the product into an alcohol insoluble compound, and partly because the initial greenish-black colour was poorly visible under the microscope. In the performance of the reaction several complicating factors were found. The incubating medium had to be freshly prepared on the day of use and immediately prior to use it had to be filtered in an incubator at 37°. Campbell (1949) found that unless a very fine filter was used (Whatman No. 40 paper), and double filtration carried out, trouble was still experienced with a fine suspension of ferric hydroxyquinoline, which tended to precipitate all over the slide. The original authors observed that diffusion of the ferric hydroxyquinoline reaction product caused a precipitate which covered the whole surface of the slide, but this was overcome, according to their statements, by saturating the incubating medium with the reaction product before starting. The incubation time was given as 3–18 hours and for most tissues the longer period was required. Since 8-hydroxyquinoline glucuronide is rapidly hydrolysed by spleen β-glucuronidase *in vitro*, and in the coupling azo dye method described above, the long incubation necessary with this method was due to inhibition of the enzyme by ferric salts. Many other metallic ions were tried by the original authors as capture reagents for 8-hydroxyquinoline but only ferric ions gave a satisfactory result.

This method is now of historical interest only. Because it was employed for a very substantial proportion of all studies on β-glucuronidase carried out between 1952 and 1962, it is still necessary to consider it here in some detail.

A critical study of the ferric hydroxyquinoline method was made by Burton and Pearse (1952). Initial experiences with the method were disappointing since it appeared capricious and at best capable of giving only very coarse localization of enzyme activity. Despite saturation of the substrate solution with ferric hydroxyquinoline, as recommended by the original authors, surface precipitation was found to be troublesome. The preparations appeared dirty as though the precipitate had been deposited at random. The whole aspect of the Friedenwald-Becker method was altered by the studies of Fishman and Baker (1956). A capricious and uncertain technique was converted into one of reasonable reliability which could apparently be used with confidence for studies in the applied histochemistry of the glucuronidases. (See Plate XVIIa, opposite).

Modifications of the ferric hydroxyquinoline method were used by a number of workers between 1956 and 1962. Hayashi and his associates (1955, 1956–57) studied changes in the distribution of uterine β-glucuronidase during the oestrus cycle in the rat and the localization of β-glucuronidase in the digestive gland of the Roman snail was studied by Billett and McGee-Russell (1955). Later (1956) these authors studied the distribution of the enzyme in newt liver. They found, especially in the latter case, that the precipitate was coarse and macrocrystalline after 1–4 hours' incubation and considered that the final stage of conversion to Prussian blue was unnecessary and dangerous. Billett and McGee-Russell also noted that saccharate (5 mM) failed to inhibit the enzyme in newt liver, and in mouse liver and kidney, when the pH was below 4, although it did so in other mammalian tissues and in snail digestive gland. They felt that the normal concentration of Fe^{3+} ions in the medium was inhibitory to the enzyme and that by reacting with Fe^{3+} saccharate might cause an apparent activation. It was concluded that the ferric hydroxyquinoline method could not give a quantitative appreciation of enzyme activity since Fe^{3+} adsorption might vary greatly between different tissues and cells. My conclusion, recorded in the 2nd Edition of this book (1960) was that "the necessary critical studies of the method have yet to be made".

The necessary critical studies were made by Janigan and Pearse (1962) and these showed that staining with the ferric hydroxyquinoline method was due

PLATE XVII

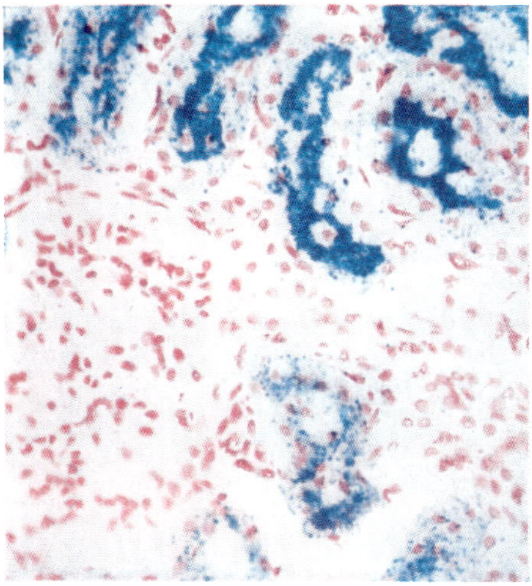

XVIIa. Rat kidney (10 μ frozen section). β-glucuronidase in proximal (above) and distal (lower right) convoluted tubules. Modified Friedenwald and Becker method. × 270.

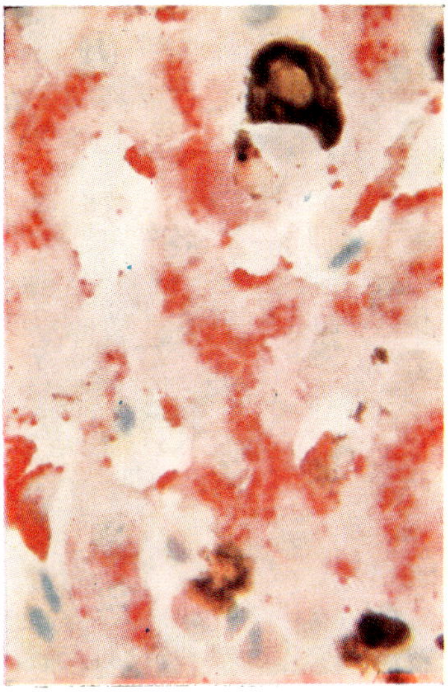

XVIIb. Newt (*Triton cristatus*) liver. Fixed in cold formalin. Cryostat section (6 μ). Pericanalicular distribution of β-glucuronidase shown by the Naphthol ASBI–HPR technique. Enzyme activity is also seen in the Kupfer cells in the sinusoids. × 410.

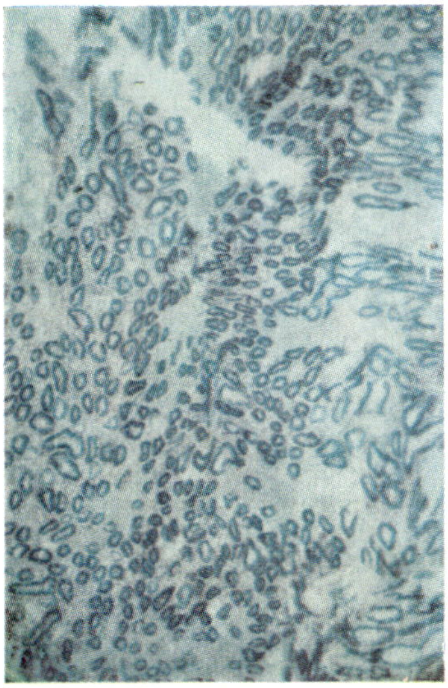

XVIIc. Rat kidney, 8 μ cold formalin-fixed section. Shows distribution of β-glucuronidase by the post-coupling method. × 100.

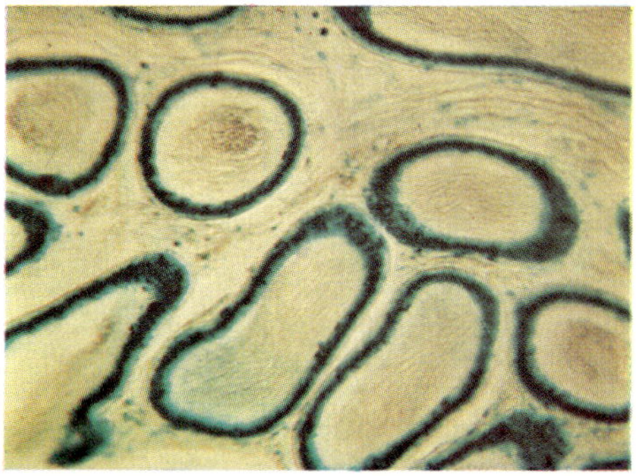

XVIId. Rat epididymis. Fixed in cold 3 per cent buffered glutaraldehyde. Shows localization of β-galactosidase by an indigogenic technique. × 150.

[*To face p.* 814

to the uptake of ferric iron from the complex equilibrium mixture of the incubating medium. This uptake was shown to be influenced by a number of factors but, essentially, it was dependant on the nature of the tissue. The recorded conclusion was that the staining product of the method was uninfluenced by the presence or absence of the enzyme. This conclusion was not quite accurate, as it stood, since the experimental work had shown that *any* change in the equilibrium of the reaction mixture would produce a deposit. While enzyme activity might thus be one of the factors producing the necessary change, the product was not necessarily, or even usually, related to enzyme localization.

Very great exception was taken to the conclusions recorded by Janigan and Pearse (see Fishman, Goldman and Green, 1964) but they were nevertheless supported by a number of independent observers. Studying the adrenal cortex in corticosteroid treated rats Bots *et al.* (1965), for example, found that substitution of the glucuronide by the equivalent of 8-hydroxyquinoline failed to affect their results at all.

While some aspects of the Fishman-Baker technique still lacked an acceptable explanation, from 1964 onwards this method ceased to be employed even by those who supported its specificity with such vigour.

Simultaneous Coupling Azo Dye Methods

Naphthyl Glucuronides. Seligman *et al.* (1949) reported attempts to localize *β*-glucuronidase using the above principle and 1- or 2-naphthyl glucuronide as substrate. Although these authors stated that hydrolysis of 1-naphthyl glucuronide was too slow to allow the establishment of a histochemical technique, I found that it was possible to obtain coloured precipitates, using this substrate at pH 5·0 and 37° for 4–16 hours, with a variety of diazonium salts. In no case was the result satisfactory from the histochemical point of view. Seligman reported a similar lack of success with 2-napththyl glucuronide. The formula for 1-naphthyl glucuronide, which appears below, indicates the general structure of the phenolic glucuronides used in biochemistry and histochemistry.

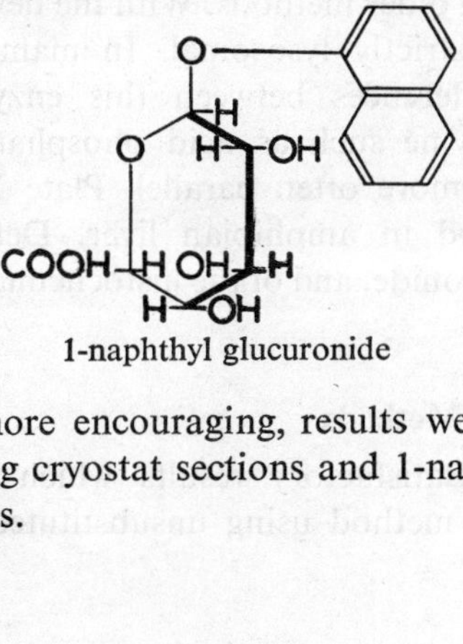

1-naphthyl glucuronide

Contrary, and far more encouraging, results were obtained by Bulmer (1967), using free-floating cryostat sections and 1-naphthyl glucuronide and Fast blue BB or RR salts.

8-Hydroxyquinoline Glucuronide. Burton and Pearse (1952) reported the largely unsuccessful use of this substrate in a coupling azo dye technique with a variety of diazotates. Further experiments showed, however, that the diazotate of 4-benzoylamino-2 : 5-dimethoxyaniline was particularly stable at pH 5·0 and that it could be used at this pH for prolonged incubation at 37° or even at 60° for some hours. Using approximately 5 mM-8-hydroxyquinoline glucuronide in 0·1 M-acetate buffer as substrate, cold formalin-fixed frozen sections of various animal tissues were incubated for 2–4 hours at 37° in the presence of 1 mg./ml. of the diazonium salt. A bright orange-red granular precipitate, and occasionally smaller amounts of a larger flocculent precipitate, were observed in tissues known to contain β-glucuronidase, or shown to do so by an alternative method (see below). Sections incubated in the presence of 110 mM-saccharate invariably showed no precipitate of either kind.

Using 8-hydroxyquinoline-β-D-glucuronide at a concentration of 2 mM, with Fast black salt K at 3mM, Rath and Otto (1966) evolved a successful simultaneous coupling technique. They applied this to a variety of tissues and obtained results which agreed with those recorded in the literature. Using the same substrate, with Fast blue RR or BB salts, or with Fast violet LB, Bulmer (1967) obtained excellent results in ovary and placenta. His localizations were identical with those given by the Naphthol AS-BI glucuronide technique (see below), and altogether different from those of the ferric hydroxyquinoline technique. These results supported the views of Janigan and Pearse, recorded on page 814.

Naphthol AS Glucuronides. The first successful synthesis of a substrate in this class, the β-glucuronide of Naphthol AS-LC, was carried out by Pugh and Walker (1961) who demonstrated that this substrate, in a simultaneous coupling technique with Fast Garnet GBC, provided greatly improved enzyme localization. The method commonly used at the present time was developed by Hayashi *et al.* (1964) and the substrate employed was Naphthol AS-BI β-D-glucuronide, in conjunction with hexazotized pararosanilin. The application of the technique was described by Hayashi (1964) who reported that the localization of β-glucuronidase was "quite different" from that obtained by workers using older methods. With the new method the distribution of β-glucuronidase is strictly lysosomal. In mammalian tissues there are often considerable differences between this enzyme and an alternative lysosomal marker enzyme such as acid phosphatase. In non-mammalian tissues the results are more often parallel. Plate XVIIb illustrates the use of the Hayashi method in amphibian liver. Details of the synthesis of Naphthol AS-BI glucuronide, and of the histochemical procedure, are given in Appendix 18, p. 1322.

Post-coupling Azo Dye Methods

Because of the unsatisfactory results which they obtained with the simultaneous coupling method using unsubstituted naphthol glucuronides,

Seligman *et al.* (1949) made use of the post-coupling azo dye principle. For this purpose they used the glucuronide of 6-bromo-2-naphthol, made biosynthetically by administration of the naphthol compound to rabbits. Unlike the majority of the glucuronides, this particular example can be obtained in an impure state by continuous ether extraction of the acid urine, followed by solution of the dried extract in hot water and its precipitation from this as the *p*-toluidine salt. Seligman, Tsou, Rutenburg and Cohen (1954), however, showed that the yield obtained was very poor and biosynthesis is thus unsuitable for the preparation of 6-bromo-2-naphthyl glucuronide. They prepared the compound in the pure state by oxidation of the β-D-glucopyranoside with oxygen in the presence of a platinum black catalyst.

Following their initial description the post-coupling azo dye methods for β-glucuronidase were little used in applied histochemistry, partly on account of difficulties in obtaining the substrate. In my hands, though the overall picture resembled that given by the modified Friedenwald-Becker technique (see above), localization was never as sharp as it is with the latter. (See Plate XVIIc, p. 814). A comparative study of the ferric hydroxyquinoline and post-coupling techniques which was made by Mori *et al.* (1962) revealed a number of differences, but a post-coupling technique using AS-BI glucuronide and Fast dark blue R was applied to skeletal muscle by Fishman (1964) to provide evidence supporting the specificity of the ferric hydroxyquinoline technique.

It is clear, in retrospect, that criticisms of the post-coupling technique made by Burton and Pearse (1952) were entirely justified.

Methods based on Indoxyl Glucuronides

In the second edition of this book it was noted that, although no reports had appeared in the literature, a number of indoxyl glycosides had been synthesized and used in histochemical techniques. It was noted also, that although little was to be expected from the use of unsubstituted indoxyl glucuronides their halogen substituted equivalents had great promise as histochemical substrates.

After long delay, due to difficulties in the synthetic prodedure, Pearson *et al.* (1967) succeeded in producing 5-bromo-4-chloro-indoxyl β-D-glucopyruroniside which they applied to the localization of β-glucuronidase. It was reported that the result was similar to that obtained with both the post-coupling technique (Seligman *et al.*, 1954) and with the Fishman-Baker (1956) method. No oxidation catalyst was employed (see Chapter 17, p. 779) and some of the results were somewhat surprising. Negative results, for instance, were given by the characteristic (lysosomal) sites of acid hydrolase activity in the pericanalicular zones of rat liver and by all sites in the cerebrum and cerebellum. Comparative studies may ultimately reveal the cause of these differences, which is not at present apparent.

Technical details of the indigogenic method of Pearson *et al.* (1967) are given in Appendix 18, p. 1323, but the modified method of Lojda (1970a), given on p. 1324, should always be employed.

A comparison between the indigogenic and azo dye techniques was made by Lojda (1970), using substrates for β-glucosidase and sections of intestine and kidney. He found that a striking improvement in localization was produced by the use of a ferri-ferrocyanide oxidation catalyst. Assays carried out on homogenates of rat intestine showed that only 26 per cent inhibition was caused, at a concentration of 0·1 mM. The localizations obtained by Pearson *et al.*, were shown to be the product of peroxidase activity. The results of Lojda's study indicated the superiority of the indigogenic methods, provided that an oxidation catalyst was used.

Glucoside Hydrolases. II Other Glycosidases and Disaccharidases

Biochemical Characteristics. From the point of view of histochemists the biochemistry of the glycosidases (other than β-glucuronidase) presents a somewhat confused picture. There is apparently a multiplicity of enzymes for each glycosidic link but, on the other hand, most members of the group show considerable lack of specificity. Many glycosidases are also transglycosylases and, as with other types of enzymes having both kinds of activity, the actual *in vivo* function is seldom clear.

Much of the available biochemical information deals with enzymes from plants, molds and bacteria. These do not concern the majority of histochemists. Scattered information is available on the main characteristics of the mammalian and invertebrate glycosidases. The brief survey which follows refers solely to those glycosidases which are likely to be of interest to histochemists. Despite considerable overlap they may be subdivided into four groups: *glucosidases, galactosidases, glucosaminidases* and *disaccharidases.*

Glucosidases. An α-D-glucosidase located in the brush border of the intestine was described by Miller and Crane (1961). This enzyme, an intestinal maltase, had been found by Larner and Gillespie (1956) to be inhibited by Tris. It is also inhibited by a number of polyols, especially erythritol (Kelemen and Whelan, 1966). The latter has been indicated (Auricchio *et al.*, 1968) as a competitive inhibitor for all mammalian glycosidases. At pH 4·5 it inhibits acid α-(1→4)-glucosidase while in the region of pH 7·5 it inhibits intestinal maltase.

Competitive inhibition of rat jejunal glycosidases, using four halogen-substituted indoxyl substrates under histochemical conditions, was investigated by Esterley (1967). The observed pH optima were: glucosidase and galactosidase, 5·4, fucosidase, 6·2, and glucuronidase, 4·8. Inhibition was strongest when both substrate and inhibitor were derived from the same sugar. The results are given in Table 63.

The cellular distribution of β-glucosidase has been studied by several

TABLE 63

Inhibition of Jejunal Glycosidases

(after Esterley, 1967)

Inhibitor	Galactosidase	Fucosidase	Glucuronidase	Glucosidase
Galactose	2	0	0	0
Fucose	0	1	0	0
Glucuronic acid	0	0	3	0
Glucose	0	0	0	0
Galactonolactone	3	1	0	0
Fuconolactone	1	3	0	0
Glucuronolactone	0	0	2	0
Saccharolactone	0	0	4	0
Gluconolactone	0	1	0	3
Gluconic acid	0	0	0	3
Saccharic acid	0	0	4	0
Mucic acid	0	0	2	1
Galacturonic acid	0	0	2	0
Polygalacturonic acid	0	0	2	3
Xylose	0	0	0	1
Lactose	1	0	0	0
Cellobiose	0	1	0	0
Melibiose	0	0	0	1

0. No inhibition.
1. Inhibition at 2% only.
2. Complete Inhibition at 2%, partial at 1%.
3. Complete Inhibition at 1%, partial at 0·1%.
4. Complete Inhibition at 0·1%, partial at 0·01%.

groups of authors. Price and Dance (1967), using rat kidney, found a soluble β-glucosidase which was identical with the soluble fraction of the β-galactosidase of this tissue. The same enzyme is active as an α-L-arabinosidase (Robinson *et al.*, 1967) and also as a β-D-xylosidase. It is therefore non-specific with regard to the C-4 and C-6 configuration of the glycoside. All three activities were found by Price and Dance to be inhibited by glucono-(1→5)-lactone. Studies carried out by Beck and Tappel (1968) showed that rat liver β-glucosidase was a lysosomal enzyme.

Almond emulsin β-glucosidase and β-galactosidase were shown by Heyworth and Walker (1962) to be identical and Conchie *et al.* (1967) observed that *Patella* β-glucosidase was associated with β-D-fucosidase activity. In rat epididymis they found no significant β-glucosidase activity.

Galactosidase. β-galactosidases occur widely in plant emulsin, in several types of *Aspergillus*, and in many microorganisms. A high activity is observed in mammalian intestine and in the albumen gland of the snail (*Helix pomatia* and *Helix aspersa*). According to McMahon *et al.* (1957) the storage polysaccharide in this gland is galactogen. In rat kidney Robinson *et al.* (1967)

found two *β*-galactosidases, one fast-moving and the other slow-moving on gel electrophoresis. The fast component contained the total *β*-glucosidase activity of the sample and inhibition studies indicated a single enzyme with a pH optimum at 5·5 to 6·0. The slow-moving component was a *β*-galactosidase with a pH optimum at 3·7. According to Barnett (1965) *Helix aspersa* *β*-galactosidase has a pH optimum at 2·2 in sodium phosphate or potassium chloride with either lactose or *o*-nitrophenyl *β*-D-galactoside as substrate. Previously two types of active centre had been proposed for *β*-galactosidase by Wallenfels and Malhotra (1961), with pH optima at 7 and 5 respectively.

Langley and Jevons (1967) identified two intestinal *β*-galactosidases one of which hydrolysed lactose but not 6-bromo-2-naphthyl-*β*-D-galactoside. The other hydrolysed lactose but not the artificial substrate. Hsia *et al.* (1966) found two galactosidases in ox liver, one of which was considered to be concerned with the degradation of glycoprotein carbohydrate. In *Patella* *β*-fucosidase and *β*-galactosidase are separate (Levvy and McAllan, 1963a) but in mammalian tissues they are both the same (Levvy and McAllan, 1963b).

In human tissues *β*-galactosidase may have some importance in relation to the accumulation of breakdown products of glycolipids in storage diseases such as Gaucher's disease (Phillipart and Menkes, 1964). Others (Brady *et al.*, 1965) have attributed the storage of gluco-cerebroside in this disorder to lack of glucosidase.

N-Acetyl-β-glucosaminidase. The occurrence of *N*-acetyl-*β*-glucosaminidase in mammalian tissues was described by Watanabe (1936a and b) and the properties of the enzyme have also been recorded by East, Madinaveitia and Todd (1941) who used partially purified preparations from ox liver and testis. The distinction of this enzyme from hyaluronidase was made by Hahn (1945) and Linker, Meyer and Weissmann (1955) have now shown that the oligosaccharides produced from hyaluronic acid by hyaluronidase are further degraded by *β*-glucuronidase and N-acetyl-*β*-glucosaminidase, in the manner illustrated below:

Polysaccharide chain
↓ polysaccharases
(hyaluronidase, heparinase)
Oligosaccharide fragments
↓ oligosaccharases
(glucosaminidase, galactosaminidase
glucuronidase)
Sugar units.

Further studies were made by Pugh, Leaback and Walker (1957) who showed that the enzyme is inhibited by acetate, acetamide and *N*-acetylglucosamine, all three behaving as typical competitive inhibitors. In 0·05 M-citrate buffer optimum activity was found at pH 4·3.

Investigating the intracellular localization of the enzyme Price and Dance (1967) found that the glucosaminidase activity of rat kidney was bimodal. The greater part, however, was lysosomal. From human spleen Robinson and Stirling (1968) extracted two components with acetylglucosaminidase activity, one an acidic the other a basic protein. The first, which they called the 'a' form, contained a number of sialic acid residues and was found in the supernatant as well as in the lysosomal compartment. The 'b' form was wholly lysosomal. Both components had identical Km values for different substrates and the same inhibition pattern.

Disaccharidases. The principal disaccharidases of histochemical interest are those of the intestinal tract. The oldest known enzyme in this group, invertase, has long been studied as the chief sugar-splitting enzyme of the yeast cell (Berthelot, 1860; Sörensen, 1909). A long series of investigations by Dahlqvist and his associates (Dahlqvist and Borgström, 1961; Dahlqvist, 1962a; Dahlqvist, Bull and Gustafson, 1965; Dahlqvist and Thomson, 1965) have outlined the importance of intestinal invertases and maltases.

The localization of invertase (β-fructofuranosidase) has been the subject of controversy at the biochemical level, Miller and Crane (1961) suggested that it was confined to the microvillous zone while Ugolev *et al.* (1964) considered that it was only on the outer surface of the enterocyte. Doell and Kretchmer (1962) found that the main activity was in the microsomal fraction while other authorities regard the lysosomes as the most important site of activity (Sellinger *et al.*, 1960).

Several attempts have been made to classify the intestinal disaccharidases. Semenza and Auricchio (1962) and Semenza *et al.* (1965) distinguished two invertases and five maltases, mainly on the basis of molecular size. Dalqvist (1962b) introduced a nomenclature, based on heat stability, which distinguished four enzymes: maltase Ia (isomaltase), maltase Ib (invertase), maltase II, and maltase III. Studying maltase, isomaltase and invertase activities in human jejunum and ileum by means of column chromatography Dalqvist and Telenius (1969) concluded that at the present time a more rational system of classification was impossible. The heat stability of some glycosidases is truly remarkable. The β-galactosidase of rat epididymis resists 120° for 16 hours, according to Raunio (1968). A detailed study of the activity of lactase, sucrase and maltase, at different levels in the intestine, was carried out by Newcomer and McGill (1966). Normally, low levels were found in duodenum and ileum with peak activity in the jejunum.

Methods for β-glucosidase (E.C. 3.2.1.21)

Seligman and his co-workers (Cohen *et al.*, 1952b) suggested that the preparation of the β-glucoside of 6-bromo-2-naphthol would enable a histochemical method for the localization of β-glucosidase to be evolved. This substrate was used in the same manner as in the method for β-galactosidase (below), except that fresh sections were essential. Localization is obtainable on the histological level in very active tissues such as cockroach colleterial glands (Fig. 171, p. 787). The tanning reaction in the oothecae of *Blatta* and *Periplaneta* depends on the liberation of free protocatechuic acid from its β-glucoside, by a β-glucosidase from the right colleterial gland (Brunet and Kent, 1955). Histochemically it can be shown that both glands possess a β-glucosidase but that on the right side is much more powerful than that on the left. It is possible that the latter exerts a synthetic effect *in vivo*.

An alternative to the post-coupling azo dye method recorded above was introduced by Pearson *et al.* (1961), using an indigogenic substrate, 3-(5-bromoindoxyl)-β-D-glucopyranoside. This method was tested by Lojda (1970a), as reported above. In his hands it gave excellent results.

Methods for α-glucosidase (E.C. 3.2.1.20)

An essentially similar method to the one recorded above for β-glucosidase was used by Rutenburg, Lang, Goldberg, and Rutenburg (1958a) for the demonstration of α-glucosidase in the small intestine and kidney of the rat. In addition to the standard post-coupling technique it was observed that hydrolysis of the substrate was sufficiently rapid for enzyme activity to be demonstrated by a simultaneous coupling technique.

A survey of the distribution of α-glucosidase in mammalian tissues, using the post-coupling technique with 6-bromo-2-naphthyl α-D-glucopyranoside as substrate, was reported by Rutenburg *et al.* (1960). Formalin-fixed sections were incubated at 26°, 37° and 50° for up to 24 hours. Positive results were obtained only with duodenum, jejunum and kidney.

Methods for β-galactosidase (E.C. 3.2.1.23)

Coupling Azo Dye. A post-coupling azo dye method for β-galactosidase was evolved by Seligman and his associates (Cohen *et al.*, 1952a) using 6-bromo-2-naphthol-β-D-galactoside as substrate.

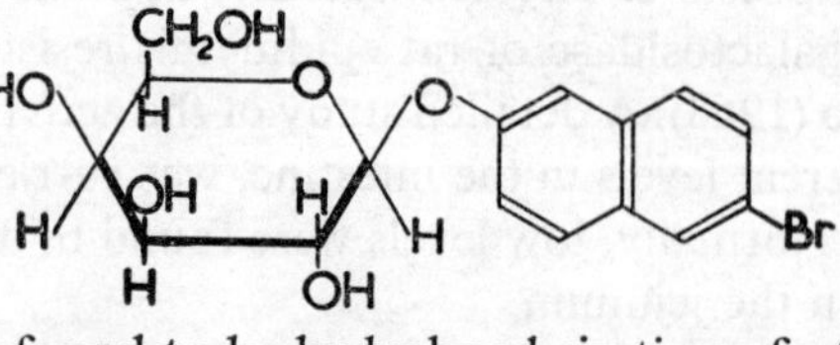

This compound was found to be hydrolysed six times faster than 2-naphthyl galactoside, by the enzyme present in tissues such as rat kidney, and a method for *in vitro* estimation was based on its use.

In the production of a histochemical method a number of difficulties were encountered and reported. Enzyme activity was initially low and inhibition by diazonium salts prevented the establishment of a simultaneous coupling technique. Acetone-fixed paraffin sections contained little enzyme and formalin-fixed frozen sections less than 25 per cent, so that fresh frozen sections had to be employed. An attempt was made to prevent diffusion of the enzyme, which was observed to be considerable, by incubation in the presence of 1 M-NaCl. In the method finally evolved the mounted sections were incubated for 2 hours at 37° and pH 4·95 with the substrate and 1 M-NaCl. They were then washed and coupled with diazotized *o*-dianisidine at pH 7·5, with the production of a bluish-purple final product.

A later report by Rutenburg *et al.* (1958b) gave details of the further development of the method. A return was made to the use of cold formalin-fixed sections and these were incubated, free-floating, for 2–4 hours at 37°. After coupling with the diazonium salt (Fast blue B was still recommended) the final product was a purplish-blue pigment except where dissolved in lipid. Here it appeared red. In rat tissues a high activity of the enzyme was found in kidney, liver, intestine, testis and epididymis and a lesser activity in salivary glands, pancreas, thymus, trachea, lung, spleen and lymphoid tissue.

Details of this method are given in Appendix 18, p. 1323. The results are illustrated in Plate XVIIIa, p. 824.

Indigogenic. A series of indolyl galactosides was synthesized by Pearson, Wolf and Vasquez (1963) of which the best for histochemical purposes proved to be 5-bromo-4-chloroindol-3-yl *β*-D-galactopyranoside. Using this substrate, *β*-galactosidase was found to be widely distributed in mammalian tissues, especially in reticulum cells in spleen, lymph nodes, and intestinal lamina propria. It was also found, as expected, in kidney and intestinal mucosa, and in liver, adrenal cortex and testis. Details of this technique will be found in Appendix 18, p. 1323 and the type of result obtainable is illustrated in Plate XVIId, p. 814. This result was not achieved, however, without adding an oxidation catalyst.

Method for α-galactosidase (E.C. 3.2.1.22)

The synthesis of 6-bromo-2-naphthyl α-D galactopyranoside (Appendix 18, p. 1324) was carried out by Monis, Tsou and Seligman (1963) and the substrate was used in a post-coupling technique at pH 5·0 with formalin-fixed tissues. Strong reactions were observed in Brunner's glands of the duodenum, in megakaryocytes, gastric parietal cells, thyroid follicular cells and, especially, in the parathyroid gland.

Methods for β-acetylaminodeoxyglucosidases (E.C. 3.2.1.30)

A histochemical method for acetyl-*β*-glucosaminidase was developed by Pugh and Walker (1958, 1961a) which was based on the hydrolysis of α-Naphthyl-*N*-acetyl-*β*-D-glucosaminide at pH 4·5 and on the simultaneous capture of the free naphthol by a suitable diazonium salt. Fast Garnet GBC (18) was found suitable for this purpose. Localization was histological rather than cytological (Fig. 172, p. 787), since diffusion of enzyme was very difficult to prevent. Cold formalin-fixed tissues were employed and the incubating medium contained methyl cellulose or PVP in order to reduce diffusion of lyo-enzyme to a minimum. The final reaction product was lipid soluble and the preparations were therefore not absolutely stable.

Later Pugh and Walker (1961b) described briefly the synthesis of the acetyl-β-glucosaminide of naphthoic-2-hydroxy-3-(2′,5′-dimethoxy-4′-chloro-anilide) (Naphthol AS-LC), and its employment as substrate in a simultaneous coupling method using Fast Garnet GBC. Further improvements, and the development of what is now the standard technique, were due to Hayashi (1965). This author synthesized the acetyl-β-glucosaminide of Naphthol AS-BI and used this substrate at pH 5·2 with hexazotized pararosanilin in a simultaneous coupling technique.

The HPR technique is always preferable to the slightly easier variation using Fast Garnet GBC and it is given in full in Appendix 18, p. 1325, together with details of the synthesis of the substrate. A very precise localization of the enzyme in discrete granules is obtained in both mammalian and non-mammalian tissues (Plate XVIIIb, p. 824), which may confidently be accepted as lysosomal. A comparative study of the localization of three lysosomal enzymes (acid phosphatase, β-glucuronidase and acetyl-β-glucosaminidase) was carried out by Hayashi (1967). Pronounced differences were noted, especially between the first two and the last, in the majority of tissues.

Methods for Fucosidase and Xylosidase

Indogenic substrates for these two enzymes were prepared by Esterley, Standen and Pearson (1967, 1968) and used on formalin or glutaraldehyde-fixed tissues. Fucosidase was restricted to the jejunal mucosa while xylosidase was strongest in liver and in developing cartilage. Inhibitor studies were carried out which supported the view that specific enzymes were being demonstrated. This method was tested by Lojda and Kraml (1971).

Coupled Oxidation Methods for Disaccharidases (E.C. 3.2.1.26 and 28)

A multi-step method for disaccharides, using disaccharide substrates (sucrose, lactose, maltose, isomaltose), was introduced by Dahlqvist and Brun (1962). The reaction sequence is shown in Fig. 173.

Glucose, released by the initial hydrolysis is oxidized by exogenous glucose oxidase and the reduced enzyme transfers electrons to a tetrazolium salt *via* phenazine methosulphate (PMS), producing an insoluble formazan dye *in situ.* No formazan is formed in absence of PMS (see Chapter 20, p. 916) since the reduced flavoenzyme cannot transfer electrons directly to tetrazolium

PLATE XVIII

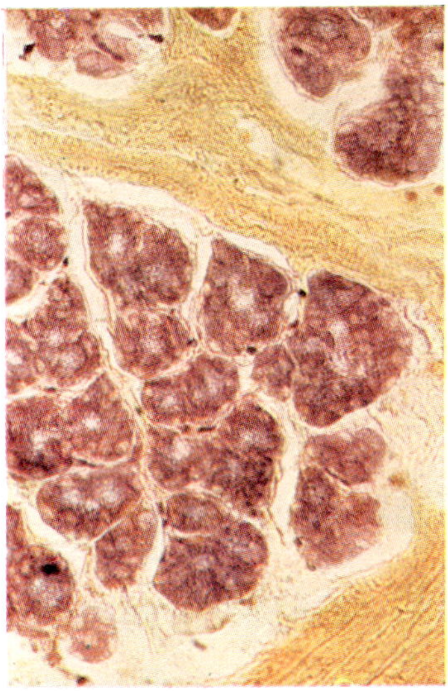

XVIIIa. Rat duodenum, fixed in cold formalin. Shows the localization of β-galactosidase in Brunner's glands, using a simultaneous coupling azo dye technique. × 300.

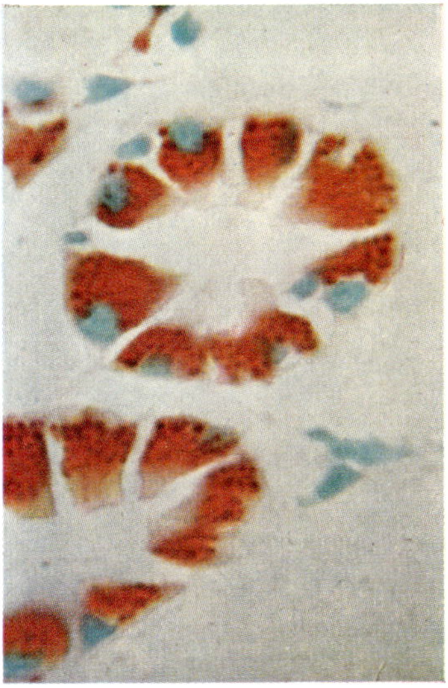

XVIIIb. Newt (*Triton cristatus*) kidney. Fixed in cold formol-calcium. Localization of *N*-acetyl-β-glucosaminidase using Naphthol AS-LC glucoside and HPR. The enzyme is predominantly lysosomal but there is some diffusion of the final reaction product. × 390.

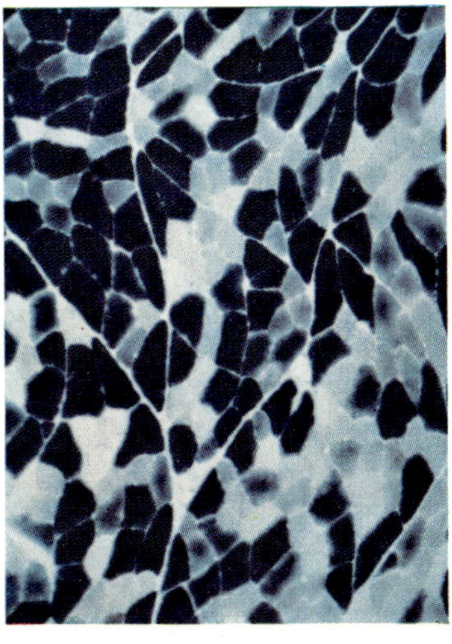

XVIIIc. Hamster skeletal muscle. Fresh frozen, 6 μ, cryostat section. Shows reaction for phosphorylase A + B with different intensities in the three fibre types. × 70.

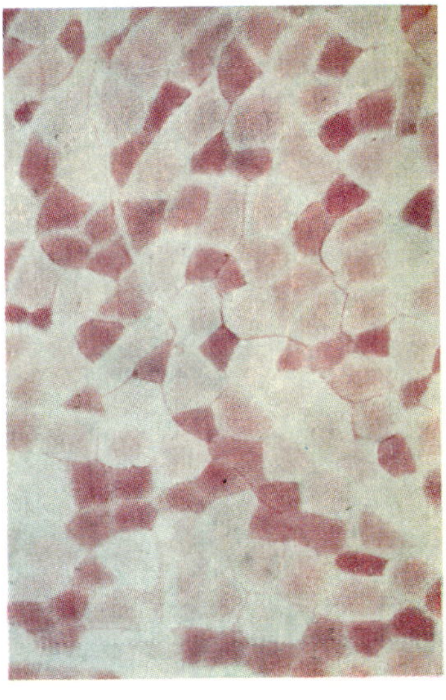

XVIIId. Rat skeletal muscle. Fresh frozen, 6 μ, cryostat section. Reaction for UDP glucose transglycosylase. The newly formed glycogen reaction product has been stained by the PAS method. × 90.

[*To face p.* 824

salts. If any of the reactants are omitted from the medium there is a negative result.

The coupled oxidation method was employed by Jos *et al.* (1967) for localization of disaccharidases in peroral biopsy specimens from human jejunum and ileum. Tetranitro-BT was used in place of the original Nitro-BT, resulting in a finer formazan precipitate and hence better localization. Activity was restricted to the intestinal brush border. The original method was used by Grossman and Sacktor (1968) for the demonstration of trehalase in

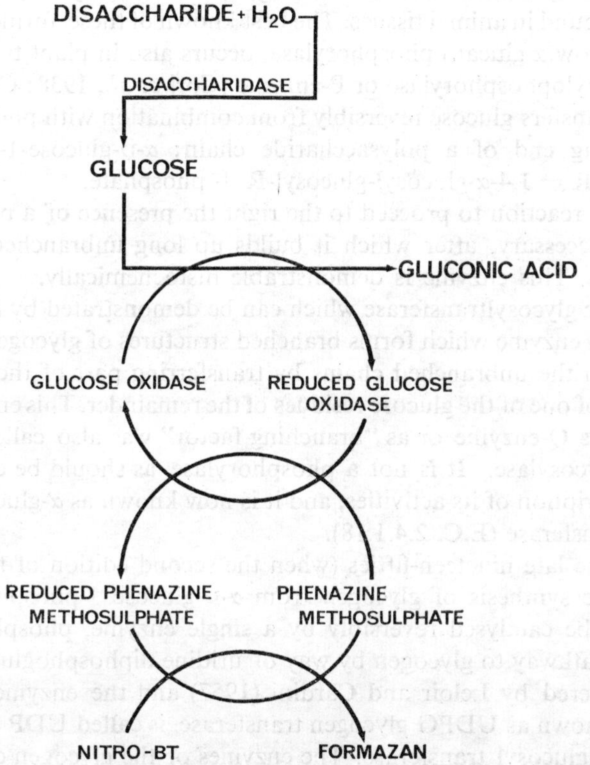

FIG. 173. Reaction sequence in the coupled oxidation procedure.

human and mouse kidney. Highest activity was recorded in the proximal convoluted tubules.

Details of the coupled oxidation technique are given in Appendix 18, p. 1326. Essentially these follow the original instructions, as modified by Lojda (1965) whose critical studies indicated the necessity for a number of changes. Lojda concluded that, because of diffusion of the primary product (glucose), the coupled oxidation method allowed localization on the histological or, at best, the cellular level. No conclusions as to intracellular localization were to be derived from its use.

Using a new (filter paper) technique, with more glucose oxidase and tetra-Nitro BT, Jos *et al.* (1967) reported improved localization of disaccharidases in peroral intestinal biopsies. Later, however, Jos (1968) indicated that the method was still far from satisfactory, principally on account of false localizations due to non-specific staining by the tetrazoles.

The Glycosyltransferases

The glycosyltransferases form a large group of enzymes, only a few of which are found in animal tissues. The best known of these, formerly known as glucosan (now α-glucan) phosphorylase, occurs also in plant tissues where it is called amylophosphorylase or P-enzyme (Cori *et al.*, 1938; Cori and Cori, 1943). It transfers glucose reversibly from combination with phosphate to the non-reducing end of a polysaccharide chain: α-D-glucose-1-phosphate + D-glucosyl-R $\rightleftharpoons$ 1,4-α-glucosyl-glucosyl-R + phosphate.

For the reaction to proceed to the right the presence of a polysaccharide starter is necessary, after which it builds up long unbranched chains with 1,4-linkages. This enzyme is demonstrable histochemically.

Another glycosyltransferase which can be demonstrated by histochemical means is an enzyme which forms branched structures of glycogen and amylopectin from the unbranched chains by transferring part of the chain to the 6-position of one of the glucose residues of the remainder. This enzyme, known in plants as Q-enzyme or as "branching factor" was also called amylo-1,4-1,6-transglycosylase. It is not a phosphorylase, as should be clear from the above description of its activities, and it is now known as α-glucan branching glycosyltransferase (E.C. 2.4.1.18).

Until the late nineteen-fifties (when the second edition of this book was written) the synthesis of glycogen from α-D-glucose 1-phosphate was considered to be catalysed reversibly by a single enzyme, phosphorylase. The synthetic pathway to glycogen by way of uridine diphosphoglucose (UDPG) was discovered by Leloir and Cardini (1957) and the enzyme responsible, formerly known as UDPG glycogen transferase, is called UDP glucose-α-1,4-glucan α-4-glucosyl transferase. The enzymes of the glycogen cycle are conveniently shown diagrammatically, as in Fig. 174.

The two enzymes (1) and (2) together constitute a reaction sequence from glucose 1-phosphate to glycogen with pyrophosphate as product (Villar-Palasi and Larner, 1958). Enzyme (2) is often simply called glycogen synthetase although (1) and (2) together might more reasonably be described in this way.

The reactions catalysed by (1) and (3) can proceed in either direction, a fact which is made use of in the histochemical reaction for phosphorylase. The synthetic activity of the latter produces orthophosphate and this fact has also been utilized for the histochemical demonstration of the enzyme. For this activity phosphorylase requires a primer or starter (usually glycogen), as mentioned above. Glycogen synthetase also requires a polysaccharide primer and, in addition, glucose 6-phosphate (Leloir and Cardini, 1957; Leloir and Goldenberg, 1960). The enzyme in rat liver is activated up to 15-fold by this ester or by the 6-phosphates of fructose, glucosamine or galactose.

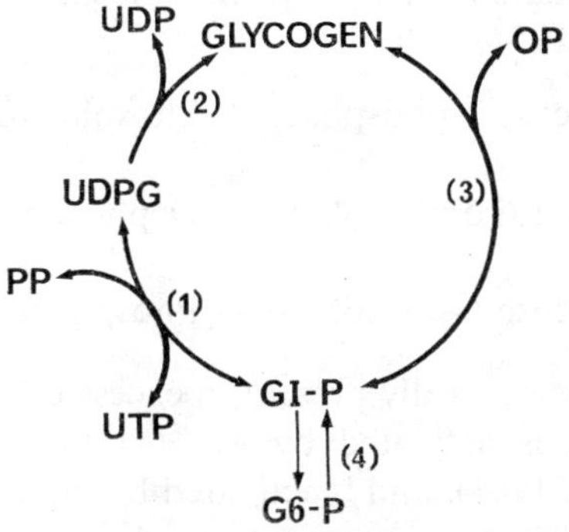

FIG. 174. The Glycogen Cycle.

(1) UDPG pyrophosphorylase (E.C. 2.7.7.9).
(2) UDPG α-glucan glucosyltransferase (E.C. 2.4.1.11).
(3) α-glucan phosphorylase (E.C. 2.4.1.1).
(4) Phosphoglucomutase (E.C. 2.7.5.1).

UTP = Uridine triphosphate PP = Pyrophosphate
UDP = Uridine diphosphate OP = Orthophosphate

Glucan phosphorylase. A survey of the molecular properties and transformations of α-glucan phosphorylase in animal tissues was presented by Krebs and Fischer (1962). Phosphorylase exists in two interconvertible forms. One, the A-form, exhibits 60–70 per cent of its full activity in the absence of adenylic acid while the other, the B-form, has an absolute requirement for the nucleotide. Phosphorylase B, which has a mol. wt. exactly half that of the A-form, is converted into the latter, in the presence of Mg^{2+} ions and ATP, by the enzyme phosphorylase kinase.

$$2 \text{ Phosphorylase B} + 4 \text{ ATP} \rightarrow \text{Phosphorylase A} + 4 \text{ ADP}$$

This process is greatly accelerated in the presence of cyclic 3′,5′-AMP. In tissue sections the kinase is present in adequate amounts and conversion takes place after addition of Mg^{2+} ions and ATP.

Phosphorylase A is converted to the inactive form by an enzyme called phosphorylase phosphatase (Cori & Green, 1943) which was formerly designated P-R enzyme, as it was considered to remove a prosthetic group. After recognition that the B to A reaction involved phosphorylation Krebs and Fischer (1956) showed that the conversion A to B was accompanied by release of inorganic phosphate. The specificity of phosphorylase phosphatase for its substrate is extremely high. It cannot be demonstrated histochemically.

In resting muscle phosphorylase is present predominantly in the inactive B-form, thus allowing glycogen to accumulate under the influence of enzymes of the UDPG pathway. During exercise activation occurs, providing the muscle with an adequate supply of hexose phosphate as an energy source.

Phosphoglucomutase. This widely distributed enzyme catalyses the reversible interconversion of glucose 1-phosphate and glucose 6-phosphate in two steps through the alternate formation of the phospho and dephosphoenzyme (Najjar and Pullman, 1954).

$$\text{Glucose 1-phosphate} + \text{Phosphoenzyme}$$
$$\rightleftharpoons$$
$$\text{Glucose 1,6-diphosphate} + \text{Dephosphoenzyme}$$
$$\rightleftharpoons$$
$$\text{Glucose 6-phosphate} + \text{Phosphoenzyme}$$

Phosphoglucomutase is generally present in excess over the other enzymes in the glycogen cycle. It is activated by Mg^{2+} ions, and also by imidazole (Robinson and Najjar, 1961), and inactivated by fluoride (Najjar, 1948).

Methods for α-Glucan Phosphorylase (E.C. 2.4.1.1)

The first histochemical reaction for phosphorylase was described by Yin and Sun (1947) who demonstrated the presence of the enzyme in plant tissues by catalysing the production of starch from glucose-1-phosphate. Two years later Cobb (1949) applied the method to fresh sections of cartilage and was able to demonstrate the formation of glycogen in the very active cells of this tissue. In Cobb's procedure pre-existing glycogen was first removed from the sections with amylase and, following incubation, the newly-formed glycogen was rendered visible with an alcoholic PAS reaction. Further development of the method was carried out by Goldberg, Wade and Jones (1952) who used the PAS reaction, Best's carmine, and also the iodine reaction for staining enzymically produced glycogen. As an alternative approach they endeavoured to demonstrate the sites of released phosphate by a modification of Gomori's acid phosphatase reaction, hydrolysis of the glucose-1-phosphate by tissue acid phosphatase being prevented by inclusion of NaF in the medium.

The phosphorylase method was placed on a sure footing by the studies of Takeuchi and Kuriaki (1955) and Takeuchi, Higashi and Watanuki (1955) who first introduced the use of activators (insulin and adenosine-5-phosphate) and primer (glycogen). Takeuchi and his associates tested several methods for staining glycogen formed in the reaction and they concluded that iodine was the most satisfactory although permanent preparations could not be obtained. The newly-formed glycogen was observed to stain bluish-black with iodine and it was thus easily distinguished from the mahogany-coloured deposits of pre-existing glycogen.

In his second paper (Takeuchi *et al.*, 1955) Takeuchi explained the colour differences as being due to the fact that the newly-formed polysaccharide was of low molecular weight and might be either amylose, amylopectin or perhaps a slightly-branched glycogen. In a third paper (1958) Takeuchi concluded that his method was capable of demonstrating not only phosphorylase but also amylo-1,4-1,6-transglycosylase. The same conclusion appears to have been reached independently by Ellis and Montagna (1958). It has now been established that an unbranched polysaccharide of the amylose type is the reaction product of the phosphorylase reaction and that this stains blue or blue-black with iodine. If the second enzyme is present and active, a branched polysaccharide (glycogen) can be formed subsequently from the amylose when the chain reaches a certain critical length. This is somewhere between 42 and 100 glucose units according to Hassid (1945).

Factors Influencing Techniques. With the realisation that histochemical demonstration of the two types of phosphorylase was possible, and that they could be shown separately or together, a number of modifications to the original technique were introduced by different groups of authors. Most of these followed established biochemical practice and were designed to alter the conditions of the reaction in favour of one or other of the phosphorylases. Other improvements were introduced specifically to increase the efficiency of the reaction by preservation of the enzymes, or of the product, or both, *in situ.*

Blocking Reagents. In order to distinguish clearly the activity of the two forms of the enzyme it is absolutely necessary to stabilize them in the tissue sections or smears and to prevent the occurrence of interconversion during incubation. The two principal reagents employed for this purpose are NaF and EDTA. The former inhibits the activity of phosphorylase phosphatase (Sutherland, 1951) and the latter phosphorylase kinase (Krebs and Fischer, 1955). Sodium fluoride was used by Guha and Wegmann (1959) and both reagents were employed by Godlewski (1960, 1963), who described the action of EDTA as a stabilization of phosphorylase A. According to Buell and Hansen (1961) EDTA has a directly protective effect on phosphorylase A, as well as preventing the conversion of phosphorylase B. A critical study of the effect of EDTA, and other chelators, on phosphorylase activity was carried out by Hori (1966a). These investigations showed that EDTA greatly stimulated the synthesis of glycogen by tissue sections. In the case of rat liver, where the result was negative unless EDTA was present in the medium, Hori suggested that the mechanism involved prevention of destruction of the newly-formed product by α-amylase (E.C. 3.2.1.1).

Activators. For histochemical media the first reagents in this category were insulin and AMP (Takeuchi *et al.*, 1955) but their use of insulin was soon discontinued. Various nucleotides, however, have continued to be employed and considerable arguments have arisen with regard to their several merits and demerits. In the case of liver phosphorylase Hadjiolov and Dancheva (1959) found that 5′-GMP, 5′-UMP and 5′-CMP produced 20 per cent of the activation brought about by 5′-AMP. This was not so with the muscle enzyme. The use of ATP and Mg^{2+} ions was advocated by Guha and Wegmann (1959,

1960a and b) in order to activate phosphorylase kinase and convert the B-form into the active A-form. Their medium was considered to demonstrate total phosphorylase but Guha and Wegmann (1961) added AMP, in addition to ATP, whenever "total" activity was required. For the demonstration of liver phosphorylase Grillo (1961) added to his medium either AMP or cyclic 3′,-5′-AMP, together with either adrenalin or glucagon. In their studies on nucleotide activation of phosphorylase in rat brain under normal and anoxic conditions Ibrahim and Castellani (1968) obtained their most intense reactions when the medium contained AMP. A descending order of activation was obtained with cyclic AMP, ATP + Mg^{2+}, and ADP. Ibrahim and Castellani interpreted their results on the basis of the findings of Breckenridge and Norman (1962, 1965) which concerned the highly sensitive adenyl cyclase system. They emphasized the great sensitivity of phosphorylase to anoxia and the need to process experimental (and control) tissues with great rapidity

Primers. Differences of opinion have long existed concerning the addition of glycogen to the histochemical medium for phosphorylase, although the use of a primer accords with biochemical precept (Cori *et al.*, 1955). Guha and Wegmann (1961, 1965), Godlewski (1963), and Hori (1966) all omitted the primer from their media but most workers, with Eränkö and Palkama (1961) apparently considered that extraneous glycogen, though not indispensable, nevertheless increased the amount of new glycogen formed. Meijer (1968, 1969) showed conclusively that added glycogen could not serve as glucosyl acceptor in the histochemical technique, probably on account of its high mol. wt. (2′500′000 to 5′000′000). It is clear that the histochemical demonstration of phosphorylase has always depended on the presence of intrinsic glycogen.

Several glycogen metabolites, such as γ-dextrins, phosphorylase limit dextrin and β-amylase limit dextrin, and also soluble starch, which can act as glucosyl acceptors in biochemical systems all fail to act under histochemical conditions. A number of dextrans (Vol 1, p. 297), however, were shown by Meijer (1968) to function successfully. The best results were obtained with dextrans having an average mol. wt. between 40′000 and 80′000. In his second paper Meijer (1969) indicated that only linear, unbranched, dextrans were useful and that the affinity of phosphorylase for the non-reducing end group of the dextran molecule increased with the increasing mol. wt. of the acceptor. He concluded that, especially in anoxic and ischaemic tissues (such as cardiac infarcts), histochemical demonstration of phosphorylase required the use of high mol. wt. unbranched dextran as acceptor. Although Meijer employed dextrans up to mol. wt. 2′000′000 it appears that, for most purposes, it is sufficient to employ 200′000 dextran.

Stabilizing Additives. Since phosphorylase is a soluble enzyme, and since the low mol. wt. unbranched glycogen product is diffusible, the original authors (Takeuchi, 1958; Takeuchi and Kuriaki, 1955) recommended the addition to the medium of 20 per cent ethanol. Eränkö and Palkama (1961)

modified the original technique by adding PVP (polyvinyl pyrrolidone) at a concentration of 7·5 per cent. At the same time they increased the concentration of glucose-1-phosphate by a factor of 10. Following their lead, PVP has been used by most workers in concentrations up to 15 per cent. Using an inverted section technique, with the medium in a small trough, Korsgaard and Wulff (1967) obtained positive results in low glycogen tissues only with the highest PVP concentrations.

Final Demonstration of the Product. Essentially only two methods are employed. The first, and still the most popular, is staining with iodine (Plate XVIIIc, p. 824). The second is to employ the PAS reaction (Plate XVIIId), after suitable fixation of the product. Although the orthophosphate liberated in the reaction can be demonstrated by a Gomori-type procedure (Pb-S), this alternative is mainly of academic interest. Lindberg and Palkama (1970) showed that phosphorylase was strongly inhibited by the concentration of Pb^{2+} ions (4·2mM) employed by Hori (1966b) to demonstrate its activity at the ultrastructural level. They found, moreover, that even when a lower concentration of Pb^{2+} (2·7mM) was used, the localization of phosphorylase failed to correspond to that shown by alternative methods and concluded that metal precipitation techniques could not be used to demonstrate phosphorylase.

Iodine staining distinguishes newly-formed glycogen (blue-black) from intrinsic glycogen (mahogany), the PAS reaction does not but it has the advantage of providing permanent sections. It can be used with confidence even after the use of dextran as primer since Meijer (1968) provided evidence that non-specific staining due to adsorbed dextran did not occur. Permanent iodine-staining methods were described by Sawyer *et al.* (1965), and by Smith and Perkins (1967), but the majority of workers are still content to photograph their preparations and discard them.

Accepting most of the conclusions recorded above, two methods for phosphorylase are given in Appendix 18. One, the more or less conventional technique for use without extraneous acceptor, can be employed in high intrinsic glycogen tissues. The other, essentially derived from Meijer's observations, should be used in all other cases and, for research projects, in parallel with the first method. In the case of blood films and smears a suitable method is that of Wulff and Sørensen (1966). The results of the application of the conventional technique (for phosphorylase A and B) are shown in Plate XVIIIc, p. 824.

Method for α-Glucan-branching Glycosyltransferase (E.C. 2.4.1.18)

Takeuchi established a series of tests to distinguish the activity of this enzyme from that of the associated α-glucan phosphorylase. The chief methods of distinction between the two enzymes were (1) the inclusion of 20 per cent ethanol in the incubating medium and (2) post-digestion of the reaction products with α- and β-amylase. In the presence of alcohol, activity of the transglycosylase is strongly inhibited. After treatment with α-amylase both amylose and glycogen are digested and the subsequent iodine reaction becomes negative. With β-amylase only amylose disappears and a mahogany stain with iodine then indicates sites of branching enzyme activity. Table 64, below, summarizes Takeuchi's observations on differences between the two enzymes.

TABLE 64

Distinction between glucan phosphorylase and transglycosylase.

Reaction	Phosphorylase (Amylose producing)	Transglycosylase (Amylopectin or glycogen)
Iodine	Blue	Mahogany
PAS	Magenta	Magenta
α-Amylase	Complete digestion	Complete digestion
β-Amylase	Complete digestion	Unaltered
Ethanol (20 per cent)	Moderate Inhibition	Strong Inhibition
Mercuric chloride (10^{+5}M)	Moderate Inhibition	Complete Inhibition

Method for α-Glucan-UDP Glucosyltransferase (E.C.2.4.1.11)

Following the description by Leloir and Cardini (1957) of the second pathway for glycogen synthesis, by way of uridine diphosphate glucose (UDPG), Takeuchi and Glenner (1960, 1961) were able to demonstrate the activity of this enzyme histochemically. Its activity may be written simply:

$$\text{UDP glucose} + (\text{glycogen})_n = \text{UDP} + (\text{glycogen})_{n+1}$$

This equation indicates the same need for a primer as in the case of α-glucan phosphorylase. Glycogen was added to the medium by the original authors, and retained by those who followed. In view of the discussions (p. 830) on the role of added glycogen in the phosphorylase reaction it must be assumed, as indicated by Brauss and Sasse (1968), that the presence and level of intrinsic glycogen in the cell is the factor controlling the histochemical result. Since dextran cannot act as primer in place of glycogen, it cannot be used in the same manner as in Meijer's phosphorylase technique. Glucose-6-phosphate and EDTA are usually incorporated in the medium as activators, following the original authors' practice. Alcohol, which Takeuchi and Glenner (1961) found to be inhibitory in all except the lowest concentrations, is nevertheless still habitually added to the medium at between 4 and 5 per cent.

Søvik *et al.* (1966) showed that UDP glucosyl transferase occurs, like phosphorylase, in two forms. The "I" form is active in the absence of activators while the "D" form requires glucose-6-phosphate. Conversion of D to I form is catalysed by a transferase phosphatase. The latter is inhibited by added glycogen, inhibition being proportional to the concentration (Huijing *et al.*, 1969).

Although one would expect the product of UDPG transferase activity to be an unbranched amylose (staining blue with iodine) the newly formed glycogen stains violet or reddish-purple under these conditions. No clear explanation for this observation is forthcoming. Usually it is demonstrated by the PAS technique (Plate XVIIId, p. 824).

The method is available in several different modifications. Sasse (1966) proposed to make distinction between intrinsic and newly-formed glycogen by extracting the former, before staining (but after incubation), with a 30-minute hydrolysis in 10 per cent H_2SO_4. Extracted and control sections were both stained by thc PAS reaction. A very considerable improvement, in the amount of newly formed glycogen which is deposited in the tissues, is produced by the addition to the incubating medium of PVP in the same concentration as in the method for α-glucan phosphorylase. Technical details of two methods for α-glucan-UDP glucosyltransferase are given in Appendix 18, p. 1328.

Method for Purine Nucleoside Phosphorylase (E.C. 2.4.2.1).

A histochemical procedure for the demonstration of this enzyme was described by Kishi (1969). Its specificity has not yet been determined but the general reaction catalysed is as follows:

Purine nucleoside + orthophosphate ⇌ α-D-ribose-1-phosphate + purine

In the histochemical reaction inosine is employed as substrate and this is converted to hypoxanthine. The latter is oxidized, by endogenous and added xanthine oxidase, to uric acid. This two-stage reaction is coupled to the reduction of Nitro-BT (see Chapter 19, p. 874):

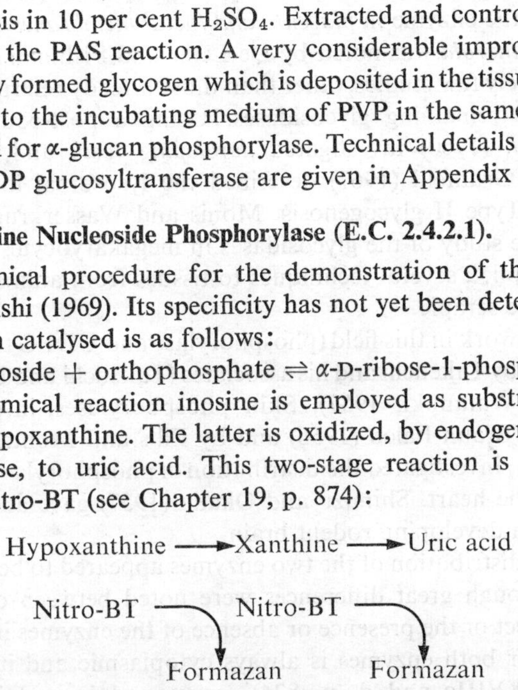

Fixation was found to be an absolute necessity and a number of alternatives were suggested, including 60 per cent cold acetone, 80 per cent cold ethanol, and 100 per cent cold acetone. The last of these provided the best localization, with a fixation time of one minute. The method, which is given in Appendix 18, p. 1329, is necessarily subject to the same objections as any of the multi-step techniques. A negative substrate-free control indicates the primary responsibility of the enzyme for the final reaction product. It does not provide control for localization.

Applications of Glycoside Hydrolase and Transferase Techniques

The greater proportion of studies on β-glucuronidase before 1963 employed the ferric hydroxyquinoline technique. In some cases localization was histologically accurate and conclusions warrantable (e.g. Cabrini and Schajowicz, 1960; Cabrini and Carranza, 1960). In spite of clearly expressed doubts on the validity of the reaction Bots (1964) and Bots *et al.* (1965) used it in their work on the adrenal cortex. More modern studies on β-glucuronidase localization have used exclusively the azo dye technique. An overall study of the enzyme in rat tissues was carried out by Hayashi (1964) and Watanabe and Fishman (1964) continued earlier studies on the *uterus and vagina.* The localization of β-glucuronidase in *bone marrow* cells was studied by Lorbacher *et al.* (1967) and conditions in the *stomach* before and after ACTH administration were investigated by Kozlowska (1968). Christie (1968) recorded the distribution of the enzyme in *placenta and foetal membranes* and the bimodal distribution in *thyroid* was noted by De Lellis and Fishman (1968).

Comparatively few studies, other than those already referred to, have been devoted to the remaining glycosidases. Using a post-coupling technique Yoshioka *et al.* (1960) investigated *developing teeth* and, more recently, Nitowsky and Grunfeld (1967) described the lysosomal localization of α-glycosidase in Type II glycogenosis. Monis and Wasserkrug (1967) carried out an extensive study of the glycosidases in megakaryocytes and Mackenzie *et al.* (1968) applied several techniques to tissues from animals infected with the virus disease scrapie.

The earliest work in this field (phosphorylase and UDPG transglycosylase) was carried out by Takeuchi and his associates (Takeuchi and Kinoshita, 1956; Sugino, 1957; Watanuki, 1957). Skin phosphorylases, in particular, were investigated by Braun-Falco (1956) and by Ellis and Montagna (1958), and Carbonell (1955) investigated the distribution of phosphorylase in the conducting system of the heart. Shimizu and Okada (1957) gave an account of the latter enzyme in developing rodent brain.

The overall distribution of the two enzymes appeared to be quite similar in most tissues though great differences were noted between one species and another in respect of the presence or absence of the enzymes in a given tissue. The reaction for both enzymes is always cytoplasmic and it is strongest in muscle (Plate XVIIIc and d, p. 824), nerve and in certain epithelia. In mammalian kidney the strongest activity is found usually in the collecting tubules (rabbit) or in the distal convoluted tubules in the rat (Macpherson and Pearse, 1957; Pearse and Macpherson, 1958). For the most part cells which show strong phosphorylase and transglycosylase activity are those which normally contain glycogen. If utilization is high, however, this relationship may not be observed. In human skin Ellis and Montagna (1958) found high activity of phosphorylase in the stratum basale and of both enzymes in the stratum spinosum. The highest activity was found in children and the lowest in older subjects.

The phosphorylase technique continued to find applications. A series of papers by Godlewski (1961, 1963) and Godlewski and Penar (1961) described the localization of the enzyme in *uterine cervix*, *breast*, *ascites tumour*, and in *liver* cells in tumour-bearing animals. Hadjiolov and Dancheva (1957, 1958) had already studied phosphorylase levels in experimental liver cancer in rats, finding a lower activity in neoplastic cells. The distribution of both phosphorylase and transglycosylase was investigated in the *uterus* by Bo and Smith (1964a and b) and Fagundes and Cohen (1965) discovered a diurnal variation in liver phosphorylase (dependence on the nutritional state) which was later studied by Egli (1967). Studies on brain included those of Gentshev (1967) on the cerebellum.

Muscle studies have naturally been substantial. Vallyathan *et al.* (1964) and Vallyathan and George (1964) investigated changes in experimental disuse atrophy and after forced exercise and Strugalska-Cynowska (1967) found an incomplete lack of phosphorylase in the myopathic muscle of Type II glycogenosis. She regarded the disorder as due to disturbance of phosphorylase kinase rather than phosphorylase deficiency.

Following the original studies on UDPG transglycosylase a small number of applications have been recorded. Hess and Pearse (1961) pointed out that the direct relationship between this enzyme and phosphorylase was not maintained in a given fibre type in skeletal muscle. Their observations were essentially the same as those of Nene and George (1965) and they were supported by the results of Bocek and Beatty (1966) who used parallel histochemical and assay techniques. Important studies on UDPG transglycosylase in embryo tissues were carried out by Grillo and Ozone (1962) and by Grillo *et al.* (1964).

It is evident that both techniques (phosphorylase and UDPG transglycosylase) have attained reasonable status in terms of reproducibility. Both require very strict control in respect of intrinsic glycogen levels and they are best used regularly, rather than sporadically, for diagnostic and experimental purposes.

REFERENCES

ANDERSON, F. B., and LEABACK, D. H. (1961). *Tetrahedron*, **12**, 236.
AURICCHIO, F., BRUNI, C. B., and SICA, V. (1968). *Biochem. J.* **108**, 161.
BAJUSZ, E., and JASMIN, G. (1964). *Experientia*, **20**, 373.
BARNETT, J. E. G. (1965). *Biochem. J.* **96**, 72P.
BECK, C., and TAPPEL, A. L. (1968). *Biochim. biophys. Acta*, **151**, 159.
BECKER, B., and FRIEDENWALD, J. S. (1949). *Archs Biochem.*, **22**, 101.
BERNFELD, P., GUARINO, A. J., and FISHMAN, W. H. (1950). *Science, N.Y.*, **112**, 653.
BERTHELOT, M. (1860). *C.r. Seanc. Acad. Sci., Paris*, **50**, 980.
BILLETT, F. (1957). *Biochem. J.*, **67**, 463.
BILLETT, F., and MCGEE-RUSSELL, S. M. (1956). *Q. J. microsc. Sci.*, **96**, 35.
BILLETT, F., and MCGEE-RUSSELL, S. M. (1956). *Q. J. microsc. Sci.*, **97**, 155.
BILLING, B. H., COLE, P. G., and LATHE, G. H. (1957). *Biochem. J.*, **65**, 774.
BILSKI, R., and GODLEWSKI, H. G. (1965). *Folia Histochem. Cytochem.*, **3**, 225.
BO, W. J., and SMITH, M. (1964a). *J. Histochem. Cytochem.*, **12**, 393.
BO, W. J., and SMITH, M. (1964b). *Anat. Rec.*, **148**, 503.
BOCEK, R. M., and BEATTY, C. H. (1966). *J. Histochem. Cytochem.*, **14**, 549.
BOTS, G. Th. A. M. (1964). "Enzymhistochemie van de Bijnierschors van de Rat." Thesis, Leiden.

Bots, G. Th. A. M., Nagy, K., and Willighagen, R. G. J. (1965). *Histochemie*, **5**, 244.
Brady, R. O., Kanfer, J., and Shapiro, D. (1965). *J. biol. Chem.*, **240**, 39.
Braun-Falco, O. (1956). *Arch. klin. exp. Derm.*, **202**, 163.
Brauss, E., and Sasse, D. (1968). *Histochemie*, **14**, 260.
Breckenridge, B. M., and Norman, J. H. (1962). *J. Neurochem.*, **9**, 383.
Breckenridge, B. M., and Norman, J. H. (1965). *J. Neurochem.*, **12**, 51.
Brunet, P. C. J., and Kent, P. W. (1955). *Proc. R. Soc. Med.*, **144**, 259.
Buell, M. V., and Hansen, R. E. (1961). *J. biol. Chem.*, **236**, 1991.
Bulmer, D. (1967). *J. R. microsc. Soc.*, **86**, 385.
Burton, J. F., and Pearse, A. G. E. (1952). *Brit. J. exp. Path.*, **33**, 1.
Cabrini, R. L., and Carranza, F. A., jr. (1960). *Arch. oral Biol.*, **1**, 28.
Cabrini, R. L., and Schajowicz, F. (1960). *Rev. Ortoped. Traumatol. Lat.-Amer.* **5**, 193.
Campbell, J. G. (1949). *Ibid.*, **30**, 548.
Carbonell, L. M. (1955). *J. Histochem. Cytochem.*, **3**, 419.
Christie, G. A. (1968). *Histochemie*, **12**, 189.
Cobb, J. (1949). *Anat. Rec.*, **103**, 531.
Cohen, R. B., Rutenburg, S. H., Tsou, K.-C., Woodbury, M. A., and Seligman, A. M. (1952a). *J. biol. Chem.*, **195**, 607.
Cohen, R. B., Tsou, K.-C., Rutenburg, S. H., and Seligman, A. M. (1952b). *Ibid.*, **195**, 239.
Conchie, J., Gelman, A. L., and Levvy, G. A. (1967). *Biochem. J.*, **103**, 609.
Conchie, J., and Levvy, G. A. (1957). *Biochem. J.*, **65**, 389.
Cori, G. T., Colowick, S. P., and Cori, C. F. (1938). *J. biol. chem.*, **123**, 375.
Cori, G. T., and Cori, C. F. (1943). *Ibid.*, **151**, 57.
Cori, G. T., and Green, A. A. (1943). *J. biol. Chem.*, **151**, 21.
Cori, G. T., Illingworth, B., and Keller, P. J. (1955). In "Methods in Enzymology." Eds. Colowick, S. P., and Kaplan, N. O. Vol. I, p. 200, Acad. Press, New York.
Dahlqvist, A. (1962a). *Gastroenterology*, **43**, 694.
Dahlqvist, A. (1962b). *J. clin. Invest.*, **41**, 463.
Dahlqvist, A., and Borgström, R. (1961). *Biochem. J.*, **81**, 411.
Dahlqvist, A., Bull, B., and Gustafson, B. E. (1965). *Archs Biochem.*, **101**, 150.
Dahlqvist, A., and Brun, A. (1962). *J. Histochem. Cytochem.*, **10**, 294.
Dahlqvist, A., and Telenius, U. (1969). *Biochem. J.*, **111**, 139.
Dahlqvist, A., and Thomson, D. L. (1965). *Archs Biochem.*, **109**, 159.
de Duve, C., and Beaufay, H. (1959). *Biochem. J.*, **73**, 610.
de Duve, C., Wattiaux, R., and Baudhuin, P. (1962). *Adv. Enzymol.*, **24**, 291.
De Lellis, R. A., and Fishman, W. H. (1968). *Histochemie*, **13**, 1.
Doell, R. G., and Kretchmer, K. (1962). *Biochim. biophys. Acta*, **62**, 353.
Dodgson, K. S., Lewis, J. I. M., and Spencer, B. (1953). *Biochem. J.*, **55**, 253.
East, M. E., Madinaveitia, J., and Todd, A. R. (1941). *Biochem. J.*, **35**, 872.
Egli, K. (1967). *Histochemie*, **8**, 164.
Ellis, R. A., and Montagna, W. (1958). *J. Histochem. Cytochem.*, **6**, 201.
Eränkö, O., and Palkama, A. (1961). *J. Histochem. Cytochem.*, **9**, 585.
Esterley, J. R. (1967). *Science, N.Y.*, **216**, 821.
Esterley, J. R., Standen, A. C., and Pearson, B. (1967). *J. Histochem. Cytochem.*, **15**, 470.
Esterley, J. R., Standen, A. C., and Pearson, B. (1968). *J. Histochem. Cytochem.*, **16**, 489.
Fabbrini, A., Re, M., and Spera, G. (1968). *Experientia*, **24**, 789.
Fagundes, L. A., and Cohen, R. B. (1965). *J. Histochem. Cytochem.*, **13**, 553.
Findlay, J., Levvy, G. A., and Marsh, C. A. (1958). *Biochem. J.*, **69**, 467.
Fishman, W. H. (1940). *J. biol Chem.*, **136**, 229.
Fishman, W. H. (1950). "The Enzymes." Vol. I, Part 1, p. 635.
Fishman, W. H. (1964). *J. Histochem. Cytochem.*, **12**, 306.
Fishman, W. H., and Anylan, A. J. (1947). *Science, N.Y.*, **106**, 66.
Fishman, W. H., Anlyan, A. J., and Gordon, E. (1947). *Cancer Res.*, **7**, 808.

FISHMAN, W. H., and BAKER, J. R. (1956). *J. Histochem. Cytochem.*, **4**, 570.
FISHMAN, W. H., and BIGELOW, R. (1950). *J. natn. Cancer Inst.*, **10**, 1115.
FISHMAN, W. H., and DE LELLIS, R. A. (1967). *Nature, Lond.*, **213**, 457.
FISHMAN, W. H., GOLDMAN, S. S., and GREEN, S. (1964). *J. Histochem. Cytochem.*, **12**, 239.
FISHMAN, W. H., and GREEN, S. (1957). *J. biol. Chem.*, **225**, 435.
FRIEDENWALD, J. S., and BECKER, B. (1948). *J. cell. comp. Physiol.*, **31**, 303.
FRIESEN, A. J. D., ALLEN, G., and VALADARES, J. R. E. (1967). *Science, N.Y.*, **155**, 1108.
GENTSHEV, T. (1967). *Brain Res.*, **5**, 350.
GODLEWSKI, H. G. (1960). *Bull. Acad. pol. sci.*, **8**, 441.
GODLEWSKI, H. G. (1961). *Acta histochem.*, **11**, 58.
GODLEWSKI, H. G. (1963). *Folia Histochem. Cytochem.*, **1**, 231.
GODLEWSKI, H. G. (1963). *J. Histochem. Cytochem.*, **11**, 108.
GODLEWSKI, H. G., and PENAR, B. (1961). *Bull. Acad. pol. sci.*, **9**, 195.
GOLDBERG, B., WADE, O. R., and JONES, H. W., jr. (1952). *J. natn. Cancer Inst.*, **13**, 543.
GREENBAUM, A. L., SLATER, T. F., and WANG, D. Y. (1965). *Biochem. J.*, **97**, 518.
GRILLO, T. A. I. (1960). *J. Histochem. Cytochem.*, **8**, 322.
GRILLO, T. A. I. (1961). *J. Histochem. Cytochem.*, **9**, 386.
GRILLO, T. A. I., OKUNO, G., PRICE, S., and FOÁ, P. P. (1964). *J. Histochem Cytochem.*, **12**, 275.
GRILLO, T. A. I., and OZONE, K. (1962). *Nature, Lond.*, **195**, 902.
GROSSMAN, I. W., and SACKTOR, B. (1968). *Science, N.Y.*, **161**, 571.
GUHA, S., and WEGMANN, R. (1959). *Ann. Histochem.*, **4**, 103.
GUHA, S., and WEGMANN, R. (1960a). *Ann. Histochem.*, **5**, 1.
GUHA, S., and WEGMANN, R. (1960b). *Bull. Soc. Chim. biol.*, **42**, 115.
GUHA, S., and WEGMANN, R. (1961). *J. Histochem. Cytochem.*, **9**, 454.
GUHA, S., and WEGMANN, R. (1965). *J. Histochem. Cytochem.*, **13**, 138.
HADJIOLOV, A. A., and DANCHEVA, K. I. (1957). *C.r. Acad. Bulg. Sci.*, **10**, 488.
HADJIOLOV, A. A., and DANCHEVA, K. I. (1958). *Nature, Lond.*, **181**, 547.
HADJIOLOV, A. A., and DANCHEVA, K. I. (1959). *Bull. Soc. Chim. biol.*, **41**, 707.
HAHN, L. (1945). *Ark. Kemi. Min. Geol. A.*, **21**, No. 1.
HASSID, W. Z. (1945). *Fedn. Proc.*, **4**, 227.
HAYASHI, M. (1964). *J. Histochem. Cytochem.*, **12**, 659.
HAYASHI, M. (1965). *J. Histochem. Cytochem.*, **13**, 355.
HAYASHI, M. (1967). *J. Histochem. Cytochem.*, **15**, 83.
HAYASHI, M., NAKAJIMA, Y., and FISHMAN, W. H. (1964). *J. Histochem. Cytochem.*, **12**, 293.
HAYASHI, M., SIMODA, K., OGATA, K., TAKAMORI, T., SHIRAOGAWA, T., SATAKE, T., UTIDA, M., SIMONAGAYOSI, S., and KAWASE, O. (1955). *Kumamoto Med. J.*, **8**, 233.
HAYASHI, M., SIMODA, K., OGATA, K., TAKAMORI, T., SHIRAOGAWA, T., and KAWASE, O. (1956–57). *Acta histochem.*, **3**, 277.
HERVEG, J. P., BECKER, C., and DE VISSCHER, M. (1966). *Biochem. J.*, **100**, 540.
HESS, R., and PEARSE, A. G. E. (1961). *Proc. Soc. exp. Biol. Med.*, **107**, 569.
HEYWORTH, R., and WALKER, P. G. (1962). *Biochem. J.*, **83**, 331.
HOLLINGER, D. M., and ROSSITER, R. J. (1952). *Biochem. J.*, **52**, 659.
HORI, S. H. (1966a). *J. Histochem. Cytochem.*, **14**, 501.
HORI, S. H. (1966b). *Stain Tech.*, **41**, 91.
HSIA, D. Y. Y., MAHLER, M., SEMENZA, G., and PRADER, A. (1966). *Biochim. biophys. Acta*, **113**, 390.
HUXLEY, J. S. (1924). *Nature, Lond.*, **114**, 895.
HUIJING, F., NUTTALL, F. Q., VILLAR-PALASI, C., and LARNER, J. (1969). *Biochim. biophys. Acta*, **177**, 204.
IBRAHIM, M. Z. M., and CASTELLANI, P. (1968). *Histochemie*, **16**, 9.
IDE, H., GREEN, S., KATO, K., and FISHMAN, W. H. (1968). *Biochem. J.*, **106**, 341.
JANIGAN, D. T., and PEARSE, A. G. E. (1962). *J. Histochem. Cytochem.*, **10**, 719.
JOS, J., FRÉZED, J., REY, J., and LAMY, M. (1967). *Nature, Lond.*, **213**, 516.

Jos, J. (1968). Proc. IIIrd. Int. Cong. Histochem. Cytochem. New York, Springer. p. 120.
Karunairatnam, M. C., and Levvy, G. A. (1949). *Biochem. J.*, **44**, 599.
Keleman, M. V., and Whelan, W. J. (1966). *Biochem. J.*, **100**, 5P.
Keller, M., and Tanka, D. (1965). *Acta histochem.*, **22**, 73.
Kerr, L. M. H., Campbell, J. G., and Levvy, G. A. (1949). *Ibid.*, **44**, 487.
Kishi, K. (1969). *Bull. Tokyo Med. Dent. Univ.*, **16**, 289.
Korsgaard, B., and Wulff, H. R. (1967). *Acta path. microbiol. scand.*, **70**, 236.
Kozlowska, K. (1968). *Acta histochem.*, **30**, 380.
Krebs, E. G., and Fischer, E. H. (1955). *J. biol. Chem.*, **216**, 113.
Krebs, E. G., and Fischer, E. H. (1956). *Biochim. biophys. Acta*, **20**, 150.
Krebs, E. G., and Fischer, E. H. (1962). *Adv. Enzymol.*, **24**, 263.
Laasonen, L., and Palkama, A. (1967). *Ann. Med. exp. Fenn.*, **45**, 307.
Langley, T. J., and Jevons, F. R. (1967). *Biochem. J.*, **104**, 31P.
Larner, J., and Gillespie, R. E. (1956). *J. biol. Chem.*, **223**, 709.
Leaback, D. H., and Walker, P. G. (1957). *J. Chem. Soc.*, 4754.
Leloir, L. F., and Cardini, C. E. (1957). *J. Am. chem. Soc.*, **79**, 6340.
Leloir, L. F., and Goldenberg, S. H. (1960). *J. biol. Chem.*, **235**, 919.
Levvy, G. A. (1947). *Nature, Lond.*, **160**, 54.
Levvy, G. A., (1948). *Biochem. J.*, **42**, 2.
Levvy, G. A. (1952). *Ibid.*, **52**, 464.
Levvy, G. A. (1954). *Ibid.*, **58**, 462.
Levvy, G. A., Hay, A. J., and Marsh, C. A. (1957). *Ibid.*, **65**, 203.
Levvy, G. A., Kerr, L. M. H., and Campbell, J. G. (1948). *Ibid.*, **42**, 462.
Levvy, G. A., and Marsh, C. A. (1959). *Advanc. Carbohydrate Chem.*, **14**, 381.
Levvy, G. A., McAllan, A. (1963a). *Biochem. J.*, **87**, 206.
Levvy, G. A., and McAllan, A. (1963b). *Biochem. J.*, **87**, 361.
Levvy, G. A., McAllan, A., and Marsh, C. A. (1958). *Ibid.*, **69**, 22.
Lindberg, L.-A., and Palkama, A. (1970). *Ann Med. exp. Fenn.*, **48**, 67.
Linker, A., Meyer, K., and Weissmann, B. (1955). *J. biol. Chem.*, **213**, 237.
Lojda, Z. (1965). *Histochemie*, **5**, 339.
Lojda, Z. (1970a). *Histochemie*, **22**, 347.
Lojda, Z. (1970b). *Histochemie*, **23**, 266.
Lojda, Z., and Kraml, I. (1971). *Histochemie*, **25**, 195.
Lorbacher, P., Yam, L. T., and Mitus, W. J. (1967). *J. Histochem. Cytochem.*, **15**, 680.
MacKenzie, A., Millson, G. C., and Wilson, A. M. (1968). *J. comp. Path*, **78**, 43.
McMahon, P., Brand, T., and Nolan, M. D. (1957). *J. cell. comp. Physiol.*, **50**, 219.
MacPherson, C. R., and Pearse, A. G. E. (1957). *Br. med. Bull.*, **13**, 19.
Meijer, A. E. F. H. (1968). *Histochemie*, **12**, 244.
Meijer, A. E. F. H. (1969). *Histochemie*, **16**, 134.
Miller, D., and Crane, R. K. (1961). *Biochim. biophys. Acta*, **52**, 293.
Mills, G. T. (1948). *Biochem. J.*, **43**, 125.
Mills, G. T., Paul, J., and Smith, E. E. B. (1953a). *Biochem. J.*, **53**, 232.
Mills, G. T., Paul, J., and Smith, E. E. B. (1953b). *Ibid.*, **53**, 245.
Mills, G. T., and Smith, E. E. B. (1951). *Science, N.Y.*, **114**, 690.
Monis, B., and Wasserkrug, H. (1967). *Histochemie*, **10**, 362.
Mori, M., Tanaka, S., and Azumagawa, H. (1962). *Arch. Histol. jap.*, **22**, 421.
Monis, B., Tsou, K.-C., and Seligman, A. M. (1963). *J. Histochem. Cytochem.*, **11**, 653.
Najjar, V. A. (1948). *J. biol. Chem.*, **175**, 281.
Najjar, V. A., and Pullman, M. E. (1954). *Science, N.Y.*, **119**, 631.
Nene, R. V., and George, J. C. (1965). *J. Anim. Morph. Physiol.*, **12**, 85.
Newcomer, A. D., and McGill, D. B. (1966). *Gastroenterology*, **51**, 481.
Nitowsky, H. M., and Grunfeld, A. (1967). *J. Lab. clin. Med.*, **69**, 472.
Oshima, G. (1934). *J. Biochem., Tokyo*, **20**, 361.
Oshima, G. (1936). *Ibid.*, **23**, 305.
Pearse, A. G. E., and MacPherson, C. R. (1958). *J. Path. Bact.*, **75**, 75, 69.

PEARSON, B., ANDREWS, M., and GROSE, F. (1961). *Proc. Soc. exp. Biol. Med.*, **108**, 619.
PEARSON, B., STANDEN, A. C., and ESTERLEY, J. R. (1967). *Lab. Invest.*, **17**, 217.
PEARSON, B., WOLF, P. and VASQUEZ, J. (1963). *Lab. Invest.*, **12**, 1249.
PELLEGRINO, C., and VILLANI, G. (1956). *Biochem. J.*, **62**, 235.
PELLEGRINO, C., and VILLANI, G. (1957). *Ibid.*, **65**, 599.
PENAR, B. (1963). *Folia histochem. cytochem.*, **1**, 198.
PHILLIPART, M., and MENKES, J. (1964). *Biochem. biophys. Res. Commun.*, **15**, 551.
PRICE, R. G., and DANCE, N. (1967). *Biochem. J.* **105**, 877.
PUGH, D., LEABACK, D. H., and WALKER, P. G. (1957). *Biochem. J.*, **65**, 464.
PUGH, D. and WALKER, P. G. (1958). *Proc. IV. Int. Cong. Biochem. Vienna*, 4–55.
PUGH, D., and WALKER, P. G. (1961a). *J. Histochem. Cytochem.*, **9**, 242.
PUGH, D., and WALKER, P. G. (1961b). *J. Histochem. Cytochem.*, **9**, 105.
QUAGLINO, D., and HAYHOE, F. G. J. (1962). *Nature, Lond.*, **194**, 929.
RATH, F. W., and OTTO, L. (1966). *Acta Histochem.*, **25**, 355.
RAUNIO, V. (1968). *Acta path. microbiol. scand.*, Suppl. 195.
ROBINSON, D., PRICE, R. G., and DANCE, N. (1967). *Biochem. J.* **102**, 525.
ROBINSON, D., and STIRLING, J. L. (1968). *Biochem. J.*, **107**, 321.
ROBINSON, J. P., and NAJJAR, V. A. (1961). *Fedn. Proc.*, **20**, 90.
RUTENBURG, A. M., GOLDBERG, J. A., RUTENBURG, S. H., and LANG, R. T. (1960). *J. Histochem. Cytochem.*, **8**, 268.
RUTENBURG, A. M., LANG, R., GOLBERG, J. A., and RUTENBURG, S. H. (1958a). *J. Histochem. Cytochem.*, **6**, 396.
RUTENBURG, A. M., RUTENBURG, S. H., MONIS, B., TEAGUE, R., and SELIGMAN, A. M. (1958a). *J. Histochem. Cytochem.*, **6**, 122.
SASAI, Y. (1965). *Tohoku J. exp. Med.*, **85**, 160.
SASSE, D. (1966). *Histochemie*, **7**, 39.
SAWYER, D., SIE, H., and FISHMAN, W. H. (1965). *J. Histochem. Cytochem.*, **13**, 605.
SCHRIEB, D. (1963). In "Ciba Found. Symp.: Lysosomes; Eds. de Reuck, A. V. S. and Cameron, M. P." London, J. & A. Churchill, p. 264.
SELIGMAN, A. M., and NACHLAS, M. M. (1950). *J. clin. Invest.*, **29**, 31.
SELIGMAN, A. M., NACHLAS, M. M., MANHEIMER, L. H., FRIEDMAN, O. M. and WOLF, G. (1949). *Ann. Surg.*, **130**, 333.
SELIGMAN, A. M., TSOU, K.-C., RUTENBURG, S. H., and COHEN, R. B. (1954). *J. Histochem. Cytochem.*, **2**, 209.
SELLINGER, O. Z., BEAUFAU, H., JACQUES, P., DOYEN, A., and DE DUVE, C. (1960). *Biochem. J.*, **74**, 450.
SEMENZA, G., and AURICCHIO, S. (1962), *Biochim. biophys. Acta*, **65**, 173.
SEMENZA, G., AURICCHIO, S., and RUBINO, A. (1965). *Biochim. biophys. Acta*, **96**, 487.
SHIBKO, S., and TAPPEL, A. L. (1965). *Biochem. J.*, **95**, 731.
SHIMIZU, N., and OKADA, M. (1957). *J. Histochem. Cytochem.*, **5**, 459.
SMITH, A. A., and PERKINS, E. M. jr. (1967). *Folia Histochem. Cytochem.*, **5**, 323.
SMITH, A. A., PERKINS, E. M., and MACHIDA, H. (1966). *Stain Technol.* **41**, 346.
SMITH, E. E. B., and MILLS, G. T. (1953). *Biochem. J.*, **54**, 164.
SÖRENSEN, S. P. L. (1909). *Biochem. Z.*, **21**, 131.
SØVIK, Ø., ØYE, I., ROSELL-PEREZ, M. (1966). *Biochim. biophys. Acta*, **124**, 26.
SPENCER, B., and WILLIAMS, R. T. (1951). *Biochem. J.*, **48**, 537.
STRUGALSKA-CYNOWSKA, M. (1967). *Folia Histochem. Cytochem.*, **5**, 151.
SUGINO, M. (1957). *Igaku Kenkyu* (*Acta med.*), **27**, 2704.
SUTHERLAND, E. W. (1951). *Ann. N.Y. Acad. Sci.*, **54**, 693.
TAKEUCHI, T. (1958). *J. Histochem. Cytochem.*, **6**, 208.
TAKEUCHI, T., and GLENNER, G. G. (1960). *J. Histochem. Cytochem.*, **8**, 227.
TAKEUCHI, T., and GLENNER, G. G. (1961). *J. Histochem. Cytochem.*, **9**, 304.
TAKEUCHI, T., HIGASHI, K., and WATANUKI, S. (1955). *J. Histochem. Cytochem.*, **3**, 485.
TAKEUCHI, T., and KURIAKI, H. (1955). *J. Histochem. Cytochem.*, **3**, 153.
TAKEUCHI, T., and KINOSHITA, K. (1956). *Blood*, **11**, 375.
UGOLEV, A. M., JESUITOVA, N. N., and DE LACY, P. (1964). *Nature, Lond.*, **203**, 879.

VALLYATHAN, N. V., CHERIAN, K. M., and GEORGE, J. C. (1964). *J. Histochem. Cytochem.*, **12**, 721.
VALLYATHAN, N. V., and GEORGE, J. C. (1964). *J. Anim. Morph. Physiol.*, **11**, 186.
VAN LANCKER, J. L. (1964). *Fed. Proc.*, **23**, 1050.
VILLAR-PALASI, C., and LARNER J. (1958). *Biochim. biophys. Acta*, **30**, 449.
WALLENFELS, K., and MALHOTRA, O. M. (1961). *Advanc. Carbohydrate Chem.*, **16**, 239.
WATANABE, K. (1936a). *J. Biochem., Tokyo*, **24**, 297.
WATANABE, K. (1936b). *Ibid.*, **24**, 315.
WATANABE, K., and FISHMAN, W. H. (1964). *J. Histochem. Cytochem.*, **12**, 908.
WATANUKI, S. (1957). *Igaku Kenkyu* (*Acta med.*), **27**, 1059.
WEBER, R. (1957). *Experientia*, **13**, 153.
WELLS, G. C., SANDERSON, K. V., and MCCABE, I. M. (1961). *Br. J. Derm.*, **73**, 337.
WILLS, E. D., and WORMALL, A. (1950). *Biochem. J.*, **47**, 158.
WOESSNER, J. F., jr. (1965). *Biochem. J.*, **97**, 855.
WULFF, H. R., and SØRENSEN, M. (1966). *Acta haemat. Basel*, **35**, 304.
YIN, H. C., and SUN, C. N. (1947). *Science, N.Y.*, **105**, 650.
YOSHIOKA, W., MORI, M., MIZUSHIMA, T., and DEGUCHI, S. (1960). *Arch. histol. jap.*, **20**, 529.

CHAPTER 19

OXIDOREDUCTASES I
(Oxidases and Peroxidases)

The histochemistry of the oxidases still lags behind their biochemistry although the position has certainly improved since 1960. The enzymes we have to consider are *cytochrome oxidase*, *peroxidase*, *catalase*, *catechol oxidase*, *monoamine oxidase*, *diamine oxidase*, *urate oxidase*, D-*amino-acid oxidase* and *xanthine oxidase*. The first three are haemoprotein, or heme-copper-protein enzymes, the next four are copper proteins and the last two flavoproteins. The chapter is therefore divided into three convenient sections, haemoproteins, copper proteins, and flavoproteins.

The Haemoprotein Oxidases

Cytochrome Oxidase and the Succinic Oxidase System

The respiratory pigments now known as cytochromes were discovered by MacMunn (1886) but his observations were belittled by Levy (1889) and by Hoppe-Seyler (1890) who maintained that the pigments were haemochromogens derived from haemoglobin. After confirming MacMunn's findings Keilin (1925) applied the name cytochromes to the pigments concerned and in a series of papers (Keilin, 1929, 1930; Keilin and Hartree, 1937, 1938, 1939, 1949) he described the nature of the cytochrome-cytochrome oxidase system. An entirely independent series of observations by Warburg (Warburg and Negelein, 1929, 1931) resulted in the description of an iron-containing enzyme called *Atmungsferment* which played an important role in aerobic cellular respiration.

According to the nature of the haemoprosthetic group the cytochromes are divided into 3 groups, *a*, *b*, and *c*, and these are the main components of the normal aerobic oxidation system of cells. The whole system has been called the succinic oxidase system and it may be written descriptively as follows:

$$\text{Succinate} \xrightarrow[\text{dehydrogenase}]{\text{succinic}} \text{cytochrome b} \rightarrow \text{factor} \rightarrow \text{cytochrome c} \rightarrow \text{cytochromes a} \rightarrow a_3 \rightarrow O_2$$

$$\text{cytochrome b} \downarrow \text{methylene blue}$$

The proteins to the left of cytochrome c. were collectively known as the succinic dehydrogenase system and those to the right of it as cytochrome oxidase (Singer, Kearney and Massey, 1956), but otherwise, the term cytochrome oxidase was regarded as synonymous with cytochrome a_3 (Keilin and Hartree, 1939; Ball and Cooper, 1952; Smith and Stotz; 1954; Stotz *et al.*,

1956). According to Dixon and Webb (1964) the precise relations between cytochromes a and a_3 are still uncertain, but since the two cannot be separated, at the present time, it is convenient to regard the complex as the oxidase. Stotz *et al.*, purified cytochrome oxidase and found that it consisted of 33·2 per cent lipid (14·3 per cent phospholipid). In mammalian cells the enzyme is entirely intramitochondrial and it is closely bound to the structure of the mitochondria. It is now known to be a heme-copper-protein consisting of 2 heme groups and two copper atoms bound to a protein of about 150,000 to 200,000 mol. wt. The porphyrin ring of the heme group possesses a long alkyl side chain which determines the lipid character of the prosthetic group (Lemberg, 1969).

Lemberg's comprehensive review should be consulted by those further interested in the composition of the enzyme, and for details of the various theories of the mechanism by which it achieves the reduction of oxygen to water. The most popular theory remains that of Keilin and Hartree ($O_2 \rightarrow a_3 \rightarrow a \rightarrow c \rightarrow c_1$).

The oxidation-reduction potentials of cytochrome oxidase have been elucidated mainly on the α-band (due 80 per cent to cytochrome a). At pH 7·4 the mid-point potential of a number of measurements made in different laboratories is + 278–290 mV.

Distribution of the Succinic Oxidase System. The components of the system are distributed widely in the cells of various organisms, often in quantities too small to be demonstrated by histochemical means. The amount of cytochrome oxidase-cytochrome c. which is present in animal tissues depends on the oxidative capacity of the organ concerned. Thus, in the rat, heart muscle and kidney are two of the richest sources. Malignant tumours have often been reported to possess a low cytochrome content (Elliot and Grieg, 1938; Schneider and Potter, 1943a), but this is not always below the normal figure for the tissue of origin. Succinic dehydrogenase is also distributed widely in normal tissues and a number of studies have been made, particularly in the rat, by Meyer and his collaborators (Meyer *et al.*, 1947; McShan *et al.*, 1946; McShan and Meyer, 1946). Since a rise in succinic oxidase activity has been observed in tissues in which the increased respiratory rate is part of a general increase in activity, a large number of studies have been made of malignant and precancerous tissues (e.g. Schneider and Potter, 1943a and b; Slack, 1943; Carruthers and Suntzeff, 1947; Gallico, 1947; Hoch-Ligeti, 1947a and b). Most of these have indicated a gradual increase in one or other of the components of the system with the onset of the malignant change. It is because of these studies that interest in the histochemical demonstration of the succinic oxidase system has increased in recent years.

Inhibitors of Cytochrome Oxidase. The oxidation of reduced cytochrome c. by cytochrome oxidase *in vitro* was found by Keilin (1936) to be inhibited by cyanide or by sodium azide (NaN_3) but *in vivo*, the results were found to be complicated by the presence of other enzyme systems. H_2S also causes inhibition of the enzyme, as does carbon monoxide in the absence of light, by forming complexes with the iron present in the molecule. Takamori *et al.* (1960) showed that a number of aldehyde reagents, such as hydrazine, phenylhydrazine, bisulphite and hydroxylamine would inhibit cytochrome oxidase at mM concentrations. Takamori *et al.* (1962) reported inhibition by basic compounds which included salmine and polylysine. Person and Fine (1960) had already shown that basic proteins (protamine sulphate, histone, lysozyme, ribonuclease) were potent inhibitors of heart muscle cytochrome oxidase. Inhibition of this type was completely reversed in the presence of strongly anionic polyglucose sulphate.

Indophenol Oxidase. It has been known for over 30 years that an enzyme is present in the majority of tissues which catalyses the oxidative reaction between α-naphthol and dimethyl-*p*-phenylenediamine, the so-called "Nadi" reaction, to form indophenol blue. The enzyme responsible was originally

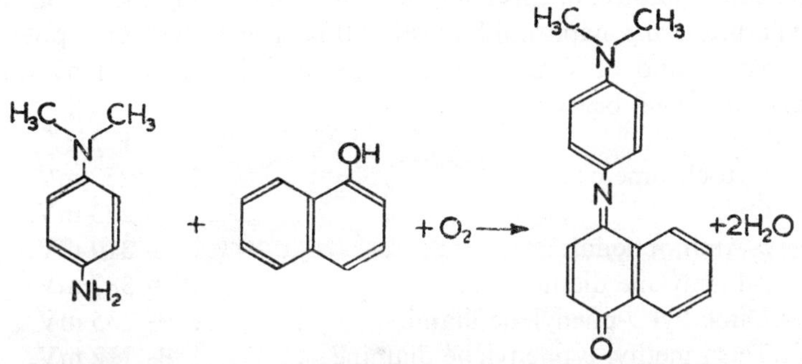

Dimethyl-*p*-phenylenediamine α-naphthol Indophenol blue

known as indophenol oxidase. In 1938, however, Keilin and Hartree showed that indophenol oxidase would catalyse the oxidation of reduced cytochrome, and that cytochrome oxidase would produce indophenol blue in the "Nadi" reaction. They concluded as the result of many experiments, that indophenol oxidase, *Atmungsferment*, and cytochrome oxidase were one and the same enzyme. It was therefore evident that the "Nadi" reaction was capable of demonstrating the presence of cytochrome oxidase and its substrate, reduced cytochrome c.

We now know, from the work of Keilin and Hartree, that the "Nadi" reaction really proceeds as follows:

α-naphthol + dimethyl-*p*-phenylenediamine + 4 cytochrome c. $\xrightarrow{\text{spontaneous}}$ Indophenol blue + 4 reduced cytochrome c.

And 4 reduced cytochrome c. + O_2 $\xrightarrow{\text{cytochrome oxidase}}$ 4 cytochrome c. + H_2O.

The α-naphthol of the "Nadi" reagent acts as a mild reducing agent and the cytochrome oxidase part of the succinic oxidase system can be demonstrated *in vitro* if the usual reducing (dehydrogenase) system is replaced by any mild reducing agent capable of acting as a hydrogen donor. Examples are *p*-phenylenediamine or hydroquinone. The oxidation of substituted phenylenediamines *via* the cytochrome system was studied by Borei and Björklund (1953). These authors found that oxidation of dimethyl-*p*-phenylenediamine by the system was only very slightly increased by the addition of cytochrome c. They therefore considered the possibility that the endogenous cytochrome c. of their preparations, mediating between oxidase and substrate, might not be sufficient to account for the total oxidation observed. It was suggested that the substrate might be directly attacked by the oxidase.

In biological systems the initial oxidation of dimethyl-*p*-phenylenediamine is rapid but the rate soon falls to zero on account of toxicity of the reaction products. Autoxidation of this compound is rapid, due to metal ion catalysis, and the addition of α-naphthol intensifies this process. The O-R potentials of cytochrome c., and of some of the compounds investigated by Borei and Björklund are given below:

	E°*p*H7
Cytochrome c.	+ 253 mV.
Quinol	+ 273 mV.
p-Aminophenol	+ 319 mV.
p-Phenylene diamine	+ 387 mV.
Dimethyl-*p*-phenylene diamine	+ 335 mV.
Tetra methyl-*p*-phenylene diamine	+ 222 mV.

The O-R potential of cytochrome c. is thus seen to be lower than any of the usual reductants but this does not prevent cytochrome c. from acting as an effective carrier in oxidizing various compounds since a substance of higher O-R potential can continue to reduce one of lower potential if the oxidized state of the former is unstable.

The Nadi Reactions

Historical Survey

Two methods were given by Lison (1936) under the heading of "Nadi-oxidase" reactions, and both were described as indicating the presence of phenoloxidases. The first, stable, or M-Nadi oxidase reaction was evolved by Schultze (1909a and b) and performed on formalin-fixed frozen sections, while the second, labile, or G-Nadi oxidase reaction (Gräff, 1916) was performed on fresh tissues. The first method was largely employed in haematological research since the granules of cells of the myeloid series were found to be invariably positive, as well as the granules of serous salivary glands and lachrymal glands in human tissues. Permanent preparations could be obtained by treating the blue granules with Lugol's iodine, turning them brown in the process, and mounting in an aqueous medium. The M-Nadi oxidase method cannot be considered as an indicator of the cytochrome oxidase-cytochrome c. complex, since this is totally inactivated by formalin. It was at one time considered to be due to the presence of a monophenol oxidase since with this group of enzymes some activity remains after treatment with formalin (see catechol oxidase methods, below). More recent work, described later in this chapter, attributes M-Nadi oxidase activity to peroxidase. A reaction of similar type was described by Loele (1912, 1913) using formalin-fixed material and an alkaline solution of α-naphthol. Bluish-violet granules were produced in certain cells, notably those of the myeloid leucocyte series, and the enzyme responsible was characterized as an α-naphthol oxidase. Alternatively it has been suggested that it may be due solely to the presence of fatty peroxides of the type shown by the Dam reaction (Chapter 26, p. 1078). Takamatsu and Hirai (1968) investigated the stable Nadi reaction of myeloid leucocytes. They showed that myeloid granules contained an acetone-soluble compound which was essential for a positive reaction. They identified it as a naphthoquinone, probably bound to the phospholipid of the granules. This compound is probably identical with the substance responsible for the positive hydroquinone-tetrazolium reduction method (see Chapter 23, p. 1001 and vol. 1, p. 429). Neither the M-Nadi oxidase reaction nor the Loele method are of use in diagnostic enzyme histochemistry.

Of greater significance is the second (G-Nadi oxidase) method, which is certainly capable of indicating the presence of cytochrome oxidase-cytochrome c. With this method Gräff showed that the pH of the incubating medium was of great importance, and that the optimum value varied with the type of cell under consideration. He observed that with animal tissues the optimum pH was between 7·8 and 8·2, while in the case of plant tissues it was between 3·4 and 5·9. Objections to the specificity of the "Nadi" reaction in histochemistry are many and varied. It was suggested that since indophenol blue is formed from the initial reagents in the presence of molecular oxygen alone, the reaction could be specific. Oxidation by air occurs very slowly, however, and this cannot be a significant factor. A more reasonable objection was that indophenol blue could be used as a stain for lipids, and a certain similarity between the results obtained with the usual lipid stains and those produced by the "Nadi" reaction was early observed. Not all lipid inclusions are shown by the latter method, however, and Lison maintained that whereas the "true reaction" gave a strong blue colour, the "false reaction" of lipids was pale and of a lilac hue. Hollande (1924), however, showed that the identical appearances of the "Nadi" reaction could be obtained by using indophenol blue as a stain. He believed that the granular localization of the oxidase reaction was secondary and that it represented staining of the granules by the dye produced in the cytoplasm of the cell. The granules, in fact, were not oxidase-positive but *indophenophile*. As a corollary, Hollande suggested that in the absence of granules the reaction remained diffuse and invisible under the microscope. Prenant (1924, 1925) was able to reproduce Holland's findings but he disagreed with that author's interpretations, maintaining that the oxidases were not diffuse but attached to the granulations. More modern work agrees substantially with Prenant's views in locating the succinic oxidase systems predominantly in particles of the size and character of mitochondria (Hogeboom, Schneider and Palade, 1948). Indophenol blue is insoluble in water so that diffusion away from the site of production, other than possibly into lipid substances, might be expected to be small.

No attempts were made to check the specificity of the histochemical "Nadi" reaction for cytochrome oxidase, by the use of specific inhibitors, until the work of Moog (1943). It is known that the succinic oxidase system is instantaneously and reversibly inhibited by cyanides, sodium azide or hydrogen sulphide, on account of their combination with the trivalent Fe of cytochrome oxidase. Moog used 0·005M-sodium azide, treating fresh tissues with this reagent in saline at pH 5·8 for a short period and then transferring the material to freshly prepared "Nadi" reagent containing 0·005M-azide. Inhibition was complete. Using the "Nadi" reaction without any inhibitor Moog found that identical results were obtained at pH 5·8 and at pH 7·2. Since the coloured oxidation products of both the "Nadi" reagent and benzidine are easily reduced to colourless substances if the reducing systems of the cell are intact, it is not always clear that a negative result is really due to lack of oxidase. If the reducing systems are blocked, or otherwise inactivated, a negative result more clearly indicates lack of oxidase and Moog therefore incubated control sections in the presence of phenylurethane which would block the reducing systems. She observed no difference between these sections and those incubated in the absence of phenylurethane. Modifications of the reaction were also made by Perner (1952) who described the deposition of indophenol blue in onion skin cells as microsomal. Numerous attempts to make the reaction permanent were all more or less unsuccessful.

Experiments carried out by Person and Fine (1961) showed clearly that the Keilin-Hartree heart muscle preparation oxidized only the diamine component of the Nadi reagents. No oxidation of α-naphthol occurred. Oxidation of the diamine produced a stable free radical and polymers with characteristic absorption spectra. These were not indophenol blue. Both the free radical and the polymers reacted rapidly with α-naphthol in a non-enzymatic oxidative coupling. Person and Fine emphasized that α-naphthol should not be used in methods for the localization of aerobic oxidases. This is because the rapid non-enzymatic coupling of α-naphthol and the diamine free radical exerts a mass action effect which substantially increases the auto-oxidation rate of the diamine.

In tissues rich in cytochromes auto-oxidation is probably not significant, provided that care is taken to exclude the presence of catalytic amounts of copper. In tissues with low terminal respiratory enzyme activity accurate interpretation of the results of the Nadi reaction is not possible. Although newer methods have replaced the latter the original reagents and their method of preparation are given in Appendix 19, p. 1330, on account of their historical interest.

Methods for Cytochrome Oxidase (E.C. 1.9.3.1)

Amine Free Radical Coupling Methods. Studies on the Nadi reaction carried out by Nachlas, Crawford, Goldstein and Seligman (1958) constituted a considerable advance in the histochemistry of cytochrome oxidase. These authors replaced the dimethyl-*p*-phenylenediamine of the original reaction with a new compound, 4-amino-*N*,*N*-dimethylnaphthylamine (ADN). The new "Nadi" reaction, shown in the equation below, gave rise to an insoluble pigment which the authors called indonaphthol purple.

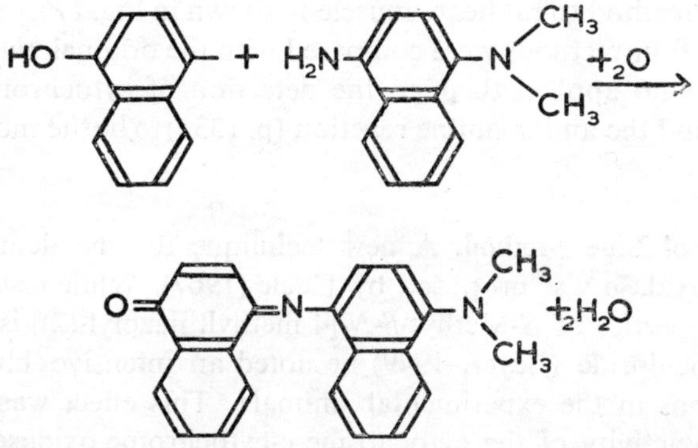

A number of naphthols were tested as substitutes for α-naphthol but none were satisfactory from the histochemical point of view. Both the rate of auto-oxidation, and of oxidation catalysed by added cytochrome c., are slower with the new reagent than with the original "Nadi" reagents but the overall histochemical result (Fig. 175, p. 848) is far more satisfactory. From this point of view the properties of indonaphthol purple are much better than those of indophenol blue. In particular, crystallization is no longer observed and the pigment is stable for a considerable period. Post-fixation of the preparations in formalin must be avoided, however, since this reagent causes definite fading of the pigment. Indonaphthol purple has two outstanding deficiencies; it is not substantive for protein, and it remains lipid soluble like indophenol blue. It is deposited in the tissues in the form of small granules and these appear to be intramitochondrial especially when protective media have been employed. The granules have a tendency to coalesce during storage and it is advisable to photograph the fresh preparations if a permanent record is required of the finer localizations of cytochrome oxidase. Details of the method ADN appear in Appendix 19, p. 1330, although it has been superseded by improved techniques (see below).

A series of alternative cytochrome oxidase methods was proposed by

Burstone (1959, 1960, 1961). These can be divided into four main groups, depending on the type of coupler employed. The four groups are (1) *Naphthol-amine* (2) *Quinoline-amine* (3) *Quinone-amine* and (4) *Amine-amine.*

The reagents selected by Burstone were designed to provide insoluble dyestaff polymers capable of metal chelation. Initially (1960) Burstone used a wide selection of amines to provide the free radical amine for coupling but later (1961) he used only one, *p*-aminodiphenylamine (*N*-phenyl-*p*-phenyl-amine), and the following couplers: 8-amino-1,2,3,4-tetrahydroquinoline, 5-nitro-, and 4-nitro-1-naphthylamine, 8-hydroxy-1,4-naphthoquinone and 1-hydroxy-2-naphthoic acid.

The nature of the final products of Burstone's cytochrome oxidase methods has not been determined. Examples of one of each of the four types of reaction are given in Appendix 19 and the result of the application of the amine-amine method to rat heart muscle is shown in Fig. 176, opposite. Three of Burstone's four methods were compared with the original Nadi reaction by Reiss (1967) who applied them to the detection of cytochrome oxidase in fungi. He found the amine-amine reaction (p. 1331) to be the most effective of the four.

The Thiazol Blue Method. A new technique for the demonstration of cytochrome oxidase was proposed by Thiele (1967). While testing the tuber-culostatic properties of *S*-Methyl-*β*-*N*-[4-methylthiazolyl-(2)]-isothiosemicar-bazide hydrochloride (Beyer, 1949) he noted an intensive blue staining in different organs in the experimental animals. This effect was shown to be dependent on activity of the cytochrome c-cytochrome oxidase system. The blue thiazol dyes from the original substrate, and from several closely related compounds, were not markedly crystalline. They were insoluble in water and in lipids.

Auto-oxidation was found to be a considerable problem at physiological pH levels but the 4-ethyl, 4-*n*-propyl and 4-isopropyl isomers were less susceptible than the original compound, in this respect. Under histochemical conditions the incubation period is short and auto-oxidation can be kept at a low level. Details are given in Appendix 19. The average redox potential of the above-mentioned compounds is in the region of + 220 mV.

Methods Using Polymerizing Diamines. Seligman *et al.* (1967) described the synthesis of a series of *p*-substituted aromatic diamines for use in the Nadi reaction. The best of these reagents, *N*-benzyl-*p*-phenylene diamine (BPDA) condensed with *α*-naphthol to produce a blue indoaniline dye which was found to be osmiophilic. None of the compounds was capable of self polymerization and all gave rise to indoanilines which were deposited in the tissues in droplet form. A search for reagents which would give rise to a non-particulate product was initiated, stimulated by the discovery (Seligman *et al.* 1968) that 3, 3′-diaminobenzidine (DAB) could be used in the Nadi reaction to

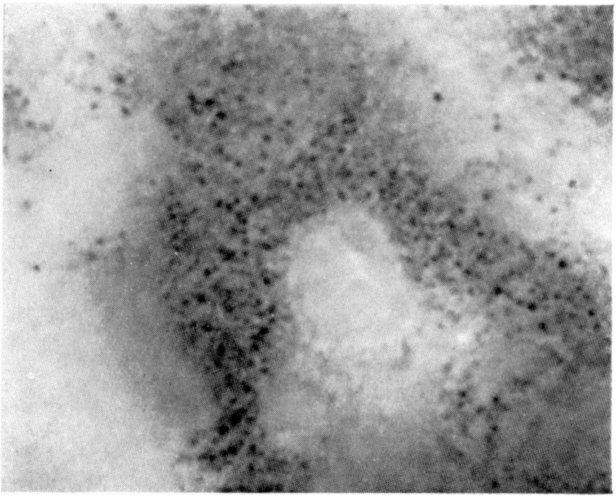

FIG. 175. High power view of single tubule. Cytochrome oxidase appears in the form of (intramitochondrial) dots. ADN method. × 2224.

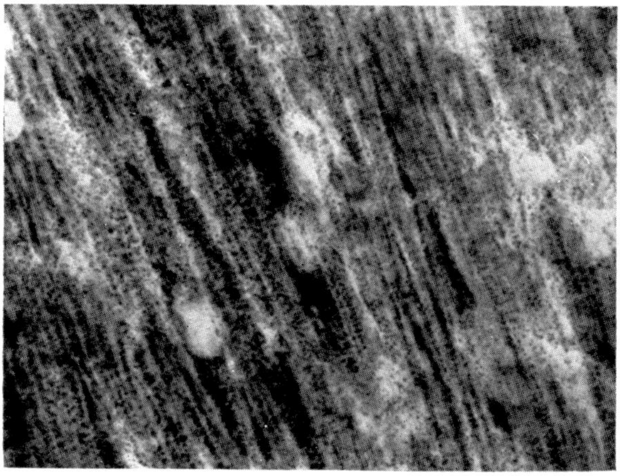

FIG. 176. Rat heart muscle. Granular distribution of cytochrome oxidase in sarcosomes. Burstone's method. × 775.

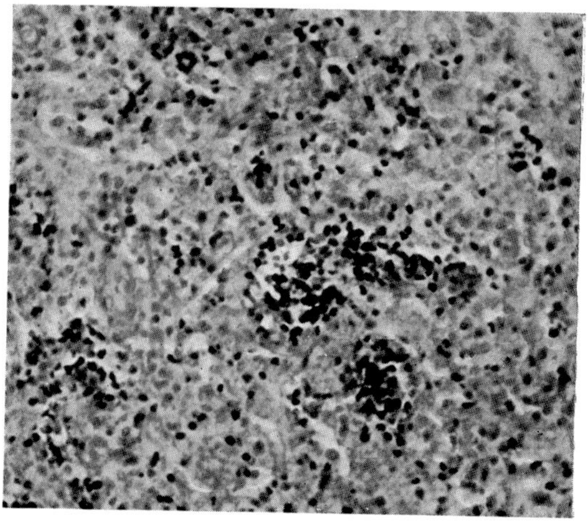

FIG. 178. Human spleen stained by the leuco-Patent Blue method for haemoglobin. × 190.

give rise to an insoluble compound, by oxidative polymerization (A) and oxidative cyclization (B).

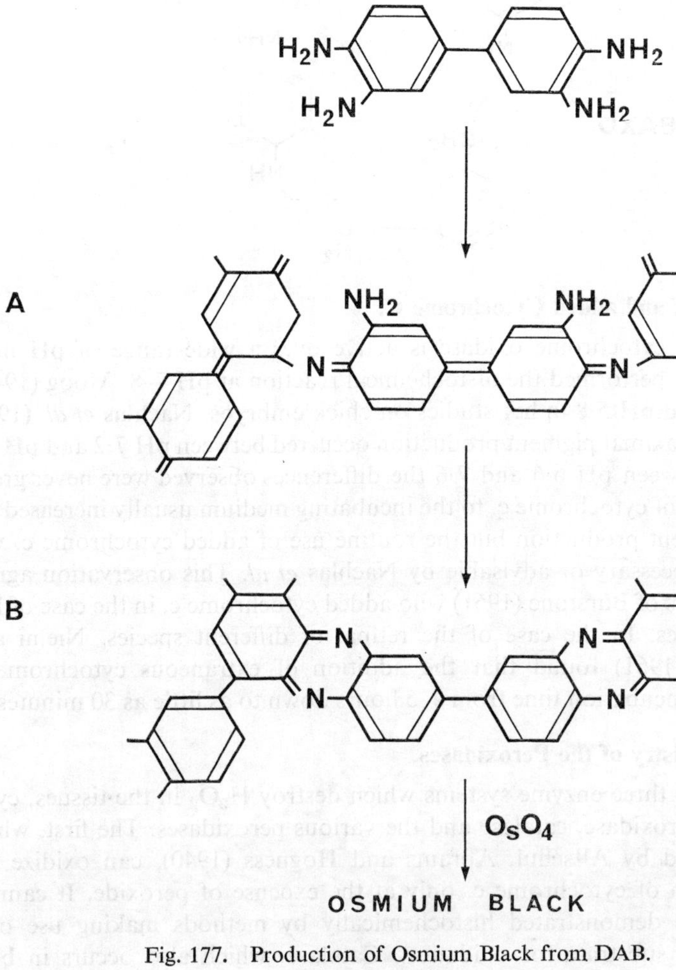

Fig. 177. Production of Osmium Black from DAB.

Treatment of the product with OsO_4 gave rise to an electron dense, coloured product which was termed osmium black.

As a result of the foregoing work, a series of compounds capable of self-polymerization was synthesized by Plapinger *et al.* (1968). The most useful of the new reagents (BAXD) was *N*,*N'*-bis (*p*-aminophenyl)-1,3-xylylene-diamine, whose structural formula is shown on p. 850. The method for the demonstration of cytochrome oxidase, at optical microscopic level, using BAXD or DAB, is given in Appendix 19, p. 1332.

BAXD

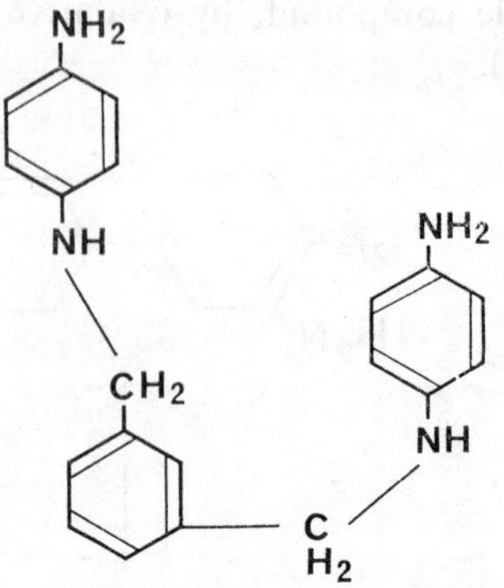

Effects of pH and added Cytochrome C.

Although cytochrome oxidase is active over a wide range of pH most workers have performed the histochemical reaction at pH 7–8. Moog (1943), however, used pH 5·8 in her studies on chick embryos. Nachlas *et al.* (1958) found that maximal pigment production occurred between pH 7·2 and pH 7·4 although between pH 6·6 and 7·6 the differences observed were never great.

Addition of cytochrome c. to the incubating medium usually increased the rate of pigment production but the routine use of added cytochrome c. was not found necessary or advisable by Nachlas *et al.* This observation agrees with the views of Burstone (1961) who added cytochrome c. in the case of low activity tissues. In the case of the retina, of different species, Niemi and Merenmies (1961) found that the addition of extraneous cytochrome c. reduced the incubation time from 4–5 hours down to as little as 30 minutes.

The Biochemistry of the Peroxidases

There are three enzyme systems which destroy H_2O_2 in the tissues, cytochrome c.-peroxidase, catalase and the various peroxidases. The first, which was described by Altschul, Abrams and Hogness (1940), can oxidize the reduced form of cytochrome c. only at the expense of peroxide. It cannot, therefore, be demonstrated histochemically by methods making use of a chromogenic substrate for oxidation. Catalase, which also occurs in both animal and plant tissues, splits H_2O_2 into water and molecular oxygen. Like cytochrome c. peroxidase, it cannot be shown histochemically with the usual chromogenic substrates since H_2O_2 acts both as substrate and acceptor. The peroxidases, which can be demonstrated by means of chromogenic substrates, are weak or lacking in most animal tissues though they occur in moderate amounts in mammary gland and in the granules of leucocytes of the myeloid series. The mammalian peroxidases differ from those occurring in plants in that the haem compound they contain is green in colour, and they are strictly classified as verdoperoxidases. Most haem and haematin compounds exhibit weak peroxidase activity, a property made use of histochemically for the demonstration of haemoglobin in the red blood cells (see below). Peroxidases catalyse reactions of the type $AH_2 + H_2O_2 \rightarrow A + 2H_2O$, and the peroxide-peroxidase system thus causes the rapid oxidation of substances which peroxide alone oxidizes extremely slowly. The need for the presence of H_2O_2 by peroxidases was made use of by Agner (1941), who demonstrated that the "Nadi" reaction of leucocyte granules could be abolished by the addition of catalase, indicating that their "Nadi oxidase" was, in fact, a peroxidase.

The status of the peroxidases has been discussed in a comprehensive review by Saunders (1957), and in a more recent treatise by Paul (1963) who concluded his relevant section with the words "a specific histochemical assay able to give permanently stained sections would be of much value for the elucidation of biological functions of peroxidases". Three enzymes in this class have now been obtained in crystalline form. They are the protoheme peroxidase of horse radish, and the two iron-containing peroxidases of myeloid leucocytes and milk. There are many other peroxidases but at the present time it seems that no detailed classification is possible. The peroxidase system is capable of oxidizing a variety of amines to coloured compounds and it will also oxidize phenolic methyl groups via aldehyde to quinone (Booth and Saunders, 1950). Hughes and Saunders (1954) have shown, furthermore, that the system is capable of breaking carbon-halogen bonds, including the C-F bond. During the oxidation of *p*-iodoaniline the iodine atom is eliminated and then oxidized to I_2, which can then iodinate a suitable acceptor. This process of transiodination is of considerable importance in relation to the iodination of tyrosine compounds in thyroid metabolism (see below).

Interest in the peroxidases has mainly been confined to the study of the leucocyte granules and this has been mainly the province of the haematologist (Yamamoto *et al.*, 1960; Archer and Broome, 1963; Inagaki, 1963; Kaplow, 1965; Lojda, 1967). Some more general interest was raised by the finding of peroxidase activity in the thyroid acini and colloid by Dempsey (1944), confirmed by de Robertis and Grasso (1946). Since peroxidase catalyses the liberation of iodine from inorganic iodides by hydrogen peroxide, its presence in the thyroid would serve a useful function. Both Dempsey, and the other authors quoted, observed that the reaction was inhibited by low concentrations of thiourea and that much larger concentrations of thiourea were required to inhibit the peroxidase activity of red blood cells. Glock (1944) criticized the specificity of Dempsey's findings on the ground that the concentration of thiourea necessary to inhibit peroxidase activity in the horse thyroid was of the order which inhibited red cell activity, and Randall (1946) showed that thiols did not inhibit peroxidase at all, but simply reduced the coloured products formed by the peroxidase-catalysed reactions.

Peroxidases can act as oxidases, in the presence of certain cofactors. This interrelationship was discussed, from the histochemical point of view, by Fujita (1959) and by Fujita *et al.* (1960). Experimental work reported by Okun *et al.* (1970a and b) has provided evidence that in the presence of

DOPA as cofactor mammalian peroxidase can catalyse the oxidation of L-tyrosine to melanin.

Histochemical Methods for Peroxidases (E.C. 1.11.1.7)

Few of the many oxidizable substrates for the peroxidase system give rise to products sufficiently deeply coloured to be visible in tissue sections under the microscope. Five groups of substances, in particular, have been used in the histochemical demonstration of peroxidases. These are (1) benzidine and its derivatives (2) phenols and naphthols (3) leuco-dyes (4) indoles and (5) amino-carbazoles.

Benzidine Reactions

These reactions depend on the oxidation of benzidine by the peroxide-peroxidase system to a blue or brown product. The formula reproduced below was given by Lison (1936, p. 166).

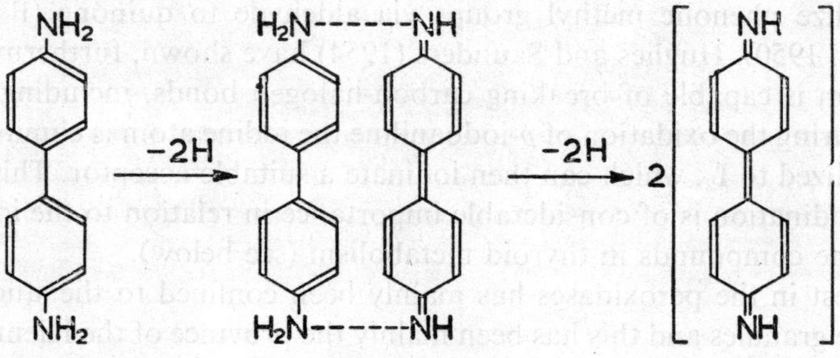

Colourless Blue Brown

The intermediate product, benzidine blue, is unstable and is oxidized to the brown compound without enzymic intervention. The reaction can take place over a wide pH range and no attempt is usually made to control the pH by means of buffers. Some authors recommend acidification of the benzidine-peroxide medium and state that if an acid solution is used the resulting colour is predominantly blue rather than brown. If the explanation given by Lison is correct this indicates that the stability of the intermediate compound is increased. Some authors, for instance Garcia Blanco and Grisolia (1946) made a histochemical distinction between pseudoperoxidase and true peroxidase reactions, separating red cell from other cytoplasmic peroxidases; others made a similar distinction by treatment with heat, since the pseudo-peroxidases are relatively heat stable. Although it may be necessary, as in the case of the thyroid acinar cells for instance, to establish that the positive reaction is not due to haemoglobin, the peroxidase of the latter should not be distinguished as pseudoperoxidase. According to Tu *et al.* (1968) the peroxidative activity of both horse heart cytochrome c. and of horse radish peroxidase are inhibited by low concentrations of CN^- whereas only the peroxidase is inhibited by fluoride ion.

The optimal concentration of H_2O_2 for the peroxidase reaction is between 0·015 and 0·03 per cent (1 or 2 drops of 3 per cent H_2O_2 in 10 ml. medium). Higher concentrations may inhibit the reaction by destroying the enzyme. Certain substances have been found to intensify the reaction and Mancini and Villamil (1950) described the action of molybdate in this respect as that of a catalyser. The exact meaning of the term, in this context, is uncertain. Ettori (1949), using *in vitro* methods for the estimation of peroxidase, showed that at pH 5·9, the rate of enzyme activity was increased by increasing the concentration of phosphate buffer up to an optimum level of 0·07M. In the presence of phosphate buffer the harmful effect of excess H_2O_2 is considerably reduced. Improvements in the technique of the benzidine-peroxidase reaction were reported by Mitsui and Ikeda (1951) and by van Duijn (1955). These improvements were based on the observation that if a high salt concentration was maintained the blue intermediate compound (above) precipitated in the form of stable crystals. Suitable conditions were established by Mitsui and Ikeda with nickel ammonium sulphate and by van Duijn with the addition of 5 per cent ammonium chloride to the incubating medium. The Japanese authors reported that their mixture was stable for months; van Duijn's must be freshly prepared since after 30 minutes benzidine blue begins to appear spontaneously. This can be controlled to some extent by addition of EDTA. The improved technique of van Duijn is given in Appendix 19, p. 1334.

A further improvement in the benzidine peroxidase reaction, for use with fresh cryostat sections, was introduced by Wachstein and Meisel (1964) who

TABLE 65

List of Benzidine-peroxide Methods

Author	Material	Fixative or Mordant
Graham (1918) . . .	Smears	Formalin-alcohol
Goodpasture (1919) . . .	Smears and sections	Formalin (optional)
McJunkin (1922) . . .	Paraffin sections	Formalin
Prenant (1924) . . .	Fresh tissues	None
Sato *et al.* (1925) . . .	Smears	Copper sulphate
Loele (1927) . . .	Smears	Various
Washburn (1928) . . .	Smears	None
Armitage (1939) . . .	Smears	Formalin-alcohol
Mancini and Celani (1940) .	Smears and sections	Ammonium molybdate
de Robertis and Grasso (1946)	Frozen sections	Ammonium molybdate
Cazal and Roure (1947) .	Smears	Formalin-alcohol-$CuSO_4$
Villamil and Mancini (1947)	Frozen sections	Ammonium molybdate
Bonati and Franchi (1948) .	Smears	None
Mitsui and Ikeda (1951) .	Smears	Nickel ammonium-SO_4
van Duijn (1955) . . .	Frozen sections	None
Avers and Grimm (1959) .	Sections	Nitroprusside
Straus (1964) . . .	Sections	Nitroprusside
Wachstein and Meisel (1964)	Sections	Ferricyanide

incubated their tissues in 6 per cent benzidine in 25 per cent ethanol containing 3 times the usual concentration of H_2O_2. Incubation was carried out at 4° for 5–30 minutes. Details of the method are given in Appendix 19.

Methods for air-dried fixed or unfixed smears of blood and bone marrow are dealt with in text books of haematology, and some of them appear in Table 65, which is a list of the various benzidine-peroxide methods, applicable both to smears and to sections. The only methods for which technical details are given in Appendix 18 are the methods of de Robertis and Grasso (1946), of van Duijn (1955), and of Wachstein and Meisel (1964).

Benzidine Substitutes. Until recently the possibility of using alternatives to benzidine was little explored in histochemical practice. On account of its carcinogenic qualities this compound has been replaced by *o*-tolidine in tests for blood in urine or faeces, and *o*-tolidine can be used in place of benzidine as a reagent for the demonstration of copper-containing pigments in the marine invertebrates (Chapter 28, p. 1146). Another alternative is mesidine, the chemistry of whose oxidation by a peroxide-peroxidase system was explored by Chapman and Saunders (1941). The oxidation product, 2,6-dimethylbenzoquinone-4-(2′,4′,6′-trimethylanil), was described as purple in colour and poorly soluble in water. In histochemical systems mesidine gives rise to diffuse purple staining of the tissues so that its solubility is probably too great.

Diaminobenzidine. The currently most popular benzidine substitute, 3,3′-diaminobenzidine, was introduced into histochemical use by Graham and Karnovsky (1966). This compound (DAB) has already received mention (p. 849) as a hydrogen donor in the Nadi reaction. It was presumed to act in the same manner in the peroxidase reaction, that is to say the oxidation product was presumed to condense to form an osmiophilic indamine polymer. Although the main use of DAB has been for the electron cytochemical identification of microbodies (see also Chapter 32, p. 1286) it can be employed, as shown by Fahimi (1968), to show the same structures at the level of the optical microscope.

The reaction of the microbodies, whose protein content is 40 per cent catalase, is inhibited by the specific catalase inhibitor 1,2,4-triazole (Margoliash and Novogrodsky, 1958). It appears that under histochemical conditions it is the peroxidatic activity of catalase which oxidises DAB. According to de Duve and Baudhuin (1966) the functional role of microbody catalase is likely to be concerned also with its peroxidatic effect.

Investigating the properties of DAB in relation to heme enzymes Hirai (1968) showed that the oxidized radical had a strong affinity for these, forming specific bonds with cytochrome c, catalase and haemoglobin *in vitro*. Using the oxidized radical in alkaline solution Hirai was able to stain not only microbodies but also mitochondrial cristae, erythrocytes and leucocyte granules. Cyanide and azide prevented staining of mitochondria and microbodies but not of leucocyte granules. The catalase inhibitor 1,2,4-triazole prevented the uptake of oxidized DAB by microbodies but *not* by mitochondria. Hirai concluded that DAB could best be used, for heme enzyme histochemistry at optical and electron microscopic levels, in conjunction with specific inhibitors. It was recommended by Goldfischer (1967), at 3× the usual concentration, for the localization of myoglobin in skeletal muscle. Used in this way non-specific reactions are prominent but, nevertheless, it can be assumed that the glutaraldehyde-resistant peroxidase of muscle is myoglobin.

Reactions Based on Naphthols and Phenols.

α-Naphthol has been used less often than benzidine for histochemical peroxidase reactions. It gives a red colour on oxidation, probably due to the formation of quinones.

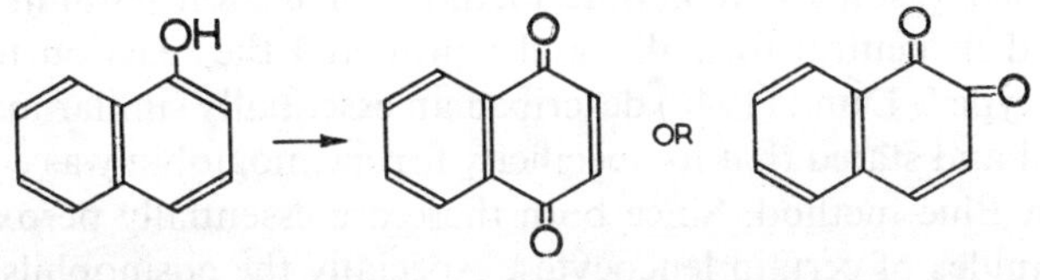

The two principal methods using this substance are those of Graham (1916) and of Ritter and Oleson (1947). The former was designed for use on fresh unfixed blood smears and the latter, which is given in Appendix 19, employs thin blocks of unfixed tissue which are subsequently fixed and embedded in paraffin. It is not very satisfactory, in practice, and the benzidine methods are to be preferred. It is probable, in fact, that the α-naphthol methods demonstrate only the so-called stable sudanophilia of Lillie and Burtner (1953).

The specificity of the α-naphthol reaction is in any case doubtful, as shown by Gomori (1953), since a reaction can be obtained in the absence of added H_2O_2. The possibility that fatty peroxides might take part in the reaction is made unlikely by the fact that if the reaction product is removed the reaction can be repeated. Using horse eosinophil leucocytes Vercauteren and Blondé (1954) were unable to demonstrate the participation of fatty peroxides in the α-naphthol or benzidine reactions. On the whole it seems that the α-naphthol reaction is not enzymic in nature and that it depends on the presence of undetermined oxidation products in solutions of α-naphthol.

Pyrogallol has been advocated as a hydrogen donor for peroxidase systems (Jensen, 1955; Doxey, 1962), particularly in plants. In his work on fungi, however, Reiss (1967) was unable to produce any reaction with pyrogallol-H_2O_2 media.

Leuco-dye Reactions

Lison (1936) introduced the use of acid fuchsin and other dyes, reduced by means of nascent hydrogen, as indicators for the peroxide peroxidase systems.

These zinc-leucos, as he called them, were produced by boiling the dye with powdered zinc and acetic acid until the colour of the original solution was discharged. Although Lison suggested that preparations made by the above methods were both beautiful and stable, they have not been much used by other workers. Fautrez and Lambert (1937) published a method for the demonstration of haemoglobin in fixed tissue sections using leuco-Cyanol (xylene-Cyanol FF), but they did not apparently consider that the method could be used as an indicator of peroxidase activity. Lison (1938) introduced a leuco-Patent Blue technique for the identification of haemoglobin casts and globules in the kidneys of frogs injected with haemoglobin solutions. He used a formalin-ferricyanide fixative and frozen sections. Patent Blue is a triphenylmethane dye which closely resembles Cyanol except that it possesses four instead of two ethyl groups and no methyl group. Dunn and Thompson (1946) modified Lison's Patent Blue method for use on paraffin sections of material fixed in neutral formalin and considered the reaction to be of the "peroxidase type". Dunn (1946) described an essentially similar method using leuco-Cyanol and stated that its specificity for haemoglobin was equal to that of the Patent Blue method. Since both these are essentially peroxidase reactions the granules of certain leucocytes, especially the eosinophils, stain blue by either. Fig. 178 (p. 848) shows red cells in splenic sinusoids demonstrated by the leuco-Patent Blue method, using formalin fixed tissue.

The use of thioindoxyls (3-hydroxythionaphthenes) as hydrogen donors in oxidase systems, introduced by Pearse (1954), has not been fully explored. The halogen-substituted thioindoxyls give rise to brightly coloured, insoluble and stable pigments on oxidation which compare favourably with the reaction products of most oxidase reactions reported in this chapter. Thioindoxyls are oxidized by the granules of mammalian eosinophils and by cells of the myelocyte series, especially those present in certain types of leukaemia. The reaction in cells from a case of myeloid leukaemia (chloroma) is shown in Plate XIXa, p. 868. It takes place between pH 6·0 and 9·2 and between 20° and 65° and, in formalin-fixed tissues, proceeds without the addition of H_2O_2. It has been postulated that the reaction is due to a peroxide-peroxidase complex which can be inhibited by ferricyanide, catalase and by excess H_2O_2. A possible source of peroxide in formalin-fixed tissues, which might be responsible for the formation of the complex, is methyl hydroperoxide ($CH_3.O.OH$). This is present in most commercial samples of formalin.

Indole Reagents

In a series of experiments designed to elucidate the relationship between peroxidase and the oxidation of dihydroxyphenylalanine (DOPA) van der Ploeg and van Duijn (1964a and b) showed that, under the influence of peroxidase and H_2O_2, one of the intermediates, 5,6-dihydroxyindole, was rapidly converted into melanin.

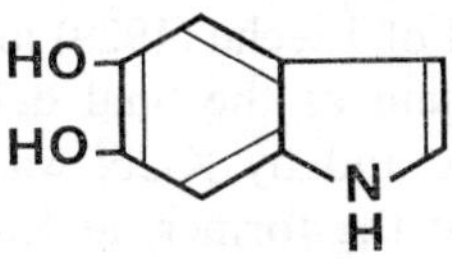

5,6-Dihydroxyindole

The synthesis of 5,6-dihydroxyindole was eventually carried out according to the method of Beer (1948) and the product was used as a substrate for the histochemical demonstration of endogenous and exogenous peroxidase. Although 5,6-dihydroxyindole is not commercially available the method is given in Appendix 19 since it provides an interesting alternative to the other peroxidase systems described in this chapter. Its potential usefulness for electron cytochemistry remains to be determined.

Carbazole Reagents

After its introduction into histochemical practice by Burstone (1960), as one of a series of amine reagents for the demonstration of cytochrome oxidase, 3-amino-9-ethylcarbazole was employed by Graham *et al.* (1965) as a reagent for peroxidase.

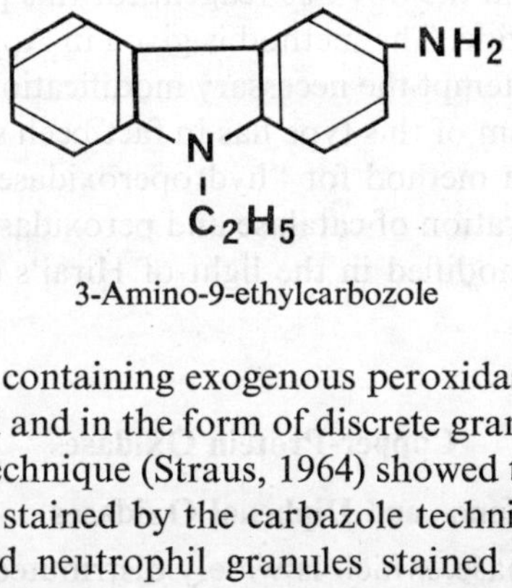

3-Amino-9-ethylcarbozole

Using fixed tissues containing exogenous peroxidase, the reaction product was observed to be red and in the form of discrete granules. Comparison with a sensitive benzidine technique (Straus, 1964) showed that some sites revealed by the latter were not stained by the carbazole technique. With fresh tissues mast cell granules and neutrophil granules stained intensively, as did the cytoplasm of a small proportion of Kupfer cells (rat liver). Once again, the sensitivity of the reaction was less than that of the benzidine technique.

The carbazole method is given in Appendix 19, p. 1338. It should always be used in conjunction with an alternative procedure.

Histochemical Method for Catalase (E.C. 1.11.1.6)

Many unsuccessful attempts have been made to produce a histochemical reaction for catalase. Only one report of a successful attempt has so far appeared in the literature (Nishiyama and Kobayashi, 1953). Their method is based on the observation by Stern (1936) that acetaldehyde is produced by the action of catalase on monoethyl hydrogen peroxide. The authors synthesized this substrate by the method of Rieche (1929) and used it in a histochemical procedure with Schiff's reagent as the final demonstrator of the aldehyde. Inhibition by hydroxylamine and by KCN was considered to indicate the specificity of the method but the former, at least, would be more likely to exert its effect by blocking pre-existing aldehydes. Some type of aldehyde blocking procedure is an essential prerequisite of methods of this type. The reaction was thought to take place as follows:

$$C_2H_5OOH \xrightarrow{\text{Catalase}} CH_3CHO + H_2O$$

$$+$$

$$\text{Schiff's Reagent}$$

$$\downarrow$$

$$\text{Red Product}$$

There are so many basic objections to the above principle that the use of the method, as it stands, cannot be advised. The primary reaction product is freely soluble and no evidence has been produced to show that the final red dye is not soluble also. The introduction of an effective capture reagent is absolutely necessary and the obvious reagent for this purpose is 2-hydroxy-3-napththoic acid hydrazide. The method is given in Appendix 19 so that those who feel disposed to attempt the necessary modifications can proceed without hindrance. A mechanism of this type has in fact been suggested by Watanabe (1958) as the basis of a method for "hydroperoxidase". This author regards the histochemical separation of catalase and peroxidase as unrealistic but this conclusion has to be modified in the light of Hirai's (1967) observations on the use of inhibitors.

Copper-Protein Oxidases

The Biochemistry of Mono- and Diphenol Oxidases

The enzyme tyrosinase, which is widely distributed in plants and animals has, for many years, been considered responsible for the conversion of tyrosine into melanin. Principally on account of the histochemical work of Bloch (1917), the precursor of melanin in animal tissues was considered to be dihydroxyphenylalanine (DOPA) and the enzyme concerned with its oxidation was called DOPA-oxidase. In the decade following Bloch's publication of his original method various objections to its specificity were put forward. Przibram (1922) considered that DOPA might be transformed into melanin by two processes, first, by a simple oxidation in alkaline solution, and secondly, by enzyme action at a neutral pH. He disagreed with Bloch's contention that DOPA-oxidase did not act on tyrosine but agreed that the cells demonstrated by the DOPA reaction contained an enzyme of some sort. Hasebroek (1921) investigated the cause of pigmentation in moths and showed that in the egg or larval stage only DOPA could be oxidized to a black pigment, but that in the chrysalis or in the adult moth both DOPA and tyrosine could be so dealt with. These views were substantiated by the findings of Lerner *et al.* (1949) that tyrosinase caused only slow oxidation of tyrosine unless a small amount of DOPA was present, when the speed of oxidation approached that of DOPA itself. Although the majority of authors agreed that enzyme action was responsible for the production of pigment from DOPA, considerable doubt was felt in identifying DOPA-oxidase with the enzyme of normal melanin production, since its substrate could not be demonstrated in mammalian tissues. The only difference between tyrosine and DOPA is the presence of a second hydroxyl group on the benzene ring and it was considered that the oxidation of tyrosine proceeded in the following manner. It is easy to see that the effect of the second hydroxyl group of DOPA is to facilitate oxidation at that locus.

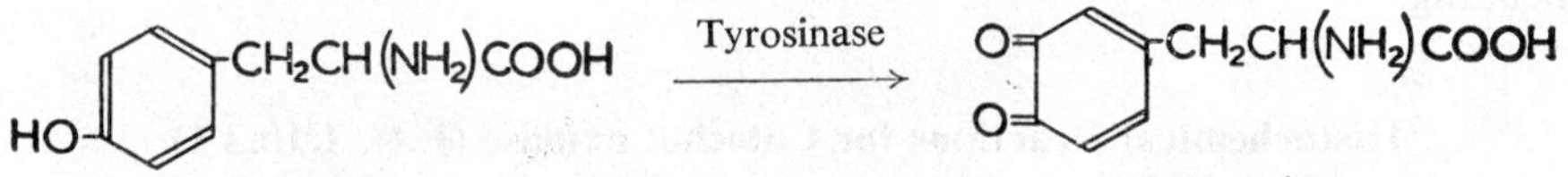

Tyrosine DOPA quinone

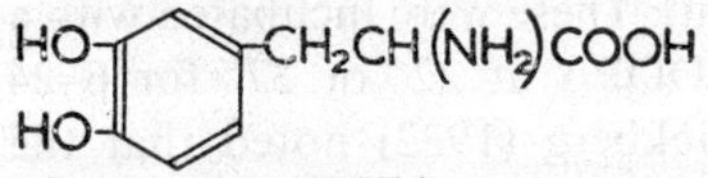

DOPA

If the views of Raper (1926) and of Evans and Raper (1937) are correct, DOPA is to be regarded as the first stage of the oxidation of tyrosine by tyrosinase. These views were based on work on the plant enzyme and they cannot necessarily be applied to animal tissues. Hogeboom and Adams (1942) first showed that tyrosinase as well as DOPA-oxidase could be demonstrated in mouse melanoma and Lerner *et al.* (1949) established the identity of these two enzymes in the same mammalian tissue (Harding-Passey mouse melanoma). These authors were not able to separate tyrosinase and DOPA-oxidase activity, although it was possible to prepare fractions which oxidized tyrosine only very slowly. They concluded that a single enzyme or enzyme-complex was involved in the oxidation of both substrates to melanin and this view is illustrated diagrammatically below.

$$\text{Tyrosine} \xrightarrow[\text{slow}]{} \text{DOPA} \xrightarrow[\text{fast}]{} \text{Intermediates} \longrightarrow \text{Melanin}$$

(Tyrosinase acting on both the slow and the fast step)

In view of this evidence it became customary to regard the enzyme demonstrated by the DOPA reaction as "tyrosinase" and to assume that it was capable of acting either as a monophenol oxidase or as an *o*-diphenol oxidase. According to Kertész (1952) the so-called tyrosinase was really a very complicated system composed of (1) an enzyme with copper as its prosthetic group, specific for *o*-diphenols; (2) an *o*-diphenol and (3) certain free metallic ions. The complexity of this system is increased by the interdependence of its components since the quantity of *o*-diphenol (or *o*-quinone) depends on the amount of enzyme and on the quality and quantity of the metallic ions present. Kertész considered that the only justification for retention of the term "tyrosinase" was an historical one.

The work of Okun and his associates (Okun *et al.*, 1969, 1970a,b,c) has indicated that peroxidase initiates melanin synthesis by oxidizing tyrosine to DOPA and DOPA quinone. In this system tyrosinase acts mainly as a DOPA oxidase, as originally conceived by Bloch. The agreed terminology (I.U.B.C. Enzyme Commission) for the enzyme, which will probably continue to be called DOPA-oxidase by histochemists, is *o*-diphenol: O_2 oxidoreductase (trivial name Catechol oxidase). The following section therefore bears this heading.

Histochemical Reactions for Catechol oxidase (E.C. 1.10.3.1)

The original method of Bloch (1917) employed agar-embedded fresh tissues from which thick frozen sections were cut. These were incubated with a freshly prepared dilute aqueous solution of DOPA at 22° or 37° for 6–24 hours. Laidlaw (1932) and Laidlaw and Blackberg (1932) noted that the substrate solution need not be fresh but could be stored at 4° and used until a red colour developed. These authors modified the original method by stabilizing the pH of the substrate solution at pH 7·4 with phosphate buffer. Until this time no particular notice had been taken of the effect of pH on the reaction, but Sharlet *et al.* (1942) noted that raising the pH increased and lowering the pH decreased the rate of the reaction; Lerner *et al.* (1949) emphasized, as was already known, that at an alkaline pH DOPA is oxidized in the presence of oxygen without the need for a catalyst. Possibly a pH of 6·8, which these authors employed for their *in vitro* estimations, should be the highest used in the histochemical DOPA reaction. Becker, Praver and Thatcher (1935) introduced a modified paraffin DOPA technique which has many advantages over other methods and is reliable and easy to use. The details given in Appendix 19, p. 1338, are those of a modification applicable to formalin-fixed frozen sections. The results are shown in Plate XIXb, p. 868.

Laidlaw cautioned against leaving sections in the incubating medium after the latter had become brown. Becker, on the other hand, recommended that both solution and block should be brownish-black at the end of incubation. A positive DOPA reaction consists of darkening of the entire protoplasm of the cells (melanoblasts in the case of skin) which contain the enzyme. Preformed melanin retains its original colour. Becker considered that only a positive reaction was significant and that with a negative one it might be impossible to determine whether the reaction was really negative or whether it had failed to occur because of some undetermined factor in the technique.

If this was true, the usefulness of the reaction in applied histochemistry would be diminished, but I consider his view to be unnecessarily cautious.

The degree of accuracy of localization achieved by the method can still be criticized and, as with all enzyme techniques, we must question whether diffusion of the enzyme or of the reaction products, or both, occurs. A proportion of the enzyme is contained in intracellular cytoplasmic granules from which it might be expected not to diffuse easily and the insolubility of the final product (melanin) cannot be questioned. The intermediate oxidation products such as DOPA-quinone, however, may well be sufficiently soluble to diffuse from their site of production. In practice it is found that with unfixed material some diffusion does occur and this is prevented by short (4 hours) fixation in formalin. Thus the diffusion effect may be due either to diffusion of the enzyme or to diffusion of protein-bound intermediate reaction products, since both these might be expected to be abolished by a protein fixative. Diffusion of the free intermediate reaction product can still occur with fixed tissues and for this the explanation may be that it is no longer bound to proteins in the immediate vicinity of its production, due to blockage of their binding groups by formalin.

The histochemical interpretation of the DOPA reaction was critically reviewed by van Duijn (1953, 1957a and b) who carried out a number of experiments to test its nature. He concluded that many of Bloch's so-called inactivations were due to loss of a soluble factor from unfixed sections. The effect of KCN was also fully considered by van Duijn who pointed out that in their original report Bloch and Schaaf (1925) stated that pretreatment with 5mM-KCN, buffered to neutrality, allowed a maximal DOPA reaction to occur. This statement has been repeatedly misquoted in the literature. In his experiments van Duijn found that concentrations of cyanide higher than 8mM inactivated the system. He concluded (1953) that the DOPA factor could not be related to any known enzyme and his later studies (1957a and b) have indicated the same conclusion. Possible relationships between the peroxidase system and the DOPA-oxidase system were suggested, since the pattern of inactivation of both systems was similar in the case of leucocytes and erythrocytes. The DOPA reaction of these cells has usually been considered to be due to a phenol oxidase but van Duijn's results suggested that it was partly due to their action in trapping coloured autoxidation products from the incubating medium.

An alternative interpretation, proposed later by van der Ploeg and van Duijn (1964b), was that during the classical histochemical DOPA reaction 5,6-dihydroxyindole is formed, either by enzyme activity or by auto-oxidation, and this compound is then converted by the phenol oxidase complex and molecular O_2 to indole-5, 6-quinone. The latter then rapidly polymerizes to melanin.

Observing that the enzyme responsible for the DOPA reaction is active in the absence of free O_2 and in an acid medium, unlike most monophenol and diphenol oxidases Radaeli (1963) proposed that it should be called DOPA quinonase. Using an acid medium (pH 6·2 to 6·8) he observed the production of melanin in the nuclei of the epidermal melanoblasts of human skin.

Histochemical Applications. Improvements in the technique of the DOPA reaction were reported by Radaeli (1953) and Rappaport (1955) described a semiquantitative application to freeze-dried skin. Kukita and Fitzpatrick (1955) used an entirely different method for demonstrating tyrosinase depending on autoradiography and radioactive tyrosine. Foster and Cook (1954) carried out a study of tyrosinase in mouse skin and Comstock and Wynne (1957) discussed the detection of the DOPA factor in pigmented tissues. The phenol oxidase systems of human melanomas were studied by Kertész (1954) and Fitzpatrick (1952) described the presence of an inhibited tyrosinase in normal skin, which was activated by irradiation. In junctional naevi he found a partially inhibited system and in melanomas, both pigmented and amelanotic, a fully active system. Riley (1967) succeeded in demonstrating the "tyrosinase" of melanocytes using 5,6-dihydroxyindole as substrate.

The non-specific "DOPA-oxidase" of neutrophil and eosinophil leucocytes and myelocytes has not been the object of much attention. Vercauteren (1951) demonstrated the presence in horse eosinophil leucocytes of an enzyme system capable of oxidizing DOPA and he showed that this was inhibited by 10mM-KCN. Similarly the DOPA reaction of erythrocytes has produced little comment although many workers (e.g. Miescher, 1923; Moncorps, 1924; Bloch and Peck, 1930; Meirowsky, 1940) noted it in passing. Erythrocytes and eosinophils from horse blood were used by van Duijn (1957b) in his inactivation studies. As mentioned above, he produced strong evidence of the identity of the DOPA factor with peroxidase or peroxidase-like activity in both cell types. Extensive tests were carried out by Lillie (1956) on the nature of phenol oxidative activities in skin. He found that the so-called DOPA (tyrosinase) system was localized at the presumed sites of cutaneous melanogenesis and that oxidation of *p*-aminodimethylaniline was situated in hair medulla and sheath cells. Activity against catechol, pyrogallol and haematoxylin was found in keratohyalin and trichohyalin granules. Although the DOPA-oxidase reaction was inhibited by 1mM-cyanide, sulphide or azide, the other activities withstood ten times these concentrations of inhibitor. Autoxidation was not completely excluded as responsible for some of the observed effects, especially those resulting in the staining of keratohyalin by polyphenol dyes like haematoxylin.

The specific nature of the DOPA-oxidase reaction of peroxidase-positive cells was thrown into question by the work of Okun and his associates (Okun, 1967a,b, Okun *et al.*, 1967). These authors showed that normal and neoplastic melanocytes, and also mast cells, were peroxidase-positive. Further work indicated that mast cells, in tissue culture, were able to synthesize "melanin" (Okun *et al.*, 1967) and that they possessed a tyrosinase activity which could be demonstrated using tyrosine as substrate, in the presence of small amounts of L-DOPA as a "cofactor" (Okun *et al.*, 1969). This activity was later shown to be due to peroxidase rather than tyrosinase Okun *et al.*, 1970a). This finding suggested that the first two stages in the conversion of tyrosine to melanin (tyrosine → DOPA and DOPA to DOPA quinone) were catalysed by peroxidase. If this is so, we can begin to comprehend some of the results obtained by earlier workers in the field.

A modified DOPA reaction, designed specifically to demonstrate the phenol oxidase activity of the granules of blood basophil cells, was described by Inagaki (1965), and three variants were devised by Rodriguez and McGavran (1969), suitable for frozen sections, paraffin sections and electron microscopy respectively. Their first two methods are described in Appendix 19, p. 1338. The tyrosine-DOPA reaction (Okun *et al.*, 1969) is also given in the Appendix.

The Biochemistry of Monoamine Oxidases

These group specific enzymes occur in a variety of animal tissues in small concentrations. They act on primary, secondary and tertiary mono-amines with the formation of either lower amines or aldehydes. On primary amines their action may be described as follows:

$$R.CH_2NH_2 + H_2O + O_2 \rightarrow R.CHO + NH_3 + H_2O_2$$

Two substrates upon which monoamine oxidase acts particularly rapidly are tyramine and adrenalin, and it is inhibited by similar monoamines in which there is a substituted methyl group in the α position, e.g. benzedrine.

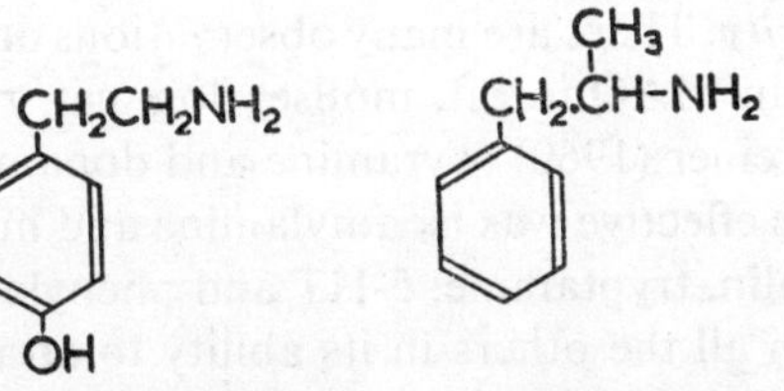

Tyramine Benzedrine

Most animal possess a high amine oxidase activity in liver and kidney, rat kidney being an exception in this respect (Blaschko *et al.*, 1937; Pugh and Quastel, 1937). Langemann (1951) reported high activity in the ox adrenal medulla and (1944) a moderate amount in human brain and heart muscle. According to Zeller (1951) an inverse relationship was always observed between amine oxidase and cholinesterase, especially in the brain.

A comprehensive review of the biochemistry of amine oxidase was presented by Blaschko (1963) who preferred to avoid the term monoamine oxidase (MAO) because the enzyme can attack several diamines. The I.U.B.C. Enzyme Commission, however, called the enzyme monoamine: O_2 oxidoreductase (trivial name monoamine oxidase) and it is by this name that it is known to most histochemists.

Inhibitor Studies. Zeller *et al.* (1955) reported that MAO contained Cu^{2+} as an essential cofactor and this was confirmed by Yamada and Yasunobu (1962). Yasunobu *et al.* (1968) found that the purified enzyme was inhibited by a number of metal chelators. In their studies 0·3 mM cuprizone produced 77 per cent inhibition, 0·3 mM neocuproine 32 per cent and 3 mM 8-hydroxyquinoline 82 per cent. Gabay and Valcourt (1968) found similar effects with rabbit liver mitochondrial MAO but in this case they noted a dual effect with the two chelators *o*-phenanthroline and neocuproine. At low concentrations (100 µM) there was some activation but at 400 µM and higher inhibition was observed. Inhibition of their purified MAO by *p*-chloromercuribenzoate (PCMB) was almost 100 per cent at 10 µM. This result confirms earlier studies which indicated that MAO was a sulphydryl enzyme and agrees with observations made by Tipton (1968) who found 100 per cent inhibition of pig brain mitochondrial MAO after preincubation with 5 µM PCMB. It would appear that this inhibitor should be used for choice in histochemical studies rather than the more usually employed pharmaceutical compounds. Of the latter the most popular has been iproniazid (Marsalid) (Spector *et al.*, 1960). In studies carried out by Tipton (1968) this compound produced 70 per cent inhibition at 0·25 mM.

Histochemical studies of inhibition by *β*-phenylisopropyl hydrazine were carried out by Mustakallio *et al.* (1961), and inhibition by isocarboxazide was recorded by Schwartz (1962). The use of glyceryl trinitrate was described by Ogawa *et al.* (1967) but this observation is of clinical rather than histochemical interest.

Substrate Specificity. There are many observations on this point. A comprehensive study of brain MAO in rat, mouse, dog, cat, rabbit, guinea-pig and man was made by Weiner (1960). Tyramine and dopamine were most readily oxidized. Slightly less effective was isoamylamine and much less effective were adrenalin, noradrenalin, tryptamine, 5-HT and phenylethylamine. The rabbit enzyme differed from all the others in its ability to attack *n*-amylamine.

The important functions carried out by MAO in the animal body probably include the detoxication of poisonous amines produced in the intestine by bacterial decarboxylation of amino-acids, and a major part of the destruction of adrenalin, at adrenergic nerve endings and elsewhere. Several groups of workers (Blaschko and Philpot, 1953; Blaschko and Hellmann, 1953; Titus and Udenfriend 1954; Udenfriend *et al.*, 1956; Sjoerdsma *et al.*, 1955) showed that serotonin (5-HT) was deaminated to 4-hydroxyindole acetaldehyde by MAO and this compound was subsequently excreted in the urine as 5-hydroxyindole acetic acid.

The exact pathways by which amines are oxidized in mammalian and plant tissues are not fully understood. Considering the factors which might interfere with the biochemical assay of MAO activity Creasey (1956) suggested that a number of other enzymes might play an auxiliary role in the process. These included aldehyde oxidase, cytochrome oxidase, catalase and peroxidase. His scheme is shown in Fig. 179, below.

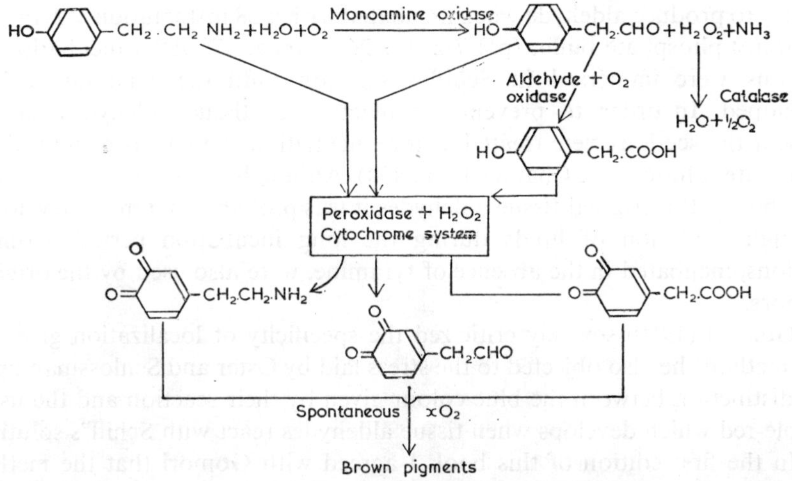

FIG. 179. Possible Oxidation Pathways for Tyramine. (After Creasey.)

In the case of the washed mitochondrial suspensions which he used Creasey found no evidence that cytochrome oxidase played any part in the oxidation of tyramine and he could not demonstrate the presence of any aldehyde oxidase activity. Since this is a soluble enzyme it is probable that it was removed during the washing process. In the absence of a suitable substrate no peroxidase activity occurred but if a small quantity of ethanol was added a coupled oxidation took place, utilizing H_2O_2 produced in the primary reaction. In view of these findings Creasey considered that spontaneous aerial oxidation of *p*-hydroxyphenyl acetaldehyde could account for the extra O_2-uptake observed in manometric studies. It was found possible to prevent this in several ways. Creasey was unable to prepare *p*-hydroxyphenyl acetaldehyde but Clarke and Mann (1957) successfully isolated 3-indolylacetaldehyde produced by the action of plant amine oxidase on tryptamine. Their work showed that previous failure to identify this oxidation product was due to secondary oxidation catalysed by other enzymes. In the presence of added H_2O_2 they found that both tryptamine and 3-indolyl acetaldehyde were oxidized by peroxidase. These observations are obviously of considerable histochemical importance.

An alternative pathway for the breakdown of catecholamines is provided by the enzyme catechol-*O*-methyltransferase (E.C. 2.1.1.a). No histochemical method exists for this enzyme and for most tissues its importance relative to MAO has not been determined.

Histochemical Methods for Monoamine Oxidase (E.C. 1.4.3.4.)

The first histochemical method for MAO was that of Oster and Schlossman (1942). This depended on the ability of fresh frozen sections of guinea-pig kidney to produce aldehyde when incubated with a 28 mM solution of tyramine in 60 mM-phosphate buffer (pH 7·2) for 24 hours at 37°. After incubation the sections were immersed in Schiff's solution until the maximum colour developed. In order to prevent the reaction of tissue aldehydes already present the sections were treated, before incubation, with a 2 per cent sodium bisulphite solution (see Chapter 13, p. 454). Although this manoeuvre successfully bound the original tissue aldehydes it was probable that more developed through oxidation of lipids during the long incubation period. Control sections, incubated in the absence of tyramine, were also used by the original authors.

Gomori (1950) severely criticized the specificity of localization given by this method; he also objected to the stress laid by Oster and Schlossman upon the distinction between the blue colour given by their reaction and the usual purple-red which develops when tissue aldehydes react with Schiff's solution.

In the first edition of this book I agreed with Gomori that the method could not localize the sites of enzyme activity. The reaction product was soluble in water and diffused freely in the incubating medium. It behaved then like any other aldehyde under similar conditions and became attached to polar groups such as NH_2, NH and SH in the tissue proteins; since the pH of the medium was above 6, such attachment was probably mainly to NH_2 groups. The presence of a blue colour in the sections did not therefore indicate the localization of amine oxidase, but only some of the sites of attachment of the reaction product to protein.

In view of these observations I suggested that a simultaneous coupling method might be designed which would trap the aldehyde at the moment of its release to form an insoluble precipitate *in situ.* Attempts which I made to achieve this object by using 3 mM-tyramine, at pH 7·4, and 37°, in the presence of 5 mM-2-hydroxy-3-naphthoic acid hydrazide, were not entirely successful. Final treatment with *o*-dianisidine at pH 7·4, however, revealed a bluish-purple deposit in tissues containing amine oxidase which was largely absent from control tissues incubated in the absence of tyramine. Tissues were pretreated with hydroxylamine hydrochloride (Chapter 13, p. 456) to bind pre-existing aldehydes and sections thus treated gave a negative reaction when incubated in the absence of substrate.

Development of Naphthoic Hydrazide Methods

A considerably improved method, using the naphthoic acid hydrazide capture reagent, was developed by Koelle and Valk (1954). Instead of hydroxylamine they used 10 mM-hydrazine as their aldehyde-blocking reagent although it was observed that in many tissues no blocking reagent was necessary. In place of tyramine they employed tryptamine as substrate and found a considerable increase in the intensity and sharpness of staining. A third modification was the incorporation of 20 per cent sodium sulphate into the medium, which was saturated with respect to naphthoic acid hydrazide. Throughout the incubation period oxygen was bubbled through the medium. Control sections were incubated in the absence of substrate and in the presence of the specific MAO inhibitor 1-isonicotinyl-2-isopropyl hydrazine (Marsilid, 1 mM). The latter completely abolished the histochemical reaction.

While the Koelle-Valk procedure was fairly satisfactory for the demonstration of MAO in the majority of mammalian tissues it was not possible to use the method on central nervous tissues or on tissues containing much oxidizable lipid. In these a pseudoplasmal reaction with the hydrazide resulted in deeply stained control sections.

The Koelle-Valk procedure was employed by Eder (1957) in an extensive study of the distribution of MAO in guinea-pig tissues. Directions given in Appendix 19, p. 1339, are derived mainly from the original paper of Koelle and Valk, with few modifications. Even in tissues with low MAO activity it is possible to obtain a positive reaction. Fig. 180, p. 869, shows the result obtained in a thin (8μ) section of rat kidney. This organ is known to possess a low MAO content. The kinetics of the capture reaction are unknown and it is probable that first order conditions (see Chapter 14, p. 484) are not attained. The solubility of naphthoic acid hydrazide is low but substitution of its more soluble hydrochloride by Glenner, Burtner and Brown (1957) produced relatively little improvement. It did, however, permit the omission of sodium sulphate from the incubating medium and the period of incubation was consequently lowered to one hour.

Methods Based on Pigment Formation

The original technique using the principle of pigment formation was that of Blaschko and Hellmann (1953) who incubated thick (25μ) fresh frozen sections with tryptamine hydrochloride at pH 7·4. A dark-brown pigment was formed after one hour's incubation. The authors noted that accelerated pigment formation occurred if H_2O_2 was added to the medium and that *sec*-octyl alcohol, which inhibits MAO, made no difference to the amount of pigment produced. A slight modification of Blaschko and Hellmann's method was used by Arioka and Tanimukai (1957) in a study of MAO activity in mouse brain. They used 5-hydroxytryptamine as substrate and heavy metals in trace amounts as additional catalysts. Their controls (heated sections; substrate-free medium) showed no pigment formation.

In the illustrations provided by Arioka and Tanimukai pigment deposition, described as in the capillary walls, appeared rather to involve their contained red cells. The peroxidase activity of haemoglobin is a factor which complicates a number of histochemical reactions for enzymes (cf. indoxyl esterase, Chapter 17, p. 781). Accentuation of pigment formation after addition of H_2O_2, observed by a number of workers, is also to be regarded as indicating secondary oxidation by peroxidases.

In many cases the site of activity of the secondary enzyme system may be closely related to the site of MAO activity and in such cases the final localization may well be that of MAO itself. Methods for MAO based on pigment formation have been used infrequently in applied histochemical studies, particularly since the development of reliable methods, described in the section below.

Methods Based on Reduction of Tetrazolium Salts

The first use of this principle for the demonstration of MAO was reported by Francis (1953) who incubated 40μ sections of guinea-pig kidney with tyramine at pH 7·4 in the presence of 0·1 per cent neotetrazolium chloride. After 2–3 hours at 37° a purple deposit of formazan occurred and this was regarded as being due to MAO activity since negative control sections were obtained if the substrate was omitted. The reaction was also observed to be inhibited by octyl alcohol and was presumably due to reduction of the tetrazolium salt by *p*-hydroxyphenyl acetaldehyde. In the hands of most workers the Francis method was unsatisfactory. By substituting tryptamine for tyramine and by using a more easily reducible tetrazolium salt Glenner *et al.* (1957) were able to obtain much better results in suitable tissues after one hour's incubation. In their experiments total inhibition was produced by preincubation either with Marsilid (1-isonicotinyl-2-isopropyl hydrazine) or with phenylhydrazine. Octyl alcohol produced no inhibition unless included in the medium, when partial inhibition occurred. Formazan pigment was produced under histochemical conditions from iodonitrotetrazolium (INT) and from nitro-blue tetrazolium (Nitro-BT) but not from neotetrazolium or blue tetrazolium. A glance at the table of redox potentials (Chapter 20, p. 884) will indicate the reason for this observation and for the failure of the Francis method in the hands of most workers. Since the formazan from INT is macrocrystalline Nitro-BT was usually employed in the histochemical performance of the MAO reaction. Although Wohlrab (1961) successfully employed methylthiazolyl tetrazolium (MTT) for the demonstration of MAO, the redox potential of this salt (see p. 885) is clearly suboptimal. On account of its higher redox potential, Tetranitro-BT provides even better localization of the enzyme than either MTT or Nitro-BT. It has therefore become the reagent of choice. Thybusch and Woohsmann (1966) using Nitro-BT, tested the effect of added phenazine methosulphate (0·05 to 0·3 mg/ml.). Higher concentrations gave strong inhibition. Lallemant and Baron (1967) indicated that TNBT considerably reduced the activity of MAO in sheep brain mitochondria, so that high concentrations must clearly be avoided.

PLATE XIX

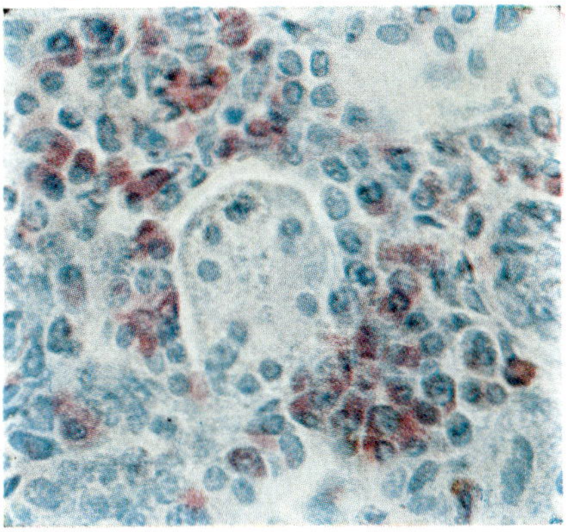

XIXa. Human kidney. Case of myeloid leukaemia. All the myeloid cells contain bright red thioindigo, indicating presence of an "oxidase" system. 4-Methyl-6-chlorothionaphthene—Haemalum. × 420.

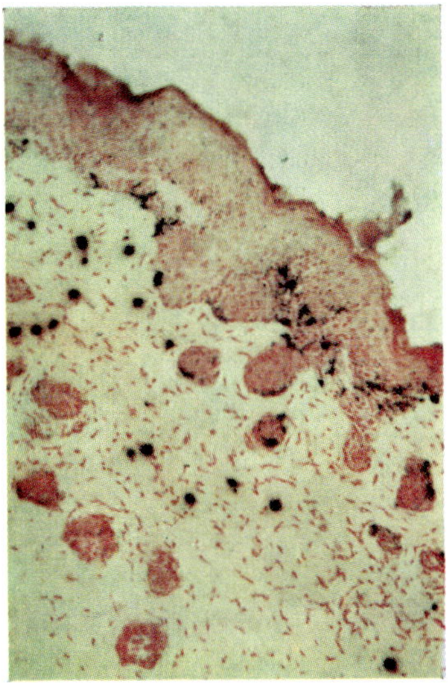

XIXb. Mouse skin. Fresh frozen, 6 μ, cryostat section. DOPA oxidase reaction, counterstained with carmalum. Shows enzyme activity in melanocytes in the epidermis and also in dermal melanophores. × 120.

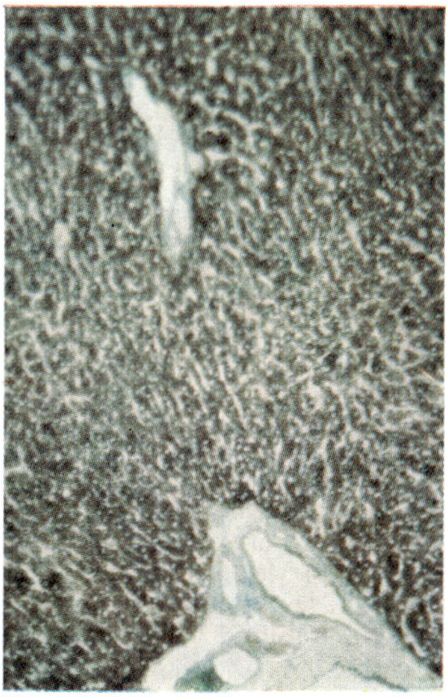

XIXc. Rat liver. Fresh frozen, 8 μ, cryostat section. Shows monoamine oxidase (MAO) activity, using tryptamine HCl as substrate (TNBT). × 70.

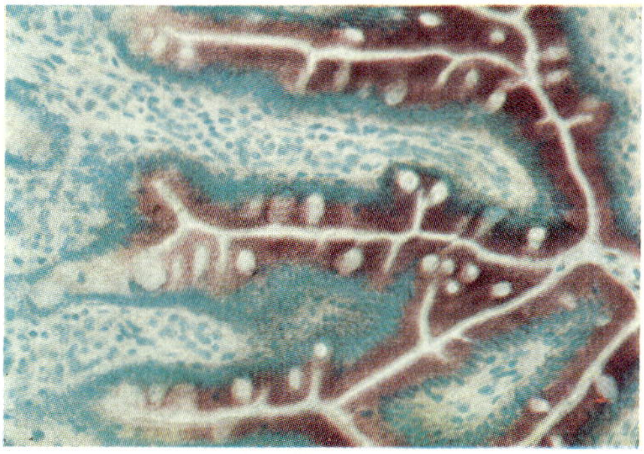

XIXd. Human jejunum (Capsule biopsy). Fresh frozen cryostat section (6 μ). Shows xanthine oxidase activity in the enterocytes. Methyl green. × 172.

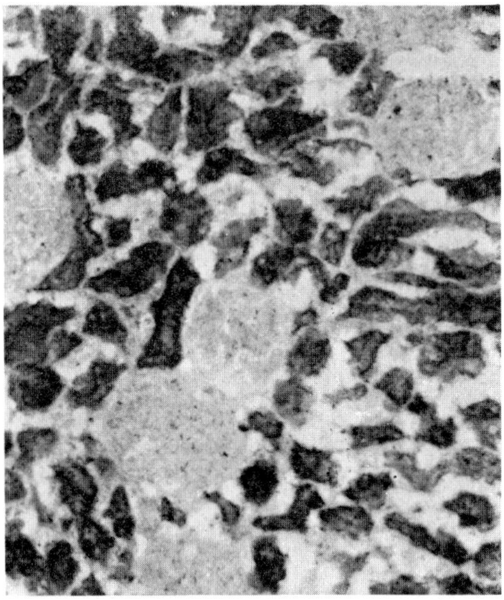

FIG. 180. Rat kidney. Distribution of monoamine oxidase by the naphthoic acid hydrazide method. × 136.

FIG. 186. Rat kidney. Succinate dehydrogenase reaction leading to production of formazan from MTT. Finely crystalline at first, crystal growth occurs rapidly, leading to the patterns shown. × 300.

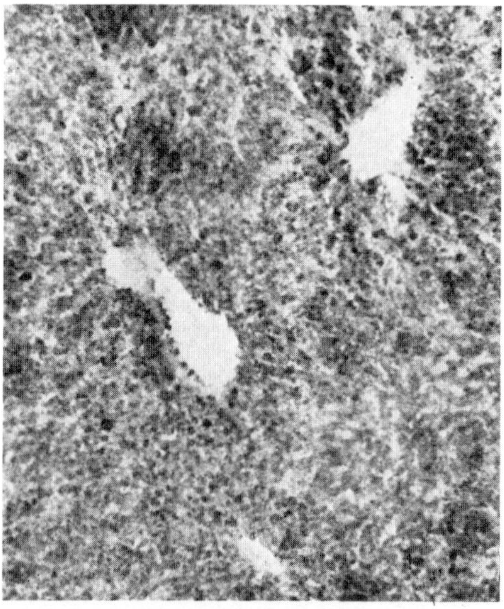

FIG. 192. Rat liver. Deposits of formazan throughout the liver lobule indicate strong activity of "nothing dehydrogenase" at pH 9·0. Nitro-BT. × 470.

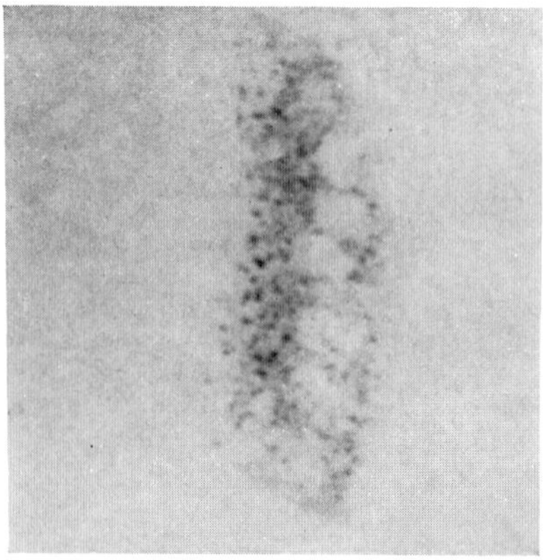

FIG. 197. Rat kidney. Cold microtome section. Shows intramitochondrial activity of "glucose-6-phosphate dehydrogenase" in the cells of the macula densa. MTT-cobalt method. × 1400.

The full tetrazolium method of choice is given in Appendix 19, p. 1340 and the type of result obtainable is shown in Plate XIXc, p. 868.

The mechanism by which tetrazolium salts are reduced in the MAO reaction was investigated by Weissbach *et al.* (1953). Their studies showed that *in vitro* INT was reduced only when an indole amine was oxidized by the MAO of rat liver mitochondria. Other substrates for MAO did not give rise to products capable of reducing tetrazolium salts. Glenner *et al.* (1957) considered the possibility that electrons from the oxidized substrate might be transferred to the tetrazolium salt *via* the diaphorase system of the mitochondria. There was no evidence to support this hypothesis, however, and Glenner *et al.* (1960) were able to show that the indole carbonyl primary reaction product could reduce the tetrazolium salt directly.

Method Based on Coupled Peroxidation

Taking advantage of the fact that one of the products of MAO activity is H_2O_2 (see Fig. 173) Graham and Karnovsky (1965b) adapted their method for uricase (p. 872) to demonstrate the enzyme by coupled peroxidation. In the presence of exogenous peroxidase, added 3-amino-9-ethylcarbozole is oxidized to an insoluble red product by H_2O_2 from the MAO-catalyzed oxidation of tryptamine. Details of the method appear in Appendix 19. Although, as the authors themselves indicated, the tetrazolium method remains. the most satisfactory one for general use the coupled peroxidation method may be used as a control in case of doubtful localizations. Furthermore, although tryptamine may be a useful substrate for guinea-pig kidney MAO, for most species and most tissues tyramine should be more effective.

Applied Histochemistry of MAO

A substantial proportion of all histochemical work on MAO refers to its activity in the brain of various species. Early studies were those of Shimizu *et al.* (1959) and Hashimoto *et al.* (1962). The second of these forms a valuable and comprehensive survey of MAO in the autonomic regions of the brain. Other brain studies include those of Duckett and Pearse (1967), who described a transient MAO-containing system of cells in the developing human foetal cortex at 2–3 months, and of Robinson (1968) who studied the developing brain stem. He found that while at birth only the locus coeruleus and the nucleus ambiguus were active, by the tenth day all the nuclei possessed strong MAO activity. This was regarded as sufficient evidence to warrant the inclusion of MAO in the list of enzymes subscribing to the caudal-rostral concept of brain maturation.

A biochemical study having some bearing on histochemical work was carried out by Tipton and Dawson (1968) who found that MAO levels in the hypothalamus and frontal cortex exceeded those of the corpus striatum, and greatly exceeded those of the thalamus, medulla and pons. Using a quantitative histochemical technique, Tyrer *et al.* (1968) found highest MAO levels in the thalamus and they observed that these appeared to be associated particularly with structures subserving olfactovisual and olfactosomatic functions.

In the case of the structures of the eye one of the original studies was that of Eränkö *et al.* (1961) and later work in this field was carried out by Lukáš and Čech (1966) and by Mustakallio (1967). The second of these three studies compared the distribution of MAO with that of adrenergic fibres (Chapter 27, p. 1109) in various ocular tissues. The enzyme was found to be both extra and intraneuronal. In a study of the carotid body in rabbit, cat and dog, Thybusch (1968) found only weak activity in the glomas caroticus of the last two species. It is often supposed that amine-storing, and amine and amine-precursor handling, cells must necessarily possess strong MAO activity. In fact, the converse is true.

The intestinal enzyme has been investigated relatively infrequently. Two studies by Penttilä (1968) and Penttilä and Mustakallio (1968) record the total lack of correlation between MAO levels and distribution, and those of 5-HT. Alterations in MAO levels with increasing age were noted by Studer *et al.* (1964) who found that the liver enzyme was relatively constant but that in cardiac muscle the level rose steadily with increasing age. Changes in the enzyme in the developing liver, brain and kidney of the rat were also reported by Kuzuya and Nagatsu (1969), and differences in the distribution of the enzyme in the hearts of different species were recorded by Müller and Pearse (1965).

Thyroid MAO, a follicular cell enzyme, was observed by Hopsu and Karinkantha (1962) to change *pari passu* with the increase in follicular cell height induced by thiouracil. There was no absolute increase. In the skin strong MAO activity has often been recorded (Hashimoto *et al.*, 1963) but Scheidegger (1967) noted a strong pseudo MAO activity in epidermal tissues. His observations must be noted by all those working in this field where it would appear that Graham and Karnovsky's coupled peroxidation technique should provide an acceptable control.

Methods for Diamine Oxidase (E.C. 1.4.3.6)

More usually called histaminase, this enzyme (DAO) might well be excluded from any rigidly correct treatise on histochemistry since no definite method for its localization exists (1969). There are, however, at least three possibilities for its demonstration *in situ* and these will be discussed below.

Histaminase activities vary widely from one species to another (Zeller, 1963) but high levels of the enzyme are usually found in kidney, intestine, liver, lung, skin and, particularly, in placenta. No DAO activity is found in mammalian brain. Inhibitors of DAO include semicarbazide, alkyl mono- and diguanidines, mono- and diamidines, hydroxylamine, methylene blue and pyocyanin. A comprehensive study carried out by Blaschko *et al.* (1951) indicated that diaminoguanidine derivatives of the type

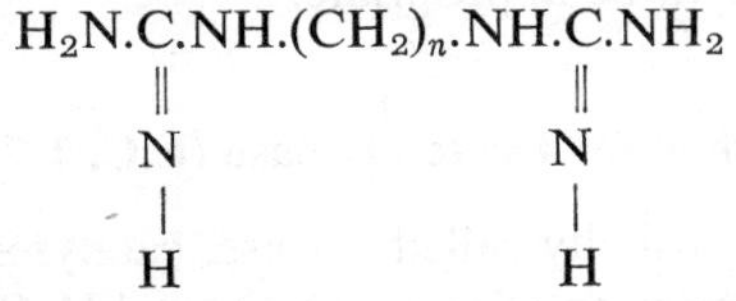

were powerful inhibitors of DAO and that their inhibitory effect was greatest with the lowest *n* values, falling to zero at $n = 18$.

Using a coupled peroxidation microassay technique based on the reaction

$$R - CH_2 - CH_2 - NH_2 + O_2 + H_2O \rightarrow R.CH_2.CHO + NH_3 + H_2O_2$$

Gunther and Glick (1967) determined the distribition of histaminase in human placenta and uterus. They employed histamine as substrate with *o*-dianisidine as the chromogenic reagent and exogenous horseradish peroxidase, as did Møller and Ottolenghi (1966). This work indicates the feasibility of a coupled peroxidation technique for *in situ* localization of the enzyme.

An alternative possibility for the development of a histochemical reaction is provided by the observations of Cavallini *et al.* (1957) who developed an assay for DAO based on the oxidation of cystamine, at pH 5·7, to cystaldimine (1, 2-dehydrothiamorpholine)

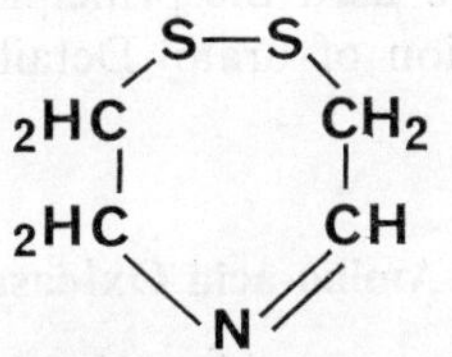

This compound combines spontaneously and specifically with *o*-aminobenzaldehyde to give yellow or orange dihydroquinazolinium derivatives:

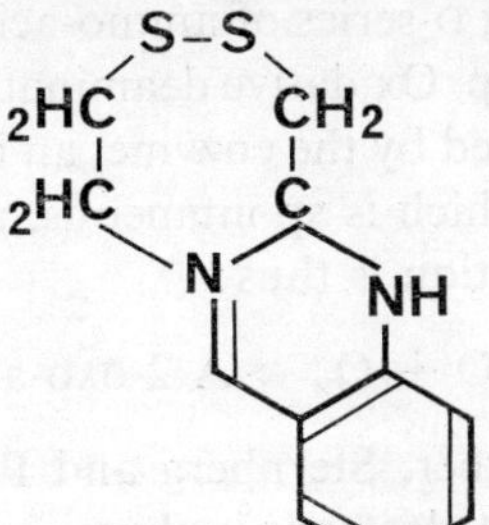

From these data, in principle, the possibility exists for the development of a fluorescent method for DAO.

A third possibility would be by utilization of the principle of decolorization of indigo disulphonate by coupled oxidation, as employed by Zeller (1940) in his assay technique for DAO. A substrate film method (see chapter 24, p. 1009) would appear to be appropriate.

Method for Urate Oxidase (E.C. 1.7.3.3)

This enzyme, more usually called uricase, catalyses the oxidation of uric acid, by molecular oxygen, to allantoin, CO_2 and H_2O_2 (Schittenhelm, 1965; Wiechowski, 1907; Batelli and Stern, 1909). Biochemical investigations by de Duve and his colleagues (Beaufay *et al.*, 1959; de Duve *et al.*, 1960) have shown that uricase is one of a spectrum of oxidases (including catalase and D-amino-acid oxidase) associated with intracellular organelles known to electron microscopists as microbodies. Its properties were reviewed by Mahler (1963), whose method (Mahler *et al.*, 1955) is the standard one for preparation of the enzyme.

Although uricase is a copper protein it is inhibited by Cu^{2+} ions at a concentration of 20μM, and by other metals (Keilin and Hartree, 1936). The specific competitive inhibitor 2,6,8-trichloropurine, described by Mahler (1963), can be used in histochemical practice. Sections are preincubated for 15 minutes in buffered inhibitor (pH 8·0), at a concentration of 2 mM, before incubation in the substrate mixture.

The development of a histochemical reaction for uricase is due to Graham and Karnovsky (1965a) who used the principle of coupled peroxidation to utilize the H_2O_2 by oxidation of urate. Details are given in Appendix 19, p. 1341.

Method for D-Amino-acid Oxidase (E.C. 1.4.3.3)

This enzyme is classified as an oxidase since its natural hydrogen acceptor is molecular oxygen, which is reduced to hydrogen peroxide. It is able to use methylene blue as an acceptor, however. The enzyme acts only on members of the non-naturally occurring D-series of amino-acids but is able to attack most of the members of this group. Oxidative deamination takes place in two stages. In the first, which is catalysed by the enzyme, an imino-acid and H_2O_2 are the products. In the second, which is spontaneous, an α-keto-acid and NH_3 are produced. The overall reaction is thus:

$$\text{A D-amino-acid} + H_2O + O_2 \rightarrow \text{A 2-oxo-acid} + NH_3 + H_2O_2$$

It was reported by Farber, Sternberg and Pearce (1958) that phenazine methosulphate could accept electrons in the D-amino-acid oxidase system and to pass them on to suitable tetrazolium salt such as Nitro-BT. In rat kidney the enzyme was observed to be concentrated in the second portion of the proximal convoluted tubule. The critical factor in the histochemical performance of this reaction appeared to be the phenazine methosulphate (PMS). With some samples no reaction whatsoever could be obtained. Dangers inherent in reactions of this kind were outlined by Pennington (1961) who showed that iodonitrotetrazolium (INT) was reduced non-enzymically in the presence of PMS in the presence of heated human serum. The latter presumably provided the necessary electrons.

The feasibility of a coupled peroxidation reaction for the histochemical demonstration of D-amino-acid oxidase was investigated by Graham and Karnovsky (1965a) who employed the same reagent used in their method for uricase. Results were described as inconsistent even when the reaction was carried out under high concentration of O_2. There was a pronounced tendency to diffuse cytoplasmic staining, as with the tetrazolium technique, and this was ascribed to the solubility of the enzyme.

All coupled peroxidation reactions should be controlled by omission of exogenous peroxidase. Positive reactions will then occur only where endogenous peroxidases are present, as in leucocyte granules. Control should also be applied, wherever possible, by the use of specific inhibitors. Finally, the possibility of non-enzymic oxidation of the chromogen must always be examined. In rabbit liver and kidney, for instance, Graham and Karnovsky found that 3-amino-9-ethylcarbazole was oxidized by some unknown mechanism. In this tissue, therefore, they were unable to demonstrate either uricase or D-amino-acid oxidase.

Methods for Xanthine Oxidase (E.C. 1.2.3.2)

This enzyme catalyses the oxidation, by molecular O_2, of the two purines hypoxanthine and xanthine to uric acid. It also oxidises other purines, pterins and various aldehydes, having thus a very low specificity. Oxidation of xanthine proceeds as follows:

$$\text{Xanthine} + H_2O + O_2 \rightarrow \text{Urate} + H_2O_2$$

According to Morell (1952) the E° of the reaction is + 0·355 mV.

The enzyme is a high molecular weight protein which is firmly bound to FAD, molybdenum and iron but is nevertheless very soluble. The precise way in which the flavin and the metals function is still unknown (Bray, 1963). In the presence of oxygen H_2O_2 is formed in the reaction but under anaerobic conditions the enzyme complex can reduce cytochrome c. and dyes such as methylene blue.

Xanthine oxidase is inhibited by SH inhibitors (Green and O'Brien, 1967), by flavin inhibitors and by cyanide. High concentrations of calcium ions, which cause dissociation of FAD, also bring about inhibition. A new class of more specific inhibitors was isolated by Fridovich (1968) from guanidinium salts. The most effective compound found to be ammeline (2,4-diamino-6-hydroxy-*s*-triazine). This competitive inhibitor was effective, at pH 7·5 in phosphate buffer, at a concentration of 16 μM.

The first histochemical method for xanthine oxidase was described by Bourne (1953) who used 30 mM xanthine as substrate, in phosphate buffer at pH 8·2 with neotetrazolium as acceptor. A similar technique was used by Villela (1955) and by Villela and Affonso (1955) but, in my hands, it proved unsatisfactory. Bourne's method was critically examined by Wohlrab (1961) who tested 13 tetrazolium salts at varying pH levels and substrate concentrations. He concluded that the reaction was unspecific and could not be used for xanthine oxidase localization.

Further attempts made by Sackler (1966) to demonstrate the enzyme by Bourne's method were unsuccessful. By using glutaraldehyde-fixed frozen sections in place of fresh sections, however, and Nitro-BT in phosphate buffer at pH 7·4, however, she succeeded in obtaining positive reactions in rat tissues which correlated well with biochemical assays. A low level of activity in rat liver was held responsible for the negative histochemical reaction in this tissue. Assays of xanthine oxidase in various tissues of cat, dog, sheep and cow were recorded by al-Khalidi and Chaglassian (1965). Highest levels were in duodenum and liver and, in dog and cow, in the lung.

Details of a suitable method are given in Appendix 19, p. 1341 and the type of result obtainable in human jejunal biopsies is shown in Plate XIXd, p. 868. In a case of xanthinuria investigated in my laboratory the absence of xanthine oxidase from the jejunal mucosa was shown by this method. Control cases gave the reaction shown in Plate XIXd.

REFERENCES

AGNER, K. (1941). *Acta physiol. scand.*, **2**, Suppl., 8.
AL-KHALIDI, U. A. S., and CHAGLASSIAN, T. H. (1965). *Biochem. J.*, **97**, 318.
ALTSCHUL, A. M., ABRAMS, R., and HOGNESS, T. R. (1940). *J. biol. Chem.*, **136**, 777.
ARCHER, R. K., and BROOME, J. (1963). *Acta haemat. Basel*, **29**, 147.
ARIOKA, I., and TANIMUKAI, H. (1957). *J. Neurochem.*, **1**, 311.
ARMITAGE, F. L. (1939). *J. Path. Bact.*, **49**, 579.
AVERS, C. J., and GRIMM, R. B. (1959). *J. exp. Bot.*, **10**, 341.
AXELROD, J., and LERNER, A. B. (1963). *Biochem. biophys. Acta*, **71**, 650.
BALL, E. G., and COOPER, O. (1952). *J. biol. Chem.*, **198**, 629.
BATELLI, F., and STERN, L. (1909). *Biochem. Z.*, **19**, 219.
BEAUFAY, H., BENDALL, D. S., BAUDHUIN, P., and DE DUVE, C. (1959). *Biochem. J.*, **73**, 623.
BECKER, S. W., PRAVER, L. L., and THATCHER, H. (1935). *Archs Derm. Syph.*, **31**, 190.
BEER, C. S. (1948). *J. chem. Soc.*, 2223.
BENNETT, H. S., and LUFT, J. H. (1959). *J. biophys. biochem. Cytol.*, **6**, 113.
BEYER, H. (1949). *Ber.*, **82**, 143.
BLASCHKO, H. (1963). In "The Enzymes." Eds. Boyer, P. O., Lardy, H., and Myrbäck, K. Acad. Press, New York, Vol. 8, p. 337.
BLASCHKO, H., FASTIER, F. N., and WADJA, I. (1951). *Biochem. J.*, **49**, 250.
BLASCHKO, H., RICHTER, D., and SCHLOSSMAN, H. (1937). *J. Physiol.*, **90**, 1.
BLASCHKO, H., and HELLMANN, K. (1953). *J. Physiol.*, **122**, 419.
BLASCHKO, H., and PHILPOT, F. J. (1953). *J. Physiol.*, **122**, 403.
BLOCH, B. (1917). *Archs Derm. Syph., Wien.*, **124**, 129.
BLOCH, B., and PECK, S. M. (1930). *Fol. Haematol.*, **41**, 166.

BLOCH, B., and SCHAAF, F. (1925). *Biochem. Z.*, **162**, 181.
BONATI, F., and FRANCHI, F. (1948). *Ateneo parmense*, **19**, 129.
BOOTH, H., and SAUNDERS, B. C. (1950). *Nature, Lond.*, **165**, 567.
BOREI, H. G., and BJÖRKLUND, U. (1953). *Biochem. J.*, **54**, 357.
BOURNE, G. H. (1953). *Nature, Lond.*, **172**, 193.
BRAY, R. C. (1963). In "The Enzymes." Eds. Boyer, P. D., Lardy, H., and Myrbäck K. Acad. Press, New York, Vol. 7, p. 533.
BURSTONE, M. S. (1959). *J. Histochem. Cytochem.*, **7**, 112.
BURSTONE, M. S. (1960). *J. Histochem. Cytochem.*, **8**, 63.
BURSTONE, M. S. (1961). *J. Histochem. Cytochem.*, **9**, 59.
CARRUTHERS, C., and SUNTZEFF, V. (1947). *Cancer Res.*, **7**, 9.
CAVALLINI, O., DI MARCO, C., and MONDOVI, B. (1957). *Biochem. biophys. Acta*, **24** 353.
CAZAL, P., and ROURE, A. (1947). *Rév. Haematol.*, **2**, 535.
CHAPMAN, N. B., and SAUNDERS, B. C. (1941). *J. chem. Soc.*, 496.
CLARKE, A. J., and MANN, P. J. G. (1957). *Biochem. J.*, **65**, 763.
COMSTOCK, E. G., and WYNNE, E. S. (1957). *Am. J. clin. Path.*, **28**, 264.
CREASEY, N. H. (1956). *Biochem. J.*, **64**, 178.
DE DUVE, C., and BAUDHUIN, P. (1966). *Physiol. Rev.*, **46**, 323.
DE DUVE, C., BEAUFAY, H., JACQUES, P., RACHMAN-LI, Y., SELLINGER, O. Z., WATTAIUX, R., and CONINCK, S. (1960). *Biochem. biophys. Acta*, **40**, 186.
DEMPSEY, E. W. (1944). *Endocrinology*, **34**, 27.
DE ROBERTIS, E., and GRASSO, R. (1946). *Endocrinology*, **38**, 137.
DIXON, M., and WEBB, E. C. (1964). "Enzymes." 2nd Edit. p. 391. Longmans, London.
DOXEY, D. L. (1962). *Stain Technol.*, **37**, 367.
DUCKETT, S., and PEARSE, A. G. E. (1967). *Revue can. Biol.*, **26**, 173.
DUNN, R. C. (1946). *Archs Path.*, **41**, 676.
DUNN, R. C., and THOMPSON, E. C. (1946). *Stain Technol.*, **21**, 65.
EDER, M. (1957). *Beitr. path. Anat.*, **117**, 343.
ELLIOT, K. A. C., and GREIG, M. E. (1938). *Biochem. J.*, **32**, 1407.
ERÄNKÖ, O., NIEMI, M., and MERENMIES, U. (1961). In "The Structure of the Eye." Ed. G. K. Smelser, Acad. Press, New York, p. 159.
ETTORI, J. (1949). *Ibid.*, **44**, 35.
EVANS, W. C., and RAPER, H. S. (1937). *Ibid.*, **31**, 2155.
FAHIMI, H. D. (1968). *J. Histochem. Cytochem.*, **16**, 547.
FARBER, E., STERNBERG, W. H., and PEARCE, N. A. M. (1958). *J. Histochem. Cytochem.*, **6**, 389.
FAUTREZ, J., and LAMBERT, P. P. (1937). *Bull. Histol. Tech. micr.*, **14**, 29.
FITZPATRICK, T. B. (1952). *Archs Derm. Syph.*, **65**, 379.
FOSTER, M., and COOK, R., jr. (1954). *Proc. Soc. exp. Biol. Med.*, **85**, 120.
FRANCIS, C. M. (1953). *Nature, Lond.*, **171**, 701.
FRIDOVICH, I. (1968). *Biochemistry*, **4**, 1098.
FUJITA, S. (1959). *Acta haemat. jap.*, **22**, 931.
FUJITA, S., KOMORI, Y., YAMAMOTO, Y., MASUI, N., SUWA, K., and TAKEOKA, O. (1960). *Acta haemat. jap.*, **23**, 78.
GABAY, S., and VALCOURT, A. J. (1968). *Biochem. biophys. Acta*, **159**, 440.
GALLICO, E. (1947). *Tumori*, **21**, 139.
GARCIA BLANCO, J., and GRISOLIA, S. (1946). *Ann Inst. med. Exper., Valenci*, **3**, 169.
GLENNER, G. G., BURTNER, H. J., and BROWN, G. W., jr. (1957). *J. Histochem Cytochem.*, **5**, 591.
GLENNER, G. G., WEISSBACH, H., and REDFIELD, B. G. (1960). *J. Histochem. Cytochem.*, **8**, 258.
GLOCK, G. E. (1944). *Nature, Lond.*, **154**, 460.
GOLDFISCHER, S. (1967). *J. Cell Biol.*, **34**, 398.
GOODPASTURE, F. W. (1919). *J. Lab. clin. Med.*, **4**, 442.
GOMORI, G. (1950). *Ann. N.Y. Acad. Sci.*, **50**, 968.
GOMORI, G. (1953). *J. Histochem. Cytochem.*, **1**, 486.
GRÄFF, S. (1916). *Zbl. Path.*, **27**, 318.

GRAHAM, G. S. (1916). *J. med. Res.*, **30**, 231.
GRAHAM, G. S. (1918). *Ibid.*, **39**, 15.
GRAHAM, R. C., jr., and KARNOVSKY, M. J. (1965a). *J. Histochem. Cytochem.*, **13**. 448.
GRAHAM, R. C., jr., and KARNOVSKY, M. J. (1965b). *J. Histochem. Cytochem.*, **13**. 604.
GRAHAM, R. C., jr., and KARNOVSKY, M. J. (1966). *J. Histochem. Cytochem.*, **14**. 291.
GRAHAM, R. C., LUNDHOLM, U., and KARNOVSKY, M. J. (1965). *J. Histochem. Cytochem.*, **13**, 150.
GREEN, R. C., and O'BRIEN, P. J. (1967). *Biochem. J.*, **105**, 585.
GUNTHER, R. E., and GLICK, D. (1967). *J. Histochem. Cytochem.*, **15**, 431.
HARLEY-MASON, J., and BULOCK, J. D. (1950). *Nature, Lond.*, **166**, 1036.
HASEBROEK, K. A. (1921). *Fermentforsch.*, **5**, 1.
HASHIMOTO, K., OGAWA, K., and LEVER, K. F. (1963). *J. invest. Derm.*, **41**, 81.
HASHIMOTO, P. H., MAEDA, T., TORII, K., and SHIMIZU, N. (1962). *Med. J. Osaka Univ.*, **12**, 425.
HIRAI, K. (1967). *Acta histochem. cytochem.*, **1**, 43.
HOCH-LIGETI, C. (1947a). *Nature, Lond.*, **159**, 786.
HOCH-LIGETI, C. (1947b). *Cancer Res.*, **7**, 148.
HOGEBOOM, G. H., and ADAMS, M. H. (1942). *J. biol. Chem.*, **145**, 273.
HOGEBOOM, G. H., SCHNEIDER, W. C., and PALADE, G. E. (1948). *Ibid.*, **172**, 619.
HOLLANDE, A. C. (1924). *Bull. Histol. Tech. micr.*, **1**, 422.
HOPPE-SEYLER, F. (1890). *Hoppe-Seyler's Z. physiol. Chem.*, **14**, 106.
HOPSU, V. K., and KARINKANTHA, H. (1962). *Ann. Med. exp. Biol. Fenn.*, **40**, 1.
HUGHES, D., and SAUNDERS, B. C. (1954). *J. chem. Soc.*, 4630.
INAGAKI, S. (1963). *Acta haemat. jap.*, **26**, 544.
INAGAKI, S. (1965). *Acta haemat. jap.*, **28**, 125.
JENSEN, W. A. (1955). *Plant Physiol.*, **30**, 426.
JENSEN, W. A. (1962). "Botanical Histochemistry, Principles and Practice." Freeman, San Francisco.
KAPLOW, L. S. (1965). *Blood*, **26**, 215.
KEILIN, D. (1925). *Proc. R. Soc. B.*, **98**, 312.
KEILIN, D. (1929). *Ibid.*, **104**, 206.
KEILIN, D. (1930). *Ibid.*, **106**, 418.
KEILIN, D. (1936). *Ibid.*, **121**, 165.
KEILIN, D., and HARTREE, E. F. (1936). *Ibid.*, **119**, 114.
KEILIN, D., and HARTREE, E. F. (1937). *Ibid.*, **122**, 298.
KEILIN, D., and HARTREE, E. F. (1938). *Ibid.*, **125**, 171.
KEILIN, D., and HARTREE, E. F. (1939). *Ibid.*, **127**, 167.
KEILIN, D., and HARTREE, E. F. (1940). *Ibid.*, **129**, 227.
KEILIN, D., and HARTREE, E. F. (1949). *Biochem. J.*, **44**, 205.
KERTÉSZ, D. (1952). *Biochim. biophys. Acta*, **9**, 170.
KERTÉSZ, D. (1954). *J. natn. Cancer Inst.*, **14**, 1081.
KOELLE, G. B., and VALK, A. de T., jr. (1954). *J. Physiol.*, **126**, 434.
KUKITA, A., and FITZPATRICK, T. B. (1955). *Science, N.Y.*, **121**, 893.
KUZUYA, H., and NAGATSU, T. (1969). *J. Neurochem.*, **16**, 123.
LAIDLAW, G. F. (1932). *Anat. Rec.*, **53**, 399.
LAIDLAW, G. F., and BLACKBERG, S. N. (1932). *Am. J. Path.*, **8**, 491.
LALLEMANT, C., and BARON, C. (1967). *Experientia*, **23**, 262.
LANGEMANN, H. (1944). *Helv. physiol. pharmac. Acta*, **2**, 367.
LANGEMANN, H. (1951). *Br. J. Pharmac. Chemother.*, **6**, 318.
LEMBERG, M. R. (1969). *Physiol. Rev.*, **49**, 48.
LERNER, A. B., FITZPATRICK, T. B., CALKINS, E., and SUMMERSON, W. H. (1949). *J. biol. Chem.*, **178**, 185.
LEVY, L. (1889). *Hoppe-Seyler's. Z. physiol. Chem.*, **13**, 309.
LILLIE, R. D. (1956). *J. Histochem. Cytochem.*, **4**, 318.
LILLIE, R. D., and BURTNER, H. J. (1953). *J. Histochem. Cytochem.*, **1**, 8.
LISON, L. (1936). "Histochemie Animale," Paris.

LISON, L. (1938). *Beitr. path. Anat.*, **101**, 94.
LOELE, W. (1912). *Fol. Haematol.*, **14**, 26.
LOELE, W. (1913). *Ergebn. Path.*, **16**, 26.
LOELE, W. (1927). *Virchows Arch.*, **265**, 827.
LODJA, Z. (1967). *Československ. Patol.*, **3**, 31.
LUKÁŠ, Z., and ČECH, S. (1966). *Acta histochem.*, **25**, 133.
MAHLER, H. R. (1963). In "The Enzymes," Eds. Boyer, P. D., Lardy, H., and Myrbäck, K. Acad. Press, New York, Vol. 8, p. 285.
MAHLER, H. R., HÜBSCHER, G., and BAUM, H. (1955), *J. biol. Chem.*, **216**, 625.
MARGOLIASH, E., and NOVOGRODSKY, A. (1958). *Biochem. J.*, **68**, 468.
MACMUNN, C. A. (1886). *Phil. Trans.*, **177**, 267.
MANCINI, R. E., and CELANI, R. B. (1940). *Rev. Soc. argent. de Biol.*, **16**, 616.
MANCINI, R. E., and VILLAMIL, M. F. (1950). *J. natn. Cancer Inst.*, **10**, 1379.
MCJUNKIN, F. A. (1922). *Anat. Rec.*, **24**, 67.
MCSHAN, W. H., and MEYER, R. K. (1946). *Archs Biochem.*, **9**, 165.
MCSHAN, W. H., ERWAY, W. F., and MEYER, R. K. (1946). *Archs Biochem. Biophys.*, **9**, 69.
MEIROWSKY, E. (1940). *Br. J. Derm. Syph.*, **52**, 205.
MEYER, R. K., SOUKUP, S. W., MCSHAN, W. H., and BIDDULPH, C. (1947). *Endocrinology*, **41**, 35.
MIESCHER, G. (1923). *Arch. mikrosk. Anat.*, **97**, 326.
MITSUI, T., and IKEDA, S. (1951). *Okajimas Fol. Anat. Jap.*, **23**, 331.
MØLLER, K. M., and OTTOLENGHI, P. (1966). *C.r. Trav. Lab. Carlsberg*, **35**, 369.
MONCORPS, C. (1924). *Münch. med. Woch.*, **71**, 1019.
MOOG, F. (1943). *J. cell. comp. Physiol.*, **22**, 223.
MORELL, D. B. (1952). *Biochem. J.*, **51**, 657.
MÜLLER, E., and PEARSE, A. G. E. (1965). *Br. Heart J.*, **27**, 116.
MUSTAKALLIO, A. (1967). *Acta ophth. (Kobh)*. Suppl 93.
MUSTAKALLIO, K. K., LEVONEN, E., and RAEKALLIO, J. (1961). *Science, N.Y.*, **134**, 344.
NACHLAS, M. M., CRAWFORD, D. T., GOLDSTEIN, T. P., and SELIGMAN, A. M. (1958). *J. Histochem. Cytochem.*, **6**, 445.
NIEMI, M., and MERENMIES, E. (1961). *J. Neurochem.*, **6**, 206.
NISHIYAMA, Y., and KOBAYASHI, I. (1953). *Acta path. Jap.*, **3**, 133.
OGAWA, K., GUDBJARNASON, S., and BING, R. J., (1967). *J. Pharmac. exp. Therap.*, **155**, 449.
OKUN, M. R. (1967a). *J. invest. Derm.*, **48**, 424.
OKUN, M. R. (1967b). *J. invest. Derm.*, **48**, 461.
OKUN, M. R., GROB, P., EDELSTEIN, L., and LEVER, W. (1967). *Hautarzt*, **18**, 489.
OKUN, M. R., EDELSTEIN, L., NIEBAUER, G., and HAMADA, G. (1969). *J. invest. Derm.*, **53**, 39.
OKUN, M. R., EDELSTEIN, L., OR, N., HAMADA, G., and DONNELAN, B. (1970a). *J. invest. Derm.*, **55**, 1.
OKUN, M. R., EDELSTEIN, L., OR, N., HAMADA, G., and DONNELAN, B. (1970b). *Life Sciences II*, **9**, 491.
OKUN, M. R., EDELSTEIN, K., HAMADA, G., OR, N., BLUMENTHAL, G., BURNETT, J. and DONNELAN, B. (1970c). Proc. VIIth Int. Pigment Cell Conf. Ed. V. Riley Appleton, New York.
OSTER, K. A., and SCHLOSSMAN, N. C. (1942). *J. cell. comp. Physiol.*, **20**, 373.
PAUL, K. G. (1963). In "The Enzymes." Eds. Boyer, P. D., Lardy, H., and Myrbäck K. Acad. Press, New York, Vol. 8, p. 227.
PEARSE, A. G. E. (1954). *J. Histochem. Cytochem.*, **2**, 475.
PENNINGTON, R. J. (1961). *Biochem. J.*, **80**, 649.
PENTTILÄ, A. (1968). *Acta physiol. scand.*, **73**, 121.
PENTTILÄ, A., and MUSTAKALLIO, A. (1968). *Acta histochem.*, **31**, 1.
PERNER, E. S. (1952). *Biol. Zbl.*, **71**, 43.
PERSON, P., and FINE, A. (1960). *Science, N.Y.*, **132**, 43.
PERSON, P., and FINE, A. (1961). *J. Histochem. Cytochem.*, **9**, 190.

PLAPINGER, R. E., LINUS, S. L., KAWASHIMA, T., DEB, C., and SELIGMAN, A. M. (1968). *Histochemie*, **14**, 1.
PRENANT, M. (1924). *Bull. Hist.*, **1**, 499.
PRENANT, M. (1925). *Ibid.*, **2**, 329.
PRZIBRAM, H. (1922). *Biochem. Z.*, **127**, 126.
PUGH, C. E. M., and QUASTEL, J. H. (1937). *Biochem. J.*, **31**, 2306.
RADAELI, G. (1953). *Archs Derm. Syph.*, **68**, 668.
RADAELI, G. (1963). *Boll. Accad. Med., Genoa*, p. 1.
RANDALL, L. O. (1946). *J. biol. Chem.*, **164**, 521.
RAPER, H. S. (1926). *Biochem. J.*, **20**, 735.
RAPPAPORT, B. Z. (1955). *Archs Path.*, **60**, 444.
REISS, J. (1967). *Histochemie*, **9**, 281.
RIECHE, A. (1929). *Chem. Ber.*, **62**, 223.
RILEY, P. A. (1967). *Nature, Lond.*, **213**, 190.
RITTER, H. B., and OLESON, J. J. (1947). *Archs Path.*, **43**, 330.
ROBINSON, N. (1968). *J. Neurochem.*, **15**, 1151.
RODRIGUEZ, H. A., and MCGAVRAN, M. H. (1969). *Am. J. clin. Path.*, **52**, 219.
SACKLER, M. L. (1966). *J. Histochem. Cytochem.*, **14**, 326.
SATO, A., and YASHIMATO, S. (1925). *Am. J. Dis. Child*, **29**, 301.
SAUNDERS, B. C. (1957). *R. Inst. Chem.*, Lecture No. 1.
SCHEIDEGGER, J. P. (1967). *Arch. klin. exp. Derm.*, **229**, 182.
SCHITTENHELM, A. (1905). *Z. phys. Chem.*, **46**, 354.
SCHNEIDER, W. C., and POTTER, V. R. (1943a). *J. biol. Chem.*, **149**, 217.
SCHNEIDER, W. C., and POTTER, V. R. (1943b). *Cancer Res.*, **3**, 353.
SCHULTZE, W. H. (1909a). *Verh. dt. path. Ges.*, **13**, quoted by Lison, L., 1936.
SCHULTZ, W. H. (1909b). *Beitr. path. Anat.*, **45**, quoted by Lison, L. 1936.
SCHWARTZ, M. A. (1962). *J. Pharmac. exp. Ther.*, **135**, 1.
SELIGMAN, A. M., KARNOVSKY, M. J., WASSERKRUG, H. L., and HANKER, J. S. (1968). *J. cell. Biol.*, **38**, 1.
SELIGMAN, A. M., PLAPINGER, R. E., WASSERKRUG, H. L., DEB, C., and HANKER, J. S. (1967). *J. cell. Biol.*, **34**, 787.
SHARLET, H., SACHS, W., SIMS, C. F., and SALZER, B. H. (1942). *Archs Derm. Syph.*, **45**, 103.
SHIMIZU, N., MORIKAWA, M., and OKADA, Z. (1959). *Z. mikrosk.-Anat.*, **49**, 389.
SINGER, T. P., KEARNEY, E. B., and MASSEY, V. (1956). In "Enzymes, Units of Biological Structure and Function." Ed. O. H. Gaebler. Acad. Press, New York.
SJOERDSMA, A., SMITH, T. E., STEVENSON, T. D., and UDENFRIEND, S. (1955). *Proc. Soc. exp. biol. Med.*, **89**, 36.
SLACK, J. (1943). *J. natn. Cancer Inst.*, **3**, 389.
SMITH, L., and STOTZ, E. (1954). *J. biol. Chem.*, **209**, 819.
SPECTOR, S., SHORE, P. A., and BRODIE, B. B. (1960). *J. Pharm. exp. Ther.*, **128**, 15.
SPURR, A. R. (1954). *Stain Technol.*, **29**, 301.
STERN, K. G. (1936). *J. biol. Chem.*, **114**, 473.
STOTZ, E., MORRISON, M., and MARINETTI, T. (1956). In "Enzymes, Units of Biological Structure and Function." Ed. O. H. Gaebler, Acad. Press, New York.
STRAUS, W. (1964). *J. Histochem. Cytochem.*, **12**, 462.
STUDER, A., BAUMGARTNER, H. R., and REBER, K. (1964). *Histochemie*, **4**, 43.
TAKAMATSU, H., and HIRAI, K. (1968). *Acta histochem. cytochem.*, **1**, 1.
TAKAMORI, S., SEKUZU, I., and OKUNUKI, K. (1960). *J. Biochem., Tokyo*, **48**, 569.
TAKAMORI, S., WADA, K., SEKUZU, I., and OKUNUKI, K. (1962). *Nature, Lond.*, **195**, 456.
THIELE, H.-J. (1967). *Acta histochem.*, **26**, 373.
THYBUSCH, D. (1968). *Acta histochem.*, **30**, 192.
THYBUSCH, D., and WOOHSMANN, H. (1966). *Histochemie*, **7**, 103.
TIPTON, K. F. (1968). *Europ. J. Biochem.*, **4**, 103.
TIPTON, K. F., and DAWSON, A. P. (1968). *Biochem. J.*, **108**, 95.
TITUS, E., and UDENFRIEND, S. (1954). *Fed. Proc.*, **13**, 411.
TU, A. T., REINOSA, J. A., and HSIAO, Y. Y. (1968). *Experientia*, **24**, 219.
TYRER, J. H., EADIE, M. J., and KUKUMS, J. R. (1968). *Brain*, **91**, 507.

UDENFRIEND, S., TITUS, E., WEISSBACH, H., and PETERSON, R. E. (1956). *J. biol. Chem.*, **219**, 335.
VAN DER PLOEG, M., and VAN DUIJN, P. (1964a). *J. R. microsc. Soc.*, **83**, 405.
VAN DER PLOEG, M., and VAN DUIJN, P. (1964b). *J. R. microsc. Soc.*, **83**, 415.
VAN DUIJN, P. (1953). *J. Histochem. Cytochem.* **1**, 143.
VAN DUIJN, P. (1955). *Recl. Trav. chim. Pays-Bas Belg.*, **74**, 771.
VAN DUIJN, P. (1957a). *Acta physiol. néerl.*, **5**, 413.
VAN DUIJN, P. (1957b). *Ibid.*, **5**, 428.
VERCAUTEREN, R. (1951). *Enzymologia*, **14**, 340.
VERCAUTEREN, R., and BLONDÉ, C. (1954). *Enzymologia*, **16**, 371.
VILLAMIL, M. F., and MANCINI, R. E. (1947). *Rec. Soc. argent. Biol.*, **23**, 215.
VILLELA, G. G. (1955). *Nature, Lond.*, **175**, 208.
VILLELA, G. G., and AFFONSO, O. R. (1955). *Nature, Lond.*, **175**, 1087.
WACHSTEIN, M., and MEISEL, E. (1964). *J. Histochem. Cytochem.*, **12**, 538.
WARBURG, O., and NEGELEIN, E. (1929). *Biochem. Z.*, **214**, 64.
WARBURG, O., and NEGELEIN, E. (1931). *Biochem. Z.*, **233**, 486.
WASHBURN, A. H. (1928). *J. Lab. clin. Med.*, **14**, 246.
WATANABE, Y. (1958). *Jap. J. med. Sci. Biol.*, **11**, 191.
WEINER, N. (1960). *Archs Biochem. Biophys.*, **91**, 182.
WEISSBACH, H., REDFIELD, B. G., GLENNER, G. G., and MITOMA, C. (1957). *J. Histochem. Cytochem.*, **5**, 601.
WIECHOWSKI, D. (1907). *Beitr. chem. physiol. Path.*, **9**, 295.
WOHLRAB, F. (1961). *Histochemie*, **2**, 163.
WOHLRAB, F. (1961). *Histochemie*, **2**, 359.
YAMADA, H., and YASUNOBU, K. T. (1962). *J. biol. Chem.*, **237**, 3077.
YAMAMOTO, Y., KOMORI, Y., MORIKUBO, T., TEGOSHI, M., TAKEBE, M., and FUJITA, S. (1960). *J. Kyoto. Pref. Univ. Med.*, **68**, 227.
YASUNOBU, K. T., IGAUE, I., and GOMES, B. (1968). *Advanc. Pharmacol.*, **6A**, 43.
ZELLER, E. A. (1940). *Helv. chim. Acta*, **23**, 1502.
ZELLER, E. A. (1951). "The Enzymes." Eds. Sumner, J. B., and Myrbäck, K. Acad. Press, New York, Vol. II, pp. 11 and 536.
ZELLER, E. A. (1963). In "The Enzymes." Eds. Boyer, P. D., Lardy, H., and Myrbäck K. Acad. Press, New York, Vol. 8, p. 313.
ZELLER, E. A., BARSKY, J., and BERMAN, E. R. (1955). *J. biol. Chem.*, **214**, 267.

CHAPTER 20

PRINCIPLES OF OXIDOREDUCTASE HISTOCHEMISTRY

THERE are a number of important differences between reactions for hydrolytic enzymes and those which demonstrate the various oxidoreductase systems. This chapter is therefore devoted to the principles on which the modern histochemistry of the dehydrogenases and diaphorases is based. These are of such importance in applied histochemistry that no further excuse is necessary. Nevertheless, much of what has already been stated in Chapter 14 is relevant here and reference will be made back to this chapter at appropriate points.

The chief difference between dehydrogenase methods and hydrolytic enzyme methods is that whereas with the latter the natural substrate is rarely employed, in the dehydrogenase methods its use is generally obligatory. All the factors taking part in the reaction can be those which are concerned with the reaction *in vivo* except for the presence of a tetrazolium salt. Our primary concern therefore is with the nature of the tetrazolium reaction and with the characteristics of tetrazolium salts and their formazans. After this we have to consider the kinetics of the various dehydrogenase reactions and the participation of the various components of the respiratory chain.

The Development and Early Use of Tetrazolium Salts

The first tetrazolium salt, triphenyl tetrazolium chloride (TTC) was prepared by von Pechmann and Runge in 1894 as a pale yellow, water-soluble, crystalline powder. Kuhn and Jerchel (1941) found that a number of colourless tetrazolium salts were reduced to coloured compounds by plant tissues according to the equation given below.

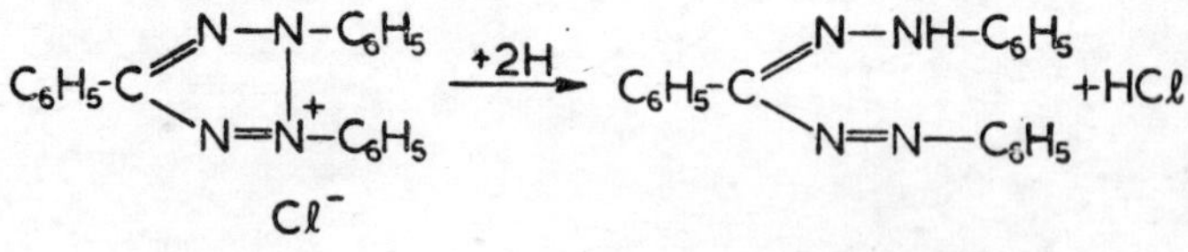

Tetrazolium salt (colourless) — Formazan (red)

It will be observed that one molecule of monotetrazole requires 1 molecule (2 atoms) of hydrogen for opening the tetrazole ring. Only 1 atom of hydrogen is contained in the formazan, however, since the other is required for combination with the chloride anion to produce HCl.

Kuhn and Jerchel noted that the production of formazan from tetrazole was not due to the usual reducing substances present in the cells (glutathione, ascorbic acid, cysteine) because these substances would not reduce tetrazolium salts below pH 9·0. The characteristic reduction took place, in plant tissues, at pH 7·2. Mattson, Jensen and Dutcher (1947) showed that reduction of TTC was not due to reducing sugars, since these were ineffective below pH 11·0, and Lakon (1942) used TTC as a test of the viability of seeds by observing the production of the red formazan in viable tissues. When tissues heated at 80° for short periods were found to be inactive, it was realized that some enzyme system in the tissues was responsible for the effects observed and Mattson *et al.* suggested that the reduction of TTC was due to dehydrogenase systems requiring Coenzymes I or II. According to Kun and Abood (1949), however, TTC revealed the presence of succinic dehydrogenase, which does not need a coenzyme, while Dufrenoy and Pratt (1948) suggested that the sites of reduction of tetrazolium salts were also the sites of reaction of phosphate ions. The effect of various inhibitors on the reduction of TTC by normal plant tissues was studied by Roberts (1951) who claimed that the formazan pattern of TTC was identical with that obtained with the Bennett sulphydryl reagent (Chapter 6, p. 143).

It soon became clear, however, that tetrazolium salts were acting as electron acceptors in enzyme-catalysed oxidations in the same way as methylene blue. A histochemical method for succinic dehydrogenase depending on the conversion of the latter to leuco-methylene blue was indeed proposed by Semenoff (1935). Diffusibility of the unreduced dye made localization by this method impossible. No doubt the method could be resurrected, through the agency of a solid medium technique (p. 923), if there were any special indications for its use.

A number of authors used TTC and related compounds in the study of animal tissues (e.g. Straus, Cheronis and Straus, 1948; Antopol, Glaubach and Goldman, 1948; Black and Kleiner, 1949; Black, Opler and Speer, 1950; Seligman, Gofstein and Rutenburg, 1949). Histochemically the difficulty of obtaining a positive reaction with TTC, and the weak colour and diffusibility of its formazan in the tissues, were serious objections to its continued use. Efforts were therefore made to overcome these objections by the synthesis of new compounds and of these three gave results which were sufficiently improved to carry dehydrogenase histochemistry forward on the next stage of its development. These compounds were 2,2′-(*p*-diphenylene)-bis-(3,5-diphenyl) tetrazolium chloride, known as *neotetrazolium* or NT; the methoxy derivative of NT, called *blue tetrazolium* or BT; and 2-(*p*-iodophenyl)-3-(*p*-nitrophenyl)-5-phenyl tetrazolium chloride or INT. The first was used by Antopol *et al.* (1948) and the second by Rutenburg, Gofstein and Seligman (1950). The monotetrazolium salt INT was produced by Atkinson *et al.* (1950), and this compound was found to be much more easily reduced than any other salt available at that time. The two ditetrazolium salts, however, were found to give better histochemical results and almost all the work in the field of dehydrogenase histochemistry between 1950 and 1958 was carried out with either NT or BT.

The reduction of BT takes place in the same way as the reduction of TTC except that a blue diformazan is the final product instead of a red monoformazan. The equation below indicates complete reduction; partial reduction can occur at one end of the molecule only, giving a monoformazan which is reddish purple and more soluble than the product of complete reduction (see below, p. 893). Three other tetrazolium salts which have been used in dehydrogenase histochemistry since 1957 are 2,2′-di-*p*-nitro-phenyl-5,5′-diphenyl-3,3′-(3,3′-dimethoxy-4,4′-biphenylene) ditetrazolium chloride (Nitro-BT) (Tsou *et*

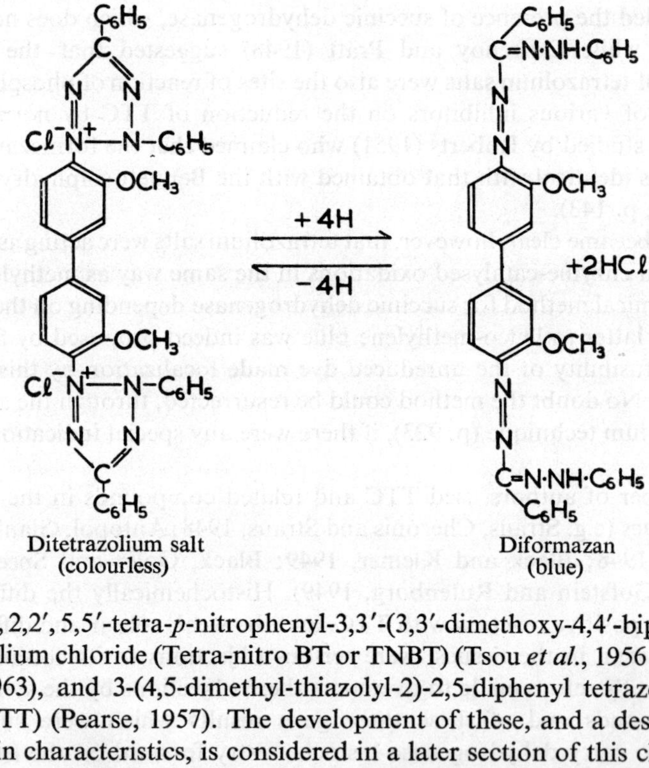

Ditetrazolium salt (colourless)

Diformazan (blue)

al., 1956), 2,2′, 5,5′-tetra-*p*-nitrophenyl-3,3′-(3,3′-dimethoxy-4,4′-biphenylene)-ditetrazolium chloride (Tetra-nitro BT or TNBT) (Tsou *et al.*, 1956; Rosa and Tsou, 1963), and 3-(4,5-dimethyl-thiazolyl-2)-2,5-diphenyl tetrazolium bromide (MTT) (Pearse, 1957). The development of these, and a description of their main characteristics, is considered in a later section of this chapter.

A new ditetrazole was synthesized by Jones (1969) expressly for quantitative histochemistry. In this compound, 2,2′-di-(3-nitrophenyl)-5,5′-dimethyl-3,3′-(4,4′-biphenylene) ditetrazolium chloride (Yellow Tetrazolium, YT), the methyl substituents shift the absorption maximum (p. 891) towards the UV while the nitrophenyl substituents facilitate absorption on to tissue proteins (p. 887) and also reducibility. YT is completely soluble in pyridine and therefore easily removed from tissue sections for photometric estimations.

Three interesting monotetrazoles, each having at least one nitro group in one of the N-phenyl rings, were synthesized by Seidler and Kunde (1969a).

These were 2,3,5-tri(*p*-nitrophenyl)-tetrazolium bromide (TNTTC), 2-phenyl-3-(3-methoxy-4-phenyl)-5-(*p*-nitrophenyl)-tetrazolium chloride (Half Nitro-BT), and 2,5-di (*p*-nitrophenyl)-3-(3-methoxy-4-phenyl)-tetrazolium chloride (Half TNBT). In histochemical systems the last of these was observed to give results superior to those of the ditetrazole (TNBT) and of Nitro-BT.

Physical and Chemical Characteristics of the Tetrazolium Salts

Redox Potentials. The function of the tetrazolium salt in dehydrogenase histochemistry is to act as acceptor of electrons from the oxidized substrate or, more usually, from some intermediate electron carrier. These electrons, if not intercepted, would normally pass by way of a succession of carriers to molecular oxygen. One of the main characteristics of the tetrazolium salt, which determines whether or not it will be suitable for histochemical purposes, is the ease with which it is reduced. This characteristic is the redox potential (E°) which is synonymous with the oxidizing capacity and is expressed in volts with reference to the normal hydrogen electrode. It can be measured by means of a potentiometer or recorded polarographically.

The earliest polarographic studies of the redox potential of TTC were those of Jerchel and Möhle (1944). Further work on mono- and ditetrazoles was carried out by Ried and Wilk (1953), Jambor (1954), Campbell and Kane (1956) and by Kivalo and Mustakallio (1956). These last authors included BT and NT in their survey. More recent studies are those of Jerchel *et al.* (1958) and Jambor (1960).

Polarograms of tetrazoles and tetrazolium salts exhibit a number of reduction waves corresponding to the production of formazan, hydrazidine and amidrazone in successive stages. The first two waves to occur, which are often called the α and β waves, were considered by Campbell and Kane to be associated with the opening of the tetrazolium ring. The α wave was believed to be characteristic of reduction under special conditions where the product is absorbed on to the mercury surface below the dropping mercury electrode. Since this condition is not present in the biological systems with which we are concerned the α wave, when present, can certainly be ignored. Although the β wave involves reduction beyond the formazan stage Campbell and Kane considered that the half wave potential of this wave could be regarded as a measure of the ease with which a tetrazolium salt is reduced to its formazan. The half wave potential of the β wave is thus the parameter with which we are here concerned.

With the exception of the last, the β wave half wave potentials shown in Table 66 were obtained with a Tinsley Recording Polarograph and dropping mercury electrode. The tetrazolium salts were dissolved in 0·1 M-phosphate buffer at a concentration of 100 µM. They were the purest available samples but the figures should nevertheless be regarded as accurate only for comparative purposes.

TABLE 66

Half Wave Potentials of Tetrazolium Salts and Tellurite

Oxidant	E ½, 22°, pH 7·2, mV.
Tellurite	−950
TTC	−490
NT	−170
BT	−160
MTT	−110
INT	− 90
Nitro-BT	− 50
TNBT	Below − 50

Tellurite was used by Wachstein (1955) as an indicator of so-called endogenous dehydrogenase activity in sections of human kidney. It was also used as an indicator of succinic dehydrogenase activity in bacteria (Mudd, Takeya and Henderson, 1956; van Iterson and Leene, 1964) and in mammalian

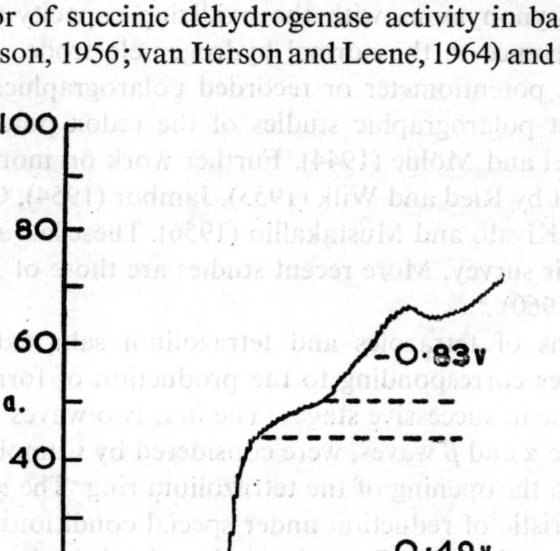

FIG. 181. Polarogram of Reduction of TTC.

tissues (Barrnett and Palade, 1957). With the latter the incubation period was very long indeed and it is perhaps surprising that any result at all was obtained.

When we come to the tetrazolium salts we see (Fig. 181) that the half wave potential for TTC is −490 mV.

This agrees closely with the result calculated from the traces for pH 7·6 given in their paper by Kivalo and Mustakallio but differs from the figures for pH 6·7 given by Campbell and Kane, which are approximately 100 mV higher. The figure recorded for TTC is well below those for NT and BT, which are practically identical. In histochemical practice, however, NT is always observed to be more easily reduced than BT and on this account it has been preferred by many workers. In this instance, therefore, the half wave potential fails to reflect accurately the differences observed in biological systems. Nevertheless it seems to be significant that the four most easily reduced tetrazolium salts in common use (INT, MTT, Nitro-BT,TNBT) have half wave potentials which are significantly higher than those of the other salts.

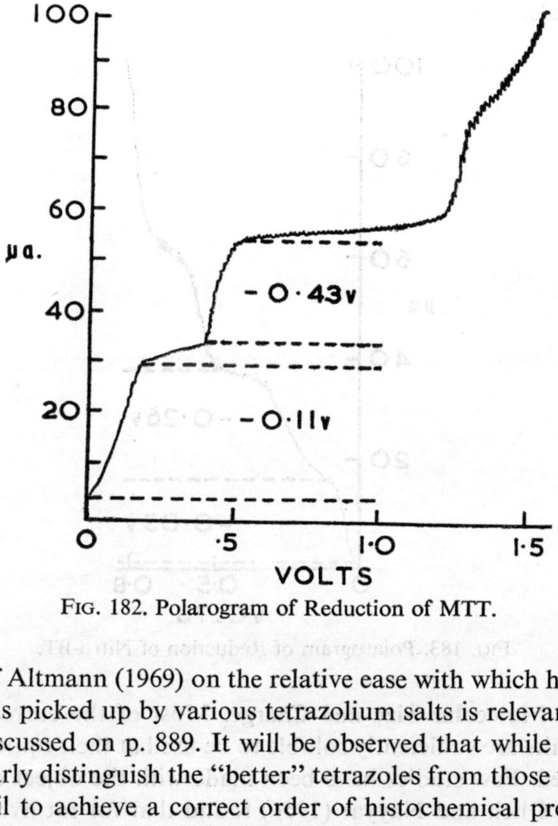

FIG. 182. Polarogram of Reduction of MTT.

The studies of Altmann (1969) on the relative ease with which hydrogen from the substrate is picked up by various tetrazolium salts is relevant here. These studies are discussed on p. 889. It will be observed that while the half-wave potentials clearly distinguish the "better" tetrazoles from those which are less good, they fail to achieve a correct order of histochemical precedence.

Figures 182 and 183 are reproductions of polarographic traces obtained with MTT and Nitro-BT respectively. The comparative steepness of the curve given by Nitro-BT is especially notable and, in practice, this salt appeared to be considerably more easily reduced than MTT. The critical work of Altmann (1969), referred to above and on p. 890, indicates that the superiority of Nitro-BT was certainly apparent rather than real.

The important question of the relationship between the half wave potential of the β wave for a given tetrazolium salt, and the E° of the reaction giving rise to the electrons accepted by this salt, is discussed in a later section of this chapter. The concept that the half wave potential of a tetrazolium salt can be too high for it to be employed safely as an electron acceptor in dehydrogenase histochemistry must be reconsidered.

It appears that, with few exceptions, the apparent reducibility (i.e. the amount of coloured product appearing in the section) cannot be used as a reliable criterion. Where the hydrogen donor is neither substrate nor intermediate carrier but itself a reducing compound, such as the aldehyde oxidation product of tryptamine in the MAO reaction (p. 866), then the order of reducibility seems more closely to approach the theoretical order derived from redox potentials.

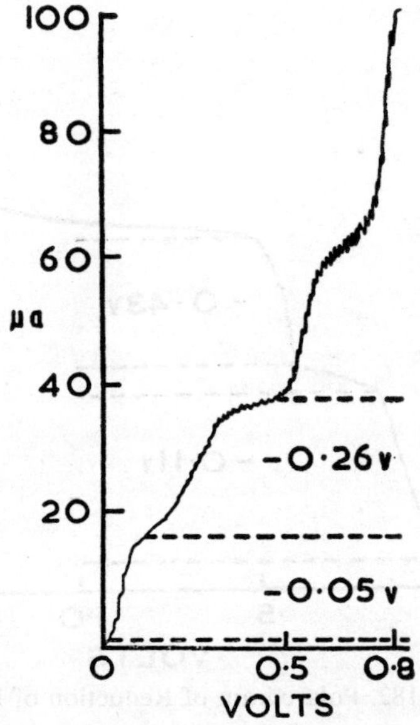

FIG. 183. Polarogram of Reduction of Nitro-BT.

Solubility, Molecular Size and Charge. Most of the currently employed tetrazolium salts are sufficiently soluble to be used at their optimal concentration if required. Few studies have been made with the object of determining the latter but Glick and Nayyar (1956) found that for rat liver sections and INT the optimal concentration of the tetrazolium salt was between 4 and 6 mM., or 2 to 3 mg/ml. in the incubating medium. This figure is well above that usually employed in dehydrogenase histochemistry.

The molecular size of all tetrazolium salts so far employed is sufficiently small for the diffusion constant to be high enough theoretically to allow rapid penetration into the tissues (see Chapter 14, p. 483). In practice, however, the *intact* membrane of the mitochondrion constitutes a barrier to the penetration of the tetrazolium salt so that the larger ditetrazolium salts may penetrate slowly or not at all. A low concentration of tetrazolium salt at the site of electron transfer may thus be the rate limiting factor even if the concentration in the medium is perfectly adequate.

In fully protected media (Chapter 14, p. 479 *et seq.*) the penetration of Nitro-BT into the mitochondrion is reduced to a low level and the amount of formazan is correspondingly small. It is necessary, in fact, for the mitochondrial membrane to be damaged before an adequate result can be obtained with Nitro-BT in the case of enzyme systems localized within this organelle.

Substantivity. Another factor which may seriously interfere with penetration of the tetrazolium salts is the possession of strong polar groups and/or substantivity for protein. Wattenberg (1954) found that when sections were incubated in the presence of low concentrations of a tetrazolium salt the latter was firmly bound to protein. This was shown by the fact that, after thorough washing, the application of a reducing agent produced a diffuse deposit of formazan throughout the section. Substantivity of the tetrazolium salt is thus generally an undesirable quality from the histochemical point of view. The part played by substituent nitro groups was investigated by Pearse and Hess (1961) who showed that (a) planar conformation of the molecule (as in ditetrazoles and in monotetrazoles with large benzeneazo substituents in the 3-position) and (b), polar groupings attached to the 2-position of the tetrazole ring conferred the property of substantivity, particularly for lipoprotein. The second property was considered to be the most important of the two.

A slightly different view was expressed by Seidler and Kunde (1969b) whose tests, using a technique essentially similar to that of Pearse and Hess (1961), indicated that linear aromatic substituents on N(3) of the tetrazole, or increase of linearity by doubling the molecule, were more important than polar groups in the 2-position. Their (1969a) half Nitro-BT was only slightly less substantive than Nitro-BT itself. Wohlrab and Fuchs (1967) investigated the non-enzymic staining of fresh and fixed cryostat sections by TNBT. They concluded that this ditetrazole was adsorbed non-specifically on to lipoprotein membranes and that non-specific localizations of enzyme activity might result therefrom. Hitzeman (1963), on the contrary, failed to obtain any evidence whatsoever of ditetrazole (Nitro-BT) substantivity. Her observations are not explicable on any scientific basis. An important contribution to the problem was the work of Brooke and Engel (1966) on selective binding of Nitro-BT to skeletal muscle fibres.

Lipid Solubility. Lipid solubility may be of value if it accelerates the rate of penetration through lipid or lipoprotein barriers. Measurements of the lipid solubility of tetrazolium salts do not appear to have been made; MTT is certainly lipid soluble but Nitro-BT and TNBT are not. The disadvantages of the last two with regard to penetration are certainly offset by the fact that progressive formazan production in lipid droplets does not occur. This process, when it occurs, is due to reduction of adsorbed tetrazolium by the autoxidation products of unsaturated fats.

Light Sensitivity. Many tetrazolium salts are sensitive to light and undergo colour changes when left in the light for short periods. In the case of TTC two compounds are produced, one being the formazan and the other, 2,3-(2,2′-diphenylene)-5-phenyl tetrazolium chloride, an oxidation product. It is fortunate that the tetrazolium salts most used in histochemistry are fairly stable to light and the thiazolyl tetrazolium salts (MTT) are exceptionally stable.

Carter and Roppel (1967) found that "conventional" Nitro-BT staining for glucose-6-phosphate dehydrogenase was greatly enhanced when, following incubation, the section was exposed to UV light from a mercury vapour lamp. The authors admitted that some negative areas in their control sections were rendered positive by irradiation. Since there is every likelihood of UV-catalysed reduction of adsorbed tetrazoles, I suggest that post-irradiation be not employed in dehydrogenase studies.

Concentration Effects and Linearity. Some interesting work has been carried out on the effect of different concentrations of tetrazolium salts on the kinetics of the dehydrogenase reactions. It had always seemed probable that inhibition of enzyme activity was a factor which had to be taken into account. It was shown by Lettré and Albrecht (1943), Jerchel and Fisher (1949), and by Rutenburg *et al.* (1950), that tetrazolium salts were highly toxic to fibroblasts in tissue culture and to intact animals. Pearse, Scarpelli and Hess (1960) compared the values for oxidation of succinate, β-hydroxybutyrate, and DPNH and TPNH by isolated submitochondrial particles, obtained by estimation of formazan and by measurement of the oxygen uptake. The formazan values were about one-tenth of the values obtained from oxygen uptake. The most likely explanation of these findings was an inhibition of enzyme activity.

Studying the effect of varying the concentration of the tetrazolium salt on succinate-NT reductase Jones (1968a) found that when the concentration was raised above 0·54 mM enzyme activity was sharply inhibited. Glick and Nayyar (1956) had indeed reported an optimum tetrazole level of 4-6 mM at pH 7·4, using a succinate-INT reductase system, and Nachlas *et al.* (1960) indicated an optimal INT concentration of 0·87 mM for the same system in rat liver homogenates. For succinate-BT reductase Kuwabara and Cogan (1959) found optima at 0·34 and 0·69 mM, observing that at maximum concentration (saturation) of the ditetrazole enzyme activity was abolished.

In the case of neotetrazolium Slater (1959) demonstrated that formazan production was not proportional to enzyme activity and Eadie *et al.* (1970), using "mock tissue preparations" made from liver homogenates, showed that the time course of formazan production was not linear with Nitro-BT or INT. With MTT linearity was achieved over a period of 80 minutes. These results are shown in Fig. 184, below.

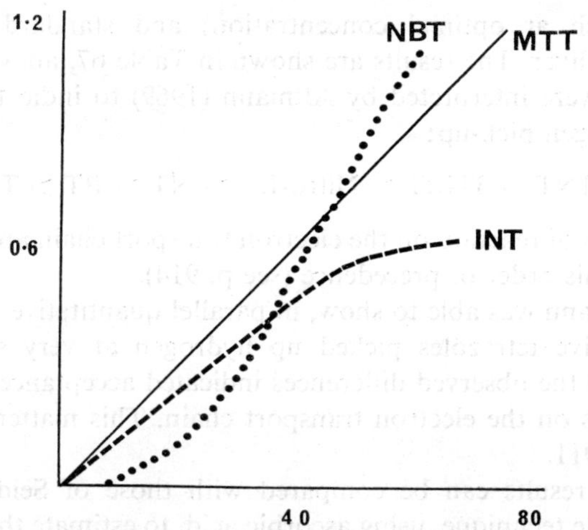

FIG. 184. Time Course of Formazan Production when NBT, MTT and INT are reduced by succinate dehydrogenase in "mock" tissue made from liver homogenate.
Y axis Formazan yield (Eq $\times 10^{-8}$)
X axis Time in minutes.

In view of increasing interest in the measurement of enzyme activity in tissue preparations, and even in single cells, these results must clearly be taken seriously. Whichever tetrazole is used for quantitative studies it will be necessary to show that formazan production is linear over the period of incubation which is employed.

Relative Order of Hydrogen Pick-up by Tetrazoles. Studies carried out by Altmann (1969) were designed to determine the relative order of pick-up of hydrogen by mono- and ditetrazoles. Incubation was carried out for the 6-phosphogluconate dehydrogenase system using media containing pairs of

TABLE 67

Relative Dehydrogenase Activities

Tetrazole Pairs	Ratios of 6-PGDH Activity
MTT/INT	3 : 1
INT/TNBT	2 : 1
INT/NBT	3 : 1
TNBT/NBT	5 : 1
NBT/BT	Nitro-BT formazan only
INT/BT	INT formazan only
BT/NT	1 : 1
BT/TV	2 : 1
NT/TTC	8 : 1
NT/TV	3 : 1
TV/TTC	1 : 1

tetrazoles (each at optimal concentration) and standard fresh cryostat sections of rat liver. The results are shown in Table 67, above.

These results were interpreted by Altmann (1969) to indicate the following order of hydrogen pick-up:

$$\text{MTT} > \text{INT} > \text{TNBT} > \text{Nitro-BT} > \text{NT} > \text{BT} > \text{TV} > \text{TTC}$$

The actual sites of reaction on the electron transport chain are not, of course, indicated by this order of precedence (see p. 914).

Since Altmann was able to show, in parallel quantitative studies, that the four most active tetrazoles picked up hydrogen at very similar rates he concluded that the observed differences indicated acceptance of hydrogen at different points on the electron transport chain. This matter is further considered on p. 911.

Altmann's results can be compared with those of Seidler (1968) who devised a simple technique, using ascorbic acid, to estimate the reducibility of a series of tetrazoles. His experiments indicated that this property was increased by interaction with other molecules found in the medium and that these could stabilize the half-reduction stage of ditetrazoles. Seidler's order of reduction was as follows:

$$\text{INT} = \text{MTT} > \text{TNBT} > \text{Nitro-BT} > \text{BT} > \text{TV} > \text{TTC}$$

He found, however, that some newer monotetrazoles, having 2 or 3 nitro groups in the molecule (Seidler and Kunde, 1969), were even more easily reduced. The most active was 2,3,5-tri(*p*-nitrophenyl) tetrazolium bromide (TNTTC). Seidler and Kunde concluded that in some tissues the results obtained with di- or trinitro-monotetrazoles were better than those obtained with Nitro-BT.

It is impossible to reduce all these conclusions to a single recommendation. The use of tetrazoles must clearly be accompanied by a better understanding of the mechanisms involved and a more critical attitude, especially if quantitative or even semi-quantitative studies are proposed.

Physical and Chemical Characteristics of the Formazans

Colour, Size and Shape of the Deposit. Formazans are intensely coloured compounds giving shades which vary from red to deep purplish or brownish black. They also show brilliant reflex colours. This property was used by Fleischhauer (1958) who examined sections containing NT formazan by polarized light. The golden reflex of the crystals made them easily distinguishable under the microscope from other (non-crystalline) reaction products deposited in the tissues. Table 67 gives details of the absorption spectra of formazans from tetrazolium salts which have been or are being used in histochemistry. Some of these have been derived from the literature and the remainder are measurements carried out in my own laboratory.

TABLE 68

Absorption Spectra of Formazans

Formazan from	Absorption Maxima (nm.)			
	λmax	εmax	λmax	εmax
TTC . . .	290	10,000	405	30,900
INT . . .	500	—	675‖	11,000
NT* . . .	567	40,000	587	14,400
BT . . .	—	—	606	10,400
Nitro-BT . .	355	37,900	574§	55,000
TNBT . . .	330	—	730‖	—
MTT . . .	560	51,000	—	—
MTT-Co† . .	560	20,000	660	20,000
5-MTT . .	550	31,000	—	—
5-MTT-Co† .	550	44,000	675	26,000
4-MTT . .	430	13,000	447	41,000
4-MTT-Co† .	436	39,000	—	—
YT‡ . . .	—	—	455	54,050

* From Jones (1968b).
† Cobalt-chelated; see subsection on chelation below.
‡ From Jones (1969).
§ From Oda (1961).
‖ From Altmann (1969); in alkaline DMF.

Absorption with two peaks, in the red and violet bands, gives the formazans their typical colour and strong absorption in the red band is characteristic of all formazans suitable for histochemical purposes. There is also strong absorption in the ultra violet at about 250 to 350 nm but little use is made of this property. These features are shown in Fig. 185, below, which is the

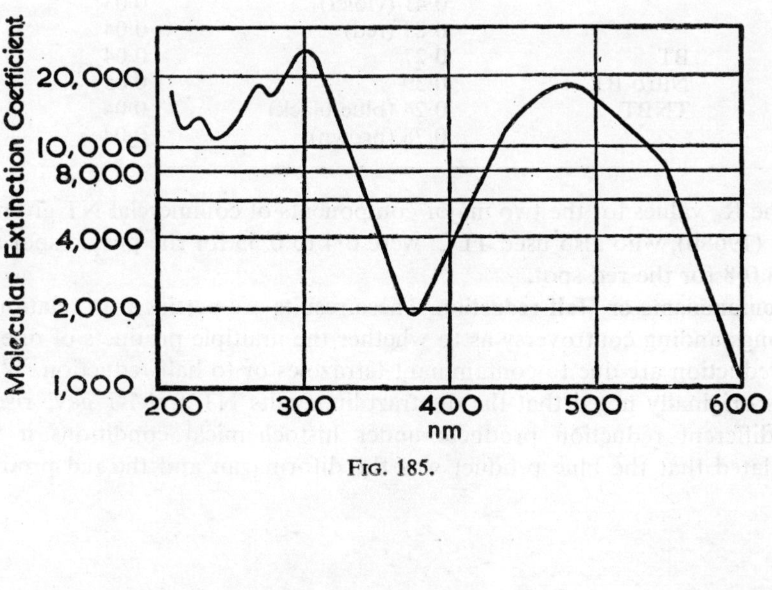

FIG. 185.

absorption spectrum of TTC. The high molecular extinctions of the diformazans are especially notable, particularly the diformazan from Nitro-BT, but the lower extinctions of the cobalt-chelated formazans from the 4,5-dimethyl and 5-dimethylthiazolyl tetrazolium salts are compensated for by the higher wavelengths at which maximal absorption occurs. When viewed by eye these cobalt-formazans appear black although they are actually deep purple in solution in organic solvents. For the detection of succinate dehydrogenase in the rabbit sperm mid-piece Young and Edwards (1963) used TNBT and examined their preparations by UV light from a condensed spark derived from rotating disc electrodes of cadmium or magnesium. They found a selective absorption peak at 330 nm.

Thin Layer Chromatography (TLC). Tyrer *et al.* (1969) used TLC on 20 × 20 cm glass plates, coated with a 0·25 mm. layer of Merck Silica Gel G, and dried at 100° for 2 hours. Ascending chromatography was carried out with *n*-butanol-water-glacial acetic acid (78 : 17 : 5). After the run the plates were dried and developed with alkaline ascorbate or ammonium sulphide. Some of their results are shown in Table 69, below.

TABLE 69

R_F Values of Tetrazolium Salts

Compound	R_F Values of Tetrazolium Salts	± S.D.
TTC	0·60	0·03
INT	0·77	0·05
MTT	0·54	0·04
TV	0·79	0·03
NT	0·32 (violet, major component)	0·04
	0·24 (red)	0·04
	0·41 (violet)	0·03
	0·83 (red)	0·04
BT	0·27	0·04
Nitro-BT	0·39	0·05
TNBT	0·24 (blue-black)	0·04
	0·78 (brown)	0·04

The R_F values for the two major components of commercial NT given by Jones (1968b), who also used TLC, were 0·4 to 0·55 for the purple spot and 0·7 to 0·8 for the red spot.

Contaminants or Half-reduction. These results are pertinent in relation to the longstanding controversy as to whether the multiple products of ditetrazole reduction are due to contaminant tetrazoles or to half reduction. When it was originally noted that the ditetrazolium salts NT and BT gave rise to two different reduction products under histochemical conditions it was postulated that the blue product was the diformazan and the red product,

PLATE XX

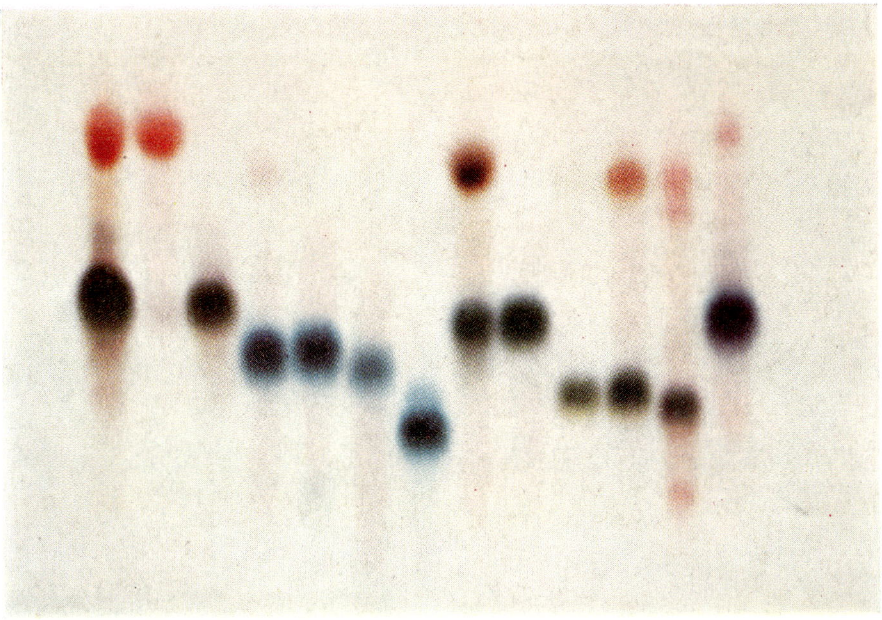

1 3
2 4 8 9 13
5 6 7 7 10 11 12

Thin-layer chromatogram of bistetrazolium salts run with pentanol-formic acid-water (J. Chromatog. *39*, 336, 1969). The dry plate has been sprayed with alkaline ascorbate to reduce the salts to coloured formazans. In the following explanation, salts may be referred to in terms of coloured spots for convenience. Except for (2) and (3), all samples were obtained commercially.

(1) Neotetrazolium chloride.

(2) 2-(4-biphenylyl)-3,5-diphenyl tetrazolium chloride, the usual impurity associated with neotetrazolium chloride.

(3) Purified neotetrazolium chloride.

(4) Blue tetrazolium, showing the main blue spot and a rapidly-moving lilac contaminant.

(5)–(7) Various derivatives of blue tetrazolium (p-anisyl, piperonyl and veratryl). Although these particular samples are free from major contaminants, the salts themselves are of no value in the cytochemical demonstration of oxidative systems.

(8) Nitroblue tetrazolium, heavily contaminated by a rapidly-moving spot which is dirty red in colour.

(9) A pure sample of nitroblue tetrazolium.

(10) and (11) Samples of tetranitroblue tetrazolium of differing purity.

(12) *m*-Nitroviolet tetrazolium. In addition to the rapidly-moving red contaminants, a slowly-moving impurity is also visible, lagging behind the bis-salt.

(13) Violet tetrazolium (*o*-tolidine). Samples of this material may contain both rapidly and slowly-moving contaminants.

PLATE XXI

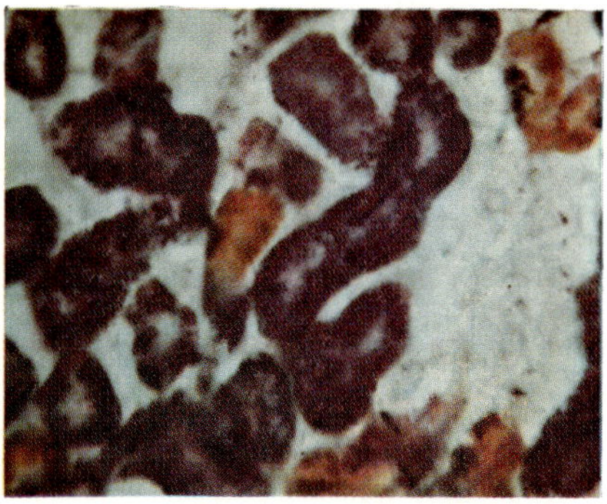

XXIa. Rat kidney. Succinate dehydrogenase reaction using INT as acceptor. Two different coloured formazan products are visible in the tubular cells. × 640.

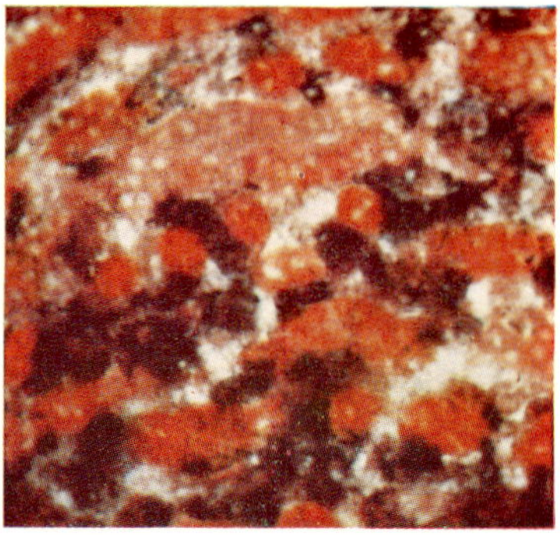

XXIb. Rat kidney. NADH diaphorase reaction using BT as acceptor. Two different formazan products are present. × 480.

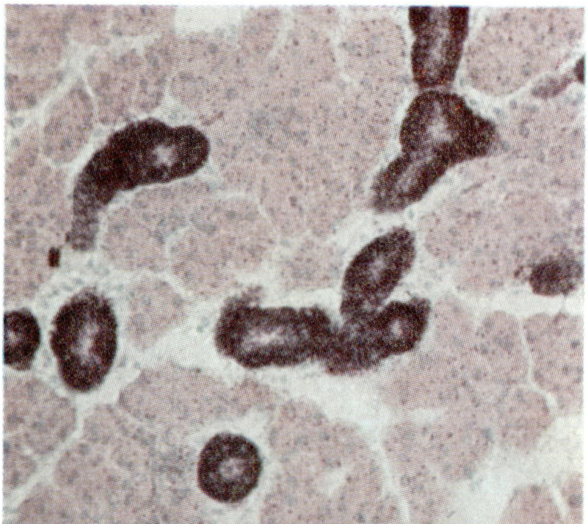

XXIc. Rat salivary gland. Shows high activity in duct lining cells but poor intracellular localization. Succinate dehydrogenase (Blue tetrazolium). × 200.

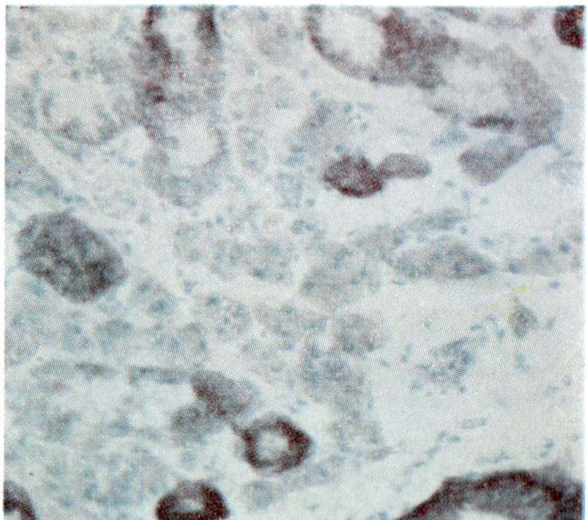

XXId. Rat salivary gland. Great improvement in intracellular localization produced by using a salt giving rise to a substantive formazan. Succinate dehydrogenase (Nitro-blue tetrazolium). × 200.

occurring in areas of lower enzyme activity, was a monoformazan produced by reduction at one end of the molecule only. Later, however, Burtner, Bahn and Longley (1956) observed that in the case of NT the red derivative was probably the formazan from a contaminating monotetrazole. This speculation was confirmed by TLC studies carried out on NT by Okui *et al.* (1963) in which a major contaminant was found in commercial samples, and in samples synthesized by the authors. This contaminant, 2,5-diphenyl-3-diphenylene tetrazolium chloride (DDTC), was responsible for the broad absorption peak (500-560 nm) observed for NT. After removal of the contaminant by column chromatography, however, the resulting pure NT still gave red and blue reaction products. Reduction was found to occur in two stages and the red half-reduction product of the first stage had an absorption peak at 530 nm. It could be produced either by short incubation with succinate and mitochondria, or by electrolytic reduction at a constant potential of −450 mV. Okui *et al.* concluded that the reduction of NT by mitochondria occurred in step-wise fashion and, using the procedure of Ball and Rodkey (1950), they observed two OR potentials at −200 and −420 mV.

Further evidence of the contamination of NT samples by monotetrazoles was obtained by Jones (1968) who described a simple non-chromatographic method for the purification of NT and identified the chief contaminant as 2-(4-biphenyl-)-3,5-diphenyl tetrazolium chloride (BDTC). In Plate XX, p. 892, are shown the thin layer chromatograms of a number of bistetrazolium salts. These indicate the degree of contamination usually experienced.

Monotetrazolium salts like TTC and INT are popularly supposed to have only one product. With many samples of INT, however, two formazans, a brick-red and a brownish purple one, are deposited in tissue sections (Plate XXIa). This may be due, of course, to a contaminating monotetrazole though the distribution of the two formazans suggests that it may be due to the presence of two isomers. Little information is available concerning the red and blue products from BT (see Plate XXIb) but Gosztonyi *et al.* (1963), using paper chromatography and the Bush 'A' and BJ systems as developers, noted two blue formazans and a red intermediate from pure BT.

In the case of Nitro-BT the position is not entirely clear. All samples of this reagent at present available give rise to two reaction products, the bluish purple diformazan and a red compound. Most authorities consider the latter to be a monoformazan from a monotetrazole contaminant. Chromatographic and spectrophotometric studies carried out by Gabler, Wendler and Schmidt (1970) indicated, clearly, that the red product of Nitro-BT reduction was due to half reduction.

Despite repeated purifications Tsou and Su (1962) continued to find two components in their samples of TNBT. On chromatograms, after reduction, the colour of the two spots obtained was identical, and different from that given by the monotetrazole contaminant. The authors suggested that tetrazoles containing the biphenylene group exist in 2 geometric isomers. There was thus still reasonable doubt as to the true state of affairs concerning the reduction products of Nitro-BT and TNBT.

Further Evidence on the Nature of Tetrazole Reduction Products. It was suggested by Ashley *et al.* (1953) that monotetrazoles were reduced by way of a free radical stage and proof of the existence of a free radical intermediate, in the case of TTC, was provided by Maender and Russell (1966). Further confirmation came from the work of Deguchi and Takagi (1967) and the nature of the intermediate was deduced by Neugebauer (1968). Chromatographically homogeneous Nitro-BT (in buffer) was reduced with ascorbic acid by Eadie (1969) who found that when the resulting dark precipitate was washed and extracted with ethanol, it could be resolved into a blue precipitate and a red solution. When dissolved in dimethylformamide (DMF), and reduced by the progressive addition of ascorbic acid in DMF, the solution became first red, then blue. Much later a blue precipitate formed. It is thus evident that Nitro-BT reduces through (a) a red, ethanol-soluble product and (b) a blue DMF-soluble product. The latter condenses to (c) a blue DMF-insoluble product. Absorption maxima were (a) 525 nm (b) 575 nm (c) 605 nm. The postulated stages of reduction are shown opposite:

The red intermediate could be either IV or VI.

While it is therefore certain that at least two alternative mechanisms exist (isomers; intermediates or half-reduction) it has not yet been shown conclusively that either is responsible for the production of different colours under histochemical conditions.

Crystal Size. The monoformazans as a class are easily crystallized and they are usually deposited in the tissues in crystalline form. Many of the diformazans (e.g. those from NT and BT) are also deposited in tissue sections as crystals. The size of these varies from long needles 5 × 1·5μ (INT) to short rods averaging 1·5 × 1μ (BT). The crystal of NT diformazan is a longer rod measuring 2·5 × 1·5μ. Crystalline products are not acceptable in dehydrogenase histochemistry since they make accurate localization impossible. The newer methods, which usually give rise to non-crystalline formazans, afford very much finer definition (see Plates XXIc and XXId).

The metal formazans from thiazolyl tetrazolium salts can still form large crystals and, if the mitochondria in which they have been deposited are for any reason allowed to rupture (see Chapter 14, p. 479; Protection of Mitochondria), diffusion of the formazan through the mounting medium allows aggregation and crystallization to take place. In the case of Nitro-BT the position is complicated. Its diformazan cannot be crystallized from solutions and initially crystals are not formed in the tissues. If, however, the sections are dehydrated in alcohols, cleared in xylene and mounted in a synthetic medium, aggregates and crystals of the diformazan or of a diformazan-protein complex are commonly observed. These, unfortunately, bear a close resemblance in size and shape to mitochondria but can be produced in model systems in which mitochondria are not present. If there is any probability of

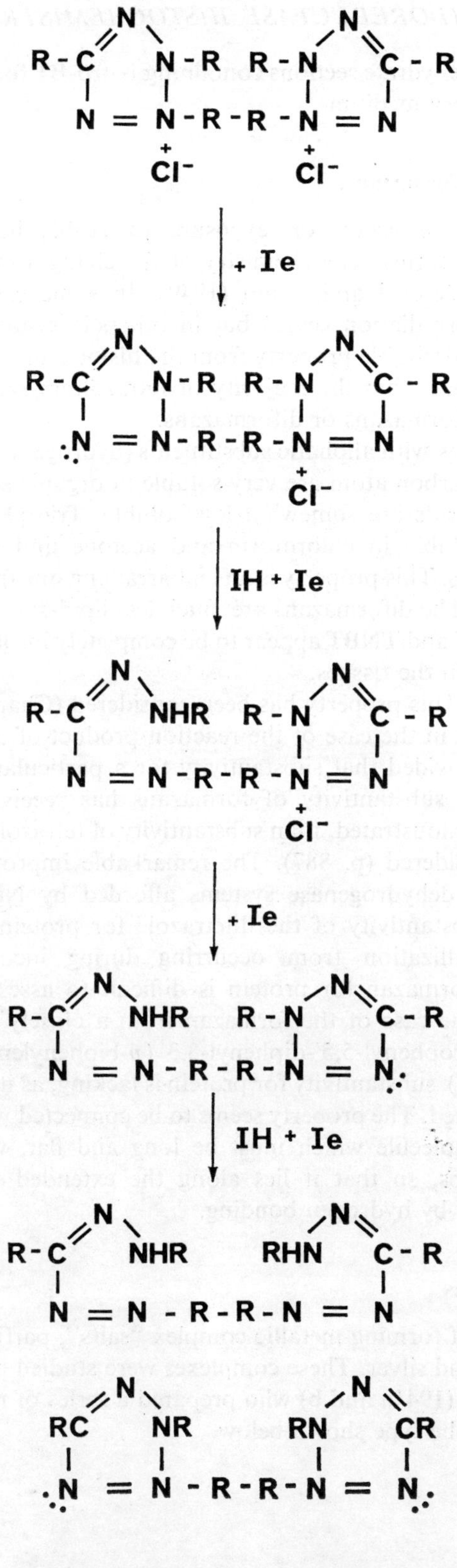

I

II

III

IV

V

VI

confusion, it is better not to dehydrate sections containing Nitro-BT formazan but to mount them in a watery medium.

Other Characteristics of the Formazans.

Light Sensitivity. Many formazans on exposure to visible light are changed into red and yellow forms. The chemistry of the changes involved was described by Hausser, Jerchel and Kuhn (1949). In some cases the reaction is reversible when irradiation ceases but in others it is not. Light sensitivity is obviously an undesirable property from the histochemical point of view and it is thus fortunate that the majority of tetrazolium salts now employed give rise to stable formazans or diformazans.

Lipid Solubility. Formazans with aliphatic substituents (hydrogen, methyl, carboxyl) attached to the 3-carbon atom are very soluble in organic solvents and those with aryl substituents are somewhat less soluble. Triaryl monoformazans are especially soluble in chloroform and acetone and also in triglycerides and phosphatides. This property is an embarrassing one from the histochemical point of view. The diformazans are much less lipid-soluble and the formazans from Nitro-BT and TNBT appear to be completely insoluble in those lipids normally found in the tissues.

Substantivity for Protein. This property has been considered (Chapter 14, p. 490) to be a desirable one in the case of the reaction product of a histochemical enzyme method provided that substantivity for a particular tissue component is avoided. The substantivity of formazans has received less attention, being less easily demonstrated, than substantivity of tetrazoles. The latter has already been considered (p. 887). The remarkable improvement in apparent localization of dehydrogenase systems afforded by Nitro-BT was clearly largely due to substantivity of the ditetrazole for protein which effectively prevented crystallization from occurring during incubation. Substantivity of Nitro-BT formazan for protein is difficult to assess. It is interesting to note that in the case of the formazan from a closely related tetrazolium salt, 2,2′-di-*p*-nitrophenyl-5,5′-diphenyl-3,3′ (*p*-biphenylene) ditetrazolium chloride (Nitro-NT), substantivity for protein is lacking, as it is with all monoformazans so far tested. The property seems to be connected with the nature of the diformazan molecule which must be long and flat, without interfering substituent groups, so that it lies along the extended protein molecule to be attached to it by hydrogen bonding.

Metal Chelation of Formazans

Formazans are capable of forming metallic complex "salts", particularly with copper, nickel, cobalt and silver. These complexes were studied particularly by Hunter and Roberts (1941a and b) who prepared a series of metallic derivatives of formazans of the type shown below.

In these the imino-hydrogen atom of the formazan is replaced by one equivalent of the metal. Chelation with copper, nickel and cobalt took place only in hot acetone solutions, under reflux, and the products were stable, black pigments, insoluble in water, ethanol or light petroleum. Despite their high mol. wt., however, they were remarkably soluble in other organic solvents.

A search for formazans which would chelate under the milder conditions experienced in histochemical reactions (Pearse, 1957) resulted in the observation that chelation readily occurred with the formazans from *N*-thiazolyl-substituted tetrazolium salts (Beyer and Pyl, 1954). When thiazolyl tetrazolium salts were used as electron acceptors in dehydrogenase reactions in the absence of a chelator (Co^{2+}, Ni^{2+}, Cu^{2+}) the formazan deposits rapidly

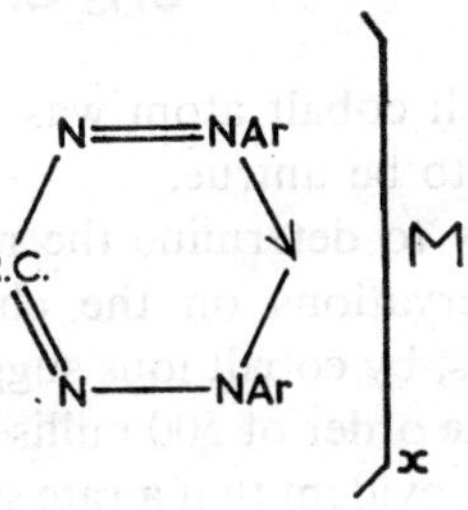

aggregated and crystallized (Fig. 186 p. 869). In the presence of suitable metallic ions, on the other hand, the deposits were found to be non-crystalline (Plate XXIIa, p. 968). Thus a capture reaction was established which prevented crystallization and which presumably also prevented diffusion of the free formazan. These observations formed the basis of the monotetrazolium-metal chelation methods for a variety of dehydrogenases and diaphorases. These are described in Chapter 21 and in Appendix 21.

In my original paper on the subject (1957) a phenomenon called "escape" was described in which, after shorter or longer periods in storage, the dot-like formazan deposits began to crystallize out. The word "escape" suggested, though this was not intended, that the metal complex could break down to liberate the free formazan. There was never any direct evidence of this and further research showed that breakdown of the complex is not responsible. Crystallization occurs if the metal-formazan is able to diffuse in the mounting medium (due to its solubility in organic solvents this has to be an aqueous medium) and this it can do if the mitochondria which contain it are damaged so that their membranes become permeable.

The mechanism by which chelation of thiazolyl formazans takes place cannot be explained on the basis of the formula used by Hunter and Roberts.

It is probable that a single cobalt atom is bound by two molecules of formazan as shown in the formula below:

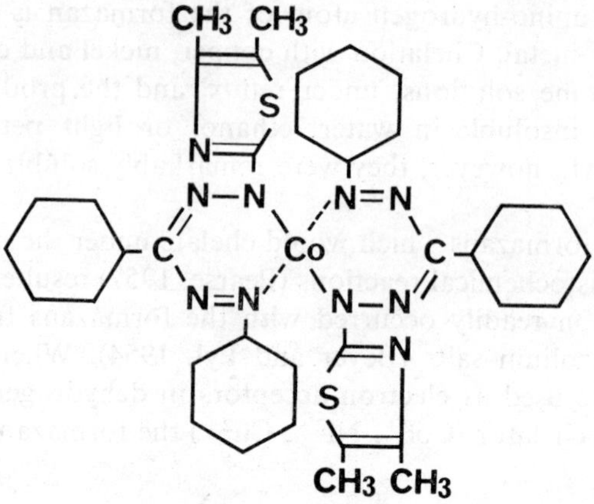

Earlier suggestions that each cobalt atom was bound by a single formazan molecule have been shown to be untrue.

It is obviously necessary to determine the rate at which chelation takes place. Initial *in vitro* observations on the chelation of MTT formazan, dissolved in organic solvents, by cobalt ions suggested that it was a relatively slow process (half time in the order of 500 millisecs). Using the data provided *in extenso* in Chapter 14, it is evident that a rate such as this could not possibly provide an efficient capture reaction. It was therefore postulated that in aqueous media and under biological conditions a different rate obtains. Evidence from polarographic studies suggested that this was in fact the case and that cobalt chelation of the nascent formazan is much faster than with the formazan dissolved in organic solvents. It is clearly sufficiently rapid to account for the observed improvement in localization.

Objections to the Chelation Principle. So far only one important disadvantage attaches to the monotetrazolium-cobalt chelation methods. This is due to the fact that they cannot be employed when, for any reason, the presence of cyanide in the incubating medium is desirable. Cyanide may be required either in order to block the electron pathway to oxygen or, more importantly, as a trapping agent for carbonyl products resulting from the oxidation of various substrates. As will be seen in Chapter 21, dehydrogenase techniques in which cyanide is present in the medium must at present be carried out with a non-chelating technique, preferably using Nitro-BT.

The reason why Co^{2+} ions cannot be included in any medium containing CN^- and an electron acceptor (tetrazolium salt) is that an unstable cobalt cyanide complex is formed $[Co(CN)_6]^{4-}$, as described by Keller and Parry (1956), and this rapidly reduces the tetrazolium salt to its formazan.

Extensions of the Chelation Principle. Alternative possibilities for metal chelation of formazans under histochemical conditions were suggested by the use of copper complexes of formazans with *ortho*-hydroxyl and carboxyl groups on the *N*-phenyl rings as dyes (Wizinger and Biro, 1949; Wizinger and

Herzog, 1953). I explored this possibility but found that tetrazolium salts containing these groups possessed redox potentials which were too low to allow them to function as electron acceptors for dehydrogenase systems. The production of copper complexes of *o*-carboxyformazyl-2-quinoline has been described by Seyhan (1952, 1953). Tetrazolium salts from such formazans have not been used in biological systems.

Physiochemical Characteristics of Dehydrogenase Reactions

Kinetics. The dehydrogenase systems which we can demonstrate histochemically in a satisfactory manner are those in which the equilibrium point of the reaction lies well over to the right, that is towards oxidation of the substrate, or in which the equilibrium can easily be shifted to this side. This fact can be illustrated with reference to alcohol dehydrogenase.

$$\text{Alcohol} \underset{\text{DPNH}}{\overset{\text{DPN}^+}{\rightleftharpoons}} \text{Acetaldehyde}$$

In the case of yeast alcohol dehydrogenase the Michaelis constant (Km) for alcohol is $2{\cdot}4 \times 10^{-2}$ whereas for acetaldehyde it is $1{\cdot}1 \times 10^{-4}$ (Racker, 1955). The equilibrium of this system thus lies very far to the left, that is to say the natural tendency of the enzyme is to reduce acetaldehyde to ethanol.

The Michaelis Constant. According to Michaelis and Menten (1913) substrate concentration was to be regarded as an important factor controlling the velocity of an enzyme reaction. The velocity (v) is related to the substrate concentration (s) by the Michaelis-Menten equation

$$v = \frac{V}{1 + \frac{Km}{s}}$$

where V is the maximum velocity of the reaction when the enzyme is saturated with substrate. The term Km is the Michaelis constant. If v is plotted against s, as in Fig. 187, below, the curve is a section of a rectangular hyperbola.

If we substitute $v = V/2$ then $s = Km$ and Km is thus the halfway point of the experimental curve or, in other words, the concentration of substrate required to give half the maximum velocity. It has been pointed out by Dixon and Webb (1964), whose book should be consulted for the derivation of the Michaelis-Menten equation, that confusion arises through the original use of the term Michaelis constant to mean the dissociation constant of the enzyme-substrate complex. For this they now propose the use of the term "substrate constant", denoted by K_s, and the restriction of the term "Michaelis constant" to the definition given above. The relationship between Km and Ks can be made more clear by some additional information.

Enzyme catalysed reactions can be expressed in the following manner:

$$E + S \underset{k_2}{\overset{k_1}{\rightleftharpoons}} ES$$

and after this essential combination of enzyme (E) with substrate (S),

$$ES\ (+\ \text{any second reactant}) \xrightarrow{k_3} E + \text{products.}$$

$$Km = \frac{k_2 + k_3}{k_1}$$

The dissociation constant (K_s) for the enzyme-substrate complex (ES) is equal to the Michaelis constant (K_m) only when k_3 is very small compared to

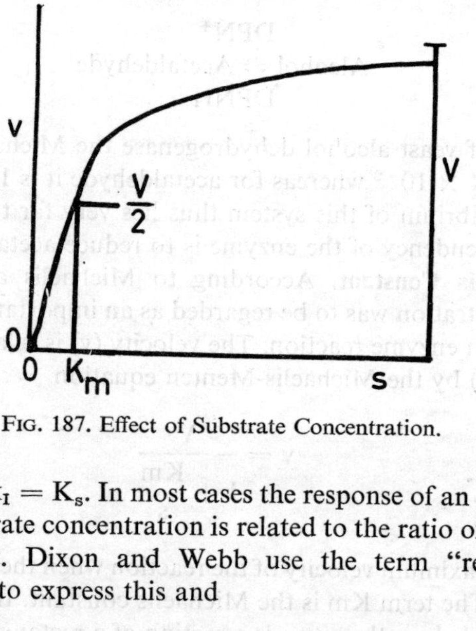

FIG. 187. Effect of Substrate Concentration.

k_2; then $K_m = k_2/k_1 = K_s$. In most cases the response of an enzyme reaction to changes in substrate concentration is related to the ratio of the latter to the Michaelis constant. Dixon and Webb use the term "relative substrate concentration" (σ) to express this and

$$\sigma = \frac{s}{Km}$$

They also define the term "relative velocity" as

$$\phi = \frac{v}{V}$$

and write the Michaelis equation as

$$\phi = \frac{\sigma}{1 + \sigma}$$

It therefore follows that when the substrate concentration is high $\phi = 1$ and when it is low $\phi = \sigma$, as should be evident from the curve shown in Fig. 188, below.

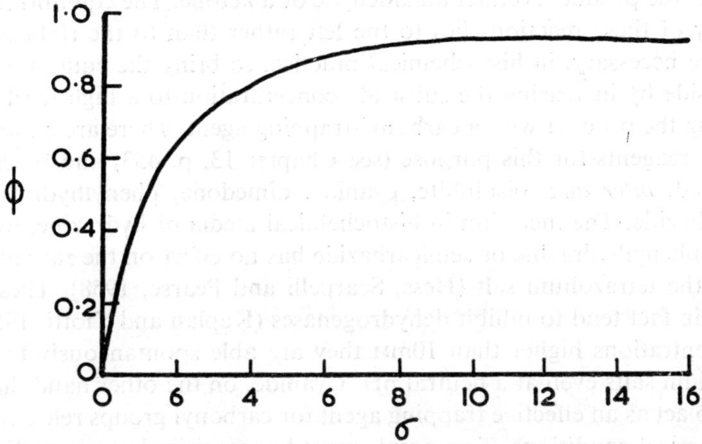

FIG. 188. Normalized Substrate Concentration Curve.

Other forms in which the Michaelis equation can be plotted are given by Dixon and Webb.

In applied dehydrogenase histochemistry the velocity of the forward reaction, and thus the Michaelis constant for the substrate of this reaction, is of paramount importance. The higher the Km value the lower the affinity of enzyme for substrate and the higher the concentration of substrate necessary to achieve a given forward velocity. In the hydrolytic enzyme field a low velocity may be quite acceptable and may indeed result in more accurate localization in cases where the capture reaction is relatively inefficient. With the dehydrogenases, however, we require the maximum possible velocity. This is partly due to the fact that prolonged incubation times damage the mitochondria and partly to the fact that when oxidation of the substrate falls to a low level alternative electron acceptors present in the tissues may absorb the entire supply of electrons, leaving the tetrazolium salt unreduced.

If we know the Michaelis constant for both sides of a dehydrogenase reaction the probability of the reaction proceeding in the required direction can be estimated. Following this the incubating medium can be designed to include the substrate at a sufficiently high concentration, provided that its solubility allows this.

Referring once again to the example of alcohol dehydrogenase we noted that biochemists, who are more interested in the oxidation of alcohol than in the reduction of acetaldehyde, introduce aldehyde-trapping reagents into their media. These combine rapidly with acetaldehyde and remove it from the system which is thus prevented from coming to an equilibrium. The demand for a continuous supply of acetaldehyde then ensures that the reaction runs to the right.

Trapping Reagents. In a substantial number of important dehydrogenase reactions the product is either an aldehyde or a ketone. The equilibrium point of many of these reactions lies to the left rather than to the right and it is therefore necessary, in histochemical practice, to bring the equilibrium over to this side by increasing the substrate concentration to a high level and by removing the product with a carbonyl-trapping agent. There are a number of possible reagents for this purpose (see Chapter 13, p. 453) and biochemists have used, *inter alia*, bisulphite, cyanide, dimedone, phenylhydrazine and semicarbazide. The inclusion in histochemical media of hydrazine, hydroxylamine, phenylhydrazine or semicarbazide has no effect on the rate of reduction of the tetrazolium salt (Hess, Scarpelli and Pearse, 1958). These compounds in fact tend to inhibit dehydrogenases (Kaplan and Ciotti, 1953) and at concentrations higher than 10mM they are able spontaneously to reduce tetrazolium salts even at a neutral pH. Cyanide, on the other hand, has been found to act as an effective trapping agent for carbonyl groups released under histochemical conditions. Two points must be mentioned in connection with its use, however. First, as noted above, cobalt chelation techniques cannot be used in the presence of cyanide though it may be possible to use other metals if these are satisfactory chelators. Secondly, cyanide has an additional important effect in blocking the electron transport system of the cell. This is more fully considered later in this chapter.

pH Optima. The effect of pH changes on the equilibria of reactions requiring pyridine nucleotide coenzymes is often profound, as in the case of alcohol dehydrogenases and small changes may alter the Km value significantly. In dehydrogenase histochemistry we cannot always work at the pH optimum of the enzyme system we are studying, particularly if this is in the alkaline range. This is especially true in the case of the pyridine nucleotide linked dehydrogenases where the so-called "nothing dehydrogenase" activity rises sharply above pH 7·0. This is described in the next section of this chapter. If, therefore, a DPN or TPN-dependent dehydrogenase has a narrow peak of activity on the alkaline side of neutrality, as has aldehyde dehydrogenase, we are unlikely to be able to localize it accurately with any of the existing methods. We are limited also in the case of a series of enzymes like the steroid dehydrogenases where, according to Marcus and Talalay (1956) the forward and backward velocities of the reaction are equal at pH 7·0.

Activation and Inhibition by Other Factors. In histochemical practice the effect of pH is not confined to altered kinetics of the enzyme reaction. As shown by Raaflaub (1953b) there is also an effect on the mitochondria. These are stable at pH 6·0 in isotonic solutions of non-electrolytes but at pH 7·4 and above very considerable swelling takes place. When this occurs the velocity of oxidation of various substrates is altered. Succinate is oxidized much faster by swollen mitochondria (Raaflaub, 1953a) but citrate, fumarate, α-ketoglutarate, malate and pyruvate are oxidized more slowly. This observation was confirmed histochemically by Scarpelli and Pearse (1958) who found that the *velocity* of the succinic dehydrogenase reaction, but not of the DPNH-diaphorase reaction, was markedly increased in swollen mitochondria. Slightly different results were obtained by Ziegler and Linnane (1958) who found that the oxidation of iso-citrate, aconitate, malate, lactate and β-hydroxybutyrate by ox heart mitochondria was in all cases increased following damage to the mitochondrial membrane.

As already noted in Chapter 14, pp. 479–480, mitochondria can be protected against non-osmotic types of swelling by the addition of ATP. The mechanism by which this acts is not clear but Leuthardt and Bruttin (1952) have shown that it is not used up in the process. Ernster and Low (1955) and Raaflaub (1956) both suggest that ATP may function by chelating intramitochondrial calcium. It is now considered that mitochondrial damage leads to release of intramitochondrial calcium and that this impairs the process of oxidative phosphorylation. This in turn results in failure to synthesize ATP, with a consequent fall in intramitochondrial ATP leading to mitochondrial swelling.

Another factor which affects the stability of mitochondria is the presence or absence of Mg^{2+} ions, which can be replaced to some extent by Mn^{2+} or Co^{2+}. The mechanism by which magnesium ions exert their protective effect is at present unexplained although Ernster and Löw postulate that they bind ATP to the mitochondrion, thus preventing its loss. It appears to have been a fortunate accident that the best chelating ion (cobalt) for the monotetrazolium salt metal-chelation techniques should have been one which protected mitochondrial morphology.

The effect of the incubating medium on the mitochondria should always be given the fullest consideration in designing reactions for dehydrogenase histochemistry. The use of hypertonic non-electrolyte media is absolutely essential for accurate localization of dehydrogenases and especially is this so in the case of soluble or partially soluble enzyme systems. The fullest possible protection is afforded by the use of 7·5 per cent polyvinyl pyrrolidone and 5 mM-magnesium ions in the incubating medium. Additional ATP is not usually required.

The Role of Coenzymes

The specific coenzymes act as carriers in the passage of electrons and hydrogen from the original substrate to molecular oxygen. Those with which we are here concerned are Coenzymes I and II, which are now called nicotinamide-adenine dinucleotide and nicotinamide-adenine dinucleotide phosphate (NAD and NADP)*. The two coenzymes have been described by a number of different names:

* Officially adopted by the International Union of Biochemistry (see Report of the Commission on Enzymes of the I.U.B.C., Pergamon Press, 1961).

Coenzyme I (Co I)	Coenzyme II (Co II)
Cozymase	Phosphocozymase
Diphosphopyridine nucleotide (DPN)	Triphosphopyridine nucleotide (TPN)
NAD	NADP

NAD and NADP can be regarded firstly as coenzymes in the various dehydrogenase reactions and secondly as substrates for further enzyme systems. Using the alcohol dehydrogenase system once more as an example

$$\text{Alcohol} + \text{NAD} \overset{\text{dehydrogenase}}{\rightleftharpoons} \text{Acetaldehyde} + \text{NADH} + \text{H}^+$$

The coenzyme is readily and reversibly reduced by the dehydrogenase and the mechanism of this reduction was elucidated by Fisher *et al.* (1953), using deuterium-labelled substrates. It had already been shown by Warburg and his associates (Warburg, Christian and Griese, 1935; Warburg and Christian, 1936) that the active part of the molecule was the nicotinamide moiety and that this underwent reversible reduction and oxidation. Fisher *et al.* showed that when deuterium-labelled alcohol was used the reduced coenzyme on the right of the equation contained deuterium, as did the acetaldehyde. If the deuterium-containing coenzyme was re-oxidized in a dehydrogenase-catalysed system, as for instance by lactate dehydrogenase, the product (pyruvate) contained the deuterium. This showed that the process was, in fact, a direct transfer of hydrogen from substrate to coenzyme.

The reduced forms of Coenzymes I and II are not oxidized at significant rates by O_2, by dyes such as methylene blue, by tetrazolium salts or by cytochrome c. They are normally oxidized by cytochromes c. or b_5 in the presence of their respective cytochrome reductases, and by methylene blue and tetrazolium salts in the presence of the appropriate diaphorase (tetrazolium reductase).

Diaphorases. An enzyme which catalysed the oxidation of reduced Coenzyme I (NADH) by methylene blue or cytochrome c. was independently discovered by Green, Dewan and Leloir (1937) and by von Euler and Hellström (1938). Shortly afterwards Straub and his co-workers succeeded in purifying the enzyme (Straub, 1939; Corran, Green and Straub, 1939) which was characterized as a flavoprotein and referred to as DPN diaphorase. At the same time Adler *et al.* (1939) concluded that the oxidation of NADPH was carried out by a separate diaphorase since the oxidation of NADPH and NADH by the tissues proceeded at very different rates. Further investigations carried out by Racker (1955) firmly established the part played by the two diaphorases in the electron transport system of the cell. According to Mahler (1955) NADH diaphorase prepared from heart muscle by the method of Straub was a degraded form of DPNH cytochrome c. reductase, differing from the latter in having lost its iron atoms and hence its capacity for reducing cytochrome c. A highly purified diaphorase produced by Savage (1957) was found to be completely inactive with cytochrome c. as electron acceptor and it was impossible to convert it to the reductase by addition of Fe^{2+} or Fe^{3+} ions.

Further work by Massey (1958) on the possible physiological role of NADH diaphorase suggested that the function of the enzyme was to couple the oxidation of NADH and the reduction of cytochrome c, employing a small molecular weight electron carrier as intermediate between the large molecules of cytochrome c. and diaphorase. Out of a large number of redox systems tested by Massey α-lipoic acid was found to act as an efficient coupling agent. This substance had been crystallized by Reed *et al.* (1951) and called α-lipoic acid to distinguish it from a modified product created by the process of extraction. Its structure, and that of its reduced form, are shown below:

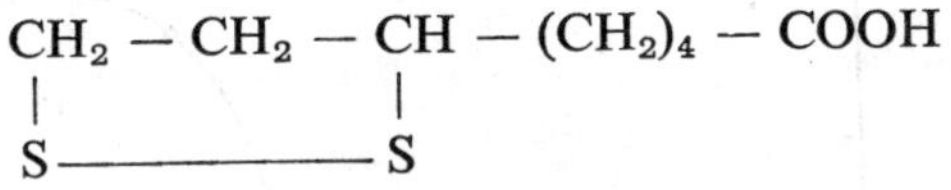

Oxidized lipoic acid

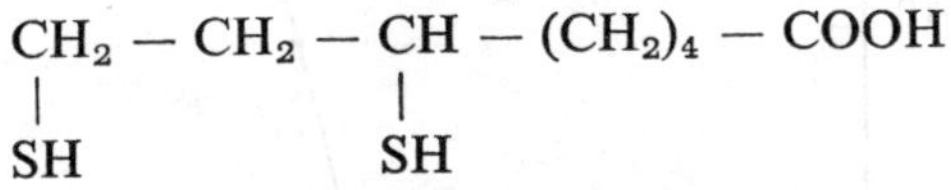

Reduced lipoic acid

The problem was finally resolved when lipoic acid (or lipoamide) dehydrogenase was identified by Massey (1960), and by Searls and Sanadi (1961), with Straub's flavoprotein diaphorase.

The physiological function of the enzyme (E.C. 1.6.4.3) has been shown to be the oxidation of reduced lipoate at the expense of NAD:

$$\text{Lipoate (SH)}_2 - \text{NAD}^+ \rightleftharpoons \text{Lipoate S}_2 + \text{NADH} - \text{H}^+$$

Reduced lipoate is produced from the oxidized form when pyruvate and 2-oxoglutarate are oxidized in the presence of their respective dehydrogenases and thiamine pyrophosphate. These mechanisms for the oxidation of keto-acids can be regarded collectively as a system.

Other pathways for the oxidation of reduced coenzymes have been described. In extracts of brain homogenates Levine *et al.* (1960) described two particulate diaphorases catalysing the oxidation of both NADH and NADPH by a number of electron acceptors of which menadione was the most active. Martius and his associates had already described a series of quinone reductases (Martius, 1954, 1958; Martius and Strufe, 1954; Martius and Märki, 1957). The majority of these enzymes were extramitochondrial and all showed extreme sensitivity to dicoumarol.

A flavoenzyme derived from the soluble fraction of rat liver homogenates was isolated and investigated by Ernster *et al.* (1960, 1962) and by Conover and Ernster (1962). The enzyme, named DT diaphorase to indicate its reaction with both D and TPNH, catalysed the oxidation of both coenzymes by artificial electron acceptors. Again, menadione was one of the most active, being exceeded only by some of the benzoquinones. Ferricyanide and dichlorophenol indophenol could act as acceptors but tetrazolium salts could not. Some of the quinones could catalyse the transfer of electrons between DT diaphorase and cytochrome c. Menadione and 1,4-naphthoquinone were the most active of these.

Attempting to localize the particulate and soluble reductase activities in brain tissue Hess and Pearse (1963) found that tetrazolium reductase activity was present only in the mitochondrial fraction of brain homogenates prepared in 0·25 M sucrose. When menadione was used as electron acceptor, quinone

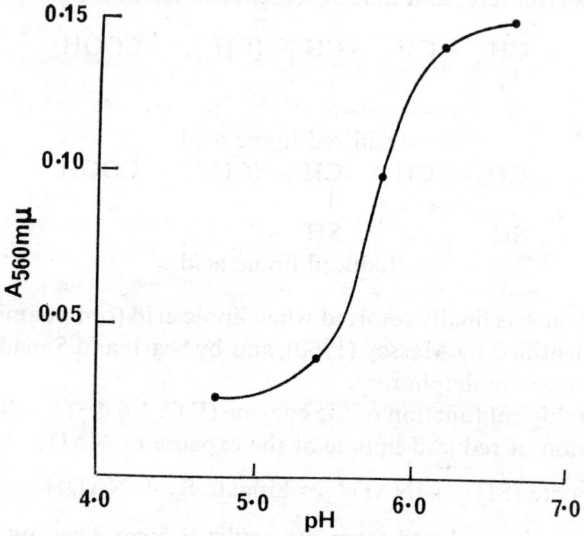

FIG. 189. pH-dependence of reduction of monotetrazole by reduced menadione. 0·2 μmole MTT was dissolved in 3 ml. 0·06 M phosphate buffer containing 20% acetone to which was added 0·15 μmole reduced menadione. After 60 sec the formazan formed was extracted in 4 ml. ethylacetate and measured at 560 nm.

reductase activity was found in both mitochondrial and cytoplasmic fractions of the homogenate. Kinetic studies showed that both diaphorases were sensitive to dicoumarol, 50 per cent inhibition being achieved at concentrations of 0·8 μM and 0·02 μM respectively. Both were also sensitive to amytal (20 μM and 2 mM, respectively) whereas the mitochondrial tetrazolium reductase was amytal insensitive.

Hess and Pearse (1963) showed that MTT could trap electrons from reduced menadione, in competition with autoxidation, and that this effect was pH dependent, as shown in Fig. 189, above.

The biochemical function of the quinone reductases is still uncertain and their histochemical significance similarly remains to be elucidated. It is clear that they are quite distinct from Straub's diaphorase (α-lipoamide dehydrogenase) on the one hand, and from the various cytochrome c. reductases on the other.

Histochemistry of the Diaphorases (Tetrazolium Reductases). The histochemical localization of NADH and NADPH diaphorase was first accomplished by Farber *et al.* (1956). These authors used an exogenous dehydrogenase together with its substrate and oxidized Coenzyme I (NAD) in order to generate NADH as substrate for the intrinsic diaphorase of the tissues. They were able to demonstrate NADPH diaphorase without exogenous dehydrogenase, however, by supplying a substrate for endogenous dehydrogenases. Subsequently Nachlas and his co-workers (Nachlas, Walker and Seligman, 1958a) developed a method based on the same principle using Nitro-BT in place of the older BT or NT. This method provided an improved localization of diaphorase but did not demonstrate all sites of NAD diaphorase activity unless exogenous enzyme was added. The method of Scarpelli. Hess and Pearse (1958), employed NADH and NADPH as substrates. It did not require exogenous dehydrogenase for the optimal demonstration of diaphorase activity and thus avoided complications due to spontaneous reduction of tetrazolium salt in the incubating medium when the exogenous dehydrogenase was contaminated with diaphorase. Racker (1955) showed that this was the case with alcohol dehydrogenase.

Further Considerations. The reduced coenzyme can certainly be considered as substrate for the diaphorase:

$$\text{NAD(NADP)H} + \text{tetrazole} \rightleftharpoons \text{NAD(NADP)}^+ + \text{Formazan}$$

Since the reaction as it stands is not reversible the equilibrium should lie to the right (formazans are not re-oxidized to tetrazoles in biological systems). The oxidation of NADH or NADPH by diaphorase should not therefore be a limiting factor in deciding the overall rate of a pyridine nucleotide-linked dehydrogenase system in histochemical practice. If, however, the NAD^+ is removed by some process (see below) the supply must be maintained either by the presence of sufficient NADH (diaphorase reaction) or of sufficient NAD for conversion to NADH (dehydrogenase reactions).

Arguments for Use of High Coenzyme Levels

Despite the apparent preservation of mitochondrial morphology achieved by the use of protective media the freezing and thawing of the normal cold microtome (cryostat) procedure results in damage to the mitochondrial membrane and in the release of soluble co-factors (Porter *et al.*, 1953). Endogenous coenzyme levels are too low to allow histochemical dehydrogenase reactions (coenzyme-dependent) to take place at all and exogenous NAD or NADP must be added. Hess, Scarpelli and Pearse (1958) found that the optimal reduction of tetrazolium salt occurred when the co-enzyme was present in the same concentration as the substrate (0·1M) although good results were obtained by using one-tenth of this amount (0·01M, or 6 mg/ml.).

These figures are much higher than those used by other workers. For example, Nachlas, Walker and Seligman (1958a and b) used 0·5 mg/ml. of NAD and 0·3 mg/ml. of NADP in their methods for NAD and NADP-linked enzyme systems.

There are many reasons why high coenzyme levels should be maintained in histochemical media and many factors operate under histochemical conditions to reduce the amount of available NAD or NADP to a low level. Some of these are of serious concern if incubation times are prolonged.

A quantitative study of the NAD(NADP)-tetrazolium reductase systems in rat liver was carried out by Kalina (1968). Both systems were found to be heterogeneous, consisting of at least two enzymes, one with a high and the other with a low affinity for reduced coenzyme. Within the liver lobule the two systems had completely different localizations. The high affinity system was found to be saturated with respect to NADH at a concentration of 65 μM, and the low affinity system at 1 mM. Similar studies carried out by Johnson (1965) produced different localizations of NADH tetrazolium reductase with 17 μM and 0·7 mM concentrations of coenzyme. At the lower level a positive reaction was obtained only in a few sites, particularly in ganglion cells and in vascular smooth muscle. McCabe *et al.* (1965) obtained evidence, from quantitative studies using NT, that NADH tetrazolium reductase is the rate-limiting step in several coenzyme-linked dehydrogenase reactions.

Binding of Coenzymes. The binding of coenzyme by the dehydrogenase (Straub, 1940) does not normally lead to deficiency because the enzyme is present in relatively low concentration. Coenzyme-linked dehydrogenases can be divided into "alcohol and aldehyde" varieties and fundamental differences exist between these two classes. Those which can be classified as "alcohol" dehydrogenases oxidize primary alcohols to aldehydes, secondary alcohols to ketones, primary amines to ketones and ammonia and hemi-acetals to lactones. Examples in these subgroups are alcohol dehydrogenase, lactic dehydrogenase, glutamic dehydrogenase and glucose-6-phosphate dehydrogenase. San Pietro and his associates (van Eys, San Pietro and Kaplan, 1958), however, have shown that the mechanism of action of NAD and NADP-dependent dehydrogenases requires the binding of two moles of NAD per molecule of enzyme. This is shown below in Fig. 190 (I) where the enzyme is represented by a hatched bar with two sulphydryl atoms, each of which binds a molecule of the coenzyme.

These diagrams represent the following equations:

$$\text{E} + \text{NAD} + \text{NAD} \longrightarrow \text{E-NAD}_2 \text{ (I)}$$

$$\text{E-NAD}_2 + \text{S} \longrightarrow \text{E-NAD-NADS} - \text{H}^+ \text{ (II)}$$

$$\text{E-NAD-NADS} \longrightarrow \text{E-NAD-NADH} + \text{P (III)}$$

We see that the enzyme binds two molecules of coenzyme and is then able to

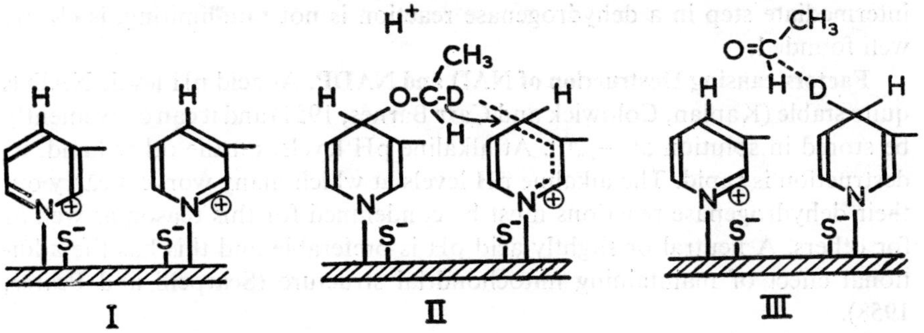

FIG. 190.

activate the substrate, deutero-ethanol in the diagrams in Fig. 190, with loss of one hydrogen atom and the transfer of the other to the second coenzyme molecule. After release of the product (P) the enzyme-coenzyme complex is thus partly reduced. It is regenerated as follows:

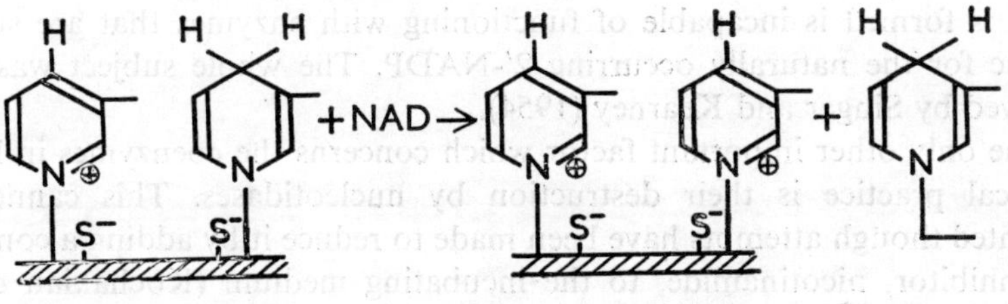

IV

FIG. 191.

This diagram (Fig. 191) represents the following equation,

$$\text{E-NAD-NADH} + \text{NAD} \longrightarrow \text{E-NAD}_2 + \text{NADH (IV)}$$

which is the normal process by which the enzyme-coenzyme is oxidized to its original state. Two further reactions may occur and these can be written:

$$\text{E-NAD-NADH} + \text{S} \longrightarrow \text{E-NADS-NADH} + \text{H}^+ \text{ (V)}$$

and

$$\text{E-NAD-NADS} + \text{S} \longrightarrow \text{E-NADS}_2 + \text{H}^+ \text{ (VI)}$$

Both these reactions will result in inactivation of the enzyme and competition will occur between reactions IV and V and also between reactions III and VI. It is evident from the above equations that, even if NAD were not destroyed in the tissues but regenerated *in toto* for further use in reaction IV, it would still be necessary to maintain a high concentration of coenzyme in histochemical media in order to satisfy the kinetics of reactions I and IV. From reactions I and II it is clear that *the affinity for the substrate depends on the NAD concentration.* The warning given by Cascarano and Zweifach (1959) that one must make certain, by providing excess NAD or NADP, that the intermediate step in a dehydrogenase reaction is not rate-limiting, is clearly well founded.

Factors causing Destruction of NAD and NADP. At acid pH levels NAD is quite stable (Kaplan, Colowick and Carr Barnes, 1951) and it can conveniently be stored in solution at − 20°. At alkaline pH levels, on the other hand, its destruction is rapid. The alkaline pH levels at which many workers carry out their dehydrogenase reactions must be condemned for this reason as well as for others. A neutral or slightly acid pH is preferable and this has the additional effect of maintaining mitochondrial structure (Scarpelli and Pearse, 1958).

The reduced coenzymes (NADH and NADPH) show the opposite characteristics, being unstable in acid but relatively stable in alkaline solutions. The marked inhibitory effect of high hydrogen ion concentration on the oxidation of NADPH by its diaphorase may well be due to isomeric conversion of the adenylic acid ribose moiety of NADPH since Shuster and Kaplan (1955) have shown that the monophosphate group of adenylic acid ribose migrates from the 2′ to the 3′ position at acid pH levels and the 3′-NADP which is formed is incapable of functioning with enzymes that are strictly specific for the naturally occurring 2′-NADP. The whole subject was ably reviewed by Singer and Kearney (1954).

The only other important factor which concerns the coenzymes in histochemical practice is their destruction by nucleotidases. This cannot be prevented though attempts have been made to reduce it by adding a competitive inhibitor, nicotinamide, to the incubating medium (Kochakian *et al.*, 1957). Under histochemical conditions this substance appears to exert little effect and far better results are achieved by maintaining high coenzyme levels in the incubating medium, despite the expense involved in this procedure.

Effect of High Coenzyme Levels at Alkaline pH. Provided that the pH of the incubating medium is maintained at 7·0 or below no untoward circumstances are observed in tissue sections. Above pH 7·0, however, an effect occurs which has been referred to as "nothing dehydrogenase" using the loose but descriptive term employed by Racker (1955). It was first noted by Hopkins and Dixon (1922) that well-washed tissue residue possessed marked reducing capacity and Racker and his associates found that well-dialysed protein fractions obtained from various tissues could catalyse the reduction of NAD or NADP in the absence of any substrate. They assumed that this activity was connected with the presence of sulphydryl groups. Zimmermann and Pearse (1959), while endeavouring to demonstrate a specific sorbitol dehydrogenase, observed with both Nitro-BT and MTT the production of formazan in systems containing NAD (or NADP), buffer and tissue section only. This effect (Fig. 192, p. 869) increased progressively from pH 7·0 up to pH 9·0 and decreased thereafter. It was not demonstrable at pH 6·5. "Nothing dehydrogenase" was inactivated by a number of sulphydryl inhibitors, incorporated in the incubating medium at a concentration of 0·1M. Among those tested were *N*-ethyl maleimide, indoacetate and *p*-chloromercuribenzoic acid. Preincubation of sections with 0·1M-*N*-ethyl maleimide for one hour at 37° also completely abolished the reaction. The substitution of NADH for NAD produced a positive reaction in the presence of any of the inhibitors, indicating lack of interference with diaphorase activity. Other considerations made it likely that protein-bound SH was responsible but the participation of smaller molecules like glutathione could not be excluded.

While studying a number of different dehydrogenases from mouse tissues, separable by starch gel electrophoresis, Koen and Shaw (1964) found one which they were unable to identify. This enzyme, originally described as DH-1, was the major site of nothing dehydrogenase activity although a few faint reactions were obtained at sites coinciding with lactate dehydrogenase bands. In a subsequent paper Shaw and Koen (1965) identified DH-1 as alcohol dehydrogenase. They confirmed the SH-dependence of its nothing dehydrogenase activity and concluded that the latter must occur at the active site of the enzyme although no mechanism to explain the observed hydrogen transfer has yet been advanced.

In histochemical practice nothing dehydrogenase activity is not usually a significant source of error but it must still be considered in dehydrogenase reactions where a pH above 7·4 is employed, particularly if incubation is prolonged and even more if an electron transfer intermediate such as phenazine methosulphate (PMS) is employed (see Ogawa and Shinonaga, 1961).

The Respiratory Chain.

Sites of Electron Transfer. The efficiency with which tetrazolium salts accept electrons is a matter of some concern in dehydrogenase histochemistry and it is important, at the same time, to know the precise point on the chain at which tetrazoles, and other acceptors, accept electrons. The natural tendency, provided that all the factors in the respiratory chain are present and functioning normally is for electrons to be transferred successively through the various stages of the chain to molecular oxygen. The whole scheme was referred to by Lundegårdh (1958) as the electron ladder, though he illustrated the passage of electrons as a stream passing over a series of waterfalls, since successive stages involve a progressive rise in redox potential.

It will be realized that the dehydrogenase is the first member of a chain of enzymes which transfer H atoms or electrons from the substrate to molecular oxygen. An oxidase, on the other hand, is an enzyme whose hydrogen acceptor is molecular oxygen. It is usually named after the hydrogen donor.

In histochemical practice there are two principal ways in which attempts are made to channel the flow of electrons into the tetrazolium trap, so that the maximum amount of formazan is produced. The first is by blocking the electron chain so that the maximum amount of formazan is produced, and the second is by the use of alternative (intermediate) electron acceptors which are capable of reversible oxidation by tetrazolium salts.

The electron pathway has been the subject of a large number of diagrammatic representations. The one reproduced in Fig. 193, below, shows the salient features of the chain as it concerns the histochemist. In this diagram the various blocking agents appear in squares and the points at which exogenous acceptors are able to take up hydrogen atoms and electrons are indicated by circles. A second diagram (Fig. 194) indicates the sites of electron

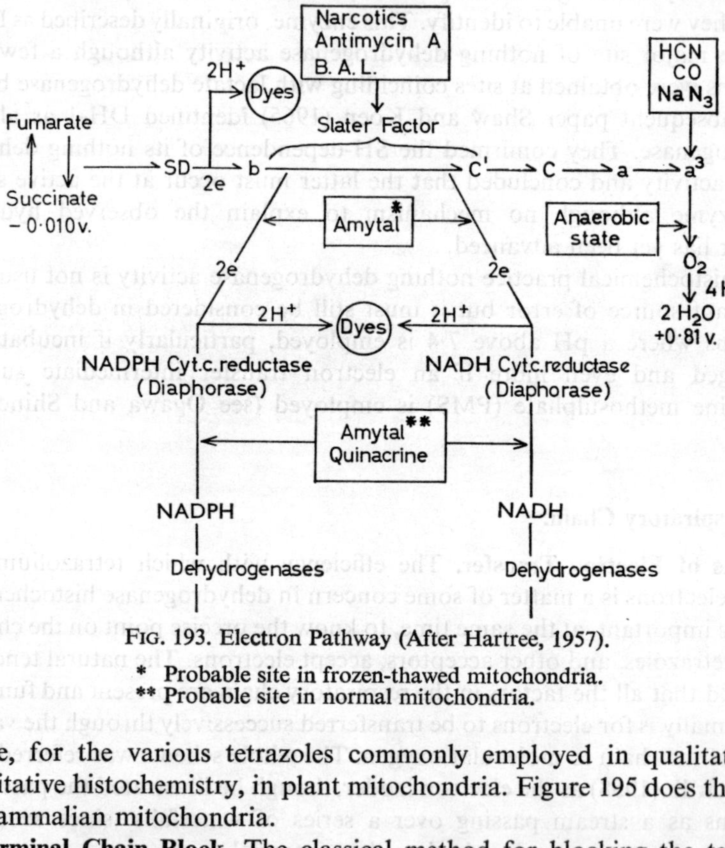

FIG. 193. Electron Pathway (After Hartree, 1957).
* Probable site in frozen-thawed mitochondria.
** Probable site in normal mitochondria.

uptake, for the various tetrazoles commonly employed in qualitative or quantitative histochemistry, in plant mitochondria. Figure 195 does the same for mammalian mitochondria.

Terminal Chain Block. The classical method for blocking the terminal portion of the respiratory chain is by incubation under anaerobic conditions. Formerly much employed, the advent of more efficient tetrazoles has rendered the manoeuvre unnecessary, with rare exceptions. Azide and cyanide, both of which block the terminal part of the chain at cytochrome a_3, are similarly seldom used in modern practice.

Mid-chain Block. Reference to Fig. 193 shows that a number of respiratory inhibitors have been described which act at points in the respiratory chain on the oxygen side of the flavoprotein carrier, whatever the identity of the latter. In particular, the work of Ernster, Jalling, Löw and Lindberg (1955) and of Jalling, Lindberg and Ernster (1955) suggested that amytal was a most efficient inhibitor for all NAD-linked systems. It was found that 1·8 mM-amytal completely blocked all NAD-linked oxidation, leaving succinate oxidation completely unaffected. Out of a large number of compounds tested Scarpelli, Hess and Pearse (1958) found that amytal was the most efficient, increasing the rate of reduction of the tetrazolium salt in the NADH diaphorase reaction so that 5 minutes' incubation was adequate. Azide was only slightly effective.

Although according to Chance (Chance and Williams, 1956; Chance, 1956) amytal was observed to block electron transfer in liver mitochondria at a point between NADH and flavoprotein (i.e. at the site marked ** in Fig. 193) the rapid rate of tetrazolium reduction by tissues in the presence of reduced pyridine nucleotide and amytal suggests that the site of amytal effect in tissue sections is beyond flavoprotein (* in Fig. 193). Recently Estabrook (1957) has shown a similar site of amytal inhibition in the case of the NADH cytochrome c. reductase system and he suggested that the difference between these findings and those of Chance might indicate structural alteration of the electron transfer chain due to damage to the mitochondria during preparation. The results obtained by Chance and Baltscheffsky (1958), in heart muscle sarcosomes have confirmed the original findings made on liver mitochondria, that the site of amytal block is between NADH and flavoprotein. Electron transport from non-NAD-linked substrates, such as succinate, was not found to be inhibited by amytal. The results obtained histochemically may possibly be explained on similar lines to those put forward by Estabrook since freezing and thawing are known to damage the mitochondrial membrane. The structural labilization of mitochondria by pre-treatment with Ca^{2+} or in the absence of Mg^{2+}, leading to decreased sensitivity to amytal blocking of respiration, was extensively discussed by Ernster (1956).

As an alternative to amytal a number of investigators have used rotenone, which blocks at, or close to, the amytal site (Ernster *et al.*, 1963).

A number of papers dealing with sites of electron transfer to tetrazoles have been published. Most of these have used information based on antimycin block. According to Potter and Reif (1952) this occurs between cytochromes b and c. Nachlas *et al.* (1960) tested six tetrazoles in respect of electron transfer from the succinoxidase system. They found that unsubstituted ditetrazoles received most of their electrons from the region of cytochrome c. Nitro-substituted mono- and ditetrazoles (INT, Nitro-BT), however, were able to pick up at the level of cytochrome b, or from the flavoprotein-cytochrome b complex.

Oda and Okazaki (1958) had already indicated, on the basis of antimycin block, that Nitro-BT could accept at the level of the dehydrogenase and Slater *et al.* (1963) confirmed this finding, showing that the reaction of Nitro-BT with the respiratory chain was virtually insensitive to antimycin A. They found, on the other hand, that MTT reacted at two sites of roughly equal importance. One, in the region of cytochrome c, was sensitive to antimycin while the other was reacting at the level of cytochrome b or ubiquinone (UQ) (Slater, 1963). Kalina and Palmer (1968), using isolated plant mitochondria found that MTT reacted entirely at the UQ level. They used blocking by 4,5-dichloro-2-trifluoromethyl benzimidazole (DCTFB) (Jones and Watson, 1967), which acts effectively between the flavoprotein dehydrogenase and UQ, to show that both Nitro-BT and TNBT accepted electrons from the former.

The scheme derived by Kalina and Palmer from their studies is shown below.

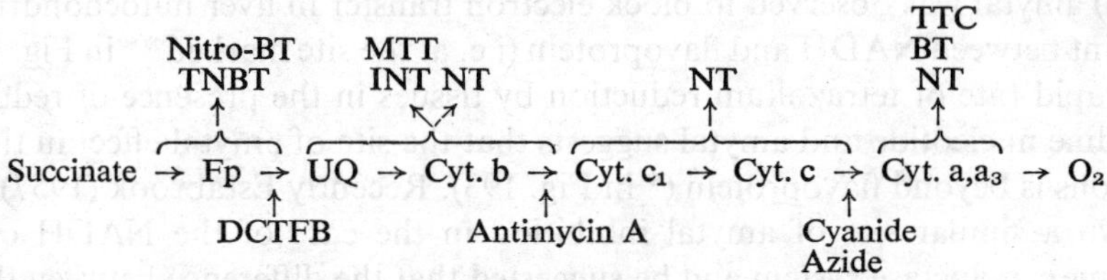

FIG. 194. Sites of Electron Transfer in Plant Mitochondria.

The situation with respect to mammalian mitochondria is somewhat different. Figure 195, which is taken mainly from the results of Slater *et al.* (1963) indicates the main points of difference

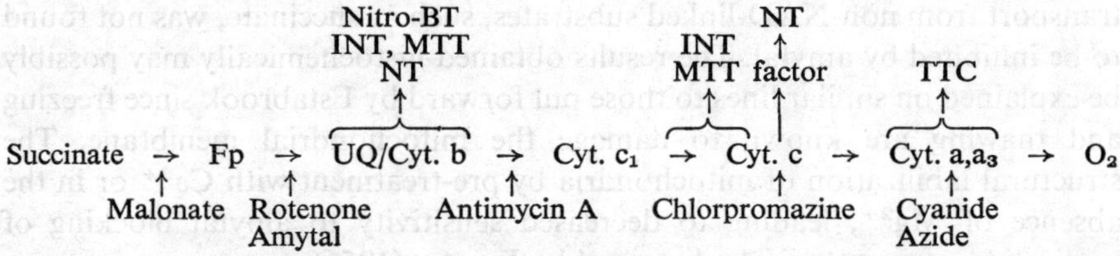

FIG. 195. Sites of Electron Transfer in Animal Mitochondria.

Competitive Electron Acceptors. The results obtained by Baltscheffsky (1957) add additional support to the histochemical findings reported above. Using the NADH oxidase system this author found that the inhibition of respiration (measured by O_2 uptake) by 2 mM-amytal or 2·4 μM-antimycin could be offset, though not entirely reversed, by the addition to the medium of 10 μM-cytochrome c. This suggested that the site of amytal action was beyond flavoprotein and that it probably involved cytochrome c. Baltscheffsky's results are shown in Fig. 196, below. The Baltscheffsky effect can be demonstrated histochemically in the NADH diaphorase reaction though, of course, the reverse effect is found. That is to say, the increased formazan production brought about by effective amytal blocking of the respiratory chain below the point at which electrons pass to the tetrazolium salt is offset by the alternative electron pathway provided by the exogenous cytochrome c. That such pathways can be produced is demonstrated by the results recorded above. Whether they exist *in vivo* or in tissue preparations is a question that requires an answer. Certainly naturally-occurring substances like quinones (e.g. ubiquinone; Morton *et al.*, 1958; Fahmy *et al.*, 1958) can transfer electrons from flavoproteins to cytochrome c, by-passing intermediate stages of the respiratory chain. If such substances are present in tissue preparations electrons may escape by this path. Combined mid-chain and terminal chain block, as employed in systems using Nitro-BT, amytal, and cyanide as trapping agent, would be expected to overcome any escape of electrons

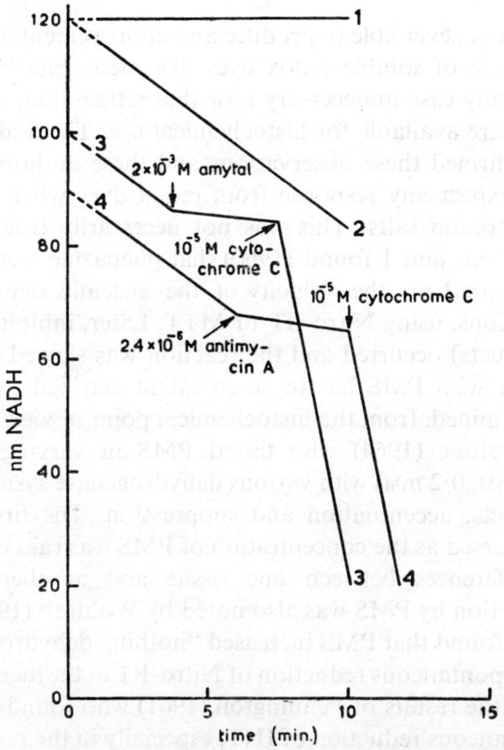

FIG. 196. Effect of Amytal and Antimycin on NADH oxidase.

induced by mechanisms of this type. It seems likely, therefore, that unknown pathways exist which prevent us from approaching the desired figure of 100 per cent efficiency for the trapping of electrons by the tetrazolium salt.

Intermediate Acceptors. The use of soluble redox compounds or dyes to act as intermediate acceptors between diaphorases and tetrazolium salts was introduced by Farber and Louviere (1956) who employed methylene blue (redox potential 0·011v.), together with the relatively inefficient BT (redox potential − 0·17v.) which was in general use at that time. Several other dyes were found to enhance formazan production. Among these were thionin (+ 0·06v.) and pyocyanin (− 0·034v.). Another suggestion made by Farber and Louviere followed the observation by Singer and Kearney (1954a) that phenazine methosulphate could reversibly accept electrons directly from succinic dehydrogenase. Later Singer, Kearney and Massey (1957) suggested that phenazine (and ferricyanide) could accept electrons directly from the Fe^{2+} ions of the iron-flavoprotein whereas the other dyes, including methylene blue and tetrazolium salts, could accept only from the flavin part of the molecule. Dyes of this variety were thus relatively inefficient and Singer believed that the natural acceptor could take electrons from Fe^{2+}, as does phenazine.

Personally I was never able to produce any improvement in the diaphorase reaction by the use of soluble redox dyes like methylene blue and thionin and these are in any case unnecessary now that tetrazolium salts with higher redox potentials are available for histochemical use. The findings of Nachlas *et al.* (1958) confirmed these observations and these authors suggested that one would not expect any response from redox dyes when using the more favourable tetrazolium salts. This was not necessarily true in the light of Singer's observations and I found (1960) that phenazine methosulphate was initially able to speed up the velocity of the succinic dehydrogenase and diaphorase reactions, using Nitro-BT or MTT. Later, inhibition (possibly by breakdown products) occurred and the reaction was slowed and stopped.

Further work with PMS has to some extent clarified the position. The situation was examined, from the histochemical point of view, by Hashimoto, Kaluza and Burstone (1964) who tested PMS in varying concentrations (2 mM, mM, 0·4 mM, 0·2 mM) with various dehydrogenase systems. They noted two distinct effects, accentuation and suppression. The first of these was progressively reversed as the concentration of PMS was raised, but there were considerable differences between one tissue and another. Inhibition of formazan production by PMS was also noted by Wohlrab (1963). Hashimoto *et al.* (1964) also found that PMS increased "nothing dehydrogenase" activity and produced a spontaneous reduction of Nitro-BT in the incubating medium. This agrees with the results of Pennington (1961) who found that 5 mM PMS caused the spontaneous reduction of INT, especially in the presence of heated serum. In the case of skeletal muscle van Wighe *et al.* (1963) showed that, because of the low level of NADH diaphorase in the white (Type II) fibres, their full lactate and α-glycerophosphate dehydrogenase activities could not be revealed without added PMS. Results with PMS, though more closely approximate to biochemical data, were less clean than those obtained in its absence. Due to transfer of electrons in solution instead of *in situ*, formazan precipitates appeared on the surface of the sections.

Although Roberts *et al.* recommended the use of PMS unreservedly for the demonstration of succinate dehydrogenase activity in plant tissues the majority of those working with mammalian tissues use PMS only to allow the demonstration of activities (such as choline oxidase) which cannot otherwise be revealed.

Menadione is the only other intermediate in common use (for demonstration of so-called mitochondrial α-glycerophosphate dehydrogenase, uridine diphosphoglucose dehydrogenase and L-gulonolactone oxidase). It is sometimes employed to increase the production of formazan by the SD complex but, as shown by Hashimoto *et al.* (1964) it produces at the same time a considerable increase in "nothing dehydrogenase" activity. Concentrations up to 4 mM can be obtained with the use of small amounts of acetone as solvent.

Redox Potentials of Oxidative Step and Acceptor. The relationship between the redox potential of the acceptor (tetrazolium salt) and that of the stage in the respiratory chain (or other reaction) at which the electrons are acquired, merits brief consideration. The redox potential of the enzyme-bound NAD-NADH system was found by Theorell and Bonnichsen (1951) to be −210 mV. and NADH can be oxidized by the cytochrome b_5 of liver microsomes (−120 mV.). According to Burton and Wilson (1953) the E° of NADP is −324 mV. at pH 7·0 and 25°. Since the redox potential of the newer tetrazoles is even higher than this there is no *a priori* reason why electrons should not flow from NADH to these compounds, as they do in practice through the medium of the diaphorase. The NAD-linked lipoic acid dehydrogenase system has a very low potential, −420 mV. according to Reed (1953), and Massey (1958) has recently shown that a diaphorase extracted from pig heart muscle behaves as a powerful lipoic acid dehydrogenase, as mentioned above. Almost all the electron acceptors used in dehydrogenase histochemistry have potentials higher than this.

We are thus confronted with no essential difficulty from the histochemical point of view provided that our tetrazolium salt has a higher redox potential than the system from which it is to accept electrons, and provided that the overall kinetics of the reaction can be adjusted to induce the reaction to proceed to the right.

REFERENCES

ADLER, E., VON EULER, H., and GÜNTHER, G. (1939). *Nature, Lond.*, **143**, 641.

ALTMANN, F. P. (1969). *Histochemie*, **19**, 363.

ANTOPOL, W., GLAUBACH, S., and GOLDMAN, L. (1948). *Publ. Hlth. Rep. Wash.*, **63**, 1231.

ASHLEY, J. N., DAVIS, B. M., NINEHAM, A. W., and SLACK, R. (1953). *J. chem. Soc.*, **4**, 3881.

ATKINSON, E., MELVIN, S., and FOX, S. W. (1950). *Science, N.Y.*, **111**, 385.

BALL, E. G., and RODKEY, F. L. (1950). *J. biol. Chem.*, **182**, 17.

BALTSCHEFFSKY, H. (1957). *Exp. Cell Res.*, **13**, 630.

BARRNETT, R. J., and PALADE, G. E. (1957). *J. biophys. biochem. Cytol.*, **3**, 577.

BEYER, H., and PYL, T. (1954). *Chem. Ber.*, **87**, 1505.

BLACK, M. M., and KLEINER, I. S. (1949). *Science, N.Y.*, **110**, 660.

BLACK, M. M., OPLER, S. R., and SPEER, F. D. (1950). *Am. J. Path.*, **26**, 1097.

BROOKE, M. H., and ENGLE, W. K. (1966). *Neurology*, **16**, 799.

BURTNER, H. J., BAHN, R. C., and LONGLEY, J. B. (1956). *J. Histochem. Cytochem.*, **5**, 127.

BURTON, K., and WILSON, T. H. (1953). *Biochem. J.*, **54**, 86.

CAMPBELL, H., and KANE, P. O. (1956). *J. chem. Soc.*, 3130.
CARTER, S. D., and ROPPEL, R. M. (1967). *Stain Technol.*, **42**, 163.
CASCARANO, J., and ZWEIFACH, B. W. (1959). *J. biophys. biochem. Cytol.*, **5**, 309.
CHANCE, B. (1956). In "Enzymes: Units of Biological Structure and Function." Ed. O. H. Gaebler, Acad. Press, New York.
CHANCE, B., and BALTSCHEFFSKY, M. (1958). *Biochem. J.*, **68**, 283.
CHANCE, B., and WILLIAMS, G. R. (1956). *Adv. Enzymol.*, **17**, 65.
CONOVER, T. E., and ERNSTER, L. (1962). *Biochim. biophys. Acta*, **58**, 189.
CORRAN, H. S., GREEN, D. E., and STRAUB, F. B. (1939). *Biochem. J.*, **33**, 793.
DEGUCHI, Y., and TAKAGI, Y. (1967). *Tetrahedron Letters*, No. **33**, 3179.
DIXON, M., and WEBB, E. C. (1964). "Enzymes," 2nd Edn. Longmans, London.
DUFRENOY, J., and PRATT, R. (1948). *Am. J. Bot.*, **35**, 333.
EADIE, M. J. (1969). Ph.D. Thesis, University of Queensland.
EADIE, M. J., TYRER, J. H., KUKUMS, J. R., and HOOPER, W. D. (1970). *Histochemie*, **21**, 170.
ERNSTER, L. (1956). *Exp Cell Res.*, **10**, 721.
ERNSTER, L., DALLNER, G., and AZZONE, G. F. (1963). *J. biol. Chem.*, **238**, 1124.
ERNSTER, L., DANIELSON, L., and LJUNGGREN, M. (1962). *Biochim. biophys. Acta*, **58**, 171.
ERNSTER, L., JALLING, O., LÖW, H., and LINDBERG, O. (1955). *Ibid.*, Suppl. **3**, 124.
ERNSTER, L., LJUNGGREN, M., and DANIELSON, L. (1960). *Biochem. biophys. Res. Commun.*, **2**, 88.
ERNSTER, L., and LÖW, H. (1955). *Ibid.*, Suppl. **3**, 133.
ESTABROOK, R. N. (1957). *J. biol. Chem.*, **227**, 1093.
FAHMY, N. I., HEMMING, F. W., MORTON, R. A., PATERSON, J. Y. F., and PENNOCK, J. F. (1958). *Biochem. J.*, **70**, IP.
FARBER, E., and LOUVIERE, C. D. (1956). *J. Histochem. Cytochem.*, **4**, 347.
FARBER, E., STERNBERG, W. H., and DUNLAP, C. E. (1956). *Ibid.*, **4**, 254.
FISHER, H. F., CONN, E. E., VENNESLAND, B., and WESTHEIMER, F. H. (1953). *J. biol. Chem.*, **202**, 687.
FLEISCHHAUER, K. (1958). *Histochemie*, **1**, 119.
GABLER, W., WENDLER, D., and SCHMIDT, W. (1970). *Acta histochemica*, **36**, 295.
GLICK, D., and NAYYAR, S. M. (1956). *J. Histochem. Cytochem.*, **4**, 389.
GOSZTONYI, T., MARTON, J., KENSENY, V., and VECSEI, P. (1963). *J. Chromat.*, **10**, 29.
GREEN, D. E., DEWAN, J. G., and LELOIR, L. F. (1937). *Biochem. J.*, **31**, 934.
HARTREE, E. F. (1957). *Adv. Enzymol.*, **18**, 1.
HASHIMOTO, T., KALUZA, J. S., and BURSTONE, M. S. (1964). *J. Histochem. Cytochem.*, **12**, 797.
HAUSSER, I., JERCHEL, D., and KUHN, R. (1949). *Chem. Ber.*, **82**, 515.
HESS, R., and PEARSE, A. G. E. (1961). *Expl. Cell Research*, **25**, 187.
HESS, R., and PEARSE, A. G. E. (1963). *Biochem. biophys. Acta*, **71**, 285.
HESS, R., SCARPELLI, D. G., and PEARSE, A. G. E. (1958). *J. biophys. biochem. Cytol.*, **4**, 753.
HITZEMAN, J. W. (1963). *J. Histochem. Cytochem.*, **11**, 62.
HOPKINS, F. G., and DIXON, M. (1922). *J. biol. Chem.*, **54**, 527.
HUNTER, L., and ROBERTS, C. B. (1941a). *J. chem. Soc.*, 820.
HUNTER, L., and ROBERTS, C. B. (1941b). *Ibid.*, 823.
JALLING, O., LINDBERG, O., and ERNSTER, L. (1955). *Acta chem. scand.*, **9**, 198.
JAMBOR, B. (1954). *Acta chim., Acad. Sci. Hung.*, **4**, 55.
JAMBOR, B. (1960). *Tetrazoliumsalze in der Biologie. Fischer Verlag, Jena.*
JERCHEL, D., and FISCHER, H. (1949). *Liebigs Ann.*, **563**, 200.
JERCHEL, D., GEYER, H., and HOLTKAMP, H. (1958). *Ann. Chem.*, **613**, 137.
JERCHEL, D., and MÖHLE, W. (1944). *Ber. dt. chem. Ges.*, **77**, 591.
JOHNSON, A. B. (1965). *J. Histochem. Cytochem.*, **13**, 583.
JONES, G. R. N. (1968a). *Expl. Cell Res.*, **49**, 251.
JONES, G. R. N. (1968b). *Histoch. J.*, **1**, 59.
JONES, G. R. N. (1969). *Histochemie*, **18**, 164.
JONES, O. T. G., and WATSON, W. A. (1967). *Biochem. J.*, **102**, 564.

KALINA, M. (1968). *Lab. Invest.*, **18**, 278.
KALINA, M., and PALMER, J. M. (1968). *Histochemie*, **14**, 366.
KAPLAN, N. O., and CIOTTI, M. M. (1953). *J. biol. Chem.*, **201**, 785.
KAPLAN, N. O., COLOWICK, S. P., and CARR BARNES, C. (1951). *Ibid.*, **191**, 461.
KELLER, R. N., and PARRY, R. W. (1956). In "Chemistry of the Co-ordination Compounds." Ed. J. C. Bailar, Reinhold, Baltimore.
KIVALO, P., and MUSTAKALLIO, K. K. (1956). *Suomen Kemi. B.*, **29**, 154.
KOCHAKIAN, C. D., CARROLL, B. R., and UHR, B. (1957). *J. biol. Chem.*, **224**, 811.
KOEN, A. L., and SHAW, C. R. (1964). *Biochem. biophys. Res. Commun.*, **15**, 92.
KUHN, R., and JERCHEL, D. (1941). *Ber. dt. chem. Ges.*, **74**B, 941.
KUN, E., and ABOOD, L. G. (1949). *Science, N.Y.*, **109**, 144.
KUWABARA, T., and COGAN, D. G. (1959). *J. Histochem. Cytochem.*, **7**, 329.
LAKON, G. (1942). *Ber. dt. botan. Ges.*, **60**, 299 and 434.
LETTRÉ, H., and ALBRECHT, M. (1943). *Hoppe-Seyler's Z. physiol. Chem.*, **279**, 206.
LEUTHARDT, F., and BRUTTIN, H. (1952). *Helv. chim. Acta*, **35**, 464.
LEVINE, W., GIUDITTA, A., ENGLARD, S., and STRECKER, H. J. (1960). *J. Neurochem.*, **6**, 28.
LUNDEGÅRDH, H. (1958). *Biochim. biophys. Acta*, **27**, 653.
MAENDER, O. W., and RUSSELL, G. A. (1966). *J. org. Chem.*, **31**, 442.
MAHLER, H. R. (1955). *Adv. Enzymol.*, **17**, 233.
MARCUS, P. I., and TALALAY, P. (1956). *J. biol. Chem.*, **218**, 661.
MARTIUS, C. (1954). *Biochem. Z.*, **326**, 26.
MARTIUS, C. (1958). *Dt. med. Wschr.*, **83**, 1701.
MARTIUS, C., and MÄRKI, F. (1957). *Biochem. Z.*, **329**, 450.
MARTIUS, C., and STRUFE, R. (1954). *Biochem. Z.*, **326**, 24.
MASSEY, V. (1958). *Biochim. biophys. Acta*, **30**, 205.
MASSEY, V. (1960). *Biochim. biophys. Acta*, **37**, 314.
MASSEY, V. (1963). In "The Enzymes." Ed. P. D. Boyer, H. Lardy, and K. Myrbäck, 2nd Edn, Acad. Press, New York, p. 275.
MATTSON, A. M., JENSEN, C. O., and DUTCHER, R. A. (1947). *Science, N.Y.*, **106**, 294.
MCCABE, M., MAPLE, A. J., and JONES, G. R. N. (1965). *J. Histochem. Cytochem.*, **13**, 541.
MICHAELIS, L., and MENTEN, M. L. (1913). *Biochem. Z.*, **49**, 333.
MORTON, R. A., WILSON, G. M., LOWE, J. S., and LEAT, W. M. F. (1958). *Biochem. J.*, **68**, 16P.
MUDD, S. TAKEYA, K., and HENDERSON, H. J. (1956). *J. Bact.*, **72**, 767.
NACHLAS, M. M., MARGULIES, S. I., and SELIGMAN, A. M. (1960). *J. biol. Chem.*, **235**, 499.
NACHLAS, M. M., MARGULIES, S. I., and SELIGMAN, A. M. (1960). *J. biol. Chem.* **235**, 2739.
NACHLAS, M. M., WALKER, W. H., and SELIGMAN, A. M. (1958a). *J. biophys. biochem. Cytol.*, **4**, 29.
NACHLAS, M. M., WALKER, W. H., and SELIGMAN, A. M. (1958b). *Ibid.*, **4**, 169.
NEUGEBAUER, F. A. (1968). *Tetrahedron Letters*, No. **17**, 2129.
ODA, T. (1961). *J. Histochem. Cytochem.*, **9**, 605.
ODA, T., and OKAZAKI, H. (1958). *Acta med. Okayama*, **12**, 193.
OGAWA, K., and SHINONAGA, V. (1961). *Acta Anat. Nippon*, **36**, 550.
OKUI, S., SUZAKI, Y. MOMOSE, K., and OGAMO, A. (1963). *J. Biochem.*, **52**, 500.
PEARSE, A. G. E. (1957). *J. Histochem. Cytochem.*, **5**, 515.
PEARSE, A. G. E., and HESS, R. (1961). *Experientia*, **17**, 136.
PEARSE, A. G. E., SCARPELLI, D. G., and HESS, R. (1960). *J. biophys. biochem. Cytol.*, **7**, 243.
PECHMANN, H. VON, and RUNGE, P. (1894). *Ber. dt. chem. Ges.*, **27**, 2920.
PENNINGTON, R. J. (1961). *Biochem. J.*, **80**, 649.
PORTER, V. S., DEMING, N. P., WRIGHT, R. C., and SCOTT, E. M. (1953). *J. biol. Chem.*, **205**, 883.
POTTER, V. R., and REIF, A. E. (1952). *J. biol. Chem.*, **194**, 287.
RAAFLAUB, J. (1953a). *Helv. physiol. Acta*, **11**, 142.

RAAFLAUB, J. (1953b). *Ibid.*, **11**, 157.
RAAFLAUB, J. (1956). *Ibid.*, **14**, 304.
RACKER, E. (1955). *Physiol. Rev.*, **35**, 1.
REED, L. J. (1953). *Ibid.*, **33**, 544.
REED, L. J., DE BUSK, B. G., GUNSALUS, I. C., and HORNBERGER, C. S. (1951). *Science, N.Y.*, **114**, 93.
RIED, W., and WILK, M. (1953). *Liebigs. Ann.*, **581**, 49.
ROBERTS, L. W. (1951). *Science, N.Y.*, **113**, 692.
ROBERTS, L. W., BABA, S., and URBAN, K. (1966). *Plant, Cell Physiol.*, **7**, 177.
ROSA, C., and TSOU, K.-C. (1961). *Nature, Lond.*, **192**, 990.
ROSA, C. G., and TSOU, K.-C. (1963). *J. Cell Biol.*, **16**, 445.
RUTENBURG, A. M., GOFSTEIN, R., and SELIGMAN, A. M. (1950). *Cancer Res.*, **10**, 113.
SAVAGE, N. (1957). *Biochem. J.*, **67**, 146.
SCARPELLI, D. G., HESS, R., and PEARSE, A. G. E. (1958). *J. biophys. biochem. Cytol.*, **4**, 747.
SCARPELLI, D. G. and PEARSE, A. G. E. (1958). *J. Histochem. Cytochem.*, **6**, 369.
SEARLS, R. L., and SANADI, D. R. (1961). *J. biol. Chem.*, **236**, 580.
SEIDLER, E. (1968). *Z. med. Labortechnik.*, **9**, 241.
SEIDLER, E., and KUNDE, D. (1969a). *Acta histochem.*, **32**, 142.
SEIDLER, E., and KUNDE, D. (1969b). *Histochemie*, **17**, 256.
SELIGMAN, A. M., GOFSTEIN, R., and RUTENBURG, A. M. (1949). *Cancer Res.*, **9**, 366.
SEMENOFF, W. E. (1935). *Z. Zellforsch. mikrosk. Anat.*, **22**, 305.
SEYHAN, M. (1952). *Rev. faculté sci. univ. Istanbul*, **17A**, 182.
SEYHAN, M. (1953). *Chem. Abstracts*, **47**, 12390.
SHAW, C. R., and KOEN, A L. (1965). *J. Histochem. Cytochem.*, **13**, 431.
SHUSTER, L., and KAPLAN, N. O. (1955). *J. biol. Chem.*, **215**, 183.
SINGER, T. P., and KEARNEY, E. B. (1954a). *Adv. Enzymol.*, **15**, 79.
SINGER, T. P. and KEARNEY, E. B. (1954b). *Biochim. biophys. Acta*, **15**, 151.
SINGER, T. P., KEARNEY, E. B., and MASSEY, V. (1957). *Adv. Enzymol.*, **18**, 65.
SLATER, T. F. (1959). *Biochem. J.*, **73**, 314.
SLATER, T. F. (1963). *Biochim. biophys. Acta*, **77**, 365.
SLATER, T. F., SAWYER, B., and STRÄULI, U. (1963). *Biochim. biophys. Acta*, **77**, 383.
STRAUB, F. B. (1939). *Biochem. J.*, **33**, 787.
STRAUB, F. B. (1940). *Ibid.*, **34**, 483.
STRAUS, F. H., CHERONIS, N. O., and STRAUS, E. (1948). *Science, N.Y.*, **108**, 113.
TALESARA, C. L., and BLANCHAER, M. C. (1967). *Can. J. Biochem.*, **45**, 299.
THEORELL, H., and BONNICHSEN, R. (1951). *Acta chem. scand.*, **5**, 1105.
TSOU, K.-C., CHENG, C. S., NACHLAS, M. M., and SELIGMAN, A. M. (1956). *J. Am. chem. Soc.*, **78**, 6139.
TSOU, K.-C., and SU, H. C. F. (1962). *J. Histochem. Cytochem.*, **10**, 689.
TYRER, J. H., EADIE, M. J., and HOOPER, W. D. (1969). *J. Chromat.*, **39**, 312.
VAN EYS, J., SAN PIETRO, A., and KAPLAN, N. O. (1958). *Science, N.Y.*, **127**, 1443.
VAN ITERSON, W., and LEENE, W. (1964). *J. Cell Biol.*, **20**, 361.
VAN WIJHE, M., BLANCHAER, M. C., and JACYK, W. R. (1963). *J. Histochem. Cytochem.*, **11**, 505.
VON EULER, H., and HELLSTRØM, H. (1938). *Z. physiol. Chem.*, **252**, 31.
WACHSTEIN, M. (1955). *J. Histochem. Cytochem.*, **3**, 246.
WARBURG, O., and CHRISTIAN, W. (1936). *Biochem. Z.*, **287**, 291.
WARBURG, O., CHRISTIAN, W., and GRIESE, A. (1935). *Ibid.*, **282**, 157.
WATTENBERG, L. W. (1954). *Proc. Soc. exp. Biol., Med.*, **87**, 535.
WIZINGER, R., and BIRO, V. (1949). *Helv. chim. Acta*, **32**, 901.
WIZINGER, R., and HERZOG, H. (1953). *Ibid.*, **36**, 531.
WOHLRAB, F. (1963). *Histochemie*, **3**, 286.
WOHLRAB, F., and FUCHS, U. (1967). *Histochemie*, **9**, 256.
YOUNG, M. R., and EDWARDS, R. G. (1963). *J. Histochem. Cytochem.*, **11**, 444.
ZIEGLER, D. M., and LINNANE, A. W. (1958). *Biochim. biophys. Acta*, **30**, 53.
ZIMMERMANN, H., and PEARSE, A. G. E. (1959). *J. Histochem. Cytochem.*, **7**, 271.

CHAPTER 21

OXIDOREDUCTASES II
(Diaphorases and Dehydrogenases)

Historical Introduction

THE first stage of dehydrogenase histochemistry began, as stated in Chapter 20, with the use of monotetrazolium salts as electron acceptors for known or unknown dehydrogenase systems. These salts (TTC and INT) were relatively useless from the histochemical point of view, providing a poorly coloured diffusible end product in one case and a macrocrystalline one in the other. During the second stage of dehydrogenase histochemistry a great expansion of ideas took place when the techniques of the science were improved by the advent of Neotetrazolium (NT) and Blue tetrazolium (BT). Almost all the applied dehydrogenase histochemistry reported in the literature between 1950 and 1958 was carried out with one or other of these two salts.

The third stage began in 1957, with the advent of the newer tetrazolium salts, still in use today, and with the adoption of new principles for their use. In place of relatively crude localization at the histological level intramitochondrial identification became possible and, in place of two diaphorases and a single dehydrogenase, the demonstration of a whole spectrum of dehydrogenases. The total number now exceeds twenty. Much of the pioneer work in applied dehydrogenase histochemistry requires revision and the list of early studies which is given below is provided as a base from which to study the tremendous advances which have been made in the last decade.

Applied Dehydrogenase Histochemistry

General Studies and Reviews. Seligman and Rutenburg (1951); Shelton and Schneider (1952); Padykula (1952); Rutenburg, Wolman and Seligman (1953); Malaty and Bourne (1953); Rosa and Velardo (1954); Novikoff (1955); Baar (1957); Ogawa (1958). *Central Nervous System and Ganglia.* Leduc and Wislocki (1952); Mustakallio (1954a); Ortmann (1957); Blasius and Zimmermann (1958). *Neoplastic Tissues.* Goddard and Seligman (1953); Hsu and Hoch-Ligeti (1953); Pearson and Defendi (1955). *Thyroid.* Goddard and Seligman (1952); Kuusisto and Telkkä (1955). *Kidney.* Mustakallio and Telkkä (1953); Bourne and Malaty (1953); Wachstein and Meisel (1954); Wachstein (1955); Telkkä and Mustakallio (1955); Mustakallio and Telkkä (1955); Reale and Luzzatto (1956); Pearse and Macpherson (1958); Rudolph and Scholl (1958). *Embryos.* Rossi, Pescetto and Reale (1954, 1957). *Liver and Gall-Bladder.* Mustakallio (1954b); Mustakallio and Saikkonen (1955); Schumacher (1957). *Skin and Appendages.* Rogers (1953); Formisano and Montagna (1954); Montagna and Formisano (1955);

Portugalov (1955); Mustakallio (1956); Argyris (1956a and b); Cruickshank, Hershey and Lewis (1958). *Cells in Blood or Tissue Culture.* Hughes (1956); Marcuse (1957a and b); Vercauteren (1957). *Blood Vessels.* Fried and Zweifach (1955). *Muscle.* Malaty and Bourne (1953); Beckett and Bourne (1958); Nachmias and Padykula (1958); Buño and Germino (1958). *Heart Muscle.* Bourne (1953); Cooper (1955); Wachstein and Meisel (1955); Cooper and Copenhaver (1957). *Female Genitalia.* Foraker, Denham and Johnston (1953); Foraker and Denham (1953); Rosa and Velardo (1958). *Bacteria.* Davis *et al.* (1953); Mustakallio and Jännes (1954); Davis and Mudd (1957).

Developments in Dehydrogenase Histochemistry

The end of the second stage development was marked by the pioneer work of Farber and his associates (Farber, Sternberg and Dunlap, 1956), using NT and BT as their hydrogen acceptors. Farber developed the thesis that three systems only could be demonstrated by the use of tetrazolium salts. These were succinate dehydrogenase, NADH diaphorase and NADPH diaphorase. This argument was based on results obtained with rat kidney, where only two different formazan patterns were obtained when systems for five different coenzyme-linked dehydrogenases were employed. These were the patterns of the two diaphorases. While admitting that the final pathway for tetrazolium reduction in coenzyme-linked systems was the diaphorase (tetrazolium reductase) Nachlas *et al.* (1958c) and Hess *et al.* (1958) concluded that distinct formazan patterns were produced by the different dehydrogenase systems. In particular (Hess and Pearse, 1959) it was consistently shown that the reaction for glucose-6-phosphate dehydrogenase, in mammalian kidney, produced formazan deposits mainly restricted to the specialized zone of the macula densa (Fig. 197, p. 869).

Further serious objections to the "single pattern" theory were put forward by Cascarano and Zweifach (1959) who found that the formazan patterns from a single tetrazolium salt, accepting electrons from various substrates for coenzyme-linked dehydrogenases, were clearly not restricted to the pattern of the appropriate diaphorase. Somewhat later came the realization that all the coenzyme-linked dehydrogenases, with the possible exception of β-hydroxybutyrate, were extremely soluble and that they diffused rapidly into the incubating medium. Fahimi and Amarasingham (1964), for instance, found that 80 per cent of white skeletal muscle lactate dehydrogenase (LDH) diffused into the medium within 10 minutes and Kalina and Gahan (1965) showed that both glucose-6-phosphate and 6-phosphogluconate dehydrogenases were totally lost from tissue sections within one minute. Upon these disclosures the pendulum swung rapidly back in favour of the "single pattern" hypothesis.

Experiments carried out by Kalina and Gahan (1965), and by Kalina *et al.* (1965) indicated conclusively that oxidation of substrate in the medium by solubilized enzyme was followed by reduction of the coenzyme and the subsequent transfer of electrons to the tissue-bound tetrazole by the appropriate reductase. In order to avoid the manifest false localization produced by soluble dehydrogenases two methods were employed. These were minimal fixation, and solid gel media.

Minimal Fixation. This principle was first employed by Walker and Seligman (1961) for the bound enzyme succinate dehydrogenase although Novikoff and Masek (1958) had already shown that LDH could withstand formol-calcium fixation. Lojda (1965), using rat tissues, found that after 10 minutes fixation in cold formol-calcium succinate, malate, isocitrate and β-hydroxybutyrate dehydrogenases were well preserved. After 24 hours' fixation lactate and malate dehydrogenases were still active. My agreement with the principle of minimum fixation constitutes an absolute withdrawal from the position which I adopted in 1960 (2nd Edition, p. 581).

Gel Media. As an alternative to fixation Fahimi and Amarasingham (1964) employed a solid gel medium, in the form of a thin film coated on glass slides. Incubation was carried out by applying the coverslip-mounted section to the surface of the gel. This method was modified by Benitez and Fischer (1964) who used 3·25 per cent gelatin in place of the original concentration (2·5 per cent). Observing that the glass-mounted medium failed to achieve proper opposition to the section, they employed a plastic film base instead. This modification was used by Kalina and Gahan (1968), with an even higher gelatin concentration (5 per cent). Their technique is described in Appendix 21, p. 1346. An acrylamide gel medium was tested by Lehrer *et al.* (1965) who observed no significant difference in formazan localization as between liquid and gel media, unless the latter contained phenazine methosulphate. Neither fixation, nor solid media methods, have proved very popular and the majority of workers still prefer to avoid them.

Soluble and Insoluble Dehydrogenases and Diaphorases. There are, fortunately, a number of "bound" enzymes which scarcely diffuse at all into the medium, even over long periods of time, provided that the mitochondria are relatively undamaged. There are others which diffuse only slowly and which can be regarded as insoluble with short incubation periods. In the first class are succinate dehydrogenase, NADH and NADPH diaphorases, and α-glycerophosphate dehydrogenase. In the second class are β-hydroxybutyrate and glutamate dehydrogenases.

"Short" formalin fixation, in the manner described by Walker and Seligman (1963), preserves *in situ*, in addition to succinate and mitochondrial α-glycerophosphate dehydrogenases, a proportion of the NAD-linked lactate, malate, isocitrate and α-glycerophosphate dehydrogenases. The soluble NADP-linked dehydrogenases, glucose-6-phosphate and 6-phosphogluconate, are very sensitive to fixation but the former survives brief fixation in dilute glutaraldehyde.

Using 0·6 per cent glutaraldehyde at 20° for one minute Flitney (1966) showed that a considerable degree of fixation was achieved, as judged by the preservation of fluorescein-labelled albumin incorporated in a gelatin gel. Succinate and mitochondrial α-glycerophosphate, as well as glucose-6-phosphate dehydrogenases will tolerate this degree of fixation. The two diaphorases of course, as indicated by McAlpine (1965), can withstand long fixation in cold formalin and prolonged storage in gum sucrose medium (Vol 1, p. 602).

Choice of Tetrazolium Salt. A great deal has been written on this subject but, unfortunately, very little of the experimental work has been truly objective or free from bias. A notable exception is the review by Lojda (1965). A considerable body of evidence was presented by myself and my associates (Hess *et al.*, 1958a and b; Scarpelli and Pearse, 1958a and b; Pearse and Scarpelli, 1958) which indicated that the formazan deposits from MTT, in the presence of cobalt, were intramitochondrial and, furthermore, that in damaged mitochondria the rate of formazan production, and the size of the final product, were sufficiently greater. This phenomenon allowed accurate appraisal of the damage and the process was termed "mitochondrial assay". "By proper choice of material" according to Novikoff *et al.* (1960) it was "possible to put such claims for intramitochondrial localizations to critical test". Using *fixed tissues*, with essentially a direct Nitro-BT stain (see Hess and Pearse, 1961; Brooke and Engel, 1966), Novikoff and his associates demonstrated the swollen mitochondria of anoxic rat kidney. They were unable, using MTT in a similar NADH diaphorase system on their fixed tissue, to distinguish normal from anoxic mitochondria and were moved, therefore, to dismiss this tetrazolium salt as "not reliable for such intracellular localization". A precisely similar view was expressed by Lojda (1965) who noted that the demonstration of intramitochondrial MTT formazan at the E.M. level (see Figs. 198, 199) did not influence this conclusion.

It is quite true that the cobalt-formazan from MTT is deposited in the same spherical (dot-like) form whether the ultimate electron donor is intramitochondrial or not. If the final donor is known to be bound to the mitochondrion then MTT formazan demonstrates the intracellular localization of functioning mitochondria as well as any other tetrazole. Being non-substantive (see Chapter 20, p. 887) it does not provide a predetermined picture of the sites of tetrazole binding.

The choice of tetrazole in modern dehydrogenase histochemistry, which still lies between one or other of the two substantive ditetrazoles (Nitro-BT, TNBT) and the non-substantive monotetrazole (MTT), is to be dictated by the considerations outlined at length in Chapter 20.

The Flavoprotein Diaphorases

The precise identity of the two diaphorases has not yet been determined and it is not possible to assign to them any definite E.C. numbers. An excellent review by Hatefi (1963) outlines the complicated relationships between

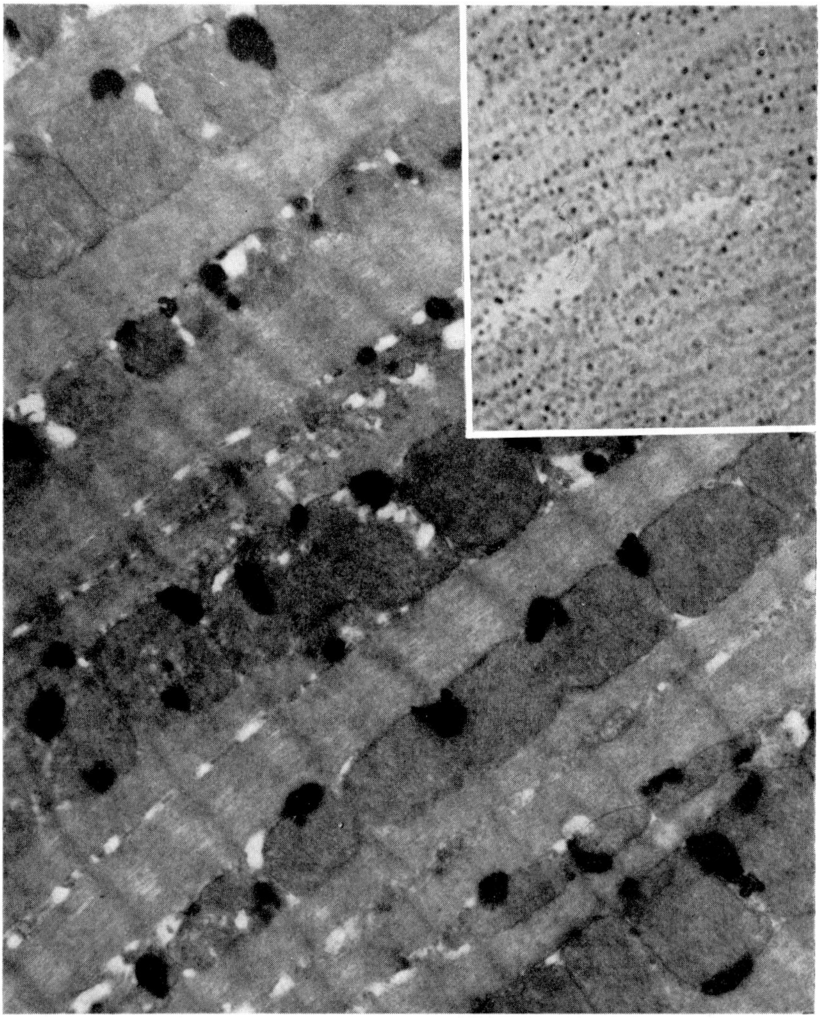

FIGS. 198 (inset) and 199.

FIG. 198. Rat heart muscle. Cryostat section (5 μ). After performing the UQ reaction, with MTT-cobalt, the metal formazan reaction product appears as intramitochondrial dots. × 1800.

FIG. 199. Tissue processed as for Fig. 198 and subsequently for electron microscopy. The electron opaque final product is substantially intramitochondrial. × 15′000.

the various functional flavoproteins and the factors which are associated with them as components of different systems.

NADH Diaphorase. NADH dehydrogenating flavoproteins, containing either FAD or FMN, react with a large number of electron acceptors. Among the latter are ferricyanide, methylene blue, 2,6-dichlorophenol-indophenol, and tetrazolium salts. Their relationship to NADH-cytochrome c reductase (E.C. 1.6.2.1) is probably a close one.

From the NADH dehydrogenase isolated by Ziegler *et al.* (1959), which contained FAD as its flavin, butanol treatment yielded a preparation with properties similar to those of Straub's (1939) diaphorase. According to Massey (1958, 1959, 1960) this diaphorase and his lipoate dehydrogenase (syn: lipoamide, lipoyl, dihydrothioctyl, or dihydrolipoyl dehydrogenase) were to be regarded as two functional aspects of the same enzyme, which was also a component of the α-ketoglutarate (syn: 2-oxoglutarate) dehydrogenase complex.

At the present time many histochemists regard NADH diaphorase and lipoamide dehydrogenase as identical. Such identity cannot be presumed on the present evidence and the two must be regarded as related but distinct.

NADPH Diaphorase. The old yellow enzyme of Warburg and Christian (1932), whose flavin component is FMN (riboflavin-5-phosphate) possesses both diaphorase and cytochrome c reductase activities. The histochemically demonstrable diaphorase may be related to old yellow enzyme (E.C. 1.6.99.1) or to the reductase (E.C. 1.6.2.3) but, at present, no definite identification can be made.

Methods for NADH and NADPH Diaphorases and Associated Enzymes

In the original method (Farber *et al.*, 1956a) both diaphorases were demonstrated by maintaining a supply of reduced coenzyme either through the activity of an exogenous enzyme-substrate system which was added to the medium, together with the oxidized coenzyme (NAD or NADP), or through the activity of an endogenous dehydrogenase whose substrate was provided. For NADH diaphorase, using endogenous enzyme, malate and glutamate were added but better results were obtained when ethanol and exogenous alcohol dehydrogenase were employed. In the case of NADPH diaphorase an exogenous enzyme was not used. Instead, either glucose-6-phosphate or a mixture of citrate and malate were added as substrates for endogenous enzymes. The addition of cysteine or thioglycollate was found essential and the presence of a sulphydryl compound was regarded as an absolute requirement for the three NADP-linked enzymes studied by Farber and his co-workers. At the pH employed cysteine and thioglycollate can produce non-enzymic reduction even of tetrazolium salts with relatively low redox potentials like NT and BT. They must never be used with tetrazolium salts having higher potentials, like MTT, Nitro-BT and TNBT.

With the advent of Nitro-BT a great improvement in the localization of the diaphorases became possible and Nachlas, Walker and Seligman (1958a) demonstrated NADH diaphorase with this salt and a system containing lactate and lactic dehydrogenase at pH 7·4. Later (1958b) these authors applied their findings to the demonstration of NADPH diaphorase. They considered two alternative procedures: (1) the inclusion in the medium of two substrates for endogenous NADP linked dehydrogenases and (2), the addition of an exogenous NADP linked dehydrogenase and its substrate (isocitrate and isocitrate dehydrogenase). Since this enzyme was not easily available the first alternative was employed. Experiments were performed with four substrates in various combinations. The mixtures which resulted in a summation of enzymic activity were isocitrate plus malate, isocitrate plus glucose-6-phosphate and 6-phosphogluconate plus malate. An unexplained but understandable finding was the failure to note summation with the combination of glucose-6-phosphate with malate and of glucose-6-phosphate plus 6-phosphogluconate. The isocitrate-malate combination was recommended as giving the best results. The localization of NADPH diaphorase afforded by the above techniques appeared to the authors to be non-mitochondrial in contrast to the mitochondrial localization which they obtained with NADH diaphorase.

Reduced Coenzymes as Substrates. The use of NADH and NADPH as substrates for their respective diaphorases was first suggested by Scarpelli, Hess and Pearse (1958). Since neither reaction required the addition of exogenous dehydrogenases, or of substrates for endogenous systems, these techniques offered distinct advantages over those previously described. The two reduced substrate techniques, which have now become standard practice, are described in Appendix 21, p. 1342.

Dihydrolipoate as Substrate. Two papers appeared in 1964 (Diculescu *et al.*, Balogh), both of which described the localization of the specific lipoamide dehydrogenase (E.C. 1.6.4.3) using in the one case the sodium salt of dihydrolipoic acid and in the other dihydrolipoamide as substrate. At the pH of incubation used by Diculescu *et al.* (7·2) Balogh observed non-specific, non-enzymic, reduction of Nitro-BT which commenced as low as pH 6·2. He therefore carried out his incubation at pH 6·0 and observed, in a few tissues, distinct differences between the formazan distribution produced by the standard NADH diaphorase reaction and the reaction for lipoamide dehydrogenase.

The presumptive pathways of this enzyme are:

$$\text{NADH} + \text{oxidized lipoamide} \rightleftharpoons \text{NAD} + \text{dihydrolipoamide}$$

and thus the histochemical reaction proceeds in the opposite direction to the standard NADH diaphorase reaction. The final pathway is still by way of the diaphorase but any difference between the two reactions should be significant.

Pyruvate and Oxoglutarate Dehydrogenases. Two further enzyme systems are associated with lipoic acid. These are pyruvate lipoate oxidoreductase

(E.C. 1.2.4.1) and 2-oxoglutarate lipoate oxidoreductase (E.C. 1.2.4.2). These catalyse the following reactions:

$$\text{Pyruvate} + \text{oxidized lipoate} \rightarrow \text{6-S-acetylhydrolipoate} + CO_2$$

$$\text{2-Oxoglutarate} + \text{oxidized lipoate} \rightarrow \text{6-S-succinylhydrolipoate} + CO_2$$

Both require thiamine pyrophosphate as cofactor and should probably be regarded as belonging to larger systems.

Lipoate, or lipoamide is a widely distributed hydrogen carrier which occurs in the tissues mainly bound to the lysine residues of protein by peptide bonds. The formula of oxidized lipoic acid is given below:

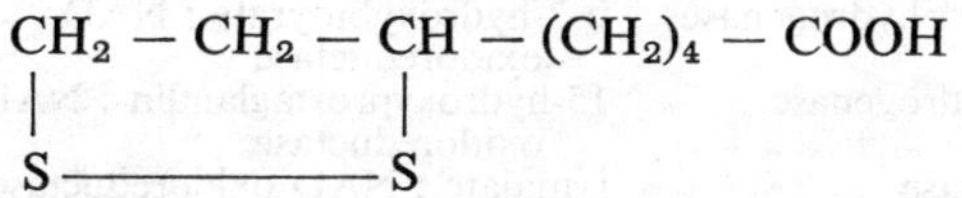

Two papers describing the histochemical localization of pyruvate and oxoglutarate oxidases have appeared in the literature. The first, by Ferguson (1966), was closely followed by a report from Coleman and Nienhuis (1966). Both authors used pyruvate and oxoglutarate as substrates, with CoA and NAD as cofactors, and incubated their tissues at pH 7·0 to 7·4. Both methods suffer from three serious deficiencies; (1) coenzyme A reduces tetrazoles spontaneously at the pH levels employed; (2) in the presence of excess pyruvate the NADH formed by oxidation of dihydrolipoate (the latter produced by conversion of 6-S-acetylhydrolipoate and CoA to acetyl CoA and hydrolipoate) will be reoxidized by LDH as well as by its diaphorase and, (3) unless the hydrolipoate and the NADH diaphorase are closely bound on the same tissue component serious loss of electrons must take place.

One can have little faith in the representation of the final formazan precipitate as the localization of either oxidase.

Oxidoreductases and Methods for Oxidoreductases

Having discussed, out of turn, the histochemistry of the diaphorases, and of some possibly closely associated flavoprotein enzymes, it is necessary to proceed to the remaining dehydrogenases. Table 70, below, is a list of those oxidoreductases for which histochemical methods have been described. It includes enzymes already presented in this chapter, and some of the oxidases dealt with in Chapter 19. In the text which follows each enzyme will be dealt with in the order in which it appears in the Table.

Alcohol Dehydrogenase (E.C. 1.1.1.1)

Alcohol dehydrogenases are widely distributed in plant and animal tissues. They catalyse the reaction:

$$R.CH_2OH + NAD \rightleftharpoons R.CHO + NADH + H^+$$

TABLE 70

Histochemical or Trivial Name	Systematic Name	E.C. Number
Alcohol dehydrogenase	Alcohol : NAD oxidoreductase	1.1.1.1
α-Glycerophosphate dehydrogenase	L-glycerol-3-phosphate : NAD oxidoreductase	1.1.1.8
Xylitol dehydrogenase	Xylitol : NAD oxidoreductase	1.1.1.9
Xylitol dehydrogenase	Xylitol : NADP oxidoreductase	1.1.1.10
Sorbitol dehydrogenase	L-iditol : NAD oxidoreductase	1.1.1.14 & 15
Glucuronate dehydrogenase	L-gulonate : NADP oxidoreductase	1.1.1.19
Gulonolactone oxidase	L-gulonolactone : NADP oxidoreductase	1.1.1.20
UDPG dehydrogenase	UDP glucose : NAD oxidoreductase	1.1.1.22
Lactate dehydrogenase	L-lactate : NAD oxidoreductase	1.1.1.27
β-hydroxybutyrate dehydrogenase	D-3-hydroxybutyrate : NAD oxidoreductase	1.1.1.30
Prostaglandin dehydrogenase	15-hydroxyprostaglandin : NAD oxidoreductase	Not allocated
Malate dehydrogenase	L-malate : NAD oxidoreductase	1.1.1.37
Malic enzyme	L-malate : NADP oxidoreductase	1.1.1.40
Isocitrate dehydrogenase	L_s-isocitrate : NAD oxidoreductase	1.1.1.41
Isocitrate dehydrogenase	L_s-isocitrate : NADP oxidoreductase	1.1.1.42
6-Phosphogluconate dehydrogenase	6-phospho-D-gluconate : NADP oxidoreductase	1.1.1.44
Glucose-6-phosphate dehydrogenase	D-glucose-6-phosphate : NADP oxidoreductase	1.1.1.49
3-α-hydroxysteroid dehydrogenase	3-α-hydroxysteroid : NAD(P) oxidoreductase	1.1.1.50
3-β-hydroxysteroid dehydrogenase	3-β-hydroxysteroid : NAD(P) oxidoreductase	1.1.1.51
17-β-hydroxysteroid dehydrogenase	17-β-hydroxysteroid : NAD(P) oxidoreductase	1.1.1.51
α-Glycerophosphate dehydrogenase (mitochondrial)	L-glycerol-3-phosphate : (acceptor) oxidoreductase	1.1.2.1
Choline dehydrogenase	Choline (acceptor) oxidoreductase	1.1.99.1
Aldehyde dehydrogenase	Aldehyde : NAD oxidoreductase	1.2.1.3
Betaine aldehyde dehydrogenase	Betaine aldehyde : NAD oxidoreductase	1.2.1.8
Glyceraldehyde phosphate dehydrogenase	D-glyceraldehyde-3-phosphate : NAD oxidoreductase	1.2.1.12
Xanthine oxidase	Xanthine : O_2 oxidoreductase	1.2.3.2
Pyruvate dehydrogenase	Pyruvate : lipoate oxidoreductase	1.2.4.1
Oxoglutarate dehydrogenase	2-oxoglutarate : lipoate oxidoreductase	1.2.4.2
Dihydro-orotate dehydrogenase	4,5-L-dihydro-orotate : (acceptor) oxidoreductase	1.3.3.1
Succinate dehydrogenase	Succinate : (acceptor) oxidoreductase	1.3.99.1
Glutamate dehydrogenase	L-glutamate : NAD oxidoreductase	1.4.1.2
L-Aminoacid dehydrogenase	L-aminoacid : NAD oxidoreductase	1.4.3.2
D-Aminoacid oxidase	D-aminoacid : O_2 oxidoreductase	1.4.3.3
Monoamine oxidase	Monoamine : O_2 oxidoreductase	1.4.3.4
Diamine oxidase	Diamine : O_2 oxidoreductase	1.4.3.6
Proline dehydrogenase	L-proline : NAD 2-oxidoreductase	1.5.1.1
Dihydrofolate reductase	5,6,7,8-tetrahydrofolate : NADP oxidoreductase	1.5.1.3
Glutathione reductase	NADPH : glutathione oxidoreductase	1.6.4.2
Lipoamide dehydrogenase	NADH : lipoamide oxidoreductase	1.6.4.3
NADPH diaphorase	NADPH : (acceptor) oxidoreductase	1.6.99.1 (possibly)
Urate oxidase	Urate : O_2 oxidoreductase	1.7.3.3
Cytochrome oxidase	Cytochrome c : O_2 oxidoreductase	1.9.3.1
Catechol oxidase	*o*-diphenol : O_2 oxidoreductase	1.10.3.1
Catalase	H_2O_2 : H_2O_2 oxidoreductase	1.11.1.6
Peroxidase	(Donor) : H_2O_2 oxidoreductase	1.11.1.7

Many alcohol dehydrogenases, such as the enzyme from mammalian liver, possess a very broad specificity for alcohols, aldehydes and ketones. A variety of primary and *sec*-alcohols, both aliphatic and aromatic, but not *tert*-alcohols, isopropanol or steroid alcohols, are oxidized. Coenzyme specificity is not absolute but, with the liver enzyme, NAD is 100 times more active than NADP.

Inhibitors. Liver alcohol dehydrogenase is a sulphydryl enzyme and it is thus inhibited by SH reagents (*p*-mercuribenzoate, iodoacetate, *N*-ethyl maleimide) and by heavy metals. The essential role of enzyme-bound Zn^{2+} is indicated by its inhibition by organic and inorganic complex-forming agents. Among these are *o*-phenanthroline (Vallee *et al.*, 1959), α,α'-dipyridyl (Plane and Theorell, 1961), and 8-hydroxyquinoline, EDTA and diethyldithiocarbamate (Vallee and Hoch, 1957).

The enzyme is quite easily demonstrated histochemically but in many tissues its activity is relatively weak. Parietal cells and salivary duct cells (rat) form an exception. The most interesting application of the alcohol dehydrogenase reaction is in the retina where, as Wald (1956) has shown, the following reaction takes place.

$$\text{Vitamin A} \underset{\text{NADH}}{\overset{\substack{\text{alcohol dehydrogenase} \\ \text{NAD} +}}{\rightleftarrows}} \text{Retinene} \xrightarrow{\text{opsin}} \text{Rhodopsin}$$

Intense alcohol dehydrogenase activity can be demonstrated in the retina of rats kept in the dark for short periods (18–24 hours). If vitamin A is used as substrate the reaction runs to the right, if hydroxylamine (or opsin) is present to trap the product. If, as in histochemical usage, alcohol is employed as substrate cyanide forms an efficient trapping agent.

NAD-linked α-Glycerophosphate Dehydrogenase (E.C. 1.1.1.8)

This *soluble* NAD-linked enzyme is the cytoplasmic component of the α-glycerophosphate shuttle (Zebe *et al.*, 1957; Boxer and Devlin, 1961). It catalyses the reaction:

$$\alpha\text{-glycerophosphate} + \text{NAD} \rightleftharpoons \text{dihydroxyacetone phosphate} + \text{NADH} + \text{H}^+$$

The equilibrium lies far to the left of this equation. The enzyme was isolated and crystallized from rabbit muscle by Baranowski (1949), by whose name it was formerly known. It was extensively studied by Bücher and his associates (Bücher and Klingenberg, 1958; Zebe, 1960), and has been shown to be absolutely specific for L-glycerol-1-phosphate.

Inhibitors and Activators. The enzyme is inhibited by sulphydryl reagents. At a concentration of 0.1 mM, *p*-chloromercuribenzoate produces 100 per cent inhibition, as does *N*-ethyl maleimide at the same concentration. In the presence of 5·4 mM EDTA at pH 7·6, activity of the enzyme is increased by 30–40 per cent.

Histochemical demonstration of NAD-linked α-glycerophosphate dehydrogenase follows the standard coenzyme-linked formula (Appendix 21, p. 1344).

Xylitol Dehydrogenases (E.C. 1.1.1.9 and 10)

These two enzymes are considered to play an important role in the so-called glucuronic acid xylulose cycle (Hollmann, 1959a and b, 1964). The NADP-linked enzyme catalyses the oxidation of L-xylulose to xylitol:

$$\text{L-xylulose} + \text{NADPH} + \text{H}^+ \rightleftharpoons \text{Xylitol} + \text{NAD}$$

and the NAD-linked enzyme thereupon converts the product back to D-xylulose:

$$\text{xylitol} + \text{NAD} \rightleftharpoons \text{D-xylulose} + \text{NADH} + \text{H}^+$$

Both enzymes have been reported to be bound to mitochondria and unfixed cryostat sections were used by Stiller and Gorski (1965) in a histochemical method employing either NAD or NADP as coenzyme.

Details are given in Appendix 21, p. 1345.

D and L-Sorbitol Dehydrogenases (E.C. 1.1.1.14 and 15)

These *soluble* NAD-linked enzymes catalyse the reaction:

$$\text{D or L-Sorbitol} + \text{NAD} \rightleftharpoons \text{Fructose} + \text{NADH} + \text{H}^+$$

The same enzymes catalyse the reduction of other polyols and, in general, resemble other polyol dehydrogenases acting on pentitol, xylitol, arabitol, iditol, galactitol and inositol (McCorkindale and Edson, 1954). High concentrations of sorbitol have been found in semen and spermatozoa (Blakley, 1951; Williams-Ashman *et al.*, 1957; King and Mann, 1959), and sorbitol dehydrogenase is particularly active in sperm and in prostate and seminal vesicles.

Early histochemical studies on this enzyme (Zimmermann and Pearse, 1959) resulted in the description of "nothing dehydrogenase" (Chapter 20, p. 910) but were otherwise unsuccessful. Cohen (1961) using fresh cryostat sections briefly post-fixed in formalin was able to demonstrate xylitol and sorbitol dehydrogenases. Although the final pathway is, as usual, *via* the diaphorase Cohen considered that his method probably gave accurate localization of a single enzyme acting on both substrates.

Using a high tonicity fixative containing formalin, ethanol and Tris at pH 7·0, followed by a high tonicity incubating medium containing D-sorbitol, NAD, EDTA, KCN and Tris buffer at pH 8·8, Johnson (1965) also succeeded in localizing the enzyme. Her reported localizations differed, however, to some extent from those of Cohen. In a later report the same author (Johnson, 1967) described technical improvements, carried out by omitting cyanide and including PMS, which resulted in a considerable decrease of false reactions. It was suggested that even at low NAD levels NADH was produced too rapidly by the dehydrogenase for efficient trapping by tissue-bound NADH diaphorase, hence the improvement manifest with PMS.

A method for sorbitol dehydrogenase was reported by Stiller and Gorski (1966). These authors used unfixed cryostat sections and a medium containing sodium azide, at pH 7·2 in order to avoid "nothing dehydrogenase" effects. These are evidently much reduced after fixation and the method of Johnson (1967) is therefore to be preferred. Details are given in Appendix 21, p. 1345.

L-Gulonate Dehydrogenase (E.C. 1.1.1.19)

The first stage in the biosynthesis of L-ascorbic acid from D-glucuronate is its reduction to L-gulonate by the *soluble* NADP-linked glucuronate reductase:

$$\text{D-glucuronate} + \text{NADPH} + \text{H}^+ \rightleftharpoons \text{L-gulonate} + \text{NADP}$$

The reversibility of this equation allowed Balogh (1965) to develop a histochemical method coupled to tetrazolium reduction. A more convenient name for the enzyme is L-gulonate dehydrogenase rather than glucuronate reductase.

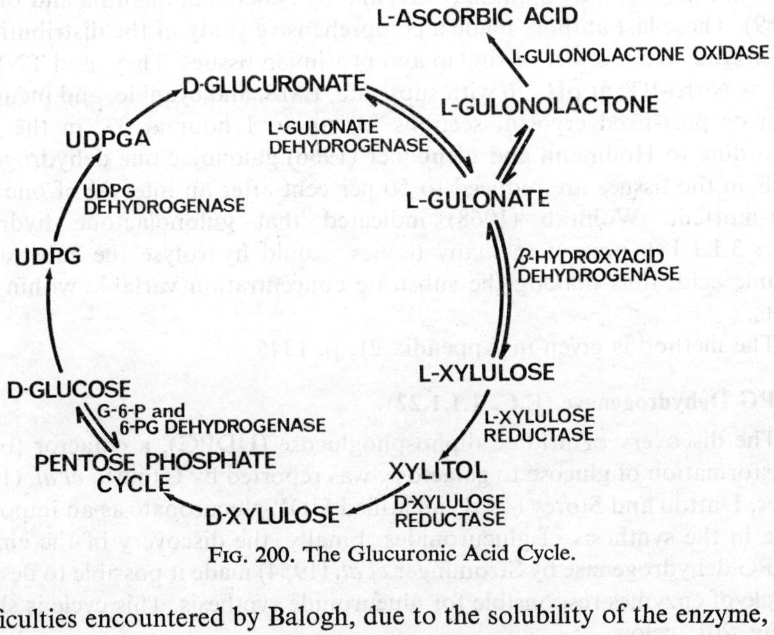

FIG. 200. The Glucuronic Acid Cycle.

Difficulties encountered by Balogh, due to the solubility of the enzyme, were partly overcome by the addition of PVP and EDTA to the medium, the composition of which is given in the appendix.

In Figure 200, above, are set forth the various stages of the glucuronic acid cycle and the enzymes concerned with these stages (Grollman and Lehninger, 1957; York *et al.*, 1961; Ashwell *et al.*, 1961).

In addition to the enzyme described in this section, histochemical methods exist for several, if not most, of the other enzymes of the cycle. These are UDPG and β-hydroxy acid dehydrogenases, the xylulose reductases, the two pentose cycle enzymes, and gulonolactone oxidase (glucuronolactone reductase).

L-Gulono-γ-lactone oxidase (E.C. 1.1.1.20)

This enzyme, as can be seen from Fig. 200, lies on the final pathway for the synthesis of ascorbic acid. It is a thiol-containing flavoprotein, inactivated by organic mercurials and other SH inhibitors (Chatterjee *et al.*, 1960; 1961; Bublitz, 1961; Caputto *et al.*, 1961). The enzyme is lacking in guinea-pig, monkeys and man but is present in the majority of mammals. In plant tissues (Mapson and Breslow, 1958) it is tightly bound in the mitochondria and when solubilized it reacts only with cytochrome c or with PMS.

$$\text{L-gulono-}\gamma\text{-lactone} + \text{NADP} \rightleftharpoons \text{D-glucurono-}\gamma\text{-lactone} + \text{NADPH} + \text{H}^+$$

It was first demonstrated histochemically by Cohen (1961) who used thick fresh sections of rat and frog tissues, incubated at pH 7·4 with L-gulonolactone, Nitro-BT and menadione. Further studies of the enzyme were made by Wohlrab (1968), by Shnitka (1969) and by Nakajima, Shantha and Bourne (1969). These last authors made a comprehensive study of the distribution of the enzyme in a number of simian and prosimian tissues. They used TNBT as well as Nitro-BT, at pH 7·6 with substrate, PMS, and cyanide, and incubated fresh or post-fixed cryostat sections for up to 1 hour at 37° in the dark. According to Holimann and Neubauer (1966) gulonolactone dehydrogenase levels in the tissues are reduced to 50 per cent after an interval of one hour post-mortem. Wohlrab (1968) indicated that gulonolactone hydrolase (E.C. 3.1.1.18), present in many tissues, would hydrolyse the substrate to gulonic acid, thus making the substrate concentration variable within wide limits.

The method is given in Appendix 21, p. 1346.

UDPG Dehydrogenase (E.C. 1.1.1.22)

The discovery of uridine diphosphoglucose (UDPG), a cofactor for the transformation of glucose to galactose, was reported by Caputto *et al.* (1950). Later, Dutton and Storey (1951) identified UDP-glucuronate as an important stage in the synthesis of glucuronides. Finally, the discovery of the enzyme UDPG dehydrogenase by Strominger *et al.* (1954) made it possible to describe a cycle of enzymes responsible for glucuronide synthesis. This cycle is shown in Fig. 201, below.

The *soluble* enzyme with which we are concerned is thus responsible for the irreversible two-step dehydrogenation of UDPG at the carbon-6 of the glucose.

$$\text{UDP glucose} + 2\text{NAD} + H_2O \rightarrow \text{UDP glucuronate} + 2\text{NADH} + 2H^+$$

Being a sulphydryl enzyme it is inhibited by the usual range of inhibitors.

The possibility exists that, as with some other coenzyme-linked dehydrogenases, a metal such as Zn^{2+} is also concerned. The histochemical reaction was described by Balogh & Cohen (1961), and also by Hess and Pearse (1961). The first authors used thick fresh sections, and Nitro-BT at pH 8·4 while the second authors employed thin sections, with MTT and menadione, at pH 7·4 in order to avoid the pronounced "nothing dehydrogenase" effect. Balogh and Cohen found high enzyme activity in cartilage while Hess and Pearse found activity only in liver where the former authors could not overcome non-specific

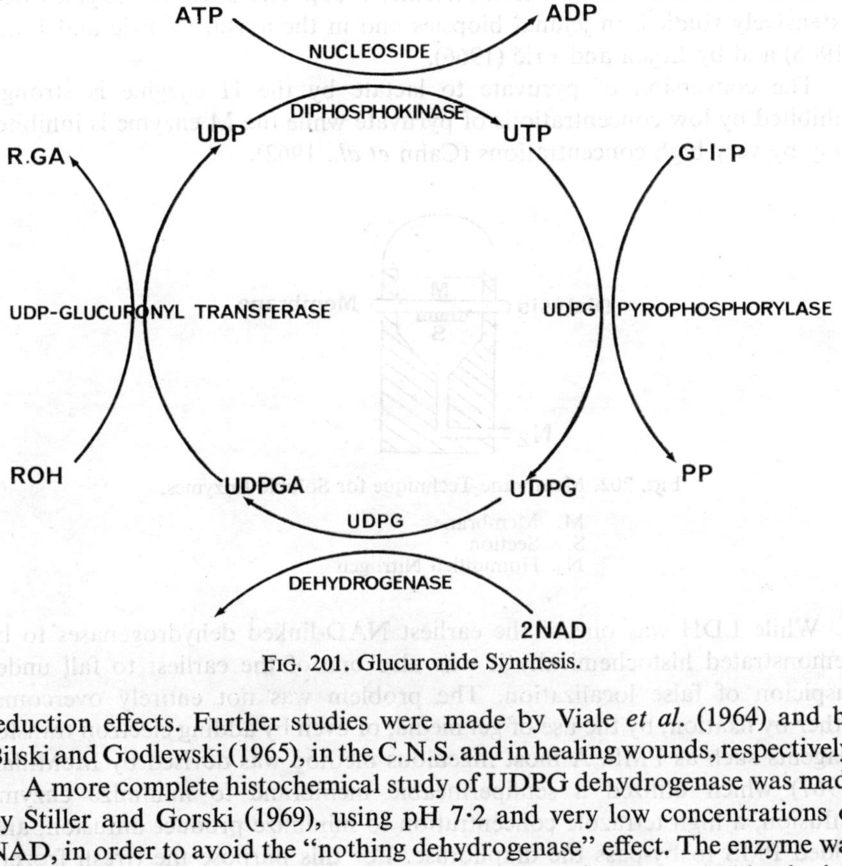

FIG. 201. Glucuronide Synthesis.

reduction effects. Further studies were made by Viale *et al.* (1964) and by Bilski and Godlewski (1965), in the C.N.S. and in healing wounds, respectively.

A more complete histochemical study of UDPG dehydrogenase was made by Stiller and Gorski (1969), using pH 7·2 and very low concentrations of NAD, in order to avoid the "nothing dehydrogenase" effect. The enzyme was found to be widely distributed but it was emphasized that only histological localizations could be provided. The technique is described in the Appendix (p. 1346).

L-Lactate Dehydrogenase (E.C. 1.1.1.27)

This ubiquitous *soluble* enzyme is part of the glycolytic system and catalyses the reaction.

$$\text{Lactate} + \text{NAD} \rightleftharpoons \text{Pyruvate} + \text{NADH} + \text{H}^+$$

As first indicated by Markert and Møller (1959) lactate dehydrogenase exists in multiple molecular forms, described as isoenzymes or isozymes (Plagemann *et al.*, 1960a and b). The various forms are made up of four subunits of two parent molecules which are designated H and M. The pure H type of LDH, occurring mainly in heart muscle, is composed of 4 H subunits (H_4) while the pure M type from skeletal muscle consists of 4 M subunits (M_4). Three molecular hybrids also exist (H_3M; H_2M_2; and HM_3). Thus five bands of LDH can be distinguished electrophoretically and these are numbered 1–5, from the fastest anodal fraction (Wieme, 1959). The LDH isoenzymes were extensively studied, in jejunal biopsies and in the aorta, by Frič and Lojda (1966) and by Lojda and Frič (1966).

The conversion of pyruvate to lactate by the H enzyme is strongly inhibited by low concentrations of pyruvate while the M enzyme is inhibited only by very high concentrations (Cahn *et al.*, 1962).

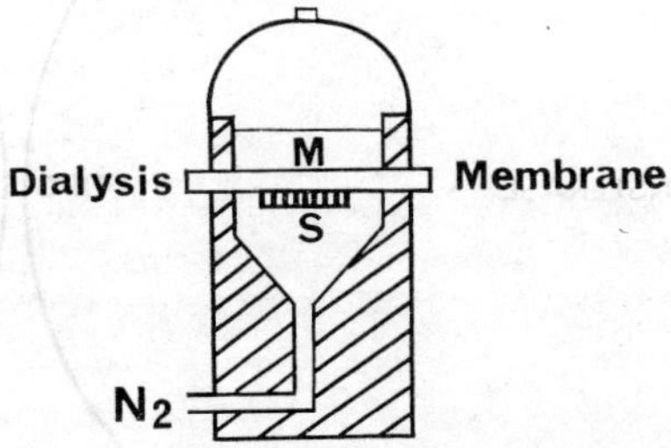

FIG. 202. Membrane Technique for Soluble Enzymes.

M. Membrane
S. Section
N_2. Humidified Nitrogen

While LDH was one of the earliest NAD-linked dehydrogenases to be demonstrated histochemically it was also one of the earliest to fall under suspicion of false localization. The problem was not entirely overcome, either by fixation, by the use of gel media, or even by adding electron transfer reagents such as PMS. A most ingenious method was devised by McMillan (1967) which utilized a semipermeable membrane to minimize enzyme diffusion, a high tetrazole concentration to minimize product diffusion, and added PMS to by-pass the diaphorase. For this purpose the (fresh frozen, cryostat) section was supported on the under surface of a dialysis membrane (Visking dialysis tubing; pore diameter 480 nm, thickness 30 μm) which formed the floor of a reaction chamber. Provision was made for the passage of humidified nitrogen through the system in order to maintain anaerobic conditions. A diagrammatic representation of McMillan's device is shown above in Fig. 202. In a footnote to his paper it is indicated that sharper localizations are obtainable by the use of freeze-dried sections.

Using the above principle McMillan was able to separate the units of the renal parenchyma in terms of H and M isozymes. He also divided the muscle fibres of rat gastrocnemius on the basis of LDH subunits, showing that the red (Type I) fibres contained the pyruvate sensitive H-isozyme which could withstand concentrations of urea (4M) which inactivated the pyruvate insensitive M-isoenzyme. The LDH isozymes of skeletal muscle had previously been demonstrated, in a similar fashion, by Brody and Engel (1964) who showed, in agreement with the combined histochemical and electrophoretic studies of van Wijhe *et al.* (1964), that isozymes I to V were progressively more sensitive to inactivation by urea. Further work on these lines by Jacobsen (1969) resulted in the development of a PVA technique for *in situ* demonstration of LDH isoenzymes. This technique is given in Appendix 21, p. 1347.

Actions of Phenazine Methosulphate. An important by-product of the investigations on LDH carried out by McMillan (1967) was the demonstration that the effect of cyanide on a histochemical system containing PMS could be duplicated by maintaining strictly anaerobic conditions (as was always done in the early days of dehydrogenase histochemistry). McMillan indicated that the effect of cyanide was clearly to prevent the oxidation of reduced PMS by atmospheric oxygen, catalysed by cytochrome oxidase. He suggested that this conclusion might be applicable to other dehydrogenase reactions.

The implications of the work reported above have not yet been afforded full recognition by the majority of applied dehydrogenase histochemists. It is clear that when these implications are fully recognized, soluble dehydrogenase studies using "conventional" techniques will no longer be made.

3-Hydroxybutyrate Dehydrogenase (E.C. 1.1.1.30)

This reaction is responsible for the reaction:

$$\text{D-3-hydroxybutyrate} + \text{NAD} \rightleftharpoons \text{acetoacetate} + \text{NADH} + \text{H}^+$$

and for the oxidation of other 3-hydroxymonocarboxylic acids. The pH optimum of the forward reaction is 8·1. For the back reaction it is 7·0. It is usually recorded as a bound mitochondrial enzyme (Schäfer and Lamprecht, 1961) but, as noted earlier in this chapter, it appears to be slowly soluble in histochemical systems. Most tissues, in the rat at least, contain very low levels of hydroxybutyrate dehydrogenase, but the enzyme is reversibly inactivated by aerobic conditions (Wise and Lehninger, 1962). It is activated also by thiol compounds and protected from inactivation by NAD and by ATP, *in vitro* but not in the tissues apparently. The standard type of incubation medium (Appendix 21, p. 1342) provides satisfactory localization of the enzyme.

15-Hydroxyprostanoate Dehydrogenase.

This enzyme, which is commonly called prostaglandin dehydrogenase, was first isolated from lung tissue by Änggård and Samuelsson (1966), and further described by these same authors (1967) and by Nakano *et al.* (1969). It acts upon the 15-carbon hydroxyl which is regarded as essential for the biological activity of the prostaglandins. Below is shown the formula of 11, 15-dihydroxy-9-oxaprosta-5, 13-dienoic acid (PGE_2) which is the main endogenous prostaglandin of rabbit kidney medulla (Lee *et al.*, 1967).

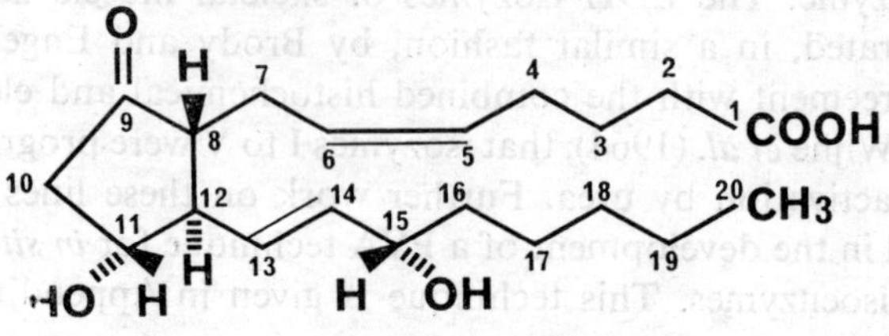

PGE_2

The enzyme is NAD-linked and shows a high degree of specificity for substrates of the prostaglandin series.

A histochemical method for prostaglandin dehydrogenase was evolved by Nissen and Andersen (1968, 1969) using PGE_1 (11, 15-dihydroxy-9-oxoprost-13-enoic acid) or PGE_2 as substrates at very low concentrations (15µM). Interference from "nothing dehydrogenase" was not marked and the most marked activity of the enzyme, in rat kidney, was in the thick ascending limb of the loop of Henle. Intense activity also occurred in the arteries and arterioles of the renal cortex. This observation agrees with the known vascular effects of prostaglandins.

Technical details are given in Appendix 21, p. 1348.

Malate Dehydrogenases (E.C. 1.1.1.37 and 1.1.1.40)

The first of these two enzymes, the NAD-linked dehydrogenase whose substrate is L-malic acid, is histochemically much more important than the second. The latter, originally known as malic enzyme, is NADP-linked and is a decarboxylating enzyme.

The two equations are:

$$\text{L-malate} + \text{NAD} \rightleftharpoons \text{oxaloacetate} + \text{NADH} + \text{H}^+$$

and

$$\text{L-malate} + \text{NADP} \rightleftharpoons \text{pyruvate} + \text{NADPH} + \text{H}^+ + \text{CO}_2$$

NAD-linked malate dehydrogenase is to be regarded as a *soluble* enzyme although biochemically a fraction is found to be bound to the mitochondria. Its specificity for L-malate is not absolute; tartrate, oxaloglycolate, tartronate and α-hydroxyglutarate are also oxidized. The substrate specificity of malic enzyme is higher and the specificity for NADP is absolute. The oxalacetate decarboxylation reaction, which requires Mn^{2+} or Mg^{2+} (10 µM), has its optimum at pH 4·5. Other metal ions are less effective, except Co^{2+} which at 5mM activates twice as much as Mn^{2+}. The two enzymes are to be regarded as *soluble*. They are not therefore commonly studied by applied histochemists. If required, the standard procedures can be used but the resulting localizations must be regarded with reserve.

Isocitrate Dehydrogenases (E.C. 1.1.1.41 and 1.1.1.42)

The reactions catalysed by these two enzymes show that both are decarboxylating.

L_s-isocitrate + NADP(P) ⇌ 2-oxoglutarate + NAD(P)H + H^+ + CO_2

The NADP-linked enzyme is substantially *soluble*, being 80 per cent soluble and 12 per cent mitochondrial bound according to Hogeboom and Schneider (1950). This member of the pair has been regarded as the most important of the two in mammalian tissues, both histochemically and biochemically. Possibly the NAD-linked enzyme should be reinvestigated. Ernster and Navazio (1956) reported that it was a mitochondrial enzyme in rat liver. As with the oxaloacetate decarboxylating reaction of malate dehydrogenase the oxalosuccinate decarboxylation reaction of isocitrate dehydrogenase requires Mg^{2+} or Mn^{2+}.

In plant tissues the NAD-linked isocitric dehydrogenase is more important than the other. It has been shown to be mitochondrial bound. Standard type techniques can be employed for the isocitrate dehydrogenases. The NAD-linked enzyme in mammalian tissues, and the NADP-linked enzyme in plant tissues, which can be regarded as bound enzymes, should be reinvestigated histochemically and the role of metallic ions determined.

Phosphogluconate Dehydrogenase (E.C. 1.1.1.43) and Glucose-6-Phosphate Dehydrogenase (E.C. 1.1.1.49)

These two enzymes, belonging to the glucose-6-phosphate oxidation system (pentose cycle) are well considered together. Both are *soluble* NADP-linked enzymes. The second, G-6-PD, catalyses the first stage of oxidation:

D-glucose-6-phosphate + NADP ⇌
D-glucono-δ-lactone-6-phosphate + NADPH + H^+

There are a considerable number of reports in the histochemical literature describing the localization of G-6-PD, and to a lesser extent of 6-PGD, using standard types of incubation media. For example, Hess and Pearse (1959) found a sharp rise in the enzyme in the macula densa of rat kidney after the application of a Goldblatt clamp to the renal artery. Cohen (1961) and Cohen and Crawford (1962) investigated G-6-PD in adrenal cortex, observing a great increase in sodium deprivation. A complete survey of rat tissues for G-6-PD activity was carried out by Rudolph and Klein (1964). These and many other investigations were carried out in the full knowledge of the soluble nature of the enzyme. In probably the majority of cases the localization afforded was similar to, or identical with, that of the diaphorase (NADPH tetrazolium reductase). In kidney, however, this was not the case and since even after fixation in glutaraldehyde it is possible to obtain localization of G-6-PD in the macula densa, it must be supposed that special conditions exist in this site —possibly bound enzyme.

Methods employing additional electron transfer agents such as menadione or PMS have also been used for pentose cycle enzymes. Hess and Pearse (1961) showed that, if menadione was used in the medium, the soluble quinone reductase of brain tissue could transfer electrons from the substrate (G-6-P) to a tetrazolium salt (MTT). Under such conditions distinction could be made between the localization of the bound NADPH reductase and the specific dehydrogenase. This is clearly shown in Figs. 203 and 204, p. 942.

Future work with the soluble pentose cycle enzymes, using conventional methods, will be obliged to rely on fixation and additional electron transfer. In only a few cases, as with the demonstration of G-6-PD deficiency and mosaicism in red blood cells (Fairbanks and Lampe, 1968), will unfixed cells or tissues be used successfully. Greater accuracy will probably follow the application of the membrane technique (p. 934).

Steroid Dehydrogenases (E.C. 1.1.1.50 and 1.1.1.51)

For the majority of biologically-trained histochemists the complexities of steroid dehydrogenase histochemistry are made worse by difficulties with the nomenclature of steroid compounds. The position has been clarified by the publication of the "Revised Tentative Rules for Nomenclature of Steroids", produced by the joint IUPAC-IUB Commission on Biochemical Nomenclature and published in Biochimica et Biophysica Acta (**164**, 453, 1968) and in the Biochemical Journal (**113**, 5, 1969). The necessarily abbreviated introduction to steroid histochemistry, which appears below, conforms to the Revised List of Rules.

Steroid Nomenclature. Steroids are numbered and lettered as shown in the formula below:

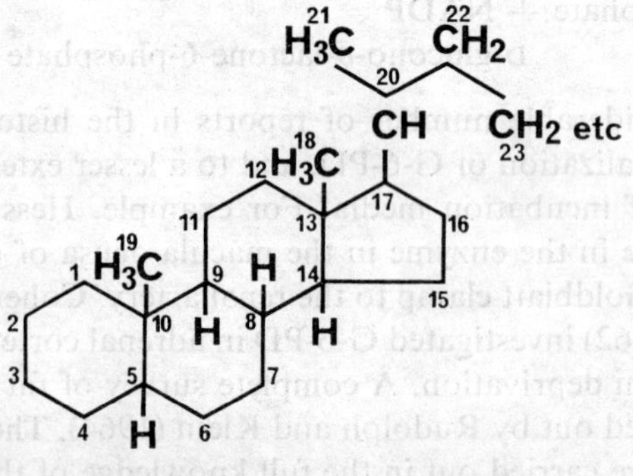

When the rings of a steroid are denoted as projections on to the plane of the paper the formula normally appears as shown below:

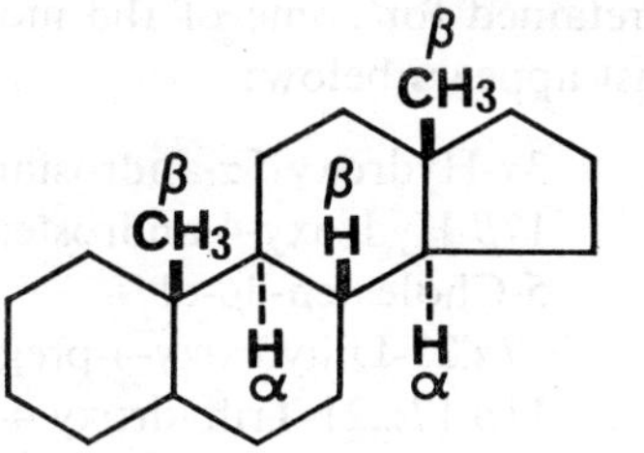

An atom or group attached to a ring depicted as in the above orientation is termed α (alpha) if it lies below the plane of the paper or β (beta) if it lies above the plane of the paper. The dotted lines represent the α configuration and the thick solid lines the β configuration (8β, 9α, 10β, 13β, 14α).

Perspective stereochemical representations are sometimes used and these make much clearer the very important difference between 5α-steroids and 5β-steroids.

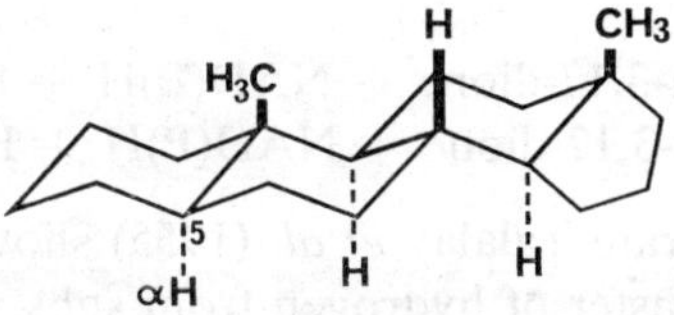

A 5α-Steroid

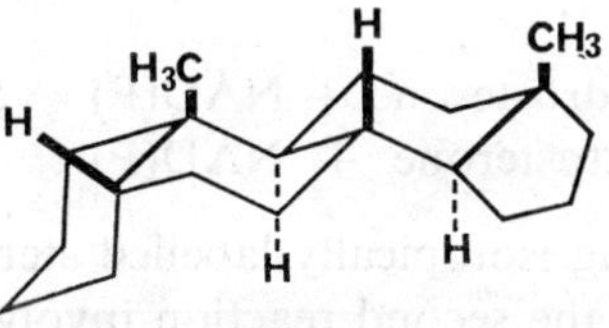

A 5β-Steroid

The parent tetracyclic hydrocarbon without methyl groups at C-10 and C-13, and without a side chain at C-17 is called *gonane*. The hydrocarbon with a methyl group at C-13 but without a methyl group at C-10 or a side chain at C-17 is called *estrane*. Hydrocarbons with methyl groups at both C-10 and C-13 and a side chain at C-17 are named for the latter (R), prefixed by an indication of the configuration at C-5 (α or β).

$R = H$	Androstane
$R = C_2H_5$	Pregnane
$R = CH(CH_3)CH_2CH_2CH_3$	Cholane
$R = CH(CH_3)CH_2CH_2CH_2(CH_3)_2$	Cholestane

Unsaturation and substituents are noted by the usual methods of organic chemistry, unsaturation being indicated by changing ane to ene.

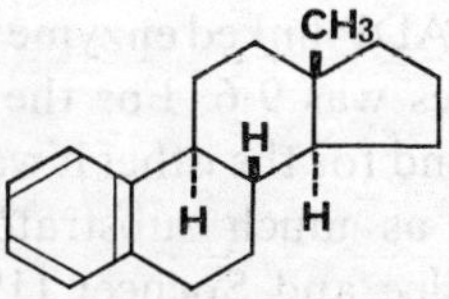

1,3,5(10)-Estratriene

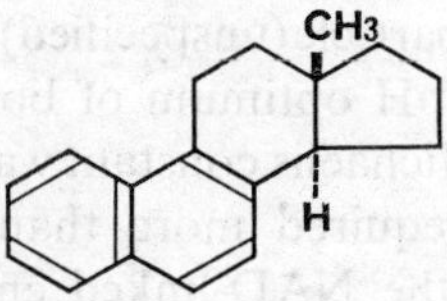

1,3,5(10),6,8-Estrapentene

Trivial names are retained for some of the most important steroids, of which an incomplete list appears below:

Androsterone	3α-Hydroxy-5α-androstan-17-one
Testosterone	17β-Hydroxy-4-androsten-3-one
Cholesterol	5-Cholesten-3β-ol
Cortisone	17α,21-Dihydroxy-4-pregnene-3,11,20-trione
Cortisol	11β,17α,21-Trihydroxy-4-pregnene-3,20-dione
Estrone	3-Hydroxy-1,3,5(10-estratrien-17-one)
Estradiol-17β	1,3,5(10)-Estratriene-3,17β-diol
DHA (dehydroepi-androsterone).	3β-hydroxy-5-androsten-17-one
Progesterone	4-Pregnene-3,20-dione

The two steroid dehydrogenases listed by the Enzyme Commission have broad substrate specificities and act on a wide range of 3α, 3β and 17β hydroxysteroids. NAD or NADP are essential cofactors, in reactions of the type shown below.

$$\text{Androsterone} + \text{NAD(P)} \rightleftharpoons 5\alpha\text{-androstan-3,17-dione} + \text{NAD(P)H} + \text{H}^+$$
$$\text{Testosterone} + \text{NAD(P)} \rightleftharpoons \text{4-androstene-3,17-dione} + \text{NAD(P)H} + \text{H}^+$$

Using isotopically labelled steroid as substrate Talalay *et al.* (1955) showed that the second reaction involves direct transfer of hydrogen from substrate to NAD. According to Talalay and Marcus (1956) the substrate constant (Ks) for dehydroepiandrosterone (DHA) with NAD is 11 μM, whereas for testosterone it is 55 μM suggesting that this last compound should be an efficient substrate in histochemical practice.

It was suggested by Hurlock and Talalay (1958) that the steroid dehydrogenases might function as pyridine nucleotide transhydrogenases and they considered them as hydrogen carriers, or coenzymes, capable of undergoing alternate oxidation and reduction by transfer of H between NAD and NADP. Factors which favour this hypothesis are the low substrate constants for both oxidized and reduced forms of the steroid and the ready reversibility of the reaction under suitable conditions. Further evidence for the participation of steroid hormones in the enzymatic transfer of hydrogen was presented by Talalay and Williams-Ashman (1960) and Baron *et al.* (1963) ascribed this activity to 3α-hydroxy-steroid dehydrogenase.

Endahl and his collaborators (1960) separated two 17β-hydroxysteroid dehydrogenases from guinea-pig liver. The NAD-linked enzyme was associated with a particle (unspecified) but the NADP-linked enzyme was completely soluble. The pH optimum of both enzymes was 9·6. For the NADP-linked enzyme the Michaelis constant was 60 μM; and for the other it was 8·3 μM. Thus the former required more than 4 times as much substrate, for maximal response, as the NAD-linked enzyme. Villee and Spencer (1960) separated the two enzymes in both liver and kidney. The NAD and NADP-linked 17β-hydroxysteroid dehydrogenases of dog prostate were investigated by Hussein and Kochakian (1968) who found that both were microsomal enzymes with pH maxima above 10.

The 3-β-ol dehydrogenase was first described by Samuels *et al.* (1951), and further characterized by Samuels and Helmreich (1956). Rat liver 3β-hydroxysteroid dehydrogenase was investigated by Rubin and Strecker (1961) who found that they could separate particulate-bound and soluble activities, which had different ratios in the two sexes.

Steroid metabolism in human proliferative endometrium *in vitro* was studied by Sweat *et al.* (1967), and by Bryson and Sweat (1967), who found high levels of conversion of estradiol-17β to estrone but no side chain cleavage of 17α-hydroxyprogesterone. This work was extended by Collins *et al.* (1969) who developed a computerized technique for the quantitative determination of radiometabolites from incubation studies. Seven steroid substrates were incubated with human endometrial tissue for 2 hours without cofactors and for 4 hours with cofactors (ATP, NAD, NADP, NADH, NADPH). All the substrates were extensively metabolized to steroids containing the same number of carbon atoms. Most transformations were to the reduced forms. For example, progesterone was transformed into 5α-pregnanedione and 20 α-dihydroprogesterone, the latter indicating 20α-hydrogenase activity. A relatively high proportion of dehydroepiandrosterone (DHA) was converted into androstenedione (3β-hydroxy dehydrogenase). Estrone and estradiol-17β were interconvertible, the reaction favouring the formation of estrone (17β-hydroxy dehydrogenase).

1,3,5(10)-Estratriene-3,17β-diol $\rightleftharpoons$ 3-Hydroxy-1,3,5(10)-estratrien-17-one

The above findings, and those of similar biochemical studies on tissues other than endometrium, must be used as yardsticks for evaluation of the results of histochemical investigations using different substrates.

3β-ol Steroid Dehydrogenase. Since the original description of a histochemical technique for 3β-SDH by Wattenberg (1958) a large number of papers have been published on the localization of this, and many other, steroid dehydrogenases. The 3β-ol enzyme was studied *inter alia* by Levy, Deane and Rubin (1959), Allen (1960), Dawson *et al.* (1961), Cavallero and Chiappino (1962), Lobel, Deane and Romney (1962), Pearson *et al.* (1964), Baillie and Griffith (1964), Goldberg *et al.* (1964), Tapia Freses *et al.* (1965) and Bergman *et al.* (1966).

The majority of workers followed the original technique using dehydroepiandrosterone (DHA) as substrate, dissolved in acetone. The usual alternative substrate was 5-pregnenolone. Some workers used propylene glycol in place of acetone as an alternative solvent but investigations carried out by Hardonk (1965) showed clearly that propylene glycol and isopropanol could act as substrates for a "secondary alcohol dehydrogenase" present in Leydig cells of the testis, theca cells of the ovary, and the zona reticularis of the adrenal cortex. A later study by Burkl and Slezak-Klemencic (1967) indicated that in some phases of the oestrus cycle in the rat ovary there was almost a reciprocal relationship between the levels of 3-β-ol steroid dehydrogenase and "secondary alcohol dehydrogenase". The use of dimethyl formamide in place of both acetone and propylene glycol is to be recommended.

Goldberg *et al.* (1964) carried out an extensive trial of substrate specificity for the 3β-SDH of human ovary and testis. They found seven additional 3β-ol steroids suitably active as histochemical substrates. The presence of a double bond in the A ring (4) was found to confer greater activity than a double bond in ring B (5) but the presence of double bond(s) was not necessary for activity and fully saturated compounds were oxidized (e.g. androstane-3β-ol,17α-one). A most important finding was that a suitable steroid substrate for one tissue could inhibit the dehydrogenase reaction in other tissues. DHA, for instance, was a good substrate for corpus luteum and ovary but an inhibitor for testis.

17β-ol Steroid Dehydrogenase. Soluble and insoluble, NAD and NADP-linked 17β-SDH enzymes were described in mammalian liver and kidney by Villee and Spencer (1960) and by Aoshima and Kochakian (1963). In porcine ovary Bjersing and Carstensen (1964, 1967) described strong activity of the enzyme and the work of Sweat *et al.* (1967) and of Collins *et al.* (1969), already quoted, refers to a similar activity in endometrium.

The first histochemical demonstration of 17β-SDH was by Pearson and Grose (1959) followed by Kellogg and Glenner (1960) who described the localization of the enzyme in human placenta. The same organ was studied by Koide and Mitsudo (1965). In rat ovary the 17β-ol enzyme activity is usually much weaker than that of the 3β-ol dehydrogenase, even when superovulation has been induced and ovarian growth stimulated by treatment with FSH (pregnant mares serum gonadotrophin) and ICSH (human chorionic gonadotrophin). The strong activity of 3β-ol steroid dehydrogenase in the very large corpora lutea which result is illustrated in Fig. 205, opposite.

A method suitable for steroid dehydrogenases, especially 3β-ol and 17β-ol is given in Appendix 21, p. 1348.

20 α-hydroxysteroid dehydrogenase. Following the description of an enzyme in rat ovary responsible for the conversion of progesterone to 4-pregnen-20α-ol-3-one by Wiest (1959), Balogh (1964) described its histochemical localization using the reverse reaction which is NADP-linked. Differences in the levels of activity in different corpora lutea did not reflect differences, which were absent, in the levels of NADPH diaphorase, the penultimate electron acceptor.

3 α-hydroxysteroid dehydrogenase. This *soluble* enzyme was identified in a number of different tissues by Marcus and Talalay (1956) and by Hamm, Kochakian and Carroll (1956). Further investigations (Rubin, 1957; Hurlock

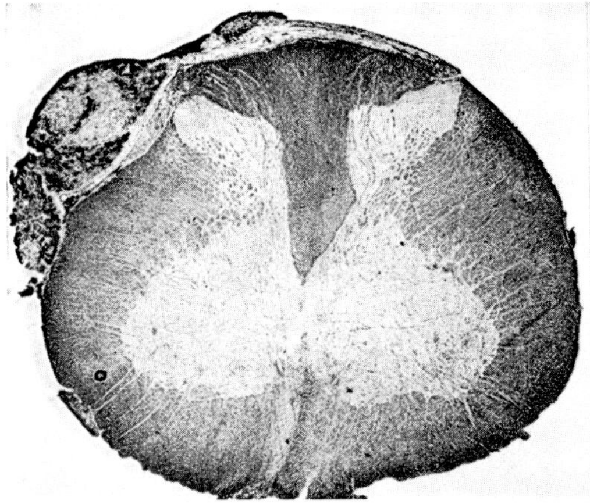

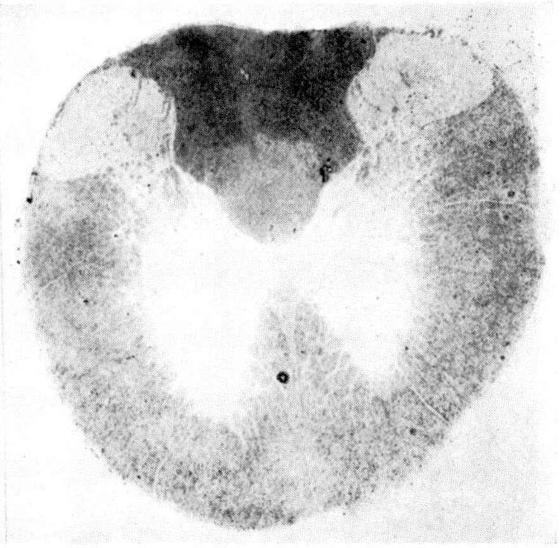

FIG. 203.

FIG. 204.

FIG. 203. Rabbit cervical cord. Shows high NADH-menadione reductase activity in white matter and spinal ganglion (upper left). MTT-Cobalt, × 14.

FIG. 204. Rabbit cervical cord. High G-6-P dehydrogenase restricted to the posterior columns. MTT-Cobalt, × 14.

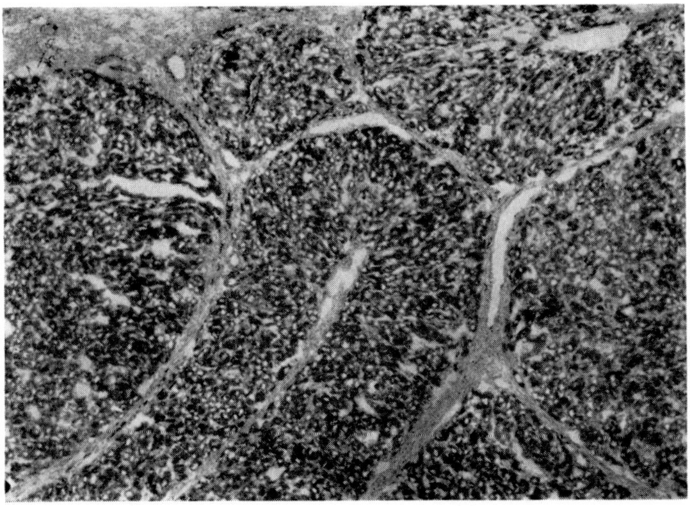

FIG. 205. Rat ovary, corpus luteum. Cryostat section (5 μ). Shows very strong reaction for 17 β-ol steroid dehydrogenase. MTT-Cobalt, × 86.

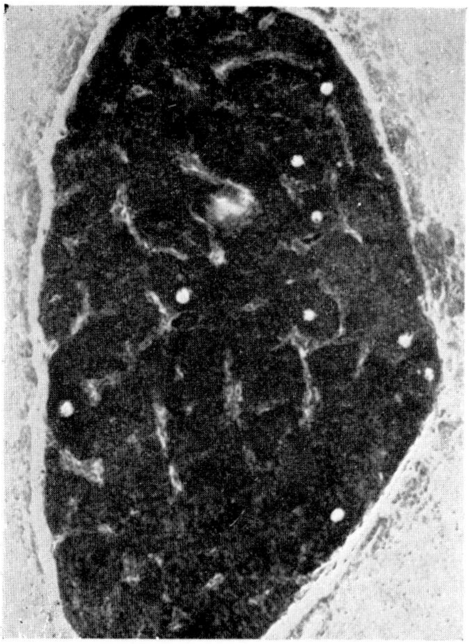

FIG. 207. Rat parathyroid gland. Embedded in liver and cut on cold microtome. Distribution of leucine aminopeptidase using L-leucyl-β-naphthylamide and Fast blue B salt. (Postchelated with copper.) Note diffusion at edge of gland. × 120.

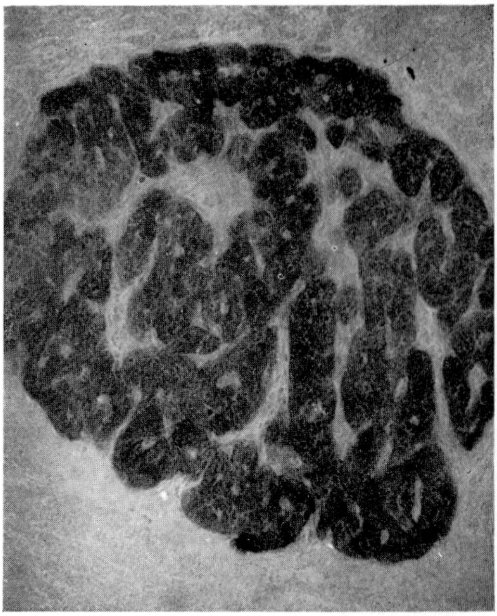

FIG. 208. As Fig. 207. Distribution of leucine aminopeptidase using L-leucyl-4-methoxy-β-naphthylamide and Fast blue B salt. (Postchelated with copper.) Note absence of diffusion. × 120.

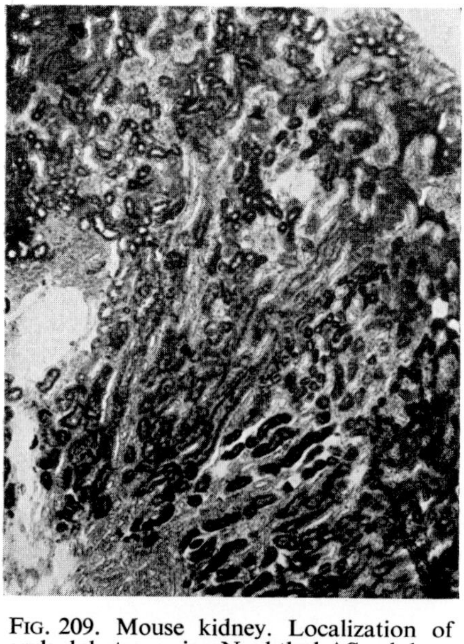

FIG. 209. Mouse kidney. Localization of arylsulphatase using Naphthol AS sulphate and Fast red TR salt. × 47.

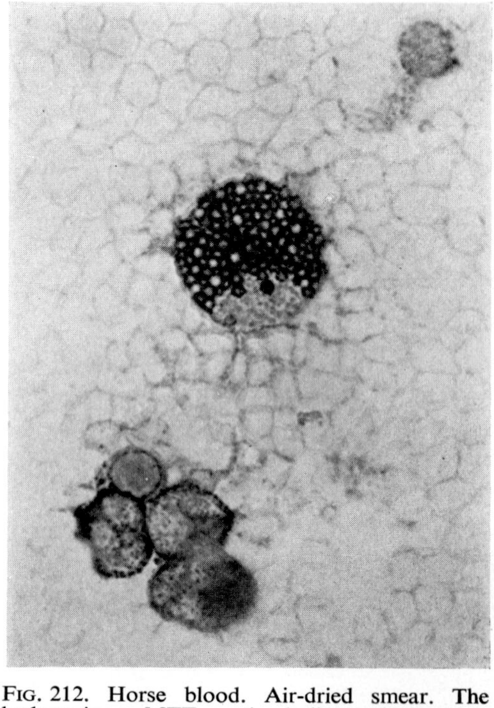

FIG. 212. Horse blood. Air-dried smear. The hydroquinone-MTT reaction indicates the presence of redox compounds (UQ or tocopherols) in neutrophil granules (below, left) and in the outer rims of the granules in an eusinophil (centre). × 1000.

and Talalay, 1958, 1959) showed that it was particularly important in the intermediary metabolism of C-19 steroids. Particulate and soluble fractions of rat ventral prostate were shown by Gore and Baron (1965) to possess 3α-hydroxysteroid dehydrogenase activity. The enzyme is active with either NAD or NADP as cofactor. Balogh (1966) evolved a histochemical method, using androsterone as substrate and high levels of PVP in the medium, in order to control, or at least diminish, enzyme diffusion.

Other steroid dehydrogenases. Many steroids, other than those specifically mentioned above, have been reported as (successful) substrates for enzymes in tissue sections. In three comprehensive papers Baxter *et al.* (1966), and Baillie, Calman, Ferguson and Hart (1965, 1966), described 6β, 11α, 12α, 16α, 16β, 17α, 20β, 21 and 24 hydroxysteroid dehydrogenases (usually NAD-linked). Some of these were dealt with, before or subsequently, in separate papers. In particular, 11β-hydroxysteroids were shown to be oxidized very strongly by salivary gland ducts (Ferguson, 1967) and many of the other substrates were oxidized by human foetal membranes (Hart, 1966).

While some authorities doubt the specificity of all histochemically demonstrable steroid dehydrogenases, with the exception of the 3β-ol and 17β-ol enzymes, others believe that the considerable degree of agreement between biochemical and histochemical findings favours a more liberal interpretation. If the participation of "secondary alcohol dehydrogenase" is avoided (by using DMF as solvent for the steroid substrate), if the final formazan product notably exceeds the control (nothing dehydrogenase) level, and if the final pattern is distinguishable from that of the related diaphorase, I consider that some confidence can be placed in the specificity of the dehydrogenase concerned. It must be added, however, that the precise localization of soluble steroid dehydrogenases is unlikely to be afforded by standard type incubation procedures.

Without doubt many other specific steroid dehydrogenases will prove to be demonstrable by histochemical techniques. The NAD and NADP-linked microsomal steroid ring dehydrogenase, found in mammalian liver and described by Breuer and Mittermayer (1963), has apparently all the required properties. The preferred substrate equilin is oxidized to equilenin (Mittermayer *et al.*, 1963) and there are two maxima, at pH 5·0 and pH 8·4.

Some recent papers dealing with specific localizations of specific steroid dehydrogenases are listed below:

Genital Ridge: Baillie *et al.*, 1966. *Rat Ovary*: Bratt *et al.*, 1968; Pupkin *et al.*, 1966. *Monkey Ovary*: Dennis and Thomas, 1967. *Testis*: Steinberger *et al.*, 1966. *Adrenal Cortex*: Pearson *et al.*, 1964. *Placenta and Foetal Membranes*: Christie, 1968; Botte *et al.*, 1968; Ferguson and Christie, 1967. *Interrenal Tissue and Stannius Corpuscles*: Chieffi and Botte, 1963. *Endometrium*: Brandau *et al.*, 1969.

α-Glycerophosphate Dehydrogenase (E.C. 1.1.2.1)

This mitochondrial-bound flavoprotein enzyme (α-GPD) is the other half of the α-glycerophosphate shuttle (see p. 929). It oxidises L-glycerol-3-phosphate to dihydroxyacetone phosphate.

L-glycerol-3-phosphate + acceptor ⇌
dihydroxyacetone phosphate + reduced acceptor

According to Singer (1963) the Enzyme Commission full name for the enzyme (L-glycerol-3-phosphate : cytochrome c oxidoreductase) is inaccurate since it does not reduce cytochrome c. Green (1936) found that the enzyme in particulate form reduced methylene blue and Tung *et al.* (1952) found that this activity was retained after extraction of the enzyme with deoxycholate. Ling *et al.* (1957) showed that after solubilization with deoxycholate and trypsin their preparation of the enzyme would transfer electrons to 2,6-dichlorophenol-indophenol (DCIP) and Ringler and Singer (1959), who solubilized the enzyme with phospholipase A, found that it could also reduce PMS and ferricyanide. Later Szarkowska and Drabikowska (1964) found that the same preparation reduced exogenous coenzyme Q-6 and Dawson and Thorne (1969a), using a preparation solubilized with Triton X-100, reduced exogenous Q-10 also.

The kinetics of α-GPD with a variety of electron acceptors, including those mentioned above and also menadione, were investigated by Dawson and Thorne (1969b). They found that in all cases the reaction mechanism involved a free modified enzyme intermediate (E′).

$$E + S1 \rightleftharpoons ES1 \rightleftharpoons E' + P1$$
$$E' + S2 \rightleftharpoons E'S2 \rightleftharpoons E + P2$$

where S1 is α-glycerophosphate and S2 the intermediate acceptor.

The Michaelis constants for (DL) glycerol-3-phosphate (K*m*1) and for the various acceptors (K*m*2), derived by extrapolation from secondary plots from the initial Lineweaver-Burk plots, are shown below in Table 71.

TABLE 71

Kinetic Constants for Various Electron Acceptors

Electron Acceptor	K*m*1 (mM)	K*m*2 (μM)
DCIP	10·0	125
Q-10	6·2	50
Q-6	9·4	110
Menadione	8·6	53

With some intermediate electron acceptors the maximum velocity of the reaction (V) and K*m*1 (glycerophosphate) were independent of the nature of the acceptor but in the case of Q–10 and menadione the constant k for the forward reaction $E'S2 \xrightarrow{k} E + P2$ was sufficiently small to make a significant contribution to the values of *Km*1 and V. The reason for this was not clear but Dawson and Thorne considered that the natural acceptor for electrons from glycerophosphate is probably mitochondrial Q–10. This agrees with the observations of Ringler and Singer (1959) on brain mitochondria where they were able to show competition between succinate and α-glycerophosphate dehydrogenases for a common factor on the respiratory chain. Since menadione is the preferred (obligate) histochemical acceptor the findings of Dawson and Thorne are of great interest and significance for us.

The histochemical demonstration of mitochondrial α-GPD was made possible by the discovery of Wattenberg and Leong (1960) that menadione could transfer electrons from the reduced flavoprotein enzyme to tetrazolium salts. Extensive studies to elucidate the mechanism of electron transfer were made by Hess and Pearse (1961a,b,c) using isolated mitochondria from mammalian and invertebrate tissues of various kinds. These studies were carried out in parallel with histochemical studies on the same tissues. With isolated rat kidney mitochondria a 4 to 5-fold increase in α-GPD activity was induced by 100 μM menadione. This activity was not affected by amytal (5 mM) or by actinomycin A (10 μg/ml.) but 10 μM dicoumarol reduced it to $\frac{1}{3}$ of the control value obtained with substrate alone. The second paper of the series (Hess and Pearse, 1961b) reported studies on mitochondria isolated from the 3 main types of skeletal muscle in the locust *Schistocerca gregaria.* Menadione was found the only effective intermediate acceptor, methylene blue and PMS were equally ineffective The optimum concentration, however, was 20 times that required for the same degree of activation in mammalian mitochondria.

Mitochondrial α-GPD can be demonstrated histochemically in mammalian tissues, with a concentration of menadione in the medium obtainable without the addition of solvents such as acetone and with the standard dehydrogenase technique (Appendix 21, p. 1342).

Choline Dehydrogenase (E.C. 1.1.99.1)

This mitochondrial enzyme was first described by Mann and Quastel (1937) and the whole system responsible for the oxidation of choline to betaine aldehyde, and of the latter to betaine by betaine aldehyde dehydrogenase, was recorded by Mann *et al.* (1938). The reaction we are considering is

$$\text{Choline} + \text{acceptor} \rightleftharpoons \text{betaine aldehyde} + \text{reduced acceptor}$$

The nature of the natural acceptor was for some years uncertain. It is now clear that NAD and NADP are not involved but rather some component of the respiratory chain which is considered to be one of the cytochromes. Two artificial acceptors were found by Rendina and Singer (1959) who considered PMS to be better than its only competitor DCIP. The organization of the respiratory chain in choline dehydrogenase was further investigated by Kimura *et al.* (1960) and by Packer *et al.* (1960).

Very little histochemical work has been done on this enzyme which was first demonstrated by Farber *et al.* (1958), using PMS as intermediate acceptor. Chang and Hori (1961) reported its demonstration in cryostat sections post-fixed by freeze-substitution in acetone at −70°. Only phosphate buffer was found to be effective and no formazan was produced in the absence of PMS. On the contrary, no cofactor or intermediate was found necessary by Glogner and Gössner (1962) or by Guha and Wegmann (1963) who used very similar incubating media containing Nitro-BT in phosphate buffer, at pH 7·5 and 6·9 respectively.

The essentially similar 2-hydroglutarate dehydrogenase (E.C. 1.1.99.2), which can use pyocyanine and PMS as acceptors, should be demonstrable histochemically by a standard technique.

Aldehyde Dehydrogenase (E.C. 1.2.1.3)

This widely distributed, soluble, mitochondrial enzyme has a very low substrate specificity. It was first described by Racker (1949) and later (1955) the same author noted the relevant facts in a comprehensive review. The enzyme oxidizes many aldehydes, including formaldehyde, acetaldehyde, glycolaldehyde, propionaldehyde, butyraldehyde, isovaleraldehyde, and a variety of aromatic aldehydes. Recently (Packer and Greville, 1969) it has been reported that glutaraldehyde in low concentration (2 mM) can also act as a substrate. At higher concentrations (1 per cent and above) it is a very efficient fixative which traps the configurational states of mitochondria (Deamer *et al.*, 1967).

The enzyme, which has an absolute requirement for NAD, catalyses the following reaction:

$$\text{Aldehyde} + \text{NAD} + H_2O \rightleftharpoons \text{Acid} + \text{NADH} + H^+$$

Distinguishing features of aldehyde dehydrogenase are hard to find. It is inhibited by the usual SH reagents, by low concentrations of arsenite (50 μM), and by its own substrates at relatively low concentrations. This last effect has been ascribed to blocking of an essential SH group (Nirenberg and Jacoby, 1960). An essential requirement for the intact enzyme complex in isolated mitochondria is Mg^{2+} and total inhibition is produced by rotenone, amytal, antimycin A and cyanide.

Histochemically aldehyde dehydrogenase has not attracted much attention. Two papers by Gabler (Bertolini *et al.*, 1966; Gabler, 1969) describe the localization of the enzyme in liver, kidney, cornea, salivary gland ducts, intestine and testis. The method described in the first paper used acetaldehyde and NAD with Nitro-BT and Mg^{2+} in phosphate buffer at pH 7·0. The authors noted that Tris, through its amino groups, formed azomethines (Vol. 1, p. 160) with the carbonyl groups of the substrate. In the second paper a new medium was employed, containing benzaldehyde as substrate and 10 per cent of propylene glycol. It is to be feared that the improved results can largely be ascribed to secondary alcohol dehydrogenase (see p. 941) but possibly the substitution of DMF for propylene glycol would still keep the substrate in solution and avoid the above complication. Gabler's medium is given in Appendix 21, p. 1349, modified as suggested.

Betaine Aldehyde Dehydrogenase (E.C. 1.2.1.8)

As already described, the oxidation of choline takes place in two stages, the second catalysed by this NAD-linked, mitochondrial enzyme. A method for its demonstration was described by Wohlrab (1965), using Nitro-BT, cyanide or azide, NAD and substrate (20 mM), under anaerobic conditions, in Tris buffer at pH 7·2. The full details of the medium are given in Appendix 21.

Glyceraldehyde Phosphate Dehydrogenase (E.C. 1.2.1.12)

This *soluble* enzyme is capable of carrying out a number of different reactions. Usually it catalyses the oxidation of glyceraldehyde-3-phosphate in a reaction which is coupled with phosphorylation:

$$\text{D-glyceraldehyde-3-phosphate} + \text{orthophosphate} + \text{NAD} \rightarrow \text{D-1, 3-diphosphoglyceric acid} + \text{NADH} + \text{H}^+$$

The affinity of the enzyme for NAD is high (see p. 909) and the usual crystalline enzyme contains two molecules of the coenzyme. Substrate affinity is also high and the need for phosphate absolute. As a sulphydryl enzyme GAPD is inhibited by iodoacetate and other SH inhibitors and also by heavy metals.

Histochemically this enzyme was first demonstrated by Himmelhoch and Karnovsky (1961) using substrate prepared from the monobarium salt of glyceraldehyde-3-phosphate diethylacetal. Phosphate buffer at pH 7·2 was employed with EDTA as activator. A similar technique was used by Fasske and Steins (1963) who studied fresh cryostat sections, sometimes briefly postfixed in acetone, from 45 different tissues. They prepared their substrate by hydrolysis with Dowex-50 resin, followed by distillation of the free alcohol.

Mention must be made of the use of an immunofluorescence technique for localizing GAPD by Emmart and his colleagues (Emmart *et al.*, 1962, 1963a and b). These papers describe the localization in kidney, invertebrate muscle and in mammalian myoblast cultures from chicken, mouse and man.

The original method is described in Appendix 21, p. 1350.

Dihydro-orotate Dehydrogenase (E.C. 1.3.3.1)

This flavoprotein enzyme, which in mammalian tissues is bound to the mitochondria, catalyses the oxidation of dihydro-orotic acid to orotic acid:

$$\text{4,5-L-dihydro-orotate} + O_2 \rightarrow \text{Orotate} + H_2O_2$$

The enzyme is unusual in containing 1 molecule of FAD and one of FMN, each associated with an Fe atom. It can react with NAD and therefore can act as a DHO oxidase, as an NAD-linked DHO dehydrogenase, or as an NADH oxidase (Dixon and Webb, 1964). We are concerned with its function as a DHO oxidase, in mammalian tissues, and not with the DHO dehydrogenase extracted from *Zymobacterium oroticum* and crystallized by Friedmann and Vennesland (1960). Mammalian DHO dehydrogenase was investigated in rat liver homogenates by Wu and Wilson (1956) and in human blood cells by Smith and Baker (1959, 1960). The enzyme is able to transfer electrons to methylene blue (Graves and Vennesland, 1957) and to DCIP, PMS, ferricyanide and tetrazolium salts, in the absence of oxygen (Miller *et al.*, 1968). According to these last authors DHO dehydrogenase is a complex containing non-heme iron, flavin, ubiquinone and lipids. The purified complex, free from cytochrome components, was incapable of reducing O_2. Ubiquinone-30 (350 μM), and 1,4-benzoquinone (170 μM), both greatly stimulated the reaction of the complex with dyes and it was therefore regarded by Miller *et al.* as dihydro-orotate-ubiquinone reductase. The possible improvement of the histochemical reaction with Q.30 or, perhaps, with 1,4-benzoquinone, should be investigated.

DHO dehydrogenase activity was demonstrated histochemically, in various organs and tissues from the rat, by Cohen (1962), who found that the enzyme did not require NAD and that the reaction was unaffected by the addition of menadione or cyanide. He was obliged to add ureidosuccinic acid to the medium, presumably to inhibit the hydrolysis of the substrate (DHO) by dihydro-orotase which converts it reversibly into ureidosuccinate. Improved localization was obtained by the addition of PVP suggesting, unexpectedly, that the enzyme was partly soluble. In a later paper Jones and Cohen (1963) studied DHO dehydrogenase in livers damaged by viral infection. More recently Castoldi *et al.* (1968) demonstrated the enzyme in human blood and bone marrow cells. These authors found that both NAD and ureidosuccinic acid were essential for the reaction, that menadione had no effect, but that PMS greatly increased formazan production from both MTT and Nitro-BT. Leukaemic cells showed increased enzyme activity, in agreement with biochemical findings (Smith and Baker, 1959).

I consider that DHO dehydrogenase is an important enzyme which should be studied further, if possible with improved methods and biochemical control. For this reason the methods of Cohen (1962) and of Castoldi *et al.* (1968) are both given in the Appendix.

Succinate Dehydrogenase (E.C. 1.3.99.1)

This bound mitochondrial, flavoprotein, enzyme is the most venerable (Thunberg, 1909; Battelli and Stern, 1910), and still perhaps the most important, of all the dehydrogenases with which we are concerned.
It catalyses the reaction:

$$\text{Succinate} + \text{acceptor} \rightleftharpoons \text{fumarate} + \text{reduced acceptor.}$$

The Krebs' Cycle. It was Szent-Gyorgyi who first suggested that succinate and fumarate acted as a carrier system for hydrogen removed in the aerobic oxidation of carbohydrate. Later, he discovered that malate and oxaloacetate could act in the same way. Unfortunately, when the whole scheme was tested with isolated components by Green and his associates the succinate system failed to couple with the NAD-linked malate system.

Resolution of this conundrum came when Krebs discovered the part played by α-ketoglutarate and citrate in the aerobic respiration of muscle and when, finally, he synthesized the whole chain into the cycle which bears his name by the finding that pyruvate and oxaloacetate could react to form a 7-carbon compound from which citrate and α-ketoglutarate could be formed once more.

Krebs' Cycle Enzymes. It can be seen from Fig. 206 that three of the dehydrogenase systems which we can demonstrate are main components of

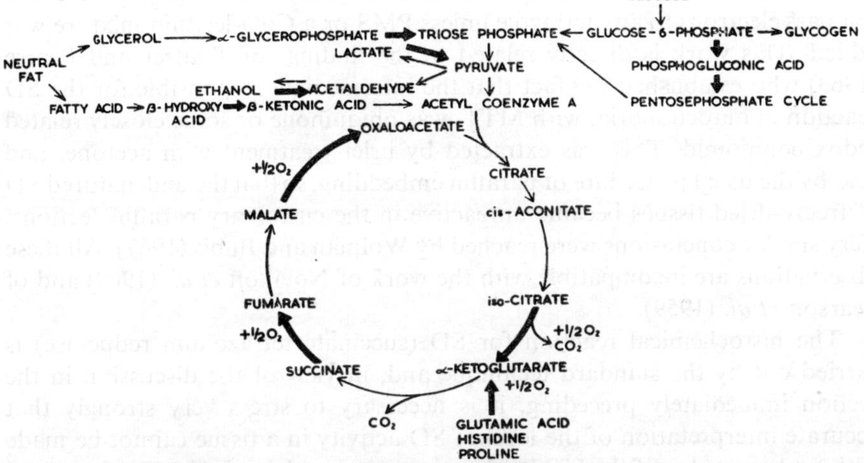

FIG. 206. Krebs' Cycle Dehydrogenases.

the tricarboxylic acid cycle. These are succinate, malate and isocitrate dehydrogenases. The latter is the NADP-linked enzyme which is found, in homogenization studies, mainly in the supernatant. In Fig. 206, above, the dehydrogenases whose activities we can at present demonstrate are shown by broad arrows. Stages which are marked by thin arrows in the Krebs' cycle, the pentose cycle (hexose monophosphate shunt), and in other oxidative pathways, cannot yet be demonstrated histochemically.

The term succinate dehydrogenase should be used to refer only to the flavoprotein part of the succinoxidase complex. Under normal conditions the passage of electrons from succinate to oxygen probably proceeds by way of ubiquinone, cytochrome b and cytochrome c_1 (Slater, 1958; Dixon and Webb, 1964). The enzyme reacts hardly at all with methylene blue or DCIP, and only slowly with ferricyanide. As first shown by Kearney and Singer (1956), it reacts rapidly with PMS in isolated preparations. It must be realised that this compound crosses the intact mitochondrial membrane only to a very limited extent.

The reaction of tetrazoles with the respiratory chain has been discussed at some length in Chapter 20. Some salient points may be restated here. The majority of tetrazolium salts are capable of accepting electrons from several different factors and there are pronounced differences between animal and plant tissues (Gahan and Kalina, 1968). The tetrazoles NT, MTT, INT, NBT and TNBT, are able to accept electrons from the flavoprotein-CoQ region. In plant mitochondria this is the only region involved with all except NT whereas in animal tissues NT, MTT, and INT accept from the cytochrome c. region and TTC, BT and NT from cytochrome oxidase as well.

The supposition that Nitro-BT and TNBT accept electrons directly from the flavoprotein dehydrogenase is not borne out by the investigations of Horwitz *et al.* (1967) who showed that acetone-extracted mitochondria could not pass electrons to any tetrazole unless PMS or a CoQ-lecithin mixture was added. This work is directly related to the findings of Tranzer and Pearse (1963) who established the fact that the intermediate responsible for the SD reaction in mitochondria, with MTT, was ubiquinone or some closely related redox compound. This was extracted by brief treatment with acetone, and also by the usual procedure of paraffin embedding, so that the undenatured SD of freeze-dried tissues became unreactive in the customary paraffin sections. Very similar conclusions were reached by Wolman and Bubis (1967). All these observations are incompatible with the work of Novikoff *et al.* (1960) and of Pearson *et al.* (1959).

The histochemical reaction for SD (succinate tetrazolium reductase) is carried out by the standard technique and, in view of the discussion in the section immediately preceding, it is necessary to stress very strongly that accurate interpretation of the level of SD activity in a tissue cannot be made without knowledge of the UQ level in the mitochondria of the cells concerned (see Chapter 23, p. 1001). The use of succinyl-lecithin as an alternative substrate for SD (Abdulla *et al.*, 1968) has interesting possibilities, particularly if it can act as a vehicle for UQ-30, since it should be capable of rapid penetration through the outer mitochondrial membrane.

Glutamate Dehydrogenases (E.C. 1.4.1.2 and 1.4.1.3)

These two mitochondrial enzymes (GDH), utilizing NAD and NADP respectively as cofactors, are to be regarded as bound, provided that the mitochondria are intact. They catalyse the reaction:

$$\text{L-glutamate} + H_2O + \text{NAD(P)} \rightleftharpoons \text{2-oxoglutarate} + NH_3 + \text{NAD(P)H} + H^+$$

As can be seen, both are deaminating enzymes. Their substrate specificity is not high but glutamate is oxidized far more readily than the next most active substrate (norvaline) (Struck and Sizer, 1960). The pH optimum of the two enzymes is said to be in the region of 8 to 8·5, with glutamate, and as high as 9·8 with leucine as substrate. Using ox liver GDH Engel and Dalziel (1969) found that NAD and NADP were utilized with the same efficiency under all the conditions studied. The rate was observed to be smaller, however, with NADP, especially at pH 7·0 with high substrate and coenzyme concentrations.

These observations are not completely in accord with histochemical practice. Only one serious attempt has been made, to my knowledge, to assess the effect on the two GDH's (as well as 12 other dehydrogenases) of alterations in concentration of substrate, coenzymes, buffer, hydrogen ion and tetrazole. I refer to the work of Braun-Falco and Petzoldt (1965), which also considered the effect of cyanide and of Mg^{2+} ions. These authors found that, provided the substrate concentration was high (100–250 mM) and the coenzyme concentration likewise (NAD 0·9 to 1·2 mM; NADP 0·8 mM) both enzymes gave equally good, strong, reactions in fresh skin sections. "Activation" of GDH by high NAD levels was observed, in Tris buffer, by Frieden (1959) and Engel and Dalziel (1969) found similar "activation" with NAD (up to 2 mM) and also with NADP (up to 1 mM) in both phosphate and Tris buffers.

Braun-Falco and Petzoldt found that cyanide was without effect but Mg^{2+} ions (5 mM) and high buffer concentrations (50 mM) increased the level of formazan production. Finally, they observed that for the full demonstration of NAD-linked GDH a very high tetrazolium concentration was required (1·85 mM for Nitro-BT). This is at least twice as high as that usually employed.

The standard method for dehydrogenases should be modified, for both GDH's therefore, in accordance with the above recorded facts.

L-Aminoacid Dehydrogenase (E.C. 1.4.1.5)

This enzyme, usually called L-aminoacid oxidase, is a mitochondrial flavoprotein containing FMN. It catalyses the following reaction:

$$\text{L-aminoacid} + H_2O + O_2 \rightarrow \text{2-oxoacid} + NH_3 + H_2O_2$$

There have been few histochemical investigations on this enzyme. Wohlrab (1965a and b, 1966) and Reiss (1967) used L-proline as substrate and investigated mammalian kidney and fungi, respectively. Castellano *et al.* (1969) used L-leucine and rat kidney, and described the enzyme which they demonstrated as L-aminoacid tetrazolium reductase (E.C. 1.4.3.2).

Describing a histochemical method for an enzyme also described as L-aminoacid tetrazolium reductase, Diculescu *et al.* (1970) described their finding that the addition of a small amount of L-glutamate to a medium, containing another L-aminoacid as substrate, produced positive results in a number of cases. Specifically, they used L-isoleucine to show activity in liver and L-tyrosine and L-methionine for kidney. For cardiac and skeletal muscle L-phenylalanine give the best results. The level of L-glutamate required was

7·5 mM which can be seen by reference to p. 951 to be below the level required to give any result at all. The reaction is thought to take place by a two stage process involving a transdeamination:

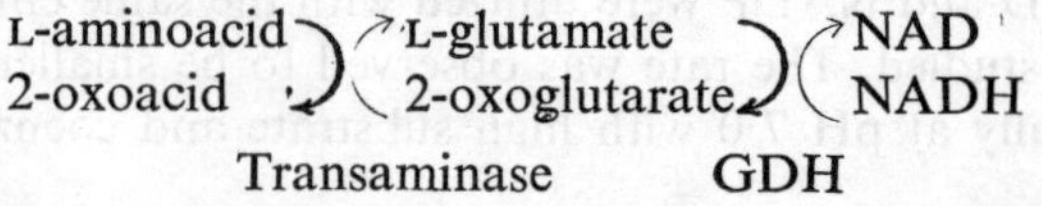

so that as long as there is some 2-oxoglutarate (formed from L-glutamate by GDH) in the medium the conversion of substrate to the corresponding oxoacid continues. This process maintains the conversion of NAD to NADH and thus the final production of formazan. As already mentioned, the enzyme system responsible for the results obtained by Diculescu *et al* was regarded by them as an L-aminoacid tetrazolium reductase but not as L-aminoacid oxidase. If their views on the mechanism of their method are correct it demonstrates the *presence* of the aminotransferase (transaminase) and the localization of GDH *via* the diaphorase.

Wohlrab (1965a) regarded his activity as probably a specific proline dehydrogenase (E.C. 1.5.1.1) but this interpretation is open to question. The use of PMS is fully justified in methods seeking to demonstrate the oxidase rather than the tetrazolium reductase since Marcus and Feeley (1962) reported that this acceptor (1·4 μM) could stimulate the oxidation of a number of aminoacids by L-aminoacid oxidase. It was considered to act by reoxidizing the reduced enzyme-flavin complex.

Dihydrofolate Reductase (E.C. 1.5.1.3)

This enzyme (5,6,7,8-tetrahydrofolate : NADP oxidoreductase) catalyses the reduction of folate to tetrahydrofolate, in two stages according to Futterman (1957), Osborn and Huennekens (1958) and Zakrzewski and Nichol (1958), as follows:

$$\text{Folate} + \text{NADPH}_2 \overset{\text{DHFR}}{\rightleftharpoons} \text{dihydrofolate} + \text{NADP}$$

$$\text{Dihydrofolate} + \text{NADPH}_2 \overset{\text{DHFR}}{\rightleftharpoons} \text{tetrahydrofolate} + \text{NADP}$$

The existence of one, or two, enzymes is still debated. However, in view of the fact that methotrexate (amethopterin) inhibits both stages of reduction at identical concentration (Werkheiser *et al.*, 1962), and that dihydrofolate reduction is completely inhibited by folate (Morales and Greenberg, 1964), a single enzyme seems to be more likely.

Dihydrofolate reductase activity is found to be high in some tissues and cells possessing large numbers of mitochondria but in others (cardiac and skeletal muscle) the enzyme levels are low. There is apparently no activity in the nervous system.

A histochemical method for the demonstration of dihydrofolate reductase was developed by Onicescu *et al.* (1970), using folate as substrate and NADP and NAD, together, as coenzymes. The precise sequence of events, in the histochemical procedure, remains unclear. Control media, both substrate-free and methotrexate-containing, gave negative results and parallel assays, using a spectrophotometric technique, indicated substantial conversion of folate to tetrahydrofolate in the histochemical medium. For these reasons it is possible to support the author's contention that their method demonstrates dihydrofolate reductase.

It is interesting to note that no reaction was obtained in fresh cryostat sections but only in those which were post-fixed in glutaraldehyde or some other aldehyde fixative. The enzyme involved may well be a mitochondrial one but it is certainly in the class of soluble enzymes.

The method is described in Appendix 21, p. 1351. It has not, at the time of writing, been tested in my laboratory.

Glutathione Reductase (E.C. 1.6.4.2)

This *soluble* enzyme produces GSH from GSSG and NADPH (Conn and Vennesland, 1951; Mapson and Goddard, 1951; Massey and Williams, 1965). Methods for its determination, by spectrophotometric assay of the consumption of NADPH in tissue sections, were applied by Malmgren and Sylvén (1960) and by Schor and Glick (1968). A method for the histochemical demonstration of the enzyme was described by Reiss (1967) who used the reverse reaction and *Saccharomyces cerevisiae*.

TABLE 72

Histochemically Demonstrable Dehydrogenases. Significance Table.

Metabolic Process	Dehydrogenase E.C. Number
Nucleotide Synthesis	1.3.3.1; 1.2.1.14
Urea Formation	1.4.1.3; 1.1.1.37
Pentose Shunt	
NADPH Supply	1.1.1.49; 1.1.1.43
Kreb's Cycle	1.1.1.37; 1.3.99.1
	1.2.4.2; 1.1.1.42
Glucuronic Acid Cycle	1.1.1.9; 1.1.1.10
Glycolysis	1.1.1.1; 1.2.1.12
	1.1.1.27; 1.6.4.3
Synthesis of CoA-linked Acyls	1.6.4.3
Glycerophosphate Shuttle	1.1.2.1; 1.1.1.8
Steroid Metabolism	
Pyridine Nucleotide-Transhydrogenation	1.1.1.50; 1.1.1.51
Lipid Synthesis	1.1.1.40

APPLIED DEHYDROGENASE HISTOCHEMISTRY

From a very large number of papers in this field the small selection given below is provided mainly to help those who may be entering the field for the first time. A full bibliography of applied dehydrogenase histochemistry up to the year 1964 can be found in *Acta Histochemica* (Vols. **24**, pp. 1–284 and **29**, pp. 1–292).

Skeletal Muscle. Magasanik and Natochin (1963); Ogata and Mori (1964a and b); Natochin and Magasanik (1964); van Wijhe *et al.* (1964); Diculescu *et al.* (1964); Tsukamoto and Mori (1966).

Cardiac Muscle. Diculescu *et al.* (1969).

Kidney. Mustakallio *et al.* (1960) (Diuretics); Natochin and Krestinskaia (1960); Hess and Regoli (1964) (Juxtaglomerular Apparatus); Wenk (1968).

Skin. Braun-Falco and Petzoldt (1964). Chouchkov (1969); Michael (1965).

Ovary. Kern-Bontke (1964); Taki *et al.* (1966): Okano *et al.* (1966).

Blood Cells. Balogh and Cohen (1961); Marcuse and Cochran (1961).

Spermatozoa. Edwards and Valentine (1963).

Tumours. Raikhlin (1961); Presnov *et al.* (1962); Mori *et al.* (1964); Goldman *et al.* (1964). See also collected papers entitled "Histochemical Methods in Diagnostic Oncology". Edited by H. A. Krachskogo, Istdatelstvo Medicina Moscow, 1968.

Blood Vessels. Lojda and Felt (1960); Lojda and Frič (1966).

Retina. Niemi and Merenmies (1961); Wolff (1969).

G.I. Tract. Niemi *et al.* (1960); Baxter-Grillo (1969).

Fungi. Reiss (1967a and b).

Endocrines. Balogh and Cohen (1961) (parathyroid); Cohen and Wolfe (1963) (pancreas); Sivaram and Sharma (1966) (Adrenal Cortex); Mori (1967) (Steroid-producing organs and tumours); Hopwood (1968) (Adrenal).

Testis. Ito (1966); Meyer (1968).

Apocrines. Tremblay (1968).

Central and Peripheral N.S. Friede (1961); Thomas (1963); Manocha and Bourne (1967); Duckett and Pearse (1967, 1968a and b); Labedsky and Lierse (1968); Fischer *et al.* (1968a and b).

Cervical Smears. Blonk (1969).

REFERENCES

ABDULLA, Y. H., ADAMS, C. W. M., and BAYLISS, O. B. (1968). *J. Histochem. Cytochem.*, **16**, 58.
ALLEN, J. M. (1960). *Anat. Rec.*, **137**, 57.
ÄNGGÅRD, E., and SAMUELSSON, B. (1966). *Ark. Kemi.*, **25**, 293.
ÄNGGÅRD, E., and SAMUELSSON, B. (1967). In "Prostaglandins. Proc. II Nobel Symp. Stockholm. Eds: S. Bergström, and B. Samuelsson. Stockholm, Almqvist and Wiksell, p. 97.
AOSHIMA, Y., and KOCHAKIAN, C. D. (1963). *Endocrinology*, **72**, 106.
ARGYRIS, T. S. (1956a). *Anat. Rec.*, **125**, 105.
ARGYRIS, T. S. (1956b). *Ibid.*, **126**, 1.
ASHWELL, G., KANFER, J., SMILEY, J. D., and BURNS, J. J. (1961). *Ann. N.Y. Acad. Sci.*, **92**, 105.
BAAR, H. S. (1957). *Schweiz, Z. allg. Path. Bakt.*, **20**, 164.
BAILLIE, A. H., CALMAN, K. C., FERGUSON, M. M., and HART, D. MCK. (1965) *J. Endocr.*, **32**, 337.
BAILLIE, A. H., CALMAN, K. C., FERGUSON, M. M., and HART, D. MCK. (1966) *J. Endocr.*, **34**, 1.
BAILLIE, A. H., FERGUSON, M. M., and HART, D. MCK. (1966). *J. clin. Endocr.*, **26**, 738.

BAILLIE, A. H., and GRIFFITH, K. (1964). *J. Endocr.*, **29**, 9.
BALOGH, K. (1964). *J. Histochem. Cytochem.*, **12**, 404.
BALOGH, K. (1964). *J. Histochem. Cytochem.*, **12**, 485.
BALOGH, K. (1964). *J. Histochem. Cytochem.*, **12**, 670.
BALOGH, K. (1965). *J. Histochem. Cytochem.*, **13**, 533.
BALOGH, K. (1966). *J. Histochem. Cytochem.*, **14**, 77.
BALOGH, K., and COHEN, R. B. (1961). *Blood*, **17**, 491.
BALOGH, K., and COHEN, R. B. (1961). *Lab. Invest.*, **10**, 354.
BALOGH, K., and COHEN, R. B. (1961). *Nature, Lond.*, **192**, 1199.
BARANOWSKI, T. (1949). *J. biol. Chem.*, **180**, 535.
BARON, D. N., GORE, M. B. R., PIETRUSZKO, R. and WILLIAMS, D. C. (1963). *Biochem. J.*, **88**, 18.
BATTELLI, F. and STERN, L. (1910). *Biochem. Z.*, **30**, 172.
BAXTER, A. O., BAILLIE, A. H., and FERGUSON, M. M. (1966). *Histochemie*, **7**, 218.
BAXTER-GRILLO, D. L. (1969). *Histochemie*, **19**, 129.
BECKETT, E. B., and BOURNE, G. H. (1958). *Acta anat.*, **33**, 289.
BENITEZ, L., and FISCHER, R. (1964). *J. Histochem. Cytochem.*, **12**, 858.
BERGMAN, S., BJERSING, L., and NILSSON, O. (1966). *Acta path. microbiol. scand.*, **68**, 461.
BERTOLINI, R., HAUPTMANN, S., and GABLER, W. (1966). *Histochemie*, **7**, 330.
BILSKI, R., and GODLEWSKI, H. G. (1965). *Folia histochem.*, **3**, 225.
BJERSING, L., and CARSTENSEN, H. (1964). *Biochem. biophys. Acta*, **86**, 639.
BJERSING, L., and CARSTENSEN, H. (1967). *J. Reprod. Fert.*, **14**, 101.
BLAKLEY, R. L. (1951). *Biochem. J.*, **49**, 257.
BLASIUS, W., and ZIMMERMANN, H. (1958). *Acta Histochem.*, **5**, 283.
BLONK, D. I. (1969). *Acta cytol.*, **13**, 264.
BOTTE, V., TRAMONTANA, S., and CHIEFFI, G. (1968). *J. Endocr.*, **40**, 189.
BOURNE, G. H. (1953). *Nature, Lond.*, **172**, 588.
BOURNE, G. H., and MALATY, H. A. (1953). *J. Physiol., Lond.*, **122**, 178.
BOXER, G. E., and DEVLIN, T. M. (1961), *Science, N.Y.*, **134**, 1495.
BRANDAU, H., BRANDAU, L., and LUH, W. (1969). *Arch. Gynäkol.*, **208**, 138.
BRATT, H., PUPKIN, M., LLOYD, C. W., WEISZ, J., and BALOGH, K. (1968). *Endocrinology*, **83**, 329.
BRAUN-FALCO, O., and PETZOLDT, D. (1964). *Arch. klin. exp. Dermatol.*, **220**, 455.
BRAUN-FALCO, O., and PETZOLDT, D. (1965). *Arch. klin. exp. Dermatol.*, **223**, 620.
BREUER, H., and MITTERMAYER, Ch. (1963). *Biochem. J.*, **86**, 12P.
BRODY, I. A., and ENGEL, W. K. (1969). *J. Histochem. Cytochem.*, **12**, 687.
BROOKE, M. H., and ENGEL, W. K. (1966). *Neurology*, **16**, 799.
BRYSON, M. J., and SWEAT, M. L. (1967). *Endocrinology*, **81**, 729.
BUBLITZ, C. (1961). *Biochim. biophys. Acta*, **48**, 61.
BÜCHER, Th., and KLINGENBERG, M. (1958). *Angew. Chem.*, **70**, 552.
BUÑO, W., and GERMINO, N. I. (1958). *Acta anat.*, **33**, 161.
BURKL, W., and SLEZAK-KLEMENCIC, E. (1967). *Acta histochem.*, **28**, 100.
CAHN, R. D., KAPLAN, N. O., LEVINE, L., and ZWILLING, E. (1962). *Science, N.Y.*, **136**, 962.
CAPUTTO, R., LELOIR, R. F., CARDINI, C. E., and PALADINI, A. C. (1950). *J. biol. Chem.*, **184**, 333.
CAPUTTO, R., TRUCCO, R. E., and KITABCHI, A. E. (1961). *Ann. N.Y. Acad. Sci.*, **92**, 79.
CASCARANO, J., and ZWEIFACH, B. W. (1959). *J. biophys. biochem. Cytol.*, **5**, 309.
CASTELLANO, M. A., GERMINO, N. I., DE HARO, N. B., and GERARD, G. (1969). *Histochemie*, **18**, 277.
CASTOLDI, G. L., YAM, L. T., and MITUS, W. J. (1968). *Acta haemat.*, **39**, 203.
CAVALLERO, C., and CHIAPPINO, G. (1962). *Experientia*, **18**, 119.
CHANG, J. P., and HORI, S. H. (1961). *Ann. Histochim.*, **6**, 419.
CHATTERJEE, I. B., CHATTERJEE, G. C., GHOSH, N. C., GHOSH, J. J., and GUHA, B. C. (1960). *Biochem. J.*, **74**, 193.
CHATTERJEE, I. B., KAR, N. C., GHOSH, N. C., and GUHA, B. C. (1961). *Ann. N.Y. Acad. Sci.*, **92**, 36.

CHIEFFI, G., and BOTTE, V. (1963). *Nature, Lond.*, **200**, 793.
CHOUCHKOV, H. (1969). *Histochemie*, **17**, 368.
CHRISTIE, G. A. (1968). *J. Endocr.*, **40**, 285.
COHEN, R. B. (1961). *Endocrinology*, **68**, 710.
COHEN, R. B. (1961). *Lab. Invest.*, **10**, 459.
COHEN, R. B. (1961). *Proc. Soc. exp. Biol. Med.*, **106**, 309.
COHEN, R. B. (1962). *Lab. Invest.*, **11**, 531.
COHEN, R. B., and CRAWFORD, J. D. (1962). *Endocrinology*, **70**, 288.
COHEN, R. B., and WOLFE, H. J. (1963). *J. Histochem. Cytochem.* **11**, 288.
COLEMAN, R. F., and NIENHUIS, A. W. (1966). *J. Histochem. Cytochem.*, **14**, 811.
COLLINS, W. P., MANSFIELD, M. D., BRIDGES, C. E., and SOMMERVILLE, I. F. (1969). *Biochem. J.*, **113**, 399.
CONN, E. C., and VENNESLAND, B. (1951). *J. biol. Chem.*, **192**, 17.
COOPER, W. G. (1955). *Anat. Rec.*, **123**, 103.
COOPER, W. G., and COPENHAVER, W. M. (1957). *Ibid.*, **126**, 503.
CRUICKSHANK, C. N. D., HERSHEY, F. B., and LEWIS, C. (1958). *J. invest. Derm.* **30**, 33.
DAVIS, J. C., and MUDD, S. (1957). *J. Histochem. Cytochem.*, **5**, 254.
DAVIS, J. C., WINTERSCHEID, L. C., HARTMAN, P. E., and MUDD, S. (1953). *Ibid.*, **1**, 123.
DAWSON, I. M. D., PRYSE-DAVIES, J., and SNAPE, Y. (1961). *J. Path. Bact.*, **81**, 1.
DAWSON, A. P., and THORNE, C. J. R. (1969a). *Biochem. J.*, **111**, 27.
DAWSON, A. P., and THORNE, C. J. R. (1969b). *Biochem. J.*, **114**, 35.
DEAMER, D. W., UTSUMI, K., and PACKER, L. (1967). *Archs Biochem. Biophys.*, **121**, 641.
DENNIS, P. M., and THOMAS, G. H. (1967). *J. Endocrinol.*, **39**, 459.
DICULESCU, I., DRAGOMIR, C. T., CONDRESCU-GUIDI, M., and POPESCU, M. (1970). *Histochemie*, **21**, 9.
DICULESCU, I., ONICESCO, D., and BARBULESCO, N. (1964). *Naturwissenschaften*, **22**, 537.
DICULESCU, I., ONICULESCU, I., ONICESCU, D., CUIDA, I., and BOBIA, C. (1969). *Histochem. J.*, **1**, 205.
DICULESCU, I., ONICESCU, D., and MISCHIU, L. (1964). *J. Histochem. Cytochem.*, **12**, 145.
DIXON, M., and WEBB, E. C. (1964). "Enzymes," 2nd ed. Longmans, London.
DUCKETT, S., and PEARSE, A. G. E. (1967). *Histochemie*, **8**, 334.
DUCKETT, S., and PEARSE, A. G. E. (1968a). *J. Anat.*, **102**, 183.
DUCKETT, S., and PEARSE, A. G. E. (1968b). *Folia Histochem. Cytochem.*, **6**, 447.
DUTTON, G. J., and STOREY, I. D. E. (1951). *Biochem. J.*, **48**, P29.
EDWARDS, R. G., and VALENTINE, R. C. (1963). *Expl Cell Res.*, **31**, 508.
EMMART, E. W., KOMINZ, D. R., and MIGUEL, J. (1963a). *J. Histochem. Cytochem.*, **11**, 207.
EMMART, E. W., SCHIMKE, R. T., SPICER, S. S., and TURNER, W. A. (1963b). *Expl Cell Res.*, **30**, 460.
EMMART, E. W., SPICER, S. S., TURNER, W. A., and HENSON, J. G. (1962). *Expl Cell Res.*, **26**, 87.
ENDAHL, G. L., KOCHAKIAN, C. O., and HAMM, D. (1960). *J. biol. Chem.*, **235**, 2792.
ENGEL, P. C., and DALZIEL, K. (1969). *Biochem. J.*, **115**, 621.
ERNSTER, L., and NAVAZIO, F. (1956). *Expl Cell Res.*, **11**, 483.
FAHIMI, H. D., and AMARASINGHAM, C. R. (1964). *J. Cell Biol.*, **22**, 29.
FAIRBANKS, V. F., and LAMPE, L. T. (1968). *Blood*, **31**, 589.
FARBER, E., STERNBERG, W. H., and DUNLAP, E. C. (1956a). *J. Histochem. Cytochem.* 4, 254.
FARBER, E., STERNBERG, W. H., and DUNLAP, C. E. (1956b). *Ibid.*, **4**, 284.
FARBER E., STERNBERG, W. H., and PEARCE, N. A. M. (1958). *J. Histochem. Cytochem.*, **6**, 389.
FASSKE, E., and STEINS, I. (1963). *Acta histochem.*, **16**, 302.
FERGUSON, M. M. (1966). *Histochemie*, **6**, 185.
FERGUSON, M. M. (1967). *Histochemie*, **9**, 269.

FERGUSON, M. M., and CHRISTIE, G. A. (1967). *J. Endocr.*, **38**, 291.
FISCHER, J., HOLUBÁŘ, J., and MALÍK, V. (1968a). *Acta histochem.*, **31**, 296.
FISCHER, J., HOLUBÁŘ, J., and MALÍK, V. (1968b). *Acta histochem.*, **31**, 305.
FLITNEY, F. W. (1966). *J. R. microsc. Soc.*, **85**, 353.
FORAKER, A. G., and DENHAM, S. W. (1953). *Surgery Gynec. Obstet.*, **96**, 259.
FORAKER, A. G., DENHAM, S. W., and JOHNSTON, M. H. (1953). *Archs Path.*, **55**, 147.
FORMISANO, V. R., and MONTAGNA, W. (1954). *Anat. Rec.*, **120**, 893.
FRIČ, P., and LOJDA, Z. (1966). *Gastroenterologia*, **106**, 65.
FRIED, G. H., and ZWEIFACH, B. W. (1955). *Anat. Rec.*, **121**, 97.
FRIEDE, R. L. (1961). *J. Neurochem.*, **8**, 17.
FRIEDEN, C. (1959). *J. biol. Chem.*, **234**, 2891.
FRIEDMANN, H. C., and VENNESLAND, B. (1960). *J. biol. Chem.*, **235**, 1526.
FUTTERMAN, S. (1957). *J. biol. Chem.*, **228**, 1031.
GABLER, W. (1969). *Acta histochem.*, **33**, 323.
GAHAN, P. B., and KALINA, M. (1968). *Histochemie*, **14**, 81.
GLOGNER, P., and GÖSSNER, W. (1962). *Histochemie*, **2**, 435.
GODDARD, J. W., and SELIGMAN, A. M. (1952). *Anat. Rec.*, **112**, 543.
GODDARD, J. W., and SELIGMAN, A. M. (1953). *Cancer*, **6**, 385.
GOLDBERG, B., JONES, G. E. S., and BORKOWF, H. J. (1964). *J. Histochem. Cytochem.* **12**, 880.
GOLDMAN, R. D., KAPLAN, N. O., and HALL, T. C. (1964). *Cancer Res.*, **24**, 389.
GORE, M. B. R., and BARON, D. N. (1965). *J. Endocr.*, **33**, 353.
GRAVES, J. L., and VENNESLAND, B. (1957). *J. biol. Chem.*, **226**, 306.
GREEN, D. E. (1936). *Biochem. J.*, **30**, 629.
GROLLMAN, A. P., and LEHNINGER, A. L. (1957). *Archs. Biochem. Biophys.*, **69**, 458.
GUHA, S., and WEGMANN, R. (1963). *Nature, Lond.*, **200**, 1218.
HAMM, D. I., KOCHAKIAN, C. D., and CARROLL, B. R. (1956). *Proc. Soc. exp. Biol. Med.*, **93**, 493.
HARDONK, M. J. (1965). *Histochemie*, **4**, 563.
HARDONK, M. J. (1965). *Histochemie*, **5**, 234.
HART, D. McK. (1966). *Histochemie*, **6**, 17.
HATEFI, Y. (1963). In "The Enzymes". Eds. P. D. Boyer, H. Lardy, and K. Myrbäck, Acad. Press, New York. Vol. 7, p. 495.
HESS, R., and PEARSE, A. G. E. (1959). *Br. J. exp. Path.* **40**, 243.
HESS, R., and PEARSE, A. G. E. (1961a). *Proc. Soc. exp. Biol. Med.*, **106**, 895.
HESS, R., and PEARSE, A. G. E. (1961b). *Enzymol. biol. clin.*, **1**, 15.
HESS, R., and PEARSE, A. G. E. (1961c). *Nature, Lond.*, **191**, 718.
HESS, R., and PEARSE, A. G. E. (1961). *Experientia*, **17**, 136.
HESS, R., and PEARSE, A. G. E. (1961). *Experientia*, **17**, 317.
HESS, R., and REGOLI, D. (1964). *Br. J. exp. Path.*, **45**, 666.
HESS, R., SCARPELLI, D. G., and PEARSE, A. G. E. (1958a). *Nature, Lond.*, **181**, 1531.
HESS, R., SCARPELLI, D. G., and PEARSE, A. G. E (1958b). *J. biophys. biochem. Cytol.*, **4**, 753.
HIMMELHOCH, S. R., and KARNOVSKY, M. J. (1961a). *J. biophys. biochem. Cytol.*, **9**, 573.
HIMMELHOCH, S. R., and KARNOVSKY, M. J. (1961b). *J. biophys. biochem. Cytol.*, **9**, 893.
HOGEBOOM, G. H., and SCHNEIDER, W. C. (1950). *J. biol. Chem.*, **186**, 417.
HOLIMANN, S., and NEUBAUER, J. (1966). *Klin. Woch.*, **44**, 722.
HOLLMANN, S. (1959a). *Z. physiol. Chem.*, **317**, 193.
HOLLMANN, S. (1959b). *Klin. Wschr.*, **37**, 727.
HOLLMANN, S. (1964). *Acta Histochem.*, **17**, Suppl. 4.
HOPWOOD, D. (1968). *Histochemie*, **16**, 85.
HORWITZ, C., BENITEZ, L., and BRAY, M. (1967). *J. Histochem. Cytochem.*, **15**, 216.
HSU, Y. T., and HOCH-LIGETI, C. (1953). *Am. J. Path.*, **29**, 105.
HUGHES, A. (1956). *Quart. J. micr. Sci.*, **97**, 165.
HURLOCK, B., and TALALAY, P. (1958). *J. biol. Chem.*, **233**, 886.
HURLOCK, B., and TALALAY, P. (1959). *Archs. Biochem. Biophys.*, **80**, 468.
HUSSEIN, K. A., and KOCHAKIAN, C. D. (1968). *Acta endocr.*, **59**, 459.

ITO, M. (1966). *Radiat. Res.*, **28**, 266.
JACOBSEN, N. O. (1969). *Histochemie*, **20**, 250.
JARRETT, A. (1962). *J. Histochem. Cytochem.*, **10**, 400.
JOHNSON, A. B. (1965). *J. Histochem. Cytochem.*, **13**, 583.
JOHNSON, A. B. (1967). *J. Histochem. Cytochem.*, **15**, 207.
JOHNSON, A. B., and STRECKER, H. J. (1963). *J. biol. Chem.*, **237**, 1876.
JONES, W. A., and COHEN, R. B. (1963). *Lab. Invest.*, **12**, 802.
KALINA, M., and GAHAN, P. B. (1965). *Histochemie*, **5**, 430.
KALINA, M., and GAHAN, P. B. (1968). *Stain Technol.*, **43**, 51.
KALINA, M., GAHAN, P. B., and JONES, G. R. N. (1965). *Nature, Lond.*, **207**, 647.
KEARNEY, E. B., and SINGER, T. P. (1956). *J. biol. Chem.*, **219**, 963.
KELLOGG, D. A., and GLENNER, G. G. (1960). *Nature, Lond.*, **187**, 763.
KERN-BONTKE, E. (1963). *Histochemie*, **4**, 56.
KIMURA, A. T., SINGER, T. P., and LUSTY, C. J. (1960). *Biochim. biophys. Acta*, **44**, 284.
KING, E. J. (1932). *Biochem. J.*, **26**, 292.
KING, T. E., and MANN, T. (1959). *Proc. R. Soc.*, **151**, 226.
KOIDE, S. S., and MITSUDO, S. M. (1965). *Endocrinology*, **76**, 403.
KUUSISTO, A. N., and TELKKÄ, A. (1955). *Acta endocr.*, **18**, 141.
LABEDSKY, L., and LIERSE, W. (1968). *Histochemie*, **12**, 130.
LEDUC, E. H., and WISLOCKI, G. B. (1952). *J. comp. Neurol.*, **97**, 241.
LEE, J. B., CROWSHAW, K., TAKMAN, B. H., ATTREP, K. A., and GOUGOUTAS, J. Z. (1967). *Biochem. J.*, **105**, 1251.
LEHRER, G. M., SILIDES, D., and FARMER, M. F. (1965). *J. Histochem. Cytochem.*, **13**, 700.
LEVY, H., DEANE, H. W., and RUBIN, B. L. (1959). *Endocrinology*, **65**, 932.
LING, K. H., WU, S. H., TING, S. M., and TUNG, T. C. (1957). *Proc. Int. Symp. Enzyme, Chem., Tokyo and Kyoto*, Pergamon, London, p. 260.
LOBEL, B. L., DEANE, H. W., and ROMNEY, S. L. (1962). *Am. J. Obstet. Gynec.*, **83**, 295.
LOJDA, Z. (1965). *Folia morphol.*, **13**, 84.
LOJDA, Z., and FELT, V. (1960). *Experientia*, **16**, 514.
LOJDA, Z., and FRIČ, P. (1966). *J. Atheroscler. Res.*, **6**, 264.
LOJDA, Z., and FRIČ, P. (1970). In "FEBS Symposium", Vol. **18**, p. 185.
MAGASANIK, L. G., and NATOCHIN, J. V. (1963). *Cytologia*, **6**, 236.
MALATAY, H. A., and BOURNE, G. H. (1953). *Nature, Lond.*, **171**, 295.
MALMGREN, H., and SYLVÉN, B. (1960). *Cancer Res.*, **20**, 204.
MANN, P. J. G., and QUASTEL, J. H. (1937). *Biochem. J.*, **31**, 869.
MANN, P. J. G., WOODWARD, H. E., and QUASTEL, J. H. (1938). *Biochem. J.*, **32**, 1025.
MANOCHA, S. L., and BOURNE, G. H. (1967). *Histochemie*, **9**, 300.
MAPSON, L. W., and BRESLOW, E. (1958). *Biochem., J.*, **68**, 395.
MAPSON, L. W., and GODDARD, D. R. (1951). *Biochem. J.*, **49**, 592.
MARCUS, A., and FEELEY, J. (1962). *Biochim. biophys. Acta*, **59**, 398.
MARCUS, P. I., and TALALAY, P. (1956). *J. biol. Chem.*, **218**, 661.
MARCUSE, P. M. (1957a). *Lab. Invest.*, **6**, 133.
MARCUSE, P. M. (1957b). *Am. J. clin. Path.*, **28**, 539.
MARCUSE, P. M., and COCHRAN, J. (1961). *Blood*, **17**, 738.
MARKERT, C. L., and MØLLER, F. (1959). *Proc. natn. Acad. Sci., U.S.A.*, **45**, 753.
MASSEY, V. (1958). *Biochim. biophys. Acta*, **30**, 205.
MASSEY, V. (1959). *Biochem. J.*, **72**, 71.
MASSEY, V. (1960). *Biochim. biophys. Acta*, **37**, 314.
MASSEY, V., and WILLIAMS, C. H. (1965). *J. biol. Chem.*, **240**, 4470.
MCALPINE, J. C. (1965). *J. Histochem., Cytochem.*, **13**, 296.
MCCORKINDALE, J., and EDSON, N. L. (1954). *Biochem. J.*, **57**, 518.
MCMILLAN, P. J. (1967). *J. Histochem. Cytochem.*, **15**, 21.
MEYER, W., (1968). *Acta histochem.*, **29**, 376.
MICHAEL, J. C. Im. (1965). *J. Histochem. Cytochem.*, **13**, 668.
MILLER, R. W., KERR, C. T., and CURRY, J. R. (1968). *Can. J. Biochem.*, **46**, 1099.

MITTERMAYER, Ch., BREURER, H., and DISCHERL, W. (1963). *Acta Endocrinol.*, **43**, 195.
MONTAGNA, W., and FORMISANO, V. R. (1955). *Anat. Rec.*, **122**, 65.
MORALES, D. R., and GREENBERG, D. M. (1964). *Biochim. biophys. Acta*, **85**, 360.
MORI, M. (1967). *Arch. histol. jap.*, **28**, 45.
MORI, M., MIZUSHIMA, T., OKA, R., OKAMOTO, Y., and TAKITA, Y. (1964). *Oral Surg.*, **17**, 352.
MUSTAKALLIO, K. K. (1954a). *Ann. med. exp. biol. Fenn.*, **32**, 175.
MUSTAKALLIO, K. K. (1954b). *Science, N.Y.*, **119**, 881.
MUSTAKALLIO, K. K. (1956). *Acta derm.-venereol.*, **36**, 279.
MUSTAKALLIO, K. K., and JÄNNES, L. (1954). *Expl Cell Res.*, **7**, 595.
MUSTAKALLIO, K. K., RAEKALLIO, E., and RAEKALLIO, J. (1960) *Ann. Med. exp. Fenn.*, **38**, 21.
MUSTAKALLIO, K. K., and SAIKKONEN, J. I. (1955). *Acta path. Scand.*, **37**, 398.
MUSTAKALLIO, K. K., and TELKKÄ, A. (1953). *Science, N.Y.*, **118**, 320.
MUSTAKALLIO, K. K., and TELKKÄ, A. (1955). *Ann. med. exp. biol. Fenn.* Suppl. 13, **33**, 1.
NACHLAS, M. M., WALKER, D. G., and SELIGMAN, A. M. (1958a). *J. biophys. biochem. Cytol.*, **4**, 29.
NACHLAS, M. M., WALKER, D. G., and SELIGMAN, A. M. (1958b) *Ibid.*, **4**, 169.
NACHLAS, M. M., WALKER, D. G., and SELIGMAN, A. M. (1958c). *J. biophys. biochem. Cytol.*, **4**, 467.
NACHMIAS, V. T., and PADYKULA, H. A. (1958). *J. biophys. biochem. Cytol.*, **4**, 47.
NAKAJIMA, Y., SHANTHA, T. R., and BOURNE, G. H. (1969). *Histochemie*, **18**, 293.
NAKANO, J., ÄNGGÅRD, E., and SAMUELSSON, B. (1969). *Europ. J. Biochem.*, **11**, 386.
NATOCHIN, J. V., and KRESTINSKAIA, T. V. (1960). *Physiol., J., U.S.S.R.*, **47**, 388.
NATOCHIN, J. V., and MAGASANIK, L. G. (1964). *J. Comp. Biol., U.S.S.R.*, **25**, 128.
NIEMI, M., and MERENMIES, E. (1961). *J. Neurochem.*, **6**, 200.
NIEMI, M., SIURALA, M., and SUNDBERG, M. (1960). *Acta path. microbiol. scand.*, **48**, 323.
NIRENBERG, M., and JACOBY, W. B. (1960). *Proc. natn. Acad. Sci., U.S.A.*, **46**, 206.
NISSEN, H. M., and ANDERSEN, H. (1968). *Histochemie*, **14**, 189.
NOVIKOFF, A. B. (1955). In "Analytical Cytology", Ed. R. C. Mellors, McGraw-Hill, New York.
NOVIKOFF, A. B., and MASEK, B. (1958). *J. Histochem. Cytochem.*, **6**, 217.
NOVIKOFF, A. B., SHIN, W., and DRUCKER, J. (1960). *J. Histochem. Cytochem.*, **8**, 37.
OGATA, T., and MORI, M. (1964a). *J. Histochem. Cytochem.*, **12**, 171.
OGATA, T., and MORI, M. (1964b). *J. Histochem. Cytochem.*, **12**, 183.
OGAWA, Y. (1958). *J. pract. Pharmacol. (Yakkyoku)*, **9**, 433.
OKANO, K., MATSUMOTO, K., KOIZUMI, T., MIZUSHIMA, T., and MORI, M. (1966). *Acta histochem.*, **25**, 12.
ONICESCU, D., TAŞCĂ, L., and TRUTZA, E. (1970). *Histochem. J.*, **2**, 289.
ORTMANN, R. (1957). *Acta Histochem.*, **4**, 158.
OSBORN, M. J., and HEUNNEKENS, F. M. (1958). *J. biol. Chem.*, **233**, 969.
PACKER, L., ESTABROOK, R. W., SINGER, T. P., and KIMURA, T. (1960). *J. biol. Chem.*, **235**, 535.
PACKER, L., and GREVILLE, G. D. (1969). *FEBS Letters*, **3**, 112.
PADYKULA, H. A. (1952). *Am. J. Anat.*, **91**, 107.
PEARSE, A. G. E. (1958). *J. clin. Path.*, **11**, 520.
PEARSE, A. G. E., and MACPHERSON, C. R. (1958). *J. Path. Bact.*, **75**, 69.
PEARSON, B., and DEFENDI, V. (1955). *Cancer Res.*, **15**, 593.
PEARSON, B., and GROSE, F. (1959). *Proc. Soc. exp. Biol. Med.*, **100**, 636.
PEARSON, B., GROSE, F., and GREEN, R. (1959). *Lab. Invest.*, **8**, 29.
PEARSON, B., WOLF, P., GROSE, F., and ANDREWS, M. (1964). *Am. J. clin. Path.*, **41**, 3.
PLAGEMANN, P. G. W., GREGORY, K. F., and WRÓBLEWSKI, F. (1960a). *J. biol. Chem.*, **235**, 2282.
PLAGEMANN, P. G. W., GREGORY, K. F., and WRÓBLEWSKI, F. (1960b). *J. biol. Chem.*, **235**, 2288.

PLANE, R. A., and THEORELL, H. (1961). *Acta chem. scand.*, **15**, 1866.
PORTUGALOV, V. (1955). *Doklady Akad. Nauk.*, **103**, 157.
PRESNOV, M. A., ABBASOV, A. T., and PESTOVA, E. A. (1962). *Invest. Oncol.*, **8**, 36.
PUPKIN, M., BRATT, H., WEISZ, J., LLOYD, C. W., and BALOGH, K., (1966). *Endocrinology*, **79**, 316.
RACKER, E. (1949). *J. biol. Chem.*, **177**, 883.
RACKER, E. (1955). *Physiol. Rev.*, **35**, 1.
RAIKHLIN, N. T. (1961). *Invest. Oncol.*, **7**, 41.
REALE, E., and LUZZATTO, L. (1956). *Acta Histochem.*, **3**, 138.
REISS, J. (1967a). *Arch. Mikrobiol.*, **57**, 285.
REISS, J. (1967b). *Arch. Mikrobiol.*, **57**, 307.
REISS, J. (1967). *J. Histochem. Cytochem.*, **15**, 273.
RENDINA, G. R., and SINGER, T. P. (1959). *J. biol. Chem.*, **234**, 1605.
RINGLER, R. L., and SINGER, T. P. (1959). *J. biol. Chem.*, **234**, 2211.
ROGERS, G. E. (1953). *Q. J. microsc. Sci.*, **94**, 253.
ROSA, C. G., and VELARDO, J. T. (1954). *J. Histochem. Cytochem.*, **2**, 110.
ROSA, C. G., and VELARDO, J. T. (1958). *Nature, Lond.*, **181**, 348.
ROSSI, F., PESCETTO, G., and REALE, E. (1954). *Boll. Soc. ital. Biol. Sper.*, **30**, 1045.
ROSSI, F., PESCETTO, G., and REALE, E. (1957). *J. Histochem. Cytochem.*, **5**, 221.
RUBIN, B. L. (1957). *J. biol. Chem.*, **227**, 917.
RUBIN, B. L., and STRECKER, H. J. (1961). *Endocrinology*, **69**, 257.
RUDOLPH, G., and KLEIN, H. J. (1964). *Histochemie*, **4**, 238.
RUDOLPH, G., and SCHOLL, O. (1958). *Beitr. path., Anat.*, **119**, 13.
RUTENBURG, A. M., WOLMAN, M., and SELIGMAN, A. M. (1953). *J. Histochem. Cytochem.*, **1**, 66.
SAMUELS, L. T., and HELMREICH, M. L. (1956). *Endocrinology*, **58**, 435.
SAMUELS, L. T., HELMREICH, M. L., LASATER, M. B., and REICH, H. (1951). *Science, N.Y.*, **113**, 490.
SASAI, Y., TANAHASHI, Y., and FUJIYAMA, T. (1969). *Tohoku. J. exp. Med.*, **98**, 99.
SCARPELLI, D. G., HESS, R., and PEARSE, A. G. E. (1958). *J. biophys. biochem. Cytol.*, **4**, 747.
SCARPELLI, D. G., and PEARSE, A. G. E. (1958a). *Anat. Rec.*, **132**, 133.
SCARPELLI, D. G., and PEARSE, A. G. E. (1958). *J. Histochem. Cytochem.*, **6**, 369.
SCHÄFER, G., and LAMPRECHT, W. (1961). *Z. phys. Chem.*, **326**, 259.
SCHOR, N. A., and GLICK, D. (1968). *J. Histochem. Cytochem.*, **16**, 185.
SCHUMACHER, H. H. (1957). *Science, N.Y.*, **125**, 501.
SELIGMAN, A. M., and RUTENBURG, A. M. (1951). *Science, N.Y.*, **113**, 317.
SHELTON, E., and SCHNEIDER, W. C. (1952). *Anat. Rec.*, **112**, 61.
SHNITKA, T. K. (1969). *J. Histochem. Cytcohem.*, **17**, 198.
SINGER, T. P. (1963). In "The Enzymes", Eds. P. D. Boyer, H. Lardy, and K. Myrbäck, Vol. **7**, pp. 356–359.
SIVARAM, S., and SHARMA, D. R. (1966). *Nature, Lond.*, **211**, 855.
SLATER, E. C. (1958). *Adv. Enzymol.*, **20**, 146.
SMITH, L. H. jr., and BAKER, F. A. (1959). *J. clin. Invest.*, **38**, 798.
SMITH, L. H. jr., and BAKER, F. A. (1960). *J. clin. Invest.*, **39**, 15.
STEINBERGER, E., STEINBERGER, A., and VILAR, O. (1966). *Endocrinology*, **79**, 405.
STILLER, D., and GORSKI, J. (1965). *Histochemie*, **5**, 407.
STILLER, D., and GORSKI, J. (1966). *Histochemie*, **6**, 156.
STILLER, D., and GORSKI, J. (1969). *Acta histochem.*, **32**, 356.
STRAUB, F. B. (1939). *Enzymologia*, **9**, 148.
STROMINGER, J. L., KALCKER, H. M., AXELROD, J., and MAXWELL, E. S. (1954). *J. Am. chem. Soc.*, **76**, 6411.
STRUCK, J., jr., and SIZER, I. W. (1960). *Archs Biochem. Biophys.*, **86**, 260.
SWEAT, M. L., BRYSON, M. J., and YOUNG, R. B. (1967). *Endocrinology*, **81**, 167.
SZARKOWSKA, L., and DRABIKOWSKA, A. K. (1964). *Life Sci.*, **2**, 519.
TAKI, I., HAMANAKA, N., and MORI, M. (1966). *Am. J. Obstet. Gynec.*, **96**, 388.
TALALAY, P., LOEWUS, F. A., and VENNESLAND, B. (1955). *J. biol. Chem.*, **212**, 801.
TALALAY, P., and MARCUS, P. I. (1956). *J. biol. Chem.*, **218**, 675.
TALALAY, P., and WILLIAMS-ASHMAN, H. G. (1960). *Recent Prog. Horm., Res.*, **16**, 1.

TAPIA FRESES, A., DIAZ, J., and FALEN, J. (1965). *Revta. ibér. Endocr.*, **12**, 5.
TELKKÄ, A., and MUSTAKALLIO, K. K. (1955). *Science, N.Y.*, **121**, 146.
THOMAS, E. (1963). *Acta neuropath.*, **2**, 231.
THUNBERG, T. (1909). *Skand. Arch. physiol.*, **22**, 430.
TRANZER, J.-P., and PEARSE, A. G. E. (1963). *Nature, Lond.*, **199**, 1063.
TREMBLAY, G. (1968). *J. invest. Derm.*, **50**, 238.
TSUKAMOTO, S., and MORI, M. (1966). *Arch. histol. jap.*, **26**, 329.
TUNG, T. C., ANDERSON, L., and LARDY, H. A. (1952). *Archs Biochem. Biophys.*, **40**, 194.
TYLER, W. S., and PEARSE, A. G. E. (1965). *Thorax*, **20**, 149.
VALLEE, B. L., and HOCH, F. L. (1957). *J. biol. Chem.*, **225**, 185.
VALLEE, B. L., WILLIAMS, R. J. P., and HOCH, F. L. (1959). *J. biol. Chem.*, **234**, 2621.
VAN WIJHE, M., BLANCHAER, M. C., and STUBBS, S. St. G. (1964). *J. Histochem. Cytochem.*, **12**, 608.
VERCAUTEREN, R. (1957). *Enzymologia*, **18**, 97.
VIALE, G. L., ANDREUSSI, L., ZACCHEO, D., and VIALE, E. (1964). *Acta histochem.*, Suppl., **4**, 103.
VILLEE C. A., and SPENCER, J. M. (1960). *J. biol. Chem.*, **235**, 3615.
WACHSTEIN, M., and MEISEL, E. (1954). *Am. J. Path.*, **30**, 147.
WACHSTEIN, M., and MEISEL, E. (1955). *Am. J. Path.*, **31**, 353.
WALKER, D. G., and SELIGMAN, A. M. (1963). *J. cell Biol.*, **16**, 455.
WARBURG, O., and CHRISTIAN, W. (1932). *Biochem. Z.*, **254**, 438.
WATTENBERG, L. W. (1958). *J. Histochem. Cytochem.*, **6**, 225.
WATTENBERG, L. W., and LEONG, J. L. (1960). *J. Histochem. Cytochem.*, **8**, 296.
WENK, H. (1968). *Histochemie*, **12**, 120.
WERKHEISER, W. C., ZAKRZEWSKI, S. F., and NICHOL, C. A. (1962). *J. Pharmac. exp. Ther.*, **137**, 162.
WIEME, R. J. (1959). "Studies on Agar Gel Electrophoresis". Arscia N.V. Brussels.
WIEST, W. G. (1959). *J. biol. Chem.*, **234**, 3115.
WILLIAMS-ASHMAN, H. G. (1953). *Cancer Res.*, **13**, 721.
WILLIAMS-ASHMAN, H. G., BANKS, J., and WOLFSON, S. K., jr. (1957). *Archs Biochem. Biophys.*, **72**, 485.
WISE, J. B., and LEHNINGER, A. I. (1962). *J. biol. Chem.*, **237**, 1363.
WOHLRAB, F. (1965). *Histochemie*, **4**, 360.
WOHLRAB, F. (1965a). *Histochemie*, **5**, 311.
WOHLRAB, F. (1965b). *Naturwissenschaften*, **52**, 211.
WOHLRAB, F. (1966). *Acta biol. med. germ.*, **16**, 210.
WOHLRAB, F. (1968). *Proc. III Int. Cong. Histo and Cytochemistry, N.Y.* p. 294.
WOLFF, H. H. (1969). *Histochemie*, **17**, 11.
WOLMAN, M., and BUBIS, J. J. (1967). *J. Histochem. Cytochem.*, **15**, 79.
WU, R., and WILSON, D. W. (1956). *J. biol. Chem.*, **223**, 195.
YORK, J. L., GROLLMAN, A. P., and BUBLITZ, C. (1961). *Biochim. biophys., Acta*, **47**, 298.
ZAKRZEWSKI, S. F., and NICHOL, C. A. (1958). *Biochim. biophys. Acta*, **27**, 425.
ZEBE, E. (1960). *Biochem. Z.*, **332**, 328.
ZEBE, E., DELRÜCK, A., and BÜCHER, Th. (1957). *Angew. Chem.*, **59**, 56.
ZIEGLER, D. M., GREEN, D. E., and DOEG, K. A. (1959). *J. biol. Chem.*, **234**, 1916.
ZIMMERMANN, H., and PEARSE, A. G. E. (1959). *J. Histochem. Cytochem.*, **7**, 271.

CHAPTER 22

PEPTIDASES

There is very little correlation between the biochemistry of the peptidases and their histochemistry. In neither case has a wholly satisfactory classification been evolved. Because there is virtually no correspondence between any histochemical peptidase and any biochemically established enzyme possessing an E.C. number, it is probably best, for the time being to use acceptable, histochemical trivial names.

Biochemistry of the Peptidases

Current biochemical classifications of the peptidases are based on the nature of their active sites (Hartley, 1960). A possible classification is as follows:

Thiol peptidases (papain, ficin, bromelain, some cathepsins).
Metal peptidases (carboxy, amino and dipeptidases).
Acid peptidases (pepsin, rennin).
Serine peptidases (chymotrypsin, trypsin, elastase, thrombin, plasmin).

The peptidases are further subdivided into exopeptidases, which hydrolyse only terminal peptides, and endopeptidases which can split peptide bonds within the protein molecule. Exopeptidases include carboxypeptidases A and B, which require an ionized terminal carboxyl group in order to exhibit activity, and aminopeptidases such as leucine aminopeptidases, which require an ionized terminal amino group.

Naphthyl amidases. In addition to the well defined peptidases described above there exists a broad group of enzymes known as arylamidases which exhibit specific activity against aminoacyl-β-naphthylamides. The list of known arylamidases, which appears below in Table 73, was compiled by Brecher and Suszkin (1969). For comparison the characteristics of the specific leucine aminopeptidase (Smith and Spackman, 1955) are included.

There are sufficient differences between the various arylamidases recorded above to justify the conclusion that they are distinct enzymes. Nevertheless, their similarities allow one to suppose that they have similar or identical active sites in spite of structural and conformational differences.

Leucine Aminopeptidase. The occurence of this enzyme in the tissues, and some of its properties, were described by Berger and Johnson (1940) and also by Johnson and Berger (1942), Fruton (1946), and Smith (1948, 1951). The properties of the pig kidney enzyme, isolated by Spackman *et al.* (1955) were described by Smith and Spackman (1955). According to these authors it had broad specificity for leucyl and other compounds, was activated by Mn^{2+}

and Mg^{2+} ions and inactivated by EDTA and citrate. These properties had previously been recorded by Berger and Johnson (1939) and by Smith (1948).

Extracts of pig intestine split leucinamide, leucyl glycine and leucylglycyl glycine at almost equal rates but during the process of purification (Smith and Bergmann, 1944) different activities towards the three compounds were observed to appear.

Cathepsins. The properties of these enzymes, recorded by Fruton (1960) and reviewed by Barrett (1969), are so diverse that their inclusion under a single comprehensive term is perhaps unwarranted. They are named in alphabetical sequence and the list at present comprises A, B, C, D and E. The first four occur in lysosomes (see Vol. 1, p. 550) and the last is possibly a derivative of cathepsin D. Cathepsins B, D and E are endopeptidases, and A and C are exopeptidases. All are intracellular acid peptidases.

The properties of bovine spleen *Cathepsin A* were set forth by Lichtenstein and Fruton (1960). It cleaves the glutamyl-tyrosyl bond in carbobenzoxy-L-glutamyl-tyrosine at an optimal pH or 5·7, does not require activation by SH compounds, and is not inhibited by iodoacetate.

The thiol-dependent *Cathepsin B*, has been demonstrated in a large variety of mammalian tissues, most of which show low activity. It was partially purified from ox spleen (200-fold) by Greenbaum and Fruton (1957) and more completely by Snellman (1969), from the lysosomal-mitochondrial pellet of calf liver. The observed mol. wt. of the enzyme is below 50,000; it has a thiol group in its active centre, requires activation by a chelating agent and a reducing agent (EDTA and cysteine, for instance), and is inhibited by iodoacetate and phenylhydrazine but not by DFP. Both peptide and ester bonds are cleaved. The recorded Km for the preferential substrate (α-*N*-benzoyl-L-arginine ethyl ester; BAEE), at pH 6·0, is 0·0175, and for α-*N*-benzoyl-L-arginine naphthylamide (BANA) it is 0·02. Other substrates, such as L-arginine β-naphthylamide (ANA) and L-leucine β-naphthylamide (LNA) are hydrolysed effectively at pH 7·0. BANA, at pH 5·0 was regarded by Blackwood and Mandl (1961) as the preferred substrate for the enzyme and, using benzoyl-L-arginine amide (BAA) and BANA, Ali *et al.* (1967) characterized the predominant cathepsin of cartilage as cathepsin B. Because both enzymes attack bonds to the same amino acid residues in a protein, cathepsin B is regarded as a trypsin-like enzyme.

The third enzyme in this group, *Cathepsin C* (E.C.3.4.4.9) was first extracted from pig kidney by Gutmann and Fruton (1948). Further purification was reported by De La Haba *et al.* (1959), Planta and Gruber (1961), and by Metrione *et al.* (1966). It is a thiol enzyme, being completely inhibited by iodoacetate, activated by SH compounds and CN^-, and unaffected by DFP (Fruton and Mycek, 1956). Its substrate specificities were determined by Tallan *et al.* (1952), Wiggans *et al.* (1954), Izumiya and Fruton (1956) and by Planta *et al.* (1964), as well as by Fruton and Mycek (1956). The enzyme acts specifically on derivatives of dipeptides composed of two amino acid

TABLE 73

Characteristics of LAP and Arylamidases

Enzyme	Source	Substrate	Substrate Specificity	pH Opt.	Activ: Agents	Inhibitors	Ref.
LAP	Pig kidney	L-Leu(NH_2)	—	7·8–9·3	Mg^{2+}, Mn^{2+}	—	Smith and Spackman (1955)
NA	Rat Brain	L-Leu BNA DL-AlaBNA DL-PheBNA	—	7·0	—	IAA, PCMB DFP, Mn^{2+}, 60°	Adams & Glenner (1962)
NA	Ascites tumour	L-Leu BNA	—	6·5, 7·0	Mg^{2+}	Mn^{2+}, Co^{2+}	Patterson *et al.* (1963)
NA	Human liver	L-Leu BNA	Ala-, Phe-, Leu-, His-.	7·0	—	Co^{2+}, PCMB Puromycin	Behal *et al.* (1965)
NA	Human intestine	L-Leu BNA	Ala-, Phe-, Leu-, His-.	6·0	—	Mg^{2+}, Zn^{2+}, Mn^{2+}, PCMB	Behal *et al.* (1965)
NA	Human pancreas	L-Leu BNA	Ala-, Phe-, Leu-.	7·0	—	Mg^{2+}, Mn^{2+}, Zn^{2+}, PCMB Puromycin	Behal *et al.* (1965)
NA	Human liver	L-Leu BNA	Ala-, Met-, Phe-, Leu-, Gly-.	6·5	Co^{2+} EDTA-Zn^{2+}	Zn^{2+}, Mg^{2+}, Mn^{2+}, PCMB EDTA	Smith *et al.* (1965)
AP-B	Cat and guinea-pig liver	L-Lys BNA L-Arg BNA	Arg-, Leu-.	7·0	Cl^-	Hg^{2+}, Pb^{2+}, Co^{2+}, PCMB MB.	Hopsu *et al.* (1966a,b)
NA	Rat liver	L-Lys BNA L-Arg BNA	Arg-, Leu-.	7·0	Cl^-	Mg^{2+}, Zn^{2+}, Mn^{2+}, Hq^{2+}, Pb^{2+}. Ni^{2+}, PCMB, IAA,	Hopsu *et al.* (1966c)

	kidney		Trp-, Leu-, Arg-, Thr-, Tyr-, Lys-, Ser-, Asp-, His-, Val-, Pro.			PCMB.	*et al.* (1967)
NA	Bovine ant. pituitary	L-Arg PNA	Arg-PNA	6·5 to 8·5	Thiols	Di and tripeptides, polylysine polyarginine protamine histones	Ellis & Perry (1966)
NA	Bovine ant. pituitary	L-Lys BNA L-Lys PNA Arg-PNA etc.	Lys-PNA	6·5–8·5	Thiols Mn^{2+} Co^{2+} Zn^{2+}	EDTA, MEA. Urea, NEM, Guanidine PCMB. Di and tripeptides polylysine etc.	Tjeder (1966)
NA	Bovine brain	L-Ala BNA	Ala-, Arg-, Leu-, Phe-, Lys-, Val-, Pro-BNA Lys-BNA Leu-BNA	7·5	DTT, ME Mn^{2+} Mg^{2+}	Co^{2+}, Zu^{2+} PCMB NEM 60°	Brecher and Suszkin (1969)

LAP, Leucine aminopeptidase
NA, Naphthylamidase
AP-B, Aminopeptidase-B
ME, 2-mercaptoethanol
MEA, 2-mercaptoethanolamine
MB, methylene blue
IAA, iodoacetate
PNA, *p*-nitroanilide
DTT, dithiothreitol

residues: $NH_2(CHR)CO\text{-}NH(CHR')CO\text{---}X$, where X may be an amide or an ester. Substrates in which R′ is the side-chain of L-phenylalanine or L-tyrosine are preferentially hydrolysed, indicating a resemblance to chymotrypsin.

It is probable that cathepsin C has limited activity *in vivo* as a protease but that it acts specifically for the hydrolysis, and transfer, of N-terminal dipeptides with a penultimate aromatic amino acid and a small unsubstituted terminal amino acid as, for example, Gly-Phe-NH_2 (Fujii and Fruton, 1958).

In addition to its sulphydryl requirement, cathepsin C has been shown by McDonald *et al.* (1966) to have an absolute requirement for chloride ion. They observed a 10-fold increase in the rate of hydrolysis of Gly-Phe-NH_2, and other substrates, as between acetic acid and hydrochloric acid buffers, at pH 5·3 in the presence of 5 mM 2-mercaptoethylamine. McDonald *et al.* (1966) suggested that although the fortuitous inclusion of Cl^- in most media had allowed cathepsin C to be demonstrated successfully, nevertheless reports of negative activity would require reassessment.

An enzyme characterized as cathepsin C was partially purified from pig kidney by Hopsu-Havu and Rintola (1968). This enzyme would not split amino acid naphthylamides but hydrolysed naphthylamides of alanyl-alanine, leucyl-leucine and glycyl-phenylalanine, yielding a dipeptide and ammonia. Optimal activity occurred between pH 5·0 and 6·0 and the mol. wt. was 134,000, as determined by gel filtration. When alanyl-alanine naphthylamide is used as a histochemical substrate for rat kidney, in the presence of a diazonium salt, an intense reaction is obtainable. This is not due to cathepsin C, but partly to microsomal aminopeptidase (Vanha-Perttula *et al.*, 1966) and partly to microsomal dipeptide naphthylamidase (Hopsu-Havu *et al.*, 1968).

A protease with a neutral isoelectric point and a mol. wt. of 58,000 was isolated by Press *et al.* (1960) from bovine spleen. This enzyme, *Cathepsin D* (E.C.3.4.4.23) has been recorded in many other tissues including rat brain (Marks and Lajtha, 1965) and thyroid gland (Kress *et al.*, 1966). It has been identified as the lysosomal acid proteinase of rabbit liver by Barrett (1967) who recorded a mol. wt. of between 50,000 and 52,000 and optimal activity against haemoglobin at pH 3·2. This particular cathepsin D was not inhibited by iodoacetamide (1 mM), *p*-chloromercuribenzoate (0·1 mM), or EDTA, and it was insensitive to polypeptide proteinase inhibitors from bovine parotid gland (Trasylol-Bayer) and soybean. It was competitively inhibited by 5 mM 3-phenylpyruvic acid and non-competitively by indolyl-3-pyruvic acid (Geratz, 1965). Preincubation with dithiothreitol (1 mM), at pH 8·0 and 4°, completely inactivated the enzyme, presumably by breaking a disulphide bond essential for the maintenance of the active configuration of the enzyme.

Barrett (1967) considered, in agreement with the views of de Duve *et al.* (1962) and Woessner (1965), that cathepsin D should be regarded as the characteristic cathepsin of the lysosomes. Purified cathepsin D preparations do not hydrolyse α-*N*-benzoyl-L-arginine amide or leucine β-naphthylamide (Barrett, 1970).

Cathepsin E, the least well characterized of the group was described by Lapresle and Webb (1962) as having its maximal activity at pH 2·5, with serum albumin as substrate.

Development of Histochemical Methods for Naphthylamidases

The first histochemical substrates for peptidase activity were produced by Gomori (1954a and b). Following the observation by Greenstein and his associates (Birnbaum *et al.*, 1952; Rao *et al.*, 1952) that chloroacetyl amino acids were hydrolysed more rapidly than the corresponding acetyl or glyceryl derivatives Gomori (1954a) tested the chloroacetylnaphthylamines as histochemical substrates. He found that these were very rapidly hydrolysed by cold acetone-fixed tissues and that they could form the basis for a histochemical method, using Garnet GBC as trapping agent for the free naphthylamines. Later (1954b) Gomori prepared glycyl-α and β-naphthylamines, and α and β-alanyl-β-naphthylamines, as chromogenic substrates for aminopeptidases. Once again he used cold acetone-fixed tissues and Garnet GBC and observed that the localization of activity was similar to that obtained with the chloroacetyl naphthylamines.

A new chromogenic substrate for aminopeptidases was synthesized, independently, by Green *et al.* (1955) and by Folk and Burstone (1955). Directly continuing the studies initiated by Gomori, and recorded above, Burstone and Folk (1956) made histochemical comparison of various substrates. They found that considerable improvement in the histochemical localization of aminopeptidases could be obtained with either L-leucyl-β-naphthylamide or DL-alanyl-β-naphthylamide. Marked inhibition was observed with Cu^{2+}, Pb^{2+} and Ca^{2+} ions, which are known inhibitors of leucine aminopeptidase, and activation was produced by the addition of cyanide (3 mM) to the incubating medium. No activation occurred, however, with Mg^{2+} or Mn^{2+}. The authors presumed that this last finding was due to the fact that in freeze-dried sections, which they preferred and recommended, these metal ions were already present in optimal concentration.

The results of the application of Burstone's technique are satisfactory in most cases. The red dye produced in freeze-dried tissues by coupling with Garnet GBC is orange coloured in fresh frozen sections and, in the case of the latter, large crystals are rapidly formed.

A method basically very similar to Burstone's was produced by Nachlas, Crawford and Seligman (1957) who also employed L-leucyl-β-naphthylamide as substrate. Since the main specificity requirements of the enzyme were provided by the leucine residue and the group linked at the leucine carboxyl was known to exert only a minor influence these authors considered that the substrate they used would be hydrolysed by leucine aminopeptidase (E.C.3.4.1.1).

pH Optimum. Nachlas and his colleagues noted that the histochemical reaction for the so-called LAP could not be carried out at the pH optimum of the enzyme (pH 8–9·5) which has a broad range of activity from pH 6·5 to 9·5. Folk and Burstone (1955) found an optimum pH activity, under histochemical conditions, at pH 7·0 to 7·4, while Green *et al.* (1955) observed an optimum at pH 6·8. Using the same substrate *in vitro* the optimum pH was 8·0.

An explanation for these findings was considered to lie in the inhibitory activity of the diazonium salts and possibly their accompanying metal ions. Nachlas, Crawford and Seligman (1957) noted that Fast blue B produced 60 per cent inhibition at pH 6·5 and 100 per cent at pH 8·0, while Fast Garnet GBC caused only 30 per cent inhibition at both pH levels.

Choice of Diazonium Salt. In spite of the apparent advantage of GBC over Blue B salt the former had three severe disadvantages which made its use, except with freeze-dried sections, impossible. The first of these was that although *in vitro* the coupling rate appeared to be faster than that of Fast blue B, in tissue sections it appeared to be slower. Diffusion of the free β-naphthylamine was apparently greater. The second disadvantage of Garnet GBC was that the azo dye which it produced gave large crystals in a very short time except, as mentioned above, in freeze-dried sections. The third disadvantage was that metal chelation of the azo dye from Garnet GBC could not be obtained.

Metal Chelation. The long molecule produced from β-naphthylamine with the dicoupler Fast blue B is moderately substantive for protein and crystallization is thus prevented. When Cu^{2+} ions are added, after incubation, the azo dye from β-naphthylamine and blue B is able to chelate 2 atoms of the metal per molecule of dye. This results in the formation of a stable copper chelate which is deep bluish-purple in colour compared with the red of the unchelated dye.

Other diazonium salts containing a methoxy group *ortho* to the diazonium were also tested (Fast red B and Fast black K salts). These, being monocouplers, produced azo dyes which chelated only one atom of copper per molecule and which were far less substantive. No diazonium salt was found which coupled rapidly enough to prevent diffusion entirely. Details of this method are given in Appendix 22, p. 1353, and the result is illustrated in Fig. 207, p. 943, and in Plate XXIIb. When this method is employed the enzyme activity which is demonstrated should be described as leucine naphthylamidase.

An alternative substrate for the histochemical demonstration of the so-called LAP was synthesized by Rosenblatt, Nachlas and Seligman (1958). This substrate, L-leucyl-4-methoxy-β-naphthylamide, was designed to couple more rapidly with the diazonium salt and thus to prevent the diffusion artifacts from which both of the older methods suffered. A preliminary report of the histochemical use of this substrate was given by Monis, Nachlas, and

PLATE XXII

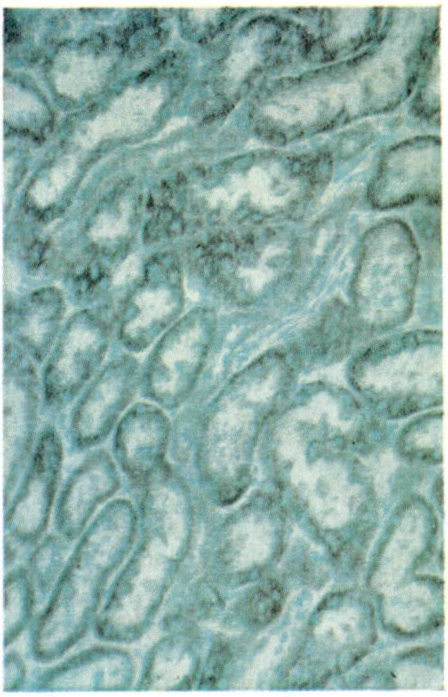

XXIIa. Rat kidney. Succinate dehydrogenase reaction using the MTT-cobalt system. Counter stained with methyl green. × 490.

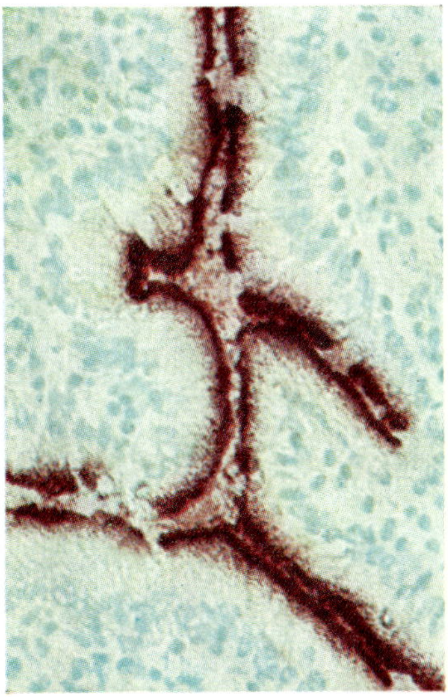

XXIIb. Rat duodenum. Leucyl naphthylamidase reaction. Shows intense enzyme activity in the brush border. Methyl green. × 260.

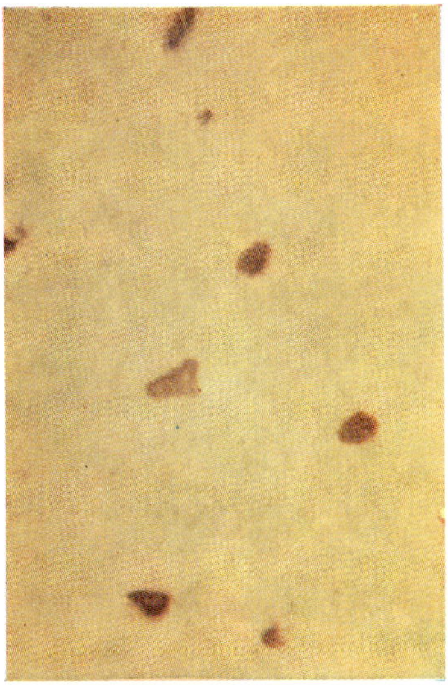

XXIIc. Rat mast cells. Amidase reaction using the phenyl propionoxy ester of Naphthol AS as substrate and Fast Garnet GBC salt. × 450.

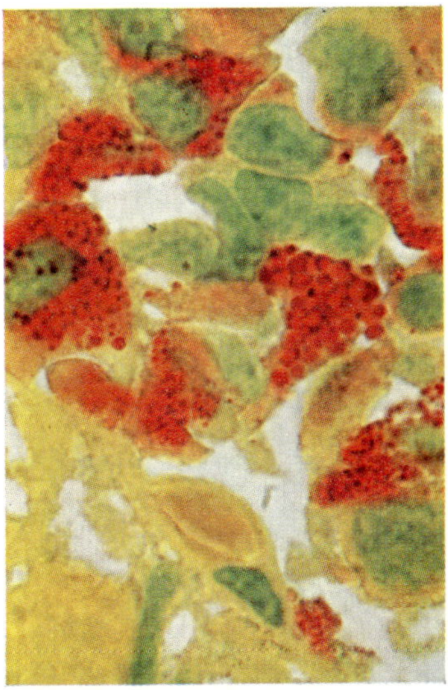

XXIId. Newt (*Triton cristatus*) Pituitary gland. Arylsulphatase reaction strongly positive in the globular basophils. Naphthol AS-BI sulphate and HPR. × 1220.

Seligman (1959) and a fuller description by Nachlas, Rosenblatt, Monis and Seligman (1960).

The diazonium salt of choice remained Fast blue B and copper chelation resulted in a stable pigment which was no longer lipid soluble. The red colour of the metal-free dye was in this case, however, substantially unaltered. The result is illustrated, in the case of the rat parathyroid gland, in Fig. 208, p. 943. This should be compared with Fig. 207, which illustrates the use of the older method on the same tissue. The degree of improvement is obviously very considerable but the older substrate continues to be the more popular of the two.

Indigogenic Methods for LAP

An indigogenic substrate, prepared by Pearson *et al.* (1963) was designed as a specific substrate for leucine aminopeptidase. This compound, DL-*N*-(5-bromoindol-3-yl) leucinamide hydrobromide, was observed to be hydrolysed by an enzyme with optimal activity between pH 7·5 and 8·0, which was activated by 0·6 mM Mg^{2+}, 0·2 mM Cu^{2+} and Mn^{2+}, and by Co^{2+} at 60 μM.

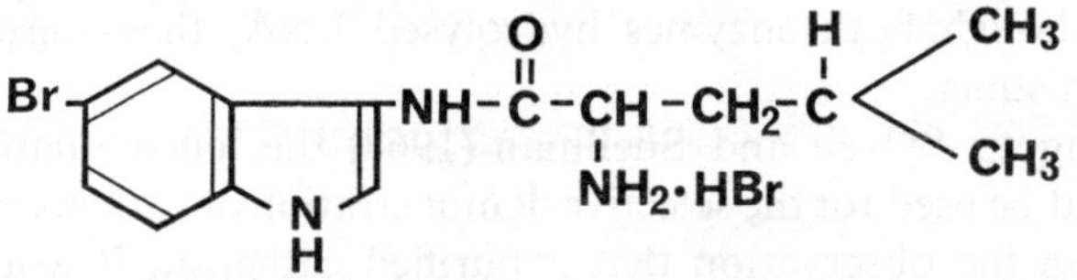

Tested against a number of proteolytic enzymes, it was hydrolysed only by a purified leucine aminopeptidase and by an impure emulsin which was presumed to contain the latter enzyme as an impurity. In the histochemical medium no oxidation catalyst was included but the Cu^{2+} ions, added as activator at 1·5 mM, presumably acted in this respect also.

Very little work has been carried out with indigogenic peptide substrates, which theoretically might be expected to have a high degree of specificity, but significant and interesting differences were recorded by Pearson *et al.* when they compared their method with standard naphthpylamidase techniques. Opinion must be reserved, however, until further proof of specificity for LAP is obtained. Technical details of the method are given in Appendix 22, p. 1354.

Interpretation of the so-called LAP Methods

As noted above, when substrates such as the leucyl naphthylamides were first used it was generally believed that the enzyme chiefly responsible for their hydrolysis was leucine aminopeptidase (LAP). Subsequently, however, a number of enzymes were described, having no LAP activity, which could hydrolyse a variety of amino acid naphthylamides (as indicated on p. 966).

Conclusive evidence was offered, to show that LAP was not significantly responsible for the hydrolysis of aminoacyl naphthylamides by tissue sections (Patterson *et al.*, 1963; Hanson *et al.*, 1963). The complexity of the situation is such that enzymes demonstrated histochemically with amino acid naphthylamides should for the present be described by their common names, according to the I.U.B.C. Enzyme Commission rules, unless an equivalent biochemical common name is available.

In a long series of papers Sylvén (Sylvén and Bois, 1962, 1963; Sylvén and Snellman, 1964, 1968; Sylvén and Bois-Svensson, 1964) indicated that the hydrolysis of LNA could be due to any or all of the following: (1) A metal-dependent group of naphthylamidases (2) A leucinamide-splitting, Mn^{2+}-activated, peptidase regarded as LAP (3) α-Chymotrypsin (4) Prolinase (5) Carboxypeptidase B and (6) Cathepsin B. The naphthylamidases were reported to be heterogeneous with respect to the effect of added cysteine + EDTA. A number of authors have reported purification or partial purification of these enzymes (Vanha-Perttula and Hopsu, 1965; Vanha-Perttula *et al.*, 1966; Ono *et al.*, 1968; Nagatsu *et al.*, 1968; Behal and Story, 1969). A particle-bound Co^{2+}-activated naphthylamidase was isolated from rat kidney by Felgenhauer and Glenner (1966), together with a soluble cysteine-activated enzyme. Although both enzymes hydrolysed LNA, they were regarded as completely distinct.

According to Sylvén and Snellman (1968) the leucine naphthylamidase method could be used for the selective demonstration of cathepsin B. This view was based on the observation that a purified cathepsin B would hydrolyse LNA at a pH as low as 5·5, at which level no hydrolysis by aminopeptidases could be observed (Sylvén and Snellman, 1968). Additional means for differentiation were considered to be provided by inhibitors and activators. A list of these, with appropriate concentrations appears in Appendix 22. Opposition to Sylvén's views was recorded by Barrett and Poole (1969) who found that while LNAse activity was powerfully inhibited by both hyamine and puromycin, cathepsins B and D were unaffected. Total inhibition of the histochemical reaction by these two inhibitors was considered to exclude cathepsin B as responsible. It was stated, furthermore, that two distinct forms of the latter enzyme, purified by Otto (1967), both failed to hydrolyse LNA.

We must conclude that for the present the lysosomal LNAses, shown by the use of LNA as substrate, do not represent any known lysosomal peptidase.

Histochemistry of the LNAse Reaction

Leucine naphthylamidase, abbreviated if required to LNAse, has been the object of a considerable number of studies. Most of these were carried out under the old name (LAP). Early studies were carried out by Burstone (1956) who described the distribution of proteolytic enzymes in human neoplasms with particular reference to the high stromal activity found in the vicinity of tumours. His evidence was not entirely in agreement with that produced by Sylvén and Malmgren (1955), who used biochemical methods. Braun-Falco (1957) used the histochemical technique and suggested that aminopeptidase activity might be connected with the invasiveness of tumours. This author (Braun-Falco and Salfeld, 1959) drew attention to the well-known fact that the mast cells of many species contain a strong aminopeptidase activity and it is of interest that Benditt (1956), using chloroacetyl-2-hydroxy-3-naphthoic acid as substrate, found a strong chymotrypsin-like activity in mast cells. The activity of LAP in the production of parathyroid hormone was considered by Pearse and Tremblay (1958) who demonstrated a presumptive direct relationship between the two. In the inactive parathyroid gland of rats treated with dihydrotachysterol aminopeptidase activity fell to less than half the resting value for the normal gland (estimated biochemically). In rats maintained on a low calcium diet there was a marked increase in enzyme activity. It was therefore assumed that the enzyme was closely connected with the production of parathyroid hormone and that the peptide hormone, or its carrier protein, might well be found to possess a terminal leucyl group. Investigations on foetal rat parathyroids by Hansson and Svalander (1967) confirmed these findings and indicated that the glands functioned from the 16th day of gestation. Niemi and Ikonen (1960) found high aminopeptidase activity in rat pineal gland and suggested that the gland secretes some agent of a protein nature.

Willighagen and Planteydt (1959), using Burstone's method, observed a strong aminopeptidase in human gastric and bile duct tumours which was absent from other types of tumour. According to these authors normal human stomach contained no AP. In the rat, however, strong activity is present in the basal portion of the gastric glands (Nachlas, Crawford and Seligman, 1957). Willighagen and Planteydt made the suggestion that examination of lymph nodes, or other tissues, by the AP method might enable one to pick out malignant cells of gastric origin. Their suggestion was an interesting one and a full tumour survey afforded considerable support so that the AP method is used by diagnostic pathologists particularly to distinguish malignant cells of gastrointestinal origin metastatic in lymph nodes. Studies on the AP reaction of histiocytes in damaged muscle fibres were carried out by Jasmin and Gareau (1961) and fibroblasts in healing wounds were noted by Raekallio (1960, 1961) to possess high activity. The stromal reaction in neoplasia was investigated by Monis *et al.* (1959) and additional evidence was provided by Pearse *et al.* (1962). Further work on these lines was reported by Monis (1963) and also by Sylvén and Bois-Svensson (1964). AP activity was reported in the spleen by Korhonen and Ruponen (1962), in blood platelets by Balogh (1963), and in ischaemic necrosis of the kidney by Nagel and Willig (1964). The activity of aminopeptidases during limb bud differentiation in the chick embryo was described by Mottet (1967) and Raekallio and Mäkinen (1967, 1969) made further studies on wound healing. Interesting results on

AP activity in rabbit epididymis were obtained by Hellströmm and Nicander (1967). Adams and Glenner (1962) gave a comprehensive account of aminopeptidase activity in the central and peripheral nervous system.

A number of studies have been devoted to aminopeptidase (naphthylamidase) activity in the placenta of different animals, including man. An early report by Hopsu *et al.* (1961) was followed by reports from Wielenga and Willighagen (1962), Wachstein *et al.* (1963), Kleiner and Wilkin (1963), Curzen (1964), Vollrath (1965), and Christie (1967). Using LAN, as well as the more specific substrate described by Semm and Waidl (1962), Seelig and Roemheld (1969) claimed that the aminopeptidase activities of trophoblast and X-cells in human placenta were those of oxytocinase. Malignant neoplasms of the female genitalia were examined by Ishihara *et al.* (1968) and Grzycki and Rzeszowska (1969) found high activity of naphthylamidase in the mucosa of the gall-bladder in *Rana esculenta.*

Using LNA as substrate Idahl and Täljedal (1968) carried out biochemical studies on isolated pancreatic islets from obese-hyperglycaemic mice. They found high LNA-splitting activity, separable into two distinct bands by disc electrophoresis. Finding that pretreatment with Triton X 100 was necessary to mobilize full activity, they expressed the opinion that their LNA-splitting activity was lysosomal in nature, agreeing with the views of Sylvén and Bois-Svensson (1964) and Sylvén and Lippi (1965). These views were extended by Niemi and Sylvén (1969) who indicated that the naphthylamidase reaction, carried out at pH 5·5, could demonstrate both lysosomes and autolytic vacuoles. They recommended the method as a diagnostic tool for the demonstration of cellular injury and autophagy.

Other Peptidases and Substrates

Enzymes described in the sections which follow are not necessarily demonstrable histochemically, using the artificial substrates towards which they show hydrolytic activity. When they are so demonstrable the fact is made clear. A very large amount of work has been carried out on the purification and characterization of aminopeptidases but is is still seldom possible to identify a given histochemical peptidase activity with a single known enzyme.

Aminopeptidase B. This enzyme was identified, in several mammalian tissues, by Hopsu, Kantonen and Glenner (1964) and purified by Hopsu, Mäkinen and Glenner (1966a and b) from rat liver. It hydrolyses only naphthylamidases of the basic amino acids arginine and lysine, is activated by chloride ions, and strongly inhibited by Hg^{2+} and Cu^{2+} ions (20 μM). Later studies by Mäkinen (1968) indicated activation by cysteine and inhibition by N-ethyl maleimide, both of the rat liver enzyme and an identical enzyme from human foetal liver.

Dipeptide Naphthylamidases. The chromogenic substrate glycyl-DL-prolyl-*β*-naphthylamide was synthesized by Hopsu-Havu and Glenner (1966) and tested against a number of commercial proteases. The greatest hydrolytic activity was shown by acylase I. An enzyme closely resembling the latter was demonstrated in rat liver and, histochemically, in rat kidney using Tris-HCl buffer at pH 7·0 and Fast Garnet GBC salt.

This enzyme was purified 600-fold from rat liver by Hopsu-Havu and Sarimo (1967). In further studies carried out by Hopsu-Havu and Eckfors (1969) the localization of the enzyme was demonstrated both by a simultaneous coupling technique and also by indirect immunofluorescence.

γ-Glutamyl transpeptidase. The γ-peptide linkage in the tripeptide γ-L-glutamyl-L-cysteinylglycine (GSH) is not hydrolysed by pepsin, papain or by the usual aminopeptidases. Enzymic hydrolysis of the γ-peptide linkage was first shown by Woodward (1939) and later by Binkley and Nakamura (1948) and by Binkley and Olson (1951), who called the enzyme glutathionase. Hanes, Hird and Isherwood (1952) described it as γ-glutamyl transpeptidase, the name by which it is still commonly known. The hydrolysis product (glutamic acid) may be released but, if a suitable amine acceptor is present, it is considered that the major part is transferred thereto. The transpeptidase is thus the more significant of the two activities. According to Jacyszyn and Laursen (1968) the enzyme is to be regarded as heterogeneous.

Histochemical demonstration of the enzyme was first achieved by Albert *et al.* (1961) using γ-L-glutamyl-α-naphthylamide and by Glenner and Folk (1961a), using the β-naphthylamide. Both groups of authors, whose work was augmented by the later studies of Glenner, Folk and McMillan (1962) and of Albert *et al.* (1964), considered with Hird and Springell (1954) that transpeptidation was of paramount importance. Both added glycylglycine (20 mmole/ml.) to their media.

Noting that at the usual pH of the reaction (6·7 to 8·5) the coupling rate of 2-naphthylamine with the diazonium salt was too slow to prevent diffusion artifacts, Rutenburg *et al.* (1969) synthesized γ-glutamyl-4-methoxy-2-naphthylamide. They used this substrate with Fast Blue BBN salt at pH 7·4. They found a 240 per cent activation with glycylglycine and 130 per cent with methionine. Cu^{2+} and Nl^{2+} (mM) produced 100 per cent inhibition.

All three methods for the enzyme are described in the Appendix (p. 1355).

α-L-Glutamyl peptidase. A peptidase with specificity for N-terminal dicarboxylic acid substrates (aminopeptidase A) was found to be capable of splitting both α-L-aspartyl and α-glutamyl-β-naphthylamides. The $Ca^{2\pm}$ activated α-L-glutamyl peptidase activity of rat kidney microsomes was investigated by Hess (1965) who found that the enzyme was apparently closely related to angiotensinase A. No transpeptidation could be observed with α-L-glutamyl arylamide hydrolysis.

Histochemical demonstration of this enzyme was obtained by Glenner and Folk (1961b), using a method similar to their own for γ-glutamyl transpeptidase, but omitting the amine acceptor, with a pH of 6·2 and 10 mM Ca^{2+} as activator.

Trypsin-like Enzymes. The chromogenic trypsin substrate N^{α}-benzoyl-DL-arginine-β-naphthylamide (BANA) was synthesized by Riedel and Wünsch (1959) and Glenner and Cohen (1960) reported its histochemical application with the demonstration, in the mast cells of certain species, of a trypsin-like amidase activity. Later Glenner, Hopsu and Cohen (1962) reported the synthesis of naphthol-AS and ASD ε-aminocaproate esters which they found were hydrolysed by a trypsin-like esterase present in the mast cells of several species. In a subsequent paper (1964) these same authors reported the synthesis of the β-naphthylamides of N^{α}-carbobenzyloxy-L-arginine and N^{α}-benzoyl-L-arginine, and of the methyl ester of N^{α}-cbz-L-arginine. They compared these, and BANA, as substrates for trypsin, and for the trypsin-like enzymes of mast cells, finding activity against the methyl ester higher than against any other substrate. The human mast cell enzyme was strongly inhibited by ethyl ε-aminocaproate and it was considered to be substrate homospecific with trypsin.

A large number of amides of arginine, synthesized by Nachlas, Plapinger and Seligman (1964) were subjected to hydrolysis by trypsin. Chain length was found to be important, the optimum length being shown by a compound containing 4 amides (pentalysine) or 3 amides and one urethane (cbz-triarginyl naphthylamide). Aromatic amides were much more susceptible to hydrolysis than simple amides. The three most active substrates N^{α}-cbz-diglycyl-L-arginine-2-naphthylamide, β-carboxy-propionyl-L-arginyl-2-naphthylamide, and N^{α}-cbz-L-triarginyl-2-naphthylamide, were hydrolysed 220, 147 and 113 times as readily as BANA. Methods for the preparation of amino acid naphthylamides were fully described by Glenner, Cohen and Folk (1965).

Chymotrypsin-like Enzymes. Using the synthetic substrate 3-chloroacetoxy-2-naphthoic acid anilide Gomori (1953) demonstrated a hydrolytic "amidase" in mast cells. Subsequently Benditt and Arase (1959) showed that this enzyme possessed a striking substrate homospecificity with α-chymotrypsin, being also coequally inhibited by varying concentrations of DFP. Later Lagunoff and Benditt (1961) obtained superior histochemical demonstration of the chymotrypsin-like esterases with alternative substrates, of which the best was the phenyl propionoxy ester of Naphthol AS. The appearance of the mast cell enzyme, using this technique (recorded in Appendix 22) is shown in Plate XXIIc, p. 968.

A biochemical and histochemical survey of enzymes hydrolysing acyl naphthylamides was carried out by Hopsu and Glenner (1964), primarily in order to ascertain the cause of their limited success in demonstrating naphthylamidases by histochemical techniques. One of the main causes was found to be the pronounced inactivation caused by diazonium salts and this was partly overcome by lowering the pH of the incubating medium as far as possible. Another cause of failure was the extreme solubility of the acyl naphthylamidases. This was partly overcome by the addition of Hg^{2+} ions

(0·2 mM) which precipitated the enzyme. Mercuric ions also brought about an apparent reversal of diazonium salt inhibition.

Hopsu, Santti and Glenner (1966) found that an enzyme in guinea-pig liver, capable of hydrolysing chloroacetyl-α-naphthylamide, could also hydrolyse *N*-formyl-L-kynurenine, that is to say it behaved as a kynurenine formamidase. The authors considered that histochemically demonstrable hydrolysis of chloroacetyl-α-naphthylamide could almost totally be attributed to this enzyme and hence could serve as an indicator of sites of tryptophan catabolism. Since kynurenine formamidase is a soluble enzyme, prefixation is a prerequisite for its demonstration.

The Silver Protein Method for Proteases

An entirely new principle was invoked by Takamatsu *et al.* (1963) in developing their method for proteinases in general. The substrate, silver proteinate, is split by proteases in the tissue sections and the released silver ions combine with bromine ions in the medium to produce insoluble AgBr. This is converted to metallic silver by reduction with hydroquinone.

The specificity of the reaction for proteases, which was not claimed by the original authors or by their successors Yamada and Ofugi (1968), depends on a number of factors. The precise nature of the silver-protein binding in the silver proteinate is unknown but it is necessary to be certain that ionic silver can be released from it by proteolytic activity. This is easily demonstrated *in vitro*. Simple acidification of the medium indeed produces a precipitate of AgBr. Thus the suggestions of Yamada and Ofugi (1968) that a linkage with protein-SH is the most probable one cannot be sustained. It is more likely that the silver atoms are held in complexes by the protein side-chains and that an intact tertiary structure is necessary to maintain these complexes. There would thus be little interference with enzyme action and breakage of peptide bonds within the protein chain by endopeptidase activity would break the tertiary structure and release silver ions into the medium. Exopeptidases might be less active in this respect.

It is necessary, furthermore, to show that the released Ag^+ ions combine rapidly with Br^-, and not with tissue-bound SH, for example, with phosphate groups in DNA and RNA, or with lipid aldehydes.

The original authors used acetate buffer at pH 6·4 or Tris buffer at pH 7·4. With the latter soluble silver complexes can be formed and I consider that Tris, and similar buffers should not be used.

As yet, no extensive critical studies of the silver proteinate method have been undertaken but the method clearly warrants further study and it can be recommended, as given in Appendix 22, p. 1358, as a starting point for further studies. Using different specific inhibitors and activators it should be possible to develop the method, both for optical and electron microscopy, so as to permit localization of specific peptidases.

Peptidases and Esterases and their Relationships

A number of proteolytic enzymes are capable of splitting esters as well as peptides (Neurath and Schwert, 1950). Peptidases like chymotrypsin and pepsin (but not trypsin) show a marked preference for links involving aromatic amino-acid residues and for both the best substrates are esters, especially those of *N*-acyl-tyrosine. Cathepsin C is one of the intracellular proteolytic enzymes which act on proteins or synthetic amino-acid esters at acid pH levels. A new classification of the animal cathepsins was proposed by Tallan, Jones and Fruton (1952) which was based on their substrate specificities and cathepsin C was recorded as homospecific with chymotrypsin.

An organophosphorus-resistant indoxyl esterase was described in rat kidney and other tissues by Hess and Pearse (1958). Its properties closely resembled those of cathepsin C. This enzyme was found in macrophages, Kupffer cells, islets of Langerhans, and in the epithelium of the pancreatic ducts, as well as in the kidney. It was similar to but not identical with the non-specific esterase described in pituitary mucoid cells by Pearse (1956) and in rat brain and rat thyroid (Pepler and Pearse, 1957a and b) as well as in human argentaffinomas (carcinoids) by Pearse and Pepler (1957).

Myers *et al.* (1955) suggested that the E600-resistant and heat-resistant esterase of rat pancreas might be related to a proteolytic enzyme, different from trypsin or chymotrypsin. The fact that the latter is able to hydrolyse *o*-acetylindoxyl supported the assumption that cathepsin-like activity might be responsible for part of the histochemically observed hydrolysis of indoxyl esters. The rat kidney enzyme was observed by Hess and Pearse (1958) to be heat-resistant, optimally active at pH 5·3 and to be activated by ascorbic acid, cysteine and glutathione.

Phenylmercuric ions activated the enzyme under histochemical conditions as well as *in vitro* and this suggested a similarity with the "C-esterase" described by Bergmann, Segal and Rimon (1957). Strong inhibition by β-phenylpropionic acid suggested a chymotrypsin-like activity. Strong activation by detergents (Tweens 60 and 80) resembled the effect of these compounds on carboxypeptidases and a similar activation by pretreatment with trypsin at pH 7·8 suggested that part of the enzyme was present as an inactive precursor.

These early views were subjected to considerable criticism, in the matter of specific relationships between esterolytic activity and any known protease. In particular, Vanha-Perttula and Hopsu (1965) separated a number of soluble proteolytic and esterolytic enzymes from rat adenohypophysis by column chromatography and showed that the enzyme hydrolysing indoxyl esters differed from those which split BANA (cathepsin B) and glycylphenylalanylamide acetate (cathepsin C). A subsequent paper by Vanha-Perttula *et al.* (1965) provided absolute proof of the non-identity of indoxyl esterase and cathepsin C. As indicated in Chapter 17, p. 765, there is now considerable evidence to show that the E600-resistant lysosomal indoxyl esterase is identical with "acid lipase". The identity of the non-lysosomal E600-labile non-specific esterase/cholinesterase of the R-type mucoid cells (Chapter 8, p. 240) of the anterior pituitary gland remains to be elucidated. Vanha-Perttula and Hopsu (1965), however, separated a non-specific cholinesterase from their pituitary preparations which also hydrolysed trifluoroacetyl-α-naphthylamide (i.e. possessed peptidase activity).

Several alkaline proteases, which have been purified and characterized, are capable of hydrolysing 5-bromoindoxyl acetate at rates which vary between 1/100th and 1/50th of the highest rate for the preferred substrate. One such protease from rat submandibular gland was described by Riekkinen *et al.* (1966).

As was made abundantly clear in a lucid paper by Hopsu, Santti and Glenner (1965), variability from species to species and from organ to organ in respect of esterases, and acyl or aryl naphthylamidases, is such that the relationships between enzymes hydrolysing trifluoroacetyl-α-naphthylamide, naphthol-AS acetate, chloroacetyl-α-naphthylamide, and 5-bromoindoxyl acetate, are not easily resolved.

REFERENCES

Adams, C. W. M., and Glenner, C. G. (1962). *J. Neurochem.*, **3**, 233.

Albert, Z., Orlowska, J., Orlowski, M., and Szewczuk, A. (1964). *Acta histochem.*, **18**, 78.

Albert, Z., Orlowski, M., and Szewczuk, A. (1961). *Nature, Lond.*, **191**, 767.

Ali, S. Y., Evans, L., Stainthorpe, E., and Lack, C. H. (1967). *Biochem. J.*, **105**, 549.

Balogh, K. (1963). *Nature, Lond.*, **199**, 1196.

Barrett, A. J. (1967). *Biochem. J.*, **104**, 601.

Barrett, A. J. (1969). In "Lysosomes in Biology and Pathology," Eds. J. T. Dingle and H. B. Fell, Nth Holland, Amsterdam, Vol. 2.

Barrett, A. J. (1970). *Biochem. J.* **117**, 601.

Barrett, A. J., and Poole, A. R. (1969). *Nature, Lond.*, **224**, 279.

Behal, F. J., Asserson, B., Dawson, F., and Hardman, J. (1965). *Archs Biochem., Biophys.*, **111**, 335.

Behal, F. J., and Story, M. N. (1969). *Archs Biochem. Biophys.*, **131**, 74.

Benditt, E. P. (1956). *Fedn. Proc. Fedn. Am. Socs. exp. Biol.*, **15**, 507.

Benditt, E. P., and Arase, M. (1959). *J. exp. Med.*, **110**, 451.

Berger, J., and Johnson, M. J. (1939). *J. biol. Chem.*, **130**, 641.

Berger, J., and Johnson, M. J. (1940). *Ibid.*, **133**, 157.

Bergmann, F., Segal, R., and Rimon, S. (1957). *Biochem. J.*, **67**, 481.

Binkley, F., and Nakamura, N. (1948). *J. biol. Chem.*, **173**, 411.

Binkley, F., and Olson, C. K. (1951). *J. biol. Chem.*, **188**, 451.

Birnbaum, S. M., Levintow, L., Kinglsey, R. R., and Greenstein, J. P. (1952). *Ibid.*, **194**, 455.

Blackwood, C., and Mandl, I. (1961). *Analyt. Biochem.*, **2**, 370.

Braun-Falco, O. (1957). *Klin. Wschr.*, **35**, 50.

Braun-Falco, O., and Salfeld, K. (1959). *Nature, Lond.*, **183**, 51.

Brecher, A. S., and Suszkin, J. B. (1969). *Biochem. J.*, **112**, 335.

Burstone, M. S. (1956). *J. natn. Cancer Inst.*, **16**, 1149.

Burstone, M. S. (1962). "Enzyme Histochemistry and its application in the study of neoplasma." Acad. Press, New York, p. 395.

Burstone, M. S., and Folk, J. E. (1956). *J. Histochem. Cytochem.*, **4**, 217.
Burstone, M. S., and Weisburger, E. K. (1961). *J. Histochem. Cytochem.*, **9**, 712.
Christie, G. A. (1967). *Histochemie*, **10**, 272.
Curzen, P. (1954). *J. obstet. Gynaec. Br. Commonw.*, **71**, 388.
de Duve, C., Wattiaux, R., and Baudhuin, P. (1962). *Adv. Enzymol.*, **24**, 291.
de la Haba, G., Cammarata, P. S., and Timasheff, S. N. (1959). *J. biol. Chem.*, **234**, 316.
Ellis, S., and Perry, M. (1966). *J. biol. Chem.*, **241**, 3679.
Enerbäck, L., and Hansson, C. G. (1968). *J. Histochem. Cytochem.*, **16**, 441.
Felgenhauer, K., and Glenner, G. G. (1966). *J. Histochem., Cytochem.*, **14**, 401.
Folk, J. E., and Burstone, M. S. (1955). *Proc. Soc. exp. Biol. Med.*, **89**, 473.
Fruton, J. S. (1946). *J. biol. Chem.*, **166**, 721.
Fruton, J. S. (1960). In "The Enzymes," Eds. P. D. Boyer, H. Lardy, and K. Myrbäck, Acad. Press, New York, Vol. 4, p. 233.
Fruton, J. S., and Mycek, M. J. (1956). *Archs Biochem. Biophys.*, **65**, 11.
Fujii, S., and Fruton, J. S. (1958). *J. biol. Chem.*, **230**, 1.
Geratz, J. D. (1965). *Archs Biochem. Biophys.*, **111**, 134.
Glenner, G. G., Burstone, M. S., and Meyer, D. B. (1959). *J. natn. Cancer Inst.*, **23**, 857.
Glenner, G. G., and Cohen, L. A. (1960). *Nature, Lond.*, **185**, 846.
Glenner, G. G., Cohen, L. A., and Folk, J. E. (1965). *J. Histochem. Cytochem.* **13**, 57.
Glenner, G. G., and Folk, J. E. (1961a). *Nature, Lond.*, **192**, 338.
Glenner, G. G., and Folk, J. E. (1961b). *Nature, Lond.*, **192**, 339.
Glenner, G. G., Folk, J. E., and McMillan, P. J. (1962). *J. Histochem. Cytochem.*, 10, 481.
Glenner, G. G., Hopsu, V. K., and Cohen, L. A. (1962). *J. Histochem. Cytochem.*, **10**, 109.
Glenner, G. G., Hopsu, V. K., and Cohen, L. A. (1964). *J. Histochem. Cytochem.*, **12**, 545.
Glenner, G. G., McMillan, P. J., and Folk, J. E. (1962). *Nature, Lond.*, **194**, 867.
Gomori, G. (1953). *J. Histochem. Cytochem.*, **1**, 469.
Gomori, G. (1954a). *Proc. Soc. exp. Biol. Med.*, **85**, 570.
Gomori, G. (1954b). *Ibid.*, **87**, 559.
Green, M. N., Tsou, K.-C., Bressler, R., and Seligman, A. M. (1955). *Archs Biochem. Biophys.*, **57**, 458.
Greenbaum, L. M., and Fruton, J. S. (1957). *J. biol. Chem.*, **226**, 173.
Grzycki, S., and Rzeszowska, G. (1969). *Acta histochem.*, **32**, 351.
Gutmann, H. R., and Fruton, J. S. (1948). *J. biol. Chem.*, **174**, 851.
Hanes, C. S., Hird, F. J. R., and Isherwood, F. A. (1952). *Biochem. J.*, **51**, 25.
Hanson, H., Bohley, P., and Mannsfeldt, H. G. (1963). *Clin. chim. Acta*, **8**, 555.
Hanson, H. H., Huetter, H. J., Mannsfeldt, H. G., Kretschmer, H., and Sohr. C. (1967). *Hoppe-Seyl. Z.*, **348**, 680.
Hansson, C. G., and Svalander, C. (1967). *Acta endocr.*, **55**, 256.
Hartley, B. S. (1960). *A. Rev. Biochem.*, **29**, 45.
Hellströmm, B., and Nicander, L. A. (1967). *Life Sci.*, **6**, 1749.
Hess, R. (1965). *Biochim. biophys. Acta*, **99**, 316.
Hess, R., and Pearse, A. G. E. (1958). *Br. J. exp. Path.*, **39**, 292.
Hird, F. J. R., and Springell, P. H. (1954). *Biochem. biophys. Acta*, **15**, 31.
Hopsu, V. K., and Glenner, G. G. (1964). *J. Histochem. Cytochem.*, **12**, 674.
Hopsu, V. K., Kantonen, U.-M., and Glenner, G. G. (1964). *Life Sci.*, **3**, 1449.
Hopsu, V. K., Mäkinen, K. K., and Glenner, G. G. (1966a). *Acta chem. scand.* **20**, 1225.
Hopsu, V. K., Mäkinen, K. K., and Glenner, G. G. (1966b). *Acta chem. scand.*, **20**, 1932.
Hopsu, V. K., Mäkinen, K. K., and Glenner, G. G. (1966a). *Archs Biochem. Biophys.*, **114**, 557.
Hopsu, V. K., Mäkinen, K. K., and Glenner, G. G. (1966b). *Archs Biochem. Biophys.*, **114**, 567.

HOPSU, V. K., RUPONEN, S., and TALANT, S. (1961). *Acta histochem.*, **12**, 305.
HOPSU, V. K., SANTTI, R., and GLENNER, G. G. (1965). *J. Histochem. Cytochem.*, **13**, 117.
HOPSU, V. K., SANTTI, R., and GLENNER, G. G. (1966). *J. Histochem. Cytochem.*, **14**, 653.
HOPSU-HAVU, V. K., and ECKFORS, T. O. (1969). *Histochemie*, **17**, 30.
HOPSU-HAVU, V. K., and GLENNER, G. G. (1966). *Histochemie*, **7**, 197.
HOPSU-HAVU, V. K., and RINTOLA, P. *Histochem. J.*, **1**, 1.
HOPSU-HAVU, V. K., RINTOLA, P., and GLENNER, G. G. (1968). *Acta chem. scand.*, **22**, 299.
HOPSU-HAVU, V. K., and SARIMO, S. R. (1967). *Hoppe-Seyler's Z. physiol. Chem.*, **348**, 1540.
IDAHL, L.Å. and TÄLJEDAL, I.-B. (1968). *Biochem. J.*, **106**, 161.
ISHIHARA, M., YAMADA, H., CHAYA, I., and NOBATA, G. (1968). *Am. J. Obstet. Gynecol.*, **102**, 525.
IZUMIYA, N., and FRUTON, J. S. (1956). *J. biol. Chem.*, **218**, 59.
JACYSZYN, K., and LAURSEN, T. (1968). *Clin. chim. Acta*, **19**, 345.
JASMIN, G., and GAREAU, R. (1961). *Br. J. exp. Path.*, **42**, 592.
JOHNSON, M. J., and BERGER, J. (1942). *Adv. Enzymol.*, **2**, 69.
KLEINER, N., and WILKIN, P. (1963). *Gynéc. Onstét.*, **62**, 191.
KORHONEN, L. K., and RUPONEN, S. (1962). *Experientia*, **18**, 364.
KRESS, L. F., PEANASKY, R. J., and KLITGUARD, H. M. (1966). *Biochim. biophys. Acta*, **113**, 375.
LAGUNOFF, D., and BENDITT, E. P. (1961). *Nature, Lond.*, **192**, 1198.
LAPRESLE, C., and WEBB, T. (1962). *Biochem. J.*, **84**, 455.
LICHTENSTEIN, N., and FRUTON, J. S. (1960). *Proc. natn. Acad. Sci. U.S.A.*, **46**, 787.
MÄKINEN, K. K. (1968). *Archs Biochim. Biophys.*, **126**, 803.
MARKS, N., and LAJTHA, A. (1965). *Biochem. J.*, **97**, 74.
MCDONALD, J. K., REILLY, T. J., ZEITMAN, B. B., and ELLIS, S. (1966). *Biochem. biophys. Res. Commun.*, **25**, 771.
METRIONE, R. M., NEVES, A. G., and FRUTON, J. S. (1966). *Biochemistry*, **5**, 1597.
MONIS, B. (1963). *Am. J. Path.*, **42**, 301.
MONIS, B., NACHLAS, M. M., and SELIGMAN, A. M. (1959). *Cancer*, **12**, 601.
MONIS, B., WASSERKRUG, H., and SELIGMAN, A. M. (1965). *J. Histochem. Cytochem.*, **13**, 503.
MOTTET, K. N. (1967). *J. exp. Zool.*, **165**, 279.
MYERS, D. K., SCHOTTE, A., BOER, H., and BORSJE-BAKKER, H. (1955). *Biochem. J.*, **61**, 521.
NACHLAS, M. M., CRAWFORD, D. T., and SELIGMAN, A. M. (1957). *J. Histochem. Cytochem.*, **5**, 264.
NACHLAS, M. M., ROSENBLATT, D. H., MONIS, B., and SELIGMAN, A. M. (1960). *J. biophys. biochem. Cytol.*, **7**, 261.
NACHLAS, M. M., PLAPINGER, R. E., and SELIGMAN, A. M., (1964). *Archs Biochem. Biophys.*, **108**, 260.
NAGATSU, I., NAGATSU, T., and YAMAMOTO, T. (1968). *Experientia*, **24**, 347.
NAGEL, W., and WILLIG, F. (1964). *Nature, Lond.*, **201**, 617.
NEURATH, H., and SCHWERT, G. W. (1950). *Chem. Rev.*, **46**, 69.
NIEMI, M., and IKONEN, M. (1960). *Nature, Lond.*, **185**, 928.
NIEMI, M., and SYLVÉN, B. (1969). *Histochemie*, **18**, 40.
ONO, T., ETO, K., and ARAKAWA, K. (1968). *Clin. Chim. Acta*, **19**, 257.
OTTO, K. (1967). *Hoppe-Seyler's Z. phsyiol. Chem.*, **348**, 1449.
PATTERSON, E. K., HSIAO, S.-H., and KEPPEL, A. (1963). *J. biol. Chem.*, **238**, 3611.
PEARSE, A. G. E. (1956). *J. Path. Bact.*, **72**, 471.
PEARSE, A. G. E., HESS, R., and SNYDER, R. (1962). Henry Ford Hosp., Symposium. Little Brown, Boston. p. 657.
PEARSE, A. G. E., and TREMBLAY, G. (1958). *Nature, Lond.*, **181**, 1532.
PEARSON, B., WOLF, P., and ANDREWS, M. (1963). *Lab. Invest.*, **12**, 712.
PEPLER, W. J., and PEARSE, A. G. E. (1957a). *J. Neurochem.*, **1**, 193.
PEPLER, W. J., and PEARSE, A. G. E. (1957b). *Br. J. exp. Path.*, **38**, 221.

PLANTA, R. J., GORTER, J., and GRUBER, M. (1964). *Biochim. biophys., Acta*, **89**, 511.
PLANTA, R. J., and GRUBER, M. (1961). *Biochim. biophys. Acta*, **53**, 445.
PRESS, W. M., PORTER, R. R., and CEBRA, J. (1960). *Biochem. J.*, **74**, 501.
RAEKALLIO, J. (1960). *Nature, Lond.*, **188**, 234.
RAEKALLIO, J. (1961). *Ann. med. exp. biol. Fenn.*, **39**, Suppl.
RAEKALLIO, J., and MÄKINEN, P.-L. (1967). *Nature, Lond.*, **213**, 1037.
RAEKALLIO, J., and MÄKINEN, P.-L. (1969). *Experientia*, **25**, 929.
RAO, K. R., BIRNBAUM, S. M., KINGSLEY, R. R., and GREENSTEIN, J. P. (1952). *J. biol. Chem.*, **198**, 507.
RIEDEL, A., and WÜNSCH, E. (1959). *Hoppe-Seyler's Z. phsyiol. Chem.*, **316**, 61.
RIEKKINEN, P. J., EKFORS, T. O., and HOPSU, V. K. (1966). *Biochim. Biophys. Acta*, **118**, 604.
ROSENBLATT, D. H., NACHLAS, M. M., and SELIGMAN, A. M. (1958). *J. Am. chem. Soc.*, **80**, 2463.
RUTENBURG, A. M., KIM, H., FISCHBEIN, J. W., HANKER, J. S., WASSERKRUG, H. L., and SELIGMAN, A. M. (1969). *J. Histochem. Cytochem.*, **17**, 517.
SEELIG, H.-P., and ROEMBELD, R. (1969). *J. Histochem. Cytochem.*, **18**, 30.
SEMM, K., and WAIDL, E. (1962). *Z. Geburtsh. Gynäkol.*, **158**, 165.
SMITH, E. E., KAUFMAN, J. I., and RUTENBURG, A. M. (1965). *J. biol. Chem.*, **240**, 1718.
SMITH, E. L. (1948). *J. biol. Chem.*, **173**, 553.
SMITH, E. L. (1951). *Adv. Enzymol.*, **12**, 191.
SMITH, E. L., and BERGMANN, M. (1944). *J. biol. Chem.*, **153**, 627.
SMITH, E. L., and SPACKMAN, D. H. (1955). *J. biol. Chem.*, **212**, 271.
SNELLMAN. O. (1969). *Biochem. J.*, **114**, 673.
STACKMAN, D. H., SMITH, E. L., and BROWN, D. M. (1955). *J. biol. Chem.*, **212**, 255.
SYLVÉN, B., and BOIS, I. (1962). *Histochemie*, **3**, 65.
SYLVÉN, B., and BOIS, I. (1964). *Histochemie*, **3**, 341.
SYLVÉN, B., and BOIS-SVENSSON, I. (1964). *Histochemie*, **4**, 135.
SYLVÉN, B., and LIPPI, U. (1965). *Expl Cell Res.*, **40**, 145.
SYLVÉN, B., and MALMGREN, H. (1955). *Expl Cell Res.*, **8**, 575.
SYLVÉN, B., and SNELLMAN, O. (1964). *Histochemie*, **3**, 484.
SYLVÉN, B., and SNELLMAN, O. (1968). *Histochemie*, **12**, 240.
TAKAMATSU, H., ITOH, M., MIZOGUCHI, K., and IKETANI, J. (1963). *Proc. jap. Histochem. Ass.* 4th A.G.M., p. 146.
TALLAN, H. H., JONES, M. E., and FRUTON, J. S. (1952). *J. biol. Chem.*, **194**, 793.
TJEDER, A. (1966). *Acta chem. scand.*, **20**, 1442.
VANHA-PERTTULA, T., and HOPSU, V. K. (1965). *Histochemie*, **4**, 372.
VANHA-PERTTULA, T. P. J., and HOPSU, V. K. (1965). *Ann. Med. exp. Fenn.*, **43**, 32.
VANHA-PERTTULA, T. P. J., HOPSU, V. K., and GLENNER G. G. (1966). *J. Histochem. Cytochem.*, **14**, 314.
VANHA-PERTTULA, T., HOPSU, V. K., SONNINEN, V., and GLENNER, G. G. (1965). *Histochemie*, **5**, 170.
VOLLRATH, L. (1965). *Histochemie*, **4**, 397.
WACHSTEIN, M., MEAGHER, J. G., and OITIZ, J. (1963). *Am. J. Obstet. Gynecol.*, **87**, 13.
WIGGANS, D. S., WINITZ, M., and FRUTON, J. S. (1954). *Yale J. Biol. Med.*, **27**, 11.
WIELENGA, G., and WILLIGHAGEN, R. G. J. (1962). *Am. J. Obstet. Gynec.*, **84**, 1059.
WILLIGHAGEN, R. G. J., and PLANTEYDT, H. T. (1959). *Nature, Lond.*, **723**, 263.
WOESSNER, J. F. (1965). *Int. Rev. conn. Tiss. Res.*, **3**, 201.
WOODWARD, G. E. (1939). *Biochem. J.*, **33**, 1171.
YAMADA, M., and OFUJI, S. (1968). *J. invest. Dermat.*, **50**, 231.

CHAPTER 23

OTHER ENZYMES AND CO-ENZYMES

There are a number of enzymes, for which histochemical techniques have been devised, which do not fit into any of the foregoing chapters. These are dealt with in Chapter 23, not in order of importance but in the order in which the groups to which they belong are listed in the Enzyme Commission List.

First, therefore, come the *carbamoyl transferases* (aspartate and ornithine carbamoyltransferases), followed by an *acyltransferase* (choline acetyltransferase), and then by the *aminotransferases* (aspartate amino-transferase, ornithine-ketoacid aminotransferase), *phosphotransferases* (hexokinase, glucokinase, creatine kinase, phosphoglucomutase), *sulphuric ester hydrolases* (aryl sulphatase), *C-N hydrolases* (glutaminase), *carboxylases* (glutamate decarboxylase), *hydrolases* (carbonic anhydrase, aconitase), *carbon-sulphur lyases* (cysteine desulphurase), and *isomerases* (UDPG epimerase).

TRANSFERASES

Method for Aspartate Carbamoyltransferase (E.C.2.1.3.2)

A procedure for the demonstration of this enzyme, at the ultrastructural level, was devised by Spors and Merker (1969). The reaction catalysed is as follows:

Carbamoylphosphate + L-aspartate ⇌ orthophosphate + *N*-carbamoyl-L-aspartate

and the method is based on the capture of the released orthophosphate by lead ions incorporated in the medium. There is therefore no reason why the method should not be employed at optical microscope level. Details are given in the Appendix, for this reason.

Method for Ornthine Carbamoyltransferase (E.C.2.1.3.3)

Merker and Spors (1969) reported the development of a method for this enzyme on the same basis as for the preceding enzyme. This followed earlier reports by Mizutani (1967a and b, 1968) on its optical and electron microscopical localization, using a similar principle. Mizutani and Fujita (1968) found the enzyme to be strictly mitochondrial, in rat liver, but Merker and Spors indicated that there were two sites, mitochondria and endoplasmic reticulum.

This enzyme is soluble and some degree of fixation is therefore required. Cold buffered formalin was found by Merker and Spors to give optimal preservation but Mizutani and Fugita used glutaraldehyde perfusion and found that the pH optimum was between 6 and 7. Their method is given in Appendix 23, p. 1359.

Methods for Choline Acetyltransferase (E.C.2.3.1.6)

This enzyme, properly known as acetyl-CoA : choline *O*-acetyltransferase, catalyses the following reaction:

$$\text{Acetyl-CoA} + \text{choline} \rightleftharpoons \text{CoA} + O\text{-acetylcholine}$$

All the acetyltransferases belonging to sub-group 2.3.1 transfer acyl groups to and from coenzyme A. The methods described below, therefore, which depend on trapping the free CoA with lead ions, by virtue of its terminal SH group, can be extended to the demonstration of other acetyltransferases. The structure of coenzyme A, which is shown below,

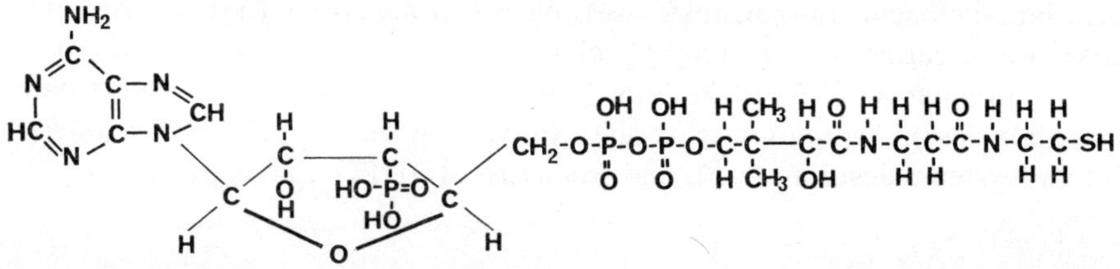

corresponds to 3′-phosphoadenosine diphosphate-pantoyl-β-alanyl-cysteamine.

A histochemical method for choline acetyltransferase was described by Burt (1969, 1970) which was based on the precipitation of CoA, released from acetyl-CoA by enzyme activity, with Pb^{2+} ions. The incubation medium contained choline (4·5 mM), acetyl-CoA (0·14 mM) and Pb^{2+} (1·8 mM), in 67 mM HEBES buffer at pH 6·0. The reaction product, converted to PbS with yellow ammonium sulphide, was limited to the cell body and proximal processes of the neurones.

A similar method was proposed by Kása *et al.* (1970) and details were given by Hebb *et al.* (1970). Thin sections of tissue were incubated in the presence of acetyl-CoA, choline and lead nitrate. The primary reaction product was converted by treatment with sodium sulphide at pH 7·5 to lead sulphide.

Critical for the establishment of the specificity of the method is the demonstration that other reactions leading to the release of CoA from acetyl-CoA are not contributing significantly. That is to say, the reaction must be shown to be dependent on the presence of choline. With his original (1969) method Burt observed non-specific hydrolysis at a level as high as 50 per cent. Subsequently (1970) he added the AChE inhibitor phospholine iodide with beneficial effects. Kása *et al.* stressed the role of AChE in the hydrolysis of acetyl-CoA and therefore preincubated their sections with eserine (0·1 mM) before incubation in the presence of this inhibitor (0·13 mM).

Even under these conditions there was still considerable non-specific hydrolysis, as shown by *in vitro* studies. This was attributed to non-specific esterase activity. When the latter was inhibited by DFP (1·0 mM) 90 per cent of non-specific hydrolysis, *in vitro*, was prevented. The dependence of the histochemical reaction on choline now became significant for the first time. Further studies showed that 1·0 mM DFP did not inhibit choline acetyltransferase.

Details of the method are given in Appendix 23, p. 1360.

Methods for Aspartate Aminotransferase (E.C.2.6.1.1)

This enzyme, which catalyses the reaction:

$$\text{L-Aspartate} + \text{2-oxoglutarate} \rightleftharpoons \text{oxaloacetate} + \text{L-glutamate}$$

was formerly known as glutamic-oxaloacetic transaminase (GOT). It was first demonstrated by Braunstein and Kritzmann (1937), and subsequently by Cohen (1940). High levels of activity have been shown in many organs and tissues, including kidney, brain and heart muscle. Two methods have been described for its histochemical demonstration. The first, by Lee (1968), was based on the precipitation of oxaloacetate by Pb^{2+} ions, and their subsequent conversion to lead sulphide in the usual manner.

Studies on the effects of lead, and of various fixatives, on the activity of GOT were described by Lee and Torack (1968a and b). In the presence of aspartic acid, as a chelating agent, and at a slightly alkaline pH, they observed (1968a) that the solubility of the lead salt of 2-oxoglutarate was at least 100 times greater than that of the oxaloacetate, which precipitated at a concentration of about 0·1 mM. The original method used cryostat sections post-fixed in cold acetone for 2–24 hours but later, in view of the observation by Garcia-Hermandez and Kun (1957) that acetone produced only a reversible precipitation of GOT, various aldehyde fixatives were tested. Brief treatment with dilute (1 per cent) glutaraldehyde for 1–5 minutes was found to give good localization of the enzyme and an additional 10–30 minutes in 4 per cent formaldehyde increased tissue preservation without further inhibiting the enzyme. Inhibition by glutaraldehyde, however, was sufficiently severe for the authors to recommend cold acetone fixation for the demonstration of low activity sites in the tissues.

In consideration of the report by Krebs (1942) on the non-enzymic decomposition of oxaloacetate in the presence of polyvalent metal ions Lee and Torack (1968b) studied the effect of Pb^{2+} ions on the reaction by a spectrophotometric method. They showed that non-enzymic decarboxylation of oxaloacetate was insignificant at slightly alkaline pH levels. They observed, further, that 3 mM Pb^{2+} did not produce serious inhibition of enzyme activity.

Technical details of Lee's method are given in Appendix 23, p. 1360.

The second method for the histochemical localization of GOT was described by Kishino (1968). This was based on the colorimetric method of

Babson *et al.* (1962) for the assay of serum GOT. This method, in turn, was based on earlier chemical studies (see Rodd, Chemistry of the Carbon Compounds III, pp. 361–2) showing that diazonium salts react with compounds containing activated methylene groups, to form "mixed" azo compounds tautomeric with phenylhydrazines.

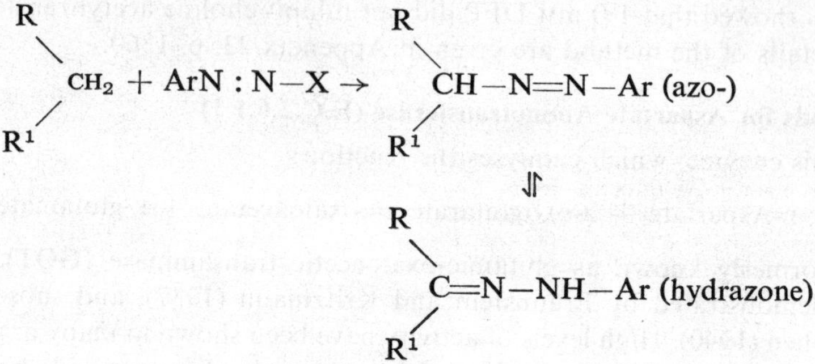

If the active-methylene compound is a *β*-keto acid, e.g. acetoacetic or oxaloacetic, decarboxylation occurs, and a *second* diazo molecule can react to give the formazan derivative:

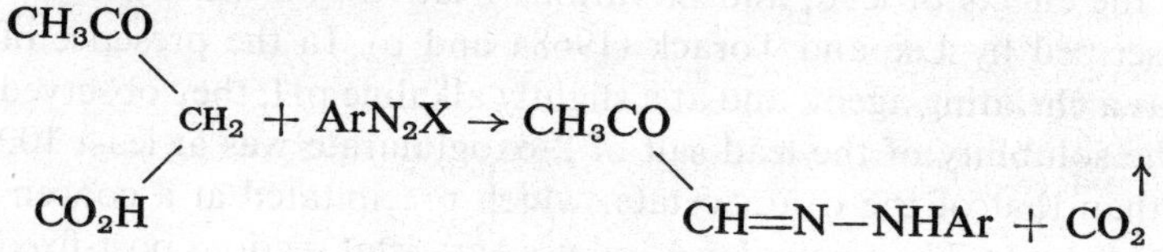

. . . (A)

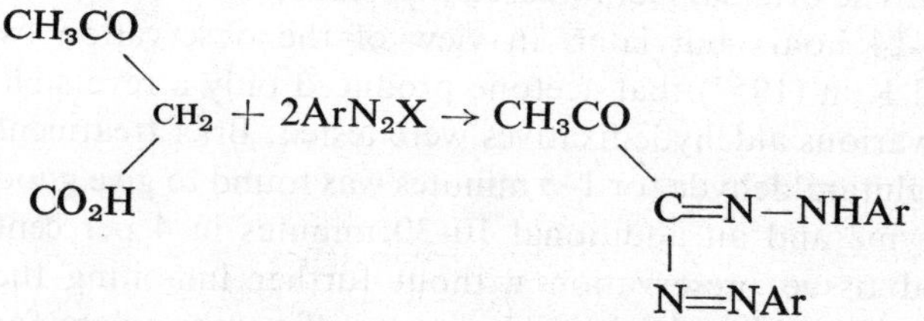

. . . . . (B)

A *third* diazo molecule can also react, with elimination of the acetyl group:

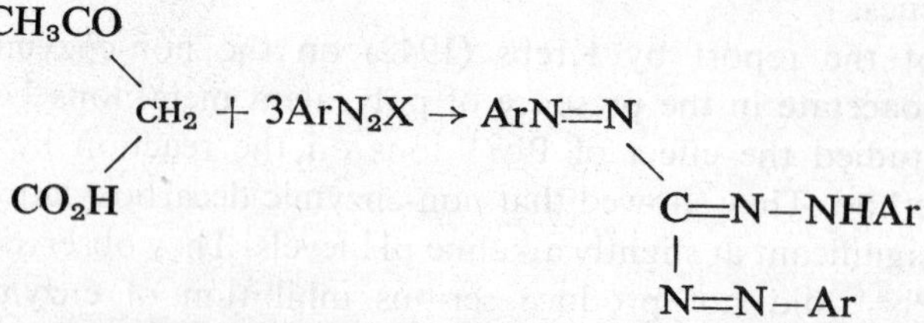

. . . . . (C)

These reactions occur with *benzene diazonium hydroxide* in *mild alkali*.

Babson *et al.* (1962) used 6-benzamido-4-methoxy-*m*-toluidine diazonium chloride, which was observed to couple rapidly with oxaloacetate at pH 7·4 to give a light-insensitive, stable, coloured compound. Serum was incubated for 20 minutes, with aspartic and 2-oxoglutaric acids, at 37° and pH 7·4. After addition of the diazonium salt, there followed a further incubation for 10 minutes. The coloured complex precipitated at high concentrations and PVP was included in the medium to prevent this. A high reagent blank was observed. After 30 minutes incubation the reagent blank was over 50 per cent of the specific value. Decker and Rau (1963) used Babson *et al's* diazonium compound to localize GOT on electropherograms (starch gel). They did not preincubate but included the diazo salt in the medium and incubated for 10–20 minutes at pH 7·4. Since the prosthetic group of GOT is pyridoxal 5-phosphate, Decker and Rau added this compound:

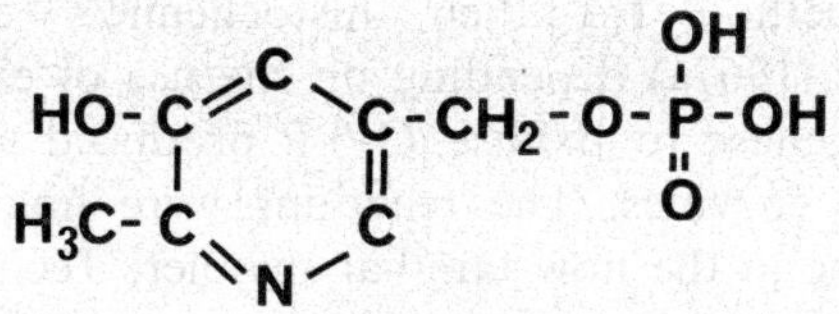

to their medium. In their controls no colour developed in the absence of aspartic acid.

Kishino's reaction is essentially that of Decker and Rau, omitting the pyridoxal phosphate. He used Azoene Fast Violet B and obtained a red precipitate convertible to reddish-brown by post-treatment with $CuSO_4$. A similar compound, Azoene Fast red salt (C.I. 37151) was used by Morgenstern *et al.* (1966) for spectrophotometric assay of serum GOT.

Kishino carried out a wide range of experiments, using assay techniques, to determine the optimal staining conditions. He found optimal preservation of enzyme activity after fixation in cold 12·5 per cent hydroxyadipaldehyde at pH 7·4 for 3 hours. Incubation was carried out for 1 hour at 37° and pH 7·4. Omission of aspartate reduced the reaction to a minimum. The method is described in Appendix 23, p. 1360, although tests carried out in my laboratory produced no staining specifically attributable to enzyme activity. Theoretical objections to the method are, first, that the solubility product of the FRP may never be exceeded under histochemical conditions. This would result in diffusion and false localization. A second objection is that the high reagent blank observed under biochemical conditions must necessarily occur in the histochemical medium. It has not been shown that this factor can be ignored.

Method for Ornithine-Ketoacid Aminotransferase (E.C.2.6.1.13)

This soluble enzyme, L-Ornithine : 2-oxoacid aminotransferase, catalyses the following reaction:

L-Ornithine +2 oxoglutarate $\rightleftharpoons$ L-glutamate + L-glutamate γ-semialdehyde

It is believed to play a role in regulating the relationship between the urea and tricarboxylic cycles in animal tissues.

A multistep method for its histochemical demonstration was described by Kishino (1967) which depended on oxidation of the glutamate PRP by exogenous glutamate dehydrogenase, in the presence of NAD and Nitro-BT, thus using the tetrazolium reductase final pathway.

Method for Hexokinase and Glucokinase (E.C.2.7.1.1 and 2)

These soluble enzymes, ATP : D-hexose phosphotransferase and ATP : D-glucose-6-phosphotransferase catalyse the transfer of the terminal phosphate group of ATP to a D-hexose with the liberation of one hydrogen ion. Glucokinase can only catalyse the phosphorylation of glucose while hexokinase can act on several different sugars (Walker, 1963; Walker and Rao, 1964).

A multistep method for their histochemical demonstration was described by Meijer (1967a) depending on the use of exogenous glucose-6-phosphate dehydrogenase to oxidise G-6-P produced from glucose by the activity of the two enzymes. The reactions were linked through NADP tetrazolium reductase in the now familiar manner. Technical details appear in the Appendix.

Methods for Creatine Kinase (E.C.2.7.3.2)

ATP : Creatine phosphotransferase, to give the enzyme its full name, transfers phosphate from ATP to creatine and from phosphocreatine to ADP:

$$\text{ATP} + \text{creatine} \rightleftharpoons \text{ADP} + \text{phosphocreatine}$$

Two distinct histochemical methods have been proposed for its demonstration in the tissues. The first, due to Hori (1966) was designed especially for the demonstration of creatine kinase in skeletal muscle. The observation that ATPase but no ADPase activity could be found in muscle suggested the possibility of demonstrating the activity of creatine kinase by a metal precipitation technique. Incubation of tissues with ADP or phosphocreatine alone produced no result. With the two together, endogenous ATPase acted on the ATP produced by endogenous creatine kinase activity. In the presence of Pb^{2+} ions a precipitate was produced (see Vol. 1, p. 528).

While this method certainly demonstrated the presence of creatine kinase, the localization of the FRP was certainly that of ATPase.

An alternative approach was used by Sjövall (1967). This author designed a multistep method depending on no less than 4 enzymes for the development of the final reaction product. The reactions were as follows:

$$\text{Phosphocreatine} + \text{ADP} \xrightarrow{\text{creatine kinase}} \text{Creatine} + \text{ATP}$$
$$\text{ATP} + \text{Glucose} \xrightarrow{\text{hexokinase}} \text{G-6-P} + \text{ADP}$$
$$\text{G-6-P} + \text{NADP} \xrightarrow{\text{G-6-PDH}} \text{6-phosphogluconate} + \text{NADPH}_2$$
$$\text{NADPH}_2 + \text{PMS} \longrightarrow \text{NADP} + \text{reduced PMS}$$
$$\text{Reduced PMS} + \text{Nitro BT} \xrightarrow{\text{Tet. reductase}} \text{Formazan} + \text{PMS}$$

It will be observed that the method incorporates within itself the multistep method for hexokinase described in the foregoing section, with the additional complication of PMS, and that the final pathway is by way of the tetrazolium reductase. Localization of the FRP may, fortuitously, be the same as that of the initiating enzyme (creatine kinase).

Method for Glucose Phosphomutase (E.C.2.7.5.1)

A multistep method for this enzyme has already been described in Vol. 1 (p. 538) and the method is given in Appendix 15, p. 727.

The enzyme catalyses the reversible reaction:

D-glucose 1,6-diphosphate + D-glucose-1-phosphate ⇌ D-glucose-6-phosphate + D-glucose 1,6-diphosphate.

It is activated by Mg^{2+} ions and by pretreatment with EDTA, L-histidine or L-cysteine. It is specifically inhibited by beryllium ions (Thomas and Aldridge, 1966).

An improved histochemical technique, making use of these facts, was reported by Yano (1968). Cryostat sections post-fixed in cold acetone for 5–10 minutes were used, and an activating (imidazole) buffer was used in the preparation of the incubating medium. Details are given in the Appendix.

Yano proposed that three control procedures must be carried out to ensure that the true localization of enzyme activity was being demonstrated. These were (1) medium lacking substrate (glucose-1-phosphate); (2) medium lacking glucose-6-phosphate dehydrogenase and (3) medium lacking both substrate and dehydrogenase. A positive reaction was obtained in the case of control 2 and this was explained as being due to endogenous glucose-6-phosphate dehydrogenase. Once again proof of the primary responsibility of first enzyme in the series (the phosphoglucomutase) is all that the controls provide.

Hydrolases

The Sulphatases

There are five known types of sulphatase in animal and plant tissues and these are known as arylsulphatase, sterol sulphatase, glucosulphatase, chondrosulphatase and myrosulphatase. The gluco- and chondrosulphatases are not present in the tissues of higher animals (Fromageot, 1950), and although myrosulphatase has been found in liver, muscle, kidney and other organs of the horse and rabbit, the only histochemical attempt at its demonstration was that of Peche (1913), until two methods were evolved by Ohara and Kurata (1950). These authors trapped the inorganic sulphate with lead ions and converted the product subsequently to lead sulphide. The natural substrate of this enzyme, sinigrin,

$$\begin{array}{c} C_6H_{11}O_5SC = N.C_3H_5 \\ | \\ O\ SO_2OK \end{array}$$

obviously lends itself to a procedure of the Gomori type, precipitating the released $KHSO_4$ with barium salts, for instance. The only other sulphatase for which histochemical methods have been employed is arylsulphatase. Attempts by Dohlman and Friedenwald (1955) to demonstrate chondrosulphatase in mammalian tissues, with heparin of chondroitin sulphate as substrates, were completely unsuccessful.

This enzyme was described by Neuberg and Kurono (1923) in takadiastase and it also occurs in *Murex*, where it is responsible for the hydrolysis of potassium indoxyl sulphate. According to Huggins and Smith (1947) arylsulphatase is found in rat tumours in far higher concentrations than in normal tissues. A transmissible rat sarcoma, for instance, contained 0·8–1·7 units of enzyme per mg. of tissue, while the connective tissue of origin contained 0·2–0·02 units per mg. only. Many substrates have been used for the chemical assay of arylsulphatases. These are potassium *p*-nitrophenylsulphate, used by Huggins and Smith, potassium indoxylsulphate used by Neuberg and Wagner (1925), nitrocatechol sulphate, potassium *p*-acetylphenyl sulphate, and potassium 2-hydroxy-5-nitrophenyl sulphate (Dodgson *et al.*, 1955; Baum *et al.*, 1958). The principles involved in the use of these formed the basis of two different histochemical methods.

A very considerable amount of biochemical work has been done on the arylsulphatases of mammalian tissues, chiefly by two groups of workers. The complex nature of arylsulphatase in ox liver was discussed by Roy (1953) who found that two fractions could be extracted from homogenates. One, having a pH optimum of 4·7 in acetate buffer, had an optimal substrate concentration of 3 mM (nitrocatechol sulphate) whereas the second was most active at pH 5·7 and showed an anomalous response to variations in substrate concentration. Dodgson, Spencer and Thomas (1953), using *p*-acetylphenyl sulphate as substrate found that for rat liver sulphatase the pH optimum was 7·2 in acetate buffer. Activity in the case of the male rat was significantly higher than in the female.

In a later paper in this series Dodgson, Spencer and Wynn (1956) divided the arylsulphatase activity of ox, rat and human livers into three distinct enzymes, as originally suggested by Roy (1953, 1954). Two of these enzymes (the sulphatases A and B of Roy) were readily solubilized and found, originally, to be associated with the mitochondrial fraction. The third (C) was considered to be microsomal in origin.

Subsequently it was shown that both A and B were lysosomal enzymes (Viala and Gianetto, 1955) while C was definitely localized in the endoplasmic reticulum (Dodgson, Spencer and Thomas, 1954; Dodgson, Rose and Spencer, 1957; Roy, 1960a; Savant *et al.*, 1964). Since this enzyme was found to be firmly bound to the insoluble membrane component of the reticulum, Milsom, Rose and Dodgson (1968) suggested its use as a microsomal marker enzyme.

A survey of the distribution of arylsulphatase in the tissues of seven mammals was made by Rutenburg and Seligman (1956) but no distinctions could be made. In his earliest work Roy (1953, 1954) indicated that sulphatases A and B hydrolysed nitrocatechol sulphate rapidly but nitrophenyl sulphate hardly at all and later (1960b) he re-emphasized these points. The third enzyme, arylsulphatase C, was found by Roy (1956b) to hydrolyse nitrophenyl sulphate very rapidly. Similar observations were made by Dodgson, Spencer and Thomas (1954), and by Dodgson *et al.* (1955), with subsequent extension to the newer substrate, potassium 2-hydroxy-5-nitrophenyl sulphate, for which A and B show high activity and affinity.

For arylsulphatases A, B and C the pH values at which optimum catalytic activity is observed are 4·5, 5·5 and 7·5, respectively (Roy, 1953; Dodgson and Spencer, 1957). Arylsulphatase C is inhibited by cyanide and sulphite anions whereas A and B are inhibited by sulphate and phosphate (Gregory and Robbins, 1960). Arylsulphatase B was found by Webb and Morrow (1960) to be activated by chloride ions but elsewhere it has been claimed that it is inhibited by chloride, especially where nitrocatechol is used as substrate. Roy (1958) found by paper electrophoresis that arylsulphatase A moves towards the anode and arylsulphatase B toward the cathode.

Possible Functions of Arylsulphatases

There are few indications of the possible functions of the various arylsulphatases. Sulphatase C was found to act as a steroid sulphatase and Roy (1956a) showed that this enzyme could synthesize dehydroepiandrosterone sulphate. Like the sulphatase described by Dodgson *et al.* (1953) it is much stronger in the male than in the female.

Working with invertebrates, Suzuki, Takahashi and Egami (1957) showed that the sulphated polysaccharide of the mucous gland of *Triton nodiferus* (*Charonia lampas*) could be sulphated by a system consisting of an arylsulphate, an arylsulphatase, a carbohydrate factor and at least one unknown factor. The physiological significance of this system is not clear, but, if similar systems exist in other animals, there should be present arylsulphates whose specific function is to act as sulphate donors. Dodgson, Rose and Tudball (1959) had considered that tyrosine-*O*-sulphate might act in this way, but their studies showed that it was not attacked by sulphatases B and C and only slowly by sulphatase A. The search for an arylsulphate sulphate donor continues therefore.

The part played by arylsulphatases A and B in the functioning of lysosomes has not yet been explained although Austin *et al.* (1964), Austin, Armstrong and Shearer (1965) and Mehl and Jatzkewitz (1965) indicated that arylsulphatase A was involved in the degradation of cerebroside sulphate esters. Low levels of activity were found in brain, liver and kidney from cases of metachromatic leucodystrophy and Langelaan (1969) showed that blood leucocytes, in this condition, contained less than 10 per cent of normal control values. Porter *et al.* (1969) demonstrated arylsulphatase A deficiency in skin fibroblast cultures from cases of this disorder. Further work by Mehl and Jatzkewitz (1968) clearly showed that cerebroside-3-sulphate was a natural substrate of mammalian arylsulphatase A. The limpet enzyme was devoid of activity for cerebroside sulphates.

Histochemically, as will be seen, we are still behind the biochemists in the matter of the sulphatases. Until we can make absolutely clear distinction between the three types of enzyme it is useless to expect much progress in the interpretation of results.

Methods for Arylsulphatases (E.C.3.1.6.1)

Methods Using Naphthyl Sulphates

Seligman *et al.* (1949) endeavoured to produce a simultaneous coupling azo dye method for the demonstration of sulphatases using α- and β-naphthyl sulphuric acid esters. The former were hydrolysed exceedingly slowly and were, therefore, histochemically unsuitable. With the latter, although hydrolysis occurred readily enough, the authors found that coupling with a diazonium salt at pH 4·5 would not occur. Since the optimum pH for arylsulphatases had been found by Huggins and Smith (1947) to be in the region of pH 6·12, and since there are several diazonium salts which will couple with naphthols at this pH, I endeavoured to produce a simultaneous coupling azo dye method, returning to the principle abandoned by Seligman *et al.* (1949). Using 10 mM potassium 2-naphthyl sulphate as substrate, maintained at pH 6·12 with 0·5 M sodium acetate-acetic acid buffer, and incubating cold formalin-fixed frozen sections of rat tissues for as long as 24 hours at 37°, no azo dye precipitate was obtained in any case. The diazonium salt usually employed, the diazotate of 4-benzoylamino-2 : 5-dimethoxyaniline, had been shown to be extremely stable at pH 6·12 and it certainly coupled with β-naphthol at this pH.

Post-coupling Azo Dye Method

Seligman and his colleagues subsequently evolved a post-coupling method using potassium 6-bromo-2-naphthyl sulphate as substrate and incubating at pH 4·5–6·0 at 37° for a suitable period. After washing, the final coupling with diazotized *o*-dianisidine was performed at pH 9·0, to give a blue insoluble azo dye. Seligman, Nachlas and Cohen (1950) made brief mention of an alternative substrate for sulphatases (potassium 6-benzoyl-2-naphthyl sulphate) with which they used a similar post-coupling technique. The method was fully described by Rutenburg, Cohen and Seligman (1952), who used the benzoyl compound with formalin-fixed or fresh frozen sections at pH 6·1. The incubation period was 2–3 hours with active tissues and subsequent coupling with Fast blue B produced red (monocoupling) and blue (dicoupling) products. In my hands the use of this method gave rise to red rather than blue products and since these were soluble in lipids the general localization of the enzyme was poor. Studies in applied histochemistry of the sulphatases have been made with post-coupling techniques using either 6-bromo or 6-benzoyl-2-naphthyl sulphate. Using the former, for instance, Hayashi *et al.* (1957) found cyclical changes in the surface layers of rat endometrium. These methods are unsatisfactory, nevertheless details are given in Appendix 23, p. 1362, for those who may wish to compare the method with those using alternative naphtholic substrates.

Naphthol AS-sulphate Method

Using Naphthol AS-sulphate synthesized by the method of Rath and Feess (1954), Goessner (1958) was able to demonstrate the localization of arylsulphatase at pH 6·2 with Fast red TR salt (9) in a simultaneous coupling method. The results he obtained, which are illustrated in Fig. 209, p. 943, in the case of mouse kidney, indicated that further developments in the same direction would be likely to result in improved localization of the enzyme.

Napthol AS and AS-D sulphates were used by Woohsmann and Brosowski (1964), with a variety of stable diazotates, and Woohsmann and Hartrodt (1964) synthesized a number of AS sulphates for use with the same technique. Using hexazonium pararosanilin (Vol. 1, pp. 518, 565) in place of Fast red LTR Wächtler and Pearse (1966) obtained greatly improved results, with strict lysosomal localization of enzyme in amphibian tissues (Plate XXIId, p. 968). With mammalian tissues the results were less satisfactory and Katchburian *et al.* (1967) therefore added Triton X (Final conc: 0·1 to 1·0 per cent) to their medium with a view to altering permeability with the lysosomal method. Only a weak and diffuse reaction was obtained, however. The intact mammalian lysosomal membrane is presumably impermeable to AS sulphates, as indicated by the work of Lloyd (1969) who employed the release of nitrocatechol sulphatase activity as an indicator of lysosomal membrane changes.

A modified Naphthol AS method is given in Appendix 23, p. 1362. It is recommended for non-mammalian tissues only and the type of result obtainable is shown in Plate XXIId, p. 968.

8-Hydroxyquinoline sulphate Method

Following the use of 8-hydroxyquinoline sulphate by Ohara and Kurata (1952), and of this and other aryl sulphate esters by Hopsu *et al.* (1965) in a method using barium salts as precipitating reagents, Woohsmann and Hartrodt (1967) employed 8-hydroxyquinoline sulphate and hexazotized pararosanilin in a method designed for optical microscopy. This method, which is described in Appendix 23, p. 1364, gives lysosomal localization in mammalian as well as non-mammalian tissues.

Metal Precipitation Techniques

The most reliable technique for the demonstration of aryl sulphatases A and B, in mammalian and non-mammalian tissues, was developed by Goldfischer (1965) primarily for their demonstration at the E.M. level. The results are excellent at optical microscopic level also.

The substrate employed was nitrocatechol sulphate and the tissues were first subjected to prolonged immersion fixation in cold formalin or glutaraldehyde. Again primarily for E.M. demonstration of aryl sulphatase Hopsu-Havu *et al.* (1967) carried out a number of experiments designed to establish the optimum conditions for a metal precipitation technique. They considered the effects of fixation, capturing ions, substrate concentration and pH. Perfusion fixation, with cold buffered 5 per cent glutaraldehyde (pH 7·4), followed by 1–2 hours post-fixation of small blocks of tissues, caused up to 22 per cent inhibition or aryl sulphatase activity and the optimum pH was found to be 5·3. A rather high optimum substrate concentration (7 mM) was indicated for unfixed kidney homogenates using nitrocatechol sulphate.

Employing the medium described by Hopsu-Havu *et al.* (1967) Abraham *et al.* (1967) studied the aryl sulphatases of rat heart muscle lysosomes. They were able to abolish staining by pretreatment of fixed frozen sections in 1 per cent Triton X at 22° for 1 hour, before incubation. A similar technique, with free floating sections, was used by Abraham (1967). Details of a modified Goldfischer technique for optical microscopy are given in Appendix 23, p. 1363, and the result is illustrated in Plate XXIIIa, p. 1363.

A metal precipitation technique, employing radioactive sulphate, was used by Kawiak *et al.* (1964) for the histochemical localization of aryl sulphatase C in cartilage. The substrate was *p*-nitrophenyl [^{35}S] sulphate. Barium acetate was used to precipitate the free ^{35}S-sulphate and slices of cartilage were incubated for one hour at 37°. Subsequently the tissues were embedded in paraffin and sections were then processed by emulsion autoradiography, using Kodak AR-10 emulsion. The results showed a random distribution of ^{35}S throughout the cartilage cells. The principle of this method is of interest, the practice a good deal less so.

C-N Hydrolases

The many enzymes in this group act specifically on C-N bonds other than peptide. Those which act on linear amides do so with the production of a carboxylic acid and ammonia. Three important representatives of the group are asparaginase, glutaminase and urease. The enzymic reactions of glutamic acid are shown in Fig. 210, below.

The glutamate enzyme systems control the amount of 2-oxoglutarate available, and thus the operating rate of the tricarboxylic acid cycle.

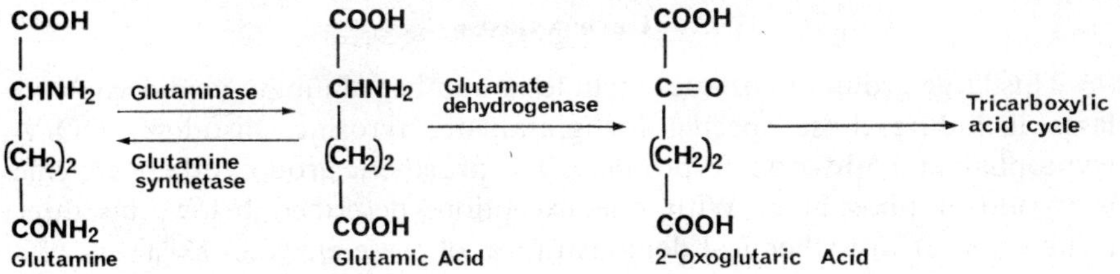

FIG. 210. Enzymatic Relations of Glutamic Acid.

Methods for Glutaminase (E.C.3.5.1.2)

Observing the important role of this enzyme in the production of ammonia by the kidney and in regulating the pH of the urine Karnovsky and Himmelhoch (1961) evolved a multistep histochemical technique for its demonstration. Their reaction depended on the oxidation of glutamate, produced by enzyme activity, by exogenous glutamate dehydrogenase. This second reaction was coupled to reduction of a tetrazole (Nitro-BT) and, as in other methods of this type, the final pathway was by way of NADH diaphorase. Controls carried out by the original authors indicated that localization was not affected by diffusion of the products. The reaction, which was inhibited by cyanide, was carried out at pH 7·9 which is dangerously high (see Chapter 20, p. 910).

An alternative method was described by Potts (1965) depending on the hydrolysis of the artificial substrate α-naphthyl-5-glutamine, and a simultaneous coupling azo dye method.

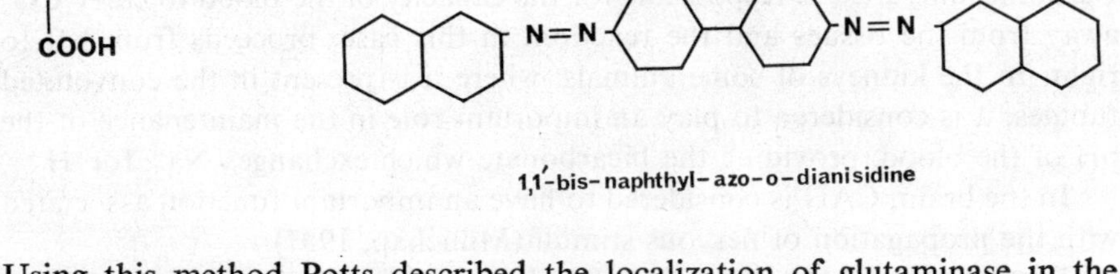

Using this method Potts described the localization of glutaminase in the retina, finding the enzyme in the inner and outer reticular layers rather than, as would be expected, in the bipolar cell layer.

Further reports on the specificity of the method, which I have not tested, may be expected.

Carboxylases

This large group of enzymes includes a number of amino-acid decarboxylases including those specific for glutamate, tyrosine, histidine, DOPA, tryptophan and 5-hydroxytryptophan. The prosthetic group of these enzymes is pyridoxal phosphate. With one exception, described below, no direct techniques for histochemical demonstration of these enzymes exist.

Method for Glutamate decarboxylase (E.C.4.1.1.15)

The reaction catalysed by this enzyme is as follows:

$$\text{L-glutamate} \rightleftharpoons \text{4-aminobutyrate} + CO_2$$

A histochemical method for its detection was described by Higashi *et al.* (1960) which involved the capture of the released CO_2 by means of Ba^{2+} ions. Exogenous carbonic anhydrase was added to the medium in order to speed up the hydration of CO_2.

The technique, which is an interesting one, is described in Appendix 23, p. 1364. I have not tested it, however, and cannot vouch for the accuracy of localization afforded.

Hydrolases

Carbonic Anhydrase (Carbonate hydrolase)

This enzyme, later shown to be a zinc protein, was discovered by Meldrum and Roughton (1932, 1933) in red blood cells. Its characteristics, in so far as they were known after the first twenty years' work on the enzyme, were described by Roughton and Clark (1951).

By the action of the enzyme carbonic acid is split to yield CO_2 and H_2O and, in the reverse direction, CO_2 is hydrated to carbonic acid.

$$O=C=O + H{-}O{-}H \underset{\text{CAH}}{\rightleftharpoons} O=C\begin{matrix}OH\\OH\end{matrix} \underset{\text{spontaneous}}{\rightleftharpoons} O=C\begin{matrix}O^-\\OH\end{matrix} + H^+$$

Carbonic anhydrase is responsible for the capacity of the blood to carry CO_2 away from the tissues and the reaction, in this case, proceeds from left to right. In the kidneys of some animals, where it is present in the convoluted tubules, it is considered to play an important role in the maintenance of the pH of the blood, providing the bicarbonate which exchanges Na^+ for H^+.

In the brain, CAH is considered to have an important function associated with the propagation of nervous stimuli (Millichap, 1957).

Erythrocyte CAH occurs, in several species, in the form of 2 or more isoenzymes (Lindskog, 1960; Rickli and Edsall, 1962; Tashian, 1965; Duff and Coleman, 1966; Furth, 1968; McIntosh, 1969; the three red cell isoenzymes described by McIntosh (1969) were shown to be closely similar in most respects. In regard to the hydration of CO_2, however, they differed very greatly. In recent years CAH has been found to catalyse a number of reactions in addition to the hydration of CO_2 and the dehydration of carbonic acid. These activities include the hydrolysis of esters, such as *p*-nitrophenyl acetate (Tashian *et al.*, 1964; Pocker and Stone, 1965; Armstrong *et al.*, 1966; Verpoorte *et al.*, 1967) and the hydration of aldehydes, such as acetaldehyde (Pocker and Meany, 1965, 1967).

Specific inhibition is produced by many aromatic sulphonamides (Mann and Keilin, 1940; Armstrong *et al.*, 1966; Whitney *et al.*, 1967; Maren *et al.*, 1960) and monovalent anions (Pocker and Stone, 1965; Kernohan, 1965; Lindskog, 1966).

A very comprehensive review of the chemistry, physiology and inhibition of CAH was provided by Maren (1967), containing most of the modern developments in the field.

Method for Carbonic Anhydrase (E.C.4.2.1.1)

The first published histochemical technique for CAH was that of Kurata (1953) who incubated cold acetone-fixed tissue slices in a bicarbonate mixture containing $MnCl_2$. After washing, the tissue was embedded in paraffin wax. Subsequently sections were cut and treated with potassium periodate in order to render visible the precipitates of manganese carbonate. Kurata suggested that calcium, cobalt, or nickel could all be employed in place of manganese.

Most workers were unable to obtain satisfactory results with the original method of Kurata but modifications using the cobalt system were employed by Goebel and Puchtler (1954) and by Braun-Falco and Rathjens (1955a). In the modification introduced by the latter authors a buffered bicarbonate solution containing cobalt ions was used and the reaction taking place in the tissues was supposedly:

$$Co(HCO_3)_2 \overset{\text{CAH}}{\rightleftharpoons} CoCO_3 + H_2O + CO_2$$

The precipitated cobalt carbonate was subsequently converted to the sulphide and thus rendered visible. The results, in my hands, were always very dirty with fine and coarse amorphous precipitates covering the entire section at times. Although the specificity of the method was supported by the observed inhibition with cyanide, azide or sulphide positive reactions could be obtained in inactivated control sections and in tissues incubated in the absence of bicarbonate. The specific CAH inhibitor Diamox (2-acetylamino-1,3,4-thiadiazol-5-sulphonamide) was used by both Puchtler and Ranniger (1955) and by Braun-Falco and Rathjens (1955b), who reported that no inhibition of the reaction occurred. The probable explanation for this last observation was the insolubility of Diamox in water. Häusler (1958), using the soluble sodium salt (Diamox parenteral-Lederle) reported inhibition both of the tissue reaction and also of the activity of a purified CAH (Cartase-Schering), adsorbed on filter paper. Modifications of the cobalt method which were introduced by Häusler successfully eliminated the troublesome precipitates associated with the older methods. This was achieved by dissolving the bicarbonate-cobalt in sodium sulphate and adding 50 mM H_2SO_4. The strongest reaction in rat tissues was in the pars recta of the proximal tubules.

The mechanism of the reaction was explained by Häusler as follows:

$$2\,HCO_3^- \rightleftharpoons CO_3^{2-} + H_2CO_3$$

$$H_2CO_3 \overset{CAH}{\rightleftharpoons} H_2O + CO_2$$

The CO_2, produced by the second, enzyme-catalysed reaction, escapes and the equilibrium of the first reaction, which is spontaneous, shifts to the right. Large numbers of CO_3^{2-} ions are produced and these are precipitated by the Co^{2+} ions present in the medium.

The question of the specificity of the histochemical reaction has been taken up by a number of workers. After testing sections from kidney and pancreas in seven mammalian species Fand, Levine and Erwin (1958, 1959) suggested that the final localization of CoS might be connected with the presence of zinc. This hypothesis arose from the observation that in the pancreas staining by the modified Kurata method occurred in the islets whereas, according to biochemical evidence, pancreatic CAH is situated in the acini. Using a Coujard slide technique Fand and her associates found that zinc, manganese, cadmium and other divalent metal ions gave strongly positive staining. A positive correlation with dithizone staining for zinc (Chapter 28, p. 1155) was obtained in the islets but kidney tissue was not stained by this reaction. It was concluded that the Kurata method did not demonstrate CAH.

In my 2nd Edition (1960) I agreed with Fand *et al.* in concluding that the Kurata method did not demonstrate the localization of CAH. Opinion was reserved on the subject of the Häusler modification. Using the latter Mustakallio *et al.* (1960) carried out tests for specificity using various metal salts in gelatine. They concluded that staining was mostly non-enzymic in character although they were able to inhibit the reaction to some extent by the *in vivo* administration of Diamox.

The specificity of the Häusler modification for CAH was "uncontested" acccording to Bleyl (1964) who nevertheless recommended the use of chelating agents, such as dithizone, in order to exclude non-specific precipitation due to the presence of Zn^{2+} or other bivalent metals. Complete inactivation of the enzyme, and inhibition of the reaction, was obtained by Diamantstein and Schlüns (1964) and by Korhonen *et al.* (1964), both using Diamox (4 mM). The first mentioned authors obtained very strong reactions in the uterus of the laying hen, agreeing with the views of Gutowska and Mitchell (1945) on the role of CAH in production of the egg shell. Korhonen and Korhonen (1965) showed that the levels of CAH in electrophoretic strips corresponded to the levels of histochemical staining. This they considered to confirm the

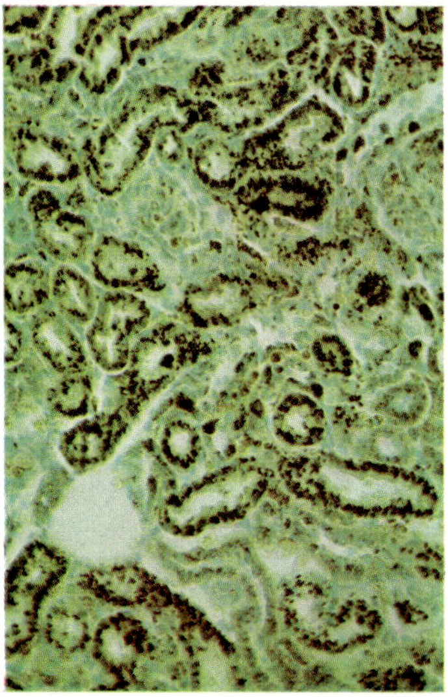

XXIIIa. Rat kidney. Arylsulphatase reaction using nitrocatechol sulphate as substrate. 2 hours after Imferon injection large numbers of phagosomes develop. These contain the enzyme as do pre-existing lysosomes. × 170.

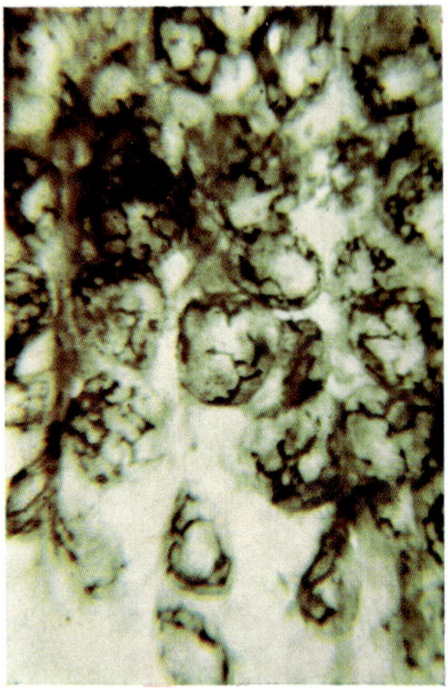

XXIIIb. Rat stomach. Localization of carbonic anhydrase by the modified Hausler technique. The enzyme is restricted to the canalicular systems of the parietal cells. × 250.

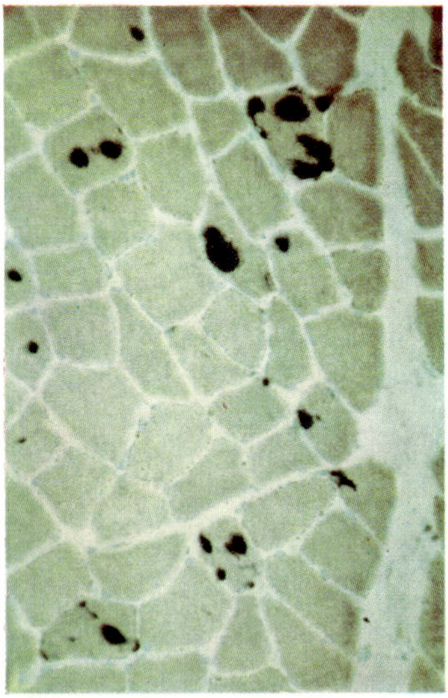

XXIIIc. Human skeletal muscle. From a case of myopathy with tubular aggregates. A strong reaction for ubiquinone, or similar redox compounds, is present in the aggregates. × 170.

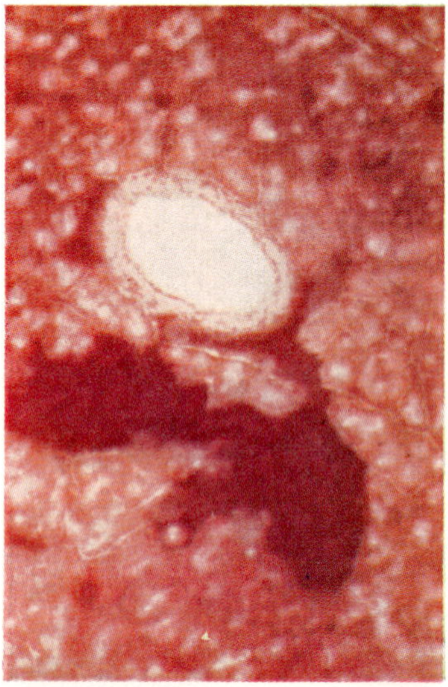

XXIIId. Rat pancreas. Substrate film technique for amylase. Demonstrated with a non-separation technique and Lugol's iodine staining. Two negative reacting islets are present. Duct epithelium and acinar cells show high activity. × 400.

[*To face p.* 997

specificity of the reaction, which was further modified by Hansson (1967) who substituted dipotassium phosphate for sodium sulphate and found that the incubation period was considerably shortened.

Errors in Häusler's reasoning were pointed out by Hansson (1967). These concerned mainly the effect of the release of CO_2 into the air which Häusler supposed would *increase* the concentration of CO_3^- ions in the medium. It is clear that a decrease would, in fact, result. Hansson repeated Häusler's control experiment showing that high levels of CO_2 over the medium totally inhibited the reaction. Thus loss of CO_2 is the essential component of the reaction, driving the equilibrium to the right and resulting in a consumption of HCO_3^- and CO_3^- and the production of OH^-. The whole reaction should therefore be written:

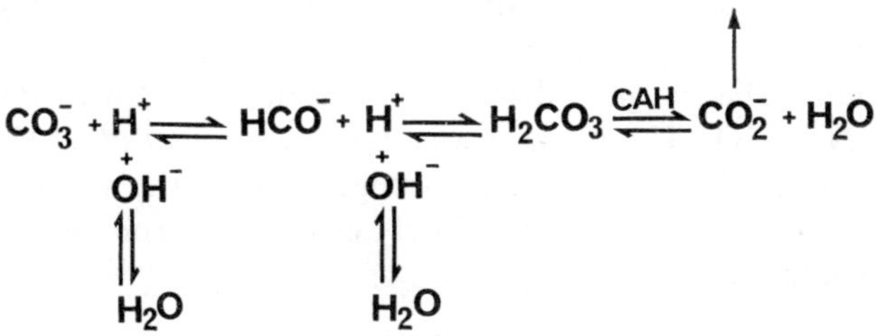

According to Hansson, in the presence of excess CO^{2+} ions the concentration of OH^- will eventually exceed the solubility product of cobalt hydroxide (10^{-16}). With a millimolar concentration of cobalt this occurs at about pH 7·5. The primary reaction product of Häusler's method is thus probably cobalt hydroxide and not the carbonate.

The matter of fixation was considered by Korhonen and Korhonen (1965), by Hansson (1967, 1968) by Ishizaki (1969) and, specifically, by Hyyppä (1968). This last author confirmed by an assay method the superiority of acetone as a preservative of CAH. Formaldehyde and glutaraldehyde inactivated the enzyme strongly, as found also by Ishizaki (1969), but 10 per cent hydroxyadipaldehyde preserved both enzyme and morphology provided that small blocks only were employed. The majority of workers continue to use acetone in preference to alternative fixatives.

Two methods for CAH are described in Appendix 23. The first is a modified Häusler technique used by Cross (1970) and the second a solid medium method described by Meijer and Bloem (1969). The results of the first method are shown in Plate XXIIIb, p. opposite.

Method for Aconitase (E.C.4.2.1.3)

A histochemical technique for this enzyme, now described as citrate hydrolase, was evolved by Yamada *et al.* (1962) and used by Mizushima (1964) in a study of enzymes in developing teeth. The medium contained *cis-*

aconitate as substrate, together with manganese chloride, NADP and Nitro-BT. No other studies are recorded, using this technique.

Carbon-Sulphur Lysases

Method for Cysteine Desulphurase (E.C.4.4.1.1)

The only representative in this section is the enzyme now known as cysteine desulphhydrase (L-cysteine hydrogensulphide lysase) which catalyses the following reaction:

$$\text{L-cysteine} + H_2O \rightleftharpoons \text{pyruvate} + NH_3 + H_2S$$

The histochemical technique was designed by Jarrett (1962) and was based on the reduction of a tetrazolium salt (BT or Nitro-BT) by free H_2S at pH 7·6 in a medium containing cysteine, pyridoxine 5-phosphate as activator, and $MgSO_4$. The incubation period was 18–24 hours at 37° with BT but only 4 hours with Nitro-BT. The method was used by Sasai *et al.* (1969) in a study of enzymes in the skin in chronic pemphigus and Darier's disease.

If this method is used it is absolutely essential to run substrate-free and tissue-free controls.

Isomerases

Method for Hydroxyproline-2-epimerase (E.C.5.1.1.a)

This enzyme, which was demonstrated by Adams and Norton (1964) in *Pseudomonas sp.*, catalyses the interconversion of L-hydroxyproline to D-allohydroxyproline, acting on carbon-2.

A method for its histochemical demonstration was described by Onicescu (1967). Unfixed cryostat sections were used and the medium contained L-hydroxyproline (20 mM) NAD (2 mM), and either GSSG (5 mM) or UDP (1. mM) as coenzymes, Nitro-BT and $MgCl_2$ at pH 7·2. The need for two coenzymes is a peculiar feature and it was assumed by the author that the first transfer from the substrate was to GSSG or UDP, and the second transfer to NAD. The epimerase was thus classified as an oxido-reductase. The highest levels of enzyme were found in Type I skeletal muscle fibres but it must be remembered that these contain high levels of NADH diaphorase which provides the final pathway in methods of this type. Details of the method are given in Appendix 23, p. 1366.

Method for UDPGlucose-4-Epimerase (E.C.5.1.3.2)

This enzyme, discovered by Leloir (1951) catalyses the interconversion of UDPGlucose and UDPGalactose by an inversion at C_4. It is thus part of the so-called galactose inverting system. A method essentially similar to the previous one was developed by Diculescu *et al.* (1968). The same basic assumption was made, that electron transfer is a part of the process of epimerization. According to Maxwell (1961) the mechanism involves oxidation-reduction, hydration-dehydration, cleavage between C_3 and C_4, followed by recondensation but Gabriel (1966) considered that a keto-enol mechanism was more likely.

A simple coenzyme (NAD) was found to be sufficient for the histochemical reaction and high activity was found in fibrocytes, also in liver, kidney and skeletal but not cardiac muscle. Either UDPGlucose or UDPGalactose could be used as substrate, as indicated by Maxwell (1961) but certain striking differences were noted. The renal collecting tubules, for instance, were unstained when UDPGlucose was used but strongly positive with UDPGalactose.

The specificity of this reaction depends on the demonstration that UDPG dehydrogenase (Chapter 21, p. 932) is inactive. Details of the technique appear in Appendix 23, p. 1366.

Method for UDPGlucuronate-4-Epimerase (E.C.5.1.3.c)

In two successive papers Diculescu and Onicescu (1965, 1966) described the development of a histochemical technique for this enzyme. The technique is essentially similar to the foregoing one, using UDPGlucuronate (0·01 μM) as substrate. This very low concentration was found optimal and a single coenzyme (NAD) was adequate. The strongest reactions were found in connective tissue cells and a curious finding was the absence of any reaction when UDPGlucose was substituted for the glucuronate. This fact is not in agreement with subsequent findings by Diculescu *et al.* (1968) but could, perhaps, be explained on the basis of substrate concentration (6 mM for UDPGlucose, 0·01 μM for UDPGlucuronate).

Methods for Glucose phosphate Isomerase (E.C.5.3.1.9)

This soluble enzyme catalyses the conversion of D-fructofuranose-6-phosphate to D-glucopyranose-6-phosphate. This is an important activity in the regulation of carbohydrate metabolism since it determines the amount of D-glucose-6-phosphate which enters the glycolytic pathway. The enzyme is competitively inhibited by intermediates of the pentose phosphate shunt, such as 6-phosphogluconate (Kahani *et al.*, 1960) and erythrose-4-phosphate (Grazi *et al.*, 1960).

The multistep histochemical method depends on the coupled oxidation of the primary reaction product by exogenous glucose-6-phosphate dehydrogenase at pH 7, with NADP as coenzyme, in the presence of Nitro-BT. The final pathway is dependent on endogenous NADPH-tetrazolium oxidoreductase (diaphorase). Methods based on the above principle were described, independently, by Meijer and Bloem (1969) and by Orchardson and McGadey (1970). The observed distribution of enzyme closely parallels that of glucose-6-phosphate dehydrogenase with which it is metabolically associated.

Details of the method are given in Appendix 23, p. 1367.

Coenzymes

As pointed out by Dixon and Webb (1964), it is not always easy to distinguish between a prosthetic group, forming part of the enzyme, a coenzyme distinct but forming part of the catalytic mechanism, and a substrate acting purely as a reactant in the enzyme-catalysed reaction. The haems, flavins, biotin and pyridoxal phosphate are to be considered as prosthetic groups while NAD, NADP and CoA are considered as carrier substrates. While the presence, or absence of a coenzyme can sometimes be deduced from the results of a histochemical reaction in which it forms an essential component, only one group of coenzymes can be demonstrated independently of their total reaction mechanism. These are the respiratory quinones (Vitamin K, ubiquinone, plastoquinone). The most important of these is the second.

Coenzyme Q, Ubiquinone

The name ubiquinone was recommended by the Enzyme Commission for the compound variously known as coenzyme Q, mitoquinone, SA, Q_{275}, and many other names. It is one of a large group of lipid-soluble, water-insoluble quinones which act as intermediate hydrogen carriers in respiratory systems. Ubiquinone was discovered by Morton (1953, 1954) and his associates (Morton *et al.*, 1957), and by Crane *et al.* (1957). The functions of the whole group of quinones have been fully considered in two recent publications (CIBA Foundation "Symposium on Quinones in Electron Transport", 1961; Proceedings of the "International Symposium on Recent Advances in research on Vitamin K and Related Quinones", 1966). The structural formula of the ubiquinones is given below:

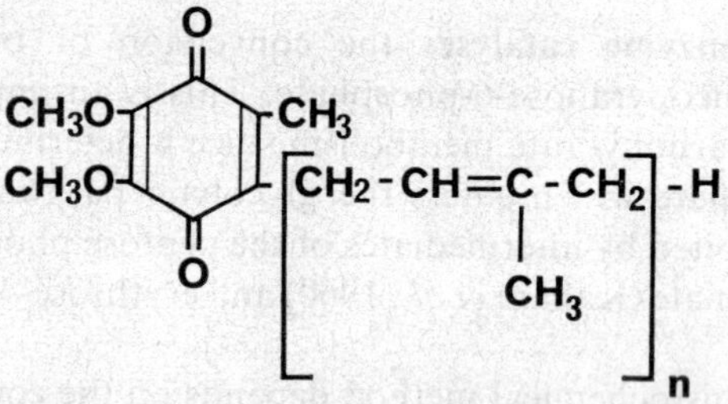

The number 'n' refers to the repeating side chain. In most animals n = 10, so that the polyisopentenoid side chain is $C_{50}H_{81}$, but in the rat Q_9 predominates.

The physiological functions of ubiquinone depend on its reduction and reoxidation (quinone-quinol-quinone). It is found in the mitochondria, where it occurs to the extent of 2–3 mg. per gram of protein. Related quinones are found in chloroplasts and in bacterial chromatophores. Extraction of mitochondria with acetone containing 4 per cent of water removes their ubiquinone, as does treatment with other lipid solvents, so that they can no longer oxidise succinate.

While studying the problem of the inactivation of succinate dehydrogenase in freeze-dried, paraffin-embedded blocks Tranzer and Pearse (1963) found that in dried cryostat sections, extracted with low melting-point waxes, succinate dehydrogenase could be reactivated by addition to the wax of a small quantity of hydroquinone.

A number of different hypotheses were considered before the conclusion was reached that the whole reaction, in which the final stage was the reduction of a tetrazolium salt, depended on the presence of ubiquinones or closely related compounds.

Method for Ubiquinones

The final method developed by Tranzer and Pearse (1963) is given in Appendix 23, p. 1367. Cryostat sections are incubated for short periods in a medium containing the tetrazolium salt (MTT), Co^{2+} ions, hydroquinone and catalase, buffered to pH 7·4 with Tris. The type of result obtainable is shown in Plate XXIIIc, p. 997.

The hydroquinone solution must be freshly prepared, and in the absence of catalase spontaneous reduction occurs. This is attributed to the following reactions:

$$\tfrac{1}{2}O_2 + H_2O \rightarrow H_2O_2$$
$$\text{Hydroquinone} + H_2O_2 \rightarrow \text{Quinone} + H_2O + (O) - (2H)$$
$$\text{MTT} + (2H) \rightarrow \text{MTT formazan}$$
$$H_2O + (O) \rightarrow H_2O_2$$

In the presence of catalase hydrogen peroxide cannot exist and the reaction series shown above is therefore prevented.

Alternative electron donors and acceptors were tested. Of the former only *p*-phenylene diamine gave a histochemical reaction. Of the alternative tetrazoles only tetra-Nitro-BT gave an immediate reaction. Treatment of sections with absolute ethanol for one hour completely abolished the reaction and absolute acetone produced a slight reduction. Acetone containing 4–10 per cent of water, however, produces complete abolition. Further tests for the specificity of the reaction were carried out and it was found that in addition to the ubiquinones a positive reaction was given by tocopherols and by some naphthoquinones, including Vitamin K. These results are shown in Fig. 211, below.

The probability that other compounds able to react with donors and acceptors of electrons would produce a positive Hq-MTT reaction has not yet fully been explored. The strongly positive reaction in the outer segments of the photoreceptors of the mammalian and avian retina (Pearse, 1965) was originally considered likely to be due to retinol. The latter, however, fails to give a reaction *in vitro*. Among numerous other biologically active substances

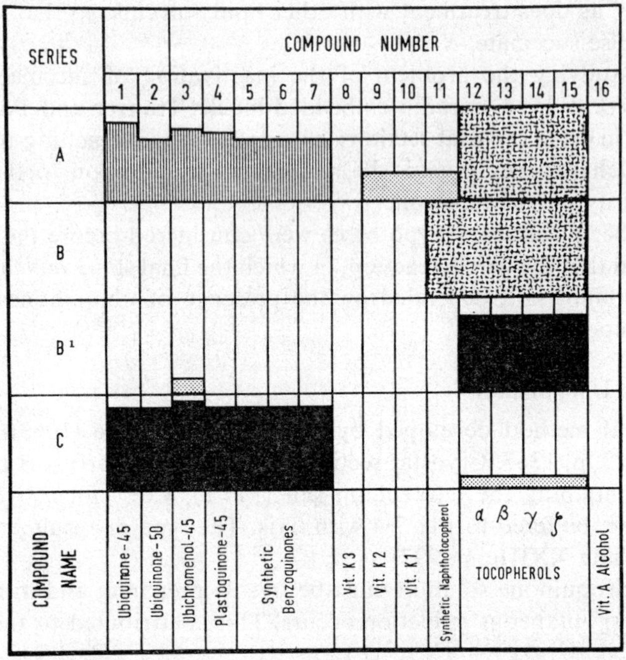

FIG. 211. Hydroquinone-MTT reactions of various compounds.

Series A. Incubation 30 mins at 37° in MTT/HQ/CO^{2+}/Catalase pH6·5.
Series B. Incubation 40 mins at 37° in TNBT/Tris-Maleate pH8.
Series B¹. Incubation 45 mins at 37° in MTT/HQ/CO^{2+}/Catalase pH4.7.
Series C. Incubation 90 mins at 32° in 1% $FeC\lambda_3$. Washed and incubated thereafter as Series A.

tested only riboflavin gave a positive result. Dilley and McConnell (1970) found 0·4 nmole α-tocopherol per mg. protein in bovine retinal mitochondria. This relatively large amount could certainly account for the positive Hq-MTT reaction.

The full mechanism of the Hq-MTT reaction is explained on the basis of three separate couples:

$$\text{(a) Hydroquinone} \underset{+2e}{\overset{-2e}{\rightleftharpoons}} \text{Quinone } (E_0 = 0{\cdot}68\text{V};\ E_0\text{pH } 7{\cdot}3 = 0{\cdot}273\text{V})$$

$$\text{(b) Ubiquinol} \underset{+2e}{\overset{-2e}{\rightleftharpoons}} \text{Ubiquinone } (E_0 = 0{\cdot}542\text{V})$$

$$\text{(c) Formazan} \underset{+2e}{\overset{-2e}{\rightleftharpoons}} \text{Tetrazolium } (E\tfrac{1}{2}\ \text{pH } 7{\cdot}2 = 0{\cdot}11\text{V})$$

The first two reactions are reversible in aqueous media and in organic solvents. The third reaction is reversible in organic solvents but will not proceed from left to right in aqueous media where the formazan is insoluble.

If the two couples (a) and (c) are in a condition of reversibility, and if the concentrations of their oxidized and reduced forms are correct, electrons should be transferred from (c) to (a). In aqueous media (c) is not reversible and in the Hq-MTT medium there is an excess of the reduced form of (a) and of the oxidized form of (c). This should favour the passage of electrons from (a) to (c). That such a transfer does not occur spontaneously may be explained by the close agreement between the two redox potentials. The role of couple (b) is thus that of an intermediate or catalyst.

The cytochemical result is not dependent solely on the different redox potentials of the tetrazolium salts but on their lipid solubility. The superior results with MTT are to be ascribed to its lipid solubility since it is in the lipid phase that the reaction occurs.

Alernative Mechanisms for the Hq-MTT Reaction

An essentially identical reaction was described by Carmichael (1963) but the mechanism was considered on an entirely different basis. Like Tranzer and Pearse (1963), Carmichael excluded an enzymic mechanism but he suggested that the reduction of the tetrazole was due to phospholipid, specifically to the bases (choline, ethanolamine, serine). This explanation was supported, indirectly, by the histochemical studies of de Pascale and Pulido (1966) and by Pulido (1966). The latter author found that a positive reaction was obtainable in the lipoprotein bands on electrophoretic separation of different sera and precipitation with specific anti-lipoprotein sera.

In later papers (Carmichael and Mander, 1967; Carmichael, 1968) explanations of the mechanism of the reaction diverged considerably from the initial premise. In the first of these two papers it was suggested that in dehydrated paraffin sections the Hq-MTT reaction was due to the catalytic action of tissue NH_2 groups, exposed by dehydration and thermal shortening, as well as by the nitrogen groupings of phospholipid. In the second paper the opinion was expressed that the reaction could be used as "a reference parameter in the numerical expression of dehydrogenase activities in tissue sections".

There have been few critical studies by independent observers on the specificity of the Hq-MTT reactions. One such study carried out by Seidler and Steinmann (1969) compared the effects of different tetrazolium salts in the presence or absence of Co^{2+} ions. They concluded that the intensity of the reaction was proportional to the reducibility (redox potential) of the tetrazole employed. They considered, furthermore, that Co^{2+} ions exerted a positive effect independently of their function as chelators for MTT formazan. No opinion was offered on the question of the effector substance. An unintentional study of the specificity of the reaction was made by Hirano and Ogawa (1969) who employed it in an endeavour to demonstrate tissue NH_2 groups at the E.M. level. They substituted Tetranitro-BT for MTT or, alternatively, used ferricyanide as their electron acceptor. The electron opaque reaction product, in either case, was confined solely to the mitochondria (the mitochondrial matrix according to the authors). This is clearly not the localization of available tissue NH_2 groups but, in the tissue selected (cardiac muscle), it could be the predominant localization of lipoprotein. In this still somewhat unlikely event one would at least expect the reaction product to stain the mitochondrial membranes rather than the matrix. The localization observed could be the localization of ubiquinone or, alternatively, of members of the tocopherol series.

An interesting and pertinent observation was made by Takamatsu and Hirai (1967, 1968) who attributed the Nadi reaction (Chapter 19, p. 845) of myeloid leucocytes to a 96 per cent acetone-soluble "naphthoquinone-like substance" which had absorption peaks at 256 and 325 nm. The strong Hq-MTT reaction of leucocyte granules, particularly noted by Tranzer and Pearse (1963) and demonstrated also by Hernandez and Martinez (1967) and by Puera *et al.* (1969), is illustrated in Fig. 212, p. 943. The UV absorption spectra recorded by Takamatsu and Hirai are closer to those of vitamin K_1 than to those of the ubiquinones and K_1 could thus be the compound responsible for both the Nadi and Hq-MTT reactions of leucocyte granules.

I conclude that the Hq-MTT reaction is essentially a simple one which can have no intrinsically high order of specificity for any particular compound unless this specificity is conferred by collateral evidence or by control experiments. That is to say, either biochemical checks, or differential extraction techniques, are an essential part of the reaction. Since all phospholipids, and presumably phospholipid membranes, are capable of dissolving ubiquinones, naphthoquinones, tocopherols and other lipid-soluble redox compounds, it is hardly surprising that the localization of formazans in the Hq-MTT reaction is exclusively associated with lipid, though certainly not exclusively with mitochondria. Most commercial samples of lipids contain lipid-soluble redox compounds as contaminants.

The first mechanism postulated by Carmichael (1963), hydrogen bond formation between the nitrogen of the phospholipid base and the hydroquinone, is not necessary. The failure of the red cell membrane to stain was explained by Carmichael on the basis of its molecular architecture. The failure of lipids and lipoproteins in brain sections requires a more luminous explanation.

While, therefore, the potential of the Hq-MTT reaction remains high, and the number of potential determinants higher still, the latter can be reduced to relatively few by means of control reactions. No one should be deterred from using the reaction by uncertainty as to the precise interpretation of its results and it is particularly important that further studies should be made at the level of ultrastructural cytochemistry.

REFERENCES

ABRAHAM, R. (1967). *Histochemie*, **11**, 129.
ABRAHAM, R., MORRIS, M., and SMITH, J. (1967). *J. Histochem. Cytochem.*, **15**, 596.
ADAMS, E., and NORTON, I. L. (1964). *J. biol. Chem.*, **239**, 1525.
ARMSTRONG, J. M., MYERS, D. V., VERPOORTE, J. A., and EDSALL, J. T. (1966). *J. biol. Chem.*, **241**, 5137.
AUSTIN, J., ARMSTRONG, D., and SHEARER, L. (1965). *Archs Neurol.*, **13**, 593.
AUSTIN, J., MCAFEE, D., ARMSTRONG, D., O'ROURKE, M., SHEARER, L., and BACHHAWAT, B. (1964). *Biochem. J.*, **93**, 15C.
BABSON, A. L., SHAPIRO, P. O., WILLIAMS, P. A. R., and PHILLIPS, G. E. (1962). *Clin. chim. Acta*, **7**, 199.
BAUM, H., DODGSON, K. S., and SPENCER, B. (1958). *Biochem. J.*, **69**, 567.
BLEYL, U. (1964). *Histochemie*, **4**, 286.
BRAUN-FALCO, O., and RATHJENS, B. (1955a). *Archs klin. exp. Dermatol.*, **201**, 73.
BRAUN-FALCO, O., and RATHJENS, B. (1955b). *Acta Histochemica*, **2**, 39.
BRAUN-FALCO, O., and SALFELD, K. (1959). *Nature, Lond.*, **183**, 51.
BRAUNSTEIN, A. A., and KRITZMANN, M. G. (1937). *Enzymologia*, **2**, 129.
BURT, A. M. (1969). *Anat. Rec.*, **163**, 162.
BURT, A. M. (1970). *J. Histochem. Cytochem.*, **18**, 408.
CARMICHAEL, G. G. (1963). *J. Histochem. Cytochem.*, **11**, 738.
CARMICHAEL, G. G. (1968). *Histochemie*, **16**, 299.
CARMICHAEL, G. G., and MANDER, S. T. K. (1967). *J. Histochem. Cytochem.*, **15**, 404.
COHEN, P. P. (1940). *J. biol. Chem.*, **136**, 565.
CRANE, F. L., HATEFI, Y., LESTER, R. L., and WIDMER, C. (1957). *Biochim. biophys. Acta*, **25**, 220.
CROSS, S. A. M. (1970). *Histochemie*, **22**, 219.
DECKER, L. E., and RAU, E. M. (1963). *Proc. Soc. exp. Biol. Med.*, **112**, 144.
DE PASCALE, A., and PULIDO, E. (1966). *Boll. Soc. ital. Biol. Sper.*, **42**, 1244.
DIAMANTSTEIN, T., and SCHLÜNS, J. (1964). *Acta histochem.*, **19**, 296.
DICULESCU, I., and ONICESCU, D. (1965). *Histochemie*, **5**, 450.
DICULESCU, I., and ONICESCU, D. (1966). *Acta histochem.*, **25**, 242.
DICULESCU, I., ONICESCU, D., SZÉGLI, G., and DUMITRESCU, A. (1968). *Histochemie*, **14**, 143.
DILLEY, R. A., and MCCONNELL, D. G. (1970). *J. Membrane Biol.*, **2**, 317.
DIXON, M., and WEBB, E. C. (1964). "Enzymes". 2nd. Edit. Longmans, London, p. 450.
DODGSON, K. S., ROSE, F. A., and TUDBALL, N. (1959). *Biochem. J.*, **71**, 10.
DODGSON, K. S., ROSE, F. A., and SPENCER, B. (1957). *Biochem. J.*, **66**, 357.
DODGSON, K. S., and SPENCER, B. (1957). *Meth. biochem. Analysis*, **4**, 211.
DODGSON, K. S., SPENCER, B., and THOMAS, J. (1953). *Biochem. J.*, **53**, 452.
DODGSON, K. S., SPENCER, B., and THOMAS, J. (1954). *Biochem. J.*, **56**, 177.
DODGSON, K. S., SPENCER, B., and THOMAS, J. (1955). *Biochem. J.*, **59**, 29.
DODGSON, K. S., SPENCER, B., and WYNN, C. H. (1956). *Ibid.*, **62**, 500.
DOHLMAN, C. H., and FRIEDENWALD, J. S. (1955). *J. Histochem. Cytochem.*, **3**, 492.
DUFF, T. A., and COLEMAN, J. E. (1966). *Biochemistry*, **5**, 2009.
FAND, S. B., LEVINE, H. J., and ERWIN, H. L. (1958). *J. biol. Chem.*, **6**, 92.
FAND, S. B., LEVINE, H. J., and ERWIN, H. L. (1959). *Ibid.*, **7**, 27.
FURTH, A. J. (1968). *J. biol. Chem.*, **243**, 4832.
GABRIEL, L. O. (1966). *J. biol. Chem.*, **241**, 924.
GARCIA-HERMANDEZ, M., and KUN, E. (1957). *Biochim. biophys. Acta*, **24**, 78.
GOEBEL, A., and PUCHTLER, H. (1954). *Naturwiss*, **41**, 53.
GOLDFISCHER, S. (1965). *J. Histochem. Cytochem.*, **13**, 520.
GRAZI, E., DE FLORA, A., and PONTREMOLI, S. (1960). *Biochem. biophys. Res. Commun.*, **2**, 121.
GREGORY, J. D., and ROBBINS, P. W. (1960). *A. Rev. Biochem.*, **29**, 347.

Gutowska, M. S., and Mitchell, C. A. (1945). *Poult. Sci.*, **24**, 159.
Hansson, H. P.-J. (1967). *Histochemie*, **11**, 112.
Hansson, H. P.-J. (1968). *Acta. physiol. scand.*, **73**, 427.
Häusler, G. (1958). *Histochemie*, **1**, 29.
Hayashi, M., Simoda, K., Ogata, K., Takamori, T., Shiraogawa, T., and Kawase, O. (1957). *Acta Histochem*, **3**, 277.
Hebb, C. O., Kása, and Mann, S. P. (1970). *J. Physiol.* **208**, 1P.
Hernandez, F., and Martinez de Morentin, J. (1967). *Stain Technol.*, **42**, 241.
Higashi, H., Okawa, K., and Takamatsu, H. (1960). *Proc. jap. histochem. Soc.*, p. 80.
Hirano, H., and Ogawa, K. (1969). *Acta histochem. cytochem.*, **2**, 112.
Hopsu, V. K., Arstila, A. U., and Glenner, G. G. (1965). *J. Histochem. Cytochem.*, **13**, 711.
Hopsu-Havu, V. K., Arstila, A. U. Helminen, H. J., Kalimo, H. O., and Glenner, G. G. (1967). *Histochemie*, **8**, 54.
Hori, S. (1966). *Histochemie*, **7**, 297.
Huggins, C., and Smith, D. R. (1947). *J. biol. Chem.*, **170**, 391.
Hyyppä, M. (1968). *Histochemie*, **12**, 184.
Ishizaki, M. (1969). *Tohoku, J. exp. Med.*, **99**, 315.
Jarrett, A. (1962). *J. Histochem. Cytochem.*, **10**, 400.
Kahani, S. E., Lowry, O. H., Schulz, D. W., Passonneau, J. V., and Crawford, E. J. (1960). *J. biol. Chem.*, **235**, 2178.
Karnovsky, M. J., and Himmelhoch, S. R. (1961). *Am. J. Physiol.*, **201**, 786.
Kása, P., Mann, S. P., and Hebb, C. (1970). *Nature, Lond.*, **226**, 812.
Katchburian, E., Katchburian, A. V., and Pearse, A. G. E. (1967). *J. Anat.*, **101**, 783.
Kawiak, J., Sawicki, W., and Miks, B. (1964). *Acta histochem.*, **19**, 184.
Kernohan, J. C. (1965). *Biochim. biophys. Acta*, **96**, 304.
Kishino, Y. (1967). *Proc. jap. Histoc. Assoc.* VII & VIIIth A.G.M., p. 50.
Kishino, Y. (1968). *Acta histochem. cytochem. japon.*, **1**, 31.
Korhonen, L. K., and Korhonen, E. (1965). *Histochemie*, **2**, 279.
Korhonen, L. K., Näätänen, E., and Hyyppä, M. (1964). *Acta histochem.*, **18**, 336.
Krebs, H. A. (1942). *Biochem. J.*, **36**, 303.
Kurata, Y. (1953). *Stain Technol.*, **28**, 231.
Lee, S. H. (1968). *Am. J. clin. Path.*, **49**, 568.
Lee, S. H., and Torack, R. M. (1968a). *Histochemie*, **12**, 341.
Lee, S. H., and Torack, R. M. (1968b). *J. Histochem. Cytochem.*, **16**, 181.
Leloir, L. F. (1951). *Archs Biochem.*, **33**, 186.
Lindskog, S. (1960). *Biochim. biophys. Acta*, **39**, 218.
Lindskog, S. (1966). *Biochemistry*, **5**, 2641.
Lloyd, J. B. (1969). *Biochem. J.*, **115**, 703.
Mann, T., and Keilin, D. (1940). *Nature, Lond.*, **146**, 164.
Maren, T. H. (1967). *Physiol. Rev.*, **47**, 595.
Maren, T. H., Parcell, A. L., and Malik, M. N. (1960). *J. Pharmac. exp. Ther.*, **130**, 389.
Maxwell, E. S. (1961). In "The Enzymes". Eds. P. Boyer, H. Lardy, and K. Myrbäck, Acad. Press, New York, Vol. 5, p. 443.
McIntosh, J. E. A. (1969). *Biochem. J.*, **114**, 463.
Mehl, E., and Jatzkewitz, H. (1965). *Biochem. biophys. Res. Commun.*, **19**, 407.
Mehl, E., and Jatzkewitz, H. (1968). *Biochim. biophys. Acta.* **151**, 619.
Meijer, A. E. F. H. (1967a). *Acta histochem.*, **28**, 286.
Meijer, A. E. F. H. (1967b). *Histochemie*, **8**, 248.
Meijer, A. E. F. H., and Bloem, J. H. (1966). *Acta histochem.*, **25**, 239.
Meijer, A. E. F. H., and Bloem, J. H. (1969). *Acta histochem.*, **32**, 110.
Meldrum, N. U., and Roughton, F. J. W. (1932). *J. Physiol.*, **75**, 12.
Meldrum, N. U., and Roughton, F. J. W. (1933). *Ibid.*, **80**, 113.
Merker, H.-J., and Spors, S. (1969). *Histochemie*, **17**, 83.
Millichap, J. G. (1957). *Proc. Soc. exo. Biol. Med.*, **96**, 125.
Milsom, D. W., Rose, F. A., and Dodgson, K. S. (1968). *Biochem. J.*, **109**, 40P.

MIZUSHIMA, T. (1964). *Histochemie*, **3**, 538.
MIZUTANI, A. (1967a). *Proc. jap. Histochem. Assoc.*, **1**, p. 51.
MIZUTANI, A. (1967b). *J. Histochem. Cytochem.*, **15**, 603.
MIZUTANI, A. (1968). *J. Histochem. Cytochem.*, **16**, 172.
MIZUTANI, A., and FUJITA, H. (1968). *Acta histochem. cytochem.*, **1**, 79.
MORGENSTERN, S., OKLANDER, M., AUERBACH, J., KAUFMAN, J., and KLEIN, B. (1966). *Clin. Chem.*, **12**, 95.
MORTON, R. A. (1953). *Biochem. J.*, **55**, 786.
MORTON, R. A. (1954). *Biochem. J.*, **57**, 595.
MORTON, R. A. (1961). In "CIBA Foundation Symposium on Quinones in Electron Transport". Eds. G. E. W. Wolstenholme, and M. J. O'Connor & A. Churchill, London.
MORTON, R. A. (1965). *Endeavour*, **24**, 81.
MORTON, R. A. (1966). *Vitam. and Horm.*, **24**, 293.
MORTON, R. A., WILSON, G. M., LOWE, J. S., and LEAT, W. M. F. (1957). *Chem. and Med.*, **1649**.
MUSTAKALLIO, K. K., RAEKALLIO, J., and RAEKALLIO, E. (1960). *Annls. Med. exp. Biol. Fenn.*, **38**, 247.
NEUBERG, C., and KURONO, K. (1923). *Biochem. Z.*, **140**, 295.
NEUBERG, C., and WAGNER, J. (1925). *Ibid.*, **161**, 492.
NYMAN, P.-O. (1961). *Biochim. biophys. Acta*, **52**, 1.
OHARA, M., and KURATA, J. (1950). *Igaku to Seibutsugaku*, **16**, 213.
OHARA, M., and KURATA, J. (1952). *Igaku to Seibutsugaku*, **25**, 215.
ONICESCU, D. (1967). *Histochemie*, **11**, 180.
ORCHARDSON, R., and MCGADEY, J. (1970). *Histochemie*, **22**, 136.
PEARSE, A. G. E. (1965). *Nature, Lond.*, **205**, 708.
PECHE, K. (1913). *Ber. Dtsch. Bot. Ges.*, **31**, 458.
POCKER, Y., and MEANY, J. E. (1965). *J. Amer. chem. Soc.*, **87**, 1809.
POCKER, Y., and MEANY, J. E. (1967). *Biochemistry*, **4**, 2535.
POCKER, Y., and STONE, J. T. (1965). *J. Am. chem. Soc.*, **87**, 5479.
POCKER, Y., and STONE, J. T. (1967). *Biochemistry*, **6**, 239.
PORTER, M. T., FLUHARTY, A. L., and KIHARA, H. (1969). *Proc. natn. Acad. Sci., U.S.A.*, **62**, 887.
POTTS, A. M. (1965). In "Biochemistry of the Retina". Ed. C. N. Graymore, Acad. Press, London, p. 155.
PUERA, B., NIR, E., and EFRATI, P. (1969). *Acta heamat.*, **41**, 296.
PULIDO, E. (1966). *Riv. istochem.*, **12**, 413.
RATH, H., and FEESS, E. (1954). *Melliand Textlber.*, **35**, 267.
RICKLI, E. E., and EDSALL, J. T. (1962). *J. biol. Chem.*, **237**, PC 258.
ROUGHTON, F. J. W., and CLARK, A. M. (1951). In "The Enzymes", Ed. J. B. Sumner, and K. Myrbäck, Acad. Press, New York. Vol. I, p. 1250.
ROY, A. B. (1953). *Biochem. J* , **53**, 12.
ROY, A. B. (1954). *Ibid.*, **57**, 465.
ROY, A. B. (1956a). *Biochem. J.*, **63**, 294.
ROY, A. B. (1956b). *Biochem. J.*, **64**, 651.
ROY, A. B. (1958). *Biochem. J.*, **68**, 519.
ROY, A. B. (1960a). *Biochem. J.*, **77**, 380.
ROY, A. B. (1960b). *Adv. Enzymol.*, **22**, 205.
RUTENBURG, A. M., COHEN, R. B., and SELIGMAN, A. M. (1952). *Science, N.Y.*, **116**, 539.
RUTENBURG, A. M., and SELIGMAN, A. M. (1956). *Archs Biochem. Biophys.*, **60**, 198.
SASAI, Y., TANAHASHI, Y., and FUJIYAMA, T. (1969). *Tohuku J. exp. Med.*, **98**, 99.
SAVANT, P. L., SHIBKO, S., KUMTA, U. S., and TAPPEL, A. L. (1964). *Biochim. biophys. Acta.*, **85**, 82.
SEIDLER, E., and STEINMANN, C. (1969). *Acta histochem.*, **34**, 378.
SELIGMAN, A. M., NACHLAS, M. M., MANHEIMER, L. H., FRIEDMAN, O. M. and WOLF, G. (1949). *Ann. Surg.*, **130**, 333.
SELIGMAN, A. M., NACHLAS, M. M., and COHEN, R. B. (1950). *Cancer Res.*, **10**, 240.
SJÖVALL, K. (1967). *Histochemie*, **10**, 336.

SPORS, S., and MERKER, H.-J. (1969). *Histochemie*, **17**, 78.
SUZUKI, S., TAKAHASI, N., and EGAMI, F. (1957). *Biochim. biophys. Acta*, **24**, 444.
TAKAMATSU, H., and HIRAI, K. (1967). *Proc. jap. Histochem. Assoc.*, 7/8th A.G.M., p. 47.
TAKAMATSU, H., and HIRAI, K. (1968). *Acta histochem. cytochem.*, **1**, 1.
TASHIAN, R. E. (1965). *Am. J. hum. Genet.*, **17**, 257.
TASHIAN, R. E., DOUGLAS, D. P., and YU, Y. L. (1964). *Biochem. biophys. Res. Commun.*, **14**, 256.
THOMAS, M., and ALDRIDGE, W. N. (1966). *Biochem. J.*, **98**, 94.
TRANZER, J.-P., and PEARSE, A. G. E. (1963). *Nature, Lond.*, **199**, 1063.
VERPOORTE, J. A., MEHTA, S., and EDSALL, J. T. (1967). *J. biol. Chem.*, **242**, 4221.
VIALA, R., and GIANETTO, R. (1955). *Can. J. Biochem. Physiol.*, **33**, 839.
WÄCHTLER, K., and PEARSE, A. G. E. (1966). *Zeit. Zellforsch.*, **69**, 326.
WALKER, D. G. (1963). *Biochim. biophys. Acta*, **77**, 209.
WALKER, D. G., and RAO, S. (1964). *Biochem. J.*, **90**, 360.
WEBB, E. C., and MORROW, P. F. W. (1960). *Biochim. biophys. Acta*, **39**, 542.
WHITNEY, P. L., FÖLSCH, G., NYMAN, P.-O., and MALMSTRÖM, B. G. (1967). *J. biol. Chem.*, **242**, 4206.
WOOHSMANN, H., and BROSOWSKI, K.-H. (1964). *Acta histochem.*, **18**, 179.
WOOHSMANN, H., and HARTRODT, W. (1964). *Histochemie*, **4**, 336.
WOOHSMANN, H., and HARTRODT, W. (1967). *Histochemie*, **11**, 81.
YAMADA, Y., ABE, T., and SHIMIZU, N. (1962). *Archs histol. jap.*, **23**, 167.
YANO, Y. (1968). *Acta histochem. cytochem. japon.*, **1**, 186.

CHAPTER 24

SUBSTRATE FILM TECHNIQUES

The principle of these techniques is essentially simple (Daoust, 1965, 1968). Fresh tissue sections are placed in contact with a thin film of the substrate and, after a period of exposure, the unreacted substrate is detected in the film by a staining or histochemical technique. The unstained portions of the film correspond to regions of enzyme activity in the overlying tissue sections.

The substrate film principle has been used to demonstrate hydrolytic enzymes in three main classes, and a single oxidoreductase:

PROTEASES	Peptidases, Cathepsins, Plasmin.
NUCLEASES	DNases, RNases.
GLYCOSIDASES	Amylases, Hyaluronidases, Lysozyme.
OXIDOREDUCTASES	Catalase.

Essentially there are two main types of substrate film method. In the first, the substrate itself forms the film while in the second the substrate is incorporated in an inert material, inert that is as far as the relevant enzyme is concerned.

Substances like starch and gelatin, acting as substrates, can obviously be prepared as films. Others, like DNA and RNA, must be included in a supporting medium. In the foreseeable future substrates will not simply be mixtures with their carrier. They will be attached by firm chemical linkages to an enzymatically inert part of the molecule. Specific peptides, for instance, will be attached by treatment with cross-linking reagents (e.g. carbodiimide, glutaraldehyde).

The basic requirements for substrate film techniques are (1), the substrate must be insoluble (or immovable) in water, (2), it must be accessible to the enzyme, (3), its product, or at least the chromogenic portion thereof, must be water soluble and/or capable of removal by aqueous washing and, finally (4), it must be possible to show that any observed modification in the film is due solely to enzyme activity on the substrate.

In practical terms there are two main divisions of substrate film technology. After incubation the tissue section may be left in contact with the film or, alternatively, it may be separated therefrom and the two halves of the process then stained separately. The sequence of procedures involved in the second of these two options is shown in Fig. 213, p. 1010.

This illustration refers to the technique for deoxyribonuclease, discussed on p. 1012. With suitable adjustment of the captions it can be made to fit any of the separation techniques described in this chapter.

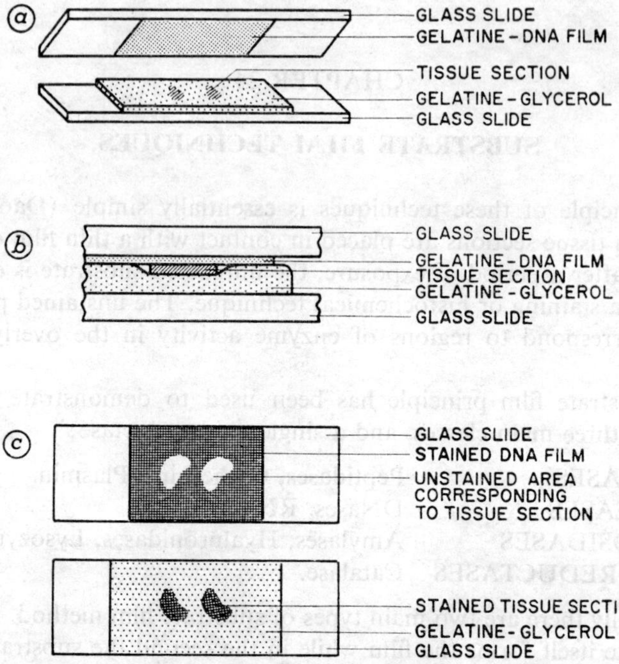

FIG. 213. Essential Steps of the Separation Technique.

(a) Materials in their relative positions before exposure.
(b) The same elements during exposure.
(c) The film and tissue sections after separation and staining.

Methods for Proteases

A general technique for proteolytic activity, developed by Adams and Tuqan (1961a) from earlier work by Adams, Fernand and Schnieden (1958), was based on the incubation of fixed frozen sections on blackened photographic plates. Digestion of the gelatin layer by proteases left clear areas in the film which could be correlated easily with structures in the overlying tissue section (non-separation technique).

This method was used by Adams and Bayliss (1961) to demonstrate peripheral nerve cathepsins and by Adams and Tuqan (1961b) to show proteinase activity in Wallerian degeneration. Distinction of cathepsin from other proteolytic activity was made by showing that the enzyme concerned was pH-dependent and thermostable.

An essentially similar method was used by Bélanger and Migicovsky (1963) to demonstrate proteolysis in bone sections after parathyroid hormone administration. Activity was related especially to mature osteocytes.

There have been a number of improvements in the gelatin technique for proteases. Films coloured with an azo dye technique were used by Cunningham (1967) as the basis of a non-separation technique which gave better localization than that obtainable with blackened photographic films. Using a variety of inhibitors Cunningham showed that the enzyme in kidney sections which digested his cross-linked substrate closely resembled cathepsin D.

In another elegant adaptation of the original principle Fratello (1968) placed his tissue sections on unexposed reversible colour film (Ferrania 3M, daylight type DIA 28). This type of film consists of a nitrocellulose backing on which three layers of gelatin are superimposed. The first, furthest from the tissue section, is a blue (cyan) layer about 6 μ thick; the second is a magenta layer 4·5 μ thick and the third yellow and 5·5 μ thick.

The three minus primary coloured layers are progressively and sequentially digested by proteases. After stopping the reaction, the mounted preparation is viewed by transmitted light. The undigested film appears brown; digestion of the yellow layer produces purple (magenta-cyan) while if the yellow and

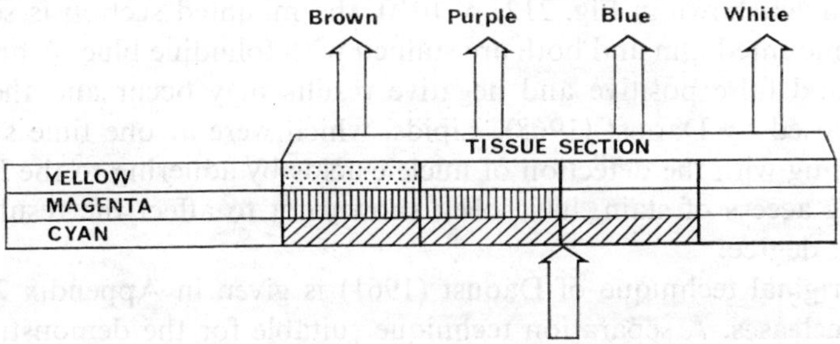

FIG. 214. Colour Film Technique. Effects of progressive digestion.

magenta layers are digested the result is blue. If all three layers are eroded the result is white.

An advantage of the colour film technique is that the gelatin layers are much thinner than the single sensitized silver layer of the black and white film. There is thus a considerable increase in sensitivity. With a silver film technique Fratello was able to demonstrate proteases in adult pigeon stomach but not in the developing digestive tract of the embryo. With the colour technique these enzymes could be shown in developing tissues.

All three protease methods are described in Appendix 24, p. 1369.

Fibrinolysin Techniques

A method of fibrinolysis autography was developed by Todd (1958, 1959) who incubated fresh tissue sections in contact with fibrin films. The technique, in essentially unaltered form, was used by Kwaan and Astrup (1963, 1967). In principle it depends on the activation of plasminogen to plasmin by an activator present in the tissues. Plasminogen is present in bovine fibrinogen and its plasmin successor digests the fibrin of the film. Specifically, therefore, it is plasminogen activator which is demonstrated by Todd's method, and not an endogenous protease.

The activator is widespread in mammalian tissues, especially in the endothelium of blood vessels (Todd, 1964a and b). It has also been shown in myometrium (but not in placenta) by Beller *et al.* (1962), in mast cells by Kwaan (1964) and in skin by Haustein (1966, 1969), and Turner and Ryan (1969).

The full method, as modified by these last authors, is given in Appendix 24, p. 1370.

Methods for Nucleases

Deoxyribonucleases. The preparation of suitable substrate films for the demonstration of this enzyme was pioneered by Daoust (1959, 1961) and by Daoust and Amano (1961). Fresh cryostat sections are placed on top of formalin-fixed 5 per cent gelatin films containing 0·2 per cent DNA. After incubation, as shown in Fig. 212, p. 1010, the mounted section is separated from the mounted film and both are stained with toluidine blue. A number of artifacts and false positive and negative results may occur and these were fully disclosed by Daoust (1968). Lipids, which were at one time suspected of interfering with the detection of nucleic acids by adhering to the film, and preventing access of stain, have been shown not to affect the result to any significant degree.

The original technique of Daoust (1961) is given in Appendix 24.

Ribonucleases. A separation technique suitable for the demonstration of this group of enzymes was described by Daoust and Amano (1960) and the method, in a modified non-separation form, was used by Mayner and Ackerman (1962, 1963) for a general survey of the localization of ribonucleases in tissues from various mammalian species. After making a critical survey of all steps of the method Daoust (1966) introduced a modified method incorporating a comparative procedure first suggested by Sierakowska and Shugar (1961). This involved flushing the section off the film after incubation and using the serial section, suitably stained, for checking the localization of the enzyme. In his modified (1966) technique Daoust recommended post-fixation of tissue section and underlying film with formaldehyde before staining. It was pointed out by Daoust and Durocher (1969), however, that in brain tissues such post-fixation produced a false positive reaction upon subsequent staining with toluidine blue. They considered that formaldehyde treatment resulted in the diffusion into the film of an unidentified, possibly lipid, component which, by becoming bound to RNA, prevented it subsequent staining with the dye.

The separation technique of Daoust (1966) is given in Appendix 24, p. 1372.

Methods for Amylases

Using a starch film as substrate Szemplinska, Sierakowska and Shugar (1962) introduced a separation technique for the demonstration of amylases.

Their report was closely followed by the description of a similar technique, independently, by Tremblay (1963). A non-separation method was evolved by Shear and Pearse (1963).

In their original technique Szemplinska *et al.* used a gelatin film containing starch while the substrate alone was used by Tremblay, and also by Shear and Pearse. The latter authors mounted their sections directly on the starch film but the other two placed their sections on a cushion of gelatin or gelatin-glycerol. For staining the residual substrate Szemplinska *et al.* and Shear and Pearse used Lugol's iodine while Tremblay used the PAS routine. To obviate tissue component staining by the latter routine, with the non-separation procedure, Shear and Pearse used starch films pre-oxidized with periodic acid. Subsequently these were developed with Schiff's reagent. Denker (1970), however, found that periodate-oxidized starch films could undergo spontaneous hydrolysis at slightly alkaline pH levels and this source of artifact must be taken into account if the modified technique is employed.

The different procedures outlined above have been subjected to numerous modifications. In Appendix 24 only the modified technique of Tremblay and Charest (1968) is described. This combines the best features of all preceding methods. The type of result obtainable is illustrated in Plate XXIIId, p. 997.

Method for Hyaluronidase

A method capable of demonstrating the tissue localization of hyaluronidase was first developed by Szemplinska *et al.* (1962), who used either sodium hyaluronate or chondroitin sulphate as substrate. Both these compounds were incorporated in gelatin films. An essentially similar technique for hyaluronidase was proposed by McCombs and White (1968), who were unaware of the existence of previous papers on the subject. Their technique gives excellent results and details are given in Appendix 24.

Method for Lysozyme

The successful development of a method for lysozyme was reported by Ghoos and Vantrappen (1970). The authors used a homogeneous film of chitin (see Vol. 1, p. 297) as substrate since this material had been shown by Dahlqvist *et al.* (1966) and by Gottschalk (1966) to be degraded by the enzyme. They showed that whereas chitin is not stained by Alcian blue (Vol. **1**, p. 344) its digestion by lysozyme results in products having pronounced alcianophilia. Ghoos and Vantrappen were able to demonstrate enzyme activity in leucocytes since, in this case, stainable breakdown products were retained within the cells. With this technique the development of a positive result must depend on *in situ* binding of the alcianophilic products. It remains to be seen whether lysozyme can be demonstrated in non-phagocytic cells by the present technique, as recorded in Appendix 24, p. 1374.

Method for Catalase

Following the development of a method for the identification of catalase in starch gels (Hale and Renwick, 1961) the method was adapted by Hale (1965) for the demonstration of this enzyme in tissue sections. In principle this technique is simple. Tissue sections, mounted on thin starch films on microscope slides, are immersed in dilute H_2O_2. The latter saturates the starch film except where it is destroyed by catalase in its passage through the overlying section. When the film (and section) are subsequently immersed in a solution of potassium iodide the dark blue colour of iodine-starch is produced, except where catalase has destroyed the peroxide. The location of catalase thus appears white on a blue background.

Details of this method, the first to use the enzymically inert supporting film as a chromogenic reagent, are given in Appendix 24.

REFERENCES

ADAMS, C. W. M., and BAYLISS, O. B. (1961). *J. Histochem. Cytochem.*, **9**, 473.

ADAMS, C. W. M., FERNAND, V. S. V., and SCHNIEDEN, H. (1958). *Brit. J. exp. Path.*, **39**, 393.

ADAMS, C. W. M., and TUQAN, N. A. (1961a). *J. Histochem. Cytochem.*, **9**, 469.

ADAMS, C. W. M., and TUQAN, N. A. (1961b). *J. Neurochem.*, **6**, 334.

BÉLANGER, L. F., and MIGICOVSKY, B. B. (1963). *J. Histochem. Cytochem.*, **11**, 734.

BELLER, F. K., GOESSNER, W., and HERRSCHLEIN, H. J. (1962). *Obstet. Gynec.*, **20**, 117.

CUNNINGHAM, L. (1967). *J. Histochem. Cytochem.*, **15**, 292.

DAHLQVIST, F., JAO, L., and RAFTERY, M. (1966). *Proc. nat. Acad. Sci. U.S.A.*, **56**, 26.

DAOUST, R. (1957). *Exp. Cell Res.*, **12**, 203.

DAOUST, R. (1959). *Exp. Cell Res.* Suppl., 7, p. 40.

DAOUST, R. (1961). *Gen. Cytochem. Methods*, **2**, 153.

DAOUST, R. (1965). *Int. Rev. Cytol.*, **18**, 191.

DAOUST, R. (1966). *J. Histochem. Cytochem.*, **14**, 254.

DAOUST, R. (1968). *J. Histochem. Cytochem.*, **16**, 540.

DAOUST, R., and AMANO, H. (1960). *J. Histochem. Cytochem.*, **8**, 131.

DAOUST, R., and AMANO, H. (1961). *Exp. Cell Res.*, **24**, 559.

DAOUST, R., and DUROCHER, J. G. (1969). *J. Histochem. Cytochem.*, **17**, 350.

DENKER, H. W. (1970). *Histochemie*, **21**, 17.

FRATELLO, B. (1968). *Stain Technol.*, **43**, 125.

GHOOS, Y., and VANTRAPPEN, G. (1970). *Histochem. J.*, **2**, 11.

GOTTSCHALK, A. (1966). In "Glycoproteins". Ed. Gottschalk, A. Elsevier, Amsterdam, p. 25.

HALE, A. J. (1965). *J. roy. micr. Soc.*, **84**, 323.

HALE, A. J., and RENWICK, J. H. (1961). *Biochem., J.*, **80**, 49P.

HAUSTEIN, U. F. (1966). *Arch. klin. exp. Dermat.*, **226**, 83.

HAUSTEIN, U. F. (1969). *Arch. klin. exp. Dermat.*, **234**, 182.

KWAAN, H. C. (1964). *Am. J. clin. Path.*, **41**, 604.

KWAAN, H. C., and ASTRUP, T. (1963). *Archs Path.*, **76**, 595.

KWAAN, H. C., and ASTRUP, T. (1967). *Lab. Invest.*, **17**, 140.

MAYNER, D. A., and ACKERMAN, G. A. (1962). *J. Histochem. Cytochem.*, **10**, 687.

MAYNER, D. A., and ACKERMAN, G. A. (1963). *J. Histochem. Cytochem.*, **11**, 573.

McCOMBS, H. L., and WHITE, H. J. (1968). *Am. J. clin. Path.*, **49**, 68.

SHEAR, M., and PEARSE, A. G. E. (1963). *Exp. Cell Res.*, **32**, 174.

SIERAKOWSKA, H., and SHUGAR, D. (1961). *Acta Biochim. pol.*, **8**, 427.

Szemplinska, H., Sierakowska, H., and Shugar, D. (1962). *Acta Biochim. pol.*, **9**, 239.
Todd, A. S. (1958). *Nature, Lond.*, **181**, 495.
Todd, A. S. (1959). *J. Path. Bact.*, **78**, 281.
Todd, A. S. (1964a). *J. clin. Path.*, **17**, 324.
Todd, A. S. (1964b). *Brit. med. Bull.*, **20**, 210.
Todd, A. S., and Nunn, A. (1967). "Atti del I Int. Symp. Tissue Factors in the Homeostasis of the Coagulation-Fibrinolysis System". Maggio, Florence, p. 57.
Tremblay, G. (1963). *J. Histochem. Cytochem.*, **11**, 202.
Tremblay, G., and Charest, J. (1968). *J. Histochem. Cytochem.*, **16**, 147.
Turner, R. H., and Ryan, T. J. (1969). *Trans. St. John's Hosp. Dermat. Soc.*, **55**, 212.

CHAPTER 25

ENZYMES AS HISTOCHEMICAL REAGENTS

Introduction

IT will be appreciated, from what has already been stated in Chapter 1, that the use of enzymes for what was originally called "enzymal analysis" is nearly as old as the science of histochemistry. Many of the modern techniques in this field are essentially those used by the older experimenters except that their crude extracts have been replaced by more or less purified crystalline enzymes. The use of these represents a very great technical advance, but a crystalline enzyme is not necessarily pure, nor is its activity necessarily specific. Danielli (1946, 1947) rightly stressed these points and maintained that there was no known method for establishing the specificity of any enzyme for a single substrate or group of substrates. A third point to which attention was drawn by this author was the importance of accessibility of the substrate in the customary tissue preparations. He claimed that a monomolecular layer of protein covering a nucleic acid would prevent access of the specific nuclease, for instance. It is clear that this particular objection is not of much importance in practice, since the results of enzyme action on intact cells in smears, and on similar cells in tissue sections, do not differ appreciably.

In addition to its purity, a large number of factors are capable of modifying the results obtained when an enzyme is allowed to react on a tissue preparation. Some of these affect the enzyme itself and others the substrate. Kaufmann, Gay and McDonald (1950) stressed the fact that the use of enzymes for accurate histochemical analysis demands control of all the variables capable of modifying the reaction, those affecting the process of hydrolysis and those affecting subsequent staining alike. The most important factors in the first class are the concentration of the enzyme, the time and temperature of incubation, the pH at which it is carried out, the presence of electrolytes, activators and inhibitors, and the stability of the enzyme under these conditions. In the second class the most important considerations are the fixative used to preserve the tissues and the stain finally employed to demonstrate the removal of substrate by the enzyme.

Until quite recently it was usual for enzymes to be dissolved in aqueous buffer solutions at their optimum pH and at a concentration more or less empirically determined. The investigations of Stowell and Zorzoli (1947), which are more fully referred to below, showed that the presence of electrolytes could exert a marked influence on the amount of material removed by enzyme action (in this case ribonuclease). Effects of this kind can be avoided by employing solutions of enzyme in glass-distilled water. The pH of these is in the region of neutrality, though often slightly on the acid side. The temperature of incubation is now usually maintained at 37°, which is optimal for most of the enzymes with which we are concerned, though in some cases lower temperatures are quite adequate and in one case the optimum is much higher. The time of incubation is varied more or less empirically but mere lengthening of the incubation period does not suffice to produce the desired result if inactivation of the enzyme is rapid. Such is the case in some instances referred to below.

Inhibitors, often essential in histochemical localization of enzymes, have not been used in analytical procedures, although the inactivation of undesirable members of enzyme mixtures might be obtained in this way. The presence of natural inhibitors in the tissue sections, however, may modify the reaction considerably. In this category mercury and chromium (derived from the fixative) are important. Activators other than inorganic ions are seldom employed in analysis with enzymes and a low concentration of the necessary ion can usually be supplied without affecting the action of the incubating medium on the substrate other than by true activation of the enzyme.

The effect of the fixative employed is in all cases of paramount importance. In the first place it is necessary that the concentration and localization of the substrate shall be as perfectly preserved as possible, and morphological considerations have therefore influenced the choice of fixative more than they would have done if a strictly histochemical view had been maintained. Secondly, it is necessary that no blocking effect should prevent access of the enzyme to its substrate in the tissues and that, granted full access, no inhibitory effect should modify its action. Thirdly, the substrate should be so preserved that the reaction product is entirely free to diffuse from the tissues. At first sight it would appear that freeze-drying offers the best chance of satisfying these criteria, but in many cases the use of unfixed material is entirely prohibited by the fact that the substrate present in the tissues is partially or entirely soluble in water, unless preserved by fixation in the histological sense. Moreover, in many cases, the precipitant fixatives alcohol and acetone, and mixtures containing them like Carnoy, fail to preserve the substrate from partial solution during the preparation of sections and they may allow its complete solution in buffer or water controls so that specific enzyme action is not demonstrable. In other cases alcoholic precipitation is satisfactory and leaves the substrate insoluble in buffers and in water. Nevertheless we are faced with the fact that choice of fixative has usually been dictated by morphological considerations and salts of mercury and chromium, and even osmium tetroxide, have been used for preserving substrates for analysis by means of enzymes.

The chief considerations governing the choice of method for demonstrating the substrate before and after enzyme action are a relatively high specificity of the chosen stain for the substrate, and as close a relationship as possible between the colour developed and the amount of substrate present.

If an ordinary staining technique is used it is essential that differentiation be avoided. Progressive staining should be carried out in very dilute solutions for a long period (24 hours) at a fairly constant temperature. The sections should then be examined after processes causing as little loss of stain as possible. Mounting in water is often the only satisfactory method since dehydration in alcohol may bring about violent differentiation and marked loss of stain. On the whole, it is preferable to take the more arduous step of employing a true histochemical reaction for some convenient grouping in the substrate, provided that the reaction chosen is free from the suspicion of removing substrate on its own account.

This general survey of the principles of the use of enzymes as histochemical reagents shows that methods based upon them are not entirely satisfactory. Even with numerous controls, it appears that the results achieved will be open to criticism. In the remainder of this chapter individual methods are considered separately, taking first those which are most used in modern histochemistry and which are, therefore, more important than the others. Three groups of enzymes are outstanding in this respect, the *nucleases*, the *hyaluronidases*, and the *neuraminidases* (*sialidases*). After these come two further groups of lesser importance, the *amylases* and the proteolytic enzymes *trypsin* and *pepsin*. Many other enzymes have been employed in analytical techniques, amongst them various *elastases*, the *collagenases*, the *pectinases*, *lipases and lecithinases*, and finally enzyme batteries such as *Clostridium welchii* filtrates and snail stomach fluid; these last will be accorded brief mention here in view of their limited use.

The Nucleases

Ribonuclease

Nucleoproteins consist of basic proteins linked by salt-like union to nucleic acids. The protein part of the molecule is acted on by the proteolytic enzymes but it is not a substrate for the nucleases which we are considering here. It is thought that the nucleic acids are, in fact, broken down by a succession of enzymes, and these are divided into three groups, nucleases or nucleodepolymerases, nucleotidases (nucleophosphatases) and nucleosidases. The first of these, which splits the nucleic acids into their component nucleotides, was at one time regarded as a pyrophosphatase (Baldwin, 1949) although Greenstein (1944) regarded the nucleases solely as depolymerases. Later, Brown and Todd (1955a and b) showed that crystalline bovine pancreatic ribonuclease degraded ribopolynucleotides in two stages. The first of these consists in the cleavage of the phosphodiester bond between the 3′ and 5′ positions of the ribose constituents in the RNA chain with the formation of oligonucleotides terminating in 2′, 3′-cyclic phosphate derivatives. Subsequently, these terminal groups are split off and hydrolyzed to nucleoside-3′-phosphates.

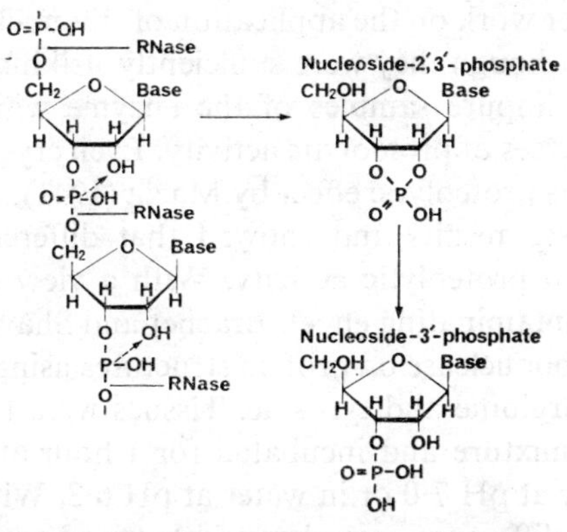

FIG. 215. Action of Ribonucleases.

The characteristics of the pancreatic ribonucleases, and their biological functions, were reviewed comprehensively by Barnard (1969).

The second group of enzymes dephosphorylates the nucleotides to give nucleosides, which are glycosides of nitrogenous bases, and the third group breaks these down into bases and sugars. In histochemical analytical procedures we are only concerned with the first type of nuclease, whether its substrate be ribonucleic acid (RNA) as in the case we are considering, or deoxyribosenucleic acid (DNA) as in a following section.

The Use of Ribonuclease in Histochemistry. In 1913 van Herwerden showed that the basophilia of the chromidial granules in the sea-urchin oocyte, now known to be due to RNA, could be reversed by means of a crude extract of spleen, described as a nuclease. Little attention was given to this work. Ribonuclease was later described by Jones (1920) but no histochemical use was made of it until, following the observations of Dubos and MacLeod (1937) on the presence of an enzyme in purulent exudates which could remove the Gram-positive layer of pneumococci, Dubos and Thompson (1938) extracted from the same source an enzyme which specifically attacked RNA. Kunitz (1940) isolated the enzyme in crystalline form from ox pancreas and a great deal of the subsequent histochemical use of ribonuclease directly followed from his work. His purified enzyme was a soluble protein of the albumin type with a molecular weight of 15,000. It had its isoelectric point and its pH optimum at 7·7 and its optimum temperature at 65°. As a natural corollary, it possessed a high degree of thermostability and could withstand a period of 3 minutes in a water-bath at 90°, with trifling loss of activity. This property is made use of (see below) in order to free the enzyme from traces of proteolytic activity which remain in the majority of samples. Ribonuclease from different sources may vary in its properties (Ledoux and Brachet, 1955) and Kaplan and Heppel (1956) described a ribonuclease from spleen and liver which differed considerably from pancreatic ribonuclease.

Much of the earlier work on the application of ribonuclease to fixed tissues gave results which, although they were sufficiently striking, were complicated by the fact that the impure samples of the enzyme which were employed possessed varying degrees of proteolytic activity. Even crystalline ribonuclease was found to have this proteolytic effect by Mazia (1941), and later Schneider (1946) confirmed these results and showed that different preparations of ribonuclease varied in proteolytic activity. With a view to determining the importance of this contaminating effect, Brachet and Shaver (1948) tested the effect of crystalline ribonuclease on protein structures using the tests described by Serra (1946) for arginine and tyrosine. Tissues were fixed in an alcohol-formalin-acetic acid mixture and incubated for 1 hour at 52° with 1 mg/ml. ribonuclease in buffer at pH 7·0 or in water at pH 6·3. With the arginine and tyrosine methods no difference was observed between treated and untreated sections. It is certain, however, that neither of these methods could give a sufficiently sensitive qualitative appreciation for the conclusions derived from the above results to be considered valid.

The earliest histochemical work on the detection of ribonucleic acids was that of Brachet (1940a and b), who showed that the cytoplasmic basophilia of cells fixed in Helly's fluid could be removed by treatment with a crude preparation of ribonuclease. He incubated sections for 1 hour at 65°—70° and subsequently stained them by the methyl-green pyronin method (see Chapter 9, p. 268). Painter and Taylor (1942) incubated tissues for several hours with crystalline ribonuclease and stained them by the same method. Gersh and Bodian (1943) used crystalline ribonuclease at a concentration of 1 mg/ml. in veronal-acetate buffer at pH 6·75, incubating for 5 hours and subsequently staining with toluidine blue. They showed the basophilia of Nissl's granules in nerve cells was entirely lost after this treatment. Dempsey and Wislocki (1945) incubated sections fixed in Zenker-formol or Zenker-acetic for ½–3 hours at 60° and stained them by the eosin-methylene blue method. Dempsey and Singer (1946) used a similar method in investigating the basophilia of thyroid colloid and they used Mallory's trichrome stain and the Heidenhain azocarmine modification in addition to an eosin-methylene blue method. Deane (1946) used a 1 mg/ml. solution of ribonuclease, at pH 6·75 for 3 hours at 60°, on liver sections fixed in Zenker-formol and stained these with eosin-methylene blue. Stowell (1946) used the same strength of enzyme in McIlvaine's citric acid-disodium phosphate buffer at pH 7 and White (1947) used a 1 mg/ml. solution of a pancreatic extract whose proteolytic activity was first destroyed by boiling for 3 minutes in acid solution before buffering with a veronal-acetate mixture at pH 6·75. Sanders (1946) used a similar product in his studies on cytoplasmic nucleotides, destroying its proteolytic activity by heating at pH 6·75 and 80° for 10 minutes. Using his own technique and methyl green-pyronin staining, White studied the cytoplasmic basophilia of human marrow cells fixed in Heidenhain's Susa fixative, and by a similar method (Pearse, 1949) I studied the basophilia of plasma cells (human) and lymphocytes (guinea-pig) engaged in the production of Russell bodies and Kurloff bodies respectively.

In the majority of the above experiments control sections were incubated in water or buffer solutions for the same time and at the same temperature as the test sections. Stowell and Zorzoli (1947) drew attention to the fact that electrolytes in solution removed a variable amount of RNA from control sections and observed that even distilled water had this effect at 60° and above. It became customary, therefore, to use ribonuclease in glass-distilled water at 1 mg/ml. and 37°. Under these conditions the incubation period, although it varied considerably with the fixative, was about 1–3 hours. Using onion root tips fixed in acetic-alcohol, and incubating for 2 hours in the manner suggested above, Kaufmann, Gay and McDonald (1950) demonstrated an increased affinity of the chromosomes for acid dyes such as acid fuchsin. This was considered to be due to the release of a basic, tryptophan-containing, protein by removal of the nucleic acid previously attached to it. Using fixation by means of Flemming's or Navashin's fluids, Kaufmann, McDonald and Gay (1948) stained plant material with the safranin-gentian violet-orange G triple stain. They found that metaphase and anaphase chromosomes were bright red, whereas the chromonemata of late telophase and early prophase were violet. Ribonuclease was observed to remove the red-staining component, so that the metaphase and anaphase chromosomes appeared violet after hydrolysis. These findings showed that the crystalline ribonuclease used was free from protease activity, since the increased acidophilia of the chromosomes was not demonstrable, after even brief incubation, if protease was present as a contaminant.

The Specificity of Ribonuclease for RNA. In spite of earlier suggestions (Stowell and Zorzoli, 1947; Pollister *et al.*, 1950) that DNA was also affected by ribonuclease, later evidence indicated that this was not the case. McDonald and Kaufmann (1954) found that the reduction in Feulgen stainability of chromatin after ribonuclease was due to an intracellular deoxyribonuclease which hydrolysed DNA only after removal of RNA. Crystalline ribonuclease can be resolved into 4 enzymically active components, RNases I, II, III and IV (Hakim, 1960). This author showed that RNases I and II liberated more uridylic and cytidylic acids from yeast RNA and that they possessed greater activity towards pyrimidine cyclic mononucleotides than RNase III and IV. Synthetic activities due to these two last fractions were shown to be similar to those exhibited by the minor component of the enzyme, the so-called ribonuclease B. The major component, ribonuclease A, was found by Hirs *et al.* (1956) to contain 124 amino acid residues per mole. The structure included 4 intramolecular S-S bridges. Since crystalline ribonuclease is now readily obtainable, substitutes such as heated human saliva should no longer be required.

The Influence of Fixation on the Result and its Interpretation. Provided that the fixative employed does not cause loss of RNA, and does not render it impervious to the action of ribonuclease, it is possible to vary the incubation time so that removal of RNA from the tissues is complete. The question arises, however, as to how much the physical structure of the ribosenucleoprotein may alter its digestibility and how much alteration in this structure can be produced by the various fixatives. All authors have agreed in finding that fixatives containing dichromates in acid solution (chromic acid) interfere dramatically with the enzymic removal of RNA. Such fixatives as Zenker, Zenker-formalin, Flemming and Navashin are now never used, therefore, for ribonuclease studies. Inferior results which were reported using Bouin, Schaudinn and methyl and ethyl alcohols in the absence of acetic acid, were largely due to imperfect preservation, or to residual tissue RNase activity. Carnoy and Susa were strongly recommended by some authors (e.g. Kaufmann *et al.*, and White) and as strongly criticized by others (e.g. Stowell and Zorzoli). It was generally agreed, however, that formalin was a satisfactory fixative for RNA, and that it could be used alone, or in combination with precipitant fixatives which did not interfere with subsequent enzyme action. Although mercuric salts inhibit ribonuclease, the amount of mercury remaining in sections after treatment with iodine and thiosulphate is not sufficient to cause significant delay in removal of RNA by the enzyme. Carnoy, Susa, and formol-mercury mixtures were, therefore, recommended for fixation of tissue blocks and the first two for material in smears.

A comprehensive study of the optimum conditions for RNase analysis, especially with respect to fixation and conditions of incubation, was made by Amano (1962). This author recommended fixation in *freshly prepared* Carnoy's solution, for at least 24 hours, followed by incubation in 1 mg/ml. crystalline RNase in distilled water, for 4 hours at 40°. Shorter fixation was found to permit loss of RNA from control, water-incubated, sections. There are still some discrepancies between these results and those of Lagerstedt (1957) and of Hartlieb *et al.* (1956).

Each of the three popular methods of demonstrating the action of ribonuclease has its adherents. A buffered eosin-methylene blue technique may be most suitable for widely distributed RNA, that of thyroid colloid, for instance. For marrow smears and most cytological work a buffered methyl green-pyronin stain gives the best results. The simplest, and often most effective method is simple and brief staining in dilute aqueous solutions of toluidine blue. The latter was used by Love and Rabotti (1963), in the form of the TBM method (see Chapter 9, p. 272), for their studies on the digestion by nucleases of the 9 types of ribonucleoprotein demonstrated by the method. Using DNase (see below) as well as RNase, they divided these 9 types into 4 distinct groups. The first group, resistant to RNase but sensitive to DNase, included the two types of chromosomal ribonucleoprotein while the second, sensitive to both nucleases, comprised both amorphous parachromatin, and the nucleolini. The third group, which was digested only by RNase, included both types of cytoplasmic and granular ribonucleoprotein while the fourth group, resistant to both enzymes, constituted the pars amorpha of the nucleolus. Studies on RNA in preneoplastic areas of rat liver, following feeding with 4-dimethylaminoazobenzene, were carried out by Brière (1970). This author showed that cytoplasmic RNA in hyperbasophilic foci, and in frank hepatomas, was susceptible to removal by short treatment with RNase in low concentration (1 to 50 μg. per ml. buffer at pH 6·0). Cytoplasmic RNA in surrounding tissues was unaffected. It was suggested that this "sensitive" RNA was either an excess of normal (messenger) RNA or, alternatively, an excess of altered (? submethylated) RNA.

Deoxyribonuclease

Much of what has been written above concerning ribonuclease refers equally to deoxyribonuclease (DNase), and it will not be repeated. The earliest histochemical work, using the latter enzyme, was carried out with preparations such as those described by Fischer *et al.* (1941) and by McCarty (1946). Although these contained no appreciable amount of ribonuclease they were contaminated with varying amounts of protease. Later experimenters used a purer preparation, devised by Kunitz (1948), which was free from both impurities. The crystalline enzyme is activated by magnesium ions (optimum concentration 3 mM) and by manganese ions. The magnesium-activated enzyme is inhibited by 0·01M-citrate but manganese activation is unaltered. The enzyme is readily inactivated by heat and has its pH optimum in the range 6·8 to 8·2. Its isoelectric point is 5·1. Chargaff and Shapiro (1955) suggested that in the case of the pancreatic enzyme magnesium ions played a double part. They proposed that after activating the enzyme, and causing the initial stages of degradation of DNA, magnesium ions were able to bring about additional, non-enzymic, breakdown of the initial reaction products.

Earlier results of the cytochemical application of purified DNase were not uniformly satisfactory. Catcheside and Holmes (1947) used a Fischer-type spleen DNase, and a McCarty pancreas DNase, and produced complete abolition of basophilia and the Feulgen reaction in alcohol-fixed beanroot chromosomes by means of the latter. The splenic enzyme would achieve such results only after previous incubation of the material with ribonuclease. The authors believed that these results were due partly to impurities in their enzyme preparations and partly to the non-specific depolymerizing action of protein solutions. They advised caution in interpreting experimental results obtained with DNase. Brachet and Shaver (1948) used DNase (Fischer) at a concentration of 10 mg/ml. and reported that the abolition of Feulgen staining and basophilia in alcohol-fixed chromosomes was "almost instantaneous" at room temperatures, provided that magnesium ions were present as activators. They used non-activated enzyme solutions for control sections. Using Ponceau 2R in acid solution as advocated by Hydén (1943) for the cytochemical detection of basic proteins of the histone type, these authors found a diminution in staining of the chromosomes after DNase which equalled the fall in basophilia demonstrated with toluidine blue. They remarked that this indicated either that Ponceau 2R stained DNA or else that DNase removed histones. The latter result was excluded by performing the arginine reaction of Serra, after DNase, and finding no diminution of staining intensity. In view of the known proteolytic activity of their preparations of DNase of the Fischer type, it was difficult to understand this last result unless the particular Ponceau employed was capable of staining DNA in an acid medium. Brachet and Shaver found that their samples of Ponceau would in fact stain in this manner, although other authors (Hydén) could not stain histones with Ponceau 2R.

Vercauteren (1950a) found that a purified DNase (McCarty) whose high activity against DNA *in vitro* was confirmed by a viscosimetric technique, failed to remove the basophilic material from chromosomes. Later (1950b) this author reported successful experiments using a crystalline DNase of Kunitz type. In these he confirmed the specificity of methyl green for DNA, already claimed by Pollister and Leuchtenberger (1949). He concluded that the DNase technique was dangerous as well as useful and that its results should always be compared with those afforded by other methods. A commercially available streptococcal deoxyribonuclease (Varidase, Streptodornase-Streptokinase, Lederle) was used by Jackson and Dessau (1955). These authors found that DNA was incompletely removed from sections treated with enzyme solutions containing less than 600 units per ml. Above this concentration complete removal was obtained but the authors did not investigate the specificity of the reaction. Kurnick (1955) reported that he obtained excellent results with a crystalline pancreatic deoxyribonuclease and in an earlier paper (1954) he discussed the mechanism of DNA depolymerization by heat and showed that relatively mild treatment, insufficient to change its elementary composition, produced a reduction in affinity for deoxyribonuclease. This change was closely related to the accompanying fall in affinity for methyl green.

A study of the action of dilute neutral or acid DNase on living fibroblast cultures was made by Chèvremont *et al.* (1960a and b). After acid DNase, Feulgen-positive spherules appeared in the cytoplasm of the cells. Using acetic-ethanol fixed cells Kasten (1964) noted that degradation of nuclear DNA began within 15 minutes after initial exposure to pancreatic DNase at 37°. Thereafter unusual phase-light granules appeared in the cytoplasm and these gave a positive fluorescent Feulgen test (Vol 1, p. 265). They stained yellow or green (fluorescent) with acridine orange (Chapter 9, p. 265 and Chapter 29, p. 1186). After 90 minutes' treatment the granules decreased in number and disappeared. Bacterial nucleoproteins were studied by Tonew (1970) using 1 per cent basic fuchsin in 5 per cent acetic-ethanol, before and after DNase extractions.

Crystalline DNase I, derived from bovine pancreas, is now freely available but, according to Laskowski (1961) many preparations of purified DNases still contain some RNase activity. The pancreatic DNases are divided into 2 types, DNase I and DNase II, according to whether their predominant products are 5′-monoesters (I) or 3′-monoesters (II). DNases from other sources do not necessarily fit precisely into this classification but Group C Streptoccal DNase produces, mainly 5′-monoesters and, like DNase I, it has its pH optimum at 7·0, is activated by Mg^{2+} ions and inactivated by versene. Studies carried out by Stone and Burton (1962) an optimum concentration of 0·6 mM for Mg^{2+} with streptodornase and an additional activation by Ca^{2+}. With both cations at 6 mM the reaction was approximately 8 times faster than with magnesium alone. High levels of Mg^{2+} inhibited the enzyme; at 2·5 mM activity was reduced to 15 per cent. Their results suggest that depolymerization of DNA by streptodornase involves two metal-binding sites, one having a specific requirement for Mg^{2+} and the other for Ca^{2+}.

Despite often expressed doubts concerning the specificity of DNase for DNA it is clear, from the work of Amano (1962) and Love and Rabotti (1963) that under optimum histochemical conditions only DNA is removed. The first author found Carnoy to be the fixative of choice and, depite Stone and Burton's (1962) findings, he was able to show by quantitative autoradiography that loss of RNA could be prevented by raising the concentration of Mg^{2+} to 0·2 M. Removal of DNA was not inhibited.

Interpretation of the results of nuclease histochemistry is not difficult but combined DNase/RNase extractions such as those used by Love and Rabotti (1963) are more complicated. These authors considered that their results could be explained on the basis of DNA/RNA hybridization, a phenomenon first described by Hall and Spiegelman (1961). Further work on these lines, with artificial hybrids, is obviously necessary.

Hyaluronidases

In recent years a number of enzymes have been obtained from various sources which are capable of depolymerizing and hydrolysing samples of hyaluronic acid and these are, therefore, referred to as hyaluronidases. A list of the sources of hyaluronidase appears in Table 74.

The history of the hyaluronidases begins with the description of the so-called "spreading factor," by Duran-Reynals (1928, 1929) and by McClean (1930, 1931), which was held to be responsible for the rate of diffusion of fluids in living tissues. The identity of this factor was discovered by Chain and Duthie (1939, 1940), who showed that the hyaluronidase activity of testicular extracts paralleled the activity of the "spreading factor". Meyer *et al.* (1936, 1937) had previously found, in autolysates of a rough Type II pneumococcus, an enzyme capable of hydrolysing hyaluronic acid. This substrate had been isolated shortly before by Meyer and Palmer (1936) from vitreous humour and umbilical cord. It has now been shown to be a protein-free linear heteropolysaccharide with repeating 3-*O*-(β-D-glucuronido)-*N*-acetyl-D-glucosamine units linked by β(1→4) hexosaminidic bonds.

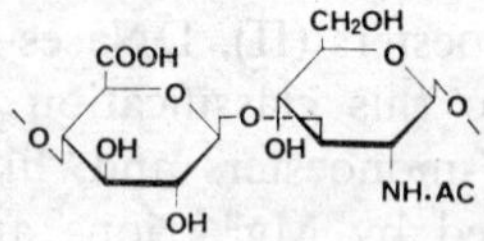

Basic disaccharide unit of hyaluronic acid (Meyer, *et al.*, 1961).

The average molecular weight of hyaluronic acid is between 200′000 and 500′000, and its solutions are characterized by high viscosity. Assuming a regular repeating unit (disaccharide) the distance between adjacent carboxyl groups would be 10Å.

TABLE 74

Sources of Hyaluronidase

Source	References
Testis	Duran-Reynals (1928); McClean (1930)
Skin	Claude and Duran-Reynals (1934); Meyer *et al.* (1941)
Leech extract	Claude (1939)
Snake venoms	Chain and Duthie (1940); Favilli (1940)
Streptococci	Duran-Reynals (1933); Crowley (1944)
Staphylococci	Duran-Reynals (1933)
Pneumococci	McClean (1936); Meyer, Dubos and Smyth (1937)
Cl. welchii, septicum, œdematiens	Meyer, Hobby, Chaffee and Dawson (1940); McClean, Rogers and Williams (1943)
Schistosoma	Levine *et al.* (1948)

Biochemistry of the Hyaluronidases

As the result of prolonged studies on the component mucopolysaccharides of mesenchymal tissues Meyer and Rapport (1951) were able to subdivide them into five types. These were hyaluronic acid, hyaluronosulphate, and chondroitin sulphates of three types, designated A, B and C. As substrate for hyaluronidase the second (hyaluronosulphate) which was found only in cornea, does not interest us very greatly, although it is rapidly depolymerized by both testis and pneumococcal hyaluronidase. The other types, however, are of widespread occurrence. Hyaluronic acid was found by Meyer and Rapport in synovial fluid, skin and umbilical cord. Chondroitin sulphate (Ch S-A) occurred exclusively in hyaline cartilage, Ch S-B in heart valves, aorta and tendon and Ch S-C in all the above sites except synovial fluid. The effects noted when these four compounds were tested *in vitro* with testis and pneumococcal hyaluronidases are condensed into tabular form below.

The data given in Table 75 repeat the statement of Meyer *et al.* (1941) that the testicular enzyme affects sulphate-containing and sulphate-free mucopolysaccharides while the bacterial enzyme affects particularly the latter. They would seem to suggest that by using hyaluronidase from the two sources mentioned we might be able to distinguish in the tissues four types of mucopolysaccharide-protein complex. This is not the case for two reasons. First, the above findings are based on hydrolysis and not on depolymerization, and the two functions do not run parallel. Secondly, they are based on material easily accessible to the enzyme. Whatever the theoretical considerations, the results given in the Table are not reproducible in the case of tissue sections, at least in those fixed in the usual formalin-containing or alcoholic fixatives. In the case of hyaline cartilage, for instance, neither the testicular nor the

TABLE 75

Actions of Hyaluronidases on Various Substrates

Substrate	Source	Hyaluronidase	
		Testis	Pneumococcal
Hyaluronic acid	Synovium Skin Umbilical cord	Labile	Labile
Ch S–A	Hyaline-cartilage	Labile	Labile
Ch S–B	Tendon Aorta Skin	Stable	Stable
Ch S–C	Tendon Aorta Umbilical cord	Labile	Stable

bacterial enzyme can remove the mucopolysaccharide component. The metachromatic substance of the aorta, the major part of which is composed of the hyaluronidase-stable ChS-B form according to Meyer and Rapport, behaves in a similarly paradoxical manner, resisting the bacterial enzyme as it should, but often being totally removed by testis extracts. The behaviour of the metachromatic substance(s) of skin, synovium and umbilical cord, which resists neither type of enzyme, is therefore histochemically correct. The metachromatic substance found in healing wounds has been stated by Bunting and White (1950) to be susceptible to testicular but not to streptococcal hyaluronidase. Penney and Balfour (1949), and Campani and Reggianini (1950) also removed the metachromatic substance of non-scorbutic wounds by means of the testicular enzyme. If unfixed tissues are used for histochemical tests with hyaluronidases the solubility of many of the mucopolysaccharide-protein complexes in water has to be taken into account. No difference may be recorded between removal by the enzyme and removal by the control solution.

The pH Optima of the Hyaluronidases. The pH optimum for hyaluronidase varies not only according to the source of the enzyme but also according to the method by which it is estimated (see below).

Thus McClean (1943) found that the optimum pH values for the viscosity-reducing activities of testicular and *Clostridium welchii* hyaluronidases were 7·0 and 6·0 respectively. Rogers (1948) found the maximum viscosity-reducing activity of streptococcal hyaluronidase at pH 5·5 and of staphylococcal hyaluronidase at pH 6·6. As measured by release of reducing sugars the pH optima of the bacterial enzymes were, for the streptococcal variety, 5·6 and 7·1 (double optimum), for the staphylococcal variety 5·6 and for the *Cl. welchii* variety 5·7. According to Meyer *et al.* (1941) the optimum hydrolytic pH for testis hyaluronidase is 4·4 and for the pneumococcal enzyme, pH 5·6–5·8. These figures show that there is a wide variation in the optimum pH of all except *Cl. welchii* hyaluronidase depending on the method of assay which is employed.

Other factors which affect the activity of the hyaluronidases are also important from the histochemical point of view. The addition of salts, especially sodium chloride, has been observed to increase the activity of the enzyme from bacterial and mammalian tissue sources, and this affects the viscosity-reducing far more than the hydrolytic activity. For testis hyaluronidase Madinaveitia and Quibell (1941) found the optimal concentration of NaCl to be between 0·07 and 0·17 M.

The Two Types of Hyaluronidase Activity. The action of hyaluronidases on hyaluronic acid is of two distinct forms (Meyer *et al.*, 1941), first a depolymerization which is rapid in action and, second, a slower hydrolytic reaction which liberates acetylglucosamine and glucuronic acid. Hydrolysis is estimated by testing the substrate solution for an increase in reducing substances, depolymerization by calculating the reduction in viscosity of the solution. Meyer (1947), in an extensive review of hyaluronic acid and hyaluronidase, suggested that two enzymes were necessary to break down the two types of glycosidic linkage in hyaluronic acid, one of which belongs to the *N*-acetyl-glucosamine and the other to the glucuronic acid (see formula on p. 1026). The existence of two enzymes was also suggested by the observation that pneumococcal hyaluronidase would almost completely hydrolyse the substrate into its component monosaccharides, while testicular hyaluronidase hydrolysed only 50 per cent in a comparable period. The latter, however, reduced the viscosity of substrate solution much more rapidly than the bacterial enzyme.

A similar conclusion to Meyer's was arrived at by Hahn (1945a and b, 1946a and b). Working with aqueous extracts of leech heads (1945a) he found that these caused only a 26 per cent hydrolysis of hyaluronic acid, as measured by increase in reducing substances, while at the same time the reduction in viscosity was equal to that given by the best testicular preparations. In the case of the latter, he suggested that there were two enzymes present, one degrading hyaluronate to a disaccharide, lowering the viscosity and liberating half the total amount of reducing sugars, and the other hydrolysing the resulting *N*-acetylglucosamine glucuronide to its constituent monosaccharides. Rogers (1946a and b) showed that when purified hyaluronic acid was partially hydrolysed by enzymes prepared from three different sources, until about 50 per cent of the theoretical amount of reducing sugar had been liberated, the size of the remaining units depended on the source of the enzyme. This suggested that the hyaluronidase enzyme system was even more complex than Meyer and Hahn believed. Rogers explained this observation by assuming that several enzymes successively degrade the polysaccharide to smaller and smaller units. He found that the optimal conditions of pH and salt concentration differed according to the enzyme preparation employed. Madinaveitia *et al.* (1940) considered that the hydrolytic and diffusing activities of testicular extracts were probably due to the presence of different entities, on account of large discrepancies between the results of two different techniques for estimating the enzyme activity. McClean and Hale (1941) and McClean (1943), however, pointed out that these authors had performed the two tests under widely different conditions. Rogers (1948) measured the effects of streptococcal and staphylococcal hyaluronidases, by viscosity-reduction and reducing sugar methods, under identical conditions. The differences amounted to more than 100 times the standard error of the method and he concluded that the two methods did not measure the activity of one and the same enzyme.

As far as the aminopolysaccharase activities of the hyaluronidases are concerned there is general agreement that the testicular enzyme degrades hyaluronic acid and chondroitin sulphate (Types A and C) mainly to tetrasaccharides although disaccharides and a few higher oligosaccharides may be formed (Weissmann *et al.*, 1954; Schütte and Greiling, 1955; Hoffman *et al.*, 1956). Bacterial hyaluronidase, on the other hand, degrades hyaluronic acid to yield mainly a disaccharide (Schütte and Greiling, 1955; Linker *et al.*, 1956). Chondroitin sulphate A can be attacked by the bacterial enzyme if its ester sulphate groups are first removed by mild acid hydrolysis. Meyer *et al.* (1956) confirm that testicular hyaluronidase is without action on chondroitin sulphate B.

The mechanism of action of the hyaluronidases was discussed byLudowieg *et al.* (1961). These authors found that when the degradation of hyaluronic acid by bacterial hyaluronidase was carried out in the presence of H_2O^{18} no O^{18} was incorporated into the products, in contrast with other known glycosidases, including testicular hyaluronidase. They presumed that the reaction catalyzed by the bacterial enzyme was an elimination resulting from an attack on the bond between O and the C-4 of the uronic acid.

Histochemistry of the Hyaluronidases

Enzymes Available. Although the hyaluronidases are widespread in nature, in their application as histochemical reagents we are concerned only with two main sources, bacterial culture filtrates and mammalian testis extracts. The former are divided into (1) coccal filtrates and (2) clostridial filtrates. We have to consider the practical suitability of hyaluronidases from each of the available sources and the substrate or substrates in the tissues upon which they act.

From the first point of view there is no doubt about the greater convenience of testis extract, available commercially as a water-soluble powder of controlled potency, by comparison with the bacterial filtrates which must be freshly prepared in the first instance, although they are remarkably stable once prepared, even at room temperature. If the crude filtrates are used a number of enzymes besides hyaluronidase will certainly be present. *Clostridium welchii* Type A filtrates for instance, which have frequently been used as a source of hyaluronidase, contain also a so-called collagenase (κ-toxin), a lecithinase (η-toxin), and other lipolytic factors (lipases or esterases). Streptococcal and pneumococcal filtrates suffer from the same disadvantages, although the presence of other enzymes does not always matter. Whether their presence is important or not depends on which particular histochemical test is used for appreciation of hyaluronidase action on tissue sections (see below). Even testis extracts, in the crude form in which they were formerly used for enzyme analysis, contained other enzymes besides hyaluronidase. Meyer and Linker (1951) reported that such extracts contained a β-glucuronidase and Houck and Pearce (1957) divided the mucolytic enzyme activities of ammonium sulphate-precipitated fractions of bovine testis into four groups. These were hyaluronidase, β-glucuronidase, β-hexosaminidase and β-galactosaminidase. The method of Dorfman (1955) has usually been employed for the preparation of testis hyaluronidase for use in biochemical studies. The β-glucuronidase, however, is separable from the hyaluronidase fraction and in any case it is not present in the purified commercial preparations of testis which are used to-day.

The sole advantage of the bacterial (streptococcal or pneumococcal) enzymes, recently stressed by Davies (1952), is their specificity for the single substrate hyaluronic acid.

Staining Methods. The difficulties discussed above are accentuated by the fact that two very different methods have been employed for demonstrating the hyaluronidase substrates. The first method makes use of the metachromatic properties of the acid mucopolysaccharides, and toluidine blue is the principal dye used for this purpose. The second is the periodic acid-Schiff (PAS) technique. This was the sole method used by Gersh and Catchpole (1949) in a comprehensive study of ground substances and basement membranes. Before we proceed further it is necessary to emphasize the profound difference between the substances demonstrated by the two methods. According to Gersh and Catchpole the most characteristic component of the microscopically non-fibrillar ground substance is a glycoprotein (i.e. a protein containing a carbohydrate moiety as an integral part of its structure) and basement membrane is regarded as a closely related substance. The metachromatic component of ground substance, however, with which the pathologist is familiar as a bluish haematoxiphil material, is an acid mucopolysaccharide (hyaluronic acid or chondroitin sulphate). Neither of these stains convincingly by the PAS method. Davies (1952) found that while an undegraded hyaluronic acid complex and a commercial preparation of potassium hyaluronate gave vigorous reactions in spot tests with the PAS reaction, an almost pure, sulphur-free, potassium hyaluronate gave a negative result. With toluidine blue all three produced metachromasia, labile to streptococcal hyaluronidase, but the PAS reaction of the first two was resistant to the enzyme.

Both the metachromatic mucopolysaccharide-protein component and the water-soluble glycoprotein component of Gersh and Catchpole are substrates for hyaluronidase but with this difference. The former was removed by short (3 hours) treatment with the enzyme, as evidenced by reversal of metachromatic staining. The latter was only affected by prolonged treatment (20 hours or more at 37°), as shown by reversal of the positive PAS reaction.

Applications to Pathology. Most of the work so far reported in the pathological literature, in which hyaluronidase has been employed, describes the accumulation in the tissues of substances believed to be acid mucopolysaccharides. These have usually been demonstrated by metachromatic techniques, and hyaluronidases from various sources have been used to effect their removal, usually from conventionally-fixed paraffin-embedded material. The majority of workers have used testicular hyaluronidase and a lesser number streptococcal and clostridial filtrates. One report of the use of leech hyaluronidase has appeared (Negri and Weber, 1948), but this enzyme was found to be inactive against all the metachromatic ground substances tested. These authors reported that their crude testis extract was "extremely active against the chromotropic substance of blood vessel walls, causing total disappearance of the metachromatic coloration with toluidine blue". This finding is at variance with those of some other workers. Altschuler and Angevine (1949), for instance, using both crude and purified testis extracts, observed no activity against the metachromatic substance in arteries. Using relatively high concentrations of the testicular enzyme in saline, I find that arterial metachromasia is usually reversible by short treatment, in the case of both frozen and paraffin sections.

The earliest studies in the histochemical field were those of Mannozzi Torini (1942) who used a testicular extract which he described as a meso-mucinase, to reverse the metachromasia of the ground substance of umbilical cord. Since this example, similar methods have been applied to the study of mucopolysaccharides in various normal and pathological tissues. Hyaluronidase-labile substances have been described, for instance, in rheumatoid nodules (Altschuler and Angevine, 1949; Fawns and Landells, 1953), in healing wounds (Penney and Balfour, 1949; Bunting and White, 1950), in

Aschoff bodies (Bunting, 1950), in lupus erythematosus (Klemperer, 1948), in generalized and circumscribed myxoedema (Pearce and Watson, 1949; Watson and Pearce, 1950), in synovium (Asboe-Hanson, 1950), and in the cock's comb after treatment of the bird with pituitary hormones (Ludwig and Boas, 1950). The individual mucopolysaccharides of the human aorta at various ages were characterized by Zugibe (1962) using testicular and streptococcal hyaluronidases, and Alcian blue for demonstration of their substrates. His conclusion was that chondroitin sulphate-B levels increased with age.

Conclusions. The conclusions which have been drawn by various authors as to the exact diagnostic significance of hyaluronidase-lability in a metachromatic mucopolysaccharide have tended to vary. Some have only made use of the observation to confirm the mucopolysaccharide nature of the substance investigated, and this is a valid if only partial conclusion if short incubation times are employed. Others have identified the labile substance with hyaluronic acid, an interpretation which may sometimes be true if testis enzyme is used, or wholly true, as in the case of streptococcal filtrates. At this present stage of our knowledge of the nature of the substrates and actions of the hyaluronidases few conclusions can be considered accurate. I consider that metachromatic material, in fixed tissues embedded in paraffin, whose metachromasia is reversible by 1–3 hours' treatment with a purified testis extract (Rondase-Evans, Hyalase-Benger): (1) is of mucopolysaccharide nature, and (2) is either chondroitin sulphate of Meyer's types A or C, or hyaluronic acid itself, or a mixture of any or all of these. Meyer himself (1947) was of the opinion that hyaluronic acid alone would not give rise to metachromasia in tissue sections, while chondroitin sulphates would do so. It is not possible to agree entirely with this view for the hyaluronidase-labile metachromasia of synovial tissues is apparently due to hyaluronic acid, the sole mucopolysaccharide constituent of synovium.

Chondroitinase

Certain strains of *Proteus vulgaris* were shown by Dodgson *et al.* (1957) and by Dodgson and Lloyd (1958) to produce two enzymes, a chondroitinase and a chondrosulphatase, which together degraded cartilage chondroitin sulphate with the release of reducing substances and sulphuric acid respectively. The chondroitinase, however, could act independently of the sulphatase. Cartilage chondroitin sulphate is mainly type A with small amounts of type C; type B is not present. The activity of their chondroitinase towards the latter was not described by Dodgson and his associates and the enzyme has not been used histochemically. Hoffman *et al.* (1957) described the production of adaptive enzymes in *Flavobacterium* which would hydrolyse chondroitin sulphates A and C at high rates. Bacteria grown with β-heparin (identical with chondroitin sulphate B) as their carbon source produced an enzyme able to break the hexosaminidic bands to produce unsaturated uronides. These results were applied histochemically by Zugibe (1962).

Neuraminidases (Sialidases)

Spicer and Warren (1960) described the use of neuraminidases, from three sources, for the histochemical distinction of sialic acid-containing mucoproteins (Vol 1, p. 300). Of these three sources, *Clostridium perfringens*, *Vibrio cholerae*, and influenza virus (RDE, receptor destroying enzyme of Burnett and Stone, 1947), the second provides the variety most easily obtainable commercially. Supplied in ampoules containing 500 units/ml. of activity *Vibrio* culture filtrates exhibit minimal activity of other enzymes although traces of protease, peptidase, and α and β-glucosidase can be detected. A very complete study of influenza virus neuraminidases was made by Wilson and Rafelson (1967) which included work on inhibition, and on activation particularly by bivalent cations.

Neuraminidase from *Vibrio* cultures was used by Quintarelli *et al.* (1964), in conjunction with Alcian blue staining, by Geyer *et al.* (1964), by Quintarelli and Dellovo (1965), and also by Leppi (1968), despite difficulties due to inefficient penetration of enzyme into the tissues and the lack of adequate controls.

The histochemical use of neuraminidase was subjected to rigorous testing by Schmitz-Moormann (1969) who showed, as was to be expected, that the greater part of tissue-bound sialic acid is removed when tissues are methylated in acid solutions (or by any treatment of fixed sections with acid). He showed, furthermore, that enzyme-free buffer removed a considerable amount of the substrate and concluded that histochemical detection of neuraminic acid can be made only if a significant difference be shown (by Alcian blue staining) between the effect of enzyme-free and enzyme-containing buffer.

Details are given in Appendix 25.

Other Polysaccharidases

Amylases

The enzyme amylase, as ptyalin in saliva, was extensively used for the digestion of glycogen in tissue sections following its introduction for this purpose by Bauer (1933). Later the use of solutions of malt diastase, advocated by Lillie and Greco (1947), largely superseded the older practice. The preponderating hydrolytic enzyme of both ptyalin and malt diastase is an α-amylase, belonging to the group which liquefy starch and glycogens in contradistinction to the more slowly acting β-amylases which exert only a saccharolytic activity against these substrates.

Neither of the two sources gives a pure α-amylase; both are contaminated with other enzymes which may lead to various side effects. Lillie (1949) observed that malt diastase solutions (often) had the property of destroying the metachromasia of nucleus pulposus and of cartilage matrix. Both the action on metachromasia and the associated ribonuclease action were to some extent dependent on the fixative used. They were both hindered by aqueous fixatives of the acid variety and by dichromate. It was not possible to obtain the metachromasia-destroying enzyme or the ribonuclease principle in separate fractions but the three enzymes could be distinguished by their differing resistance to certain procedures (Lillie, Greco and Laskey, 1949). The ribonuclease, characteristically, resisted 90° at pH 3–4, the metachromasia-destroying enzyme resisted 60° at pH 3–4 and the amylase was destroyed by 60° at pH 6·0. Furthermore, it was observed that certain large Gram-positive bacilli present in the rabbit caecum retained their Gram-positive staining, and their positive PAS reaction after 4 hours' digestion with 0·1 per cent malt diastase at 37°. When stained by means of Lillie's (1948) buffered azure-eosin technique, however, their staining changed from blue to pink. Precisely what action of diastase this indicates is not certain. It draws attention once more to the principle of using different final reactions to demonstrate the effect of any enzyme on tissue structures.

The above-mentioned side effects of ptyalin and diastase do not in any way influence the interpretation of results achieved by the use of saliva or malt diastase for removing glycogen from sections. With diastase it is unnecessary to use the enzyme either in buffer or saline, both of which have the tendency to increase its ribonuclease activity. Solutions in glass-distilled water remove glycogen adequately, even within as short a period as 30 minutes at room temperature, and if they are thus used little or no change will be produced in the tissue sections by the other enzymes present. The opinion expressed in Chapter 10 (p. 363), that glycogen cannot be digested from resin-embedded sections, must be reviewed in the light of the report by Rosa and Johnson (1967) indicating that this can certainly be achieved.

The method of using either ptyalin or diastase is given in Appendix 25. The choice of methods for demonstrating their glycogenolytic effect was discussed in Chapter 10, and the methods for this have been given in the Appendix to that chapter.

Debranching Enzyme.

Substitution of this enzyme (Diazyme) for the amylases or diastase has already been noted (Chapter 10, p. 363).

Pectinases

Biochemical Aspects. Enzymes of the pectinase group were applied to acetone-fixed paraffin-embedded animal tissues by McManus and Saunders (1950). This group of enzymes was first described in 1898 in barley malt and they have since been shown to be widely distributed in fruits and to be formed by a number of bacteria and fungi. There are two main enzymes in the mixtures described as pectinase and the proportion of each varies widely, depending on the source.

The first enzyme, pectinesterase (PE), is particularly plentiful in plant sources such as tomato, potato leaves and tobacco. According to Lineweaver and Jansen (1951), PE hydrolyses methyl ester groups in polygalacturonides, if they are adjacent to free carboxyl groups, as at the × in the diagram.

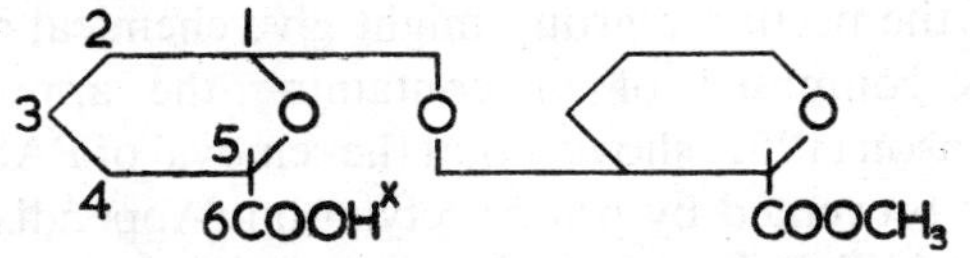

Methyl esters of polygalacturonic acid occur in long chains in the wide range of fruit pectins and these are the natural substrates for the de-esterifying action of PE. The enzyme is apparently highly specific (MacDonnell *et al.*, 1950) since it will not act on galacturonides but only on polygalacturonides. The second enzyme, polygalacturonase (PG), seldom occurs in higher plants but is widely distributed in bacteria and fungi. It also occurs in the animal kingdom having been found in snail-stomach fluid (Holden *et al.*, 1950). The usual commercial source is *Aspergillus niger* and the enzyme, poly-α 1, 4-galacturonide glycanohydrolase, acts on de-esterified portions of polygalacturonide chains, breaking the glycosidic links to give products of varying molecular weight.

Histochemical Aspects. From the histochemical point of view the situation would be much simpler than it is if the two enzymes described above were the only ones present in the various commercial and other extracts used under the comprehensive term pectinase. Unfortunately, a variety of different polysaccharidases may be present.

McManus and Saunders (1950) investigated the activity of several pectinase preparations, and also a preparation which they described as a β-glucuronidase, using 0·4 per cent solutions at pH 4 and 37° for 48 hours in each case. After incubation and brief washing, their sections were stained with haematoxylin and eosin and by the periodic acid-Schiff method. The authors noted that a loss of PAS-positive material, mucin, glycogen, splenic reticulin and the ground substance of cartilage, occurred after the use of a preparation called pectinase (Nutritional Biochemical Company). A similar loss was observed after the use of polygalacturonase (Jansen and MacDonnell, 1945), and the same effect, to a lesser extent, occurred after the use of Pectinol O (Rohm and Hass). Pectin esterase (Rohm and Hass) was noted to increase rather than decrease the PAS reaction of the tissues listed. β-glucuronidase was without action, while diastase at pH 6·8 removed "everything" from the sections.

The authors divided the actions of their preparation called pectinase, which must presumably have been a polygalacturonase, into two groups. The first was a morphological action, removing such materials as "hyaline" while leaving elastica and nuclei untouched, which brought about a kind of microdissection on the slide. The second was a purely histochemical action. Their "morphological" action was, in fact, the employment of enzymal analysis in the old sense where the enzyme is used as a macerating agent. They believed that the evidence that PAS-positive materials were carbohydrates was enhanced by their removal of pectinase, and concluded that, in the future, pure enzymes of the pectinase group might give chemical information about tissue structures composed of or containing the appropriate substrate. McManus and Cason (1951) showed that the removal of PAS-positive material by pectinase was prevented by prior acetylation (Appendix 10, p. 660). This was regarded as additional evidence in favour of the action of pectinase on a carbohydrate substrate. The results obtained by McManus and Saunders in acetone-fixed tissues were duplicated by Cambel (1952) with ethanol-fixed mucin-containing tissues, and by Eidinger and Ghosh (1956) using acetone. The latter authors agreed that pectinesterase intensified the PAS reaction of collagenous tissues but found this effect with pectinase also. Eidinger and Ghosh also examined the effect of the two enzymes on tissues fixed in Bouin, Helly and Carnoy's fluids. In this case, they noted little loss of staining but an intensification with both enzymes in a number of sites which were usually PAS-negative or only weakly positive. Boiled enzyme solutions did not have any effect on removal of tissue components, but some of the increased staining was observed to be due to adsorption of carbohydrate components of the enzyme on to tissue proteins. This was prevented by prior acetylation. They considered that an unknown mucinase, present in commercial pectinase preparations, must be responsible for removing carbohydrate components from acetone-fixed tissues.

These propositions must be examined. First of all, we can omit the further consideration of pectinesterase for two reasons. This enzyme should not remove any naturally-occurring substrate from animal tissues, and McManus and Saunders observed that it indeed had no such action on the tissues which they used. When we come to examine the case of polygalacturonase we have to decide two points, the interpretation of the removal of PAS-positive substances from animal tissues by crude extracts containing the enzyme, and the probable interpretation if pure polygalacturonase is employed with similar results.

According to Lineweaver *et al.* (1949), PG is highly specific by comparison with the simple glycosidases from which it can be differentiated by its failure to hydrolyse methyl-α-D-galacturonic acid. Neither crude nor purified PG attacked gum ghatti, pneumococcus Type I polysaccharide, heparin or hyaluronic acid. The first two of these contain 12 and 60 per cent respectively, of galacturonic acid, while the other two are glucuronides. The polymannuronide alginic acid, occasionally found in sections from operation sites where calcium alginate has been used for haemostasis and packing, is similarly impervious to crude or purified PG. The crude preparations, however, contain enzymes which can hydrolyse starch, maltose, sucrose, carboxymethylcellulose, inulin and other glycosides. It is probable that some of these enzymes act upon the glycosidic linkages of hyaluronic acid and chondroitin sulphate. These are both glucuronides but differ in that the former contains *N*-acetylglucosamine and *β*-glucosidic linkages while the latter possesses *N*-acetylgalactosamine and *β*-galactosidic linkages.

Now that a relatively pure *Aspergillus* PG is available, if this is found to remove PAS-positive substances from animal tissues, it is necessary to conclude either that it is not a pure enzyme at all (if it is true that no substrate for PG is normally present in such tissues), or that PG acts on linkages in polysaccharides which have not yet been described. For the animal tissue histochemist, therefore, the use of either crude or purified preparations of pectinase is unlikely at present to yield any information of an exact nature, and if the component being studied is, in fact, removed by a pectinase preparation the picture may well be complicated instead of the reverse. In spite of these strictures, details of the use of pectinase are given in the Appendix to this chapter.

β-glucosidases

There are two convenient sources of enzymes belonging to the class of *β*-glucosidases; these are the so-called emulsins, derived from *Aspergillus oryzae* (taka-diastase) and from the digestive tracts of various snails, especially *Helix pomatia*. The term "emulsin" means a crude or partially purified mixture of *β*-glucosidases derived from any source, and there are numerous other enzymes present in both the above-mentioned examples. Neither has been much used in histochemical research but the second, under the heading of snail-stomach fluid, has perhaps received more attention than the first. It is the reddish-brown liquid from the crop of a starved snail and is available commercially as "snail digestive juice". More than thirty enzymes have been detected in this fluid (Holden and Tracey, 1950) and of this number two-thirds were classified as carbohydrases.

Snail-stomach fluid was recommended by Fabergé (1945) as a reagent for use in plant cytology and its use for plant materials was discussed by Chambers (1955). It was used by Gersh (1949) in his studies on the Golgi apparatus in mammalian duodenal cells. He found that 1 hour's incubation reduced the PAS reaction of the apparatus to vanishing point. This finding Gersh interpreted as indicating the presence of glycoprotein since he regarded the chief action of snail-stomach fluid as a depolymerization of polysaccharides. Gersh and Catchpole (1949) also used the fluid and found that, with 1 hour's incubation, it had no effect on the glycoproteins of ground substance and basement membranes which they were studying. They rightly ignored this result. We are faced with the question of deciding what substrates in animal tissues are attacked by the battery of enzymes present in snail-stomach fluid.

Holden and Tracey have shown that hyaluronic acid and alginic acid are resistant to snail-stomach fluid and there is no reason to suppose that chondroitin sulphate would behave differently. Presumably the prominent arylsulphatase present in the fluid would not affect this substrate. Glycogen is certainly attacked, but this substance is easily distinguishable, and fatty materials may be broken down by a lipase which is present. These are also distinguishable, by parallel tests, from the purely protein constituents of any given tissue. It is probable that the views of Gersh are correct and that any reduction of the PAS-positive reaction of a tissue structure by snail-stomach fluid can be taken to indicate the presence of carbohydrate.

The use of snail-stomach fluid is scarcely warranted in the ordinary diagnostic histochemistry of animal tissues. In the case of plant tissues it may be useful as a source of cellulase.

Acetylglucosaminases

A number of crude extracts containing chitinases have been described (Zechmeister and Tóth, 1939) and enzymes of this type are present in bacteria, molds and *Actinomyces*. The preparation of a much purer chitinase was described by Jeuniaux (1957). This enzyme, obtained from *Streptomyces* sp. had an activity 70 times higher than that of the original product. Purified chitinases have obvious uses in the histochemical study of developing chitins and for investigating the large number of reputedly chitinous tissue components which fail to react in a satisfactory manner to histochemical tests.

The Proteinases

Trypsin, Chymotrypsin and Pepsin

The three main proteolytic enzymes are here dealt with in a single section although their histochemical actions are not entirely similar. All three were long ago prepared in crystalline form, pepsin by Northrop (1930), chymotrypsin by Kunitz and Northrop (1935), and trypsin by McDonald and Kunitz (1946). They are available in a high degree of purity. These three enzymes are all described as endopeptidases, since they can act only upon the peptide links of terminal amino-acids in the protein chain, like the exopeptidases, but also on similar links within the chain. Their specificities for these internal links are somewhat different and it is worthwhile recording them although, at present, their translation into differential histochemical diagnosis is not possible.

Trypsin acts upon peptide links incorporating the carboxyl (COOH) groups of arginine and lysine, chymotrypsin on those formed from the carboxyl group of an aromatic amino-acid, and pepsin on the link between the α-carboxyl of a dicarboxylic amino-acid (e.g. aspartic, glutamic) with the amino group of an aromatic amino-acid. Because of these different specificities mixtures of the three enzymes will break down a greater number of peptide links than any one of the three acting separately.

A histochemical interpretation of the actions of trypsin and pepsin was suggested by the work of Mazia *et al.* (1947) and of Kaufmann *et al.* (1950) on the protein structure of the chromosomes. The former authors showed that trypsin, as expected, was able to digest both histone (basic) and non-histone proteins although they believed that the main histochemical action of trypsin was attributable to the digestion of peptide links in basic amino-acids. These would be more numerous in histones than in more acid proteins. Kaufmann *et al.* agreed substantially with these findings and observed no final difference in sections treated with trypsin and those treated with chymotrypsin. They showed, furthermore, that in the absence of electrolytes solutions of either enzyme (0·1 mg/ml. in water at pH 6) failed to digest proteins sufficiently to cause their removal from the sections. (They did, however, have an effect in increasing the affinity of the chromosomes for pyronin which was attributed to some degradation of ribonucleoprotein.) If these trypsin-digested sections were subsequently placed in buffer or sodium chloride solutions, full digestion of their protein components took place and the protein could then be completely removed by washing with water. Kaufmann *et al.* therefore declared that the histochemical action of trypsin was largely attributable to the removal by water of the enzymic digestion products after their further degradation by various salts.

Using salivary gland chromosomes as his test object Mazia (1941) showed that solutions of pepsin in HCl caused only shrinkage and no impairment of structure. He believed that this shrinkage was due to removal of a "matrix protein" containing many acidic groups. Mirsky (1943), however, suggested that it was equally likely to be due to the breakdown of histones to histopeptones. The problem resolved itself into whether histones as well as non-histones were removed by treatment with pepsin. Mazia *et al.* (1947) endeavoured to answer this question. They found that pepsin would not digest histones or combined histones (nucleohistones) in solution, but that it would digest more acidic proteins under similar conditions. These findings, unfortunately, are not exactly applicable to the proteins of fixed tissue sections since wide variations can be caused by the chemical and physical assaults suffered during fixation and embedding (see Chapter 5). Kaufmann *et al.* believed that the principal action of pepsin, recognizable histochemically, was the removal of tryptophan-containing proteins of the non-histone variety. Their findings in the case of chromosomes (onion root) suggested that such proteins were present in considerable amounts, and this agreed with the views of Pollister and Leuchtenberger (1949) though not with the values derived from chemical analysis of isolated nuclei (which ought to be more reliable).

Pepsin and trypsin differ in their action on native collagen fibres in the test-tube, only pepsin having any marked digestive effect at temperatures below 40°, but fixed collagen in tissue sections is digestible by both enzymes though relatively faster by pepsin. The difference in the digestibility of reticulin and collagen, already referred to in Chapter 8, is not of much practical importance since distinction between the two is more readily made by means of silver methods of staining. Tryptic digestion has been used for the purpose of indicating whether a tissue component is or is not composed of protein. Gersh (1949) thus recorded the protein nature of the Golgi apparatus in mammalian duodenal cells and Day (1947) established the fact that the amorphous ground substance between collagen fibres also contained protein. We may conclude from all the above studies that trypsin and pepsin can be used histochemically either for the simple confirmation of the presence of protein or for the more important purpose of obtaining information upon the nature of the protein in a given structure.

Benditt and French (1953), in a study of the effects of various enzyme preparations on cartilage considered that the reduction or abolition of a specific staining method after treatment of sections with protease-containing solutions might be due to four activities: (1) depolymerization (specific or non-specific), (2) proteolysis, (3) masking of reactive groups by protein binding, and (4) displacement without depolymerization. The presence of traces of proteolytic activity in many depolymerase preparations was considered to be the most common source of confusion. The other contingencies mentioned by Benditt and French should be borne in mind, however. The elastolytic effect of pepsin was discussed by Fisher *et al.* (1960). These authors showed that three different commercial pepsins had an equally marked effect on alcohol or formalin-fixed elastic fibres in aorta or skin. Two hours' incubation at 37° in a 0·1 per cent solution of the enzyme in 0·1N-HCl rendered the fibres non-fluorescent and unstainable with aldehyde fuchsin or with orcein (Chapter 8, p. 225). Affinity for iron haematoxylin (Verhoeff) was lost much earlier.

Pronase, a commercial protease from *Streptomyces griseus*, was shown by Nomoto *et al*, (1960) to contain several enzymes. Purification by ion exchange chromatography (Trop and Birk, 1970) gave four proteolytically active fractions. One of these, Fraction D, contained an apparently homogeneous endopeptidase, inhibited by DFP, which adsorbed to and hydrolysed elastin. Earlier studies by Morihara and Tsuzuki (1967) had indicated that there was no significant difference in side-chain specificity between a number of bacterial proteinases possessing and not possessing elastolytic activity. They showed that complex formation with elastin was the function required for elastolysis.

Papain

From the mixture of proteolytic enzymes found in crude extracts of the latex of *Carica papaya* (paw-paw), which is called papain, two peptidases have been isolated in crystalline form. Besides these two a number of other enzymes exist in the crude extract and this makes the use of the latter as a histochemical reagent a source of difficulty in interpretation of results. Nevertheless, the biochemical work referred to below suggested the possibility of using even the crude extract in diagnostic histochemical procedures. Middlebrook and Phillips (1941) showed that the keratin of wool fibres was rapidly attacked by papain in the presence of sodium bisulphite at pH 6·7 and this action was further investigated by Blackburn (1950) who discussed the whole question of keratins and enzymic digestion. While intact keratins are resistant to the action of proteolytic enzymes such as trypsin, pepsin and papain, keratins which have been reduced—i.e. in which the disulphide groups have been converted into thiols—are readily attacked. Blackburn stated that sodium bisulphite at pH 6 or 7 markedly increased the digestion of keratin by trypsin, but in this case, and after digestion of wool by papain, histological examination of the residues indicated that the cortical cells were still resistant. This finding appeared to limit the usefulness of the above for histochemical application. A crystalline keratinase isolated from *Streptomyces frondiae* was shown by Nickerson *et al.* (1963) to have the property of solubilizing more than one third of the weight of native wool. The enzyme appeared to be homogeneous by electrophoretic and ultracentrifugal analysis and exhibited maximum against wool at its isoelectric point, approximately pH 9 (Nickerson and Durand, 1963).

Blumberg and Ogston (1958) used a crystalline preparation of papain for removing protein from the protein complexes of hyaluronic acid. They suggested that in rheumatoid arthritis the protein part of the complex is abnormal. The effect of papain on the staining properties of bone and teeth was investigated by Irving (1960) and its effect on isolated amyloid fibrils was studied by Kim *et al.* (1969). These last authors used also pronase, Nagarse and trypsin but found that even after 2 weeks incubation they could identify amyloid fibrils by electron microscopy (see Chapter 11, p. 391). Bromelin, the proteolytic enzyme from pineapple, was found by Thomas and Partridge (1960) to digest elastic tissue completely.

Elastase

An elastolytic enzyme was discovered in bacterial culture filtrates by Eijkman (1904). The most active organism in this respect was *Ps. pyocyanea.* The preparation from pancreas of an enzyme which specifically attacked elastic fibres was described by Baló and Banga (1949, 1950). These authors used their preparation to digest elastic fibres in histological sections of aorta and they showed that the enzyme had no action upon collagen fibres from tendo Achillis, although it caused complete removal of elastic tissue. The first step in this process was described as a swelling, followed by breaking of the fibres into small particles, and subsequently by their complete solution. Since no amino-acids were liberated from elastin when this was acted upon by the enzyme the authors considered that its elastolytic activity depended on the conversion of the "long rod-shaped" molecules of elastic into molecules of a globular type. The pH optimum of the enzyme was in the region of 10·3.

Banga and Schuler (1953) concluded that elastin was a glycoprotein since they were able to demonstrate carbohydrate in elastic tissue after depolymerization by elastase. Pepler and Brandt (1954) compared the activity of a bacterial chondrosulphatase obtained from *Ps. pyocyanea* and from *P. vulgaris* with bovine pancreatic elastase. With short incubation both types of enzyme had an identical chondrosulpholytic effect and with longer incubation an identical elastolytic effect. They maintained, therefore, that elastase could be regarded as a chondrosulphatase. Banga and Baló (1956) later regarded elastase as inhomogeneous. Their preparations contained a mucolytic enzyme in addition to the elastase which acted on serum mucoproteins and on the carbohydrate-containing protein of native and fixed collagen. Lewis *et al.* (1956) described the preparation of a purified elastase, together with its properties, and Tennant *et al.* (1956) described the application of such a preparation to experimental atherosclerosis.

Pancreatic elastase was differentiated from other pancreatic proteinases by Lewis *et al.* (1959) and the isolation, purification and properties of bacterial elastase was described by Mandl and Cohen (1960). Assays for the enzyme were developed by Lamy *et al.* (1961) and the selective inhibition of elastolytic and proteolytic properties of elastase were described by Walford and Kickhöfen (1962). Elastase produced by ringworm fungi was investigated by Rippon (1967) and an elastase from *Staphylococcus epidermitis* was described by Varadi and Saqueton (1968).

Naughton and Sanger (1961) prepared a purified elastase, by CM-cellulose column chromatography, which split peptide bonds adjacent to neutral amino acids. The enzyme preparation produced by this method was shown by Gertler and Hofmann (1967) to be inhomogeneous. Using DEAE-cellulose and DEAE-Sephadex, Narayanan and Anwar (1969) produced a porcine pancreatic elastase, free from trypsin and chymotrypsin activity. One of the proteolytic peaks removed by their procedure was shown to be carboxypeptidase A and it was suggested that this enzyme was responsible for the appearance of free amino acids when crystalline elastase was used to digest elastin (Ling and Anwar, 1966).

Histochemical applications of elastase have been relatively infrequent. Fullmer (1960), using the crystalline enzyme (Worthington) in borate buffer at pH 9, was able to remove mucopolysaccharides entirely from tooth pulp and Adams and Bayliss (1962) compared elastase and pepsin, in respect of their activity towards arterial elastica. Hassler and Herbertson (1962) also studied the removal of arterial elastica, in formalin-fixed tissues, by electron microscopy, as did Ebel and Fontaine (1963), by a number of techniques.

Collagenase and Reticulinase

After the toxin found in *Clostridium welchii* Type A filtrates had been shown by Macfarlane and MacLennan (1945) to produce necrosis in rabbit muscle, the associated histological picture was described by Robb-Smith

(1945). This author observed the disappearance of the reticulin membrane of the muscle fibre and the intermysial connective tissue and ascribed these to the action of a collagenase. Dresner and Schubert (1955) applied a purified clostridial collagenase of this type to a number of collagen and reticulin preparations *in vitro*. All were readily hydrolysed by the enzyme with the release of hydroxyproline-containing peptides. Bidwell (1949) reported her findings on the nature of the collagenase activity of *Cl. welchii* filtrates. She suggested that a second collagenase could be produced in the filtrates either by heating to 50° for 10 minutes or by altering the pH to 9·0 or 10·0. This second enzyme was considered to be derived from an inactive precursor, or from the original enzyme. Many authors disagreed with Robb-Smith both in respect of his observations and of his interpretation of these. Gersh and Catchpole (1949), for instance, used a purified κ-toxin (collagenase) and found it inactive against purified collagen or against collagen in tissue sections; it was capable of removing the glycoproteins of the ground substance, however. They concluded that the term collagenase was inappropriate. Aikat and Dible (1950), using Type A filtrates, could find no evidence of collagenase activity (i.e. destruction of collagen) but they did find strong evidence of what was thought to be a hyaluronidase effect in that the collagen bundles tended to be broken down into their constituent reticulin fibrils. It was believed that the cement substance binding the fibrils together contained hyaluronic acid.

Robb-Smith (1953) reviewed the literature on proteolytic enzyme systems derived from clostridial filtrates. Some of these were certainly collagenases and gelatinases, broadly speaking. Ogle and Logan (1956), for instance, showed that the "gelatinase" of *Cl. histolyticum* contained at least two metal-containing peptidases, inactivated by versene, in addition to many others. My conclusion, in the first edition of this book, that clostridial enzymes could not be regarded as collagenases was certainly an overstatement.

Collagenase or, as it is now termed, clostridiopeptidase A was very fully purified by Mandl *et al.* (1958), using continuous electrophoresis. It was shown to require the presence of some metal, probably calcium, for full activity. Its specificity was further characterized by Grassmann *et al.* (1959) and the general status of the collagenases and elastases was reviewed by Mandl (1961).

Once again, it is necessary to observe that relatively few histochemical studies using collagenase have appeared in the literature. The effect on its activity of a number of fixatives was studied by Green (1960). He showed that collagen was impervious to the enzyme after fixation in Bouin, Zenker, formalin, acrolein, osmium tetroxide, formol-sublimate and Susa. Ethanol and Carnoy, on the other hand, did not prevent digestion and the latter was recommended as the fixative of choice. A comprehensive study of the effect of purified *Cl. perfringens* and *histolyticum* collagenases on the human auricular appendage was made by Beletskaya *et al.* (1966). They found no effects on other tissue components such as mucopolysaccharides, elastic fibres or nuclei. An analysis of the structural stability of biopolymers by Velican (1968) included the use of a number of enzymes as histochemical reagents. These comprised collagenase, sialidase, lysozyme, hyaluronidase, pectinase, pepsin, trypsin, and papain. The suitability of collagenase treatment for the radio-autographic identification of newly synthesized collagen, labelled with ^{3}H-glycine or ^{3}H-proline, was investigated by Carneiro and Leblond (1966). As a result of their studies on Carnoy-fixed bones and teeth from young rats and adult mice they concluded that the collagen nature of the labelled protein could be established by this method.

Lipase and Lecithinase

Lipases have found little use in histochemistry but a small amount of work has been done with the various lecithinases. The usual substrate for these is lecithin but evidence was given by Macfarlane (1942) that sphingomyelins were also split. There are four types of lecithinase (A, B, C and D). The first liberates one only of the two esterified fatty acids to give a product known as lysolecithin. To this product the snake venoms owe some of their toxic character. Lecithinase B liberates the two fatty acids and lecithinase C liberates choline only, leaving a phosphatidic acid. Lecithinase D splits lecithin into two fragments, a diglyceride and phosphoryl choline (see Chapter 17, p. 788).

The α-toxin of *Cl. welchii* owes its activity to the presence of a lecithinase, as shown originally by Macfarlane and Knight (1941) who found that when α-toxin split lecithin into phosphoryl choline and a diglyceride the presence of Ca or Mg ions was necessary.

Lecithinase C activity is also apparently present in *Cl. welchii* α-toxin and this was used by Nygaard *et al.* (1954) in their studies on the effect of these enzymes on mitochondria. Morrison and Zamecnik (1950) used snake venom and bacterial lecithases on tissues and found that fat (demonstrable with Oil red O) appeared after incubation with the snake venoms only.

Only unfixed tissues are suitable for lecithinase studies and *Cl. welchii* α-toxin is the reagent of choice. It is usual to incubate thin blocks at 37° for 12–18 hours and then to fix the tissues in formalin before cutting sections and using fat stains. Materials capable of taking up the solvent fat stains, such as Oil red O, are produced by the action of lecithinase on a variety of tissues. Interpretation of the results, which often have a rather dirty appearance, is far from easy.

The action of a lecithinase (phospholipase) A from the dried venom of the cottonmouth moccasin *Agkistrodon piscivorus piscivorus* on ovolecithin was described by Long and Penny (1957). Its activity is stimulated by Ca^{2+} ions between 40 and 80 μM when the substrate concentration is 1·3 to 3·3 mM. and EDTA (ethylenediamine tetraacetic acid) acts, not unnaturally, as a potent inhibitor of the activity of this type of lecithinase. Davidson and Long (1957) carried out a survey of plant tissues in search of a strong source of phospholipase D, capable of liberating choline from the naturally occurring phospholipids. The highest activity was present in the leaves of the Savoy cabbage and in Brussels sprouts. An associated phosphodiesterase activity could be separated from the lecithinase activity by adsorption on $Ca_3(PO_4)_2$ gel.

REFERENCES

Adams, C. W. M., and Bayliss, O. B. (1962). *J. Histochem. Cytochem.*, **10**, 222.
Aikat, B., and Dible, J. H. (1950). Unpublished observations.
Altschuler, C. H., and Angevine, D. M. (1949). *Amer. J. Path.*, **25**, 1061.
Amano, M. (1962). *J. Histochem. Cytochem.*, **10**, 204.
Asboe-Hanson, G. (1950). *Ann. rheum. Dis.*, **9**, 149.
Baldwin, E. (1949). "Dynamic Aspects of Biochemistry." 2nd Ed., Cambridge Univ. Press, p. 85.
Baló, J., and Banga, I. (1949). *Schweiz. Z. Path. Bakt.*, **12**, 350
Baló, J., and Banga, I. (1950). *Biochem. J.*, **46**, 384.
Banga, I., and Baló, J. (1956), *Nature, Lond.*, **178**, 310.
Banga, I., and Schuler, D. (1953). *Acta phys. Acad. Sci. Hung.*, **4**, 13.
Barnard, E. A. (1969). *Ann. Rev. Biochem.*, **38**, 677.
Bauer, K. (1933). *Z. mikr.-anat. Forsch.*, **33**, 143.
Beletskaya, L. V., Larina, I. A., and Levdikova, G. A. (1966). *Arkiv. Patol. Moscow*, **1**, 80.
Benditt, E. P., French, J. E. (1953). *J. Histochem. Cytochem.*, **1**, 315.
Bidwell, O. (1949). *Biochem. J.*, **44**, 28.
Blackburn, S. (1950). *Ibid.*, **47**, 443.
Blumberg, B. S., and Ogston, A. G. (1958). *Biochem. J.*, **68**, 183.
Brachet, J. (1940a). *C. R. Soc. Biol., Paris*, **133**, 88.
Brachet, J. (1940b). *Ibid.*, **133**, 90.
Brachet, J., and Shaver, J. R. (1948). *Stain Tech.*, **23**, 177.
Brière, N. (1970). *J. Histochem. Cytochem.*, **18**, 498.
Brown, D. M., and Todd, A. R. (1955a). *Ann. Rev. Biochem.*, **24**, 311.
Brown, D. M., and Todd, A. R. (1955b). In "The Nucleic Acids" Eds. Chargaff, E. and Davidson, J. N. Acad. Press, New York, Vol. 1, p. 409.
Bunting, H. (1950). *Ann. N. Y. Acad. Sci.*, **52**, 977.
Bunting, H., and White, R. F. (1950). *Arch. Path.*, **49**, 590.
Burnett, F. M., and Stone, J. D. (1947). *Australian J. exp. Biol. med. Sci.*, **25**, 227.
Cambel, P. (1952). *Science*, **115**, 596.
Campani, M., and Reggianini, O. (1950). *J. Path. Bact.*, **62**, 563.
Carneiro, J., and Leblond, C. P. (1966). *J. Histochem. Cytochem.*, **14**, 334.
Catcheside, D. G., and Holmes, B. (1947). *Symp. Soc. exp. Biol.*, **1**, 225.
Chain, E., and Duthie, E. S. (1939). *Nature, Lond.*, **144**, 977.
Chain, E., and Duthie, E. S. (1940). *Brit. J. exp. Path.*, **21**, 324.
Chambers, T. C. (1955). *Nature, Lond.*, **175**, 215.
Chargaff, E., and Shapiro, H. S. (1955). *Exp. Cell. Research, Suppl.*, **3**, 64.
Chèvremont, M., Baekeland, E., and Chèvremont-Comhaire, S. (1960a). *Bull. Acad. roy. méd. Belg.*, **25**, 367.
Chèvremont, M., Baekeland, E., and Chèvremont-Comhaire, S. (1960b). *Bull. Acad. roy. méd. Belg.*, **25**, 579.
Claude, A. (1939). *J. exp. Med.*, **66**, 353.
Claude, A., and Duran-Reynals, F. (1934). *Ibid.*, **60**, 457.
Crowley, N. (1944). *J. Path. Bact.*, **56**, 27.
Danielli, J. F. (1946). *J. exp. Biol.*, **22**, 110.
Danielli, J. F. (1947). *Symp. Soc. exp. Biol.*, **1**, 101.
Davidson, F. M., and Long, C. (1957). *Biochem. J.*, **66**, 23P.
Davies, D. V. (1952). *Stain Tech.*, **27**, 67.
Day, T. D. (1947). *J. Path. Bact.*, **59**, 567.
Deane, H. W. (1946). *Amer. J. Anat.*, **78**, 227.
Dempsey, E. W., and Singer, M. (1946). *Endocrinology*, **38**, 270.

DEMPSEY, E. W., and WISLOCKI, G. B. (1945). *Amer. J. Anat.*, **76,** 277.
DODGSON, K. S., LLOYD, A. G., and SPENCER, B. (1957). *Biochem. J.*, **65,** 131.
DODGSON, K. S., and LLOYD, A. G. (1958). *Biochem. J.*, **68,** 88.
DORFMAN, A. (1955). In "Methods in Enzymology," Vol. I., p. 166. Academic Press Inc., N. Y.
DRESNER, E., and SCHUBERT, M. (1955). *J. Histochem. Cytochem.*, **3,** 360.
DUBOS, R. J., and MACLEOD, C. M. (1937). *Proc. Soc. exp. Biol., N.Y.*, **36,** 696.
DUBOS, R. J., and THOMPSON, R. H. S. (1938). *J. biol. Chem.*, **124,** 501.
DURAN-REYNALS, F. (1928). *C.R. Soc. Biol., Paris*, **99,** 6.
DURAN-REYNALS, F. (1929). *J. exp. Med.*, **50,** 327.
DURAN-REYNALS, F. (1933). *Ibid.*, **58,** 161.
EBEL, A., and FONTAINE, R. (1963). *Path. Biol.*, **11,** 885.
EIDINGER, D., and GHOSH, A. (1956). *J. Histochem. Cytochem.*, **4,** 200.
EIJKMAN, C. (1903–4). *Zbl. Bakt.*, **35,** 1.
FABERGÉ, A. C. (1945). *Stain Tech.*, **20,** 1.
FAVILLI, G. (1940). *Nature, Lond.*, **145,** 866.
FAWNS, H. T., and LANDELLS, J. W. (1953). *Ann. Rheum. Dis.*, **12,** 105.
FISCHER, F. G., BÖTTGER, I., and LEHMANN-ECHTERNACHT, H. (1941). *Z. physiol. Chem.*, **276,** 271.
FISHER, E. R., ROSENTHAL, T. B., and LANSING, A. I. (1960). *J. Histochem. Cytochem.*, **8,** 102.
FULLMER, H. M. (1960). *J. Histochem. Cytochem.*, **8,** 113.
GERSH, I. (1949). *Arch. Path.*, **47,** 99.
GERSH, I., and BODIAN, D. (1943). *Biol. Symp.*, **10,** 163.
GERSH, I., and CATCHPOLE, H. R. (1949). *Amer. J. Anat.*, **85,** 457.
GERTLER, A., and HOFMANN, T. (1967). *J. biol. Chem.*, **242,** 2522.
GEYER, G., FEUERSTEIN, H., and GRUNDIG, E. (1964). *Acta histochem.*, **18,** 186.
GRASSMANN, W., HÖRSMANN, H., NORDWIG, A., and WÜNSCH, E. (1959). Hoppe-Seyler's *Z. physiol. Chem.*, **316,** 287.
GREEN, J. A. (1960). *Stain Technol.*, **35,** 273.
GREENSTEIN, J. P. (1944). *Adv. Prot. Chem.*, **1,** 209.
HAHN, L. (1945a). *Ark. Kemi. Min. Geol.*, **19**A, No. 33.
HAHN, L. (1945b). *Ibid.*, **21**A, No. 1.
HAHN, L. (1946a). *Ibid.*, **22**A, No. 1.
HAHN, L. (1946b). *Ibid.*, **22**A, No. 2.
HAKIM, A. A. (1960). *Enzymologia*, **22,** 73.
HALL, B. D., and SPEIGELMAN, S. (1961). *Proc. natn. Acad. Sci. U.S.A.*, **17,** 137.
HARTLIEB, J., DIEFENBACH, H., and SANDRITTER, W. (1956). *Acta histochem.*, **2,** 196.
HASSLER, O., and HERBERTSON, S. (1962). *Acta path. scand.*, **55,** 14.
HIRS, C. H.W., STEIN, W. H., and MOORE, S. (1956). *J. biol. Chem.*, **219,** 623.
HOFFMAN, P., MEYER, K., and LINKER, A., (1956). *J. biol. Chem.*, **219,** 653.
HOFFMAN, P., LINKER, A., SAMPSON, P., and MEYER, K., and KORN, E. D. (1957). *Biochim. Biophys. Acta*, **25,** 658.
HOLDEN, M., and TRACEY, M. V. (1950). *Biochem. J.*, **47,** 407.
HOUCK, J. C., and PEARCE, R. H. (1957). *Biochim. Biophys. Acta*, **25,** 555.
HYDÉN, H. (1943). *Acta Physiol. scand.*, **6,** Suppl., 17.
IRVING, J. T. (1960). *Archs oral Biol.*, **2,** 241.
JACKSON, B., and DESSAU, F. I. (1955). *Stain Tech.*, **30,** 9.
JANSEN, E. F., and MACDONNELL, L. R. (1945). *Arch. Biochem.*, **8,** 97.
JEUNIAUX, C. (1957). *Biochem. J.*, **66,** 29P.
JONES, W. (1920). *Amer. J. Physiol.*, **52,** 243.
KAPLAN, H. S., and HEPPEL, L. A. (1956). *J. biol. Chem.*, **222,** 907.
KASTEN, F. H. (1964). *J. Cell. Biol.*, **23,** 48A.
KAUFMANN, B. P., GAY, H., and MCDONALD, M. R. (1950). *Cold Spr. Harb. Symp. quant. Biol.*, **14,** 85.
KAUFMANN, B. P., MCDONALD, M. R., and GAY, H. (1948). *Nature, Lond.*, **162,** 814.
KIM, I. C., FRANZBLAU, C., SHIRAHAMA, T., and COHEN, A. S. (1969). *Biochim. biophys Acta*, **181,** 465.
KLEMPERER, P. (1948). *Ann. intern. Med.*, **28,** 1.

KUGLER, J. H. (1965). M.Sc. Thesis, University of Sheffield.
KUNITZ, M. (1940). *J. gen. Physiol.*, **24,** 15.
KUNITZ, M. (1948). *Science*, **108,** 19.
KUNITZ, M., and NORTHROP, J. H. (1935). *J. gen. Physiol.*, **18,** 433.
KURNICK, N. B. (1954). *J. Amer. Chem. Soc.*, **76,** 417.
KURNICK, N. B. (1955). *Int. Rev. Cytol.*, **4,** 221.
LAGERSTEDT, S. (1957). *Zeit. Zellforsch.*, **45,** 472.
LAMY, F., CRAIG, C. P., and TAUBER, S. (1961). *J. biol. Chem.*, **236,** 86.
LASKOWSKI, M. (1961). In "The Enzymes" Eds. Boyer, P. D., Lardy, H., and Myrbäck, K. Acad. Press, New York, Vol. 5, p. 123.
LEDOUX, L., and BRACHET, J. (1955). *Biochim. Biophys. Acta*, **16,** 290.
LEPPI, T. J. (1968). *Histochemie*, **15,** 68.
LEVINE, M. D., GARZOLI, R. F., KURTZ, R. E., and KILLOUGH, J. H. (1948). *J. Parasitol.*, **34,** 158.
LEWIS, U. J., WILLIAMS, D. E., and BRINK, N. G. (1956). *J. biol. Chem.*, **222,** 705.
LEWIS, U. J., WILLIAMS, D. E., and BRINK, N. G. (1959). *J. biol. Chem.*, **234,** 2304.
LILLIE, R. D. (1948). "Histopathologic Technic." Philadelphia, p. 82.
LILLIE, R. D. (1949). *Anat. Rec.*, **103,** 611.
LILLIE, R. D., and GRECO, J. (1947). *Stain Tech.*, **22,** 67.
LILLIE, R. D., GRECO, J., and LASKEY, A. (1949). *Anat. Rec.*, **103,** 635.
LINEWEAVER, H., and JANSEN, E. F. (1951). *Advanc. Enzymol.*, **11,** 267.
LINEWEAVER, H., JANG, R., and JANSEN, E. F. (1949). *Arch. Biochem.*, **20,** 137.
LING, V., and ANWAR, R. A. (1966). *Biochem. biophys. Res. Commun.*, **24,** 593.
LINKER, A., MEYER, K., and HOFFMAN, P. (1956). *J. biol. Chem.*, **219,** 613.
LONG, C., and PENNY, I. F. (1957). *Biochem. J.*, **65,** 382.
LOVE, R., and RABOTTI, G. (1963). *J. Histochem. Cytochem.*, **11,** 603.
LUDOWIEG, J., VENNESLAND, B., and DORFMAN, A. (1961). *J. biol. Chem.*, **236,** 333.
LUDWIG, A. W., and BOAS, N. F. (1950). *Endocrinology*, **46,** 291.
MADINAVEITIA, J., and QUIBELL, T. H. H. (1941). *Biochem. J.*, **35,** 453.
MADINAVEITIA, J., TODD, A. R., BACHARACH, A. L., and CHANCE, M. R. A. (1940). *Nature, Lond.*, **146,** 197.
MACDONNELL, L. R., JANG, R., JANSEN, E. F., and LINEWEAVER, H. (1950). *Arch. Biochem.*, **28,** 260.
MACFARLANE, M. G. (1942). *Biochem. J.*, **36,** *Proc. Biochem. Soc.* p. 111.
MACFARLANE, M. G., and KNIGHT, B. C. J. G. (1941). *Biochem. J.*, **35,** 884.
MACFARLANE, R. G., and MACLENNAN, J. D. (1945). *Lancet*, **2,** 328.
MANDL, I. (1961). *Adv. Enzymol.*, **23,** 163.
MANDL, I., and COHEN, B. B. (1960). *Archs Biochem.*, **91,** 47.
MANDL, I., ZIPPER, H., and FERGUSON, L. T. (1958). *Archs Biochem. Biophys.*, **74,** 465.
MANNOZZI TORINI, M. (1942) *Boll. Soc. ital. Biol.*, **17,** 153.
MAZIA, D. (1941). *Cold Spr. Harb. Symp. quant. Biol.*, **9,** 40.
MAZIA, D., HAYASHI, T., and YUDOWITCH, K. (1947). *Ibid.*, **12,** 122.
MCCARTY, M. (1946). *J. gen. Physiol.*, **29,** 123.
MCCLEAN, D. (1930). *J. Path. Bact.*, **33,** 1045.
MCCLEAN, D. (1931). *Ibid.*, **34,** 459.
MCCLEAN, D. (1936). *Ibid.*, **42,** 477.
MCCLEAN, D. (1943). *Biochem. J.*, **37,** 169.
MCCLEAN, D., and HALE, C. W. (1941). *Ibid.*, **35,** 159.
MCCLEAN, D., ROGERS, H. J., and WILLIAMS, B. W. (1943). *Lancet*, **1,** 246 and 355.
MCDONALD, M. R. (1948). *J. gen. Physiol.*, **32,** 39.
MCDONALD, M. R., and KUNITZ, M. (1946). *Ibid.*, **29,** 155.
MCDONALD, M. R., and KAUFMANN, B. P. (1954). *J. Histochem. Cytochem.*, **2,** 387.
MCMANUS, J. F. A., and CASON, J. E. (1951). *Arch. Biochem.*, **34,** 293.
MCMANUS, J. F. A., and SAUNDERS, J. C. (1950). *Science*, **111,** 204.
MEYER, K. (1947). *Physiol. Rev.*, **27,** 335.
MEYER, K., CHAFFEE, E., HOBBY, G. L., and DAWSON, M. H. (1941). *J. exp. Med.*, **73,** 309.

MEYER, K., DAVIDSON, E., LINKER, A., and HOFFMAN, P. (1956). *Biochim. Biophys. Acta*, **21**, 506.
MEYER, K., DUBOS, R., and SMYTH, E. M. (1936). *Proc. Soc. exp. Biol., N.Y.*, **34**, 816.
MEYER, K., DUBOS, R., and SMYTH, E. M. (1937). *J. biol. Chem.*, **118**, 71.
MEYER, K., HOBBY, G. L., CHAFFEE, E., and DAWSON, M. H. (1940). *J. exp. Med.*, **71**, 137.
MEYER, K., HOFFMAN, P., and LINKER, A. (1961). In "The Enzymes" Eds. Boyer, P. D., Lardy, H., Myrbäck, K. Acad. Press, New York, Vol. 4, p. 447.
MEYER, K., and LINKER, A. (1951). *Fed. Proc.*, **10**, 223.
MEYER, K., and PALMER, J. W. (1936). *J. biol. Chem.*, **114**, 689.
MEYER, K., and RAPPORT, M. (1951). *Science*, **113**, 596.
MEYER, KURT., and FELLIG, J. (1950). *Experientia*, **6**, 186.
MIDDLEBROOK, W. R., and PHILLIPS, H. (1941). *J. Soc. Dye Col., Bradford*, **57**, 137.
MIRSKY, A. E. (1943). *Advanc. Enzymol.*, **3**, 1.
MORIHARA, K., and TSUZUKI, H. (1967). *Archs Biochem.*, **120**, 68.
MORISON, L. R., and ZAMECNIK, P. C. (1950). *Arch. Neurol. Psychiat.*, **63**, 367.
NARAYANAN, A. S., and ANWAR, R. A. (1969). *Biochem. J.*, **114**, 11.
NAUGHTON, M. A., and SANGER, F. (1961). *Biochem. J.*, **78**, 156.
NEGRI, L., and WEBER, G. (1948). *Arch.* "De Vecchi" *Anat. Patol.*, **2**, 947.
NICKERSON, W. J., and DURAND, S. C. (1963). *Biochim. biophys. Acta*, **77**, 87.
NICKERSON, W. J., NOVAL, J. J., and ROBISON, R. S. (1963). *Biochim. biophys Acta*, **77**, 73.
NOMOTO, M., NARASHASHI, Y., and MURAKAMI, J. (1960). *J. Biochem., Tokyo*, **48**, 593.
NORTHROP, J. H. (1920). *J. gen. Physiol.*, **13**, 739.
NYGAARD, A. P., DIANZANI, M. U., and BAHR, G. F. (1954). *Exp. Cell Res.*, **6**, 453.
OGLE, J. D., and LOGAN, M. A. (1956). *Fed. Proc.*, **15**, 323.
PAINTER, T. S., and TAYLOR, A. N. (1942). *Proc. Nat. Acad. Sci., Wash.*, **28**, 311.
PEARCE, R. H., and WATSON, E. M. (1949). *Canad. J. Res., Sect. E.*, **27**, 43.
PEARSE, A. G. E. (1949). *J. clin. Path.*, **2**, 81.
PENNEY, J. R., and BALFOUR, B. M. (1949). *J. Path. Bact.*, **61**, 171.
PEPLER, W. J., and BRANDT, E. A. (1954). *Brit. J. exp. Path.*, **35**, 41.
POLLISTER, A. W., FLAX, M., HIMES, M., and LEUCHTENBERGER, C. (1950). *J. nat. Cancer Inst.*, **10**, 1349.
POLLISTER, A. W., and LEUCHTENBERGER, C. (1949). *Proc. Nat. Acad. Sci., Wash.*, **35**, 66.
QUINTARELLI, G., and DELLOVO, M. C. (1965). *Histochemie*, **5**, 197.
QUINTARELLI, G., SCOTT, J. E., and DELLOVO, M. C. (1964). *Histochemie*, **4**, 91.
RIPPON, J. W. (1967). *Science, N.Y.*, **157**, 947.
ROBB-SMITH, A. H. T. (1945). *Lancet*, **2**, 362.
ROBB-SMITH, A. H. T. (1953). In "Nature and Structure of Collagen." Edit. Randall, J. T. Acad. Press Inc., N.Y., p. 14.
ROGERS, H. J. (1946a). *Biochem. J.*, **40**, 583.
ROGERS, H. J. (1946b). *Ibid.*, **40**, 782.
ROGERS, H. J. (1948). *Ibid.*, **42**, 633.
ROSA, F., and JOHNSON, F. B. (1967). *J. Histochem. Cytochem.*, **15**, 14.
SANDERS, F. (1946). *Quart. J. micr. Sci.*, **87**, 203.
SCHMITZ-MOORMANN, P. (1969). *Histochemie*, **20**, 78.
SCHNEIDER, W. C. (1946). *J. biol. Chem.*, **164**, 241.
SCHÜTTE, E., and GREILING, H. (1955). *Hoppe-Seyl. Z.*, **302**, 55.
SERRA, J. A. (1946). *Stain Tech.*, **21**, 5.
SPICER, S. S., and WARREN, L. (1960). *J. Histochem. Cytochem.*, **8**, 135.
STONE, A. B., and BURTON, K. (1962). *Biochem. J.*, **83**, 492.
STOWELL, R. E. (1946). *Cancer Res.*, **6**, 426.
STOWELL, R. E., and ZORZOLI, A. (1947). *Stain Tech.*, **22**, 51.
TENNANT, D. M., ZANETTI, M. E., OTT, W. H., KURON, G. W., and SIEGEL, H. (1956). *Science*, **124**, 588.
THOMAS, J., and PARTRIDGE, S. M. (1960). *Biochem. J.*, **74**, 600.

TONEW, E. (1970). *Acta histochem.*, **35,** 186.
TROP, M., and BIRK, Y. (1970). *Biochem. J.*, **116,** 19.
VAN HERWERDEN, M. A. (1913a). *Arch. Zellforsch.*, **10,** 431.
VAN HERWERDEN, M. A. (1913b). *Arch. int. Physiol.*, **13,** 359.
VARADI, D. P., and SAQUETON, A. C. (1968). *Nature, London,* **218,** 468.
VELICAN, C. (1968). *Acta histochem.*, **31,** 261.
VERCAUTEREN, R. (1950a). *Nature, Lond.*, **165,** 603.
VERCAUTEREN, R. (1950b). *Bull. Soc. Chim. biol.*, **32,** 473.
VERHOEFF, F. H. (1908). *J. Am. med. Assoc.*, **50,** 876.
WALFORD, R. L., and KICKHÖFEN, B. (1962). *Archs Biochem. Biophys.*, **98,** 191.
WATSON, E. M., and PEARCE, R. H. (1950). *Ann. N.Y. Acad. Sci.*, **52,** 1004.
WEISSMANN, B., MEYER, K., SAMPSON, P., and LINKER, A. (1954). *J. biol. Chem.*, **208,** 223.
WHITE, J. C. (1947). *J. Path. Bact.*, **59,** 223.
WILSON, V. W. JR., and RAFELSON, M. E. JR. (1967). *Biochim. biophys. Acta*, **146,** 160.
ZECHMEISTER, L., and TÓTH, G. (1939). *Naturwiss.*, **27,** 367.
ZUGIBE, F. T. (1962). *J. Histochem. Cytochem.*, **10,** 448.

CHAPTER 26

PIGMENTS AND PIGMENT PRECURSORS

Introduction

In this chapter the following substances will be considered: melanins, adrenochromes, noradrenochromes, echinochrome, arenichrome, ochronosis pigment, lipo- and haemo-fuscins, lipochromes, haemoglobin and porphyrins, haemosiderin and bilirubin or haematoidin. Theoretically, this arrangement is indefensible. Most members of the group could more logically be considered elsewhere: thus melanin and haemoglobin should appear in Chapter 6 as chromoproteins, and the lipid pigments might be dealt with in Chapter 12. Haemosiderin might logically be considered in Chapter 28, where the methods for its demonstration are described. In practice, however, histologists and cytologists very often find pigmented substances in the sections they are examining and wish to identify them. This chapter collects together all the important naturally occurring pigmented substances, or their precursors, and considers their histochemical identification. The practical value of this procedure outweighs any theoretical disadvantages. Some degree of overlap with the subject matter of Chapter 27 (biogenic amines) is inevitable especially where, as so often, these are themselves pigment precursors.

The chapter falls naturally into three divisions: (A) Phenolic (*Tyrosine-Tryptophan*) *pigments.* (B) *Haem pigments.* (C) *Lipid-pigments.* The majority of the substances to be dealt with can be classified as one of these three. A few appear to be hybrids between two groups or classes. These will be dealt with under the heading of their most prominent constituent. The intestinal pseudomelanins are considered in the section on lipofuscins (p. 1085) as also are the "lipomelanins" of the Dubin-Johnson syndrome.

The Tyrosine Group

The Melanins

The word melanin is derived from the Greek μέλας, meaning black, but it is used to refer to a group of pigments which may be yellow, brown, black or even violet. The process by which the melanin pigments are thought to be elaborated from tyrosine or tyrosine-containing precursors has been described, in part, in Chapter 19, in the section dealing with catechol oxidase (tyrosinase, DOPA oxidase). It is not known whether tyrosine itself is the precursor of melanin in the mammalian organism but, at all events, brown or black pigments can be produced *in vitro* by the oxidation of dihydroxyphenyl compounds and the subsequent polymerization of the products into substances of high molecular weight. The biochemical basis of human melanin pigmentation was discussed by Fitzpatrick and Lerner (1954) and Mason (1955) reviewed the matter from the enzymological point of view. The association of melanin with protein is an important one which is responsible for much of the difficulty in extracting the pigment from the tissues. It is also responsible for differences observed between natural and synthetic melanins (Stein, 1954). Gortner (1912) described a melanoprotein in wool and Greenstein *et al.* (1940) obtained a water-soluble melanin-containing pseudoglobulin from S91 mouse melanoma. Mason *et al.* (1947) subjected this water-soluble pseudoglobulin, and also melanin granules from the same source, to electron microscopy. They found that the two were identical and consisted of regular round or rod-shaped bodies, 0·2 μ in average diameter, which were aggregated in ammonium sulphate and redispersed by its dialytic removal. This is to say, they behaved as a pseudoglobulin. It is still possible that melano-proteins may be formed by the oxidation of tyrosyl radicals in a protein chain, rather than from free tyrosine with later combination with protein. In this context it is important to note that Sizer (1947, 1948) reported that mushroom tyrosinase could oxidize tyrosine present in the peptide chain of proteins. However, mammalian tyrosinase probably cannot act in this manner since it cannot oxidize tyrosine if the amino group is combined with acetyl or formyl radicals.

The most acceptable explanations at present centre on the enzymic or non-enzymic conversion of 3,4-dihydroxyphenylalanine, 3,4-dihydroxyphenylethylamine and 5,6-dihydroxyindole to melanin (Swan, 1963). In the enzymic conversion of DOPA to melanin it is presumed that dopachrome is an intermediate (λ max 305 and 475nm)

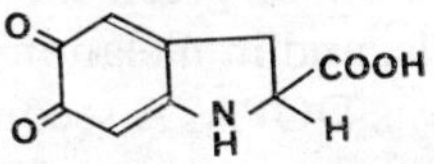

and that this is slowly converted into 5,6-dihydroxyindole

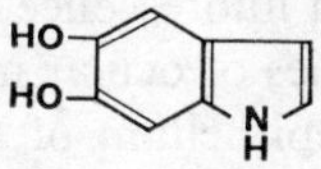

and, thence, to indole-5,6-quinone

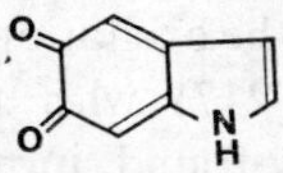

In the case of the melanin of *Sepia officinalis* Nicolaus *et al.* (1959), Piatelli and Nicolaus (1961) and Piatelli *et al.* (1962) obtained confirmation of the indole-quinone theory. Although Mason (1947) suggested that indole-5,6-quinone was the purple melanochrome obtained by oxidation of 5,6-dihydroxyindole Beer *et al.* (1954) indicated that deeply coloured melanochromes were more likely to be composed of linked quinonoid indoles. Bu'Lock (1960) obtained spectroscopic evidence that the first pigment derived from the oxidation of 5,6-dihydroxyindole was 5,6-dihydroxyindolylindole-5,6-quinone and Swan (1963) suggested that melanin derived from the enzymic oxidation of 3,4-dihydroxyphenylethylamine might be a copolymer of indole-5,6-quinone and a dopachrome analogue lacking the carboxyl group.

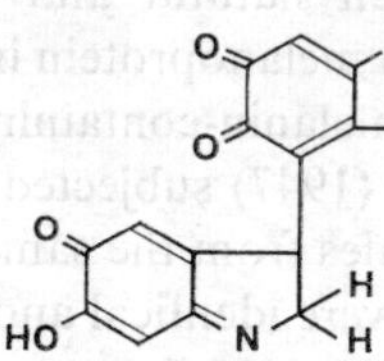

The characteristics of Swan's presumptive melanin precursor (formula above) do not appear to have been recorded but Bu'Lock found that his intermediate pigment had an absorption maximum at 530 nm and that its reduced form exhibited a characteristic fluorescence. Further stages were observed to have flatter absorption maxima (around 540 nm) and no fluorescent reduced forms could be extracted by ethyl acetate.

This observation brings us to an important development affecting the diagnostic pathology of the melanomas. In a fluorescence microscopic study of three such tumours Baroni (1933) found a yellow fluorescence not only in the basal cell layer of the epidermis, but also in the tumour cells themselves. Much later Falck *et al.* (1965, 1966a and b), using freeze-dried, formaldehyde vapour-fixed material (Falck *et al.*, 1962), fluorescence microscopy, and a yellow barrier filter, found yellow or green fluorescence in normal melanocytes, in junctional naevus cells, and in melanomas. They considered that the main responsible compound was DOPA. A second, unknown, substance with a fluorescence emission at 500 nm was later found (Falck *et al.*, 1966c). In melanomas of the Syrian golden hamster Olivecrona and Rorsman (1966) found no formaldehyde-induced fluorescence (FIF) but Ehinger *et al.* (1967) found such fluorescence in a series of ocular melanomas as did Winckler and Turner (1969) in the pigment epithelium of the developing guinea-pig eye. This they also attributed to DOPA.

Using a microspectrofluorimeter developed by Pearse and Rost (1968), Rost and Polak (1969) compared the excitation and emission spectra of DOPA models (shown in Figs. 216 and 217) with those of melanocytes and of melanoma and naevus cells. They found, in most cases, unimodal excitation and emission curves with peaks at about 440 and 490 nm, respectively. In one section, bimodal curves with peaks at about 400/440 and 490/520 nm were observed. The fluorescence excitation curves were not altered by exposure to HCl vapour (Björklund *et al.*, 1968). The DOPA model gave excitation and emission maxima at about 420 and 490 nm and the excitation maximum was shifted to 390 nm after short exposure to hydrochloric acid. It may be con-

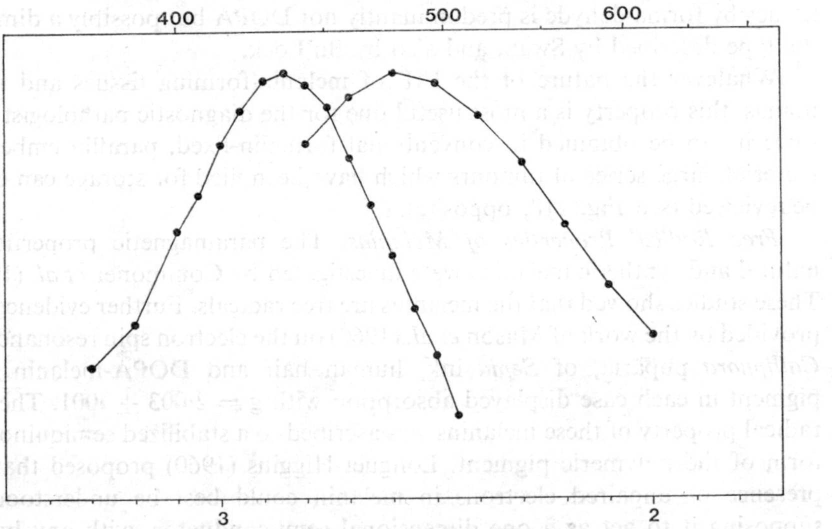

FIG. 216. Excitation (left) and emission spectra of FIF in a malignant melanoma cell. Abscissa: quantum energy in eV (lower scale) wavelength in nm (upper scale). Ordinates: Relative quanta emitted per quanta of excitation (excitation spectra) or relative quanta per unit quantum energy interval (emission spectra). Arbitrary units.

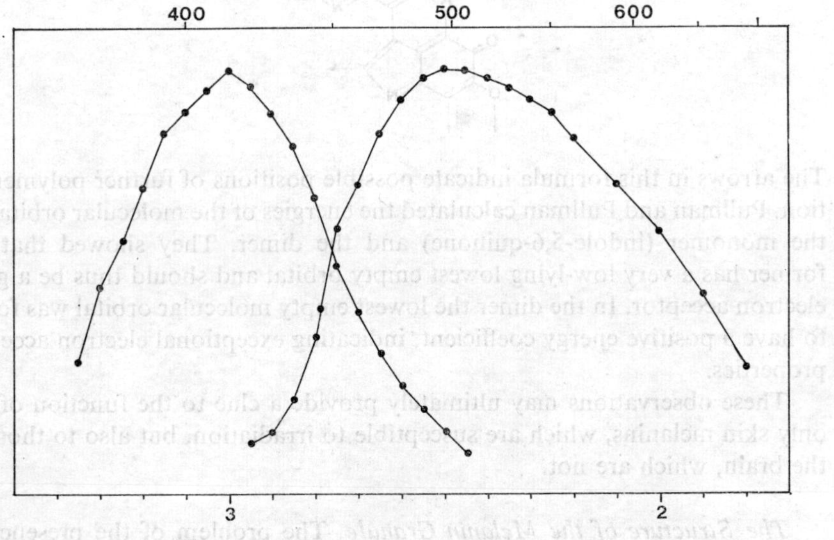

FIG. 217. Excitation and emission spectra of FIF of DOPA model. Scales as Fig. 216.

cluded that the fluorescent compound retained (induced) in melanin-forming tissues by formaldehyde is predominantly not DOPA but possibly a dimer of the type described by Swan, and also by Bu'Lock.

Whatever the nature of the FIF of melanin-forming tissues and melanomas, this property is a most useful one for the diagnostic pathologist and, since it can be obtained in conventional formalin-fixed, paraffin embedded material, large series of tumours which have been filed for storage can easily be reviewed (see Fig. 218, opposite).

Free Radical Properties of Melanins. The paramagnetic properties of natural and synthetic melanins were investigated by Commoner *et al.* (1954). These studies showed that the melanins are free radicals. Further evidence was provided by the work of Mason *et al.* (1960) on the electron spin resonances of *Calliphora* puparia, of *Sepia* ink, human hair and DOPA-melanin. The pigment in each case displayed absorption with $g = 2{\cdot}003 \pm {\cdot}001$. The free radical property of these melanins was ascribed to a stabilized semiquinonoid form of the polymeric pigment. Longuet-Higgins (1960) proposed that the presence of unpaired electrons in melanin could best be understood by supposing it to act as a one dimensional semi-conductor, with any bound protons acting as electron traps. A formula for DOPA melanin polymer (dimer) was suggested by Pullman and Pullman (1961).

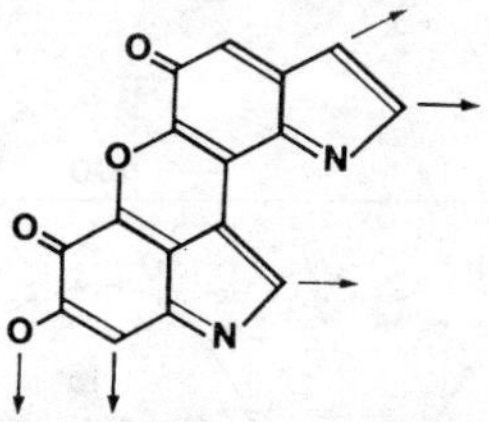

The arrows in this formula indicate possible positions of further polymerization. Pullman and Pullman calculated the energies of the molecular orbitals in the monomer (indole-5,6-quinone) and the dimer. They showed that the former has a very low-lying lowest empty orbital and should thus be a good electron acceptor. In the dimer the lowest empty molecular orbital was found to have a positive energy coefficient, indicating exceptional electron acceptor properties.

These observations may ultimately provide a clue to the function of not only skin melanins, which are susceptible to irradiation, but also to those of the brain, which are not.

The Structure of the Melanin Granule. The problem of the presence or absence of a carrier, in the form of a protein nucleus to the melanin granule was considered extensively by Lignac and his associates (Lignac *et al.*, 1952; Lignac, 1954) who concluded that there was a residuum remaining after bleaching which can be so regarded. The problem of the development of the

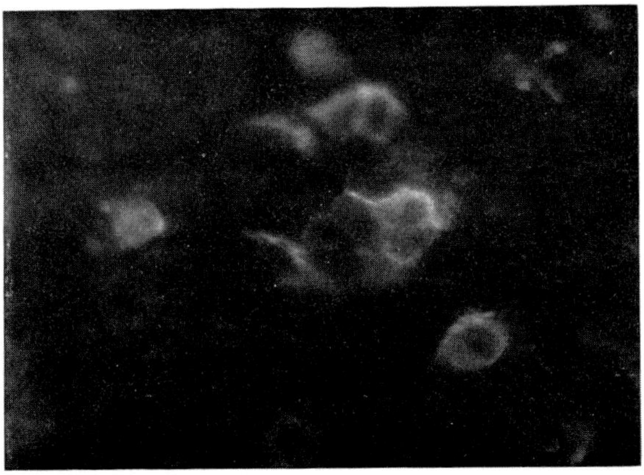

FIG. 218. Formalin-fixed paraffin section. Shows FIF in some of the cells of an amelanotic malignant melanoma. × 470.

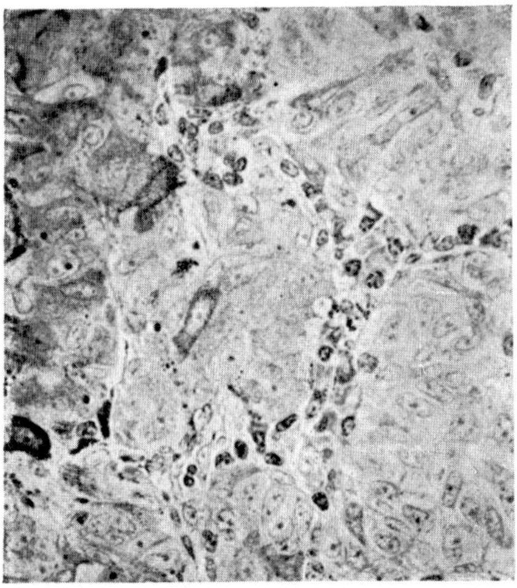

FIG. 219. Benign naevus-cell melanoma. Melanin granules appear black. Masson-Fontana stain. × 370.

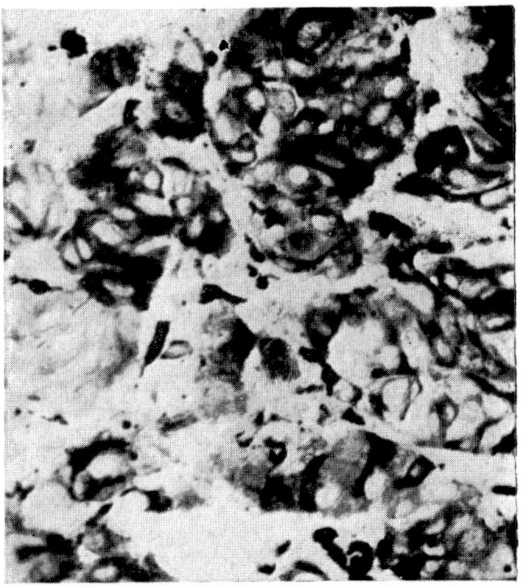

FIG. 220. For comparison with Fig. 219. A similar field in a similar section stained by Schmorl's method. The degree of staining is much greater. × 370.

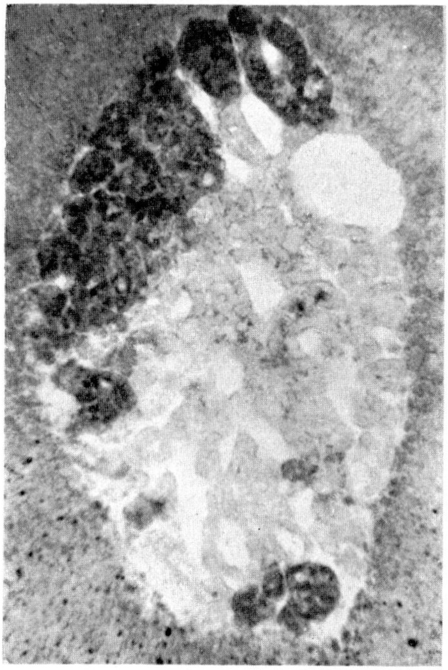

FIG. 221. Hamster adrenal gland. Iodate reaction shows brown pigment from noradrenalin in islets in the medulla. × 60.

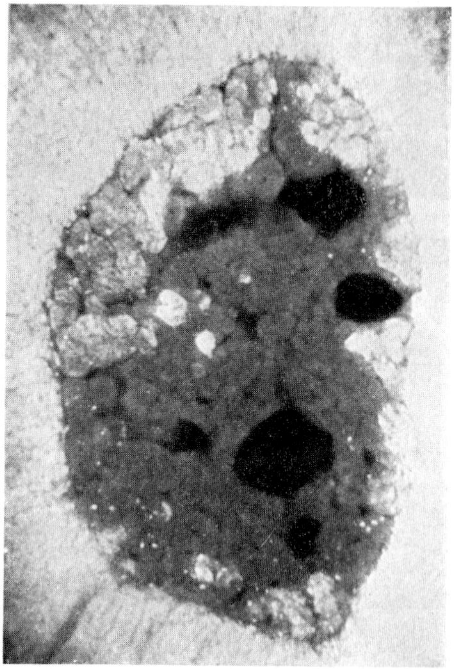

FIG. 222. As FIG. 221 but contralateral adrenal. Formol calcium—induced fluorescence in peripherally situated islets in the adrenal medulla. × 60.

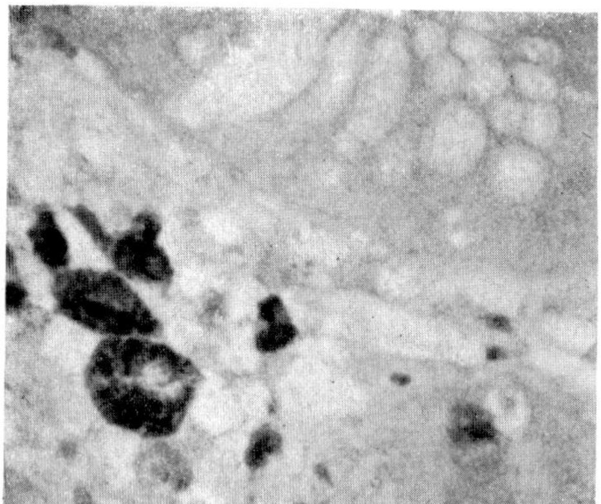

FIG. 225. Human appendix (paraffin section). Showing, in the lamina propria (*left*) strongly staining "pseudomelanin" granules and also (*right*) argentaffin granules. Schmorl's reaction. × 520.

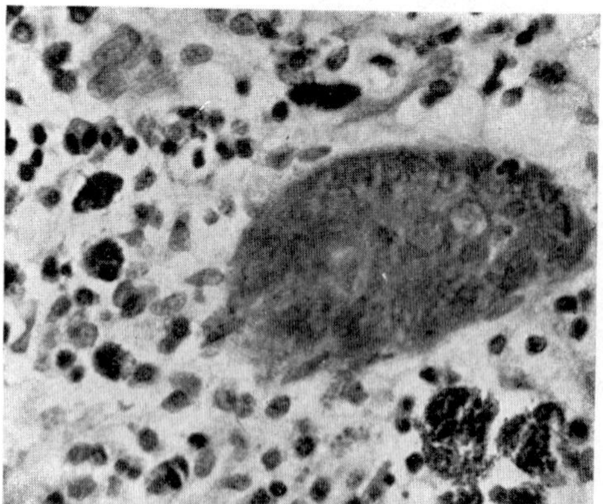

FIG. 226. As FIG. 225. "Pseudomelanin" granules in macrophages in the lamina propria showing various degrees of positive staining. Periodic acid-Schiff, haemalum. × 370.

melanin granule from mitochondria was examined by Stein (1955) by means of comparative chemical studies. He showed that the differences between the two were so considerable that it is unlikely that melanin granules are in fact derived from mitochondria.

Electron microscopical studies of the melanocyte carried out by Birkbeck, Mercer and Barnicot (1956) indicated that melanin granules appeared first as hollow granules in which, subsequently, a "tenuous" material appeared in the form of a folded lamella. Their findings were confirmed and extended by those of Wellings and Siegel (1959, 1963) and by Seiji *et al.* (1961, 1963). It is now agreed that there are 3 stages in the formation of the melanin granule. In the first stage tyrosine-rich polypeptides are synthesized in the endoplasmic reticulum and in the second stage they are transferred *via* the Golgi region into vesicles. These constitute the *premelanosomes*, having an inner structure of consecutive sheets or lamellae, closely resembling the internal organization of the chloroplast. The protein tyrosinase, also synthesized in the endoplasmic reticulum, is thought to be added at the second stage. In the third stage melanin synthesis begins and melanin begins to accumulate. The granule still contains tyrosinase and is described as a *melanosome*. The final product, with the completion of melanization, contains no enzyme and is described as the melanin granule.

Evidence for the production of melanin in the nuclei was presented by Meirowsky and Freeman (1951), by Redaelli (1961, 1965), and by Santini (1964) but, generally speaking, the views expressed by these authors have been unfavourably received. The work of Okun *et al.* (1970), however, suggests that the nuclear "DOPA oxidase" activity noted in chick retinal pigment cells by Redaelli (1961, 1965), and by McCurdy (1969) in neural crest melanoblasts of the California newt *Taricha torosa*, is possibly peroxidase-dependent. Okun (1967) had already shown that melanocytes possessed peroxidase activity and he and his colleagues postulated an important role for this enzyme in the initiation of melanogenesis *in vivo*.

The mitochondrial origin of melanosomes (du Buy *et al.*, 1963) is supported by few. As Moyer (1963) stated, the problem is no longer to work out the origin of the melanosomes but to determine how and why the polypeptides from which they are formed come to aggregate in the first place.

Histochemical Characteristics of the Melanins

Recognition of melanin in the skin or the eye is not in practice a matter of difficulty. Difficulties arise with unidentified brown pigments in, for instance, the bowel wall or in secondary tumours of uncertain origin. It is here that the few positive reactions of melanin, usually regarded as typical, are in fact a source of confusion since they are shared with other types of pigment (see Table 77, p. 1090).

Solubilities. The most striking physical characteristic of the melanins, from the point of view of the histologist, is their complete insolubility in organic solvents or in anything which is not markedly destructive towards the tissue preparations which contain them. Lea (1945) stated that melanin was soluble in ethylene chlorohydrin (2-chloroethanol), but this solvent and others mentioned by Mason (1948a) are said to be satisfactory only after the pigment has been separated from its protein component. Taft (1949) treated fixed and unfixed sections of skin with ethylene chlorohydrin and with pyridine, at 22° and 66°, for periods of 15 minutes to 24 hours. He could find no diminution in the amount of melanin in any of the sections. It is often stated that melanin granules are soluble in strong alkali (NaOH, KOH) if this is applied for a sufficient period of time, and that lipofuscin is not. This feature is not much used in histochemical diagnosis, although it might be if an unknown pigment was thought to be either melanin or one of the lipofuscins. (The solubility in fat solvents of the latter, even when unfixed, is not comparable to that of glycerol esters or phospholipids, so that no useful purpose is served by testing with various extractives.) In making a survey of the histochemical properties of *Amphiuma* melanins Helmy and Hack (1964) observed that pigment granules isolated from liver were insoluble in hot glacial acetic acid and hot pyridine. They were moderately soluble, however, in hot 2·5 N-NaOH.

Bleaching. The second characteristic of the melanins is the facility with which they are bleached by strong oxidizing agents such as H_2O_2, acid chlorate, $KMnO_4$ or $FeCl_3$, although the process is always a slow one. The blacker the melanin the slower its decolorization. Figge (1939, 1940), reported that sodium hydrosulphite ($Na_2S_2O_4$ dithionite) or ascorbic acid would convert the black melanin from the ink sac of the squid, or that produced by the *in vitro* oxidation of DOPA, to a light tan colour. On addition of potassium ferricyanide the black colour was restored. The brown, yellow and finally colourless melanin produced in tissue sections by any of the four strong oxidizing agents mentioned above cannot be restored by ferricyanide. After decolorization, melanin granules in human skin and melanomata cannot be shown to contain tyrosine either by Millon's or by the coupled tetrazonium reaction. They can be demonstrated, however, by means of silver reactions employing a reducing agent (argyrophilia). At least 24 hours' treatment is necessary to decolorize melanin as a rule; 10 per cent H_2O_2 is a most convenient and satisfactory reagent in my experience (see Appendix 26, p. 1381), but stronger solutions often have to be employed. As a rapid bleach for retinal melanin Curd (1968) proposed a dilute H_2O_2-ammonia reagent and this gives a satisfactory result in less than 6 hours. Performic and peracetic acids also bleach most melanins very rapidly (Lillie, 1956, 1965) and the second, being commercially available is the more convenient of the two. Melanin in paraffin sections is more easily bleached than melanin in frozen sections, from identical sites. A table giving the bleaching characteristics of a number of melanins, derived from Lillie (1957), will be found in Appendix 26, p. 1381.

Reduction of Silver Solutions. Another characteristic of the melanins is their ability to reduce solutions of ammoniacal silver nitrate to metallic silver. This property was first made use of for their demonstration in tissues by Bizzozero (1908). Masson's (1914) method for their demonstration, which also depends on this property, has been universally employed for the purpose, in many modifications. Melanins are also blackened by acid silver nitrate solutions, as observed by Giroud *et al.* (1934) and by Clara (1942). This phenomenon was considered by Lillie *et al.* (1957) who showed that at pH 4·0 relatively few substances reduce $AgNO_3$ in 10 minutes or less. Of naturally occurring substances likely to be present in tissue sections only two, ascorbic acid and *o*-diphenols, performed in this manner. Gomori's (1946) hexamine-silver method, recommended for the demonstration of melanin by Lennox (1949), and the more usual Masson-Fontana technique are equally efficient. Details of both appear in Appendix 26. (See Fig. 219).

Besides being argentaffin, melanin granules are also argyrophil, using this word to mean that silver impregnation methods employing a reducing agent colour them black. This reaction is not of diagnostic value.

Basophilia. Many of the basic dyes colour melanin strongly, especially methylene blue and toluidine blue in acid solutions, but this property is relatively useless since the lipid pigments are often just as basophilic when stained with the dyes mentioned. Lillie (1955) made some tests of this property and observed that the basophilia of melanin was a true basophilia and not a false one as had been supposed by Hueck (1912). Cutaneous, ocular and pial melanins can be stained green with thionin at pH 1·0 but the locus coeruleus and substantia nigra pigments do not stain below pH 3, and only weakly at this pH. According to Coupland and Heath (1961) all melanin granules stain with azure dyes after formalin fixation but only dermal melanins after formol-dichromate. Spectroscopic studies carried out on substantia nigra pigment by Van Woert *et al.* (1967) indicated that it was not the same as DOPA-melanin but probably derived from one or more of the neurotropic amines present in the cells. The only other tissue components which stain are those containing sulphonic acid groups or ester sulphates, Lillie suggested that these were responsible for the basophilia of melanin. Sulphur has long been known to be present in melanin and it is possible that oxidation of S-S groups to SO_3H can occur under natural conditions.

Studies carried out by Okun (1965) appeared to confirm Lillie's views in demonstrating sulphate groups in melanins. Ruddell (1969) observed that NaOCl oxidation of melanins produced anions which were assumed to be sulphate.

The reduction of ferricyanide to ferrocyanide and the production of Prussian blue in the presence of ferric salts (the so-called Schmorl's reaction, see below), is shared by the lipid pigments and by argentaffin cell granules. One can nearly always say that a pigment insoluble in anything which leaves the tissues intact, but soluble in N–NaOH, which is bleached by strong oxidizers in less than 48 hours, and which reduces ammoniacal silver nitrate, belongs to the class of melanins. In my opinion Schmorl's reaction is superior to the ammoniacal silver techniques for demonstrating melanin, especially the finer grades of melanin and diffusely distributed "melanin precursors", which do not reduce silver salts. Figs. 219 and 220 illustrate the difference between the two techniques, in serial sections of the same tumour.

Ferrous Ion Uptake. A reaction originally thought to be specific for melanin was described by Lillie (1957a) who found that melanins formed complexes with Fe^{++} which could be demonstrated by means of potassium ferricyanide (Turnbull blue reaction). Ferrous ion uptake, demonstrated in this way, differs from Schmorl's reaction in that there is no reduction of Fe^{+++} to Fe^{++}. As Lillie pointed out, *preliminary* treatment of melanin with $FeCl_3$ results in a positive Turnbull blue reaction when ferricyanide is applied later. This is not true of other components which give a positive Schmorl's reaction. With these it is necessary to apply the reagents simultaneously. Thus it appears that some part of the melanin complex is able to chelate Fe^{++} and Lillie suggested that this was an *o*-quinhydrone configuration. Details of this technique are given in Appendix 26, p. 1380. Its specificity for melanin is not very high since enterochromaffin and bile pigments also react.

The PAS Reaction. McManus (1949) stated that melanin under certain conditions contained carbohydrate, an interpretation based on the occasionally positive PAS reaction which it gave. This reaction, in my experience (1960), was only positive with melanin in melanophages (phagocytes) and always negative in epidermal melanoblasts and epithelial cells. The presence of a PAS-positive component in melanin was reported also by Sacchi (1962), who considered that this was probably quite distinct from the component responsible for argentaffinity. According to Lillie (1969) cutaneous and ocular melanins are PAS-negative while neuromelanins are coloured brownish-red. Extraction for 24 hours at 60° in dry pyridine failed to reverse this reaction, making unlikely the hypothesis of an unsaturated lipid component. It remains possible that mild oxidation of some melanins may produce aldehydes from the pigment polymer itself.

Trichoxanthin

This name was proposed by Lillie (1957b) for the yellow granular pigment occurring in the hair medulla of the paler varieties of guinea-pig. Its histochemical characteristics strongly resemble those of the melanins and Lillie suggested that, like these pigments, trichoxanthin contains quinhydrone groupings.

Arenichrome

A green granular pigment in the epithelial cells of *Arenicola marina* L. was named arenichrome by Lignac (1945) who described its histological and histochemical characteristics. This pigment had already been noted by Fauvel (1899) under the title "lipochrome jaune". This author described the solubility of the pigment in alcohol, its preservation by alkali, and its conversion by the influence of light, especially at acid pH levels, into a brownish-black insoluble pigment. Lignac, agreeing with these findings, regarded arenichrome as a coloured premelanin, capable of acting as a "mesocatalyst" or respiratory carrier pigment. Later studies by Van Duijn (1952) resulted in the extraction and crystallization of arenichrome as its potassium salt ($C_{21}H_{15}S_2K_3O_{14}$). The presence of sulphur in the form of strongly acid groups, presumably SO_3H, provides a point of similarity to the melanins. The absence of nitrogen, however, must place arenichrome in an entirely different category. Further histochemical studies on arenichrome have not been carried out.

Echinochromes

A reddish pigment, echinochrome, characteristic of many species of Echinoids, was described in *Arbacia* by McMunn (1885) and more modern work on this (Thomson, 1962) revealed several different forms. A red pigment was found in the radial nerve of *Diadema antillarum* by Millott and Okumura (1968). It was soluble in acidified diethyl ether and had an absorption maximum at 465 nm, corresponding to that of echinochrome A. The formula of the latter is shown below:

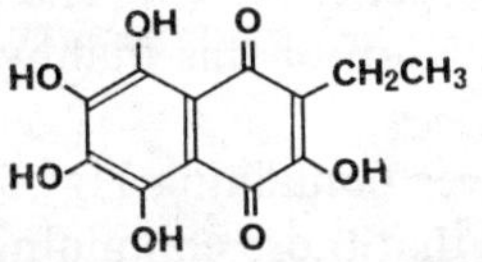

It is a substituted naphthazarin (dihydroxy-α-naphthoquinone).

The echinochrome pigments of the axial organ in *Diadema*, *Arbacia*, and *Paracentrotus* were investigated cytochemically by Millott and Vevers (1968) who found, in agreement with the observations of Millott (1957), that the amoebocytes contained a red pigment which, after liberation, changed to orange, brown, and finally to black. Much of the chemistry and function of the echinochromes remains to be elucidated.

Adrenochromes and Noradrenochromes

Older Views on the Chromaffin Reaction. The observation that certain cells in the adrenal medulla contained dark brown material in granular form, after fixation with dichromate or chromic acid, was made by Henle in 1865. The chromaffin reaction, as it has long been called, was attributed to adrenalin by Meulon in 1905. Verne (1923) showed that not only adrenalin, but also other aromatic compounds containing two OH or NH_2 groups *ortho* or *para* to one another, would give the brown colour on reaction with dichromate. The experiments of Ogata and Ogata (1923), who demonstrated the presence of chromium in the chromaffin granules, seemed to indicate that the reaction depended on the reduction of dichromate to a yellow or brown compound. Gérard, Cordier and Lison (1930) proved, however, that it was due to an oxidation of the reacting substance by the dichromate. They were able to obtain the typical result by fixation in a fluid containing the mild oxidizing agent potassium iodate. Lison (1936) pointed out that these findings increased the specificity of the reaction which, in the former hypothesis, could be due to any reducing substance. He considered that the reacting substance was converted into a quinone which then coupled with a molecule of the original substance to give a deeply coloured quinhydrone.

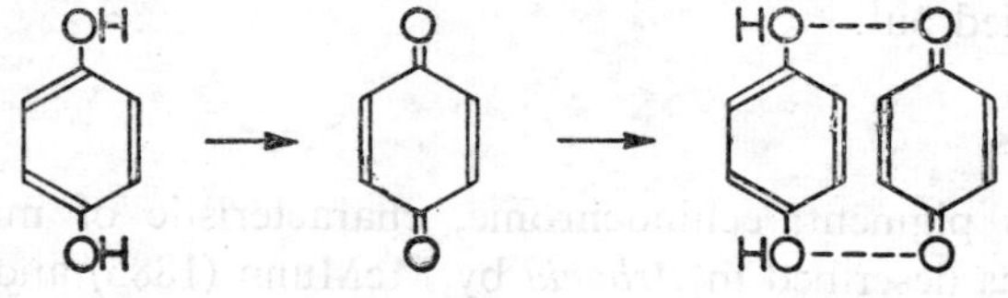

Colourless Pale yellow Brown

From this reasoning Lison put forward four criteria of chromaffinity: (1) All substances which belong to the series of aromatic photographic developers are chromaffin. (2) The reagents giving the chromaffin reaction are oxidants which do not act too energetically. (3) The final coloured product is a quinhydrone. (4) Decolorization of this quinhydrone is produced by strong oxidation.

Lison stated that fixatives containing $HgCl_2$ as well as dichromate gave only inconstant results and that those containing chromic acid generally gave negative results. He recommended the use of formalin containing potassium iodate, since this avoided the pseudo-chromaffin reaction given by substances capable of absorbing chromium salts. Other authors recommend a simple formalin-dichromate solution. Lillie (1948) could not obtain the chromaffin reactions with Lison's fixative, and he stated that as little as one hour's preliminary treatment with formalin would prevent production of the reaction by dichromate.

It is sometimes helpful, though never diagnostic, to stain the chromaffin granules, subsequent to fixation with dichromate, by some method like that of Wiesel (given by Roulet, 1948). This depends on staining with aniline blue or toluidine blue followed by nuclear staining with safranin; chromaffin cells appear green, other cells blue. Alternatively, modifications of Schmorl's Giemsa method can be employed to give a similar final effect. Sevki's (1934) modification of Schmorl's method (Appendix 26, p. 1382) was successfully employed by Blacklock *et al.* (1947) to demonstrate chromaffin granules in cases of phaeochromocytoma. This method, originally designed for use with non-chromated tissues, gives a pink colour with chromaffin cells under such conditions. In the case of chromated tissues, however, the colour is green.

Chromaffin granules and the PAS reaction. Lillie (1948) mentioned, but did not explain, the fact that chromaffin granules stain red by the PAS technique. I could not produce a positive PAS reaction in the chromaffin granules of either human or dog adrenals (Pearse, 1953). The cytoplasm of the chromaffin cells, however, stained deeply by this method when an aqueous solution of HIO_4 was employed as the oxidizing agent. It is theoretically not impossible for noradrenochrome to react with Schiff's reagent. Adrenochrome should not react since it is a tertiary amine. Model experiments, and work on acid dichromate-fixed adrenal medullary tissue from *Columba livida* and *Phalacrocorax niger*, carried out by Mukherjee (1966) showed that only noradrenalin would give a positive PAS reaction after chromic acid oxidation. This was attributed to cleavage of the $CHOH\text{-}CH_2NH$ grouping in the 2–3 position by periodic acid.

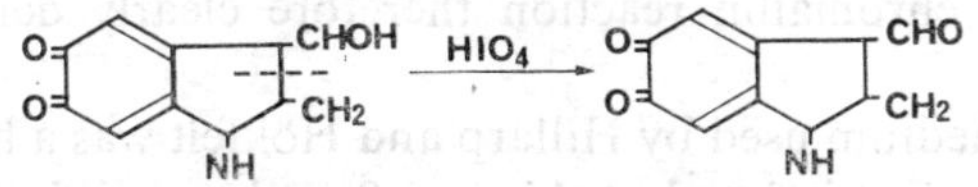

Modern Views on the Adrenochromes. More modern work on the pigments derived from the oxidation of adrenalin led to modifications of the views put forward by Lison although it did not significantly alter the histochemical significance of the reaction. Green and Richter (1937) produced a deep red compound (adrenochrome) by the oxidation of adrenalin with catechol oxidase. They advanced the following equation for this reaction.

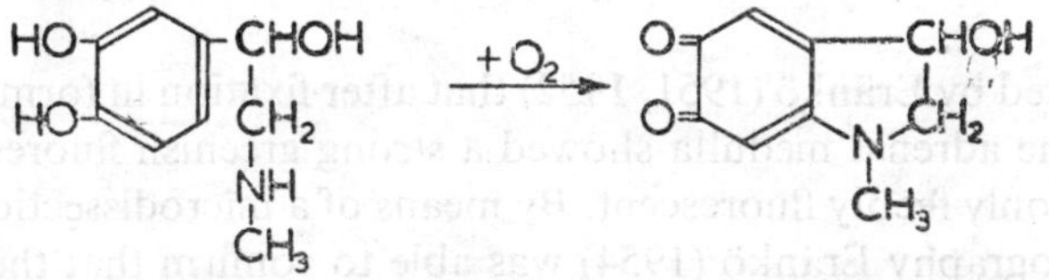

Adrenalin Adrenochrome

Reduction of the product, with H_2S or SO_2, gave a colourless leucoadrenochrome which reacted with ferric chloride to give the typical green colour of a catechol. A non-oxidative condensation mechanism by which two molecules of adrenochrome combine in acid solution to form a dimer was proposed by Bu'Lock (1961).

Richter and Blaschko (1937) obtained a red iodo-adrenochrome by treating adrenalin with iodate in acid solution. This finding suggested that the success of Lison's iodate method depended on the presence of sufficient acidity in the formalin employed. Using iodate-formalin solutions I found (1953) that "chromaffin" was produced between pH 5·0 and 6·8 but that the amount decreased if the iodate-containing formalin was more acid or more alkaline than this. Lillie's failure to produce chromaffin granules with the iodate-formalin fixative was therefore perhaps explicable on the basis of insufficient acidity, since the formalin he used was buffered. The reaction product of Lison's method was presumably iodo, rather than simple, adrenochrome. This reacts essentially in the same way as the substance produced by dichromate oxidation, but is less strongly reducing towards ferricyanide and silver solutions.

Distinction between Adrenalin and Noradrenalin. In a series of papers (1953, 1954, 1955) Hillarp and Hökfelt described a method for the distinction of noradrenalin from adrenalin and its application to the adrenal medulla in a number of different species. These authors found that when oxidation was carried out with potassium iodate at pH 5–6 noradrenalin was converted into a dark pigment within a few minutes while adrenalin, on the other hand, required about 24 hours. Oxidation with chromate or dichromate rapidly produced brown insoluble pigments from both adrenalin and noradrenalin and the classical chromaffin reaction therefore clearly demonstrates both these substances.

The original medium used by Hillarp and Hökfelt was a buffered solution containing 2·5 per cent iodate but this was found unsatisfactory both by the authors and by other workers (e.g. Eränkö, 1955a). A 10 per cent solution of iodate (unbuffered) was therefore substituted. Only thin tissue slices can be used for the method, which is given in full in Appendix 26, p. 1382. Frozen sections or freeze-dried sections are quite useless. The localization of noradrenalin afforded by this method is histologically satisfactory. Within the cell, however, diffusion takes place so that the nucleus is stained brown. Application of the technique to phaeochromocytomas was described by Ghislandi *et al.* (1955).

It was noted by Eränkö (1951, 1952) that after fixation in formalin some of the cells of the adrenal medulla showed a strong greenish fluorescence while the rest were only feebly fluorescent. By means of a microdissection technique and chromatography Eränkö (1954) was able to confirm that the fluorescing islets contained noradrenalin and later (1955b and c) he showed that there was an excellent correlation between noradrenalin content and fluorescence in seven species of mammals. *In vitro* experiments showed that noradrenalin and formalin formed an insoluble fluorescent compound but that adrenalin did not react in this way. Comparative studies made by Eränkö (1955a, 1956) of the fluorescent and iodate methods for the demonstration of noradrenalin led to the conclusion that the two reactions gave positive results in the same types of medullary cell and that either method might therefore be employed with confidence. Comparison between the two methods is illustrated in Figs. 221 and 222. In practice, the iodate method appeared to be easier and could be used on fresh frozen sections. One or other of the fluorogenic methods described in Chapter 27 and in Appendix 27, p. 1394, will probably be preferred.

Further developments in this field followed the use of glutaraldehyde as a fixative (for E.M. studies). Wood and Barrnett (1963, 1964) found that noradrenalin granules were selectively oxidized by dichromate at pH 4·1, whereas at pH 6·5 both noradrenalin (NA) and adrenalin (A) granules became electron opaque. Coupland, Pyper and Hopwood (1964) followed glutaraldehyde fixation with OsO_4 as a method of distinction between the two types of granule, at optical and electron microscope levels. A similar technique using silver carbonate as the final reagent was evolved by Tramezzani *et al.* (1964a and b) who observed that their yellow, glutaraldehyde-fixed NA granules rapidly reduced this reagent. No reaction occurred in A cells and the authors considered that this was due to non-fixation of adrenalin and its consequent escape from the tissues during fixation and washing. There would appear to be some discrepancy between their findings and those of Wood and Barrnett (1964).

The method of Tramezzani *et al.* (1964b), which is given in Appendix 26, p. 1380, was originally recommended for optical and electron microscopy. Later studies (Cannata *et al.*, 1968) indicated that its specificity was wider than at first appreciated. A positive reaction is obtained with NA, dopamine, melanin, and lipofuscin. Adrenalin and 5-HT were negative at optical microscope level but with long exposure (30 minutes) fine silver deposits could be seen in the polymorphic EC cell granules by electron microscopy. The deposits in the case of dopamine granules were also too small to be visible by optical microscopy. The type of result obtained at this level, in NA cells of chick adrenal medulla, is shown in Plate XXIVa, p. 1075. An alternative method, described by Jones (1967), was also based on glutaraldehyde-dichromate and silver intensification. The author could distinguish NA and A cells in paraffin sections of the adrenal medulla of both new-born and adult rats.

Other Methods for Adrenalin. Coujard and Coujard-Champy (1942) described the use of an osmium tetroxide-potassium iodide mixture for demonstrating adrenalin in sympathetic nerves and adrenal medulla. This method which was also used by Champy and Hatem (1955) was criticized by Lison (1953), and also by Hillarp (1959), as unspecific. Lison stated that although the mixture was reduced particularly strongly by phenols it reacted in tissue sections with many other types of compound.

Relationship of Chromaffin Granules to Mitochondria. Histologists long supposed that adrenalin in the chromaffin cells was present in the form of granules. Blaschko and his associates (Blaschko and Welch, 1953; Blaschko, Hagen and Welch, 1955) made a particular study of this problem and found that adrenalin was certainly contained in the large granule fraction of sucrose homogenates. From these granules it was easily released by acetolysis, high temperatures and by electrolytes. Feldberg (1940) and Hillarp and Nilson (1954) both described the release of adrenalin after treatment of the granules with lysolecithin and the latter authors described the liberation of both adrenalin and noradrenalin by hypotonic solutions of electrolytes and non-electrolytes and by lecithinase A.

All these observations suggested that the adrenalin-containing granules possessed some kind of membrane and Hillarp *et al.* (1954) considered that they were probably mitochondria. E.M. studies however, particularly those carried out with glutaraldehyde fixation, revealed that both A and NA granules are true secretion granules, formed in the Golgi region, and covered only by a single membrane.

Ochronosis Pigment

Pigmentation of the tissues of sufferers from the hereditary enzyme defect causing alkaptonuria was termed ochronosis by Virchow (1866) long before the aetiology of the condition was understood. Alkaptonuria was recognized as an inborn error of metabolism by Garrod (1909) who proposed that it was due to the absence of the enzyme homogentisic acid oxidase. In this event oxidation to maleylacetoacetic acid of metabolic homogentisic acid cannot take place.

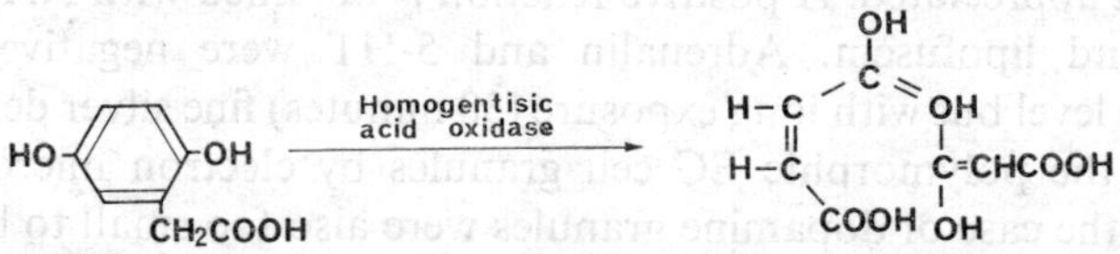

It accumulates particularly in the connective tissues where it is oxidized, presumably by reactions resembling those involved in the synthesis of melanin from dopaquinone (see p. 1051), to a melanin-like pigment. The question was investigated by Zannoni *et al.* (1969) and the whole subject of ochronosis was reviewed by La Du and Zannoni (1969).

As observed by Virchow (1866) the colour of the pigment is yellowish-brown. Its histochemical characteristics are ill defined. It has been said to reduce ferricyanide (Schmorl-positive), but not silver nitrate, and to remain unbleached after H_2O_2. Most of the reported staining characteristics are shared with the melanins but, if verified, its yellow autofluorescence should distinguish it clearly.

Tryptophan Pigments and the Indole Reaction

Little attention has been given to the histochemistry of the red pigments derived from tryptophan, which can be extracted from the urine and which may also occur in the tissues, if only in the form of intermediates. They are better considered at this point than in Chapter 6, into which they do not fit readily because they are not demonstrable by the indole reaction.

The tryptophan pigments are of two classes, the indigoids, formed by the oxidation of free tryptophan, and the Hopkins-Cole pigments, which are formed from free or peptide-linked tryptophan by condensation with an aldehyde and subsequent oxidation. Fearon and Boggart (1950) obtained a fluorescent red pigment from tryptophan by oxidation with iodate in acid solution, which they called tryptochrome. This was found to be a very unreactive substance, insoluble in water and dilute alkalis but soluble in organic solvents and especially in chloroform. If a pigment such as tryptochrome was formed in the body it might be expected to be stored in solution in lipids as are the fat-soluble carotenoid pigments. No reactions have been described by which tryptochrome can be demonstrated histochemically although it has a characteristic fluorescence spectrum.

The Haem Group

Haemoglobin

This histochemistry of the haemoglobins is divisible into two parts, that of the globins, already dealt with in Chapter 6, and that of the haematins, which we must consider here. Whether the globin is present or not makes little difference in the tests which we apply for the haematin part of the molecule. As Lison has stated, the histological demonstration of haemoglobin is not easy and distinction between the various types can only be made spectroscopically or by absorption spectrography. The former method is satisfactory when applied to pigments *in vitro* but not when applied to tissue sections.

As far as chemical methods are concerned, with one dubious exception, we have only a choice of modifications of the peroxidase reaction, and these are equally positive with the native globins (reduced and oxy-haemoglobin, met- and sulphaemoglobin), denatured globins (haemochromogen and parahaematin), and with the globin-free compounds (haematin, haem and haemin). Peroxidase reactions are negative with the porphyrins, haematoidin and haemosiderin. In the past the most commonly employed methods were those using benzidine and fixed rather than fresh material. Lison (1936, p. 249) used a formalin-lead acetate mixture in order to convert haemoglobin into haematin, but he also recommended the method of Slonimski and Lapinski which involved fixation in ferricyanide-formalin for 24 hours. This converted the haemoglobin into methaemoglobin. Most workers preferred to employ formalin fixation and some modification of the benzidine-nitroprusside methods of Lepehne (1919) and Pickworth (1934). As an alternative to these procedures some used the leuco-Patent Blue method of Lison (1938) or modifications such as those of Dunn (1946) and Dunn and Thompson (1946), with the original leuco-dye or leuco-Cyanol (see Chapter 19, p. 856, and Appendix 19, p. 1355). It is possible to stain haemoglobin by means of Perls' reaction if sections are treated for a short period with 10 per cent H_2O_2 before application of the iron reagent. This method gives only a bluish-green tinge to the red cells and it has no important practical application. Similar treatment with H_2O_2 will sometimes reveal haemosiderin iron which has not proved reactive when treated with the HCl of Perls' reagent alone.

With all these methods the primary objective has usually been the clear demonstration of red blood cells by virtue of their haemoglobin content. For this purpose there is really very little to choose between the peroxidase techniques and simpler staining methods for haemoglobin such as Lendrum's

(1949) Kiton-red—almond green. Methods revealing the lipoprotein envelope of the red cell give equally good results (i.e. Sudan black B or Baker's acid haematein method, both after proper fixation). Haemoglobin outside red cells is much less easily demonstrated since none of the available reactions possess a sufficiently high degree of specificity. It is possible to distinguish between tissue peroxidases and haemoglobin and its derivatives, since the former are heat labile and destroyed by warming the slides at 90° (Lison recommended 180°) for 10–20 minutes, while haemoglobin is not. There are, however, certain other thermostable tissue peroxidases which appear during the process of autolysis. Benzidine methods give a positive reaction with these, in cell nuclei for instance, but Lison stated that his zinc-leuco method did not do so. In practice these objections are not found to carry much weight and the development of a "pseudoperoxidase" reaction can be taken to indicate the presence of haemoglobin or haemochromogen. Gomori (1951) considered the zinc-leuco methods to be superior to all others for this purpose but noted (1952) that only intact haemoglobin, and not its protein-free derivatives, would stain properly by these methods.

While working with 3,3′-diaminobenzidine (DAB) as a peroxidase reagent Hirai (1968) observed that one of its coloured intermediate oxidation products, which he regarded as a free radical although its structure remained undetermined, possessed specific affinity for heme proteins. The reagent was used as a stain, visible by optical or electron microscopy, to demonstrate catalase-rich microbodies as well as erythrocytes, inner mitochondrial membranes and leucocyte cytoplasmic particles. The reaction occurred anaerobically, and after heating for 30 minutes at 120°. It was inhibited, in some components, by mM KCN or NaN_3. Hirai suggested that cytochrome c, catalase, and haemoglobin were the principal heme enzymes demonstrated. Details of the reaction, which promises to be a useful one, are given in Appendix 26.

A method for the detection and localization of intracellular haemoproteins such as haemoglobin, peroxidases, catalase and cytochrome c, was described by Granick and Levere (1965). This depends on the treatment of cells, usually in smears or monolayer cultures, with perchloric acid in the presence of 0·2 M mercaptoethylamine, followed by UV irradiation (365–440 nm). This procedure converts the hemes to porphyrins which fluoresce red (Emission maxima, 597–600 and 651–653 nm; see below). It cannot be used to distinguish the heme of haemoglobin from that of cytochrome c, since the porphyrins formed are in each case identical. The method, which is undoubtedly the forerunner of a number of histo-photochemical techniques, is described in Appendix 26, p. 1387.

Myoglobin

This heme-protein respiratory pigment is located in striated muscle fibres where it is responsible for their red colour. Several attempts have been made to provide histochemical localization, usually by means of the benzidine-peroxidase reaction (Drews and Engel, 1961; Chinoy, 1963). More recently the DAB method has been used (James, 1968; Morita *et al.*, 1969, 1970). Glutaraldehyde-fixed, free-floating cryostat sections are used and the results are very satisfactory.

The Porphyrins

These substances, which normally only occur in small quantities in most animals, are considered to be precursors of the haem part of haemoglobin. They possess the same tetrapyrrole nucleus as haem but lack the central atom of iron. Of the large number of possible tetrapyrroles only two types occur in nature, the so-called Series I and Series III porphyrins. Those of Series III are the direct precursors of haem and those of Series I are regarded as useless by-products of haemoglobin synthesis. In the foetus, where this process is vigorous, porphyrins of both series are present in greater amounts than in the adult but, while those of Series III are utilized at once, those of Series I accumulate and are demonstrable in the bone marrow and bone, and also in the meconium. In the adult increased amounts of porphyrin (Series I) are excreted in various anaemias and in the congenital or acquired porphyrias they may be present in large amounts in the tissues. The condition giving rise to the so-called pink-tooth cattle was described by Chu & Chu (1962).

Unfortunately, although the chemistry of the porphyrins is fascinating (see Lemberg and Legge, 1949; Margoliash, 1961; Goodwin, 1968), their histochemistry is unsatisfactory and disappointing. The only method so far employed for their demonstration makes use of the red and orange fluorescence given by their solutions in ultra-violet light. Lison (1936, p. 255) described this fluorescence as secondary, since in his view it was produced only after treatment of the porphyrins with appropriate solvents. He noted that in alkaline media ultra-violet light rapidly destroyed the specific fluorescence. Since reddish fluorescence is produced by other substances besides the porphyrins (e.g. carotenoids, lipofuscins, oxidized cytochrome), it is necessary to make spectroscopic (or better spectrographic) examination in order to be certain about their presence in the tissues. Spectroscopic analysis was applied by Sjöstrand (1946) to fluorescent substances in freeze-dried material mounted in inert media, and the freeze-drying technique is particularly suited to such a purpose. Histochemical studies of the porphyrins were made by Borst and Königsdorfer (1929), in a case of congenital porphyrinuria, by Ellinger (1938, 1940) and Dobriner and Rhoads (1940), and by Grafflin (1942) on the Harderian gland of the albino rat. This last tissue is the classical material for most experimental studies in porphyrin histochemistry and a survey of four species of *Gerbillidae* was made by Arvy and de Lerma (1961). These authors found that the fluorescent material in the Harderian glands, studied in fresh frozen sections, gave emission spectra characteristic of protoporphyrins. Gillman *et al.* (1945a and b) reported brief investigations of porphyrins present in the liver in cases of pellagra, using a fluorescence technique with fixed frozen sections. They described the fluorescence as intensely scarlet red, localized to hepatic cells and intensified by acids and alkalis. The red fluorescence noted in erythrocytes in cases of anaemia by Seggel (1940) has been attributed to the presence of porphyrins. With this last case forming the exception to the rule, it should be made clear that the most that can be achieved at present is a particularly poor localization of porphyrins as a diffuse pinkish-red fluorescence in tissues where they are present in sufficient amount. Since porphyrins as a class are lipid-soluble, when intracellular they are invariably associated with lipid bodies. Figge *et al.* (1953) observed that porphyrins had a tendency to accumulate especially in neoplastic, embryonic and traumatized tissues and also in normal lymph nodes. Later Figge (1959) investigated the red fluorescence of the fore-stomach of rats and mice fed on a chlorophyll-rich diet. Part of the fluorescence was attributed to breakdown products such as phylloerythrins and part to porphyrins secreted by the Harderian gland. Klüver (1955) found that the white matter of normal brain, exposed to near ultra-violet irradiation, showed fluorescence with a maximum at 625nm. He was able to confirm that this was due to the presence of coproporphyrin.

A technique by which the porphyrins react *in situ* with a specific, or relatively specific, reagent is badly needed. Porphyrins can form complexes with metallic ions such as Fe^{++} or Cu^{+} but this property cannot be used as the basis of a histochemical test since these ions also become attached to a wide variety of active groups in the tissues.

Haemosiderins

The term siderosis was first used by Neumann (1888), together with the name haemosiderin. The haemosiderin pigments occur in the form of yellow to brown granules which are normally always intracellular. They contain iron in an easily unmasked condition and have long been considered to be composed of a protein framework combined with ferric iron in the form of the hydroxide, $Fe(OH)_3$. It was shown by Rich (1924) that histiocytes in tissue culture could break down the haemoglobin of phagocytosed red cells into an iron-containing haemosiderin and an iron-free haematoidin. The time scale of events was established by Muir and Niven (1935) who injected blood subcutaneously into various animals. As early as 24 hours after injection iron-containing pigment was observed in tissue histiocytes and, in rats and mice, haematoidin appeared on the 7th day. In rabbits haematoidin was not visible at any stage. Strassmann (1944) observed histiocytes containing haemosiderin 33 hours after intra-tracheal introduction of blood into rabbits. The same author (1949) showed histiocytes with haemosiderin granules 48–72 hours after experimental brain injury in mice. In man, a similar occurrence has been noted on the third day (Hammes, 1944), and on the fifth day (Baggenstoss *et al.*, 1943) after ventricular puncture. The work of Rich and of Strassmann indicated that, following haemorrhage, haematoidin appears much later in histiocytes than haemosiderin, probably not until 10–14 days afterwards.

Fresh haemosiderin is insoluble in alkalis but soluble in strong acid solutions; after fixation with formalin, however, it is slowly soluble in dilute acids, best in oxalic, then sulphuric, then nitric, formic and hydrochloric acids (Lillie, 1939). Fixatives that contain acids but no formalin may completely remove haemosiderin or so alter it that a positive reaction for iron is no longer obtainable. Lillie (1954) gave an excellent description of the haemosiderins and divided them into three groups, all of which are easily demonstrated by means of Perls' acid ferrocyanide reaction. In his first group he placed those haemosiderins which gave a diffuse light blue staining when Perls' reaction was applied. He described these as showing initially a "diffuse brownish staining of cytoplasm". Very often, however, a diffuse Perls' reaction is given by connective tissues in the vicinity of old as well as recent haemorrhage, and these are not coloured at all in unstained sections. While the iron may possibly be present here in the form of haemosiderin, it is equally possible that it may simply be combined with certain acid groups of the protein as is, for instance, the colloidal iron in Hale's method for acid mucopolysaccharides. It is certainly more rapidly removed by treatment with dilute (5 per cent) oxalic acid than is the iron in haemosiderin granules. The haemosiderins in Lillie's second and third groups both occur in granular form but they are distinguished by the fact that one stains slightly with basic dyes (especially thiazines) while the other has no affinity for these.

The constitution of the material remaining in the haemosiderin granule after removal of its iron was investigated by Goessner (1953) and by Gedigk and Strauss (1954). Goessner used 5 per cent oxalic acid but Gedigk and Strauss preferred 20 per cent HCl on account of the more rapid and complete extraction of iron which it allowed. Both authors found that the carrier substance was a polysaccharide-containing protein giving a strong tetrazonium reaction after benzoylation and being strongly PAS-positive. Gedigk and Strauss also studied the development of haemosiderin after subcutaneous injection of colloidal iron in mice. They found that the basophilia of the carrier substance progressively increased, together with the iron-binding capacity, and that in the later stages (14–84 days after injection) some lipid was also present. The relationship of the iron-binding protein to apoferritin has to be considered, as has the relationship of haemosiderin to ferritin. The reactions of the latter are dealt with in Chapter 28, p. 1132. Microspectrophotometry of haemosiderin granules carried out by Wells and Wolken (1962) revealed protoporphyrin absorption peaks, indicating the origin of the granules by degradation of red cells.

The electron microscope studies of Farrant (1954) and Richter (1957, 1958) threw much light on the nature of the haemosiderin granule and the observations of the latter author indicated that ferritin was one of the constituents of haemosiderin. Richter observed that haemosiderin was formed within intracellular organelles and that these might resemble mitochondria (c.f. Gillman and Gillman, 1945). He preferred, however, to reserve judgement on their actual identity and to call them siderosomes. These views were opposed by Sturgeon and Shoden (1969) who considered that the relationship of haemosiderin to ferritin was still ambiguous.

Histochemistry of the Haemosiderins. The histochemistry of the haemosiderins is closely bound up with the demonstration of inorganic iron and methods for demonstrating the latter are considered in Chapter 28, p. 1130, and in Appendix 28, p. 1402.

Differential Diagnosis of Haemosiderin. Theoretically haemosiderin might have to be distinguished from any of the pigments already mentioned in this chapter and from haematoidin and the lipofuscins (see below). No difficulty is to be expected in making the diagnosis except where members of the lipid series of pigments contain iron. This is not a very frequent occurrence and when it occurs, usually in the lower range of oxidation, it seldom affects all the pigment in a particular location. If any doubt is felt about the nature of an iron-containing pigment associated with undoubted lipofuscin an attempt must be made to stain the granules by one or either of the lipofuscin reactions (PAS and Schmorl's reaction are especially useful), comparing the control slide with a serial slide stained by Perls' method.

Haemochromatosis

This condition, which has had many definitions, was divided by MacDonald (1964, 1969) into two types. The primary type occurs when excess iron is derived from the diet, as in liver or pancreatic disease, chronic anaemia or nutritional imbalance. The secondary type is due to parenterally administered iron or blood. In experimental haemochromatosis MacDonald *et al.* (1968) showed that iron was deposited not only in the normal storage tissues but in the pancreas and heart muscle, and in the parenchyma of other non-storage tissues, as it is in the human disorder. The major part of the iron-containing pigment in haemochromatosis is haemosiderin.

Bile Pigments. Haematoidin and Bilirubin

Historical Perspective. Haematoidin was the name given by Virchow (1847) to the extracellular yellowish-brown crystals found in old haemorrhagic foci. The relationship between this pigment and bilirubin remained uncertain until Fischer and Reindel (1923) showed that a sample of haematoidin from a cyst of the liver was identical with bilirubin. Rich and Bumstead (1925) obtained the same result with haematoidin crystals from other sources. Histochemically, therefore, the two substances were long considered together as "bile pigments" whereas morphologically it was still reasonable to distinguish between them. The histological characteristics of the haematoidins, and their formation from blood and bile, were considered by Lignac (1923) in a comprehensive paper.

Since the work of Rich (1924) we know that haematoidin, like haemosiderin, is formed intracellularly in phagocytes which have ingested red blood corpuscles. Unlike haemosiderin, however, it does not remain there. Normally it is transferred into the blood stream and excreted by the liver but under certain circumstances, such as in old infarcts, its removal is prevented and it remains in the form of yellowish-brown crystals and amorphous granules, lying always extracellularly. Bilirubin appears in the gall bladder and bile ducts usually as bright orange-yellow amorphous masses and in the liver of obstructive jaundice in the form of bile thrombi.

Chemistry of the Bile Pigments. The bile pigments are a group of coloured substances, usually yellow, orange or green, which have a tetrapyrrole structure

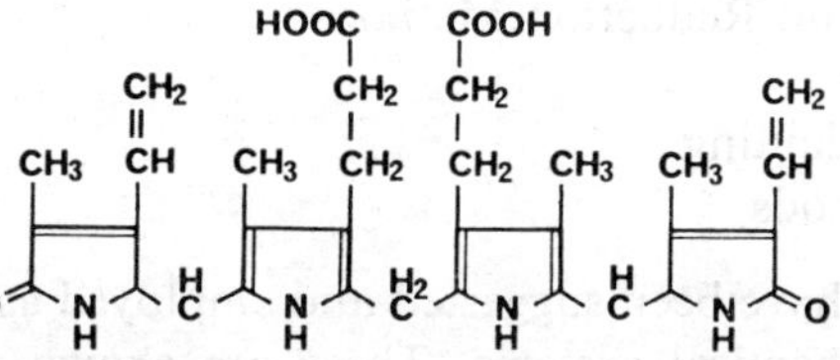

the four rings of which are linked by α-methyne groups (Gray *et al.*, 1958). The two most important compounds are bilirubin and biliverdin and the former occurs in the tissues in two different forms, known as unconjugated and conjugated.

The difference between bilirubin and the bilirubin-like pigment excreted in the urine was first recorded by van den Bergh and Müller (1916) who observed that bilirubin and the serum pigment of patients with haemolytic jaundice required ethanol before they would couple with diazotized sulphanilic acid. The pigment of bile and from the sera of patients with obstructive jaundice, on the other hand, would react directly. Cole and Lathe (1953) separated the direct and indirect types of pigment chromatographically and Cole *et al.* (1954) showed that the direct pigment had two components (Pigments I and II). Both of these were water-soluble and they appeared in the urine in conditions such as obstructive jaundice and hepatitis where they formed the main bile pigments of the serum. Bilirubin, on the other hand, is insoluble in water and in conditions such as acholuric jaundice, where it is the main bile pigment of the blood, the urine remains free of bile. Pigment II is the chief pigment of human bile and when coupled with benzenediazonium chloride it forms azo pigment B whereas bilirubin itself forms azo pigment A. Billing and Lathe (1956) and Billing *et al.* (1957) later showed that pigment B was the ester glucuronide of pigment A. It is probable that these ester glucuronides are formed only in the liver, by the action of the enzyme glucuronyl transferase (Billing and Lathe, 1958).

Although both Pigment I and Pigment II are unstable they were identified by Billing *et al.* (1957) as the mono- and diglucuronide of bilirubin, respectively. The existence of Pigment I remained open to question and Nosslin (1960) proposed that it was a labile equimolar complex of bilirubin diglucuronide and free bilirubin. Ostrow and Murphy (1970) showed clearly that a true bilirubin monoglucuronide, not a complex, was present in normal rat bile.

Following the work of Billing, Cole and Lathe (1957) it finally became clear that the "direct" reaction was due to the water-soluble bilirubin glucuronides and the "indirect" reaction to the insoluble, lipophilic unconjugated bilirubin. The action of alcohol, and other solvents, as "accelerators" was identified as a simple dissolving of the water-insoluble pigment.

Histochemistry of the Bile Pigments. Methods for the demonstration of bilirubin in the tissues can be divided into four groups:

(1) Oxidation and Reduction Methods
(2) Basophilia
(3) Metal Mordanting
(4) Diazo Methods

Many compounds have been suggested and employed as oxidizing agents for bilirubin in histochemical systems. These are shown in Table 76, below.

TABLE 76. *Oxidation Reactions*

Author and Date	Reagent(s)
Gmelin: 1826	HNO_3/HNO_2
Stein: 1935	Lugol's Iodine
Okamoto: 1952	HNO_3/HCl
Gomori: 1952	HNO_2
Lillie: 1954	H_2O_2
Glenner: 1957	Dichromate
Kutlík: 1957	ferric chloride/acetic
Hall: 1960	Fouchet reagent
Leibnitz: 1964	Iodine/CH_3OH
Lillie, Pizzolato: 1967	Bromine/CCl_4

The oldest oxidation method for bilirubin and haematoidin is Gmelin's reaction (Tiedemann and Gmelin, 1826), usually carried out histochemically by the inclusion of concentrated HNO_3 under a coverslip and rapid examination under the microscope. A positive reaction is shown by a spreading greenish, and later blue or purple coloration. According to Lemberg and Legge (1949) the Gmelin reaction depends on the dehydrogenation of bilirubin to the greenish-blue bilatriene (verdin), and the green stage of the reaction is due to mixtures of bilatriene and unoxidized bilirubin. Similarly the blue stage is also due to a mixture, this time of bilatriene with violet pigments (purpurins) produced by its further oxidation to biladienediols. Finally the blue and violet compounds are oxidized to yellow choletelins. In the histochemical test, besides giving a display of colours, the original pigment is progressively dissolved. A successful Gmelin's reaction is completely diagnostic but a negative one does not exclude bile pigments, especially if the other tests are positive, since the reaction is apt to be capricious. With large amorphous masses of bilirubin it invariably works well but with small granules of haematoidin it is often negative. Crystalline haematoidin performs excellently.

Stein's (1935) test is supposed to depend on the oxidation of the pigment to green biliverdin by means of a solution of iodine. It works better with small granules of pigment than with the large masses and in my hands produces, at best, a darkening of the original golden yellow pigment to an olive-brown colour. The fact that this is not reversed by removal of the excess iodine with thiosulphate favours an oxidative mechanism for the change in colour but the possibility of some kind of combination with iodine cannot be ruled out. A negative Stein's reaction does not exclude the presence of bile pigment but a positive one, although iodine treatment may produce some darkening in pigments of the lipid series, is usually to be regarded as strong evidence in favour of its presence.

Oxidation methods for the demonstration of bilirubin and biliverdin were studied particularly thoroughly by Kutlík (1957) who tested chromates, permanganate, iron alum, ferric sulphate, ferricyanide, peracetic and performic acids, alcoholic benzoquinone, H_2O_2, chlorate, periodic acid and nitrate, *inter alia.* Most of these reagents, in practice, were found to have undesirable qualities but, in addition to iodine, trivalent iron as ferric sulphate or iron alum gave reasonably controlled oxidation to green biliverdin. A further stage of oxidation to the reddish-violet bilipurpurine was best carried out with ferricyanide in alkaline solution at pH 7·2. These oxidation methods were regarded by the author as specific for bilirubin. They were negative with protein-bound bilirubin, kernicterus pigment, haematoidin and bilirubin-calcium compounds.

Contrary to common practice Glenner (1957) recommended the use of unfixed cryostat sections, post-fixed in 1·5 per cent potassium dichromate in buffer at pH 2·2. Leibnitz (1964), on the other hand, used alcohol-fixed paraffin sections, and a non-aqueous solvent for his oxidizing agent. Formalin-fixed paraffin sections were used by Lillie and Pizzolato (1967), with bromine in carbon tetrachloride as oxidant. Formalin fixation is not known to interfere with oxidation but Glenner (1957) considered that small intracellular granules might be dissolved by this process.

It is generally concluded that only conjugated bilirubin is transformed into biliverdin by oxidation techniques. Lillie and Pizzolato (1967), however, indicated that with their technique haematoidin (in infarcts) could be made to give a positive reaction by repeating the whole procedure if necessary. The green phase of bilirubin oxidation can be reversed to yellow-brown (Lillie and Pizzolato, 1968) by a number of reducing agents, including acid bisulphite and sodium dithionite. Walker *et al.* (1970) distinguished between bilirubin and haematoidin by first reducing in dithionite (0·1 M, 1–2 days) and then re-oxidizing in a modified Stein medium. Colourless bilirubin, produced by the first reagent, was restored to its green phase. Haematoidin remained yellow through both stages. Both bilirubin and haematoidin reduce ferric ferricyanide, the former somewhat more rapidly (3 hours) than the latter (12 hours), and also diammine silver (Appendix 26, p. 1379).

The property of staining strongly with methylene blue was shown by Reinhold and Fowler (1947) to be due to the formation of a compound of methylene blue with bilirubin, and this compound was successfully isolated. Unfortunately, from the histochemical point of view, nearly all the pigments we have to consider in this chapter share this property and stain strongly with methylene blue and other basic dyes. A positive test is therefore useless, but a pigment which does not stain with methylene blue is unlikely to be bilirubin or haematoidin.

Observing the 0·2 per cent chromic acid bleached bile pigments completely in 2–4 hours at 24° Lillie and Pizzolato (1969a) showed that after as little as 4–6 hours chromation these pigments were coloured blue-black by acetic-haematoxylin. Di- and trivalent metal salts (Cr^{3+}, Cu^{2+}, Fe^{3+}, Fe^{2+}, Sn^{2+}) could be substituted for hexavalent chromium. The mechanism of the reaction is not fully understood. It differs, apparently, from the known capacity of bile pigments to form complexes with Zn^{2+}, many of which are fluorescent. The zinc-bilirubin complex is not fluorescent but, according to O'Carra (1962), this complex, formed by treatment with zinc acetate in chloroform/methanol, is oxidized by iodine in $CHCl_3$ much more rapidly than the uncomplexed pigment. Oxidation of haematoidin should be accelerated by such treatment.

The first recorded histochemical diazo reaction for bilirubin was by Daddi (1933) but the salt he used, p-nitrodiazobenzene, has not since been employed. I myself (2nd Edition, 1960) reported negative results with both bilirubin and haematoidin in fixed tissues at alkaline or acid pH levels although Gedigk and Gries (1952) had shown that bilirubin would couple directly in alkali. With unfixed cryostat sections, however, and freshly diazotized *p*-nitroaniline, a rose-red colour rapidly developed in liver sections from cases of cirrhosis and biliary obstruction. Barka and Anderson (1963) stained bilirubin deposits, in fresh or formalin-fixed frozen sections, dark brown, using hexazotized pararosanilin at pH 6·5. Despite the foregoing observations diazo reactions were seldom employed for histochemical identification of bile pigments until Raia (1965, 1967) developed a suitable van den Bergh technique, based on the serum bilirubin technique of Rand and di Pasqua (1962), using diazotized 2,4-dichloroaniline. To demonstrate conjugated bilirubin the reagent was used in aqueous solution while for total (conjugated and unconjugated) bilirubin an "accelerator" mixture was added. The results of the direct reaction are

PLATE XXIV

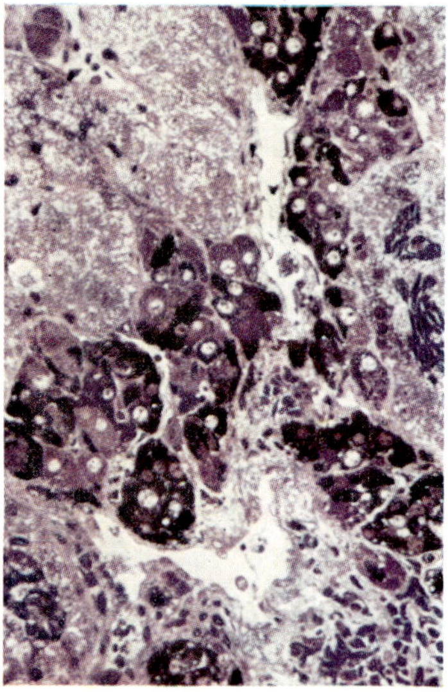

XXIVa. Resin-embedded 1 μ section. Chick adrenal medulla. Stained with toluidine blue and the Masson–Fontana silver method. Noradrenalin cells reduce alkaline silver solutions and appear black while adrenalin cells (purple) do not. × 250.

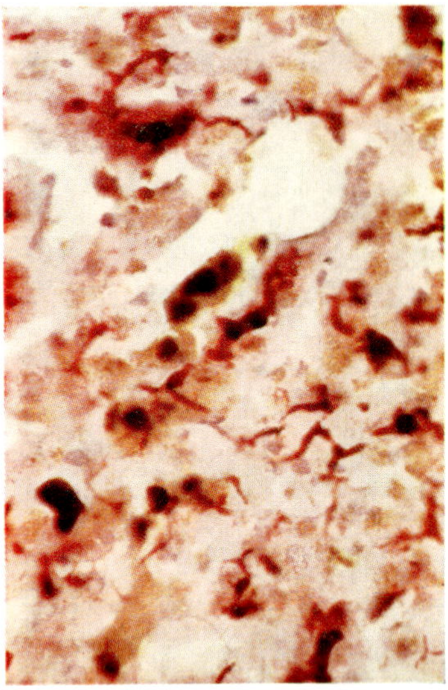

XXIVb. Human liver, obstructive jaundice (needle biopsy). Formol-calcium fixed, cryostat section. Direct histochemical Van den Bergh reaction. Shows conjugated bilirubin in the dilated bile canaliculi, stained deep red. Unconjugated bilirubin masses, in Kupfer cells and hepatocytes are unstained (brown). × 400.

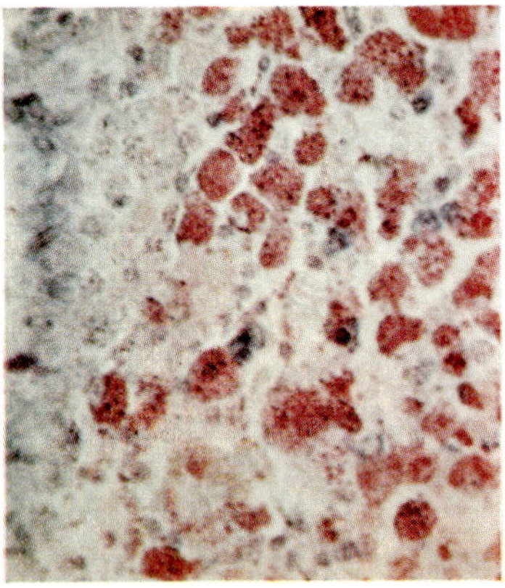

XXIVc. Human fallopian tube. A group of large cells is present, each filled with a granular material (ceroid) which is strongly acid-fast. Long Ziehl–Neelsen stain. × 170.

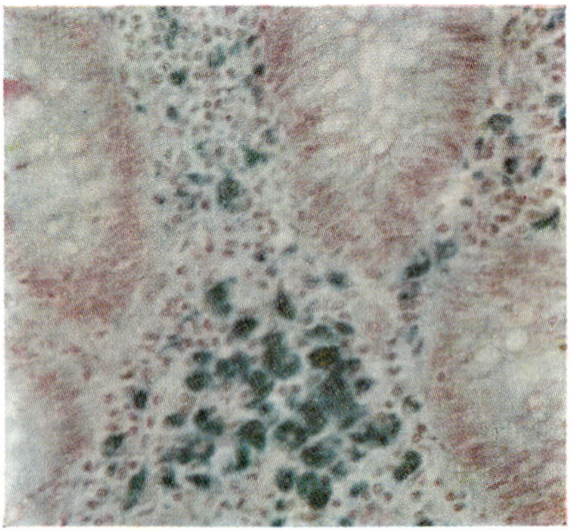

XXIVd. Human vermiform appendix. Numerous macrophages in the lamina propria contain "pseudomelanin" pigment. Schmorl's ferric-ferricyanide-Carmalum. × 95.

[*To face p.* 1075

shown in Plate XXIVb, opposite, and the two methods are described in Appendix 26.

A successful diazo reaction for conjugated bilirubin was subsequently developed by Desmet *et al.* (1968) following the demonstration by Van Roy and Heirwegh (1968) that the diazo salt of ethyl anthranilate (EA) would react selectively with this variety of pigment. Desmet and his colleagues used fresh or cold formalin-fixed cryostat sections. Agar films containing various bilirubins showed the high specificity of the EA technique for the conjugated form. Details of the reaction are given in the Appendix. With the Raia technique, in order to be certain of the identity of the reacting pigment as solely or predominantly conjugated bilirubin, it is necessary to keep the diazotization period as short as possible (Raia, 1970). This is because unconjugated bilirubin always takes part, to some extent, in the direct phase. Using the EA procedure Lillie and Pizzolato (1969b) found that more consistent results could be achieved by using a modified Claus diazotization procedure. Both this and the normal method used by Lillie are given in Appendix 26, p. 1392.

The properties of azobilirubins and azohaematoidins were investigated fully by Lillie and Pizzolato (1970). Both were resistant to acid treatment but were gradually bleached by fresh dithionite. It was concluded that azoreactive bile pigments are protein-bound.

Haematins (Formalin, Malarial and Bilharzial Pigment)

Acid formalin pigment is a dark brown or black substance occurring in minute rhomboidal crystals and granules. It is produced in tissues rich in blood when these are fixed in formalin at an acid pH and is regarded as a haematin derivative. Lillie and Hershberger (1947a and b) and Hershberger and Lillie (1947) give excellent descriptions of its genesis and properties. It is easily distinguished histochemically by means of several characteristics, being doubly refractile, easily bleached by 3 per cent H_2O_2 (30 minutes) and more easily bleached by 99 per cent formic acid. It is extremely insoluble in strong acids but is extracted by alkalis, especially in alcoholic solutions. Saturated alcoholic picric acid removes it instantly. As emphasized by Lillie it is better practice to prevent its formation rather than to allow it to form and subsequently to remove it. In most cases formalin pigment does not form if the pH of the fixing bath is maintained above pH 6·0. The presence of mercuric chloride, potassium dichromate, or picric acid, also inhibits its formation. According to Waldrop *et al.* (1969) the last of these can be used, as a saturated solution in absolute or 95 per cent ethanol, to remove all acid haematin type pigments. A period of 15 hours, or more, is required.

Malarial pigment is also regarded as a haematin and its histochemical properties are apparently identical with those of formalin pigment. Lillie (1948) discusses reported differences but considers their validity doubtful. Johnson *et al.* (1954) found that the histochemical reactions of the haematin pigment of schistosomiasis were identical with those of malarial pigment. On the other hand, Sawada *et al.* (1956) indicated that the pigment produced in mice by *Schistosoma japonicum* was analogous to melanin. A quantitative study of schistosomal pigments in marine hepatic bilharziasis was carried out by Ramadan and Michael (1969). Using a spectrophotometric technique, with haemin as a standard, they found a progressive increase up to 90 days after the initial infection.

Aposiderin

This term is used to describe a brown granular iron-negative pigment which is not doubly refractile and which is extremely resistant to extraction. It is apparently formed from haemosiderin by the action of acid fixatives which either remove the iron or render it inactive to Perls' and other tests. Its chief importance lies in the distinction between it and the lipofuscins of medium and high grades of oxidation. The best way out of the difficulty is to avoid the formation of aposiderin by the use of neutral or buffered neutral formalin. The former is usually adequate but it must be remembered that with some tissues the pH of the formalin solution, initially perhaps 7·4, may fall to 6·0 or lower. If distinction must be made, Schmorl's reaction is probably more reliable for the purpose, being negative with haemosiderin and aposiderin.

The Lipid Pigments

All the pigmented substances discussed under this heading are derived mainly or partly from lipid precursors. They constitute a large class to which many names have been applied and, on the whole, the various subdivisions are characterized by rather indeterminate differences in reactivity with a number of histochemical tests. Most of the iron-free members of the class are considered to belong to the so-called wear and tear pigments (syn. brown atrophy, abnützungspigment, lipofuscin), others have been described as chromolipoid, haemofuscin, cytolipochrome, lipochrome and ceroid. The iron-containing lipid pigments are sometimes regarded as variants of haemosiderin and sometimes separately as cytosiderins.

The following members of the class of lipid pigments are described in this chapter: *lipofuscins*, *haemofuscin* and *cytolipochrome*, *ceroid*, *pseudomelanosis pigment*, *iron-containing lipid pigment* and finally *lipochrome* and its contained carotenoids. No single histochemical property characterizes all these substances, and in many of them it is difficult or impossible to demonstrate lipid by means of the usual tests.

Lipofuscins

These pigments are derived from a number of lipid or lipoprotein sources and the process by which they are produced is at least partly an oxidation. Bensley (1947) described a red pigment occurring in the mitochondria and submicroscopic particles of guinea-pig liver cells. He compared this pigment with those produced by the autoxidation of phospholipids and unsaturated fats and found a close resemblance. Bensley suggested that the colour of the lipid pigments was due to the polymerization of aldehydes produced by oxidation. In this way the original easily soluble lipid, colourless unless it contains dissolved carotenoids, is oxidized to a very insoluble, often highly coloured product. The process varies with the nature of the original lipid and with the cell in which it takes place. It also varies with the extent to which protein forms part of the parent substance and on whether the constituent amino-acids of the protein are subsequently oxidized. Chemical studies carried out by Heidenreich and Siebert (1955) showed that the lipid content of an

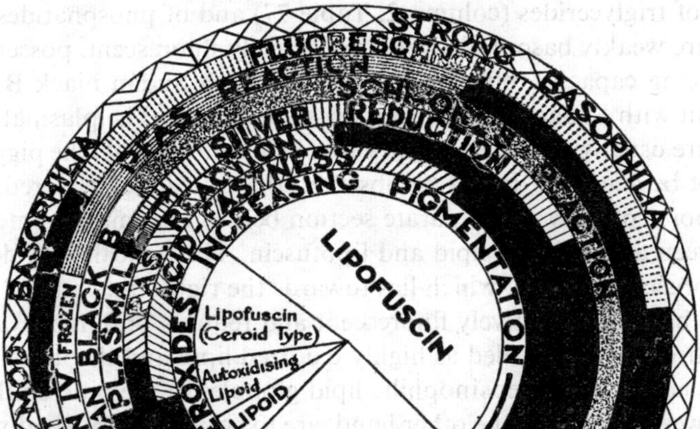

FIG. 223

isolated lipofuscin (cardiac lipochrome) was 20 per cent of the dry weight. The nitrogen content of the fat-free residue was 14·7 per cent and hydrolysates revealed the presence of a number of amino-acids. Later studies (Siebert *et al.*, 1962) indicated a content of 50 per cent protein while the remaining 30 per cent was described as a "melaninhaltige, schwarzer Rückstand". The lipofuscins are the same as the chromolipoids of Ciaccio (see Chapter 12) which he regarded, similarly, as oxidation products of lipids. His chromolipoids, however, retained their sudanophilia in almost undiminished form, although they became gradually more and more insoluble in fat solvents. They differed, moreover, in not reducing alkaline silver solutions. The term chromolipoid is short and accurate and it deserves almost equally with lipofuscin to be used for the description of the whole class of pigments which we are considering here. But because the more advanced stages of the pigments are no longer lipids in the accepted sense, it is probably better to call them fuscins, with the prefix lipo- to suggest their origin.

Views on the Histogenesis of Lipofuscins. My own views of the natural history of the lipofuscins can be expressed diagrammatically in terms of the histochemical tests which indicate their development (Fig. 223). At the lower left-hand edge of the diagram the lipid precursors of the lipid pigments are found in their original form. As oxidation begins their chemical and physical characteristics begin to alter and as it continues the characteristics of fat are replaced by those of partly oxidized fat. Pigmentation characteristically, but not inevitably, rises with the progress of oxidation; some of the early reactions of oxidizing fat become negative and others become positive in their place. Thus, at the lower left-hand edge of the scale are the early products of the oxidation of triglycerides (column 2, Table 77) and of phosphatides (column 3). These are weakly basophilic, yet often actively fluorescent, possessing only weak reducing capacity, but strongly positive with Sudan black B and to a lesser extent with Sudan IV and other fat-soluble dyes. The plasmal and PAS reactions are usually positive and fatty peroxides are present; the pigment may or may not be acid-fast. It will be observed that ceroid, considered from the standard point of view in a separate section below, becomes an intermediate stage between autoxidizing lipid and lipofuscin. The typical, well-developed, brown lipofuscin pigment, which lies towards the right-hand end of the scale, is strongly basophilic, actively fluorescent and reduces both ferricyanide and silver solutions. It is regarded as highly oxidized lipid and the usual fat stains are always negative. The eosinophilic lipid globules in the pink epithelium of apocrine sweat glands, on the other hand, are to be regarded as hardly oxidized at all.

The point at which the lipid propigment becomes lipofuscin is not exactly determinable. It may logically be taken as the point at which a definite yellow colour becomes visible, remembering that in many cases a brownish-yellow colour is all that develops, although all the tests at the higher end of the scale become positive. Alternatively, it may be taken as the point at which the peroxide reaction of Dam (Glavind *et al.*, 1949) becomes negative. This reaction was shown by Dam and Granados (1945) to be due to oxidation products of highly unsaturated fatty acids and it disappeared during the production of lipofuscin pigment at a relatively early stage. The process of lipid peroxidation was investigated, in the tissues of vitamin E-deficient rats, by Pritchard and Singh (1960) and quantitative studies of lipid peroxide formation by normal mitochondria were made by Thiele and Huff (1960). These authors found that ascorbic acid increased peroxide formation 10–20 fold and that the process was independent of pH, from 5·6 to 8·4. Inhibition was produced by Mn^{2+}, versene, and adrenalin.

The histochemical method (see Appendix 26, p. 1385) was developed for the demonstration of peroxides appearing in lipid material during oxidation to acid fast pigment. The authors of the method had observed that, under suitable conditions, peroxidized fats and fatty acids reacted with a number of oxidizable substances (benzidine, leuco-malachite green and leucodichlorophenol-indophenol) to give coloured compounds. These reactions were accelerated by mild acidity and by heat and they were catalyzed by haemin. In the method finally evolved fresh or formalin-fixed frozen sections were treated with a solution of leuco-3:5-dichloro-4:4′-dihydroxyphenylenediamine in the presence of haemin for 3–5 minutes, when the substrate became oxidized to a red compound at sites containing peroxide groups. The specificity of this method is regarded as high and the localization of the dye within a variable proportion of fat cells, and sometimes in central zones only, seems to indicate that the explanation offered by the authors for the mechanism of the reaction is a true one.

In the second edition of his book on "Spot Tests" Feigl (1954) gave several reactions for the oxidation products of unsaturated fatty acids. As an alternative to the Dam reaction he suggested the use of 2,7-diaminofluorene, with haemin as catalyst, for the detection of organic peroxides. This reaction I found (1960) to be quite sensitive but the greenish-blue product was diffusible and localization poor. According to Feigl (1954) hydroxy fatty acids give a red colour with *sym* diphenylcarbazide in organic solvents and this reaction is sometimes weakly positive in sections containing oxidizing fats. A third reaction, this time for epihydrinaldehydes, is also occasionally positive in ceroid-type lipofuscins and in cardiac "lipochrome" if fresh sections are cut and stored for varying periods. This reaction depends on the development of a red colour when the aldehyde condenses with phloroglucinol in the presence of HCl.

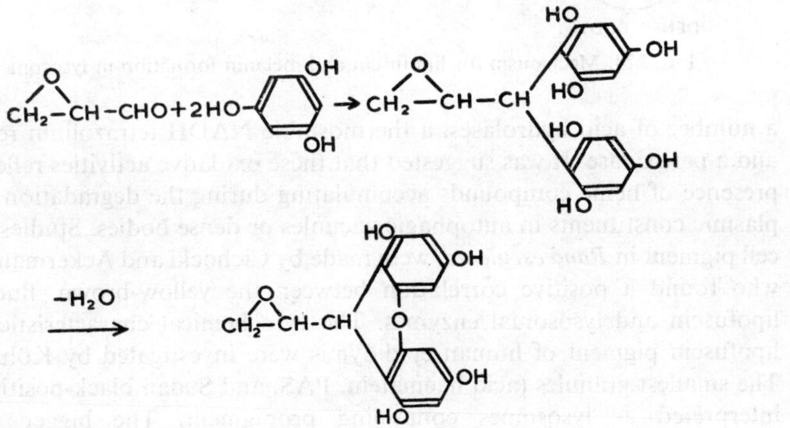

Epihydrinaldehydes are volatile and cannot be expected to be present in large quantities in oxidizing fats. A negative reaction is therefore without significance, as indeed are negative reactions for organic peroxides and hydroxy fatty acids.

Enzymes in Lipofuscins. The presence of a non-specific esterase capable of hydrolysing Naphthol AS acetates was recorded by Gomori (1955) in various types of lipofuscin. I observed a similar non-specific esterase, especially in the lipofuscins of the nervous system, using indoxyl acetate methods (Pearse, 1955). A more extensive study was carried out by Gedigk and Bontke (1956) who found strong AS-type esterases and also acid phosphatases in a variety of lipid pigments. Gomori doubted the accuracy of enzyme localization in lipofuscins but I considered the presence of these enzymes as certain (1960).

Further work has clearly established a relationship between lipofuscin granules and lysosomes. The connection between human hepatocellular pigments and lysosomes was established by Novikoff and Essner (1960), and liver cell pigment granules were shown by Goldfischer *et al.* (1966) to contain

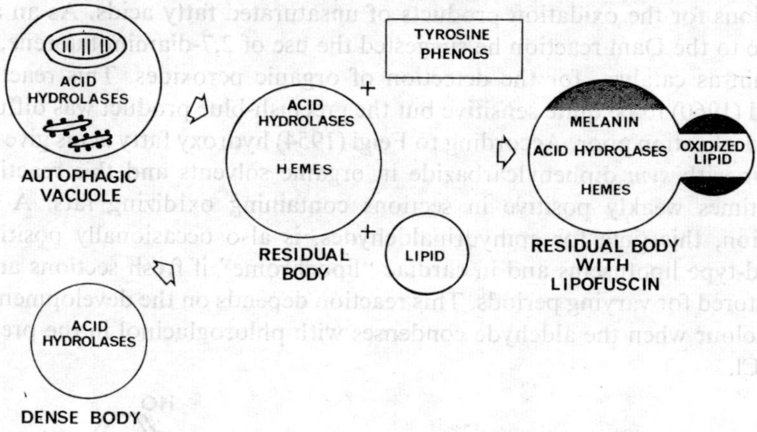

FIG. 224. Mechanism for lipofuscin and melanin formation in lysosomes.

a number of acid hydrolases, a thermostable NADH tetrazolium reductase, and a peroxidase. It was suggested that these oxidative activities reflected the presence of heme compounds accumulating during the degradation of cytoplasmic constituents in autophagic vacuoles or dense bodies. Studies on liver cell pigment in *Rana esculenta.* were made by Cichocki and Ackermann (1967) who found a positive correlation between the yellow-brown, fluorescent, lipofuscin and lysosomal enzymes. The cytochemical characteristics of the lipofuscin pigment of human epididymis were investigated by Köhl (1968). The smallest granules (acid haematein, PAS, and Sudan black-positive) were interpreted as lysosomes containing propigment. The bigger granules (Schmorl-positive, basophilic, acid fast) were characterized as lipofuscin. They could equally well be called ceroid.

The concept of lysosomal involvement in lipofuscin production is an important one which merits the fullest consideration. The diagram shown above (Fig. 224), modified from Goldfischer *et al.* (1966), was designed to illustrate the concept and to explain the finding of melanin or "melanins" in association with lipid pigment production in lysosomes.

It is supposed that heme compounds possessing oxidase or peroxidase activity accumulate within residual bodies. These hemes may be derived from mitochondria, or from fragments of endoplasmic reticulum, degraded within autophagic vacuoles. Unsaturated lipids and metabolites of tyrosine, adrenalin and noradrenalin, which are incorporated into such bodies, are subsequently oxidized to lipofuscin and "melanin".

Significance of the Various Tests. At present we have little idea of the significance of the various tests which are positive in the case of the lipofuscins. Their solubility in fat solvents diminishes as their ability to dissolve fat stains diminishes, reflecting a progressive physical and chemical alteration. The number of acid groups available for combination with the basic dyes begins to rise almost as soon as oxidation commences and at this stage the plasmal reaction may become positive (see Chapter 13, p. 460). It remains positive only for a short period although the comparable PFAS reaction persists for a somewhat longer period. The reason for the positive PAS reaction only in the intermediate stages of the oxidation of lipids is far from clear. It certainly does not mean, as many authors have concluded, that the lipid concerned belongs to the carbohydrate-containing glycolipid (cerebroside) group. We know (Chapter 12, p. 437) that aldehyde can be produced from non-carbohydrate-containing unsaturated phosphatides by the action of periodic acid, and we also know that such aldehydes cannot be produced from all phosphatide-containing tissue components by this method. Until much more work has been done upon the problem we can only say that, at a certain intermediate stage in the oxidation of lipids, Schiff-positive groups are revealed by brief oxidation with HIO_4.

At an early stage in the oxidative process reducing groups are formed which give a positive Schmorl's reaction and a weaker reduction of alkaline silver solutions. Beyond a certain stage of oxidation the latter effect becomes no longer noticeable but the reduction of ferricyanide increases in intensity and seldom drops to the point of extinction. Adams (1956) considered that the pigment produced in lipofuscins by the ferric-ferricyanide reaction was not Prussian blue but the green pigment ferroso-ferric ferricyanide (Lange, 1937) which is known as Prussian green. Lillie believed, however, that the green colour was a mixture of the yellow or brown of the pigment with Prussian blue. Details of Schmorl's reaction given in Appendix 26, p. 1383, have been modified in the light of Adams' work.

The quality of acid-fastness depends to some extent on the way in which the test is performed and on the interpretation of what is acid-fast. Berg (1953) showed that the acid-fastness of ceroid was dye specific. It is obtainable only when basic fuchsin is employed. I therefore use a long Ziehl-Neelsen method (Appendix 26, p. 1385) and regard the presence of any redness as positive. It is naturally less easy to distinguish this in the more highly coloured pigments so that, in the diagram, acid-fastness is indicated as falling off towards the right-hand end of the scale. A chemical explanation for this property was suggested by the work of Reeves and Anderson (1937) on the chemistry of the lipids of the tubercle bacillus. Acid-fastness was thought to signify the presence of unsaturated fatty acids of high molecular weight. The mechanism of the acid fast reaction was investigated by Lartique and Fite (1962). They concluded, in agreement with Lamanna (1946) that the action of phenol could only be to increase the lipid solubility of the dye employed. Acid fastness should not be used to distinguish between lipofuscin and ceroid pigment (see below).

A method for distinguishing melanins from lipofuscins was evolved by Hueck (1912) which depended on staining with Nile blue sulphate and subsequent treatment with H_2O_2 to bleach the melanins. This method was modified by Lillie (1956a) and followed by acetone extraction which leaves the melanins stained dark green and the lipofuscins colourless. Lillie (1956b) investigated the mechanism of Nile blue staining of lipofuscins and found that two types of staining could occur. The first type he attributed to oil solubility and this occurred promptly with dilute solutions of Nile blue at pH levels down to 0·85 or lower. The process was rapidly reversed by acetone or alcohol extractions and could be repeated indefinitely. The second type of staining he attributed to formation of salt linkages. This took place with a number of basic dyes but was extinguished in every case at pH 3·0.

An indophenol method for demonstrating lipofuscins was developed by Alpert *et al.* (1960) which gives positive (red) staining with most lipofuscins, with the notable exception of cardiac lipochrome. The mechanism of the reaction is presumed to involve preferential solubility of the red, non-ionized form of the dye (2,6-dichlorophenolindophenol) in the pigment granules. Details of the method are given in Appendix 26.

Haemofuscin and Cytolipochrome

The name haemofuscin was given by von Recklinghausen (1899) to a yellow, granular, iron-free pigment occurring in cases of haemochromatosis together with very much larger amounts of haemosiderin. Since by the majority of tests it reacted in the same way as lipofuscin, it was regarded as such by Hueck (1912). Mallory (1938), who described a basic fuchsin staining method for haemofuscin, suggested that this pigment could be distinguished by its solubility in 5 per cent H_2O_2. I do not find the iron-free pigment of haemochromatosis, or the recognized lipofuscins, to be soluble in this reagent. There is not sufficient histochemical evidence to permit the retention of a separate fuscin pigment, derived from red cells, under the title haemofuscin. Where red cell envelopes contribute to the formation of a lipid pigment, this has at all stages the characteristics of a lipofuscin. As such it should be regarded and described.

Gillman and Gillman (1945) described as cytolipochrome a pigment having the characteristics of haemofuscin, which they found in the livers of pellagrins. This designation is entirely to be condemned. There is nothing in the nature of this pigment, as described, to warrant the invention of a new term, even if it had the merit of histochemical accuracy.

Ceroid

This pigment was first described, in the cirrhotic livers of animals maintained on inadequate diets, by Lillie *et al.* (1941, 1942) and by Edwards and White (1941). Edwards and Dalton (1942) later described a similar pigment occurring in the livers of mice treated with carbon tetrachloride. Wolf and Pappenheimer (1945) and Pappenheimer and Victor (1946) described the occurrence of a pigment, which they related to ceroid by virtue of its acid-fastness, in a wide variety of human tissues. They suggested that its development was due to deficiency of vitamin E. Endicott and Lillie (1944) established the main characteristics of ceroid pigment and the means of distinguishing it from other pigments such as haemofuscin. It was described as occurring in yellow globules 1–20 μ in diameter, usually in phagocytes, and as rims of basophilic, acid-fast material around large fat globules. It stained well with fat-soluble dyes in both frozen and paraffin sections. Lillie noted that the property of acid-fastness was shared by some of the lipofuscins, or as we should say, was shared by the lipofuscins at some stages of their development. This, of course, means that it cannot be used to distinguish between the two pigments. Lillie, however, maintained that distinction could be made by the negative reaction of ceroid with Schmorl's test. He did not (1948) mention the PAS reaction but Lee (1950) observed that ceroid in rat and mouse livers, produced by a variety of means, was variably positive. He observed in respect of this reaction that sometimes only a part of the pigment was stained and concluded that "although the pigment studied exhibited some similarity in acid-fast and fat-staining properties the variation in staining with iron reaction, periodic-acid-leucofuchsin, Schiff reaction, basic fuchsin and pyronin-methyl green indicated that ceroid may be a variable mixture of substances". McManus (1949) stated that ceroid gave a positive reaction for carbohydrate components by the Hotchkiss (PAS) and Feulgen reactions, "with as well as without acid hydrolysis". This last statement requires clarification since both reactions, without acid hydrolysis, are merely pseudoplasmal reactions. Artificial "chromolipoids", produced by injection of neutralized cod-liver oil into adult rats, were studied by Wolman and Shoshan (1960). These pigment inclusions were found to stain by the Feulgen reaction for DNA. This component was excluded, however, by spectrophotometry indicating clearly the presence of plasmalogens (Chapter 13, p. 460). According to Wolman and Zaidel (1962) high O_2 tension favours the formation of "chromolipoids".

All lipids in an early stage of oxidation are likely to react as does ceroid. The PAS reaction, when positive, does not necessarily indicate the presence of carbohydrate either in ceroid or in lipofuscin, but the results obtained with blocking techniques suggest that some at least of the positive groups are due to the presence of polysaccharide. In the case of cardiac "lipochrome", however, which has some relationship to ceroid, the PAS reaction requires long oxidation to develop any intensity and this suggests (see Chapter 10, p. 311) that non-polysaccharide groups are responsible for the positive reaction.

My observations (1960) agreed entirely with those of Lee in the matter of the variability of the pigment by various histochemical tests. I concluded, however, not only that ceroid is a mixture of substances but that it represents a typical lipofuscin in an early stage of oxidation. The acid-fast rims of large fat globules, which Lillie regarded as ceroid, occur in many other tissues besides liver. In particular, the lipid rims around the vacuoles of basophils in the human anterior pituitary are acid-fast, as are the PAS-positive vesicles described by me in the pituitary chromophobes (1948, 1949). Since these are not pigmented they are not to be regarded as lipofuscin but they can probably be regarded as ceroid. Alpert (1953), in a paper describing the hormonal induction of deposition of ceroid pigment in the mouse, summarized the now firmly held belief that it is a fatty acid derivative perhaps due to the failure of the normal cell mechanisms for the breakdown of unsaturated fats. Alpert believed, and I agreed (1960), that the vacuoles which are so often seen in paraffin sections of ceroid masses represent part of the fat which has not undergone sufficient polymerization to be insoluble in fat solvents. In this material fatty peroxides can be demonstrated if cold microtome sections are used.

A high proportion of the lipoprotein envelopes of red cells, after fixation in mild oxidant fixatives such as Müller or Regaud, are strongly acid-fast. The substance responsible is not ceroid in the usual sense though it may well be chemically related to ceroid in possessing the same acid-fast groupings. Further confirmation of these views can be derived from the experimental production of a ceroid-like substance from erythrocytes, in rats, by Hartroft (1951), and from the work of Casselman (1951) on the histochemical properties of a similar artificially produced "ceroid". Acid-fast yellowish-brown autofluorescent pigments were observed to develop in Vitamin E-deficient rats and mink by Mason and Hartsough (1951) and in pigs by Robinson and Coey (1951). Tappel (1955) was able to produce yellowish-brown copolymers of ceroid type by the *in vitro* oxidation of unsaturated fats in the presence of proteins, especially haemoglobin. The term ceroid originally applied only to the particular material in animal livers, produced by experimental means. Naturally occurring pigments with similar characteristics, in man and other animals, can be referred to as lipofuscins (ceroid type), or as lipofuscins (early), or more simply as ceroid. This last term is now sanctified by common usage. Ceroid has been reported in atheroma (Burt, 1952), in the coronary arteries (Schornagel, 1956), in fluorocytes in the liver (Schmidt, 1953), and in haemochromatosis (Weber, 1954). It has also been described in a series of experimental studies by various authors (Williams and Aronsohn, 1953;

Klein and Johnson, 1954; Wood and Yasutaki, 1956). Its occurrence in jejunal mucosa was described by Gresham *et al.*, (1958) and in placenta by Ishizaki (1960). Ceroid is common in the Fallopian tube, as shown in Plate XXIVc, p. 1075. Katchburian *et al.* (1961) described a ceroid-like pigment in the livers of the new-world monkey *Alouatta fusca,* and Bednář and his colleagues (1964) described a condition of "vascular pseudohaemophilia" in albinos which was associated with ceroid laden macrophages in the bone-marrow. In the nervous system of vitamin E-deficient monkeys Einarson and Telford (1960) found large amounts of a typical ceroid pigment, especially concentrated in the cell bodies of the neurones of the anterior horn. Changes were also described in the skeletal muscles where they had already been described, in vitamin E-deficient rats by Gedigk and Fischer (1959). "Lipoid pigments" were also studied in ganglion cells in human brain by Stammler (1959) and in the pars nervosa of the pituitary by Müller (1963). Maeda *et al.* (1968) described the strongly positive reaction of ceroid by the Okamoto method (Vol 1, p. 424). They attributed this to its unsaturated C-C bonds since it was prevented by prior treatment with peracetic acid.

The formation of ceroid pigment in subcutaneous implants in mice was noted by Winnail and Williams (1968), and a pigment described as ceroid was found in the gastrointestinal smooth muscle of persons in the Thai-Lao ethnic group by Nye and Chittayasothorn (1967).

Pseudomelanosis Pigment

At one time there was a great deal of interest in the nature of this peculiar pigment, found almost exclusively in macrophages in the lamina propria of the large intestine and appendix. Hueck (1912), Henschen and Bergstrand (1913) and Dalldorf (1927) regarded the pigment as intermediate between melanin and lipofuscin, while Pick (1911), Lubarsch and Borchardt (1929) and Pick and Brahn (1930) considered it as a true melanin. Pick's final theory that aromatic protein degradation products, absorbed from the large intestine, were converted into melanin within the connective tissues by a tyrosinase-like enzyme, was accepted by Stewart and Hickman (1931). Speare (1951) revived the "anthracene" purgative theory of pseudomelanosis by reporting a series of cases which, without exception, were associated with chronic medication with *cascara sagrada.* This substance contains anthraquinone glucuronides. Some German authors, apparently on the basis of a reported positive reaction by both Prussian and Turnbull blue tests (Appendix 28, p. 1402), have considered pseudomelanin to be a haemosiderin, altered by the action of H_2S. Lillie *et al.* (1964) could find no evidence to support this theory in their studies on guinea-pig enterosiderosis. All the pigment, in those examples of pseudomelanosis of the colon and appendix which I studied (1953) while investigating this particular problem, was negative by both the Turnbull and Prussian blue reactions and I agreed with Stewart and Hickman's decision that the pigment is not derived from haemoglobin. Piringer-Kuchinka (1952) considered pseudomelanin as an "exogenous pigment" and a similar view was taken by Hieronymi (1954). The condition has to be distinguished from the so-called intestinal lipofuscinosis, described as brown bowel syndrome by Toffler, Hukill and Spiro (1963). Earlier authors (Pappenheimer and Victor, 1946; Adlersberg and Schein, 1947; Tverdy *et al.*, 1949; Gresham *et al.*, 1958; Richards, 1959; and Fullerton, 1960) had noted the deposition of lipid pigments in the smooth muscle, especially of the small intestine, in various conditions, such as hepatic cirrhosis, pancreatic mucoviscidosis and Whipple's disease (Vol 1, p. 393). Conditions giving rise to pigmentation of the intestine were reviewed by Fisher (1969), and an account of lipofuscinosis in the human gastrointestinal tract was given by Fox (1967). Pseudomelanosis pigment is shown in Figs. 225 and 226, and also in Plate XXIVd, p. 1075 where its reaction with Schmorl's ferric-ferricyanide is recorded.

Iron-containing Pigments

This is not the point at which to consider the differential diagnosis of the haemosiderins; it is necessary, however, to mention the occasional occurrence in collections of lipofuscin pigment, of almost identical but non-fluorescent, iron-containing, granules. It is possible to regard these as true haemosiderin but, more probably, they are due to the combination of iron salts with the lipofuscin at some stage in its formation.

Lipochromes

By definition these are lipids containing in solution coloured hydrocarbons of the widely occurring carotenoid series. The presence of dissolved carotenoids is usually obvious enough, in fresh tissues, because of the brilliant yellows and reds which they impart. Some examples are the corpora lutea of women and the corpora lutea and corpora rubra of cows, the ordinary body fats of many animals and the adrenal cortical lipids. The carotenoids are soluble in light petroleum, benzene, xylene, chloroform, ether, alcohol, acetone, etc., and they are largely dissolved out during the preparation of paraffin sections. For this reason we are not usually concerned with lipochromes except in fresh or formalin-fixed frozen sections and in freeze dried material. Carotenoids combined with protein may remain sufficiently firmly attached to withstand the process of paraffin embedding, but little is known of the occurrence of this particular type of chromoprotein in mammalian tissues.

The Carotenoid Pigments. Whatever the nature of the lipid part of the lipochrome, and this may vary within wide limits, the diagnosis of the whole depends on the demonstration of carotenoids. The methods available for this purpose are discussed below. According to Karrer and Jucker (1950), approximately eighty natural carotenoids are now known. They are all closely related chemically and belong to the class of polyenes (polyisoprenes) whose chief characteristic is the possession of a large number of conjugated double bonds. Below is illustrated the formula of half the molecule of β-carotene; the other half being the mirror image of that given. The whole molecule consists of eight isoprene residues linked together, not all in the same way. The formula of isoprene is also given.

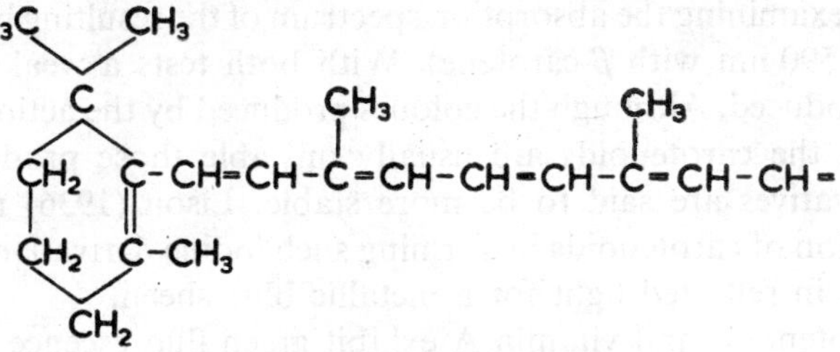

β-carotene (½ molecule)

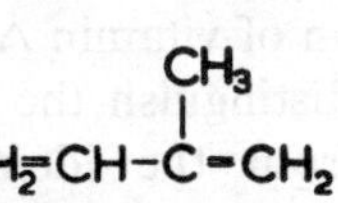

Isoprene

As far as animals are concerned, the carotenoid pigments are derived entirely from the diet and, originally, from vegetable sources. They are partially absorbed from the intestine and partially excreted unchanged. Three carotenoids in particular have been recorded as occurring in human serum (Willstaedt and Lindquist, 1936); these are carotene, lycopene and xanthophyll. The same three have been identified in human liver (Zechmeister and Tuzson, 1936). They are known to act as sources of vitamin A, as first suggested by Steenbock *et al.* (1921); later Karrer *et al.* (1931, 1933) showed that β-carotene could be converted by the uptake of water into two molecules of vitamin A. Carotenoids also play a part in the process of vision. Rhodopsin (visual purple) was considered by Wald (1935) to be a carotenoid-protein complex, and retinene, which also occurs in the eye, is the aldehyde of vitamin A.

Since the carotenoids are extremely soluble in fat solvents they are seldom found in paraffin sections except where they occur as chromoproteins. When dissolved in lipids, however, they are insoluble in water, glycerol, formalin, and dilute acids and alkalis; they are relatively insoluble in ethanol and methanol. In fresh or formalin-fixed frozen, or in freeze-dried sections they remain *in situ.* Since β-carotene and the other carotenoids are autoxidized in the presence of atmospheric oxygen to colourless compounds, material for histochemical tests should be examined as soon as possible. Judging by the progressive reduction in colour of the adrenal cortical lipids, autoxidation also occurs in formalin.

The Histochemistry of the Carotenoids. Carotenoids have been demonstrated histochemically by means of certain colour reactions. They give green, blue or violet colours with a variety of strong acids such as sulphuric, hydrochloric, phosphoric, perchloric, formic, and trichloroacetic, and blue colours with the chlorides of antimony and arsenic. In practice it is usual to carry out the reaction with sulphuric acid or the Carr-Price test with antimony chloride in chloroform. Neither reaction is specific though they both might well be made more so by examining the absorption spectrum of the resulting blue compound (maximum 590 nm with β-carotene). With both tests a weak and transient colour is produced. Although the colours produced by the action of antimony chloride on the carotenoids are usually unstable those produced with the iodine derivatives are said to be more stable. Lison (1936) mentioned the demonstration of carotenoids by forming such iodine derivatives and examining sections in reflected light for a metallic blue sheen.

The carotenoids and vitamin A exhibit green fluorescence in ultra-violet light and this property was utilized in the extensive researches of Popper (1941, 1944) on the distribution of vitamin A in human tissues. According to this author it is possible to distinguish the green fluorescence of carotene, which fades very slowly, from the identically coloured, rapidly fading fluorescence of vitamin A. Ball and Morton (1949) described the reactions of the latter with sulphuric and phosphoric acids. With conc. sulphuric acid at 0° a blue compound was produced (absorption maximum 620 nm) which quickly faded to a series of red compounds (absorption maxima 465, 520 and 580 nm). At the lower temperature fading of the blue component was much less rapid than at room temperature. Ball *et al.* (1949) gave similar data about retinene and, in addition, they described the combination of this substance with amines. The reaction was usually slow but with the aromatic amines coloured compounds were quickly produced. Aniline (in excess in acid solution) gave a red compound (absorption maxima 490–435 nm).

In routine histochemistry the carotenoid pigments are of very little importance though in research their interest is obviously greater. The results obtained with the usual reactions (Carr-Price and conc. sulphuric acid) are disappointing and the knowledge gained in biochemical investigations of the type quoted above has not yet found histochemical application.

Dubin-Johnson Pigment

The fact that this entity has to be considered separately is an indication of the uncertainty surrounding its identity and origin. The condition known as Dubin-Johnson syndrome was described by the eponymous authors Dubin and Johnson (1954) as a chronic idiopathic jaundice with deposition, in liver cells only, of a pigment of unknown constitution. Subsequently this was regarded by some authors as a "lipomelanin" and by others as a lipochrome. Though, originally, he considered the pigment to be a mesobilirubin Dubin (1958) was content to identify it as a lipochrome.

The pigment has few reliably constant features. It is basophilic, weakly PAS-positive, argentaffin and, usually, autofluorescent. Until extraction and biochemical characterization has been achieved the present uncertainty cannot be resolved. The present status of Dubin-Johnson pigment was well described, with an abundant bibliography, by Barone *et al.* (1969).

A condition occurring in a mutant strain of Corriedale sheep, giving rise to hepatic pigmentation resembling that seen in Dubin-Johnson syndrome, was described by Cornelius *et al.* (1965). In sheep, the condition is complicated by a fatal photosensitivity due to an excess of chlorophyll-derived phylloerythrin, which is inadequately excreted. The precise relationship of the two conditions has not been established. A review of the whole syndrome was presented by Schillinger (1967).

The Differential Diagnosis of Tissue Pigments

Table 77 shows the reactions of various types of lipid pigment and pro-pigment and, for comparison, the reactions of melanin, haemosiderin, haematoidin, argentaffin cell granules, adrenal chromaffin granules, Dubin-Johnson pigment and of "pseudomelanin" in the lamina propria of the colon and appendix. The pigments in the first ten columns are all of lipid origin and in the remaining columns appear the other pigments dealt with in this chapter. The table may therefore be used as a reference sheet of reactions in the diagnostic histochemistry of mammalian pigments.

At first sight, the results appear to show only a general similarity between all the lipo-pigments and a number of minor points of variance. A most important observation, which cannot be shown clearly in a bald record of responses to the various tests, is that in any one situation the individual granules of lipo-pigment often differ both in appearance and reaction. This point is best illustrated by example. In the posterior pituitary gland of man there accumulate numerous intracellular pigment granules varying in size from 0·2 to 1 μ, and in colour from yellow to brown. Similar pigment granules occur in many other sites listed in the table, notably liver, adrenal and ganglion cells, and they also occur in numerous sites not listed. In the case of the neurohypophysis about half of the granules, mainly the larger and less pigmented ones, can be stained with Sudan black B in paraffin sections; they are PAS-positive, reduce ferricyanide (Schmorl) and alkaline silver solutions, and are only weakly basophilic. The remainder, consisting of the smaller and more pigmented half of the granules, stain strongly with basic dyes and are acid-fast; they are less strongly reducing towards ferricyanide and silver solutions, and they are Sudan black and PAS-negative. In my opinion these differences reflect not the existence of two different types of lipid pigment, but of two main varieties of the same pigment, lipofuscin.

The pigments listed in the first three columns of the table are shown by their content of peroxides to be in an early stage of autoxidation, and the positive plasmal reaction indicates that acetals are among the reaction products of this process. All three can be shown with fat-soluble dyes to contain at least a proportion of lipid. The pigments in columns 4–8 inclusive usually still

TABLE 77

Comparative Histochemistry of Pigments

TEST OR PROPERTY	LIPO-PIGMENTS									OTHER PIGMENTS (FOR COMPARISON)						
	Apocrine gland globules	Perirenal fat (early oxidation)	Lipid in macrophages	Heart muscle "lipochrome"	Posterior pituitary	Ganglion cells	Liver cells	Adrenal (reticularis)	Ceroid liver	Dubin-Johnson Pigment	Melanin (melanoma)	Haemosiderin	Haematoidin	Argentaffin cell granules	Pseudomelanosis (appendix)	Chromaffin (adrenal medulla)
	(1)	(2)	(3)	(4)	(5)	(6)	(7)	(8)	(9)	(10)	(11)	(12)	(13)	(14)	(15)	(16)
Colour	Pale yellow	Pale yellow	Yell. to brown	Gold-brown	L. to D. brown	Yell. to D. brown	Light brown	Gold-brown	Yellow-brown	Yell. to blk.	Brown to Black	Yell. to brown	Yellow-brown	Pale yellow	Yell. to brown	Brown
Eosinophilia	+	±	−	−	−	−	−	−	−	−	−	−	−	+	−	−
Basophilia	−	+	+	weak +	+ v.	weak + to +	+	+	+	+	+	weak + to −	−	−	weak + to +	weak +
Acid-fast (long Z-N)	−	−	weak +	+ v.	+ v.	+ v.	+	+	+	+ v.	occ. +	−	−	−	−	−
Sudan black	+	+	+ to −	±	±	±	±	±	+	Grey	−	−	−	−	−	−
Oil Red O (frozen)	mod. +	mod. +	mainly −	mod. +	usually −	−	−	−	weak +	−	−	−	−	−	−	−
PFAS	+	+	+	+	+ v.	+ v.	+ v.	+ −	+ v.	−	−	−	−	−	±	−
PAS	−	−	±	−	±	±	±	±	±	−	−	−	−	−	±	−
Plasmal	part +	part +	part +	−	−	−	−	−	−	−	occ. + −	−	−	−	−	−
Alkaline silver	weak +	weak +	weak +	−	weak +	weak +	weak ±	−	weak +	weak +	+	−	−	+	±	+
Schmorl	weak +	weak +	weak +	weak + to −	+ to weak +	str. +	+	+	±	+	+	−	−	+	±	str. +
Gmelin	−	−	−	−	−	−	−	−	−	−	−	−	+	−	−	−
H_2O_2 (bleaching)	−	−	−	48 +	48 +	48 +	48 +	48 +	−	−	24 to 48	−	−	−	6–8 +	1–2 +
Diazo (alkaline)	−	−	−	−	−	−	−	−	−	+ v.	−	−	−	+	Deep yell. brn.	Deep yell. brn.
Diazo (acid)	−	−	−	−	−	−	−	−	−	−	−	−	−	+	Yell.brn.	Yell. brn.
Gibbs'	−	−	−	−	−	−	−	−	−	+ v.	−	−	−	+	−	−
Fluorescence (U.V.)	Gold-yellow	Brown-gold	Yellow	Red-brown	Green-yellow	Gold-yellow	Dull brown	Weak to nil	Gold-brown	Yellow	−	−	−	White-yellow	Gold-yellow	−
Chrome haematoxylin	−	−	−	+	+	+	+	+	−	+	−	−	−	−	−	−
Peroxide (Dam)	+	+	+	−	−	−	−	−	−	−	−	−	−	−	−	−

The following abbreviations are used: yell., yellow; wh., white; brn., brown; blk., black; L., light; D., dark; v., varies; mod., moderate; str., strong. In the case of bleaching by 3 per cent hydrogen peroxide the numbers refer to the approximate time (hours) of treatment necessary to decolorize the darkest form of the particular pigment listed. The reference 48 + means more than 48 hours. Where the symbols ± or + − occur in the Table, this does not mean the pigment as a whole may be either positive or negative but that at the same site it is partly positive and partly negative; the upper symbol indicates predominance. Similarly, the symbol + v. means that the pigment is usually entirely positive but that some granules in the same situation in the section may be negative.

contain lipid demonstrable by this means and the presence of unsaturated lipids in nearly all of them is revealed by the positive PAS and PFAS reactions. The behaviour of heart muscle "lipochrome" (column 4) is anomalous and this material cannot be fitted into the same precise group (of intermediate lipofuscin) as the others.

The reactions of "ceroid" pigment show that it can conveniently be considered as an early lipofuscin in the stage of autoxidation, and that none of the tests performed can be used to distinguish it from other members of this group. This does not mean that the pigments in columns 4–10 have identical origins and identity; only that we are at present incapable of demonstrating any differences which may exist between them.

Of the pigments in the comparison group (columns 11–16) only one resembles the lipo-pigments in its reactions. This is the so-called "pseudomelanin" found in the intestinal wall. Where two symbols or descriptions appear in column 15 the upper one refers in each case to the pigment situated at the mucosal border of the lamina propria and the lower symbol to the pigment in cells deeply situated just under the muscularis mucosae. I assume that the former is the younger and the latter the more mature pigment. The newer material occurs in the form of yellow granules which are brilliantly fluorescent, strongly reducing towards both ferricyanide (Fig. 225) and alkaline silver solutions, weakly basophilic, strongly PAS-positive (Fig. 226) and positive also by the PFAS reaction. The older pigment is darker in colour sometimes approaching black, less strongly fluorescent, and much less strongly reducing. The oldest pigment is PAS and PFAS-negative.

These characteristics are those of a lipofuscin which contains lipid, possibly small in amount and combined as lipoprotein, oxidized sufficiently at all stages to give negative results with the Sudan dyes. The results do not suggest that the pigment is a true melanin since, in those critical positive reactions which it shares with lipofuscin, melanin never shows the same range of variability. The rapid bleaching of the intestinal pigment with H_2O_2 also suggests that it is not a true melanin, but only one reaction gives any indication that it may not be a true lipofuscin. This is the diazo reaction which invariably increases the colour of the granules to a deep yellow, and even to a reddish-brown in some instances. This is in contrast to the behaviour of true lipofuscin granules whose colour, at all stages, is unaffected by diazonium salts or their breakdown products. Chromaffin granules, on the other hand, give a reaction almost identical with that given by pseudomelanin. Even if the increase of colour is due not to coupling with the diazonium salt but to a strong affinity for its coloured breakdown products in alkaline solution, the similarity between pseudomelanin and chromaffin suggests that Pick's hypothesis of aromatic breakdown products offers the best explanation in this case. Staining with a diazotate (inactivated by treatment at 60° and pH 9·2 for 3 hours) scarcely affects either pseudomelanin or chromaffin granules but this control experiment does not exclude the possibility that intermediate breakdown products of the diazonium salt may be responsible. I, therefore, regard pseudomelanin as belonging to the class of lipofuscins but containing an additional element derived from lipid-soluble phenolic oxidation products.

The reactions recorded for the remaining pigments (columns 11, 12, 13, 14 and 16) show that no difficulty is experienced in recognizing the two belonging to the haem group (haemosiderin and haematoidin) and in distinguishing them from the others in the table. The tyrosine tryptophan group of pigments

TABLE 78

Diagnostic Schedule for Organic Pigments

	SCHMORL POSITIVE	SCHMORL NEGATIVE
PERLS' POSITIVE	Lipoprotein-bound iron (Lipofuscin)	HAEMOSIDERIN
	MELANIN Fe^{2+} uptake, H_2O_2 bleaching, Masson-Fontana +, Alk. tet. −, non-fluorescent	CEROID Oil Red +, Acid fast, fluorescent, Ald. fuchsin +.
	PSEUDOMELANIN PAS +, Alk. tet +, weak fluorescence	HAEMOGLOBIN Benzidine or leuco dye reaction.
	LIPOFUSCIN PAS +, fluorescent, Alk. tet. +, Sudan black + (CEROID)	HAEMATIN Formalin Pigment } Bleaching, Malaria Pigment } birefringence
PERLS' NEGATIVE	BILIRUBIN (Conj) Direct Diazo +, Gmelin +, Stein + Alk. tet. +, Diammine silver.	
	BILIRUBIN (Unconj) = HAEMATOIDIN Indirect Diazo +, Gmelin −, Stein −.	APOSIDERIN Produced by acid fixatives PAS +, Isotropic.
	NORADRENOCHROME (Glutaraldehyde-fixed) PAS +, Silver +.	
	ENTEROCHROMAFFIN (5-HT/HCHO) Fluorescent, PAS −, Diazo +	

(columns 11, 14 and 16), on the contrary, is shown to possess a number of reactions in common with the lipopigments. Those reactions usually thought to be diagnostic for one or another of the pigments (e.g. Schmorl's reaction for lipofuscin and the alkaline silver reaction for melanin) are obviously unspecific. In each case, however, there is at least one negative or positive reaction to distinguish the members of the tyrosine group from each other and from the lipo-pigments in the middle and upper ranges of oxidation. Melanin, except in melanophages, never gives a positive PAS or PFAS reaction and this feature, and the shorter time required for bleaching, distinguish it from the lipofuscins. It can also be distinguished by the alkaline tetrazolium reaction which is positive only with lipofuscins and by methylene blue extinction (MBE). Melanins are basophilic down to and sometimes below pH 1·2. Lipofuscins seldom bind methylene blue below pH 1·8. Chromaffin granules are distinguished from the latter, and from melanin, by an extremely short bleaching time, and by the yellow-brown colour which they give with the diazo reaction. Argentaffin granules give a true positive diazo reaction (red) and a positive Gibbs' reaction, and they differ on this account from the other members of the group, and from lipofuscins.

In spite of the above criteria, however, the histochemical diagnosis of pigmented substances in tissue sections is not always easy or straightforward. It will not be so until some of the many gaps in our knowledge of the mechanism of pigment formation are closed. The schedule laid down in Table 78, below, was designed mainly for pathologists, to enable them to distinguish by means of two simple tests (Perls', Schmorl) four groups of pigments. By the application of a further test, in some favourable instances, or of a small battery of tests in others, they can usually pin-point the nature of their pigment without too much difficulty.

REFERENCES

ADAMS, C. W. M. (1956). *J. Histochem. Cytochem.*, **4**, 23.

ADLERSBERG, D., and SCHIEN, J. (1947). *J. Am. med. Ass.*, **134**, 1459.

ALPERT, M. (1953). *Anat. Rec.*, **116**, 469.

ALPERT, M., JACOBOWITZ, D., and MARKS, B. H. (1960). *J. Histochem. Cytochem.*, **8**, 153.

ARVY, L., and DE LERMA, B. (1961). *C. R. Acad. Sci. Paris*, **253**, 1012.

BAGGENSTOSS, A. H., KERNOHAN, J. W., and DRAPEWSKI, J. F. (1943). *Am. J. clin. Path.*, **13**, 333.

BAKER, J. R. (1946). *Q. J. microsc. Sci.*, **87**, 441.

BALL, S., COLLINS, F. D., DALVI, P. D., and MORTON, R. A. (1949). *Biochem. J.*, **45**, 304.

BALL, S., and MORTON, R. A. (1949). *Ibid.*, **45**, 298.

BARKA, T., and ANDERSON, P. J. (1963). "Histochemistry, Theory, Practice and Bibliography". Hoeber, New York, p. 187.

BARONE, P., INFERRERA, C., and CARROZZA, G. (1969). In "Pigments in Pathology". Ed. M. Wolman, Acad. Press, N.Y. p. 307.

BARONI, B. (1933). *Archs ital. Dermatol.*, **9**, 543.

BEDNÁŘ, B., HEŘMANSKY, F., and LOJDA, Z. (1964). *Am. J. Path.*, **45**, 283.

BEER, R. J. S., BROADHURST, T., and ROBERTSON, A. (1954). *J. chem. Soc.*, 1947.

BENSLEY, R. R. (1947). *Anat. Rec.*, **98**, 609.

BERG, J. W. (1953). *J. Histochem. Cytochem.*, **1**, 436.

BILLING, B. H., COLE, P. G., and LATHE, G. H. (1957). *Biochem. J.*, **65**, 774.

BILLING, B. H., and LATHE, G. H. (1956). *Biochem. J.*, **63**, 6p.

BILLING, B. H., and LATHE, G. H. (1958). *Am. J. Med.*, **24**, 111.

BIRBECK, M. S. C., MERCER, E. H., and BARNICOT, N. A. (1956). *Expl. Cell Res.*, **10**, 505.

BIZZOZERO, E. (1908). *Münch. med. Wschr.*, **55**, 2140.

BJÖRKLUND, A., EHINGER, B., and FALCK, B. (1968). *Acta physiol. scand.*, **72**, 253

BLACKLOCK, J. W. S., FERGUSON, J. W., MACK, W. S., SHAFER, J., and SYMINGTON T. (1947). *Br. J. Surg.*, **35**, 179.

BLASCHKO, H., and WELCH, A. D. (1953). *Archs exp. Path. Pharmacol.*, **219**, **17.**

BLASCHKO, H., HAGEN, P., and WELCH, A. D. (1955). *J. Physiol.*, **129,** 27.
BORST, M., and KÖNIGSDORFER, H., JR. (1929). "Untersuchungen über Porphyrie mit besonderer Berüchsichtigung der Porphyria congenita". Leipzig.
BU'LOCK, J. D. (1960). *Archs Biochem. Biophys.*, **21,** 189.
BU'LOCK, J. H. (1961). *J. chem. Soc.*, 52.
BURT, R. C. (1952). *Am. J. Clin. Path.*, **22,** 135.
BURTNER, J., and LILLIE, R. D. (1949). *Stain Technol.*, **94,** 225.
CANNATA, M. A., CHIOCCIO, S. R., and TRAMEZZANI, J. H. (1968). *Histochemie*, **12,** 253.
CASSELMAN, W. G. B. (1951). *J. exp. Med.*, **94,** 549.
CHAMPY, C., and HATEM, S. (1955). *Bull. Histol. Tech. micr.*, **5,** 95.
CHINOY, N. J. (1963). *J. Animal Morphol. Physiol.*, **10,** 74.
CHU, T. C., and CHU, E. J.-H. (1962). *Biochem. J.*, **83,** 318.
CICHOCKI, T., and ACKERMANN, J. (1967). *Folia histochem. cytochemi.*, **5,** 145.
CLARA, M. (1942). *Z. mikrosk-anat. Forsch.*, **52,** 359.
COLE, P. G., and LATHE, G. H. (1953), *J. clin. Path.*, **6,** 99.
COLE, P. G., LATHE, G. H., and BILLING, B. H. (1954). *Biochem. J.*, **57, 514.**
COMMONER, B., TOWNSEND, J., and PAKE, G. E. (1954). *Nature, Lond.*, **174,** 689.
CORNELIUS, C. E., ARIAS, I. M., and OSBURN, B. I. (1965). *J. Am. vet. med. Ass.*, **146,** 709.
COUJARD, R., and COUJARD-CHAMPY, C. (1942). *Archs Zool. exp. gen.*, **89,** 47.
COUPLAND, R. E., and HEATH, I. D. (1961). *J. Endocrinol.*, **22,** 59.
COUPLAND, R. E., PYPER, A. S., and HOPWOOD, D. (1964). *Nature, Lond.*, **201,** 1240.
CURD, M. R. (1968). *Stain Technol.*, **43,** 42.
DADDI, C. (1933). *Riv. clin. Med.*, **34,** 78.
DALLDORF, G. J. G. (1927). *Beitr. path. Anat.*, **54,** 68.
DAM, H., and GRANADOS, H. (1945). *Science, N.Y.*, **102,** 327.
DESMET, V. J., NULLENS, A.-M., DE GROOTE, J., and HEIRWEGH, K. P. M. (1968). *J. Histochem. Cytochem.*, **16,** 419.
DOBRINER, K., and RHOADS, C. P. (1940). *Physiol. Rev.*, **20,** 416.
DREWS, G. A., and ENGEL, W. K. (1961). *J. Histochem. Cytochem.*, **9,** 206.
DUBIN, I. N. (1958). *Am. J. Med.*, **24,** 268.
DUBIN, I. N., and JOHNSON, F. B. (1954). *Medicine*, **33,** 155.
DU BUY, H. G., SHOWACRE, J. L., and HESSELBACH, M. L. (1963). *Ann. N.Y. Acad. Sci.*, **100,** 569.
DUNN, R. C. (1946). *Archs Path.*, **41,** 676.
DUNN, R. C., and THOMPSON, E. C. (1946). *Stain Technol.*, **21,** 65.
EDWARDS, J. E., and DALTON, A. J. (1942). *J. natn. Cancer Inst.*, **3,** 19.
EDWARDS, J. E., and WHITE, J. (1941). *Ibid.*, **2,** 157.
EHINEGR, B., OLIVECRONA, H., and RORSMAN, H. (1967). *Acta path. microbiol. scand.*, **69,** 179.
EHRLICH, P. (1883). *Zent. klin. Med.*, **4,** 721.
EINARSON, L., and TELFORD, I. A. (1960). *Anat. Skrift. Univ. Aarhus*, **3,** 5.
ELLINGER, P. (1938). *Biochem. J.*, **32,** 376.
ELLINGER, P. (1940). *Biol. Rev., Trans. Camb. Phil. Soc.*, **15,** 323.
ENDICOTT, K. M., and LILLIE, R. D. (1944). *Am. J. Path.*, **20,** 149.
ERÄNKÖ, O. (1951). *Acta physiol. Scand.*, **25,** Suppl. 89, p. 22.
ERÄNKÖ, O. (1952). *Acta anat.*, **16,** Suppl. 17.
ERÄNKÖ, O. (1954). *Ann. med. exp. biol. Fenn.*, **32,** 392.
ERÄNKÖ, O. (1955a). *Endocrinology*, **57,** 363.
ERÄNKÖ, O. (1955b). *Nature, Lond.*, **175,** 88.
ERÄNKÖ, O. (1955c). *Ann. med. exp. biol. Fenn.*, **33,** 278.
ERÄNKÖ, O. (1956). *J. Histochem. Cytochem.*, **4,** 11.
ESSNER, E., and NOVIKOFF, A. B. (1960). *J. ultrastruct. Res.*, **3,** 374.
FALCK, B. (1962). *Acta physiol. scand.*, **56,** Suppl. 197.
FALCK, B., JACOBSSON, S., OLIVECRONA, H., OLSEN, G., RORSMAN, H., and ROSENGREN, E. (1966a). *Acta derm. venereol. Stockh.*, **46,** 65.
FALCK, B., JACOBSSON, S., OLIVECRONA, H., and RORSMAN, H. (1965). *Science, N.Y.*, **149,** 439.

FALCK, B., JACOBSSON, S., OLIVECRONA, H., and RORSMAN, H. (1966b). *Archs Dermatol.*, **94,** 363.
FALCK, B., ROSENGREN, A. M., and ROSENGREN, E. (1966c). *Commun. Dept. Anat. Univ. Lund. No.* 5.
FARRANT, J. L. (1954). *Biochim. Biophys. Acta*, **13,** 569.
FAUVEL, P. (1899). *C. R. Acad. Sci., Paris*, **129,** 1273.
FEARON, W. R., and BOGGART, W. A. (1950). *Biochem. J.*, **46,** 62.
FEIGL, F. (1954). "Spot Tests", Vol. 2, Elsevier, Amsterdam, p. 352.
FELDBERG, W. (1940). *J. Physiol.*, **99,** 104.
FIGGE, F. H. J. (1939). *Proc. Soc. exp. Biol. Med.*, **41,** 127.
FIGGE, F. H. J. (1940). *Ibid.*, **44,** 293.
FIGGE, F. H. J. (1959). *J. Histochem. Cytochem.*, **7,** 257.
FIGGE, F. H. J., PECK, G. C., and MCCAFFERTY, R. E. (1953). *J. Histochem. Cytochem.*, **1,** 385.
FISCHER, H., and REINDEL, F. (1923). *Z. physiol. Chem.*, 127, 294.
FISHER, E. R. (1969). In "Pigments in Pathology" Ed. M. Wolman, Acad. Press, New York, p. 489.
FITZPATRICK, T. B., and LERNER, A. B. (1954). *Archs Dermatol. Syph.*, **69,** 135.
FOX, B. (1967). *J. clin. Path.*, **20,** 806.
FULLERTON, P. M. (1960). *Brit. Med. J.*, **1,** 249.
GARROD, A. E. (1909). "Inborn Errors of Metabolism". Hodder and Stoughton, London.
GEDIGK, P., and BONTKE, E. (1956). *Z. Zellforsch.*, **44,** 495.
GEDIGK, P., and FISCHER, R. (1959). *Virchows Archiv.*, **332,** 431.
GEDIGK, P., and GRIES, G. (1952). *Hoppe-Seyl. Z.*, **298,** 261.
GEDIGK, P., and STRAUSS, G. (1954). *Virchows Arch.*, **326,** 172.
GÉRARD, P., CORDIER, R., and LISON, L. (1930). *Bull. Histol. Tech. micr.*, **7,** 133.
GHISLANDI, E., GUIDOTTI, G., and CAVALCA, L. (1955). *Virchows Arch.*, **327,** 92.
GHOSH, R., and MEDDA, A. K. (1968). *Acta histochem.*, **30,** 247.
GILLMAN, J., and GILLMAN, T. (1945). *Archs Path.*, **40,** 239.
GILLMAN, J., GILLMAN, T., and BRENNER, S. (1945a). *Nature, Lond.*, **156,** 689.
GILLMAN, J., GILLMAN, T., and BRENNER, S. (1945b). *S. Afr. J. med. Sci.*, **10,** 67.
GIROUD, A., LEBLOND, C. P., and RABINOWICZ, M. (1934). *C. R. Soc. Biol., Paris*, **115,** 1088.
GLAVIND, J., GRANADOS, H., HARTMANN, S., and DAM, H. (1949). *Experientia*, **5,** 84.
GLENNER, G. G. (1957). *Am. J. clin. Path.*, **27,** 1.
GOESSNER, W. (1953). *Virchows Arch.*, **323,** 685.
GOLDFISCHER, S., VILLAVERDE, H., and FORSCHIRM, R. (1966). *J. Histochem. Cytochem.*, **14,** 641.
GOMORI, G. (1941). *Archs Path.*, **32,** 169.
GOMORI, G. (1946). *Am. J. clin. Path.*, **10,** 177.
GOMORI, G. (1951). *Modern Methods in Medical Research*, **4,** 3.
GOMORI, G. (1952). "Microscopic Histochemistry, Principles and Practice". *Univ. Chicago Press.*
GOMORI, G. (1955). *J. Histochem. Cytochem.*, **3,** 479.
GOODWIN, T. W. (1968). "Porphyrins and Related Compounds". Biochem. Soc. Symposia No. 28, Acad, Press, London.
GORTNER, R. A. (1912). *Proc. Soc. exp. Biol. Med.*, **9,** 120.
GRAFFLIN, A. L. (1942). *Am. J. Anat.*, **71,** 43.
GRANICK, S., and LEVERE, R. D. (1965). *J. cell Biol.*, **26,** 167.
GRAY, C. H., NICHOLSON, D. C., and NICOLAUS, R. A. (1968). *Nature, Lond.*, **181,** 183.
GREEN, D. E., and RICHTER, D. (1937). *Biochem. J.*, **31,** 596.
GREENSTEIN, J. P., TURNER, F. C., and JENRETTE, W. V. (1940). *J. natn. Cancer Inst.* **1,** 377.
GRESHAM, G. A., CRUICKSHANK, J. G. and VALENTINE, J. C. (1958). *Nature, Lond.*, **181,** 538.
HALL, M. J. (1960). *Am. J. clin. Path.*, **34,** 313.
HAMMES E. M., JR. (1944). *Archs Neurol. Psychiat.*, **52,** 505.

HARTROFT, W. S. (1951). *Science, N.Y.*, **113,** 673.
HEIDENREICH, O., and SIEBERT, G. (1955). *Virchows Arch.*, **327,** 112.
HELMY, F. M., and HACK, M. H. (1964). *Acta histochem.*, **18,** 194.
HENLE, J. (1865). *Z. rationelle Med.* **24,** 143.
HENSCHEN, F., and BERGSTRAND, H. (1913). *Beitr. path. Anat.*, **56,** 103.
HERSHBERGER, L. R., and LILLIE, R. D. (1947). *Bill. int. Ass. med. Mus.*, **27,** 162.
HIERONYMI, G. (1954). *Z. allg. Path. path. Anat.*, **91,** 428.
HILLARP, N-Å. (1959). *Acta. anat. Basel*, **38,** 379.
HILLARP, N-Å., and HÖKFELT, B. (1953). *Acta physiol. Scand.*, **30,** 55.
HILLARP, N-Å., and HÖKFELT, B. (1954). *Endocrinology*, **55,** 255.
HILLARP, N-Å., and HÖKFELT, B. (1955). *J. Histochem. Cytochem.*, **3,** 1.
HILLARP, N-Å., HÖKFELT, B., and NILSON, B. (1954). *Acta anat.*, **21,** 155.
HILLARP, N-Å., and NILSON, B. (1954). *Acta physiol. Scand.*, **31,** Suppl. 113, p. 79.
HIRAI, K. (1968). *Acta histochem. cytochem.*, **1,** 43.
HUECK, W. (1912). *Beitr. path. Anat.*, **54,** 68.
ISHIZAKI, Y. (1960). *Am. J. Obstet. Gynec.*, **80,** 245.
JAMES, N. T. (1968). *Nature, Lond.*, **219,** 1174.
JOHNSON, F. B., HAMILTON, P. K. and GRIDLEY, M. F. (1954). *J. Histochem. Cytochem.*, **2,** 481.
JONES, P. A. (1967). *Stain Technol.*, **42,** 1.
KARRER, P., MORF, R., and SCHÖFF, K. (1931). *Helv. chim. Acta*, **14,** 1036 and 1431.
KARRER, P., MORF, R., and SCHÖFF, K. (1933). *Ibid.*, **16,** 557.
KARRER, P., and JUCKER, C. (1950). "Carotenoids". Elsevier, New York.
KATCHBURIAN, E., DA SILVA SASSO, W., and DELLA SERRA, O. (1961). *Rev. brasil. Biol.*, **21,** 131.
KLEIN, P. D., and JOHNSON, R. M. (1954). *Archs Biochem. Biophys.*, **48,** 380.
KLÜVER, H. (1955). In "Biochemistry of the Developing Nervous System", Acad. Press, N.Y., p. 137.
KÖHL, W. (1968). *Histochemie*, **16,** 236.
KUTLÍK, I. E. (1957). *Acta Histochem.*, **4,** 141.
LA DU, B. N., and ZANNONI, V. G. (1969). In "Pigments in Pathology". Ed. M. Wolman, Acad. Press, New York, p. 465.
LAMMANNA, C. (1946). *J. Bact.*, **52,** 99.
LANGE, N. A. (1937). "Handbook of Chemistry", Handbook Publishers Inc., p. 148.
LARTIGUE, D. J., and FITE, G. L. (1962). *J. Histochem. Cytochem.*, **10,** 611.
LEA, A. J. (1945). *Nature, Lond.*, **156,** 478.
LEA, A. J. (1949). *Archs Path.*, **47,** 211.
LEE, C. S. (1950). *J. natn Cancer Inst.*, **11,** 339.
LEIBNITZ, H. L. (1964). *Acta histochem.*, **17,** 73.
LEMBERG, R., and LEGGE, J. W. (1949). "Haematin Compounds and Bile Pigments". New York.
LENDRUM, A. C. (1949). *J. Path. Bact.*, **61,** 443.
LENNOX, B. (1949). *Ibid.*, **61,** 587.
LEPEHNE, G. (1919). *Beitr. path. Anat.*, **65,** 163.
LIGNAC, G. O. E. (1923). *Virchows Arch.*, **243,** 18.
LIGNAC, G. O. E. (1945). *Proc. Koninkl. Akad. Wet. Amsterdam*, **48,** 406.
LIGNAC, G. O. E. (1954). *J. Path. Bact.*, **68,** 273.
LIGNAC, G. O. E., VAN DUIJN, P., SCHILLING, P. H. M., and ZWIJSEN, H. J. C. (1952). *J. Path. Bact.*, **64,** 675.
LILLIE, R. D. (1939). *Am. J. clin. Path.*, **15,** 225.
LILLIE, R. D. (1948). "Histopathological Technic". Blakiston, N. Y.
LILLIE, R. D. (1954). "Histopathologic Technic and Practical Histochemistry". 1st Ed., Blakiston, N.Y.
LILLIE, R. D. (1955). *J. Histochem. Cytochem.*, **3,** 453.
LILLIE, R. D. (1956a). *Stain Technol.*, **31,** 151.
LILLIE, R. D. (1956b). *J. Histochem. Cytochem.*, **4,** 377.
LILLIE, R. D. (1957a). *J. Histochem. Cytochem.*, **5,** 346.
LILLIE, R. D. (1957b). *Archs Path.*, **64,** 100.
LILLIE, R. D. (1957c). *J. Histochem. Cytochem.*, **5,** 346.

LILLIE, R. D. (1965). "Histopathologic Technic and Practical Histochemistry", 3rd Edit. McGraw-Hill, New York.
LILLIE, R. D. (1969). In "Pigments in Pathology". Ed. M. Wolman, Acad. Press, New York, p. 327.
LILLIE, R. D., ASHBURN, L. L., SEBRELL, W. H. J., DAFT, F. S., and LOWRY, J. V. (1942). *Publ. Hlth. Rep. Wash.*, **57**, 502.
LILLIE, R. D., DAFT, F. S., and SEBRELL, W. H. J. (1941). *Publ. Hlth. Rep. Wash.*, **56**, 1255.
LILLIE, R. D., GEER, J. C., and GUTIÉRREZ, A. (1964). *J. Histochem. Cytochem.*, **12**, 715.
LILLIE, R. D., GRECO HENSON, J. P., and BURTNER, H. C. J. (1957). *J. Histochem. Cytochem.*, **5**, 311.
LILLIE, R. D., and HERSHBERGER, L. R. (1947a). *Bull. int. Ass. med. Mus.*, **27**, 136.
LILLIE, R. D., and HERSHBERGER, L. R. (1947b). *Ibid.*, **27**, 145.
LILLIE, R. D., and PIZZOLATO, P. (1967). *J. Histochem. Cytochem.*, **15**, 600.
LILLIE, R. D., and PIZZOLATO, P. (1968). *J. Histochem. Cytochem.*, **16**, 17.
LILLIE, R. D., and PIZZOLATO, P. (1969a). *J. Histochem. Cytochem.*, **17**, 467.
LILLIE, R. D., and PIZZOLATO, P. (1969b). *J. Histochem. Cytochem.*, **17**, 738.
LILLIE, R. D., and PIZZOLATO, P. (1970). *J. Histochem. Cytochem.*, **18**, 75.
LISON, L. (1936). "Histochimie Animale". Gautier Villars, Paris.
LISON, L. (1938). *Beitr. path. Anat.*, **101**, 94.
LISON, L. (1953). "Histochimie et Cytochimie Animales", Gautier-Villars, Paris, p. 425.
LONGUET-HIGGINS, H. C. (1960). *Archs Biochem. Biophys.*, **86**, 231.
LUBARSCH, O., and BORCHARDT, H. (1929). In "Handbuch der Spez. path. Anat. u. Hist". Vol. 4, Berlin.
MACDONALD, R. A. (1964). "Hemochromatosis and Hemosiderosis". Thomas, Springfield.
MACDONALD, R. A. (1969). In "Pigments and Pathology". Ed. M. Wolman, Acad. Press, New York, p. 115.
MACDONALD, R. A., ENDO, H., and PECHET, G. S. (1968). *Archs Path.*, **85**, 366.
MALLORY, F. B. (1938). "Pathologic Technic". Philadelphia.
MARGOLIASH, E. (1961). *Ann. Rev. Biochem.*, **30**, 549.
MASON, H. S. (1948a). In "Biology of Melanomas", New York, p. 399.
MASON, H. S. (1948b). *J. biol. Chem.*, **172**, 83.
MASON, H. S. (1955). *Adv. Enzymol.*, **16**, 105.
MASON, H. S., INGRAM, D. J. E., and ALLEN, B. (1960). *Archs Biochem. Biophys.*, **86**, 225.
MASON, H. S., KAHLER, H., MACCARDLE, R. K., and DALTON, A. J. (1947). *Proc. Soc. exp. Biol. Med.*, **66**, 421.
MASON, J. S. (1947). *J. biol. Chem.*, **168**, 433.
MASON, K. E., HARTSOUGH, G. R. (1951). *J. Am. Vet. Med. Assoc.*, **69**, 72.
MASSON, M. P. (1914). *C. R. Acad. Sci., Paris*, **158**, 57.
MCCURDY, H. M. (1969). *J. Cell Biol.*, **43**, 220.
MCMANUS, J. F. A. (1949). *Am. J. Path.*, **25**, 800.
MCMUNN, C. A. (1885). *Quart. J. micr. Sci.*, **25**, 469.
MAEDA, R., IHARA, N., and KAINAZAWA, K. (1968). Proc. III Int. Conf. N.Y. p. 164.
MEIROWSKY, E., and FREEMAN, L. W. (1951). *J. invest. Dermatol.*, **16**, 527.
MEULON, A. (1905). Quoted by Lison, 1936.
MILLOTT, N. (1957). *Proc. zool. Soc. London*, **129**, 263.
MILLOTT, N., and OKUMURA, H. (1968). *Nature, Lond.*, **217**, 92.
MILLOTT, N., and VEVERS, H. G. (1968). *Phil. Trans. roy. Soc., London Series B*, **253**, 201.
MORITA, S., CASSENS, R. G., and BRISKEY, E. J. (1969). *Stain Technol.*, **44**, 283.
MORITA, S., CASSENS, R. G., and BRISKEY, E. J. (1970). *J. Histochem. Cytochem.*, **18**, 364.
MOYER, F. H. (1963). *Ann. N.Y. Acad. Sci.*, **100**, 584.
MUIR, R., and NIVEN, J. S. F. (1935). *J. Path. Bact.*, **41**, 183.
MUKHERJEE, M. (1966). D.Sc. Thesis. University of Calcutta.

MÜLLER, W. (1963). *Naturwiss.*, **50,** 445.
NEUMANN, E. (1888). *Virchows Archiv.*, **111,** 25.
NICOLAUS, R. A., PIATTELLI, M., and NARNI, G. (1959). Tetrahedron Letters, No. 21, 14.
NOSSLIN, B. (1960). *Scand. J. clin. Lan. Invest.*, **12,** Suppl., 49, p. 1.
NOVIKOFF, A. B., and ESSNER, E. (1960). *Am. J. Med.*, **29,** 102.
NYE, S. W., and CHITTAYASOTHORN, K. (1967). *Am. J. Path.*, **51,** 287.
O'CARRA, P. (1962). *Nature, Lond.*, **199,** 899.
OGATA, T., and OGATA, A. (1923). *Beitr. path. Anat.*, **71,** 376.
OKAMOTO, K. (1952). *Hyogo. J. med. Sci.*, **1,** 89.
OKUN, M. R. (1965). *J. invest. Derm.*, **44,** 285.
OKUN, M. R. (1967). *J. invest. Derm.*, **48,** 461.
OKUN, M. R., EDELSTEIN, L. M., OR, N., HAMADA, G., DONNELLAN, B., and LEVER, W. F. (1970). *Histochemie*, **23,** 295.
OLIVECRONA, H., and RORSMAN, H. (1966). *Acta derm. venercol., Stockh.*, **46,** 403.
OSTROW, J. D. and MURPHY, N. H. (1970). *Biochem. J.*, **120,** 311.
PAPPENHEIMER, A. M., and VICTOR, J. (1946). *Am. J. Path.*, **22,** 395.
PEARSE, A. G. E. (1948). *Nature, Lond.*, **162,** 451.
PEARSE, A. G. E. (1949). *J. Path. Bact.*, **61,** 195.
PEARSE, A. G. E. (1953). "Histochemistry, Theoretical and Applied". Vol. 1. J. & A. Churchill, London.
PEARSE, A. G. E. (1955). *Proc. 2nd Int. Cong. Neuropath. Excerpta Med., Amsterdam*, p. 601.
PEARSE, A. G. E., and ROST, F. W. D. (1968). *J. Microsc.*, **89,** 321.
PIATTELLI, M., FATTORUSSO, E., MAGNO, S., and NICHOLAUS, R. A. (1962). *Tetrahedron*, **18,** 941.
PIATTELLI, M., and NICHOLAUS, R. A. (1961). *Tetrahedron*, **15,** 66.
PICK, L. (1911). *Berl. klin. Wschr.*, **48,** 840 and 884.
PICK, L., and BRAHN, B. (1930). *Virchows Arch.*, **175,** 37.
PICKWORTH, F. A. (1934). *J. Anat.*, **69,** 62.
PIRINGER-KUCHINKA, L. (1952). *Virchows Arch.* **322,** 433.
POPPER, H. (1941). *Archs Path.*, **31,** 766.
POPPER, H. (1944). *Physiol. Rev.*, **24,** 205.
PRITCHARD, E. T., and SINGH, H. (1960). *Biochem. biophys. Res. Commun.*, **2,** 184.
PULLMAN, A., and PULLMAN, B. (1961). *Biochim. biophys. Acta*, **54,** 384.
RAIA, S. (1965). *Nature, Lond.*, **205,** 304.
RAIA, S. (1967). *Ph.D. Thesis, Univ. of London.*
RAIA, S. (1970). *J. Histochem. Cytochem.*, **18,** 153.
RAMADAN, M. A., and MICHAEL, A. I. (1969). *Acta histochem.*, **34,** 238.
RAND, R., and DI PASQUA, A. (1962). *Clin. Chem.*, **8,** 570.
REDAELLI, G. (1961). *Acad. Med.*, **76,** 113.
REDAELLI, G. (1965). *Att. Soc. ital. Patol.*, p. 817.
REEVES, R. E., and ANDERSON, R. J. (1937). *J. Am. chem. Soc.*, **59,** 858.
REINHOLD, J. C., and FOWLER, C. B. (1947). *J. biol. Chem.*, **167,** 401.
RICH, A. R. (1924). *Bull. Johns Hopk. Hosp.*, **35,** 415.
RICH, A. R., and BUMSTEAD, J. H. (1925). *Ibid.*, **36,** 225.
RICHARDS, W. C. D. (1959). *Lancet*, **1,** 683.
RICHTER, D., and BLASCHKO, H. (1937). *J. chem. Soc.*, p. 601.
RICHTER, G. W. (1957). *J. exp. Med.*, **106,** 203.
RICHTER, G. W. (1958). *J. biophys. biochem. Cytol.*, **4,** 55.
ROBINSON, K. L., and COEY, W. E. (1951). *Nature, Lond.*, **168,** 997.
ROST, F. W. D., and POLAK, J. M. (1969). *Virchows Arch. Abt. A. Path. Anat.*, **347,** 321.
ROST, F. W. D., POLAK, J. M. and PEARSE, A. G. E. (1969). *Virchows Arch. Abt. B. Zellpath.*, **4,** 93.
ROULET, F. (1948). "Methoden der Pathologischen Histologie". Vienna, p. 234.
RUDDELL, C. L. (1969). *Histochemie*, **19,** 319.
SACCHI, S. (1962). *G. ital. Derm.*, **1,** 3.
SANTINI, D. (1964). *Anthologica med. santoriana*, **73,** 3.

SAWADA, T., HARA, K., TAKAGI, K., NAGAZAWA, Y., and OKA, S. (1956). *Am. J. trop. Med. Hyg.*, **5**, 847.
SCHILLINGER, H. (1967). *Med. Klinik*, **62**, 161.
SCHMIDT, R. (1953). *Virchows Arch.*, **323**, 123.
SCHORNAGEL, H. E. (1956). *J. Path. Bact.*, **72**, 267.
SEGGEL, K. A. (1940). *Ergebn. inn. Med. Kinderh.*, **58**, 582.
SEIJI, M., FITZPATRICK, T. B., and BIRBECK, M. S. C. (1961). *J. invest. Derm.*, **36**, 343.
SEIJI, M., FITZPATRICK, T. B., SIMPSON, R. T., and BIRBECK, M. S. C. (1963). *Nature, Lond.*, **197**, 1082.
SEVKI, K. (1934). *Virchows Arch.*, **294**, 65.
SIEBERT, G., DIEZEL, P. B., JAHR, K., KRUG, E., SCHMITT, A., GRÜNBERGER, E. and BOTTKE, I. (1962). *Histochemie*, **3**, 17.
SIZER, I. W. (1947). *J. biol. Chem.*, **169**, 303.
SIZER, I. W. (1948). *Science, N.Y.*, **108**, 335.
SJÖSTRAND, F. (1946). *Nature, Lond.*, **157**, 348.
SPEARE, G. S. (1951). *Am. J. Surg.*, **82**, 631.
STAMMLER, A. (1959). *Virchows Arch.*, **332**, 347.
STEENBOCK, H., SELL, M. T., NELSON, E. M., and BUELL, M. V. (1921). *J. biol. Chem.*, **46**, xxxii.
STEIN, J. (1935). *C. R. Soc. Biol., Paris*, **120**, 1136.
STEIN, W. D. (1954). *Nature, Lond.*, **174**, 601.
STEIN, W. D. (1955). *Ibid.*, **175**, 256.
STEWART, M. J., and HICKMAN, E. M. (1931). *J. Path. Bact.*, **34**, 61.
STRASSMANN, G. (1944). *Archs Path.*, **38**, 76.
STRASSMANN, G. (1949). *Ibid.*, **47**, 205.
STURGEON, P., and SHODEN, A. (1969). In "Pigments in Pathology". Ed. M. Wolman. Acad. Press, New York, p. 93.
SWAN, G. A. (1963). *Ann. N.Y. Acad. Sci.*, **100**, 1005.
TAFT, E. B. (1949). *Nature, Lond.*, **164**, 1133.
TAPPEL, A. L. (1955). *Archs Biochem. Biophys.*, **54**, 266.
THIELE, E. H., and HUFF, J. W. (1960). *Arch. Biochem.*, **88**, 203.
THOMSON, R. H. (1962). In "Comparative Biochemistry", Vol. 3, "Quinones, Structure and Distribution". Eds. Florkin, M., and Mason, H. S., Acad. Press, New York.
TIEDEMANN, F., and GMELIN, L. (1826). "Die Verdauung nach Versuchen". K. Gross, Heidelberg and Leipzig, Vol. I, p. 89.
TOFFLER, A. H., HUKILL, P. B., and SPIRO, H. M. (1963). *Ann. intern. Med.*, **58**, 872.
TRAMEZZANI, J. H., CHIOCCHIO, S., and WASSERMANN, G. F. (1964a). *Acta physiol. latinoamer.*, **14**, 122.
TRAMEZZANI, J. H., CHIOCCHIO, S., and WASSERMANN, G. F. (1964b). *J. Histochem. Cytochem.*, **12**, 890.
TVERDY, G., FROEHLICH, A. L., and FIERENS, B. (1949). *Acta gastroenterol. belg.*, **12**, 221.
VAN DEN BERGH, A. A. H., and MÜLLER, P. (1916). *Biochem. Z.*, **77**, 90,
VAN DUIJN, P. (1952). *Rec. Trav. chim. Pays-Bas.*, **71**, 585.
VAN ROY, F., and HEIRWEGH, K. P. M. (1968). *Biochem. J.*, **107**, 507.
VAN WOERT, M. H. PRASAD, K. N., and BORG, D. C. (1967). *J. Neurochem.*, **14**, 707.
VERNE, J. (1923). *Bull. Soc. Chim. biol.*, **5**, 227.
VIRCHOW, R. (1847). *Virchows Arch.*, **1**, 379.
VIRCHOW, R. (1866). *Arch. path. anat. physiol.*, **37**, 212.
VON RECKLINGHAUSEN, R. (1899). *Versamml. Gesell. Natürf. Arzte.*, **62**, 324.
WALD, G. (1935). *J. gen. Physiol.*, **19**, 351.
WALDROP, F. S., PUCHTLER, H., and TERRY, M. S. (1969). *Stain Technol.*, **44**, 279.
WALKER, D., PIZZOLATO, P., and LILLIE, R. D. (1970). *J. Histochem. Cytochem.*, **18**, 367.
WEBER, G. (1954). *Arch. de Vecchi*, **20**, 913.
WELLINGS, S. R., and SIEGEL, B. V. (1959). *J. ultrastruct. Res.*, **3**, 147.
WELLINGS, S. R., and SIEGEL, B. V. (1963). *Ann. N.Y. Acad. Sci.*, **100**, 548.

WELLS, C. L., and WOLKEN, J. J. (1962). *Nature, Lond.*, **193,** 977.
WILLIAMS, W. L., and ARONSOHN, R. B. (1953). *Fed. Proc.*, **12,** 157.
WILLSTAEDT, H., and LINDQUIST, T. (1936). *Z. physiol. Chem.*, **240,** 10.
WINCKLER, J., and TURNER, H.-B., (1969). *Histochemie*, **19,** 272.
WINNAIL, D. S., and WILLIAMS, W. L. (1968). *Anat. Rec.*, **160,** 759.
WOLF, A., and PAPPENHEIMER, A. M. (1945). *J. Neuropath. exp. Neurol.*, **4,** 402.
WOLMAN, M., and SHOSHAN, S. (1960). *Histochemie*, **2,** 69.
WOLMAN, M., and ZAIDEL, L. (1962). *Experientia*, **18,** 323.
WOOD, E. M., and YASUTAKI, W. T. (1956). *Am. J. Path.*, **32,** 591.
WOOD, J. G., and BARRNETT, R. J. (1963). *Anat. Rec.*, **145,** 301.
WOOD, J. G., and BARRNETT, R. J. (1964). *J. Histochem. Cytochem.*, **12,** 197.
ZANNONI, V. G., LOMTEVAS, N., and GOLDFINGER, S. (1969). *Biochim. biophys. Acta*, **177,** 94.
ZECHMEISTER, L., and TUZSON, P. (1936). *Z. physiol. Chem.*, **239,** 147.

CHAPTER 27

BIOGENIC AMINES

Introduction

FORMERLY included under the heading of Pigments, with which many of them were closely associated as precursors, the biogenic amines are now considered in a separate chapter. A degree of overlap with Chapter 26 is inevitable, as when the condensation products, or the oxidized condensation products, of an amine are themselves pigments. The enormous expansion which has taken place, since 1962, in the applied histochemistry of the biogenic amines is due solely to the development of the hot formaldehyde vapour technique, by Falck and his associates (Falck, 1962; Falck *et al.*, 1962).

The bibliography of formaldehyde-induced fluorescence (FIF; Eränkö, 1967) is already so vast as to require a specialized volume of its own. In this chapter, therefore, I propose to deal very fully with the chemistry and technology of the FIF method, and of other specific methods for individual amines. On the other hand, only selected examples of FIF will be considered. These, but not the foregoing text, will be biased by my own particular interests.

The list of endogenous amines demonstrable by FIF, or by alternative condensation methods, is not a closed one. Undoubtedly many presently unidentified amines are detected. Well characterized amines which are included here are 5-hydroxytryptamine (5-HT), tryptamine, tyramine, noradrenalin (NA), adrenalin (A), dopamine (DA), leptodactyline and histamine. A number of these amines have been demonstrated successfully after their administration and uptake by cells or tissues. Two amines can be demonstrated by FIF after administration and uptake in the form of the precursor amino acid (DOPA or 5-hydroxytryptophan) followed by decarboxylation, and storage in the form of the amine (dopamine, or 5-HT). "Uptake and Decarboxylation" will be dealt with in a separate section (see p. 1111). Over 25 compounds are known to give FIF in tissues or models. All, except histamine and 5-hydroxyindole acetic acid, are aryl ethylamines.

Early History of FIF

All the early work, leading up to the concept of formaldehyde-induced fluorescence, was centred on the granules of the Kultschitsky cells. In the first edition of this book these cells were described as argentaffin cells. This title was changed, in the second edition, in favour of the term enterochromaffin (E.C.) cells largely in acknowledgement of the great contributions towards an understanding of their functions made by Erspamer, Vialli, and others of the Italian school. A very considerable volume of work by Erspamer and his associates culminated in the identification by Erspamer and Asero (1952) of the "hormone of the enterochromaffin cells" as 5-hydroxytryptamine. This they obtained in the form of its picrate from extracts of the salivary glands of *Octopus vulgaris* (which contain chromaffin cells). The formula of 5-hydroxytryptamine is given below:

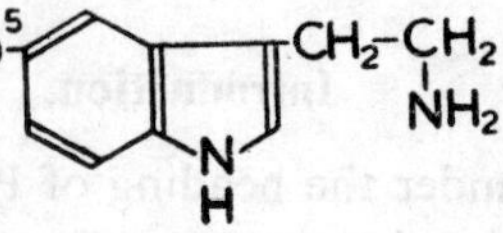

It is seen to be a derivative of 3(β-aminoethyl) indole, (tryptamine), the latter being produced from tryptophan in the body in the same way as tyramine is produced from tyrosine. This hormone the authors called enteramine. The history and histology of the E.C. cells was comprehensively reviewed by Vialli (1965) in the Handbook of Experimental Pharmacology and his chapter in this work should be consulted by all who are interested in the function of the cells and the nature of their granules. The first mention of the yellow autofluorescence shown by the E.C. cells when formalin-fixed tissues are exposed to ultra-violet light has long been attributed to Erös (1932). This author ascribed the fluorescence which he noted to autofluorescent lipids and it was probably these (lipofuscins) which he was in fact observing. A green fluorescence was observed in some of the cells in the adrenal medulla, following formalin fixation, by Eränkö (1951, 1952). These observations led to the development of formaldehyde-condensation as a histochemical technique for noradrenalin (Eränkö, 1955).

Seeking identification of the fluorescence of the E.C. cells, wrongly attributed to Erös, Jacobson (1939) measured the fluorescence spectra and ultra-violet absorption spectra of sections of a human carcinoid tumour. The fluorescence spectra were shown to have two bands, one with a peak at 610 nm, the other at 550 nm. The U.V. absorption maximum was at 270 nm. In a second paper (1946) this author claimed that the fluorescence spectrum of an ethylene glycol monoethyl ether extract of a carcinoid tumour was almost the same as that given by a solution of xanthopterin in this solvent. From his spectrographic results and those of certain histochemical tests (notably the murexide test) Jacobson suggested that the granules contained a pteridine compound. Xanthopterin and leucopterin, the yellow and white pigments of insects, are the two most important members of this class. Gomori (1948), however, tested 2-amino-4-hydroxypteridine-6-carboxylic acid by the five major histochemical reactions for E.C. granules and in no case did he obtain a positive result. Personally, I could not succeed in demonstrating a positive murexide test in the granules from mammalian sources and Jacobson's views were never substantiated.

The fluorescence of freeze-dried rat duodenal E.C. cells was obtained by Barter and Pearse (1953, 1955) only after exposure of the sections to formaldehyde vapour. Substances suggested by various authors as possible precursors of the fluorescent compound were tested *in vitro*. Of these only 5-HT was found to be non-fluorescent before and fluorescent after treatment with formaldehyde. On account of their findings Barter and Pearse first postulated that a harmaline derivative was present. Later they proposed that a fully conjugated β-carboline (Jepson and Stevens, 1953) was the type of substance actually present in formalin-fixed E.C. cells and that this was therefore responsible for the classical reactions of the granules. This was elaborated in the 2nd Edition of this book (1960) by the suggestion that a Mannich reaction took place with the active hydrogen atom in position 2, to form a cyclic compound (I) and that this could subsequently oxidize to the fully conjugated β-carboline derivative (II).

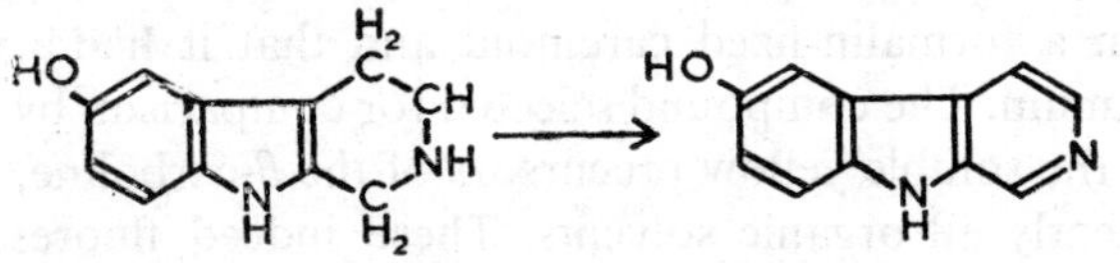

I II

This relatively simple theory was complicated by the fact that active hydrogen atoms are present also at positions 4 and 6 in the 5-HT molecule. Opportunity for the occurrence of polymers was thus considered to be present. Possibly these are actually formed *in vitro* and even perhaps in E.C. cells subjected to treatment with formaldehyde. On the other hand, in the case of the E.C. cells, the reaction of intermediates with protein might well occur.

The ninhydrin reaction. A specific test for tryptamines unsubstituted at position 2 and either of the nitrogen atoms was described by Jepson and Stevens (1953). A bright blue-green fluorescence is given by 5-hydroxytryptamine, on filter paper, after treatment with 0·2 per cent ninhydrin in acetone containing 10 per cent v/v acetic acid, drying and heating to 100° for 2-3 minutes. This reaction, with freeze-dried sections of duodenum, gave negative results. These could have been due to quenching effects from the blue ninhydrin-amino-acid reaction product but more probably they were due to the fact that adsorption of the compound on filter paper is necessary for a positive reaction to occur.

There were some considerable objections to the hypothesis of Barter and Pearse (1955). The first of these was that the *in vitro* condensation of 5-HT with HCHO is slow and markedly pH dependent. In the tissues it appeared to be more rapid and relatively independent of pH. I believed (1960) that this discrepancy would ultimately be resolved when the nature of the structure responsible for binding the product within the E.C. cell was known. The role of associated protein was subsequently elucidated by the work of Corrodi and Jonsson (1965a), which is described in a succeeding section of this chapter.

A second objection was raised by the work of Christie (1955a). This author stated that in unfixed sections of freeze-dried guinea-pig duodenum the E.C. cells fluoresced brightly yellow when irradiated by UVL at 275 nm. The effect of formalin was to diminish the colour and to change it in the direction of orange. Repeated attempts to reproduce these findings in the guinea-pig were entirely without success. The activation maximum for 5-HT was recorded as 295 nm (Bowman *et al.*, 1955) and its fluorescence maximum was said to be 330 nm except in strongly acid solution (Udenfriend *et al.*, 1955) where 5-HT and other 5-hydroxyindoles fluoresced at 550 nm. It was clear, therefore, that if a yellow fluorescent compound was present in unfixed E.C. cells it was not 5-HT.

A third objection to the hypothesis of Barter and Pearse came from further studies by Jacobson (1958) who claimed to have shown that the fluorescence of the 5-HT-formaldehyde compound was much weaker than that of the material extracted from a formalin-fixed carcinoid and that it had a slightly lower emission maximum. The compound selected for comparison by Jacobson was clearly one of the soluble yellow precursors of the β-carboline, which itself is insoluble in nearly all organic solvents. These indeed fluoresce weakly by comparison with the final product. The investigations of Corrodi and Jonsson (1965b) finally disposed of any remaining objections to the β-carboline hypothesis and the way was made clear for more important studies on the role of 5-HT in the E.C. cells.

Chemistry of FIF

The original studies of Corrodi and Hillarp (1963, 1964) were concerned solely with the fluorescent products of primary and secondary catecholamines, when these were treated in a dried protein layer with formaldehyde gas. These, and subsequent, studies (Corrodi and Jonsson, 1965a and b; Corrodi *et al.*, 1966) showed that when a primary catecholamine such as DA or NA (I) is so treated a very rapid condensation takes place, leading to the formation of a 6,7-dihydroxy-1,2,3,4-tetrahydroisoquinoline (II).

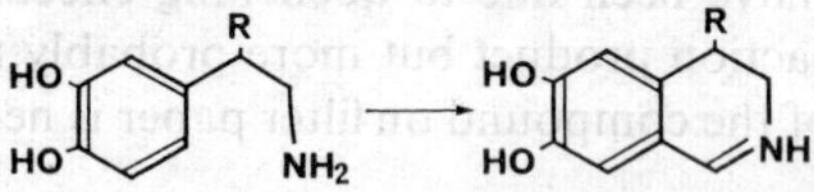

I II

R is H or OH, respectively, for dopamine and noradrenalin.

This is known in organic chemistry as a Pictet-Spengler reaction. A further stage is necessary for the development of a fluorescent product. In this stage the tetrahydroisoquinoline is dehydrogenated to the corresponding 3,4-dihydroisoquinoline in a reaction which is facilitated by the presence of protein.

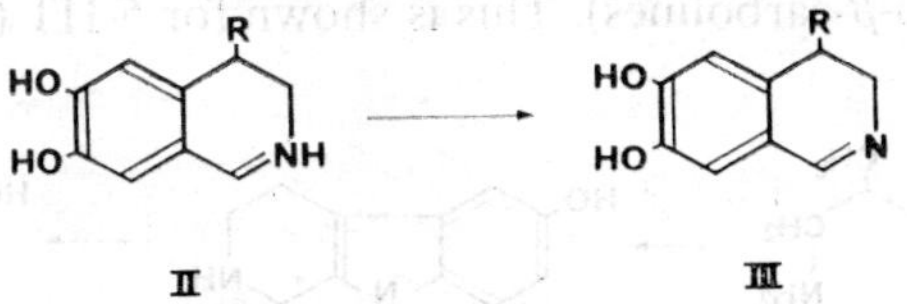

II III

The dihydroisoquinolines (III) exist in a pH-dependent equilibrium with their tautomeric quinoidal form (IV) which is the compound responsible for the fluorescence observed under histochemical conditions (Emax, 480 nm).

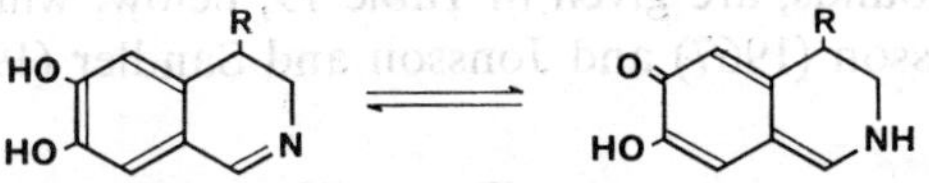

III IV

Secondary catecholamines such as adrenalin (A), when enclosed in a protein layer react with formaldehyde in an essentially similar manner but the dehydrogenation step (II to III) requires much more severe reaction conditions (Jonsson, 1967). According to Corrodi and Hillarp (1964) this is due to the formation of a quaternary 3,4-dihydroisoquinoline.

The effect of pH on the intensity of fluorescence given by 3,4-dihydroisoquinolines was measured by Corrodi *et al.* (1966) and their findings, in the case of the NA derivative, are shown in Fig. 227, below.

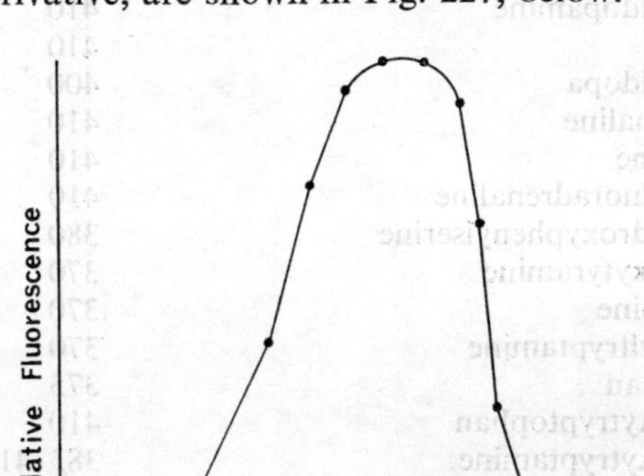

FIG. 227. Effect of pH on fluorescence intensity of the condensation product of noradrenalin and formaldehyde in dried protein.

Using methods essentially similar to those described above Corrodi and Jonsson (1965b) confirmed that tryptamines such as 5-HT, in association with protein, could undergo essentially similar condensation reactions with the production of fluorescent 3,4-dihydro-β-carbolines by way of intermediates

(1,2,3,4-tetrahydro-β-carbolines). This is shown for 5-HT (I) in the equations below:

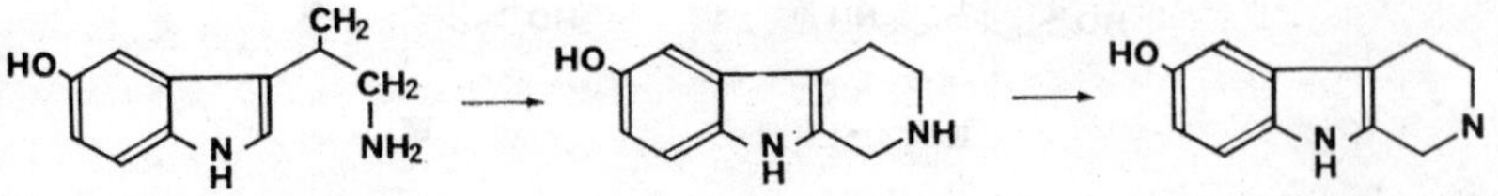

I II III

The fluorescence characteristics of a number of biogenic amines, precursors and related compounds, are given in Table 79, below, which is substantially derived from Jonsson (1967) and Jonsson and Sandler (1969).

TABLE 79

Formaldehyde-induced Fluorescence Characteristics of Amines and Precursors

No.	Compound giving FIF	Excitation max. (nm.)	Emission max. (nm.)
1	m-tyramine	385	415/510
2	m-hydroxyamphetamine	385	415/510
3	metaraminol	385	415/510
4	dopamine	410	480
5	α-methyldopamine	410	480
6	DOPA	410	490
7	α-methyldopa	400	480
8	noradrenaline	410	480
9	adrenaline	410	480
10	α-methylnoradrenaline	410	480
11	3,4-dihydroxyphenylserine	380	470
12	3-methoxytyramine	370	470
13	Tryptamine	370	490
14	N-methyltryptamine	370	490
15	tryptophan	375	490
16	5-hydroxytryptophan	410	525
17	5-hydroxytryptamine	385–410	520–540
18	6-hydroxytryptamine	400	505
19	5,6-dihydroxytryptamine	400	500
20	5-methoxytryptamine	380	525
21	α-methyl-5-hydroxytryptamine	410	520

These results indicate that the formaldehyde condensation products of all the catecholamines, their α-methylated analogues, and the corresponding amino acids, in dried protein layers, have almost identical fluorescence characteristics. The tryptamines and corresponding amino acids showed a relatively weak fluorescence by comparison with that of the catecholamines, after identical treatment. The β-carboline products were found, moreover, to be very sensitive to irradiation at the maximal activation wavelength, by comparison with catecholamine products. The fading characteristics of the β-carbolines in tissue components seem to be variable and Jonsson's (1967) suggestion that these can be used as a means of distinction is not acceptable. His alternative (microspectrofluorimetry), combined with automatic recording of fading (Pearse and Rost, 1969), is the only acceptable alternative.

Alternatives to Formaldehyde

Fluorophores similar to those produced by formaldehyde can be obtained by treating tissues with other aldehydes, or with carboxy acids. Isoquinolines were first synthesized from phenylethylamines by condensation with a carboxy acid and dehydration (Bischler and Napieralski, 1893) and only later by condensation with aldehydes, such as formaldehyde (Pictet and Spengler, 1911). The possibility of using acetaldehyde as an alternative to formaldehyde was indicated by Corrodi and Hillarp (1964). The histochemical application of the Bischler-Napieralski syntheses was investigated by Rost and Ewen (1971), who exposed freeze-dried tissues to vapours of acetaldehyde, glutaraldehyde, and formic and acetic acids. Vapours of acetaldehyde, glutaraldehyde, and acetic acid were found to be capable of forming fluorophores with substances giving FIF. In some cases, acetaldehyde and glutaraldehyde vapours were more effective than formaldehyde. These techniques are described in Appendix 27, p. 1400.

Fluorescence Techniques for Arylethylamines

Formalin Techniques. The procedure for the demonstration of NA which was developed by Eränkö (1955) is now largely of historical interest. Nevertheless, in order to maintain proper historical perspective, details are given in Appendix 27 and the result is illustrated in Fig. 222, p. 1055.

Freeze-dry, Hot Formaldehyde Vapour Techniques. These are all elaborations of the hot vapour FIF method of Falck *et al.* (1962). Many, or perhaps most, of the modifications were introduced by Falck and his associates but the standard procedure remains that of Falck and Owman (1965). Details of this are given in Appendix 27. The method involves freeze-drying (see Chapter 3), followed by treatment of the block with formaldehyde gas, at 60°, 70°, or 80°. The formaldehyde is derived from paraformaldehyde powder which releases the gas when heated.

Humidity Control. A considerable mystique has grown up around procedures developed to ensure that the correct degree of humidity is present in the warm chamber containing the paraformaldehyde. Excess humidity causes diffusion of the fluorescent product. If the level of water vapour is too low the condensation reactions take place with reduced efficiency. The critical effect of the water content of the paraformaldehyde powder (the sole source other than ambient air) was stressed by Falck and Owman (1965), who developed a procedure by which a standard quantity of paraformaldehyde (5 g) was enclosed with the freeze-dried blocks in a standard volume container (1 litre). Before use the powder was equilibrated for 5–7 days (later changed to 10 days, as recommended by Hamberger *et al.*, 1965, and Hamberger, 1967) over sulphuric acid at a relative humidity of 50–70 per cent, at 22° in a desiccator. This humidity was achieved with about 500 g/litre conc. H_2SO_4 (see Table 94, Appendix 27). The uptake of water by dried paraformaldehyde, under these conditions, was followed by Hamberger (1967) who showed that equilibrium was not reached until after 25 days, as shown in Fig. 228.

There is no doubt that the optimal conditions of humidity, temperature and time of exposure vary for different tissues, different species and different arylethylamines. In countries where the mean humidity is around 65 per cent it is possible for most purposes to use paraformaldehyde directly from the

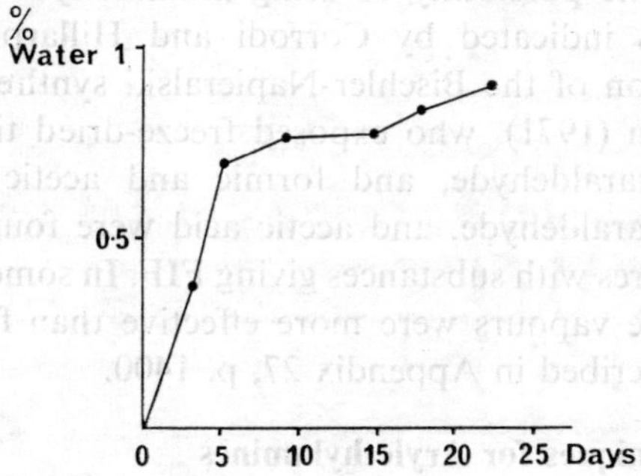

FIG. 228. Water content of paraformaldehyde equilibrated at 70 per cent relative humidity and 21–24°.

bottle (freshly for each exposure). Measurements made, with the apparatus illustrated in Vol. 1, p. 51, indicated that with the closed chamber at 60–70° the mean humidity of 28 per cent could be raised or lowered only with considerable effort. Results achieved at this level were totally satisfactory for most mammalian tissues. The temperature of 80° used by Falck and Owman (1965) is not required, except for the demonstration of adrenalin and for some other amines in the nervous system. For countries where the mean humidity is too low, or too high, to provide optimal chamber humidity, equilibration of paraformaldehyde must be carried out. The "rule of 110", suggested by Eränkö (1967) can be employed with advantage. This rule states that the sum of the percentage relative humidity of the *air with which the powder has been equilibrated*, added to the temperature of exposure in degrees Centigrade, should be 110.

A variation developed by Fuxe and Jonsson (1967) involved treatment of the dried blocks first with HCHO vapour at low humidity (70 per cent equilibrated paraformaldehyde) for an hour, followed by HCHO vapour at

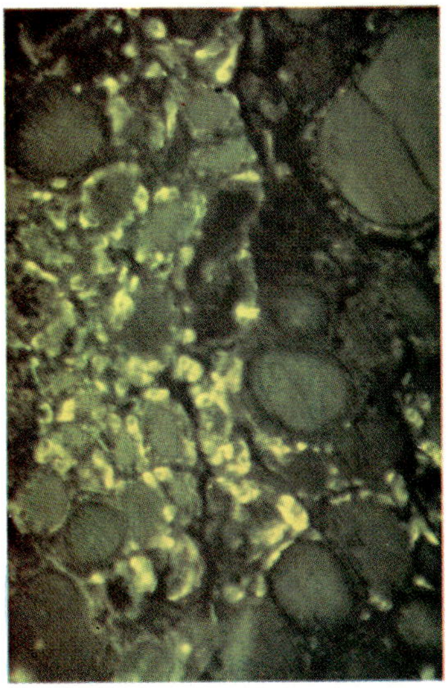

XXVa. Mouse thyroid gland. Prepared by freeze-drying and hot formaldehyde vapour fixation to demonstrate FIF. Animal killed 4 hours after receiving an injection of L-5-Hydroxytryptophan. Shows yellow (5-HT) fluorescence in the C cells and green (noradrenalin) fluorescence in adrenergic nerve endings. × 170.

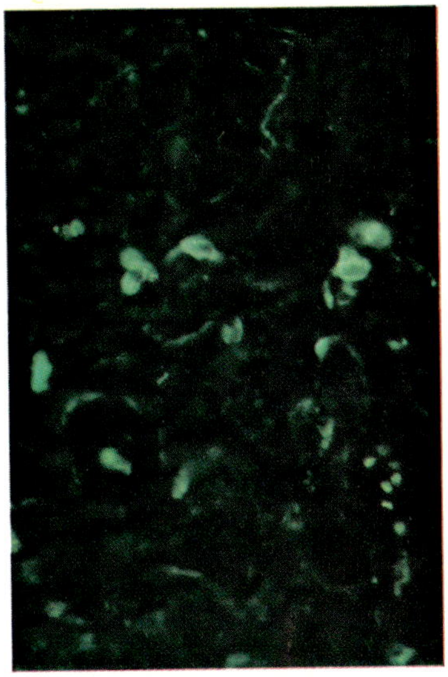

XXVb. Tree-shrew (*Tupaia glis*) thyroid gland. Prepared as XXVa, two hours after injection of L-DOPA. Shows green (dopamine) fluorescence in the C cells. × 256.

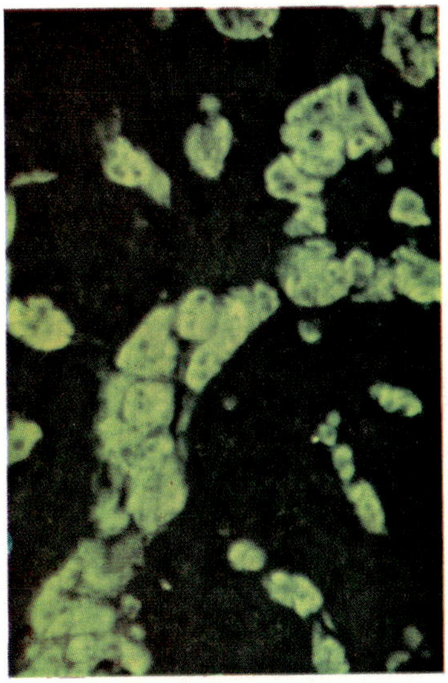

XXVc. Chick adrenal. Prepared as XXVa and b, two hours after injection of L-DOPA. Shows intense yellow fluorescence in both adrenalin and noradrenalin cells in the medulla. × 170.

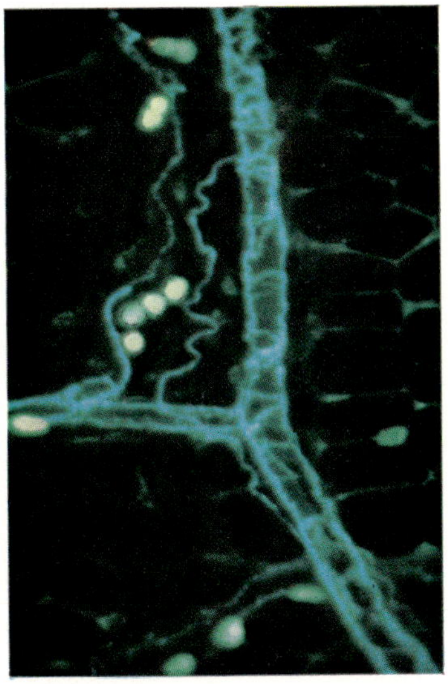

XXVd. Rat mesentery (Stretch preparation). Freeze-dried and treated with hot formaldehyde vapour to demonstrate FIF. Use of the Ploem illuminator gives true colour rendering of 5-HT FIF (yellow) in mast cells and NA-FIF (blue-green) in adrenergic endings. × 150.

[*To face p.* 1108

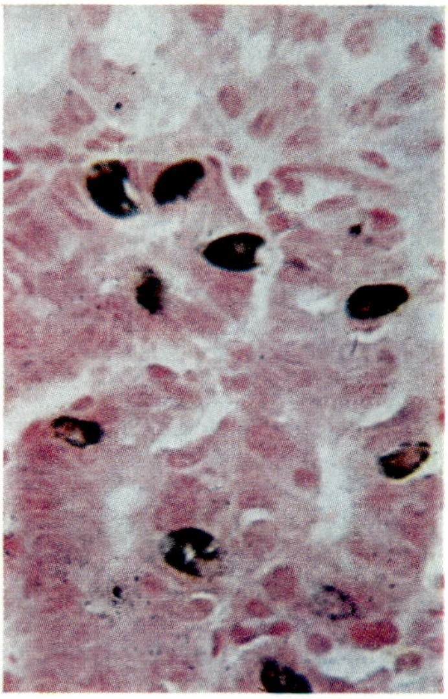

XXVIa. Guinea-pig duodenum. Formaldehyde-fixed cryostat section. Enterochromaffin cells reduce alkaline silver solutions (Masson–Fontana method) and appear black. Counterstained with carmalum. × 790.

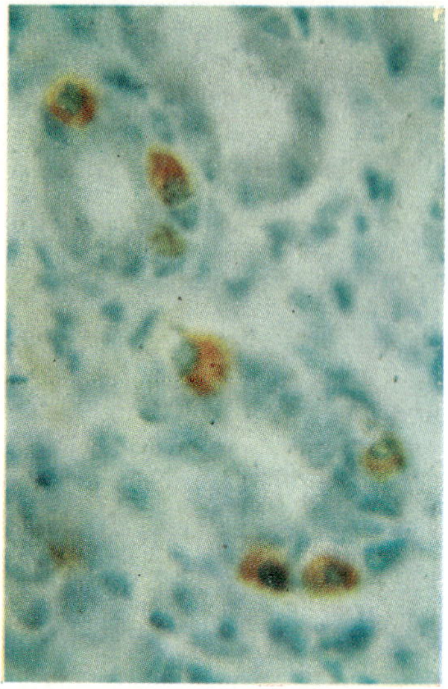

XXVIb. Guinea-pig duodenum. Formaldehyde-fixed cryostat section. Diazo reaction (hexazotized pararosanilin) shows yellow EC cells. Nuclei blue (haemalum) × 1100.

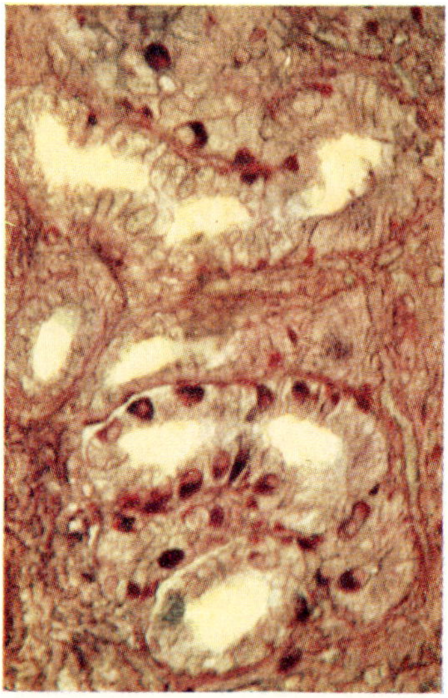

XXVIc. Human duodenum. Fixed in glutaraldehyde-picric acid. Acid hydrolysis for 1 hour at 60° in 0·1 N-HCl, followed by toluidine blue (pH 5·0). Masked metachromasia of endocrine polypeptide (APUD) cells in basal glands.

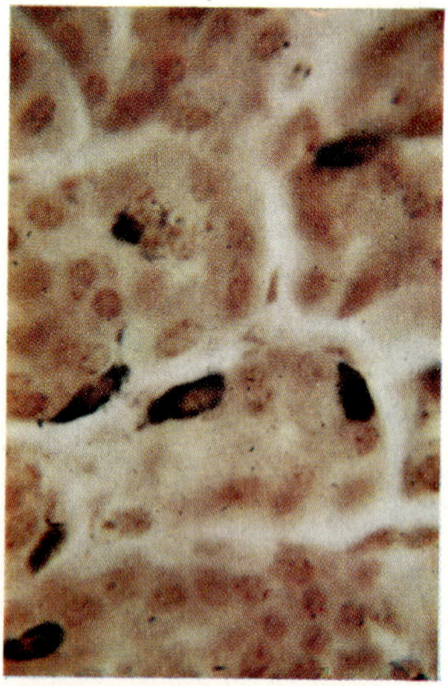

XXVId. Mouse stomach (Cardia). Formaldehyde-fixed paraffin section. Bodian silver reaction shows argyrophil endocrine polypeptide cells in the glands. × 920.

high humidity (97·5 per cent equilibrated paraformaldehyde) for 1–2 hours. This procedure was claimed to produce improved results, especially in nerve terminals containing 5-HT.

Embedding, Sectioning and Mounting. Normally, vacuum embedding in degassed paraffin wax (48–52°) is employed although, if thin sections are required, epoxy resins can be used as for electron microscopy. The solubility of some fluorophores, such as the adrenalin derivative, in xylene prevents the use of conventional embedding techniques. From certain fibre structures fluorophores are also extracted by hot paraffin wax, according to Björklund and Falck (1968), who developed a modified technique to reduce exposure to hot wax to a minimum. Details are given in Appendix 27.

Mounting is usually performed by covering the sections, mounted on warm slides, with liquid paraffin. Further warming dissolves the wax. Xylene and Entellan (Merck) are used as alternatives.

The results of the standard FIF procedure are shown in Plate XXV p. 1108.

Fresh Frozen Section Techniques. The demand for simultaneous demonstration of adrenergic fibres (by FIF) and cholinergic fibres (by ChEsterase techniques) produced a number of methods using fresh sections (Hamberger and Norburg, 1964). Cryostat sections are either mounted on slides, and immediately treated with hot HCHO vapour, or first allowed to dry for several hours in the cryostat at −20°. In a technique developed by Heene (1968) cryostat sections were freeze-dried in a special container mounted on the top module of a thermoelectric dryer. Subsequently they were exposed to vapour from *dry* paraformaldehyde at 80°.

A modification introduced by Laties *et al.* (1967) employed cryostat sections of perfusion-fixed tissues. Details are given in Appendix 27, p. 1395. A fresh slice technique developed by Ehinger *et al.* (1969) is also given in the Appendix.

Distinction Techniques. It is often necessary to distinguish, in a single section, between two closely associated fluorophores or, more simply, to identify one fluorophore from another having similar fluorescence characteristics.

Filters. The simplest methods are those based on selection of suitable filters. Distinction between catecholamine fluorescence and 5-HT fluorescence is often quite easily made by eye, using standard colour filter systems (Plate XXVa and b). The eye responds to different fluorescence intensities, however, with an altered colour sense (intense DA or NA fluorescence, which should be green, appears yellow; Plate XXVc). It may be necessary, therefore, to use interference filters, with a sharp cut-off at specific wavelengths, to provide narrow band irradiation of the fluorophores (see Chapter 29, p. 1199). These procedures have been described by Angelakos (1964) and, more extensively, by Ploem (1969). The type of result obtainable is shown in Plate XXVd.

Microspectrofluorimetry After HCl Treatment. The provision of excitation and emission spectra of fluorophores by analytical microspectrofluorimetry provides valuable information but cannot, of course, distinguish fluorophores with identical characteristics such as NA and DA. *In vitro* experiments carried out by Corrodi and Jonsson (1965) showed that vigorous treatment with HCl or $SOCl_2$ could convert the 4,6,7-trihydroxy-3,4-dihydroisoquinoline from NA (II) into a 6,7-dihydroxyisoquinoline (III). The 6,7-dihydroxy-3,4-dihydroisoquinoline from DA (I) cannot react in this way.

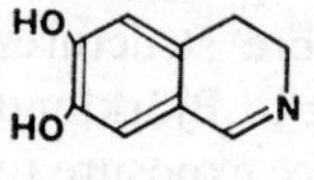

I

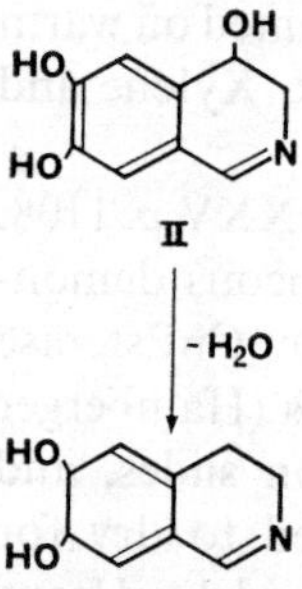

II → III

The fluorophore (III) has spectral characteristics different from those of compounds I & II. Thus, by HCl treatment, compound II (excit. max. 370 nm) is converted to compound III (excit. max. 330 nm). The DA fluorophore (I) remains unaffected so that a progressive change in excitation maximum from 370 nm to 330 nm indicates the presence of NA. The excitation maximum of the fluorophore from 5-hydroxydopamine changes from 380 nm to 345 nm after acid treatment (Ehinger and Falck, 1969).

The original *in vitro* results have been confirmed for tissue components by Björklund *et al.* (1968a and b) and in my own laboratory using the instrument described by Pearse and Rost (1969). Treatment with thionyl chloride has not been used successfully on tissue sections. Details of the HCl procedure, which requires considerably less rigorous conditions than originally suggested, are given in Appendix 27.

Borohydride Reduction. As a means of distinguishing the fluorescence of specific catecholamine or 5-HT fluorophores from non-specific autofluorescence in tissue sections Corrodi *et al.* (1964) introduced the technique of borohydride reduction. This reagent rapidly reduces fluorescent 3,4-dihydroisoquinolines and β-carbolines to the equivalent non-fluorescent tetrahydro-iso-derivatives. The original fluorescence is restored by a further period of treatment with hot HCHO vapour. Borohydride reduction was investigated by Vialli and Fantin (1966) who found that the fluorophores in E.C. cells of *Octopus vulgaris* and *Eledone moschata* and in the chromaffin cells of *Calliactus parasitica* were not always completely reduced by the process.

A modification of the original borohydride technique was suggested by Mukherji *et al.* (1966) who found that periodic acid could cleave the C-C bond at the 3,4-position of the NA fluorophore (Formula II, above) but had no effect on those of A or DA which lack the —CH.OH — CH_2NH grouping. The authors proposed the borohydride-periodic acid sequence as a method of distinguishing NA from DA. The fluorophore of the latter was restored fully by HCHO vapour after borohydride reduction and periodic acid oxidation while that of NA was not. Details are given in Appendix 27.

Augmentation of Fluorophores

As already stated, the FIF technique can be used for the demonstration of exogenous fluorogenic amines, administered in various ways. An important tool for increasing the demonstrability of a variety of tissue components is the augmentation of a feeble or absent FIF by causing the particular component involved to take up either its normal amine, or some other. This process can take place *in vivo* or *in vitro*. As an alternative, accumulation of the endogenous product may be brought about in some other manner. Dahlström and Fuxe (1964), for instance, showed that crushing of adrenergic nerves brought about an accumulation of NA on the side of the neuron above the site of crushing. Other workers have produced increased levels of normal fluorogenic amines by administration of specific MAO inhibitors such as malamide (Carlsson, 1964; Fuxe, 1965). These techniques are particularly important in studies of nerve fibres and of the central nervous system (see Clarke *et al.*, 1969; Ehinger and Falck, 1970; Sachs, 1970; Jonsson and Sachs, 1970).

Precursor Uptake and Decarboxylation

A different process is involved in the uptake by different cells of the amino acid precursors of DA and 5-HT. Both L and D-isomers of DOPA and 5-HTP have been employed, either by injection or by an *in vitro* technique first described by Håkanson *et al.* (1969). Original studies on the general distribution of parenterally administered ^{3}H-5-HTP were made by Ritzén *et al.* (1965), and by Gershon and Ross (1966), and a specific study of the uptake of 5-HTP by endocrine polypeptide cells of the so-called APUD series was made by me (Pearse, 1966). This study, employing the FIF technique, was based on the presumption that uptake of the amino acid was followed by decarboxylation to the amine, and by storage of the latter. Contemporary studies by Larson *et al.* (1966) showed that both DA and L-DOPA were taken up by mouse thyroid parafollicular, C, cells and Håkanson *et al.* (1967) showed that the so-called enterochromaffin-like cells of the gastric mucosa could take up both D and DL-DOPA whereas in most other cells only the L-isomer was handled.

My own earlier studies were followed by others using both L-DOPA and 5-HTP (Pearse and Carvalheira, 1967). These were administered to pregnant rodents and, after crossing the placenta, both were shown to be taken up and stored by the C cells in the ultimobranchial component of the thyroid gland. Later this work was extended to show uptake and presumptive decarboxylation by the other endocrine cells of the APUD series (Pearse, 1968), and these aspects have been considered by Dawson (1970) in a comprehensive review of the endocrine polypeptide cells of the gastrointestinal tract. These cells share a number of cytochemical and ultrastructural characteristics in addition to those described above. The initial letters APUD are derived from their fluorogenic properties (**A**mine and **A**mine **P**recursor **U**ptake and **D**ecarboxylation).

For most of the cells concerned the presence of a specific or non-specific decarboxylase has not been demonstrated directly. It is possible, however, to demonstrate the presence of arylethylamines corresponding to the two amino acids by silver reduction techniques (argentaffinity, Chapter 26, p. 1063) and, in the case of 5-HT, if sufficient conversion and storage of this amine occurs, it can be shown by the xanthydrol technique (p. 616) after glutaraldehyde fixation or by the alkaline diazo reaction (p. 1114) after formaldehyde fixation. These reactions would be equally positive if the compounds stored were the precursor amino acids but there is reason to suppose that the latter are not amenable to storage in the specific granules of endocrine polypeptide cells.

A comprehensive study of the uptake and turnover of L-DOPA and related compounds by the exocrine pancreas was made by Alm *et al.* (1969). These authors found FIF in the zymogen granules only after L-DOPA or DA and they suggested that conversion of the former to DA preceded its uptake by the granules. Certainly dopamine, produced in this manner *in vivo* by the widely distributed decarboxylases of the tissues, is taken up by a very large number of tissue protein components but especially by those having high levels of reactive carboxyl side chains (glutamic and aspartic acids). The undoubted uptake of DA by the parietal cells of rat stomach, demonstrated by Penttilä and Hirvonen (1969) was not shown to be due to uptake and decarboxylation of DOPA. Since the latter is an amino acid it is possible that it may be handled as such by cells outside the APUD series (and related neurons in the C.N.S.). In my experience conditions can always be arranged so that specific uptake and decarboxylation by endocrine cells is the predominant source of observable FIF.

In the mammalian pituitary gland the cells which take up and decarboxylate amine precursors are the two cell types which already contain a fluorogenic amine or amines. These are the MSH-producing cells of the pars intermedia (when present) and the R-type mucoid cells (see Vol. 1, p. 240) of the pars distalis which presumably produce ACTH. Basing their views solely on the colour and fading characteristics of the autofluorescence of the pituitary gland of rat, dog and pig, Pearse and McGregor (1964) and Pearse (1966) identified the stored amine as 5-HT. Dahlström and Fuxe (1966), on the other hand, considered that it was a primary catecholamine. Björklund and Falck (1969) using cat pituitary gland, observed a yellow fluorescence whose microspectrofluorimetric characteristics distinguished it from both 5-HT and the catecholamines. They suggested that the compound responsible was tryptamine.

Techniques for Individual Amines

5-HT and the Enterochromaffin Granules. Five major histochemical tests give insight into the chemical nature of the granules. These are the *argentaffin*, *chromaffin*, *diazonium*, *indophenol*, and *Gibbs'* reactions. A few subsidiary tests and characteristics are also of sufficient interest to be mentioned here. Their property of reducing ammoniacal silver solutions to metallic silver (argentaffin reaction) is still most commonly employed by histologists for demonstration of the granules. The most popular techniques are those of Masson-Hamperl, Masson-Fontana and the later Gomori hexamine-silver method; the last two are given in Appendix 26 p. 1379 and the type of result obtainable is shown in Plate XXVIa, p. 1109. The argentaffin reaction is of low chemical specificity, as pointed out by Lison, who considered that it was given particularly strongly by dihydric and polyhydric phenols, aminophenols and polyamines, and less strongly by many other substances, so that for practical purposes in histochemistry a positive reaction suggests the presence of phenols. Cordier (1927), Hamperl (1932), Lison (1936) and Gomori (1948) all strongly stressed the fundamental difference between the argentaffin reaction and silver impregnation methods, in that the former is brought about by the reducing capacity of the tissue component itself and the latter by the addition of an extraneous reducer. It is still necessary to stress this point although it is entirely justifiable to diagnose argentaffin cell tumours (carcinoids) on the basis of granules demonstrated solely by impregnation techniques employing a reducer. The position is not as simple as it might be because, if tissues containing argentaffin cells are fixed in alcoholic fixatives, so that the specific granules are absent, the cells which contained the precursor substance can still be demonstrated by a reducing impregnation technique. That is to say, the cells are still argyrophil though they are no longer argentaffin. If the argyrophil method is applied to formalin-fixed tissue, intestine for instance, the morphology and position of the resulting argyrophil cells closely resembles that of the argentaffin cells. Numerically, however, they exceed the latter by a considerable amount. Hamperl (1951, 1952) believed that the argyrophil but non-argentaffin cells were precursors of the argyrophil *and* argentaffin (true argentaffin) cells. Sharples (1945a and b) observed that the Bodian (reducing) silver technique often demonstrated large numbers of argyrophil cells in the deeper layers of the stomach mucosa, and Gomori (1948) stated that the majority of these cells, though scarcely recognizable by the usual (non-reducing) argentaffin techniques, were well shown by his hexamine-silver method. The same cells were demonstrated by the other specific reactions mentioned in this chapter but in shades so pale that they were only recognizable with difficulty. This observation seemed to accord well with Hamperl's hypothesis but Clara (1957) believed that only a proportion of the argyrophil basal granulated cells could be regarded as precursors of the E.C. cells. Certainly anyone who is investigating the endocrine polypeptide (APUD) cells of the gastrointestinal tract from either the morphological or the histochemical point of view, cannot overlook the value of the argyrophil reactions in demonstrating them. A great deal of new work on this subject has been carried out in the last 5 years. Reference to this is made in a later section on the Applied Histochemistry of the E.C. cells.

Returning to the histochemistry of the granules, it is not necessary to mention the *chromaffin reaction* in this context, other than briefly. It was considered by Lison to be a specific reaction solely exhibited by the dihydric and polyhydric phenols, aminophenols and polyamines. It has not been used for the diagnosis of carcinoid tumours and is seldom employed for the demonstration of the specific granules in intestinal cells. Nevertheless a positive chromaffin reaction can be obtained either with iodate or chromate after formalin fixation, since the specific substance of the E.C. cells is preserved by the latter.

The *diazonium reaction* demonstrates the granules particularly clearly (Plate XXVIb) and the resulting colour varies with the particular *diazonium* salt which is employed. Technical details of the method, and considerations of the various salts which can be used, appear in Appendix 27, p. 1396. Lison considered that the diazonium reaction, in alkaline solution, was given by all phenolic substances on condition that the hydroxyl was unsubstituted and that at least one of the positions *ortho* and *para* to the hydroxyl group was also free.

The *indophenol reaction* (Lison, 1931) depends on the production of a blue or green indophenol dye when a phenol combines in the presence of an oxidizing agent with dimethyl*para*phenylenediamine, as shown in the equation below.

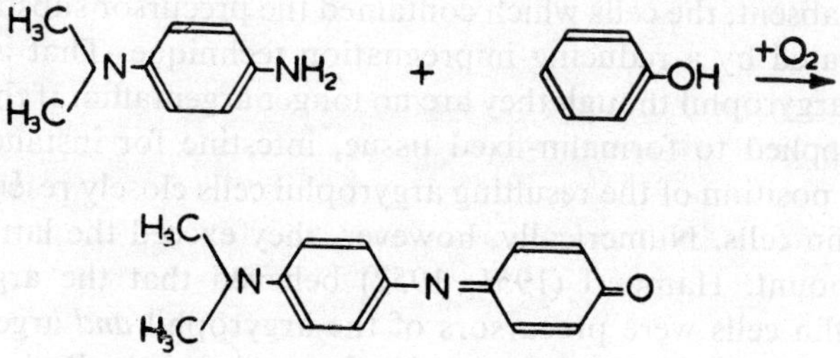

According to the author, the technique was difficult in application and, in practice, the colour developed is too pale to be of much use in diagnosis.

The fifth and last of those main reactions which we have been considering is Gibb's reaction (Gibbs, 1926). This is based on the formation of an indophenol dye when 2,6-dichloroquinonechloroimine reacts with a phenol in alkaline solution.

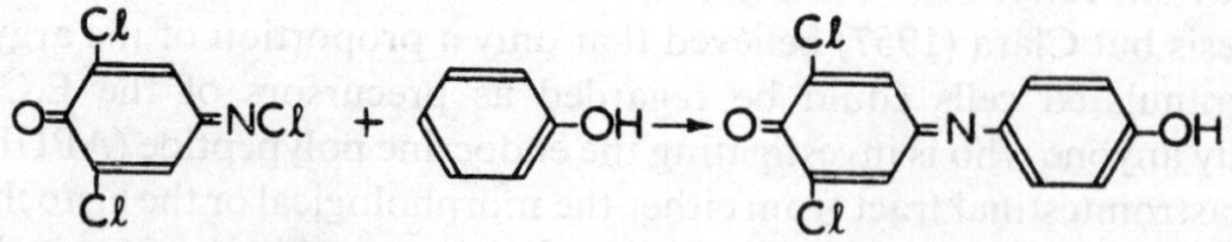

The final colour depends on the phenol with which the quinoneimine is coupled. With phenol itself, or α- and β-naphthol, it is blue, and with resorcinol reddish-purple. Coupling does not occur with catechol or with any of the *para*-substituted phenols. The use of this reaction in the particular instance of the argentaffin granules was first described by Gomori (1948) and the black or brownish-black colour developed in tissue sections is very satisfactory.

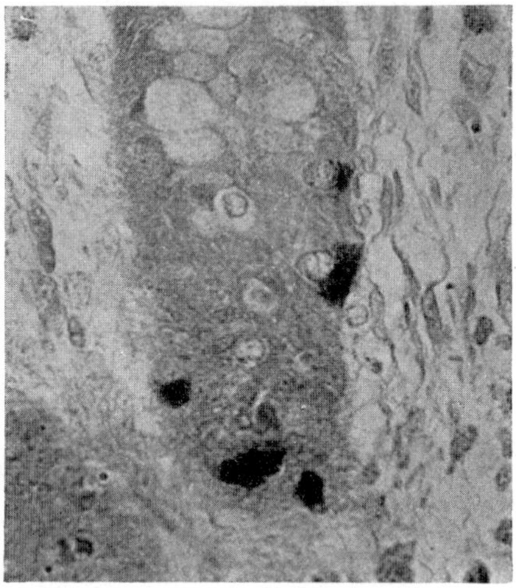

FIG. 229. Paraffin section (5·5 μ) of human small intestine. Argentaffin granules are stained black. Gibbs' reaction, carmalum. × 460.

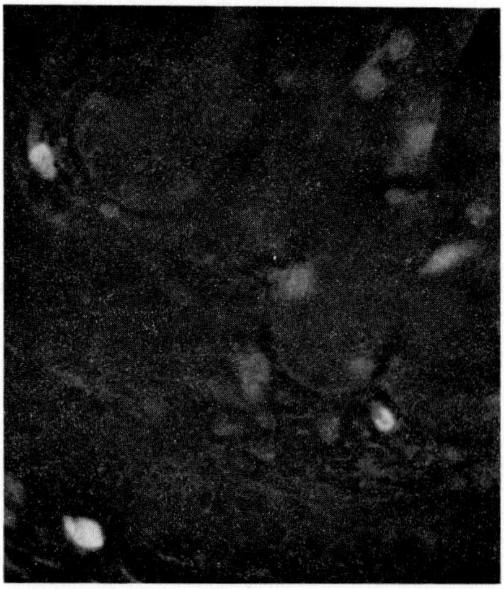

FIG. 230. Rat stomach (Fundus). OPT reaction for histamine. Shows strong (yellow) fluorescence of mast cells in the submucosa and weaker (blue) fluorescence in some endocrine cells in the fundic glands. × 610.

Fig. 229 illustrates the results of its application to material from the human intestine.

Other Histochemical Tests. Four lesser characteristics of the E.C. granules may well be described at this point. The first is the *Vulpian reaction* (Vulpian, 1856) which is given, according to Lison, by all *ortho*diphenols (catechols). It depends on the development of a green colour when a very weak solution of ferric chloride is applied to tissues containing reactive phenols. Vulpian himself used the reaction to demonstrate the substance subsequently identified as adrenalin in the adrenal medulla, and it was observed by Jonnard (1934) to give a weak reaction with the argentaffin granules.

In a large proportion of the E.C. cells of the human intestine the granules have sufficient reducing capacity to give a strong blue colour with *Schmorl's ferric-ferricyanide reaction.* This reaction is of no use in elucidating the exact chemical nature of the granules, since it indicates only the presence of a substance capable of reducing ferricyanide to ferrocyanide. The property is shared, *inter alia*, by the melanins, the lipofuscins (below), sulphydryl groups (Chapter 6, p. 146), and probably by aldehydes such as plasmal (Chapter 13). It is necessary to remember these facts when attempting the differential diagnosis of obscure pigment granules having reducing properties towards a number of reagents. Fig. 225 shows argentaffin granules in the human appendix stained by Schmorl's reaction.

The fact that the E.C. granules give a positive acid haematein reaction was recorded by Christie (1954) who showed also that the reaction was negative after pyridine extraction. In my hands, using duodenum from guinea-pig, rat and man, the reaction was always a weak one but definitely positive in a proportion of the E.C. cells in all three species. In a number of carcinoid tumours only an occasional positive cell was recorded. The interpretation of a positive acid haematein reaction (with negative pyridine-extracted control) is not usually difficult and the presence of phospholipid can be confirmed by a number of alternative reactions. I failed consistently to demonstrate the E.C. granules by means of any other phospholipid reaction. Moreover, it can be shown that the condensation product of 5-HT and formaldehyde, after postchroming as in Baker's method, can be stained with acid haematein. This condensation product, whose use was suggested by the work of Erspamer and his associates referred to below (p. 1118), and the phenolic material in the fixed E.C. cells, are both soluble in hot pyridine so that in this case the control reaction has no differential significance. It is reasonable to suppose that the water-soluble phenol present in the E.C. cells occurs in bound form and one of the most likely forms of binding would be with lipid or phospholipid. Christie's interesting hypothesis cannot be sustained, however, unless some corroborative evidence of the presence of phospholipid in the granules can be produced.

The last of the four lesser characteristics referred to is the so-called alkaline thioindoxyl reaction of Pearse (1956). Originally developed to provide evidence that the compound stored in the E.C. granules was 5-HT and not an *ortho*- or *meta*diphenol, the reaction proved useful, especially in the hands of Italian workers (Vialli and Quaroni, 1956; Vialli, 1962, 1965). It is based on the coupling in alkaline solution of an unsubstituted or substituted thioindoxyl (hydroxythionaphthene) (I) with the quinoneimine (II) formed by mild oxidation of the *β*-carboline present in formaldehyde-fixed E.C. cells, to give reddish or reddish-brown thioindigoid dyes (III).

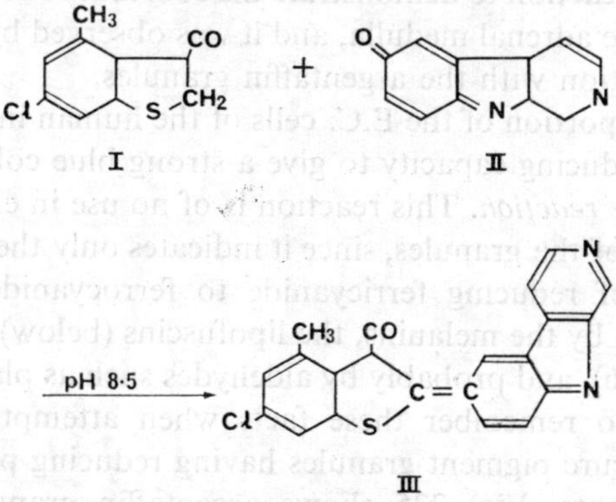

The dyestuff (III) formed from 6-chloro-4-methyl-thionaphthene is insoluble in chloroform whereas thioindigos formed by oxidation of thioindoxyls dissolve easily in this solvent. This indicates that simple oxidation of the *β*-carboline is not responsible for the reaction.

The thioindoxyl reaction cannot demonstrate resorcinols because these do not form quinones on oxidation and, although they can act in the tautomeric carbonyl form, they fail to condense with thioindoxyls at an alkaline pH. At acid pH levels they can be persuaded to do so. Details of the reaction are given in Appendix 27, p. 1397.

Investigations carried out by Martin *et al.* (1960) on the metal-catalysed oxidation of 5-hydroxyindoles indicated that cupric ions were most active in this respect. Vanadium (as *ortho*vanadate, VO^{3-}) was also effective. These observations offer an explanation for the relative failure of the thioindoxyl reaction after fixation in analytical reagent formaldehyde. Commercial formalin contains trace metals including Fe^{3+} and Cu^{2+}. It is probably these which catalyse initial oxidation of the *β*-carboline to the quinoneimine.

The Ninhydrin Reaction. A specific test for tryptamines was described by Jepson and Stevens (1953) involving condensation with ninhydrin to produce brightly fluorescent products. The reaction was applied to freeze-dried (unfixed) rat duodenum by Barter & Pearse (1955) but no fluorescent products developed. It was applied to (wet) formalin-fixed paraffin sections by Holcenberg and Benditt (1961) who found that, with the *β*-carboline product of 5-HT, ninhydrin gave a brilliant orange-red colour. This reaction has been little used in the applied histochemistry of tryptamines.

Indole Reactions. The controversy regarding the negative indole reactions of formaldehyde-fixed E.C. cells was resolved by Solcia and his associates (Solcia *et al.*, 1966; Solcia and Sampietro, 1967) who demonstrated, simply and clearly, that while the formaldehyde-fixed granules would not react their glutaraldehyde-fixed equivalents did so readily. They used the DMAB reaction (p. 615), and the post-coupled benzylidene and xanthydrol methods (p. 616). The application of this last method to glutaraldehyde-fixed E.C. cells has proved to be a useful means of distinction.

This work was amplified by Solcia *et al.* (1969) who provided evidence that ring closure to the fluorescent β-carboline did not occur with glutaraldehyde fixation. They failed to obtain fluorescence, but did obtain positive indole reactions indicating, supposedly, that the 2(α) carbon of the amine was still free to react. On the other hand, they obtained a grey-black reaction product with the alkaline thioindoxyl method. This indicated that the 5-HT—glutaraldehyde condensation product remains susceptible to conversion to the quinonemine by mild oxidation and that the latter can then react with thionaphthenes to produce thioindigoid dyes.

Historical Considerations. Cordier and Lison (1930) made a thorough study of the nature of the E.C. granules using the first four major reactions detailed above and also Vulpian's reaction. From the diazo and indoreactions they concluded the presence of a phenolic group and, from the stability to alkali of the colour developed in the former reaction, that coupling must have occurred in the *ortho* position. This meant that the *para* position was not free since coupling invariably occurs at that point if it is. From the chromaffin and argentaffin reactions, in addition, they concluded that the granules contained an *ortho* or *para*-polyphenol, and with the aid of the Vulpian reaction the structure was imagined as an *ortho*-diphenol (catechol) of the type

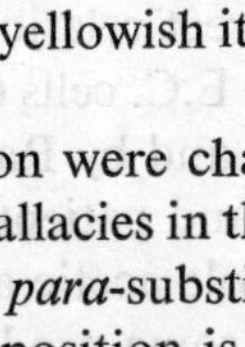

carrying a side-chain in the *para* position. Because the shades of the dyes formed in the diazo reaction were yellowish it was assumed that the side chain was relatively simple.

The ideas of Cordier and Lison were challenged by Gomori (1948) who pointed out that there were three fallacies in their reasoning. First, the positive indophenol reaction was against a *para*-substituted phenol since these dyes do not usually form unless the *para* position is free. Secondly, catechol and its derivatives were far stronger reducing agents than argentaffin granules and, thirdly, there were numerous exceptions to the rule that the colour of the diazo dye obtained from *ortho*-coupled phenols is stable to alkali, while that obtained from *para*-coupled phenols is not. From the result of the Gibbs reaction Gomori excluded the possibility of the granules containing catechols or any *para*-substituted phenol, since with these the reaction was negative. In *in vitro* experiments he showed that catechol or hydroquinone would reduce alkaline silver solutions in a few minutes while resorcinol and phloroglucinol took several hours. Furthermore, short treatment with iodate or dichromate would abolish the silver-reducing power of the first two substances leaving that of the other two, like that of the granules, unchanged. Finally, Gomori showed that the shades produced by the diazo reaction with resorcinol were the same as those given by the granules, and he therefore suggested that these contained a derivative of *meta*diphenol (resorcinol) rather than *ortho*-diphenol (catechol), as previously believed. Following the work of Erspamer and Asero (1952), described in the next paragraph, Gomori (1954) retested the above-mentioned compounds, and also 5-HT, with various diazonium salts. He considered that the colours given by the E.C. granules resembled those given by resorcinol more closely than those given by 5-HT. The resorcinol hypothesis of Gomori was supported, to some extent, by the results obtained by Lillie, Burtner and Henson (1953) with their diazo-safranin method. It was opposed by Barter and Pearse (1953, 1955) and by the result of the alkaline thioindoxyl reaction (Pearse, 1956).

These historical considerations were brought up to date by Lillie (1961) and his conclusions, that the presence of 5-HT in normal E.C. cells in demonstrable amounts was contraindicated and that the "azo-positive phenol" in the cells was a distinct substance, can be considered correct or incorrect according to whether one is dealing with formaldehyde-fixed, or unfixed, E.C. cells. The conclusions of Vialli and Traverso (1960) and of Ghiringelli (1961), both based on extensive studies of the various reactions of the E.C. cells, were that these were all compatible with the presence of substantial amounts of 5-HT.

The Biological Importance of 5-HT in Mammals. It is generally agreed that 5-HT is a substance of great physiological importance although its exact function under normal conditions has yet to be determined. It is known to be formed by hydroxylation of tryptophan to 5-hydroxytryptophan (5-HTP), and by the subsequent decarboxylation of this compound (Udenfriend *et al.*, 1953). The enzyme responsible for this second stage is present in most mammalian tissues and, according to Gaddum and Giarman (1956), the amount was found to have a direct relationship to their 5-HT content.

Metastasizing tumours of the E.C. cells (carcinoids, argentaffinomas) were shown by Lembeck (1953, 1954) and by Ratzenhofer and Lembeck (1954) to contain large amounts of 5-HT and patients with these tumours excrete very large quantities of 5-hydroxyindole acetic acid (5-HIAA), the normal breakdown product. The clinical syndrome produced by argentaffinomas was first recognized by Waldenström and his colleagues (Pernow and Waldenström, 1954) and it is notable that no effects due to the action of 5-HT on the nervous system are recorded. This is because 5-HT cannot penetrate the blood brain barrier although 5-HTP does so readily (Udenfriend *et al.*, 1956). In a case of argentaffinoma with renal metastases reported by Smith *et al.* (1957) large amounts of 5-HTP and 5-HT, as well as some 5-HIAA, were excreted in the urine. These findings, together with other evidence, suggested to Dalgleish and Dutton (1957) that 5-HTP was formed at some central locus and converted elsewhere into 5-HT, and that this locus was very possibly the E.C. cell.

Subsequently, tryptophan hydroxylase activity has been demonstrated in a variety of tissues including brain (Gal *et al.*, 1966) and pineal gland (Lovenberg *et al.*, 1967). These authors demonstrated it also in carcinoid tumours. Although neoplastic mast cells contained the enzyme (Levine *et al.*, 1964) Slorach and Uvnäs (1968) found that normal rat mast cells could not hydroxylate either tryptophan or tyrosine.

Applied Histochemistry of the E.C. Cells

The distribution of the E.C. cells and the corresponding levels of 5-HT in the gastrointestinal tract continues to be an important subject. A survey carried out by Penttilä (1966) on the mammalian duodenum showed a very close correlation, and later (1967) this author found, in the guinea-pig, a highly significant correlation between E.C. cell numbers and 5-HT content from stomach to rectum. Investigations carried out by Thompson and Campbell (1967) also revealed a similar state of affairs in the rat although there were wide distributional differences between the two species.

The division of the E.C. cells into argentaffin and argyrophil, still used by some, perpetuates earlier views (Erspamer, 1939; Hamperl, 1952) that the two states represented functional stages of a single cell type. More recent E.M. studies (Ratzenhofer, 1966a and b) and cytochemical studies (Penttilä, 1966, 1967) supported this view. Using the technique of masked metachromasia (Plate XXVIc) Solcia and Sampietro (1965) clearly distinguished argyrophil from E.C. cells and Carvalheira *et al.* (1968), using a spectrum of cytochemical tests, described a cholinesterase-rich, argyrophil APUD cell in the gastrointestinal tract of 4 mammalian species which was functionally quite distinct from the E.C. cell. The large number of techniques for demonstrating argyrophilia is a reflection of the need for precise conditions, differing significantly from cell to cell and species to species. The older Bodian technique, still in constant use in many laboratories, is illustrated in Plate XXVId, p. 1109.

Further cytochemical and ultrastructural studies (Capella *et al.*, 1969; Vassallo *et al.*, 1969; Pearse *et al.*, 1970) have shown that there are at least 6 (more probably 7) types of endocrine polypeptide (APUD) cell in the mammalian gastrointestinal tract. The relationship of all these cells to each other is close and the possibility of interconversion must still be entertained. No evidence for the existence of a cell intermediate between E.C. and any other cell type has yet been found, unless it be the E.C.-like cell of the fundic glands of the stomach.

Methods for Tryptamine. This amine reacts with formaldehyde to give a fluorescent product but the amount of fluorescence developed is much less than that produced by equimolar quantities of noradrenalin or dopamine (Ritzén, 1967; Jonsson, 1967). Observing that the high fluorescence yields of catecholamines with the FIF method were due to the fact that the necessary dehydrogenation could take place in the dry state Björklund *et al.* (1968) investigated the reaction with tryptamine in order to establish optimal conditions. According to Hess and Udenfriend (1959) tryptamine (I) reacts with formaldehyde to form a non-fluorescent tetrahydronorharman (II).

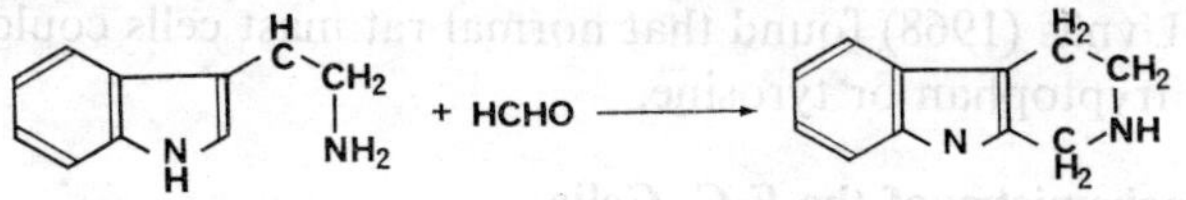

I II

In a second stage the tetrahydronorharman is oxidized to a variety of fluorescent end products (III).

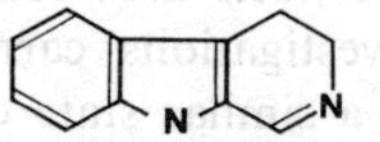

IIIa
3, 4-dihydronorharman

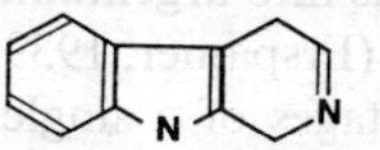

IIIb
1, 4-dihydronorharman

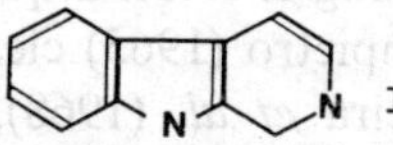

IIIc
1, 2-dihydronorharman

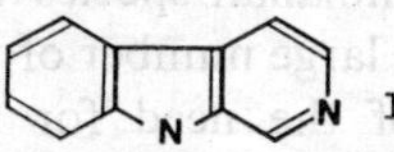

IIId
Norharman

By introducing ozone into the formaldehyde gas chamber Björklund *et al.* (1968) were able to increase the fluorescence yield of tryptamine in models 13- to 18-fold. They considered this to be a specific oxidation of the initial condensation product (II) to the fluorophores (IIIa to d). Details of the reaction are given in Appendix 27, p. 1399.

Model experiments by Björklund and Stenevi (1970) showed that formaldehyde condensation with tryptamine would proceed at a greatly increased rate in the presence of a trace of HCl gas. A 20–200-fold increase in fluorescence was obtained. The acid-catalysed formaldehyde reaction was specific for indolylethylamines and 3-hydroxy or 3-methoxy phenylethylamines.

Methods for Catecholamines, A, NA, DA. Special histochemical methods for adrenalin and noradrenalin have been developed on the basis of the trihydroxyindole (THI) method and on the ethylenediamine (ED) condensation reaction. The latter was first described by Wallerstein *et al.* (1947) and the procedure was developed by Weil-Malherbe and Bone (1952). According to Weil-Malherbe (1960) NA reacts with 2 molecules of ED as shown below

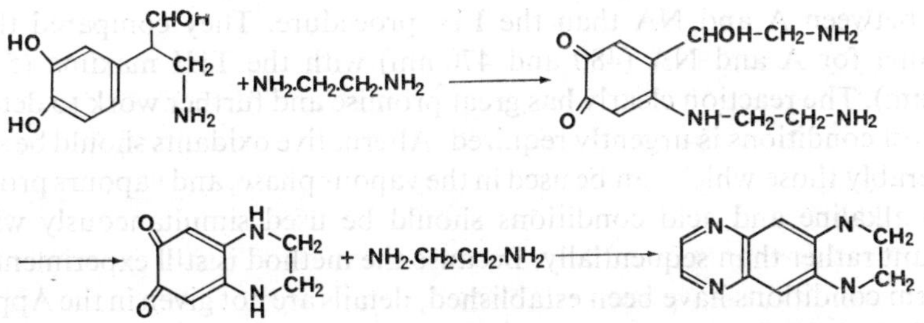

The reaction takes place in two stages and there are at least two products. The major fluorescent product has its fluorescence maximum at 485 nm. Adrenalin, on the other hand, reacts with only one molecule of ED to give a single final product (Udenfriend, 1964) whose formula has been given as

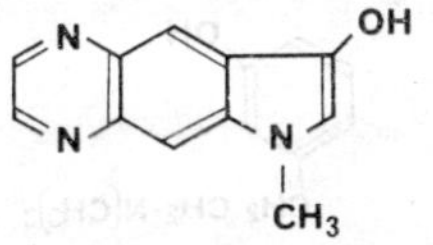

This compound fluoresces maximally at 525 nm and clear differentiation between the two catecholamines is thus afforded.

Histochemical methods based on the reactions described above were reported by Angelakos and King (1968), using ED either dissolved in isopropanol or as its vapour. Their fluorescence maxima were recorded as 470 nm (NA), 500 nm (A) and 490 nm (DA).

The trihydroxyindole method depends on oxidative conversion of adrenalin to the highly fluorescent compound adrenolutin:

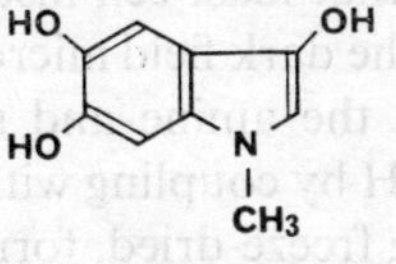

As described by Heller *et al.* (1949) the product was unstable but further refinements (Price and Price, 1957; von Euler and Lishajko, 1959), using either ferricyanide or iodine as oxidizing agent, made the use of the method appropriate for tissue assay of adrenalin. Weil-Malherbe and Bigelow (1968) described further improvements in the fluorometric estimation of A and NA by the THI method.

A histochemical technique based on the above was developed by Angelakos and King (1967), using whole mount preparations stretched on slides. These were placed in a closed Coplin jar and exposed to iodine vapour for 5 to 10 minutes. They were then exposed to ammonia vapour for 50–60 seconds. The authors stressed that the times of exposure were critical and my own attempts to reproduce their results were not very successful. King and Angelakos (1968), however, recorded their opinion that the THI method provided better distinction between A and NA than the FIF procedure. They compared the FIF maxima for A and NA (480 and 470 nm) with the THI maxima (515 and 480 nm). The reaction clearly has great promise and further work to determine the best conditions is urgently required. Alternative oxidants should be sought, preferably those which can be used in the vapour phase, and vapours producing both alkaline and acid conditions should be used simultaneously with the oxidant rather than sequentially. Because the method is still experimental and no firm conditions have been established, details are not given in the Appendix.

Method for Leptodactyline. The skin of the South American amphibian *Leptodactylus pentadactylus labyrinthicus* contains a large number of biologically active amines which include histamine and related imidazolealkylamines (Erspamer *et al.*, 1963). According to Erspamer (1959) and Vialli (1966) the tertiary amine leptodactyline is also present, and treatment with formaldehyde

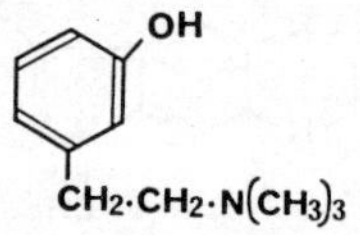

vapour produces a fluorescent product having two peaks at 552 and 562 nm. The presumptive formaldehyde condensation product *in vitro* was observed to fluoresce maximally at 497 nm with a second peak at 570 nm.

Methods for Histamine. The earliest description of a histochemical method for histamine (H) was that of Schauer and Werle (1959) who obtained a coloured azo dye product by the reaction of diazotized sulphanilic acid with tissue histamine post-fixed in Reineke salt (tetrathiocyano-diammonochromic acid). I have not been able to find further references to this method but Whur and Gracie (1967) reported that mast cell histamine, precipitated by Reineke salt, became luminescent in the dark field microscope. This property belongs to the granules rather than to the amine and should be further investigated. Further attempts to localize H by coupling with diazonium salts were made by Lagunoff *et al.* (1961). Using freeze-dried, formaldehyde vapour-fixed tissues, and diazotized *p*-bromoaniline, an orange coloured product was obtained. It was postulated that the reacting compound in the tissues was the presumptive tetrahydropyridoiminazole from H and formaldehyde. The weight of present evidence suggests that it was rather the *β*-carboline from 5-HT.

The OPT Method for Histamine. A sensitive fluorometric procedure for histamine was described by Shore *et al.* (1959), based on the condensation of the amine with *ortho*-phthalaldehyde (OPT). The initial condensation product is considered to have the structure shown in the equation below:

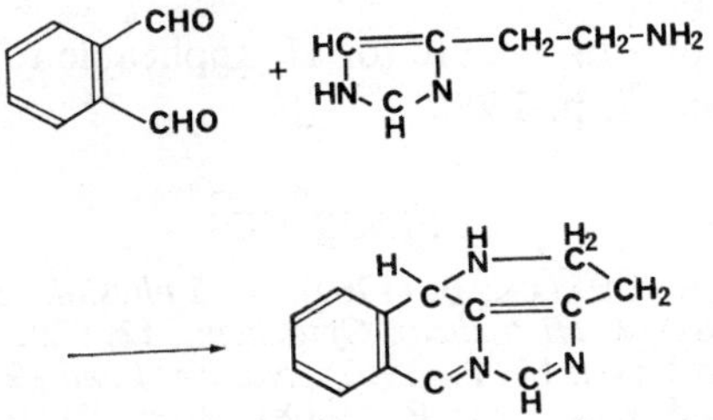

It is probable, however, that rearrangement and oxidation take place and that the final fluorophore or fluorophores differ from the product described above.

From this reaction Juhlin and Shelley (1966) developed a histochemical method for H which they applied to cell suspensions or fresh tissue sections. Later Shelley and Öhman (1968) described a modification of this procedure for fixed tissues, Shelley *et al.* (1968) applied it to freeze-dried, paraffin sections and Enerbäck (1969) to Carnoy-fixed paraffin sections. All three groups used OPT dissolved in ethylbenzene. Independently Ehinger and Thunberg (1967) developed a histochemical method on the same basis as the foregoing one and their method was amplified and extended by Ehinger *et al.* (1968).

Methods using OPT in liquid media give positive results with mast cells (mast cell histamine) but not with H in other situations. The vapour phase method of Ehinger and Thunberg gives bright yellow fluorescence with mast cells and a paler blue or bluish fluorescence with the histamine-containing cells of rat gastric mucosa. These results are illustrated in Fig, 230, p. 1115. In my experience the blue fluorescence is unstable and fades rapidly. The yellow fluorescence is more stable.

The microspectrofluorimetric characteristics of the fluorophore(s) from H and OPT were investigated by Håkanson *et al.* (1970) in model systems. Histamine was used in a silica gel thin layer or in dried albumin droplets. These authors found a concentration-dependent shift of the fluorescence maximum from yellow (high concentration) to blue (low concentration). The excitation maxima were 365 nm (blue fluorescence) and 405 nm (yellow fluorescence). These results suggest that at least two fluorophores are formed by the condensation of H and OPT.

There are few reports of the demonstration of H other than in mast cells or in rat gastric mucosa but Håkanson *et al.* (1968) observed a positive reaction in pituitary gland and in the endocrine pancreas, while Öhman and Shelley (1968) reported OPT fluorescence in photoreceptor cells and Öhman and Shelley (1969) found it in the renal vasa recta. The yellow fluorescence of pancreatic alpha (α_2) granules in fresh tissue sections, after treatment with OPT, was described by Takaya (1970). This observation brings the specificity of the OPT reaction for histamine into question since the fluorescent staining reaction of OPT is not dependent on the presence of any known biogenic amine. Further elucidation of the mechanisms of the several reactions must be awaited.

Technical details of a procedure for H, applicable to freeze-dried tissues, are given in Appendix 27, p. 1400.

REFERENCES

ALM, P., EHINGER, B., and FALCK, B. (1969). *Acta physiol. scand.*, **76**, 121.
ANGELAKOS, E. T. (1964). *J. Histochem. Cytochem.*, **12**, 929.
ANGELAKOS, E. T., and KING, M. P. (1967). *Nature, Lond.*, **213**, 391.
ANGELAKOS, E. T., and KING, M. P. (1968). Proc. III Int. Cong. Histochem. Cytochem. Springer, New York, p. 7.
BARTER, R., and PEARSE, A. G. E. (1953). *Nature, Lond.*, **172**, 810.
BARTER, R., and PEARSE, A. G. E. (1955). *J. Path. Bact.*, **69**, 25.
BENDITT, E. P., and WONG, R. L. (1957). *J. exp. Med.*, **105**, 509.
BISCHLER, A., and NAPIERALSKI, B. (1893). *Ber. dt. chem. Ges.*, **26**, 1903.
BJÖRKLUND, A., EHINGER, B., and FALCK, B. (1968a). *Acta physiol. scand.*, **72**, 253.
BJÖRKLUND, A., EHINGER, B., and FALCK, B. (1968b). *J. Histochem. Cytochem.*, **16**, 263.
BJÖRKLUND, A., and FALCK, B. (1968). *J. Histochem. Cytochem.*, **16**, 717.
BJÖRKLUND, A., and FALCK, B. (1969). *Z. Zellforsch.*, **93**, 254.
BJÖRKLUND, A., FALCK, B., and HÅKANSON, R. (1968). *Acta physiol. scand.*, Suppl. 318.
BJÖRKLUND, A., and STENEVI, U. (1970). *J. Histochem. Cytochem.*, **18**, 794.
BLACK, W. C. (1968). *Lab. Invest.*, **19**, 473.
BOWMAN, R. L., CAULFIELD, P. A., and UDENFRIEND, S. (1955). *Science, N.Y.*, **122**, 32.
CAPELLA, C., SOLCIA, E., and VASSALLO, G. (1969). *Archs histol. japon.*, **30**, 479.
CARLSSON, A. (1964). *Prog. Brain Res.*, **8**, 9.
CARVALHEIRA, A. F., WELSCH, U., and PEARSE, A. G. E. (1968). *Histochemie*, **14**, 33.
CHRISTIE, A. C. (1954). *Nature, Lond.*, **173**, 589.
CHRISTIE, A. C. (1955a). *Quart. J. micr. Sci.*, **96**, 289.
CHRISTIE, A. C. (1955b). *Quart. J. micr. Sci.*, **96**, 295.
CLARA, M. (1957). *Acta neurovegetativa*, **16**, 294.
CLARKE, D. E., JONES, C. J., and LINLEY, P. A. (1969). *Br. J. Pharmac.*, **37**, 1.
CORDIER, R. (1927). *Bull. Histol. Tech. micr.*, **4**, 161.
CORDIER, R., and LISON, L. (1930). *Ibid.*, **7**, 140.
CORRODI, H., and HILLARP, N. Å. (1963). *Helvetia chim. Acta*, **46**, 2425.
CORRODI, H., and HILLARP, N. Å. (1964). *Helvetia chim. Acta*, **47**, 911.
CORRODI, H., HILLARP, N. Å., and JONSSON, G. (1964). *J. Histochem. Cytochem.*, **12**, 582.
CORRODI, H., and JONSSON, G. (1965a). *J. Histochem. Cytochem.*, **13**, 484.
CORRODI, H., and JONSSON, G. (1965b). *Acta histochem.*, **22**, 247.
CORRODI, H., and JONSSON, G. (1967). *J. Histochem. Cytochem.*, **15**, 65.
CORRODI, H., JONSSON, G., and MALMFORS, T. (1966). *Acta histochem.*, **25**, 367.
DAHLSTRÖM, A., and FUXE, K. (1964). *Z. Zellforsch.*, **62**, 602.
DAHLSTRÖM, A., and FUXE, K. (1966). *Acta endocrinol. (Kbh.)*, **51**, 301.
DALGLEISH, C. E., and DUTTON, R. W. (1957). *Br. J. Cancer*, **11**, 296.
DAWSON, I. (1970). *Histochem. J.*, **2**, 527.
EHINGER, B., and FALCK, B. (1969). *Histochemie*, **18**, 1.
EHINGER, B., and FALCK, B. (1970). *Acta physiol. scand.*, **78**, 132.
EHINGER, B., FALCK, B., and STENEVI, U. (1969). *J. Histochem. Cytochem.*, **17**, 351.
EHINGER, B., HÅKANSON, R., OWMAN, C., and SPORRONG, B. (1968). *Biochem. Pharmac.*, **17**, 1997.
EHINGER, B., and THUNBERG, R. (1967). *Expl Cell Res.*, **47**, 116.
ENERBÄCK, L. (1969). *J. Histochem. Cytochem.*, **17**, 757.

Eränkö, O. (1951). *Acta physiol. scand.*, **25**, Suppl. 89.
Eränkö, O. (1952). *Acta anat.*, **16**, Suppl. 17.
Eränkö, O. (1955). *Acta endocrinol.*, **18**, 174.
Eränkö, O. (1967). *J. roy. micr. Soc.*, **87**, 259.
Erös, G. (1932). *Zbl. allg. Path. Anat.*, **54**, 385.
Erspamer, V. (1939). *Z. Anat. Entwickl. Gesch.*, **109**, 586.
Erspamer, V. (1959). *Archs Biochem. Biophys.*, **82**, 431.
Erspamer, V., and Asero, B. (1952). *Nature, Lond.*, **169**, 800.
Erspamer, V., and Boretti, G. (1950). *Experientia*, **6**, 348.
Erspamer, V., Vitali, T., Roseghini, M., and Cei, J. M. (1963). *Experientia*, **19**, 346.
Falck, B. (1962). *Acta physiol. Scand.*, **56**, Suppl. 197.
Falck, B., Hillarp, N. Å., Thieme, G., and Torp, A. (1962). *J. Histochem. Cytochem.*, **10**, 348.
Falck, B., and Owman, C. (1965). *Acta Univ. Lund.*, Sect., II, No. 7.
Fuxe, K. (1965). *Z. Zellforsch.*, **65**, 573.
Fuxe, K., and Jonsson, G. (1967). *Histochemie*, **11**, 161.
Gaddum, J. H., and Giarman, N. J. (1956). *Br. J. Pharmac.*, **11**, 88.
Gal, E. M., Armstrong, J. C., and Ginsberg, B. (1966). *J. Neurochem.*, **13**, 643.
Gershon, M. D., and Ross, L. L. (1966). *J. Physiol.*, **186**, 477.
Ghiringhelli, F. (1961). *Riv. Istochim.*, **7**, 185.
Gibbs, H. D. (1926). *Chem. Rev.*, **3**, 291.
Glenner, G. G., and Lillie, R. D. (1957). *J. Histochem. Cytochem.*, **5**, 279.
Gomori, G. (1948). *Archs Path.*, **45**, 48.
Gomori, G. (1954). *J. Histochem. Cytochem.*, **2**, 50.
Grutzner, P., and Menzel, H. (1879). *Pflug. Archs ges. Physiol.*, **20**, 395.
Håkanson, R., Juhlin, L., Owman, C., and Sporrong, B. (1970). *J. Histochem. Cytochem.*, **18**, 93.
Håkanson, R., Lilja, B., and Owman, C. (1967). *Europ. J. Pharmac.*, I, 188.
Håkanson, R., Lilja, B., and Owman, C. (1969). *Histochemie*, **18**, 74.
Håkanson, R., Owman, C., and Sporrong, B. (1968). Proc. III Int. Cong., Histochem. Cytochem. Springer, New York, p. 89.
Hamberger, B. (1967). *Acta physiol. scand.*, Suppl. 295, p. 1.
Hamberger, B., Malmfors, T., and Sachs, C. (1965). *J. Histochem. Cytochem.*, **13**, 147.
Hamberger, B., and Norberg, K.-A. (1964). *J. Histochem. Cytochem.*, **12**, 48.
Hamperl, H. (1932). *Virchows Archs*, **286**, 811.
Hamperl, H. (1951). Personal communication.
Hamperl, H. (1952). *Virchows Archs*, **321**, 482.
Heene, R. (1968). *Histochemie*, **14**, 324.
Heller, J. H., Setlow, R. B., and Mylor, E. (1949). *Am. J. Physiol.*, **161**, 268.
Hess, S. M., and Udenfriend, S. (1959). *J. Pharmac. exp. Ther.*, **127**, 175.
Holcenberg, J., and Benditt, E. P. (1961). *Lab. Invest.*, **10**, 144.
Jacobson, W. (1939). *J. Path. Bact.*, **49**, 1.
Jacobson, W. (1946). *Biochem. J.*, **40**, 3.
Jacobson, W. (1958). "5-Hydroxytryptamine". Ed. G. P. Lewis. Pergamon, London, pp. 38 and 158.
Jepson, J. B., and Stevens, B. J. (1953). *Nature, Lond.*, **172**, 772.
Jonnard, R. (1934). *J. Physiol. Path. gen.*, **32**, 731.
Jonsson, G. (1967). *Histochemie*, **8**, 288.
Jonsson, G., and Sachs, C. (1970). *Europ. J. Pharmac.*, **8**, 58.
Jonsson, G., and Sandler, M. (1969). *Histochemie*, **17**, 207.
Juhlin, L., and Shelley, W. B. (1966). *J. Histochem. Cytochem.*, **14**, 525.
King, M. J., and Angelakos, E. T. (1968). Proc. III Int. Cong. Histochem. Cytochem. Springer, New York, p. 128.
Lagunoff, D., Phillips, M., and Benditt, E. P. (1961). *J. Histochem. Cytochem.*, **9**, 534.
Larson, B., Owman, C., and Sundler, F. (1966). *Endocrinology*, **78**, 1109.
Laties, A. M., Lund, R., and Jacobowitz, D. (1967). *J. Histochem. Cytochem.*, **15**, 535.

LEMBECK, F. (1953). *Nature, Lond.*, **172,** 910.
LEMBECK, F. (1954). *Archs exp. Path. Pharmakol.*, **221,** 50.
LEVINE, R. J., LOVENBERG, W., and SJOERDSMA, A. (1964). *Biochem. Pharmac.*, **13,** 1283.
LILLIE, R. D. (1956). *J. Histochem. Cytochem.*, **4,** 118.
LILLIE, R. D. (1961). *J. Histochem. Cytochem.*, **9,** 184.
LILLIE, R. D., BURTNER, H. J., and HENSON, J. P. G. (1953). *J. Histochem. Cytochem.*, **1,** 154.
LISON, L. (1931). *Archs Biol.*, Paris, **41,** 343.
LISON, L. (1936). "Histochimie Animale". Gautier-Villars, Paris.
LOVENBERG, W., JEQUIER, E., and SJOERDSMA, A. (1967). *Science, N.Y.*, **155,** 217.
MANHEIMER, L. H., and SELIGMAN, A. M. (1949). *J. natl Cancer Inst.*, **9,** 181.
MARTIN, G. M., BENDITT, E. P., and ERIKSEN, N. (1960). *Nature, Lond.*, **186,** 884.
MENTEN, M. L., JUNGE, J., and GREEN, M. H. (1944). *J. biol. Chem.*, **153,** 471.
MUKHERJI, M., RAY, A. K., and SEN, P. B. (1966). *J. Histochem. Cytochem.*, **14,** 479.
NUSSBAUM, M. (1879). *Archs mikr. Anat.*, **16,** 532.
ÖHMAN, S., and SHELLEY, W. B. (1968). *Nature, Lond.*, **220,** 378.
ÖHMAN, S., and SHELLEY, W. B. (1969). *Life Sci.*, **8,** 27.
PEARSE, A. G. E. (1956). *Riv. Istochim.*, **2,** 103.
PEARSE, A. G. E. (1960). "Histochemistry, Theoretical and Applied". 2nd ed. J. & A. Churchill, London.
PEARSE, A. G. E. (1966). *Nature, Lond.*, **211,** 598.
PEARSE, A. G. E. (1966). In "Immunological Properties of Protein Hormones". Acad. Press, New York, p. 7.
PEARSE, A. G. E. (1968). *Proc. roy. Soc. B.*, **170,** 71.
PEARSE, A. G. E., and CARVALHEIRA, A. F. (1967). *Nature, Lond.*, **214,** 929.
PEARSE, A. G. E., COULLING, I., WEAVERS, B., and FRIESEN, S. (1970). *Gut*, **11,** 649.
PEARSE, A. G. E., and MCGREGOR, M. M. (1964). *Rep. Br. Europ. Cancer Camp.*, **2,** 665.
PEARSE, A. G. E., and PEPLER, W. J. (1957). *Nature, Lond.*, **179,** 589.
PEARSE, A. G. E., and ROST, F. W. D. (1969). *J. Microsc.*, **89,** 321.
PENTTILÄ, A. (1966). *Acta physiol. scand.*, **69,** Suppl. 281.
PENTTILÄ, A. (1967). *Histochemie*, **11,** 185.
PENTTILÄ, A., and HIRVONEN, J. (1969). *Scand. J. Gastroenterol.*, **4,** 489.
PERNOW, B., and WALDENSTRÖM, J. (1954). *Lancet*, ii, 951.
PICTET, A., and SPENGLER, T. (1911). *Ber. dt. chem. Ges.*, **44,** 2030.
PLETSCHER, A., SHORE, P. A., and BRODIE, B. B. (1955). *Science, N.Y.*, **122,** 374.
PLETSCHER, A., SHORE, P. A., and BRODIE, B. B. (1956). *J. Pharmac. exp. Ther.*, **116,** 84.
PLOEM, J. S. (1969). *Leitz. Mitt. Wiss. Tech.*, **4,** 225.
PRICE, H. L., and PRICE, M. L. (1957). *J. lab. clin. Med.*, **50,** 769.
RATZENHOFER, M. (1966a). *Mikroscopie*, **21,** 57.
RATZENHOFER, M. (1966b). *Klin. Wschr.*, **44,** 109.
RATZENHOFER, M., and LEMBECK, F. (1954). *Z. Krebsforsch.*, **60,** 169.
RITZÉN, M. (1967). *Exp. Cell Res.*, **44,** 505.
RITZÉN, M., HAMMARSTRÖM, L., and ULLBERG, S. (1965). *Biochem. Pharmac.*, **14,** 313.
ROST., F. W. D., and EWEN, S. W. B. (1971). *Histochem. J.*, **3,** 207.
SACHS, C. (1970). *Acta physiol. scand.* Suppl. 341.
SCHAUER, A., and WERLE, E. (1959). *Z. ges. exp. Med.*, **131,** 100.
SHARPLES, W. (1945a). *Anat. Rec.*, **91,** 107.
SHARPLES, W. (1945b). *Ibid.*, **91,** 237.
SHELLEY, W. B., and ÖHMAN, S. (1968). *Histochemie*, **15,** 287.
SHELLEY, W. B., ÖHMAN, S., and PARNES, H. M. (1968). *J. Histochem. Cytochem.*, **16,** 433.
SHORE, P. A., BURKHALTER, A., and COHN, V. H. JR. (1959). *J. Pharmac. ex. Ther.*, **127,** 182.
SLORACH, S. A., and UVNÄS, B. (1968). *Acta physiol. scand.*, **73,** 457.

SMITH, A. N., NYHUS, L. M., DALGLIESH, C. E., DUTTON, R. W., LENNOX, B., and MACFARLANE, P. S. (1957). *Scot. med. J.*, **2**, 24.
SOLCIA, E., and SAMPIETRO, R. (1965). *Riv. Istochim. norm. patol.*, **11**, 265.
SOLCIA, E., and SAMPIETRO, R. (1967). *Nature, Lond.*, **214**, 196.
SOLCIA, E., SAMPIETRO, R., and CAPELLA, C. (1969). *Histochemie*, **17**, 273.
SOLCIA, E., SAMPIETRO, R., and VASSALLO, G. (1966). *J. Histochem. Cytochem.*, **14**, 691.
TAKAYA, K. (1970). *J. Histochem. Cytochem.*, **18**, 178.
THOMPSON, J. H., and CAMPBELL, L. B. (1967). *Experientia*, **23**, 825.
UDENFRIEND, S. (1962). "Fluorescence Assay in Biology and Medicine". Acad. Press, New York, p. 152.
UDENFRIEND, S. (1964). "Fluorescence Assay in Biology and Medicine". Acad. Press, New York, p. 152.
UDENFRIEND, S., BOGDANSKI, D. F., and WEISSBACH, H. (1955). *Science, N.Y.*, **122**, 972.
UDENFRIEND, S., BOGDANSKI, D. F., and WEISSBACH, H. (1956). *Fed. Proc.*, **15**, 493.
UDENFRIEND, S., CLARK, C. T., and TITUS, E. (1953). *J. Am. chem. Soc.*, **75**, 501.
VASSALLO, G., SOLCIA, E., and CAPELLA, C. (1969). *Z. Zellforsch.*, **98**, 333.
VIALLI, M. (1962). *Riv. Anat. patol. Oncol.*, **21**, 392.
VIALLI, M. (1965). In "Handbook of Experimental Pharmacology". Vol. 19, p. 1, Springer, Berlin.
VIALLI, M. (1966). *Acta histochem.*, **23**, 163.
VIALLI, M., and FANTIN, A. M. B. (1966). *Acta histochem.*, **23**, 171.
VIALLI, M., and QUARONI, E. (1956a). *Riv. Istochim.*, **2**, 213.
VIALLI, M., and QUARONI, E. (1956b). *Ibid.*, **2**, 111.
VIALLI, M., and TRAVERSO, G. (1960). *Istituto Lombardo* (*Rend. Sci. B.*)., **24**, 167.
VON EULER, U. S., and LISHAJKO, F. (1959). *Acta physiol. scand.*, **45**, 122.
VULPIAN, A. (1856). *C. R. Acad. Sci., Paris*, **43**, 663.
WALLERSTEIN, J. S., ALBA, R. T., and HALE, M. G. (1947). *Biochim. biophys. Acta*, **1**, 175.
WEIL-MALHERBE, H. (1960). *Biochim. biophys. Acta*, **40**, 351.
WEIL-MALHERBE, H., and BIGELOW, L. B. (1968). *Analyt. Biochem.*, **22**, 321.
WEIL-MALHERBE, H., and BONE, A. D. (1952). *Biochem. J.*, **51**, 311.
WHUR, P., and GRACIE, M. (1967). *Experientia*, **15**, 655.
ZBINDEN, G., PLETSCHER, A., and STUDER, A. (1957). *Schweiz med. Wschr.*, **87**, 629.

CHAPTER 28

INORGANIC CONSTITUENTS AND FOREIGN SUBSTANCES

Introduction

This chapter is divided into two sections. The major part concerns the localization of inorganic constituents of the tissues, and of inorganic materials introduced into the body by one means or another. The second, lesser, part deals with the localization of a number of exogenous organic compounds, introduced into the organism as drugs or for other reasons, and of a few endogenous products which do not fit readily into any other chapter.

From the point of view of the routine histochemist, iron and calcium are the two most important inorganic constituents of tissue sections, with phosphate and carbonate closely following. This is because satisfactory and simple methods exist for their demonstration. From the physiological point of view, sodium, potassium and chloride are far more important than the constituents mentioned above but for these the ordinary methods of histochemistry have in the past been largely unsatisfactory.

Numerous chemical techniques exist for the localization in the tissues of other metallic ions, e.g. gold, silver, platinum, mercury, lead, copper, nickel, aluminium, zinc, palladium, uranium, bismuth and arsenic. With most, if not all, of these metals it is necessary first to introduce them into the tissues, and their subsequent "specific" demonstration depends largely on the fact that one knows what to expect. There are three alternative ways in which the problem of localizing inorganic ions in the tissues may be undertaken. These are (1) micro-incineration, (2) autoradiography and (3) electron probe analysis. The first is capable of giving accurate information about the total distribution of certain minerals in tissue sections but, unless emission histospectrography is used in conjunction, only calcium, magnesium, iron and silicon can be distinguished. The amount of information which can be obtained from the simple micro-incineration of a tissue section is limited, although by the addition of histospectrographic techniques such as those employed by Policard (1933) and by Scott and Williams (1935), and of analytical electron microscopy (Scott and Packer, 1939), quantitative information about very small quantities of minerals becomes available. An excellent review by Horning (1951) appeared in Bourne's *Cytology and Cell Physiology*, in which the status of micro-incineration in modern histochemistry was fully discussed. The method itself was dealt with by Horning, and also by Glick (1949).

The applications of autoradiography are discussed in Chapter 30. This method cannot give information about minerals of the type given by micro-incineration but only about the handling in the tissues of various elements introduced in radioactive form. This kind of information, being functional and not static, is likely to be of greater importance than the other. The use of the electron probe is discussed in Chapter 32 and in this chapter, therefore, only the ordinary histochemical methods for inorganic elements will be discussed.

Histochemistry of Inorganic Cations

Lakes and Chelates

Many histochemical methods for inorganic elements, especially metals, depend on the formation of coloured compounds with coloured or colourless reagents. In the pigment and dye industry the word lake (Sanskrit, *laksha*, a hundred thousand) has been used very loosely, but in histological practice it usually means an adsorption compound formed by union of a dye with a metal or metal salt. Lake-forming metal salts, when used as histological reagents, are referred to as mordants (Latin, *mordere*, to bite) from an old idea that they attack the material being dyed before causing the dye to adhere. In this chapter the word lake has been used to mean an adsorption compound of metal salt with dye, particularly when the nature of their union is a simple electrovalent salt linkage, or when it is unknown. The term chelate, on the other hand, is used to refer to non-ionic metallic linkages of the types illustrated below:

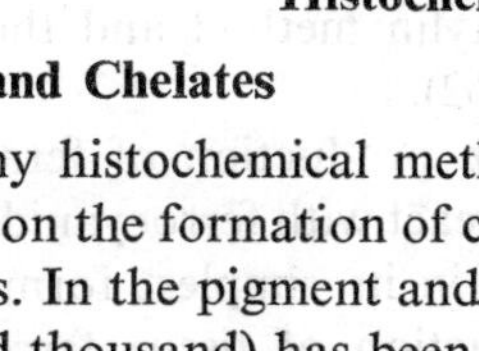

1, Nitroso—2—naphthol

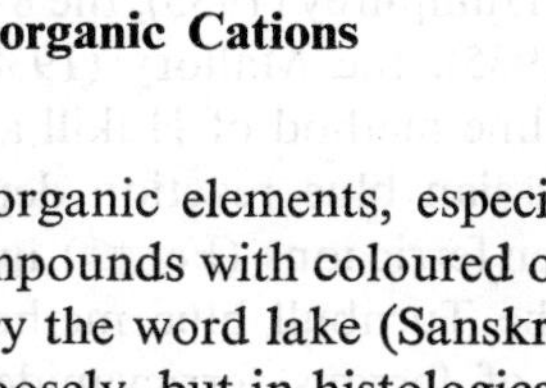

Anthraquinone

where the metal is held by strong covalent (co-ordinate) linkages (35–100 k.cal. per mole). A chelator, therefore, is a compound capable of forming a metal chelate, like those shown above, where M may be any of a number of metals and where the atoms of the metal may be bound in a ratio of 1 : 1 of the chelator or in some other ratio. For a molecule to function as a chelator it must (1) possess at least 2 atoms which can attach to the metal, and (2) the atoms must be joined so that their valence bonds are at their natural angle. (The natural angle of the C atom = 108, i.e. the angles of a pentagon.) The most usual atoms which form the electron donors of a chelator are O, N, and S.

The literature on colour lakes was thoroughly reviewed by Blumenthal (1946) and also by Venkataraman (1952) in his two volumes on the "Chemistry of Synthetic Dyes." The whole subject of the coordination compounds was covered by Bailar (1956) whose book should be consulted by those interested in the theoretical aspects of chelation.

Methods for Inorganic Iron

There are three classical methods for this purpose: the Prussian blue (Perls, 1867), Turnbull blue (Schmeltzer, 1933), and iron sulphide (Quinke, 1868) methods. Four other tests which are available are the dinitrosoresorcinol methods of Humphrey (1935), the 8-hydroxyquinoline method of Thomas and Lavalloy (1935), the Mallory (1938) haematoxylin method and the bathophenanthroline method of Hukill and Putt (1962).

The Prussian blue reaction depends on the production of ferric ferrocyanide when ferric ions (Fe^{+++}) in the tissues react with ferrocyanide in acid solution. The Turnbull blue method depends, in its simplest form, on the production of ferrous ferricyanide by the reaction of tissue ferrous ions (Fe^{++}) with ferricyanide. Since, however, iron in the tissues is almost invariably present in ferric form, the Turnbull blue reaction is usually performed after the manner of Schmeltzer (Tirmann-Schmeltzer), that is to say, following conversion of Fe^{+++} to Fe^{++} with $(NH_4)_2S$. The product of Perls reaction, Prussian blue, formed from iron (III) and hexacyanoferrate (II), is identical with Turnbull's blue, formed from iron (II) and hexacyanoferrate III).

The three best reviews on the demonstration of inorganic iron in the tissues are those of Gomori (1936), Lillie (1948) and Bunting (1949). The last two authors agreed that neutral 10 per cent. formalin was the best fixative for the purpose and that by its use more iron is preserved, with less diffusion, than by other and especially by alcoholic fixatives. Bunting confirmed the earlier observations of Gomori that the particles of iron demonstrated by the Turnbull blue method were larger than those produced by Quinke's method. Both were larger and more distorted than the Prussian blue particles. This last reaction, moreover, is recorded as having the highest sensitivity of the three methods (0·002 γ of Fe^{+++}, according to Lison, 1953; 0·1 γ of Fe^{+++} according to Feigl, 1954) and it certainly reveals more iron than the other two.

Treatment with dilute acid is an essential preliminary in the performance of the Prussian blue reaction since by this means the ferric ion is liberated from unreactive loose combinations with protein, such as haemosiderin. Iron combined in the tissues as haemoglobin (masked iron) is more strongly bound and cannot be released by acid treatment. The application of 100 vol. H_2O_2 (30 per cent) for a short period makes some of this masked iron available for Perls' reaction. In performing the reaction the two reagents are invariably employed simultaneously, since, if treatment with ferrocyanide follows treatment with acid, diffusion of iron is considerable. Various investigators have used different concentrations of acid and ferrocyanide. Gomori (1936) used equal parts of 20 per cent HCl and 10 per cent $Na_4Fe(CN)_6$; Lison (1936) used equal parts of 2 per cent (v/v) HCl and 2 per cent (w/v) ferrocyanide, and Lillie (1948) equal parts of 5 per cent acetic acid and 2 per cent $K_4Fe(CN)_6$, at a temperature of 80°. Bunting (1949) varied the concentrations of both acid and ferrocyanide and obtained the best results with those employed by Lison. Because of its high degree of sensitivity and specificity the Prussian blue method of Perls, using the concentrations advised by Lison, is the method of choice for the histochemical demonstration of haemosiderin iron and other protein-bound forms. The employment of Perls' reaction following micro-incineration, and celloidinization of the ash, was proposed as a sensitive method for the demonstration of very small quantities of iron by Fenton *et al.* (1964).

The Quinke method gives a green or brownish-black colour to the large particulate deposits of haemosiderin but the finer particulate and the non-particulate deposits are not distinguishable by this method. Much more sensitive is the $(NH_4)_2S$-dinitrosoresorcinol method of Humphrey, which depends on the formation of a greenish-black ferrous iron-lake. This method works well and may be employed if, for any reason, complete freedom from iron-containing reagents is necessary, as it might be in a critical histochemical investigation. The Humphrey method used to have the advantage of permanency over the Prussian blue reaction, for mounting in Canada balsam, but since the introduction of synthetic mounting media this advantage has largely disappeared.

The Tirmann-Schmeltzer reaction is not recommended. Lillie has shown that it demonstrates only a small portion of the available iron present. If sulphide-treated sections are placed in acid ferrocyanide instead of in acid ferricyanide, a perfectly good Prussian blue reaction occurs.

The three remaining methods depend, like that of Humphrey, on the formation of a coloured iron-lake. With the Thomas and Lavalloy method I was (1953, 1960) completely unable to obtain a satisfactory result and, indeed, the rationale of the method requires explanation. If ferric iron is released from combination with protein by the 4 per cent acetic acid of their hydroxyquinoline solution, the ferric hydroxyquinoline complex so formed may not be present in sufficient concentration for its solubility product to be exceeded locally. Thus a precipitate may not occur. Thomas and Lavalloy sought to establish conditions suitable for precipitation, after the initial reaction was ended, by the addition of 25 per cent ammonium hydroxide. In my experience this produces a diffuse precipitate of hydroxyquinoline together with a few particles of the ferric salt. The latter are not related to iron-containing structures and one would scarcely expect them to be. Glick (1949) suggested that the ferric hydroxyquinoline produced by this method might be identified by its characteristic fluorescence, but a precipitate of this substance in the tissues does not give a satisfactory fluorescence.

Mallory's method for iron is sensitive but of low specificity, since a blue-black haematoxylin lake is formed with many other metals, including calcium. It is possible to exclude most of these without much difficulty and the method may still be useful, on occasion, for inorganic iron. Haemosiderin usually stains only brown, but if it is first treated with $(NH_4)_2S$ the blue-black lake is produced, The interstitial and connective tissue varieties of diffuse "haemosiderin" give the blue-black colour without sulphide treatment. The bathophenanthroline method (Hukill and Putt, 1962) produces a red coloured dye with ferrous iron in the pH range 3–7. The reagent itself (4,7-diphenyl-1,10-phenanthroline) is poorly soluble in water, in which the iron chelate is insoluble. Both reagent and chelate are soluble in ethanol and in most lipid solvents. As the chelate is formed only with Fe^{2+} it is necessary to convert tissue Fe^{3+} either before staining or by the incorporation of a reducing agent in the medium. In practice the second alternative has been found effective. The method is described in Appendix 28.

Other chelating reagents have been used as sensitive analytical reagents for Fe^{2+}. Collins *et al.* (1959), and Diehl *et al.* (1960) used 2,4,6-tri-(2′-pyridyl)-1,3,5-triazine which forms an insoluble red chelate with Fe^{2+}.

Ferritin

This soluble iron-containing protein was first isolated from horse spleen in 1937 by Laufberger, who showed that it contained 17–23 per cent of iron. It consists (Granick, 1951) of a homogeneous protein called apoferritin (mol. wt. 460,000) and an iron-containing part said to be present in the form $(FeOOH)_8.(FeOPO_3H_2)$. This iron can be removed from the protein by the action of reducing agents such as hydrosulphite. Both apoferritin and ferritin are precipitated in crystalline form by cadmium sulphate (Granick, 1946), and this forms the basis of one histochemical method for their demonstration (see below). Using rabbits, Granick (1943) prepared anti-horse ferritin antibodies which precipitated both ferritin and apoferritin and, by means of the precipitin reaction, he showed that ferritin was present in spleen, bone marrow, liver, testis, kidney, pancreas, ovary and lymph nodes. These organs are listed in order of diminishing ferritin content. In the case of the dog Mazur and Schorr (1950) calculated the γ of ferritin N per gram of wet tissue as follows: liver, 97; spleen, 94; bone marrow, 16; kidney cortex, 17; pancreas, 6; skeletal muscle, 4.

Ferritin is now regarded as the main storage form of iron, having thus taken the place of haemosiderin, which is to be considered as the storage of iron in excess. It is formed in the mucosa of the gastro-intestinal tract when iron is given by mouth and in other tissues when, for instance, iron is set free by haemorrhage or haemolysis. In general, tissues which have a high haemosiderin content contain large amounts of ferritin. In the placenta iron is transported from mother to foetus against a concentration gradient, and Vosburgh and Flexner (1950) have suggested that a special transport mechanism exists. Nylander (1953) succeeded in demonstrating a soluble form of ferric iron in the villous portion of the visceral yolk sac membrane in the rat by means of the Prussian blue reaction on fresh tissues. In the visceral yolk sac, and in the allantoic placenta, he was able to demonstrate the typical isotropic, octahedral yellowish-brown crystals after incubation in cadmium sulphate. This technique was also used by Diezel (1955) to demonstrate the presence of ferritin in the globus pallidus of the human brain. Richter (1960) used three different methods of fractionating haemosiderin granules and demonstrated, by electron microscopy, the presence in the granules of dense particles corresponding to the iron hydroxide micelles of ferritin. He considered that an abnormal cellular metabolic pathway of ferritin was implicated in the pathogenesis of haemochromatosis and transfusional haemosiderosis.

Although it is strictly a method for demonstrating the protein part of ferritin, it is probable that the cadmium sulphate technique, together with the Prussian blue reaction, will continue to be used for the localization of ferritin. For this reason it has been considered in this chapter and details are given in Appendix 28, p. 1403.

Methods for Calcium

These methods are divisible into two classes: those which are more or less specific for calcium, and those which really indicate the phosphate or carbonate with which the calcium is often combined. Calcium occurs in the animal body in a number of different forms. It may be present as a soluble salt (chloride, sulphate or lactate) in the blood serum, in an ionized or ionizable form in combination with serum or other proteins, in masked form, or in an insoluble form as deposits of phosphate or carbonate, or of the two together.

Soluble Forms. These have been demonstrated by treating fresh tissues with soluble salts of oxalic or tartaric acids, the insoluble calcium salt being deposited *in situ* in crystalline form. According to Lison (1936a, p. 156), the sensitivity of these methods was high; he gave a figure of $0{\cdot}06\gamma$ for the oxalate method. Crétin's method, described below, can also be used for the demonstration of soluble calcium salts in fresh tissues. Rabl (1923, 1926) fixed thin slices of tissue in a mixture of equal parts of 4 per cent ammonium oxalate and neutral formalin or in formalin saturated with ammonium oxalate in the cold. Phosphates and carbonates of calcium did not react by this method, according to its author, but Freudenberg (1926) maintained that the prolonged action of oxalate could cause these insoluble salts to react. Lison cautioned against too strict an interpretation of the localization of soluble calcium by Rabl's method.

Ionized or Ionizable Forms. Calcium in these forms is identifiable by two methods whose specificity is high. In the first, the section is placed in 40 per cent alcohol and treated with a single drop of 3 per cent sulphuric acid. A coverslip is then applied and the section observed under the microscope. In a short while coarse crystals of gypsum ($CaSO_4 . 2H_2O$) are formed. The localization of the crystals is broadly over the area in which the calcium was previously situated, but definition on the cytochemical scale is impossible. Lillie (1954, p. 266) recommended as an alternative that 10 per cent oxalic acid should be drawn under the coverslip by means of blotting-paper applied to the opposite edge. Crystals of calcium oxalate which formed were easily recognizable, but localization was no more accurate by this than by the gypsum method. Turchini (1924) employed a solution of iodic acid in gelatine, which he applied to sections with the idea of forming insoluble calcium iodate, diffusion being kept to a minimum by the presence of gelatine. This process was designed for, and used upon, micro-incinerated specimens and it seems that it is not applicable to fresh or fixed tissues. Godlewski (1960) recommended micro-incineration followed by spraying of the spodogram with H_2SO_4.

Masked Calcium. Calcium in this form is demonstrable by micro-incineration, applied either to fresh frozen sections, or to sections of fixed tissue or, preferably, to freeze-dried material. It is not proposed to deal with the technique in this book.

Insoluble Forms. The methods described below are principally applicable to insoluble deposits of calcium phosphate and carbonate, though some have been employed for the other forms. A method depending on the formation of a blue calcium lake was *Crétin's gallic acid-metaldehyde method* (Crétin, 1922, 1923, 1924) and according to Lison (1936a), this had a high degree of sensitivity. Other metals were stated to give distinguishable colours with the Crétin reagent (barium, cadmium and strontium bluish-green; iron dark brown; and magnesium, zinc, lead and silicon various shades of yellow). Lillie (1954) continued to give details of the method but stated that he had not, in fact, tried it. Bunting (1951), in a paper on the histochemical analysis of mineral deposits, found the Crétin reaction to be useless, and I agreed entirely with this view. Gomori (1952, p. 33) regarded the Crétin test as too capricious to be dependable and Lison (1953, p. 162) no longer recommended it.

A second method making use of the formation of a calcium lake is the *Roehl and Leutert haematoxylin test*. Sections are treated for a few minutes in a concentrated alcoholic solution of either haematoxylin or haematein, and deposits of calcium are coloured bluish-black. This method we customarily employ in principle when staining sections with Ehrlich's haematoxylin and eosin. Its specificity is obviously very low. It serves primarily to draw attention to the presence of calcium deposits but may be used more critically for the demonstration of these if a control section is stained identically after brief treatment (1–2 minutes) with dilute acetic (5 per cent) or nitric (0·25 per cent) acids. The blue-black calcium deposits of the test section will be absent in the case of the control. This reaction must not be used for critical histochemistry.

A third group of methods depending on the formation of calcium lakes employs various compounds related to alizarin. *Alizarin-purpurin methods.* The dyes used are all derivatives of anthraquinone. Grandis and Mainini (1900, 1902) used purpurin as a saturated alcoholic solution applied to alcohol-fixed frozen sections. Crétin (1924) and Cameron (1930) used alizarin sulphonic acid in the form of its sodium salt, Bourne (1943) used either this salt or 1,2,5,8-tetrahydroxyanthraquinone for staining the calcium salts of bone *in vivo*. Langeron's (1934) method for staining calcium deposits in paraffin sections employed sodium alizarin sulphonate also. The early alizarin-purpurin methods possessed a low degree of sensitivity for calcium and their specificity was not high. Aluminium salts may give brilliant scarlet red lakes which are more satisfactory than the calcium lakes, often described as red but usually appearing in various shades of orange. Chromium, barium, strontium and iron also give coloured lakes with various anthraquinone derivatives, those of the last-mentioned being deep violet-black in colour.

A more sensitive technique using alizarin red S was evolved by Dahl (1952), who stressed the importance of pH in the performance of the test. Table 80, below, summarizes his findings in respect of the alizarin complexes of various elements. The sensitivity of his alizarin method was stated by Dahl to be greater when used histochemically than in spot tests. With most

TABLE 80

Solubility of Alizarin Complexes

Salt	pH range (max. ppt.)	Colour
$CaCl_2$ . . .	4·28–6·75	Reddish-orange
$FeCl_2$ . . .	5·62–8·05	Deep purple
$Fe(NO_3)_3$ . .	5·73–6·25	Purplish-black
$Ba(C_2H_3O_2)_2$ . .	4·08–8·05	Reddish-orange-magenta
$SrCO_3$ . . .	4·5 –8·05	Reddish-orange-magenta
$Be(NO_3)_2$ . .	5·73–8·05	Magenta
$CdCl_2$ . . .	5·11–5·78	Reddish-orange or magenta
$La(C_2H_3O_2)_3$ . .	5·73–8·05	Cherry to light purple
$Pb(C_2H_3O_2)_2$ . .	4·08–4·5	Reddish-orange to purple
$Hg_2(NO_3)_2$ . .	4·5 –5·62	Purple

inorganic methods it is my impression that the opposite is the case. For those who wish to try it this method is given in Appendix 28, p. 1404. Alizarin red S was used by Wiener and Wolman (1965), following micro-incineration, to demonstrate especially lipid-bound calcium in cells and nerve fibres. The mechanism of alizarin staining was investigated by Puchtler *et al.* (1969). They found that salt formation between calcium (deposits) and alizarin or alizarin red S was markedly sensitive to pH. Alizarin itself stained well only at pH 12 while the optimum for alizarin red S was around pH 9. This observation, which is clearly accurate, does not correspond, as might be expected, with Dahl's figures for maximum precipitation.

Other lake-forming, or chelating, compounds have been brought into histochemical use for demonstrating calcium deposits. Kaufmann and Adams (1954) used a saturated solution of murexide (ammonium purpurate) in 0·1 N KCN, and McGee-Russell (1955) described the use of the anthraquinone dye Nuclear fast red. This he proposed to call calcium red because of the ambiguity attached to the former name on account of the German equivalent,

Kernechtrot, which is used for two very different classes of dye. The calcium red method (Appendix 28, p. 1404) works very well in the case of many invertebrate tissues, but, in my hands, it is less reliable for calcium deposits in mammalian tissues where diffuse staining of the tissues usually results. Since the calcium-calcium red lake is water-soluble the shortest possible staining time should be employed.

Many members of the phthalocyanin class of dyes will stain calcium deposits (Pearse, 1955), and of these Durazol fast blue 8G and Alcian blue 8GS (I.C.I. Ltd.) are particularly useful. The former was employed to stain calcium in the gill plates of *Anodonta cygnea*, illustrated in Figs. 231 and 232, p. 1150. Because of the effect on lake formation of the different physical states in which calcium occurs in the tissues it is often worth while to use a battery of phthalocyanins or a water-soluble followed by a spirit-soluble one (see Appendix 28, p. 1404).

Kaufmann and Adams refer to the use of Eriochrome Black T as a stain for tissue calcium. This dye forms complexes with Ca and Mg, which differ in colour from the tissues stained by the dye, and the nature of these complexes has been described by Young and Sweet (1955). I had little success with this method, however, and cannot recommend it.

A new complexing reagent for calcium was developed by Close and West (1960). Described as calcichrome this compound,

[cyclo-tris-7-(1-azo-8-hydroxynaphthalene-3 : 6-disulphonic acid)],

whose formula is shown below, gives a red colour with Ca^{2+} in alkaline solution.

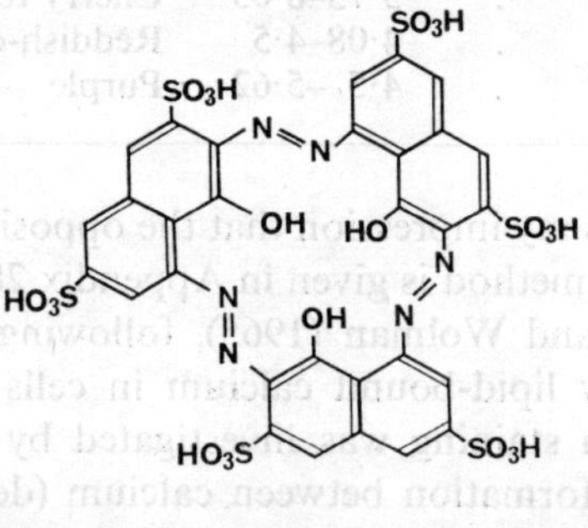

Its application to the histochemical demonstration of calcium was proposed by Ogawa (1964). I have no experience of the method but the sensitivity of the reagent and the lack of interference from magnesium ions suggests that it should be particularly suitable for invertebrate tissues.

The dinaphthylphenylmethane dye Naphthochrome green B (or G), which was used by Denz (1949) to stain deposits of beryllium (*q.v.*) in the tissues, forms a reddish-brown complex with calcium salts. In mercury-fixed tissues, however, calcium deposits are stained green. These are easily distinguishable from those of beryllium (pale apple green) or aluminium (rich dark green) by the fact that they are removed, before or after staining, by treatment with dilute acids. I consider that after mercury fixation Denz's method (using 0·5 per cent aqueous Naphthochrome green at pH ± 7·2 instead of the original pH 5·0), merits serious consideration as a routine method for demonstrating calcium salts. In my hands it is one of the best of the chelation techniques, having a coloured product visible in ordinary light (see Fig. 233, p. 1150).

The GBHA Method. Following the synthesis by Bayer (1957) of glyoxal-bis-(2-hydroxyanil) and its use by him and also by Goldstein and Stark-Mayer (1958) as a complexing reagent for Ca, Ba, Sr, Cd, Ca, Co and Ni ions, Kashiwa and Atkinson (1963) evolved a successful GBHA method for calcium in the tissues. The formula of GBHA (I), given below, suggests that the mechanism of calcium binding is by chelation (II), as shown.

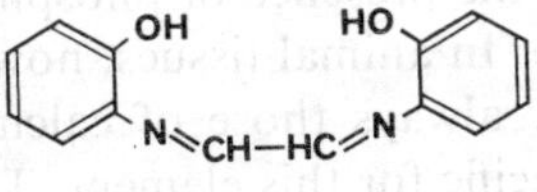

I

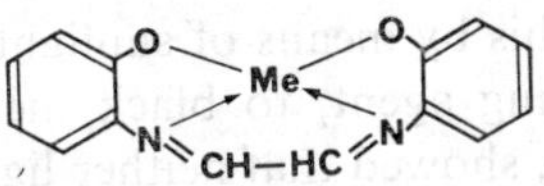

II

The specificity of the histochemical reaction depends on the use of a differentiating alcoholic carbonate-KCN solution. The GBHA complexes of Ba, Sr, Cd, Cu, Co and Ni are decolorized but the calcium complex is stable. The reaction is certainly a sensitive one but it is difficult to achieve the quality of results indicated, and illustrated, by the original authors. Fading is a notable feature, especially in less strongly stained areas. The mechanism of fading, using GBHA in a direct spectrophotometric technique, was investigated by Linstrom and Milligan (1967). Technical details of the method are given in Appendix 28, p. 1406. It was employed by Bots and Hardonk (1965) specifically for the demonstration of Ca^{2+} and Mg^{2+} in so-called pseudo-calcium deposits in brain.

Fluorescence Method for Calcium and other Metals

According to Feigl (1954) alcoholic solutions of 3,5,7,2′,4′-pentahydroxy-flavanol (morin) react with salts of aluminium, beryllium, indium, gallium, thorium, scandium and zirconium to give fluorescent complexes or adsorption compounds. Aluminium and beryllium (*q.v.*) are the only two metals in this list whose salts are at all likely to be found as foreign substances in mammalian tissues. I have observed, however, that calcium deposits as well as calcium in bone and cartilage give an intense whitish-green fluorescence when treated with morin in acetic-alcohol solution. This observation forms the basis of the fluorescent reaction for calcium described in Appendix 28, p. 1407.

Distinction between the fluorescence of the morin-calcium complex and those of other metals might possibly be made on the difference in colour (fluorescence spectrum). The aluminium complex, for instance, gives a much greener fluorescence than does the calcium lake (see p. 1141). The morin reaction in calcified cartilage from the bronchial tree of the rat is shown in Fig. 234, p. 1150.

The sensitivity of this method, histochemically, is higher than that of any of the visible light chelation methods, so that very small deposits of calcium phosphate or carbonate can easily be distinguished. Tissues containing calcium salts in non-particulate form also give the characteristic fluorescence.

Substitution Methods for Calcium. It is customary in routine histology to use the *silver method of von Kóssa* (1901) for the demonstration of calcium deposits. This depends on the presence of phosphate and carbonate and not on the presence of calcium. In animal tissues, however, insoluble phosphates and carbonates are nearly always those of calcium and the test is usually regarded as sufficiently specific for this element. The original principle of the method was the conversion of either salt into its silver equivalent and the subsequent reduction of this by means of sunlight or ultra-violet light, or by means of a strong reducing agent, to black metallic silver. Pizzolato and McCrory (1962), however, showed that neither light, nor reducing agent, was necessary to bring about the second stage of the reaction. None of the many substances which reduce silver nitrate to metallic silver is likely to cause confusion under histochemical conditions, except perhaps uric acid or urates. Gomori (1951) recommended the prior removal of these by treating sections with saturated aqueous lithium carbonate.

Oxalosis. Crystals of calcium oxalate were first demonstrated in urine by Donné (1839) and deposits in the kidneys were first recognized by Lepoutre (1925) and later described by Laas (1941) and Vischer (1947). The term oxalosis was invented by Chou and Donohue (1952) since when it has become established. Papers by Archer *et al.* (1957), Pyrah *et al.* (1959), Chisholm and Heard (1962), and Broyer and Berger (1968) may be consulted.

The reaction of calcium oxalate with the von Kóssa method was long supposed to be variable and uncertain. As an alternative Johnson (1956), Johnson and Pani (1962) and Wolman and Goldring (1962) suggested conversion of oxalate to carbonate by heating to 450°. The subsequent identification was originally made by addition of acid and detection of the resulting CO_2 bubbles. Wolman and Goldring, however, compared serial sections treated with the von Kóssa technique, one without and one following incineration. Oxalate crystals stained only after incineration. Noting that "pure" calcium oxalate crystals, produced in rats by the administration of ammonium oxalate or ethylene glycol, failed to stain by the von Kóssa technique Pizzolato (1964) developed a silver nitrate-hydrogen peroxide method which gave a black reaction product. Prior removal of phosphate and carbonate with dilute acid left only the birefringent oxalate crystals free to

TABLE 81

In Vitro *Tests on Reagents for Metals*

METAL	Rubeanic Acid	*o*-Tolidine	Benzidine	Rhodizonate	Diethyldithiocarbamate	Dithizone	DMABR *	Naphthochrome green	Acid solochrome cyanine	Solochrome azurine	Magneson	Titan yellow	SALT
Be^{2}								**Green**	**Purple**				$BeCl_2$
Mg^{2}											**Blue**	**Red**	$MgCl_2$
Al^{3}				Yellow		Orange		**Deep green**	**Rose-pink**	**Purple**			$Al(OH)_3$
Ca^{3}					Gold	Grey-green				Purple			$CaCl_2$
V^{5}				Yellow				Grey	Pink	Pink			V_2O_5
Mn^{2}					Brown	Light brown	+			Red			$Mn(NO_3)_2$
Fe^{2}	Yellow			**Blue-black**	**Black**	Light brown	+	Green		Blue			$FeSO_4$
Fe^{3}		Blue	Dark blue	**Blue-black**	**Black**	Pink	(+)	Green		Blue			$Fe_2(SO_4)_3$
Co^{2}	Yellow				Lime-green	Pink	+	Blue-grey		Purple			$Co(NO_3)_2$
Ni^{2}	**Blue**				Yellow	Brown-red		Green		Reddish-blue			$NiCl_2$
Cu^{1}	**Olive**	Light blue			Yellow-brown	Brown	++ **Red**	Green		Blue			$CuCl_2$†
Cu^{2}	**Olive**	**Dark blue**	**Blue-black**	Yellow	Yellow-brown	Brown	+ Brown	Green		Blue			$CuSo_4$
Zn^{2}						**Deep red**		Grey-green		Purple			$ZnSo_4$
Sr^{2}				**Red**						Pale blue			$SrCl_2$
Sn^{2}					Yellow	Pink	+			Orange			$SnCl_2$‡
Ba^{2}				**Red**		Brown		Red		Orange			$Ba(OH)_2$
Hg^{2}						Red	++ Purple	**Green**		Purple			$HgCl_2$
Pb^{2}	Yellow			**Purple**			++ Purple	Black					$Pb(NO_3)_2$
Bi^{3}						Brown-red				Orange			$BiCl_3$

* Dimethylaminobenzylidene rhodanine.
† Dissolved in NH_4OH and evaporated to dryness.
‡ Dissolved in 5 N–HCl.

react. Further use of this method was described by Pizzolato and Pizzolato (1966) and its application to plant tissues by Silver and Price (1969). The method is described in Appendix 28, p. 1405.

A different procedure was suggested by Yasue (1969) who treated calcium oxalate deposits, after removal of phosphate/carbonate, with 5 per cent $AgNO_3$, followed by rubeanic acid. The excellent results obtained with this method justify its inclusion in the Appendix.

Other substitution methods for calcium salts are not used for the demonstration of deposits. The cobalt and yellow ammonium sulphide method, which is used for revealing calcium orthophosphate in the alkaline phosphatase technique (Chapter 15), gives poor results on older calcium deposits. The method of Macallum, which involved conversion of the deposits to sulphate followed by treatment successively with lead acetate and yellow $(NH_4)_2S$, was shown by Lison (1936a) to be completely useless.

Methods for Various Metals

Inorganic histochemistry is probably the most neglected division of the science. The development of worthwhile techniques is slow and new methods are derived exclusively from the methods of chemical assay. The remainder of this section of the chapter is concerned with techniques for a variety of metal ions and for a few anions. Few of them find even occasional employment in applied histochemistry.

Methods for Aluminium

This element is not normally present in mammalian tissues, but powdered aluminium has been introduced into the lungs as a preventative for silicosis (Denny, Robson and Irwin, 1937). After such therapy it is present as a mixture of metal and metal hydroxide. Hydroxides of aluminium are known to form adsorption complexes with a variety of compounds, alizarin, quinalizarin, morin and aluminon among them, and the latter (ammonium aurine tricarboxylic acid) was described by Hammett and Sottery (1925) as a specific reagent for aluminium. Denny *et al.* (1939) first described the histochemical use of the *aluminon method*, and a modification of this was used by King, Harrison, Mohanty and Nagelschmidt (1955) for the demonstration of aluminium (hydroxide) in rat lung. The reagent is used in solution as the ammonium salt and differentiation is performed with ammonium carbonate. Lakes formed by the combination of aurine tricarboxylic acid and metals other than aluminium or beryllium are dissolved by ammonium ions at pH 7·2, and therefore by ammonium carbonate. In the absence of excess dye the beryllium lake is unstable so that the method becomes specific for aluminium salts. This method works reasonably well (Fig. 235, p. 1150), though it sometimes fails inexplicably when aluminium is present in the tissues in large amounts by chemical assay.

PLATE XXVII

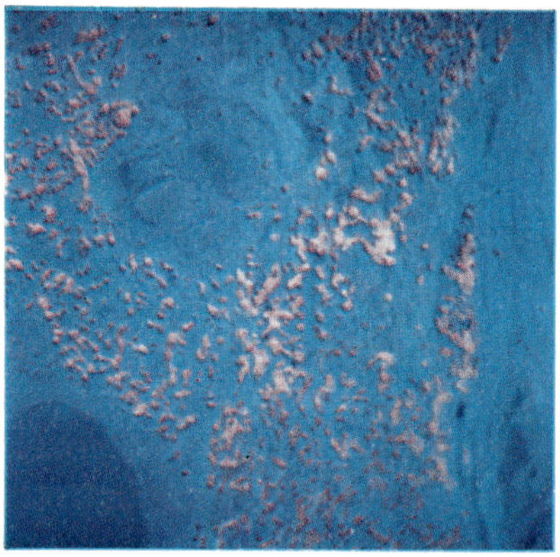

XXVIIa. Rat lung containing partly solubilized deposits of aluminium dust. These show strawberry-pink fluorescence. Solochrome dark blue BS. × 385.

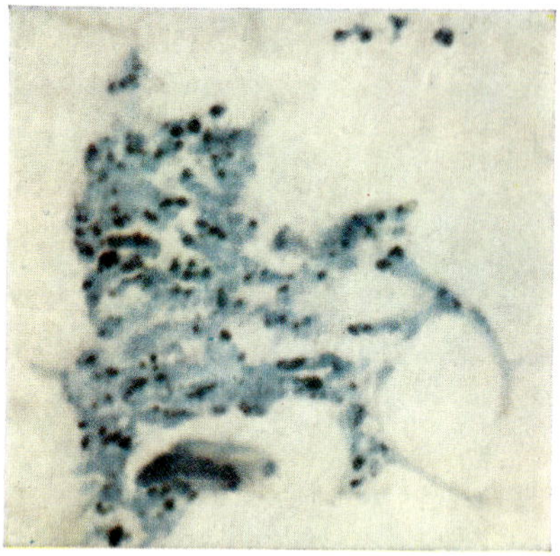

XXVIIb. Rat lung containing similar deposits. Solochrome azurine. × 250.

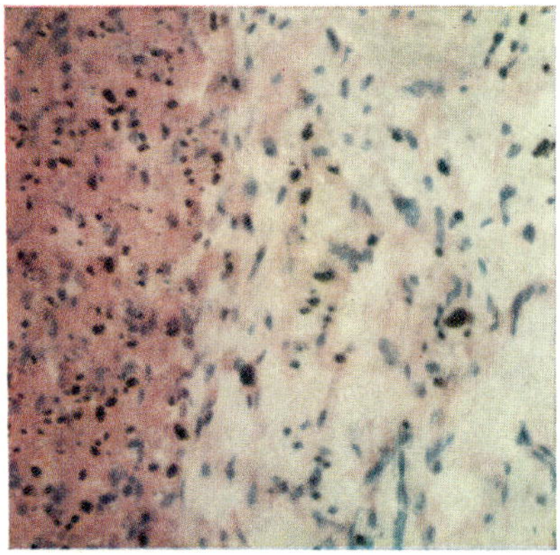

XXVIIc. Rat lung as in Plates XXVIIa and b. Acid Solochrome cyanine. × 300.

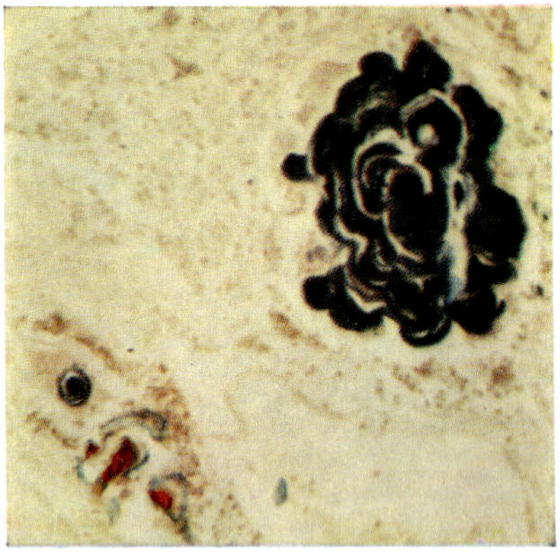

XXVIId. Human lymph node. Crystalline doubly refractile beryllium deposit surrounded by zone where a soluble beryllium salt is combined with protein. Alkaline Solochrome azurine. × 200.

[*To face p.* 1141

A second method which may be used to demonstrate aluminium hydroxide histochemically is the *morin method* already described for calcium. According to Feigl (1954) an inner complex salt is formed and this gives an intense greenish fluorescence. The identification of aluminium dust in lungs is more easily made by this method than by the aluminon method described above and distinction from calcium salts (by acid pretreatment) is also easy.

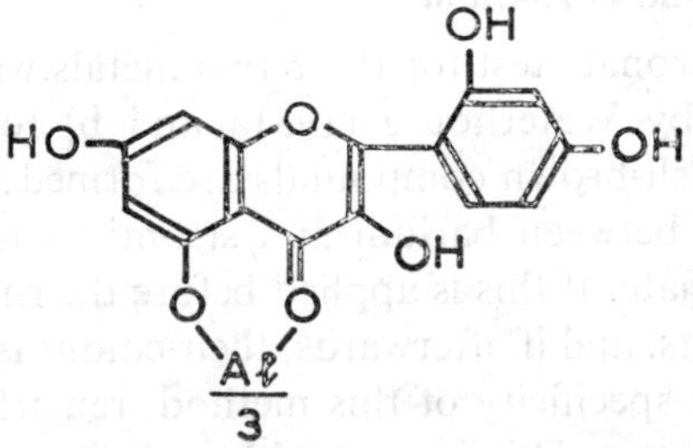

The *naphthochrome green method* (at acid or alkaline pH levels) gives a rich green lake with aluminium hydroxide. When, as with aluminium dust inflation experiments, the nature of the metal introduced into the tissues is known, this method is far more satisfactory than the more specific aluminon method.

A fourth method for aluminium salts, whose sensitivity and specificity appears to be high, is the *Solochrome blue-black R method.* In acid alcoholic solution this dye forms reddish chelates with salts of aluminium and calcium. The latter are first dissolved and then reprecipitated as the calcium-solochrome lake, which is insoluble in dilute acids. The new precipitate is diffuse, however, and in place of, say, a crystal of calcium phosphate measuring $2 \times 2\,\mu$ there is now a circular patch of amorphous granular deposit 100–200 μ in diameter. Moreover, while these granules do not fluoresce, the Solochrome-aluminium chelate fluoresces brilliantly in shades varying from yellowish-orange to strawberry-pink (Plate XXVIIa, opposite). These stand out clearly on a pale blue background. Other Solochrome dyes give coloured, non-fluorescent complexes with aluminium hydroxide. Methods employing these dyes are given in Appendix 28, pp. 1406 and 1408, and they are illustrated in Plates XXVIIb and XXVIIc.

Methods for Arsenates and Arsenites

The method most commonly used for the histochemical demonstration of As^{+++} and As^{++++} is that of Castel (1936), in which both arsenites and arsenates are precipitated as their cupric salts by fixing the tissues in a medium containing $CuSO_4$. The resulting green granular precipitate is visible microscopically. I have no experience of this method, but alternatives would seem to be worth trying. The method of Bettendorf (1870) is mentioned by Feigl (1954) as highly specific. It is based on the principle that ter- and quinquevalent arsenic compounds are reduced to the metal by warming with stannous chloride in 35 per cent HCl. Arsenic appears as a brownish-black precipitate. A test for As^{+++} described by this same author depends on the development of a red colour when arsenites are warmed with 0·5 per cent *n*-ethyl-*o*-hydroquinoline (kairine) and 1 per cent ferric chloride in strong HCl solution. Arsenates are previously reduced with hydroxylamine sulphate.

Method for Barium and Strontium

The sodium rhodizonate test for these two metals was described by Feigl (1924) and adapted by Waterhouse (1951a and b) for histochemical use. Intensely red or reddish-brown compounds are formed; calcium salts do not react and distinction between barium and strontium is made by treatment with potassium chromate. If this is applied before the rhodizonate it prevents staining of barium salts, and if afterwards, their colour is discharged. Gomori (1952) questioned the specificity of this method, remarking that Hg and Pb also give red rhodizonates. This is true (although from neutral solutions Pb gives a violet salt), but distinction of this sort is unlikely to be required histochemically. What is more usually called for is the localization of some already chemically determined metal. In practice deposits of barium sulphate, in tissues or on filter paper, give an almost black colour after rhodizonate treatment (Appendix 28, p. 1410), and this is converted to a fine scarlet red by treatment with strong mineral acids. Ferric and ferrous salts give a black colour with rhodizonate (see Table 81, p. 1139) and some forms of tissue iron react in this way.

Methods for Beryllium

The only histochemical method so far employed for salts of beryllium is that of Denz (1949) using *Naphthochrome green*, already described in this chapter as a reagent for calcium and aluminium salts (*q.v.*). This method was

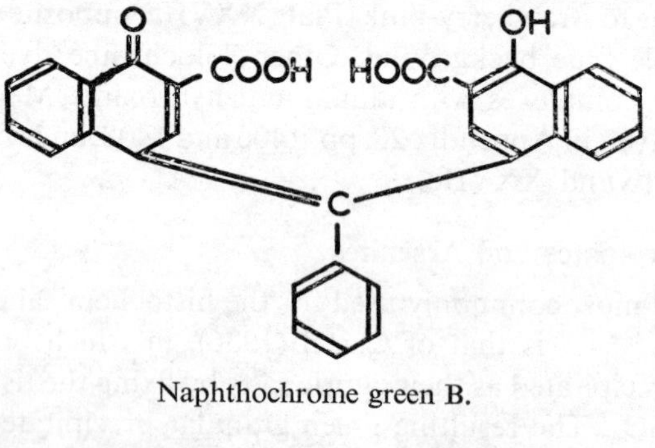

Naphthochrome green B.

used successfully by several authors. Firket (1953), for instance, demonstrated Be in cells in tissue culture, and Cheng (1956) localized the metal in the Kupfer cells in the liver of experimental berylliosis. The specificity of the test for beryllium depends on exclusion of Ca and Al salts, and this is done by performing the reaction at pH 5·0 when only very high concentrations of aluminium will react, and calcium salts will dissolve.

Many types of beryllium deposits appear dark blue to almost black in haematoxylin and eosin preparations. This is due to the formation of a haematoxylin lake which cannot be regarded as a specific test although it may draw attention to the presence of the deposits.

Since many beryllium salts fluoresce it is always worth while to examine an unstained section in ultra-violet light. The deposits often emit a characteristic bluish-white fluorescence, and this can be quenched by several different chelators such as alkaline quinalizarin, for instance. The *morin (fluorescent) method* for beryllium salts was described by Zermatten (1933), and according to Feigl (1954) a yellowish-green fluorescent compound is formed when alcohol or acetone solutions of morin come into contact with $Be(OH)_2$. If the deposits are autofluorescent this method is not applicable although a slight increase in fluorescence can be observed after performing the morin reaction in such cases.

A third method for beryllium is the *alkaline quinalizarin method* (Appendix 28, p. 1409). This colours the deposits a deep purple (Fig. 236, p. 1150), which changes to brown in acid solution. Distinction from the neodymium, praseodymium, cerium, lanthanum, thorium and zirconium lakes of quinalizarin is unlikely to be required in histochemical practice. The magnesium lake is blue, but can be distinguished from the beryllium lake by its stability to bromine in the presence of sodium hydroxide. This destroys the colour of the beryllium lake, while bromine in the presence of ammonia destroys only the magnesium lake. Beryllium deposits can usually be rapidly and specifically demonstrated with 0·5 per cent aqueous *Solochrome azurine*, which stains them deep blue (Plate XXVIId, p. 1141). Differentiation from Ca and Al is easy (see Appendix 28, p. 1409).

Method for Bismuth

This metal is found in the tissues for an indefinite period, after its injection for therapeutic purposes, in the form of brown or black granules. These are identifiable as bismuth by the Komaya-Christeller method (Komaya, 1925; Christeller, 1926) or by a modification described by Castel (1935). The latter's brucine method was further modified by Wachstein and Zak (1946), and their method appears in Appendix 28, p. 1410. I have no experience of its use. The demonstration of bismuth in the internal organs was described by Roncoroni (1962).

Methods for Cobalt and Nickel

Cobalt salts can be demonstrated in quite small amounts in the tissues by conversion to the brown sulphide. This is done in many histochemical enzyme reactions where the precipitate contains cobalt. Other chemical methods for this metal (α-nitroso-β-naphthol, rubeanic acid, ammonium thiocyanate) were used as histochemical techniques by Voigt, Bog, and Wolf (1954). They found that only the sulphide method gave accurate localization in the tissues.

Nickel salts were shown by Wells (1956) to have a considerable affinity for certain keratins, in fixed or unfixed state. These keratins can take up in a few hours, from solutions containing 1 p.p.m. of nickel, a sufficient amount of the metal to give a mauve to bluish colour with rubeanic acid (dithiooxamide). The dimethylglyoxime method can also be used to demonstrate nickel salts taken up by keratin in this way.

Methods for Copper

Although copper is present in traces in most vertebrate tissues the amount is too small to be detected histochemically by any of the existing methods. Increased amounts are said to be present in foetal and cirrhotic livers (Okamoto and Utamura, 1938) and in the liver of hepato-lenticular degeneration relatively enormous amounts (up to 30 mg./100 grams) occur. Such figures are at least ten times higher than the highest figures for normal tissues. Enzymes of the mono- and polyphenol oxidase classes are copper proteins, and the tyrosinase of the wall of the ink sac in the octopus and squid is especially rich in copper. The respiratory proteins of *Mollusca* and *Arthropoda* contain copper in place of the iron of haemoglobin, and in some marine species the total copper content of the organism reaches a high level. Echave (1941) includes a number of estimations on such materials in his monograph on the biological use of copper.

Most of the older methods for copper used in histochemistry lacked both specificity and sensitivity and were therefore abandoned. These methods included the haematoxylin-lake method of Mallory and Parker (1939), and the same authors' methylene blue technique. A dilute haematoxylin method was used by Mendel and Bradley (1905) in their studies on the distribution of copper and zinc in the liver of the gastropod *Sycotypus canaliculatus* (Figshell) collected from Long Island Sound. They used $(NH_4)_2S$, and a potassium ferrocyanide method, with equally good results. Lison (1953) condemned most of these methods, but continued to support the use of the potassium ferrocyanide test as a specific reaction. Its sensitivity is certainly not high enough to demonstrate physiological or pathological levels of copper in mammalian tissues. As their final method Mendel and Bradley used a concentrated solution of hydrobromic acid containing free bromine, which gives a violet colour with copper salts. They found the reaction "brilliant and very delicate", but the coloured compound was observed to diffuse rapidly through the tissues and thereafter to fade. Lison regarded this reaction as highly specific. Although its sensitivity in spot tests (0·15λ Cu) is high enough, Feigl (1954) pointed out that Ag, Hg, Fe and Cl interfere with the reaction. In histochemical practice only artificially high levels of Cu^{++} can be demonstrated in mammalian tissues.

There remain to be discussed seven more modern methods. The first, and probably the most important of these methods in mammalian histochemistry, employs *rubeanic acid* (*dithiooxamide*). This compound has already been mentioned as giving a coloured chelate compound with nickel salts. It was first used successfully for the histochemical localization of copper by Okamoto and his associates (1938, 1939), and it gives a deep greenish-black colour with salts of this metal. Feigl (1954) regarded this as due to formation of an inner complex salt of the di-imido form of rubeanic acid.

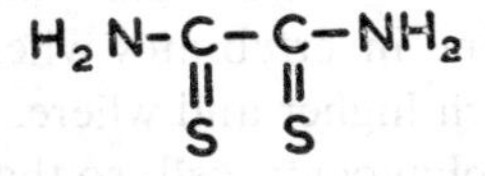

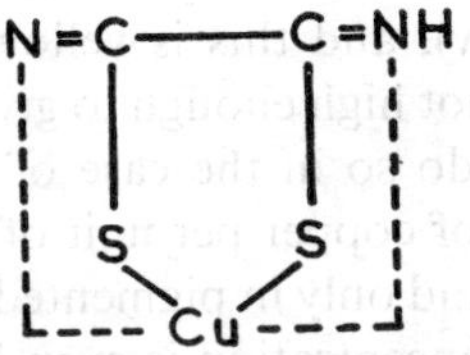

In the alkaline solution used in the test only copper, cobalt and nickel salts give a positive reaction, but the colours (greenish-black, yellowish-brown and mauve) are easily distinguishable. Moreover, if the medium contains ethanol and acetate ions Co and Ni rubeanates are soluble. Fig. 237 shows the result in the case of the copper pigment of *Triton rubicundus*.

Another method was described at the same time as the above by Okamoto *et al.* which made use of *p-dimethylaminobenzylidene rhodanine* (Fig. 238). This compound reacts to form a reddish-brown precipitate. It is more usually employed for the demonstration of mercury and silver (*q.v.*). The third method suggested by Okamoto *et al.*, which does not yield permanent preparations, is the *diphenylcarbazide* method. This compound has the following structure:

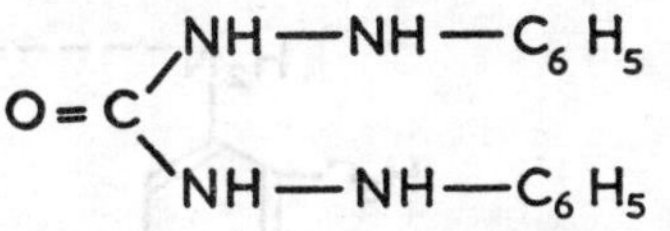

and it reacts with copper salts in neutral alcoholic solutions to give a reddish-violet colour. In strongly acid solutions only hexavalent chromium and molybdenum react.

Waterhouse (1945a, b) made a number of studies on the distribution of copper in the sheep blow-fly *Lucilia cuprina*, and for these he employed a fourth compound, *sodium diethyldithiocarbamate* (Fig. 239), which has the structure:

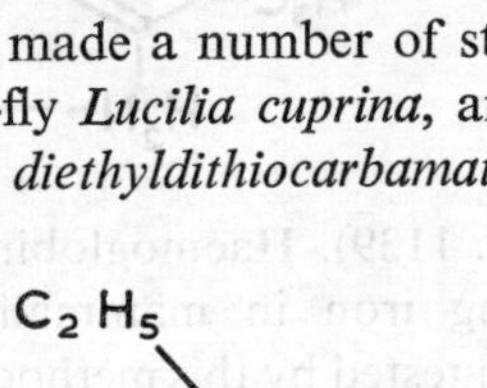

It was originally used as a copper reagent in the form of its zinc salt by Clarke and Hermance (1937). Whichever salt is used the yellowish-brown cuprous derivative is precipitated. The *diphenylthiocarbazone* method gives coloured precipitates with a number of metals, including Ag, Cu, Ni, Cd and Zn. In neutral or ammoniacal solution it gives a yellowish-brown precipitate with copper salts, but it has not found much employment in mammalian histochemistry.

Cupric ions form an inner complex salt with *dithizone* (see section on Zinc, below), and this is yellowish-brown in colour. The sensitivity of this method is not high enough to give a positive result in most mammalian tissues, but it will do so in the case of certain marine invertebrates where the concentration of copper per unit of tissue is much higher and where, in fact, the metal is found only in pigmented granules in phagocytic cells so that the histochemical concentration is even higher.

The final group of methods may be described as *oxidation catalyst reactions.* Copper salts are known to react with *benzidine* in the presence of alkali cyanides to give benzidine blue, and Feigl (1954) recommended the substitution of 0-*tolidine* for benzidine in order to increase the sensitivity of the reaction. Both these methods depend on the fact that the oxidation potential of Cu^{2+} salts is raised when Cu^{+} ions are removed by the formation of insoluble cyanides. Like benzidine blue (see p. 852), tolidine blue is an unstable compound easily reduced to a colourless product or oxidized to a brown one.

Interference with the specificity of the tolidine reaction for copper might be expected to come from salts of silver, mercury, iron, thallium, cerium and gold, but, in histochemical practice, iron is the only one likely to be of

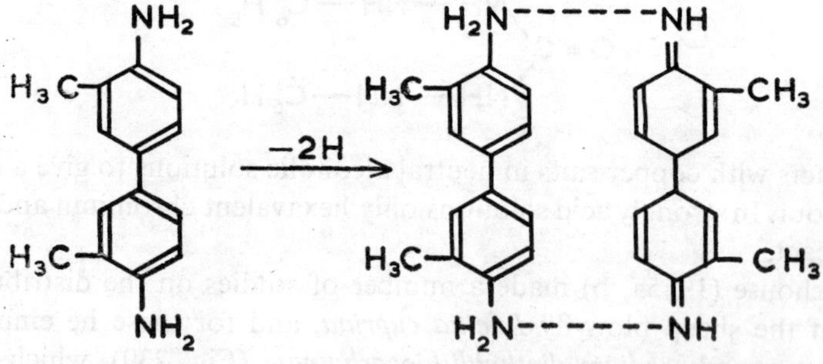

importance (see Table 81, p. 1139). Haemoglobin, haemosiderin, and other forms of naturally-occurring iron in mammalian tissues, however, give entirely negative results when tested by this method. When used for the demonstration of copper in invertebrate tissues the tolidine reaction shows a higher degree of sensitivity than any of the other methods listed above. Copper-containing pigments in the perivascular phagocytes in the hepatopancreas of *Triton rubicundus* are well shown by its use, but in mammalian tissues I have not yet succeeded in obtaining a positive result, even in material known to contain copper (natural) in excessive amounts.

Unreactive Copper

It is thought that Cu^{2+} or Cu^{+} may be present in some tissues in an unreactive form from which they must first be released before they can react. Gomori (1952) recommended the use of hydrogen peroxide for this purpose, and oxidizing agents of various types are commonly used. They are not altogether effective and Uzman (1956) used concentrated HCl, applied to sections in the form of its vapour (see Appendix 28, p. 1411). Chelators of various types may possibly release bound copper, as BAL (2,3-dimercapto-propanol) does *in vivo*, but even if successful their chief use would be to provide negative controls for methods depending on the formation of coloured chelates.

The Applied Histochemistry of Copper

Workers using mammalian tissues have shown a strong preference for the rubeanic acid method, and it was the only one with which I could demonstrate copper in the liver and kidney in Kinnier-Wilson's disease (hepato-lenticular degeneration). Brand and Takats (1951) were able to obtain a positive result also in the characteristic Kayser-Fleischer rings in Descemet's membrane (horn-hautringes). Working with tissues from a case of this disorder Green (1955), however, obtained positive results equally well by the Mallory (haematoxylin), rubeanic acid, diphenylcarbazide and diethyldithiocarbamate methods. Using fresh as well as formalin-fixed tissues from a similar case, Uzman (1956) obtained excellent results with a modification of the Okamoto-Utamura rubeanate method. He regarded the original method as deficient in two major respects as a histochemical reaction. First, he considered that the use of watery solutions might lead to diffusion of copper, and secondly, he believed that failure to use copper-free materials could cause serious artifacts to occur. Unfortunately Uzman did not consider any of the other available methods for copper. Howell (1959) used this method to demonstrate copper deposits in experimental rats, and also in hepatolenticular degeneration, and Schnabel and Nisch (1961) demonstrated the metal in liver and cornea, in this condition, with diethlydithiocarbamate, rubeanic acid and dimethylamino-benzylidine rhodanine.

Discounting for the moment the *o*-tolidine oxidation method, which has an *in vitro* sensitivity of $0{\cdot}003\gamma$ Cu^{++}, histochemical comparison of the remaining methods suggests that rubeanic acid (sensitivity $0{\cdot}006\gamma$) is definitely superior to dithizone ($0{\cdot}1\gamma$), diphenyldithiocarbamate ($0{\cdot}002\gamma$), and dimethyl-aminobenzylidene rhodanine ($0{\cdot}6\gamma$), for demonstrating naturally-occurring copper. In the case of tissue sections cut from blocks fixed in copper sulphate, and containing varying amounts of copper, I found these differences harder to detect. All three methods (dithizone excepted) appeared to become positive at much the same level of copper although the colour of the rubeanate was the strongest. In these artificially prepared tissues the sensitivity of the *o*-tolidine reactions was higher than that of any of the other methods.

In insect tissues the diethyldithiocarbamate method was preferred to the rubeanic acid method by Waterhouse (1951a and b), who failed to obtain a positive result at all with the latter. It was preferred also by Poulsen and Bowen (1952) for their studies on copper metabolism in *Drosophila* larvae. According to Clarke and Hermance (1937) the sensitivity of zinc diethyldithiocarbamate for cuprous ions *in vitro* is 0·002γ, and it seems likely that copper in insect tissues is present in cuprous form rather than in the cupric form which is usual in mammalian tissues. If this is true it would account for the invariable failure of the dimethylaminobenzylidene rhodanine method in the latter. This compound reacts only with cuprous salts, or with cupric salts, in inorganic combinations, after their reduction by sulphurous acid,

At least three new reagents for copper have been described in the literature. The first, bis-cyclohexanone oxalyldihydrazone (Peterson and Bollier, 1955), used in alcoholic solution, was said to be twice as sensitive as diethyldithiocarbamate. The second method, described by Gehauf and Goldenson (1955), made use of a dye called Zolon red. This forms red chelate compounds with silver and with cuprous salts. A third compound, 1-(2-thiazolylazo)-2-naphthol, was shown by Wada and Nakagawa (1966) to form a 1 : 1 complex with Cu^{2+} at pH 0·5. Interference from other metals, especially Ni^{2+} and Co^{2+} could be expected to occur.

Methods for Gold

Gold salts are sometimes introduced into the tissues therapeutically either in the form of radioactive or inert compounds. Two methods have been recommended for demonstration of the latter. The first is the method of Christeller (1927), which works well in the case of inorganic gold but gives negative results where the metal is part of an organic molecule. This method depends on the reduction of the gold salts to purple of Cassius with stannous chloride in strong HCl, but in practice reduction to metallic gold (brown or black) also occurs. The degree of localization afforded by this method is not likely to be precise, as pointed out by Gérard and Cordier (1932), since solution of the original gold salt necessarily precedes the precipitation of purple of Cassius.

A second method was described by Elftman and Elftman (1945). It depends on the bleaching of all pigmented substances (except silver and gold) by means of the prolonged action of 30 per cent hydrogen peroxide.

Although water-soluble Au^{+++} salts produce a purple colour with dimethylaminobenzylidene rhodanine this method does not seem to have achieved any histochemical success.

Methods for Lead

A number of dyes form coloured lakes with lead salts or $Pb(OH)_2$, among them haematoxylin, gallocyanin and carminic acid. Mallory's (1938) method employed a freshly prepared haematoxylin solution, and lead salts were said to stain bluish-grey. These lake methods are neither sensitive nor specific and their use is not recommended.

For many years the most popular method for demonstrating Pb in tissues was the *chromate method* (Crétin, 1929; Frankenberger, 1921), in which the tissues are fixed in neutral potassium dichromate (Regaud) or, if already fixed, are treated with this mixture subsequently. Lead is found in the form of opaque yellow crystals.

In modern histochemistry the most popular method has been the so-called sulphide-silver method of Timm (Timm, 1936, 1958; Timm and Arnold, 1958; Timm and Neth, 1959). Tissues are fixed, by perfusion if necessary, in absolute or 70 per cent alcohol saturated with H_2S. This transforms lead salts into the brownish-black PbS. Since this is insoluble in dilute and moderately strong acids, bones in which it has been deposited can be decalcified. The specificity of the sulphide-silver method is low. In addition to Pb, the reaction is positive with Au, Ag, Pt, Fe, Cd, Cu, Co, Ni, Zn and Hg. The possibility of differentiation between these is very limited but the successful extension of the method to the electron microscope (Chapter 32, p. 1290) has increased its popularity despite this lack of specificity. Technical details are given in Appendix 28.

The sulphide-silver method was used by Riecken (1963) for the demonstration of Pb in human bones and the type of result obtainable is shown in Plate XXVIIIa, p. 1180. Unidentified heavy metals were demonstrated by Müller (1964) in the Golgi region of many cells in tissues from animals poisoned with H_2S.

A Carnoy-sodium sulphide fixative developed by Kodousek (1963) gives excellent results when combined with the sulphide-silver technique. Differences between paraffin and cryostat sections, found by Brunk and Sköld (1967), were attributed by them to oxidation of the metal sulphides to sulphate, sulphite or hydroxide during dehydration and embedding. Some reversal could be effected by treating the undeparaffinized sections with H_2S but the authors strongly recommended the use of cryostat sections.

Coloured chelate reactions. Lillie (1954), following up Molnar's (1952) suggestion to use sodium rhodizonate for the demonstration of the lead-precipitated reaction products of enzyme activity, suggested this method for the demonstration of lead deposits in general. From the information given by Feigl (1954) it would appear that sodium rhodizonate is more likely to give positive histochemical results than dithizone or any other metal chelating agents, since it is capable of reacting even with the very slightly soluble PbS, $PbSO_4$ and $PbCrO_4$.

The type of chelate compound formed is illustrated below:

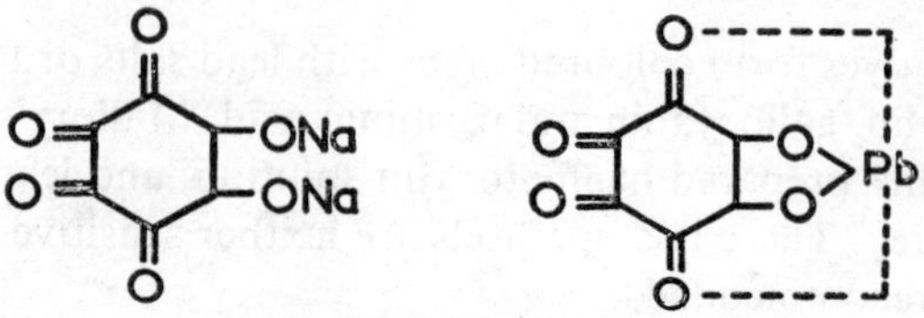

Sodium rhodizonate

In practice I found that small concentrations of an insoluble lead salt ($PbSO_4$) in the tissues could readily be demonstrated by the rhodizonate method.

Methods for Magnesium

There are three histochemical methods for magnesium, two of which, *titan yellow* and *magneson* are of high specificity. The third, the alkaline quinalizarin reaction, has already been mentioned as a method for beryllium (p. 1143).

The titan yellow reaction was introduced into botanical histochemistry by Broda (1939), but it is equally applicable to invertebrate and mammalian tissues. A striking flame red colour is produced but, unfortunately, no method is known by which this can be stabilized and it fades within 24 hours. Magnesium-containing deposits in the mantle region of *Limnaea* are shown in Fig. 240, stained by this method.

The magneson method depends on the affinity for magnesium salts of certain azo dyestuffs of which *p*-nitrobenzene-azo-1-naphthol is the best known. This dye is reddish in colour and it is adsorbed by magnesium salts without change in colour. When the medium is made strongly alkaline (pH 12) by the NaOH the colour changes to bright blue. This is also unstable in tissue due to the difficulty of maintaining so high a pH. Staining by this method gives precise and clear localization of magnesium deposits, as indicated by Fig. 241, p. 1151. Both methods are given in detail in the Appendix to this chapter (p. 1414).

Methods for Mercury

Early histochemical methods for this metal depended on the conversion of its salts either to black HgS or to metallic Hg. The older *sulphide methods* in which H_2S was applied in the fixative were abandoned when the work of Lombardo (1906, 1908) showed that H_2S, when added to the fixative, produced a whole spectrum of pigments from organic as well as inorganic precursors. The more modern sulphide methods employ treatment with H_2S after a variety of fixatives. Deposits produced by experimental mercury poisoning were demonstrated in rat kidney by Wöckel *et al.* (1961), using the sulphide-silver method of Timm.

Reduction methods are still used and stannous chloride remains a popular agent for this purpose. It was used by Debenedetti (1925) and by Hand, Edwards and Caley (1943) to reduce both mercurous and mercuric forms. The latter authors also used thioglycolic acid for reducing the mercurous form only.

Diphenylcarbazide and diphenylcarbazone methods were introduced to histochemical use by Brandino (1927), who used the first compound in alcoholic solution. His method, according to Lison (1953), showed only mineral compounds of mercury and consequently failed to reveal mercury in organic combination. The colour developed was violet.

Feigl (1954) indicated that the second of these two compounds forms an inner complex mercury salt with the constitution I. It is possible, however, that the constitution of the complex may be as shown in II.

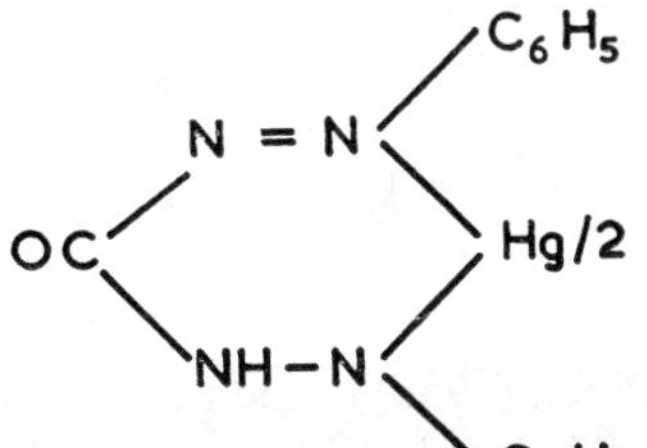

I

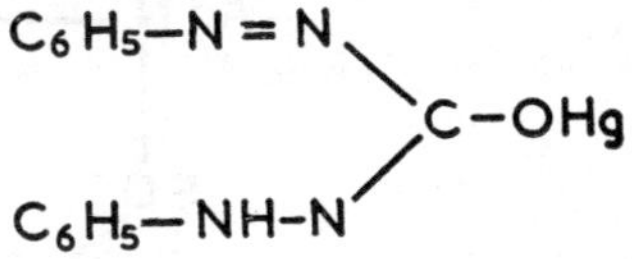

II

The specificity of the reaction for mercury depends on the pH. In neutral, or slightly acid, solutions other heavy metals (Cu, Fe, Co) react to give coloured compounds. After treatment of tissues with alcoholic mercuric nitrate, as in the phospholipid reaction of Okamoto *et al.* (1947) (see Chapter 12, p. 424), a very strong reddish-purple colour is produced in structures which retain mercury. These same tissues, however, fail to react by the dimethylamino-benzylidene rhodanine method, which *in vitro* has a sensitivity for salts of mercury not much below that of diphenylcarbazone.

Methods for Silver

When silver salts are introduced into the tissues they are rapidly transformed into brown or black deposits of unknown constitution, presumably containing metallic silver bound to some protein carrier. Seen through the skin in cases of therapeutic argyrosis the deposits appear steel-blue. They have been identified histochemically mainly by their removal or decolorization with various reagents which do not affect melanin in the same period of time. A solution of 1 per cent potassium ferricyanide in 20 per cent sodium thiosulphate was recommended by Lison or, as an alternative, 1 or 2 hours' treatment with Lugol's solution followed by treatment with thiosulphate. Yellow ammonium sulphide turns silver deposits into the black AgS, while potassium dichromate forms the brownish-red silver salt. These two salts are soluble in 5 per cent aqueous KCN.

Two alternative methods were proposed by Okamoto, Utamura and Akagi (1939), neither of which can be regarded as specific for silver. The first is the rubeanic acid method (see section on Copper, above) and the second the dimethylaminobenzylidene rhodanine method. Silver salts give a brown precipitate with neutral or alkaline solutions of rubeanic acid, but they will not react under the conditions given in Appendix 28, p. 1411, where the method is used for detection of copper.

Rhodanine and similar compounds form insoluble silver salts selectively in acid solution, the silver taking the place of H in the cyclic imine (NH) group. At alkaline pH levels a number of metals form coloured insoluble rhodanates. The silver compound, whose formula is shown below, is reddish-violet to reddish-brown in colour.

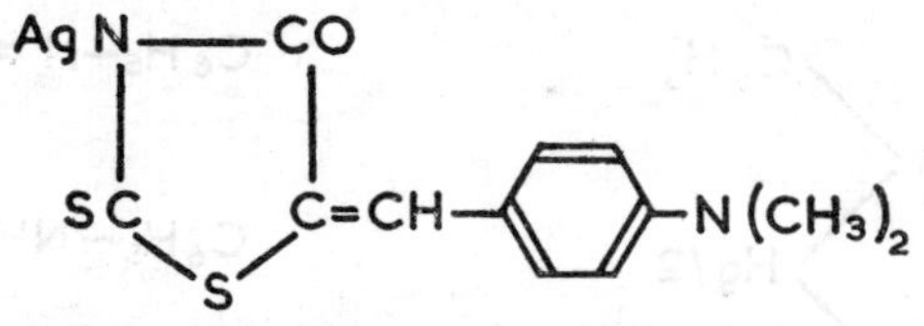

Both these methods were used by Gedigk and Pioch (1956) in their studies on metal-binding by connective tissues. They found that localization of silver (AgO) was better with the rubeanic acid method since silver rhodanate was sufficiently soluble to diffuse away from its original site.

Methods for Tin

Occasionally dust from ores containing tin is inhaled into the lungs where it may produce a type of pneumokoniosis. The chief form in which tin is mined is cassiterite (SnO_2), the other forms being mixed ones containing other metals such as stannite (Cu_3FeSnS_4) and cylindrite ($Pb_3FeSn_4Sb_2S_{14}$). The problem therefore usually resolves itself into demonstration of stannic oxide.

There are three possible methods, all mentioned by Feigl (1954) in his volume on the inorganic applications of spot tests. The first of these, the *morin method*, has already been mentioned (p. 1137) as a method for calcium and aluminium. The second (Eegriwe, 1940) was based on the fact that in the presence of strong mineral acids stannic salts form orange lakes with 1,2,7-*trihydroxyanthraquinone*, and similarly, in the third method (Kuznetsov, 1941) a blue-violet lake is formed with the red dye *anthraquinone*-1-*azo*-4-*dimethylaniline hydrochloride*. I tested these three methods on human lungs containing inhaled tin ores and on rat lungs with very large amounts of oxides of tin in the form of discrete deposits. In no case could a positive reaction be obtained.

Other tests for tin which can be employed after reduction of stannic to stannous salts are the *cacotheline method* (Gutzeit, 1929) and the *diazine green S method* (Eegriwe, 1928). With neither of these have I been able to

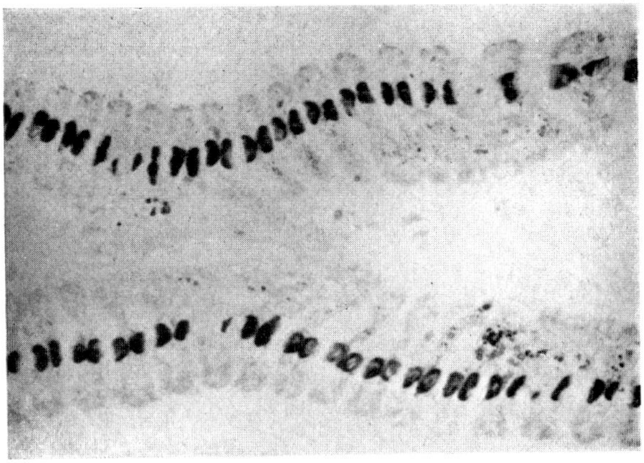

FIG. 231. *Anodonta cygnea* (fresh water mussel). Gills. In each bar is a bright blue-stained mineral deposit containing a large amount of calcium. Copper phthalocyanin method. × 130.

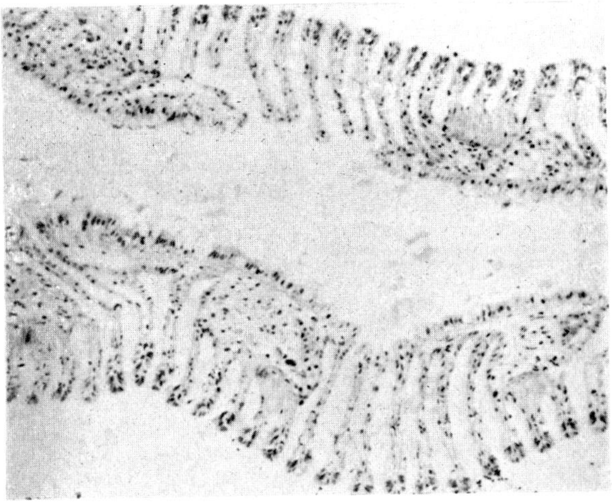

FIG. 232. As Fig. 231. Control section pretreated with 1 per cent acid. Calcium deposits dissolved out. Copper phthalocyanin—neutral red. × 130.

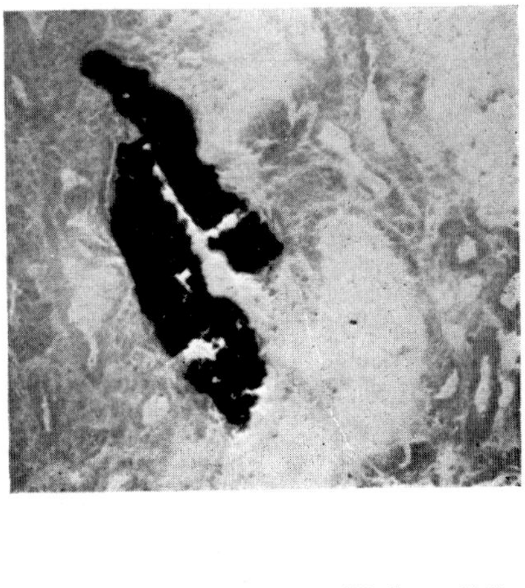

FIG. 233. Human kidney. Nephrocalcinosis. Shows large amorphous calcium phosphate deposit, stained bright green. Mercury—naphthochrome green. × 130.

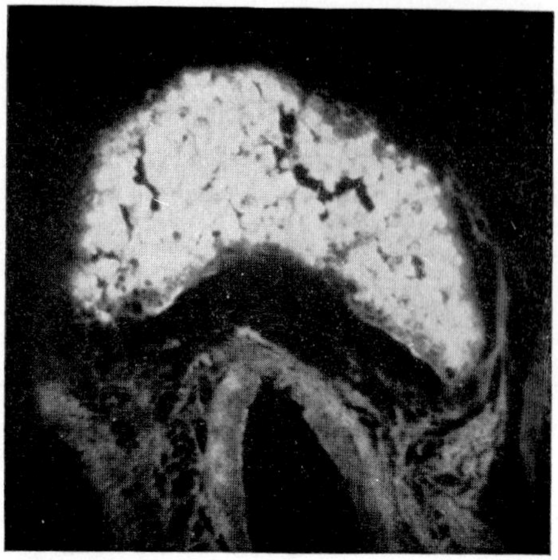

FIG. 234. Rat bronchus. Calcified cartilage shows bright whitish-green fluorescence. Morin method. × 110.

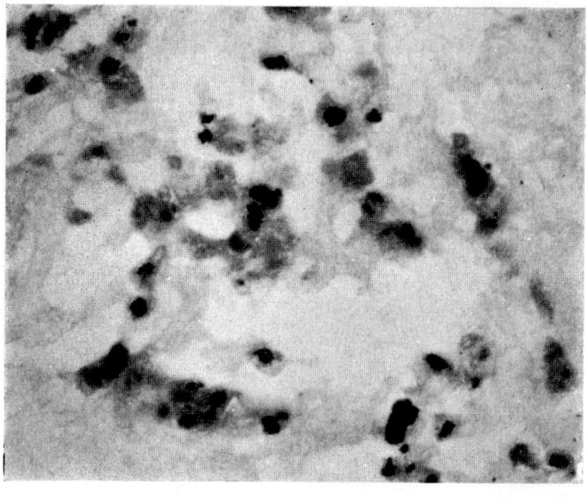

FIG. 235. Rat lung. After insufflation of metallic aluminium dust. Aluminon method. × 510.

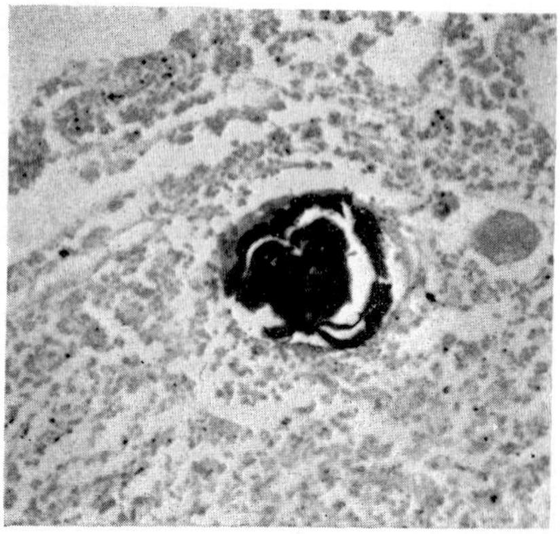

FIG. 236. Human lung. Case of berylliosis. A large deposit of beryllium stained deep purple. Alkaline quinalizarin method. × 130.

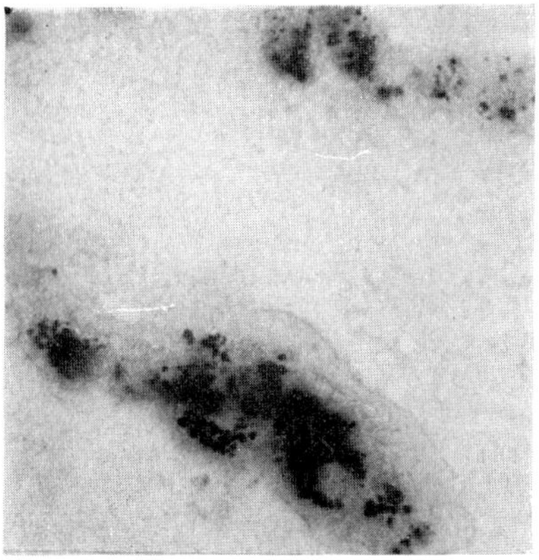

FIG. 237. *Triton rubicundus* (whelk). Copper-containing pigment granules in macrophages. Rubeanic acid method. × 510.

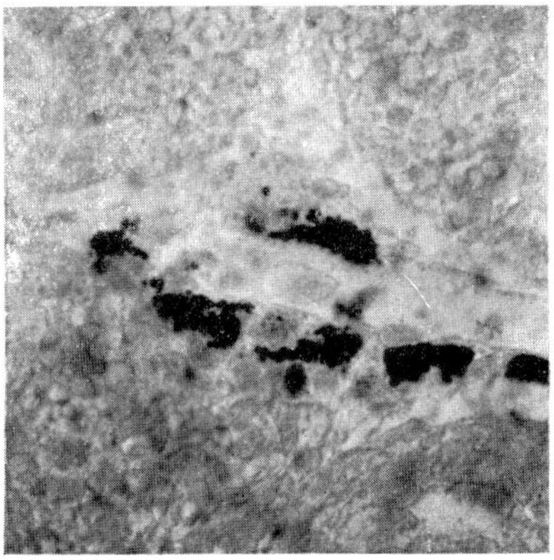

FIG. 238. As Fig. 237. Dimethylamino-benzylidene rhodanine. × 510.

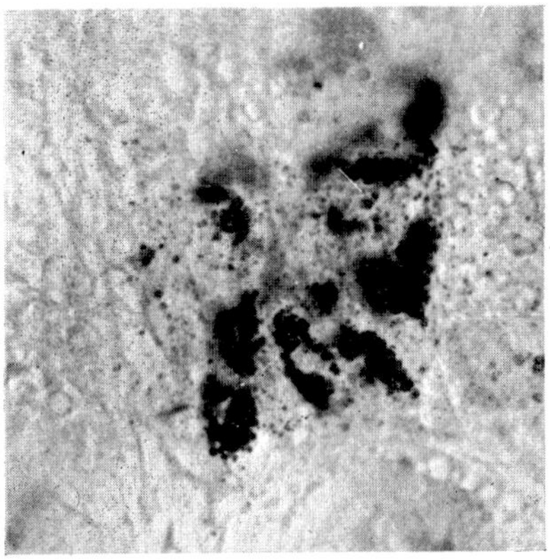

FIG. 239. As Figs. 237 and 238. Diethyldithiocarbamate. × 510.

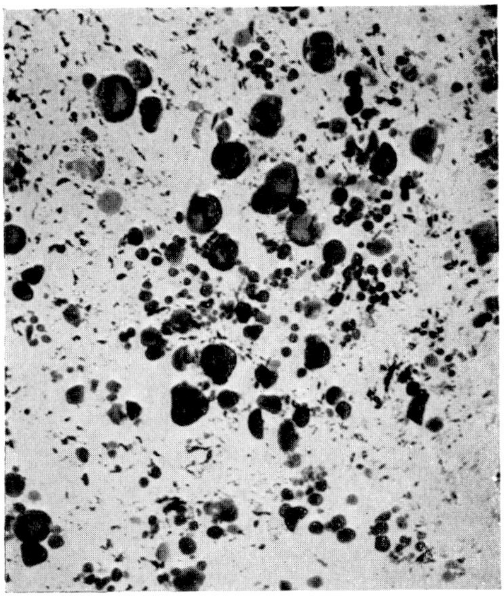

FIG. 240. *Limnaea stagnalis* (water snail). Deposits in the connective tissues containing large amounts of magnesium salts, stained bright red. Titan red method. × 120.

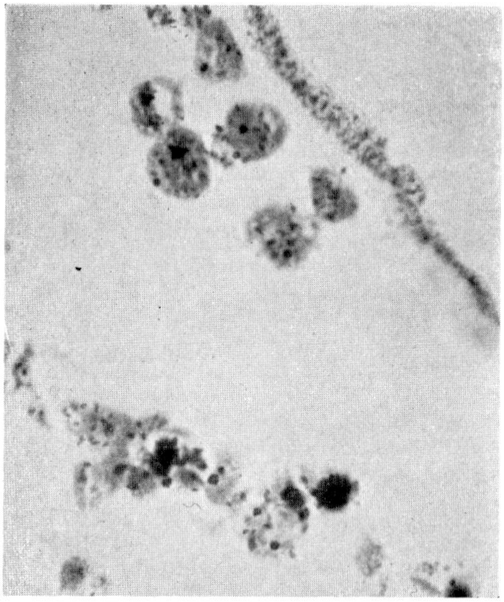

FIG. 241. As Fig. 225. Magnesium deposits stained bright blue. Magneson method. × 120.

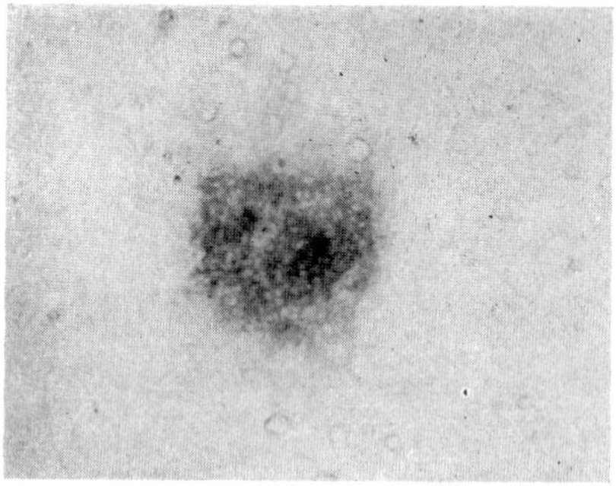

FIG. 242. Rat pancreas. Single islet of Langerhans. Zinc salts appear red. Dithizone method. × 130.

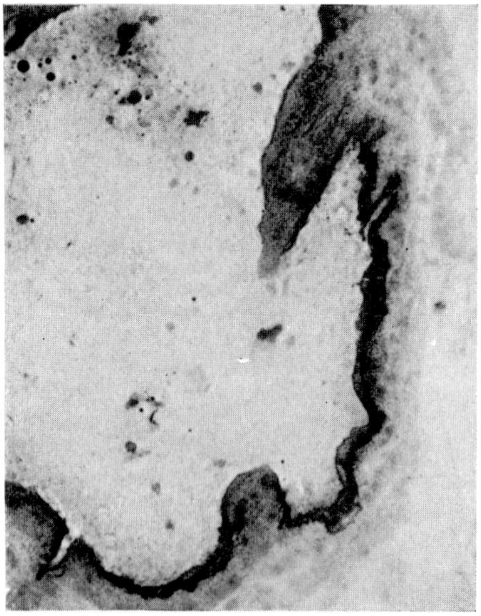

FIG. 243. Rat prostate. Dorsal lobe. Shows high content of zinc in glandular epithelium and in secretions in the lumen. Dithizone method. × 188.

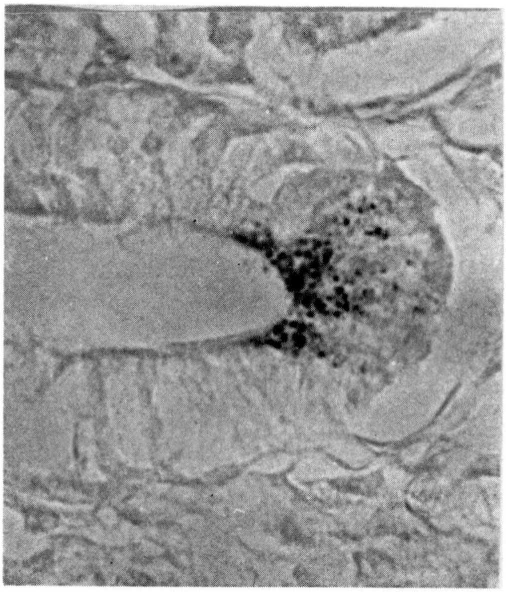

FIG. 244. Rat jejunum. Sulphide-Silver Method (Timm). Shows a positive reaction in the Paneth cell granules which is strongest in the apical part of the cells. × 1200.

demonstrate the presence of powdered tin-containing ores artificially insufflated into the lungs of rats.

After introducing soluble tin salts into rats in varying amounts Khau van Kien and Thaï-Tuong (1956) made experiments with three staining reactions in order to determine the optimum type of fixation for tin in rat tissues. They found formaldehyde vapour to be the best, closely followed by Carnoy's fluid and alcohol. Their three methods were: (1) Dithiol (toluene-3,4-dithiol), used as a 2 per cent solution in water or dilute HCl; (2) Acetone-H_2S used specifically at pH 2·0; and (3) Mayer's haematein (sic) or an old solution of haematoxylin. These authors claimed that all forms of tin could be demonstrated and that the haematein reaction was the most sensitive in practice. The dithiol reaction, on the other hand, was the most specific, but it gave a red precipitate only with tin in mineral form. I have no experience of the three methods of Khau van Kien and Thaï-Tuong, but it would appear that these demonstrate only the complexes formed by soluble tin salts with tissue proteins.

Methods for Zinc

Zinc is present in the tissues as an essential component of insulin and of the enzyme carbonic anhydrase. A purified sample of this enzyme obtained from the ox contained 0·3 per cent of zinc, equivalent to 0·2 μg. per unit of

TABLE 81a

Enzyme and Zinc Levels in Rat Prostate. (*After Fischer* et al.)

Moiety	Ventral	Dorsolateral	Dorsal	Lateral
Zn (μg./g.) . .	13·7	180	103	590
C.A. (units/mg.) .	0·11	10·5	5·1	16·0
Ac. Ph. (units/g.) . .	14·8	11·1	9·6	12·4
Alk. Ph. (units/g.) . .	53·9	21·0	16·5	23·6

enzyme (Fischer, Tikkala and Mawson, 1955). These authors, following up earlier work by Mawson and Fischer (1952a and b), showed that the amount of zinc in the dorsolateral prostate of the rat increased rapidly during infancy to reach a level of 250–300 μg. per gram at 160 days. This is ten times the amount present in other tissues. Carbonic anhydrase activity increased in the same way up to a maximum at about 200 days. In spite of a constant relationship between the two the very large amount of zinc in the prostate could not be correlated directly with the level of carbonic anhydrase. Nor were the levels of acid or alkaline phosphatase found to have any relationship to zinc levels, as the above table shows.

Assuming that rat carbonic anhydrase contains the same amount of zinc as in the ox it can be calculated that at 0·2 μg./enzyme unit the amount of zinc directly related to carbonic anhydrase activity in the dorsolateral prostate is

6·3 μg. per gram or 3·5 per cent of the total. The function of the remainder is unexplained.

Using human prostate glands Hoare, Delory and Penner (1956) determined the levels of zinc and of acid phosphatase in different physiological and pathological states. They found that the normal gland contained 601 μg./gram of tissue, while in benign hyperplasia the figure was 460 μg./gram (mean values). Although the amount of acid phosphatase could be directly related to the amount of glandular tissue present there was no such relationship in the case of zinc. In human semen Mawson and Fischer (1953) found a high concentration of zinc, and they recorded the fact that spermatozoa contain more zinc than any other human tissue.

Some other associations of zinc are worthy of note. In the melanin-pigmented ocular tissues of some fresh-water fishes Leiner and Leiner (1944) found very large amounts of the metal, and in the agranular cells of the pancreatic islets of *Cottus* Falkmer *et al.* (1964) identified zinc in appreciable amounts. Quinlan-Watson (1953) recorded a fall in aldolase level in oats and clover plants suffering from zinc deficiency. Hoch and Vallee (1952) described two zinc-protein complexes in the leucocytes and, in a later paper, Vallee, Hoch and Hughes (1954) recorded the properties of a soluble zinc-containing protein from human leucocytes. This protein had no carbonic anhydrase activity nor (see below) did it act as an alcohol dehydrogenase or as a carboxypeptidase. Vallee and his associates were among the first to record the existence of zinc-protein enzymes other than carbonic anhydrase, and Tupper, Watts and Wormall (1954), in a paper on the uptake of ^{65}Zn by avian eggs, recorded that zinc might be a specific activator for several enzymes. Hanson and Smith (1949) described the activation of carnosinase by zinc and similar activation has been reported for peptidases of various kinds (Johnson, 1941; Yudkin and Fruton, 1947; Smith, 1948). The carboxypeptidase of ox pancreas has been shown by Vallee and Neurath (1955) to be a zinc metalloprotein, containing 1,870 μg. per gram of zinc, and Vallee and Hoch (1955) have found a similar amount of the metal in crystalline yeast alcohol dehydrogenase. Zymohexase (Warburg, 1949) and some types of phosphatase (Cloetens, 1942) have also been found to be zinc-activated enzymes.

An older method for demonstrating zinc in the tissues, used by Mendel and Bradley (1905) in their studies on the hepatopancreas in *Mollusca*, depended on the treatment of paraffin sections with *sodium nitroprusside*, and afterwards with sodium or potassium sulphide. An intense purple colour indicated the presence of zinc and the reaction was regarded by Lison (1953) as sufficiently specific. The nitroprusside method was used by Sebruyns (1946) for localization of the metal in mice which had received various doses of ZnO by mouth. As an alternative an *alizarin lake method* attributed to Crétin (1933) was employed.

Zinc can be demonstrated histochemically in the islets of Langerhans (Fig. 242, p. 1151) and in the prostate (Fig. 243, p. 1151) of many species by the original method of Okamoto (1942a and b, 1943) using diphenyl thiocarbazone (dithizone). This forms a deep red inner complex salt with zinc ions, especially in alkaline solutions:

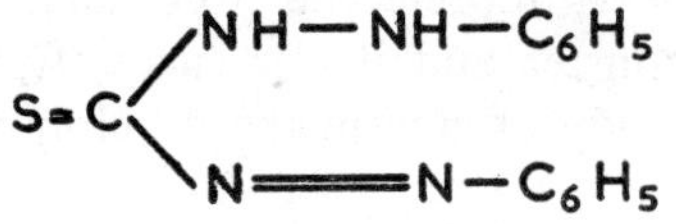

Dithizone

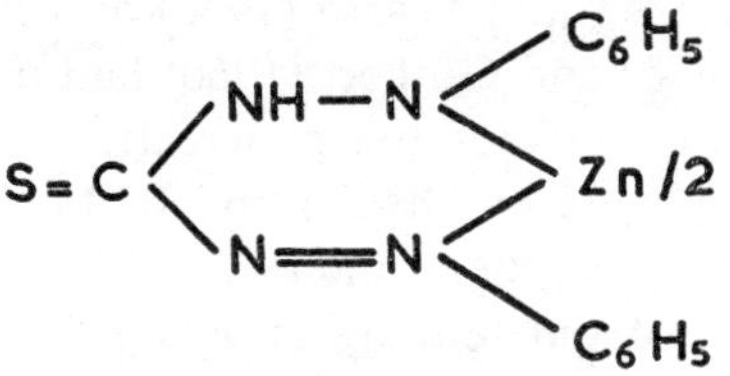

The original method was used as an intravital reaction by Wolff and Ringler (1954), who were able to demonstrate zinc in the islets of the rabbit, dog, rat, mouse and guinea-pig in this way. A similar technique was used by Logothetopoulos (1960) who found that examination of fresh-frozen sections by dark-field illumination was particularly useful.

In order to increase the specificity of the method for zinc Okamoto, Aoyama, Ibaraki, Narumi, Shibata, Kawasaki, Shibata and Komatsu (1951) used ammonium chloride (pH 8·4–8·9) and ammonium carbonate. They claimed that the former prevented both zinc and cadmium from reacting with dithizone, and the latter zinc only. Mager, McNary and Lionetti (1953) introduced for the same purpose the use of Vallee and Gibson's (1948) complex-forming buffer, which prevents the formation of metal-dithizone complexes other than zinc. Control sections were treated with dithizone in this buffer and compared with those stained with dithizone in 60 per cent acetone. The specificity of the dithizone method for zinc was examined critically by Schmidt and Rautschke (1963a, b, c, 1964).

If cold-knife or cold-microtome sections are used, and these are undoubtedly best for the demonstration of natural tissue zinc, acetone or alcohol in high concentrations in the medium causes considerable shrinkage artifact when free-floating sections are used. For freeze-dried sections and mounted cold-microtome sections, on the other hand, the acetone-dithizone reagent of Mager *et al.* is excellent. Alternatives are therefore given in Appendix 28, p. 1417, one or other of which should prove suitable for most materials.

The fact that fat in sections is stained (pink) by the dithizone method was noted by Mager and his associates, and in order to distinguish this type of non-specific staining they treated sections with a solution of versene (ethylenediamine tetra-acetic acid) before applying dithizone. Since versene is a non-specific metal chelator it was assumed that a positive result reversible by this reagent was due to metallic causes.

By the use of one or other modification of the Okamoto method zinc present in the tissues at a concentration of 50 μg./gram or higher can usually be demonstrated histochemically. Lower concentrations will be positive if the metal is localized in small intracellular components. Midorikawa and Eder (1962) demonstrated zinc with great regularity in the Paneth cells of rat, rabbit and guinea-pig intestine, but not in human intestine, using dithizone staining, intravital dithizone, and the sulphide-silver reaction. The latter gives accurate localization, restricted to the Paneth cell granules, as shown by Riecken and Pearse (1966) and illustrated in Fig. 244, p. 1151. Previously Müller and Güthert (1960) had used this method, in preference to others, in their study of the pancreatic islets in the hamster and it was successfully employed by Otsuka and Ibata (1968), in a study of zinc in the developing hippocampus of the rat.

An outstanding anomaly is the behaviour of the leucocytes which are known to contain a large amount of zinc. Vallee *et al.* (1954) found 3,000 µg./gram in their soluble protein from human leucocytes. Although Chauncey and Lionetti (1952) reported positive results in leucocytes with the dithizone technique these were later found to be reproducible only irregularly.

An alternative, fluorescence, method based on staining with 8-hydroxyquinoline was introduced by Smith *et al.* (1969). Specificity was achieved by staining at pH 8·0, at which level the complexes formed with Mg^{2+} and Ca^{2+} are unstable. This technique, which in the original authors' hands gave good results with zinc in leucocyte granules, is less satisfactory for the demonstration of the metal in other sites. The technique is given, nevertheless, in Appendix 28.

Methods for Thallium

An early technique for this metal, introduced by Barbaglia (1930) was based on the formation of insoluble yellow thallous iodide by fixation with KI in ethanol. Pritschow and Lhotka (1961) developed a method, applicable to paraffin sections, which depended on the formation of a red thallium-bismuth complex. Both methods proved unsatisfactory in respect of sensitivity and Voigt (1966) successfully used a modified sulphide-silver technique. Giusti and Fiori (1969) adapted this for use with paraffin sections and their technique is given in Appendix 28.

Methods for Potassium

The method of Macallum (1905) for the demonstration of potassium was based on the formation of an insoluble triple salt by the application of sodium cobaltinitrite to potassium salts in fresh tissue sections. The orange-yellow crystalline precipitate of sodium potassium cobaltinitrite was then converted into cobalt sulphide with yellow $(NH_4)_2S$. A large number of modifications of Macallum's method have been devised. Menten (1922) observed that the maximum precipitate was obtained after 2 hours, but preferred to stain for 15 minutes only. Rohdenburg and Geiger (1928) placed small tissue blocks in the cobaltinitrite solution for 1 hour, afterwards washing for 20 minutes and then placing the tissue in ammonium sulphide. Poppen, Green and Wrenn (1953) found that this last method was unacceptable because it was impossible to remove the large excess of precipitate in the interstitial tissues by washing.

They employed sections from heart muscle fixed in alcoholic formalin for 48 hours. Sections were cut at 7 μ and dewaxed in xylene followed by two changes of absolute alcohol. They were then transferred directly into the solution of cobaltinitrite.

The freeze-drying method of preparing tissue sections overcomes the problem of diffusion of potassium ions up to the stage at which an aqueous solution can be applied to the section. Gersh (1938) therefore evolved a modified cobaltinitrite method, for application to 15 μ freeze-dried sections, in which elaborate precautions against this diffusion were taken. Observing that potassium cobaltinitrite is soluble in water at room temperatures, but scarcely at all at 0°, he employed this temperature for the reaction. Crout and Jennings (1957) compared the original method of Macallum with that of Poppen *et al.* and with the method of Gersh. They observed that sections stained by the latter method showed a fine dense granular precipitate, evenly distributed throughout muscle cells, whereas with the other two methods the precipitate was irregularly distributed and the granules much more coarse. By means of potassium analyses carried out in parallel with the histochemical studies it was shown that all the potassium present in the tissues is precipitated by cobaltinitrite. Crout and Jennings varied Gersh's method in respect of some minor details, and these, as well as the original method, are given in Appendix 28, p. 1415.

The reaction was used on freeze-substituted rabbit brain by Collewijn (1963) who developed an alternative method in order to check his results. This second method was based on the use of sodium tetraphenylboron, $Na(C_6H_5)_4B$, first used as a sensitive analytical reagent for potassium ions. Although the author used freeze-substituted tissues also for the tetraphenylboron method I found freeze-dried sections equally suitable. The sensitivity of the method, applied to heart muscle (Pearse, 1964), appeared to be high. It was capable of accurate localization, and provided a satisfactorily dark final product. Details of the reaction are given in the Appendix.

The reaction of Carere-Comes (1938), using dipicrylamine, is quite useless. This dye (aurantia) stains basic proteins independently of their potassium content.

Methods for Sodium

The zinc uranyl acetate method (Kolthoff, 1927), which is widely used for spot tests (see Feigl, 1954), cannot be applied to fresh frozen sections with any hope of successfully localizing the sodium ions. When it is applied to such sections or to freeze-dried material and examined under the fluorescence microscope the tissues exhibit a general fluorescence, and it is impossible to say how much of this is related to sodium and how much is non-specific. No other methods for sodium have been applied histochemically.

A technique for EM localization of sodium ions, using pyroantimonate, was proposed by Komnick (1962) and by Komnick and Komnick (1963). It was used by a number of workers including Kaye and Donn (1965) and Bulger (1969). The technique was shown by Torack and La Valle (1970) to be invalid for several reasons, notably because of the coprecipitation of potassium ions. Doubtless calcium ions will also produce a precipitate.

Histochemistry of Inorganic Anions

Methods for Phosphate and Carbonate

Provided that phosphate and carbonate are present as insoluble calcium salts they can be distinguished very simply by mounting the section containing them in water and running glacial acetic acid under one edge of the coverslip while the preparation is examined under the microscope. The deposit, easily seen with the condenser slightly out of focus, dissolves quietly if it is phosphate and with evolution of bubbles of CO_2 if carbonate. A mixture of the two salts is not distinguishable, but CO_2 can be positively identified by the method of Bunting (1951) which is given in Appendix 28.

Molybdate methods can be used to demonstrate both organic and inorganic phosphates, but in the first case the conversion of the organic salts to inorganic phosphate is a necessary preliminary to their demonstration. The molybdate methods are variations of the ammonium molybdate technique of Lilienfeld and Monti (1892). A useful modification is that of Serra and Queiroz Lopes (1945) (referred to in Chapter 9, p. 252), which, like the other methods, gives a blue colour with phosphate. Although the accuracy of localization of organic phosphate by this method is very low, preformed inorganic phosphate deposits are better demonstrated. Methods for the intracellular localization of orthophosphate were critically considered by Tandler (1960) and this author (1961) described the detection of inorganic phosphate in the nucleus.

Cheng (1956) employed a modification of the method of Lowry and Lopez (1946) for demonstrating inorganic phosphate in the presence of labile phosphate esters. In his technique (see Appendix 28, p. 1416) short formalin fixation was employed, and frozen sections, but loss of inorganic phosphate into the fixative was noted to be a serious factor.

Methods for Chlorides

A large number of authors have reported work on the histochemical localization of chloride, using silver techniques exclusively. Lison (1936b) sought to avoid the gross errors of earlier methods using tissue blocks by the intravascular injection of the silver solution. He showed that silver salts could penetrate only a few cells deep in a block of tissue after 4 hours' incubation and, since chloride ions are extremely diffusible, localization was obviously impossible.

Lison (1953) was extremely critical of the silver reaction for chloride in histochemistry. He pointed out that the reaction is an indirect one, since the final product (metallic silver) is related to the total number of silver ions in the tissue and not to the number of chloride ions. If traces of silver remain in the tissues false localization will occur. These can best be avoided by washing sections (not blocks) with dilute nitric acid in place of distilled water. Chloride bromide and iodide of silver are insoluble in dilute mineral acids, phosphate and other salts are removed. According to Lison nitric acid possesses two other advantages. First, it dissolves out metallic silver precipitated in the tissues by the direct action of reducing agents, such as ascorbic acid, and secondly, it releases silver which has been adsorbed on the tissue proteins and lipoproteins.

Gersh (1938) evolved a procedure for localizing phosphate (phosphate-carbonate) and chloride in freeze-dried sections. He employed two solutions of 60 per cent silver nitrate, one containing phosphoric acid and saturated with silver chloride, the other containing non-phosphoric acid and saturated with silver phosphate and chloride. The first precipitates chloride only, the second both phosphate and chloride. The sections are examined after short treatment with ultra-violet light.

The histochemical differentiation of chloride from other ions precipitated by silver nitrate was described by Van Harreveld and Potter (1961). Their method was based on substitution-fixation at $-25°$ in 90 per cent ethanol saturated with silver nitrate. Provided this medium was acidified all of the reaction product in the tissues was regarded as reduced silver chloride.

Methods for Iodides

Inorganic iodides have not been localized successfully by histochemical methods other than those employing an autoradiographic technique (see Chapter 30, p. 1208). Gersh and Stieglitz (1933) and Lison (1936a and 1953) concluded that it was impossible to localize diffusible iodide by means of precipitant reactions. These have also been largely unsuccessful.

One method, that of Turchini (1930), may be mentioned. This author demonstrated bound iodides in the tissues of the oyster by treating them with nitrous acid (10 per cent sodium nitrite in 1 per cent sulphuric acid). This released iodine in elemental form which stained the tissues yellow. Treatment with starch solution produced a blue colour. According to Lison (1953) only weakly bound iodine can be demonstrated by this method.

McAllister (1953) described a method which gave a red fluorescence in the presence of iodides. This depended on their reaction with 1-methyl-2-mercaptoimidazole and bismuth acid sulphate, $BiH(SO_4)_3$. The possibility of adapting this reaction for histochemical use might be considered.

Recapitulation

Because of the fact that many of the metals dealt with in this chapter can be demonstrated by a variety of methods I have set down, below, a list of metals for which more than a single workable method exists. The order in which the methods are given is to some extent the order of choice, for mammalian and human pathological tissues, but it is obvious that no hard and fast rules can be laid down concerning the choice of method except in rare cases.

Aluminium

Neutral solochrome azurine (p. 1141).
Acid solochrome cyanine (p. 1141).
Naphthochrome green (p. 1141).
Morin-fluorescence (p. 1141).
Aluminon (p. 1140).
Solochrome-fluorescence (p. 1141).

Beryllium

Alkaline solochrome azurine (p. 1143).
Alkaline quinalizarin (p. 1143).
Naphthochrome green (p. 1142).
Neutral solochrome azurine (p. 1143).
Autofluorescence.
Ehrlich's haematoxylin.

Calcium

GBHA (p. 1137).
Morin-fluorescence (p. 1137).
Phthalocyanin (p. 1136).
Gypsum method (p. 1133).
Alizarin red (p. 1135).
Calcium red (p. 1135).
Alkaline quinalizarin (p. 1143).
Naphthochrome green (p. 1136).

Copper

Rubeanic acid (p. 1145).
Dethyldithiocarbamate (p. 1145).
Oxidation reactions (pp. 1146, 1147).
Dithizone (p. 1155).

Zinc

Sulphide-Silver (p. 1156).
Dithizone (p. 1155).
Nitroprusside.

Lead

Sulphide-Silver (p. 1149).
Rhodizonate (p. 1150).
Chromate (p. 1149).

Iron

Prussian blue (p. 1130).
Turnbull blue.
Dinitrosoresorcinol (p. 1131).

Nickel

Rubeanic acid (p. 1145).
Dimethylglyoxime acid (p. 1144).

Silver

Dimethylaminobenzylidene rhodanine (p. 1152).
Dithizone.

Localization of Exogenous Substances

The first two substances which we have to consider, the dextrans and polyvinyl pyrrolidone (PVP), are used as plasma substitutes but, unfortunately, they escape rapidly from the blood stream and find their way into the tissues. Polyvinyl pyrrolidone has also been employed as a vehicle for various pharmacologically active materials which must be given by subcutaneous or intramuscular injection. Carboxymethylcellulose is employed as a vehicle for contrast materials used in Radiology. Dextran sulphate is an anti-coagulant which has been used as a substitute for heparin.

A number of textile fibres may find their way into the tissues when used as surgical ligatures and the commonest of these are linen or cotton thread, silk or nylon. Other materials for which operative interference is responsible include gelfoam (fibrin foam) alginate, polyvinyl alcohol and of course talc, which produces a characteristic granuloma. Starch granules are introduced, together with talc, from the surgeon's gloves. They are usually identified only in operation specimens.

The trypanocidal drugs, Suramin and Germanin, are two examples of pharmaceutical products which can be localized in the tissues by histochemical methods after administration in the therapeutic doses. The list of substances in this category might well be made larger by including all those drugs which are fluorescent and which can be localized by the simple expedient of examining sections containing them under the fluorescence microscope. These include various acridine and quinoline derivatives, such as quinacrine (atebrin, mepacrine), some of the sulphonamides, a number of pteridines and the tetracyclines. Substances introduced for experimental purposes include all kinds of metallic dusts, powdered ores and other industrial products. Some of these have been considered earlier in the chapter, while others do not warrant consideration here.

Dextran and Dextran Sulphate

The deposition of high mol. wt. dextrans in rabbit kidneys was first described by Goldenberg, Crane and Popper (1947), but many authors subsequently failed to find any change in the kidney or other organs and no accumulation of dextran in the cells of the reticulo-endothelial (R.E.) system (Ingelman, 1947, 1949; Thorsén, 1949; Bull *et al.*, 1949). Partially hydrolysed dextrans, whose molecular weight varied from 28,000 to 400,000, were injected into rabbits by Friberg, Graf and Aberg (1951). By means of the PAS reaction these authors were able to demonstrate dextran in the glomeruli and in the cells of the convoluted tubules and loops of Henle. Excreted dextran was found in masses in the lumen of the collecting tubules. Further histochemical studies were made by Persson (1952a and b), who also used rabbits. He fixed his materials in a variety of fixatives, of which the best was found to be a mixture of equal parts of ethanol and methanol. This precipitated the dextrans very rapidly. A 3 per cent solution of sulphosalycylic acid in methanol was also found to be effective. Persson tested a number of oxidation procedures and followed these in each case by treatment with Schiff's solution. He found that periodic acid was the most satisfactory. Any substance in the tissues which was strongly PAS-positive, unstained by the Bauer-Feulgen procedure, blue with Azure A after HIO_4 and only light yellow with iodine, was regarded as dextran. Apart from the kidneys, dextran was found in a number of other sites in the body, particularly in the cells of the R.E. system.

Mowry, Longley and Millican (1952) were the first to take adequate precautions against loss of the water-soluble dextrans during the PAS routine. These authors fixed thin blocks of tissue in absolute alcohol and avoided contact with water thereafter by floating out their (paraffin) sections on warm 95 per cent alcohol and by using alcoholic periodic acid and alcoholic Schiff's reagent. Details of their method are given in Appendix 28, p. 1418. Further work with this method was reported by Mowry and Millican (1953).

A method for the demonstration of dextran in the tissues evolved by Jancsó-Gabor was described by Jancsó (1955). Methyl alcohol-fixed fresh preparations of tissues were stained with Niagara sky blue in 90 per cent methanol for 15–20 minutes. A different problem is the demonstration of dextran sulphate, also mentioned by Jancsó. For this purpose Mowry used a descending series of alcohols containing toluidine blue. This dye forms a precipitate with dextran sulphate which is insoluble in water, and it is thus possible to proceed from the alcoholic toluidine blue solution to the aqueous solution in which the maximum degree of staining can be achieved. Details of this method are given in the Appendix (p. 1418). A study of the fate of dextran in the human subject (Vickery, 1956) was based entirely on the use of Mowry's techniques.

Details of methods for the demonstration of carboxymethyl cellulose were given by Schnabel (1959, 1960).

Polyvinyl Pyrrolidone (PVP)

The storage of this substance, whose formula is given below, was described in human tissues by Nelson and Lusky (1951), Husselmann (1952), and also by Jeckeln (1952). Later studies were made by Jancsó *et al.* (1953), Jancsó (1954) and by Freiman and Gall (1955). A number of staining methods give

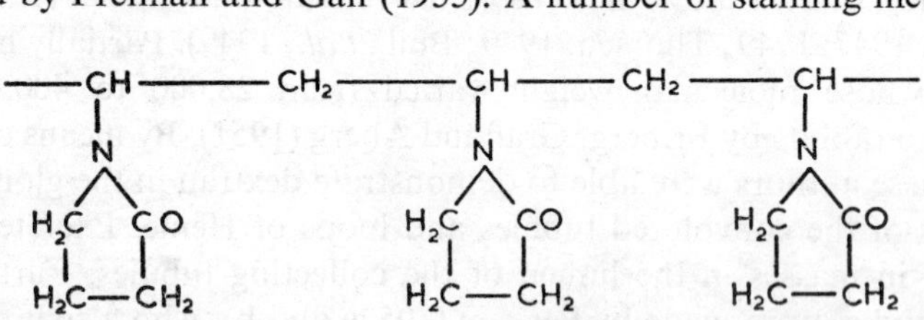

satisfactory results. The simplest is the iodine method (Jancsó, 1954), but this lacks permanence. Congo red stains PVP a deep cherry-red colour and Freiman and Gall recommend a technique based on this stain. Unfortunately Congo red staining of PVP is also impermanent, since fading takes place within a few days.

The best and most permanent method in my hands has been Jancsó's Chlorazol fast pink method. The formula of this dye is somewhat similar to that of Congo red (see Vol. 1, p. 386) and it stains PVP a pale pinkish-red colour.

A report on this method was made by Towers (1957), and Frommer (1956) described the effect of intravenous PVP on the cells of mouse liver. All the above methods are given in Appendix 28, pp. 1418, 1419.

Polyvinyl Alcohol

This polymer is less often found in human tissues than PVP, but tubes composed of a sponge-work of the compound have been used in arterial grafting. The graft stains black with Weigert's elastic stain, which should surprise nobody if its formula (below) is considered in the light of modern views (Chapter 8, p. 225) on the staining of elastic tissue. Polyvinyl alcohol also stains black with iodine and red with Chlorazol fast pink.

$$-CH_2\cdot\underset{OH}{CH}-CH_2\cdot\underset{OH}{CH}-CH_2\cdot\underset{OH}{CH}-$$

Polyvinyl alcohol

Suramin

A method for demonstrating this substance in the tissues was reported by Jancsó and Jancsó-Gabor (1953), which depended on simultaneous fixation of freshly teased tissues with methanol containing methylene blue eosinate (May-Grunwald stain). Other dyes capable of forming a precipitate with Suramin were phenosafranin, cresyl blue, and Azure B. With any of these dyes it was essential to avoid contact with water and, after a few minutes in the methanolic stain, differentiation was carried out by agitation in absolute alcohol followed by clearing in xylene and mounting in Canada balsam.

Quinacrines (Atebrin, Mepacrine)

The clinical use of mepacrine as a vermifuge, and in the treatment of lupus erythematosus, produced two papers, by Mustakallio and Saikkonen (1954) and Mustakallio (1954), describing its localization in the tissues. Provided that no solvent fixative is used, and freeze-drying is the method of choice, it is easy to demonstrate the characteristic fluorescence (Excitation maxima, 285 and 420 nm.; emission maximum, 500 nm.). In a case of unexplained "jaundice" I found high concentrations of the drug particularly in the cutaneous glands (biopsy). This is shown in Plate XXVIIIb, p. 1180.

Textile Fibres

Due to the presence of side groups which are highly polarizable, cellulose nitrate and polyacrylonitrile (Orlon) show negative birefringence in polarized light. Because of its crystalline nature, nylon suture material in tissue sections is also birefringent, as are fibres of lint or cotton thread. Polarization microscopy is therefore of little use from the diagnostic point of view. Venkataraman (1952) gives a differential table for the identification of textile fibres and some of the methods suggested therein may be found useful histochemically. Silk and nylon, he says, are soluble in conc. HCl while cotton is not. Silk and cotton, on the other hand, are soluble in ammoniacal copper solutions while nylon is not. Strong solutions (90 per cent) of phenol dissolve nylon alone. A large number of stains for the primary wall of the cotton fibre are given by Moore (1953). These include ruthenium red, since the primary wall contains pectic acid (see Chapter 10, p. 359) as well as cellulose.

Special staining techniques exist for polyamide fibres like nylon which contain a number of free carboxyl groups, together with a smaller number of amino groups and some terminal amides. The formula of a linear polyamide molecule given below suggested that the free carboxyl groups would be likely to play an important part in staining reactions.

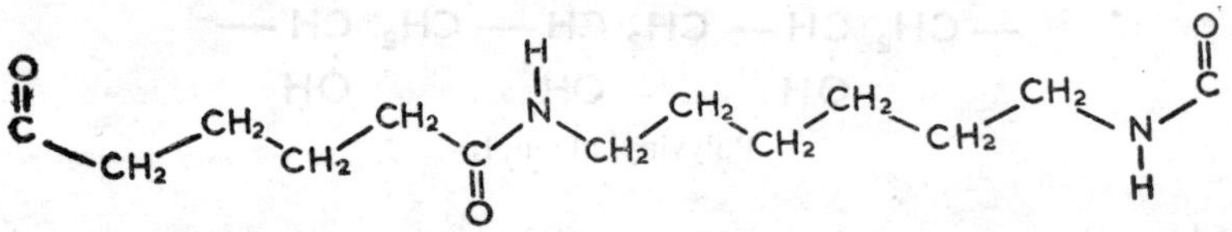

It was found, however (Vickerstaff, 1954), that the terminal amide groups are of great importance, especially in the staining of nylon by acid dyes.

Differential stains for the histochemical identification of nylon have not been investigated. Dyes of the type known as "dispersed," that is to say, water-insoluble anthraquinone dyes employed as aqueous suspensions, can be used to stain nylon fibres. Since the drawn nylon fibre is highly crystalline the dispersed dyes, combined with the amide groups, are orientated along the fibre. This produces dichroism when the stained fibre is viewed by polarized light. Many soluble dyes have been used to stain nylon fibres, but these stain cotton and silk even more readily and are thus useless for differentiation. Although the need for histochemical distinction between the various natural and synthetic fibres is not very great, a definitive study of the staining and other properties of all materials likely to be found in tissue sections would be of considerable value to histopathologists.

Miscellaneous Endogenous Substances

Urea

The only specific test for urea depends on the formation of crystals of dixanthydrol urea by fixing fresh pieces of tissue in xanthydrol in alcoholic acetic acid. Rosettes of yellowish-brown needle-shaped crystals indicate the localization of urea. According to Gomori (1952), the usefulness of this test in histochemistry was limited because of damage to the structure of the tissues by the strong acetic acid and by the amount of diffusion which occurred during the reaction. Details are nevertheless given in Appendix 28, p. 1419. Lillie (1954) modified the method by performing the first stage of the reaction for 14 days at −25°. He considered that further study was necessary before the method could be used routinely.

Feigl (1954) suggested the production of a sensitive test for urea by condensation with phenylhydrazine to form diphenylcarbazide. In the presence of certain metals (e.g. nickel) an insoluble highly coloured complex would be produced. Unfortunately it seems that the conditions for urea-phenylhydrazine condensation are unlikely to be realized histochemically.

Uric Acid and Urates

In the tissues uric acid is normally present in the form of crystalline acid sodium urate. The crystals are slightly soluble in dilute alkali, but insoluble in all ordinary reagents. Urate deposits can be distinguished from phosphate and carbonates by their solubility in dilute solutions of lithium carbonate. They possess two histochemical properties, basophilia and argentaffinity, and tests for their demonstration depend on one or the other of these. Of the two the silver methods are the most popular. Lillie (1954) gave both De Galantha's (1935) silver nitrate method and Gomori's hexamine-silver technique and this last method is reproduced in Appendix 28.

REFERENCES

ARCHER, H. E., DORMER, A. E., SCOWEN, E. F., and WATTS, R. W. E. (1957). *Clin. Sic.*, **16,** 405.

ALDRIDGE, W. N., and LIDDELL, H. F. (1948). *Analyst*, **73,** 603.

BAILAR, J. C. (1956). "Chemistry of Coordination Compounds." Reinhold, New York.

BARBAGLIA, V. (1930). *Studi Sassaresi*, **8,** 253.

BAYER, E. (1957). *Ber. dtsch. chem. Ges.*, **90,** 2325.

BETTENDORF, A. (1870). *Z. analyt. Chem.*, **9,** 105.

BLUMENTHAL, W. B. (1946). *Amer. Dyestuff Rep.*, **35,** 520.

BOTS, G. T. A. M., and HARDONK, M. J. (1965). *Psychiat. Neurol. Neurochir.*, **68,** 259.

BOURNE, G. H. (1943). *Quart. J. exp. Physiol.*, **32,** 1.

BRAND, I., and TAKATS, I. (1951). *von Graefes Arch. Ophthalmol.*, **151,** 391.

BRANDING, G. (1927). *Studi sassari*, **5,** 85.

BRODA, B. (1939). *Mikrokosmos.*, **32,** 184.

BROYER, M., and BERGER, R. (1968). *Presse med.*, **76,** 2395.

BRUNK, U., and SKÖLD, G. (1967). *Acta histochem.*, **27,** 199.

BULGER, R. E. (1969). *J. Cell Biol.*, **40,** 79.

BULL, J. P., RICKETTS, C., SQUIRE, J. R., MAYCOCK, W. D'A., SPOONER, S. J. L., MOLLISON, P. L., and PATTERSON, J. C. S. (1949). *Lancet*, **256,** 134.

BUNTING, H. (1949). *Stain Tech.*, **24,** 109.

BUNTING, H. (1951). *Arch. Path.*, **52,** 458.

CAMERON, G. R. (1930). *J. Path. Bact.*, **33,** 329.

CARERE-COMES, O. (1938). *Z. wiss. Mikr.*, **55,** 1.

CASTEL, P. (1935). *Arch. Soc. med. biol. Montpelier et Languedoc*, **16,** 453.

CASTEL, P. (1936). *Bull. Histol. Tech. Micr.*, **13,** 106.

CHAUNCEY, H. H., and LIONETTI, F. (1952). *Fed. Proc.*, **11,** 196.

CHENG, K-K. (1956). *J. Path. Bact.*, **71,** 265.

CHISHOLM, G. D., and HEARD, B. E. (1962). *Br. J. Surg.*, **50,** 78.

CHOU, L. Y., and DONOHUE, W. L. (1952). *Pediatrics*, **10,** 660.

CHRISTELLER, E. (1926). *Med. Klin.*, **22,** 619.
CHRISTELLER, E. (1927). *Verh. deutsch. path. Ges.*, **22,** 173.
CLARKE, B. L., and HERMANCE, H. W. (1937). *Indust. eng. Chem., Analytical Ed.*, **9,** 292.
CLOETENS, R. (1942). *Arch. int. Pharmacodynam.*, **68,** 419.
CLOSE, R. A., and WEST, T. S. (1960). *Talant*, **5,** 221.
COLLEWIJN, H. (1963). Thesis, University of Amsterdam.
COLLINS, P. F., DIEHL, H., and SMITH, G. F. (1959). *Analyt. Chem.*, **31,** 1862.
CRÉTIN, A. (1922). *Ass. Anat.*, **17,** 100.
CRÉTIN, A. (1923). *Ibid.*, **18,** 163.
CRÉTIN, A. (1924). *Bull. Histol. Tech. micr.*, **1,** 334.
CRÉTIN, A. (1929). *C.R. Acad. Anat. Sci., Paris*, **16,** 241.
CRÉTIN, A. (1933). *Bordeaux Chirurg.*, p. 125.
CROUT, J. R., and JENNINGS, R. B. (1957). *J. Histochem. Cytochem.*, **5,** 170.
DAHL, L. K. (1952). *Proc. Soc. exp. Biol., N.Y.*, **80,** 474.
DEBENEDETTI, S. (1925). *Arch. Sci. med. Pavia*, **47,** 261.
DE GALANTHA, A. (1935). *Amer. J. clin. Path.*, **5,** 165.
DENNY, J. J., ROBSON, W. D., and IRWIN, D. A. (1937). *Canad. Med. Assoc. J.*, **37,** 1.
DENNY, J. J., ROBSON, W. D., and IRWIN, D. A. (1939). *Ibid.*, **40,** 213.
DENZ, F. A. (1949). *Quart. J. micr. Sci.*, **90,** 317.
DIEHL, H., BUCHANAN, E. B., and SMITH, G. F. (1960). *Analyt. Chem.*, **32,** 1117.
DIEZEL, P. B. (1955). In "Biochemistry of the Developing Nervous System." Academic Press Inc., p. 145.
DONNÉ, A. (1839). *C.R. Acad. Sci., Paris*, **8,** 805.
ECHAVE, D. (1941). "El Metal Cobre en Biologia." Libreria El Ateneo, Buenos Aires.
EEGRIWE, E. (1928). *Z. anal. Chem.*, **74,** 223.
EEGRIWE, E. (1940). *Ibid.*, **120,** 81.
ELFTMAN, H., and ELFTMAN, A. G. (1945). *Stain Tech.*, **20,** 59.
FAIRHALL, L. T. (1940). *Publ. Health Bull.* (*U.S. Public Health Service*), No. 253, p. 22.
FALKMER, S., HELLMANN, B., and VOIGT, G. E. (1964). *Acta physiol. scand.*, **60,** 47.
FEIGL, F. (1924). *Mikrochemie.*, **2,** 188.
FEIGL, F. (1954). "Spot Tests, Vol. I, Inorganic Applications." Elsevier, Amsterdam.
FEIGL, F. (1954). "Spot Tests." Elsevier, Amsterdam. Vol. 2, p. 11.
FENTON, R. H., JOHNSON, F. B., and ZIMMERMAN, L. E. (1964). *J. Histochem. Cytochem.*, **12,** 153.
FIRKET, H. (1953). *C.R. Soc. Biol. Paris*, **147,** 167.
FISCHER, M. I., TIKKALA, A. O., and MAWSON, C. A. (1955). *Canad. J. Biochem.*, **33,** 181.
FRANKENBERGER, Z. (1921). *C.R. Acad. Anat. Sci., Paris*, **16,** 241.
FREIMAN, D. G., and GALL, E. A. (1955). *Amer. J. clin. Path.*, **95,** 1427.
FREUDENBERG, E. (1926). *Klin. Wschr.*, **5,** 66.
FRIBERG, U., GRAF, W., and ABERG, B. (1951). *Scand. J. clin. Lab. Invest.*, **3,** 221.
FROMMER, J. (1956). *Amer. J. Path.*, **32,** 433.
GEDIGK, P., and PIOCH, W. (1956). *Beit. path. Anat. allg. Path.*, **116,** 124.
GEHAUF, B., and GOLDENSON, J. (1955). *Analyt. Chem.*, **27,** 420.
GÉRARD, P., and CORDIER, R. (1932). *C.R. Soc. biol., Paris*, **43,** 367.
GERSH, I. (1938). *Anat. Rec.*, **70,** 311.
GERSH, I., and STIEGLITZ, E. J. (1933). *Ibid.*, **56,** 185.
GIUSTI, G. V., and FIORI, A. (1969). *Stain Technol.*, **44,** 263.
GLICK, D. (1949). "Techniques of Histo- and Cyto-Chemistry." New York.
GODLEWSKI, H. G. (1960). *J. Histochem. Cytochem.*, **8,** 122.
GOLDENBERG, M., CRANE, R. D., and POPPER, H. (1947). *Amer. J. Path.*, **17,** 939.
GOLDSTEIN, D., and STARK-MAYER, C. (1958). *Anal. chim. Acta.*, **19,** 437.
GOMORI, G. (1936). *Amer. J. Path.*, **13,** 655.
GOMORI, G. (1951). *Methods in Medical Research*, **4,** 3.
GOMORI, G. (1952). "Microscopic Histochemistry." Chicago University Press, p. 35.

Gomori, G. (1952). "Microscopic Histochemistry." Chicago Univ. Press, p. 117.
Grandis, E., and Mainini, A. (1900). *Arch. ital. Biol.*, **34**, 73.
Grandis, E., and Mainini, A. (1902). *Ibid.*, **38**, 143.
Granick, S. (1943). *J. biol. Chem.*, **149**, 157.
Granick, S. (1946). *Ibid.*, **164**, 737.
Granick, S. (1951). *Physiol. Revs.*, **31**, 489.
Green, C. L. (1955). *Amer. J. Path.*, **31**, 545.
Gutzeit, G. (1929). *Helv. chim. Acta*, **12**, 720.
Hammett, L. P., and Sottery, C. T. (1925). *J. Amer. chem. Soc.*, **47**, 142.
Hand, W. C., Edwards, B. B., and Caley, E. R. (1943). *J. lab. clin. Med.*, **28**, 1835.
Hanson, H. T., and Smith, E. L. (1949). *J. biol. Chem.*, **179**, 789.
Hoare, R., Delory, E. E., and Penner, D. W. (1956). *Cancer*, **9**, 721.
Hoch, F. L., and Vallee, B. L. (1952). *J. biol. Chem.*, **195**, 531.
Horning, E. S. (1951). In "Cytology and Cell Physiology". Oxford, p. 287.
Howell, J. S. (1959). *J. Path. Bact.*, **77**, 473.
Hukill, P. B., and Putt, F. A. (1962). *J. Histochem. Cytochem.*, **11**, 490.
Humphrey, A. A. (1935). *Arch. Path.*, **20**, 256.
Husselmann, H. (1952). *Klin. Woch.*, **30**, 801.
Ingelman, B. (1947). *Acta chem. Scand.*, **1**, 731.
Ingelman, B. (1949). *Uppsala läkarcf. förh.*, **54**, 107.
Irwin, D. A. (1955). *Arch. indust. Health*, **12**, 218.
Jancsó, N. (1954). *M. Tud. Akad. Biol. Orv. Oszt Közleményei*, **5**, 381.
Jancsó, N. (1955). "Speicherung." Akademiai Kiado, Budapest.
Jancsó, N., and Jancsó-Gabor, A. (1953). *Nature, Lond.*, **170**, 567.
Jancsó, N., Jancsó-Gabor, A., Lakos, A., and Draskóczy, P. (1953). *Acta physiol. Hung. Supp.*, **4**, p. 30.
Jeckeln, E. (1952). *Arch. Path. Anat.*, **322**, 529.
Johnson, F. B. (1956). *J. Histochem. Cytochem.*, **4**, 404.
Johnson, F. B., and Pani, K. (1962). *Arcs Path.*, **74**, 347.
Johnson, M. J. (1941). *J. biol. Chem.*, **137**, 575.
Kashiwa, H. K., and Atkinson, W. B. (1963). *J. Histochem. Cytochem.*, **11**, 258.
Kaufmann, H. E., and Adams, E. C. (1954). *Science*, **120**, 723.
Kaye, G. I., and Donn, J. D. (1965). *Science, N.Y.*, **150**, 1167.
Khau van Kien, L., and Thaï-Tuong (1956). *Ann. Histochim.*, **1**, 15.
King, E. J., Harrison, C. V., Mohanty, G. P., and Nagelschmidt, G. (1955). *J. Path. Bact.*, **69**, 81.
Kodousek, R. (1963). *Acta histochem.*, **15**, 386.
Kolthoff, I. M. (1927). *Z. wiss. Mikr.*, **55**, 1.
Komaya, G. (1925). *Arch. Dermatol. Syph.*, **149**, 277.
Komnick, H. (1962). *Protoplasma*, **55**, 414.
Komnick, H., and Komnick, U. (1963). *Z. Zellforsch.*, **60**, 163.
Kuznetsov, V. J. (1941). *Chem. Abstracts*, **35**, 3190.
Laas, E. (1941). *Frankfurt Z. Path.*, **55**, 265.
Langeron, M. (1934). "Précis de microscopie." 5th Ed., Masson, Paris.
Laufberger, V. (1937). *Bull. Soc. chim. Biol.*, **19**, 1575.
Leiner, M., and Leiner, G. (1944). *Biol. Zbl.*, **64**, 293.
Lepoutre, C. (1925). *J. Urol. méd. chir.*, **20**, 424.
Lilienfeld, L., and Monti, A. (1892). *Z. wiss. Mikr.*, **9**, 332.
Lillie, R. D. (1948). "Histopathologic Technic." Philadelphia.
Lillie, R. D. (1954). "Histopathologic Technic and Practical Histochemistry." Blakiston, New York, p. 152.
Lillie, R. D., Geer, J. C., and Gutiérrez, A. (1963). *J. Histochem. Cytochem.*, **11**, 662.
Linstrom, F., and Milligan, C. W. (1967). *Analyt. chem.*, **39**, 132.
Lison, L. (1936a). "Histochimie Animale." Gautier Villars, Paris.
Lison, L. (1936b). *C.R. Soc. biol., Paris*, **191**, 900.
Lison, L. (1953). "Histochimie et Cytochimic Animales." Gautier Villars, Paris.
Lloyd, F. E. (1925). *Flora*, **18**, 369.
Logothetopoulos, J. (1960). *Am. J. Path.*, **37**, 357.

LOMBARDO, C. (1906). *La Riforma medica*, **22**, 26.
LOMBARDO, C. (1908). *Arch. farmacol.*, **7**, 8.
LOWRY, O. H., and LOPEZ, J. A. (1946). *J. biol. Chem.*, **162**, 421.
MACALLUM, A. B. (1905). In Abderhalden's "Handbuch der biochemischen Arbeitsmethoden", Vol. II, 2, 1099.
MCALLISTER, R. A. (1953). *Nature, Lond.*, **169**, 708.
MCGEE-RUSSELL, S. M. (1955). *Nature, Lond.*, **175**, 301.
MAGER, M., MCNARY, W. F., Jr., and LIONETTI, F. (1953). *J. Histochem. Cytochem.* **1**, 493.
MALLORY, F. B. (1938). "Pathological Technique." Philadelphia.
MALLORY, F. B., and PARKER, F., Jr. (1939). *Amer. J. Path.*, **15**, 517.
MAWSON, C. A., and FISCHER, M. I. (1952a). *Canad. J. med. Sci.*, **30**, 336.
MAWSON, C. A., and FISCHER, M. I. (1952b). *Arch. Biochem. Biophys.*, **36**, 485.
MAWSON, C. A., and FISCHER, M. I. (1953). *Biochem. J.*, **55**, 696.
MAZUR, A., and SCHORR, E. (1950). *J. biol. Chem.*, **82**, 607.
MENDEL, L. B., and BRADLEY, H. C. (1905). *Amer. J. Physiol.*, **13**, 17.
MENDEL, L. B., and BRADLEY, H. C. (1905). *Ibid.*, **14**, 313.
MENTEN, M. (1922). *J. med. Res.*, **43**, 323.
MIDORIKAWA, O., and EDER, M. (1962). *Histochemie*, **2**, 444.
MOLNAR, J. (1952). *Stain Tech.*, **27**, 221.
MOORE, A. T. (1953). *Stain Tech.*, **28**, 149.
MOWRY, R. W. (1954). *J. Histochem. Cytochem.*, **2**, 470.
MOWRY, R. W., LONGLEY, J. B., and MILLICAN, R. C. (1952). *J. lab. clin. Med.*, **39**, 211.
MOWRY, R. W., and MILLICAN, R. C. (1953). *Amer. J. Path.*, **29**, 523.
MÜLLER, P. (1964). *Acta histochem.*, **17**, 118.
MÜLLER, D., and GÜTHERT, H. (1960). *Virchows Archs.*, **333**, 97.
MUSTAKALLIO, K. K. (1954). *Acta derm-venereol.*, **34**, 93.
MUSTAKALLIO, K. K., and SAIKKONEN, J. I. (1954). *Exp. Parasitol.*, **3**, 167.
NELSON, A. A., and LUSKY, L. M. (1951). *Proc. Soc. exp. Biol.*, **76**, 765.
NYLANDER, G. (1953). *Acta physiol. scand.*, **29**, Suppl. 107.
OGAWA, Y. (1964). *Odontology (japon)*, **52**, 153.
OKAMOTO, K. (1942a). *Acta Schol. med. Univ. imp. Kioto*, **32**, 99.
OKAMOTO, K. (1942b). *Trans. Soc. Path., Japan*, **32**, 99.
OKAMOTO, K. (1943). *Ibid.*, **33**, 247.
OKAMOTO, K., and UTAMURA, M. (1938). *Ibid.*, **20**, 573.
OKAMOTO, K., UTAMURA, M., and AKAGI, T. (1939). *Ibid.*, **22**, 361.
OKAMOTO, K., UTAMURA, M., and MIKAMI, G. (1938). *Ibid.*, **22**, 348.
OKAMOTO, K., SHIMAMOTO, T., SENO, M., UEDA, M., KUSUMOTO, Y., KATO, A., and SHIBATA, D. (1947). *Trans. Soc. path. jap.*, **36**, 16.
OKAMOTO, K., AOYAMA, T., IBARAKI, H., NARUMI, M., SHIBATA, Y., KAWASAKI, D. SHIBATA, A., and KOMATSU, K. (1951). *Hyogo J. Med. Sci.*, **1**, 77.
OTSUKA, N., and IBATA, Y. (1968). *Histochemie*, **12**, 357.
PEARSE, A. G. E. (1953). "Histochemistry, Theoretical and Applied." J. & A. Churchill, London.
PEARSE, A. G. E. (1955). *J. Path. Bact.*, **70**, 554.
PEARSE, A. G. E. (1960). "Histochemistry, Theoretical and Applied." 2nd Ed. J. & A. Churchill, London.
PEARSE, A. G. E. (1964). Ciba Foundation Symp., "Cardiomyopathy". Ed. Wolstenholme, G. E. W., and O'Connor, M. Churchill, London, p. 132.
PERLS, M. (1867). *Virchows Arch.*, **39**, 42.
PERSSON, B. (1952a). *Acta Soc. med. Upsal.*, **57**, 411.
PERSSON, B. (1952b). *Ibid.*, **57**, 421.
PETERSON, R. E., and BOLLIER, M. E. (1955). *Analyt. Chem.*, **27**, 1195.
PIZZOLATO, P. (1964). *J. Histochem. Cytochem.*, **12**, 333.
PIZZOLATO, P., and MCCRORY, P. (1962). *J. Histochem. Cytochem.*, **10**, 102.
PIZZOLATO, P., and PIZZOLATO, J. M. (1966). *Am. J. med. Tech.*, **32**, 247.
POLICARD, A. (1933). *Protoplasma*, **19**, 602.

Poppen, K. J., Green, D. S., and Wrenn, H. T. (1953). *J. Histochem. Cytochem.*, **1,** 160.
Poulson, D. F., and Bowen, V. T. (1952). *Exp. Cell. Research Supp.*, **2,** 161.
Pritschow, A., and Lhotka, J. F. (1961). *Stain Technol.*, **36,** 293.
Puchtler, H., Meloan, S. N., and Terry, M. S. (1969). *J. Histochem. Cytochem.*, **17,** 110.
Pyrah, L. N., Anderson, C. K., Hodgkinson, A., and Zarembski, P. M. (1959). *Br. J. Urol.*, **31,** 235.
Quinke, H. I. (1868). Referred to in *Arch. exp. Path. Pharmak., Leipzig*, **37,** 183.
Quinlan-Watson, F. (1953). *Biochem. J.*, **53,** 457.
Rabl, C. H. R. (1923). *Klin. Wschr.*, **2,** 1644.
Rabl, C. H. R. (1926). *Ibid.*, **5,** 365.
Richter, G. W. (1960). *J. exp. Med.*, **112,** 551.
Riecken, E. O. (1963). *Histochemie*, **3,** 298.
Riecken, E. O., and Pearse, A. G. E. (1966). *Gut*, **7,** 86.
Roberts, W. J. (1935). *Bull. Histol. Tech. micr.*, **12,** 400.
Rohdenburg, G. L., and Geiger, J. (1928). *Arch. Path.*, **6,** 215.
Roncoroni, G. (1962). *Riv. istochim.*, **8,** 77.
Schmeltzer, W. (1933). *Z. wiss. Mikr.*, **50,** 99.
Schmidt, R., and Rautschke, R. (1963a). *Acta histochem.*, **15,** 87.
Schmidt, R., and Rautschke, R. (1963b). *Acta histochem.*, **15,** 359.
Schmidt, R., and Rautschke, R. (1963c). *Acta histochem.*, **15,** 373.
Schmidt, R., and Rautschke, R. (1964). *Acta histochem.*, **19,** 1.
Schnabel, R. (1959). *Acta histochem.*, **8,** 50.
Schnabel, R. (1960). *Naturwiss.*, **8,** 180.
Schnabel, R., and Nisch, G. (1961). *Graefes. Archs Ophthalmol.*, **164,** 220.
Scott, G. H., and Packer, D. M. (1939). *Anat. Rec.*, **74,** 17.
Scott, G. H., and Williams, P. S. (1935). *Ibid.*, **64,** 107.
Sebruyns, M. (1946). *C.R. Soc. Biol., Paris*, **140,** 1141.
Serra, J. A., and Queiroz Lopes, A. (1945). *Port. Acta. biol.*, **1,** 111.
Silver, V. L., and Price, J. L. (1969). *Stain Technol.*, **44,** 257.
Smith, E. L. (1948). *J. biol. Chem.*, **176,** 9.
Smith, G. L., Jenkins, R. A., and Gough, J. F. (1969). *J. Histochem. Cytochem.*, **17,** 749.
Tandler, C. J. (1960). *Biochim. biophys. Acta.*, **44,** 536.
Tandler, C. J. (1961). *Exp. Cell Res.*, **25,** 246.
Thomas, J. A., and Lavalloy, J. (1935). *Bull. Histol. Tech. micr.*, **12,** 400.
Thorsén, G. (1949). *Lancet*, **256,** 132.
Timm, F. (1958). *Dtsch. Z. ges. gerichtl. Med.*, **46,** 706.
Timm, F., and Arnold, M. (1958). *Archs exp. Path. Pharmakol.*, **233,** 422.
Timm, F., and Neth, R. (1959). *Histochemie*, **1,** 403.
Timm, F. (1936). *Virchow's Arch.*, **297,** 502.
Tirmann, J. (1898). *Gorbersdorfer Veroffentl.*, **2,** 101.
Torack, R. M., and La Valle, M. (1970). *J. Histochem. Cytochem.*, **18,** 635.
Towers, R. P. (1957). *J. clin. Path.*, **10,** 175.
Tupper, R., Watts, R. W. E., and Wormall, A. (1954). *Biochem. J.*, **57,** 245.
Turchini, J. (1924). *Ass. Anat.*, **19,** quoted by Lison, 1936.
Turchini, J. (1930). *Bull. Acad. sci. Lettres, Montpellier*, **60,** 29.
Uzman, L. L. (1956). *Lab. Invest.*, **5,** 299.
Vallee, B. L., and Gibson, J. G. (1948). *J. biol. Chem.*, **176,** 435.
Vallee, B. L., and Hoch, F. L. (1955). *Proc. Nat. Acad. Sci., Wash.*, **41,** 327.
Vallee, B. L., Hoch, F. L., and Hughes, W. L., Jr. (1954). *Arch. Biochem. Biophys.*, **48,** 347.
Vallee, B. L., and Neurath, H. (1955). *J. biol. Chem.*, **217,** 253.
Van Harreveld, A., and Potter, R. L. (1961). *Stain Technol.*, **36,** 185.
Venkataraman, K. (1952). "The Chemistry of Synthetic Dyes." Academic Press, New York, Vol. 1, p. 269.
Vickerstaff, T. (1954). "The Physical Chemistry of Dyeing." Oliver and Boyd, Edinburgh.

VICKERY, A. L. (1956). *Amer. J. Path.*, **32,** 161.
VISCHER, W. (1947). *Schweiz. Z. Path.*, **10,** 286.
VOIGT, G. E. (1966). *Acta med. leg. Soc., Liège*, **19,** 17.
VOIGT, G. E., BOG, R., and WOLF, I. (1954). *Z. allg. Path. path. Anat.*, **92,** 1.
VON KÓSSA, J. (1901). *Beitr. path. Anat.*, **29,** 163.
VOSBURGH, G. I., and FLEXNER, L. B. (1950). *Amer. J. Physiol.*, **161,** 202.
WACHSTEIN, M., and ZAK, F. G. (1946). *Amer. J. Path.*, **22,** 603.
WARBURG, O. (1949). "Heavy Metal Prosthetic Groups and Enzyme Action." p. 191. Trans. A. Lawson, Oxford.
WATERHOUSE, D. F. (1945a). *Bull. Council Sci. indust. Res. Austral.*, **191,** 7.
WATERHOUSE, D. F. (1945b). *Ibid.*, **191,** 21.
WATERHOUSE, D. F. (1951a). *Nature, Lond.*, **167,** 358.
WATERHOUSE, D. F. (1951b). *Australian J. Sci. Res. B.*, **4,** 145.
WELCHER, F. J. (1947). "Organic Analytical Reagents." New York, p. 337.
WELLS, G. C. (1956). *Brit. J. Dermatol.*, **68,** 237.
WIENER, H., and WOLMAN, M. (1965). *Histochemie*, **4,** 357.
WÖCKEL, W., STEGNER, H. E., and JANISCH, W. (1961). *Virchows Archs*, **334,** 503.
WOLFF, H., and RINGLER, D. (1954). *Z. ges. exp. Med.*, **124,** 236.
WOLMAN, M., and GOLDRING, D. (1962). *J. Histochem. Cytochem.*, **10,** 505.
YASUE, T. (1969). *Acta histochem. cytochem.*, **2,** 83.
YOUNG, A., and SWEET, T. R. (1955). *Analyt. Chem.*, **27,** 418.
YUDKIN, W. H., and FRUTON, J. S. (1947). *J. biol. Chem.*, **170,** 421.
ZERMATTEN, H. L. (1933). *J. Proc. Acad. Sci., Amsterdam*, **36,** 899.

CHAPTER 29

FLUORESCENCE MICROSCOPY

Introduction

Fluorescence microscopy dates from about 1904, when Köhler reported observing fluorescence of tissues under ultraviolet irradiation in a microscope. Largely due to the awkwardness of early instrumentation, the technique was not widely used until Coons (1941) introduced the fluorescent antibody technique, with a consequent revolution in immunology. At the present time, immunofluorescence is the most widely used of all fluorescence microscopic techniques, but the further development and greater availability of suitable equipment has encouraged the greater use of fluorescence techniques generally. Development of instrumentation, of staining and other techniques, and of applications has been rapid during the past few years, so that the chapter on fluorescence microscopy in the previous edition of this book is now hopelessly out of date. Further rapid progress can confidently be expected.

The present chapter is concerned mainly with the principles of qualitative fluorescence microscopy. Quantitative fluorescence microscopy will be dealt with in Chapter 31. Many individual applications of fluorescence microscopy have already been mentioned in previous chapters, particularly immunofluorescence (Chapter 7) and the aldehyde-induced fluorescence of biogenic amines (Chapter 27).

The value of fluorescence microscopy in histochemistry derives primarily from the ability of the method to reveal substances in very low concentration. Fluorescence microscopy also extends the range of substances which can be revealed, either directly or after chemical treatment. In addition, changes in excitation and emission spectra of fluorescent substances due to binding on to a substrate can be used to obtain information about the conformation of the substrate molecules.

The only general text-book on fluorescence microscopy available at the present time appears to be that of Haitinger (1959), which consists of a reprint of the original (1938) text with additions by the editors. It deals particularly with autofluorescence and fluorochromes, and has an extensive bibliography of the early literature. The basic technique of fluorescence microscopy is described by Price and Schwartz (1956). Early applications to histochemistry have been reviewed, with an enormous bibliography, by de Lerma (1958). In immunofluorescence, the standard texts are those of Bals (1966), Goldman (1968), and Nairn (1969). Much of general interest is to be found in papers on standardization in immunofluorescence edited by Hol-

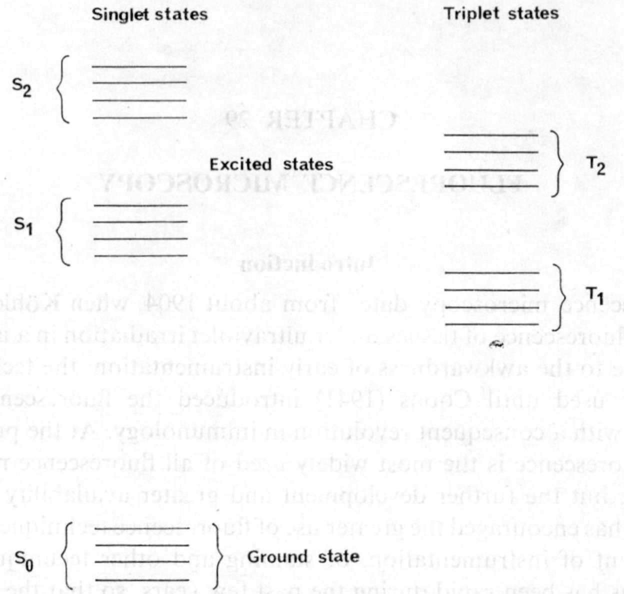

FIG. 245. Molecular energy levels. Singlet and triplet electronic states (S_0, S_1, S_2, . . .; T_1, T_2, . . .) with some vibrational levels of each electronic state.

borow (1970), and in Zelenin's (1967) book on fluorescence cytochemistry of nucleic acids.

Of bibliographic value in fluorescence are the Index Medicus section on fluorescence microscopy (at September 1967, 1130 references had been indexed; by June 1970 the number had risen to 2317, an increase of about 500

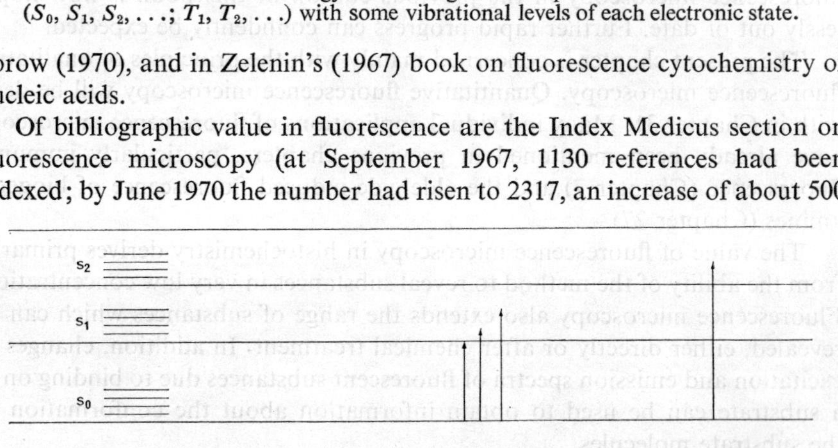

FIG. 246. Showing transitions between electron states corresponding to various points of an absorption/excitation spectrum.

per annum), the Guide to Fluorescence Literature (Passwater, 1967, 1970), and the annual reviews of fluorescence in Analytical Chemistry.

Basic principles

Essentially, fluorescence is an optical phenomenon in which light is absorbed by a substance (known as a fluorophore) and almost instantaneously re-emitted as light of a longer wavelength. The change in wavelength is known as the Stokes shift. For details of the fluorescence process, see Seliger and McElroy (1965), Hercules (1966), Udenfriend (1969), Barenboim, Domanskii and Turoverov (1969), and McCarthy and Moyer (1970).

As a result of absorption of light, the fluorophore molecules become excited, that is to say they absorb the energy of the light and their electronic state is changed to an excited state in which the energy of each molecule is

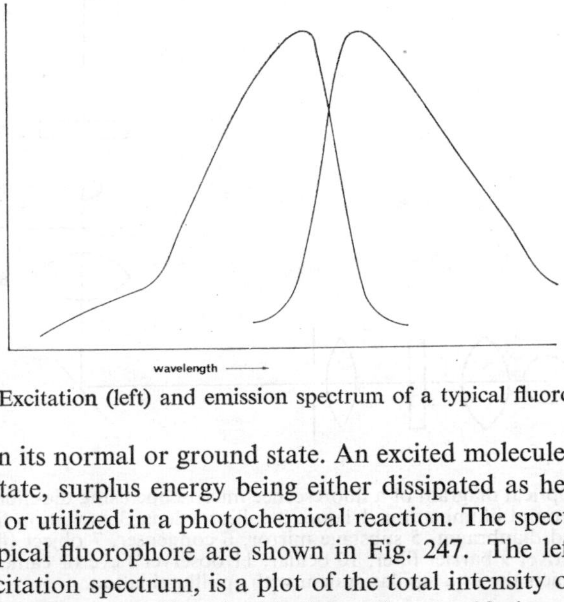

FIG. 247. Excitation (left) and emission spectrum of a typical fluorophore.

higher than in its normal or ground state. An excited molecule will return to the ground state, surplus energy being either dissipated as heat, emitted as fluorescence, or utilized in a photochemical reaction. The spectral characteristics of a typical fluorophore are shown in Fig. 247. The left-hand curve, called the excitation spectrum, is a plot of the total intensity of fluorescence when the specimen is irradiated with light of a specified wavelength. The shape of the excitation spectrum is usually substantially identical to that of the absorption spectrum. It is evident that fluorescence is best excited by irradiation of light of a wavelength corresponding as closely as possible to the peak of the excitation or absorption spectrum. The use of a still shorter wavelength results in less efficiency of fluorescence excitation and an increased risk of damage to the specimen by heating. The right-hand curve, the emission spectrum, shows a spectral distribution of the fluorescence emitted by the fluorophore. For a pure substance, the emission spectrum is independent of of the wavelength used for excitation. The fact that the emitted light is of a longer wavelength than that used for excitation is the basis for the optical system of fluorescence microscopes, as shown in Fig. 248. Such a microscope differs from a microscope used for conventional transmitted light absorption studies in that it has two filters. A primary (excitation) filter is placed somewhere between the light source and a specimen. This filter, in combination with the light source used, should provide light only over a comparatively narrow band of wavelengths corresponding to the excitation peak of the fluorophore being investigated. The other filter, the secondary or barrier filter, is intended solely to prevent light of the excitation wavelength from reaching the observer's eye, and is placed between the object and the eye. Its trans-

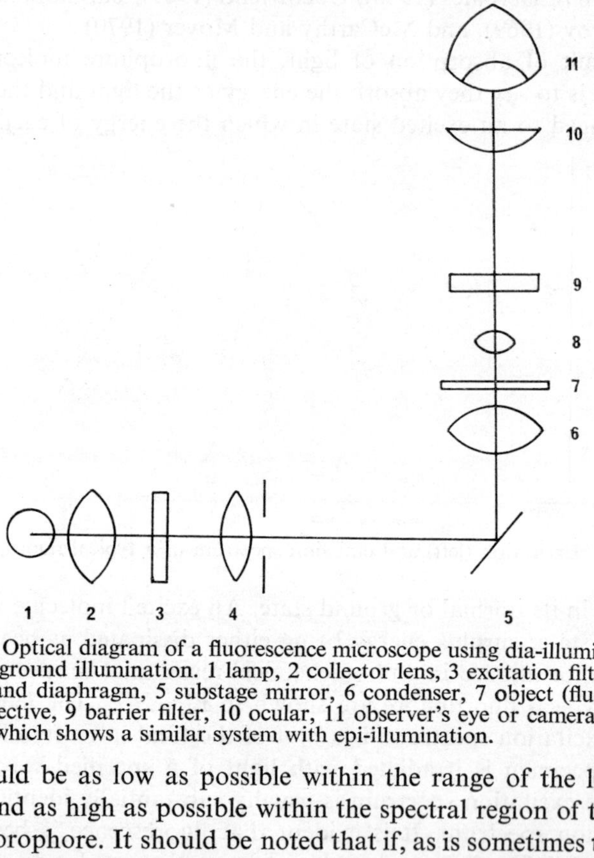

FIG. 248. Optical diagram of a fluorescence microscope using dia-illumination or dark-ground illumination. 1 lamp, 2 collector lens, 3 excitation filter, 4 field lens and diaphragm, 5 substage mirror, 6 condenser, 7 object (fluorescent), 8 objective, 9 barrier filter, 10 ocular, 11 observer's eye or camera. Cf. Fig. 254, which shows a similar system with epi-illumination.

mission should be as low as possible within the range of the light use for excitation, and as high as possible within the spectral region of the emission from the fluorophore. It should be noted that if, as is sometimes the case, the barrier filter is not absolutely colourless to visible light but has perhaps a yellow tinge, then the apparent colour of the fluorescence will be distorted by the colour of the barrier filter. For example, the formaldehyde-induced fluorescence of noradrenaline should appear a greenish-blue colour with a transparent barrier filter but it is commonly observed as a green fluorescence due to the use of a yellow barrier filter (see Plate XXVa).

The usual optical arrangement of a fluorescence microscope is such that the fluorescent specimen is viewed against a dark background although it is possible, if required, to have a background of a contrasting colour using phase contrast or some such technique. Fluorescence microscopy is capable of revealing the presence of very small quantities of fluorophore, which will stand out against the dark background. It is the extreme sensitivity of the fluorescence technique which gives it great value in immunofluorescence and for other purposes requiring the demonstration of small amounts of substances.

The apparent colour and contrast of a fluorescent specimen depends not only upon the filter system, but also upon the characteristics of the observer's eye (or photographic film). The human eye is unable to appreciate colour properly in dim light. The effect of the observer's vision upon the apparent colour of fluorescence has been discussed by Ritzén (1967) and Sander (1970). It has already been pointed out that the colour of the barrier filter may gravely distort the apparent colour of fluorescence towards longer wavelengths. The only accurate way to determine the colour of fluorescence is to measure the emission spectrum with a microspectrofluorimeter (see Chapter 31, p. 1241). The apparent contrast of a fluorescent field will also depend upon the overall brightness (Ploem, 1970a) because the dimmer parts may be below the limit of colour vision. This is of particular importance in immunofluorescence, where the intensity of background staining or the relative fluorescence of test and control preparations will appear to change according to the optical conditions. The conclusions to be drawn are simple: the optical conditions should be chosen to make the fluorescence microscopic image as bright as possible, and in important cases actual measurement (microfluorimetry) must be used.

Fluorescence fading

All stained preparations fade upon exposure to strong light, particularly ultra-violet, and fluorescent preparations are particularly subject to this trouble. Fading gives rise to difficulties with photography (see p. 1200) and with quantitation (see Chapter 31, p. 1250). The rate of fading of fluorescence depends upon several factors, varying from one fluorophore to another and according to the environment of the fluorophore and the nature of the excitation. Fluorochromes which are comparatively resistant to fading should of course be chosen when possible, particularly for quantitative studies. Fading can be minimized by careful selection of mounting medium and of excitation wavelength. Fading, in general, is increased by the presence of oxidizing agents and by excitation at an unduly short wavelength. The rate of fading increases with increasing intensity of irradiation. If, after examination, specimens are kept in the dark at a low temperature, some recovery may occur. Fading is due to photochemical reactions, i.e. a chemical reaction of the excited molecules of the fluorophore usually leading to decomposition, and to a reduction in quantum efficiency due to local heating or other causes.

Both may be reversible during storage in the dark, preferably at a low temperature.

Applications

Fluorescence techniques can be applied to all kinds of biological material, either as alternatives to other methods or for specific purposes for which only fluorescence is suitable. It has already been indicated that a major advantage of fluorescence microscopy is its very high sensitivity, making possible the demonstration of substances in very low concentrations or in particles below the resolution of a transmitted-light microscope.

Coloured preparations, such as those which have been stained with a dye, which can appropriately be examined in a conventional transmitted light microscope can often also be examined in a fluorescence microscope. However, the use of fluorescence microscopy extends the range of dyes which can be used because of the additional possibility of using dyes which absorb in the ultra-violet region, fluorescing in the visible region (e.g. Thioflavine T). Specimens which are transparent at all wavelengths can only be examined by some form of retardation microscopy (interference, phase-contrast, or polarization). Opaque objects or relatively opaque objects, such as very thick sections, can very conveniently be examined by fluorescence microscopy using an epi-illuminator.

The basic types of fluorescent material that can be studied in tissues by fluorescence microscopy are indicated in Table 82. The number of tissue components which are naturally fluorescent (autofluorescent) is small, but some substances can be converted to fluorophores by a chemical method

TABLE 82

Fluorescence Microscopy of Biological Material

1. AUTOFLUORESCENCE
   Natural fluorescence of substance(s) in tissue

2. INDUCED FLUORESCENCE
   Substance in tissue converted to fluorophore

3. DIRECT FLUOROCHROMY
   A. Without pretreatment: simple staining techniques
   B. With pretreatment: chemical reaction following by staining

4. INDIRECT FLUOROCHROMY
   Immunofluorescence

5. FLUORESCENT END-POINT REACTIONS
   Enzyme methods

(induced fluorescence), or the specimen can be "stained" with a fluorescent dye (fluorochromy). In immunofluorescence, the site of an antigen-antibody reaction is revealed by labelling one of the components of the reaction (antibody, antigen, or complement) with a fluorescent dye. Finally, enzymes present in tissue can be studied by conversion of a non-fluorescent substrate to a fluorescent product.

Autofluorescence

Unstained tissues possess, in most tissue components, some degree of fluorescence: this is known as autofluorescence or primary fluorescence. This is particularly marked in plant tissues. In animal tissues, connective tissue fibres (collagen, elastin) and lipofuscin have a strong autofluorescence. Within cells, most of the autofluorescence is believed to be due to the presence of NADH bound to a mitochondrial component or a dehydrogenase (Chance and Thorell, 1959). For review, see Haitinger (1959) and Barenboim, Domanskii and Turoverov (1969).

To most fluorescence microscopists, autofluorescence is or may be a nuisance, if it mimics some specific secondary fluorescence. In such cases, distinction can almost always be made by careful selection of optimal excitation wavelength so as to favour the desired fluorophore, together with a barrier filter which permits observation of the true colours of the fluorophores.

All proteins can be expected to fluoresce, if excited in the region 250-280 nm, due to the presence of tryptophan, tyrosine and phenylalanine. This subject is discussed by Konev (1967), and by Barenboim, Domanskii and Turoverov (1969).

Porphryin fluorescence is of diagnostic value in biopsy material from cases of porphryia (Cripps, Hawgood and Magnus, 1966; Cripps and Peters, 1967; Czitober, Schnack and Wewalka, 1968; Schmid, Schwartz and Sundberg, 1955). Porphryin fluorescence fades rapidly under irradiation; Cripps,

TABLE 82a

Strongly autofluorescent substances which may occur in tissues

(all proteins are to some extent autofluorescent)

Substance	*Fluorescence Colour*
Collagen fibres	blue-green
Elastic fibres	blue-green
Protein-bound $NADH_2$	blue
Vitamin A	red
Lipofuscin	orange
Porphyrins	red

Hawgood and Magnus (1966) found that fading was reduced by using an iodine quartz lamp rather than a mercury arc.

In lipid storage diseases, lipid droplets may be demonstrable by fluorescence microscopy; however, this is unlikely to be of diagnostic value since the fluorescence is believed to be due to non-specific substances dissolved in the lipid, rather than the lipid itself.

Foreign Substances

The list of fluorescent compounds introduced into the tissues for therapeutic or other purposes, and studied thereafter by fluorescence microscopy is not a large one. Reference has already been made in Chapter 28 to the demonstration of quinacrine (atebrin) in human skin by Mustakallio (1954), whose studies were based on earlier animal experiments by Jailer (1945). The distribution of this substance in *Taenia saginata* was studied by Mustakallio and Saikkonen (1954) after its (successful) use as a vermifuge. Lewis and Goland (1948) studied the uptake of acridine compounds by tumours in mice. Flavin compounds are readily studied by fluorescence microscopy and the trypanocidal drug Acriflavine was investigated in this way by Jancso (1932). Alin and Helander (1948) used the method to detect the distribution of *p*-amino-salicylic acid.

The fluorescence of mouse skin, after painting with benzpyrene in acetone or benzene, was examined by Beck and Peacock (1940), by Hamperl, Graffi and Langer (1942), by Doniach, Mottram and Weigert (1943), and also by Nordén (1953) who identified the benzpyrene by microspectrofluorimetry. Simpson and Cramer (1945) observed a blue fluorescence after the application of methylcholanthrene to the skin. The uptake of hydrocarbon carcinogens by lysosomes was demonstrated by Allison and Mallucci (1964).

A fluorescent pteridine compound (2-methylamino-4-amino-6, 7-diphenylpteridine) which has a selective action on myocardial muscle has been described by Paget (1957). This can be detected in the tissues by fluorescence microscopy.

The localization of antibiotics after absorption has been studied by fluorescence microscopy. The technique has been applied mainly to tetracyclines, and also to dihydrostreptomycin (Abello, 1954) and chloramphenicol (Diamond, 1963). Tetracyclines tend to localize preferentially in bone, teeth, and in soft tissues which are necrotic, inflamed, or neoplastic. The uptake by gastric tumours of tetracyclines administered orally was at one time thought to be of diagnostic value (Klinger and Katz, 1961) particularly for screening by gastric lavage, but subsequent studies have not justified early optimism, and current opinion is that the test is of dubious value (Rachlin and Harrower, 1967). Of greater current importance is the deposition of tetracyclines in bone, which can be used in the study of bone changes. The tetracyclines are deposited in bone undergoing active mineralization at the time when the tetracycline is circulating in the blood (Milch, Rall and Tobie, 1958; Urist and McLean, 1963). Tetracyclines are not deposited in bone which has already been mineralized (Frost, Villanueva and Roth, 1960) nor in unmineralized osteoid (Harris, Jackson and Jowsey, 1962). The tetracyclines are also demonstrable in osteocyte lacunae, non-growing osteons and areas of secondary mineralization, while the circulating blood level is high, but disappear from these sites when the blood level falls (Steendijk, 1964) and cannot be observed in these sites more than 72 hours after administration of the antibiotic (Frost, Villanueva and Tobie, 1960; Harris, Jackson and Jowsey, 1962).

Induced Fluorescence

Some substances, themselves non-fluorescent in the visible region, can be converted to fluorescent compounds by chemical treatment. At present, by far the most important method of this type is the induction of fluorescence with formaldehyde by condensation reactions leading to ring formation. Other reagents include glutaraldehyde, *o*-phthalaldehyde and ninhydrin. For details of these methods, see Chapter 27.

Fluorochromy

Fluorescent dyes used as stains are known as fluorochromes (Haitinger, 1934) to distinguish them from those intended to be examined by transmitted light, which are called diachromes. An incomplete list of fluorochromes, and the excitation and emission maxima of some of them, are given in Appendix 29, p. 1422. Many dyes can be used both as diachromes and as fluorochromes: examples are Congo Red, Eosin, Basic Fuchsin (pararosaniline). Most yellow, orange and red dyes are in fact fluorescent. With these dyes, fluorescence microscopy is only another way of revealing the dye, and its use does not alter the specificity of the staining. With strongly fluorescent dyes the sensitivity of the staining reaction is increased because lower concentrations of the dye can be observed. However, strongly stained regions of a specimen may appear only weakly fluorescent unless epi-illumination is used, due to concentration quenching. For these reasons, dyes are normally used in much weaker solutions for fluorochromy than for diachromy: a typical concentration of fluorochrome in a staining solution is 1:10 000, and much weaker solutions are sometimes used.

Two advantages in particular may accrue from the use of fluorescence microscopy to examine the result of a staining reaction. First, the range of possible dyes is extended to include those which are fluorescent but colourless or only weakly coloured in the visible region. Examples of such fluorochromes are Thioflavine T and BAO (Table 95, Appendix 29). The second advantage of fluorescence microscopy is that much smaller amounts of dye can be revealed by this means than by conventional absorption microscopy.

The nomenclature of fluorochromes is still in a slightly confused state. All publications concerning fluorochromes (including commercial catalogues) should specify either the chemical name of the substance (if possible) or the Colour Index Number of the Society of Dyers and Colourists. The situation is complicated by the fact that commercial preparations of fluorochromes tend to be mixtures, due to incomplete purification or to the presence of oxidation products. For details of the chemical nature of fluorochromes (and diachromes) see Harms (1965) and Lillie (1969). Impurities have been discussed by Kasten (1967) and Horobin (1969, 1970). Fluorochromes and their applications have been reviewed by Haitinger (1959) and Harms (1965). Many of them are mentioned individually in other chapters (locatable by the Index), e.g. Thioflavine T for amyloid (Chapter 11, p. 387) and fluorochrome methods for lipids (Chapter 12). Acridine Orange and Coriphosphine O are dealt with below (p. 1186).

Vital fluorochromy

Because of the high sensitivity of fluorescence methods, fluorochromes are particularly suitable for vital staining. Suitable substances include Neutral Red, Euchrysine, Congo Red, Evans Blue, Acridine Orange, and the quinacrines and tetracyclines. For a recent review of vital staining, see Barbosa and Peters (1971). Maddy (1964) described a fluorochrome, 4-acetamido-4′-isothiocyanostilbene-2,2′-disulphonic acid, for the specific labelling of the outer components of the plasma membrane. Fluorescence has been reported following vital staining of tumour cells with the non-fluorescent thiazine and oxazine dyes Azure II, Azure A, Azure I, Methylene Blue, Toluidine Blue, Thionin and Brilliant Cresyl Blue followed by ultra-violet irradiation (Bastos *et al.*, 1969).

The microfluorimetric identification of fluorochromes, taken up by living cells, will be discussed in Chapter 31.

Basic dye fluorochromes

In this section, the opportunity will be taken to review the histochemistry of basic dyes in the light of developments since publication of the previous volume, and to endeavour to clarify and correlate aspects of basophilia, the Schiff and Feulgen reactions, and the more recent discovery of masked metachromasia. There is a further excuse in that information obtained by fluorescence microscopy, particularly microspectrofluorimetry, has been responsible for some notable advances in this field.

Basic dyes, by definition, are those which ionize to carry a positive charge. They are accordingly attracted to acidic groups, carrying a negative charge,

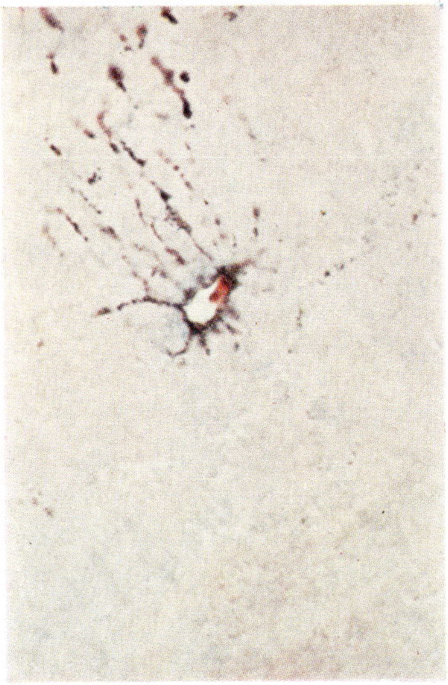

XXVIIIa. Undecalcified bone (Rat). Processed by the Timm (sulphide-silver) method, counterstained with azocarmine. Shows black granular Pb deposits in the cytoplasm of an osteocyte. × 168.

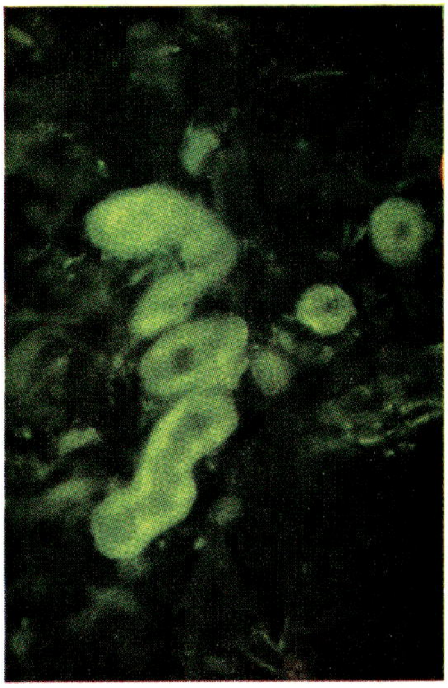

XXVIIIb. Human skin (sweat glands). Freeze-dried formaldehyde vapour-fixed. Shows autofluorescence of (self-administered) mepacrine. × 150.

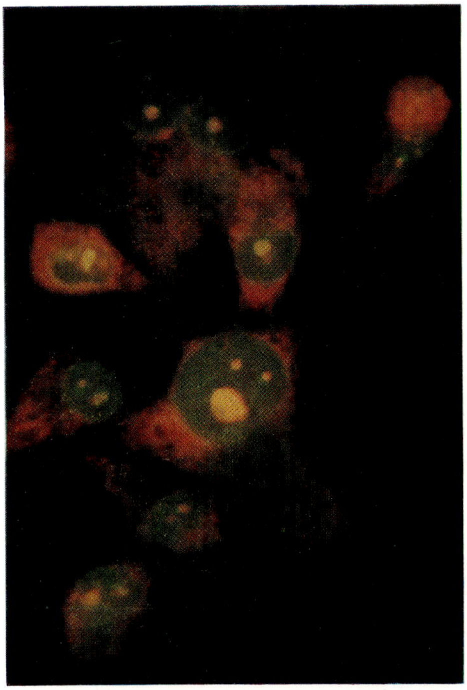

XXVIIIc. Human liver cells (culture). Stained with dilute acridine orange and examined by fluorescence microscopy. DNA, green, RNA, yellow to red. × 1200.

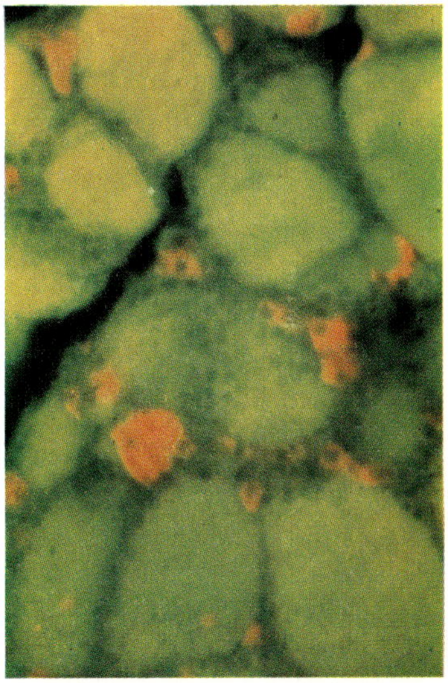

XXVIIId. Dog thyroid gland. Glutaraldehyde-picric acid fixed cryostat section. Stained with 1:10′000 Coriphosphine O in 0·1 M phosphate buffer (pH 5·0). The C cells are reddish orange, follicular cells green. × 250.

and can bind into such sites by a salt linkage. This is, probably, the main mechanism of staining by the majority of basic dyes, and therefore the basis of the phenomenon of basophilia exhibited by such acidic tissue components as nucleoproteins and acidic mucins.

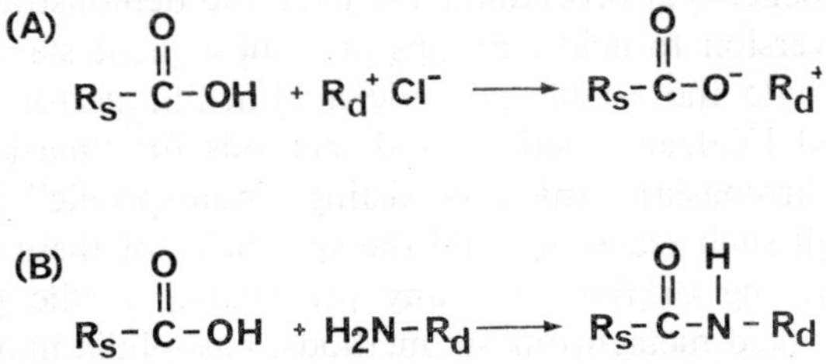

FIG. 249. Comparison of basic dye binding by salt linkage (A) and condensation reaction (B). R_s and R_d are parts of the substrate and dye respectively. Carboxyl groups are shown as the acid groups of the substrate: sulphate and phosphate groups would react similarly.

Basic dyes can of course also react by any other groups which they may possess. Those possessing primary amine groups, as many basic dyes do, have the option of reacting with acidic groups by a condensation reaction instead of by a salt linkage (see Fig. 249). Condensation reactions produce much firmer binding than salt linkages, and staining is accordingly more resistant to, e.g. removal during the processes of dehydration and mounting.

A secondary reaction of some basic dyes is polymerization, leading to a change in colour (metachromasia). This is of some importance in the study

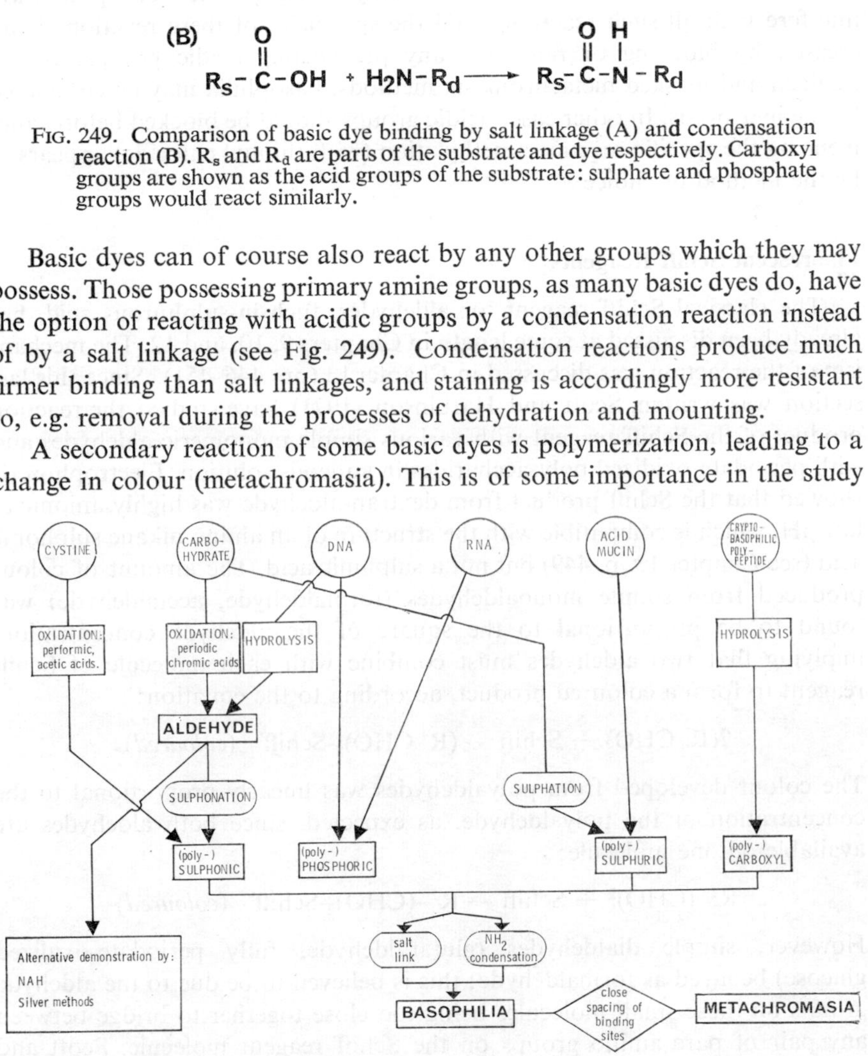

FIG. 250. Chart indicating applications of basic dyes with or without pretreatment of the substrate.

of polyanions, since metachromatic dyes not only demonstrate the presence of acidic groups (by basophilia) but may also give some indication of the distance between them, by metachromasia in regions of closely spaced binding sites. This subject is dealt with in greater detail below (p. 1185).

Several histochemical procedures exist for the demonstration of specific groups by conversion to acidic groups and subsequent staining by a basic dye. These include the Schiff, periodic acid-Schiff, performic acid-Schiff, U.V.-Schiff, and Feulgen reactions and methods for "masked" basophilia (masked metachromasia). Any pre-existing "non-specific" basophilia will interfere with all such reactions, and the specificity of these reactions is increased by blocking or removing any pre-existing acidic groups. In the Feulgen and masked metachromasia methods, basophilia may be eliminated by the hydrolysis. In other cases acidic groups should be blocked before commencing the specific reaction; methylation (with thionyl chloride) appears to be the method of choice.

Fluorescent Schiff Reagents

The classical Schiff reagent for aldehydes, fuchsin-sulphurous acid, has already been discussed at some length in Chapters 9, 10, and 12. The mechanism of the reaction was discussed in Chapter 13 (pp. 446–451). Since this last section was written, Scott and Harbinson (1971) have studied the reaction product of the Schiff reagent with various simple monomeric aldehydes and with periodate-oxidized polysaccharides in aqueous solution. Electrophoresis showed that the Schiff product from dextran-aldehyde was highly anionic at low pH, which is compatible with the structure of an amino-alkane sulphonic acid (see Chapter 13, p. 449) but not a sulphinic acid. The amount of colour produced from simple monoaldehydes (formaldehyde, acetaldehyde) was found to be proportional to the square of the aldehyde concentration, implying that two aldehydes must combine with each molecule of Schiff reagent to form a coloured product, according to the equation:

$$2(\text{R–CHO}) + \text{Schiff} \rightarrow (\text{R–CHO})_2\text{-Schiff} \quad (\textit{coloured})$$

The colour developed from polyaldehydes was linearly proportional to the concentration of the polyaldehyde, as expected, since both aldehydes are available on one molecule:

$$\text{R}^1\text{–(CHO)}^2 + \text{Schiff} \rightarrow \text{R}^1\text{–(CHO)}_2\text{-Schiff} \quad (\textit{coloured})$$

However, simple dialdehydes (glutaraldehyde, fully periodate-oxidized glucose) behaved as formaldehyde; this is believed to be due to the aldehyde groups on these small molecules being too close together to bridge between any pair of para-amino groups on the Schiff reagent molecule. Scott and Harbinson concluded that in solution, the stoichiometry of the Schiff reaction depends upon the distance apart of the aldehyde groups. The Schiff reagent must now clearly be regarded as a reagent for *pairs* of aldehyde groups, suitably spaced on the same or adjacent molecules.

It is evident that the situation in solution is not the same as in tissue sections. In model experiments in solution, the reactions are carried out in the presence of excess bisulphite, whereas in sections, in the majority of cases, the bisulphite is removed in the final washing. In solution, the bisulphite decolourizes any dye which is either free or bound to only one aldehyde. Only when the dye is combined with 2 aldehyde groups is the bisulphite unable to decolourize the dye. In histological sections the Schiff dye may bind by one aldehyde group and become recolourized in the process of washing away bisulphite. The reaction in tissue sections might be expected to be more specific if it were possible to retain bisulphite in the section.

It has been indicated (Chapter 13, p. 449) that the Schiff reaction is believed to consist of two stages: in the first, aldehyde groups are sulphonated, and in the second these sulphonic groups condense with a form of dye. It might reasonably be supposed that after the sulphonation stage, any basic dye could be used to stain the sulphonic groups, and this indeed appears to be so. Schiff-type reagents not employing Basic Fuchsin are called pseudo-Schiff reagents to distinguish them from the classical Schiff reagent, the mechanism of staining by which is believed to be different. Individual reagents are named according to the dye employed, e.g. Acriflavine-Schiff or Acriflavine-SO_2.

Schiff-type reagents for aldehydes can be prepared from at least 18 different basic fluorochromes, including those indicated in the Table in Appendix 29, p. 1422. The classical Schiff reagent, fuchsin-sulphurous acid, is itself usable for fluorescence microscopy (Prenna and Sacchi, 1964; Böhm and Sprenger, 1968), because Basic Fuchsin (mainly pararosaniline) is a red fluorochrome. Other fluorochromes favoured at present for the preparation of Schiff-type reagents include BAO (Ruch, 1966), Acriflavine, Acridine Yellow and Auramine O. For other possible dyes, see Appendix 29, p. 1422. Investigations of the suitability of various dyes have been reported by Kasten, Burton and Glover (1959), Prenna and de Paoli (1964), Ruch (1966), Prenna (1968), and Böhm and Sprenger (1968). Kasten (1959) investigated a large number of dyes, including some fluorochromes, and concluded that a primary amine group on the dye molecule is the reaction site for aldehyde binding in the presence of sulphurous acid, the structure of the remainder of the molecule being of no significance in the Schiff reaction but leading on occasion to staining of substances other than the aldehyde. However, subsequent work has shown that aminoacridines can be used as Schiff dyes without having reactive amine groups, leading Stoward (1967b) to the conclusion that the mechanism of action of the Schiff reagents is by sulphonation followed by binding of a basic dye by salt linkage. Following this to a logical conclusion, Stoward (*loc. cit.*) realized that the stain did not necessarily have to be dissolved in the sulphonation reagent, and found that better results could be achieved by sequential sulphonation followed by staining with a basic fluorochrome. To increase the specificity of the PAS reaction, Stoward eliminated staining of normally basophilic tissue components by a blocking procedure (methylation) before the periodic acid treatment. The resulting method, methylation-periodic acid-sulphonation-basophilia, is described in Appendix 29, p. 1421.

Fluorescent Schiff-type reagents can be used in all the procedures requiring a Schiff reagent, e.g. the various modifications of the periodic-acid-Schiff and Feulgen reactions. Potentially, a fluorescence Schiff method is more sensitive than a diachrome method, and also, offers the possibility of quantitation by microfluorimetry. In practice, there does not appear to be any point in using a fluorescent Schiff reagent for general purposes, but applications to quantitative histochemistry are increasing (see Chapter 31). The use of a metachromatic fluorochrome may give information about the secondary structure of PAS-positive substances. For reviews on fluorescent Schiff reagents see Kasten (1959, 1960), Ruch (1966), Stoward (1967b) and, Prenna (1968). Another fluorescent reagent for aldehydes is used in the periodic acid-salicylhydrazide reaction (Stoward, 1967c) which is described in Chapter 10, p. 324, and Appendix 10, p. 661.

Fluorochromes for chromosome identification

Selective labelling of chromosomes by fluorochromes is of considerable value. Caspersson and his colleagues (Caspersson *et al.*, 1968) originally reasoned as follows. Alkylating agents exert their biological effects by alkylation of DNA, particularly by action on the N-7 atom of guanine (see p. 249). An alkylating fluorochrome might therefore selectively accumulate in guanine-rich segments of DNA. The bifunctional alkyating agent 2-methoxy-6-chloro-9(4-bis(2-chloroethyl)amino-1-methylbutylamino) acridine, commonly known as quinacrine mustard*, was used to test this hypothesis. Quinacrine itself was used as a control, being a closely analogous fluorochrome incapable of covalent binding to DNA. In the event, it was found that both quinacrine mustard and quinacrine become bound to chromosomes, giving a diffuse fluorescence with a clear pattern of cross-striations extending across both sister chromatids. There are twin regions of particularly strong fluorescence on the Y chromosome. Comparison with the pattern of distribution of DNA along the chromosomes indicated that the striations were not due to variations in DNA content but to differences in chemical reactivity.

In man, quinacrine hydrochloride can also be used to label the Y chromosome in interphase nuclei (Pearson, Bobrow and Vosa, 1970).

The method is useful in studying the fine structure of human chromosomes and in karyotyping them. Visual and fluorometric differentiation of all human chromosomes has already been reported (Caspersson *et al.*, 1971).

* Quinacrine mustard dihydrochloride is 2-methoxy-6-chloro-9[4-bis(2-chloroethyl)-amino-1-methylbutylamino]acridine.

For further references, see Caspersson, Zech and Modest (1970), a *Lancet* editorial (1971) and Caspersson *et al.* (1971).

At the time of writing, rapid progress is being made in this field. The method is given in Appendix 29, p. 1423.

The fluorochromation of DNA by Acridine orange is dealt with below, in the section on metachromasia.

Fluorescence metachromasia

Some fluorochromes are metachromatic, i.e. fluorescence with more than one colour. Metachromasia has already been discussed, mainly in relation to diachromes (Chapter 10, p. 330). With fluorochromes, as in the case of diachromes, a change from orthochromasia to metachromasia involves a shift in the excitation (absorption) peak towards shorter wavelengths, and a decrease in the molar extinction coefficient at the absorption maximum; in addition, there is a corresponding shift of the emission spectrum towards

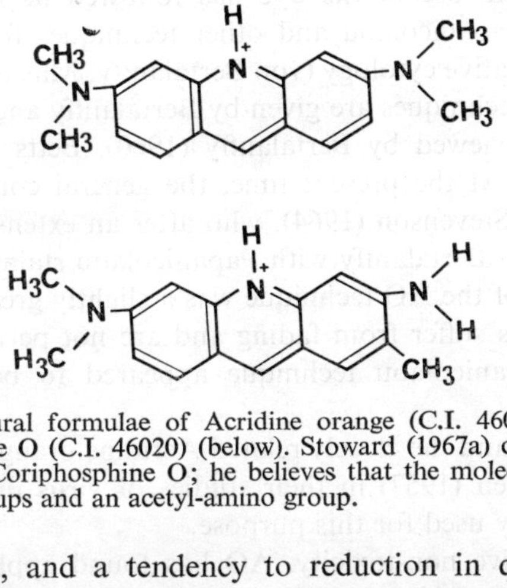

FIG. 251. Structural formulae of Acridine orange (C.I. 46005) (above), and Coriphosphine O (C.I. 46020) (below). Stoward (1967a) disagrees with this formula for Coriphosphine O; he believes that the molecule also contains two ethyl groups and an acetyl-amino group.

longer wavelengths, and a tendency to reduction in quantum efficiency. Fluorescence intensity is decreased, because of reductions in absorption and in quantum efficiency. In short, the fluorescence emission colour changes to one of a longer wavelength, and the brightness of fluorescence decreases.

Fluorescence metachromasia is due to the formation of dimers and polymers, as a result of close aggregation of dye molecules. Metachromasia can be taken as indicative of a high local concentration of the dye. Staining techniques for metachromasia are therefore sensitive to small variations in technique, since under-staining will prevent metachromasia and overstaining is likely to produce a preparation which is metachromatic all over.

All metachromatic fluorochromes known at present are acridines. The most widely used are Acridine orange (C.I. 46005) and Coriphosphine O (C.I. 46020), of which the structural formulae are given in Fig. 251. In both of these dyes, the orthochromatic form fluoresces green, the metachromatic form red. These dyes appear to function essentially as basic dyes. Their particular usefulness compared to other basic dyes lies in the fact that both the number of dye binding sites and the distance between them are indicated by the fluorescence intensity and the degree of metachromasia respectively.

Acridine orange

Acridine orange AO; C.I. 46005) is an acridine dye which has been widely used as a basic, metachromatic fluorochrome since its introduction by Bukatsch and Haitinger (1940) and Strugger (1940). As a superficial rule of thumb, one can say that staining results in green fluorescence from DNA, and red fluorescence from RNA, denatured DNA, and acid polysaccharides (Plate XXVIIIc, p. 1180). It was initially advocated by Strugger (1940, 1949) for the differentiation of living and dead cells which gave green and red fluorescence respectively. This technique, however, was found to be unreliable. The most widespread use of the dye has followed its introduction as an alternative to the Papanicolaou and other techniques for the diagnosis of malignancy in exfoliative cytology (von Bertalanffy, Masin and Masin, 1956). Details of staining techniques are given by Bertalanffy and Nagy (1962). The subject has been reviewed by Bertalanffy (1960), Betts *et al.* (1962), Sani (1964), and others. At the present time, the general consensus of opinion appears to support Stevenson (1964), who after an extensive comparison of the technique of von Bertalanffy with Papanicolaou staining concluded that the sole advantage of the AO technique was a slightly greater speed; against this, its preparations suffer from fading and are not permanent, and, most important, the Papanicolaou technique appeared to be somewhat more accurate.

Acridine orange and the closely related Acridine orange-red were used by Armstrong and Niven (1957) in their studies of virus inclusions: Acridine orange is now widely used for this purpose.

Because of relative non-toxicity, AO has found applications as a vital fluorochrome (Schümmelfeder, 1948; Lewis and Goland, 1948; Debruyn, Robertson and Farr, 1950; Armstrong, 1956; Contier, 1956). Toxicity is however relative, since AO has been found to exercise a narcotic action on protozoal cells, to inhibit tumour growth (Lewis and Goland, 1948), and to be mutagenic (Brenner *et al.*, 1961; see also Lerman, 1963).

Despite disappointments in applications as a routine fluorochrome, AO has undoubted applications as a histochemical reagent for the investigation of nucleic acids. This subject has been reviewed by Rigler (1966), Zelenin (1967), and Kasten (1967).

AO and Coriphosphine O are essentially basic dyes, binding to acidic groups (carboxyl, phosphoric, sulphuric, sulphonic) by salt linkages. When binding to polyanions, dye molecules bound to adjacent acidic groups may be brought into sufficiently close proximity for dye-dye interaction to occur, with the formation of dimers or polymers and consequent metachromasia. Both AO and Coriphosphine O fluoresce green when in the orthochromatic (monomeric) form, and red in the metachromatic forms. The majority of the staining reactions of these dyes can be explained on the basis of this simple mechanism, particularly in the cases of staining of acid polysaccharides (Stoward, 1967a), and in pseudo-Schiff reagents (Stoward, 1967b).

The staining of nucleic acids by aminoacridines has received much attention. It is probable that the staining of RNA occurs in the simple manner described above. The staining of DNA, however, is more complicated. The negatively charged phosphate groups have been suggested as binding sites, on theoretical grounds (Michaelis, 1947; Bradley and Wolf, 1959), and also on the experimental evidence that Hg^{++} ions, which are assumed to bind preferably to the heterocyclic bases of the DNA helix have no, or very slight, influence (compared to other bivalent ions) on the formation of AO-nucleic acid complexes (Koudelka, Kleinwächtler and Blazicek, 1964; Sivarama Sastry and Gordon, 1964). Furthermore the dissociation velocity of the DNA-proflavine complex during filtration on Sephadex is greatly increased by high ionic strength, suggesting a dye-DNA interaction by electrostatic forces (Liersch and Hartmann, 1964). On the other hand there is evidence that the binding rate of Acriflavine (another acridine dye) varies with the base composition of DNA (Tubbs, Ditmars and van Winkel, 1964; Gersch and Jordan, 1965), indicating a reaction with the purine and pyrimidine bases as well.

Lerman (1961) found that the combination in solution of DNA with small amounts of Acridine orange, Acridine, or Proflavine resulted in markedly increased viscosity and a diminution of the sedimentation coefficient of the DNA. These changes were regarded as contrary to those expected on the basis of simple electrostatic binding. Characteristic changes were also found in the X-ray diffraction patterns of fibres of the complex with Proflavine, suggesting considerable modification of the helical structure of the DNA. Lerman concluded that the DNA molecule is modified, the double helix stretching and untwisting slightly to allow the acridine molecules to be intercalated between neighbouring base pairs inside the two polynucleotide chains. This hypothesis was supported by further studies (Luzzati, Masson, and Lerman, 1961; Lerman, 1963, 1964), and is now generally accepted.

Rigler (1966) determined the ratio of AO to nucleic acid phosphorus by microfluorimetry, and found that for DNA the ratio was about 1:6, indicating one AO molecule for every 3 base pairs. For an artificial random coil polyU the ratio was about 1:1·5, indicating that almost every nucleotide unit bound an AO molecule. From these results, Rigler concluded that when binding to the double stranded helix of DNA, AO cations are intercalated after approximately every third base pair. Their dimethylamino groups are in contact with the negatively charged PO_4^- groups, and the acridine ring is in contact with the purine and pyrimidine rings of the upper and lower bases of a pair, thereby constituting a stable complex by means of ionic as well as dipole-dipole bonds (Pritchard, Blake and Peacocke, 1966). Under these circumstances, the distance between dye molecules is great enough to prevent dye-dye interaction, so that the fluorescence characteristics are those of the Acridine orange monomer, i.e. green emission (525 nm). On the other hand, when binding to a single-stranded random coil (as in the polyU), AO is bound to almost every nucleotide unit by its PO_4^- group, and flexing of the random coil permits adjacent dye molecules to approach near enough for dye-dye interaction and polymer formation, leading to a metachromatic (red) fluorescence. This latter arrangement appears to occur in cytoplasmic RNA.

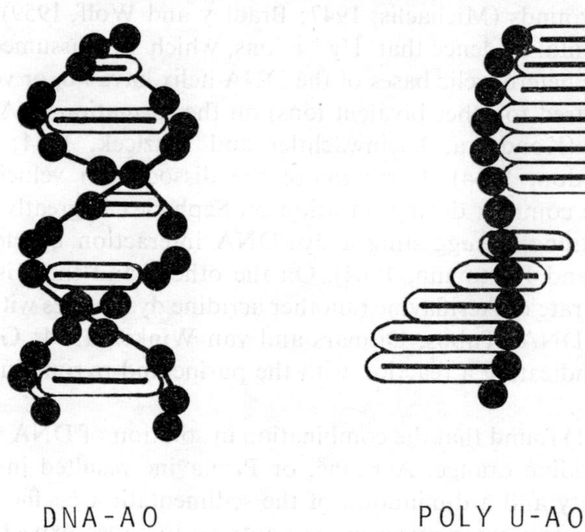

FIG. 252. Schematic molecular configurations of nucleic acid-AO complexes with fixed-coil (left) and random-coil (right) chains (From Rigler, 1966).

Fluorescence metachromasia is therefore dependent upon the secondary structure of the nucleic acid molecule, metachromasia being permitted by a random coil conformation but not by a highly ordered structure.

The position of the aminoacridine molecule, inside the helix, renders it less accessible: Lerman (1963) found that the amino groups of aminoacridines were much less accessible to nitrosating reagents when the dye was bound to DNA, than in free solution or when bound to other polyanions. Displacement of the base pairs by intercalation of aminoacridine molecules has been suggested as a likely mechanism for mutagenesis, by leading to recombination errors (Lerman, 1963).

The rigid double helix of DNA can be disorganized to a random coil configuration by heat denaturation (Marmur and Doty, 1959), and as might be expected this leads to metachromatic staining with AO. Metachromatic staining by AO of DNA also occurs after acid hydrolysis (Schümmelfeder, Ebschner and Krogh, 1957). Structural differences in DNA of heterochromatin and euchromatin of interphase nuclei (Sandritter and Böhm, 1964) are reflected in greater resistance of the latter to acid-induced metachromasia (Roschlau, 1965).

Masked metachromasia

Feulgen-type hydrolysis followed by staining with one of a number of basic metachromatic dyes results in specific metachromatic staining of certain types of endocrine cells. The reaction is believed to be due to the presence of a protein or polypeptide with a high concentration of side chain acidic groups. This subject belongs properly to Chapter 6, but has developed since the publication of Volume 1; the present volume gives an opportunity to bring the subject up to date.

Details of the methods are given in Appendix 29, p. 1421. In principle, suitably fixed material is subjected to hydrolysis in a hot solution of mineral acid followed by staining with a basic dye, which may be metachromatic or a fluorochrome or both. Dyes which have been used include Toluidine blue (Manocchio, 1960, 1964; Solcia and Sampietro, 1965a, b), Azure A, Methylene Blue, Safranin, Pseudoisocyanin, and Alcian Blue (Solcia, Vassallo and Capella, 1968), and Coriphosphine O (Bussolati, Rost and Pearse, 1969; Rost and Maunder, 1971). Of these dyes, the first three are metachromatic diachromes, Pseudoisocyanin is a red fluorochrome, and Coriphosphine O (Plate XXVIIId, p. 1180) is a metachromatic fluorochrome (green → red). Alcian Blue, alone of these dyes, gives a satisfactory permanent preparation but cannot be used in the presence of acidic mucosubstances. Descriptions of the development and application of these methods are given by Solcia, Vassallo and Capella (1968), Bussolati, Rost and Pearse (1969), and Rost and Maunder (1971). Studies carried out by Manocchio (1960, 1964), and subsequently by Solcia and Sampietro (1965a, b), indicated that metachromasia or increased basophilia, demonstrated with Toluidine Blue at pH 5, could be induced in a variety of cell types by methylation of suitably fixed tissue sections with methanolic solutions of hydrochloric acid. Saponification, or treatment with alkali, was also noted to be capable of inducing basophilia under some circumstances. It was soon realized that this "masked basophilia", perhaps often better termed "masked metachromasia", was not due to methylation but to an accompanying hydrolysis by the hot mineral acid. Thereafter, a standard Feulgen-type hydrolysis was employed (0·2 N hydrochloric acid at 60°, or 0·5 to 1 per cent at 65°), for up to 18 hours.

Among the cells originally or subsequently described as giving a positive reaction were the argyrophil or G cells of the pyloric antrum (Solcia, Vassallo and Sampietro, 1967; Carvalheira, Welsch and Pearse, 1968), the argyrophil δ cells of the pancreatic islets (Cavallero, Solcia and Sampietro, 1967), and the (sometimes) argyrophil thyroid parafollicular or C cells (Solcia and Sampietro, 1965a). A comprehensive synthesis of these results, with new data on fixation and on suitable dyes for recording masked basophilia, has been presented by Solcia, Vassallo and Capella (1968).

Most of the cell types that exhibit masked metachromasia are included in a cytochemically, and embryologically, related series of endocrine cells which share the common property of polypeptide hormone production (Pearse, 1966, 1968). These are now called APUD cells, a term derived from the first letters of three of their six common cytochemical characteristics.

The basophilia and metachromasia are undoubtedly due to the unmasking of polycarboxyl groups in the side-chains of polypeptides or proteins. The unmasking by acid hydrolysis is believed to be due to a combination of at least two processes. First, pre-existing basophilia is removed by extraction of cytoplasmic RNA and acid polysaccharides, and to a lesser extent of DNA. A second effect, which is obviously important in some cases at least, is the conversion of carboxamido groups. Cleavage of carboxamido groups in proteins by acid hydrolysis has been described by Chibnall *et al.* (1958) and by Hill (1965). If there is a sufficient density of side-chain carboxyls in a given protein, or polypeptide, chain, and if the pH of staining is such that these carboxyls are in the dissociated form, the result should be recordable cytochemically as basophilia or metachromasia, depending on the basic dye employed.

Two of the endocrine polypeptide or protein materials which occur in the cells of the APUD series would be expected to give metachromatic staining without prior acid hydrolysis. These are gastrin (–$\overset{6}{\mathrm{Glu}}$-$\overset{7}{\mathrm{Glu}}$-$\overset{8}{\mathrm{Glu}}$-$\overset{9}{\mathrm{Glu}}$-$\overset{10}{\mathrm{Glu}}$) and chromogranin (26 per cent w/w glutamic acid residues). In some animals (Solcia and Sampietro, 1969a) the presumptive gastrin-secreting or G cells of the stomach have been recorded as giving Toluidine Blue metachromasia without acid hydrolysis. We did not find this in our own investigations, (Bussolati, Rost and Pearse, 1969), where minimal hydrolysis was always necessary. In all animals tested, the A and NA cells of the adrenal medulla were found to give metachromasia at pH 5·0 with both Toluidine Blue and Coriphosphine. With both dyes the level of metachromasia was greatly increased by acid hydrolysis.

In the case of some other polypeptide hormones the possible reacting sequence contains carboxamido rather than carboxyl groups (e.g. Calcitonin $\overset{15}{\mathrm{Asn}}$-$\overset{16}{\mathrm{Leu}}$-$\overset{17}{\mathrm{Asn}}$-$\overset{18}{\mathrm{Asn}}$.) It must be emphasized, however, that we do not know the composition of the precursor proteins of the various polypeptide hormones. These may or may not possess the required density of side-chain carboxyl groups. Nevertheless, it seems reasonable to postulate that the primary structure of a protein giving masked metachromasia will include amino acid sequences providing adjacent or closely associated carboxyl or carboxamido groups, and that the latter will be converted to carboxyls by

FIG. 253. Molecular model of human calcitonin M, made with Biobits ®.

acid hydrolysis. The first and second effects of hydrolysis are thus (1) removal of RNA and (2), the conversion of carboxamido groups to carboxyls.

A third effect of acid hydrolysis we regard as more significant than the other two, in terms of interpretation of the metachromatic reaction. The association of fluorescence metachromasia with random-coil conformation (Rigler, 1966) has been described above (p. 1187). The acidic protein, chromogranin, has been shown by Smith and Winkler (1967) to behave as a random-coil polymer. It is perhaps significant that this protein, and gelatin which has been shown by Gouinlock *et al.* (1955) to possess the random-coil conformation, both give strong masked metachromasia. Collagen, which is regarded as a triple helix composed of three polypeptide chains, joined by inter- and intra-molecular cross-links (Francois and Glimcher, 1967) does not react in this way. Simple polyanion chains, such as polyglutamate, give strong metachromasia without prior acid hydrolysis, as would be expected (Pearse, 1969).

Further information concerning the action of acid hydrolysis on secondary protein structure can be derived from the results of the two methods for carboxyl groups, applied after acid hydrolysis. The observed increase in side-chain carboxyls (Bussolati, Rost and Pearse, 1969) is almost certainly due in part to conversion of carboxamido groups. Part, however, may be due to breakage of labile interchain cross-linking peptide bonds (γ-glutamyl or β-aspartyl). The striking increase in C-terminal carboxyl groups must be due to the breakage of labile intra-chain linkages. This process might be expected not only to supply more carboxyl groups for the metachromatic reaction but also to allow some reorganization of the structure of the polypeptide. Thus the third effect of acid hydrolysis is the breaking of inter- and intra-chain linkages with consequent production of new carboxyl groups.

If the propositions recorded above are accepted, the correct interpretation of a positive result with Manocchio's metachromatic staining, with the masked basophilia of Solcia and his associates, or with our own masked fluorescence metachromasia, will be to signify the presence of a protein whose primary structure includes a sufficient density of carboxyl or carboxamido side-chains and whose secondary structure is predominantıy that of a random-coil polymer.

This concept should be elastic enough to allow the presence in the natural (unfixed) protein of a limited content of the α-helix conformation. Fixation presumably involves some degree of denaturation and the "unmasking" effect of acid can be considered, broadly speaking, as a denaturation involving helix-random coil transition. More specifically, it may be attributed to hydrolysis of labile γ-glutamyl and β-aspartyl cross-links, and to breakage of intra-chain hydrogen bonds.

It has been postulated (Pearse, 1969) that the provision of random-coil structure for the protein precursors of polypeptide hormones is a mechanism for ensuring the rapid release of the smaller active polypeptide from the larger precursor molecule. This would operate by facilitating the hydrolysis of specific linkages in the protein chain by proteolytic or esterolytic enzyme activity.

Although only primary and secondary conformations have been considered here, we can only agree with Kopple (1966), that "it is apparent that the chemical, physiological and biological properties of a protein depend as much on secondary and tertiary structure as they do on amino-acid sequence". It is possible that further microfluorimetric studies of masked metachromasia, similar to those of Rigler in relation to nucleic acids, may reveal further information concerning the conformation of these polypeptide hormones.

For routine use, the methods for masked metachromasia are of value in the diagnosis of malignant melanomas (Rost, Polak and Pearse, 1969) and other tumours of APUD cells.

Immunofluorescence

This subject has already been covered in Chapter 7. Since that chapter was written, there have been two international conferences on Standardization in Immunofluorescence (Florence, 1967; London, 1968), the proceedings of which have now been published in a single volume (Holborow, 1970). Probably the most important developments have been in the manufacture of interference filters for specific excitation of FITC, giving greater optical specificity. Advantages to be gained include reduction in autofluorescence, and increased possibility of double staining methods using two fluorochromes, e.g. FITC and TRITC, distinguished optically by appropriate filters. The wider availability of tungsten-halogen lamps, together with the further development of suitable filter systems, has placed the microscope equipment necessary for immunofluorescence within the financial reach of many more laboratories, as well as extending the possible application of immunofluorescence techniques in the field (e.g. with the portable Macarthur microscope). Quantitative methods in immunofluorescence will be referred to in Chapter 31, p. 1241.

Enzymatically-produced fluorescence

Enzyme activity in cells, living or fixed, can be studied in systems where the enzyme produces a change in fluorescence by action on a substrate or co-enzyme. Oxidation-reduction kinetics of enzyme systems involving pyridine nucleotides have been studied by making use of the fluorescence of NADH (see Chapter 31, p. 1242). Rotman and Papermaster (1966) studied the properties of cell membranes in living cells using fatty acid esters of fluorescein, which pass fairly freely into the cells and are hydrolysed to produce fluorescein, which accumulates inside the cell. Other fluorogenic esters are those of methylumbelliferone and of naphthyl derivatives. As an alternative to coupling reactions in the demonstration of phosphatases and esterases, it is possible to use the fluorescence of the actual enzymatic products: this may lead to some gain in sensitivity, in optical specificity, and in possibilities for microfluorimetric quantitation and kinetic studies. Consideration of this topic will be resumed in Chapter 31, p. 1242.

Fluorescence microscopy in the study of neoplasia

The high sensitivity of fluorescence methods makes them particularly suitable for many studies of neoplastic cells, both in experimental investigations and in diagnostic pathology.

At the experimental level, fluorescence microscopy and microfluorimetry offer many possibilities. The uptake of fluorescent carcinogens can be followed, as in the classical investigations of Nordén (1953) on the uptake of benzypyrene. Fluorescence methods offer a number of delicate probes which may be of value for demonstrating changes in neoplastic cells. Sensitive methods are available for the study of chromosomes (quinacrine mustard, see p. 1184), of plasma membrane morphology (SITS, p. 1180) and dynamic characteristics (fluorogenic esters, see Chapter 31, p. 1242), and of enzyme systems (NADH fluorescence, p. 1242; fluorogenic substrate methods, p. 1242).

For the diagnosis of tumours, Acridine orange fluorochroming (p. 1186) has had many advocates but is now little used. It is possible that the technique might be revived if certain technical disadvantages (particularly fading) were eliminated. Several studies have been made of uptake of various substances by tumours. Such substances have included tetracycline (p. 1178), oxazine and thiazine dyes (p. 1179), and certain aminoacids (Chapter 27, p. 1111). Although at present only the last of these methods is of proven diagnostic value (for apudomas), uptake methods should be worth further investigation. Other fluorescence methods of value are aldehyde-induced fluorescence (phaeochromocytomas, carcinoids, melanomas, mastocytomas), masked metachromasia (apudomas generally) and immunofluorescence (for specific secretory products).

Combined fluorescence and electron microscopy

In principle, an electron-dense fluorochrome should be capable of examination with both optical and electron microscopes, the high sensitivity of fluorescence microscopy making possible the observation of a small amount of fluorophore such as would be present in a thin section. Such a reagent would have to be either a fluorochrome containing a heavy atom, or a fluorochrome complex containing a heavy atom.

At present few fluorochromes contain atoms heavier than sulphur or chlorine, although some contain bromine (the eosins, phloxines, Rose Bengals and Methyl green) or iodine (the erythrosins). Mercurochrome 220 is a fluorescein derivative containing mercury, with staining characteristics resembling eosin.

An acriflavine-phosphotungstate complex has recently been developed as a stain for nucleic acids (Chan-Curtis, Belt and Ladoulis, 1970; Chan-Curtis, Beer and Koller, 1970).

Section preparation and mounting

The requirements for section preparation for fluorescence microscopy do not differ, in general, from those for other forms of optical microscopy. Quartz slides and cover-glasses are not required (except for examination of fluorescence excited at wavelengths below 350 nm). Non-fluorescent mounting media must of course be used: water, inorganic buffer solutions, and glycerine are all completely non-fluorescent. Of the solid non-aqueous media, in our experience, DPX is as non-fluorescent as the special commercial media ("Entellan", Merck; "Fluorolite", Lamb). During dehydration, it is necessary to avoid using solvents (alcohol etc.) which have already been used for dehydrating sections containing eosin, which is strongly fluorescent. Stoward (1967) recommends blotting dry, dehydration in 3 changes of isopropyl-alcohol, followed by 50:50 isopropyl alcohol and xylene, 3 changes of xylene, and mounting in DPX. Alternatively, sections can be dried in air and mounted direct in DPX. The mounting medium affects the rate of fading of fluorescence. Resin embedding media ("Maraglas", "Epon") are suitable for embedding freeze-dried formaldehyde-vapour fixed material, enabling very thin sections to be cut. If a fairly soft mixture is used, sections about 2 μm thick can be cut on a sledge microtome.

Instrumentation

In the previous edition, would-be fluorescence microscopists were advised to assemble their own equipment, in view of deficiencies in commercially available equipment. In the decade that has passed, there has been considerable progress in the development and availability of commercial fluorescence microscopes, now made by all the major microscope manufacturers and many of the lesser ones, so that at the present time the main problem (finance apart) is to choose from the multitude. At the time of writing, further commercial development is proceeding very rapidly, so that if any specific recommendations were made here they would certainly be out of date before appearing in print. Accordingly, only general indications will be given. Undoubtedly the most satisfactory way to choose a fluorescence microscope is to go round an exhibition of microscopes, armed with a preparation of the type which one intends to examine, and to ask to examine one's preparation under each microscope. It must be emphasized that only the most versatile (and expensive) microscopes will be capable of coping adequately with a wide range of fluorescent preparations; of the simpler instruments, one suitable for (say) immunofluorescence may be unsuitable for FIF, and vice-versa. In this respect, versatility depends upon the light source(s) and the possible filter combinations, and to some extent also upon the optical arrangement of the illumination system.

Light sources

It should be evident that the function of a light source is to provide light at a wavelength corresponding to an excitation maximum of the fluorophore. In the earliest days, following the experiments of Köhler, it was believed that true ultra-violet (UV) was required, and carbon arc lamps were used, followed by high-pressure mercury arc lamps. The latter are still in use today, and are likely to remain in wide use. The advantage of high-pressure mercury lamps is that they have strong emission at certain specific wavelengths (365, 406, 436, 537 nm) which can fairly easily be isolated by a filter to give substantially monochromatic light of high intensity. The 406 line is particularly suitable for excitation of catecholamine FIF (excitation maximum at 410 nm). Xenon arc lamps give light of a more uniform spectral distribution, and are to be preferred to mercury lamps for the UV below 365 nm and for much of the visible region. Unfortunately these lamps are at present very expensive. Other arc lamps may prove of value. The cadmium arc, with emission lines at 468, 480 and 509 nm, has been suggested for immunofluorescence with FITC (Bals and Velculescu, 1966). The zinc arc may be useful for excitation.

Much development is proceeding on the use of the tungsten-halogen ("iodine-quartz") lamp. This consists essentially of a conventional tungsten filament, surrounded by iodine vapour in a quartz envelope. The iodine has the effect of reducing vaporization of tungsten from the filament, allowing it to be run at a higher temperature than would otherwise be possible. The tungsten-halogen lamp has the advantages, compared to arc lamps, of cheapness and convenience. It is suitable for excitation in the green and blue, but completely hopeless for the UV region.

Illumination systems

There are three main possibilities for illumination of the specimen: dia-illumination by a substage bright-field condenser, oblique illumination by a substage dark-ground condenser, and epi-illumination by a beamsplitter through the objective. A fourth system, oblique illumination from above, is possible but offers no particular advantage and is technically awkward. Each of the first three has advantages and disadvantages and a good microscope should have provision for both dia- and epi-illumination. The advantages and disadvantages of the three systems are set forth in Table 83. With dark-ground illumination, because only scattered light and fluorescence enters the objective, the demands made upon the filter system are somewhat reduced. However, dark-ground condensers in general give low levels of illumination, and moreover dark-ground condensers for high-power objectives give particularly weak illumination, just when intense illumination is most wanted. Difficulties may arise with scattered light at the excitation wavelength, unless the barrier filter is fairly opaque at that wavelength in which case dia-illumination could probably be used in any case.

The optical arrangement of a fluorescence microscope employing dia-illumination is shown in Fig. 248, p. 1174, which should be compared with the corresponding arrangement for epi-illumination shown in Fig. 254, below. For Köhler illumination, an image of the source is imaged into the "exit" pupil of the objective, a field diaphragm being imaged into the object plane. Because the objective is also the condenser, focusing the objective automatically focuses the condenser; this gives good reproducibility for quantitation (see Chapter 31). A major advantage is that each objective concentrates the illumination precisely on to the field of view. For greatest brightness, the beamsplitter should be a dichroic mirror (Ploem, 1967a, 1969a, b). If the beamsplitter has 50 per cent transmission and 50 per cent reflection at all wavelengths, only 50 per cent of the excitation is reflected, and only 50 per cent of the fluorescence reaching the objective is transmitted, giving an overall reduction in intensity to 25 per cent. Theoretically, a dichroic mirror having 100 per cent reflectivity at the excitation wavelength and 100 per cent transmission (nil reflectivity) at the fluorescence wavelength would give an increase

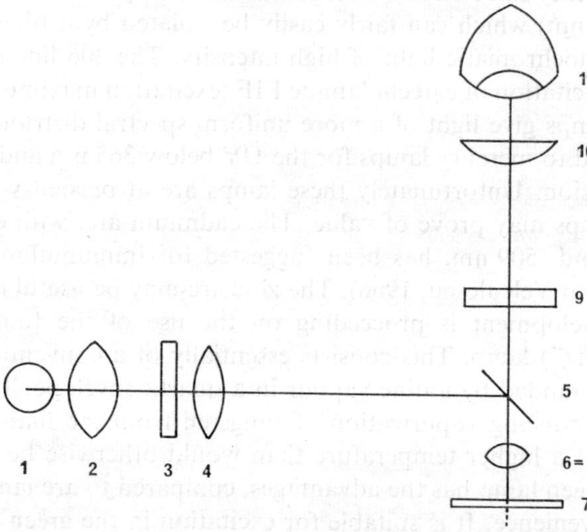

FIG. 254. Optical system of fluorescence microscope with epi-illumination (Cf. Fig. 248, p. 1174). 1 lamp, 2 collector lens, 3 excitation filter, 4 field lens and diaphragm, 5 beam-splitter, 6 objective, 7 object (fluorescent), 8 condenser, 9 barrier filter, 10 ocular, 11 observer's eye or camera.

in brightness by a factor of 4; in practice a factor of about 3 can be obtained. The dichroic mirror, since it reflects at the excitation wavelength and transmits at the fluorescence wavelength, serves also as a barrier filter.

It is interesting to calculate the theoretical changes in apparent fluorescence intensity due to changing objectives, using either epi-illumination or dia-illumination (see Table 84). In the case of dia-illumination, with a fixed condenser, the brightness of the image will be affected only by the characteristics of the objective. Ignoring light losses in the objective (which usually increase with increasing complexity of the objective), the brilliance of the final image depends upon the square of the N.A. of the objective and inversely upon the square of the magnification. In the case of epi-illumination, the objective is also the condenser, so that the brightness of the illumination of the specimen, and therefore of the fluorescence, will depend similarly upon

TABLE 83

Comparison of dia-illumination, dark-ground illumination, and epi-illumination for fluorescence microscopy.

	DIA	DG	EPI
Intensity:			
lowpower objective	high	moderate	low
highpower objective	moderate	very low	high
Reproducibility (for quantitation)	possible	difficult	absolute
compatibility with absorption and phase-contrast	yes	no	yes
simultaneous phase-contrast	awkward	impossible	easy
evenness of illumination	good	poor	good
thick or opaque objects	impossible	impossible	easy
quantitation: self-absorption	bad	bad	best

TABLE 84

Relative brilliance of observed fluorescence to be expected from a typical set of objectives, ignoring light losses due to flare etc. Comparison valid only within each column.

Objective		Relative brightness	
Magn.	NA	dia	epi
4	0·10	6·15	1
10	0·30	9·0	81
25	0·50	4·0	625
40	0·65	2·6	1785
100	1·30	1·7	28,561

the square of the numerical aperture and upon the square of the magnification (due to greater demagnification of the light source). Combining the various effects, the relationships can be summarized thus:

For dia-illumination, $I_{dia} \propto NA^2/M^2$
For epi-illumination, $I_{epi} \propto NA^4$

From the above it is evident that an ideal fluorescence microscope should be capable of both epi-illumination (for high powers) and dia-illumination (for low powers). The use of a dark-ground condenser is an unsatisfactory compromise which has hitherto been forced upon us both by the unavailability of suitable epi-illumination systems and by inadequate filter systems.

Filters

There can be no question but that the most critical feature of a fluorescence microscope is the filter combination. Both the excitation filter and the barrier filter must be correct for the fluorophore(s) being examined and in relation to each other. Wrong choice of filters for a particular application may give misleading or even erroneous results.

Systems available for monochromation are of three basic types: prism or grating monochromators, liquid filter solutions, coloured glass filters, and interference systems. Prism and grating monochromators are not really practical for general fluorescence microscopy, mainly because of expense and also because of low light transmission. Liquid filters used to be popular particularly for removing long wavelengths (heat, infra-red, red), but are much less convenient than solid filters.

Glass filters are of two types: those in which the colour is due to the presence of coloured ions in solution, and temperature-coloured glasses in which the colour is due to submicroscopic crystals whose optical effect is partly dependent upon their size. The former typically have a bell-shaped transmission curve, and can be used as excitation filters. The temperature-coloured glasses have a long-wave region of high transmission and a short-wave region of opacity, separated by a cut-off region of which the slope and position are determined during manufacture. This second type of glass is suitable for use in barrier filters.

Interference systems are of two types: transmission filters and mirrors. Interference transmission filters are suitable for excitation filters, transmitting mainly in a narrow band. Present-day interference filters do not, in general, have as high maximal transmission as glass filters and neither do they have as high opacity in regions removed from their transmission maxima; also, most interference filters have small pin-holes in the coating which of course transmit some unwanted light. The great advantage of interference filters is that they can be made for all wavelengths, whereas the number of wavelengths for which satisfactory glass filters is available is small. They are also resistant to heat. The disadvantages of interference filters are usually circumvented by using a supplementary glass filter. Mirrors with interference coatings have been used as excitation filters (Ploem, 1967b). They are now widely used as dichroic mirrors in epi-illumination systems in which they function in part as both excitation and barrier filters.

Selection of filters

The choice of filters depends primarily upon the fluorophore. The excitation filter must be chosen so as to transmit selectively at the longest-wavelength excitation peak of the fluorophore, and otherwise to be as opaque as possible. The barrier filter is then chosen so as to transmit as much as possible of the emission of the fluorophore, while blocking any light transmitted by the excitation filter.

In practice, the excitation filter will usually consist of several elements, as follows: (1) a specific narrow-band filter to select the wavelength of the excitation peak of the fluorophore, (2) a second filter to support the first, by removing any unwanted transmission, (3) a heat filter to protect the other filters and the specimen, and (4) in the case of epi-illumination, a dichroic mirror. For specific selection of the 365 nm mercury line, UG1 glass is excellent, unsurpassed in our experience by any interference filter in either maximal transmission or cut-off. For other lines, and for use with tungsten lamps, interference filters are best.

Barrier filters, at the present time, are made from "temperature-coloured" glass, and therefore are subject to some inconsistency in manufacture. With dichroic epi-illumination systems (see below) the dichroic mirror serves in part as a barrier filter. Nomenclature of barrier filters at the present time tends to be unsatisfactory, particularly among the lesser manufacturers of fluorescence equipment. The most important characteristic of a barrier filter is the position of the transition region ("cut-off") between high transmission and low transmission, and that should be stated for each filter. In the case of Schott filter glasses, the code number indicates the approximate wavelength in nm at which a 3 mm thickness has 50 per cent internal transmission. Leitz barrier filters have code numbers, prefixed by the letter K, indicating the approximate wavelength of 50 per cent transmission of the filter. Zeiss (Oberkochen, West Germany) use a different system, in which the number refers to the wavelength in units of 10 nm at which the filter has 10 per cent transmission.

When selecting a filter combination for a specific application, in most cases the best approach is to try several combinations and see which one gives the best results. The theoretical approach, however, is as follows. The excitation peak of the fluorophore is first found (e.g. from Table 96, Appendix 29, p. 1423). An excitation filter is then chosen: if a mercury lamp is being used, the filter should select the nearest line to the desired wavelength. It is next necessary to know the emission spectrum; the barrier filter must be chosen so as to transmit most of the fluorescence. As an example, consider the question of examining the formaldehyde-induced fluorescence of nor-adrenalin. In Table 79, Chapter 27, the excitation and emission maxima are given as 410 nm and 480 nm respectively. The nearest mercury line to 410 nm is the 406 nm line, and this should be selected by a suitable interference filter (e.g. Schott AL 406). The barrier filter should be chosen with its cut-off below the emission maximum of 480 nm; the Leitz K460 or K470, or the Zeiss 41 would be appropriate. The next thing is a practical trial. The interference filter will probably have to be supplemented with a glass filter; wide-band glass filters (e.g. BG12) and red-absorbing filters (e.g. BG38) can be tried. Finally a barrier filter should be chosen with cut-off at as short a wavelength as possible consistent with blocking the excitation. The selection of filters for dichroic epi-illumination systems has been discussed by Ploem (1969, 1970b)

When determining the excitation and emission spectra of fluorophores, for choosing filter combinations or for any other purpose, it is necessary to remember that the spectrum of a fluorochrome is not necessarily the same in solution as bound to a protein or whatever. Accordingly, spectra of a "pure" fluorochrome in solution can only be taken as a rough guide. Metachromatic fluorochromes present the difficulty that they have two or more excitation and emission spectra, one set of each form of the dye. For adequate simultaneous visualization of both orthochromasia and metachromasia, a moderately broad-band excitation is required (to excite both metachromatic and orthochromatic forms), while the barrier filter must be chosen on the basis of the orthochromatic form which has the shortest-wavelength fluorescence. For Acridine orange and Coriphosphine O, one possible combination is a mercury arc lamp with BG12 excitation filter (supported by BG38 and KG1 or KG3) together with a barrier filter with cut-off at about 510 nm.

Fluorescence photomicrography

Many fluorescent preparations are evanescent, and photography is necessary if a permanent record is required. Photomicrography of fluorescence presents special problems, mainly because of the low light levels involved and the fading of the fluorophore. The photography of immunofluorescence preparations has been discusssd by Schuit (1970). The photography of porphyrin fluorescence is described by Cripps, Hawgood and Magnus (1966).

Because of the dimness of the image on the photographic film, a fast film and/or long exposures are required. Fast film suffers from graininess. Long exposures (over 0·1 second) lead both to fading of the fluorophore and to reciprocity failure in the film. The term "reciprocity failure" refers to loss of proportionality between the brightness of the image and the reciprocal of the required exposure time; this failure occurs both for extremely short exposure times and for long exposure times. The sensitivity (ASA, DIN) rating of the film is usually quoted by the manufacturer on the assumption that the exposure will be of the order of 1/100 sec. However, for very long or intermittent exposures, the sensitivity of the film is reduced. The approximate speed of some current film types for various exposure times is indicated in Appendix 29, p. 1424. For colour films, the speeds given relate to the film as a whole; strictly, each of the sensitive layers has its own reciprocity characteristics, resulting in loss of correct colour balance for long exposures. In our experience, loss of colour balance is not of importance for fluorescence photomicrography, since there are other variables of greater importance—the colour distortion of the barrier filter, and the adequacy of the observer's vision in dim light.

Returning to the essential problem, it will be seen that a vicious circle can develop: weak fluorescence requires a long exposure, which results in reciprocity failure and fading, which both tend to lengthen the required exposure, leading to further reciprocity failure and fading, so forming a vicious circle.

Design considerations for equipment suitable for photomicrography of fluorescence are therefore quite different from those for conventional transmitted-light photography. First, every effort must be made to obtain as bright an image as possible on the photographic film. This requires an efficient optical system. The objective NA should be as high as possible (this may be limited by requirement for depth of field; high NA objectives with iris are convenient). An efficient barrier filter, transmitting most of the fluorescence, is important: if the cut-off of the filter is at too long a wavelength, not only will the colour of the fluorescence be distorted, but also the intensity will be greatly decreased if only the long-wavelength tail of the fluorescence is transmitted. An efficient body-tube system should send 100 per cent of the available light to the camera, although perhaps up to 10 per cent could be diverted to a photometer if required. Any light subtracted by a beamsplitter leads to an even greater increase in the exposure because of reciprocity failure and fading (see below). The magnification to the film place should be as low as possible, since the intensity of the light is inversely proportional to the square of the magnification. In practice, this means that a small film format is needed. 35 mm film is the obvious photosensitive material to use. The standard 24 × 36 mm format is not obligatory, however, and the use of still smaller formats (e.g. 24 × 24 mm, 17 × 24 mm) would give shorter exposure times.

Because of the weakness of fluorescence compared with the usual transmitted light, every effort should be made to conserve light. Objectives should be chosen for high NA, if possible. It used to be said that the simplest possible objectives (achromatics) should be used to avoid autofluorescence of the fluorite glass used in more highly-corrected systems. In our experience however, this is not a problem: we have successfully used fluorite, apochromat, flat-field, and even planapochromat objectives. It is true that light losses (due mainly to reflections) are greater in the more complicated objectives, but this may be offset by higher numerical aperture. With epi-illumination systems, the amount of light at the excitation wavelength scattered back up the microscope tube by the objective may be significant, particularly for microfluorometry. This is determined by the design of the objective and by the anti-reflection coatings (if any) of the glass surfaces.

In order to conserve light, the total magnification of the system should be kept as low as practicable. Because of the correlation of numerical aperture and magnification of objectives, in practice as much as possible of the desired magnification should be achieved by the objective, while the magnification of the ocular should be as low as possible. Additional magnifying systems, particularly zooms, should be avoided both for this reason and because of additional loss of light by reflections. It used to be said that binocular systems should be avoided, due to great light losses in the beam-splitter system of binocular tubes. With binocular tubes of good modern design, this is no longer mandatory except for the most extremely weak fluorescence.

Basically there are three ways by which the exposure can be determined: by trial and error, by measurement with a photometer followed by calculation, or by an automatic system controlled by a photometer. As has already been indicated, a system convenient and suitable for transmitted-light photography is not necessarily ideal or even possible for fluorescence. As regards trial-and-error, this method involves least capital cost and would appear to be adequate for routine photomicrography of a run of similar preparations: it may even be advantageous to give always a standard exposure, provided that the lamp remains constant, to record changes in fluorescence intensity from one preparation to another. Certainly for immunofluorescence, test and control specimens should all be photographed under identical conditions with the same exposure: this tends to be difficult with automatic cameras. Fully automatic systems in which the exposure is controlled automatically by a photometer may be convenient in some respects, but suffer from four limitations: (1) some light is lost to the photometer, which must continuously monitor the image; (2) identical exposures cannot conveniently be given for test and control preparations; (3) with spot measuring systems, the area of the field from which a reading should be taken may not coincide with the centre of the field where the measuring spot is; and (4) corrections are not made for the reciprocity characteristics of the individual film in use.

It appears that the most satisfactory arrangement for fluorescence photomicrography is one where a photometer is used to measure the brightness of the fluorescence; the exposure is calculated or read from a chart, taking into account the reciprocity failure of the film at the anticipated exposure; and the exposure of required length given, preferably controlled by an electric timing device.

The choice of film is of course determined primarily by what is commercially available. Films have improved greatly in the past decade, and are likely to improve further, so that any specific advice would be likely to be soon out of date. With black-and-white film, one must choose a compromise between high speed, giving a short exposure, or finer detail with longer exposure. In practice, any moderately fast film (100 to 400 ASA; 21 to 27 DIN) should give satisfactory results. Some types of film have bad reciprocity characteristics, and these should be avoided. For recording fluorescence in colour, the choice is more difficult. At present, colour negative film is useless unless one is prepared to print the pictures oneself, or can arrange for a standard printing: current commercial printing, tailored to the supposed needs of the worst amateurs, adjusts the overall colour of prints to a neutral grey. For colour positive film, the principle is to choose a daylight type film with as high a sensitivity as possible consistent with an acceptable degree of graininess. At the time of writing, the fastest available film (Ansco 500, 500 ASA, 29 DIN) is too grainy for general scientific use. The next fastest, Ansco 200 (200 ASA, 24 DIN) gives adequate detail for many purposes and is comparatively resistant to reciprocity failure. Probably the most widely used film for this purpose at present is High Speed Ektachrome (ASA 160, 23 DIN), which in our hands has given good results but suffers badly from reciprocity failure. Kodachrome X (64 ASA, 19 DIN) is more resistant to reciprocity failure, so that for exposures of about 100 seconds it is almost as sensitive as High Speed Ektachrome.

REFERENCES

ALLISON, A. C., and MALLUCCI, L. (1964). *Nature, Lond.*, **203**, 1024.

ARMSTRONG, J. A. (1956). *Expl Cell Res.*, **11**, 640.

ARMSTRONG, J. A., and NIVEN, J. S. F. (1957). *Nature, Lond.*, **180**, 1335.

BALS, M. G. (1966). "Immunofluorescenta". Editura Acad. RSR Bucharest.

BALS, M. G., and VELCULESCU, V. G. (1966). *Nature, Lond.*, **210**, 1073.

BARBOSA, P., and PETERS, T. M. (1971). *Histochem. J.*, **3**, 71.

BARENBOIM, G. M., DOMANSKII, A. N., and TUROVEROV, K. K. (1969). "Luminescence of Biopolymers and Cells". Plenum Press, New York and London. (Translated from Russian.)

BASTOS, A. L., MARQUES, D. V., SILVA, J. F. DE, NUNES, J. B. M., VIGARIO, J. D., and TERRINHA, A. M. (1968). *Z. Naturforsch.*, **23B**, 969.

BERTALANFFY, F. D. (1960). *Postgraduate Medicine*, **28**, 627.

BERTALANFFY, F. D., and NAGY, K. P. (1962). *Radiogr. y Fotogr. clin.*, **29**, 2.

BETTS, A., ACHENBACH, R. R., and FRIEDELL, G. H. (1962). *Archs Path.*, **73**, 86.

BÖHM, N., and SPRENGER, E. (1968). *Histochemie*, **16**, 100.

BORST, M., and KÖNIGSDÖRFER, H. (1929). "Untersuchungen über Porphyrie mi besonderer Berücksichtigung der Porphyria Congenita". Leipzig, S. Hitzel.

BRENNER, S., BARNETT, L., CRICK, F. H. C., and ORGEL, A. (1961). *J. Molec. Biol.*, **3**, 121.

BUKATSCH, F., and HAITINGER, M. (1940). *Protoplasma*, **34**, 515.

BUSSOLATI, G., ROST, F. W. D., and PEARSE, A. G. E. (1969). *Histochem. J.*, **1**, 517.

CARVALHEIRA, A. F., WELSCH, U., and PEARSE, A. G. E. (1968). *Histochemie*, **14**, 33.

CASPERSSON, T., FARBER, S., FOLEY, G. E., KUDYNOWSKI, J., MODEST, E. J., SIMONSSON, E., WAGH, U., and ZECH, L. (1968). *Expl Cell Res.*, **49**, 219.

CASPERSSON, T., LOMAKKA, G., and ZECH, L. (1971). *Hereditas*, **67**, 89.
CASPERSSON, T., ZECH, L., and MODEST, E. J. (1970). *Science, N.Y.*, **170**, 762.
CAVALLERO, C., SOLCIA, E., and SAMPIETRO, R. (1967). *Gut*, **8**, 172.
CHANCE, B., and THORELL, B. (1959). *J. Biol. Chem.*, **234**, 3044.
CHAN-CURTIS, V., BELT, W. D., and LADOULIS, C. T. (1970). *J. Histochem. Cytochem.*, **18**, 609.
CHAN-CURTIS, V., BEER, M., and KOLLER, T. (1970). *J. Histochem. Cytochem.*, **18**, 628.
CHIBNALL, A. C., HASELBACH, C., MANGAN, J. L., and REES, M. W. (1958). *Biochem. J.*, **68**, 122.
CONTIER, L. (1957). *Blut*, **3**, 185.
COONS, A. H., CREECH, H. J., and JONES, R. N. (1941). *Proc. Soc. exp. Biol. med.*, **47**, 200.
CRIPPS, D. J., HAWGOOD, R. S., and MAGNUS, I. A. (1966). *Archs Derm.*, **93**, 129.
CRIPPS, D. J., and PETERS, H. A. (1967). *Archs Derm.*, **96**, 712.
CZITOBER, H., SCHNACK, H., and WEWALKA, F. (1968). *Germ. med. Monthly*, **13**, 19.
DEBRUYN, P. P. H., ROBERTSON, R. C., and FARR, R. S. (1960). *Anat. Rec.*, **108**, 279.
DE LERMA, B. (1958). "Handbuch der Histochemie", I/1, 78. G. Fischer Verlag, Stuttgart.
DIAMOND, L. (1963). *Nature, Lond.*, **199**, 90.
FRANCOIS, C. J., and GLIMCHER, M. J. (1967). *Biochim. biophys. Acta*, **133**, 91.
FROST, H. M., VILLANUEVA, A. R., and ROTH, H. (1960). *Henry Ford Hosp. Med. Bull.* **8**, 239.
GERSCH, N. F., and JORDAN, D. O. (1965). *J. Molec. Biol.*, **13**, 138.
GOLDMAN, M. (1968). "Fluorescent Antibody Methods". Acad. Press, New York and London.
GOUINLOCK, E. V., FLORY, P. J., and SCHERAGA, H. A. (1955). *J. Polym. Sci.*, **16**, 383.
HAITINGER, M. (1934). In Arnerhalde, "Handbuch der Biol. Arbeitsmethoden". Abt. II, Teil **2**, 3307.
HAITINGER, M. (1959). "Fluoreszenz-Mikroscopie". Akademie-Verlag, Leipzig, 2nd edn. Eds. J. Eisenbrand and G. Werth.
HARMS, H. (1965). "Handbuch der Farbstoffe für die Mikroskopie". Stauffen Verlag, Kamp-Lintfort.
HARRIS, W. H., JACKSON, R. H., and JOWSEY, J. (1962). *J. Bone Jt Surg.*, *A*. **44**, 1308.
HERCULES, D. M. (1966). Theory of luminescence processes. In "Fluorescence and phosphorescence analysis", Ed. Hercules. Interscience, New York.
HILL, R. L. (1965). *Adv. Prot. Chem.*, **20**, 37.
HOLBOROW, E. J. (Ed.) (1970). "Standardization in Immunofluorescence". Blackwell, Oxford.
HOROBIN, R. W. (1969). *Histochem. J.*, **1**, 231.
JAILER, J. W. (1945). *Science, N.Y.*, **102**, 258.
KASTEN, F. H. (1959). *Histochemie*, **1**, 466.
KASTEN, F. H. (1960). *Int. Rev. Cytol.*, **10**, 1.
KASTEN, F. H. (1967). *Int. Rev. Cytol.*, **21**, 141.
KASTEN, F. H., BURTON, V., and GLOVER, P. (1959). *Nature, Lond.*, **184**, 1797.
KLINGER, J., and KATZ, R. (1967). *Am. J. dig. Dis.*, **12**, 593.
KONEV, S. V. (1967). "Fluorescence and Phosphorescence of Proteins and Nucleic Acids" Plenum Press, New York.
KOUDELKA, J., KLEINWACHTLER, V., and BLAZICEK, G. (1964). *Biochim. biophys. Acta*, **91**, 337.
KOPPLE, K. D. (1966). "Peptides and Aminoacids" Benjamin, New York.
KUYPER, C. M. A. (1962). *Histochemie*, **3**, 46.
LANCET (editorial, 1971). *Lancet*, i, 275.
LERMAN, L. S. (1961). *J. Molec. Biol.*, **3**, 18.
LERMAN, L. S. (1963). *Proc. Natn. Acad. Sci., U.S.A.*, **49**, 94.
LEWIS, M. R., and GOLAND, P. C. (1948). *Am. J. Med. Sci.*, **215**, 282.

LIERSCH, M., and HARTMANN, G. (1964). *Biochem. Z.*, **340**, 390.
LILLIE, R. D. (1969). H. J. Conn's "Biological Stains". Williams and Wilkins, Baltimore.
LUZZATI, V., MASSON, F., and LERMAN, L. S. (1961). *J. Molec. Biol.*, **3**, 634.
MADDY, A. H. (1964). *Biochem. biophys. Acta*, **88**, 390.
MANOCCHIO, I. (1960). *Zentbl. allg. Path. path., Anat.*, **101**, 1.
MANOCCHIO, I. (1964). *Archs vet. ital.*, **15**, 3.
MARMUR, J., and DOTY, P. (1959). *Nature, Lond.*, **183**, 1427.
MICHAELIS, L. (1947). *Cold Spring Harb. Symp. Quant. Biol.*, **12**, 131.
MILCH, R. A., RALL, D. P., and TOBIE, J. E. (1958). *J. Bone Jt Surg. A.*, **40**, 897.
MUSTAKALLIO, K. K. (1954). *Acta Dermato-Venereol.*, **34**, 93.
MUSTAKALLIO, K. K., and SAIKKONEN, J. I. (1954). *Expl. Parasit.*, **3**, 167.
NAIRN, R. C. (1969). "Fluorescent Protein Tracing". E. & S. Livingstone, Edinburgh and London.
NORDÉN, G. (1953). *Acta path. microbiol. scand. Suppl.* 96.
PASSWATER, R. A. (1967, 1970). "Guide to Fluorescence Literature". Plenum Press Data Division, New York.
PEARSE, A. G. E. (1966). *Vet. Rec.*, **79**, 587.
PEARSE, A. G. E. (1968). *Proc. Roy. Soc. B.*, **170**, 71.
PEARSE, A. G. E. (1969). *Nature, Lond.*, **221**, 1210.
PEARSON, P. L., BOBROW, M., BARLOW, P. B., and VOSA, C. G. (1971). *Nature, Lond.*, **231**, 326.
PEARSON, P. L., BOBROW, M., and VOSA, C. G. (1970). *Nature, Lond.*, **226**, 78.
PLOEM, J. S. (1967a). *Z. Wiss. Mikr.*, **68**, 129.
PLOEM, J. S. (1967b). M.D. Thesis, University of Amsterdam.
PLOEM, J. S. (1969a). *Archs. int. Pharmacodyn.*, **182**, 421.
PLOEM, J. S. (1969b). *Leitz. Mitt. Wiss. Techn.*, **4**, 225.
PLOEM, J. S. (1970a). In: "Holborow 1970", p. 25.
PLOEM, J. S. (1970b). "Progress in Brain Research". (In press.)
PRENNA, G. (1964). *Riv. Istoch. norm. Pat.*, **10**, 469.
PRENNA, G. (1968). *Mikroskopie*, **23**, 150.
PRENNA, G., and DE PAOLI, A. M. (1964). *Riv. Istoch. norm. Pat.*, **10**, 185.
PRENNA, G. (1968). *Mikroskopie*, **23**, 150.
PRENNA, G., and SACCHI, S. (1964). *Boll. Soc. Med.-Chir. Pavia*, **78**, 719.
PRICE, G. R., and SCHWARTZ, S. (1956). Fluorescence Microscopy. In: Oster, G., and Pollister, A. W.: "Physical Techniques in Biological Research". Vol. 3, 91. Acad. Press, New York.
PRITCHARD, N. J., BLAKE, A., and PEACOCKE, A. R. (1966). *Nature, Lond.*, **212**, 1360.
RACHLIN, L., and HARROWER, J. R. (1967). *Am. J. dig. Dis.*, **12**, 593.
RIGLER, R. jr. (1966). *Acta physiol. scand.*, **67**, Suppl. 267.
RITZÉN, M. (1967). M.D. Thesis, Stockholm.
ROSCHLAU, G. (1965). *Histochemie*, **5**, 396.
ROST, F. W. D., and MAUNDER, Caroline (1971). *Proc. R. microsc. Soc.*, **6**, 20.
ROTMAN, B., and PAPERMASTER, B. W. (1966). *Proc. Natn. Acad. Sci., U.S.A.*, **55**, 134.
RUCH, F. (1965). *Acta Histochem (Jena). Suppl.* **6**, 117.
RUCH, F. (1966). In: Wied, G. (Ed.): "Introduction to Quantitative Cytochemistry", p. 281. Acad. Press, New York and London.
SANDER, G. (1970). Quoted by Holborow (1970) Chapter 21.
SANDRITTER, W., and BÖHM, N. (1964). *Naturwiss.*, **51**, 273.
SANI, G. (1964). "Fluorescence Microscopy in the Cytodiagnosis of Cancer". Thomas, Springfield, Ill.
SCHMID, R., SCHWARTZ, S., and SUNDBERG, R. D. (1955). *Blood*, **10**, 416.
SCHUIT, Henrica R. E. (1970). In: Holborow, E. J. (ed.) "Standardization in Immunofluorescence", p. 159, Blackwell, Oxford.
SCHÜMMELFEDER, N. (1948). *Naturwiss.*, **35**, 346.
SCOTT, J. F., and HARBINSON, R. J. (1971). *Proc. roy. microsc. Soc.*, **6**, 22.

SELIGER, H. H., and McELROY, W. D. (1965). In "Light: Physical and Biological Action". Acad. Press, New York.
SIVERAMA SASTRY, K., and GORDON, M. P. (1964). *Biochim. biophys. Acta*, **91**, 406.
SOLCIA, E., and SAMPIETRO, R. (1965a). *Riv. Istochim. norm. Pat.*, **9**, 265.
SOLCIA, E., and SAMPIETRO, R. (1965b). *Z. Zellforsch. Mikrosk. Anat.*, **68**, 689.
SOLCIA, E., VASSALLO, G., and CAPELLA, C. (1968). *Stain Technol.*, **43**, 257.
STEENDIJK, R. (1964). *Acta anat. (Basel)*, **59**, 368.
STEVENSON, J. (1964). *Acta Cytol. (Balt.)*, **8**, 224.
STOWARD, P. J. (1967). *J. Roy. Micr. Soc.*, **87**, 215.
STOWARD, P. J. (1967). *J. Roy. Micr. Soc.*, **87**, 237.
STOWARD, P. J. (1967). *J. Roy. Micr. Soc.*, **87**, 247.
STRUGGER, S. (1940). *Z. naturwiss., Jena*, **73**, 97.
STRUGGER, S. (1949). In: "Fluoreszenzmikroskopie und Mikrobiologie". Schaper, Hanover.
TUBBS, R. K., DITMARS, W. E., and VAN WINKEL, Q. (1964). *J. Molec. Biol.*, **9**, 545.
UDENFRIEND, S. (1969). Fluorescence Assay in Biology and Medicine., Acad. Press, New York and London.
URIST, M. R., and McLEAN, F. C. (1963). *J. Bone Jt Surg. A.*, **45**, 1305.
VON BERTALANFFY, L., MASIN, F., and MASIN, M. (1956). *Science, N.Y.*, **124**, 1024.
WEST, S. S. (1965). *Acta histochem. Suppl.*, **6**, 135.
ZELENIN, A. V. (1967). "Fluorescence Cytochemistry of Nucleic Acids". Moscow.

CHAPTER 30

AUTORADIOGRAPHY AND ITS APPLICATIONS

Historical Introduction

The basis of autoradiography (or radioautography) is the demonstration of radioactive isotopes in tissue sections by means of their ability to reduce silver salts in a photographic plate, film or emulsion. The method is by no means a new one, having been employed as early as 1904 by Loudon. Kotzareff (1922) and Lacassagne and Lattes (1924) carried out autoradiographic studies, the latter authors using polonium in paraffin-embedded material. At a later date successful studies with radioactive lead in bones were made by Behrens and Baumann (1933a and b), and with radioactive phosphorus by Dols *et al.* (1938). In the case of radioactive iodine the studies of Hamilton *et al.* (1940) on the thyroid gland are particularly important from the historical point of view. Pecher (1942) used phosphorus for studies of bone and soft tissues and strontium in the case of bone tumours. Treadwell *et al.* (1942) also used strontium in a similar manner.

After 1940 the techniques and practice of autoradiography increased considerably in scope and achievement. A large number of radioactive elements were used and these fall naturally into two main groups.

Substances Foreign to the Tissues. Included in this category are polonium and strontium, already mentioned, and other elements such as gallium, radium, plutonium, cadmium and gold. Strontium and radium, when administered in ionic form, mostly end up in the growing ends of bone and gallium (Dudley and Maddox, 1949) was found to have a similar distribution. Endicott and Yagoda (1947) injected polonium (Radium F) in buffered physiological saline into mice and Lindenbaum and Smoler (1969) described the intracellular location of plutonium in high resolution autoradiographs. The distribution of cadmium in mouse kidney tissues was studied by Berlin *et al.* (1964). Besides these elements, administered in ionic form, a number of others have been given in combined form in substances foreign to the tissues. Axelrod and Hamilton (1947), for instance, studied the distribution of sulphur-labelled mustard gas and arsenic-labelled lewisite in the tissues of the skin and eye.

Physiological Substances. In the case of substances occurring naturally in the tissues, early studies were made either with the elements in ionic form or with various combinations into organic molecules. Bloom *et al.* (1947) injected rats with ^{14}C-labelled bicarbonate and found marked deposition in the shafts of the long bones and in liver and kidney. Armstrong *et al.* (1948) made similar investigations of ^{14}C distribution in the rat. Further details were given by Bloom (1949). Boyd *et al.* (1948) described the localization of ^{14}C

in individual blood cells, and Boyd and Levi (1950) were able to show β-particle tracks from the livers of rats injected with ^{14}C-labelled glycine. Numerous authors made studies with ^{32}P (e.g. Bélanger and Leblond, 1946; Percival and Leblond, 1948; Palm, 1948; Bulliard *et al.*, 1938; Marshak, 1941; Bayley, 1947) and these demonstrated a number of interesting points. Radioactive phosphate ions were found to exchange with non-radioactive phosphate in a large number of situations, but most rapidly, apparently, where there was a rapid rate of metabolism. They were shown to be incorporated both into phospholipids, and also into the nucleic acids (Leblond, Stevens and Bogoroch, 1948). These authors considered that normally all ^{32}P introduced into the tissues was incorporated into newly-formed nucleic acids.

Dudley and Dobyns (1949) illustrated the deposition of radiocalcium in the subperiosteal region of ribs and Berggren (1946) studied the distribution of ^{24}Na in teeth. This element was studied in ocular tissues by von Sallman *et al.* (1949). Radiopotassium was shown by Colfer and Essex (1946) to be fairly uniformly distributed through the brain of the rat, with a tendency to greatest concentration in the cerebral and cerebellar cortex, and in the basal ganglia. Howard and Pelc (1951) treated seedlings with ^{32}P and subsequently extracted the root tissues with hot HCl. After this treatment ^{32}P was present in the form of DNA-phosphorus. Using this method the authors found that DNA synthesis occurred in early interphase only, in those cells which were preparing for division. When ^{35}S was used in place of ^{32}P Howard and Pelc found a positive result in the same proportion of nuclei as gave a positive result with ^{32}P. They interpreted this finding as indicating the build-up of sulphur-containing amino-acids in parallel with nucleic acid synthesis.

In the early days of autoradiography radio-iodine (^{131}I) was probably more extensively used than any other element and this was largely due to the ease with which it becomes incorporated into thyroglobulin after being initially concentrated in the thyroid gland in inorganic form. The majority of thyroid autoradiographs illustrated in the literature showed protein-bound ^{131}I mainly confined to the colloid. Leblond and Gross (1948), however, maintained that in iodine-deficient rats incorporation of ^{131}I took place mainly in the luminal part of the lining cells of the acinus. Since all inorganic ^{131}I is removed during the processing of the tissues, this was considered to indicate the production and secretion into the acinus of new thyroglobulin. Doniach and Pelc (1949) were unable to confirm these findings and they considered that protein-bound ^{131}I was always situated in the colloid. These authors (Doniach *et al.*, 1952) suggested that ^{131}I diffused through the acinar cells and was bound to colloid within the follicle lumen. According to Doniach *et al.*, the two types of thyroid autoradiographs (*Ring* and *Blob*) represented two different types of hormone synthesis in the underactive and normally active follicles respectively.

Other interesting studies of thyroid function were made with the assistance of ^{131}I. Gorbman and Evans (1941) prepared autoradiographs of the developing thyroids of tadpoles kept in water containing ^{131}I. They were able by this means to show that storage of iodine commenced as soon as follicles appeared in the developing thyroid gland. Later Gorbman (1947) demonstrated by means of autoradiography the presence in mouse thyroids of non-functional structures which he considered, correctly, to be derived from the ultimobranchial bodies. The localization of labelled thyroid hormones in rat tissues was studied by Baserga (1959). The use of ^{131}I was not confined to studies of thyroid function. Albert *et al.* (1949), for instance, used radioactive iodo-alpha-oestradiol (non-biologically active) and found that after injection this substance was localized in the dermal connective tissues and, to a lesser extent, in liver, intestine and breast.

Incorporation of radioactive elements into organic substances produced much interesting work. Boyd *et al.* (1948) used glycine labelled with ^{14}C and showed that this substance became incorporated into liver cells, white blood cells, and into a small proportion of the circulating erythrocytes. Proteins were labelled by combination with various radioactive elements, the most popular, being the easiest to apply, was ^{131}I. This element is taken up by the tyrosine groups of the protein and the number so iodinated is subject to fairly accurate control. If only a small percentage of the available tyrosine is iodinated the properties of the original protein are not radically altered. Using such a method Warren and Dixon (1948) showed that in the guinea-pig killed by anaphylactic shock due to sensitivity to bovine gamma globulin, the antigen was concentrated not in the smooth muscle of the bronchi but in the peribronchial connective tissues. Pressman *et al.* (1949) showed by a similar means that anti-mouse kidney serum globulin, labelled with ^{131}I, could be found after injection solely in the glomeruli as far as the kidney was concerned.

Hormones labelled with ^{131}I were used by several authors for the demonstration of the supposed sites of specific activity. Jensen and Clark (1951) administered ^{131}I-labelled thyroxine to rabbits which they killed four hours after the injection. Strong ARG were obtained from sections of the neurohypophysis and tuber cinereum. Sonenberg *et al.* (1951a) showed that ^{131}I-labelled ACTH gave positive autoradiographs only in adrenals and thyroid and this same author (Sonenberg *et al.*, 1951b) demonstrated that labelled prolactin gave no significant result in resting or lactating mammary glands. In the ovary, however, positive results were obtained and these were confined to the corpora lutea. ^{131}I-labelled iodate in the iodate reaction for the demonstration of noradrenalin in the adrenal medulla Eränkö (1957) demonstrated the presence of iodine in the noradrenochrome pigment.

The greatest volume of studies continued to be in the nucleic acid field and a variety of labelled substances were employed. Considerable improvement in localization was produced by the substitution of ^{14}C-labelled adenine for ^{32}P as phosphate. Clowes (1956) used the former, together with the stripping film technique, in studying the metabolism of nucleic acids in *Zea mays* seedlings. Pelc and his associates (Pelc, 1958; Hornsey and Howard, 1956; Pelc and Howard, 1956; Pelc, 1956; Lasnitsky and Pelc, 1956) continued their studies in this field, using the same methods. Electron track autoradiographs of sections prepared by the coating technique were used by Ficq (1955) to obtain intracellular localization of ^{14}C-adenine in frogs and Ficq and Errera (1958) studied the metabolism of phenylalanine-2-^{14}C in isolated nuclei. Walker and Leblond (1958) used ^{14}C-adenine and ^{14}C-thymidine to demonstrate sites of nucleic acid synthesis in the mouse. Following treatment of the sections with ribonuclease sites of incorporation of ^{14}C-adenine were interpreted as sites of newly synthesized DNA. The chief sites of production were the epithelia of the digestive tract and other epithelia, and the germinal centres of the lymph nodes and spleen. Considerable activity was also observed in many types of connective tissue.

A truly great advance was made when tritium (^{3}H) was introduced as a radioactive label by Fitzgerald *et al.* (1951). This isotope has been used on an ever-increasing scale. Tritium-labelled adenine and thymidine, according to McQuade *et al.* (1955), was incorporated solely into DNA and by the use of labelled thymidine, therefore, selective extraction procedures could be avoided (Plaut and Mazia, 1956). This work marked the beginning of a period of concentrated studies on DNA and RNA metabolism. Some of these are afforded brief mention in the concluding section of this chapter.

A large amount of work continued to be done with the aid of ^{35}S, particularly in the use of inorganic ^{35}S and ^{35}S-methionine in studies of the metabolism of sulphur in the acid mucopolysaccharides (Boström and Odeblad, 1953, 1954; Bélanger, 1953, 1954; Davies and Young, 1954; Engfeldt *et al.*, 1954; Friberg and Ringertz, 1954; Boström *et al.*, 1954; Fajberg and Ringertz, 1954; Pelc and Glucksmann, 1955; Fitzgerald *et al.*, 1954; Dziewiatkowski, 1956; Johnson and Comar, 1957; Gracheva, 1957; Lea and Vaughan, 1957). An extensive study of the epithelial mucins in 5 species was made by Jennings and Florey (1956). These authors studied the uptake of ^{35}S, administered intravenously either as $Na_2\,^{35}SO_4$ or as ^{35}S-methionine, and they found that with the latter radioactivity was situated diffusely throughout the gastrointestinal tract without being concentrated in any particular area. After $Na_2\,^{35}SO_4$, however, they noted that certain epithelial cells of the stomach and duodenum took up the radioactive label and incorporated it into their mucopolysaccharides. In the cat (Plate XXIXA, opposite) the strongest activity was found in the goblet cells in the crypts of Lieberkühn. In the rat (Plate XXIXB) the pyloric glands were strongly active while the Brunner's glands of the duodenum were inactive. In the guinea-pig almost the opposite condition was observed.

Metallic Ions. Comparatively few studies have been carried out using isotopes other than the more or less conventional ones although many are quite suitable for autoradiographic studies (Taylor, 1956). Among these are

PLATE XXIX

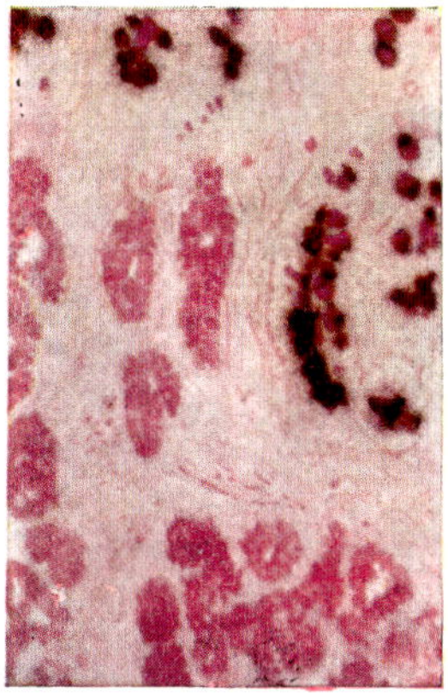

XXIXa. Cat. Brunners glands and bases of crypts of Lieberkühn. Fixed 6 hours after injection of $Na_2{}^{35}SO_4$. Strong radioactivity in goblet cells of the crypts. (Exposure 41 days.) PAS × 120.

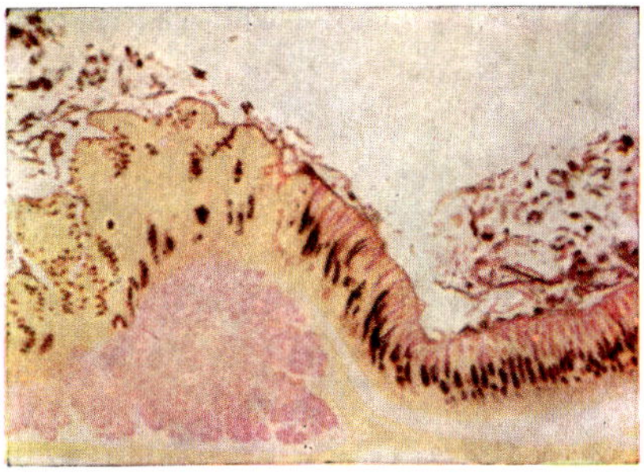

XXIXb. Rat pylorus and duodenum. Fixed 130 minutes after injection of $Na_2{}^{35}SO_4$. Note strong activity of the pyloric glands. (Exposure 22 days). PAS × 20.

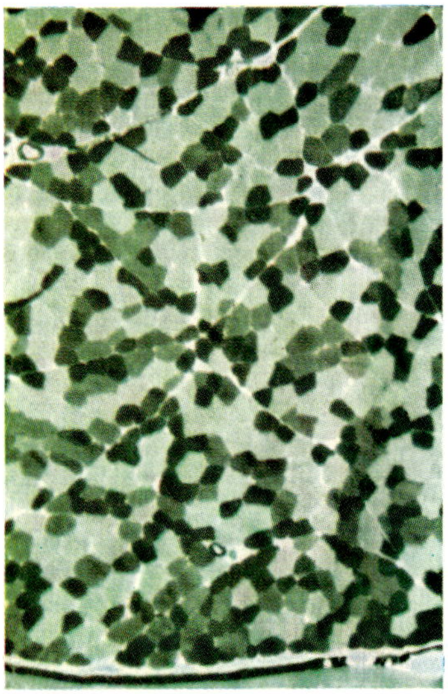

XXIXc. Human muscle biopsy. Myosin ATPase reaction in a case of thyroid carcinoma in which 3 distinct fibre types, instead of the usual two, are visible. × 56.

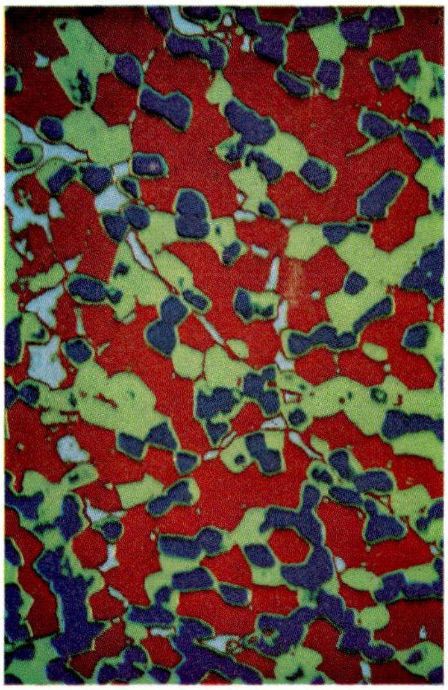

XXIXd. TV Image Conversion of part of the transparency from which XXIXc was made. Each of the fibre types is represented by a different colour. Type II, blue; Type II (intermediate), yellow; Type I, red.

^{45}Ca, ^{22}Na, ^{52}Mn, ^{54}Mn, ^{65}Zn, ^{42}K, ^{58}Co and ^{59}Fe. Radioactive calcium was used by Bélanger *et al.* (1954) in autoradiographic studies of the growth of incisor teeth in fluoride-treated pigs, and the uptake of ^{45}Ca by bone slices *in vitro* was reported by Amprino (1952). Radiopotassium was used by Morel and Guinnebault (1956) to study the pattern of tubular excretion in the rabbit kidney. Most of the specific activity was found in the cortex, with a very high exchange rate (15 per cent per minute).

Studies with radioactive zinc were made by Lowry *et al.* (1954) but these did not include autoradiography. The use of ^{65}Zn was also mentioned by Steinberg and Solomon (1949) and ^{59}Fe was employed in autoradiographic studies by Austoni (1954). The distribution of ^{65}Zn was studied by McIsaac (1955) in rat pancreas, by Daniel *et al.* (1956) in human prostate, by Ryder (1959) in skin, by Wetterdal (1958) in testis, and by Millar *et al.* (1961) in a variety of tissues.

The Techniques of Autoradiography

Physical Considerations. It is not proposed to enter into the physical aspects of autoradiography in this chapter. For theoretical considerations of these the reader is referred to excellent reviews or monographs on the subject by Heller, Axelrod and Hamilton (1950), by Gross *et al.* (1951), Doniach *et al.* (1952), and by Fitzgerald (1955), Boyd (1955), Taylor (1956), Pelc (1958), Joftes (1963), Baserga (1967), Rogers (1967), Stumpf and Roth (1968), Baserga and Malamud (1969), and Schultze (1969). The points dealt with by all the above authors include the relationship between the source of radioactivity in the tissue and the overlying emulsion, the mechanism of photographic action in the emulsion, the dose and energy of the radiation employed, the characteristics of the emulsion and the conditions of its development and fixation.

Summarizing briefly, the finest resolution is obtained by reduction of the thickness of the section, and of the width of the interspace between the section and the overlying emulsion, to a minimum. The maximum theoretical resolution with a 2 μ section, a 2 μ emulsion, and a 0·1 μ interspace has been calculated to be 2·3 μ. By reducing each of these the theoretical maximum resolution of 0·2 μ can be approached, for optical microscopic techniques. According to Baserga and Malamud (1969), for ^{3}H the usual resolution is between 0·2 and 1·0 μ whereas for ^{14}C it lies between 2 and 14 μ. For ^{32}P resolution is 5–10 μ, for ^{35}S it is 2–5 μ and for ^{125}I, 0·5 to 1·0 μ (Rogers, 1967). Resolution is decreased by prolonged exposures because the diameter of the image is thereby increased. As far as the emulsion is concerned, the larger the grain size the more β-particles are trapped and the greater the amount of energy absorbed. Emissions from radioactive material are of three types, alpha, beta and gamma particles, but the majority of isotopes normally employed are beta and gamma emitters. The greater the grain size, the greater the number of silver atoms reduced by the same amount of radiation and, therefore, the larger the grain the greater the apparent sensitivity of the emulsion when examined by the low-power microscope. At the same time the resolving power is much lower (owing to the larger image). For this reason, from the point of view of accurate localization, the use of fine grain emulsions is preferable. Some improvement in sensitivity can be achieved in these fine grain emulsions by increasing the concentration of grains to a maximum, and by sensitizing the grains in various ways.

Technical Procedures

Only the processing of soft tissues will be considered here; details of the preparation of bone and teeth for autoradiography will be found in papers by Leblond *et al.* (1950), Laude *et al.* (1949), Sognnaes *et al.* (1949), Duthie (1954), Arnold and Jee (1954a and b), Tonna and Cronkite (1958) and Kember (1960).

Fixation. Although it is possible to make autoradiographs of thick frozen sections by the contact method, high resolution autoradiography involves the preparation of thin tissue sections. For most purposes some kind of fixation is essential. In this context it must be remembered that a major cause of artifacts in autoradiography is translocation of the isotope. The choice of fixative to some extent depends on the nature of the isotope concerned and on the tissue component with which it has become incorporated if this has occurred. Protein-bound isotopes have been studied in material fixed by most of the conventional fixatives. Endicott and Yagoda (1947) used formalin and Bélanger and Leblond (1946) stated that commonly used fixatives such as Bouin and neutral formalin did not interfere with the technique. Evans (1947) used fixation, dehydration, embedding and sectioning "according to the usual histologic techniques". Doniach and Pelc (1950), however, tried formal saline, Bouin, Helly, Susa and absolute alcohol as fixatives for rat thyroids containing protein-bound ^{131}I. They found, with the first two fixatives, a peculiar patchy inhibition of the photographic process whereby parts of the film, on subsequent development, appeared devoid of silver granules. The two mercury-containing fixatives also produced an occasional inhibition and sections fixed in them suffered from the disadvantage that mercury deposits could not be removed in the usual manner with iodine and thiosulphate, in case this treatment caused removal of ^{131}I also. Doniach and Pelc therefore advised fixation in absolute alcohol until such time as the process of autoradiography allowed fine cytological localizations to be made. Quantitative autoradiography of mouse liver after injection of cytidine-^{3}H was carried out by Kopriwa and Leblond (1962), using a variety of fixatives. The results with Carnoy, Bouin and formalin were little different but practically no silver grains were present after Zenker fixation.

In the case of water-soluble isotopes a somewhat different technique is necessary. Holt *et al.* (1949) found considerable loss of ^{32}P from formalin-fixed paraffin sections floated out on water. They tested three methods in particular for avoiding this loss: (1) Freeze-dried sections embedded in paraffin. (2) Alcohol-formalin fixation (2 hours), followed by dehydration in alcohol, clearing in xylene and paraffin embedding. (3) Alcohol-formalin fixation followed by dehydration and clearing with dioxane, and paraffin embedding. Floating out was avoided in each case, and the third method was found to be most satisfactory for avoiding loss of water-soluble ^{32}P. They also froze small pieces of tissue at $-170°$ and then immersed them in absolute alcohol saturated with basic lead acetate, in order to precipitate phosphate as the lead salt. MacDonald *et al.* (1949) using a method for preparing frozen sections originally described by Linderstrøm-Lang and Mogensen (1938). This consisted of freezing the tissues in liquid nitrogen and sectioning on a freezing microtome in a cold room at $-10°$ to $-15°$.

Normally all radioactivity preserved by conventional wet fixatives is protein-bound. Unbound isotope, in whatever form it has been administered, is totally lost. This process is often described as "differential fixation".

Application of the Photographic Emulsion. *Contact Autoradiographs.* The simplest of the various methods of autoradiography is the contact method. This requires no special equipment and can easily be performed in any routine laboratory. It depends on the placing of the tissue slice or section in contact with a photographic plate and maintaining contact for an appropriate period. The results with this method are poor in that resolution is bad (50–100 μ), but the speed obtained is high if fast X-ray film is used. Dobyns and Lennon (1948) used contact autoradiography to determine the presence of active and inactive nodules in operative specimens of human thyroids and they obtained much useful information by this means. Details of the contact process are given in Appendix 30, p. 1429. An illustration of the type of result obtainable is provided by Fig. 255. This shows the technique of whole body (apposition) autoradiography, developed by Ullberg (1954, 1958) and successfully applied by Ritzén *et al.* (1965) for 5-HT and 5-HTP, by Hammarström (1966) for ascorbic acid and by Flodh (1968) for vitamin B12. The autoradiographs of water-diffusible substances in sections of whole baby rats was described by Aitken *et al.* (1968).

Mounted Autoradiographs. This method, first described by Endicott and Yagoda (1947) and later by Evans (1947), depends on the mounting of sections (5 μ thick in the original article) directly on to the photographic emulsion. Staining may be carried out before or after mounting, exposure and development.

Stripping Film Autoradiographs. This popular technique, which was developed by Pelc (1947, 1956) is based on the use of a thin photographic emulsion, on a gelatin base, which must be stripped from the plate or celluloid film backing before application. The film usually employed is a Kodak fine grain emulsion (AR 10 plates). Full details of this technique are given in Appendix 30.

With the older stripping film techniques some trouble was experienced on account of displacement of the photographic image by movement of the film. This displacement was of two types: (*a*) generalized and (*b*) localized. Generalized displacement occurred when the whole emulsion and gelatin backing became shifted relative to the underlying section and it was overcome by paying particular attention to efficient attachment of the emulsion to the slide. Bogoroch (1951) suggested the wrapping of the base and emulsion completely round the slide and emphasized the importance of sufficient preliminary soaking. For this purpose she advised a Dupanol-water mixture (Dupanol C, or lauryl sulphate) and 5 minutes immersion of the emulsion before its application to the slide. After application the preparations were thoroughly dried in a desiccator over $CaCl_2$. Localized displacement occurred with folding and swelling of the emulsion and this was overcome by efficient soaking of the type described above. Coating of slides with a film of gelatin previous to mounting of the sections (Doniach and Pelc, 1950; Berriman *et al.*, 1950) is an efficient method of avoiding both generalized and localized displacement and it will be found that displacement of both types can be entirely avoided by strict adherence to the coating and processing temperatures recommended in Appendix 30, p. 1429.

In dry climates an electrostatic discharge may be provoked by stripping the film off the plate. This can be prevented by carrying out the whole process in the darkroom, at 55–65 per cent relative humidity. An alternative technique, suggested by Mazia and Bucher (1960) involved immersion of the plate in 100 per cent ethanol, and transfer to 95 per cent ethanol for the peeling off process. The resulting strips were floated in water, emulsion side down.

Dry Mounting. For studying the incorporation of soluble isotopes special methods of dry mounting were developed. The original technique of Fitzgerald (1961) was modified by Appleton (1964). In his technique fresh cryostat sections were cut in the dark-room and picked up on cold coverslips coated previously with Kodak AR 10 stripping film. Until completion of exposure the mounted sections were kept frozen. The method was subjected to critical study by Appleton (1966) who showed that a resolution of 2–4 μ could be obtained with ^{3}H-thymidine and 9–13 μ with ^{22}NaCl. There was no significant diffusion at the optimum exposure temperature of −25°.

An alternative procedure described by Nagata *et al.* (1969) employed freeze-dried tissue blocks, infiltrated in the unfixed state with an Epon mixture by way of a resin embedding accessory (see Vol. 1, p. 596). As would be expected, ice crystal damage was the main problem but their results suggest that this theoretically excellent technique can be made to yield good results in practice. Freeze-dried tissues, embedded in silicone plastic, were used by Stirling and Kinter (1967) to study ^{3}H-galactose in hamster intestine.

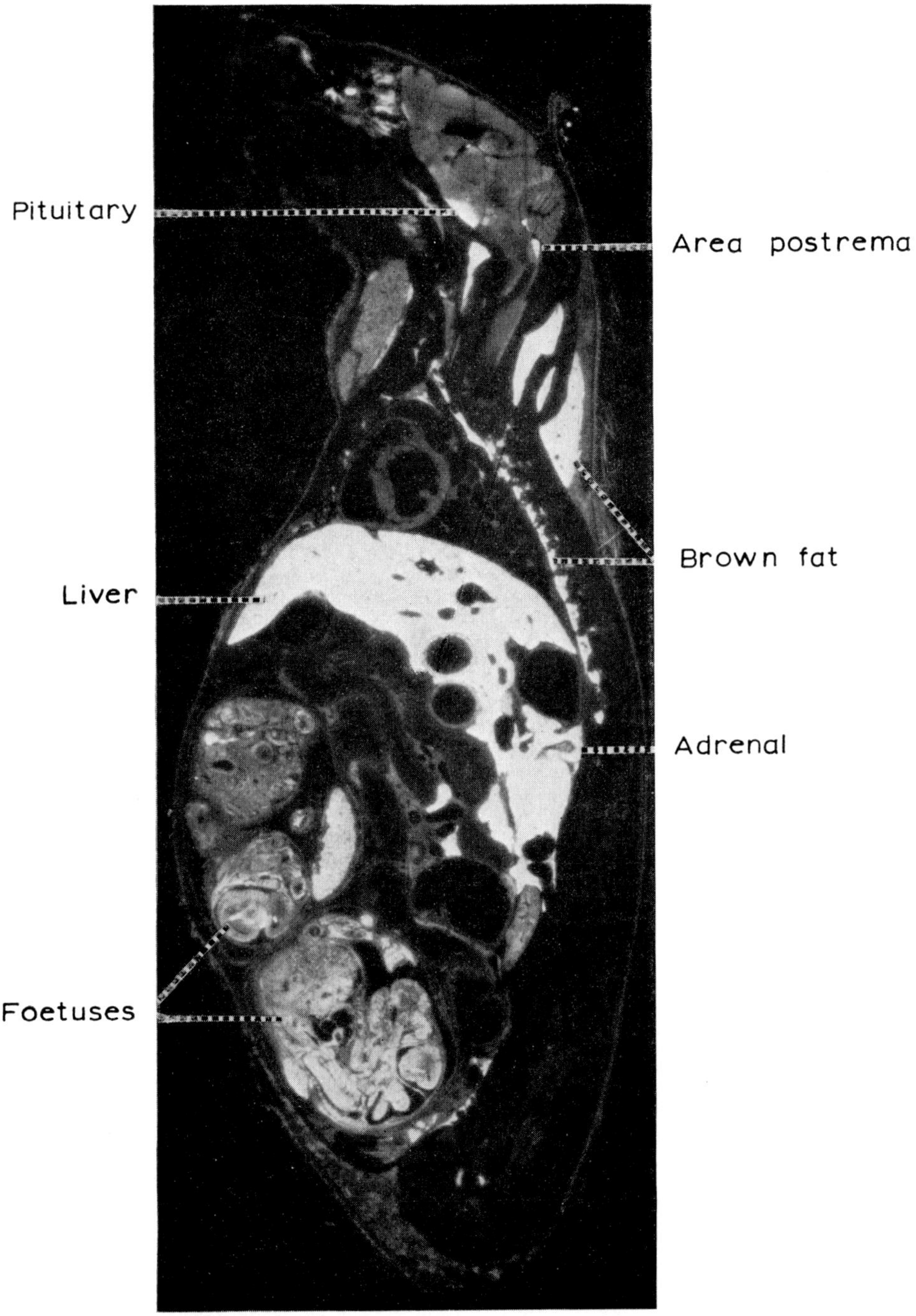

FIG. 255. Autoradiogram of ^{58}Co-vitamin B_{12} in a pregnant mouse, 4 hours after intravenous injection. High concentration of the isotope in the foetuses and placentae.

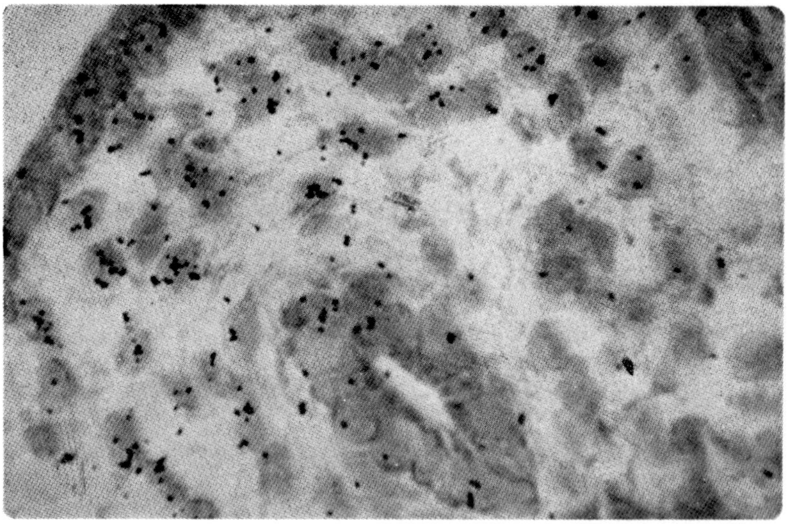

FIG. 256. Female rat, 24 days old. 2 μ freeze-dried section. Autoradiograph of uterine gland and epithelium 2 hours after subcutaneous injection of 0·63 μg of ^{3}H-oestradiol. Exposure, 54 days. Counterstained with methyl green-pyronin. × 1680.

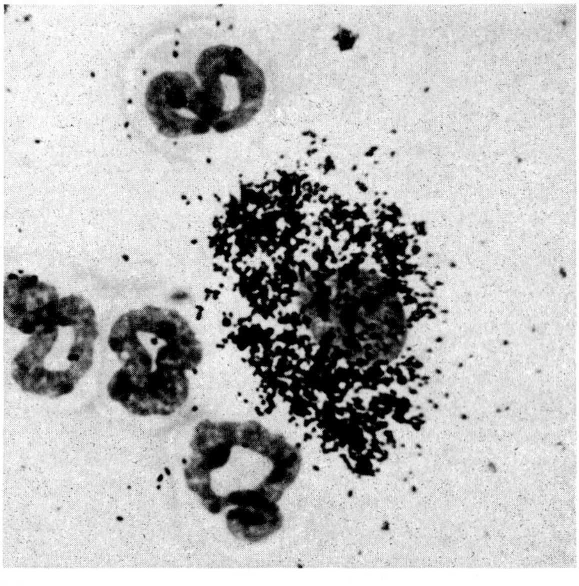

FIG. 257. Mouse. Peritoneal exudate containing blood cells and ^{3}H-antigen-carrying macrophages. (15 days after injection of antigen). Shows positive chemiotaxis by eosinophil leucocytes. Coated autoradiograph, Giemsa counterstain. × 1200.

A possible source of error was described by Stocker and Müller (1967) who found that in squash preparations stained with orcein, and covered with stripping film and exposed, silver grains appeared over cold nuclear DNA. This production of silver grains in the absence of any emitting isotope was ascribed to chemical interaction with the orcein stain.

Coated Autoradiographs. There are several types of coated autoradiograph: in the first, described by Bélanger and Leblond (1946) and by Leblond, Perceval and Gross (1948), sections mounted on slides were covered by a thin layer of photographic emulsion. This was painted on to the section by means of a brush or a glass rod, used as a roller. The second, or dipping, technique was introduced by Joftes and Warren (1955) and modified subsequently by Messier and Leblond (1957), Joftes (1963) and Prescott (1964).

In the technique most usually employed (details are given in Appendix 30) the liquid emulsion is melted in a water bath at 40°, either in undiluted or diluted form. The most popular emulsions are Kodak NTB, NTB2 or NTB3 and Ilford G5, K5, K2 or L4. The letters of the Ilford preparations refer to grain size; large (G), medium (K) and small (L). The numbers indicate sensitivity. Emulsions labelled O, for instance, record α but not β-particles while those labelled 2 record β-particles up to 50 keV. Diluted emulsions provide a thinner coating and the choice of diluted or undiluted emulsion depends on the purpose of the investigation (Falk and King, 1963). A semi-automatic instrument for dip-coating was devised by Kopriwa (1967).

A third type of coated autoradiograph was the inversion type, introduced by MacDonald *et al.* (1948) and by Bélanger (1950). With this procedure the mounted unstained tissue section, covered with celloidin, is coated with a special emulsion (Kodak matrix). After exposure and development, the preparation is immersed in water and the emulsion is lifted from the slide with a razor blade, the section adhering to its undersurface. It is then inverted and mounted, section upwards, on a clean slide. Staining can now be carried out with ease.

Dry Mounting. The dry mounting of freeze-dried frozen sections on dip-coated slides was developed as a technique for high resolution autoradiography by Stumpf and Roth (1966, 1967), and a further description of the method was given by Stumpf (1968). Full details of the method are given in Appendix 30 and the type of result obtainable is shown in Fig. 256. A similar technique was used by Sander and Attramadal (1968).

Double-labelling (*Two emulsion*) *Techniques*. A special application of the dip-coating technique was evolved by Baserga (1961) and Baserga and Nemeroff (1962). In order to distinguish between β-particles emitted from ^{14}C and ^{3}H, two layers of emulsion were applied, separated by a layer of celloidin. Because of their longer range the ^{14}C particles reached the second emulsion while those from tritium were arrested by the first emulsion. By using a tritiated precursor of DNA and a ^{14}C-labelled precursor of RNA it was possible to investigate the metabolism of these two substances simultaneously. By this procedure any pair of distinct metabolic processes may theoretically be followed at one and the same time. Many modifications of the original method have been published. Most of them use two successive dipping emulsions but some (Lennartz and Maurer, 1964; Pilgrim *et al.*, 1966) have employed stripping film for the first layer and liquid emulsion for the second.

Exposure. The exposure time for an autoradiograph depends on many factors, the most important being the type of isotope, the dose, the duration of its effect and the type of emulsion. It is common practice to calculate the exposure by making test autoradiograms using a series of sections and successive intervals. It may be necessary also to vary the exposure according to special needs. For grain counting, for instance, the lowest background. The whole process is normally carried out at 4° in a closed box containing Drierite or other water-absorbing material.

Development. The end result of this process, which can be modified within wide limits, is the production of silver grains. For most emulsions and stripping film it is customary to use Kodak D19 developer and a time of 4–6 minutes at 18°. For the production of large grains from fine grains emulsions Kopriwa and Leblond (1962) recommend Kodak D178 or D170 developers. The time of development depends basically on the thickness of the emulsion. With thick layers 20–30 minutes may be necessary at 12–14°. After development, washing in water is followed by treatment with an acid fixer and by a final thorough wash in water, provided that the sections have not been prestained.

Staining. This process can be carried out (1) before application of the emulsion or (2) after development. Stains which are applied at stage 1 must be resistant to all subsequent procedures, they must not bring about removal or translocation of the isotope and they must not cause fogging of the emulsion. Stains which are applied at stage 2 must be able to penetrate the gelatin film without causing damage and without loss of silver grains.

Prestaining procedures were described by Thurston and Joftes (1963) and Harris' haematoxylin was used by Kopriwa and Leblond (1962). The Feulgen method (Vol. 1, p. 647) was at one time popular but several authors (Lang and Maurer, 1965; Baserga and Nemeroff, 1962) showed that with tritiated thymidine substantial losses of isotope (up to 50 per cent) occurred during the phase of acid hydrolysis. A method of pre-vital staining with neutral red and trypan blue which avoids these difficulties was described by Sawicki *et al.* (1967) and this has been found to give excellent and reproducible results.

For post-staining, haematoxylin (Mayer's) and eosin continue to remain popular but many other basic dyes have been used for nuclear staining. These include basic fuchsin, toluidine blue, methylene blue, azure A and cresyl violet. Many staining mixtures or sequences have been used also, such as May-Gruenwald Giemsa (Fig. 257) methyl green-pyronin and neutral fast red/indigocarmine. Gallocyanin-chromalum was recommended by Bowie and Edmonson (1960). Since it is a progressive stain no differentiation is necessary but Stenram (1962), and also Linnartz-Niklas *et al.* (1964), described the removal of developed silver grains by this dye.

Quantitative Autoradiography

There are four types of procedure by which the amount of radioactivity in cells and tissues can be calculated:

(1) Grain Counting (Andresen *et al.*, 1952).
(2) Track Counting.
(3) Counting of Labelled Units (e.g. Nuclei).
(4) Photometric Densitometry.

These methods were very fully considered by Pelc (1957) and by Levi (1957). Excellent coverage of the subject was provided also by Schultze (1969) and by Baserga and Malamud (1969). The first two, and especially the first, are extremely laborious and the third has only limited applications. It would therefore be much preferable to employ the fourth type of method.

An automatic grain counter based on the principle of the flying-spot microscope was designed by Dudley and Pelc (1953) but this apparatus was not suitable for routine use. Levi (1957) pointed out that direct estimation of the density of grains with a densitometer was impossible, for several reasons. She recommended the use of instruments containing a cylindrical lens, which spreads the grain image into a band (Mazia, Plaut and Ellis, 1955). This method is less open to criticism than most others but still apparently has many shortcomings. An alternative instrument, designed by Gullberg (1957) used a direct recording microscope to produce a photometric trace of the density of the grains in an autoradiograph. A flying-spot scanning instrument was described by Tolles (1959) and Stubblefield (1965) used photographs at different levels combined into a single picture and with the grains converted into lines for photodensitometry. Ritzén (1967) described a method for the determination of absolute specific radioactivity in cells, and Dormer *et al.* (1966) and Dormer (1967) used a new photometric method for quantitative autoradiography. Initial difficulties due to light-scattering effects were overcome and the method is apparently capable of distinguishing between closely adjacent cells.

The problem of quantitative electron autoradiography, especially when using tritium, is less simple than at optical microscope level. This is because self absorption of tritium β-particles in suitable sections for electron microscopy is very low and variations in section thickness produce corresponding variations in grain yield. A second factor is the difficulty of producing uniform emulsion layers. Quantitative EM autoradiography was nevertheless carried out by Salpeter (1967) and by Bachmann and Salpeter (1967). An alternative approach, using an internal standard, was proposed by Flitney

(1969). He incorporated in his blocks a seam of ^{3}H-labelled araldite of known radioactivity. Comparing grain density over the tissue with that over the standard permitted accurate quantitation.

Measurement of Enzymes by Quantitative Autoradiography

The labelled inhibitor method of Ostrowski and Barnard (1961) is a logical and elegant extension of the practice of quantitative autoradiography. The method involves the use of an isotopically labelled specific, and preferably irreversible, inhibitor. This binds to the enzyme active site and provides a stoichiometric 1:1 relationship between enzyme molecules and atoms of the isotope. This principle, which merits much wider extension, has been discussed in full Barnard (1970).

Most of the early work carried out by means of the labelled inhibitor method employed ^{3}H-DFP for the quantitation of acetylcholinesterase at the motor end plates and elsewhere in muscle (Ostrowski *et al.*, 1963; Barnard and Ostrowski, 1964; Rogers *et al.*, 1964). In these studies there was no adequate separation of the various sites but this was achieved in later work (Rogers *et al.*, 1966). Both grain and track counting methods were employed and with the former it was necessary to stain the motor end plates after carrying out the autoradiographic process. This was done by using pyridine-2-aldoxime methiodide (2-PAM) to extract the labelled inhibitor and reactivate the cholinesterase. For track autoradiography it was necessary to microdissect single muscle fibres, each bearing an end plate.

Electron Microscopical Autoradiography

Extension of optical autoradiography to the electron microscope was introduced by Liquier-Milward (1956) and further developments were due to O'Brien and George (1959) who use ^{210}Po. Progress was at first very slow and no increase in resolution was obtained. Using tritium labelling and bacterial cells, together with a fine grain emulsion, van Tubergen (1961) first achieved resolutions superior to those of optical microscopic autoradiography and his method was developed further by Caro (1961, 1962) and Caro and Palade (1962). A standard technique was developed by Caro and van Tubergen (1962) and this was described fully by Caro (1964). The methods of EM autoradiography were discussed by Fromme (1964) and practical considerations dealt with more fully by the same author (Fromme, 1970). Special emulsions with very small silver halide crystals are necessary for EM autoradiography. The most commonly used examples are Ilford L4 nuclear track emulsion (halide crystal size 120–160 nm), Kodak NTE (30–50 nm), and Gevaert Scientia NUC 307 (70 nm.). The latter was assessed in relation to the others by Young and Kopriwa (1964) and a similar assessment was made by Hülser and Rajewsky (1966). A fine grain silver bromide emulsion, prepared

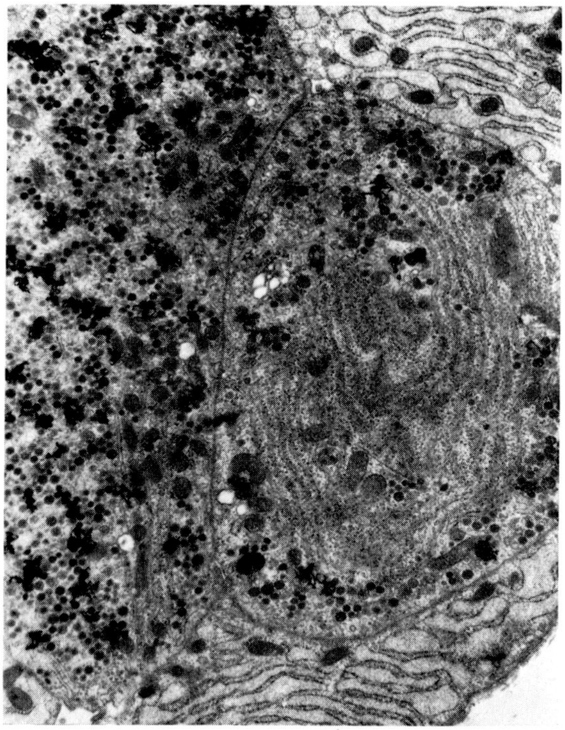

FIG. 259. Rat thyroid gland. Parts of two C cells and three follicular cells can be seen. The autoradiographic grains are localized, only in the C cells, in areas containing the specific granules. They are absent from the whorled endoplasmic reticulum. Preparation made after injection of ^{3}H-5-HTP. × 6500.

[*To face p.* 1219

by Van Kleef *et al.* (1969) was reported to have a mean diameter of less than 10 nm. Details are given in Appendix 30.

The factors involved in EM autoradiography were listed by Caro (1964). They are:

(1) The range of the β-particles in tissue, resin or gelatin.

(2) Geometry: with a monolayer of crystals and a point source of β-particles the probability of a hit on a crystal decreases as the solid angle from which it is seen by a source.

(3) The shielding of the silver halide crystals away from the source by those which are close to it (thus reducing the probability of finding an exposed grain away from the source).

The combination of these factors results in a distribution curve from which one can predict a resolution of the order of 0·1 μ (with Ilford LV emulsion). Distribution curves were calculated for various distances of source to emulsion and for various grain sizes. Such curves, for four different combinations are shown in Fig. 258, below.

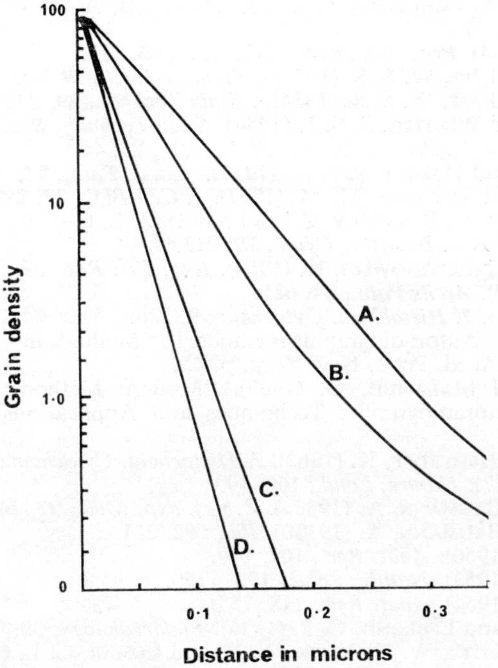

FIG. 258. Grain density distribution, calculated for a variety of situations. (A) Crystal diameter 100 nm, source 50 nm. from emusion; (B) 10 nm and 50 nm; (C) 100 nm and 10 nm; (D) 10 nm and 10 nm.

It will be observed that resolution increases if either the diameter of the silver halide crystals, or the thickness of the specimen, is reduced.

Electron microscopy thus brings the limit of resolution down from approximately 1 micron to about 0·1 micron, a ten-fold improvement. Numerous successful applications testify to the efficacy of EM autoradiography. One such application is illustrated in Fig. 259, p. 1219, which shows the localization of 5-HT, derived from ^{3}H-5-HTP by decarboxylation, in the calcitonin-containing granules of the thyroid C cells (Ericson, 1970). It is only by the use of such techniques that we can establish the precise localization of such small, potentially water-soluble, molecules at the cellular level.

REFERENCES

AITKEN, W. M., WRIGHT, E., and GRAY, H. S. (1968). *Stain Technol.*, **43**, 1.
ALBERT, S., HEARD, R. D. H., LEBLOND, C. P., and SAFFRAN, J. (1949). *J. biol. Chem.*, **177**, 247.
AMPRINO, R. (1952). *Experientia*, **8**, 380.
ANDRESEN, N., ANDRESEN, C. C., and HOLTER, H. (1952). *C. R. trav. Lab.Carlsberg, Sér. Chim.*, **28**, 189.
APPLETON, T. C. (1964). *J. roy. micr. Soc.*, **83**, 277.
APPLETON, T. C. (1966). *J. Histochem. Cytochem.*, **14**, 414.
ARMSTRONG, W. D., SCHUBERT, J., and LINDENBAUM, A. (1948). *Proc. Soc. exp. Biol., N.Y.*, **68**, 233.
ARNOLD, J. S. (1954). *Proc. Soc. exp. Biol., N.Y.*, **85**, 113.
ARNOLD, J. S., and JEE, W. S. S. (1954a). *Stain Technol.*, **29**, 49.
ARNOLD, J. S., and JEE, W. S. S. (1954b). *Stain Technol.*, **29**, 225.
ARNOLD, J. S., and WEBSTER, S. S. J. (1954). *Stain Technol.*, **29**, 225.
AUSTONI, M. E. (1954). *Proc. Soc. exp. Biol., N.Y.*, **85**, 48.
AXELROD, D. J., and HAMILTON, J. G. (1947). *Am. J. Path.*, **23**, 349.
BACHMANN, L., and SALPETER, M. M. (1967). *J. Cell Biol.*, **33**, 299.
BARKA, T., SZALAY, S., POSALAKY, Z., and KERTESZ, L. (1952). *Acta anat.*, **16**, 45.
BARNARD, E. A. (1970). *Int. Rev. Cytol.*, **29**, 213.
BARNARD, E. A., and OSTROWSKI, K. (1964). *Exp. Cell Res.*, **36**, 28.
BASERGA, R. (1959). *Archs Path.*, **68**, 621.
BASERGA, R. (1961). *J. Histochem. Cytochem.*, **9**, 586.
BASERGA, R. (1967). Autoradiographic methods. In "Methods in Cancer Research". Ed. H. Busch. Acad. Press, New York, p. 45.
BASERGA, R., and MALAMUD, D. (1969). "Modern Methods in Experimental Pathology, Autoradiography, Techniques and Applications". Hoeber, New York.
BASERGA, R., and NEMEROFF, K. (1962). *J. Histochem. Cytochem.*, **10**, 628.
BAYLEY, S. T. (1947). *Nature, Lond.*, **160**, 193.
BEHRENS, B., and BAUMANN, A. (1933a). *Z. ges. exp. Med.*, **92**, 241.
BEHRENS, B., and BAUMANN, A. (1933b). *Ibid.*, **92**, 251.
BÉLANGER, L. F. (1950). *Anat. Rec.*, **107**, 149.
BÉLANGER, L. F. (1953). *Nature, Lond.*, **172**, 1150.
BÉLANGER, L. F. (1954). *Anat. Rec.*, **118**, 755.
BÉLANGER, L. F., and LEBLOND, C. P. (1946). *Endocrinology*, **39**, 8.
BÉLANGER, L. F., LOTZ, W. E., VISEK, W. J., and COMAR, C. L. (1954). *Anat. Rec.* **119**, 53.
BERGGREN, H. (1946). *Acta radiol.*, **27**, 248.
BERLIN, M., HAMMERSTRÖM, L., and MAUNSBACH, A. B. (1964). *Acta Radiol. Ther.*, **2**, 345.
BERRIMAN, R. W., HERTZ, R. H., and STEVENS, G. W. W. (1950). *Br. J. Radiol.*, **23**, 472.
BLOOM, W. (1949). *Anat. Rec.*, **103**, Suppl. 9.
BLOOM, W., CURTIS, H. J., and MCLEAN, F. C. (1947). *Science, N.Y.*, **105**, 45.
BOGOROCH, R. (1951). *Stain Technol.*, **26**, 43.

BOSTRÖM, H., FRIBERG, U., ODEBLAD, E., and RINGERTZ, N. R. (1954). *Acta path. Scand.*, **35**, 1.
BOSTRÖM, H., and ODEBLAD, E. (1953). *Anat. Rec.*, **115**, 505.
BOSTRÖM, H., and ODEBLAD, E. (1954). *Acta physiol. Scand.*, **32**, 124.
BOWIE, D. J., and EDMONSON, V. R. (1960). *Stain Technol.*, **35**, 1.
BOYD, G. A. (1955). "Autoradiography in Biology and Medicine". Acad. Press, New York.
BOYD, G. A., CASARETT, G. W., ALTMAN, K. I., NOONAN, T. R., and SALOMON, K. (1948). *Science, N.Y.*, **108**, 529.
BOYD, G. A., and LEVI, H. (1950). *Ibid.*, **111**, 58.
BUDD, G. C. (1970). *Int. Rev. Cytol.*, **29**, 245.
BULLIARD, H., GRUNDLAND, I., and MOUSSA, A. (1938). *C. R. Acad. Sci., Paris*, **207**, 745.
CARO, L. G. (1961). *J. biophys. biochem. Cytol.*, **10**, 37.
CARO, L. G. (1962). *J. Cell Biol.*, **15**, 189.
CARO, L. G. (1964). *J. roy. micr. Soc.*, **83**, 127.
CARO, L. G., and PALADE, G. (1962). *C. R. Soc. Biol., Paris*, **155**, 1750.
CARO, L. G., and VAN TUBERGEN, R. P. (1962). *J. Cell Biol.*, **15**, 173.
CARO, L. G., VAN TUBERGEN, R. P., and KOLB, J. A. (1962). *J. Cell Biol.*, **15**, 173.
CLOWES, F. A. L. (1956). *New Phytologist*, **55**, 29.
COLFER, H. F., and ESSEX, H. E. (1946). *Proc. Soc. exp. Biol., N.Y.*, **63**, 243.
CORMACK, D. V. (1955). *Br. J. Radiol.*, **28**, 450.
DANIEL, O., HADDAD, F., PROUT, G., and WHITMORE, W. F. (1956). *Br. J. Urol.*, **28**, 271.
DAVIES, D. V., and YOUNG, L. (1954). *Nature, Lond.*, **173**, 448.
DOBYNS, B. M., and LENNON, B. (1948). *J. clin. Endocrinol.*, **8**, 732.
DOLS, M. J. L., JANSEN, B. C. P., SIZOO, G. J., and VAN DER MAAS, G. J. (1938). *Nature, Lond.*, **142**, 953.
DONIACH, I., HOWARD, A., and PELC, S. R. (1952). *Progress in Biophysics*, **3**, 1.
DONIACH, I., and PELC, S. R. (1949). *Proc. R. Soc. Med.*, **42**, 957.
DONIACH, I., and PELC, S. R. (1950). *Br. J. Radiol.*, **23**, 184.
DORMER, P. (1967). *Histochemie*, **8**, 1.
DORMER, P., BRINKMANN, W., STIEBER, A., and STICK, W. (1966). *Klin. Wschr.*, **44**, 477.
DUDLEY, R. A. (1954). *Nucleonics*, **12**, 24.
DUDLEY, R. A., and DOBYNS, B. M. (1949). *Science, N.Y.*, **109**, 327.
DUDLEY, H. C., and MADDOX, G. E. (1949). *J. Pharmac. exp. Ther.*, **96**, 224.
DUDLEY, R. A., and PELC, S. R. (1953). *Nature, Lond.*, **172**, 992.
DUTHIE, R. B. (1954). *J. Path. Bact.*, **68**, 296.
DZIEWIATKOWSKI, D. D. (1956). *J. biophys. biochem. Cytol.*, **2**, 29.
EDWARDS, L. C. (1955). *Stain Technol.*, **30**, 163.
ENDICOTT, K. M., and YAGODA, H. (1947). *Proc. Soc. exp. Biol., N.Y.*, **64**, 170.
ENGFELDT, B., ENGSTRÖM, A., and BOSTRÖM, H. (1954). *Expl Cell Res.*, **6**, 251.
ERÄNKÖ, O. (1957). *J. Histochem. Cytochem.*, **5**, 408.
ERICSON, L. E. (1970). *J. Ultrastructure Res.*, **31**, 162.
EVANS, T. C. (1947). *Proc. Soc. exp. Biol., N.Y.*, **64**, 313.
FAJBERG, V., and RINGERTZ, N. R. (1954). *Experientia*, **10**, 67.
FALK, G. J., and KING, R. C. (1963). *Radiation Res.*, **20**, 466.
FICQ, A. (1955). *Exp. Cell Res.*, **9**, 286.
FICQ, A., and ERRERA, M. (1958). *Ibid.*, **14**, 182.
FITZGERALD, P. J. (1955). In "Analytical Cytology". Ed. Mellors, R. C. McGraw-Hill, New York.
FITZGERALD, P. J. (1961). *Lab. Invest.*, **10**, 846.
FITZGERALD, P. J., EIDINOFF, M. L., KNOLL, J. E., and SIMMEL, E. B. (1951). *Science, N.Y.*, **114**, 494.
FITZGERALD, P. J., SIMMEL, E. B., WEINSTEIN, J., and MARTIN, L. (1953). *Lab. Invest.*, **2**, 181.
FITZGERALD, P. J., WEINSTEIN, J., BATCHELDER, R., and ALVIZOURI, M. (1954). *J. Histochem. Cytochem.*, **2**, 463.

FLITNEY, F. W. (1969). *J. Microsc.*, **90**, 151.
FLODH, H. (1968). *Acta radiol.*, Suppl. 284.
FRIBERG, U., and RINGERTZ, N. R. (1954). *Expl Cell Res.*, **6**, 527.
FROMME, H. G. (1964). *Z. Naturforsch.*, **196**, 851.
FROMME, H. G. (1970). *Acta histochemica*, **36**, 261.
GORBMAN, A. (1947). *Anat. Rec.*, **98**, 93.
GORBMAN, A., and EVANS, H. M. (1941). *Proc. Soc. exp. Biol., N.Y.*, **47**, 103.
GRACHEVA, N. D. (1957). *Doklady Acad. Nauk.*, **113**, 688.
GROSS, J., BOGOROCH, R., NADLER, N. J., and LEBLOND, C. P. (1951). *Am. J. Roentgenol.*, **65**, 420.
GULLBERG, J. E. (1957). *Expl Cell Res.*, Suppl., **4**, 222.
HAMILTON, J. G., SOLEY, M. H., and EICHORN, K. B. (1940). *Univ. Calif. Publ. Pharmac.*, **1**, 339.
HAMMARSTRÖM, V. (1966). *Acta physiol. scand.*, **70**, Suppl. 289.
HARRIS, J. E., SLOANE, J. F., and KING, D. T. (1950). *Nature, Lond.*, **166**, 26.
HEATLEY, N. G., JERROME, D. W., JENNINGS, M. A., and FLOREY, H. W. (1956). *Quart. J. exp. Physiol.*, **41**, 124.
HELLER, D., AXELROD, D. T., and HAMILTON, J. G. (1950). In McClung's "Handbook of Microscopical Technic". 3rd Ed., Hoeber, New York.
HOLT, M. W., COWING, R. F., and WARREN, S. (1949). *Science, N.Y.*, **110**, 328.
HORNSEY, S., and HOWARD, A. (1956). *Ann. N.Y. Acad. Sci.*, **63**, 915.
HOWARD, A., and PELC, S. R. (1951). *J. exp. Cell Res.*, **2**, 178.
HÜLSER, D. F., and RAJEWSKY, M. F. (1966). *Biophysik*, **3**, 123.
JENNINGS, M. A., and FLOREY, H. W. (1956). *Quart. J. exp. Physiol.*, **41**, 131.
JENSEN, J. M., and CLARK, D. E. (1951). *J. Lab. clin. Med.*, **38**, 663.
JOFTES, D. L. (1963). *J. Nuclear Med.*, **4**, 143.
JOFTES, D. L., and WARREN, S. (1955). *J. Biol. Phot. Assoc.*, **23**, 145.
JOHNSON, P. M., and COMAR, C. L. (1957). *J. biophys. biochem. Cytol.*, **3**, 231.
KAMINSKI, E. J. (1955). *Stain Technol.*, **30**, 139.
KEMBER, N. F. (1960). *J. Bone Joint Surg.*, **42B**, 824.
KOPRIWA, B. (1967). *J. Histochem. Cytochem.*, **14**, 923.
KOPRIWA, B., and LEBLOND, C. P. (1962). *J. Histochem. Cytochem.*, **10**, 269.
KOTZAREFF, A. (1922). *J. radiol. d'elect.*, **6**, 131.
KRAKUSIN, J. S., and JENNINGS, R. B. (1955). *Archs Path.*, **59**, 471.
LACASSAGNE, A., and LATTES, J. (1924). *C. R. Acad. Sci., Paris*, **178**, 488.
LANG, W., and MAURER, W. (1965). *Expl Cell Res.*, **39**, 1.
LASNITSKY, I., and PELC, S. R. (1956). *Expl Cell Res.*, **13**, 140.
LAUDE, P. P., JAMES, R. G., and BOYD. (1949). *Anat. Rec.*, **104**, 11.
LEA, L., and VAUGHAN, J. (1957). *Quart. J. micr. Sci.*, **98**, 369.
LEBLOND, C. P., and GROSS, J. (1948). *Endocrinology*, **43**, 306.
LEBLOND, C. P., PERCEVAL, N. L., and GROSS, J. (1948). *Proc. Soc. exp. Biol., N.Y.*, **67**, 74.
LEBLOND, C. P., STEVENS, C. E., and BOGOROCH, R. (1948). *Science, N.Y.*, **108**, 531.
LEBLOND, C. P., WILKINSON, E. W. W., BÉLANGER, L. F., and ROBICHON, J. (1950). *Am. J. Anat.*, **86**, 289.
LENNARTZ, K.-J., and MAURER, W. (1964). *Z. Zellforsch.*, **63**, 478.
LEVI, H. (1957). *Expl Cell Res.*, Suppl. **4**, 207.
LINDENBAUM, A., and SMOLER, M. H. (1969). *Science, N.Y.*, **165**, 192.
LINDERSTRØM-LANG, K., and MOGENSEN, K. R. (1938). *C. R. trav. Lab. Carlsberg, Sér. Chim.*, **23**, 27.
LINNARTZ-NIKLAS, A., HEMPEL, K., and MAURER, W. (1964). *Z. Zellforsch.*, **62**, 443.
LIQUIER-MILWARD, J. (1956). *Nature, Lond.*, **177**, 619.
LOTZ, W. E., GALLIMORE, J. C., and BOYD, G. A. (1952). *Nucleonics*, **10**, 28.
LOUDON, E. S. (1904). *Archs d'elect. med.*, **12**, 363.
LOWRY, J. R., HARRINGTON, R. V., and BALDWIN, R. R. (1954). *Science, N.Y.*, **119**, 219.
MACDONALD, H. M., COBB, J., and SOLOMON, A. K. (1948). *Science, N.Y.*, **107**, 550.
MACDONALD, H. M., COBB, J., SOLOMON, A. K., and STEINBERG, A. K. (1949). *Proc. Soc. exp. Biol., N.Y.*, **72**, 117.

Magnusson, B., and Odebald, E. (1954). *Expl Cell Res.*, **6**, 525.
Marshak, A. (1941). *J. gen. Physiol.*, **25**, 275.
Mazia, D., Plaut, W. S., and Ellis, G. W. (1955). *Expl Cell Res.*, **9**, 305.
Mazia, D., and Bucher, N. L. R. (1960). *Experientia*, **16**, 215.
McIsaac, R. J. (1955). *Endocrinology*, **57**, 571.
McQuade, H. A., Friedkin, M., and Atchison, A. (1955). *Nature, Lond.*, **175** 1038.
Messier, B., and Leblond, C. P. (1957). *Proc. Soc. exp. Biol. Med., N.Y.*, **96**, 7.
Millar, M. J., Vincent, N. R., and Mawson, C. A. (1961). *J. Histochem. Cytochem.*, **9**, 11.
Morel, F., and Guinnebault, M. (1956). *Helv. physiol. pharmac. Acta*, **14**, 255.
Nagata, T., Nawa, T., and Yokota, S. (1969). *Histochemie*, **18**, 241.
O'Brien, R. T., and George, L. A. (1959). *Nature, Lond.*, **183**, 1461.
Ostrowski, K., and Barnard, E. A. (1961). *Exp. Cell Res.*, **25**, 465.
Ostrowski, K., Barnard, E. A., Stocka, Z., and Darzynkiewicz, Z. (1963). *Expl Cell Res.*, **31**, 89.
Palm, E. (1948). *Acta ophthal.*, Suppl. 32.
Pease, D. C. (1966). *J. Ultrastructure Res.*, **14**, 379.
Pecher, C. (1942). *Univ. Calif. Publ. Pharmacol.*, **2**, 117.
Pelc, S. R. (1947). *Nature, Lond.*, **160**, 749.
Pelc, S. R. (1956). *Ibid.*, **178**, 359.
Pelc, S. R. (1957). *Expl Cell Res.*, Suppl. **4**, 231.
Pelc, S. R. (1958). In "General Cytochemical Methods". Ed. J. F. Danielli, Acad. Press, New York, Vol. I, p. 279.
Pelc, S. R., and Glücksmann, A. (1955). *Expl Cell Res.*, **8**, 336.
Pelc, S. R., and Howard, A. (1956). *Ibid.*, **11**, 128.
Percival, W. L., and Leblond, C. P. (1948). *Rev. canad. Biol.*, **7**, 217.
Pilgrim, C. Lang, W., and Maurer, W. (1966). *Expl Cell Res.*, **44**, 129.
Plaut, W., and Mazia, D. (1956). *J. biophys. biochem. Cytol.*, **2**, 573.
Prescott, D. M. (1964). "Methods in Cell Physiology". Acad. Press, New York.
Pressman, D., Hill, R. F., and Foote, F. W. (1949). *Science, N.Y.*, **109**, 65.
Ritzén, M., Hammarström, V., and Ullberg, S. (1965). *Biochem. Pharmac.*, **14**, 313.
Ritzén, M. (1967). *Expl Cell Res.*, **45**, 250.
Rogers, A. W. (1967). "Techniques of Autoradiography". Elsevier, Amsterdam.
Rogers, A. W., and Barnard, E. A. (1964). *J. Cell. Biol.*, **41**, 686.
Rogers, A. W., Darzynkiewicz, Z., Salpeter, M. M., Ostrowski, K., and Barnard, E. A. (1966). *J. Cell Biol.*, **41**, 665.
Ryder, M. L. (1959). *J. Histochem. Cytochem.*, **7**, 133.
Salpeter, M. M. (1967). *J. Cell Biol.*, **32**, 379.
Salpeter, M. M., and Bachmann, L. (1964). *J. Cell Biol.*, **22**, 469.
Sander, S., and Attramadal, A. (1968). *Acta endocrinol.*, **58**, 235.
Sawicki, W., Kieler, J., and Briand, P. (1967). *Stain Technol.*, **42**, 143.
Schultze, B. (1969). "Physical Techniques in Biological Research", Vol. III, Part B, "Autoradiography at the Cellular Level". Acad Press, New York.
Sherman, A. A., and Essman, L. J. (1954). *Science, N.Y.*, **119**, 659.
Siess, M., and Seybold, G. (1957). *Z. wiss. Mikr.*, **63**, 156.
Simmel, E. B., Fitzgerald, P. J., and Godwin, J. T. (1951). *Stain Technol.*, **26**, 25.
Simmel, E. B. (1957). *Ibid.*, **32**, 294.
Sognnaes, R. F., Shaw, J. H., Solomon, A. K., and Harvold, E. H. (1949). *Anat. Rec.*, **104**, 319.
Sonenberg, M., Keston, A. S., and Money, W. L. (1951a). *Endocrinology*, **48**, 148.
Sonenberg, M., Money, W. L., Keston, A. S., Fitzgerald, P., and Godwin, J. T. (1951b). *J. Endocrinol.*, **11**, 747.
Steedman, H. F. (1947). *Quart. J. micr. Sci.*, **88**, 123.
Steinberg, D., and Solomon, A. K. (1949). *Rev. Sci. Instrument*, **20**, 655.
Stenram, U. (1962). *Stain Technol.*, **37**, 231.
Stocker, E., and Müller, H.-A. (1967). *Histochemie*, **11**, 167.
Stirling, C. E., and Kinter, W. B. (1967). *J. Cell Biol.*, **35**, 585.

STUBBLEFIELD, E. (1965). *J. Cell Biol.*, **25**, 137.
STUMPF, W. E. (1968). In "Radioisotopes in Medicine, in vitro studies". Eds. Hayes, R. L., Godwitz, F. A., and Murphy, B. E. P. U.S. Atomic Energy Comm: p. 633.
STUMPF, W. E., and ROTH, L. J. (1966). *J. Histochem. Cytochem.*, **14**, 274.
STUMPF, W. E., and ROTH, L. J. (1967). *J. Histochem. Cytochem.*, **15**, 243.
STUMPF, W. E., and ROTH, L. J. (1968). "Advances in Tracer Methodology". Vol. 4. Ed. S. Rothchild, Phenum Press, New York, p. 122.
TAYLOR, J. H. (1956). "Physical Techniques in Biological Research". Ed. G. Oster and A. W. Pollister, Acad. Press, New York, Vol. **3**, p. 545.
THURSTON, J. M., and JOFTES, D. L. (1963). *Stain Technol.*, **38**, 231.
TOLLES, W. E. (1959). *Lab. Invest.*, **8**, 99.
TONNA, E. A., and CRONKITE, E. P. (1958). *Stain Technol.*, **33**, 255.
TREADWELL, A. DE G., LOW-BEER, B. V. A., FRIEDELL, H. L., and LAWRENCE, J. H. (1942). *Am. J. med. Sci.*, **204**, 521.
ULLBERG, S. (1954). *Acta radiol.*, Suppl. 118.
ULLBERG, S. (1958). In "Second U.N. Int. Conf. Peaceful Uses of Atomic Energy", **24**, 248.
VAN KLEEF, B. H. A., DE BOER, W. E., KOKKE, R., and WIKÉN, T. O. (1969). *Exp. Cell Res.*, **54**, 249.
VAN TUBERGEN, R. P. (1961). *J. biophys. biochem. Cytol.*, **9**, 219.
VON SALLMAN, L., EVANS, T. C., and DILLON, B. (1949). *Acta ophthal.*, **41**, 611.
WALKER, B. E., and LEBLOND, C. P. (1958). *Expl Cell Res.*, **14**, 510.
WARREN, S., and DIXON, F. J. (1948). *Am. J. med. Sci.*, **216**, 136.
WETTERDAL, B. (1958). *Acta Radiol.*, Suppl., 156.
WITTEN, V. H., and HOLMSTROM, V. (1953). *Lab. Invest.*, **2**, 368.
YOUNG, B. A., and KOPRIWA, B. M. (1964). *J. Histochem. Cytochem.*, **12**, 438.

CHAPTER 31

QUANTITATIVE HISTOCHEMISTRY

Measurement in histochemistry may be used to provide either analytical information, concerning the nature of some histochemical component of a tissue, or quantitative information on the precise amount of a given substance which is present. Examples of the former (analytical investigation) are the identification of biogenic amines by microspectrofluorimetry of aldehyde-induced fluorescence (Chapter 27), and the investigation of enzyme kinetics in tissue sections. Examples of quantitative investigations are the measurement of nuclear DNA, cytoplasmic glycogen, and the total dry mass of individual cells.

Quantitative histochemistry has grown enormously since the publication of the previous edition of this book (1960), and within the limits of this chapter only a fairly superficial review is possible. An attempt will be made at least to mention all the important aspects of the subject, to deal in somewhat greater detail with aspects not readily accessible elsewhere, and also to cover the subject matter of the chapter on "Other Physical Methods" in the 1960 edition.

The goal of quantitative histochemistry has been stated to be "the quantitative chemical analysis of the cell, its parts, and extracellular products, *in situ* and in the living state" (Glick, 1953). It must be admitted that this goal is very far from being achieved. Methods for the biochemical assay of substances of histochemical interest have become routine; histochemical localization of many substances is perhaps not so far behind; but the combination of localization and accurate quantitation still presents great difficulties except in specially favourable circumstances. It seems that an uncertainty principle is in operation, whereby perfect localization cannot usually be combined with perfect quantitation. Although combinations of imperfect localization with imperfect quantitation are possible, the accuracies of localization and quantitation have a kind of reciprocal relationship whereby any increase in one, beyond a certain limit, requires a corresponding decrease in the other.

One approach to quantitative histochemistry involves the isolation of the region of interest by microdissection, differential centrifugation or a similar technique, followed by chemical microanalysis. Such methods are capable of accurate quantitation, but with relatively imprecise localization—illustrating the uncertainty principle enunciated above. Much of this field has been pioneered by Lowry and his colleagues, who developed numerous methods for microanalysis of enzymes and other substances (e.g. Lowry and Hastings, 1942; Lowry *et al.*, 1954a, 1954b; Lowry, 1957) in material obtained by microdissection of freeze-dried tissue (Lowry, 1953). Many recent advances in this field were discussed in a series of papers edited by Dubach and Schmidt (1971). Methods of this type were ably reviewed by Glick (1963). These methods may best be described as microchemical, rather than histochemical, since the final result is not observed *in situ* through a microscope. Accordingly, despite the undoubted value of microchemical methods, no further mention will be made of them in this chapter.

The other approach may be called histochemical quantitation, in contrast to microchemical quantitation. This implies that quantitation is carried out *in situ*, using a microscope. Usually, the substance to be assayed is demonstrated by an appropriate histochemical reaction, leading to the generation or deposition of a reaction product (FRP, final reaction product) at the site of the original substance. The specimen is then examined under a microscope, the region of interest defined, and its image isolated by a diaphragm. The amount of FRP is then measured by optical means; similar measurements are made on a standard, and the results compared. In a few cases, direct optical quantitation of the substance of interest is possible, e.g. assay of nucleic acids by ultraviolet absorption, or determination of total dry mass by interferometry. The light or other electromagnetic radiation utilized in the microscope need not be limited to the visible: infra-red, ultra-violet, X-rays and electron beams have all been used.

Current methods for histochemical quantitation have been reviewed in an excellent series of papers edited by Wied (1966) and Wied and Bahr (1970). Other general publications of interest include those of Caspersson (1950, 1955), Eränkö (1955), and a series of papers in *Acta histochemica* Vol. 9 (1960). Problems relating to quantitative immunofluorescence are discussed in a series of papers edited by Holborow (1970), many of which are of interest in relation to microfluorimetry in general. For further information on the basic principles of optics and their application to biology, see Pollister (1955), Seliger and McElroy (1965), and Slayter (1970).

Optical Quantitation (Cytophotometry)

Optical quantitation of the FRP can in principle be carried out by employing any of the phenomena of absorption, retardation, reflection, and luminescence. In practice, at present, the following techniques of measurement are available:

(1) Absorptiometry (densitometry).
(2) Interferometry.
(3) Fluorimetry.
(4) Reflectiometry.

An additional method, that of point counting, is applicable mainly to quantitative autoradiography (see Chapter 30, p. 1217) and it can also be applied, manually or electronically, for determining relative areas. It is also combined with absorptiometry and interferometry in scanning methods.

Absorptiometry is suitable for quantitating any FRP which absorbs light (infra-red to ultra-violet) or X-rays. Some optical specificity can be obtained by measurement at a wavelength corresponding to an absorption peak of the FRP. Scanning or other special techniques are necessary to avoid distributional error in large inhomogeneous fields.

Interferometry is usually applied to measure the dry mass of a translucent FRP or whole cell. It cannot be used for an opaque FRP. There is no optical specificity in this method. A large inhomogeneous field can only be measured accurately by scanning. Interferometry can also be used for measurements of thickness and of refractive index.

Fluorimetry can, of course, only be applied to fluorescent FRPs. Optical specificity is high, since both the excitation and emission wavelengths can be selected. High sensitivity can be achieved. Measurements can be made of the fluorescence of particles too small to be resolved by the microscope. Distributional error is absent, so that large areas can be covered by a single measurement—which may save a lot of time. Opaque or thick preparations can be measured, using epi-illumination.

Reflectiometry can only be used to quantitate suitable particulate deposits, e.g. silver grains and formazan deposits. As in fluorimetry, there is no distributional error, and measurements can be made from non-resolvable points. Optical specificity is dependent upon the absence of reflective material other than the FRP. Accurate standardization of reflection techniques is at present difficult.

Optical Absorptiometry

Since the vast majority of histochemical reactions at present terminate in the deposition of a coloured or opaque FRP, the most obvious method for quantitation is that of measuring the optical density of the FRP at an appropriate wavelength. This section deals with absorptiometry (densitometry) using visible and ultra-violet light: X-rays are referred to on p. 1233.

The principles concerned with absorption measurements are set forth in a simple manner by Mendelsohn (1966) and by Ruthmann (1970), in other papers in the books edited by Wied (1966) and Wied and Bahr (1970), and more fully by Mellon (1950) and Lothian (1969) in two works devoted to the subject, and also by Sandritter (1958).

Absorptiometry depends upon the fact that the proportion of light absorbed by a layer of absorbing material is dependent upon the amount of the absorbing substance. The transmittance, T, of an absorbing object may be defined as the ratio of transmitted to incident light. The optical density (A) of the object is defined by the equation

$$A = -\log_{10} T$$

The Beer–Lambert law states that for monochromatic light, the optical density of a layer of absorbing material is proportional to the amount of absorbing material the light traverses, i.e.

$$A = kcl$$

where c is the concentration of material, l is the pathlength traversed by the light, and k is the absorptivity of the material—a characteristic of the absorbing material. The Beer–Lambert law is valid for many of the dyes used in histology and histochemistry, but deviations from it are frequent: many coloured substances absorb less light at high concentrations than they would if they obeyed the law. The Beer–Lambert law should never be expected to be valid in the case of stained sections (Eränkö, 1955).

One-wavelength, one-area method

In the simplest possible type of absorption cytophotometer, an image of the cell or part of the cell to be studied is allowed to fall on a photosensitive device (photomultiplier, etc.), and the electric current produced is measured with a galvanometer. A control reading is taken from an empty field, corresponding to 100 per cent transmission. Simple instruments of this type are unfortunately subject to a number of errors, of which the most important is distributional error.

In chemical analysis by absorptiometry, the solutions measured are usually clear and the substance is evenly distributed in the solvent. Problems of light refraction (at interfaces between regions of different refractive index) and of uneven distribution of absorbing substance therefore do not present themselves. Tissue sections and cell smears are another matter.

The subject of errors in absorption cytophotometry has been reviewed by Garcia (1962), Mayall and Mendelsohn (1970), and Ruthmann (1970). Errors due to refractive index differences have been discussed by Caspersson (1950), Glick, Engström and Malmström (1951), and Kiefer (1970). If the area which is illuminated is greater than the area from which measurement is made, an appreciable amount of stray light may be included in the measurement (the Schwarzchild-Villiger effect). This has been discussed by Naora (1951, 1952, 1955, 1958), Howling and Fitzgerald (1959), and Goldstein (1970). Errors due to the orientation of molecules have been considered by Commoner (1949) and by Commoner and Lipkin (1949). Errors due to light scattering in general have been considered by Barer (1952). The effect of path length variations due to too high numerical aperture of the objective and condenser have been studied by King (1959) and Pillat (1960). Errors due to thickness of the specimen exceeding the depth of the field of the objective can be reduced or eliminated by crushing the specimen by the Davies "crushing condenser" (Davies, Wilkins and Boddy, 1954; Atkin, 1970). The most important error in absorption measurements is the distributional error, which is due to the nature of the relationship between transmission and concentration of substance. Distributional error can only be avoided if the absorbing substance is homogeneously distributed, which rarely occurs in biological systems. Distributional errors can be eliminated or minimized in three main ways: by scanning, by two wavelength methods, and by photographic methods.

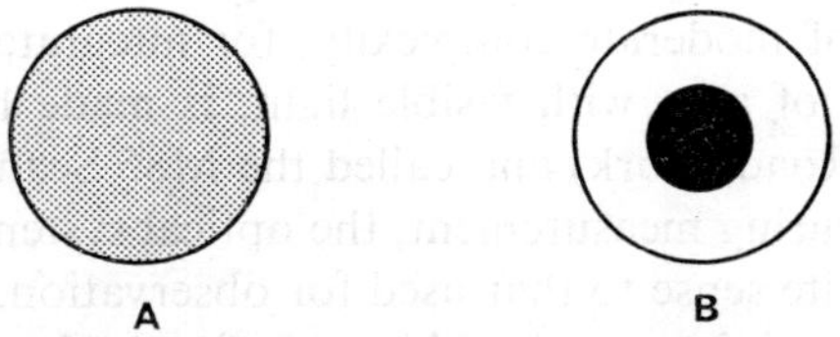

FIG. 260. Illustration of distributional error in absorbance measurements. (A) uniform field, unit area, with 50 per cent transmission, O.D. 0·3, integral of area × O.D. = 1 × 0·3 = 0·3; (B) same amount of material in 1/100 unit area, O.D. of object 30·0, O.D. of remainder of field 0·0; integral of area × O.D. = 30 × 1/100 + 0 × (99/100) = 0·3; however observed transmission = 99 per cent!

In the "plug" method, a one-wavelength measurement is made of the optical density of a region (such as the centre of a nucleus) selected to have a reasonably homogeneous distribution of chromophore. Unfortunately this method has only limited application.

If errors other than distributional error are minimized, e.g. by limiting the condenser aperture and the illuminated field, it may be possible to use the one-wavelength one-area method for approximate measurement of deposits of opaque particles, as in the Gomori-type enzyme methods (Scott, 1969).

Scanning Methods

In scanning microabsorptiometry, the field to be measured is divided up into a large number of small regions whose size is not larger than the size of the smallest inhomogeneity resolvable by the optical system employed. The transmission is measured in each of these small regions in turn, each measurement is converted to optical density, and the results integrated. Since thousands of individual measurements may be required from a single object (e.g. a cell, a nucleus, a nucleolus) the method is only practicable with an instrument which will scan the field and integrate the optical densities automatically.

The simplest integrating microdensitometer which is commercially available is that designed by Deeley (1955, 1960) and manufactured by Barr and Stroud. The operation of the instrument has been described by Atkin (1970). This instrument has been applied mainly to measurement of DNA using Feulgen staining (for references, see Atkin, 1970). Other instruments, designed by Caspersson and his group in Stockholm (Caspersson *et al.*, 1953; Caspersson and Lomakka, 1962, 1970) were the basis of the Zeiss UMSP1 (Carl

Zeiss, Oberkochen; Caspersson and Lomakka, 1970). This instrument is capable of performing integrating absorbance measurements with visible and ultra-violet light. Zeiss have recently produced an instrument which is mechanically simpler but controlled by a digital computer (Fig. 261). One important virtue of this instrument is that it is based on a good standard microscope, to which all normal accessories can be fitted. Both this feature and its computer "brain" give great versatility.

An instrument of moderate complexity, for integrated measurements of optical density and of area with visible light, is made by Vickers (Vickers Limited, Haxbury Road, York) and called the M85 Scanning Microdensitometer (Fig. 262). During measurement, the optical system of the microscope is used in the opposite sense to that used for observation, i.e. the objective is used as condenser to project a reduced image of a diaphragm on to the object. This image is made to scan the specimen by a prismatic system, measurements of optical density being made every 100 μsec. For setting up operations and for spot measurements, the image of the diaphragm can be manually positioned anywhere in the field of view. The actual measuring area, within which integration takes place, is determined by an indirect masking system. In principle, it is possible to make a special mask to fit any desired shape, such as a single cell in a tissue section. At the time of writing, the instrument has been applied to measurements of nuclear DNA and of acid phosphatase, and in a study of the effects of glare (Goldstein, 1970). A relatively simple instrument for integrated measurements in the visible region only was designed by Jansen (1961). Scanning methods have been widely used, mainly for the determination of nuclear DNA by absorptiometry of Feulgen-stained material (usually with the Deeley or Caspersson instrument) or by direct UV absorptiometry (with the Caspersson instrument). As a result of the more ready availability of suitable equipment for scanning absorptiometry, there is no doubt that its applications will multiply.

Two-wavelength methods

With these methods the absorption of the object is measured in light of two different wavelengths chosen so that, for the object concerned, the absorbance at the second wavelength is half that at the other wavelength. Any apparent change in this ratio in actual measurements will be a measure of the distributional error: the correction can be calculated. The two-wavelength method was introduced simultaneously and independently by Ornstein (1952) and Patau (1952). Patau's method, although essentially the same as Ornstein's, is the more convenient for manual operation and has been the most used of the two. Mendelsohn (1961) introduced a third method, in which the area of the photometric field is varied until the transmittances at the two wavelengths bear a definite ratio to one another. Microspectrophotometers designed specifically for the two-wavelength method were

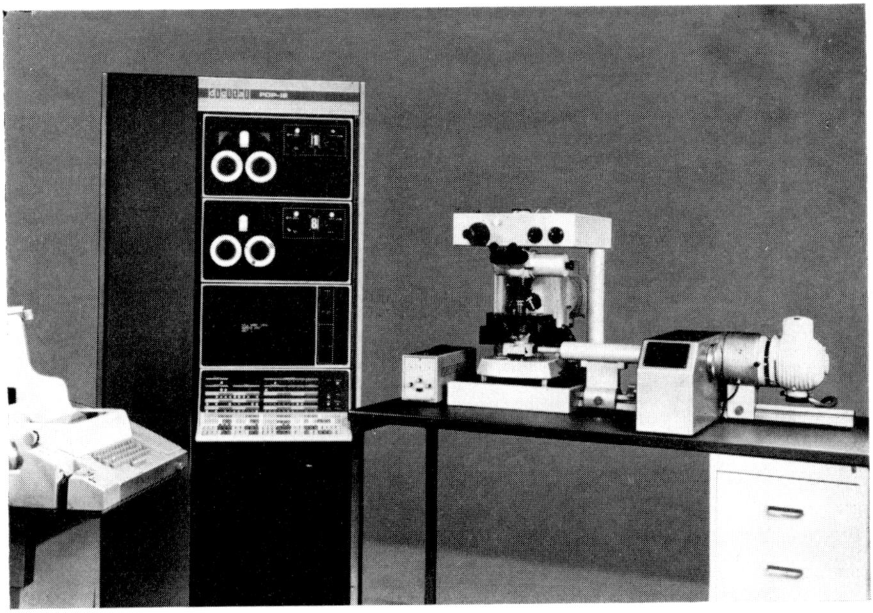

FIG. 261. The Zeiss Scanning Microscope Photometer. The main components (right to left) are: Arc lamphouse, monochromator, microscope with automatic scanning stage and photometer, digital computer (PDP 12) and Teletype. The computer controls the scanning stage and integrates density measurements in a specified area. The Teletype is used for giving instructions to the computer and for printing of results.

[*To face p.* 1230

FIG. 262. The Vickers M85 Scanning Microdensitometer.

described by van Duijn (1965) and Alpen (1966). Practical details of the technique are given by Garcia (1962) and by Rasch and Rasch (1970).

Photographic methods (Photographic colorimetry)

Caspersson (1936) used positive photographic images instead of the objects themselves for scanning densitometry, as a matter of convenience (the photographic image can be much larger). Ornstein (1952) extracted and measured the amount of silver in the positive photographic image, as a measure of the total amount of chromophore in the original object. This method takes advantage of the fact that the density of a photographic image is linearly proportional to the logarithm of the exposure, over a fairly wide range of

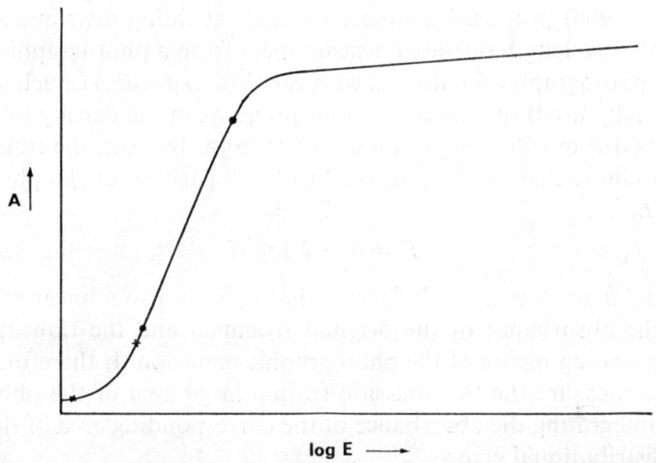

FIG. 263. The Hurter-Driffield curve of a photographic emulsion, showing the relationship between the logarithm of the exposure (E) and the resulting density (A). Photographic colorimetry by elution of silver or dye from a positive is only accurate if the linear part of the curve (between dots) is used; the Adams method employs a region on the tail of the curve (between the crosses).

density (Fig. 263). Accordingly the amount of silver present in a given area is proportional (within the linear range) to the optical density of the corresponding area in the original. The amount of silver, unlike optical density, is additive and therefore assay of the total amount of silver gives a measure of the integrated optical density of the original, i.e. of the total amount of chromophore present. This method was applied by Niemi (1958) to the determination of the haemoglobin content of erythrocytes. Ornstein, finding the classic gravimetric assay for silver unduly laborious and inaccurate, subsequently described the use of coloured photographic positives from which the dye could be eluted and measured in a colorimeter (Pollister and Ornstein, 1959). A similar conversion to a coloured positive was described by Mendelsohn (1958), who however measured the dye *in situ* by a two-wavelength method. Kelly (1962, 1966) described a modification employing commercially-developed colour films and Polacolor®. The dye-conversion and elution method was studied and described in detail by den Tonkelaar and van Duijn (1964a, b, c) who found that the reproducibility of the method, as tested on Feulgen-stained nuclei, ranged from 3 per cent to 5 per cent. By making a colour positive on paper, greatly enlarged, it is possible by this method to make measurements from irregular objects surrounded by other objects, the precise area from which measurements are to be made being determined by cutting out the required areas with scissors. The method has been applied by Gaillard, van Duijn and Schaberg (1968) to the assay of DNA in human chromosomes and parts of chromosomes, using as standard a cytochemical model system of DNA-containing cellulose films (van Duijn and Persijn, 1960).

Adams (1968) proposed a simple method, avoiding distributional error, using a one-wavelength one-area measurement from a photographic negative. For many photographic emulsions, with very low exposures (much lower than those normally used) the linear relationship between the density of the image and the logarithm of the exposure does not hold. Instead, the transmittance (T) of the film is linearly proportional to the logarithm of the photographic exposure E, i.e.

$$T = a - b \log E$$

where a and b are constants. It follows that there is also a linear relationship between the absorbance of the original specimen and the transmittance of the corresponding region of the photographic emulsion. It therefore becomes possible to measure the transmission from a large area of the photographic negative, integrating the absorbance of the corresponding area of the original without distributional error.

Wavelength

In principle, any wavelength of electromagnetic radiation can be used for absorption. For practical reasons, visible light is most commonly used, but ultra-violet, infra-red, and X-rays have also been used. Special equipment is required for non-visible light. In the case of ultra-violet and X-rays, a fluorescent screen can be used to make the image visible for orienting the specimen. Special optical systems are required for ultra-violet and infra-red. Reflecting optical systems are convenient, since these do not change focus with change of wavelength from visible to infra-red or ultra-violet. Ultra-violet optical systems require quartz instead of glass, as the latter is opaque to ultra-violet rays with a wavelength below about 350 nm. Photomultipliers with quartz windows are suitable for ultra-violet and visible ranges; a special cathode is required for sensitivity in the infra-red. The application of various wavelengths in cytochemical methods has been discussed by Caspersson (1955).

Microradiography

X-rays are absorbed by atoms, particularly those of high atomic weight. X-ray absorption can therefore be used to study tissue components containing atoms of relatively high molecular weight, e.g. calcium and iron, while the X-rays will pass comparatively freely through tissue components mainly composed of low-molecular weight atoms, e.g. hydrogen, carbon, nitrogen and oxygen. In principle, the extremely short wavelength of X-rays (c. 0·1 mm.) should make possible correspondingly fine resolution. In practice, resolution is limited by difficulties in obtaining an adequate magnifying system.

The principles of microradiography and X-ray analysis have been clearly set forth by Beneke (1966b) and in greater detail by Neumann (1958), Cosslett and Nixon (1960), Pattee, Cosslett and Nixon (1960), and Engström (1962, 1966). A bibliography for the period 1913–1963 has been compiled by Ely (1963).

Three basic techniques are at present available: contact microradiography, projection microradiography, and reflection microradiography (or true X-ray microscopy). In *contact microradiography*, the specimen is placed directly on a photographic plate. After exposure and development, the plate is examined with a normal microscope. The secondary (optical) enlargement of the microradiogram is limited by the grain of the photographic emulsion. In *Projection microradiography*, a divergent beam of X-rays from a small source is passed through the specimen, an enlarged image of which is projected onto a photographic plate at a distance. In "*reflection microradiography*" or true X-ray microscopy, an enlarged X-ray image is produced by a system of metal mirrors reflecting X-rays at a glancing angle. A very high resolution is theoretically possible by this method, but more work needs to be done on the practical application of this technique.

The most obvious applications of microradiography are to the investigation of bone and similar structures. Quantitative X-ray absorptiometry can be used in the measurement of total dry mass, providing a useful check on results obtained by interferometry and UV absorption.

X-rays can be used for identifying the atoms present, using X-ray microdiffraction and X-ray emission studies. The latter is the basis of the electron probe microanalyser (Castaing and Guinier, 1949) in which a beam of electrons is shone on to the region to be studied, X-rays being generated with wavelengths dependent upon the elements present in the irradiated area. The method has been reviewed by Hale (1966). An example of the application of the method is the study of copper deposits in Descemet's membrane in Wilson's disease, by Tousimis and Adler (1963). (See also Chapter 32, p. 1293.)

Microfluorimetry

For quantitation, microfluorimetry has several advantages as compared to microdensitometric methods: there is no distributional error, so that scanning or two-wavelength techniques are not necessary; great sensitivity can be achieved; specificity can be obtained by selection of appropriate wavelengths both for excitation and for measurement of emission; and thick or opaque objects can be examined by epi-illumination. It is also possible to measure fluorescence from particles too small to be resolved by the microscope. Despite these advantages, practical application of fluorimetry has so far been rather limited. This has probably been due in part to the fact that equipment for fluorescence microscopy is not as widely available as it should be, and partly to difficulties which have been encountered in the development of adequate standards.

Theoretical aspects of microfluorimetry have been discussed particularly by Rigler (1966), Böhm and Sprenger (1968), McCarthy and Moyer (1970), and Ruch (1970). Difluorescence (polarized fluorescence) is dealt with in the section on polarization (p. 1243).

Instrumentation for microfluorimetry

By analogy with fluorimeters and spectrofluorimeters, instruments for microfluorimetry fall into two classes: those using only filters for monochromation, and those with at least one prism or grating monochromator. It is convenient to refer to these two types as microfluorimeters and microspectrofluorimeters respectively. The practical difference between the two is that monochromators offer advantages of continuously variable wavelength

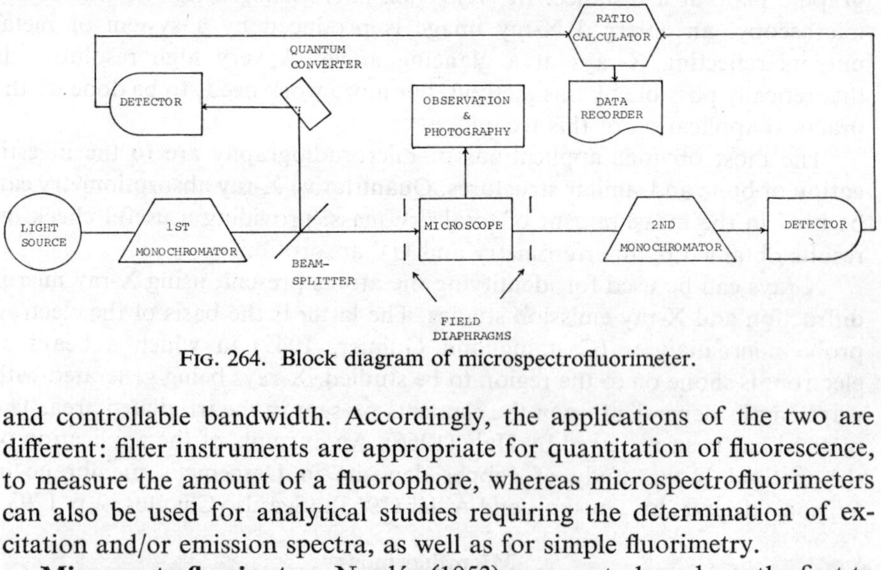

FIG. 264. Block diagram of microspectrofluorimeter.

and controllable bandwidth. Accordingly, the applications of the two are different: filter instruments are appropriate for quantitation of fluorescence, to measure the amount of a fluorophore, whereas microspectrofluorimeters can also be used for analytical studies requiring the determination of excitation and/or emission spectra, as well as for simple fluorimetry.

Microspectrofluorimeters. Nordén (1953) appears to have been the first to describe the construction and use of a microspectrofluorimeter, and his discussion of the principles involved is still worth reading. Other instruments have been described by several authors, notably Olson (1960), West (1965), Caspersson, Lomakka and Rigler (1965), Runge (1966), Thieme (1966), Pearse and Rost (1969), Parker (1969) and Van Orden (1970). The only commercially available instrument of this type at present appears to be the Leitz microspectrograph (Ruch, 1960) as modified for fluorescence (Thaer, 1966b; Björklund, Ehinger and Falck, 1968; Pearse and Rost, 1969; Sprenger and Böhm, 1971a).

The essential instrumental requirements for a microfluorimeter are as follows. (1) A light source of appropriate wavelength, (2) a good microscope with epi-illumination, (3) Field diaphragms to limit the area illuminated and to delimit the measuring field, (4) either a reference channel or a very stable light source (or both), and (5) a sufficiently sensitive photometer system.

Tungsten filament lamps are only suitable for use as light sources for excitation at wavelengths longer than about 450 nm, e.g. in the green for excitation of Pararosaniline-Schiff and FITC. A high degree of short-term electrical and mechanical stability can be achieved with such a lamp. The xenon arc lamp is at present the source of first choice for most purposes. These lamps are suitable for excitation at all visible wavelengths and the ultra-violet down to about 250 nm. Electric stability is achieved by stabilizing the power supply: this should ideally be accomplished by a constant-current DC supply. Like all arcs, xenon arcs tend to wander, although modern design of the electrode shape has reduced this tendency and mercury arc lamps are convenient for those cases where excitation at a mercury line is appropriate (e.g. noradrenalin FIF at 406 nm), but the spatial stability of mercury arcs is much inferior to that of xenon arc lamps. Other arc lamps, particularly cadmium and zinc, may prove useful.

It should hardly be necessary to state that a microfluorimeter should be based on a good microscope: the stand must be rigid, and there should be adequate provision for examining and photographing the object by transmitted light, or phase-contrast, as well as fluorescence. Epi-illumination (see Chapter 29, p. 1196, and Fig. 265) is practically essential for microfluorimetry, for the following reasons:

(1) Error due to reabsorption is reduced (Rigler, 1966).

(2) Correct focusing of the objective guarantees correct focusing of the condenser, since the two are identical—this aids standardization.

(3) Only the area being looked at need be irradiated, even for scanning.

(4) With high-power objectives such as are usually used for fluorimetry, epi-illumination can give brighter illumination—this is useful if a monochromator is used in the excitation system.

(5) The simultaneous use of phase-contrast (by dia-illumination) and fluorescence (by epi-illumination, with a restricted field) is facilitated, so that a small measuring field (fluorescence) can be accurately positioned on the desired object.

A reference channel, continuously or intermittently monitoring the in-

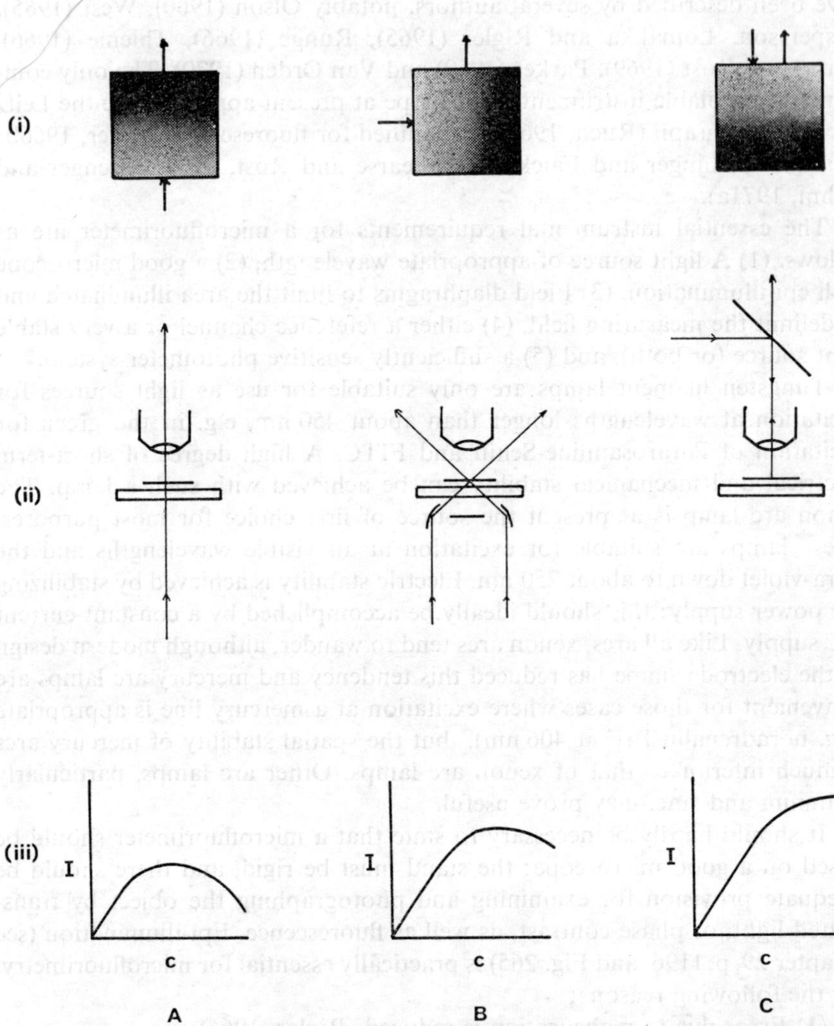

FIG. 265. The relationships between concentration of a fluorophore and the intensity of the measured fluorescence as observed with three optical arrangements. The first row (i) shows measurement with a cuvette, the second row (ii) shows how similar optical systems are obtained in a microfluorimeter, and the third row (iii) shows the relationship between concentration (c) and apparent fluorescence intensity (I). The first column (A) shows longitudinal observation with the excitation passing through the object, i.e. dia-illumination (iiA). The second column (B) shows oblique illumination, approximately equivalent to illumination by a dark-ground condenser (iiB). The third column (C) shows measurement from the same side as is illuminated, i.e. epi-illumination. The advantage of epi-illumination over dia-illumination and dark-ground illumination is obvious. (After Udenfriend (1962) and Rigler (1966).)

tensity of the light used for excitation, can be used to compensate for changes due to instability of the light source, change of wavelength, change of monochromator setting, etc. Such a reference channel has been incorporated into several instruments which have been described in the literature, e.g. by Mellors, Glassman and Papanicolaou (1952), Chance and Legallais (1959), Olson (1960), Ruch and Bosshard (1963), Caspersson, Lomakka and Rigler (1965), Gijzel (1966), Goldman (1967), Björklund, Ehinger and Falck (1968), Pearse and Rost (1969), and Parker (1969). Unfortunately this necessary feature is still not generally available commercially, although Leitz have exhibited a prototype. If a quantum converter, e.g. a cuvette of Rhodamine B solution, is used in the reference channel, it is not necessary to correct for the spectral sensitivity of the detector in the reference channel, so that corrected excitation spectra can be measured directly (Ritzén, 1967). The light intensity in the main and reference channels can be measured with either the same photomultiplier using a beam chopper (Caspersson, Lomakka and Rigler, 1965) or a separate one: in the latter case, measurements can be made in both channels simultaneously. The ratio between the two channels can be computed, depending on the nature of the output from the detector systems, by either a digital computer, a ratio chart recorder, or an analogue computer. In our experience, the first is the most accurate and the last the most convenient.

The microspectrofluorimeter shown in Fig. 266 (Rost and Pearse, 1969) is designed primarily for the measurement of fluorescence excitation and emission spectra. It is also possible to measure transmission and reflection. Briefly, the most usual arrangement of the instrument for fluorescence measurements is as follows. The light source is a xenon arc lamp (XBO 150, not visible in the illustration). The light passes through a Leitz monochromator and then through a supplementary glass filter to remove light of unwanted wavelengths, e.g. due to fluorescence of dust on optical surfaces of the monochromator. A portion of the light beam is split off by a quartz beam-splitter, and enters a reference channel which includes a Rhodamine B quantum converter (Ritzén, 1967) and a red-sensitive photomultiplier (EMI 9558). The main beam passes through field diaphragm and a shutter into an epi-illuminator, thence down through the objective on to the specimen. For excitation in the region above 360 nm, a Leitz ×95 Fluorite oil objective is usually used, otherwise a Zeiss Ultrafluar. The fluorescence of the specimen is examined with the microscope. Alternative means of illumination by a substage condenser are available, e.g. for phase-contrast. Light from a selected area of the field passes into what is in effect a second monochromator, and thence to a photomultiplier (EMI 9558QA). It is possible either to image a selected part of the field into the entrance slit of the second monochromator, or alternatively to image the exit pupil of the objective on to the entrance slit of the second monochromator; the latter method is usually preferable except with long thin specimens such as fibres.

For the simultaneous recording of fluorescence emission and of transmission at the excitation wavelength, using the epi-illumination system, light passing through the specimen is collected by the substage condenser (acting as an objective) and measured by a third photomultiplier. This arrangement has proved useful for studies of fading.

Output from the main and reference photomultipliers is measured by photon counting, i.e. by counting the current pulses in the photomultiplier circuits each due to the incidence of one photon on the photomultiplier cathode (similar pulses originating within the photomultiplier constitute the "dark current"). Pulses are counted for a preset time, usually 1 second; the result is displayed digitally and can be printed automatically on the Teletype machine (see Fig. 266). It is possible to trigger the photon counting by the opening of the shutter in the excitation beam; this means that, if the desired area of the specimen can be located by phase-contrast, a fluorescence intensity measurement can be made with only 200 msec. of irradiation, this being the shortest counting time available with this equipment. For dynamic studies, the photon counting is triggered at preset intervals by a digital clock, both the measurements and the time being recorded automatically on the Teletype.

In addition to the digital system, the count rate is displayed on meters in each channel, and an analogue signal proportional to the count is available; the ratio of these (after subtraction of dark current) is calculated by a small analogue computing device, the output of which can be used to operate a chart recorder or oscilloscope.

The photon counting system has several advantages over analogue systems: it is more sensitive, quicker in making a measurement, and allows the measurement conditions to be more easily known and adjusted to suit the conditions of each experiment. For further discussion, see Kemplay (1962), Barenboim, Domanskii and Turoverov (1969), and Rost and Pearse (1968).

The following particularly good features are to be noted in instruments which have been described in the literature:

(1) A light source with a substantially continuous spectrum, e.g. a xenon arc lamp (e.g. Caspersson, Lomakka and Rigler, 1965).

(2) A reference channel (Mellors, Glassman and Papanicolaou, 1952; Chance and Legallais, 1959; Olson, 1960; Ruch and Bosshard, 1963; Caspersson, Lomakka and Rigler, 1965; Gijzel, 1966; Goldman, 1967; Björklund, Ehinger and Falck, 1968; Pearse and Rost, 1969; Parker, 1969).

(3) Identical monochromators for excitation and emission monochromation (Caspersson, Lomakka and Rigler, 1965; Thieme, 1966) and synchronous motor drive (Thieme, 1966).

(4) Wavelength analogue output from voltage divider attached to each monochromator (Thieme, 1966).

(5) Facility for rapid spectral scanning (Olson, 1960; West, Loeser and Schoenberger, 1960; Thaer, 1966b; Björklund, Ehinger and Falck, 1968; Pearse and Rost, 1969).

FIG. 266. A Leitz microspectrofluorimeter (Pearse and Rost, 1969). The main component parts (left to right) are: Rack containing electronic system; a Teletype machine on which the data is printed automatically or by hand, producing also a punched paper tape for input to a computer; the Leitz microspectrograph, with monochromator, microscope, and spectrophotometer; and a chart recorder.

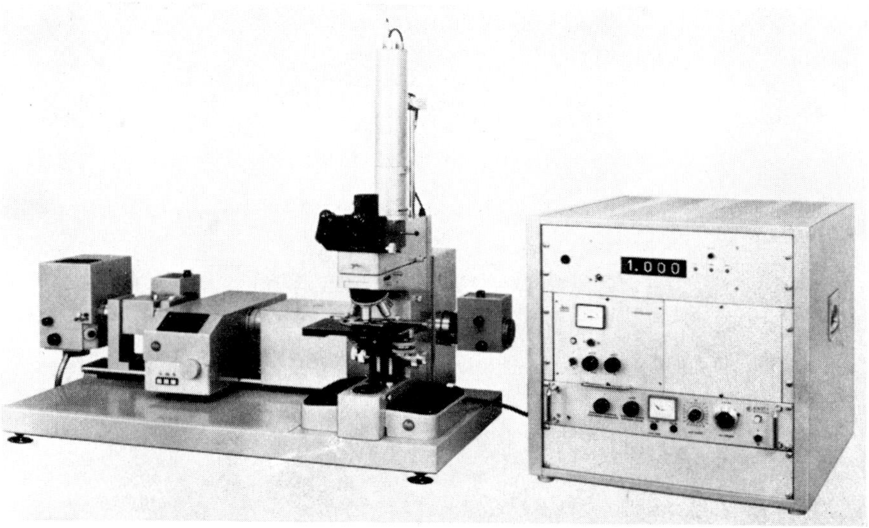

FIG. 267. A microfluorimeter by E. Leitz (Wetzlar), using a xenon arc lamp, grating monochromator, Orthoplan fluorescence microscope, and MPV photometer. Either dia- or epi-illumination can be selected by a mirror system at the rear of the microscope.

[To face p. 1238

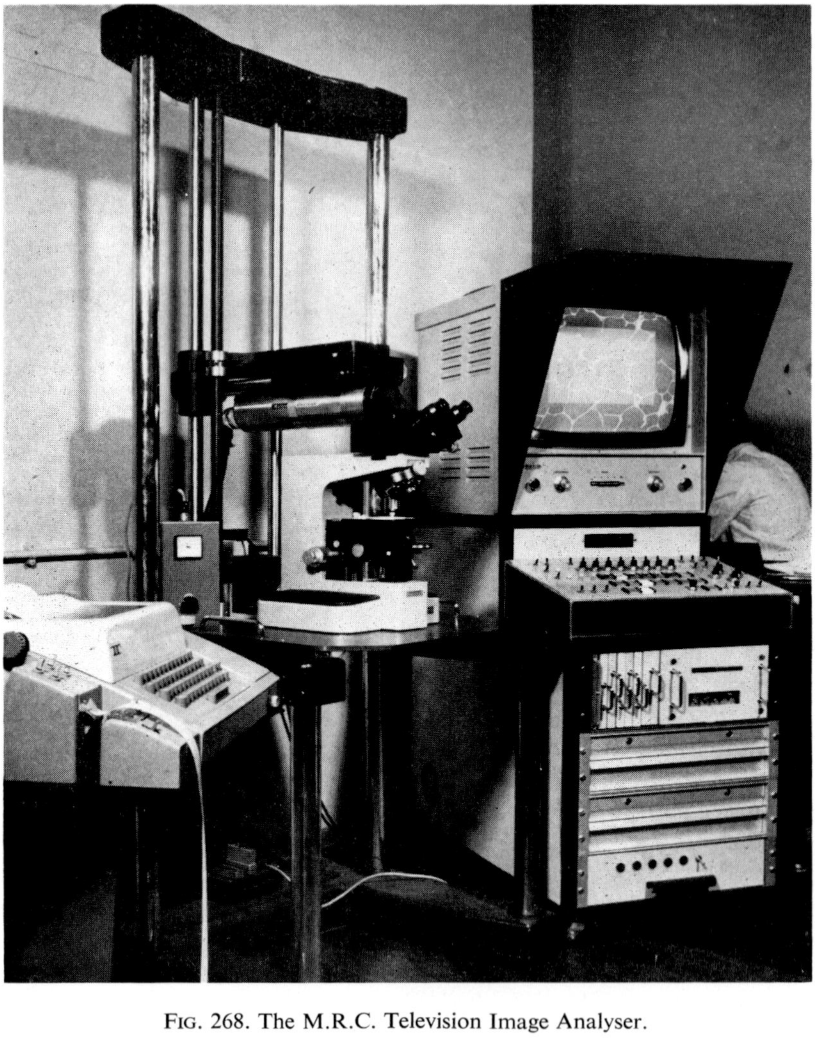

FIG. 268. The M.R.C. Television Image Analyser.

(6) Epi-illumination for excitation of fluorescence (Mellors and Silver, 1951; Böhm and Sprenger, 1968; Pearse and Rost, 1969; Parker, 1969).

(7) Photo-electric detection with electronic amplification (e.g. Mellors, Glassman and Papanicolaou, 1952), and photon counting (Pearse and Rost, 1969).

(8) Light choppers for time resolution of luminescence (Parker, 1969).

Simpler microfluorimeters. The same general principles apply to simpler instruments as to microspectrofluorimeters, the differences being mainly that filters are normally used instead of prism or grating monochromators. Such simpler instruments are more suitable for routine measurements of fluorescence intensity, e.g. for quantitation of immunofluorescence, while the more complex microspectrofluorimeters are required for analytical studies involving the measurement of excitation and emission spectra. Microfluorimeters employing filters have been reviewed by Thaer (1966a). At the time of writing, suitable instruments are available from several manufacturers. For excitation, an interference line filter can be used; Leitz (Wetzlar) and Zeiss (Oberkochen) can supply a grating or prism monochromator in the excitation system (see Fig. 267). For monochromation of the emission, an appropriate interference line filter or a conventional barrier filter is probably to be preferred for most purposes, but continuous interference filters are also available.

Standards for microfluorimetry

Microfluorimetry has to some extent been bedevilled by the absence, in most instrumentation, of a satisfactory means for measuring the intensity of the incident light. In absorptiometry, measurements are made of the incident and transmitted light intensities, and the ratio (per cent transmission) calculated; similarly, in fluorimetry the intensities of the incident and fluorescent light should be measured and the ratio calculated. There is unfortunately no name for the quantity so obtained, which should ideally be expressed as the ratio of emitted quanta to incident quanta at specified wavelengths.

Microfluorimetry has therefore come to rely exclusively on the comparison of the fluorescences of the unknown and of a standard, under stable conditions. The essential requirements for fluorescence standards are: (1) the same fluorophore as that to be measured, or at least a fluorophore with identical or very similar fluorescence excitation and emission spectra; (2) reproducibility; and (3) sufficient stability. It must be noted that the fluorescence characteristics of a fluorochrome bound to a protein are not the same as those of the unbound fluorochrome, so that a solution of a fluorochrome may not be a suitable standard for bound fluorochrome. Most early efforts to standardize fluorimetry were based on the use of a piece of fluorescent uranium glass (Rigler, 1966; Ruch, 1970) and this method is still in use. Unfortunately the fluorescence of uranium glass is not reproducible and is subject to fading. Ploem (1970b) found uranium glass adequate for short-term standardization (hours to days), and introduced the use of microdroplets of fluorochrome solution in a silicone oil as reproducible long-term standards: the diameter of the droplets is measured with an eyepiece micrometer. Sernetz and Thaer (1970) proposed the employment of standard fluorochrome solutions in capillary tubes mounted in a medium of the same refractive index as the tubes (e.g. DPX for glass tubes, glycerol for quartz). The internal diameter of the tubes is measured by interferometry, and the fluorescence is measured from a part of the tube, the length of which is measured by an eyepiece micrometer. This technique is likely to be of great value. Jongsma, Hijmans and Ploem (1971) compared uranyl glass standards with standard fluorescein solutions in microdroplets and capillary tubes, in relation to quantitation of immunofluorescence. They found that, of these alternatives, fluorescein solution (see Appendix 31) in capillary tubes was to be preferred. They proposed as a standard fluorescein unit (SFU), 10^{-3} of the fluorescence of 1 picolitre of a 50 μM fluorescein solution at pH 8·5. This pH was chosen because the fluorescence is then maximal and the influence of small pH changes is negligible. Their original paper should be consulted for a discussion of the effects of temperature, fading, and other causes of error.

Analytical microspectrofluorimetry

Determination of fluorescence excitation and emission spectra of a known or unknown fluorophore is of value in three types of investigation: the identification of an unknown fluorophore by its fluorescence characteristics, for investigating the conditions under which the fluorophore is acting (e.g. as a probe for the conformation of a substrate), and for assessing the ideal conditions for microscopy of a known fluorophore from the excitation and emission maxima.

Excitation and emission spectra can be determined with the microspectrofluorimeters, employing monochromators in both excitation and emission paths, to which reference has already been made (p. 1234). If results are to be comparable from one laboratory to another, or even between different instruments in the same laboratory, all instrumental spectra should be corrected, e.g. by the use of a standard lamp.

Excitation spectra should be plotted as relative quanta (quanta emitted per quanta incident) as ordinate; the abcissa should be a scale linear in quantum energy or wavenumber (the effect is the same) (Chapman *et al.*, 1965). It is usual to plot blue on the left, red on the right. The use of a quantum converter in the reference channel containing a strong solution of Rhodamine B (Ritzén, 1967) is becoming accepted (Björklund, Ehinger and Falck, 1968; Pearse and Rost, 1969) and may well become standard in equipment of this type. In this case, corrections still have to be made for the reflectivity of the beam-splitter, if any, giving rise to the reference channel, and for the characteristics of the epi-illuminator (especially if this is dichroic).

Emission spectra should be plotted as quanta per unit quantum energy interval (quanta per unit wavenumber interval) (Chapman *et al.*, 1965), against quantum energy or wavenumber.

Applications

Quantitation. Microfluorimetry is obviously of great significance in relation to immunofluorescence. The subject has been discussed by Goldman (1960, 1967), Mansberg and Kusnetz (1966), Pittman *et al.* (1967), and in a series of papers originally presented at international conferences in Florence and London and edited by Holborow (1970). Difficulties include standardization of the immunological reagents and of the fluorochroming; instrumental difficulties seem now to have been largely solved.

The use of a fluorescent Schiff reagent (see Chapter 29, p. 1182) in the Feulgen-Schiff and periodic acid-Schiff reactions permits microfluorimetric quantitation of DNA (Bosshard, 1964; Ruch, 1966a; Böhm and Sprenger, 1968; Böhm, Sprenger and Sandritter, 1970) and of PAS positive material such as glycogen (Yataganas, Gahrton and Thorell, 1969). The selection of a suitable fluorescent pseudo-Schiff reagent has been investigated by Prenna and Bianchi (1964b), Prenna, Zanotti and Bianchi (1964), Ruch (1966), Yataganas, Gahrton and Thorell (1969), Böhm and Sprenger (1968), and Böhm, Sprenger and Sandritter (1970). Present opinion appears to favour either the classical (pararosaniline-)Schiff reagent used as a fluorochrome, excited in the green and emitting in the red (Böhm and Sprenger, 1968) or Acriflavin (Böhm and Sprenger, 1968; Sprenger and Böhm, 1971b), or bis-aminophenyl-oxidiazole (BAO) employed by Ruch (1968, 1970). This dye is excited at 365 nm and emits in the blue.

Microfluorimetry has also been applied to the quantitation of arginine (Rosselet, 1967; Bock and Schlüter, 1971), lysine (Rosselet and Ruch, 1968), and nuclear histones and total protein (Ruch, 1970). Analytical dynamic studies by microfluorimetry are referred to below (p. 1242).

Analytical studies. Nordén (1953) applied microspectrofluorimetry to the identification of benzpyrene taken up by tissues, and laid down many of the principles to be observed in such investigations. Several groups of workers have applied microfluorimetry to the identification of biogenic amines (see Chapter 27). Spectral shifts in the fluorescence of fluorochromes bound to various substrates may give useful information about dye-substrate binding, from which information can be deduced concerning the substrate itself. The most obvious application of this principle is the measurement of fluorescence metachromasia (Rigler, 1966; Bussolati, Rost and Pearse, 1969; Rost and Maunder, 1971; see Chapter 29, p. 1185).

Kinetic studies

Fluorescence methods are particularly suitable for kinetic studies in living tissue or unfixed sections, because of the high sensitivity which can be achieved. The deleterious effects of ultra-violet irradiation must of course be minimized by using the least possible irradiation (both in intensity and duration).

Chance and co-workers have developed techniques for the study of oxidative enzyme systems in living cells by measuring changes in the fluorescence of reduced pyridine nucleotide (NADH) bound to protein, using a differential microfluorimeter (Chance and Legallais, 1959; Kohen *et al.*, 1967; Kohen, Kohen and Thorell, 1969) which was subsequently equipped with a flowchamber (Kohen, 1963) and apparatus for direct or microelectrophoretic injection (Kohen and Legallais, 1965; Kohen, Legallais and Kohen, 1966). For details of methods and applications see Chance and Thorell (1959), Chance (1962), Kohen and Kohen (1966), and Kohen, Kohen and Thorell (1968, 1969).

Rotman and Papermaster (1966) observed that living cells in a medium containing certain fluorescein esters of fatty acids became fluorescent. This was due to the non-polar ester passing through the cell membrane, and becoming hydrolysed by intracellular esterases to produce free fluorescein, which, being polar, could not readily pass the cell membrane and therefore accumulated in the cell. This phenomenon allows kinetic investigations of the properties of cell membranes (Rotman and Papermaster, 1966; Sernetz, 1969).

The use of cryostat sections of isolated living cells for enzyme kinetic studies has several potential advantages compared to the use of an enzyme in solution. For example, alkaline phosphatase (EC 3.1.3.1) can only be obtained in solution by vigorous procedures such as autolysis or extraction with *n*-butanol, since it is firmly bound to structural elements of cells in which it occurs. These extractive procedures are potentially disruptive of enzyme structure and therefore it may be doubted whether characteristics such as substrate specificity, Michaelis constants, pH optimum, or responses to activators and inhibitors when determined with the soluble enzyme truly reflect the properties of the enzyme within the living cell. Jeffree (1970) investigated the effects of pH and inhibitors upon the phosphatases of osteoclasts using a semi-quantitative azo-dye technique. Täljedal (1970) studied the apparent Km of glucose-6-phosphatase in cryostat sections, by microfluorimetry of reduced pyridine nucleotide. Rost, Nägel and Moss (1970), in preliminary experiments using α-naphthyl orthophosphate as a fluorogenic substrate for alkaline phosphatase (Moss, 1960), in cryostat sections mounted in a flow chamber in the M.R.C. Microspectrofluorimeter (Pearse and Rost, 1969), were able to obtain an apparent Km value of 1·1 mM at pH 10·1, in good agreement with the value of 0·8 mM at the same pH for the soluble enzyme. The pH dependence of the activity of the enzyme in tissue sections was also similar to that of the enzyme in solution.

Reflex Microscopy

Microscopy of biological materials by reflected light has hitherto been but little used. The equipment required is fairly simple, namely a microscope provided with an epi-illuminator. The technique can be used for the demonstration of reflecting crystals (e.g. in fish scales, and formazan deposits) and of silver grains in autoradiographs. Quantitative reflex microscopy (microreflectiometry) has been applied to autoradiography as an alternative to silver grain counting (Rogers, 1961; Dörmer *et al.*, 1966; Dörmer, 1967; Rogers, 1967). Reflex microscopy could possibly be used in periodic-acid-Schiff and Feulgen-Schiff techniques by replacing the Schiff reagent by a silver method (e.g. hexamine silver).

Polarization: Birefringence, Dichroism and Difluorescence

Polarization microscopy may give information on the submicroscopic arrangement or molecular structure of biological and other material. The properties of dichroism, birefringence, and difluorescence (described below) are to be found in optically inhomogeneous objects, in which the molecules or other submicroscopic units tend to a definite orientation rather than a random arrangement. The polarization microscope, as a means of observing orientation of tissue elements not directly visible in an optical microscope has to some extent been supplanted by the electron microscope, of which the greater resolution may enable direct visualization of detail, the presence of which could only be inferred by polarization studies. However, the polarization microscope is still a valuable tool for the biologist, particularly because of the possibility of examining fresh unfixed tissue or living cells and because of its simplicity and cheapness compared to the electron microscope.

Dichroism is the property of preferentially absorbing light polarized in a certain direction, light polarized at right angles to this direction being transmitted to a greater extent. Some crystals and amyloid are dichroic. The best known example of dichroism is Polaroid material, which consists of sheets of transparent plastic containing dichroic crystals in parallel alignment. Light which has passed through a sheet of Polaroid is substantially plane-polarized, light oscillating in other planes have been absorbed. The wavelengths at which dichroism is present depends upon the material—e.g. Polaroid is dichroic in the visible but not in the ultra-violet, while amyloid is dichroic only to green light. Dichroism is a particularly sensitive index of submicroscopic orientation, and may reveal a regular array of molecules which escapes polarization studies. Like difluorescence, but in contrast to birefringence, dichroism can be specific for a particular substance.

Dichroism can be observed in a microscope similar to a conventional polarizing microscope, but employing only either a polarizer or an analyser but not both; since the dichroic substance itself serves as analyser or polarizer depending upon the optical arrangement. In most cases, it is best to use a polarizer, since a binocular microscope can then be used. Either the polarizer or the object (if on a rotating stage) should be rotated through at least 90° while watching for changes in intensity of light transmitted through the region of the specimen thought to be dichroic. The detection of dichroism in microscopic objects has been discussed by Goldstein (1969). Ruch (1951) has described an apparatus for measuring ultra-violet dichroism of cellular structures.

Some applications of dichroism have already been described in relation to amyloid (Chapter 11, p. 386). For other applications, particularly to botany, see Ruch (1966b, c).

Birefringence is the possession of different refractive indices with respect to light polarized in different directions. Birefringence is shown by many crystals and some biological materials, particularly connective tissue fibres and the anisotropic (A) bands of striated muscle. Birefringent crystals are used in the construction of Nicol and other prisms used in the production of plane-polarized light (see Slayter, 1970). Compared to Polaroid, these prisms are more expensive, particularly in large sizes, but give more complete polarization, greater light transmission, and are available for a wider range of wavelengths.

Birefringent objects in a microscope field can be observed by their property of rotating the plane of polarization. In a simple polarizing microscope, a polarizer (e.g. a Polaroid disc) is placed between the light source and the specimen, so that the specimen is illuminated by plane polarized light. In practice, the best place for the polarizer is usually in a filter-holder immediately beneath the condenser. A second polarizer, in this case referred to as the analyser, is placed somewhere between the object and the observer's eye. If a binocular or inclined monocular microscope tube is used, for best results the analyser should be placed in the body of the microscope below the prisms of the binocular head, since reflections within the prisms will affect the polarization. The polarizer and analyser should be so placed that their polarization axes are at right angles, so that no light passes through the analyser and the microscope field appears black. Any birefringent object, if suitably oriented, will rotate the plane of the polarized light produced by the polarizer, and therefore this will appear bright against the dark background. It is necessary to be able to rotate the specimen (e.g. on a rotating stage) in order to be sure of observing birefringence, if present.

Birefringence is of two basic types: intrinsic, and form. Intrinsic birefringence is due to the molecular arrangement of the substance. Form birefringence is due to parallel orientation of sub-units of the object, e.g. in the case of Congo red staining of amyloid to the orientation of molecules bound on to bundles of parallel fibrils. Form birefringence can be induced by stress, stretching or compression resulting in the long axes of constituents becoming more or less parallel. For further details, see Oster (1955). Micropolarimetry applied to cytochemistry has been reviewed by West (1970).

Polarization microscopy in relation to amyloid has already been referred to in Chapter 11, p. 387. For other applications and further details, the reader is referred to Oster (1955), Schmidt (1958), Ruch (1966b, c), Dayan (1967) and Ruthmann (1970).

Difluorescence is the emission of polarized fluorescence. Like dichroism, it may be specific to a particular substance. Its observation requires a fluorescence microscope equipped with an analyser (Ruch, 1957, 1962). The phenomenon has been used in the study of chloroplasts (Ruch, 1957, 1966b), and may in the future find application to the investigation of proteins (cf. Konev, 1967).

Phase-Contrast and Interference Microscopy

Phase-contrast and interference microscopes render transparent objects, such as cells, visible by converting phase changes in the light passing through them into changes of wave amplitude. Regions of differing refractive index are therefore made visible as regions of differing brightness. These forms of microscopy can be used to examine living and unstained material. The principles and applications of these techniques have been ably reviewed by Ross (1967); a more mathematical discussion is given by Osterberg (1955).

Phase-contrast

Phase-contrast microscopy is sometimes useful for scanning a specimen to locate a desired area to be investigated by ultra-violet absorption or fluorimetry, the object being to minimize irradiation of the object with ultra-violet light. In such circumstances it must be remembered that if the object to be studied is visible by phase-contrast, its refractive index must differ from that of the medium and this may affect the accuracy of other measurements. For observation and photography, it is particularly easy to combine fluorescence excitation by epi-illumination with phase-contrast by transmitted light: the colour of the latter can be selected (by filters) to contrast with the colour of fluorescence.

In selecting equipment for phase-contrast, it may be useful to note that phase-contrast objectives are of two kinds: positive contrast, in which objects of high refractive index appear dark against a grey background, and negative contrast in which objects of high refractive index appear bright against a dark background. At least one manufacturer (Leitz) produces phase contrast objectives with both high-contrast and low-contrast plates: the former is more common, gives a greater contrast, and is suitable for low-contrast objects; the latter will cope with a wider range of contrasts and have smaller halos (Ross, 1967). For further details, the reviews of Barer (1959, 1966), Ross (1967), and Ruthmann (1970) should be consulted. Sources of error in the measurement of dry-mass concentration in tissue sections have been discussed by Galjaard (1967).

Interference microscopy

Interference microscopy differs from phase-contrast microscopy only in that, whereas in phase-contrast light refracted by the object is made to interfere with light which has passed through the same field, in interference microscopy the light which has passed through the object is made to interfere with a beam which has passed along a different path. The principles are explained by Hale (1958), Barer (1959, 1966), Krug, Rienitz and Schultz (1961), Richards (1961), Beneke (1966a), Ross (1967), and Ruthmann (1970). For details of instrumentation, see also Hale (1958), Dyson (1964), Lomakka (1965), Allen and Brault (1966), Smith (1967), and Carlson (1970).

Both phase-contrast and interference microscopy can be used for determining the refractive index of cellular components, by successive immersion in media of various refractive indices until one is found in which the object in question cannot be seen, i.e. its refractive index is the same as that of the medium. For details of this technique, see Ross (1967).

The main value of interference microscopy is that it permits measurements to be made of the dry mass of an object. The only alternative method is by X-ray absorption, which has been shown to give the same results when identical objects are examined by both methods (Davies, Engström and Lindström, 1953; Sandritter and Müller, 1959; Müller *et al.*, 1959; Kimball *et al.*, 1959). The measurement of total dry weight is important in histochemistry as a basis for the quantitative expression of results obtained with a variety of other techniques. Davies, Barter and Danielli (1954) in a classic paper described the measurement of alkaline phosphatase activity by interferometry of calcium phosphate deposited at the site of enzyme activity. This approach to quantitative histochemistry, as an alternative to densitometry or fluorimetry, may prove of great value, particularly since the dry mass of a transparent reaction product can be measured in this way. The method has been reviewed by Danielli (1958). Dry mass determination by interferometry has mainly been applied to studies of cell growth by measurement of nuclear dry mass, which can be compared with DNA estimations by UV absorption and Feulgen photometry.

Thickness Measurements

The precise determination of section thickness is of special importance for quantitative cytochemical methods. A microtome setting cannot be relied upon to give more than a very approximate indication of section thickness.

There are four main ways in which the thickness of microscopic objects can be measured (Lange and Engström, 1954):

(1) Focusing on lower and upper surfaces.
(2) Stereomicroradiography.
(3) Mechanical methods.
(4) Interferometry.

The first of these can be carried out with the calibrated fine adjustment of an ordinary microscope. Accuracy is increased by using an objective with a shallow depth of field, i.e. with a high N.A. At best, however, only low accuracy can be obtained: Lange and Engström (1954) found an error of at least 10 per cent for sections 5 μm thick, although Brattgård (1954) calculated the maximal error in 5 μm sections as 3 per cent.

Stereomicroradiography (Bellman, 1953) requires special apparatus and gives excellent results with thick sections (50–200 μm) but with thin sections it is quite inaccurate. Glimstedt and Håkansson (1951) introduced a mechanical method for measuring section thickness. This depended on the use of a commercially available instrument, the mikrokator, adapted by the provision of a special probe and used with the minimum possible pressure (200 mg.). Even then it is probable that the pressure of the probe on the sections will cause some deformity since 200 mg. on an area 100 × 100 μm corresponds to a pressure of 2 kg/cm^2 (Eränkö, 1955). Hallén (1956) modified the mikrokator by substituting a profile microscope for the probe. His instrument (Aktiebolaget and Johnsson, Eskilstuna, Sweden) was based on the following principle. If a flat surface is focused in a profile microscope the wire in the plane of the field diaphragm gives a sharp shadow line in the field of view. Alteration of focus up or down causes lateral displacement of this line which becomes simultaneously weaker in contrast. The lateral movement is symmetrical around its centre position with regard to motion, sharpness and contrast. A cross line in the eyepiece of the microscope indicates the position of the axis of symmetry in the field of view and under these conditions great accuracy can be achieved if focused on different levels of an object. Under ideal measuring conditions an accuracy of 0·12 μm. is possible. Hallén's instrument thus represents the best that can be achieved by the use of profile microscopy.

Interferometry is the most accurate of the available methods. The thickness of the object can be calculated from the optical path difference (measured interferometrically) between two rays, one passing through the object and the other through the mounting medium only, if the refractive index of the object is known. Unfortunately, very often the refractive index of the object is not known, and several methods have been proposed for obtaining the necessary information by two interference measurements. In Barer's double-immersion method (Barer and Dick, 1957) if the object is capable of being immersed in two media of different known refractive indices, both the thickness and refractive index of the object can be calculated from the phase changes in the two media. A similar effect is achieved by a two-wavelength method by measurements at two wavelengths, at which the immersion medium has different refractive indices (Chaubal, Lodin and Pilny, 1967). Nomarski (1967) has pointed out that the necessary information can be obtained by two measurements, one made in the usual way and the other with the object tilted relative to the optic axis of the microscope, using a universal stage. Goldstein (1967) has proposed an analogous but more convenient method which he called the "two-aperture" method, in which the first measurement is made with axial parallel light, and the second measurement with an oblique beam of light using only marginal rays from a wide-angle condenser.

The distance between the slide and cover-glass can also be determined by multiple-beam interferometry, if the upper surface of the slide and the lower surface of the coverglass are metallized with a non-toxic coating (Osterberg, 1955). The thickness of parts of the section may however be significantly less than this.

Measurement of Area

Measurements of area may be of value either for use with measurements of thickness to obtain volume measurements, or on a comparative basis—e.g. to determine the relative areas of a section occupied by different types of tissue. Four main methods are available:

(1) Planimetry.
(2) Photography.
(3) Line and point sampling.
(4) Automatic scanning.

In planimetry, the area of the relevant region of a photograph or projection of the microscope field is measured with a mechanical device, the planimeter. The method is somewhat tedious. Another method is to make a photograph, cut out the relevant area with scissors, and determine the area by weighing the cut-out and comparing with the weight of a known area. Line sampling (Uotila and Kannas, 1952) and point sampling (Chalkley, 1943; Leibnitz, 1964) are manual methods in which greater speed is attained at the expense of a reduction in accuracy. The point sampling technique has been extended and automated by using a television camera to scan the field, the parts of the field being computed electronically. In the MRC Television Image Analyser (Fig. 268), a television camera scans either a part of the microscope field or a photomicrograph (which latter can be produced in an optical, electron or X-ray microscope, and is usually more convenient). A television image is produced on a monitor in colour, in which each colour represents a particular intensity range seen by the camera (corresponding to various densities on the photomicrograph). The intensity limits for each colour can be adjusted, and up to seven colours can be used. In the example (Plate XXIXc and d) a photograph of an ATPase reaction in a muscle biopsy (c) is examined and a part of the field is reproduced on the television screen (d) in four colours corresponding to high, medium, and low staining and blank areas respectively. By electronic means, the relative areas of these colours can be determined. The instrument can also be used for counting particles and measuring the length of interfaces between areas of differing densities. Somewhat similar instruments are commercially available.

Checking and Calibration of Quantitative Optical Systems

Before any optical equipment is used for histochemical quantitation, it is necessary both to check on the accuracy of the measuring system and to calibrate it with appropriate standards: otherwise all measurements are valueless. Caspersson emphasized the importance of knowing the degree of accuracy, or rather the amount of error, attained with a particular system.

Testing for homogeneity of illumination of the measuring field is carried out, basically, by making measurements of a small object placed successively in different parts of the field. In absorptiometry this can be accomplished by, for example, repeated measurements from a Feulgen-stained nucleus placed in five or more different positions within a large measuring field (Garcia, 1962; Garcia and Iorio, 1966), or by moving a second small diaphragm to various parts of the field (van Duijn, 1965). The latter principle has been applied to fluorimetry by Böhm and Sprenger (1970) who described a "nine-position test" in which a homogeneous plate of fluorescent uranyl glass (GG21, Schott) was placed in contact with the front lens of the objective, and a small measuring aperture was moved to different parts of the field. An alternative method which we have found more convenient involves measurement from a small fluorescent particle placed in various parts of the field; fluorescent plastic "scintillator" material is suitable (e.g. the BBL fluorescence test slide*).

Model systems in which a known amount of the substance to be assayed was incorporated into polyacrilamide or cellulose films have been extensively studied by van Duijn and his colleagues, particularly in relation to alkaline or acid phosphatase in polyacrilamide (van Duijn, Pascoe, and van der Ploeg, 1967; Lojda, van der Ploeg and van Duijn, 1967; van der Ploeg and van Duijn, 1968), DNA in cellulose (van Duijn and Persijn, 1960a, b; Hardonk and van Duijn, 1964a, b) and Schiff-positive groups attached to cellulose films (Hardonk and van Duijn, 1964a). A method for the determination of the molar extinction coefficient of structure-linked chromophores in model systems was described by van Dalen, Ahsmann and van Duijn (1970).

For microspectrofluorimetry, models are usually made by incorporation of substances (e.g. amines) into serum albumin, 0·1 to 1 per cent aqueous, spread and dried on to quartz or glass slides as a thin layer (Corrodi and Jonsson, 1965) or droplets (Ritzén, 1967). In some applications, polyvinyl-

* BBL division of BioQuest, Cockeysville, Maryland 21030, U.S.A.

pyrrolidone (PVP) is to be preferred to albumin (Rost and Polak, 1969; Rost and Ewen, 1971).

General Considerations

Whatever means of quantitation are used, it is necessary to investigate carefully the precision and accuracy of the optical quantitation, the preparation of the tissue, the histochemical reaction, and the mounting of the preparation.

Mounting

The mounting medium is of some importance. The process of mounting must not remove any of the FRP. Dehydration by ethanol is particularly liable to offend in this respect: dehydration in isopropanol or by air-drying are possible alternatives (see Chapter 29, p. 1194). The refractive index of the mountant should be as close as possible to that of the specimen for absorptiometry and fluorimetry, but must be different for interferometry and for scanning by phase-contrast. Kiefer (1970) has discussed mounting media for UV microphotometry. It is probably no longer necessary to point out that a non-fluorescent medium is required for fluorimetry—DPX, glycerol, liquid paraffin, and aqueous buffer solutions are all satisfactory in this respect.

Fading

Fading of a preparation either before, during, or between measurements obviously leads to errors in quantitation. Spectral measurements (e.g. by microspectrofluorimetry) may be affected in two ways: fading during scanning of a spectrum will cause distortion of the spectrum, and if the photochemical reaction responsible for fading leads to the production of fluorescent products the actual spectrum of the preparation will change accordingly. The fading of preparations is particularly important in fluorescence methods, in which the intensity of illumination tends to be high. Comparatively few special investigations have been made of this problem, and the nature of the phenomenon is still not well understood. The problem has two aspects: fading of a preparation during measurement, due to irradiation, and fading before measurement or between successive measurements on the same preparations. In the development of any new technique for quantitation, these problems must be borne in mind and may affect the choice of dye, mounting medium, and measuring technique.

Feulgen. Most of the studies on fading to date have been in relation to the fading of Feulgen-stained material. Hillary (1939) reported fading of a Feulgen preparation of root tip squashes mounted in a corn syrup-acetic acid medium. Ris and Mirsky (1949) found no demonstrable change in one preparation 4½ months after its initial measurement. Kasten (1959), in testing a large number of dyes as possible Schiff reagents, exposed stained slides to light and heat for periods up to 24 hours and reported no obvious visual change. De la Torre and Salisbury (1962) studied the fading of Feulgen-stained bovine spermatozoa, and found a marked decolourization on re-measurement after 30–38 days, the amount of decolorization depending upon the batch of Feulgen reagent used.

Prenna and Bianchi (1964) studied the fading of Feulgen-stained material using Acriflavine pseudo-Schiff reagent. Ruch (1965) measured the fading of Auramine O and BAO used in pseudo-Schiff reagents for Feulgen staining of liver nuclei, and found BAO preferable. Böhm and Sprenger (1968) carried out similar studies, using pararosaniline, Acridine yellow, Acriflavine, and Coriphosphine (all as pseudo-Schiff fluorochromes). In the latter case, they found an initial increase in fluorescence during excitation with UV light. Changes in intensity and emission spectrum of Acriflavine pseudo-Schiff stained nuclei were further studied by Sprenger and Böhm (1971b).

Immunofluorescence. Goldman (1960) investigated the fading of the fluorescence of fluorescein solutions at various dye concentrations and light intensities. He found that fading was more rapid with higher light intensities irrespective of concentration, and that the more concentrated solutions faded more rapidly than the less concentrated ones. The slower the rate of fading, the sooner was a level reached which thereafter remained relatively constant during the time interval studied.

Other fluorochromes. West (1965), found that the metachromatic red emission of Acridine orange faded much more rapidly than the green. Thaer (1966b) and Yamada *et al.* (1966) obtained similar results. Similar findings with Coriphosphine O in the masked metachromasia reaction were reported by Bussolati, Rost and Pearse (1969).

FIF. Fading of catecholamine (CA) and tryptamine FIF was noted by Falck (1962), who later observed that the FIF of CA's is less sensitive to UV irradiation than that of 5-HT (Falck and Owman, 1965). This was confirmed by Caspersson, Hillarp and Ritzén (1966), who found the rate of fading of the fluorescence to be characteristic of the different groups of monoamines. They found that the FIF of 5-HT fading twice as rapidly as that of CA's. In concentrations of up to 5 per cent by dry weight of the protein medium, the fading was practically independent of the monoamine concentration and of the reaction time with formaldehyde. Fading curves were given for the FIF from noradrenalin (Ritzén, 1966) and 5-HT (Ritzén, 1966a). Rapid fading of FIF from *m*-tyramine, metaraminol, α-methyl-*m*-tyrosine and *m*-tyrosine was noted by Jonsson and Ritzén (1966). Fading of indolylethylamine FIFs was studied by Jonsson and Sandler (1969): 6-HT was found to give a more stable FIF than 5-HT.

The problem of obtaining spectral excitation and emission curves from rapidly fading FIF was discussed by Ritzén (1966), who stated that errors would occur if measurement took more than 5–10 seconds. Fading was said to commence 5–10 seconds after switching on the light, and to recover partially if the preparation was subsequently kept in the dark. He recommended making the measurement after 3 minutes of irradiation, as the fading rate is then less. Vialli and Prenna (1969) did not observe any latent period before fading occurred in 5-HT FIF.

For the microspectrofluorimetric identification of FIF in tissues, Ritzén (1967) gives as one criterion of identification that the fading rate under irradiation of the specimen should be shown to be characteristic of the presumptive amine.

In our own experiments, of which a preliminary report has been made (Rost and Pearse, 1968), simultaneous measurements were made of fluorescence emission and of transmission at the exciting wavelength, using the modified Leitz microspectrofluorimeter described above. Catecholamine FIF in sections of rat adrenal medulla exposed to radiation for some hours showed fading in which the emission declined more rapidly than the absorption, indicating some loss of quantum efficiency. After the same sections had been kept in the dark overnight, there was recovery of quantum efficiency, suggesting that local heating may have played part in the fading.

Measurements. Several methods have been proposed for minimizing errors due to fading during microfluorimetry. In general, the time and intensity of irradiation should be kept to a minimum (Ruch, 1970). Searching and focusing should if possible be carried out at a relatively long wavelength (e.g. green light) using phase-contrast if necessary. The actual measurement can either be made at a fixed time after commencement of irradiation (Ritzén, 1967; Ruch, 1970) or as quickly as possible (Böhm and Sprenger, 1968; Sprenger and Böhm, 1971a). If a number of measurements are made at known times after the commencement of irradiation, it may be possible to extrapolate backwards to estimate the intensity at the commencement of irradiation, i.e. before fading (Rost and Pearse, 1968).

Nordén (1953) and Ritzén (1966) pointed out that fading during recording of fluorescence spectra would cause distortion of the apparent spectrum. To obtain consistent readings, Nordén always recorded emission spectra in the same manner, beginning with shorter wavelengths. Ritzén (1967) found that reasonably consistent spectra could be obtained if the specimen was irradiated for 3 minutes before starting the measurement. Thaer (1966b), using a television system with the Leitz microspectrograph, demonstrated changes in the emission spectrum of Acridine orange fluorochromed cells occurring within the interval from 5 seconds to 45 seconds after the commencement of irradiation. Fading of emission spectra of 5-HT FIF was investigated with a similar system by Vialli and Prenna (1969).

Histochemical considerations

The ideal histochemical requirements for quantitation of a substance are as follows:

(1) The substance to be measured must be kept *in situ*, and not allowed to diffuse either into other parts of the tissue or into a "fixative" or incubation medium.

(2) A fixative, employed to prevent diffusion of the substance, should neither extract part of it ("One man's fixative is another man's extractant") nor damage its chemical reactivity to the subsequent histochemical reaction.

(3) The histochemical reaction employed to demonstrate the substance should react with all of it, giving a FRP whose amount is linearly proportional to the amount of the substance. The components of the reaction mixture should not extract or denature unreacted substance.

(4) The FRP should be quantifiable, i.e. if absorptiometry is to be employed the FRP should obey the Beer–Lambert Law.

Of course, no present-day technique completely satisfies all the above. In fixation in particular, some compromise is inevitable, particularly in enzyme assays where fixation adequate to preferent diffusion is likely to partly denature that enzyme. Loss of water-soluble substances into the fixative can be completely eliminated by the use of freeze-drying and vapour fixation, where appropriate.

For the rest, the most important thing is that the amount of the error must in all cases be known. So long as the final reading can be related to the original amount of the substance present, using model systems, it does not matter if a (known) proportion of substance is lost, or if the histochemical reaction demonstrates only a (constant) proportion of the substance.

At the present time, a major difficulty in quantitative histochemistry lies in defining the region of the specimen to be measured. The problem is easiest in the special case of isolated or isolatable cells, e.g. leucocytes in a smear of blood or bone marrow, and tissue culture cells. It is hardly surprising that the most successful quantitative histochemistry to date has been carried out on cells of these types, e.g. determination of alkaline phosphatase in leucocytes (van der Ploeg and van Duijn, 1968). It is to be hoped that applications to clinical haematology and possibly exfoliative cytology will soon become routine.

Tissue sections are another matter. There are two difficulties: variations in thickness of the section, and the problem of isolating an irregularly-shaped cell in contiguity with its neighbours. It cannot be assumed that the thickness of a section is precisely, or even quite approximately, equal to the number of microns to which the microtome is set. Nor can serial sections be absolutely guaranteed to be of identical thickness; for example every microtomist is familiar with the phenomenon of alternate thick and thin sections, which is sometimes obvious but which may pass unnoticed if of small degree. Finally, the isolation of an image of a single cell of irregular shape in a measuring diaphragm is related to the problem of fitting a square (or worse!) peg in a round (or possibly rectangular) hole. One approach to this problem is that of the new Zeiss Scanning Photometer, the computer of which can be instructed to integrate measurements within an irregular area; the conversation which is required between computer and operator is at present lengthy, but in the future a light-pen could be used to define the measuring area.

Measurements on tissue sections are dependent upon knowledge of the section thickness. There does not appear to be any really satisfactory method at present for measuring section thickness; interferometry offers the best hopes.

Comparative or statistical studies of a cell population within a single section avoid some of the difficulties of absolute quantitation, and the use of instruments like the MRC Television Image Analyser is likely to increase. Such studies are particularly applicable to muscle (Rost, 1971).

Instrumental error

Finally there is the question of instrumental error in measurement of the amount of FRP, e.g. by absorptiometry or fluorimetry. This is a purely technical problem, and is probably already the least significant source of error in quantitative histochemistry if adequate equipment is used.

Conclusions

Quantitative histochemistry for the assay of enzymes and other substances is obviously of great importance. Great progress has been made since the pioneering work of Caspersson and his colleagues. Many difficulties still remain. Most of the instrumental problems appear to have been essentially solved, at least for those laboratories which can afford to pay for adequate equipment. Purely histochemical difficulties remain: fixation, the stoichiometry of the reaction, and diffusion of reaction products. One of the outstanding problems is the difficulty of isolating the desired volume of tissue: relatively easy for isolated blood or tissue culture cells, still almost impossible in tissue sections. In view of all the doubts and difficulties which exist, it is obviously still only prudent to employ more than one method wherever possible, so that each can be a check on the others. In all cases, an appropriate standard or model system must be set up. Obviously the final justification of any quantitative histochemical method will depend upon comparison with results obtained by other means, if possible, and upon the results of calibration with proper model systems.

REFERENCES

Adams, L. R. (1968). *Acta cytologica,* **12**, 3.
Ahsmann, W. B. A. M., and Van Duijn, P. (1968). *Histochemie* **12**, 285.
Allen, R. D., and Brault, J. W. (1966). *Adv. optical and electron Microsc.*, **1**, 77.
Alpen, E. L. (1966). *J. Sci. Instrum.*, **43**, 443.
Atkin, N. B. (1970). In: Wied, G., and Bahr, G. F. "Introduction to Quantitative Cytochemistry", p. 1. Acad. Press, New York and London.

BARENBOIM, G. M., DOMANSKII, A. N., and TUROVEROV, K. K. (1969). In "Luminescence of Biopolymers and Cells". (Trans. from Russian by R. F. Chen.) Plenum Press, New York and London.
BARER, R. (1952). *J. Physiol.*, **119**, 52 P.
BARER, R. (1959). In: Mellors, R. C. (ed.) "Analytical Cytology", Vol. 2, pp. 1–60. Acad. Press, New York and London.
BARER, R. (1966). In: Pollister, A. W. (ed.) "Physical Techniques in Biological Research", Vol. IIIA, pp. 1–56 (2nd edn). Acad. Press, New York and London.
BARER, R., and DICK, D. A. T. (1957). *Expl. Cell Res. Suppl.*, **4**, 103.
BELLMAN, S. (1953). *Acta Radiol.*, Suppl. 102.
BENEKE, G. (1966a). In: Wied, G. (ed.) "Introduction to Quantitative Cytochemistry", pp. 63–92. Acad. Press, New York and London.
BENEKE, G. (1966b). In: Wied, G. (ed.) "Introduction to Quantitative Cytochemistry", p. 107. Acad. Press, New York and London.
BJÖRKLUND, A., EHINGER, B., and FALCK, B. (1968). *J. Histochem. Cytochem.*, **16**, 263.
BOCK, R., and SCHLÜTER, G. (1971). *Histochemie*, **25**, 152.
BÖHM, N., and SPRENGER, E. (1968). *Histochemie*, **16**, 100.
BÖHM, N., and SPRENGER, E. (1970). *Histochemie*, **22**, 59.
BÖHM, N., SPRENGER, E., and SANDRITTER, W. (1970). *Acta histochem.*, **35**, 324.
BOSSHARD, U. (1964). *Z. wiss. Mikr.* **65**, 391.
BRATTGÅRD, S. O. (1954). *Jl R. microsc. Soc.*, **74**, 113.
BUSSOLATI, G., ROST, F. W. D., and PEARSE, A. G. E. (1969). *Histochem. J.*, **1**, 517.
CARLSON, L. (1970). *Histochemie*, **21**, 289.
CASPERSSON, T. (1950). "Cell Growth and Cell Function. A Cytochemical Study". W. W. Norton Co., New York.
CASPERSSON, T. (1955). *Experientia*, **11**, 45.
CASPERSSON, T. (1936). *Skand. Arch. Physiol.* **73**, Suppl. 8, 1.
CASPERSSON, T., CARLSON, L., and SVENSSON, G. (1954). *Expl Cell Res.*, **7**, 601.
CASPERSSON, T., JACOBSSON, F., and LOMAKKA, G. (1951). *Expl Cell Res.*, **2**, 301.
CASPERSSON, T., JACOBSSON, F., LOMAKKA, G., SVENSSON, G., and SAFSTROM, R. (1953). *Expl Cell Res.*, **5**, 560.
CASPERSSON, T., and LOMAKKA, G. (1962). *Ann. N.Y. Acad. Sci.*, **97**, 449.
CASPERSSON, T., and LOMAKKA, G. (1970). In: Wied, G. L. and Bahr, G. F. "Introduction to Quantitative Cytochemistry", II, p. 29. Acad. Press, New York and London.
CASPERSSON, T., LOMAKKA, G., and RIGLER, R. JR. (1965). *Acta Histochem. Suppl.*, **6**, 123.
CASPERSSON, T., LOMAKKA, G., and SVENSSON, G. (1957). *Acta Histochem. Suppl.*, **4**, 9.
CASTAING, R., and GUINIER, A. (1949). Proc. Conf. Electron Microscopy, Delft, p. 60. Nijhoff, The Hague.
CHALKLEY, H. W. (1943). *J. natn. Cancer Inst.*, **4**, 47.
CHANCE, B. (1962). *Ann. N.Y. Acad. Sci.*, **97**, 431.
CHANCE, B., and LEGALLAIS, V. (1959). *Rev. Sci. Instr.*, **30**, 732.
CHANCE, B., and THORELL, B. (1959). *J. biol. Chem.*, **234**, 3044.
CHAPMAN, J. H., FORSTER, T. H., KORTUM, G., PARKER, C. A., LIPPERT, E., MELHUISH, W. H., and NEBBIA, G. (1965). *Appl. Spectr.*, **17**, 171.
CHAUBALL, K. A., LODIN, Z., and PILNY, J. (1967). *Acta histochem.*, **26**, 131.
COMMONER, B. (1950). *Science, N.Y.*, **111**, 71.
COMMONER, B., and LIPKIN, D. (1949). *Science, N.Y.*, **110**, 41.
COSSLETT, V. E., and NIXON, W. C. (1960). "X-ray Microscopy". Cambridge Univ. Press, London and New York.
DANIELLI, J. F. (1958). In: Danielli, J. F. (ed.), "General Cytochemical Methods", Vol. 1, Acad. Press, New York and London.
DAVIES, H. G., BARTER, R., and DANIELLI, J. F. (1954). *Nature, Lond.*, **173**, 1234.
DAVIES, H. G., ENGSTRÖM, A., and LINDSTRÖM, B. (1953). *Nature, Lond.*, **172**, 1041.
DAVIES, H. G., WILKINS, M. F., and BODDY, R. G. H. B. (1954). *Expl Cell Res.*, **6**, 550.

Dayan, A. D. (1967). *J. Histochem. Cytochem.*, **15**, 421.
Deeley, E. M. (1955). *J. sci. Instr.*, **32**, 263.
Deeley, E. M. (1960). *Biochem. Pharmacol.*, **4**, 104.
de la Torre, L., and Salisbury, G. W. (1962). *J. Histochem. Cytochem.*, **10**, 39.
Den Tonkelaar, E. M., and Van Duijn, P. (1964). *Histochemie*, **4**, 1.
Den Tonkelaar, E. M., and Van Duijn, P. (1964). *Histochemie*, **4**, 10.
Den Tonkelaar, E. M., and Van Duijn, P. (1964). *Histochemie*, **4**, 16.
Dörmer, P. (1967). *Histochemie*, **8**, 1.
Dörmer, P., Brinkman, W., Stieber, A., Stich, W. (1966). *Klin. Wschr.*, **9**, 477.
Dubach, U. C., and Schmidt, U. (1971). "Recent Advances in Quantitative Histo- and Cytochemistry". Huber, Berne.
Dyson, J. (1964). *Nature, Lond.*, **203**, 1300.
Ely, R. V. (1963). "Bibliography Micro X-Radiography and Analysis for the Years 1913–1962". Guildford, England.
Engström, A. (1962). "X-ray Microanalysis in Biology and Medicine". Elsevier, Amsterdam.
Engström, A. (1966). In: Pollister, A. W. (ed.), "Physical Techniques in Biological Research", Vol. IIIA, p. 87, Acad. Press, New York and London.
Eränkö, O. (1955). "Quantitative Methods in Histology and Microscopic Histochemistry". S. Karger, Basel.
Fischer, H. A., and Werner, G. (1968). *Histochemie*, **15**, 84.
Galjaard, H. (1967). *Jl. R. microsc. Soc.*, **87**, 157.
Gaillard, J. L. J., Van Duijn, P., and Schaberg, A. (1968). *Expl Cell Res.*, **53**, 417.
Garcia, A. M. (1962). *Histochemie*, **3**, 178.
Garcia, A. M., and Iorio, R. (1966). In: Wied, G. (ed.), "Introduction to Quantitative Cytochemistry", pp. 215–234. Acad. Press, New York and London.
Glick, D. (1953). *Int. Rev. Cytol.*, **2**, 447.
Glick, D. (1963). "Quantitative Chemical Techniques of Histo- and Cytochemistry". New York, Interscience.
Glick, D., Engström, A., and Malmstrom, B. G. (1951). *Science, N.Y.*, **114**, 253.
Glimstedt, G., and Håkansson, A. (1951). *Nature, Lond.*, **167**, 397.
Goldman, M. (1960). *Expl. Parasitol.*, **9**, 25.
Goldman, M. (1967). *J. Histochem. Cytochem.*, **15**, 38.
Goldstein, D. J. (1967). *Nature, Lond.*, **213**, 386.
Goldstein, D. J. (1969). *J. Microscopy*, **89**, 19.
Goldstein, D. J. (1970). *J. Microscopy*, **92**, 1.
Hale, A. J. (1958). "The Interference Microscope in Biological Research". Edinburgh, Livingstone.
Hale, A. J. (1966). In: Wied, G. L. (ed.), "Introduction to Quantitative Cytochemistry". Acad. Press, New York and London.
Hallén, O. (1956). *Acta Anat.*, **26**, Suppl. 25.
Hardonk, M. J., and Van Duijn, P. (1964). *J. Histochem. Cytochem.*, **12**, 533.
Hardonk, M. J., and Van Duijn, P. (1964). *J. Histochem. Cytochem.*, **12**, 752.
Hillary, B. B. (1939). *Stain Technol.*, **14**, 97.
Holborow, E. J. (1970). "Standardization in Immunofluorescence". Blackwell, Oxford.
Howling, D. H., and Fitzgerald, P. J. (1959). *J. biophys. biochem. Cytol.*, **6**, 313.
Jansen, M. T. (1961). *Histochemie*, **2**, 342.
Jeffree, G. M. (1970). *Histochem.*, **2**, 231.
Jongsma, A. P. M., Hijmans, W., and Ploem, J. S. (1971). *Histochemie* (in press).
Kasten, F. H. (1959). *Histochemie*, **1**, 466.
Kelly, J. W. (1962). *Science, N.Y.*, **138**, 1272.
Kelly, J. W. (1966). In: Wied, G. (ed.), "Introduction to Quantitative Cytochemistry". Acad. Press, New York and London.
Kemplay, J. R. (1962). *Electron. Engng.*, **34**, 820.
Kiefer, G. (1970). In: Wied, G. L., and Bahr, G. F. (eds.), "Introduction to Quantitative Cytochemistry". II. p. 199. Acad. Press, New York and London.

KIMBALL, R. F., CASPERSSON, T., SVENSSON, G., and CARLSON, L. (1959). *Expl Cell Res.*, **17**, 160.
KING, R. J. (1959). *Jl. R. microsc. Soc.* **77**, 99.
KOHEN, E., and KOHEN, C. (1966). *Histochemie*, **7**, 348.
KOHEN, E., KOHEN, C., NORDBERG, L., THORELL, B., and ÅKERMAN, L. (1967). Digest of the 7th Internat. Conference on Medical and Biological Engineering, p. 274. Stockholm.
KOHEN, E., KOHEN, C., and THORELL, B. (1968). *Histochemie*, **16**, 170.
KOHEN, E., KOHEN, C., and THORELL, B. (1969). *Biomed. Engng*, **4**, 544.
KOHEN, E., and LEGALLAIS, V. (1965). *Rev. Sci. Instrum.*, **36**, 1890.
KOHEN, E., LEGALLAIS, V., and KOHEN, C. (1966). *Expl. Cell Res.*, **41**, 223.
KONEV, S. V. (1967). "Fluorescence and Phosphorescence of Proteins and Nucleic Acids". (Trans. from Russian by S. Udenfriend.) Plenum Press, New York.
KRUG, W., RIENITZ, J., and SCHULTZ, G. (1961). "Beitrage zür Interferenz-Mikroskopie". Berlin, Akademie-Verlag. Trans. by Home-Dickson, J. (1964), London, Adam Hilger.
LANGE, P. W., and ENGSTRÖM, A. (1954). *Lab. Invest.*, **3**, 116.
LEIBNITZ, L. (1964). *Histochemie*, **4**, 123.
LOJDA, Z., VAN DER PLOEG, M., and VAN DUIJN, P. (1967). *Histochemie*, **11**, 13.
LOMAKKA, G. (1965). *Acta histochem. Suppl.*, **6**, 393.
LOTHIAN, G. F. (1969). "Absorption Spectrophotometry". Adam Hilger, London.
LOWRY, O. H. (1953). *J. Histochem. Cytochem.*, **1**, 420.
LOWRY, O. H. (1957). In: Colowick, S. P., and Kaplan, N. O. "Methods in Enzymology", Vol. 4, Acad. Press, New York, p. 366.
LOWRY, O. H., and HASTINGS, A. B. (1942). *J. biol. Chem.*, **143**, 257.
LOWRY, O. H., ROBERTS, N. R., LEINER, K. Y., WU, M.-L., and FARR, A. L. (1954). *J. biol. Chem.*, **207**, 1.
LOWRY, O. H., ROBERTS, N. R., WU, M.-L., HIXON, W. S., and CRAWFORD, E. J. (1954). *J. biol. Chem.*, **207**, 19.
MANSBERG, H. P., and KUSNETZ, J. (1966). *J. Histochem. Cytochem.*, **14**, 260.
MAYALL, B. H., and MENDELSOHN, M. L. (1970). In: Wied, G. L., and Bahr, G. F. (eds.), "Introduction to Quantitative Cytochemistry", II, p. 171. Acad. Press, New York and London.
McCARTHY, W. J., and MOYER, E. S. (1970). In: Wied, G. L., and Bahr, G. F. (eds.), "Introduction to Quantitative Cytochemistry", II, p. 399. Acad. Press, New York and London.
MENDELSOHN, M. L. (1958). *J. biophys. biochem. Cytol.*, **4**, 425.
MENDELSOHN, M. L. (1961). *J. biophys. biochem. Cytol.*, **11**, 509.
MENDELSOHN, M. L. (1966). In: Wied, G. L. (ed.), "Introduction to Quantitative Cytochemistry", p. 202. Acad. Press, New York and London.
MELLORS, R. C., GLASSMAN, A., and PAPANICOLAOU, G. N. (1952). *Cancer*, **5**, 458.
MELLORS, R. C., and SILVER, R. (1951). *Science, N.Y.*, **114**, 356.
MELLON, M. G. (1950). "Analytical Absorption Spectroscopy". Wiley, New York.
MOSS, D. W. (1962). Ph.D. Thesis, University of London.
MÜLLER, D., SANDRITTER, W., SCHIEMER, H. G., and ENDRES, K. (1959). *Histochemie*, **1**, 438.
NAORA, H. (1951). *Science, N.Y.*, **114**, 279.
NAORA, H. (1952). *Science, N.Y.*, **115**, 258.
NAORA, H. (1955). *Expl Cell Res.*, **8**, 259.
NAORA, H. (1958). *Handbuch der Histochemie*, **1**, 192.
NEUMANN, K. (1958). *Handbuch der Histochemie*, **1**, 339.
NIEMI, M. (1958). *Acta Anat. Suppl.* **34**, 1.
NOMARSKI, G. (1967). Personal communication, acknowledged by Goldstein (1967).
NORDÉN, G. (1953). *Acta Pathol. Microbiol. scand., Suppl.* 96.
OLSON, R. A. (1960). *Rev. Sci. Instrum.*, **31**, 844.
ORNSTEIN, L. (1952). *Lab. Invest.*, **1**, 250.
OSTER, G. (1955). In: Oster, G., and Pollister, A. W. (eds.), "Physical Techniques in Biological Research", Vol. 1, p. 439. Acad. Press, New York and London.

OSTER, G., and POLLISTER, A. W. (1955) (eds.). In "Physical Techniques in Biological Research", Vol. 1 (Optical techniques). Acad. Press, New York and London.

OSTERBERG, H. (1955). In: "Physical Techniques in Biological Research". Ed. G. Oster, and A. W. Pollister. Acad. Press, New York, Vol. 1, p. 377.

PARKER, C. A. (1969). *Analyst*, **94**, 161.

PATAU, K. (1952). *Chromosoma*, **5**, 341.

PATTEE, H. H., COSSLETT, V. E., and ENGSTRÖM, A. (1963). (eds.), "X-ray Optics and X-ray Microanalysis". Acad. Press, New York.

PEARSE, A. G. E., and ROST, F. W. D. (1969). *J. Microscopy*, **89**, 321.

PILLAT, G. (1960). *Acta histochem.*, **9**, 169.

PITTMAN, B., HEBERT, G. A., CHERRY, W. B., and TAYLOR, G. G. (1967). *J. Immunol.*, **98**, 1196.

PLOEM, J. S. (1970a). In: Holborow, E. J., "Standardization in Immunofluorescence", p. 63. Blackwell, Oxford.

PLOEM, J. S. (1970b). In: Holborow, E. J. "Standardization in Immunofluorescence", p. 137. Blackwell, Oxford.

POLLISTER, A. W. (1966) (ed.), "Physical Techniques in Biological Research". 2nd Edn. Vol. IIIA (Cells and Tissues). Acad. Press, New York and London.

POLLISTER, A. W., and ORNSTEIN, L. (1959). In: Mellors, R. C. (ed.), "Analytical Cytology", 2nd edn, p. 431. McGraw-Hill, New York.

PRENNA, G., and BIANCHI, U. A. (1964a). *Riv. Istoch. norm. pat.*, **10**, 667.

PRENNA, G., and BIANCHI, U. A. (1964b). *Riv. Istoch. norm. pat.*, **10**, 645.

PRENNA, G., ZANOTTI, L., and BIANCHI, U. A. (1964). *Riv. Istoch. norm. pat.*, **10**, 539.

RASCH, E. M., and RASCH, R. W. (1970). In: Wied, G. L., and Bahr, G. F. (eds.), "Introduction to Quantitative Histochemistry", II, p. 297. Acad. Press, New York and London.

RICHARDS, O. W. (1966). In: Wied, G. L. (ed.), "Introduction to Quantitative Cytochemistry", p. 43. Acad. Press, New York and London.

RIGLER, R. JR. (1966). *Acta physiol. scand.*, **67**, *Suppl.* 267.

RIS, H., and MIRSKY, A. E. (1949). *J. gen. Physiol.*, **33**, 125.

RITZÉN, M. (1967). M.D. Thesis, Stockholm.

ROGERS, A. W. (1961). *Expl Cell Res.*, **24**, 228.

ROGERS, A. W. (1967). "Techniques of Autoradiography". Elsevier, Amsterdam.

ROSS, K. F. A. (1967). "Phase Contrast and Interference Microscopy for Cell Biologists". Arnold, London.

ROSSELET, A. (1967). *Z. wiss. Mikr.*, **68**, 22.

ROSSELET, A., and RUCH, F. (1968). *J. Histochem. Cytochem.*, **16**, 459.

ROST, F. W. D. (1971). *J. clin. Path.*, **24**, *Suppl.* (Ass. clin. Path.) **4.**

ROST, F. W. D., and EWEN, S. W. B. (1971). *Histochem. J.*, **3**, 207.

ROST, F. W. D., and MAUNDER, C. (1971). *Proc. R. microsc. Soc.*, **6**, 20.

ROST, F. W. D., NÄGEL, L. C. A., and MOSS, D. W. (1970). *Proc. R. microsc. Soc.*, **5**, 76.

ROST, F. W. D., and PEARSE, A. G. E. (1968). Proc. 3rd Intern. Congr. Histochem. Cytochem., New York, p. 226, Springer, Berlin.

ROST, F. W. D., and PEARSE, A. G. E. (1971). *J. Microscopy*, **94**, 93.

ROST, F. W. D., and POLAK, JULIA M. (1969). *Virchows Archs A. Path. Anat.*, **347**, 321.

ROTMAN, B., and PAPERMASTER, B. W. (1966). *Proc. Natn. Acad. Sci., U.S.A.*, **55**, 134.

RUCH, F. (1951). *Expl Cell Res.*, **2**, 680.

RUCH, F. (1957). *Expl Cell Res.*, *Suppl.*, **4**, 58.

RUCH, F. (1960). *Z. wiss. Mikrosk.*, **64**, 453.

RUCH, F. (1962). *Acta histochem. Suppl.*, **2**, 175.

RUCH, F. (1965). *Acta histochem. Suppl.*, **6**, 117.

RUCH, F. (1966a). In: Wied, G. L. (ed.), "Introduction to Quantitative Cytochemistry", p. 281. Acad. Press, New York and London.

RUCH, F. (1966b). In: Wied, G. L. (ed.), "Introduction to Quantitative Cytochemistry", p. 549, Acad. Press, New York and London.
RUCH, F. (1966c). In: Pollister, A. W. (ed.), "Physical Techniques in Biological Research", 2nd edn. Vol. IIIA (Cells and Tissues), p. 57. Acad. Press, New York and London.
RUCH, F. (1970). In: Wied G. L., and Bahr, G. F. (eds.), "Introduction to Quantitative Cytochemistry", II, p. 431. Acad. Press, New York and London.
RUCH, F., and BOSSHARD, U. (1963). *Z. wiss. Mikroskop. Tech.*, **65**, 335.
RUNGE, W. J. (1966). *Science, N.Y.*, **151**, 1499.
RUTHMANN, A. (1970). "Methods in Cell Research". Bell, London.
SANDRITTER, W. (1958). *Handbuch der Histochemie*, **1**, 220.
SANDRITTER, W., and MÜLLER, D. (1959). *Experientia*, **15**, 158.
SCHMIDT, W. J. (1958). *Handbuch der Histochemie*, **1**, 171.
SELIGER, H. H., and MCELROY, W. D. (1965). Light: "Physical and Biological Action". Acad. Press, New York and London.
SERNETZ, M. (1969). Private communication.
SERNETZ, M., and THAER, A. (1970). *J. Microscopy*, **91**, 43.
SLAYTER, E. M. (1970). "Optical methods in biology", New York, Wiley-Interscience.
SMITH, F. H. (1967). *Jl. R. microsc. Soc.*, **87**, 165.
SPRENGER, E., and BÖHM, N. (1971a). *Histochemie*, **25**, 163.
SPRENGER, E., and BÖHM, N. (1971b). *Histochemie*, **25**, 177.
TÄLJEDAL, I.-B. (1970). *Histochemie*, **21**, 307.
THAER, A. (1966a). In: Wied, G. L. (ed.), "Introduction to Quantitative Cytochemistry". Acad. Press, New York and London.
THAER, A. (1966b). Haus der Technik-Vortragsveroffentlichungen, Heft 69/66, "Fernsehen in der Industrie", p. 49.
THIEME, G. A. (1966). *Acta physiol. scand.*, **67**, 514.
TOUSIMIS, A. J., and ADLER, I. (1963). *J. Histochem. Cytochem.*, **11**, 40.
UOTILA, U., and KANNAS, O. (1952). *Acta endocrinol.*, **11**, 49.
VAN DALEN, J. P. R., AHSMANN, W. B. A. M., and VAN DUIJN, P. (1970). *Histochem. J.*, **2**, 329.
VAN DUIJN, P., and VAN DER PLOEG, M. (1970). In: Wied, G. L., and Bahr, G. F. (eds.), "Introduction to Quantitative Cytochemistry". Acad. Press, New York and London.
VAN DUIJN, P. (1965). *Acta histochem. Suppl.*, **6**, 61.
VAN DUIJN, P., PASCOE, E., VAN DER PLOEG, M. (1967). *J. Histochem. Cytochem.*, **15**, 631.
VAN DUIJN, P., and PERSIJN, J.-P. (1960). *Nature, Lond.*, **186**, 805.
VAN DER PLOEG, M., and VAN DUIJN, P. (1968). *J. Histochem. Cytochem.*, **16**, 693.
VAN GIJZEL, P. (1966). *Leitz-Mitt. wiss. Techn.*, **3**, 206., *Eng. Edn.*, **1**, 174.
VAN ORDEN, L. S. (1970). *Biochem. Pharmac.*, **19**, 1105.
VIALLI, M., and PRENNA, G. (1969). *J. Histochem. Cytochem.*, **15**, 321.
WEST, S. S. (1965). *Acta histochem. Suppl.* VI, 135.
WEST, S. S., LOESER, C. N., and SCHOENBERGER, M. D. (1960). *IRE Trans. med. Electronics, ME-*7, 138.
WIED, G. L. (1966) (ed.), "Introduction to Quantitative Cytochemistry". Acad. Press, New York and London.
WIED, G., and BAHR, G. F. (1970). "Introduction to Quantitative Cytochemistry", II. Acad. Press, New York and London.
YAMADA, M., TAKAKUSU, A., YAMAMOTO, K., and IWATA, S. (1966). *Archs histol. jap.*, **27**, 387.
YATAGANAS, X., GAHRTON, G., THORELL, B. (1969). *Expl Cell Res.*, **56**, 59.

CHAPTER 32

ELECTRON HISTOCHEMISTRY (ULTRAHISTOCHEMISTRY)

In the second edition of this book (1960), this chapter was exceedingly brief. Electron histochemistry was just beginning to develop as an independent discipline and no comprehensive or critical treatise had been, or could have been, written. Now, ten years later, a number of short review articles, a comprehensive review (Scarpelli and Kanczak, 1965), and a single text on the subject (Geyer, 1969) have appeared in print. It is clear, from the number of references in the literature, that the subject now deserves monographic treatment. In a book devoted to the whole of histochemistry, however, common sense dictates an abbreviated presentation. The essential difficulty is to know what to leave out. It is safe to prophesy that, in any succeeding editions of this book, at least two of the subsections of this Chapter (Immunoelectron microscopy and Enzyme electron cytochemistry) will require separate chapters of their own. My endeavour here, therefore, has been to emphasize the principles and practice particularly of the advancing frontiers of electron histo- and cytochemistry, dealing only briefly with other sections of the art, and to provide a basic minimum of technical assistance in the Appendix. Before dealing with the actual application of the electron microscope to histochemistry it may be well to describe some of the factors with which we must necessarily be concerned. These are connected with the production of contrast in the electron microscope image (Valentine, 1958).

Contrast in the Electron Microscope Image (Principles)

The detail observable under the high magnification of the electron microscope is limited more by lack of contrast than by any lack of resolution in the microscope. The latter can now resolve objects as small as 4·5 Å, that is to say having a diameter of only a few atoms. Various techniques are used for increasing contrast in specimens by artificial means and these include treatment with various reagents and metal shadowing of the surface. The latter cannot be employed in electron histochemistry but the former, by analogy with histological practice, has come to be known as "electron staining". This term, though it is not a very accurate one, is convenient and it will probably continue to be used for this reason.

The precise physical property which makes any compound useful as an electron stain is still far from clear. Some authorities have gone so far as to claim that all substances are almost equally efficient in increasing contrast by increasing electron density. The problem is made more difficult by the fact that exact measurements of contrast in the electron image are not easy to make (Hall, 1955) and, except in rare cases of very intense staining, subjective opinions are obviously unreliable.

A simple theoretical approach to the problem of contrast was made by Zeitler and Bahr (1957) and from their formulae Valentine (1958) deduced the conditions necessary to produce a just appreciable increase in contrast in the image of any part of a specimen.

If the object under consideration covers an area of S cm.^{2} at right angles to the beam then we may deduce that an increase in contrast requires the addition of N atoms of atomic number Z uniformly over or through the object where $NZ^{4/3}/S = 5{\cdot}5 \times 10^{17}$ for a microscope working at 60 kV ($4{\cdot}7 \times 10^{17}$ for 50 kV ; $8{\cdot}25 \times 10^{17}$ for 100 kV). By rewriting this formula in simple terms one can predict the effect of adding various amounts of different substances to the specimen and one can, to some extent, reconcile conflicting views on requirements for increasing contrast. There are three useful bases on which contrast can be compared and it is important to distinguish which is relevant in any particular case.

Contrast on the Basis of Equal Thickness. This is relevant particularly when selecting a material for shadowing a specimen and for calculating the minimum thickness of deposit which is required. It has little direct concern with electron histochemistry, but the calculations made by Valentine led him to suggest that a useful histochemical technique for the electron microscope would be one which formed small deposits of some dense substance. This prediction has been most amply confirmed, as the rest of this chapter indicates.

Contrast on the Basis of Equal Weight. If an object *d* cm. thick of density ρ g./cm.$^{-3}$ can bind *p* per cent by weight of an added substance, then a detectable increase of contrast requires *p* to have a value not less than that given by $(p/100)\rho dLZ^{4/3}/A = 5{\cdot}5 \times 10^{17}$ where A is the atomic weight of the added substance and L is Avogadro's number ($6{\cdot}02 \times 10^{23}$). Taking $Z^{4/3}/A$ and the density of the specimen as unity, $p = 9 \times 10^{-5}/d$. Thus for a specimen of given thickness the percentage of an electron stain required to increase contrast appears to be independent of the nature of the stain used. For an object 0·5 μ thick such as a bacterium a 2 per cent uptake of any stain should theoretically increase the contrast sufficiently and Valentine suggested that the usual histological stains should be tested by means of the electron microscope. For a thin (200 Å) section on the other hand, the above calculations indicate that the addition of 45 per cent of stain is required to increase contrast. Calculations of this type were made by Ornstein (1957) who deduced that, since osmium is only taken up to the extent of some 10 per cent by tissues fixed in osmium tetroxide, "staining" of objects in thin sections by this means must be fictitious. On the basis of practical observations most electron microscopists have seen fit to quarrel with this conclusion.

Contrast on the Basis of Equal Numbers of Molecules. If the object to be stained is *d* cm. thick and can bind *n* molecules of stain per cm.^{3} then to obtain an increase of contrast it can be calculated that $nd(\Sigma Z^{4/3}) = 5{\cdot}5 \times 10^{17}$

where $(\Sigma Z^{4/3})$ is the effective value of $Z^{4/3}$ obtained by adding up the effective value of $Z^{4/3}$ for each atom in the molecule of the stain. If, as is often the case, the molecules of the stain are bound to the surface of membrane and not uniformly distributed, as has been assumed above, then if n molecules are bound per cm.^{2} of membrane surface, the normal to which makes an angle θ with the beam at any point, contrast will be obtained when $n(\Sigma Z^{4/3})$sec. $\theta = 5 \times 10^{17}$. When tissues are fixed with $OsO_4$the latter is bound as OsO_2 $(\Sigma Z^{4/3}$ 353) and if $\theta = 890$, as where the membrane concerned runs at right angles to the surface, then $n = 2{\cdot}5 \times 10^{13}$cm.$^{-2}$. This means that one osmium atom per 400 Å^2 of surface will increase its contrast. This figure is much more reasonable than those calculated in the preceding section and osmium staining of surface membranes may therefore be a fact.

The foregoing considerations may be summarized by saying that when substances are added to a specimen in layers of equal thickness the increase in contrast will be proportional to the density of each substance. If, however, equal weights of substances are added, the changes in contrast will be almost the same in each case. If equal numbers of molecules are added the increase in contrast will depend on the sum of $Z^{4/3}$ for each atom in the molecule, that is to say on the molecular weight.

Contrast in the Electron Microscope Image (Practice)

The problem of OsO_4 as an electron stain was considered very fully by Bahr (1954, 1955) and also by Houck and Dempsey (1954). Since it was, for many years, the standard fixative for electron microscopy, the results of its use were not usually subjected to histochemical analysis. Phosphotungstic acid has been employed extensively as an electron stain, especially for muscle (Schmitt *et al.*, 1945; Hodge, 1956), and also for collagen (Kühn *et al.*, 1958, 1960), and as a selective contrast agent for various structures (Robertson, 1955; Wohlfarth-Bottermann, 1957; Brody, 1959; Bloom and Aghajanian, 1966). Chromium compounds were studied by Low and Freeman (1956) and especially recommended for nucleic acid studies. Ferric salts were used for the same purpose by Bernstein (1956).

Potassium permanganate was first described as a fixative by Luft (1956) and it was used by Usuku (1958) in studies on elastic fibres. Other papers describing the use of $KMnO_4$ as a fixative and electron stain are those of Lawn (1960), Yasutake *et al.* (1961), Reedy (1965), and Sutton (1968). Its use as a post-fixative was described by Parsons (1961).

The most popular contrast metal is lead, introduced by Mudd and Anderson (1942) and developed by Watson (1958), who used it in strong alkaline solutions. Further work on staining with lead ions at high pH levels by Dalton and Zeigel (1960), Lever (1960), Parsons and Darden (1960), Millonig (1961), Karnovsky (1961) and Feldman (1962), culminated in the most extensively used (lead citrate) method of modern electron microscopy (Reynolds, 1963).

Contrast staining with uranyl cation (UO_2) has proved almost equally popular. This is usually applied in the form of its acetate, either in the fixative (Strugger, 1956), or, more usually, to thin sections (Watson, 1958a).

Many other metal ions have been employed as electron stains and some of these are referred to in succeeding sections of the Chapter. These sections are devoted to electron histochemistry under headings which follow, broadly, the chapter headings of Volumes 1 and 2.

Techniques of Electron Histochemistry

Protein End-Groups

Sulphydryl and Disulphide. Working with yeast cells Mundkur (1964a) found that methyl mercury acetate and mercury orange (red sulphydryl reagent), used as indicators of SH groups, gave only weak contrast in the EM. Other procedures have been used, such as silver reduction techniques, in attempts to demonstrate SH groups. Little success attended these efforts but Swift (1968, 1969) evolved a method for SS groups in hair, following the precedent set by Dobb and Sikorski (1961) who had stained keratin fibres in bulk with $AgNO_3$. He used a strongly alkaline (pH 9·2) methenamine silver and showed conclusively that the reduction of the silver salt was due to SS and not to SH. These results were confirmed for hair keratin by Thompson and Colvin (1970) who applied the method also to plant cell walls. The alkaline methenamine-silver method for SS groups is given in Appendix 32, p.1437.

Tyrosine, Tryptophan, Histidine. The coupled tetrazonium reaction (Vol. 1, p.127) was extended to the EM level by Tice and Barrnett (1965). These authors synthesized a number of diazophthalocyanins, containing Mg, Cu or Pb as the chelated metal, and used these in place of tetrazotized benzidine. Increased contrast was produced particularly in the nuclei and in the endoplasmic reticulum. The specificity of the optical microscopical technique is not matched by this method.

Carboxyl. The carbodiimide method for COOH groups developed by Geyer (1964), (see Vol. 1, p. 126), was adapted by him for ultrastructural investigations by substituting the water-soluble thiocarbohydrazide or thiosemicarbohydrazide (Hanker *et al.*, 1964) for the naphthoic acid hydrazide of the optical microscopic method. I have not tested this reaction but the strictures recorded in Chapter 6 must clearly apply.

Elastic Tissue. Though not an end-group method the electron dense staining method for elastic tissue developed by Albert and Fleischer (1970) deserves mention here. These authors extended the fluorescence technique of Winkleman and Spicer (1962) for elastic tissue, which employed tetraphenylporphine sulphonate, by complexing the latter with silver or gold. The metal chelated porphyrin derivative was used in 5 per cent aqueous solution and staining was complete in 5 minutes. While this technique is not strictly histochemical, its specificity is apparently as high as any of the current optical microscopical methods for elastic tissue (Chapter 8).

Acid and Basic Proteins. A series of osmium coordination compounds containing acidic or basic ligands was prepared by Seligman *et al.*, (1968). These water-soluble, brownish black, monomers (occasionally polymers) were used as protein stains in the manner of acid and basic dyes. Two of the most useful reagents were osmium-dimethylethylenediamine (Os-DMEDA) and osmium-tetramethylenediamine (Os-TMEDA).

Immuno-electron Microscopy

The techniques of immuno-electron microscopy, which is sometimes called ultrastructural immunocytochemistry, can localize at the sub-cellular level any antigenic component to which specific antibodies can be obtained. They thus complement the methods of immunofluorescence (Vol. 1, Chapter 7) which have become a spearhead of modern applied histochemistry. A number of excellent reviews have appeared, dealing with theoretical and practical aspects of immuno-electron microscopy (Morgan *et al.*, 1962; Pierce *et al.*, 1964; Baxandall, 1966; Sternberger, 1967; Tanaka, 1968).

Although, exceptionally, *unlabelled* antibody which has been reacted with its antigen in tissue sections or smears can be observed in the electron microscope, it is usually necessary to label it in some way with an electron opaque marker. The latter can be attached either to the crude immunoglobulin fraction or to the purified specific antibody. A prerequisite condition is that minimal interference with antibody activity should be caused by whatever technique is employed.

As pointed out by Sternberger (1967) the forces between antigen and antibody are effective only over a short range (Pauling *et al.*, 1943; Pressman *et al.*, 1948) and thus a close fit is necessary between an antibody-combining site and an antigenic determinant. This site probably contains about 4 or 5 amino acid residues (Schechter and Sela, 1965) or 6 to 7 monosaccharide units (Kabat, 1960), or 5 or more thymidylate units (Stollar *et al.*, 1962). Therefore the degree of specificity of an antigenic determinant is limited by the number of permutations possible among the units fitting into the antibody site. Additional specificity may be contributed by the different tertiary conformations possible for each of the sequences of residues composing the antigenic determinant and it is clear that *for successful immuno-electron microscopy minimal interference with the conformations of both antibody and tissue antigen are required.*

Preparation of Tissues. The problems of fixation, in general, have been dealt with in Chapter 5 and the preparation of tissues for immunofluorescence is discussed in Chapter 7 (pp. 202–203). The latter is relevant to the problems of fixation for immuno-electron microscopy. Not only fixation, but also the processes of dehydration and embedding, can destroy the antigenic determinant site. In order to avoid this complication, in the past, the majority of work was carried out with unembedded tissues (fixed or unfixed) in the form of slices or sections and the electron-opaque immunoglobulin fraction was applied directly to them. The main problem was, and is, the low degree of penetration achieved by the antibody and a secondary difficulty was, and is, the degree of non-specific absorption of antibody to the tissues. This is not reversible by washing. As an alternative, and increasingly in recent years, the application of an electron-opaque specific antibody directly to ultrathin sections has been employed. The preparation of specific antibody, prior to labelling, requires the use of immunospecific purification (Sternberger *et al.*, 1965). The pure antibody is removed from the antiserum with insolubilized specific antigen. The antibody is eluted from the washed precipitate with dilute alkali.

It has been noted that many antigens are able to withstand osmium tetroxide fixation but few will tolerate glutaraldehyde even for very short periods. This ability to survive fixation is considered to be due to its dependence more on the intact primary, and perhaps tertiary (Lloyd *et al.*, 1966), conformations of the sequences of 4 or 5 amino acids forming the antigenic determinant site, than on the inviolability of the main structure of the protein. Accumulating evidence shows that carbodiimide (Kendall, Polak and Pearse, 1971) is particularly inocuous in this respect.

Procedures developed for the preparation of electron-opaque antibody can be divided into 3 groups:

(1) Conjugation with metallo-protein
(2) Conjugation with enzyme protein
(3) Conjugation with metals.

Labelling with Metallo-protein. The first of these three methods of approach was introduced by Singer (1959) who showed that ferritin could be conjugated to an antibody molecule without loss of immunological specificity. Ferritin is a thermostable protein, containing 23 per cent of iron, with a molecular weight of 600′000 to 700′000. Only the iron-containing internal core of the molecule is visible in the electron microscope. It consists of 4 or 6 subunits, each about 20Å in size (Kerr and Muir, 1960); the overall diameter of the molecule is 100Å and that of the internal core is 50Å. The substructure of the iron core of the ferritin molecule was investigated by Towe (1969) who concluded that the ferritin micelle might have as little as one subunit. Four different conjugating reagents have been employed for linking ferritin and antibody. These are *m*-xylylene diisocyanate (XC) (Singer, 1959), toluene-2, 4-diisocyanate (TC) (Singer and Schick, 1961), *p,p′*-difluoro-*m, m′*-dinitrophenylsulphone (FNPS) (Tawde *et al.*, 1962), and bis-diazotized benzidine (BDB) (Williams and Gregory, 1967). These last authors made a comparative survey of the four conjugating techniques, by assessing their ability to form dimers of bovine serum albumin. They found the BDB was the most effective. According to Schick and Singer (1961) TC was superior to XC on account of its being a milder reagent and producing no non-covalent linkage of antibody to ferritin. In practice both XC and TC have proved popular, used in a two-stage procedure:

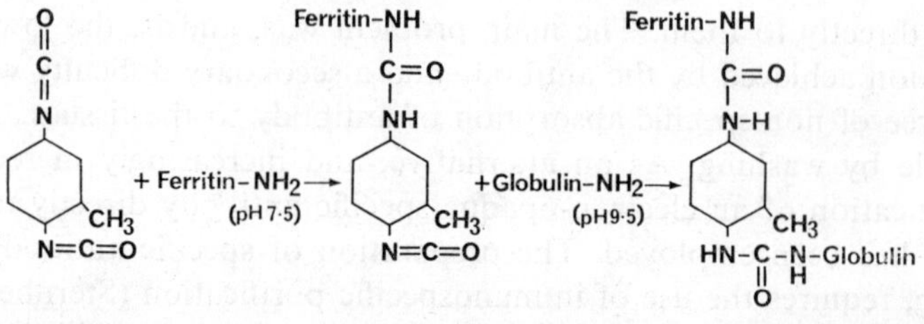

In the first stage of the reaction the diisocyanate is mixed with ferritin in phosphate buffer at pH 7·5. Ureido bonds are formed with terminal and ε-amino groups in the ferritin. At pH 7·5 the isocyanate group in position-2 is unreactive and coupling occurs only in the 4-position. Subsequently, the ferritin-diisocyanate conjugate is added to the immunoglobulin fraction in borate buffer at pH 9·5. Unreacted isocyanate groups cross-link to the globulin.

The second step of the above procedure is relatively mild and yields about 33 per cent of conjugated globulin. Most of the latter exists as a 1:1 antibody-ferritin complex. Since conjugated antibody interferes with localization by competing with labelled antibody it is necessary to purify the latter. This can be done, simply, by ultracentrifugation (Hsu *et al.*, 1963).

The advantages and disadvantages of the ferritin-labelling technique were discussed by Tanaka (1968). The most important complications are:

(1) Loss of antibody after conjugation.
(2) Strong non-specific staining.
(3) Inapplicability to resin-embedded sections.
(4) Failure of the antibody to penetrate intact cells, subcellular organelles or viruses.
(5) Lack of distinction from natural ferritin.

Retention of the specific activity of ferritin-antibody conjugate is favoured by mild reacting conditions, giving low yields, (Vogt and Kopp, 1965). The problem of non-specific staining is a more serious one. Frequently ferritin granules are found around the nuclei or other structures and it may be difficult, even with the use of control ferritin-globulin (non-antibody) to determine whether in a given site the reaction is true or false. Problems of the non-applicability of the ferritin-labelled antibody method to resin-embedded sections have not been overcome. According to Mott (1963) and Striker *et al.* (1966) ferritin conjugates are absorbed non-specifically on the embedding media commonly employed in electron microscopy. Natural ferritin occurs in two forms. Isolated ferritin granules may be scattered throughout cells, especially in nuclei and mitochondria, and accumulations of ferritin may occur in lysosomes. If ferritin-labelled conjugate is used on living cells they may take it up by pinocytosis. Failure to penetrate cell and organelle membranes, a more serious objection, has been countered by subjecting the tissues to a number of disruptive procedures, before incubation with the labelled antibody. Rifkind *et al.* (1964) cut very small pieces of tissue into

5 per cent formalin and, if necessary, ground them in a mortar. Repeated freezing and thawing was employed by Morgan *et al.*, (1961) and treatment with digitonin by Dales *et al.*, (1965).

Progressive damage to the cell is not an acceptable principle and methods which avoid it, some of which are discussed below, have begun to replace older methods such as ferritin-labelling. The method is applicable particularly to the localization of surface antigens, such as the capsular antigens of microorganisms (Rifkind *et al.*, 1964), surface antigens of cell membranes (Baxandall *et al.*, 1963) and to the localization of extracellular materials such as fibrin (White *et al.*, 1964; Wyllie, 1964; Daria Haust *et al.*, 1965). A double labelling technique (fluorescein and ferritin) enabled Chapman and Taylor (1966) to localize rheumatoid factor on intracellular granules in cells from human lymph nodes.

Labelling with Enzyme proteins. Conjugation of antibody to enzyme proteins was successfully achieved by SriRam *et al.*, (1966) and by Nakane *et al.*, (1966). This approach was not only logical (Sternberger, 1967), but also eminently useful. The principle incorporates the notion that while with an antibody containing a fixed amount of an electron opaque label contrast is proportional simply to the number of antibody molecules reacting, an enzyme label can produce an increasing amount of electron opaque reaction product as the time of reaction is increased. Already, in 1966, a number of EM enzyme methods were available (see section on Enzymes, below) and the choice fell upon peroxidase (Nakane and Pierce, 1966) and acid phosphatase (Nakane and Pierce, 1967). As coupling agents for linking enzyme to antibody SriRam *et al.*, (1966) used either *p,p'*-difluoro-*m,m'*-dinitrodiphenyl sulphone or 1-ethyl-3-(3-dimethylaminopropyl)-carbodiimide. An alternative method of conjugation, using glutaraldehyde, was employed by Avrameas and Ternynck (1969) and Avrameas (1969). This method gives very satisfactory labelling and is easy and convenient to use. Details are given in Appendix 32, p. 1437. A standard metal precipitation technique (Chapter 16, p. 552) was used to localize acid phosphatase and the DAB technique of Graham and Karnovsky (1966) was used to demonstrate peroxidase (see Chapter 19, p. 854, and Fig. 269, p. 1280).

The simultaneous localization of three different antigens (hormones) was performed by Nakane (1968) using three different peroxidase techniques. After application of the first labelled antibody, its situation was marked with the DAB reaction. The antibody was then eluted (Appendix 32, p. 1439) and the second labelled antibody applied. An α-naphthol-pyronin sequence was used for revealing this antibody. A further elution was followed by the third application and demonstration by means of 4-chloro-1-naphthol. The brown, pink, and grayish-blue products were clearly distinguishable.

Since the molecule of peroxidase is smaller than that of acid phosphatase, peroxidase-labelling came to be regarded as the method of choice. An additional factor in its favour was the low degree of interference by the natural enzyme. A strong peroxidase is an infrequent occurrence in mammalian tissues.

Exogenous cytochrome c. (mol wt 12'270) was used as an ultrastructural tracer by Karnovsky and Rice (1969), taking advantage of its ability to act as a peroxidase at low pH and high H_2O_2 levels, in the presence of chlorides, citrate and acetate. In addition to the enzymes considered above, alkaline phosphatase was employed by Leduc *et al.*, (1969) for the localization of immune globulins in plasma cells and lymphoblasts. These authors attempted to demonstrate immunoglobulins both in thin resin-embedded sections and in ultra-thin frozen sections. In these projects they were not successful. Kawarai and Nakane (1970) however, using the peroxidase label and an indirect technique, localized L.H., G.H. and prolactin in thin sections of methylacrylate-embedded rat anterior pituitary gland. In each case the hormone was found in the secretion granules, but not elsewhere.

The problem of penetration of antisera through the tissues disappears when ultra-thin sections are employed. Since even unconjugated antibody fails to penetrate freely into a normal intact cell (except by pinocytosis in cells using this mechanism) it may be assumed that ultra-thin section application of conjugate is the method of choice. This technique has its own problems however. These include interference by the embedding medium, where one is employed, and the often severe alterations which take place in the conformation of the antigenic determinant sequence, effectively altering its reactivity. Various techniques have been used to reduce, or abolish, interference caused by steric hindrance of the antigenic determinant site by Epon or methacrylate. In the case of the latter Sternberger *et al.*, (1965) recommended etching with benzene-saturated water.

A heme-protein with a mol wt of 1900 was prepared by Feder (1970) for use as an ultrastructural tracer. In this compound, known as microperoxidase, the heme molecule is bound by thioether bridges to the cysteine residues of an undecapeptide (Falk, 1964). Its structure is shown below:

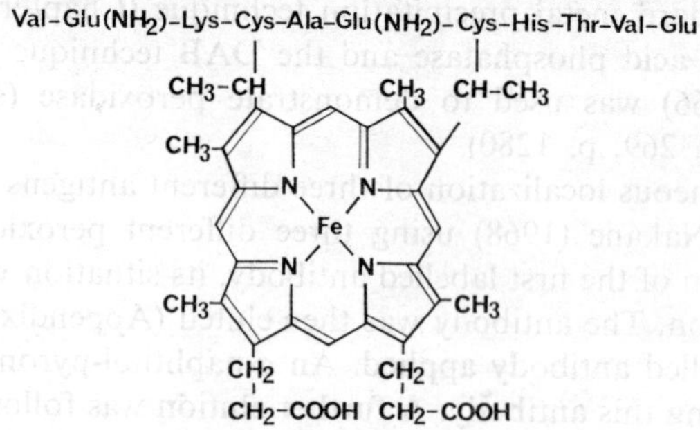

The method of production is given in Appendix 32, p. 1440 but, at the time of writing, microperoxidase has not been used in immunoelectron microscopy.

Labelling with Metals. A variety of metals, and a variety of techniques have been used in attempts to confer electron opacity on immunoglobulins.

A metal chelation method, coupling antibody to anthranilic acid and Pb^{2+} ions, was used by Koíke *et al.*, (1964) but the process destroyed 90–95 per cent of the antibody activity at a level of 7·5 atoms of the metal per mole.

Uranium. Successful labelling of antibody with uranyl acetate, using the principle of immunospecific protection, was achieved by Sternberger *et al.*, (1965). Uranium chelation destroys antibody activity but if the specific binding sites of the antibody are first occupied by antigen, a degree of protection is afforded. Subsequent elution of the antigen leaves uranium-labelled antibody. This technique was further developed by the application of the so-called "osmium-black" principle.

The use of bidentate ligands for bridging metals to osmium was introduced, for both optical and electron microscopy, by Seligman *et al.*, (1966) and Hanker *et al.*, (1966). For example, thiocarbohydrazide ($H_2N.NHCS.NH.NH_2$) was shown to bridge through one of its hydrazines, and the thiocarbonyl group, to osmium present in osmium tetroxide-fixed ultrathin sections. Using excess reagent, the second hydrazino group remained free to react with additional osmium producing a deposit of osmium-black (see below, p. 1278). In a similar reaction uranium is capable of chelation by thiocarbohydrazide (TCH) to provide the second hydrazino group for subsequent reaction with osmium.

An alternative technique, developed by Sternberger *et al.*, (1966) was based on the use of diazothioether, introduced by Seligman *et al.*, (1965). This method, immunodiazothioether-OsO_4 (immuno-DTO) suffers from technical difficulties due to instability of the reagent, even at − 70°.

Iron. The use of ferrocene for marking proteins was suggested by Franz (1967a and b, 1968a) and its application to Immunoelectron Microscopy was described by the same author (1968b). Ferrocene, bis(cyclopentadienyl)iron, was first synthesized by Miller *et al.*, (1952) and assumed to have the structure:

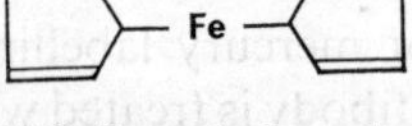

however, Wilkinson *et al.*, (1956) found that the compound was diamagnetic. This fact, taken into consideration with its reported infrared spectrum, compels the adoption of a different structure (Bailar, 1956):

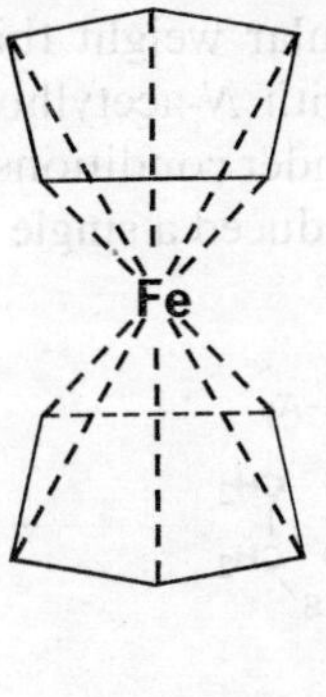

There are several different ways of coupling ferrocene to protein, as described by Franz (1968b). The reagent recommended by the latter, 3-carboxy-4-ferrocenyl phenylisothiocyanate (CFPI), was synthesized from ferrocene coupled to diazotized 2-amino-5-nitrobenzoic acid, by reduction with nascent hydrogen (catalyst, Raney nickel), and subsequent treatment with thiophosgene.

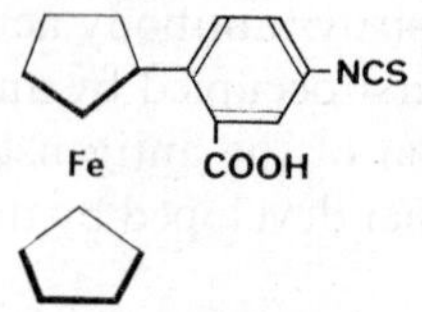

3-Carboxy-4-ferrocenylphenylisothiocyanate

For the demonstration of bacterial capsular antigens, using appropriate antisera coupled to CFPI, the method gave excellent results.

Mercury. Labelling methods using this metal, proposed and used by Pepe (1961), Pepe and Finck (1961) and by Zhdanov *et al.*, (1965, 1966), depended on diazotization of a suitable mercury-substituted aromatic compound followed by diazo coupling on to the protein. Pepe (1961) used tetra-(acetoxymercuri)arsanilic acid but Zhdanov *et al.*, (1966) found this compound to be unstable, releasing mercury in a form capable of sublimation in the electron beam. They used instead *p*-aminophenyl mercuric acetate.

The method used by the Russian authors was severely criticized by Sternberger and Donati (1966) who observed that it would produce, at best, an antibody labelled with 2 to 3 metal atoms per molecule. They suggested, furthermore, that since immunospecific protection was not employed the antibody combining sites would probably be blocked. It is clear that the group -Hg$\overline{\text{OAc}}$ is by no means inert and that special steps must be taken to prevent non-specific binding of the labelled antibody (or of labelled globulin if whole immunoglobulin fractions are used) to reactive groups, such as thiols, in the tissues.

An alternative method for mercury labelling of proteins was developed by Kendall (1965) in which antibody is treated with a thiolating reagent, which reacts with amino groups to introduce thiol groups which can then bind, for example, methylmercury in a mercaptide linkage. The methylmercurimercapto group (CH_3Hg-S-) is inert and stable: the Hg-S bond can be broken only by an excess of some low molecular weight thiol. In Kendall's original procedure, α-globulin was treated with *N*-acetylhomocysteine thiolactone (AHTL) in a silver-catalysed reaction under conditions modified from those of Benesch and Benesch (1958). This introduced a single (silver-protected) thiol group for each amino group reacted:

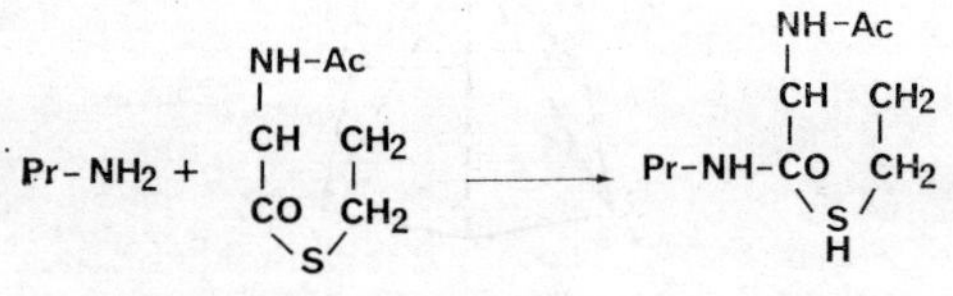

Removal of the protective silver, followed by treatment with methylmercuric hydroxide yielded the methylmercuric mercaptide. In practice, about 40 per cent of the NH_2 groups of rabbit γ-globulin could be thiolated, giving a labelled derivative containing about 26 Hg atoms per molecule, equivalent to a 3·3 per cent mass or density increase. When the globulin was from an antiserum to a haemagglutinating virus, it was found that the ability specifically to inhibit haemagglutination was reduced by less than 50 per cent on mercury labelling, i.e. affinity for the antigen was still very great. However, it is doubtful whether this relatively small density increase could be sufficient to confer visibly increased electron scattering power on the antibody.

In an extension of the same principle, Kendall (1971) applied homocysteine thiolactone (HTL) in place of AHTL. HTL contains an unblocked primary amino group, which can undergo reaction with another HTL molecule, thereby promoting introduction of more than one SH group per NH_2 reacted:

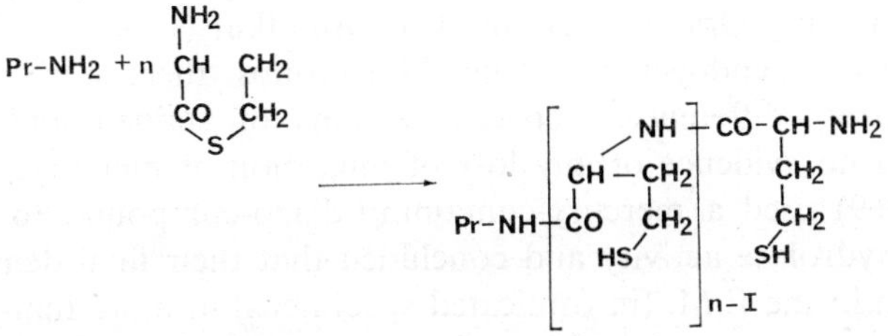

A single-stage treatment of IgG with HTL at pH 10 in the presence of methylmercuric hydroxide yields a soluble protein labelled with, typically, 10–11 per cent of mercury. Further successive treatments are possible, since proteins thiolated with HTL always contain an (α)-amino group for each ε-amino reacted (see formula above). In practice, sequential treatments appeared to promote some denaturation of the protein, although soluble derivatives bearing up to 14–15 per cent of mercury were obtained.

IgG from antisera to haemagglutinating viruses showed no loss in specific inhibiting activity when treated with HTL and methylmercuric hydroxide. Hence, modification of antibody by HTL is even less harmful than treatment with AHTL, despite the much greater amount of mercury bound. This is presumed to be because the modified lysine residues still bear an amino group after HTL treatment, as opposed to the uncharged *N*-acetylated substituent introduced by AHTL. Further qualitative evidence that antibody activity is conserved in the HTL-treated derivatives was derived, in the case of antiglobulin antibodies, from immunodiffusion, where it was shown that the mercury labelled antibody was non-precipitating, but could specifically inhibit precipitation of antigen by the homologous unlabelled antibody.

Although Kendall's mercury labelling method is founded on a sound rational basis, its actual value for localizing antigens in the E.M. still remains to be demonstrated. In view of the data presented above, it would seem that failure could only arise in three situations: (a) if the binding of mercury to antibody did not survive reaction with the antigen; (b) if the 10–14 per cent of bound mercury should still prove insufficient to scatter electrons sufficiently; and (c) if the mercury label should prove unstable in the operating conditions of the electron microscope (sublimation).

As regards (a), it is highly improbable that mercury bound as methylmercury mercaptide could be dislodged from antibody either by exposure to fixed tissue, or by engagement with the antigen. Situation (b) could almost certainly be remedied by improving reaction conditions to effect greater binding. As explained above, there is no theoretical obstacle to further reaction. Hence, only (c) deserves further consideration at this stage.

Pepe and Finck (1961) observed loss of mercury from their tetra-(acetoxymercuri)-substituted benzene label in the electron beam, which could be greatly minimized by layering the (Araldite) section with evaporated carbon. However, when methylmercuric chloride and 1-(4-chloromercuriphenylazo)-naphthol-2(Mercury Orange) were used by Mundkur (1964) as thiol-specific electron stains for endogenous tissue thiol groups, there were clearly discernible contrast differences between sections of stained and unstained material, and no evidence of any loss of migration of mercury. Smith and Fishman (1969) used a mercury-containing diazo-compound to reveal the site of acid hydrolase activity and concluded that their final dense product could be seen in the E.M. (in osmicated specimens) in more than one form. Again there was no sign of significant loss or migration of mercury. It would appear that firm predictions cannot be made as to how a particular mercury label will survive in the electron beam; certainly there can be no automatic assumption that mercury is unsuitable. Most relevant to the antibody labelling procedure described above is a very recent demonstration by Formanek and Formanek (1970) of staining of bacterial murein sacculi. Unstained sacculi are barely visible in the E.M., but they have significant contrast when stained with heavy metals by different methods, including mercaptide formation by mercurials after thiolation of their amino groups. Furthermore, this readily visible staining is brought about by the introduction of one mercury atom per sacculus subunit (mol. wt. 1500), equivalent to a mass (density) increase of 200/1500, or about 13·5 per cent. This is of the same order of density increase as that produced by Kendall's method.

It is not possible to make any final assessment of the merits and demerits of the various techniques for labelling antibody proteins, outlined above. At the present time most work is carried out with enzyme-labelled antibodies. This will continue to be the case unless some of the theoretical advantages of metal-labelling can be translated into practical advantages.

Nucleic Acids

Two principal methods (Feulgen-silver, metal staining) have been used for the positive demonstration of nucleic acids at the E.M. level, and the two standard methods of extraction (nucleases, acids) for their negative demonstration. The classical Feulgen technique for DNA, modified by the use of alkaline silver reagents (Chapter 9, p. 265) in place of Schiff's reagent, was used by Jurand *et al.*, (1958), by Bryan and Brinkley (1963) and by Bryan (1964). Adams *et al.*, (1965) used a periodic acid-silver diamine sequence for the preferential demonstration of DNA at the optical and E.M. level. The degree of localization obtained by these last authors appeared to be very satisfactory. Despite some views to the contrary (Geyer, 1969, p. 59) osmium was observed to interfere with staining and various devices were used in endeavours to overcome this. The osmium black method, using thiocarbohydrazide or thiosemicarbohydrazide, was applied to Feulgen-hydrolysed tissues by Seligman *et al.*, (1965). The method is essentially similar to the PATO and PATCO methods for carbohydrates (see below, p. 1274).

Methods based on staining with metal salts demonstrate both DNA and RNA. An early application was the use of uranyl acetate by Huxley and Zubay (1961), for contrast staining of DNA, and the use of lead hydroxide to show RNA, in tissues fixed in a variety of fixatives. The mechanisms of uranyl acetate and lead hydroxide contrast staining were discussed by Zobel and Beer (1961) and by Marinozzi and Gautier (1962). Sodium tungstate (10 per cent), adjusted with HCl to pH 5·6, was recommended as a selective stain for both DNA and RNA by Swift and Adams (1966), and Watson and Aldridge (1961, 1964) used indium ($InCl_3$), originally in water but later in acetone, for the same purpose. Bismuth was recommended by Albersheim and Killias (1963) as a stain for chromatin. Their reagent was prepared by dissolving metallic bismuth in nitric acid, diluting, adding citric acid and adjusting the pH to 7. Both RNA and DNA showed additional contrast.

A procedure permitting the localization of DNA and RNA at both optical and electron microscope levels was reported by Chan-Curtis *et al.*, (1970a). These authors prepared a complex from acriflavine (3,6,-diamino-*N*-methyl acridinium chloride) and phosphotungstic acid, which they dissolved in ethanol or buffer for use as a stain. The reagent produced fluorescent staining of DNA and RNA and strong contrast in these two materials at the E.M. level. Sulphated cerebrosides were also found to be reactive but sulphated glucosaminoglycans were not stained. Physical studies of the acriflavine-PTA complex were carried out by Chan-Curtis *et al.*, (1970b). These showed that the effect of PTA was to lower the net positive charge of the dye and to shift the absorption maximum from 452 to 446 nm.

Carbohydrates

The most important ultrastructural techniques for polysaccharides and polysaccharide-protein complexes are based on periodic acid oxidation, despite recent indications that this procedure is in no way stoichiometric. The presumptive dialdehyde oxidation product of tissue-bound *vic*-glycols (Vol. 1, p. 308) has been demonstrated in a variety of different ways. The most important of these are (1) Silver methods, (2) Thiosemicarbazide and similar methods and (3) Fluorinated arylhydrazine methods.

Silver Methenamine. The original silver methenamine technique (Gomori, 1946) was adapted for use at the E.M. level by Dettmer and Schwarz (1952–53) and, subsequently, by a large number of workers (Churg *et al.*, 1958; Marinozzi, 1961; Movat, 1961; Suzuki and Sekiyama, 1961; De Martino and Zamboni, 1967; Swift and Saxton, 1967; Hernandez and Rambourg, 1967). Useful methods were described by Pickett-Heaps (1967), for botanical materials and by Rambourg (1967) for animal tissues. The technique described by the latter author is given in Appendix 32. Interpretation of the results of silver methenamine methods is complicated by lack of specificity. Many of the authors listed above used control, non-oxidized, preparations and compared the degree of silver deposition with that obtained after pretreatment with periodic acid.

Thiocarbohydrazide. Thiocarbohydrazide (TCH) and thiosemicarbazide (TSC) were introduced into ultrastructural cytochemistry by Seligman *et al.*, (1965) as osmiophilic reagents for demonstrating aldehyde-containing macromolecules, produced by periodic acid oxidation of tissue sections. The product, described as osmium black, possessed excellent qualities for the demonstration of aldehydes at optical microscope level. It is primarily as electron opaque products, however, that the many varieties of osmium black find their chief use.

The two reagents were used for many years as blocking agents for aldehyde groups (Vol. 1, p. 456). Seligman and his coauthors pointed out that their hydrazino groups are powerful reducing agents and that they readily reduce OsO_4. The two techniques which were developed were described as periodic acid-TSC-OsO_4, or PATO (Hanker *et al.*, 1964) and periodic acid-TCH-OsO_4, or PATCO. They are described in Appendix 32 and have a much higher degree of specificity than the silver techniques described above. Using the PATO reaction McAlpine (1969) showed that the mucosubstance(s) present in experimental amyloid were localized exclusively between the fibrils, in the ground substance.

A modification of the PATO/PATCO method was introduced by Thiéry (1967) who developed his electron opaque final product with silver proteinate instead of osmium tetroxide. This technique (PA-TSC-SP) was used by Anderson and Personne (1970) in their studies on the localization of glycogen in vertebrate and invertebrate spermatozoa, and by Jewell and Saxton (1970) for the demonstration of *vic*-glycols in plant cell walls. In both cases the results were excellent and the method is strongly recommended. Details are given in Appendix 32, p. 1442.

Fluorinated Arylhydrazines. A technique for the E.M. demonstration of periodate-reactive mucosubstances was described by Bradbury and Stoward (1967). The original reagent which these authors employed was pentafluorophenylhydrazine. When this became unavailable from commercial sources

Stoward and Bradbury (1968) suggested the substitution of *p*-fluorophenylhydrazine hydrochloride. Whichever reagent is used the intermediate reaction product (hydrazone) is subjected to further treatment, first with ammonium sulphide to produce a thiol derivative, and then with OsO_4 to produce osmium black. Details of the method appear in the Appendix.

Contrast Stains. The earliest simple contrast stain for carbohydrate complexes was phosphotungstic acid (PTA), originally recommended by Pease (1966) who stated that multiple hydrogen bonds could form between PTA and polysaccharide chains. Silverman and Glick (1969) suggested, however, that the intense electron-opaque staining produced by treating aldehyde-fixed tissue sections with PTA, in aqueous acidic medium, was due to interaction with proteins. They detected no relationship to carbohydrate and Glick and Scott (1970) declared that PTA should be regarded as an anionic stain for positively charged groups, particularly of proteins. These authors denied that PTA was a stain for carbohydrates but its virtues, in this respect, were defended by Pease (1970) who stressed the pH-dependence of the reaction. He claimed that acidic PTA (acidified to below pH 1·0 with HCl) was "reasonably specific" for complex carbohydrates. The truth, as so often, lies somewhere between the two extremes.

After testing a number of heavy metals Watson (1958a and b) indicated the superiority of Pb^{2+} for staining glycogen. Lead ions have no specificity for polysaccharide, however, but after periodic acid oxidation glycogen granules are well stained with lead citrate (Perry, 1967) and they can be distinguished from other particles, such as ribosomes, by diastase digestion.

Several other contrast methods have been used for ultrastructural definition of carbohydrates. Among these are the colloidal iron technique, Alcian blue, and ruthenium red. The first of these was used by Curran *et al.*, (1965) and by Wetzel *et al.*, (1966), to demonstrate acid mucopolysaccharides at the E.M. level after initial studies by Gasic and Berwick (1963) had produced a suitable adaptation of the original Hale method (Vol. 1, p. 349). Further work with colloidal iron was carried out by Gasic *et al.*, (1968) and by Arnold and Hager (1969). Both these authors indicated that penetration was an outstanding problem and that the method was not applicable to intracellular components.

Alcian blue was used as an electron stain by Rothman (1970) to demonstrate a mucopolysaccharide on the surface of an acanthocephalan worm *Moniliformis dubius*. Thin slices of the worms (200μ thick) were stained in the dye, at pH 2·5. Processing for electron microscopy was carried out in the usual manner. A dual staining technique, using Alcian blue and ruthenium red, was reported by Yamada (1970). The second dye was recommended by Luft (1964) as an electron stain for cell surfaces and it was used by Gustafson and Pihl (1967a and b) and by Pihl *et al.*, (1968) for studies of acid mucopolysaccharides. Poor penetration of the reagent was noted by Gustafson and Pihl (1967a) and by Fowler (1970). The latter author attempted to overcome this deficiency by means of a perfusion technique. Deposits in the basal lamina of the glomerular capillaries were attributed to acidic glycoproteins. Details of a suitable ruthenium red technique are given in Appendix 32.

Lipids

As emphasized by Idelman (1965) fundamental problems are posed by the preservation of lipids for ultrastructural studies but more lipid survives the usual aldehyde fixation and resin embedding than would be expected. Quantitative studies were carried out by Cope and Williams (1967) using tritiated glyceryl oleates, taken up into peritoneal macrophages. The amount of lipid extracted by each processing stage was estimated by scintillation counting. Distribution of the label within the cells was assessed by autoradiography. Losses of neutral glycerides varied from 50 to 100 per cent and they were found to be soluble in Araldite and glycol methacrylate, as well as in ethanol, with only a slight increase in retention after OsO_4 post-fixation. Negatively stained preparations of macromolecular assemblies of lecithin, cholesterol and saponin were examined by Glauert and Lucy (1968). Information was obtained on the mode of formation of biological membranes *in vivo*.

A method for the demonstration of phosphoglyceride which was developed by Weller *et al.*, (1965) was based on the modified ferric hydroxamate-silver technique of Gallyas (1963) for phosphoglycerides (see Vol. 1, p. 426). The digitonin reaction (Vol. 1, p. 409) was used by Ökrös (1968) to demonstrate cholesterol at the E.M. level. The digitonin-cholesterol crystals formed in the reaction were osmiophilic and they could be recognized by their cylindrical structure made up of coaxial lamellae.

Enzyme Techniques

The early phases of electron enzyme cytochemistry were bedevilled by fixation problems. If morphological criteria were satisfied enzyme activity was reduced or absent and *vice versa*. The situation was transformed with the introduction into histochemistry of the dialdehyde fixatives, particularly glutaraldehyde, by Sabatini, Bensch and Barrnett (1963). It was not difficult to prophesy (Pearse, 1963), in view of the cogent data supplied by Valentine (1958), that the easily foreseeable expansion in enzyme ultracytochemistry would be based, substantially, on the metal precipitation principle.

Pioneer work in the field was carried out by Barrnett and his associates (Barrnett, 1959; 1962, 1964; Barrnett and Palade, 1957, 1959; Tice and Barrnett, 1962; Torack and Barrnett, 1963), and the worst problems were rapidly overcome. The early lead taken by the hydrolases, and particularly the phosphatases, has never been reversed. Nevertheless, the present distribution of available methods is much wider, as will be seen.

While the majority of E.M. enzyme methods are carried out on tissues prefixed in aldehyde or dialdehyde fixatives there is increasing interest in the use of unfixed materials (Zagury *et al.*, 1968; Hajós and Kerpel-Fronius, 1970), and in techniques like those presented by the first of these authors, which permit the observation of identical tissue components by both optical and electron microscopy. A very competent review of the use of aldehyde and dialdehyde fixatives for electron microscopy was given by Reale and Luciano (1970). The danger of removal of enzyme products from ultrathin sections by counterstaining and other procedures was described by Kalimo *et al.*, (1968). Lead citrate at pH 12 caused considerable loss of $PbSO_4$ and $BaSO_4$ (reaction products of aryl sulphatase) and staining with uranyl acetate removed both. Heavy metal phosphates, on the other hand, were relatively stable.

The methods of enzyme electron cytochemistry will be discussed under specific individual headings and individual techniques will be dealt with when they stress points of particular importance.

It is necessary to discuss here a principle which is particularly applicable to the demonstration of enzymes at the ultrastructural level. It is widely applicable also to the demonstration of metal ions or metallic compounds deposited in the tissues by virtue of their selective activity for particular tissue components. This principle (Hanker, Anderson and Bloom, 1971) is described as osmiophilic polymer generation.

Osmiophilic Polymer Generation. In cases where Hatchett's brown (cupric ferrocyanide) had been formed either by the activity of cholinesterases (see p. 793) or succinate dehydrogenase (p. 948) Hanker *et al.*, (1971) noted that it had been found beneficial to convert this product to the sulphide and to attach to it either osmium or silver (bridging technique). The thiocarbohydrazide procedure (described in the PATCO method, p. 1274) had also been employed, by several different groups of workers, to bridge osmium to Hatchett's brown. Three to four molecules of osmium can be bridged to each molecule of the latter. This procedure produces greatly enhanced contrast, for both optical and electron microscopy, and it allows a marked decrease in the incubation period.

As an alternative to this approach Hanker *et al.*, (1971) proposed that the metal ion captured or deposited as the result of a cytochemical reaction should be used to catalyse the generation of an osmiophilic polymer. The secondary metal, osmium, is attached to the sites where osmiophilic polymer has been bound by the action of the primary metal catalyst. A convenient monomer, from which osmiophilic polymers may be generated, is DAB, as shown in the formulation on p. 1278. By this procedure approximately 8 atoms of osmium can be bound to each molecule of Hatchett's brown.

The osmiophilic polymer generation method has been applied, successfully, to the demonstration of acid phosphatase, esterases, cholinesterases and dehydrogenases. In each case, as described in the appropriate section below, and in Appendix 32, after removal of the incubation medium with distilled water the tissues are treated with 0·5 per cent DAB in 0·05M phosphate buffer,

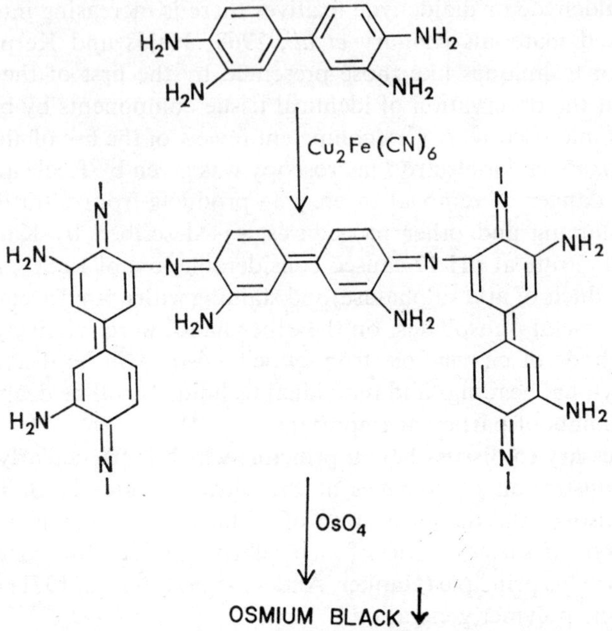

at pH 6·8 for one hour. Further rinsing is then followed by osmication, either with vapour (50–55°) or by immersion in 2 per cent unbuffered OsO_4 at 5° for 15 minutes.

Non-specific Alkaline Phosphatase. A calcium-lead variant of Gomori's (1939) method was used by Reale (1962) and this method, as well as the original calcium-cobalt variant, was used by Reale and Luciano (1967), and applied to cryostat sections. Two-step methods such as these have not given sufficiently accurate localization for critical studies. A cadmium one-stage technique, allowing incubation at pH 9, was developed by Mizutani and Barrnett (1965) but this has not proved popular. A direct lead method was described by Mölbert *et al.*, (1960) but here, in order to keep the metal in solution in Tris buffer, a chelating agent was used with disodium phenyl phosphate as substrate. The pH of incubation was 7·6 and there was therefore some opportunity for activity of other phosphatases to occur. Lead citrate at pH 9 in veronal buffer was used by Tranzer (1965) who obtained a fine precipitate in fresh tissue sections incubated without fixation after attachment to small blocks of polymerized resin (Araldite). The high concentration of Pb^{2+} ions in this medium inhibits the enzyme.

A technique employing lead nitrate at pH 9·0, with no chelating agent, was developed by Hugon and Borgers (1966). This was further used by them (1968) for studies on the absorbing cells of the duodenum. Another lead citrate method was developed by Mayahara *et al.*, (1967). This method gives reliably reproducible results, although the size of the final reaction product is too large for the finest localizations of enzyme to be made. Details of the technique are given in Appendix 32 and the type of result obtainable is shown in Fig. 270. Sources of error affecting alkaline phosphatase techniques were discussed by Reale and Luciano (1964) and the effect of thickness of the specimen was investigated by Mayahara and Ogawa (1968). These authors found that activity of the enzyme, in rat kidney tubules, could be demonstrated only when 20–40μ sections (fixed or unfixed) were used for incubation.

Various substrates have been used for the E.M. demonstration of alkaline phosphatase. Sodium-β-glycerophosphate does not always give optimum results. Oledzka-Slotwińska *et al.*, (1967) suggested the use of cytidine monophosphate, especially for demonstration of the liver enzyme.

Specific Alkaline Phosphatases (1) Myosin ATPase. This enzyme has been most extensively investigated. Early studies by Persijn *et al.*, (1961) were followed by those of Tice and Barrnett (1962), Hori (1963), Hori and Chang (1963), Man *et al.*, (1963), and Zebe (1963). Later studies were those of Zebe and Falk (1964), Sommer and Spach (1964), Tice and Smith (1965), Dorn (1966) and Somogyi and Sotonyi (1969). Considerable variation in technique can be noted, as would be expected. One author (Dorn, 1966) used thallium nitrate, in place of the more usual lead salt, for his capture reaction. Somogyi and Sotonyi (1969) replaced lead nitrate by lead citrate or cobalt chloride and they added 25mM cysteine and 50 mM sodium tartrate.

(2) Mitochondrial ATPase. Studies on this enzyme by Lazarus and Barden (1962, 1964) were complimented by those of Scarpelli and Craig (1963), whose elegant preparations showed the enzyme in the rod outer segments of the retina as well as in mitochondrial matrix, and of Ashworth *et al.*, (1963) who found the enzyme on the inner (cristal) membranes. With earlier methods using cryostat sections preservation of tissue was poor but Ashworth *et al.* (1963) claimed localization on the mitochondrial cristae. Lazarus and Vethamany (1966), using lead nitrate at pH 7·2 found the reaction product diffusely distributed in the mitochondrial matrix.

Later studies were carried out by Schultze and Wollenberger (1965), Shiose and Sears (1965), Rechardt and Kokko (1967), Grossman and Heitkamp (1968) and Ogawa and Mayahara (1969). Unfixed, isolated, intact skeletal muscle mitochondria were used by Grossman and Heitkamp (1968) and, with a lead nitrate medium, the rather large deposits of reaction product were localized in the matrix. Rechardt and Kokko (1967) obtained a similar localization in some mitochondria in the spinal cord neuropil. The same result was obtained using ITP as well as ATP as substrate. The degree of localization obtained by Ogawa and Mayahara (1969) was superior to that obtained by other workers and their method is therefore given in Appendix 32.

(3) Membrane (Na-K) ATPase. The ouabain sensitive sodium-potassium activated enzyme was successfully localized, at optical microscope level, by Palkama and Uusitalo (1968) and later Uusitalo and Palkama (1970) adapted their method for ultrastructural localization of the enzyme. It was observed that ouabain inhibition could be obtained either by intravital injection or by application to unfixed tissues. After fixation the effect could not be obtained.

Lead Ions and Nucleoside Phosphatase Histochemistry. In a long series of papers Rosenthal and his colleagues (Rosenthal *et al.*, 1966; Moses *et al.*, 1966; Ganote *et al.*, 1969; Tice, 1969; Rosenthal *et al.*, 1969) discussed the various factors involved. In the first and second papers of the series it was shown that lead ions could hydrolyse ATP, at pH 7·2 and 37°, and it was suggested that this was a source of artifact. The third communication (Rosenthal *et al.*, 1969) indicated that with low ATP or high Pb^{2+} concentrations more phosphate was bound in the tissues than could be attributed to enzyme activity. Ganote *et al.* (1969) reached the further conclusion that tissue and chemical events other than enzyme activity contributed to reaction product localization in the ATPase technique, and Tice (1969) showed that the characteristics of the enzyme inhibition by Pb^{2+} were consistent with the predicted effects of lead-ATP chelate formation. Increased concentrations of ATP in the medium produced increased solubility of lead phosphate. The objections recorded above were convincingly answered by Novikoff (1967), who pointed out *inter alia* that, while non-enzymic dephosphorylation of IDP runs parallel to that of ATP, with the former as substrate it was possible to demonstrate reaction product in the endoplasmic reticulum of kidney tubule cells while with ATP only the β-cytomembranes contained a deposit. These two systems are no more than 25 nm apart. The argument continued, however, (Moses and Rosenthal, 1967; Rosenthal *et al.*, 1970) with a final flourish from Novikoff (1970) who indicated that all the criticism poured on the method was of little consequence, a conclusion which some of us had already reached. Current users of the technique have (rightly) chosen to ignore most of the contradictory evidence while remaining properly cautious in their interpretation of results.

Thiamine Pyrophosphatase. The establishment of this enzyme as a Golgi-marker (Novikoff and Goldfischer, 1961) was followed by a number of studies at the ultrastructural level (Essner and Novikoff, 1962; Meek and Bradbury, 1963; Saito and Ogawa, 1966; Ishikawa *et al.*, 1967; Vorbrodt, 1967; Saito, 1968). In general these have confirmed the already well established conclusions drawn from the use of the technique in optical histochemistry (Vol. 1, p. 531).

Glucose-6-Phosphatase. Fine structural localization of this enzyme was first reported by Tice and Barrnett (1962), in rat liver. Excellent results were obtained by Ericsson (1966) who demonstrated the enzyme in the smooth ER of glutaraladehyde-fixed rat liver. Saito and Ogawa (1967) used a similar technique and found activity in the membranous cisternae of the ER, on the matrical side of the membrane.

Acid Phosphatases. The original Gomori (1941) technique for this enzyme (Vol. 1, p. 552) was modified for E.M. studies by Holt and Hicks (1961). It is one of the most reliable and reproducible methods of modern electron

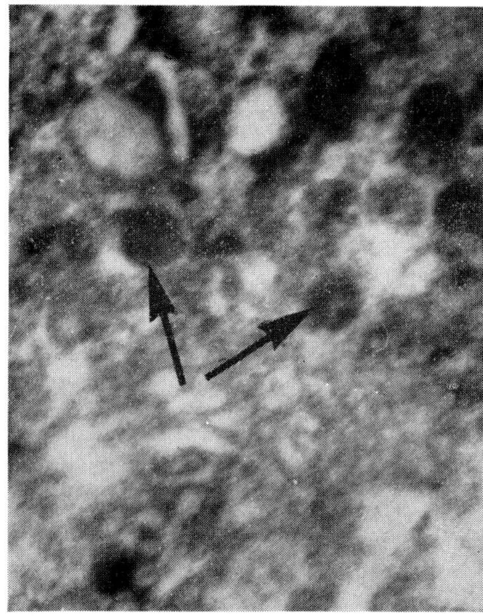

FIG. 269. Immunoelectron microscopical localization of calcitonin in the specific granules of a thyroid C cell. Peroxidase-labelled antibody technique. No counterstain has been used. Some of the granules are uniformly dense. Others (arrows) irregularly. × 45′000.

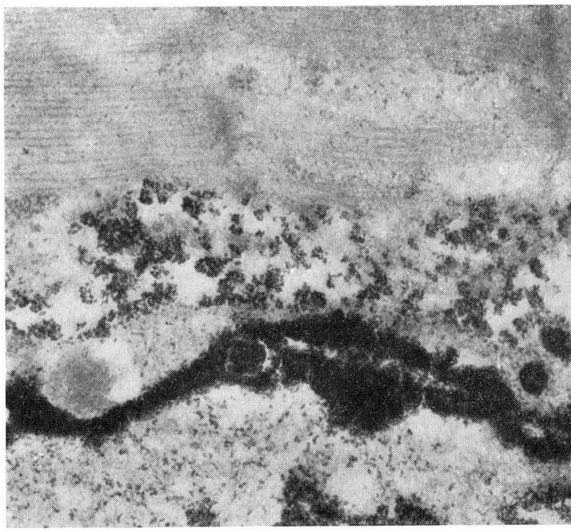

FIG. 270. Localization of alkaline phosphatase in rat atrial muscle (technique of Mayahara *et al.*, 1967). The electron opaque deposits are associated with the fine network of adrenergic nerve fibres. Clumped deposits extend up to the sarcolemma. Glycogen particles are present in the muscle fibres. × 30′000.

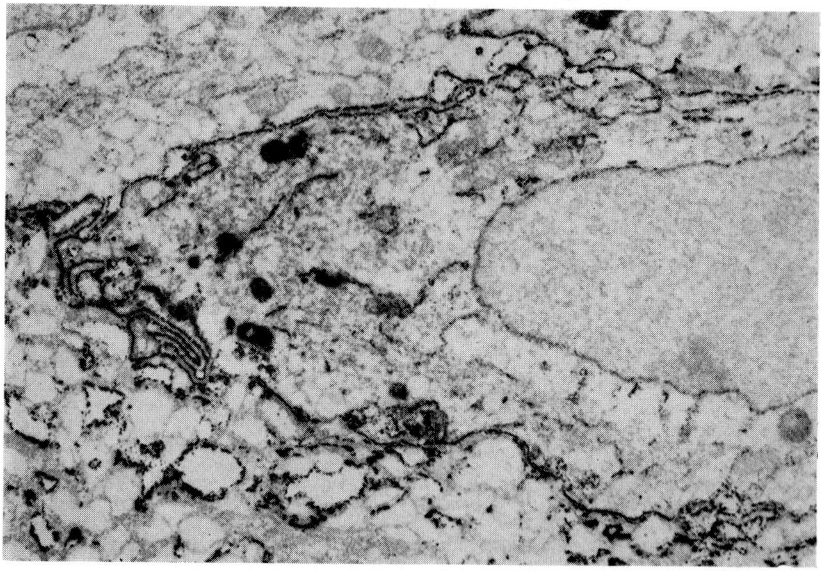

FIG. 271. Epithelial cells in mouse skin, showing acid phosphatase in endoplasmic reticulum as well as in lysosomes. Osmium bridging technique. × 14′000

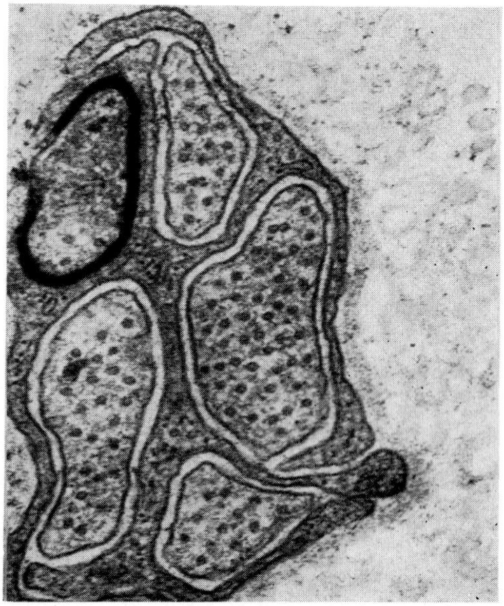

FIG. 272. Localization of non-specific cholinesterase in a single myelinated fibre in a mixed nerve. Copper-lead thiocholine technique. × 45′000.

enzyme cytochemistry but, as pointed out by Etherton and Botham (1970), it is necessary in the case of many tissues to alter both the method of fixation and the substrate. Palkama and Rechardt (1970) indicated that the results are strictly dependent on the concentration of Pb^{2+} ions in the medium. Using rat and mouse adrenal medulla they showed that with 10mM substrate (*β*- or *α*-glycerophosphate) and 3·4mM Pb^{2+} localization was mainly lysosomal. With 6·8mM Pb^{2+} nearly all the specific (amine) storage granules contained deposits. They regarded this result as non-enzymic. The technique of Holt and Hicks (1961) as modified by Etherton and Botham (1970) is given in Appendix 32. The fine localization of acid glucose-6-phosphatase activity was reported by Kazimierczak (1970). Reaction product in the ER was attributed to glucose-6-phosphatase. That which was found on the surfaces of endothelial cells, and within the glomerular basement membrane, could not be interpreted.

Following the use of a mercury-containing diazotate for E.M. demonstration of acid hydrolases by Smith and Fishman (1969), an alternative procedure was developed by Livingston *et al.* (1970) who synthesized two lead-containing diazonium chlorides, and used these in a Naphthol AS-Bl system to produce an insoluble electron opaque product. The diazonium chlorides were derived from triphenyl-*p*-aminophenethyl-lead and triphenyl-*p*-aminophenyl-lead. To judge from the illustrations accompanying the paper, the results were very satisfactory. I have had no opportunity to test the method, however.

An osmium bridging technique for acid phosphatase was described by Hanker, Yates, Clapp and Anderson (1972), using di-dicyclohexylammonium 2-naphthylthiol phosphate (DDNTP) as substrate, in the presence of cupric ions and potassium ferricyanide. The method, which is described in Appendix 32, gives precise enzyme localization, as shown in Fig. 271, opposite.

Cholinesterases and Non-specific Esterases. The literature on this group of enzymes is very considerable and a large number of methods, based on different principles, are in existence. The first application of any of the cholinesterase techniques of optical histochemistry, to the E.M. localization of the enzyme, was made by Lehrer and Ornstein (1959).

Azo-Dye Techniques. These authors took the then new diazonium salt, hexazotized pararosanilin, and employed it to capture the free 1-naphthol released from its acetate by acetylcholinesterase activity in the motor-end-plate. Their results suggested that the product had insufficient electron density and the method was criticized also on account of lack of specificity of the substrate (Bergman *et al.*, 1967). Despite these strictures the principle of the method is sound and it might well be reinvestigated with alternative naphtholic substrates and rapid-coupling electron opaque diazotates.

Thiolacetic Acid Method. The optical histochemical method (Chapter 17, p. 794) was modified by Barrnett (1962) and applied to the demonstration of specific cholinesterase in the myoneural junction. The method was used by Mori *et al.* (1964), who applied it to rat brain but it has found little general favour. Complex formation between lead ions and thiolacetic acid was investigated by Haugaard *et al.* (1965) who found that both these and aurous ions could form coordinate complexes of one metal ion and two thiolacetic acid residues. The nature of the "M-band Enzyme" originally demonstrated by Barrnett and Palade (1959), using the thiolacetic acid technique, was investigated by Karnovsky and Hug (1963). Their observations suggested that the enzyme was either an organophosphate-resistant non-specific esterase or a site readily acetylated by thiolacetic acid, essentially a non-enzymic feature.

Thiocholine-Copper-Ferrocyanide Methods. These methods were introduced into optical and electron histochemistry by Karnovsky (1964) and Karnovsky and Roots (1964). They employ thiocholine esters as substrates and the capture reaction, supposedly, is the reduction of ferricyanide ions in the medium by the free thiocholine. In the presence of Cu^{2+} the product is the coloured, electron opaque, copper ferrocyanide. In the absence of copper ions no reaction product can be demonstrated.

The thiocholine-copper-ferrocyanide technique has the merit of using the relatively specific substrates acetyl- or butyrylthiocholine. Using the direct-colouring method Eränkö *et al.* (1967) made a study of cholinesterases in rat spinal cord. They found that the large size of the reaction product, a serious defect at E.M. level, could be reduced by prolonged formaldehyde fixation. This apparently improved the otherwise poor penetration of the reagents. The enzyme was localized in the nuclear envelope, in the ER, the Golgi vesicles and around the synaptic terminals. The direct-colouring method was used also by Bell (1966) who found that omission of sucrose from the incubating medium improved penetration which, in pre-fixed unfrozen sections, was confined to a layer 10μ thick. Rat brain cholinesterases were also investigated by Shimizu and Ishii (1966), using Karnovsky's method and perfusion fixation with either 2 per cent glutaraldehyde or 4 per cent formaldehyde. The myoneural junctions in rat diaphragm were studied by Teräväinen (1967) who found acetylcholinesterase in the muscle sarcoplasm close to the post-synaptic membrane.

An osmium bridging technique was applied to the product of the Karnovsky and Roots procedure by Hanker, Dixon, Bloom and Weitsen (1972) who fixed small blocks in the glutaraldehyde-formaldehyde fixative described by Karnovsky (1965).

Thiocholine-Copper Methods. These methods were developed directly from their optical histochemical forbears by modifications of the medium designed to increase the efficiency of the capture reaction. The substitution of silver for copper was proposed by Birks and Brown (1960) but the most consistently successful modifications have been those introduced by Lewis and Shute (1964, 1966, 1969). Details are given in Appendix 32.

Thiocholine-Copper-Lead Technique. Preliminary studies by Joó *et al.*

(1965) indicated that increased electron density of the copper thiocholine product could be obtained by substituting Pb^{2+} for Cu^{2+}. Later Kása and Csíllik (1966, 1968) found that retention of both Cu^{2+} and Pb^{2+} provided still finer resolution. The constitution of their incubating medium is given in Appendix 32, and the type of result obtainable is shown in Fig. 272.

Gold-Thiocholine and Gold-Thiolacetic Methods. The substitution of aurous gold for lead as the capture reagent for the thiolacetic acid method (Koelle and Foroglou-Kerameos, 1965) provided sufficient improvement for Koelle and Gromadzki (1966) to make the same modification to the thiocholine method. When the reaction is carried out in the presence of a high concentration of phosphate buffer a fine granular precipitate, presumably gold thiocholine phosphate, marks sites of enzyme activity. Details of incubating media required for the AuThCh method are given in the Appendix and the result is shown in Fig. 273, p. 1286. A full account of the localization of cholinesterases by means of gold salt techniques was provided by Koelle, Davis and Gromadzki (1967). The authors pointed out that the chief drawback of the thiocholine methods, their reliance on quaternary ammonium substrates which penetrate poorly through cell membranes, is a factor of major importance at E.M. level.

Osmiophilic Diazoether Techniques. Three new esterase substrates which were synthesized by Bergman *et al.* (1967) were designed to provide an osmium black final reaction product. Two of the substrates 2-naphthylthiolacetate (NTA) and 2-thiolacetoxybenzanilide (TAB) were preferentially hydrolysed by acetylcholinesterase while the third 2-thiolpropionoxybenzanilide (TPB) was hydrolysed at faster rates by non-specific cholinesterases and esterases. The substrates were used, in phosphate buffer at pH 7·4, and the liberated thiols were captured (by S-coupling) by a diazonium salt (Fast blue BBN) to form yellow osmiophilic diazoethers. Subsequent exposure to OsO_4 vapour converted the latter into osmium black. In electron micrographs the final product appeared as discrete electron dense droplets, localized solely on the post-synaptic membrane.

Osmium Bridging Technique. The thiolacetoxybenzanilide substrate, described above, was used by Hanker, Yates, Clapp and Anderson (1972) in a bridging procedure based on Hatchett's brown, for the demonstration of esterases, rather than cholinesterases.

Dimethylbutylthioacetate Technique. This quaternary carbon analogue of acetylthiocholine was synthesized by Nyberg-Hansen *et al.* (1969) and used in a more or less standard thiocholine-type technique with Cu^{2+} ions in the medium. The primary precipitate was transformed into CuS by incubation with 3 per cent Na_2S in dilute HCl (pH adjusted to 6·5 to 7·0). The results in no way approach in clarity those obtainable by the gold thiocholine method but are comparable with those obtainable by other thiocholine methods.

Acetylselenocholine and Acetyl-β-Methylthiocholine Methods. The use of these two substrates in place of acetylthiocholine was reported by Kokko,

Mautner and Barrnett (1969). Using the copper-ferrocyanide principle, with tartrate instead of citrate as the chelating agent, they found the most rapid reaction with acetylselenocholine but the second substrate was more specific for acetylcholinesterase.

Indoxyl Acetate-BAXD Method. In order to provide a check for the whole range of metal precipitation techniques for cholinesterases Kawashima and Murata (1969) modified the indoxyl-HPR method of Holt and Hicks (1966), producing an azo-indoxyl method with indoxyl acetate as substrate and tetrazotized *N,N'*-bis(*p*-aminophenyl)-l, 3-xylenediamine (BAXD) as coupling agent. A final stage of post osmification produced an electron dense deposit of finely granular character. This technique, details of which are given in Appendix 32, might be improved by the use of a variety of substituted indoxyl acetate substrates. In any case it is extremely valuable, and an excellent principle in enzyme cytochemistry, to correct excessive devotion to a single type of method by the application of an alternative.

The indoxyl-BAXD method was used by Kawashima (1968) for the demonstration of non-specific esterases.

Lipases. A modification of the original Gomori (1949) method for lipase was made by Murata *et al.* (1968). In order to avoid non-specific deposition of lead the concentration of lead nitrate in the medium was considerably reduced. Applying the technique to mouse pancreas the authors found specific activity, in the form of rather large single deposits, in rough ER, Golgi lamellae, zymogen granules and secretory canaliculi. Further improvements are clearly required with this technique.

Techniques for Beta-glucuronidases. A simultaneous coupling method using naphthol AS-BI glucuronide and hexazotized pararosanilin was developed for use at the E.M. level by Hayashi *et al.* (1968). Formalin-cacodylate-sucrose fixation and frozen sections were employed. Brief pre-treatment of sections with cold ethanol prior to incubation produced marked enhancement of staining. The final product was localized, in variable amounts, in dense bodies.

A post-coupling procedure for *β*-glucuronidase was introduced by Smith and Fishman (1969). These authors incubated frozen and non-frozen sections in media containing naphthol AS-BI glucuronide as substrate, for one to several hours. After washing the sections were transferred to a post-coupling medium containing diazotized acetoxymercuric aniline. Subsequently they were reacted for 30–45 minutes with 1 per cent thiocarbohydrazide and finally with osmium tetroxide. The reaction product was predominantly lysosomal.

Phosphorylase Techniques. Demonstration of the polyglucose product of phosphorylase activity in tissue sections was first made, at the E.M. level, by Sasaki and Takeuchi (1963). Improvements in the technique were recorded by Sasaki and Takeuchi (1968) and by Sasaki and Aoki (1970). These last authors incubated non-frozen pieces of tissue (rat skeletal muscle) in a standard Takeuchi medium (Appendix 18, p. 1327) at room temperature for 30 minutes. After incubation the tissues were fixed in cacodylate-buffered glutaraldehyde and then post-fixed in osmium tetroxide. As described by Takeuchi and Sasaki (1968), it is easy to distinguish the newly synthesized polyglucose from native glycogen originally present in the muscle fibres.

Aryl Sulphatases. Techniques for this group of enzymes were evolved, practically simultaneously, by Goldfischer (1965) and by Hopsu *et al.* (1965 a and b). Both for optical and electron microscopy (see Chapter 23, p. 992) lead or barium ions provided the capture reaction and the substrates employed were either 8-hydroxyquinoline sulphate, *p*-nitrophenyl sulphate or 2-hydroxy-5-nitrophenyl sulphate. A technique using this last compound (*p*-nitrocatechol sulphate) is described in Appendix 23, p. 1363, and the adaptation for E.M. studies follows the usual pattern.

The optimum conditions for the demonstration of aryl sulphatases were worked out by Hopsu-Havu *et al.* (1967) who recommended fixation in 5 per cent cacodylate-buffered glutaraldehyde, and incubation with *p*-nitrocatechol sulphate in the presence of barium ions at pH 5·5.

The technique, which is reliable and reproducible, is given in Appendix 32.

Cytochrome Oxidase. A modification of Burstone's (1960) technique (Chapter 19, p. 847), using *N*-phenyl-*p*-phenylenediamine as substrate was used by Kerpel-Fronius and Hajós (1967) for the E.M. demonstration of cytochrome oxidase. No coupling agent was employed and the final product was first postchelated with copper sulphate and then with potassium ferrocyanide. With this method structural preservation is less good than is desirable and the product occurs in the form of relatively large droplets. An improved ultrastructural localization of the enzyme was obtained by Seligman *et al.* (1967a) who used as substrate *N*-benzyl-*p*-phenylenediamine (BPDA) and as coupler either l-naphthol, or one of a number of substituted naphthols.

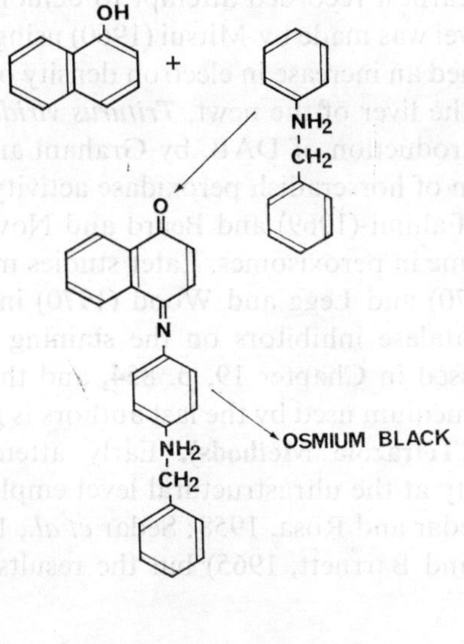

The indoaniline product of the reaction, on treatment with OsO_4 yields an electron opaque coordination polymer (osmium black). With this method sites of activity of cytochrome oxidase were revealed in the form of small droplets.

A series of *p*-substituted aromatic diamines was synthesized by Plapinger *et al.* (1968) and these were tested in the cytochrome oxidase reaction. A method was described, using *N,N'*-bis(*p*-aminophenyl)-1, 3-xylyene diamine (BAXD) as substrate, which was satisfactory at optical microscope levels only.

Non-droplet localization of cytochrome oxidase was achieved by Seligman *et al.* (1968) using a method based on the oxidative polymerization of 3,3'diaminobenzidine (DAB). The principle of the method is described in Chapter 19, p. 848 and details of the E.M. procedure are given in Appendix 32.

Monoamine Oxidase. The osmiophilic ditetrazolium salt, thiocarbamyl nitro-blue tetrazolium, was used by Boadle and Bloom (1969) for the demonstration of this enzyme at the ultrastructural level. Tissues were fixed in 2 per cent buffered paraformaldehyde, cut at 50–80μ in the cryostat, and reacted at 37° for 4 hours. After post-fixation in OsO_4 an increase in electron opacity was noted in the outer mitochondrial membranes.

DOPA Oxidase. The ultrastructural localization of DOPA oxidase, in Harding-Passey melanoma, was achieved by Novikoff *et al.* (1968) using the optical microscopical technique of Becker *et al.* (1935). The same method was used by Okun *et al.* (1970) who were able to determine that the oxidation of tyrosine to melanin in mammalian melanocytes, mast cells and eosinophils was mediated by peroxidase while the conversion of the intermediate (DOPA) to melanin was catalysed by a copper-containing tyrosinase. The formation of DOPA melanin, in Harding-Passey melanoma, is shown in Fig. 274, opposite.

Peroxidase. The earliest recorded attempt to demonstrate this enzyme at the ultrastructural level was made by Mitsui (1960) using a standard benzidine procedure. He obtained an increase in electron density of the specific granules of the leucocytes in the liver of the newt, *Triturus viridescens.*

Following the introduction of DAB, by Graham and Karnovsky (1966), for the demonstration of horseradish peroxidase activity it was used by Hirai (1968), and later by Fahimi (1969) and Beard and Novikoff (1969), to show the endogenous enzyme in peroxisomes. Later studies made by Wood (1969), Wood and Legg (1970) and Legg and Wood (1970) included investigations into the effects of catalase inhibitors on the staining of microbodies. The matter is fully discussed in Chapter 19, p. 854, and the composition of the modified incubating medium used by the last authors is given in Appendix 32.

Dehydrogenases (Tetrazole Methods). Early attempts to demonstrate dehydrogenase activity at the ultrastructural level employed either Nitro-BT or Tetra-nitro-BT (Sedar and Rosa, 1958; Sedar *et al.*, 1962; Rosa and Tsou, 1964, 1965; Ogawa and Barrnett, 1965) but the results were far from satis-

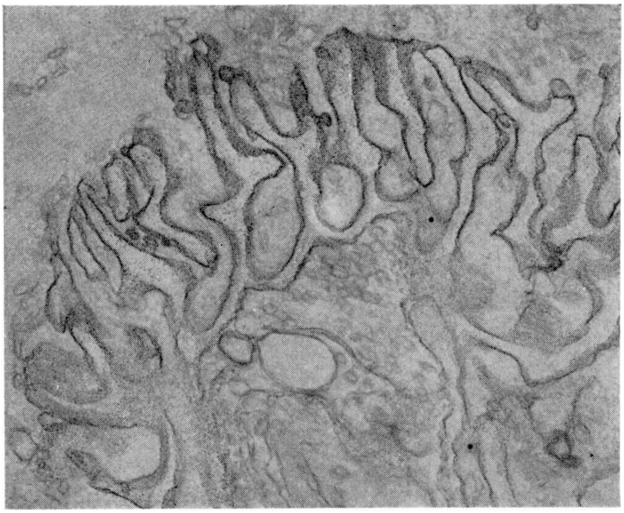

FIG. 273. Mouse intercostal muscle. Incubated for 120 minutes by the acetyldisulphide-$Na_3Au(S_2O_3)_2$ method, following fixation in 4 per cent formaldehyde-sucrose-maleate solution. The reaction product is confined to the prejunctional membrane and the postjunctional membrane and its infoldings. × 31′000.

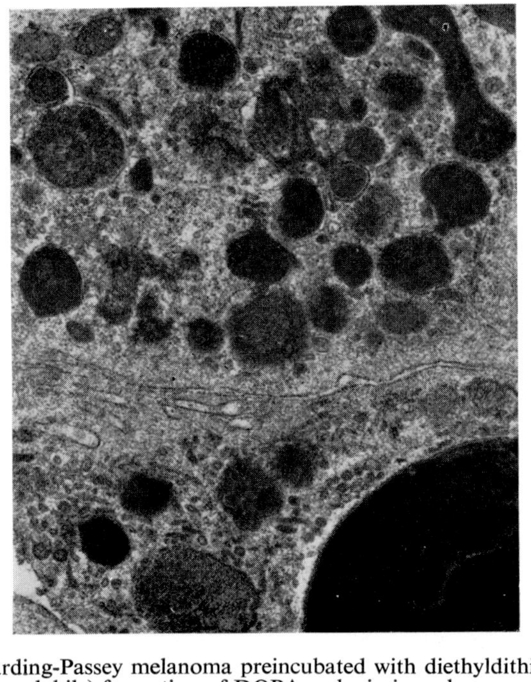

FIG. 274. Harding-Passey melanoma preincubated with diethyldithiocarbamate. Shows (catalase-labile) formation of DOPA-melanin in melanosomes and in the nucleus. Counterstain uranyl acetate and lead citrate. × 20′500.

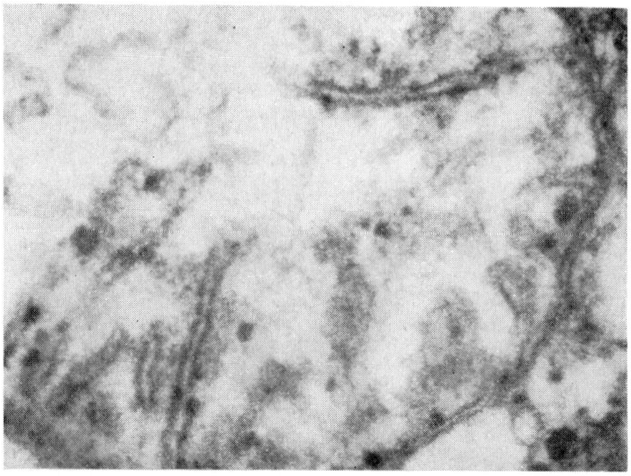

FIG. 275. Rat liver. Isolated mitochondrial preparation incubated for succinate dehydrogenase by the modified Ogawa *et al.* (1968) technique (incubation period 2 minutes). Shows part of a single mitochondrion. Reaction product, chiefly on the cristal membranes, in the form of single deposits which are often paired (i.e. on opposite sides of the cristal membrane). × 90′000.

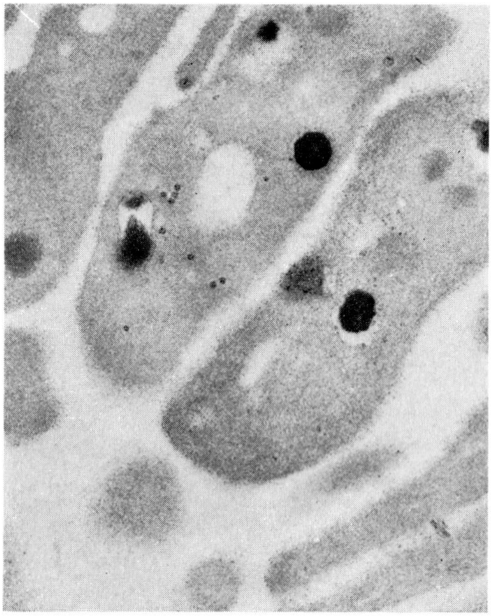

FIG. 276. Ultrastructural localization of 5-HT in blood platelets. The strongly osmiophilic 5-HT-containing organelles shown here are discharged by brief pretreatment, *in vitro*, with reserpine. × 39′000.

factory. Observing the need for a tetrazolium salt which would yield an osmiophilic formazan Seligman *et al.* (1967b) produced thiocarbamyl nitro-BT (TC-NBT). The synthesis of this compound, 2,2′-di-*p*-nitrophenyl-5,5′-di-thiocarbamylphenyl-3,3′-(3,3′-dimethoxy-4,4′-biphenylene)ditetrazolium chloride, was accomplished by starting with *p*-cyanobenzaldehyde, preparing the *p*-cyano analogue of nitro-BT, and converting this to the thiocarbamyl analogue with thioacetamide. The structure of TC-NBT is shown below:

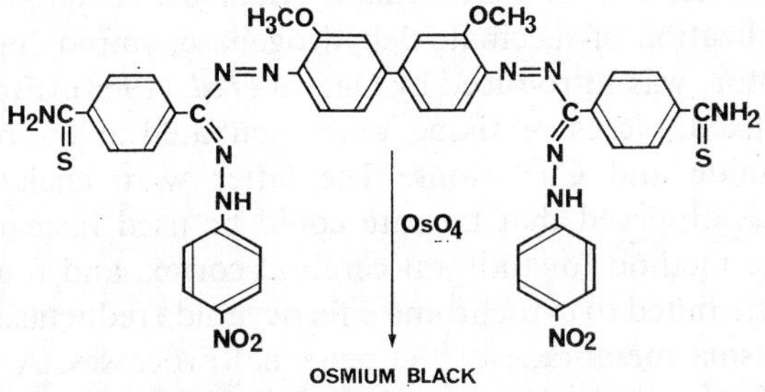

by treatment with OsO_4 this product is converted to osmium black.

The original authors found that TC-NBT would penetrate into slices of rat heart muscle to a depth of about 30μ, whether the tissue was fresh or fixed. The use, and duration, of fixation depended, of course, on the sensitivity of the particular enzyme being demonstrated. Diformazan deposits from TC-NBT were observed to be uniformly distributed along both surfaces of the mitochondrial cristae, with encroachment of the intracristal spaces. The outer limiting membranes were unstained.

A number of tetrazoles, designed to improve the localization of dehydrogenases, were synthesized by Tsou *et al.* (1968). The best of these, the osmate of 2,2′,5,5′-tetra-*p*-nitrophenyl-3,3′-stilbene ditetrazolium chloride (Os-TNST), was used with glutaraldehyde-fixed rat heart muscle which, after incubation, was treated with uranyl acetate only. The formula of Os-TNST is given below:

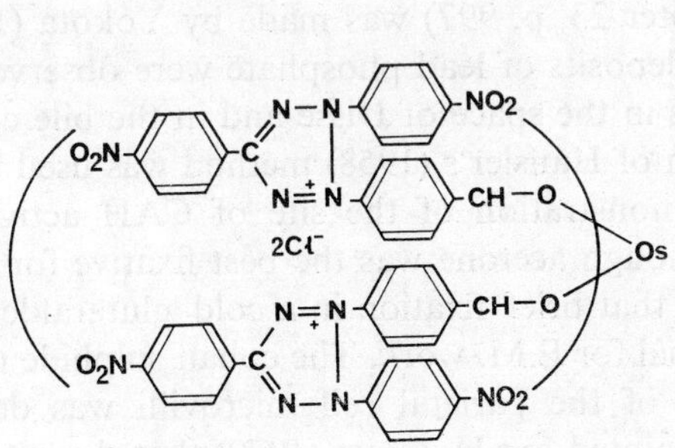

Sites of mitochondrial activity were convincingly demonstrated by the electron opaque formazan as intracristal. At times the whole intracristal space was delineated while at other times dense particular deposits were noted to be arranged in a helical manner with a distance of 10 nm between the coils.

Further observations on the use of tetrazoles (particularly TC-NBT) for mitochondrial localization of succinate dehydrogenase were made by Haydon, Smith and Seligman (1968). These authors showed that in non-incubated control slices of rat heart muscle there were mitochondria with densely stained matrices indistinguishable from those found in incubated samples. Since adjacent lightly, and densely, stained mitochondria were found it seemed likely that the differences were not due to experimental artifact.

Dehydrogenases, Copper-Ferrocyanide Methods. A method for the ultrastructural localization of succinate dehydrogenase, employing ferricyanide as electron acceptor, was introduced by Ogawa *et al.* (1968). Fresh, or hydroxyadipaldehyde-fixed, pieces of tissue were incubated in the presence of substrate ferricyanide and Cu^{2+} ions. The latter were chelated with citrate although it was observed that tartrate could be used instead. Hirano *et al.* (1968) used the method to study rat cerebral cortex, and found oxidoreductase activity, attributed to cytochrome c-ferricyanide reductase, in the synaptic region and plasma membrane of the neuronal processes. A modification of the original Ogawa method was made by Kerpel-Fronius and Hajós (1968) who showed that the choice of chelator was a crucial one. They used sodium-potassium tartrate, as did Kalina *et al.* (1969) for their studies on isolated mitochondria in different physiological states. The latter authors found that after short incubation periods (1–5 minutes) the reaction product was deposited in punctate form on the outer side of the cristal membrane (Fig. 275, p. 1287). In many cases two deposits were opposite one another, on opposite sides of the crista that is. The method of Kerpel-Fronius and Hajós (1968) is given in Appendix 32, p. 1451.

A bridging technique for lactate dehydrogenase was described by Hanker, Kusyk, Bloom and Pearse (1972). In this method, details of which are given in Appendix 32, a modified Ogawa technique is followed by treatment of the post-fixed blocks with thiocarbohydrazide and osmium tetroxide.

Carbonic Anhydrase. An adaptation of Hansson's (1967) method for this enzyme (see Chapter 23, p. 997) was made by Yokota (1969). Using mouse liver, particulate deposits of lead phosphate were observed on the microvilli of the hepatocytes in the space of Disse and in the bile canaliculi.

A modification of Hausler's (1958) method was used by Cross (1970) for the successful demonstration of the site of CAH activity in rat stomach parietal cells. Although acetone was the best fixative for optical microscope studies she found that brief fixation in a cold glutaraldehyde-formaldehyde mixture was optimal for E.M. work. The cobalt sulphide reaction product on the outer surface of the parietal cell microvilli was deposited at regular intervals. It was pointed out by Cross (1970) that the extent of hydration of the cobalt sulphate used to prepare the medium was of vital importance. Cobalt sulphate heptahydrate slowly loses water and may be converted into the monohydrate in a period of a few months. If an under-hydrated salt is used in the belief that it is the heptahydrate, gross overstaining results.

Transferases. A method for the ultrastructural demonstration of aspartate aminotransferase (E.C. 2.6.1.1.) was evolved by Papadimitriou and van Duijn (1970). Rapid inactivation of the soluble enzyme by aldehyde fixation was avoided by the addition of ketoglutarate to the fixative. The method is discussed in Chapter 23, p. 983.

The ultracytochemical localization of carnitine acetyltransferase (E.C. 2.3.1.7) was made by Higgins and Barrnett (1970).

Ligases. Fine structural localization of acetyl coenzyme A carboxylase (E.C. 6.4.1.2) in rat liver cells was demonstrated by Yates, Higgins and Barrnett (1969) using a method by which the hydrolysis of ATP, coupled to the synthesis of malonyl CoA, was demonstrated by precipitation of lead phosphate. The reaction product was closely associated with the outer surfaces of the membranes of the granular endoplasmic reticulum.

Peptidases. An osmium black technique was developed by Rutenburg *et al.* (1969) to demonstrate γ-glutamyl transpeptidase activity at the E.M. level. A new synthetic substrate, *N*-(γ-L-glutamyl)-4-methoxy-2-naphthylamide, was used in a simultaneous coupling azo dye technique with Fast blue BBN (diazotized 4′-amino-2′,5′-diethozybenzanilide) and the product was subsequently chelated with Cu^{2+} ions. Following treatment with thiocarbohydrazide and osmium, the black, electron-dense, final product was observed in the endoplasmic reticulum in the vicinity of the zymogen granules in the apical portion of pancreatic acinar cells.

An alternative osmium black technique, essentially applicable to the E.M. localization of any peptidases capable of yielding naphthylamine products from the hydrolysis of artificial histochemical substrates, was proposed by Seligman *et al.* (1970). The principle of the reaction is shown below:

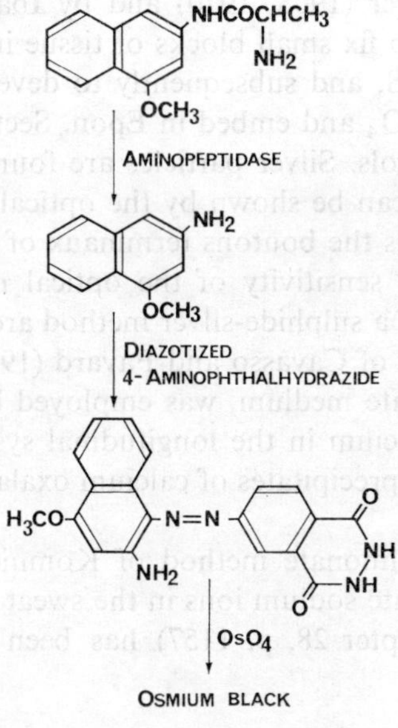

Freshly diazotized 4-aminophthalhydrazide is coupled to the product (4-methoxy-2-naphthylamine) of γ-glutamyl transpeptidase activity. The product, a lipophobic osmiophilic dye, is treated with OsO_4 to yield osmium black.

Extraction Techniques for Nucleic Acids and Proteins

Pioneer work in this field was carried out by Leduc and Bernhard (1960) and by Leduc *et al.* (1960), using fixed materials embedded for electron microscopy in a water-soluble medium. The reagents used were nucleases and HCl, for the extraction of nucleic acids, and pepsin and trypsin for various protein components. A summary of the various techniques was presented by Leduc and Bernhard (1961), who used the water-soluble polyepoxide (X133/2097) developed by Stäubli (1960). Later, glycol methacrylate, or hydroxypropyl methacrylate (Ripley and Pease, 1965) became the preferred embedding media and Douglas (1970) developed a method of perchloric acid extraction of nucleic acids applicable to glutaraldehyde and osmium-fixed materials embedded in water-insoluble epoxy resins.

Techniques for Inorganic Ions

Heavy Metals. Most of the work in this field has been concerned with the localization of heavy metals at the ultrastructural level. Various modifications of the original sulphide-silver method of Timm (1958) have been evolved for the purpose, (see Chapter 28, p. 1149). The main contributor has been Pihl (Pihl, 1967, 1968; Pihl and Falkmer, 1967) but the method has been used also by Müller and Geyer (1968, 1970) and by Ibata and Otsuka (1969). The usual procedure is to fix small blocks of tissue in glutaraldehyde (3 per cent), saturated with H_2S, and subsequently to develop them with Timm's developer, post-fix in OsO_4 and embed in Epon. Sections from these blocks are compared with controls. Silver particles are found at all the usual sites where heavy metal ions can be shown by the optical microscope technique, and in some sites, such as the boutons terminaux of the mossy fibres of the hippocampus, where the sensitivity of the optical method is insufficiently high. Technical details of a sulphide-silver method are given in Appendix 32.

Calcium. The method of Cavasso and Favard (1966), employing fixation in a glutaraldehyde-oxalate medium, was employed by Komnick (1969) for the demonstration of calcium in the longitudinal system of tubules in frog skeletal muscle. The fine precipitates of calcium oxalate, after OsO_4 fixation, are electron opaque.

Sodium. The pyroantimonate method of Komnick (1962) was used by Ochi (1968) to demonstrate sodium ions in the sweat glands of the rat's foot pad. This method (Chapter 28, p. 1157) has been severely and properly criticized.

Chloride. The principle of silver precipitation of chloride ions was extensively discussed by Komnick and Bierther (1969). These authors used a more or less standard method and produced fine silver precipitates in a variety of tissues. The precise relationship of these to pre-existing chloride ions is difficult to determine.

Localization of Biogenic Amines

A large number of techniques have been recorded for the E.M. demonstration of biogenic and exogenous amines in storage granules in cells and nerve endings. These have all been developed from existing techniques of optical histochemistry (see Chapter 27, p. 1112). While a considerable degree of specificity has often been claimed in respect of one or other of the common biogenic amines (noradrenalin, NA; Adrenalin, A; dopamine, DA; 5-hydroxytryptamine, 5-HT) the true specificity is usually much lower than this. The easiest distinction is between A and NA, as pointed out by Wood and Barrnett (1963, 1964) for the two cell types in the adrenal medulla.

Available techniques can be divided into four groups: (1) *Chromaffin*, (2) *Argentaffin*, (3) OsO_4, (4) $KMnO_4$. All are used in conjunction with or following glutaraldehyde fixation, without which only empty vesicles may be found to mark the site of amine-storage granules.

According to Wood and Barrnett (1964) the **chromaffin reaction** carried out at pH 4·1 was specific for NA while at pH 5·6 both NA and A were demonstrable. An explanation of the mechanism of the reaction was provided by Coupland and Hopwood (1966a and b). These authors (1966b) pointed out that at pH 6·5 dichromate could react with the matrix of the amine-storing vesicles.

The chromaffin reaction given by 5-HT was discussed by Wood (1967) and by Etcheverry and Zieher (1968a and b). Both authors stressed the fact that whereas the reaction is negative with catecholamines if prior fixation with formaldehyde is used, this procedure does not affect the result with 5-HT. Using spot test analysis Arnold and Hager (1968) demonstrated the presence of chromium in amine granules containing both A and NA and they correctly attributed their contrast in the E.M. to this fact.

Reactions for **argentaffinity** are carried out, invariably, after glutaraldehyde fixation. The reaction was used originally for distinction between A and NA granules by Tramezzani *et al.* (1964) and the specificity of the glutaraldehyde-silver technique was investigated fully by Cannata, Chiocchio and Tramezzani (1968). These authors showed that with the standard technique (1 minute in silver bath) a positive result was obtained with noradrenalin, dopamine, melanin, and lipofuscin while 5-HT and adrenalin gave negative results. After 30 minutes' treatment with silver carbonate 5-HT became positive. These results are shown in Table 85, below:

The **glutaraldehyde-osmium** technique was fully investigated by Coupland

TABLE 85

The Glutaraldehyde-Silver Reaction

Amine or Substance	Optical Microscope	Electron Microscope	
		1 min. Silver	30 min. Silver
Noradrenalin	+	+	+
Adrenalin	−	−	−
Dopamine	−	+	+
5-HT	−	−	+
Melanin	+	+	+
Lipofuscin	+	+	+

and Hopwood (1966) and also by Van Orden *et al.* (1966), Wood (1966) and Vitry (1967). It is clear that the product of glutaraldehyde condensation with NA, A or DA can reduce OsO_4 and that the final product (OsO_2) is retained in the storage granule. The presence of 5-HT in such granules in blood platelets is shown in Fig. 276, p. 1287.

The **permanganate** method was developed by Richardson (1966, 1968) and thoroughly tested by Hökfelt and Jonsson (1968). The original technique called for fixation in 3 per cent $KMnO_4$ but after carrying out a series of model experiments Hökfelt and Jonsson found that 0·3 per cent, at pH 7·0 and 22°, was the optimal concentration.

A large number of monoamines react rapidly with $KMnO_4$ to give coloured products and, usually after a longer period, precipitates also. The strongest reactions are with tyramine and 3-methoxytyramine but NA, DA and 5-HT also react strongly. It is presumed that the reaction product (MnO_2) is the main constituent of precipitates formed *in vitro* and that it is responsible for the observed electron opacity of amine granules after $KMnO_4$ fixation.

Specificity of the Various Reactions. It is clear that much of the so-called specificity claimed by various authors for the procedures outlined above depends on fixation and, particularly, on the failure of aldehyde fixation to immobilize adrenalin. Dichromate oxidation (the chromaffin reaction) is specific for biogenic amines but not for any single substance. The glutaraldehyde-silver method, like the others, can differentiate DA and NA from A and DA and NA from 5-HT.

In the present state of the art it is more important, as emphasized by Tranzer *et al.* (1969) to achieve optimum fixation. The procedure developed by these authors is given in Appendix 32.

Microincineration for Electron Microscopy

While earlier workers in this field used the electron beam for the purpose of incineration, later studies (Yasuzumi and Sawada, 1950) used material heated in a standard muffle furnace. Incineration of ultrathin sections for

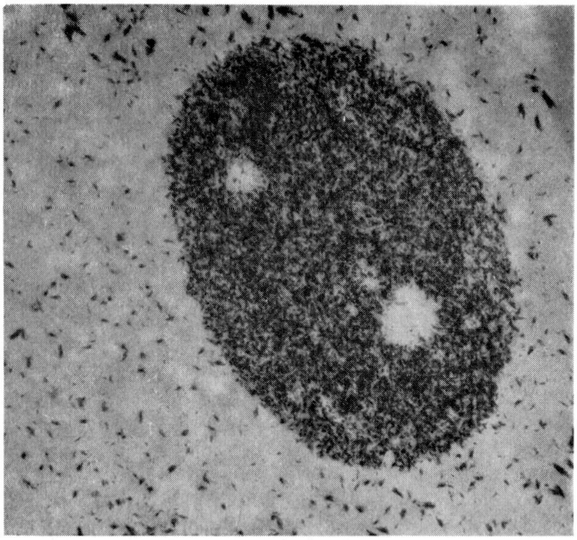

FIG. 277. Part of a maize root tip cell nucleus, fixed in lead acetate at pH 6·0. A mass of electron-opaque crystals occupies the nucleolus. Fewer, coarser crystals are present in the nucleoplasm. Unstained. × 15′000.

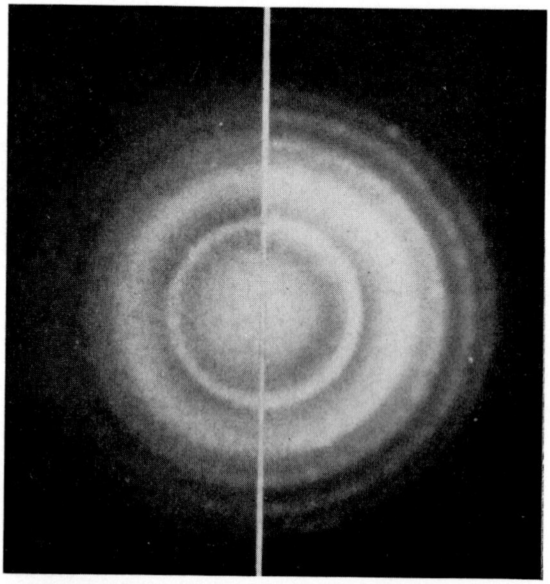

FIG. 278. X-ray diffraction patterns obtained from (left) crystals of lead orthophosphate and (right) from an area, 1 micron in diameter, in the nucleolus illustrated in Fig. 277. The patterns are identical.

E.M. studies was reported by Arnold and Sasse (1963) and Thomas (1964) examined ashed material from *Bacillus megatherium* spores. A suitable technique for microincineration and electron microscopy was developed by Boothroyd (1968) and this is applicable to mammalian tissues.

Electron-Microprobe Analysis

The design of a suitable instrument for electron probe microanalysis was published by Duncomb and Melford (1960) and an instrument of this type was used by Hale *et al.* (1967) for their studies on the distribution of calcium, phosphorus and sulphur in the human aorta. The technique depends on the fact that when tissue sections are exposed to bombardment by electrons from a fine probe (approximate diameter 200 nm) X-rays are produced which are characteristic of particular elements. The distribution of several elements can be shown in the same specimen, since the technique is not especially destructive to the tissue components. The detectability of any particular element depends on its local concentration and on the sensitivity of the instrument employed.

The electron probe microanalyser gives a relatively poor scanning electron image while producing X-ray emission in both scanned display form for elemental distribution and probe analysis. It is possible to provide an analysing attachment for the transmission electron microscope but this procedure involves loss of image resolution and specimen facilities and the probe size is limited by the origin condenser system. Probe microanalysers can be converted for transmission electron microscopy but imaging and analysis of thin specimens must still be carried out separately. The analytical electron microscope EMMA-4, produced in Britain by Associated Electrical Industries, was designed to overcome these problems and to allow simultaneous observation and probe analysis on ultrathin sections. The instrument provides diffraction facilities from areas less than 100 nm up to 400 nm, with an image resolution better than 1 nm.

The pyroantimonate method of Komnick (1962), for E.M. detection of sodium ions, was subjected to electron probe analysis by Lane and Martin (1969). Their work emphasizes the desirability of such controls when reasonable doubt exists as to the precise composition of the precipitate produced by any histochemical procedure.

Electron Diffraction Techniques

Electron diffraction has, as yet, been little used as a tool in ultrastructural cytochemistry. As a control for enzyme histochemical reactions it was proposed, and employed, by Arstila *et al.* (1966). These authors analysed the presumptive barium sulphate precipitate in ultrathin sections of tissues reacted for aryl sulphatase. They were able to confirm that the electron dense material in their cytosomes was $BaSO_4$ and hence that the observed staining was not due to non-specific binding of Ba^{2+} ions.

Working with the optical microscope Tandler (1956, 1967) found that when an aqueous solution of lead acetate was used as a fixative, massive precipitates occurred in the nucleoli of plant and animal cells. Evidence accumulated that the deposit was really a precipitate (not merely a binding of Pb^{2+} ions) and that the tissue component precipitated was a highly diffusible substance not retained by the usual aldehyde or protein-precipitant fixatives. It was considered that the precipitate consisted of lead orthophosphate (Tandler, 1961). On the other hand, Studzinski and Love (1964) considered that the lead precipitate in the nucleoli of mammalian cells could be due to the presence of an easily denatured protein.

Electron microscopical examination of the reaction product in lead acetate-fixed plant cells (Fig. 277) was shown by Tandler and Solari (1969) to be in the form of microcrystals, largely confined to the nucleolus. Electron diffraction studies were carried out on selected areas of the precipitate, with the results illustrated in Fig. 278. The diffraction pattern obtained was precisely that of lead orthophosphate.

Diffraction Analysis

In order to improve on the interpretation of structure in electron micrographs provided by the human eye Boseck and Lange (1970) constructed an apparatus based on the Frauenhofer diffraction principle which separates all structures registered on the photographic plate into elementary linear gratings with a sine-wave modulation and a certain number of periods, or line-pairs, per unit length. Testing the method on structures in electron micrographs interpreted visually as having single or double periodicity, from 3 to 5 periodic structures were revealed. This method has considerable promise for the analysis of crystalline and paracrystalline materials in specific storage granules in mammalian cells.

REFERENCES

Adams, C. W. M., Bayliss, O. B., and Weller, R. O. (1965). *J. Histochem. Cytochem.*, **13,** 694.

Albersheim, P., and Killias, U. (1963). *J. Cell Biol.*, **17,** 93.

Albert, E. N., and Fleischer, E. (1970). *J. Histochem. Cytochem.*, **18,** 697.

Anderson, W. A., and Personne, P. (1970). *J. Cell Biol.*, **44,** 29.

Arnold, M., and Hager, G. (1968). *Histochemie*, **14,** 297.

Arnold, M., and Hager, G. (1969). *Histochemie*, **17,** 312.

Arnold, M., and Sasse, D. (1963). *Acta. histochem. Suppl.* **3,** 204.

Arstila, A. U., Jaakkola, S., Kalimo, H., Helminen, H., and Hopsu-Havu, V. K. (1966). *J. Microsc.*, **5,** 777.

Ashworth, C. T., Luibel, F. J., and Stewart, S. C. (1963). *J. Cell Biol.*, **17,** 1.

Avrameas, S. (1969). *Immunochemistry*, **6,** 825.

Avrameas, S., and Ternynck, T. (1969). *Immunochemistry*, **6,** 53.

Bahr, G. F. (1954). *Expl Cell Res.*, **7,** 457.

Bahr, G. F. (1955). *Ibid.*, **9,** 277.

BAHR, G. F., and MOBERGER, G. (1954). *Ibid.*, **6**, 506.
BAILAR, J. C. jr. (1956). "The Chemistry of the Coordination Compounds", Reinhold, New York, p. 507.
BARRNETT, R. J. (1959). *Expl Cell Res.*, Suppl 7, p. 65.
BARRNETT, R. J. (1962). *J. Cell Biol.*, **12**, 247.
BARRNETT, R. J. (1964). *J. roy. micr. Soc.*, **83**, 143.
BARRNETT, R. J., and PALADE, G. E. (1957). *J. biophys. biochem. Cytol.*, **3**, 577.
BARRNETT, R. T., and PALADE, G. E. (1959). *J. biophys. biochem. Cytol.*, **6**, 163.
BAXANDALL, J. (1966). *J. Histochem. Cytochem.*, **14**, 730.
BAXANDALL, J., PERLMANN, P., and AFZELIUS, B. A. (1963). *J. roy. micr. Soc.*, **81**, 155.
BEARD, M. E., and NOVIKOFF, A. B. (1969). *J. Cell Biol.*, **42**, 501.
BECKER, S. W., PRAVER, L. L., and THATCHER, H. (1935). *Archs Dermatol.*, **31**, 190.
BEHNKE, O. (1969). *J. Histochem. Cytochem.*, **17**, 62.
BELL, C. (1966). *J. Histochem. Cytochem.*, **14**, 567.
BENESCH, R., and BENESCH, R. E. (1958). *Proc. Natn. Acad. Sci., U.S.A.*, **44**, 848.
BERGMAN, R. A., UENO, H., MORIZONO, Y., HANKER, J. S., and SELIGMAN, A. M. (1967). *Histochemie*, **11**, 1.
BERNSTEIN, M. H., (1956). *J. biophys. biochem. Cytol.*, **2**, 633.
BIRKS, R. I., and BROWN, L. M. (1960). *J. Physiol., Lond.*, **152**, 5P.
BLOOM, F. E., and AGHAJANIAN, G. K. (1966). *Science, N.Y.*, **154**, 1575.
BOADLE, M. C., and BLOOM, F. E. (1969). *J. Histochem. Cytochem.*, **17**, 331.
BOOTHROYD, B. (1968). *J. roy. micr. Soc.*, **88**, 529.
BOSECK, S., and LANGE, R. (1970). *Z. wiss. Mikr.*, **70**, 65.
BRADBURY, S., and STOWARD, P. J. (1967). *Histochemie*, **11**, 71.
BRODY, I. (1959). *J. ultrastruct. Res.*, **3**, 84.
BRYAN, J. H. D. (1964). *Quart. J. micr. Sci.*, **105**, 367.
BRYAN, J. H. D., and BRINKLEY, W. R. (1963). *J. Cell Biol.*, **19**, 10A.
BURSTONE, M. S. (1960). *J. Histochem. Cytochem.*, **8**, 63.
CAVASSO, N., and FAVARD, P. (1966). *J. Miscrosc.*, **5**, 759.
CANNATA, M. A., CHIOCCHIO, S. A., and TRAMEZZANI, J. H. (1968). *Histochemie*, **12**, 253.
CHAN-CURTIS, V., BELT, W. D., and LADOULIS, C. T. (1970a). *J. Histochem. Cytochem.*, **18**, 609.
CHAN-CURTIS, V., BEER, M., and KOLLER, T. (1970b). *J. Histochem. Cytochem.*, **18**, 628.
CHAPMAN, J. A., and TAYLOR, G. (1966). *J. roy. micr. Soc.*, **85**, 455.
CHURG, J., MAUTNER, W., and GRISHMAN, E. (1958). *J. biophys. biochem. Cytol.*, **4**, 841.
COPE, G. H., and WILLIAMS, M. A. (1967). *J. roy. micr. Soc.*, **88**, 259.
COUPLAND, R. E., and HOPWOOD, D. (1966a). *Nature, Lond.*, **209**, 590.
COUPLAND, R. E., and HOPWOOD, D. (1966b). *J. Anat.*, **100**, 227.
COUPLAND, R. E., PYPER, A., and HOPWOOD, D. (1964). *Nature, Lond.*, **201**, 1240.
CROSS, S. A. M. (1970). *Histochemie.*, **22**, 219.
CURRAN, R. C., CLARK, A. E., and LOVELL, D. (1965). *J. Anat.*, **99**, 427.
DALES, S., GOMATOS, P. J., and HSUO, K. C. (1965). *Virology*, **25**, 193.
DALTON, A. J., and ZEIGEL, R. F. (1960). *J. biophys. biochem. Cytol.*, **7**, 409.
DARIA HAUST, M., WYLLIE, J. C., and MORE, R. H. (1965). *Expl Mol. Pathol.*, **4**, 205.
DAVIS, R., and KOELLE, G. B. (1967). *J. Cell Biol.*, **34**, 157.
DE MARTINO, C., and ZAMBONI, L. (1967). *J. Ultrastruct. Res.*, **19**, 273.
DETTMER, N. I., and SCHWARZ, W. (1952–53). *Z. wiss. Microsc.*, **61**, 423.
DOBB, M. G., and SIKORSKI, J. (1961). In "Structure de la laine", *Colloq. Inst. Textil. Franc. Paris.*
DORN, A. (1966). *Acta histochemica*, **25**, 375.
DOUGLAS, W. H. J. (1970). *J. Histochem. Cytochem.*, **18**, 510.
DUNCOMB, P., and MELFORD, D. A. (1960). *Proc. Int. Symp.* "X-ray Microscopy and X-ray Microanalysis". Ed. A Engström, p. 358.
ERÄNKÖ, O., RECHARDT, L., and HÄNNINEN, L. (1967). *Histochemie*, **8**, 369.

ERICSSON, J. L. E. (1966). *J. Histochem. Cytochem.*, **14**, 361.
ESSNER, E., and NOVIKOFF, A. B. (1962). *J. Cell Biol.*, **15**, 289.
ETCHEVERRY, G. J., and ZIEHER, L. M. (1968a). *Experientia*, **24**, 593.
ETCHEVERRY, G. J., and ZIEHER, L. M. (1968b). *J. Histochem. Cytochem.*, **16**, 162.
ETHERTON, J. E., and BOTHAM, C. M. (1970). *Histochem. J.*, **2**, 507.
FAHIMI, H. D. (1969). *J. Cell Biol.*, **43**, 275.
FALK, S. (1964). In "Porphyrins and Metalloporphyrins". Elsevier, Amsterdam, p. 96.
FEDER, N. (1970). *J. Histochem. Cytochem.*, **18**, 911.
FELDMAN, D. G. (1962). *J. Cell Biol.*, **15**, 592.
FORMANEK, H., and FORMANEK, S. (1970). *Europ. J. Biochem.*, **17**, 78.
FOWLER, B. A. (1970). *Histochemie*, **22**, 155.
FRANZ, H. (1967a). *Naturwiss*, **54**, 339.
FRANZ, H. (1967b). *Z. Chem.*, **7**, 235.
FRANZ, H. (1968a). *Z. Chem.*, **7**, 427.
FRANZ, H. (1968b). *Histochemie*, **12**, 230.
GALLYAS, F. (1963). *J. Neurochem.*, **10**, 125.
GANOTE, C. E., ROSENTHAL, A. S., MOSES, H. L., and TICE, L. W. (1969). *J. Histochem. Cytochem.*, **17**, 641.
GASIC, G., and BERWICK, L. (1963). *J. Cell Biol.*, **19**, 223.
GASIC, G., BERWICK, L., and SORRENTINO, M. (1968). *Lab. Invest.*, **18**, 63.
GEYER, G. (1969). "Ultrahistochemie". "Histochemische Arbeitsvotschiften für die Elektronenkroskopie". Fischer Verlag, Jena.
GEYER, G. (1964). *Acta histochemica*, **19**, 73.
GLAUERT, A. M., and LUCY, J. A. (1968). *J. Microsc.*, **89**, 1.
GLICK, D., and SCOTT, J. E. (1970). *J. Histochem. Cytochem.*, **18**, 455.
GOLDFISCHER, S. (1965). *J. Histochem. Cytochem.*, **13**, 520.
GOMORI, G. (1939). *Proc. soc. exp. biol. Med.*, **42**, 23.
GOMORI, G. (1941). *Archs Pathol.*, **32**, 189.
GOMORI, G. (1946). *Am. J. clin. Path.*, **10**, 177.
GOMORI, G. (1949). *Proc. Soc. exp. Biol. Med.*, **72**, 697.
GOMORI, G. (1952). "Microscopic Histochemistr". Chicago Univ. Press, p. 193.
GRAHAM, R. C., and KARNOVSKY, M. J. (1966). *J. Histochem. Cytochem.*, **14**, 291.
GROSSMAN, I. W., and HEITKAMP, D. H. (1968). *J. Histochem. Cytochem.*, **16**, 645.
GUSTAFSON, G. T., and PIHL, E. (1967a). *Acta path. scand.*, **69**, 393.
GUSTAFSON, G. T., and PIHL, E. (1967b). *Nature, Lond.*, **216**, 697.
HAJÓS, F., and KERPEL-FRONIUS, S. (1970). *Histochemie*, **23**, 120.
HALE, A. J., HALL, T., and CURRAN, R. C. (1967). *J. Path. Bact.*, **93**, 1.
HALL, C. E. (1955). *J. biophys. biochem. Cytol.*, **1**, 1.
HANKER, J. S., ANDERSON, W. E., and BLOOM, F. E. (1971). *Science*, in press.
HANKER, J. S., DEB, C., WASSERKRUG, H. L., and SELIGMAN, A. M. (1966). *Science, N.Y.*, **152**, 1631.
HANKER, J. S., DIXON, A. D., BLOOM, F. E., and WEITSEN, H. A. (1972). In press.
HANKER, J. S., KUSYK, C. J., BLOOM, F. E., and PEARSE, A. G. E. (1972). In press.
HANKER, J. S., SEAMAN, A. R., WEISS, L. P., UENO, H., BERGMAN, R. A., and SELIGMAN, A. M. (1964). *Science, N.Y.*, **146**, 1039.
HANKER, J. S., YATES, P. E., CLAPP, D. H., and ANDERSON, W. E. (1972). In press.
HANSSON, H. P. J. (1967). *Histochemie*, **11**, 112.
HÄUSLER, G. (1958). *Histochemie*, **1**, 29.
HAUGAARD, N., HORN, R. S., and KOELLE, G. B. (1965). *J. Histochem. Cytochem.* **13**, 566.
HAYASHI, M., SHIRAHAMA, T., and COHEN, A. S. (1968). *J. Cell Biol.*, **26**, 289.
HAYDON, G. B., SMITH, S. O., and SELIGMAN, A. M. (1968). *J. Histochem. Cytochem.*, **15**, 752.
HERNANDEZ, W., and RAMBOURG, A. (1967). *Anat. Rec.*, **157**, 260.
HIGGINS, J. A., and BARRNETT, R. J. (1970). *J. Cell Sci.*, **6**, 29.
HIRAI, K. (1968). *Acta histochem. Cytochem.*, **1**, 43.
HIRANO, G., SAITO, T., and OGAWA, K. (1968). *Expl Cell Res.*, **51**, 259.
HODGE, A. J. (1956). *J. biophys. biochem. Cytol.*, Vol 2, Suppl., p. 131.

HÖKFELT, T., and JONSSON, G. (1968). *Histochemie*, **16**, 45.
HOLT, S. J., and HICKS, R. M. (1961). *J. biophys. biochem. Cytol.*, **11**, 47.
HOLT, S. J., and HICKS, R. M. (1966). *J. Cell Biol.*, **29**, 361.
HOPSU, V. K., ARSTILA, A., and GLENNER, G. G. (1965a). *J. Histochem. Cytochem.*, **13**, 711.
HOPSU, V. K., ARSTILA, A., and GLENNER, G. G. (1965b). *Ann. Med. exp. Fenn.*, **43**, 114.
HOPSU-HAVU, V. K., ARSTILA, A., HELMINEN, H. J., KALIMO, H. O., and GLENNER, G. C. (1967). *Histochemie*, **8**, 54.
HORI, S. H. (1963). *Histochemie*, **7**, 297.
HORI, S. H., and CHANG, J. P. (1963). *J. Histochem. Cytochem.*, **11**, 71.
HOUCK, C. E., and DEMPSEY, E. W. (1954). *Stain Technol.*, **29**, 207.
HSU, K. C., RIFKIND, R. A., and ZABRISKIE, J. B. (1963). *Science, N.Y.*, **142**, 1471.
HUGON, J., and BORGERS, M. (1966). *J. Histochem. Cytochem.*, **14**, 429.
HUGON, J., and BORGERS, M. (1968). *Histochemie*, **12**, 42.
HUXLEY, H., and ZUBAY, G. (1961). *J. biophys. biochem. Cytol.*, **11**, 273.
IBATA, Y., and OTSUKA, N. (1969). *J. Histochem. Cytochem.*, **17**, 171.
IDELMAN, S. (1965). *Histochemie*, **5**, 18.
ISHIKAWA, Y., SAITO, T., and OGAWA, K. (1967). *J. Electronmicrosc.*, **16**, 52.
JEWELL, G. G., and SAXTON, C. A. (1970). *Histochem, J.*, **2**, 17.
JOÓ, F., SÁVAY, G., and CSILLIK, B. (1965). *Acta histochem.*, **22**, 40.
JURAND, A., DEUTSCH, K., and DUNN, E. A. G. (1958). *J. roy. micr. Soc.*, **78**, 46.
KABAT, E. A. (1960). *J. Immunol.*, **84**, 82.
KALIMO, H. O., HELMINEN, H. J., ARSTILA, A. U., and HOPSU-HAVU, V. K. (1968). *Histochemie*, **14**, 123.
KALINA, M., and PEARSE, A. G. E. (1971). *Histochemie*, **26**, 1.
KALINA, M., WEAVERS, B., and PEARSE, A. G. E. (1969). *Nature, Lond.*, **221**, 479.
KARNOVSKY, M., and HUG, C. (1963). *J. Cell Biol.*, **19**, 255.
KARNOVSKY, M. J. (1961). *J. biophys. biochem. Cytol.*, **11**, 729.
KARNOVSKY, M. J. (1964). *J. Cell Biol.*, **23**, 217.
KARNOVSKY, M. J. (1965). *J. Cell Biol.*, **27**, 137A.
KARNOVSKY, M. J., and RICE, D. F. (1969). *J. Histochem. Cytochem.*, **17**, 751.
KARNOVSKY, M. J., and ROOTS, L. (1964). *J. Histochem. Cytochem.*, **12**, 219.
KÁSA, P., and CSERNOVSZKY, E. (1967). *Acta histochem.*, **28**, 274.
KÁSA, P., and CSILLIK, B. (1966). *J. Neurochem.*, **13**, 1345.
KÁSA, P., and CSILLIK, B. (1968). *Histochemie*, **12**, 175.
KAWARAI, Y., and NAKANE, P. K. (1970). *J. Histochem. Cytochem.*, **18**, 161.
KAWASHIMA, T. (1968). *Acta histochem. cytochem.*, **1**, 147.
KAWASHIMA, T., and MURATA, F. (1969). *Acta histochem. cytochem.*, **2**, 54.
KAZIMIERCZAK, J. (1970). *Proc. VII Int. Cong. E.M.* p. 551.
KENDALL, P. A. (1965). *Biochim. biophys. Acta*, **97**, 174.
KENDALL, P. A. (1971). *Biochim. biophys. Acta*, (in press).
KENDALL, P. A., and BARNARD, E. A. (1969). *Biochim. biophys. Acta*, **188**, 10.
KENDALL, P. A., POLAK, J. M., and PEARSE, A. G. E. (1971). *Experientia*, **27**, 1104.
KENDALL, P. A., POLAK, J. M., and PEARSE, A. G. E. (1972). *Experientia*, **28**, In Press.
KERPEL-FRONIUS, S., and HAJÓS, F. (1697). *Histochemie*, **10**, 216.
KERPEL-FRONIUS, S., and HAJÓS F. (1968). *Histochemie*, **14**, 343.
KERR, D. N. S., and MUIR, A. R. (1960). *J. Ultrastruct. Res.*, **3**, 313.
KOELLE, G. B., DAVIS, R., and GROMADZKI, C. G. (1967). *Ann. N.Y. Acad. Sci.*, **144**, 613.
KOELLE, G. B., and FOROGLOU-KERAMEOS, C. (1965). *Life Sci.*, **4**, 417.
KOELLE, G. B., and GROMADZKI, C. G. (1966). *J. Histochem. Cytochem.*, **14**, 443.
KOÍKE, M., NAGAYAMA, A., NAGAYAMA, H., AMAKO, K., MATSUO, T., and TAKEYA, K. (1964). *J. Electron Microsc.*, **13**, 133.
KOKKO, A., MAUTNER, H. G., and BARRNETT, R. J. (1969). *J. Histochem. Cytochem.*, **17**, 625.
KOMNICK, H. (1962). *Protoplasma*, **55**, 414.
KOMNICK, H. (1969). *Histochemie*, **18**, 24.
KOMNICK, H., and BIERTHER, M. (1969). *Histochemie*, **18**, 337.

Kühn, K., Grassmann, W., and Hofmann, U. (1958). *Naturwiss.*, **45**, 521.
Kühn, K., Grassmann, W., and Hofmann, U. (1960). *Naturwiss.*, **47**, 15.
Kul'berg, A. Y., and Azadova, N. B. (1964). *Fed. Proc. (Transl. Suppl.)* **23**, T69.
Lane, B. P., and Martin, E. (1969). *J. Histochem. Cytochem.*, **17**, 102.
Lawn, A. M. (1960). *J. biophys. biochem. Cytol.*, **7**, 197.
Lazarus, S. S., and Barden, H. (1962). *J. Histochem. Cytochem.*, **10**, 285.
Lazarus, S. S., and Barden, H. (1964). *J. Ultrastruct. Res.*, **10**, 189.
Lazarus, S. S., and Vethamany, V. G. (1966). *J. Histochem. Cytochem.*, **14**, 810.
Leduc, E., and Bernhard, W. (1960). *C.R. Acad. Sci. Paris*, **250**, 2948.
Leduc, E., and Bernhard, W. (1961). *J. Cell Biol.*, **10**, 437.
Leduc, E., and Bernhard, W. (1967). *J. Ultrastruct. Res.*, **19**, 196.
Leduc, E., Byczkowska-Smyk, W., and Bernhard, W. (1960). *C.R. Acad. Sci. Paris*, **250**, 4052.
Leduc, E. H., Scott, G. B., and Avrameas, S. (1969). *J. Histochem. Cytochem.*, **17**, 211.
Legg, P. G., and Wood, R. L. (1970). *Histochemie*, **22**, 262.
Lehrer, G. M., and Ornstein, L. (1959). *J. biophys. biochem. Cytol.*, **6**, 399.
Lever, J. D. (1960). *Nature, Lond.*, **186**, 810.
Lewis, P. R., and Shute, C. C. D. (1964). *J. Physiol., Lond.*, **175**, 5P.
Lewis, P. R., and Shute, C. C. D. (1966). *J. Cell Sci.*, **1**, 381.
Lewis, P. R., and Shute, C. C. D. (1969). *J. Microsc.*, **89**, 181.
Livingston, D. C., Fisher, S. W., Greenoak, G. E., and Maggi, V. (1970). *Histochemie*, **24**, 159.
Lloyd, K. O., Kabat, E. A., and Rosenfield, R. E. (1966). *Biochemistry*, **5**, 1502.
Low, F. N., and Freeman, J. A. (1956). *J. biophys. biochem. Cytol.*, **2**, 62G.
Luft, J. H. (1956). *J. biophys. biochem. Cytol.*, **2**, 799.
Luft, J. H. (1964). *J. Cell Biol.*, **23**, 54A.
Man, J. C. H., Beyer, I. M., and Persijn, J. P. (1963). *Histochemie*, **3**, 269.
Marinozzi, V. (1961). *J. biophys. biochem. Cytol.*, **9**, 121.
Marinozzi, V., and Gautier, A. (1962). *J. Ultrastruct. Res.*, **7**, 436.
Mayahara, H., Hirano, H., Saito, T., and Ogawa, K. (1967). *Histochemie*, **11**, 88.
Mayahara, H., and Ogawa, K. (1968). *J. Histochem. Cytochem.*, **16**, 721.
Mcalpine, J. C. (1969). *Histochemie*, **19**, 135.
Meek, G. A., and Bradbury, S. (1963). *J. Cell Biol.*, **18**, 73.
Miller, S. A., Tebboth, A., and Tremaine, J. F. (1952). *J. chem. Soc.*, 632.
Millonig, G. (1961). *J. biophys. biochem. Cytol.*, **11**, 736.
Mitsui, T. (1960). *J. biophys. biochem. Cytol.*, **7**, 251.
Mizutani, A., and Barrnett, R. J. (1965). *Nature, Lond.*, **206**, 1001.
Mölbert, E. R. G., Duspiva, F., and Deimling, O. H. (1960). *J. biophys. biochem. Cytol.*, **7**, 387.
Morgan, C., Hsu, K. C., Rifkind, A., Knox, A. W., and Rose, H. M. (1961). *J. exp. Med.*, **114**, 833.
Morgan, C., Rifkind, R. A., and Rose, H. M. (1962). *Cold Spring Harb. Symp. Quant. Biol.*, **27**, 57.
Mori, S., Maeda, T., and Shimizu, N. (1964). *Histochemie*, **4**, 65.
Moses, H. L., and Rosenthal, A. S. (1967). *J. Histochem. Cytochem.*, **15**, 354.
Moses, H. L., Rosenthal, A. S., Beaver, D. L., and Schuffman, S. S. (1966). *J. Histochem. Cytochem.*, **14**, 702.
Mott, M. (1963). *J. roy. micr. Soc.*, **81**, 159.
Movat, H. Z. (1961). *Am. J. clin. Path.*, **35**, 528.
Mudd, S., and Anderson, T. F. (1942). *J. exp. Med.*, **76**, 103.
Müller, A., and Geyer, G. (1968). *Acta histochem.*, **31**, 98.
Müller, A., and Geyer, G. (1970). *Acta histochem.*, **36**, 87.
Mundkur, B. (1964a). *Expl Cell Res.*, **34**, 155.
Mundkur, B. (1964b). *Z. Zellforsch.*, **63**, 52.
Murata, F., Yokota, S., and Nagata, T. (1968). *Histochemie*, **13**, 215.
Nakane, P. K. (1968). *J. Histochem. Cytochem.*, **16**, 557.
Nakane, P. K., and Pierce, G. B. (1966). *J. Histochem. Cytochem.*, **14**, 929.

NAKANE, P. K., and PIERCE, G. B. (1967). *J. Cell Biol.*, **33**, 307.
NAKANE, P. K., SRIRAM, J., and PIERCE, G. B. (1966). *J. Histochem. Cytochem.*, **14**, 789.
NOVIKOFF, A. B. (1967). *J. Histochem. Cytochem.*, **15**, 353.
NOVIKOFF, A. B. (1970). *J. Histochem. Cytochem.*, **18**, 916.
NOVIKOFF, A. B., ALBALA, A., and BIEMPICA, L. (1968). *J. Histochem. Cytochem.*, **16**, 299.
NOVIKOFF, A. B., and GOLDFISCHER, S. (1961). *Proc. Natn. Acad. Sci. U.S.A.*, **47**, 802.
NYBERG-HANSEN, R., RINVIK, E., AARSETH, P., and BARSTAD, J. A. B. (1969). *Histochemie*, **20**, 40.
OCHI, J. (1968). *Histochemie*, **14**, 300.
OGAWA, K., and BARRNETT, R. J. (1965). *J. Ultrastruct. Res.*, **12**, 488.
OGAWA, K., and MAYAHARA, H. (1969). *J. Histochem. Cytochem.*, **17**, 487.
OGAWA, K., SAITO, T., and MAYAHARA, H. (1968). *J. Histochem. Cytochem.*, **16**, 49.
ÖKRÖS, I. (1968). *Histochemie*, **13**, 91.
OKUN, M. R., EDELSTEIN, L. M., OR, N., HAMADA, G., DONELLAN, B., and LEVER, W. F. (1970). *Histochemie*, **23**, 295.
OLEDZKA-SLOTWÍNSKA, H., CREEMERS, J., and DESMET, V. (1967). *Histochemie*, **9**, 320.
ORNSTEIN, L. (1957). *J. biophys. biochem. Cytol.*, **3**, 809.
PALKAMA, A., and RECHARDT, L. (1970). *Acta Med. exp. Fenn.*, **48**, 77.
PALKAMA, A., and UUSITALO, R. (1968). *Scand. J. clin. Lab. Invest.*, **21**, Suppl. 101, 48.
PAPADIMITRIOU, J. M., and VAN DUIJN, P. (1970). *J. Cell Biol.*, **47**, 84.
PARSONS, D. F. (1961). *J. biophys. biochem. Cytol.*, **11**, 492.
PARSONS, D. F., and DARDEN, E. B. (1960). *J. biophys. biochem. Cytol.*, **8**, 834.
PAULING, L., CAMPBELL, D. H., and PRESSMAN, D. (1943). *Physiol. Rev.*, **23**, 203.
PEARSE, A. G. E. (1963). *J. roy. micr. Soc.*, **81**, 107.
PEASE, D. C. (1966). *J. Ultrastruct. Res.*, **15**, 555.
PEASE, D. C. (1970). *J. Histochem. Cytochem.*, **18**, 455.
PEPE, F. A. (1961). *J. biophys. biochem. Cytol.*, **11**, 515.
PEPE, F. A. (1966). *J. Cell Biol.*, **28**, 505.
PEPE, F. A., and FINCK, H. (1961). *J. biophys. biochem. Cytol.*, **11**, 521.
PERRY, M. M. (1967). *J. Cell Sci.*, **2**, 257.
PERSIJN, J. P., DAEMS, W. T., MAN, J. C. H., and MEIJER, A. E. F. H. (1961). *Histochemie*, **2**, 372.
PICKETT-HEAPS, J. D. (1967). *J. Histochem. Cytochem.*, **15**, 442.
PIERCE, G. B., SRIRAM, Y., and MIDGLEY, A. R. (1964). *Int. Rev. exp. Path.*, **3**, 1.
PIHL, E. (1967). *Histochemie*, **10**, 126.
PIHL, E. (1968). *Acta path. scand.*, **74**, 145.
PIHL, E., and FALKMER, S. (1967). *Acta histochem.*, **27**, 34.
PIHL, E., GUSTAFSON, G. T., and FALKMER, S. (1968). *Histochem. J.*, **1**, 26.
PLAPINGER, R. E., LINUS, S. L., KAWASHIMA, T., DEB, C., and SELIGMAN, A. M. (1968). *Histochemie*, **14**, 1.
PRESSMAN, D., BRYDEN, J. H., and PAULING, L. (1948). *J. Am. Chem. Soc.*, **70**, 1352.
RAMBOURG, A. (1967). *J. Histochem. Cytochem.*, **15**, 409.
REALE, E. (1962). *Expl Cell Res.*, **26**, 210.
REALE, E., and LUCIANO, L. (1964). *J. Histochem. Cytochem.*, **10**, 713.
REALE, E., and LUCIANO, L. (1967). *Histochemie*, **8**, 302.
REALE, E., and LUCIANO, L. (1970). *Histochemie*, **23**, 144.
RECHARDT, L., and KOKKO, A. (1967). *Histochemie*, **10**, 278.
REEDY, M. K. (1965). *J. Cell Biol.*, **26**, 309.
REVEL, J.-P. (1966). *J. Histochem. Cytochem.* **14**, 740.
REYNOLDS, E. S. (1963). *J. Cell Biol.*, **17**, 208.
RICHARDSON, K. C. (1966). *Nature, Lond.*, **210**, 756.
RICHARDSON, K. C. (1968). *Life Sci.*, **7**, 599.

RIFKIND, R. A., HSU, K. C., and MORGAN, C. (1964). *J. Histochem. Cytochem.*, **12,** 131.
RIPLEY, R. C., and PEASE, D. C. (1965). *J. appl. Phys.*, **36,** 2609.
ROBERTSON, J. D. (1955). *J. biophys. biochem. Cytol.*, **1,** 271.
ROSA, C.-G., and TSOU, K.-C. (1964). *J. Histochem. Cytochem.*, **12,** 23.
ROSA, C.-G., and TSOU, K.-C. (1965). *Nature, Lond.*, **206,** 103.
ROSENTHAL, A. S., MOSES, H. L., BEAVER, D. L., and SCHUFFMAN, S. F. (1966). *J. Histochem. Cytochem.*, **14,** 698.
ROSENTHAL, A. S., MOSES, H. L., and GANOTE, C. E. (1970). *J. Histochem. Cytochem.* **18,** 915.
ROSENTHAL, A. S., MOSES, H. L., TICE, L. W., and GANOTE, C. E. (1969). *J. Histochem. Cytochem.*, **17,** 608.
ROTHMAN, A. H. (1970). *Expl Cell Res.*, **58,** 177.
RUTENBURG, A. M., KIM, H., FISHBEIN, J. W., HANKER, J. S., WASSERKRUG, H. L., and SELIGMAN, A. M. (1969). *J. Histochem. Cytochem.*, **17,** 517.
SABATINI, D. D., BENSCH, K. G., and BARRNETT, R. J. (1963). *J. Cell Biol.*, **17,** 19.
SAITO, T. (1968). *J. Kansai Med. Sch.*, **20,** 17 (in Japanese).
SAITO, T., and OGAWA, K. (1966). *Archs histol. japon.*, **27,** 473.
SAITO, T., and OGAWA, K. (1967). *Okajimas Fol. anat. japon.*, **44,** 11.
SASAKI, M., and AOKI, T. (1970). *Acta Histochem. Cytochem.*, **3,** 139.
SASAKI, M., and TAKEUCHI, T. (1963). *J. Histochem. Cytochem.*, **11,** 342.
SASAKI, M., and TAKEUCHI, T. (1968). *Acta Histochem. Cytochem.*, **1,** 197.
SCARPELLI, D. G., and CRAIG, E. L. (1963). *J. Cell Biol.*, **17,** 279.
SCARPELLI, D. G., and KANCZAK, N. M. (1965). *Int. Rev. exp. Pathol.*, **4,** 55.
SCHECHTER, I., and SELA, M. (1965). *Biochim. biophys. Acta*, **104,** 298.
SCHICK, A., and SINGER, S. J. (1961). *J. Biol. Chem.*, **236,** 2477.
SCHMITT, F. O., HALL, C. E., and JAKUS, M. A. (1945). *Appl. Physiol.*, **16,** 459.
SCHULTZE, W., and WOLLENBERGER, A. (1965). *Histochemie*, **5,** 417.
SEDAR, A. W., and ROSA, C. G. (1958). *Anat. Rec.*, **130,** 371.
SEDAR, A. W., ROSA, C. G., and TSOU, K.-C. (1962). *J. Histochem. Cytochem.*, **10,** 506.
SELIGMAN, A. M., HANKER, J. S., KATZOFF, L., DEB, C., and WASSERKRUG, H. (1965). *J. Histochem. Cytochem.*, **13,** 705.
SELIGMAN, A. M., HANKER, J. S., WASSERKRUG, H., DMOCHOWSKI, H., and KATZOFF, L. (1965). *J. Histochem. Cytochem.*, **13,** 629.
SELIGMAN, A. M., KARNOVSKY, M. J., WASSERKRUG, H. L., and HANKER, J. S. (1968). *J. Cell Biol.*, **38,** 1.
SELIGMAN, A. M., PLAPLINGER, R. E., WASSERKRUG, H., DEB, C., and HANKER, J. S. (1967a). *J. Cell Biol.*, **34,** 787.
SELIGMAN, A. M., UNEO, H., MORIZONO, Y., WASSERKRUG, H. L., KATZOFF, L., and HANKER, J. S. (1967b).
SELIGMAN, A. M., SAITO, T., and PLAPINGER, R. E. (1970). *Histochemie*, **22,** 85.
SELIGMAN, A. M., WASSERKRUG, H. L., DEB, C., and HANKER, J. S. (1968). *J. Histochem. Cytochem.*, **16,** 87.
SELIGMAN, A. M., WASSERKRUG, H. L., and HANKER, J. S. (1966). *J. Cell Biol.*, **30,** 424.
SELIGMAN, A. M., WASSERKRUG, H. L., PLAPINGER, R. E., SAITO, T., and HANKER, J. S. (1970). *J. Histochem. Cytochem.*, **18,** 542.
SHIMIZU, N., and ISHII, S. (1966). *Histochemie*, **6,** 24.
SHIOSE, Y., and SEARS, M. (1965). *Invest. Ophthalmol.*, **5,** 152.
SILVERMAN, L., and GLICK, D. (1969). *J. Cell Biol.*, **40,** 761.
SINGER, S. J. (1959). *Nature, Lond.*, **183,** 1523.
SINGER, S. J., and SCHICK, A. (1961). *J. biophys. biochem. Cytol.*, **9,** 519.
SMITH, R. E., and FISHMAN, W. H. (1969). *J. Histochem. Cytochem.*, **17,** 1.
SOMMER, J. R., and SPACH, M. S. (1964). *Am. J. Path.*, **44,** 491.
SOMOGYI, E., and SOTONYI, P. (1969). *Acta histochemica*, **34,** 70.
SRIRAM, J., NAKANE, P. K., RAWLINSON, E. G., and PIERCE, G. B. (1966). *Fed. Proc.*, **25,** 732.

SRIRAM, J., TAWDE, S. S., PIERCE, G. B., and MIDGELEY, A. R. (1963). *J. Cell Biol.*, **17,** 673.
STÄUBLI, W. (1960). *C.R. Acad. Sci. Paris*, **250,** 1137.
STERNBERGER, L. A. (1967). *J. Histochem. Cytochem.*, **15,** 139.
STERNBERGER, L. A., and DONATI, E. J. (1966). *J. Histochem. Cytochem.*, **14,** 606.
STERNBERGER, L. A., DONATI, E. J., CUCULIS, J. J., and PETRALI, J. P. (1965). *Exp. Molec. Path.*, **4,** 112.
STERNBERGER, L. A., DONATI, E. J., HANKER, J. S., and SELIGMAN, A. M. (1966). *Exp. Molec. Path., Suppl.* **3,** 36.
STOLLAR, B. D., LEVINE, H., LEHRER, H. I., and VAN VUNAKIS, H. (1962). *Proc. Natn. Acad. Sci., U.S.A.*, **48,** 874.
STOWARD, P. J., and BRADBURY, S. (1968). *Histochemie*, **15,** 93.
STRIKER, G. E., DONATI, E. J., PETRALI, J. P., and STERNBERGER, L. A. (1966). *Exp. molec. Path., Suppl.* **3,** 52.
STRUGGER, S. (1956). *Naturwiss*, **43,** 357.
STUDZINSKI, G. P., and LOVE, R. (1964). *Stain Technol.*, **39,** 397.
SUTTON, J. S. (1968). *J. Ultrastruct. Res.*, **31,** 424.
SUZUKI, T., and SEKIYAMA, S. (1961). *J. Electronmicrosc.*, **10,** 36.
SWIFT, J. A. (1968). *J. roy. micr. Soc.*, **88,** 449.
SWIFT, J. A. (1969). *Histochemie*, **19,** 88.
SWIFT, H., and ADAMS, B. J. (1966). *J. Histochem. Cytochem.*, **14,** 744.
SWIFT, J. A., and SAXTON, C. A. (1967). *J. Ultrastruct. Res.*, **17,** 23.
TAKEUCHI, T., and SASAKI, M. (1968). *Acta Histochem. Cytochem.*, **1,** 63.
TANAKA, H. (1968). *Acta haematol. japon.*, **31,** 135.
TANDLER, C. J. (1956). *J. Histochem. Cytochem.*, **4,** 331.
TANDLER, C. J. (1957). *J. Histochem. Cytochem.*, **5,** 489.
TANDLER, C. J. (1961). *Expl Cell Res.*, **25,** 246.
TANDLER, C. J., and SOLARI, A. J. (1969). *J. Cell Biol.*, **41,** 91.
TAWDE, S. S., and SRIRAM, J. (1962). *Archs Biochem. Biophys.*, **97,** 429.
TERÄVÄINEN, H. (1967). *Histochemie*, **10,** 266.
THIÉRY, J. P. (1967). *J. Microsc.*, **6,** 987.
THOMAS, R. S. (1964). *J. Cell Biol.*, **23,** 113.
THOMPSON, E., and COLVIN, J. R. (1970). *J. roy. micr. Soc.*, **91,** 87.
TICE, L. W. (1969). *J. Histochem. Cytochem.*, **17,** 85.
TICE, L., and BARRNETT, R. J. (1962). *J. Histochem. Cytochem.*, **10,** 754.
TICE, L. W., and BARRNETT, R. J. (1962). *J. Cell Biol.*, **15,** 401.
TICE, L. W., and BARRNETT, R. J. (1965). *J. Cell Biol.*, **25,** 23.
TICE, L. W., and SMITH, D. S. (1965). *J. Cell Biol.*, **25,** 126.
TIMM, F. (1958). *Dtsch. Z. gerichtl. Med.*, **46,** 706.
TORACK, R. M., and BARRNETT, R. J. (1963). *J. Histochem. Cytochem.*, **11,** 763.
TOWE, K. M. (1969). *J. Microsc.*, **90,** 279.
TRAMEZZANI, J. H., CHIOCCHIO, S., and WASSERMANN, G. F. (1964). *J. Histochem. Cytochem.*, **12,** 890.
TRANZER, J.-P. (1965). *J. Microsc.*, **4,** 409.
TRANZER, J.-P., and THOENEN, H. (1967). *Experientia*, **24,** 484.
TRANZER, J.-P., THOENEN, H., SNIPES, R. L., and RICHARDS, J. G. (1969). *Prog. Brain Res.*, **31,** 34.
TSOU, K. C., GOODWIN, C. W., SEAMOND, B., and LYNM, D. (1968). *J. Histochem. Cytochem.*, **16,** 487.
USUKU, G. (1958). *Kumamoto med. J.*, **11,** 84.
UUSITALO, R., and PALKAMA, A. (1970). *Acta Med. exp. Fenn.*, **48,** 84.
VALENTINE, R. C. (1958). *Nature, Lond.*, **181,** 832.
VAN ORDEN, L. S., BLOOM, F. E., BARRNETT, R. J., and GIARMANN, N. J. (1966). *J. Pharmac. exp. Ther.*, **154,** 185.
VITRY, G. (1967). *Archs micr. Morphol. exp.*, **56,** 31.
VOGT, A., and KOPP, R. (1965). *Z. Bakteriol.*, **198,** 270.
VORBRODT, A. (1967). *Acta histochemica*, **28,** 204.
WATSON, M. L. (1958a). *J. biophys. biochem. Cytol.*, **4,** 475.
WATSON, M. L. (1958b). *J. biophys. biochem. Cytol.*, **4,** 727.

WATSON, M. L., and ALDRIDGE, W. G. (1961). *J. biophys. biochem. Cytol.*, **11,** 257.
WATSON, M. L., and ALDRIDGE, W. G. (1964). *J. Histochem. Cytochem.*, **12,** 96.
WELLER, R. O., BAYLISS, O. B., ABDULLA, Y. H., and ADAMS, C. W. M. (1965). *J. Histochem. Cytochem.*, **13,** 690.
WETZEL, M., WETZEL, B., and SPICER, S. (1966). *J. Cell Biol.*, **30,** 299.
WHITE, J. G., KRIVIT, W., and VERNIER, R. (1964). *Blood*, **24,** 443.
WILKINSON, G., COTTON, F. A., and BIRMINGHAM, J. M. (1956). *J. inorg. nucl. Chem.*, **2,** 95.
WILLIAMS, M. A., and GREGORY, D. W. (1967). *J. roy. micr. Soc.*, **86,** 397.
WINKLEMAN, J., and SPICER, S. (1962). *Stain Technol.*, **37,** 303.
WOHLFARTH-BOTTERMANN, K. E. (1957). *Naturwiss.*, **44,** 287.
WOOD, J. (1966). *Nature, Lond.*, **209,** 1131.
WOOD, J. (1967). *Anat. Rec.*, **157,** 343.
WOOD, J., and BARRNETT, R. J. (1963). *Anat. Rec.*, **145,** 301.
WOOD, J., and BARRNETT, R. J. (1964). *J. Histochem. Cytochem.*, **12,** 197.
WOOD, R. L. (1969). *Anat. Rec.*, **163,** 287.
WOOD, R. L., and LEGG, P. G. (1970). *J. Cell Biol.*, **45,** 576.
WYLLIE, J. C. (1964). *Exp. Molec. Path.*, **3,** 468.
YAMADA, K. (1970). *Histochemie*, **23,** 13.
YASUTAKE, S., MURAKAMI, M., and YOSHIDA, N. (1961). *J. Electronmicrosc.*, **10,** 233.
YASUZUMI, G., and SAWADA, H. (1950). *Cytologia, Tokyo*, **15,** 295.
YATES, R. D., HIGGINS, J. A., and BARRNETT, R. J. (1969). *J. Histochem. Cytochem.*, **17,** 379.
YOKOTA, S. (1969). *Histochemie*, **19,** 255.
ZAGURY, D., MODEL, P. G., and PAPPAS, G. D. (1968). *J. Histochem. Cytochem.*, **16,** 40.
ZAMBONI, L., and DE MARTINO, C. (1967). *J. Cell Biol.*, **35,** 148A.
ZEBE, E. (1963). *Z. naturforsch.*, **18b,** 501.
ZEBE, E., and FALK, H. (1964). *Histochemie*, **4,** 161.
ZEITLER, E., and BAHR, G. F. (1957). *Expl Cell Res.*, **12,** 44.
ZHDANOV, V. M., AZADOVA, N. B., and KULBERG, A. Y. (1965). *J. Histochem. Cytochem.*, **13,** 684.
ZHDANOV, V. M., KULBERG, A., and AZADOVA, N. (1966). *J. Histochem. Cytochem.*, **14,** 609.
ZOBEL, R. C., and BEER, M. (1961). *J. Biophys. Biochem. Cytol.*, **10,** 335.

APPENDIX 17

α-NAPHTHYL ACETATE METHODS FOR ESTERASE

(*Cold formalin, frozen sections; cold acetone, paraffin sections; cold microtome, formalin post-fixed sections*)

Standard Method. (1) Cut frozen sections 10–15 μ thick and mount on clean slides. Dry in air to ensure adherence.

(2) Incubate for 1 to 15 minutes at room temperature in the following medium: Dissolve 10 mg. α-naphthyl acetate in 0·25 ml. acetone and add 20 ml. 0·1 M-phosphate buffer (pH 7·4).* Shake thoroughly until most of the initial cloudiness disappears. Add 50–100 mg. Fast blue B salt,† shake, and filter directly on to the sections. These should be dry if frozen sections but brought to water and left wet if paraffin sections.

(3) Wash in running water, 2 minutes.

(4) Counterstain in Mayer's haemalum, 4–6 minutes.

(5) Wash in running water for at least 30 minutes.

(6) Mount in glycerine jelly.

Result. Esterase, black; nuclei, dark blue. (Lipase, AChE and ChE can also hydrolyse α-naphthyl acetate and appear black, see Chapter 17, pp. 796–799, for methods of distinguishing.)

HPR Method

Preparation of Incubation Medium. For preparation of hexazotized pararosanilin see Vol. 1, p. 732. Add to 7·5 ml. 0·2 M-phosphate buffer stock solution (Na_2HPO_4), 2·5 ml. distilled water and 0·25 ml. substrate solution (1 per cent α-Naphthyl acetate) in acetone. Shake and add 0·8 ml. of HPR.

The pH of this mixture is about 6·5, at which it can remain. It should be filtered immediately before use.

Incubation. (1) Incubate for 10–30 minutes at room temperature.

(2) Wash briefly in water.

(3) Counterstain nuclei in Mayer's haemalum for 4–6 minutes.

(4) Wash in running water for 30 minutes.

(5) Dehydrate in alcohols, clear in xylene and mount in DPX or other synthetic resin.

Result. Esterases, brick red; nuclei, blue.

THE NAPHTHOL AS ACETATE METHOD FOR ESTERASE

(after Gomori)

(*Cold formalin, frozen sections; cold acetone, paraffin sections; cold microtome, formalin post-fixed sections*)

PREPARATION OF THE SUBSTRATE

Dissolve 5 g. 2-hydroxy-3-naphthoic anilide (Naphthol AS, or Brenthol AS, I.C.I. Ltd.) in 10 ml. dry pyridine with 20 ml. acetic anhydride in addition. Heat

*Alternatively, add 1·0 ml. 1 per cent α-naphthyl acetate in acetone to 10 ml. of the buffer.

† Other diazonium salts can be used for coupling. I have tried the twenty-two salts which appear in the Table on page 711, Vol 1; the majority of these give an orange or red particulate deposit.

under a reflux condenser for 1 hour. Pour into cold water, filter off the pasty product and dry. Recrystallize from ethanol (containing a little charcoal) to obtain the acetate as a cream-coloured powder (m.p. 160°–161°).

Method. (1) Cut frozen sections at 10 μ and mount on clean slides, dry in air to ensure adherence.

(2) Add 0·1 ml. 1 per cent. Naphthol AS acetate in acetone to 10 ml. 0·05 M-phosphate buffer at pH 7·0. Shake to form a slightly turbid medium and add 10 mg. of a suitable stable diazotate.* Stir, and filter on to the dry section.

(3) Incubate at 17°–22° for 20–30 minutes.

(4) Wash in running water for 1 minute.

(5) Counterstain, according to need, with either Mayer's haemalum, 4–6 minutes, or with carmalum, 1–6 hours.

(6) Mount in glycerine jelly.

Result (Plate XVa). A particulate azo dye is deposited at sites of esterase activity. Nuclei, blue or red.

Gomori (1952) recommended the diazotate of *o*-aminoazotoluene; this gives a finely particulate orange-red precipitate. He also recommended the use of cold acetone-fixed paraffin sections and the technique given above can equally well be employed for these. The necessary incubation time is longer (1–1⅓ hours at room temperature) but, because diffusion is less likely than with the α-naphthyl acetate method, the Naphthol AS method is well suited to use with acetone-fixed paraffin sections. Naphthol AS is particularly sensitive to pH changes when coupling with diazonium salts. The best results are obtained within the range of pH 6·8–7·2.

NAPHTHOL AS-LC ACETATE METHOD FOR ESTERASE

(after Burstone)

(*Cold acetone or freeze-dried, paraffin sections; cold formalin, frozen sections*)

Preparation of the Substrate

Suspend 5 g. naphthol AS-LC in 50 ml. tetrahydrofuran and add 4 ml. acetic anhydride, and then 4 ml. of 5 N-NaOH with constant stirring until the solution becomes clear. Pour the clear supernatant into light petroleum and collect the precipitate under suction. Dry, dissolve in hot methanol, filter and recrystallize several times from cold methanol. (Naphthol AS-LC acetate; mol. wt. 399·82; m.p. 138°).

Method (paraffin sections).† Dissolve 20 mg. of the substrate in 4 ml. dimethylformamide and add 5–10 ml. ethyl cellosolve (ethylene glycol monoethyl ether) and 10 ml. 0·2 M-tris buffer (pH 7·1). Make up to 50 ml. with distilled water. Add 30 mg. of diazonium salt (Garnet GBC, or Corinth LB), shake, filter and use at room

* Of the twenty-two stable diazotates listed in the Table on page 711, Vol 1, only four give a satisfactory result. These are the diazotates of 4-benzoylamino-2:5-dimethoxyaniline, 4-chloro-*o*-anisidine, 5-chloro-*o*-toluidine, and *o*-aminoazotoluene. The first gives a steel blue precipitate of moderate size which does not obscure cellular detail, the second gives a very fine particulate dull red precipitate (Plate XVa), and the third a bright scarlet precipitate in which the particle size is too large for convenience even with low magnification. The fourth gives a fine orange-red precipitate.

† A somewhat different medium should be used for frozen sections. Burstone recommended placing 3–5 mg. of substrate in a 50-ml. flask. After adding 2 ml. dimethylformamide and 5 ml. ethyl cellosolve the volume was made up by adding distilled water and buffer (pH 7·2 to 7·4). This solution could not be filtered but 40 mg. Garnet GBC salt in 3 ml. water was filtered into it. Sections were preferably used free-floating.

temperature. The addition of a few drops of 20 per cent Tween 20 is sometimes helpful.

(2) Incubate sections for 10–40 minutes.

(3) Wash in running water.

(4) Counterstain nuclei with Mayer's haemalum, 1½ minutes.

(5) Wash in running water, mount in glycerine jelly or in 80 per cent alcoholic polyvinyl acetate.

Result. Esterases produce a bright red precipitate; nuclei blue.

NAPHTHOL AS-D ACETATE METHOD FOR ESTERASE

(Burstone, 1957, 1962)

(*Cold formalin, frozen sections and freeze-dried, paraffin embedded sections were recommended*)

Preparation of the Incubating Medium

Place approximately 5·0 mg. naphthol AS-D acetate in a 50 ml. Erlenmeyer flask and add 0·5 ml. dimethylformamide or acetone. Add 25 ml. distilled water, 25 ml. 0·2 M Tris buffer (pH 7·1) and 20–40 mg. Fast blue RR salt. Mix by shaking and filter through coarse paper into a staining jar. Check pH. If this rises above 7·1 there will be spontaneous breakdown of the substrate.

Method. (1) Incubate sections for 5 minutes to several hours (depending on type of preparation).

(2) Wash briefly in water.

(3) Counterstain nuclei in Mayer's carmalum (10 minutes), or by the Feulgen procedure (Appendix 9, p. 647.)

(4) Wash and mount in glycerine jelly oı in a water miscible synthetic mountant.

Result (Plate XVb). Esterases deep blue; nuclei in shades of red.

HEAVY-METAL METHOD FOR A-ESTERASES

(after Deimling, 1965)

(*Fresh frozen cryostat sections, postfixed in cold 6 per cent glutaraldehyde*)

Preparation of the Substrate (8-propionoxy-5-nitroquinoline)

Mix 5 g. 5-nitro-8-hydroxyquinoline with 15 ml. propionic acid anhydride in a 50 ml. Erlenmeyer flask and bring to 100° on an oil bath. Raise the temperature of the bath for a few minutes to 150° to bring everything into solution. After 10 minutes, cool the flask to −18°. Compact crystals develop very slowly. After 18 hours or more, decant the supernatant and wash the crystals with 5 ml. ether. The remainder is extracted with 20 ml. warm ether. Pool the ether solutions and cool to −18°. Wash the resulting crystals twice with cold ether (−18°). Yield: 3·6 g.

Preparation of the Incubating Medium

(1) Stock Bismuth Solution (0·05 M). Dissolve 1·525 g. bismuthyl nitrate, $(BiO)NO_3.H_2O$, mol. wt. 305·03, and 6 g. L(+)-tartaric acid in 100 ml. distilled water. Add NaOH to bring the pH to 7·0. At 4° this solution keeps for one month.

(2) Final Medium. 24 ml. Tris-HCl buffer, pH 7·6, 23 ml. distilled water, 3 ml. 0·05 M bismuth solution. Bring this solution to 17–20°. Immediately before use add 16 mg. 8-propionoxy-5-nitroquinoline dissolved in 1 ml. acetone and shake vigorously. This final solution is stable at 19° for about 40 minutes.

Method. (1) Post-fix fresh frozen cryostat sections, 7-10 μ thick in 6 per cent glutaraldehyde at 0–4°, for 20 minutes.

(2) Wash briefly in water.

(3) Transfer to the incubation medium at 17–20° until the sections show a golden yellow colour in esterase-positive regions (10–30 minutes).

(4) Wash briefly in water.

(5) Immerse for 5 minutes in dilute yellow ammonium sulphide.

(6) Wash briefly and mount in glycerine jelly.

Result. Brown bismuth sulphide is deposited in esterase-positive regions.

INDIGOGENIC METHODS FOR ESTERASES

(Holt and Withers, 1952; Holt, 1958)

(*Cold formalin, frozen sections; cold acetone, cold formalin and acetone, paraffin sections*)

PREPARATION OF THE INCUBATING MEDIUM

O-acetyl-5-bromoindoxyl*	1·3 mg.	
Ethanol	0·1 ml.	
Allow to dissolve and add:		*Final concn.*
0·1 M tris(hydroxymethyl)amino-methane/HCl buffer (pH 6–8·5)† .	2·0 ml.	0·02 M
0·05 M-potassium ferricyanide . .	1·0 ml.	5×10^{3} M
0·05 M-potassium ferrocyanide . .	1·0 ml.	5×10^{-3} M
0·1 M-calcium chloride	1·0 ml.	0·01 M
Add water to 10 ml.‡		

This medium should be freshly prepared.

Method. (1) Fix thin pieces of tissue in 5 per cent neutral formaldehyde in 1 per cent saline, at 4° for 24 hours. Alternatively use 4 per cent formaldehyde containing 1 per cent $CaCl_2$.

(2) Avoid washing in water. This causes loss of enzyme.

(3) Cut frozen sections at 10–15 μ and transfer these to the incubating medium, at 22° or 37°, for 10 seconds to 15 hours. (Development can be carried out under inspection, using the microscope for small structures. Within fairly broad limits over-incubation is not dangerous since it only increases the intensity without altering the localization.)

(4) Counterstain the nuclei, if desired, in Mayer's Carmalum, 10 minutes.

(5) Rinse briefly in water.

(6) Mount sections on clean dry slides.

(7) Either mount directly in glycerine jelly or dehydrate gently in 50 per cent, 70 per cent and 95 per cent alcohol, clear in carbol-xylene (1 part phenol, 3 parts xylene) and mount in Canada balsam.

Result (Figs. 165, 166 and Plates XVd and XVIa). A blue precipitate of indigo forms at the site of esterase activity. Nuclei (if stained), deep red.

* Use alternatively, *O*-acetyl-4-chloro-5-bromoindoxyl, 1·5 mg., or *O*-acetyl-5-bromo-6-chlorindoxyl, 3·0 mg.

† At pH 6·0 the final reaction product is diffuse whereas at pH 8·5 a droplet pattern usually results.

‡ Holt used 10 ml. 1·0 M-NaCl in place of distilled water.

INDOXYLAZO METHODS FOR ESTERASES

(after Delellis and Fishman, 1965; and others)

(*Cold formalin, frozen sections, preferable*)

PREPARATION OF TISSUES

Fix small tissue blocks for 12–24 hours in cold 4 per cent formol-calcium. Transfer to gum-sucrose (Vol. 1, p. 592) for 24 hours at 4°. Wash for 5 minutes in water, blot dry and cut cryostat sections at 5–10 μ

PREPARATION OF THE INCUBATING MEDIUM (pH 6·1)

Dissolve 1 mg. *O*-acetyl-5-bromoindoxyl (alternatives are the 4-chloro-5-bromo- and 4-chloro-5-bromo-7-methyl- substituted indoxyl acetates) in 0·1 ml. ethanol and add to 5 ml. 0·2 M Tris-HCl buffer at pH 7·4 with 4·75 ml. distilled water. Add 0·125 ml. hexazotized pararosanilin stock solution (Vol. 1, p. 732) mixed with 0·125 ml. freshly prepared 4 per cent sodium nitrite solution and allow to stand for 1 minute before adding to the rest of the medium. Finally, add 11 mg. calcium chloride and adjust the pH to 6·1 by adding sufficient 0·1 N-NaOH. Filter before use.

Method. (1) Incubate sections at 37° for 15–30 minutes.
(2) Wash well in distilled water.
(3) Counterstain nuclei with 2 per cent chloroform-washed methyl green for 5 minutes.
(4) Wash briefly in tap water.
(5) Dehydrate in alcohols, clear in xylene and mount in a suitable synthetic resin.

Result (Plate XVIb). A dark brown indoxylazo dye is deposited in sites of esterase activity. Nuclei, deep green.

PREPARATION OF INCUBATING MEDIUM (pH 8·8)

Dissolve 1 mg. *O*-acetyl-5-bromoindoxyl in 0·1 ml. ethanol and add to 9·5 ml. Tris-HCl buffer (pH 8·8). Add 11 mg. calcium chloride (optional) and 10 mg. Fast blue RR salt (Vol. 1, p. 711). Adjust pH to 8·8 with 0·1 N-NaOH. Filter before using.

Method. (1) Incubate sections at 37° for 15–30 minutes.
(2) Wash well in distilled water.
(3) Counterstain nuclei in carmalum for 5 minutes.
(4) Wash briefly in tap water.
(5) Mount in glycerine jelly.

Result. Nuclei, red; Esterases, brown.

IMPROVED SYNTHESES OF DIHALOGEN-O-ACETYLINDOXYLS

(after Kambli, 1964)

The classical Sandmeyer synthesis, producing isatins from arylamines *via* isonitrosoacetarylides, was used by Holt and Sadler (1958) for the production of dihalogen acetylindoxyls. The production of 4-chloro-5-bromoisatin in this manner is very difficult because there are many opportunities for the formation of isomeric compounds.

These difficulties can be avoided in two ways:

(A) by choosing a different principle for the production of 4-chloro-5-bromoisatin.

(B) still using the Sandmeyer synthesis, by carrying out the whole sequence, from the beginning, with a further substituent (i.e. 7-methyl or 7-methoxy) in the benzene ring.

A. Isatin Intermediate Synthesis. Oxidative breakdown of indigo and substituted indigos leads first to the corresponding isatins. Chromic acid is a suitable oxidant. Brilliant indigo 4 G (4,4′-dichloro-5,5′-dibromindigo) is the ideal starting material. An 80% yield of the isatin, of sufficient purity, can be obtained.

Details. To a suspension of 245 g. Brilliant indigo 4 G (0·5 mol.) in 2500 ml. 30% sulphuric acid containing 20 g. of a dispersing agent effective in acid medium (Dispergator SF, Ciba; Tamol NNO, BASF), add a solution containing 240 g. $Na_2Cr_2O_7,2H_2O$ (0·8 mol.) in 500 ml. sulphuric acid, over a period of 15 minutes. Warm to 80° and stir for 1 hour. The appearance of foaming can be countered by the addition of a sufficient quantity of nitrobenzene. The crude product contains some unchanged indigo which can be removed by dissolving the isatin in alkali. Crystallize from butanol as orange needles (m.p. 269–271°).

B. Synthesis of 4-chloro-5-bromo-7-methyl (or 7-methoxy) isatin. This method was suggested by the fact that all steps of the Sandmeyer synthesis lead to chemically uniform compounds if in the initial amine the *p*-position to the chlorine atom is substituted. The additional substituent is localized in the 7-position of the 4,5-dihalogen isatin. The two-step reaction begins with the conversion of 4-chloro-2-amino-1-methyl benzene (or 1-methoxybenzene) to the corresponding nitrosoacetate compounds. In the second step, 4,7-substituted isatins are formed from the iso-nitroso compounds by treatment with conc. H_2SO_4. Halogenation thus takes place exclusively in the 5-position but at different rates. With the 4-chloro-7-methyl isatin bromination in the 5-position is easy whereas, in the case of 4-chloro-7-methoxyisatin bromination takes place only in nitrobenzene in the presence of a catalyst. The formation of 4,5-dihalogen-*O*-acetylindoxyls from the isatins is carried out by the method of Villiger (1909).

Details. (1) **Isonitrosoacetarylide:** Warm a solution containing 10 g. chemically pure H_2SO_4, 246 g. (1·5 mol.) hydroxylamine sulphate and 1·7 kg. Na_2SO_4, $10H_2O$ in 8000 ml. water to 50–60°, and add 71 g. 2-Amino-4-chloro-1-methyl benzene or 79 g. 2-Amino-4-chloro-1-methoxybenzene (0·5 mol.) and 120 g. glacial acetic acid. After the addition of 96 g. (0·65 mol.) chloral hydrate and heating for 7 hours at 80–90°, the isonitrosoacetarylides can be separated by cold filtration.

(2) **Isatin:** Dissolve 213 g. 2-Isonitrosoacetyl-4-chloro-1-methyl benzene or 229 g. 2-Isonitrosoacetyl-4-chloro-1-methoxybenzene (1 mol.) in 2300 g. chemically pure sulphuric acid at 70°. After heating to 80–90° for 1 hour, cool the solution with ice-water; then warm to 70–75° for 30 minutes and filter. Dissolve the precipitate in warm alkali, filter through charcoal, and acidify with HCl.

(3) **4-Chloro-5-bromo-7-methyl isatin:** Suspend 196 g. 4-chloro-7-methyl isatin in 1100 ml. glacial acetic acid and add this suspension dropwise to a solution containing 176 g. bromine in 160 ml. glacial acetic acid and stir for 8 hours at 50°. Crystallize as brownish orange flakes from butanol (m.p. 294–295°).

(4) **4-Chloro-5-bromo-7-methoxy isatin:** Suspend 212 g. 4-chloro-7-methoxy isatin (1 mol.) and 4 g. water-free ferric chloride in 500 ml. nitrobenzene and add dropwise to a solution of 185 g. bromine in 200 ml. nitrobenzene over a period of 2 hours at 30–35°. Subsequently warm and stir at 45–50° for 7 hours. Crystallize as dark red needles from butanol, (m.p. 265–266°).

THE TWEEN METHOD FOR LIPASE

(after Gomori)

(*Cold formalin, frozen sections; cold acetone, paraffin sections*)

A large number of substrates have been employed for this method. Gomori recommended as best for most purposes either Tween 60 (Atlas Powder Co.), the stearic ester of a polymannitol, or Product 81 (Onyx Oil and Chemical Co.), the stearic ester of a polyglycol. To demonstrate his "true lipase" Gomori recommended Tween 80, the oleic ester of a polymannitol, or G-6486, a ricinoleic acid ester.

Stock Solutions. (1) Five per cent Tween 60 (Tween 80 for "true lipase").

(2) 0·5 M-tris(hydroxymethyl)aminomethane buffer pH 7·2–7·4 (Appendix 1). This substance was at one time not easily available in the U.K. In its place others employed 0·2 M-veronal acetate buffer (Mark, 1950) or bicarbonate buffer (Sneath, 1950).

(3) Ten per cent $CaCl_2$.

Substrate and buffer solutions are preserved with 0·25 per cent chloretone or with a crystal of thymol.

Incubating Medium

5 ml. buffer.
2 ml. $CaCl_2$.
2 ml. Tween solution.
40 ml. distilled water.

With a crystal of thymol included, this working solution keeps well in the refrigerator.

Method. (1) Cut frozen sections and mount on slides.

(2) After drying to ensure adherence, incubate for 3–12 hours in the above medium.

(3) Wash thoroughly in distilled water.

(4) Immerse in 1 per cent lead nitrate for 15 minutes.

(5) Wash in running water, 5 minutes.

(6) Immerse in dilute yellow ammonium sulphide, 1–2 minutes.

(7) Wash and counterstain with 1 per cent aqueous eosin, 5 minutes.

(8) Wash well, mount in glycerine jelly.

Result (Fig. 167). A brownish-black deposit indicates the presence of lipase. (In my hands satisfactory results were obtained only with sections of pancreas.)

MODIFIED TWEEN METHOD FOR LIPASE-ESTERASE

(after Martin)

(*Formalin-fixed frozen sections; cold acetone, paraffin sections*)

Preparation of Stock Solution

Dissolve 2 per cent Tween 40 in 0·2 per cent $CaCl_2$ in veronal-acetate buffer (pH 7·4). Add a crystal of thymol. Incubate at 37° for 2 days and filter through a Seitz filter.

PREPARATION OF INCUBATING MEDIUM

0·05 M-veronal acetate (pH 7·2 to 7·6)	30 ml.
2 per cent $CaCl_2$	2–3 ml.
Glycerol	10–20 ml.
Stock solution	2–4 ml.
Thymol	A small crystal

Method. Proceed as in Gomori method (above).

IMPROVED LIPASE METHOD USING TWEEN 85

(George and Iype, 1960; Bokdawala and George, 1964)

Method. (1) Cut fresh sections, 10–15 μ thick, directly into cold 6 per cent neutral formalin, *via* the catchment shute of the cryostat. Allow to fix at 4° for 4 hours.

(2) Mount sections on clean slides and allow to dry.

(3) Apply a thin coating of 1 per cent gelatin and fix for 30 minutes in cold formalin.

(4) Wash in running water for 30 minutes.

(5) Rinse with distilled water and transfer to 0·2 M borate buffer, pH 8·0, containing 2mM Versene, at 4°. Leave for up to 15 minutes.

(6) Wash in distilled water

(7) Incubate at 37° for 16 hours, together with control sections previously treated with boiling water (10 minutes) to inactivate the enzyme, in the following medium:

2 ml.	5 per cent Tween 85*
5 ml.	bicarbonate buffer (0·2 M $NaHCO_3$)
2 ml.	10 per cent $CaCl_2$
40 ml.	distilled water.

This medium should be made up and then incubated for 10 hours at 37°, filtered through Whatman 42 paper, and stored with the addition of a crystal of thymol as a preservative.

(8) Wash in distilled water.

(9) Immerse in 1 per cent Alizarin red S at pH 6·5 to 6·8 for 1 minute.

(10) Rinse in distilled water and mount in glycerine jelly.

Result. Orange red deposits indicate sites of lipase activity. Background faint pink.

NAPHTHOL AS NONANOATE METHOD FOR LIPASE

(Abe, Kramer and Seligman, 1964)

Method. (1) Fix tissues for 24 hours in Baker's calcium formalin (p. 601) at 4°.

(2) Rinse three times with cold physiological saline and infiltrate with 0·88 M sucrose solution, containing 9 g. per litre of gum acacia, for 12–24 hours at 4°.

(3) Blot blocks to remove excess medium, quench in cold (−70°) isopentane, cut 4–10 μ sections in the cryostat and transfer to cold 2 M NaCl.

* Polyoxyethylene sorbitan trioleate.

(4) Float on the following medium at 37° for 40–90 minutes (usually 60 minutes):
5 ml. 0·4 M-Tris buffer, pH 7·4.
1 ml. 2·5 per cent sodium taurocholate.
3·9 ml. distilled water.
0·1 ml. 2 per cent Naphthol AS nonanoate in dimethylacetamide.
10 mg. Fast blue BB salt.*
The turbid solution is filtered before use.
(5) Wash in distilled water for 10 minutes.
(6) Counterstain nuclei with haematoxylin.
(7) Wash and mount in 90 per cent w/v polyvinyl pyrrolidone.

Result (Plate XVId). Lipase activity is shown by a blue or red azo dye (depending on whether the diazonium or hexazonium salt is used for coupling).

METHOD FOR PHOSPHOLIPASE B

(after Ottolenghi *et al.*, 1966)

(*Cryostat sections post-fixed in cold formol-calcium*)

PREPARATION OF SOLUTIONS AND REAGENTS

Stock Tris Buffer. Dissolve 36·3 g. reagent grade Tris in 400 ml. distilled water and add 100 ml. 3N-acetic acid. Adjust to pH 6·6 by adding 3N-acetic acid and make up to 1000 ml.

Stock Incubating Solution A. Add 100 ml. stock Tris buffer to 150 ml. water. Dissolve 3 g. cobalt acetate ($4H_2O$) with stirring and add 0·1 N-NaOH to restore the original pH. Dissolve 24 g. sucrose and bring volume to 270 ml.

Preparation of Lysolecithin. Dissolve 2·5 g. lecithin in 1000 ml. ethyl ether. Add 10 mg. lyophilized *Agkistrodon piscivorus piscivorus* venom dissolved in 10 ml. 8 mM $CaCl_2$, with vigorous stirring. Continue stirring for a few minutes and allow to stand overnight. Separate precipitate by centrifuging, and wash 3 times with ethyl ether and then with acetone. Dissolve the product in hot absolute ethanol. Centrifuge and recrystallize from ethanol (Saunders and Thomas, 1958). Commercial lysolecithin may be used if available.

Lysolecithin Stock Solution. Dissolve a small quantity of lysolecithin in 50 ml. water by heating to 40°. Dialyse overnight against distilled water. Estimate organic phosphorus content of the solution and adjust to 20 μmoles/ml. Store aliquots (5 ml.) in plastic tubes at 4°. For use melt at 40° and add to incubating medium to produce final concentration of about 2 μmoles/ml.

Nuclear Counterstain. 0·1 g. Nuclear Fast red in 100 ml. of 5 per cent aluminium sulphate.

Incubating Medium. Add 5 ml. lysolecithin stock to 40–45 ml. stock solution A.

Method. (1) Incubate for 30–60 minutes.
(2) Rinse twice in 0·9 per cent NaCl.
(3) Immerse in dilute ammonium sulphide at 4° for 15 seconds.
(4) Rinse twice in 0·9 per cent NaCl.
(5) Counterstain in Nuclear Fast red, 2 minutes, and wash briefly.
(6) Dry in air and mount in a synthetic resin.

Result. Brownish-black deposits indicate activity of phospholipase B.

* Alternatively, use 0·8 ml. freshly prepared hexazonium pararosanilin (p. 732) per 10 ml. medium.

THE MYRISTOYL CHOLINE METHOD FOR CHOLINESTERASES

(after Gomori)

(*Cold formalin, frozen sections; cold acetone, paraffin sections*)

PREPARATION OF SOLUTIONS

Stock Solution 1. 0·1 M-veronal-acetate buffer pH 7·6, 50–100 ml.
0·1 M-cobaltous acetate, 30–50 ml.
Distilled water to make up 300 ml.
Add 1 mg. each of $CaCl_2$, $MgCl_2$ and $MnCl_2$.

Add a crystal of thymol and store at 4°.

Stock Solution 2. 0·02 M-myristoyl choline in distilled water.

Add a crystal of thymol and store at 4°.

For use, add 1 ml. of stock solution 2 to 50 ml. of stock solution 1, previously heated to 37°.

Method. (1) Cut frozen sections 10–15 μ thick and mount on clean slides (alternatively, bring paraffin sections to water).

(2) Incubate for 1–4 hours (frozen sections), 2–16 hours (paraffin sections), in the working solution of myristoyl choline.

(3) Wash in running water, 2 minutes.

(4) Immerse in a dilute solution of yellow ammonium sulphide, 1–2 minutes.

(5) Wash in water.

(6) Counterstain in 1 per cent aqueous eosin or 1 per cent light green.

(7) Wash in water.

(8) Mount in glycerine jelly (frozen sections) or dehydrate rapidly, clear, and mount in Canada balsam (paraffin sections).

Result (Figs 168 and 169). Sites of ChE, and possibly some non-specific esterases, appear black.

THE ACETYLTHIOCHOLINE METHOD FOR CHOLINESTERASES

(after Gomori, 1952)

(*Cold formalin, frozen sections; cold microtome, free-floating sections*)

PREPARATION OF STOCK SOLUTIONS

Copper sulphate ($CuSO_4.5H_2O$)	0·3 g.
Glycine	0·375 g.
Magnesium chloride ($MgCl_2.6H_2O$)	1·0 g.
Maleic acid	1·75 g.
NaOH (4 per cent)	30 ml.
Na_2SO_4 (40 per cent; hot saturated)	170 ml.

This solution keeps indefinitely (pH 6·0).

PREPARATION OF INCUBATING MEDIUM

Dissolve about 20 mg. acetylthiocholine iodide in a few drops of water. Add 10 ml. stock solution.

Method. (1) Incubate sections for 10–60 minutes at 37°.

(2) Rinse in 3 changes of saturated Na_2SO_4.

(3) Treat with dilute yellow ammonium sulphide, 2 minutes.*

(4) Wash briefly and mount in glycerine jelly.

* In Holmstedt's modification this step is omitted.

Result. (Plate XVIc). Sites of cholinesterase activity appear in shades of brown. This method is much simpler than the Koelle method. Some workers may still prefer the latter, however, and directions therefore appear below

THE ACETYL- AND BUTYRYLTHIOCHOLINE METHOD FOR CHOLINESTERASES

(after Koelle)

(Fresh frozen)

Preparation of Solutions

STORAGE SOLUTIONS

(1) **DFP** Distilled water . . . . . . 4·5 ml.
40 per cent Na_2SO_4 . . . . . . 9·0 ml.
1 μM-DFP . . . . . . 1·5 ml.

The 1 μM-DFP solution was prepared within 30 minutes of the time of use by serial dilution of a 0·1 M stock solution in propylene glycol.

(2) **Storage Solution C**
Distilled water . . . . . . 4·5 ml.
40 per cent Na_2SO_4 . . . . . 10·5 ml.

(3) **Storage Solution B-D**
Distilled water . . . . . . 6·0 ml.
40 per cent Na_2SO_4 . . . . . 9·0 ml.

Storage solutions are kept at 35° and used at this temperature.

INCUBATING SOLUTIONS

The following Table gives the composition of the final incubation media for localization of AChE and ChE.

Solution	Enzyme localized	Cu-Gl ml.	H_2O ml.	Mal ml.	Na_2SO_4 ml.	$MgCl_2$ ml.	CuThCh	AThCh ml.	BuThCh ml.
B	AChE	0·6	2·1	1·5	9·0	0·6	trace	1·2	—
C	ChE	0·6	0·6	1·5	10·5	0·6	trace	—	1·2
D	Control	0·4	1·4	1·0	6·0	0·4	trace	—	0·8

All solutions allowed to incubate 15 minutes after addition of CuThCh; substrates (AThCh and BuTuCh) added immediately prior to filtering and introducing slides.

The solutions from which the final incubating media are made up are composed as follows:

(1) Copper-glycine (Cu-Gl): 3·75 g. glycine, 2·5 g. $CuSO_4$, $5H_2O$, distilled water 100 ml.

(2) Maleate buffer (Mal.): 9·6 g. sodium hydrogen maleate, 52·2 ml. N-NaOH, distilled water to 100 ml.

(3) Sodium sulphate (Na_2SO_4): 40 per cent (W/v) Na_2SO_4 adjusted to pH 6·0, stored at 37°.

(4) Magnesium chloride ($MgCl_2$): 9·52 g. $MgCl_2$ in 100 ml. distilled water.

(5) Acetylthiocholine (AThCh): 23 mg. acetylthiocholine iodide, 1·2 ml. distilled water, 0·4 ml. 0·1 M-$CuSO_4$, $5H_2O$; centrifuged and the supernatant decanted for use.

(6) Butyrylthiocholine (BuThCh): 43 mg. butyrylthiocholine iodide, 1·8 ml. distilled water, 0·6 ml. 0·1 M-$CuSO_4$, $5H_2O$; centrifuged, etc., as above.

(7) Copper thiocholine (CuThCh): This is initially prepared by alkalinizing a solution of AThCh in copper-glycine solution to pH 12·0 with KOH, allowing to stand overnight at room temperature, collecting the precipitate and washing with water. Subsequently, the incubating solutions, after use, are filtered, allowed to stand at 37° for 2–4 days to permit spontaneous hydrolysis of the substrate, and the precipitate is collected and washed with water.

Method. (1) Following a minimal period of storage in the appropriate storage solutions, transfer the mounted sections into incubating solutions for 5–60 minutes at 37° (average time, 30 minutes).

(2) Transfer to rinsing solution I (20 per cent Na_2SO_4, saturated with CuThCh) for 5 minutes.

(3) Transfer to rinsing solution II (10 per cent Na_2SO_4, saturated with CuThCh) for 1 minute.

(4) Transfer to rinsing solution III (water, saturated with CuThCh) for 1 minute.

(5) Treat with CuS-saturated ammonium sulphide solution (dilute yellow ammonium sulphide treated with a few drops of $CuSO_4$, shaken and filtered) for 20 seconds.

(6) Rinse rapidly in water.

(7) Fix for 30 minutes in 10 per cent neutral formalin, saturated with CuS.

(8) Dehydrate in CuS-saturated alcohols and clear in CuS-saturated xylene.

(9) Mount in Canada balsam.

Result. A dark brown precipitate indicates sites of cholinesterase activity. In sections incubated with solution B this indicates AChE, in those incubated with solution C, ChE. Control sections treated with DFP and incubated with solution D should be blank.

MODIFIED THIOCHOLINE AND INHIBITOR METHOD

(after Holmstedt)

(*Fresh frozen, cold microtome, mounted sections*)

STOCK SOLUTIONS

(1) N-NaOH

(2) 0·1 M-$CuSO_4$

(3) 40 per cent Na_2SO_4 adjusted to pH 6·0.

(4) 0·04 mM-Mipafox.

(5) 0·5 mM-284C51.

(6) Copper glycinate (3·75 g. glycine and 2·5 g. $CuSO_4 . 5H_2O$ in 100 ml water).

(7) Maleate buffer, pH 6·0 (9·6 g. sodium hydrogen maleate with 50 ml. N-NaOH in 100 ml. water).

(8) $MgCl_2$, 9·52 per cent aqueous solution.

(9) Acetylthiocholine sulphate (45 mg. acetylthiocholine iodide with 0·8 ml. 0·1 M-$CuSO_4$ and 2·4 ml. water; precipitate discarded).

(10) Butyrylthiocholine sulphate (50 mg. butyrylthiocholine iodide treated as above).

Solutions 3, 6, 7 and 8 should be stored in an incubator at 38°.

Method. (1) Cut fresh cryostat sections 10–15 μ thick and mount on slides or coverslips.

(2) Place sections in 25 per cent Na_2SO_4 until required for use (as short a time as possible).

(3) Incubate a series of 7 sections as shown in the Table below.

TABLE 86

Stage 1		Storage			
	Slides . .	ABCD		EFG	
	Solutions . .	St_1		St_2	
Stage 2		Incubation with inhibitor (30 mins. 37°)			
	Slides . .	BD	F	C	G
	Solutions . .	I_1	I_2	I_3	I_4
Stage 3		Wash (10 mins.)			
	Slides . .	BD	F		
	Solutions . .	St_1	St_2		
Stage 4		Re-incubation with inhibitor (30 mins. 37°)			
	Slides . .	D			
	Solutions . .	I_3			
Stage 5		Incubation with substrates (5–120 mins. 37°)			
	Slides . .	AB	CD	EF	G
	Solutions . .	S_1	S_2	S_3	S_4

Solutions indicated in the above table are composed as follows:

Solution	Amounts (ml.) of Stock Solutions								Amount of Water (ml.)
	3	4	5	6	7	8	9*	10*	
St_1	9·0	—	—	—	—	—	—	—	6·0
St_2	10·5	—	—	—	—	—	—	—	4·5
I_1	9·0	1·5	—	—	—	—	—	—	4·5
I_2	10·5	1·5	—	—	—	—	—	—	3·0
I_3	9·0	—	1·5	—	—	—	—	—	5.85
I_4	10·5	—	0·15	—	—	—	—	—	4·35
S_1	9·0	—	—	0·6	1·5	0·6	1·2	—	2·1
S_2	9·0	—	0·15	0·6	1·5	0·6	1·2	—	1·95
S_3	10·5	—	—	0·6	1·5	0·6	—	1·2	0·6
S_4	10·5	—	0·15	0·6	1·5	0·6	—	1·2	0·45

*Add just before use.

(4) Examine slides without further treatment.

Result. The precipitate of copper thiocholine sulphate appears as a microcrystalline deposit visible by ordinary microscopy. Interpretation of results is given below:

TABLE 87

Slide	Treatment		Cholinesterase Indicated
	Inhibitor	Substrate	
A	—	AcThCh	AChE + BuChE
B	Mipafox	AcThCh	AChE
C	284C51	AcThCh	BuChE
D	Both	AcThCh	Nil
E	—	BuThCh	BuChE
F	Mipafox	BuThCh	Nil
G	284C51	BuThCh	BuChE

TABLE 88

INHIBITOR CONCENTRATIONS

Eserine sulphate	30 μ M	1·9 mg./100 ml.
Ethopropazine HCl	100 μ M	3·12 mg./100 ml.
284c51 (BW)	50 μ M	28 mg./100 ml.
IsoOMPA	1 μ M	0·034 mg./100 ml.
Mipafox	50 μ M	0·9 mg./100 ml.
E600	0·1 μ M	·0026 mg/100 ml.

Pb MODIFICATION OF THE THIOCHOLINE TECHNIQUE

Joó, Sávay and Csillik, 1965)

(*Frozen sections, blocks fixed in formalin mixtures*)

Method. (1) Incubate free-floating sections for 5–10 minutes at 4° in the following medium:

5 mg.	Acetylthiocholine
2·5 ml.	0·1 N Sodium acetate
0·3 ml.	0·1 N Acetic acid
0·5 ml.	0·03 M Lead nitrate

(2) Wash in distilled water.

(3) Immerse in a dilute solution of yellow ammonium sulphide for 1–2 minutes.

(4) Wash in distilled water, dehydrate in alcohols, clear and mount in a suitable synthetic resin.

Result. Black deposits indicate sites of cholinesterase activity.

SILVER INTENSIFICATION METHOD

(Henderson, 1967)

PROCEDURE

Carry out whichever copper thiocholine technique is desired. Shorten incubation time to approximately one half of that normally used. Proceed to the stage of treatment with $(NH_4)_2S$ or alternatives.

Method. (1) Wash in three changes of distilled water.

(2) Immerse in 5–10 per cent aqueous silver nitrate for 10–40 seconds. Sections should become straw-coloured.

(3) Wash in three changes of distilled water.

(4) Treat with 5 per cent sodium thiosulphate for 5 minutes.

(5) Wash in water, counterstain, dehydrate and mount as usual.

THE GOLD THIOCHOLINE METHOD

(Koelle and Gromadzki, 1966)

(*Cold formalin or cold Pearson; cryostat sections*)

PREPARATION OF PREINCUBATION, INCUBATION AND RINSING MEDIA

(1) $AuNa_3(S_2O_3)_2 . 2H_2O$, 0·1 M. Prepared by dissolving 100 mg. Sanocrysin* in 1·9 ml. water.

* Ferrosan, Medizinalfabrik, Copenhagen.

(2) AThCh or BuThCh, 0·05 M. Prepared by dissolving 46 mg. AThCh iodide or 50 mg. BuThCh iodide in 2·05 ml. water After addition of 1·15 ml. 0·1 M, $AgNO_3$ the resulting precipitate of AgI is allowed to settle and the supernatant decanted for use.

(3) Eserine salicylate and BW 284C51 are prepared as 1·0 mM stock solutions.

(4) $NaH_2PO_4 . H_2O$ and K_2HPO_4, 6 M.

(5) Acid-alcoholic $(NH_4)_2S$. 20 ml. ethanol, 6 ml. glacial acetic, 1 drop 0·1 M $AuNa_3(S_2O_3)_2$, add finally 3·6 ml. $(NH_4)_2S$ and adjust to pH 5·5.

Compositions of Media

(a) With inhibitor (optional).
(b) Stand for 40 minutes. Filter (Whatman 3) before use.
(c) Saturated with gold thiocholine phosphate.* Filtered.

TABLE 89

Solution	Stock Solutions					
	1. $AuNa_3(S_2O_3)_2$	H_2O (ml.)	2. ThCh	3. Inhibitor	4. NaH_2PO_4	K_2HPO_4
Final Concn.	4 mM		4 mM	·01 mM	4	4
Preincubation	0·4 ml.	2·9 or 2·8[a]		(0·1 ml.)	4·6 ml.	2·1 ml.
Incubation[b]	0·4 ml.	2·1 or 2·0[a]	0·8 ml[c]	(0·1 ml.)	4·6 ml.	2·1 ml.
Rinse	—	3·3			4·6 ml.	2·1 ml.

a. With inhibitor (eserine or BW284)
b. Allowed to stand until visibly colloidal (20–40 mins), then filtered (Whatman no. 3) immediately prior to use.
c. Saturated with gold thiocholine phosphate; filtered through Whatman no. 3 paper.

Method. (1) Preincubate for 15 minutes.
(2) Incubate for up to 3 hours (average 60–90 minutes).
(3) Rinse for 10 minutes.
(4) Immerse in acid-alcoholic $(NH_4)_2S$ for 10 minutes.
(5) Rinse in 8 per cent formalin in 70 per cent ethanol.
(6) Dehydrate, clear and mount in synthetic resin.

Result. Yellow-brown gold sulphide deposits show ChE activity.

THE DIRECT-COLOURING THIOCHOLINE METHOD

(Karnovsky and Roots, 1964)

(Cold formol calcium, gum-sucrose, cryostat sections)

Method. (1) Cut 5–10 μ sections and mount on gelatinized slides.
(2) Incubate at 37° for 15–120 minutes in the following medium.
5 mg. AThCh iodide or BuThCh iodide in 6·5 ml. 0·1 M buffer (pH 6·0). Acetate, phthalate, phosphate or maleate may be used.

Add with stirring:

0·5 ml.	0·1 M-sodium citrate
1·0 ml.	30 mM-$CuSO_4$
1·0 ml.	distilled water
1·0 ml.	5 mM-potassium ferricyanide.

* Obtained by allowing spontaneous hydrolysis of incubating medium for 3 days and centrifuging to collect precipitate.

When eserine sulphate is used as inhibitor it is added as 1·0 ml. of a 1 mM solution in place of the 1·0 ml. of water.

The final incubating medium is clear, greenish in colour, and stable for some hours.

(3) Rinse in distilled water.

(4) Counterstain nuclei, if required, in haematoxylin.

(5) Dehydrate, clear and mount in a suitable synthetic medium.

Result. A reddish-brown precipitate indicates ChE activity.

THIOCHOLINE-LEAD FERROCYANIDE METHOD

(Eränkö, Koelle and Räisänen, 1967)

(*Cold formalin, Krebs–Ringer–Calcium, cryostat sections*)

Preparation of Fixative

5·0 ml. formalin (40 per cent formaldehyde) in 50 ml. Krebs–Ringer–calcium, prepared as follows:

Mix in the amounts indicated:

100 ml.	0·9 per cent NaCl
4·0 ml.	1·15 per cent KCl.
1·27 ml.	1·22 per cent $CaCl_2$.
1·0 ml.	2·11 per cent KH_2PO_4.
1·0 ml.	3·13 per cent $MgCl_2.6H_2O$.
40 ml.	3·55 per cent Na_2HPO_4.
2·0 ml.	N-HCl.

Fix blocks at 0–4° for 2–4 hours, rinsing for 1–8 hours in Krebs–Ringer-calcium. Cut 10 μ cryostat sections.

Preparation of Substrate Mixture

All reagents are made up in glass-distilled water, boiled for one hour before use. They should be freshly prepared. Add the reagents in the order listed, mixing thoroughly after each addition. Tris-acetate buffer (pH 6·0) is made up by adding approximately 2 ml. 1 M-acetic acid to 10 ml. 0·2 M-Tris. The lead acetate solution should be filtered before use. The acetylthiocholine solution is made by dissolving 23 mg. acetylthiocholine iodide in 1·2 ml. water and adding 0·4 ml. 0·1 M-lead acetate to precipitate the iodide. The clear supernatant is obtained by gentle centrifugation at about 100 × g.

Distilled water	4·9 ml.
Tris-acetate buffer (0·17 M)	4·0 ml.
Lead acetate (0·1 M)	0·5 ml.
Potassium ferricyanide (0·1 M)	0·05 ml.
Acetylthiocholine (0·05 M)	0·6 ml.
Potassium ferrocyanide (0·05 M)	0·01 ml.

The last reagent is added in order to saturate the medium with lead ferrocyanide, forming a yellowish-white precipitate. After mixing, the solution is filtered (Whatman 3 paper) and cooled in ice.

Method. (1) Rinse sections in three changes of 0·17 M-Tris-acetate buffer, to remove phosphate ions.

(2) Incubate for 1–60 minutes at 0°.

(3) Rinse in three changes of distilled water at 0°.

(4) Dehydrate, clear and mount in synthetic resin (for direct observation).

(5) Alternatively, treat with dilute yellow ammonium sulphide to which a small amount of 0·1 M-lead acetate has been added (filter).

Result. Pale-yellow or brown deposits indicate ChE activity.

THE THIOLACETIC ACID METHOD FOR CHOLINESTERASE

(after Crevier and Bélanger)

(*Fresh, cold microtome sections; formalin-fixed frozen sections*)

PREPARATION OF INCUBATING MEDIUM

Dissolve 0·12 M-thiolacetic acid* (CH_3COSH) (0·75 g.) and 0·001 M-lead nitrate (33 mg.) in 83 ml. Na_2HPO_4 (0·1 M). Add 17 ml. McIlvaine's Phosphate-Citrate buffer (pH 5·6).

Method. (1) Incubate for 30–60 minutes at 22°.

(2) Wash in ice water at 4°.

(3) Counterstain nuclei, if desired, with 0·02 per cent basic fuchsin.

(4) Wash in cold water, dehydrate, clear in xylene and mount in Permount (Fisher) or other suitable synthetic medium. Any crystalline deposit will wash out in alcohol without altering the stain.

Result. A black deposit indicates sites of cholinesterase activity.

A MODIFIED Pb THIOLACETIC ACID METHOD

(Koelle, 1962)

(*Cold buffered formalin, cryostat sections*)

PREPARATION OF THE INCUBATING MEDIUM

Mix, in the order given below, the following reagents:

Distilled water	11·3 ml.
0·1 M-Citric acid	2·0 ml.
N-NaOH	0·6 ml.
2 M-Sucrose	4·4 ml.
M-$MgCl_2$	0·5 ml.
0·24 M-Ch_3COSH	1·0 ml.

This reagent is prepared by adding 2 ml. N-NaOH to 0·17 ml. CH_3COSH and adding water to 10 ml. Finally, add 0·2 ml. 0·1 M-$PbNO_3$. Adjust final pH to 5·4.

Method. As for preceding technique.

Result. As for preceding technique.

A GOLD-THIOLACETIC ACID MODIFICATION

(Koelle and Gromadzki, 1966; Koelle and Horn, 1968)

PREPARATION OF TISSUES

As for the gold thiocholine method (p. 1316).

* All samples of reagent grade thiolacetic acid do not produce equally good results. The reasons for this have been clarified by the work of Koelle and Horn (1968) (see p. 795).

PREPARATION OF STOCK REAGENTS

(1) $AuNa_3(S_2O_3)_2 . 2H_2O$, 0·1 M.

(2) $MgCl_2 . 6H_2O$, 1·0 M.

(3) Maleate buffer, 0·703 M. Dissolve 4·8 g. sodium hydrogen maleate in 26·1 ml. N-NaOH and make up to 20 ml. with distilled water.

(4) Thiolacetic acid, 0·2 M. Dissolve 0·288 ml. ThAc in 3·6 ml. N-NaOH and make up to 20 ml. with cold, freshly boiled distilled water. (Allow insoluble material to settle.)

(5) N-HCl.

(6) Inhibitors as in gold thiocholine method (p. 1317).

(7) Yellow ammonium sulphide 0·3/100.

Preparation of Preincubation and Incubation Solutions.

TABLE 90

	Stock Solutions							
Solution	1. ml.	H_2O ml.	2. ml.	3. ml.	4. ml.	5. ml.	6. ml.	7.
Preincubation	0·3	8·2, 8·1[b]	0·32	0·8	0·3	0·12	(0·1)	
Incubation	0·3	6·5, 6·4[b]	0·32	0·8	2·0	0·12	(0·1)	1 drop

Final concentrations: $AuNa_3(S_2O_3)_2$, 0·003 M
$MgCl_2$, 0·032 M
Maleate, 0·056 M
ThAc, 0·04 M

b. When inhibitor is used.

Method. (1) Preincubate sections for 15 minutes at 22°.

(2) Incubate for 10–170 minutes at 22°.

(3) Transfer to 4 per cent formaldehyde in 0·85 per cent NaCl for 10 minutes.

(4) Dehydrate, clear, and mount in synthetic resin.

Result. Yellow and brown deposits indicate sites of ChE activity.

APPENDIX 18

8-HYDROXYQUINOLINE COUPLING AZO DYE METHOD FOR BETA-GLUCURONIDASE

(*Cold formalin, frozen sections*)

Method. (1) Cut frozen sections 10–15 μ and mount on clean glass slides. Dry in air to ensure adherence. Ring the sections with paraffin wax so as to enclose a small pool of the incubating medium.

(2) To 10 ml. 3 to 5 mM-8-hydroxyquinoline glucuronide in 100 mM-acetate buffer pH 5·2 add 10 mg. of the stable diazotate of 4-benzoylamino-2: 5-dimethoxy-aniline (I.C.I. Ltd.). Shake the mixture and filter directly on to the dry slides.

(3) Incubate for 2–4 hours in a closed dish in which is placed some wet blotting paper to minimize evaporation.

(4) Wash in running water and remove the ring of wax at the same time.

(5) Counterstain in Mayer's haemalum, 4–6 minutes.

(6) Wash in running water for 30 minutes.

(7) Mount in glycerine jelly.

Result. An orange precipitate is deposited in the neighbourhood of structures containing *β*-glucuronidase.

8-HYDROXYQUINOLINE AZO DYE METHOD

(after Rath and Otto, 1966)

(*Fresh cryostat sections*)

PREPARATION OF THE MEDIUM

Dissolve 15 mg. 8-hydroxyquinoline-*β*-D-glucuronide in 1 ml. ethanol and add 4 ml. of 200 mM acetate buffer at pH 5·2. Immediately before use add 20 mg. Fast black salt K (Salt 14, p. 712, Vol. 1) in 15 ml. distilled water.

Method. (1) Incubate fresh sections (mounted on coverslips) for 2–4 hours at 37°.

(2) Wash in distilled water.

(3) Immerse in 1 per cent aqueous sulphanilic acid for 10 minutes.

(4) Wash in distilled water.

(5) Mount in glycerine jelly.

Result. A dark brown, finely granular, precipitate indicates *β*-glucuronidase.

POST-COUPLING METHOD FOR BETA-GLUCURONIDASE

(after Seligman, Tsou, Rutenburg and Cohen)

(*Cold microtome, mounted sections, post-fixed in neutral formalin*)

PREPARATION OF SUBSTRATE SOLUTION

Dissolve 2 mg. 6-bromo-2-naphthyl-*β*-D-glucopyruronoside (glucuronide) in a few drops of dimethyl formamide and add 10 ml. 100 mM. phosphate-citrate buffer (pH 4·95).

Method. (1) Incubate sections for 4–6 hours in the substrate solution at 37°.

(2) Rinse in tap water.

(3) Immerse in a solution of Fast blue B (Salt 6, p. 712, Vol. I) in cold (4°) 20 mM-phosphate buffer (pH 7·5). A strength of 1 mg. per ml. is adequate.

(4) After 2 minutes remove sections, wash in two changes of cold distilled water.

(5) Rinse in 0·1 per cent acetic acid.

(6) Mount in glycerine jelly.

Result. A blue or purple colour indicates sites of high *β*-glucuronidase activity. A red colour may indicate lower activity or the presence of lipid.

NAPHTHOL AS-BI GLUCURONIDE METHOD

(after Hayashi, *et al.*, 1964)

(*Cold formol-calcium, gum sucrose. Free floating sections*)

PREPARATION OF STOCK SOLUTIONS

A. Substrate Stock. Dissolve 28 mg. naphthol AS-BI glucuronide in 1·2 ml. 50 mM-sodium bicarbonate (420 mg. $NaHCO_3$ in 100 ml. H_2O) and bring to 100 ml. with 0·2 N-acetate buffer (pH 5·0). This stock solution keeps for several weeks at room temperature.

B. Pararosanilin Stock. Dissolve 1 g. pararosanilin hydrochloride in 20 ml. distilled water and 5 ml. conc. HCl, with gentle warming. Filter after cooling and store at room temperature.

C. Sodium Nitrite Stock. Prepare a 4 per cent solution in distilled water (freshly each week) and store at 0°.

D. Working Solution. Mix 0·3 ml. pararosanilin (B) and sodium nitrite (C) in a 20 ml. beaker. After 1 minute, add 10 ml. substrate stock solution (A). Adjust pH to 5·2 with N-NaOH. Add distilled water to a final volume of 20 ml. Filter through a No. 1 Whatman paper. (Final Concentrations: Substrate 0·25 mM, hexazonium reagent 1·8 mM).

Method. (1) Treat sections (free-floating or mounted) with cold ethanol (0°) for 2 minutes.

(2) Incubate free-floating (or mounted) sections, for 20–30 minutes at 37°, after rinsing in distilled water to remove ethanol.

(3) Rinse well in distilled water.

(4) Mount section at this point (if necessary).

(5) Counterstain nuclei in 1 per cent methyl green (chloroform-washed) for 5–10 minutes.

(6) Wash in running water.

(7) Dehydrate rapidly in alcohols, clear in xylene and mount in a suitable synthetic medium.

Result. (Plate XVIIb and d). *β*-glucuronidase red, nuclei green.

SYNTHESIS OF NAPHTHOL AS-BI-*β*-D-GLUCOSIDURONIC ACID

(Fishman *et al.*, 1964)

Dissolve 4·46 g. crystalline Naphthol AS-BI in alcoholic KOH (0·67 g. in 200 ml.). Add 4·0 g. acetobromomethyl glucuronate (methyl-1-bromo-1-deoxy-2,3,4-tri-*O*-acetyl-*α*-D-glucuronate). Keep for 1 week at 20° in a glass-stoppered vessel.

Pour the reaction mixture into 1000 ml. cold water and extract the turbid suspension with 50 ml. quantities of $CHCl_3$ (total 280 ml.). Wash the chloroform extract with three successive 100 ml. portions of 2N-KOH, then with distilled water (300 ml.). Dry over anhydrous sodium sulphate and evaporate to dryness.

Dissolve the residue in 50 ml. methanol and add hot filtered barium hydroxide (10 g. $Ba(OH)_2 . 8H_2O$ in 200 ml. H_2O). The yellow barium salt of the deacetylated product is precipitated. Collect the latter, next morning, on a Buchner filter. Wash with 50 ml. cold water and then with 50 ml. cold methanol. Next, suspend the precipitate in the funnel (slight negative pressure) with three successive 10 ml. portions of 1·25 N-HCl. This transforms the barium salt into the insoluble free acid. Wash out residual HCl with 50 ml. cold water and suspend the precipitate in 1 per cent $NaHCO_3$. Filter to remove insoluble impurity and precipitate the glucosiduronic acid from its bicarbonate solution with 1·25 N-acid. Filter and dry the white gelatinous precipitate.

Crystallize by dissolving in 75 ml. hot methanol and adding 20 ml. H_2O. On standing white needles separate out (m.p. 210°). Two further recrystallizations should be carried out (m.p. 219°).

INDIGOGENIC METHOD FOR β-GLUCURONIDASE*

(after Pearson, *et al.*, 1967)

(*Fresh cryostat: post-fix buffered glutaraldehyde*)

PREPARATION OF BUFFERED FIXATIVE

Dissolve the following salts in 1000 ml. 85 mN-NaCl: 0·2 g. KCl, 0·2 g. KH_2PO_4, 1·15 g. Na_2 HPO_4. Add glutaraldehyde to 2·5 per cent when required for use.

PREPARATION OF SUBSTRATE SOLUTION

Dissolve 8 mg. 5-Bromo-4-chloroindolyl-β-D-glucopyruroniside in 30 ml. 0·1 M-acetate buffer (pH 4·8) to which has been added 0·5 ml. 15 mM-NaCl, 0·5 ml. 15 mM-$MgCl_2$ and 8 ml. spermidine trihydrochloride.

Method. (1) Incubate in substrate solution for 2–4 hours at 37°.

(2) Rinse in water.

(3) Counterstain nuclei, if required, in carmalum (10 minutes).

(4) Wash in water. Dehydrate in alcohols, clear in xylene, and mount in a suitable synthetic medium.

Result. A fine bluish-green precipitate indicates localization of β-glucuronidase. Nuclei red.

POST-COUPLING METHOD FOR BETA-GALACTOSIDASE†

(after Rutenburg *et al.*, 1958)

(*Cold microtome, mounted sections; cold formalin, frozen sections*)

PREPARATION OF SUBSTRATE SOLUTION

Dissolve 100 mg. 6-bromo-2-naphthyl-β-D-galactopyranoside (galactoside) in 15 ml. methanol and add 200 ml. hot distilled water (70°). Allow the solution to cool and add 85 ml. phosphate-citrate buffer (pH 4·95). Prepare the buffer from equal

* Similar methods have been published, using the appropriate 5-bromo-4-chloroindoxyl glycosides, for β-glucosidase (pH 5·4), β-galactosidase (pH 5·4) and fucosidase (pH 6·2).

† A similar method can be used for β-glycosidase using 6-bromo-2-naphthyl-β-D-glycoside as substrate. Also for α-galactosidase using 6-bromo-2-naphthyl-α-D-galactopyranoside.

volumes of 0·2 M-disodium phosphate and 0·1 M-citric acid. Add a further 100 ml. distilled water. The solution is stable for 6 months at 4°.

Method. (1) Incubate sections at 37° in the above solution. Active tissues need 1–4 hours. Tissues containing only weak enzyme activity may be incubated for 8 hours or more at 25°.

(2) Wash in 3 changes of distilled water, for 3 minutes in each.

(3) Transfer to a freshly prepared solution of Fast blue B (6), 1 mg./ml., at 4° and at pH 7·4 to 7·8. Agitate gently for 3–5 minutes.

(4) Wash 3 times in cold water.

(5) Mount in glycerine jelly.

Result. (Fig. 171, Plate XVII). A blue or purple colour indicates sites of β-galactosidase activity.

POST-COUPLING METHOD FOR ALPHA-GLUCOSIDASE*

(after Rutenburg, *et al.*, 1960)

(*Cold neutral formalin, frozen sections*)

Method. (1) Dissolve 1 to 1·5 mg. 6-bromo-2-naphthyl-α-D-glucopyranoside (glucoside) in 1 ml. ethanol and add 9 ml. 0·025 M-phosphate buffer (pH 6·5).

(2) Incubate free-floating sections for 1–2 hours at 37°.

(3) Rinse in water.

(4) Treat with coupling solution (1 mg./ml. Fast blue B in phosphate buffer at pH 7·5) for 2 minutes.

(5) Wash in water and mount in glycerine jelly.

Result. Blue or purple dyes indicate sites of enzyme activity.

INDIGOGENIC METHOD FOR β-GLYCOSIDASES

(after Lojda, 1970)

(*Cold formol-calcium fixed, free-floating, frozen sections. Fresh cryostat sections*)

PREPARATION OF INCUBATION MEDIA

Dissolve 3 mg. of the appropriate indoxyl glycoside (i.e. 5-bromo-4-chloroindolyl-3-β-D-glucoside) in 0·3 ml. dimethylformamide and add 7 ml. 0·1 M citrate-phosphate buffer at pH 3·5 or 5·5. Shake well and add 0·5 ml. 50 mM potassium ferricyanide (1·65 per cent) and 0·5 ml. 50 mM potassium ferrocyanide (2·11 per cent of the crystalline form). Shake well and filter. (The medium keeps for several days at 0–4°). For most tissues pH 3·5 is suitable; pH 5·5 is optimal for the intestinal brush border enzyme.

Method. (1) Incubate at 37° for 30 minutes to 24 hours.†

(2) Wash in water.

(3) Mount sections, if required, on slides and cover with Apáthy's medium or with glycerine-cadmium chloride (Romeis, 1948).

(4) Alternatively, dehydrate in alcohols, clear in xylene, and mount in Canada balsam.

Result. (Plate XVIId, p. 814). A deep blue colour indicates sites of glycosidase activity.

* See note on p. 1323.

† Incubation times vary. Rat epididymis, 1–2 hours; kidney, 2–4 hours; liver, 6 hours; lung and brain, 6–8 hours; intestine brush border, 8 hours.

NAPHTHOL AS-BI–HPR METHOD FOR BETA-GLUCOSAMINIDASE

(after Hayashi, 1965)

(*Cold formol-calcium, green sucrose; frozen sections*)

PREPARATION OF STOCK PARAROSANILIN AND NITRITE

See p. 732.

PREPARATION OF WORKING SOLUTION

Dissolve 3 mg. Naphthol AS-BI *N*-acetyl-β-glucosaminide in 0·5 ml. ethylene glycol monomethyl ether and add 5 ml. 0·1 M-citrate buffer (pH 5·2). Immediately add 0·6 ml. freshly prepared hexazotized pararosanilin solution. Adjust pH to 5·2 with N-NaOH and dilute with distilled water to a final volume of 10 ml. Filter through a Whatman No. 1 paper.

(Final concentrations: substrate, 0·5 mM; hexazotate, 3·6 mM.)

Method. (1) Treat sections, before or after mounting on slides or coverslips, with cold absolute ethanol (0°) for 5 minutes.

(2) Rinse in distilled water at 20°.

(3) Incubate in the working solution for 15–30 minutes at 37°.

(4) Rinse in distilled water.

(5) Counterstain nuclei in 1 per cent methyl green (chloroform-washed) for 5–10 minutes.

(6) Wash in running water.

(7) Dehydrate in alcohols, clear in xylene and mount in a synthetic resin.

Result. (Plate XVIIIb). Acetyl-β-glucosaminidase activity is indicated by bright red deposits; nuclei green.

NAPTHHOL AS-BI–GARNET GBC METHOD FOR BETA-GLUCOSAMINIDASE

(after Hayashi, 1965)

(*Cold formol-calcium, gum sucrose; frozen sections*)

PREPARATION OF WORKING SOLUTION

Dissolve 3 mg. Naphthol AS-BI *N*-acetyl-β-glucosaminide in 0·5 ml. ethylene glycol monomethyl ether and add 9·5 ml. 0·1 M-citrate buffer (pH 5·2) and 10 mg. Fast Garnet GBC salt (Salt 18, Vol. 1, p. 713). Filter through Whatman No. 1 paper.

Method. (1) Proceed as in stages 1–4 of previous method, using the above working solution.

(2) Counterstain nuclei in methyl green (if required).

(3) Rinse in water and mount in glycerine jelly or in an alternative aqueous mounting medium.

Result. Acetyl-β-glucosaminidase activity is indicated by a violet precipitate. Nuclei green.

SYNTHESIS OF NAPHTHOL AS-BI *N*-ACETYL-β-GLUCOSAMINIDE

(Anderson & Leaback, 1961; Hayashi, 1965)

Dissolve 4·46 g. Naphthol AS-BI (7-bromo-3-hydroxy-2-naphth-*O*-anisidide) and 3·65 g. acetochloroglucosamine (2-acetamido-3,4,6-tri-*O*-acetyl-1-chloro-2-deoxyβ-D-glucose; Leaback and Walker, 1957), in 200 ml. cold acetone and add 50 ml. 1·6 per cent KOH. Allow the reaction mixture to stand, in a glass-stoppered vessel, for 18 hours at room temperature. After filtration, evaporate the acetone *in vacuo* and dissolve the products in 200 ml. chloroform.

Extract the chloroform layer with 3 successive 100 ml. portions of ice-cold 2 N-KOH. Wash with ice-cold water, dry with anhydrous sodium sulphate and evaporate to dryness.

Dissolve the residue in 100 ml. dry methanol, and cool in an ice-salt mixture. Pass a rapid stream of dry ammonia into the solution for 45 minutes. Keep at room temperature for 18 hours and evaporate to dryness at this temperature. Add a further 100 ml. methanol and again evaporate *in vacuo*.

Wash the crude products with 3 × 100 ml. ethyl ether and with 100 ml. acetone.

Yield 2·5 g. (43 per cent) m.p. 189–191°. After two recrystallizations from 1 : 1 methanol/water, m.p. 195–196°.

COUPLED OXIDATION METHOD FOR DISACCHARIDASES

(Dahlqvist & Brun, 1962; Lojda, 1965)

(*Fresh-cryostat, mounted, sections*)

Stock Solutions and Buffers

0·2 per cent aqueous Nitro-BT.

4 per cent solutions of sugars (sucrose, lactose monohydrate, maltose monohydrate, trehalose dihydrate, melibiose dihydrate). 0·1 M-Citrate-phosphate buffer (pH 6·0 or 6·5).

Standard Incubating Medium

6 ml. Citrate-phosphate buffer (pH 6·0 for invertase or lactase; pH 6·5 for maltase and trehalase).

1·0 mg. Glucose oxidase.

0·2 mg. Phenazine methosulphate.

To 1·5 ml. of the above add 0·5 ml. disaccharide solution and 1·0 ml. Nitro-BT solution. Finally, dilute with 2 ml. buffer, or with 2 ml. 6·5 per cent gelatin in buffer, or with 1 per cent Agar (Difco) in buffer.

In these last two cases it is necessary to heat first to 37°, or 50°, respectively; after mixing pour onto coverslips and allow to gel.

Method (Gel media). (1) Without removing coverslip from the gel medium place the medium on top of the section.

(2) Run in sufficient buffer to form a continuous layer.

(3) Incubate at room temperature, in a closed moist box, for 24 hours.

(4) Remove coverslip and wash with warm water.

(5) Mount in Apáthy's medium.

Method (Buffer Media). (1) Place a plastic or paraffin wax ring around the section (mounted on a slide).

(2) Fill the ring with medium and apply a coverslip.

(3) Incubate for 18–24 hours at 37°.

(4) Wash briefly in water.

(5) Counterstain nuclei, if required, with carmalum.

(6) Mount in a suitable aqueous medium.

Result. Disaccharidase activity produces a fine blue deposit of formazan.

STANDARD METHOD FOR α-GLUCAN PHOSPHORYLASE

(After Takeuchi & Kuriaki, 1955; Eränkö & Palkama, 1961; Ibrahim & Castellani, 1968)

(*Air-dried cryostat sections (16 μ); post-fixed in cold acetone*)

Preparation of Stock Solution

To 100 ml. 0·1 M-acetate buffer (pH 5·9) add 1 g. glucose-1-phosphate, 100 mg. adenosine-5-phosphate, 20 mg. glycogen, 1·8 g. NaF, 9 g. polyvinyl pyrrolidone (mol. wt. 24·000), and 10 drops insulin (40 i.u./ml.). Do not filter. This solution keeps several months at 0–4°.

Preparation of Incubating Medium

For active or total phosphorylase add 0·5 ml. absolute ethanol to 2·5 ml. freshly filtered stock solution. For phosphorylase + branching enzyme, use filtered stock solution only.

Method. (1) Incubate sections for 1–2 hours at 37° (phosphorylase) or for 30 minutes at 37° (phosphorylase + branching enzyme), preferably in a moist environment to prevent evaporation of the (small amount of) medium on the slide.

(2) Shake off excess medium.
(3) Wash briefly in 40 per cent ethanol and dry in air.
(4) Fix for 30 minutes in absolute ethanol and dry in air.
(5) Stain in dilute Gram's iodine (1 : 10), for 5 minutes.
(6) Mount in iodine-glycerine (1 : 5) and seal the coverslip.

Result. (Plate XVIIIc). A fine dark blue or blue-black precipitate indicates phosphorylase. Mahogany colours indicate branching enzyme activity.

(7) (If required.) Destain in absolute ethanol.
(8) Stain by the PAS routine (Vol. 1, p. 660).
(9) Dehydrate in alcohols, clear in xylene and mount in synthetic resin.

Result. Diffuse magenta staining shows all types of glycogen (preformed, amylose, amylopectin).

PERMANENT IODINE-STAINING METHODS

Keller and Tanka (1965). After incubation, without removing the medium, add acid Lugol (Lugol's solution, 2·5 parts; distilled water, 2·5 parts; 20 per cent H_2SO_4, 5 parts) in the proportion 1 : 25 of the medium. Stand for 5 minutes. Wash in distilled water and mount in iodine-glycerine.

Sawyer *et al.* (1965). Stain as usual in Gram's iodine. Transfer to cold 4 per cent formol chloral hydrate (100 ml. formaldehyde, 1 g. chloral hydrate, 900 ml. water) for 5 minutes. Counterstain for 2 minutes in 0·4 mM-methyl green in 1·0 N-acetate buffer, pH 4·8. Dehydrate rapidly in absolute tertiary butanol. Clear in xylene and mount in synthetic resin (Stable 2 months).

Smith and Perkins (1967). Stain as usual in Gram's iodine. Dehydrate in graded alcohols and mount in Histoclad* containing 1 mg. iodine per ml. (Stable 8 months).

* Clay-Adams Co., New York

PHOSPHORYLASE TECHNIQUE FOR BLOOD FILMS

(after Wulff and Sørensen, 1966)

Method. (1) Make films of capillary blood (or other smears) on clean slides. Allow to dry in air for not longer than 30 minutes.

(2) Using the medium described on p. 1327, place on a hollow slide sufficient to fill the central circular concavity.

(3) Invert the slide and film over the medium and incubate for 1 hour at 37°.

(4) Shake off medium and fix in absolute methanol for 1 minute. Dry in air.

(5) Mount in iodine glycerine.

Result. Blue-black staining indicates phosphorylase.

DEXTRAN METHOD FOR PHOSPHORYLASE

(after Meÿer, 1968)

(*Fresh cryostat sections, mounted on slides*)

PREPARATION OF INCUBATION MEDIUM

Disodium glucose-1-phosphate	100 mg.
Disodium ethylenediamine tetra acetate	50 mg.
Sodium fluoride	40 mg.
Acetate buffer 10 mM, pH 5·6	25 ml.
Dextran (Av. mol. wt. 4′0000)	2 g.

When total phosphorylase is required add:

Disodium adenosine-5-phosphate	20 mg.

Method. (1) Incubate 7 μ sections for 30 minutes at 37°.

(2) Shake off excess medium.

(3) Wash in 40 per cent ethanol.

(4) Fix for 10 minutes in absolute ethanol.

(5) Carry out the PAS routine (Vol. 1, p. 660).

(6) Dehydrate, clear and mount in synthetic resin.

Result. Phosphorylase A, B or A + B activity is shown by a magenta coloured deposit.

METHOD FOR UDPG GLYCOGEN TRANSGLYCOSYLASE

(after Takeuchi and Glenner, 1960, 1961)

(*Unfixed cryostat sections, 15–20 μ thick*)

PREPARATION OF THE INCUBATING MEDIUM

Dissolve 20 mg. UDPG, 4 mg. Glycogen and 8 mg. EDTA in 6 ml. distilled water and 4 ml. 0·2 M Tris buffer (pH 7·4). Add 0·5 ml. ethanol and 2·25 g. PVP.

Method. (1) Incubate sections, mounted on coverslips or slides, at 37° for 1–3 hours.

(2) Place in dilute Gram's iodine solution until the red-brown colour of glycogen appears.

(3) Cover sections with iodine-glycerol (Gram's iodine 1 part, glycerol, 9 parts) and seal with paraffin wax.

Alternatively,

(2) Rinse briefly in distilled water and allow to dry.

(3) Fix in 80 per cent ethanol for 5 minutes.

(4) Stain with an aqueous PAS routine (Vol. 1, p. 660).
(5) Dehydrate, clear, and mount in synthetic resin.

Result. Red-brown colours indicate glycogen (iodine method). Comparison with a control section is necessary in order to differentiate between endogenous and newly-formed glycogens.

Magenta staining (PAS) indicates glycogen (Plate XVIII d). Control section equally necessary.

METHOD FOR UDPG GLYCOGEN TRANSGLYCOSYLASE

(after Sasse, 1966)

(*Fresh cryostat sections, 20 μ thick*)

PREPARATION OF INCUBATING MEDIUM

To 14 ml. distilled water, 10 ml. 0·2 M Tris buffer (pH 7·4) and 1 ml. ethanol add 10 mg. glucose-6-phosphate, 50 mg. UDPG, 20 mg. EDTA, and 10 mg. glycogen. Adjust pH to 7·5.

Method. (1) Post-fix sections (mounted on slides or coverslips) for 10 minutes in ice-cold (4°) ethanol.
(2) Incubate for 3 hours at 37°.
(3) Treat with 10 per cent H_2SO_4 at room temperature for 30 minutes.
(4) Wash twice, for 2 minutes, in distilled water.
(5) Stain with an aqueous PAS routine (Vol. 1, p. 660).
(6) Counterstain nuclei, if required, with haemalum.
(7) Dehydrate, clear and mount in synthetic resin.

Result. "Primary" glycogen is removed by the acid hydrolysis, leaving only newly synthesized glycogen for demonstration (by the PAS reaction) as magenta staining.

METHOD FOR PURINE NUCLEOSIDE PHOSPHORYLASE

(after Kishi, 1969)

(*Cold acetone, 1 minute; post-fixed cryostat sections*)

PREPARATION OF INCUBATING MEDIUM

This medium must be freshly prepared immediately before incubation.

Inosine 5·4 mg. (20 μmoles)
0·2 M Phosphate buffer (pH 7·8), 0·4 ml.
Xanthine oxidase 0·4 ml. (1·6 units).
10 per cent gelatine 0·1 ml.
Nitro-BT (5 mg./ml.) 0·1 ml.

Mix all components, except the Nitro-BT and incubate at 37° for 5 minutes to remove any contaminating xanthine or hypoxanthine.

Add the Nitro-BT solution. No reduction should occur.

Method. (1) Post-fix 10–20 μm sections in cold acetone for 1 minute.
(2) Dry in air and apply one or two drops of the incubating medium.
(3) Incubate at 37° for 1–2 hours in a covered Petri dish containing wet blotting paper.
(4) Wash sections in warm water.
(5) Fix for 30 minutes in neutral formalin.
(6) Mount in glycerine jelly.

Result. Blue formazan deposits indicate enzyme activity.

APPENDIX 19

G-NADI REACTION FOR CYTOCHROME OXIDASE

(after Moog)

(*Fresh slices*)

Moog (1943) applied this method to whole chick embryos. It is equally applicable to thin unfixed tissue slices, or to thick sections.

PREPARATION OF REAGENTS

(1) **"Nadi Reagent."** Just before use combine equal parts of 10 mM dimethyl-*p*-phenylene diamine HCl and 10 mM α-naphthol, each in 1 per cent NaCl with 66 mM phosphate buffer at pH 5·8.

(2) **Sodium Azide.** A 5 mM solution of NaN_3 in physiological saline.

(3) **"Nadi Reagent" + Sodium Azide.** The Nadi reagent, as above, with 5 mM sodium azide.

(4) **Phenylurethane.** A 3 mM solution in physiological saline.

(5) **"Nadi Reagent" + Phenylurethane.** The Nadi reagent, as above, with 3 mM phenylurethane.

Method. (1) Cut fresh tissue slices 100–150 μ thick into Ringer at pH 7·2.

(2) Incubate with "Nadi reagent" at 37° for 2–5 minutes.

(3) Incubate control sections, after washing in sodium azide and phenylurethane solutions respectively, in "Nadi reagent"/sodium azide (3) and in "Nadi reagent"/ phenylurethane (5) for 2–5 minutes.

(4) Wash in physiological saline.

(5) Counterstain nuclei in carmalum (30 minutes) if desired.

(6) Mount in 5 per cent potassium acetate and "ring" the coverslip with paraffin wax. Permanent preparations cannot be made.

Result. Sections incubated in "Nadi reagent" or in "Nadi reagent" with phenylurethane should show a blue or bluish-violet colour in areas of cytochrome oxidase activity. Sections incubated in the presence of sodium azide should be negative.

ADN METHOD FOR CYTOCHROME OXIDASE

(after Nachlas *et al.*, 1958)

(*Cold microtome, fresh frozen sections; mounted on slides or coverslips*)

PREPARATION OF THE MEDIUM

Mix the following reagents:

0·1 M-phosphate buffer (pH 7·4)	3·0 ml.
α-Naphthol (1 mg./ml.)	4·0 ml.
*Cytochrome c. (5 mg./ml.)	3·0 ml.
Catalase (30 μg./ml.)	1·0 ml.
Immediately before use add:	
ADN (2 mg./ml.)	4·0 ml.

* If known active tissues are being examined this (expensive) component of the medium may be omitted.

Method. (1) Incubate sections for 10–30 minutes at room temperature.
(2) Rinse briefly in 0·9 per cent saline.
(3) Mount in glycerine jelly.

Result. Purple granules indicate sites of cytochrome oxidase activity.

NAPHTHOL-AMINE METHOD FOR CYTOCHROME OXIDASE

(after Burstone, 1959)

(*Fresh frozen, cryostat, sections; mounted*)

PREPARATION OF INCUBATING MEDIUM

Place 10 mg. 1-hydroxy-2-naphthoic acid and 10 mg. *p*-aminodiphenylamine* (*N*-phenyl-*p*-phenylene diamine) in a 50-ml. Erlenmeyer flask. Add 0·5 ml. reagent ethanol to dissolve the reagents. Add 35 ml. distilled water and 15 ml. "Tris" buffer (0·2 M, pH 7·4). Shake and filter into a suitable container.

Method. (1) Incubate sections at room temperature for 15–60 minutes or longer.

(2) Transfer directly to 1 per cent cobalt acetate in 10 per cent formalin. Fix for 1 hour.

(3) Wash in water and mount in glycerine jelly.

Result. (Fig. 176). Reddish brown dyes are deposited at sites of cytochrome oxidase activity. These are finely particulate and stable.

QUINOLINE-AMINE METHOD FOR CYTOCHROME OXIDASE

(after Burstone, 1961)

(*Fresh frozen, cryostat, sections; mounted*)

PREPARATION OF INCUBATING MEDIUM

Dissolve one drop of 8-amino-1,2,3,4-tetrahydroquinoline and 10 mg. *p*-aminodiphenylamine in 0·5 ml. reagent ethanol. Add 35 ml. distilled water and 15 ml. 0·2 M Tris-HCl buffer (pH 7·4). Filter and proceed as in the above method. (It is possible to dehydrate in alcohols and mount in 25% PVA.)

Result. A brownish violet product indicates enzyme activity.

AMINE-AMINE METHOD FOR CYTOCHROME OXIDASE

(after Burstone, 1961)

(*Fresh frozen, cryostat, sections; mounted*)

PREPARATION OF THE INCUBATING MEDIUM

Dissolve 10 mg. *p*-aminodiphenylamine and 10 mg. *p*-methoxy-*p*′-aminodiphenylamine (Variamine blue B base) in 0·5 ml. reagent ethanol. Add 35 ml. distilled water and 15 ml. 0·2 M Tris-HCl buffer (pH 7·4). Filter the mixture before use. In the case of tissues known to have low enzyme activity add 10–20 mg. cytochrome c and dissolve by stirring.†

Method. (1) Incubate sections for 15–180 minutes.
(2) Wash briefly in tap water.
(3) Counterstain nuclei with 1% methyl green, 2 minutes.

* Use the base *not* the hydrochloride.
† This procedure applies equally to Burstone's other cytochrome oxidase techniques.

(4) Wash briefly.
(5) Mount in a suitable watery mounting medium.

Result. A brownish-red product indicates sites of cytochrome oxidase.

NITRONAPHTHYLAMINE-AMINE METHOD FOR CYTOCHROME OXIDASE

(after Burstone, 1961)

(*Fresh frozen, cryostat, sections; mounted*)

PREPARATION OF THE INCUBATING MEDIUM

Dissolve 10 mg. *p*-aminodiphenylamine and 10 mg. of either 4-nitro or 5-nitro-1-naphthylamine in 0·5 ml. reagent ethanol and proceed as in the above method.

Result. Brownish dyes are produced from both naphthylamines but the crystals are much finer with the 5-nitro derivative.

THIAZOLYLTHIOSEMICARBAZIDE METHOD FOR CYTOCHROME OXIDASE

(after Thiele, 1967)

(*Fresh frozen, cryostat, sections; mounted*)

PREPARATION OF THE MEDIUM

Dissolve 0·5 mg. per ml. 4-isopropylthiazolyl-(2)-isothiosemicarbazide hydrochloride in 0·2 M-phosphate buffer (pH 7·0), and use immediately.

Method. (1) Incubate at 37° for 5–10 minutes.
(2) Wash briefly in water and counterstain nuclei in 1% methyl green for 2 minutes.
(3) Wash briefly in water.
(4) Dehydrate rapidly in 75% and absolute ethanol.
(5) Mount in DPX or other suitable synthetic medium.

Result. A fine blue precipitate indicates sites of cytochrome oxidase activity.

DAB METHOD FOR CYTOCHROME OXIDASE

(after Seligman *et al.*, 1968)

(*Fresh frozen, cryostat; cold formalin,* cryostat sections*)

PREPARATION OF THE INCUBATING MEDIUM

Dissolve 5 mg. 3,3′-diaminobenzidine tetrahydrochloride (DAB) in 9 ml. 50 mM-phosphate buffer (pH 7·4). Add 1 ml. catalase solution (20 μg./ml.), 10 mg. cytochrome C† and 750 mg. sucrose.

Method. (1) Incubate sections for 30–60 minutes at 22°.
(2) Wash briefly in distilled water.
(3) Dehydrate in alcohols and mount in a suitable synthetic medium.

Result. A fine brown stain indicates cytochrome oxidase activity.

* Use depolymerized paraformaldehyde 4 per cent solution, prepared by dissolving in hot water (70°) to which the necessary quantity of N-NaOH has been added to maintain neutrality. Neutralise further to pH 7·4 before use.

† Type II, Sigma.

METHOD FOR CATALASE

(Nishiyama and Kobayashi)

(Smears, fresh sections, cold acetone-fixed paraffin sections)

Method. (1) Incubate sections in a mixture of equal parts of monoethyl hydrogen peroxide* and phosphate buffer (pH 6·8) for 10–15 minutes at 22°.

(2) Immerse in Schiff's reagent, 3–5 seconds.

(3) Rinse in sodium bisulphite solution.

(4) Wash, dehydrate, clear and mount in a permanent medium.

Result. A magenta colour was attributed by the authors to catalase activity (but see Chapter 19, p. 858).

ETHYL HYDROGEN PEROXIDE METHOD FOR "HYDROPEROXIDASE"

(after Watanabe)

(Fresh frozen, cold microtome sections, mounted on slides or coverslips)

PREPARATION OF SOLUTIONS

Stabilizer. Dissolve 0·001 per cent albumin, 0·001 per cent glycogen and 0·001 per cent urea in 0·1 M-veronal acetate buffer (pH 7·0).

Coupling medium. 0·5 g. Fast blue B salt in 50 ml. of 2 per cent pyridine in water.

Substrate. 1–2 per cent ethyl hydrogen peroxide . .	10 parts
Phenylhydrazine	0·5 parts

Filter and add to 8 parts of the filtrate 1 part of phenylhydrazine and 1 part of the stabilizer solution.

The incubating medium should be freshly prepared before use, and adjusted to pH 7·0 with Na_2HPO_4.

Method. (1) Immerse sections for 1 minute in equal parts of carbon tetrachloride and light petroleum. Dry in air.

(2) Incubate for 5–10 minutes at 22° in the substrate medium.

(3) Wash in water.

(4) Treat with coupling medium for 4–6 minutes.

(5) Wash in water and fix in 10 per cent neutral formalin for 10 minutes.

(6) Mount in glycerine jelly.

Result. Sites of "hydroperoxidase" activity are shown by the presence of a reddish-brown precipitate of diformazan (see Chapter 10, p. 326).

PREPARATION OF ETHYL HYDROGEN PEROXIDE

(after Rieche and Hitz)

Add 200 ml. 40 per cent KOH, by dropping, into a mixture of 10 per cent H_2O_2 (100 vol., 30 per cent) in 100 ml. diethyl sulphate. Cool continuously with ice-salt mixture. After 30 minutes adjust the pH to 7·0 with 50 per cent H_2SO_4.

Distil off the ethyl hydrogen peroxide (95–99°).

* Prepared by method of Rieche and Hitz (1929).

THE BENZIDINE REACTION FOR PEROXIDASE

(after DeRobertis and Grasso, 1946)

(*Cold microtome, fresh frozen sections*)

Method. (1) Cut sections, 20–40 μ thick, from fresh tissues, into 0·85 per cent saline at 4°.

(2) Treat with 0·1 per cent ammonium molybdate in 0·85 per cent saline for 5 minutes.

(3) Immerse in saturated benzidine in 0·85 per cent saline to which 1 drop of 20 vol. H_2O_2 per 2 ml. has been added immediately before. Move the sections in the liquid, with a glass rod, until a blue colour appears (30–40 seconds). Leave for a total of 2 minutes.

(4) Wash in 0·85 per cent saline.

(5) Mount in glycerine jelly.

The preparations are not permanent.

Result. The presence of peroxidase is indicated by a blue precipitate.

AN IMPROVED BENZIDINE-PEROXIDASE REACTION

(after van Duijn, 1955)

(*Smears; cold microtome, fresh frozen sections*)

Solutions

(1) Saturated aqueous solution of benzidine. Dissolve 50 mg. in 200 ml. distilled water at 80°. Cool to room temperature and filter.

(2) 3 per cent H_2O_2.

(3) Saturated ammonium chloride. Dissolve 40 g. NH_4Cl in 100 ml. of hot water and cool.

(4) 5 per cent EDTA. Prepare a solution of ethylene diamine tetra-acetic acid and buffer to pH 6·0 with NaOH.

Incubating Medium

Add 1 ml. saturated ammonium chloride and 1 ml. EDTA solution to 9 ml. benzidine and add 1 drop of H_2O_2.

Method. (1) Incubate smears or sections for 5–10 minutes.

(2) Rinse briefly.

(3) Mount in PVA-Fructose medium.*

Result. Blue crystals indicate peroxidase activity. These are stable for some weeks.

* Described by Spurr (1953), this consists of the following:

Polyvinyl alcohol (low viscosity grade)	18 g.
Cadmium iodide	34 g.
Fructose	8 g.
Distilled water	40 ml.

Wash the PVA with methanol, dry. Grind in a mortar. Dissolve the cadmium iodide and add the PVA with continuous stirring. Heat, while stirring, to 75° and add the fructose After the foam has dispersed the medium is ready for use.

ALTERNATIVE BENZIDINE-PEROXIDASE METHOD

(after Wachstein & Meisel, 1964)

(*Fresh frozen, 10–20 μ, cryostat sections*)

PREPARATION OF THE INCUBATING MEDIUM

Dissolve 600 mg. sodium ferricyanide and 600 mg. benzidine in 100 ml. 25% ethanol. Cool to 4°. Filter immediately before use and add 6 drops of 30% H_2O_2.

Method. (1) Incubate sections at 4° for 5–30 minutes.

(2) Wash in distilled water.

(3) Mount in glycerine jelly.

Result. A green or blue reaction product indicates sites of peroxidase activity. In some sites (e.g. leucocytes) the reaction product is in the oxidized form (brown). Slides should be stored at 4°, in the dark.

α-NAPHTHOL REACTION FOR STABLE SUDANOPHILIA

(after Ritter and Oleson)

(*Thin tissue slices; alkaline alcohol-formalin*)

Method. (1) Fix thin slices (1–4 mm.) of tissue in alcoholic alkaline formalin (95 per cent alcohol, 90 ml. 40 per cent formalin, 10 ml. 0·1 N NaOH, 1 ml.) for not less than 24 hours.

(2) Wash in running water for 30 minutes and stand in water for 2 hours.

(3) Immerse for 24 hours in α-naphthol-H_2O_2 solution, freshly prepared (α-naphthol 1 g., 40 per cent alcohol 10 ml., 30 vol. H_2O_2 0·2 ml.).

(4) Rinse in running water.

(5) Stain for 3–24 hours in pyronin solution (pyronin 0·1 g., 40 per cent alcohol 96 ml., aniline oil 4 ml.).

(6) Rinse in alcohol.

(7) Dehydrate in alcohol, clear in benzene and embed in paraffin wax.

(8) Cut sections 6–8 μ thick and mount on albuminized slides.

(9) Bring to water and counterstain nuclei if required.

(10) Dehydrate in alcohol, clear in xylene, mount in Canada balsam.

Result. Peroxidase granules (in the neutrophil leucocytes for instance) appear bright red.

(This method makes use of the increased basophilia of the granules after treatment with α-naphthol-H_2O_2.)

LEUCO-PATENT BLUE METHOD FOR PEROXIDASE (HAEMOGLOBIN)

(after Dunn)

(*Formalin; paraffin sections*)

Fix thin blocks (3–5 mm. thick) in 10 per cent formalin buffered to pH 7·0. Embed in paraffin and section at 5–6 μ.

PREPARATION OF SOLUTIONS

Prepare leuco-patent blue solution as follows: To Patent Blue V (C.I. No. 712) (1 g. in 100 ml. distilled water) add 10 g. powdered zinc and 2·0 ml. glacial acetic acid. Boil until colourless. Cool and filter. Before use add to 10 ml. of the stock solution 2 ml. glacial acetic acid and 1 ml. 3 per cent H_2O_2.

Prepare carmine alum solution as follows: Dissolve 1 g. carmine (alum lake) in 200 ml. of water with heat. Cool and add 0·2 g. salicylic acid. Filter.

Method. (1) Bring paraffin sections to water.
(2) Stain in freshly prepared working reagent, 3–5 minutes.
(3) Wash briefly in water.
(4) Counterstain in carmine alum, 30–60 seconds (or on 0·1 per cent safranin).
(5) Dehydrate in alcohols, clear in xylene and mount in DPX.

Result. (Fig. 178). Haemoglobin stains dark blue-green. Background light pink, nuclei red (with safranin). Eosinophil and neutrophil granules stain dark blue.

DIHYDROXYINDOLE METHOD FOR PEROXIDASE

(after van der Ploeg and van Duijn, 1964b)

(*Smears; or cold formalin-fixed cryostat sections*)

SYNTHESIS OF 5,6-DIHYDROXYINDOLE

Place 1 g. (0·0033 mole) 5,6-dibenzyloxyindole* in a 500 ml. Parr shaker bottle. Add 0·3 g. 5% palladium on charcoal and 120 ml. ethyl acetate. Shake the mixture for 4 hours at room temperature under 46 p.s.i. of hydrogen. Filter the reaction mixture and evaporate under reduced pressure to a red oil. Add hexane and allow the oil to crystallize as a tan coloured solid. Recrystallize from benzene, after shaking with charcoal, by adding hexane.

Yield 0·4 g. (89 per cent) 5,6-dihydroxyindole; m.p. 142-144°. Recrystallize further, as above, to a white granular solid; m.p. 143–144°.

PREPARATION OF 5,6-DIHYDROXYINDOLE SOLUTION (alternative to above)

Harley-Mason and Bu'Lock (1950) described a method for preparing the substrate from the more readily obtainable DL-3,4-dihydroxyphenylalanine, as follows:

To a solution containing 0·5 g. DOPA and 0·84 g. $NaHCO_3$, in 100 ml. distilled water, add 3·29 g. $K_3Fe(CN)_6$. After a minute or so bubble O_2-free nitrogen through the reddish-brown solution (all further operations must be carried out under nitrogen) and add a saturated aqueous solution of zinc acetate, containing 5·59 g. A precipitate of $Zn_2Fe(CN)_6$, coloured grey by melanin products, is formed. Remove this by centrifugation and add to the yellow supernatant 5 ml. 0·25 M-citrate buffer (pH 4·5). Extract the 5,6-dihydroxyindole with two portions of 60 ml. peroxide-free ether or (Axelrod and Lerner, 1963) with ethyl acetate.

PREPARATION OF THE INCUBATING MEDIUM

Dissolve 5,6-dihydroxyindole†, at 4–7 mg. /100ml., in 25 mM-phosphate buffer (pH 7·0 to 7·3). Bubble O_2-free nitrogen through the solution for at least 10 minutes. Add H_2O_2 to a final concentration of 0·03 per cent.

Method. (1) Incubate smears or sections for 3–30 minutes.
(2) Wash, counterstain nuclei in 1 per cent methyl green.
(3) Dehydrate rapidly in alcohols, clear and mount in a synthetic medium.

Result. Dark brown melanin deposits indicate peroxidase.

* Regis Chemical Co, Chicago, Ill.

† This compound is unstable to both light and air but it can be stored in the dark, under N_2, in sealed ampoules.

DAB METHOD FOR PEROXIDASE

(after Graham and Karnovsky, 1966; Fahimi, 1968)

(*Buffered formal-glutaraldehyde, cryostat; glutaraldehyde perfusion, chopper*)

PREPARATION AND USE OF FORMAL-GLUTARALDEHYDE

Add 2 g. paraformaldehyde to 25 ml. water in a flask. Heat to 70° with constant stirring. Add slowly N-NaOH, drop by drop, until the solution becomes clear (2–3 drops). Cool under the tap. Add 5 ml. 50 per cent glutaraldehyde, 20 ml. 0·2 M-cacodylate buffer (Vol. 1, p. 586) and 25 mg. $CaCl_2$. Mix thoroughly.

Fix thin (3 mm.) slices for 5 hours at room temperature. Wash for at least 18 hours in 0·1 M-cacodylate buffer (pH 7·2). Cut 40 μ cryostat sections, or 50 μ sections on the tissue chopper.

PREPARATION OF THE INCUBATING MEDIUM

Dissolve 5 mg. 3,3′-diaminobenzidine tetrahydrochloride in 10 ml. Tris-HCl (pH 7·6) and add 0·2 ml. 1 per cent H_2O_2.

Method. (1) Incubate sections for 3–10 minutes at room temperature.

(2) Wash in 3 changes of distilled water.

(3) Post-fix for 90 minutes in 1·3 per cent OsO_4 in 0·2 M-collidine buffer* (pH 7·2–7·4), containing 5 per cent sucrose.

(4) Dehydrate in ethanol (ascending concentrations).

(5) Embed in Epon.

(6) Cut 0·5 to 1 μ sections for light microscopy and stain briefly with toluidine blue before examination.

Result. Fine brownish granules indicate the localization of peroxidase (catalase) containing microbodies. In some tissues heme-containing dense bodies of lysosomal origin may also stain. The latter, which stain with DAB mixtures from which H_2O_2 has been omitted (see Hirai's method below), usually also contain acid phosphatase.

OXIDIZED DAB STAIN FOR HEME PROTEINS

(after Hirai, 1968)

(*Cold formalin, cryostat sections. Epon embedding*)

PREPARATION OF THE MEDIUM

Dissolve 30 g. DAB and 4·3 g. sucrose in 50 ml. 0·1 M-Tris-HCl buffer (pH 8·4). Adjust pH to 7·2 by adding 1 ml. N-HCl. Keep this solution in the air, and light, at room temperature. Filter and prepare the final medium by adding 1 ml. of 0·1 M-Tris-HCl (pH 7·2).

Method. (1) Incubate 8 μ cryostat sections for 3–6 hours.†

(2) Wash in 0·25 M-sucrose.

(3) Post-fix for 2 hours in cold 1 per cent OsO_4 in Michaelis buffer at pH 7·4 (Vol. 1, p. 584) containing 0·25 M-sucrose.

(4) For optical microscopy, wash sections in 0·25 M-sucrose and mount in glycerine jelly.

* Dissolve 2·67 ml. pure *s*-collidine (2,4,6-trimethylpyridine) in 50 ml. distilled water and add 9ml. N-HCl. Dilute to 100 ml. with distilled water. The product is 100 ml. of 0·2 M-collidine buffer at pH 7·2 to 7·4 (Bennett and Luft, 1959).

† The intermediate radical (oxidized DAB) is irreversibly bound to the heme proteins cytochrome c, catalase and haemoglobin. Specific inhibition of microbody staining is produced by adding 10 mM 3-amino-1,2,4-triazole to the medium. Heme proteins (other than leucocyte granules are inhibited by mM-KCN.

(5) For optical (thick resin embedded sections) and electron microscopy, dehydrate in ethanol and propylene oxide.
(6) Embed in Epon.
(7) Cut ½ to 1 µ sections (for optical microscopy) and counterstain briefly with toluidine blue before examination.

Result. A deep brown granular appearance indicates microbodies, leucocyte granules, etc. Mitochondria are much less strongly stained.

AMINOETHYLCARBAZOLE METHOD FOR PEROXIDASE

(Graham, Lundholm and Karnovsky, 1965)
(*Cold buffered formalin-sucrose, cryostat sections*)

PREPARATION OF THE INCUBATING MEDIUM

Dissolve 2 mg. 3-amino-9-ethylcarbazole in 0·5 ml. dimethylformamide. Add 9·5 ml. 50 mM-acetate buffer (pH 5·0). Immediately before use add one drop of 3 per cent H_2O_2.

Method. (1) Fix tissues for 18 hours at 0° and transfer to cold 5 per cent sucrose for at least 24 hours.
(2) Cut 8 µ cryostat sections and mount on gelatin-coated slides or coverslips. Dry in air.
(3) Incubate for 2–5 minutes at room temperature.
(4) Wash in distilled water and mount in glycerine jelly, first counterstaining the nuclei, if required, with 1 per cent aqueous methyl green.

Result. A discrete red product indicates sites of peroxidase activity.

CATECHOL (DOPA) OXIDASE METHOD

(after Becker *et al.*, 1935; Rappaport, 1955)
(*Fresh frozen or cold formalin-fixed, cryostat; FD, paraffin sections*)

Method. (1) Incubate sections in 0·1 per cent (5·6 mM) DL-DOPA, in 0·1 M-phosphate buffer at pH 7·4, for 1 hour at 37°.
(2) Change DOPA solution and continue incubation as necessary.
(3) Wash in water.
(4) Counterstain nuclei, if required, in Mayer's carmalum for 5–10 minutes.
(5) Wash in water, dehydrate, clear, and mount in a synthetic resin.

Result. Blackish-brown granules indicate the presence of the enzyme.

CATECHOL OXIDASE METHOD FOR CRYOSTAT SECTIONS

(after Rodriguez and McGavran, 1969)
(*Fresh frozen cryostat sections, 8–20 µm thick*)

PREPARATION OF STOCK SOLUTIONS

Solution A. 0·2 M-sodium cacodylate, pH 7·42 (dissolve 42·8 g. sodium dimethyl arsenate in 500 ml. distilled water containing 9·6 ml. 1N-HCl. Dilute to 1000 ml.

Solution B. 0·2 per cent DL-DOPA in distilled water. Mix equal parts of A and B, just before use, to provide the incubating medium.

Method. (1) Post-fix sections for 1 hour at 4° in 4 per cent formaldehyde (buffered at pH 7·4 with 0·1 M cacodylate buffer made hypertonic with 0·44 M-sucrose. Alternatively use 2 per cent glutaraldehyde, buffered to pH 7·4 with 0·1 M-cacodylate buffer.

(2) Rinse sections in cold (4°) buffer for 1 minute.
(3) Incubate for 4–7 hours in the incubating medium.
(4) Wash twice in distilled water and counterstain nuclei (if required).
(5) Dehydrate, clear, and mount in a synthetic resin.

BLOCK INCUBATION METHOD FOR CATECHOL OXIDASE

(after Rodriguez and McGavran, 1969)

Method. (1) Cut tissues into small blocks ($1 \times 1 \times 0{\cdot}2$ cm^3) and fix in fixatives as in Stage 1 of the preceding method, for 3 hours at 4°.
(2) Rinse blocks in cold cacodylate buffer.
(3) Incubate blocks at 37° for 10–16 hours in DOPA solution (equal parts A and B).
(4) Change DOPA solution every 3–4 hours.
(5) Wash blocks in distilled water.
(6) Post-fix 18 hours in 10 per cent neutral formalin.
(7) Dehydrate, clear, and embed in paraffin wax.
(8) Cut sections 10–20 μm thick, and bring to water.
(9) Counterstain nuclei with nuclear fast red or carmalum.
(10) Dehydrate, clear and mount.

TYROSINE—DOPA REACTION REACTION FOR "TYROSINASE"

(after Okun *et al.*, 1969)

(*Fresh frozen, or formalin-fixed, cryostat sections*)

Method. (1) Incubate for 3 hours at 37° in the following medium:
(a) 0·1 M phosphate buffer, pH 7·4.
(b) 10 ml. phosphate buffer, pH 7·4, containing 2 mg. L-tyrosine and 0·1 or 0·2 mg. DL-DOPA.
(c) 10 ml. phosphate buffer, pH 7·4, containing 0·1 or 0·2 mg. DL-DOPA.
(d) As (b) with 1 mM sodium diethyldithiocarbamate.

Controls (a), (c) and (d) should contain sections as nearly serial as possible to the sections incubated in medium (b). Preformed pigment is seen in (a), usually very little induced pigment in (c). No new pigment should be visible in (d).

Result. Grayish-brown to black pigment in sections incubated in medium (b) absent from controls, indicates melanin synthesis initiated by peroxidase.

Some degree of cell damage is necessary for maximum development of the histochemical reaction.

NAPHTHOIC HYDRAZIDE METHOD FOR MONOAMINE OXIDASE

(after Koelle and Valk; Eder)

(*Cold microtome, fresh frozen sections; mounted*)

PREPARATION OF STOCK SOLUTIONS

(1) 0·1 M-hydrazine hydrochloride.
(2) 40 per cent Na_2SO_4 (w/v), adjusted to pH 8·6 with NaOH.
(3) 0·2 M-Na_2HPO_4 (buffer).
(4) 2-Hydroxy-3-naphthoic acid hydrazide (add to final medium).

(5) M-isonicotinyl-2-isopropyl hydrazine phosphate (Marsilid).
(6) 0·1 M-tryptamine hydrochloride.
(7) 1 N NaOH.

PREPARATION OF WORKING SOLUTIONS

(1) Preincubating medium. 3·0 ml. water, 1·5 ml. hydrazine, 3·0 ml. buffer, 7·5 ml. Na_2SO_4 (pH should be 7·6).
(2) Control pre-incubating medium. As above with 0·15 ml. Marsilid.
(3) Rinse. 4·5 ml. water, 3·0 ml. buffer and 7·5 ml. Na_2SO_4.
(4) Incubating medium. 4·35 ml. water, 0·15 ml. NaOH, 3·0 ml. buffer, 7·5 ml. Na_2SO_4. Heat to 80–90° and saturate with naphthoic acid hydrazide. Cool, filter and add 1·0 ml. tryptamine. Saturate with O_2 immediately before use.
(5) Control incubating medium. Prepare as (4) but add, finally, 0·15 ml. Marsilid.
(6) Developing medium. 300 mg. Fast blue B salt, dissolved in 10 ml. water with 5 ml. phosphate buffer, pH 7·4. Filter and use at once.

Method. (1) Immerse sections in pre-incubating medium for 1 hour at 22°.
(2) Rinse in buffered sodium sulphate medium.
(3) Blot, dry, and immerse in incubating medium for 2 hours at 37°. During incubation a stream of O_2 must be passed through the medium.
(4) Rinse in distilled water.
(5) Develop in Fast blue B solution for 3 minutes at room temperature.
(6) Rinse in distilled water.
(7) Fix in 10 per cent formalin for 1 hour.
(8) Mount in glycerine jelly.

Result. (Fig. 180). A purplish-blue colour indicates sites of MAO activity. Control sections preincubated in Marsilid and incubated in the presence of this inhibitor should show no colour. Control sections incubated in the absence of substrate should likewise show no colour.

TETRAZOLIUM METHOD FOR MONOAMINE OXIDASE

(after Glenner *et al*, 1957)

(*Cold microtome, fresh frozen sections; mounted*)

PREPARATION OF INCUBATING MEDIUM

Tryptamine hydrochloride	25 mg.
Sodium sulphate*	4 mg.
Tetranitro-blue tetrazolium	5 mg.
0·1 M-Phosphate buffer (pH 7·6)	5 ml.
Distilled water	15 ml.

Method. (1) Incubate 8–16 μ fresh cryostat sections, mounted on coverslips, for 30–45 minutes at 37°.
(2) Wash in running water for 2 minutes.
(3) Fix in 10 per cent neutral formalin for 24 hours.
(4) Mount in glycerine jelly.

Result. Brown to black formazan deposits indicate sites of MAO activity.

* Better results may be obtained in nervous tissues if sodium sulphate is substituted in the medium by $MgSO_4$ (2 drops of a 0·1 M solution).

COUPLED PEROXIDATIC METHOD FOR MAO

(Graham and Karnovsky, 1965)

(*Fresh frozen cryostat sections, post-fixed in acetone*)

Preparation of the Incubation Medium

Dissolve 2 mg. 3-amino-9-ethylcarbazole in 0·5 ml. dimethylformamide. Add 9·5 ml. 50 mM-phosphate buffer (pH 7·6). Shake and filter. Add to the filtrate 12 mg. tryptamine HCl and 10 mg. horseradish peroxidase. Shake but do not filter again.

Method. (1) Treat fresh cryostat sections with cold acetone for 15 minutes and then wash in 0·9 per cent saline.

(2) Incubate for 20–40 minutes at 37°.

(3) Wash in 0·9 per cent saline.

(4) Post-fix for 18 hours in 10 per cent formalin.

(5) Wash in water and mount in glycerine jelly.

Result. A red granular reaction product indicates MAO activity.

COUPLED PEROXIDATIC METHOD FOR URICASE

(Graham and Karnovsky, 1965)

(*Fresh frozen cryostat sections; post-fixed in cold acetone*)

Preparation of the Medium

Dissolve 3 mg. 3-amino-9-ethylcarbazole in 1 ml. dimethylformamide. Add 10 ml. 50 mM-Tris-HCl buffer (pH 8·5). Shake, allow to stand for 2–3 minutes, and filter. Add to the filtrate 3 mg. sodium urate, 5 mg. EDTA, and 3 mg. horseradish peroxidase.

Method. (1) Incubate sections in air at 37° for 30–60 minutes.

(2) Wash in 0·15 M-NaCl.

(3) Fix in 4 per cent formaldehyde for 1–2 hours.

(4) Wash in 0·15 M-NaCl and mount in glycerine jelly.

Result. A red precipitate forms at sites of uricase activity.

TETRAZOLIUM METHOD FOR XANTHINE OXIDASE

(after Sackler, 1966)

(*Buffered glutaraldehyde; cryostat sections*)

Preparation of Incubation Medium

Dissolve 10 mg. tetranitro-BT and 3·4 mg. hypoxanthine in 10 ml. 0·2 M-phosphate buffer (pH 7·4).

Method. (1) Fix tissues for 1 hour at 4°, in 2·5 per cent glutaraldehyde in 0·1 M-cacodylate buffer containing 1 per cent $CaCl_2$ (pH 7·4).

(2) Rinse in 3 changes of buffer at 4° for 1 hour.

(3) Cut cryostat sections at 8 μ and mount on albuminized coverslips. Dry in air.

(4) Incubate for 10 minutes at 37°.

(5) Rinse in distilled water.

(6) Clear in successive rinses of cellosolve and toluene.

(7) Mount in synthetic resin.

Result. Brown formazan deposits indicate xanthine oxidase activity.

APPENDIX 21

STANDARD METHOD FOR BOUND ENZYMES

(*This technique is usually applied to fresh frozen, or to briefly formalin or glutaraldehyde-prefixed, cryostat sections. Post-fixed cryostat sections are sometimes employed*)

The technique is applicable to the following enzymes:

NADH Diaphorase
NADPH Diaphorase
Succinate (SD)
β-Hydroxybutyrate (HBD)
Glutamate (GLD)
α-Glycerophosphate (α-GPD)
} Dehydrogenases

Preparation of Stock Substrate Solutions (pH adjusted to 7.0)

Enzyme	Substrate	Mol. Wt.	Final Stock Concn.	Amount Required	Vol. H_2O (ml.)	Neutralization	H_2O to (ml.)
SD	DiNa Succinate	270·12	2·5 M	6·75 g.	8	0·05 ml. N-HCl	10
HBD	Na-D-3-hydroxy-butyrate	127·09	1·0 M	1·27 g.	8	0·15 ml. N-HCl	10
GLD[s]	Na-L-glutamate (monohydrate)	187·13	1·0 M	1·87 g.	8	0·05 ml. N-HCl	10
α-GDP (Mitoc)	DiNa-glycerol-3-phosphate	315·15	1·0 M	3·15 g.	8	0·7 ml. N-HCl	10

These solutions are stored at −20°. They are stable for several months under these conditions.

PREPARATION OF STOCK 0·2 M BUFFER (pH 7·4)

To 20·7 ml. of 0·2 M Tris (hydroxymethyl) aminomethane (24·2 g./litre) add 79·3 ml. of 0·2 M-HCl.

PREPARATION OF STOCK INCUBATING SOLUTIONS

MTT (2 mg/ml.)	2·5 ml.	or	Nitro-BT (4 mg/ml.)*	2·5 ml.
Tris buffer (pH 7·4)	2·5 ml.		Tris buffer (pH 7·4)	2·5 ml.
Cobalt chloride (50 mM)	0·5 ml.		$MgCl_2$	1·0 ml.
$MgCl_2$ (5 mM)	1·0 ml.		Distilled water	3·0 ml.
Distilled water	2·5 ml.			

Check the pH of these solutions and adjust to pH 7·0 to 7·2 using stock 0·2 M-Tris or 0·2 M-HCl. These solutions, stored at −20° are stable for several months,

STOCK RESPIRATORY CHAIN INHIBITOR

Sodium azide (100 mM). Add 0·1 ml. to 1 ml. final incubating medium.

* Alternatively, if TNBT is preferred, this salt should be used at a similar high concentration.

Respiratory Inhibitor (Prepare freshly)

Potassium or Sodium cyanide (100 mM). Add 0·1 ml. to 1 ml. final incubating medium.

Preparation of Incubating Media (Volume 1 ml.)

Enzyme	Vol. Stock Soln. (Incubating)	Vol. Stock Soln. (Incubating)	Vol. Distilled Water	Coenzyme†
NADHD	0·9 ml.	—	0·1 ml.	NADH 2 mg.
NADPHD	0·9 ml.	—	0·1 ml.	NADPH 2 mg.
SD	0·9 ml.	0·1 ml.	—	—
HBD	0·9 ml.	0·1 ml.	—	NAD 2 mg.
GLD[s]	0·9 ml.	0·1 ml.	—	NAD or NADP 2 mg.
α-GPD (Mitoc)	0·9 ml.*	0·1 ml.	—	—

* Use Stock Solution saturated with menadione (Vitamin K_3) as follows: Add 5 mg. menadione (twice recrystallized from methanol) to 1 ml. stock incubating solution. Shake and incubate for 30–60 minutes at 37°. Filter before use.

† Coenzymes are added just before use. Check pH and adjust, if necessary, to pH 7·0 to 7·1 before incubating tissues.

Method. (1) Mount cryostat sections (normally 5 μm thick) on coverslips.

(2) Deliver on to each section sufficient incubating medium to cover the tissue completely (approximately 0·2 ml.). Incubate (aerobically) for 10–60 minutes.

(3) Pour off incubating medium and immerse sections in 15 per cent formol saline for 15 minutes.

(4) Wash in running tap water for 2 minutes.

(5) Counterstain nuclei in 2 per cent (chloroform-washed) methyl green for 2–5 minutes (MTT).

(6) Rinse well in distilled water.

(7) Mount in glycerine jelly (MTT) or dehydrate in alcohols, clear in xylene and mount in synthetic resin (Nitro-BT).

Result. Black (MTT) or purple (Nitro-BT) formazan deposits indicate enzyme activity. Nuclei-green (MTT). (Plate XXIIa, p. 968).

STANDARD METHOD FOR SOLUBLE ENZYMES

If this method is used without modification for the demonstration of soluble enzymes the localization of the final product may be that of the corresponding diaphorase. Whenever possible rigorous controls, or alternative methods (see below) should be used.

Preparation of Stock Solutions (pH adjusted to 7.0)

Enzyme	Substrate	Mol. Wt.	Final Stock Concn.	Amount Required	Vol. H_2O (ml.)	Neutralization	H_2O to (ml.)
MD	L-malic acid	134·09		1·34 g.		With malic acid	
	Na-L-malate	187·07	1·0 M	1·87 g.	8	0·9 ml. 40% NaOH	10
ICD	TriNa-DL-isocitrate	276·12	1·0 M	2·76 g.	8	0·9 ml. N-HCl	10
LD	Na-DL-lactate	112·07	1·0 M	1·25 ml.	—	—	10
α-GPD	DiNa-glycerol-3-phosphate	315·15	1·0 M	3·15 g.	8	0·7 ml. N-HCl	10
G-6-PD	DiNa-glucose-6-phosphate	304·11	1·0 M	3·04 g.	8	0·6 ml. N-HCl	10
6-PGD	Ba-6-phosphogluconic acid	430·37	1·0 M	4·3 g.	8	0·6 ml. N-HCl	10
ALC	Ethanol	46·07	1·0 M	0·58 ml.	8	1 drop stock Tris (pH 10·4)	10

These solutions are stored at −20°.
They are stable for several months.

Preparation of Incubating Media

Enzyme	Vol. Stock Soln. (Incubating)	Vol. Stock Soln. (Substrate)	Resp. Inhibitor	Coenzyme	PVP (optional)
MD	0·9 ml.	0·1 ml.	—	NAD 2 mg.	—
ICD	0·9 ml.	0·1 ml.	NaCN 0·1 ml.	NAD or NADP 2 mg.	—
LD	0·9 ml.	0·1 ml.	—	NAD 2 mg.	75 mg.
α-GPD	0·9 ml.*	0·1 ml.	—	NAD 2 mg.	75 mg.
G-6-PD	0·9 ml.	0·1 ml.	NaCN 0·1 ml.	NADP 2 mg.	75 mg.
6-PGD	0·9 ml.	0·1 ml.	NaCN 0·1 ml.	NADP 2 mg.	75 mg.
Alcohol	0·9 ml.	0·1 ml.	—	NAD 2 mg.	—

* Use Stock Incubating Solution made up with 0·06 M Phosphate buffer in place of 0·2 M Tris.

INTERMEDIATE ELECTRON ACCEPTOR

If it is proposed to use phenazine methosulphate (PMS) with any of the above incubating media for soluble enzymes, this compound should be added at the rate of 1 mg. per ml. of the medium. Thereafter incubation should take place in the dark.

Method. As for standard method for bound enzymes.

TECHNIQUES FOR INDIVIDUAL DEHYDROGENASES

When these are given in incomplete form (preparation of incubating medium only) all remaining procedures should be carried out as with the standard methods.

METHOD FOR XYLITOL DEHYDROGENASE

(after Stiller and Gorski, 1965)

(*Fresh cryostat sections, 5–8 μm, mounted on coverslips*)

INCUBATING MEDIUM

Xylitol (1 or 2 M)	0·1 ml.
Tris-HCl buffer (0·2 M, pH 7·2)	0·25 ml.
KCN (0·1 M)	0·1 ml.
Nitro-BT (1 mg/ml.)	0·25 ml.

Add NAD sufficient to provide a final concentration of 1 mM. Check pH and adjust to 7·0 to 7·2 with stock Tris buffer. Add distilled water to final volume of 1 ml. Incubate for 20–30 minutes only.

METHOD FOR SORBITOL DEHYDROGENASE

(after Johnson, 1956, 1967)

(*Cryostat sections. Post-fixed in special fixative*)

PREPARATION OF MEDIA. (FINAL CONCENTRATIONS)

Medium A		**Medium B**	**Medium C**
Tris-HCl (pH 8·8)	0·3 M	As A but with the addition of PMS (15 μM)	As A, but with the omission of Sorbitol and NAD, and addition of NADH 0·4 mM
Nitro-BT	0·3 mM*		
EDTA	1 mM		
KCN	0·1 mM		
D-Sorbitol	30 mM		
NAD	0·12 mM		

Method. (1) Mount fresh sections on coverslips and fix in a cold (0–4°) solution containing ethanol (50 per cent), formalin (10 per cent), Tris base (0·6 M), neutralized with acetic acid.

(2) Transfer to cold acetone, 5 minutes.
(3) Rinse three times (5 minutes each) in cold Tris buffer (pH 8·8).
(4) Incubate (3) serial sections, in each of the 3 media, for 30–60 minutes at 37°.
(5) Post-fix in formol-ethanol fixative.
(6) Wash and mount in glycerine jelly.

METHOD FOR GLUCURONATE DEHYDROGENASE

(after Balogh, 1965)

(*Fresh, 20 μm, cryostat sections*)

PREPARATION OF INCUBATING MEDIUM

Dissolve 10 mg. Nitro-BT, 20 mg. EDTA, 1 mg. NADP in 4 ml. of 0·2 M Tris-HCl buffer. Finally, pipette 2 ml. 20 mM sodium L-gulonate into the medium. (Final substrate concentration 4 mM).

* Probably sub-optimal.

PREPARATION OF SODIUM L-GULONATE

A 20 mM solution was prepared from L-gulono-γ-lactone by hydrolyzing it with an equivalent of NaOH at room temperature. To complete hydrolysis the solution was kept overnight at pH 10, then brought to pH 7·4 and stored at room temperature.

Method. (1) Incubate for 30 minutes at 37°.
(2) Fix for two hours in 10 per cent formalin.
(3) Mount in glycerine jelly.

METHOD FOR L-GULONOLACTONE OXIDASE

(after Nakajima *et al.*, 1969)

(*Fresh, 10 μm, cryostat sections, optional post-fixation in cold acetone, 80 per cent ethanol, or 10 per cent formalin*)

PREPARATION OF MEDIUM

Dissolve 40 mg. L-1,4-gulonolactone, 10 mg. Nitro-BT, and 0·13 mg. KCN in 15 ml. 0·2 M phosphate buffer (pH 7·6). Add 0·2 ml. 0·2 per cent PMS and 4·8 ml. distilled water.

Method. (1) Incubate for 10–60 minutes at 37°.
(2) Wash with 0·9 per cent NaCl.
(3) Fix in 10 per cent neutral formalin, 20 minutes.
(4) Wash with water.
(5) Mount in aqueous medium.

METHOD FOR UDGP DEHYDROGENASE

(after Stiller and Gorski, 1969)

(*Fresh frozen cryostat sections, post-fixed in acetone at 22°*)

INCUBATING MEDIUM

UDP-Glucose (5 mM)	0·1 ml.
Tris buffer (0·2 M, pH 7·2)	0·25 ml.
KCN (0·1 M)	0·1 ml.
Nitro-BT (1 mg/ml.)	0·25 ml.

Add NAD sufficient to give a final concentration of 2 mM. Check pH and adjust to pH 7·0 to 7·2 with Tris. Finally add distilled water to 1 ml. Incubate for up to 60 minutes at 37°.

GELATIN GEL MEDIUM METHOD FOR SOLUBLE DEHYDROGENASES

(after Fahimi and Amarasingham, 1964; Benitiez and Fisher, 1964)

The original technique was devised in order to demonstrate LDH in skeletal muscle. The gel method is widely applicable and the modification described here overcomes many of the disadvantages of the original.

METHOD OF PREPARATION

Take a sheet of commercial PVC (polyvinyl chloride) film (0·09 to 0·13 mm thick). Stretch this over a suitable frame or piece of glass (approximately 10 × 8 cm.) and tape the edges so that the film remains perfectly flat. Pipette 1·6 ml. of the gelatin incubation medium over the surface of the PVC film. Allow to gel, in the dark, for at least 15 minutes. Cut strips of the desired size. Apply film to tissue sections (fresh cryostat sections 2–5 μm) mounted on slides, gently and without pressure. After incubation (5–10 minutes) strip off the film with forceps or melt it away with a stream of warm water (45°).

Fix sections in neutral buffered formalin for 30 minutes. Mount in aqueous medium after counterstaining nuclei (if required).

INCUBATION MEDIUM (FOR LDH)

Prepare the following mixture in a test-tube (total volume of medium 2 ml.): 0·2 μmoles Nitro-BT; 0·5 μmoles NAD; 100 μmoles lithium lactate; 0·2 μmoles PMS. Mix these with an equal volume of 0·2 M Tris buffer (pH 7·4) containing 6·5 per cent gelatin.

METHOD FOR LDH ISOENZYMES

(after Jacobsen, 1969)

(Fresh frozen, cryostat sections, 6–12 μ, air-dried)

PREPARATION OF MEDIA

Solution A. Dissolve the following in 0·1 M Tris buffer (pH 7·2): Sodium DL-lactate 0·4 M (equivalent to L-lactate 0·2 M); Nitro-BT 8 mg/ml. (9·8 mM); NaCN 10 mM; PMS 0·2 mg/ml. (0·66 mM); PVA (mol. wt. 30′000) 0·33 g/ml.

Solution B. To 60 ml. 50 mM Tris buffer (pH 7·0) add 24 g. PVA and dissolve by warming to 90°. Adjust pH to 7·2 at 25°.

FINAL STANDARD MEDIUM (TOTAL LDH)

On a 40° water-bath mix 3·75 solution A with an equal quantity of solution B. Add NAD 1 mg/ml. (1·5 mM) immediately before incubation. The precipitate of Nitro-BT in solution A dissolves on addition of solution B.

MEDIA FOR DISTINCTION OF H AND M ISOENZYMES

(1) To the standard medium add 3·25 M Urea. The M-enzyme is almost totally inhibited.

(2) The substrate concentration (L-lactate) is varied between 5 mM and 500 mM. At the standard concentration (100 mM) and above the M enzyme is activated to a greater extent than the H enzyme (variation between different tissues).

Controls. (1) For α-Hydroxyacid oxidase. Omit NAD, otherwise incubate as for standard procedure.

(2) Total LDH. Add p-chloromercuribenzoate, 1 mM.

(3) Nothing Dehydrogenase. Incubate without lactate.

Method. (1) Fix mounted sections for 10 minutes in cold acetone (0–4°). Follow by treatment with $CHCl_3$ at −15° for 10 minutes. Wash in cold acetone and dry in air.

(2) Incubate sections for times varying according to the concentration of substrate:

5 mM	20–30 minutes
100 mM	5–15 minutes
500 mM	2–5 minutes

(3) Wash in warm 45° water.

(4) Fix for 15 minutes in neutral formol-calcium.

(5) Remove monoformazans by brief treatment with acetone.
(6) Mount in an aqueous medium.

METHOD FOR PROSTAGLANDIN DEHYDROGENASE

(after Nissen and Andersen, 1968, 1969)

(*Cold formalin—gum sucrose, 10 minutes each; 4–6 μm cryostat sections*)

INCUBATING MEDIUM

(Total Volume 0·5 ml.). Contains the following:

PGE_1 or PGE_2*	0·3 mmol.
Nitro-BT	0·075 mmol.
Phosphate buffer (ph8)	0·033 M

Just before incubation add:

NAD	0·65 mmol.

Incubate from 5–50 minutes at 37°.

METHOD FOR 3 BETA-HYDROXYSTEROID DEHYDROGENASE

(after Wattenberg, 1958)

(*Fresh frozen cryostat sections, 10–20 μm*)

INCUBATING MEDIUM

NAD	6·6 mg/ml.	0·1 ml.
$MgCl_2$	4·8 mg/ml.	0·1 ml.
Nitro-BT	2 mg/ml. in DMF	0·1 ml.
DHA	5 mg/ml. in DMF	0·1 ml.
Tris buffer	(pH 8·3, 0·2 M)	0·6 ml.

Method. (1) Remove lipids by treatment with cold acetone (−20°) for 20 minutes.
(2) Dry sections in air.
(3) Incubate at 37° for 20–30 minutes.
(4) Rinse briefly in distilled water.
(5) Fix in neutral formol saline for 30 minutes.
(6) Rinse in distilled water and treat for 2 minutes with 20 per cent ethanol.
(7) Rinse in distilled water and counterstain nuclei (if required) in 2 per cent methyl green.
(8) Wash and mount in glycerine jelly.

Result. Purple formazan deposits indicate localization of the enzymes.

METHOD FOR 3 ALPHA-HYDROXYSTEROID DEHYDROGENASE

(after Balogh, 1966)

(*Fresh frozen, 20 μm, cryostat sections*)

INCUBATING MEDIUM

Dissolve 5 mg. Nitro-BT, 5 mg. NAD, and 10 mg. EDTA in 2 ml. 0·2 M Tris-HCl buffer (pH 7·0). To this add 2 ml. 50 per cent PVP in the same buffer. Finally, add 5 mg. androsterone dissolved in 1 ml. DMF.

* Upjohn.

INCUBATION

60 minutes at 37°.

METHOD FOR CHOLINE DEHYDROGENASE

(after Glogner and Gössner, 1962)

(*Fresh frozen cryostat sections, 10 μm*)

INCUBATION MEDIUM

Choline chloride	(100 mg/ml.)	0·3 ml.
Phosphate buffer	(0·2 M, pH 7·5)	2·1 ml.
Nitro-BT	(5 mg/ml.)	0·3 ml.
Distilled water		0·3 ml.

INCUBATION

45–60 minutes at 37°.

METHOD FOR ALDEHYDE DEHYDROGENASE

(after Gabler, 1969)

(*Fresh frozen, cryostat sections, 10 μm*)

INCUBATION MEDIUM

Dissolve the following in 9 ml. of 0·06 M Phosphate buffer, saturated with benzaldehyde.

Sodium Azide	6·5 mg.
$MgCl_2.6H_2O$	10 mg.
Nitro-BT	2·5 mg.
NAD	11 mg.

Add 1·0 ml., 1,2-Propylene-glycol.*

METHOD FOR BETAINE ALDEHYDE DEHYDROGENASE

(after Wohlrab, 1965)

(*Fresh frozen, cryostat sections, 20 μm*)

INCUBATION MEDIUM

Betaine aldehyde (monomer)	(14 mg/ml.)	2·0 ml.
Nitro-BT	(5 mg/ml.)	2·5 ml.
Tris-HCl buffer	(0·2 M, pH 7·2)	2·5 ml.
KCN	(0·1 M)	2·0 ml.
Distilled water.		

Add 9·0 mg. NAD immediately before use.

INCUBATION

It is recommended that this process should be carried out under nitrogen, after bubbling nitrogen through the medium, for 30–60 minutes at 37°.

* Since this compound can act as substrate for secondary alcohol dehydrogenase it may be advisable to substitute another organic solvent, such as DMF.

METHOD FOR GLYCERALDEHYDE PHOSPHATE DEHYDROGENASE

(after Himmelhoch and Karnovsky, 1961)

(*Cryostat sections, 8–10 µm, post-fixed in cold acetone, 4°, 20 minutes*)

Preparation of Substrate

Place 250 mg. glyceraldehyde-3-phosphate diethyl-acetal (monobarium salt) in a glass centrifuge tube with 0·8 g. Dowex-50 resin (hydrogen form) and 2 ml. distilled water. The contents should be mixed thoroughly and the tube then placed for 3 minutes in a boiling water-bath. During this time a stream of oxygen-free nitrogen should be passed through the contents of the tube. After 3 minutes place the tube in a crushed ice bath and wash down its walls, and the nitrogen pipette, with a few drops of distilled water. Centrifuge for 10 minutes. Pipette off supernatant and transfer to another container. Wash the resin with 1 ml. distilled water. Collect after 10 minutes centrifugation and add to the supernatant (total 3 ml.). The glyceraldehyde-3-phosphate now present is titrated to pH 7·2 with 2·0 M-NaOH. (Usually this requires 0·6 ml.) Suitable dilutions can now be prepared for the incubating medium, and for assay.

Assay of Substrate Concentration

Take a suitable aliquot of the final substrate preparation and add NaOH to bring the concentration of the latter to 1N. Allow to stand at room temperature for 10 minutes. Titrate back to neutrality with conc. HCl and analyse the inorganic phosphate concentration by the method of King (1932). Compare this value with that of another aliquot which has not been treated with alkali. The difference is taken as a measure of the concentration of glyceraldehyde-3-phosphate.

Incubating Medium

Substrate solution (10–40 mM*)	0·3 ml.
NAD (5 mg/ml.)	0·25 ml.
Nitro-BT (5 mg/ml.)	0·2 ml.
Phosphate buffer 90·2 M, pH 7·2)	0·5 ml.
EDTA (diNa) 6 mg/ml.)	0·1 ml.
Distilled water	0·25 ml.

This medium can be stored at −20° for some months.

* Optimum concentration 36 mM.

Method. (1) Rinse sections briefly in 0·85 per cent saline.
(2) Incubate at 37 µ for 3–20 minutes.
(3) Rinse in saline.
(4) Fix for 20 minutes in formol-calcium, at 4°.
(5) Rinse in saline.
(6) Mount in glycerine jelly.

METHOD FOR DIHYDRO-OROTATE DEHYDROGENASE

(after Cohen, 1962)

(*Fresh cryostat sections, 30 µm*)

Incubating Medium

Dissolve the following in 4 ml. 0·1 M phosphate buffer (pH 8·0):

5 mg. DHO	(6 mM)
2 mg. Ureidosuccinic acid	(2·3 mM)
2·5 mg. Nitro-BT	(0·5 mM)

Bring to pH 8 with 0·1N-NaOH and to a total of 5 ml. by addition of buffer.

INCUBATION

60 minutes at 37°.

METHOD FOR DIHYDRO-OROTATE DEHYDROGENASE

(after Castoldi *et al.*, 1968)

This method is designed for use with smears.

(*Air-dried smears, fixed with acetone for 40 seconds*)

INCUBATING MEDIUM

L-Dihydro-orotic acid	1 mg/ml.	(6 mM)
Ureidosuccinic acid	0·4 mg/ml.	(23 mM)
NAD	0·15 mg/ml.	(0·22 mM)
Nitro-BT	0·5 mg/ml.	(0·55 mM)
Phosphate buffer M/15, pH 7·8	8 ml.	

Bring to pH 7·8 by titration with 0·1N-NaOH.
Bring final volume to 10 ml. by adding buffer.

INCUBATION

60 minutes at 37°.
Counterstain carmalum.
Aqueous mounting medium.

METHOD FOR L-AMINOACID TETRAZOLIUM REDUCTASE

(after Castellano *et al.*, 1969)

(*Cryostat sections, 10–16 μm, post-fixed in cold acetone for 30 seconds*)

INCUBATING MEDIUM

Dissolve in phosphate buffer (0·1 M, pH 7·6), the following:

L-leucine	10 mM
NNT (or Nitro-BT)	1 mg/ml.
PMS	1 mg/ml.

INCUBATION

15 minutes to 2 hours at 37°.
Stop reaction with 10 per cent formalin.

METHOD FOR DIHYDROFOLATE REDUCTASE

(after Onicescu *et al.*, 1970)

(*Fresh frozen cryostat sections, 8–10 μ, post-fixed in cold glutaraldehyde*)

FIXATION

The authors used 20 per cent glutaraldehyde at 4° for 30 minutes but it seems likely that a purified glutaraldehyde might be used at much lower concentrations than this, with equally good effect.

INCUBATION MEDIUM

The medium, which should be freshly prepared, contains the following:

Pteroylglutamate (sodium salt)	4 mM
NADP	5 mM
NAD	5 mM
Nitro-BT	10 mM
Phosphate buffer, pH 7·0	0·1 M

Method. (1) Incubate for 40–60 minutes at 37°.
(2) Wash in distilled water.
(3) Mount (preferably) in glycerine jelly.

Result. Purple formazan deposits indicate reductase activity.

APPENDIX 22

LNA METHOD FOR LEUCINE NAPHTHYLAMIDASE

(Nachlas, Crawford and Seligman, 1957)

(*Fresh frozen, cold microtome sections, mounted; freeze-dried paraffin-embedded sections*)

PREPARATION OF THE SUBSTRATE SOLUTION

Stock Substrate. Dissolve 8 mg. per ml. of L-leucyl-β-naphthylamide in distilled water. This solution can be stored at 0–4° for several months.

INCUBATING MEDIUM

Stock substrate solution	1 ml.
Acetate buffer (0·1 M, pH 6·5)	10 ml.
Sodium chloride (0·85 per cent)	8 ml.
Potassium cyanide (20 mM)	1 ml.
Fast blue B salt	10 mg.

Method. (1) Incubate sections for 15 minutes to 2 hours at 37°.
(2) Rinse in saline for a few seconds.
(3) Immerse in 0·1 M-cupric sulphate for 2 minutes.
(4) Rinse again in saline.
(5) Dehydrate in graded alcohols, clear in xylene and mount in Canada balsam. Alternatively sections may be mounted without dehydration in glycerine jelly or polyvinyl pyrrolidone medium.

Result. A purplish blue colour indicates the site of leucine naphthylamidase activity.

Since the coupling rate of the capture reaction is somewhat slow, as pointed out by Enerbäck and Hansson (1968) strongly electronegative tissue components such as mast cell granules may absorb, first diazonium salt and secondly the free naphthylamine. This results in false localizations of the final azo dye product.

MNA METHOD FOR AMINO ACID NAPHTHYLAMIDASES

(after Monis, Wasserkrug and Seligman, 1965)

(*Cold glutaraldehyde, cryostat sections*)

FIXATION PROCEDURE

Prepare 10 ml. of a 6-fold dilution of 25 per cent glutaraldehyde with 0·1 M phosphate buffer (pH 7·4). Cool to 4° and immerse small tissue blocks, not more than 2 mm. thick, for 1–2 hours. Blot gently, quench on cryostat tissue holder, cut 6 μm sections.

PREPARATION OF INCUBATING MEDIUM

The composition is as follows:

Phosphate buffer (0·1 M, pH 7·4)	2·5 ml.
Substrate solution	0·25 ml.
Saline, 0·85 per cent	2·0 ml.
KCN (20 mM)	0·25 ml.
Fast blue B salt	5 mg.

Alternative substrates are:

L-alanyl-4-methoxy-2-naphthylamide	8 mg/ml.
L-methionyl-4-methoxy-2-naphthylamide-HCl	9·6 mg/ml.
L-leucyl-4-methoxy-2-naphthylamide-HCl	10 mg/ml.

Method. (1) Place sections directly in the incubating medium (at room temperature). Bring to 37° and incubate for 5–30 minutes.

(2) Remove sections with a glass lifter and rinse in

(3) 0·85 per cent saline. Thence, transfer to

(4) 0·1 M $CuSO_4$.

(5) Rinse in distilled water.

(6) Mount in glycerine jelly.

Result. A bright red precipitate indicates sites of amino acid naphthylamidase activity.

ACYL NAPHTHYLAMIDASE METHOD

(after Hopsu and Glenner, 1964)

(*Fresh cryostat; brief cold formalin, cryostat sections*)

PREPARATION OF THE INCUBATING MEDIUM

Dissolve the substrate (chloroacetyl-α-naphthylamide is recommended) in dimethylformamide at 10 mg/ml.

Prepare the following medium:

Substrate in DMF	0·2 ml.
Acetate buffer, 0·1 M, pH 6·2	7·8 ml.
Mercuric chloride, 2 mM	1·0 ml.
Fast garnet GBC salt, 4 mg/ml.	1·0 ml.

Method. (1) Incubate mounted sections for 5–30 minutes at 27°.

(2) Rinse briefly in distilled water.

(3) Mount in glycerine jelly.

Result. Chelation between azo dye and Hg^{2+} ions produces a blue final product.

INDIGOGENIC METHOD FOR AMINOPEPTIDASE

(after Pearson, Wolf and Andrews, 1963)

(*Fresh cryostat sections, 6 μ, mounted on slides or coverslips*)

PREPARATION OF THE INCUBATING MEDIUM

Add the following components:

Tris-HCl buffer, 0·2 M, pH 7·7	11 ml.
Copper glycinate, 1·5 mM	0·5 ml.
Substrate*, 15 mM	0·5 ml.
Spermidine trihydrochloride	4 mg.

*PREPARATION OF STOCK SUBSTRATE SOLUTION

Add 300 mg. DL-*N*-(5-bromoindol-3-yl) leucinamide hydrobromide to 50 ml. distilled water and neutralize with 3 drops of 6N-NaOH and store at 4°. This solution keeps for 2–3 months.

Method. (1) Incubate sections for 1–4 hours at 27°.

(2) Immerse sections in 90 per cent acetone and 90 per cent *n*-butanol (3 : 1) for 15 minutes.

(3) Rinse briefly in water.

(4) Mount in glycerine jelly.

Result. A blue indigo precipitate indicates sites of aminopeptidase activity.

TABLE 91

Inhibitors and Activators for Peptidases

Compound	Final Molarity	Amount per 100 ml.	Buffer, pH	Effect: Cathepsin	Effect: AP's & NA's
$Mn_2SO_4.4H_2O$	5 mM	155 mg.	0·1 M Acetate, 5·5	Inhib.	Activ.
	10 mM	230 mg.	0·1 M Acetate, 6·5	Inhib.	Activ.
EDTA	50 mM	1·85 g.	0·1 M Acetate, 6·5	Activ.	Inhib.
EDTA + GSH	50 & 4 mM	1·85 g., 61·5 mg.	0·1 M Acetate, 6·5	Activ. +	Total Inhib.
EDTA + Cysteine-HCl	50 & 4 mM	1·85 g., 63 mg.	0·1 M Acetate, 6·5	Activ. +	Total Inhib.
Lactic acid	~20 mM	About 150 mg.	0·1 M Acetate, 6·5	Activ.	Inhib.
Lactic + GSH	20 & 4 mM	150 mg., 61·5 mg.	0·1 M Acetate, 6·5	Activ. +	Total Inhib.
BPP	50 mM	730 mg.	0·1 M Acetate, 6·5	Nil	Inhib.
BPP followed by EDTA + GSH	50 mM / 50 & 4 mM	as above	0·1 M Acetate, 6·5	Activ.	Total Inhib.
p-HMB	1 mM	44·8 mg.	0·1 M Acetate, 6·5	Inhib.	Nil
Hyamine 1622	1 mM	45·4 mg.	In Medium	Nil	−99
Puromycin	1 mM	47·15 mg.	In Medium	Nil	−89

Except where indicated these figures refer to 30 minutes preincubation, at the pH indicated, of fresh cryostat sections

BPP = β-phenylpropionate

p-HMB = *p*-hydroxymercuribenzoate

Hyamine = benzyldimethyl-2-[2-(*p*-1,3,3 tetramethylbutylphenoxy)ethoxy] ethyl ammonium chloride.

γ-GLUTAMYL TRANSPEPTIDASE METHOD I

(after Rutenberg *et al.*, 1969)

(*Fresh frozen cryostat sections, 4–8 μ, mounted on slides or coverslips*)

PREPARATION OF THE INCUBATING MEDIUM

Prepare freshly for each incubation the following:

γ-Glutamyl-4-methoxy-2-naphthylamide (2·5 mg/ml.)	1·0 ml.
Tris buffer (0·1 M), pH 7·4	5·0 ml.
Saline, 0·85 per cent	14·0 ml.
Glycylglycine	10 mg.
Fast blue BBN salt	10 mg.

Filter before use. (Final substrate concentration, 0·125 mg/ml.).

STOCK SUBSTRATE SOLUTION

Dissolve 25 mg. of the substrate in 0·5 ml. dimethylsulphoxide and 0·5 ml. 1·0 N-NaOH and add 9 ml. distilled water.

This solution is stable for 3 days at 4°.

Method. (1) Incubate sections for 3–45 minutes at 25°.
(2) Rinse briefly in 0·85 per cent saline.
(3) Transfer to 0·1 M $CuSo_4$ for 2 minutes.
(4) Rinse again in saline.
(5) Dry and mount in Apathy's medium (Vol. 1, p. 579).

Result. The bright red copper chelate of the azo dye final product indicates sites of γ-glutamyl transpeptidase activity.

γ-GLUTAMYL TRANSPEPTIDASE METHOD II

(after Glenner and Folk, 1961)

(*Fresh frozen cryostat sections, 10 μ, mounted on slides or coverslips*)

PREPARATION OF THE INCUBATING MEDIUM

Prepare the following medium freshly and filter before use:

Suspension of *N*-(γ-L-glutamyl-*β*-naphthylamide), 4 mg/ml. in acetone	0·5 ml.
Aqueous glycylglycine, 20 mg/ml.,	0·5 ml.
Tris buffer, 0·1 M, pH 7·2,	10 ml.
Fast garnet GBC salt	15 mg.

Method. (1) Incubate at room temperature for 10–60 minutes.
(2) Rinse in distilled water.
(3) Counterstain nuclei, if necessary in Mayer's haemalum, 5 minutes.
(4) Rinse in distilled water for 5 minutes.
(5) Mount in glycerine jelly.

Result. A reddish brown colour indicates sites of enzyme activity.

γ-GLUTAMYL TRANSPEPTIDASE METHOD III

(after Albert *et al.*, 1964)

(*Cold ethanol or acetone-fixed, 10 μ, paraffin sections*)

PREPARATION OF THE INCUBATING MEDIUM

Dissolve 20 mmols. of γ-L-glutamyl-*α*-naphthylamide in 6 ml. 0·9 per cent saline. Add 4 ml. 0·1 M phosphate buffer (pH 6·7). Alternatively use phosphate-citrate buffer at the same pH. Dissolve glycylglycine at the rate of 10 mmols./ml. Immediately before use add 5 mg. Fast garnet GBC salt to each 10 ml. of medium.

Method. (1) Bring sections to water, through ethanol and xylene.
(2) Incubate for 10–45 minutes at 22°.
(3) Rinse in saline for a few seconds.
(4) Immerse for 2 minutes in 0·1 M $CuSO_4$.
(5) Rinse briefly in distilled water.
(6) Counterstain in Mayer's haemalum, 2 minutes.
(7) Rinse in tap water and mount in glycerine jelly.

Result. A brownish-red colour indicates transpeptidase activity; nuclei blue.

METHOD FOR CHYMOTRYPSIN-LIKE ESTERASE

(after Lagunoff and Benditt, 1961)

(*Cold formalin, gum sucrose routine, cryostat sections, post-fix in methanol*)

PREPARATION OF THE MEDIUM

Substrate: Naphthol AS-phenylproprionoxy ester, 40 μM, in 40 per cent methanol in Tris buffer (pH 8·0).

Add 5mg. Fast garnet GBC salt per 10 ml. medium, just before use and filter.

Method. (1) Incubate for ½ to 1 hour at 22°.
(2) Wash in distilled water.
(3) Mount in PVP medium.

Result. (Plate XXII c.) A brownish-purple colour indicates esterolytic activity.

METHOD FOR TRYPSIN-LIKE ENZYMES

(after Glenner and Cohen, 1960)

(*Fresh or cold formalin-fixed, cryostat sections, 10 μ, mounted*)

PREPARATION OF THE MEDIUM

Substrate: *N*-(*a*-benzoyl-DL-arginine-*β*-naphthylamide) hydrochloride. (BANA). Dissolve 30 mg. substrate and 10 mg. Fast garnet GBC salt in 20 ml. 0·1 M Tris buffer (pH 7·0).

Method. (1) Incubate sections for 2–30 minutes at room temperature.
(2) Rinse in distilled water and transfer for 2 minutes to 0·1 M $CuSO_4$.
(3) Rinse in water for 2 minutes.
(4) Mount in glycerine jelly containing 0·05 M $CuSO_4$.

Result. Red azo dye precipitates indicate sites of enzyme activity.

OXIDATION TECHNIQUE FOR AMINOPEPTIDASE

(after Burstone and Weisburger, 1961)

Freeze-dried unfixed, or fresh cryostat sections, 8–16 μ)

PREPARATION OF THE MEDIUM

Alternative Substrates: DL-Alanyl derivatives of 3-amino-9-ethyl carbazole, 3-amino carbazole, 4-aminodiphenylamine, 8-amino-1,2,3,4, tetrahydroquinoline or 2-aminofluorene-7-sulphonic acid.

Dissolve 5 mg. substrate in 0·5 ml. ethanol and add 25 ml. distilled water and 2 ml. 50 mM potassium ferricyanide. (The first substrate is to be preferred.)

Method. (1) Incubate sections for 15–16 minutes at 37°.
(2) Wash briefly in distilled water.
(3) Immerse in 0·1 M $CuSO_4$, 2 minutes.
(4) Wash briefly and mount in PVP medium.

Result. Brownish or reddish dyes, which may be iron or copper complexes, indicate sites of aminopeptidase activity.

SILVER PROTEINATE METHOD FOR ENDOPEPTIDASES

(after Takamatsu *et al.*, 1963; Yamada and Ofugi, 1968)

(*Fresh frozen, or cold acetone-fixed, cryostat sections*)

INCUBATING MEDIUM

1 per cent aqueous silver proteinate	10 ml.
2 per cent KBr	2 ml.
0·5 M buffer (acetate or borate)	20 ml.
Distilled water	15 ml.

Method. (1) Incubate mounted sections at 37°, in the dark, for 4–6 hours.

(2) Rinse briefly in distilled water.

(3) Immerse in a dilute solution of hydroquinone (photographic developer) for 15 minutes at room temperature.

(4) Rinse in distilled water.

(5) Treat with 2 per cent sodium thiosulphate for 5 minutes.

(6) Wash well in running water.

(7) Dehydrate in alcohols, clear in xylene, mount in Canada balsam.

Result. Brown granulation indicates the sites of proteolytic (endopeptidase) activity.

APPENDIX 23

METHOD FOR ASPARTATE CARBAMOYLTRANSFERASE

(after Spors and Merker, 1969)

FIXATION

Fix small blocks in 3 per cent glutaraldehyde buffered with cacodylate, pH 7·2 (Vol. 1, p. 586). Three to five hours fixation is recommended. Wash briefly after fixation and cut sections (10–15 microns) in the cryostat.

PREPARATION OF INCUBATING MEDIUM

The reaction mixture contains the following:

L-Aspartic acid	(2·6 mM)
Di-lithium carbamoyl phosphate	(5 mM)
Lead nitrate	(3 mM)
KCl	(6 mM)
Trisodium citrate	(5 mM)
Glucose	(0·11 M)

in 50 mM Tris-maleate buffer (pH 7·2).

Method. (1) Incubate sections for 30–45 minutes at 37°.
(2) Wash briefly in distilled water.
(3) Treat with dilute yellow ammonium sulphide, 2 minutes
(4) Rinse in distilled water
(5) Counterstain nuclei, if required, in 2 per cent methyl green for 2 minutes.
(6) Wash briefly.
(7) Mount in glycerine jelly.

Result. Brown deposits of lead sulphide indicate sites of enzyme activity.

METHOD FOR ORNITHINE CARBAMOYL TRANSFERASE

(after Mizutani and Fujita, 1968)

FIXATION

The authors recommended fixation by perfusion with 3 per cent cacodylate buffered (pH 7·2) cold (4°) glutaraldehyde for 2–5 minutes. Cut 10–30 micron frozen or cryostat sections and transfer these directly into 50 mM Tris-maleate buffer at pH 7·2, containing 7·5 per cent sucrose. Leave for 1 hour to 8 days.

PREPARATION OF THE INCUBATING MEDIUM

Di-lithium carbamoyl phosphate	3 mg.
50 mM Tris-maleate (pH 6·7 to 7·0)	4 ml.
1 per cent lead nitrate	1 ml.
Distilled water	5 ml.
Sucrose	0·7 g.

Add the lead nitrate in the distilled water as a final step, with continuous stirring. Filter the slightly turbid mixture and use immediately.

Method. (1) Incubate sections for 15–30 minutes at 37°.
(2) Wash briefly in distilled water.
(3) Treat with dilute yellow ammonium sulphide.
(4) Wash in distilled water and mount in glycerine jelly.

Result. Brown PbS deposits indicate enzyme activity.

METHOD FOR CHOLINE ACETYLTRANSFERASE

(after Burt, 1970)

(*Fresh cryostat sections, 8 μ, mounted on coverslips*)

PREPARATION OF INCUBATING MEDIUM

The basic incubating medium is as follows:

Choline	(4 mM)
Acetyl CoA	(0·2 mM)
$Pb(NO_3)_2$	(1·8 mM)
Phospholine iodide	(0·1 mM)

in 25 mM HEPES (*N*-2-hydroxyethylpiperazine-*N*-2 ethane sulphonic acid) at a final pH of 6·0.

Method. (1) Rinse sections in 25 mM HEPES buffer (pH 7·0) containing 10 per cent sucrose and 2 per cent formaldehyde, for 5 minutes.

(2) Wash, for 15 minutes each, in 3 changes of the buffered sucrose medium.

(3) Transfer to incubating medium at 37° for 15–30 minutes.

(4) Wash in 3 changes of distilled water.

(5) Transfer to dilute yellow ammonium sulphide for 2–3 minutes.

(6) Wash in distilled water and mount in glycerine jelly.

Result. Brown PbS deposits indicate sites of enzyme activity.

METHOD FOR ASPARTATE AMINOTRANSFERASE

(after Lee, 1968)

FIXATION

Lee and Torack (1968) recommended brief fixation of cryostat sections with 1 per cent glutaraldehyde in place of cold acetone, used by Lee (1968). To this fixative they added 50 mM imidazole, and the pH was adjusted to 7·2 to 7·4 with normal HNO_3. After 1 to 10 minutes' fixation, the tissues were washed in four changes of distilled water. This should preferably be cold (0–4°).

PREPARATION OF THE INCUBATING MEDIUM

Dissolve 266·2 mg. L-aspartic acid and 58·4 mg. 2-oxoglutaric acid in 10 ml. distilled water and add 25 ml. 0·2 M imidazole. Adjust pH to 7·5 with 0·2 N-NaOH. Bring total volume to 50 ml. with distilled water.

Shortly before use add 50 ml. 12 mM lead nitrate, dropwise, with constant stirring. The final pH should be 7·2 to 7·4. Filter the medium if it remains cloudy.

Method. (1) Incubate sections for 30–60 minutes at room temperature (22–28°).

(2) Wash briefly in 4 changes of distilled water.

(3) Immerse in 1 per cent yellow ammonium sulphide for 1–2 minutes.

(4) Wash and mount in glycerine jelly.

Result. Lead sulphide product indicates sites of enzyme activity.

DIAZO METHOD FOR ASPARTATE AMINOTRANSFERASE

(after Kishino, 1968)

FIXATION

The author recommended fixation of small pieces of tissue in cold (0–4°) 12·5 per cent hydroxydipaldehyde, buffered with 0·1 M phosphate buffer to pH 7·4.

After fixing for 3 hours the tissues are washed briefly in water and 20 μ sections cut in the cryostat. Free floating or mounted sections may be employed.

PREPARATION OF THE INCUBATING MEDIUM

The medium, prepared freshly immediately before use, contains the following:

0·2 M Aspartic acid	1 ml.
0·1 M 2-Oxoglutaric acid	1 ml.
Fast violet B salt	3–5 mg.
0·3 per cent EDTA	0·5 ml.
0·1 M Phosphate buffer (pH 7·4)	2 ml.

Adjust pH to 7·4 with N-NaOH.

Method. (1) Incubate sections for 1 hour at 37°.
(2) Wash in distilled water.
(3) Treat with 0·1 M $CuSO_4$ for 3–5 minutes.
(4) Wash thoroughly in water.
(5) Dehydrate in alcohols, clear in xylene and mount in Canada balsam.

Result. Sites of enzyme activity are indicated by a reddish brown deposit of the formazan.

METHOD FOR HEXO- AND GLUCOKINASE

(after Meijer, 1967)

(*Fresh frozen, 7 μ, cryostat sections, mounted on slides*)

PREPARATION OF THE INCUBATING MEDIUM

This must be freshly prepared, immediately before use. The medium has the following composition:

D-glucose	30 mg.
NADP	2·5 mg.
ATP	5·5 mg.
$MgCl_2.6H_2O$	20 mg.
Nitro-BT	2·5 mg.
Imidazole buffer, 40 mM, pH 7·5	2 ml.
Gelatin (6 per cent)	3·8 ml.
G-6-P dehydrogenase (1 mg/ml.)	0·005 ml.

Adjust pH to 7·5 before use with N HNO_3.

Method. (1) Incubate at 37° for 1–2 hours.

(2) Wash briefly and fix for 30 minutes in 4 per cent neutral formaldehyde solution.

(3) Mount in glycerine jelly.

Result. Dark blue formazan granules indicate sites of activity of the two kinases.

METHOD FOR PHOSPHOGLUCOMUTASE

(after Meijer, 1967 and Yano, 1968)

(*Fresh frozen 8–10* μ *cryostat sections, mounted, dried at −20°*)

PREPARATION OF THE MEDIUM

The medium, freshly prepared before use, is made up as follows:

D-Glucose-1-phosphate	80 mg.
NADP	2·5 mg.

ATP	5 mg.
$MgCl_2$	1 mM
L-Histidine	5 mM
Nitro-BT (1 mg/ml.)	5 ml.
Imidazole buffer, pH 7·4, 40 mM	10 ml.
Gelatin, 3 per cent	5 ml.
G-6-P dehydrogenase (1 mg/ml.)	0·04 ml.
Distilled water to	25 ml.

Two control media are required. The first is a substrate (G-1-P) free medium as above. The second medium contains 1 mM beryllium sulphate ($BeSO_4.4H_2O$) as inhibitor.

Method. (1) Cryostat sections, dried at −20° for 24–48 hours, are post-fixed in cold (0–4°) acetone for 5 minutes.

(2) Incubation is carried out at 37° for 1 hour.

(3) The reaction is stopped by immersion in 10 per cent formalin for 10 minutes.

(4) Washing in water is followed by mounting in PVP.

Result. Dark blue formazan product present in the test sections and absent from the two controls, indicates phosphoglucomutase activity.

POST-COUPLING METHOD FOR ARYLSULPHATASE

(Rutenburg, Cohen and Seligman, 1952)

(*Cold formalin-fixed, frozen sections; cold microtome sections, mounted*)

Preparation of Substrate Solution

Dissolve 25 mg. potassium 6-benzoyl-2-naphthyl sulphate in 80 ml. hot 0·85 per cent sodium chloride solution. Add 20 ml. 0·5 M-acetate buffer (pH 6·1). This solution is stable and can be stored for at least 15 days. Before use this solution is made hypertonic by the addition of 2·6 g. solid sodium chloride to each 100 ml.

Method. (1) Pretreat sections, for 2–5 minutes in each, in successive solutions of sodium chloride (0·85, 1·0 and 2·0 per cent).

(2) Incubate for 2–8 hours at 37° in the substrate solution.

(3) Wash twice with cold saline (fresh tissues) or water (formalin-fixed tissues).

(4) Transfer to an ice-cold freshly prepared solution of Fast blue B salt (1 mg/ml.) in 0·05 M-phosphate buffer (pH 7·6). Agitate gently for 5 minutes.

(5) Wash three times in cold 0·85 per cent saline and then in water.

(6) Mount in glycerine jelly.

Result. A blue or purplish-blue colour indicates sites of strong arylsulphatase activity. A red or pink colour may indicate weaker activity or diffusion of the final product with solution in lipids.

NAPHTHOL AS METHOD FOR ARYLSULPHATASES A AND B

(after Woohsman and Hartrodt, 1964; Wächtler and Pearse, 1966)

Fixation

Small blocks, or whole organs if small, are fixed in cold (0–4°) 10 per cent formol-calcium (Baker, 1944), or in buffered neutral 10 per cent formalin, for 12–24 hours. Thereafter they are stored for at least 24 hours at 4° in gum sucrose (Holt, 1960). Cryostat sections, 6–10 μ thick, mounted on coverslips, are used. It was found useful to pretreat sections with a mixture of equal parts of 2 per cent formaldehyde and 1 per cent gelatin (Romeis, 1948).

PREPARATION OF THE INCUBATING MEDIUM

Solution A. Dissolve 20 mg. Naphthol AS-BI sulphate (potassium salt) in 8 ml. 0·85 per cent saline and add 2 ml. 0·2 M acetate buffer (pH 6·0).

Solution B. Prepare a stock solution containing 2 g. pararosanilin hydrochloride in 50 ml. 2N-HCl. Warm to dissolve, cool and filter.

Immediately before use add 0·3 ml. of this solution to 0·3 ml. of 4 per cent sodium nitrite. Allow to stand for 3–5 minutes before use.

Mix 10 ml. of Solution A with 0·6 ml. of solution B and add 260 mg. NaCl. Dissolve and adjust pH to 6·4 to 6·8.

Method. (1) Incubate for 30 minutes to 2 hours at 37°.
(2) Wash in distilled water.
(3) Counterstain nuclei in 2 per cent methyl green, 2 minutes.
(4) Wash in running water for 30 seconds.
(5) Dehydrate in alcohols, clear in xylene, and mount in a suitable synthetic medium.

Result (Plate XXIId). Arylsulphatase activity is indicated by red to reddish-brown deposits. Nuclei, green.

NCS METHOD FOR ARYLSULPHATASES A AND B

(after Goldfischer, 1965; Hopsu-Havu, *et al.*, 1967; Abraham, 1967)

FIXATION

This is a critical factor in the demonstration of lysosomal sulphatase. The original author (Goldfischer, 1965) recommended either perfusion-fixation with 3 per cent glutaraldehyde, followed by 3 hours' immersion fixation, or prolonged (1 week) fixation in Baker's (1946) cold formol-calcium, or overnight fixation in 2 per cent cacodylate-buffered glutaraldehyde.

Particularly for mammalian liver, Abraham (1967) fixed 1 mm. slices for 3–24 hours in two different cold (4°) fixatives:

(a) 2 or 5 per cent glutaraldehyde buffered to pH 7·4 with 0·1 M cacodylate.
(b) 4 per cent formaldehyde, similarly buffered.

After brief washing of the fixed tissues cut 8 μ cryostat sections directly (*via* the chute if provided) into 0·1 M cacodylate containing 7·5 per cent sucrose.

INCUBATING MEDIUM

Dissolve 160 mg. nitrocatechol sulphate in 4 ml. distilled water; add 12 ml. 0·1 M acetate buffer (pH 5·5) and 4 ml. 8 per cent lead nitrate. Adjust the pH to 5·5 with 0·2 M acetic acid.

Method. (1) Incubate free-floating sections for 1 hour at 37°.
(2) Rinse in distilled water.
(3) Treat with 1 per cent yellow ammonium sulphide for 2 minutes.
(4) Wash in distilled water.
(5) Mount in glycerine jelly.

Result (Plate XXIIIa). Black deposits appear in lysosomes containing arylsulphatase.

HYDROXYQUINOLINE METHOD FOR ARYLSULPHATASES

(after Woohsmann and Hartrodt, 1967)

FIXATION

Cold (4°) formol-calcium (Baker, 1946) followed by gum sucrose. Cut 8–10 μ sections directly into the incubation medium.

INCUBATION MEDIUM (Freshly prepared)

8-Hydroxyquinoline sulphate	50 mg.
NaCl	200 mg.
Veronal acetate buffer, pH 6·0	10 ml.
Hexazonium pararosanilin (fresh)	0·6 ml.

Adjust pH, to 5·8 to 6·0, with N-NaOH.

Method. (1) Incubate for 30–60 minutes at 37°.
(2) Wash in distilled water.
(3) Counterstain nuclei in 2 per cent methyl green for 2 minutes.
(4) Wash in running water.
(5) Dehydrate in alcohols, clear in xylene and mount in Canada balsam.

Result. Bright red staining indicates sites of arylsulphatase activity.

METHOD FOR GLUTAMATE DECARBOXYLASE

(after Higashi, *et al.*, 1960)

(*Fresh frozen, 20 μ, cryostat sections, mounted on coverslips*)

INCUBATION MEDIUM

Prepare three stock solutions, as follows:

Solution A. 0·4 per cent gum arabic.

Solution B. 0·8 g. $BaCl_2$ and 1·18 g. sodium L-glutamate dissolved in distilled water to 100 ml.

Adjust this solution to pH 6·6.

Solution C. 1·7 g. 2,4,5-trichlorophenol in 100 ml. ethanol.

Mix 5 ml. Solution A, 5 ml. Solution B and 0·1 ml. Solution C and add 5 mg. carbonic anhydrase and 1 mg. glutathione. The final concentrations are as follows:

$BaCl_2$	0·33 M
Sodium glutamate	70 mM
Gum arabic	0·2 per cent
2,4,5-trichlorophenol	0·86 mM
CAH	0·05 per cent
Glutathione	0·01 per cent

The addition of 3 mg. pyridoxal phosphate to the above medium intensifies the reaction considerably.

Method. (1) Incubate for 18 hours at 37° keeping sections in a stoppered chamber communicating directly with a flask containing barium hydroxide solution.
(2) Wash in distilled water.
(3) Immerse in 2 per cent lead nitrate for 5 minutes.
(4) Wash in distilled water.
(5) Immerse in yellow 1 per cent ammonium sulphide for 2 minutes.

(6) Wash in water.

(7) Dehydrate in alcohols, clear in xylene and mount in Canada balsam.

Result. Lead sulphide deposits indicate decarboxylase activity.

METHOD FOR CARBONIC ANHYDRASE

(after Häusler, 1958; Hansson, 1967; Cross, 1970)

Fixation

Two alternatives have been used. Either freeze-dried cryostat sections or cryostat sections post-fixed in cold acetone, preferably the latter, are recommended. In the second case the sections are delivered directly into the cold fixative *via* the cryostat shute. They are then fixed for 1½ to 2 hours in the same fixative, at room temperature.

Incubation Medium

Preparation of Incubating Medium

Solution A. To 1·0 ml. 0·1 M $CoSO_4$ add 6·0 ml. 0·05 M-H_2SO_4. (This solution is stable.)

Solution B. Dissolve 1·0 g. $NaHCO_3$ in 50 ml. 0·1 M-Na_2SO_4. (Freshly prepared before use.)

Before incubation pour solution B into solution A in order to avoid a momentary high concentration of cobalt ions.

If tissues with low CAH levels are under investigation the concentration of $CoSO_4$ can be raised by using a 0·125 M solution, as above.

Method. (1) Incubate floating sections for 60–120 minutes at room temperature. An improvized section lifter, made of angled paper, can be employed to ensure that each section floats on the surface of the medium.

(2) Wash in distilled water.

(3) Transfer to 1 per cent $(NH_4)_2S$ solution for 1 minute.

(4) Wash in tap water.

(5) Mount in glycerine jelly.

Result (Plate XXIIIb). A positive result is indicated by black or brown deposits of cobalt sulphide. Control sections incubated in the presence of acetazolamide (Diamox sodium salt) (1–5 mM) should show no staining.

SOLID MEDIUM METHOD FOR CARBONIC ANHYDRASE

(after Meijer and Bloem, 1966)

(*Fresh frozen, 7 μ, cryostat sections*)

Incubation Medium

Solution A. 1 part 0·1 M cobalt sulphate; 6 parts 0·05 M H_2SO_4. Stable indefinitely.

Solution B. 1 g. sodium hydrogen carbonate in 25 ml. 0·2 M sodium sulphate Prepare just before use.

Solution C. 1 g. of gelatin in 25 ml. distilled water.

Mix Solution C with 7 ml. Solution A. After mixing add 25 ml. Solution B. Pour the medium into Petri dishes and cool to 4° to produce a gel.

Method. (1) Lay the sections carefully on the surface of the medium.

(2) Warm the Petri dish to room temperature and allow incubation to proceed for 1–2 hours.

(3) Cool the Petri dish to 4° and cut out individual blocks containing single sections.

(4) Place blocks, section down, on coverslips. Apply gentle pressure.

(5) Transfer coverslips, block and section down, to distilled water at 70°.

(6) Rinse sections carefully with distilled water.

(7) Proceed as in previous method.

Result. Brown deposits of cobalt sulphide indicate enzyme activity.

METHOD FOR HYDROXYPROLINE-2-EPIMERASE

(after Onicesco, 1967)

(*Fresh frozen* (*6–8* μ) *cryostat sections, mounted*)

Preparation of Incubating Medium

The following concentrations of substrate, coenzyme and activators were found optimal by the original author:

L-hydroxyproline	20–30 mM
NAD	2 mM
Oxidized glutathione	5 mM
(alternatively) UDP	1 mM
Nitro-BT	2 mM
Phosphate buffer (pH 7·2)	0·2 M
$MgCl_2$	10 mM

Method. (1) Incubate sections for 30–45 minutes at 37°.

(2) Wash in distilled water.

(3) Fix in 10 per cent formalin.

(4) Mount in Apathy's medium.

Result. Blue formazan deposits, absent from sections incubated without substrate, indicate epimerase activity.

METHOD FOR URIDINE DIPHOSPHOGALACTOSE EPIMERASE

(after Diculescu, *et al.*, 1968)

(*Fresh frozen cryostat sections, 6–8* μ, *mounted*)

Incubation Medium

Uridine diphosphogalactose	2 mM
or Uridine diphosphoglucose	6 mM
NAD	5 mM
NaCl	40 mM
Tris-HCl buffer (pH 7·4)	0·1 M
Nitro-BT	10 mM

Method. (1) Incubate for 45 minutes at 37°.

(2) Fix in 10 per cent formalin for 30 minutes.

(3) Mount in Apathy's medium.

Result. Blue formazan precipitates indicate enzyme activity.

METHOD FOR GLUCOSE PHOSPHATE ISOMERASE

(after Meijer and Bloem, 1970)

(*Fresh frozen sections, 7 μ, mounted on slides or coverslips*)

INCUBATION MEDIUM

Prepare freshly immediately before use.

Disodium D-Fructose-6-phosphate	10 mg.
NADP (5 mg/ml.)	0·6 ml.
Nitro-BT (5 mg/ml.)	0·5 ml.
Imidazole buffer (0·15 M), pH 7·0	2 ml.
Gelatin (3 per cent)	3 ml.
G-6-P dehydrogenase (1 mg/ml.)	0·02 ml.

Adjust pH to 7·0 before use.

Method. Post-fix sections for 30 minutes in acetone at −25°.
(2) Incubate for 60 minutes at 37°.
(3) Fix in 4 per cent neutral formaldehyde for 30 minutes.
(4) Mount in glycerine jelly.

Result. Blue deposits indicate isomerase activity.

METHOD FOR UBIQUINONES (and related compounds)

(after Tranzer and Pearse, 1963)

(*Fresh frozen, 8 μ, cryostat sections, mounted*)

INCUBATION MEDIUM

MTT (1 mg/ml.)	0·25 ml.
Cobalt chloride (0·5 M)	0·05 ml.
Tris buffer (0·2 M, pH 7·4)	0·40 ml.
Hydroquinone (40 mg/ml.)	0·10 ml.
Distilled water	0·20 ml.
Catalase	2 mg.

Prepare freshly just before use.

Method. (1) Incubate for 10–30 minutes at 37°.
(2) Fix in 10 per cent formalin at 22° for 15 minutes.
(3) Wash in running water.
(4) Counterstain nuclei (if required) in 2 per cent methyl green.
(5) Wash, mount in glycerine jelly.

Result (Plate XXIIIc). Black formazan deposits indicate sites of redox activity due to ubiquinones, naphthoquinones, and related compounds.

HYDROQUINONE-FERRICYANIDE METHOD FOR UQ

(after Hirano and Ogawa, 1969)

(*Buffered formaldehyde or glutaraldehyde, cryostat sections*)

INCUBATION MEDIUM

0·1 M Phosphate buffer (pH 6·0)	6·5 ml.
0·1 M Hydroquinone	0·1 ml.

0·3 M Sodium citrate	0·5 ml.
30 mM $CuSO_4$	1·0 ml.
5 mM Potassium ferricyanide	1·0 ml.
Sucrose	0·8 g.

Method. This method was designed for application to small blocks for ultrastructural studies. It can be applied to 6–8 μ sections, for optical microscopy.

(1) Incubate for 1–2 hours at 37°.

(2) Wash in distilled water.

(3) Dehydrate in alcohols, clear in xylene and mount in a suitable synthetic medium.

Result. Brown precipitates of copper ferrocyanide indicate sites of redox activity (UQ and related quinones and tocopherols).

APPENDIX 24

GELATIN METHOD FOR PROTEASES

(after Adams and Tuqan, 1961)

PREPARATION

Fix tissues in cold (4°) 4 per cent formaldehyde-saline for 24 hours. Wash briefly in water and cut 10–15 μ cryostat sections. Mount sections directly on to a blackened photographic plate, cut to suitable size. Prepare this by exposing rapid panchromatic quarter-plate to daylight for 10–15 minutes. Develop in the usual way and harden in a thiosulphate-metabisulphite bath before drying.

INCUBATION

After mounting allow the slide to become just dry and then lightly damp the section, and a large area of the surrounding gelatin, with either 0·15 M phosphate buffer (pH 7·6) or 0·15 M acetate buffer (pH 5·0). Alternative buffer systems may be used, depending on the pH optimum of the protease in question. Specific inhibitors of individual proteases or protease groups may also be incorporated in the buffer. It is important that no free fluid be left on top of the section since this encourages diffusion of enzymes.

The preparation should now be incubated for 30–60 minutes, at 37°, in a moist chamber. A Petri dish containing wet filter paper is quite suitable. At 5-minute intervals the sections should be inspected to make sure that they have not become too dry. In such an event further buffer must be added.

After incubation the preparation is dried without washing and subsequently dehydrated in alcohol, cleared in xylene and mounted in a synthetic medium. Rapid dehydration and clearing are essential, in order to avoid crazing of the gelatin film.

Result. Protease activity is shown as a clear area where the gelatin film and its contained granules have disappeared. Histological, but not intracellular, definition is obtained.

DYED GELATIN METHOD FOR PROTEASES

(after Cunningham, 1967)

PREPARATION OF SUBSTRATE FILMS

Suspend 1·75 g. gelatin in 50 ml. distilled water and place in a refrigerator for 30 minutes. Dissolve the swollen gelatin by warming and agitating for 3–5 minutes at 56°. Use this solution within one hour.

Prepare films on standard (precleaned) microscope slides by the following procedure. Spread 0·2 ml. of gelatin solution over one half of a standard slide by tipping the latter in all directions, and by stirring and tapping with the end of the pipette. For thin films use 0·05 ml. of gelatin solution. Place the slide on a flat surface and dry slowly (12 hours). Load the dried films into staining racks and place in a freshly prepared 0·2 per cent solution of the stable diazotate Fast blue B (Vol. **1**, p. 712), in 40 mM veronal acetate buffer at pH 9·0 (Vol. **1**, p. 584). Leave for 15 minutes for thick FBB-15 films or for 45 minutes for thick FBB-45 and thin FBB-45 films. At 1–2 minute intervals detach gas bubbles from the films by drawing a metal spatula briskly across the upper edges of the slides. Wash slides in three changes of buffer, for 15 seconds in each.

Subsequently the colour of the films is intensified by treatment with a 2 per cent suspension of H-acid in veronal buffer at pH 9·0, for 5 minutes. Transfer the films through 0·1 N acetate buffer (pH 4·4), 5 minutes; 10 mM EDTA, 30 minutes; 0·1 N acetate buffer (pH 4·4), 5 minutes; distilled water, 5 minutes and, finally, a further bath of distilled water. Place slides, film surface upwards, on filter paper to dry. Remove excess, loosely bound, stain in 3 successive baths of 70 per cent ethanol, for 48, 3 and 2 hours, respectively. Dry slides once more.

Incubation of Tissue Sections

For preference cold formalin or formol-calcium fixed tissues, post-treated in gum-sucrose, are cut on the cryostat at 6–10 μ. Sections are mounted directly on top of the film which is taken from 2·5 mM phosphate buffer (pH 6·8) and blotted with filter paper to remove excess water. Incubation takes place in a wet box at 37° for up to 20 minutes. If no proteolytic activity can be demonstrated fresh frozen cryostat sections, mounted as above, should be used. According to the original author it is necessary to incubate fresh-frozen preparations in a fluid medium rather than in the moist chamber.

Incubation is terminated by fixation in acetic ethanol containing 10 per cent formalin. After one hour the film is sufficiently hard. Counterstaining of nuclei can be carried out. The Feulgen procedure is particularly recommended (see Vol. **1**, p. 647), with or without the PAS routine.

Finally, the slides are dehydrated in ethanol, cleared in xylene, and mounted in D.P.X.

Result. Clear areas in the stained film indicate sites of protease activity.

COLOUR FILM METHOD FOR PROTEASES

(after Fratello, 1968)

Preparation of Tissues

Fix tissues in cold formol-saline at 4° for 24 hours. Cut at 10–20 μ (cryostat) after washing. The cold gum sucrose routine (Vol. **1**, p. 602) may be interposed. Pick up sections directly on to a suitable colour film. The author recommended Ferrania 3M, daylight type DIA 28, film. Many other colour films are equally suitable. The picking-up procedure is facilitated by the use of a film holder, improvised according to taste. The film should be dampened with buffer, at a suitable pH, until the gelatin is soft but not wet.

Incubation

This takes place in a moist chamber at 37° (10–30 minutes). Following incubation the film is allowed to dry and, subsequently, it is mounted on a slide, on Canada balsam, and protected by a coverslip on a drop of the same medium.

Result. Progressive proteolytic digestion produces magenta, blue and white successively. Unaffected areas appear brown.

METHOD FOR FIBRINOLYSIN (PLASMINOGEN ACTIVATOR)

(after Todd and Nunn, 1967; Turner and Ryan, 1969)

Preparation of Films

The whole procedure is greatly facilitated by the construction of an apparatus illustrated below in diagrammatic form.

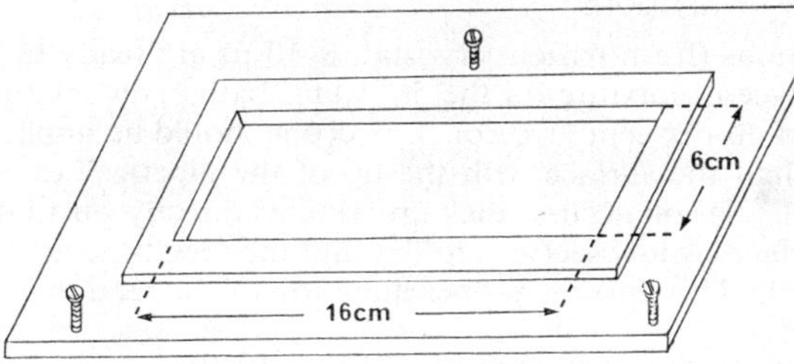

This apparatus consists of a sheet of plate glass with 3 levelling screws and a Perspex (Lucite) frame with a central aperture having the dimensions indicated. The frame is laid on the levelled plate to form a trough. The latter is lined with wettable Cellophane film,* which is moistened with veronal acetate buffer (pH 7·4) and then smoothed on to the plate with a bent glass rod.

A clotting mixture, formed by the addition of 0·1 ml. bovine thrombin† (25 units/ml. in veronal buffer) to 3·5 ml. bovine fibrinogen‡ (2 per cent solution in veronal buffer), is poured into the trough and distributed evenly with the bent glass rod. The apparatus is then left at room temperature for 30 minutes to promote stabilization of the clot. The film is then removed and cut into strips ready for use.

INCUBATION PROCEDURE

Fresh frozen cryostat sections (8 μ) are mounted on slides or coverslips. Each section is covered with a strip of the substrate film and incubated in a moist chamber at 4° for 18–24 hours. Proper adhesion of the film to the section can be promoted by the interposition of a piece of plastic foam between it and the under surface of the lid of the Petri dish. Inspection is carried out at intervals and the development of clear areas of lysis in the opaque medium is noted. When this process appears to be sufficient the sections are fixed in 10 per cent formalin for 1 hour. The cellophane is then removed and fixation continued for a further 12–18 hours. If no lysis is found in cold incubated sections, the procedure is carried out at 37°.

After fixation the preparations are rinsed in tap-water, counterstained with haematoxylin (no acid), washed in water and allowed to dry thoroughly. They are then quickly dehydrated in acetone, cleared in xylene and mounted in D.P.X. medium.

Result. Pale or clear areas in the stained fibrin film indicate regions with plasminogen activator activity.

METHOD FOR DEOXYRIBONUCLEASES
(after Daoust, 1961)

PREPARATION OF DNA FILMS

Heat on a water-bath at 56° a 1 : 1 mixture of 5 per cent gelatin and 0·2 per cent aqueous DNA. Apply one drop to the surface of a precleaned glass slide and spread rapidly with the tip of the pipette. Place the slide on a level surface, at room temperature, to allow the film to set at a uniform thickness. Allow to dry completely. Fix the film by standing overnight in 10 per cent formalin at room temperature. This procedure renders the DNA insoluble in water. Wash the fixed film in 3 successive baths of distilled water, 15 minutes each, and allow to dry once more.

* PT 400, British Cellophane Ltd.
† Topical Thrombin, Parke Davis.
‡ Bovine Plasma Fraction I, Armour.

MOUNTING OF TISSUE SECTIONS

When the sections (fresh-frozen, cryostat, 8–10 μ) are ready to be cut, prepare a warm gelatin-glycerol mixture on the 56° water-bath. This mixture consists of 7 per cent gelatin in 40 per cent glycerol. Ten drops should be applied to each slide and spread over half the surface with the tip of the pipette. Leave the slide on a level surface until gelation occurs. Pick up sections directly on to the gel from the cryostat knife surface. Allow sections to dry and then melt the gel once more in an incubator at 45–50°. This ensures proper adhesion of the section.

EXPOSURE OF FILMS TO SECTIONS (See Fig. 213, p. 1010)

Place the slide supporting the section against a slide supporting the DNA film and hold the two together by means of paper clips. Leave at room temperature for 15–90 minutes. Separate the two slides. Wash the DNA film in two changes of distilled water and allow to dry. Stand the section slide in 10 per cent formalin for 18 hours.

STAINING OF FILMS AND SECTIONS

Stain the DNA film in 0·2 per cent toluidine blue for 5 minutes. Rinse and allow to dry. Mount under a coverslip with Canada balsam.

Treat the section slide with 1 : 3 acetic ethanol for 5 minutes. Transfer directly to 0·2 per cent toluidine blue for 1–2 minutes. Rinse in water. Allow to dry and mount in Canada balsam.

Result. Comparison of the two slides, with photography where necessary, allows the localization of toluidine blue negative areas in the DNA film to be correlated with identifiable components in the tissue sections.

METHOD FOR RIBONUCLEASES

(after Daoust, 1966)

PREPARATION OF SUBSTRATE

Chill a 10 per cent aqueous solution of sodium ribonucleate to 4° and precipitate with 2 volumes of chilled absolute ethanol (4°). Collect the precipitate by centrifugation, dry *in vacuo*, and grind with a pestle in a mortar.

PREPARATION OF RNA FILMS

Prepare a 5 per cent solution of gelatin, by heating to 60°. Add an equal volume of 10 per cent RNA (purified substrate) at the same temperature. Filter at 60°.

Place 3–5 drops of the gelatin-RNA mixture on a glass slide and spread over an area of about 2·5 × 5 cm. Drain off excess by holding the slide vertically on a piece of filter paper. Dry at room temperature.

EXPOSURE OF FILMS TO SECTIONS

Cut one 10–15 μ fresh frozen cryostat section and mount directly on to the gelatin-RNA film. Cut the serial section at 6–8 μ and mount on a precleaned slide.

Place the RNA film slide in a Petri dish containing wet filter paper and incubate at 20–22° for 1 minute to 2 hours. After exposure flush the section off the surface of the film and place the latter in neutral 20 per cent formalin for 1 hour at room temperature. Wash in three changes of distilled water, 5 minutes each, and then allow to dry.

STAINING OF FILMS AND SECTIONS

Stain the RNA films for 5 minutes in 0·2 per cent toluidine blue. Wash in distilled water, dry in air and mount in Canada balsam.

Stain tissue sections in 0·1 per cent toluidine blue for 2 minutes, wash in water, dry and mount as above.

Result. Clear regions indicate RNase activity. Comparison with the stained serial section can be made when necessary. Localization is superior to that afforded by the preceding method.

MODIFIED STARCH FILM FOR AMYLASE

(Tremblay and Charèst, 1968)

PREPARATION OF STARCH FILMS

Heat a 4 per cent suspension of starch (Hydrolysed Starch, Connaught Medical Research Laboratories, Toronto) to 100°, on a water bath, for 15 minutes. Filter while hot through glass wool and gauze. Deposit 8–12 drops of this preparation on a precleaned glass slide and spread rapidly over a surface of approximately 2·5 × 4 cm. with the tip of the pipette. Place the slide vertically on a piece of filter paper and allow excess of the suspension to drain off. Allow the film to dry at room temperature and then fix overnight in methanol-acetic acid-water (5 : 1 : 5). Wash in 3 baths of distilled water and dry at room temperature.

EXPOSURE OF STARCH FILM TO TISSUE SECTIONS

Cut a single fresh frozen cryostat section, 4–8 μ thick, and mount directly on the starch film. Prior to mounting the film should be kept for 2 hours at 22° in a moist chamber. Cut the serial section, at 4 μ, on to a clean slide. Incubate the first section at 22° for 30 seconds to 5 minutes in a moist chamber. After incubation flush the section off the film with a stream of distilled water. Immerse film in the methanol-acetic fixative for 15 minutes. Stain by the PAS routine (Vol. **1**, p. 660). Wash, dry and mount in a suitable synthetic medium.

Stain the adjacent serial section as required. Dehydrate, clear and mount.

Result. Clear areas in a magenta field indicate amylase activity.

METHOD FOR HYALURONIDASES

(after McCombs and White, 1968)

PREPARATION OF SUBSTRATE FILMS

Prepare a solution of sodium hyaluronate (5 mg. in 0·8 ml. distilled water) and add to 0·4 ml. of 12 per cent gelatin at 60° on a water bath. Agitate the mixture for 2 minutes and, while still warm, use a 0·2 ml. pipette to place 1 drop in the centre of an acid-cleaned slide. The resulting round patches of substrate film average 0·5 to 1·0 cm. in diameter. Allow the slides to dry for 24 hours in dust-free, covered, slide trays. Fix in 40 per cent formaldehyde for 1 hour. Rinse thoroughly in several changes of distilled water and allow to dry for several hours before use. These preparations keep indefinitely.

EXPOSURE TO TISSUE SECTIONS

Cut fresh frozen 10 μ cryostat sections and mount directly on hyaluronate slides, leaving a small part of the hyaluronate film uncovered. Prepare a gelatin-glycerol mixture, as described on p. 577. Maintain at a temperature just hot enough to maintain this in the liquid state. Add sufficient gelatin-glycerol to cover both section and uncovered substrate film. Incubate for 3 minutes to 24 hours (optimum 5 minutes) at 25°. Gently remove the coating mixture by immersion with gentle agitation in lukewarm tap water (30–60 seconds).

Fix slides for 10 minutes in 40 per cent formaldehyde. Rinse in distilled water and stain for 3 minutes in 0·05 per cent aqueous toluidine blue buffered to pH 4 with phosphate buffer. After staining rinse in distilled water, dehydrate rapidly in absolute ethanol, clear in xylene and mount in a synthetic resin.

Result. Clear areas in the purple-violet background indicate sites of hyaluronidase activity.

METHOD FOR LYSOZYME IN LEUCOCYTES

(after Ghoos and Vantrappen, 1970)

Preparation of Substrate Film

Prepare a concentrated solution of lithium iodide (120 g. per 100 ml.), bring to 80° and add chitin, with constant swirling until excess becomes apparent. Maintain at 80° for 24 hours. If the solution crystallizes add an aliquot of water. Filter to remove solid particles. Smear the yellow-brown solution on to hot (80°) acid-cleaned microscope slides and keep at 80° for at least two hours. Cool to room temperature. The solution solidifies leaving the chitin evenly distributed in a lithium iodide network.

To eliminate the latter place the slides in a horizontal position and cover with ethanol for 20 minutes. Remove ethanol and cover the slides with 5 per cent potassium iodate in 0·2 N-H_2SO_4. This procedure results in the precipitation of iodine-chitin on the surface of the glass. Remove surplus iodine by sublimation at 80°. Further purify by a 30 minute wash in ethanol and washing in water overnight at 50°.

Exposure of Films to Leucocyte Smears

Smear buffy coat preparations on to the chitin film slides. Incubate in a moist chamber for 1 hour at 38–40°. Fix for 2 minutes in absolute methanol.

Staining Procedure

Allow slides to dry (from methanol) and stain for 8 minutes in Alcian blue 8GS (400 mg. in 100 ml. water, brought to pH 4 with acetic acid). Counterstain nuclei with Nuclear fast red (100 mg. in 100 ml. 5 per cent aluminium sulphate, pH 3·5) for 5 minutes. Rinse in water, dehydrate in alcohols, clear in xylene, and mount in D.P.X.

Result. A blue colour, concentrated in the cytoplasm especially of neutrophil leucocytes, indicates activity of the enzyme. Nuclei reddish-brown.

METHOD FOR CATALASE

(after Hale, 1965)

Preparation of Starch Film

Prepare films of hydrolysed starch, in the manner described on p. 1373, on a series of clean slides.

Exposure of Films to Sections

Fix tissues in cold 4 per cent formaldehyde (prepared freshly from paraformaldehyde) overnight. Wash and treat with gum sucrose (Vol. **1**, p. 602). Cut cryostat sections directly on to the starch films. Dry thoroughly in air, at room temperature.

Immerse slides gently in 10 mM H_2O_2 for 2 hours at 4°, in a suitable container. Remove the hydrogen peroxide with the minimum disturbance and replace by 2 per cent KI. Leave in this solution for 15–120 minutes at room temperature. Replace KI solution with distilled water. Remove slide and apply coverslip directly.

Result. Pale or clear areas indicate concentrations of catalase activity. The original author noted (with his separation technique) that certain tissue elements in the separated section were coloured blue. He suggested that these might be sites of peroxidase activity. This assumption was based on the function of KI as a substrate for oxidation to iodine by peroxidase. Subsequent reaction with starch would produce the blue colour observed.

APPENDIX 25

RIBONUCLEASE EXTRACTIONS

(*Carnoy; paraffin sections; cold microtome sections, post-fixed in acetic-ethanol*)

Method

(1) Bring paraffin sections to water.

(2) Incubate for 1 hour at 37° in a solution of the enzyme (0·5 to 1 mg/ml.)* in glass-distilled water.

(3) Wash in running water.

(4) Stain the treated and a control section by whatever method is desired. (For choice of methods for staining the sections see text, Chapter 9).

Result (see Figs. 88 and 89)

Basophilic material removed by this treatment is to be regarded as ribonucleic acid.

DEOXYRIBONUCLEASE EXTRACTIONS

(*Carnoy; paraffin sections. Acetic-ethanol, cold microtome sections*)

Method

(1) Bring paraffin sections to water.

(2) Treat sections for 3–6 hours at 37° with crystalline DNase (0·05 mg/ml.) or with streptodornase (1,000 units/ml.) in 25 mM-Veronal buffer at pH 7·5 containing 0·2 M-$MgSO_4$. Take care not to shake the enzyme solution. This hastens its destruction.

(3) Wash in water.

(4) Dehydrate in alcohol and pass to alcohol/ether.

(5) Cover with a thin film of 1 per cent celloidin.

(6) Perform the standard Feulgen reaction (p. 648).

Result

Control slides show staining of nuclear chromatin, treated slides should not do so.

HYALURONIDASE EXTRACTIONS

(*Formalin, alcohol, etc.; paraffin sections. Acetic ethanol, cold microtome sections*)

Method

(1) Bring paraffin sections to water.

(2) Incubate for 3 hours at 37° in a solution of either of the above-mentioned preparations (1 mg/ml.) in 0·85 per cent saline.

(3) Wash thoroughly in distilled water.

(4) Stain in ½ per cent aqueous toluidine blue, 20 minutes.

(5) Rinse in distilled water, mount in water and compare with a control section incubated at 37° for 3 hours in saline alone, and stained as above.

* As low as 0·1 mg/ml. is often sufficient.

Result

For interpretation of hyaluronidase-lability see text, Chapter 25, p. 1027.

TRYPSIN EXTRACTIONS

(*Carnoy; smears or paraffin sections*)

Method

(1) Bring sections to water.

(2) Incubate for 15–60 minutes at 37° in 0·05 M phosphate buffer at pH 8·9, containing 0·1 mg/ml. purified trypsin. Incubate control section in buffer without enzyme. As an alternative 0·2 mg/ml. in 50 mM-borate buffer can be employed.

(3) Wash both sections in water.

(4) Treat by whatever staining method is considered desirable (see text, Chapters 6, 8 and 9).

Result

Histone and non-histone proteins alike are degraded, and removed by the subsequent washing in water.

PEPSIN EXTRACTIONS

(*Smears or paraffin sections after Carnoy fixation*)

Method

(1) Bring sections to water.

(2) Incubate for 2–3 hours at 37° in 0·02 N HCl at pH 1·6, containing 2 mg/ml. crystalline pepsin. Incubate control section in acid only.

(3) Wash both sections in water.

(4) Treat by whatever staining method is considered desirable (see text, Chapter 25, p. 1040).

Result

Kaufmann *et al.* (1950) regarded the shrinkage observed in tissues after treatment with pepsin as due to removal, primarily, of trytophan-containing proteins.

PECTINASE DIGESTION (after McManus and Cason)

(*Acetone; paraffin sections*)

Pectinase Solutions

McManus used Pectinase (Nutritional Biochemicals Corp.) or Pectinol O (Rohm and Hass). A commercial preparation, containing pectinesterase and polygalacturonase, available in the U.K., is Pectozyme (Norman Evans and Rais Ltd.).

Dissolve 0·8 g. of the enzyme in 100 ml. of acetate buffer at pH 4·2. The buffer solution alone should be used for control sections.

Method

(1) Bring sections to distilled water via light petroleum and absolute acetone.

(2) Incubate for up to 48 hours in the enzyme solution.

(3) Wash thoroughly in tap water.

(4) Stain buffer control and test sections by the selected procedure (usually the PAS method).

Result

PAS-positive materials are removed from animal tissues, presumably by the polygalacturonase present in pectinase. This action is reversibly blocked by previous acetylation of the sections (see Appendix 6, p. 613 for details of acetylation).

NEURAMINIDASE EXTRACTIONS

(*Cold formalin, cryostat or paraffin sections, FDFV*)

Method

(1) Bring sections to water and incubate with neuraminidase (**Vibrio cholerae**) (500 units/ml.), diluted with an equal volume of 0·1 M-acetate buffer, containing approximately 1 per cent NaCl and 0·1 per cent $CaCl_2$, at pH 5·5, for 16–24 hours at 39–41°.

(2) Incubate control sections in 0·05 M buffer, as above.

(3) Wash carefully in distilled water.

(4) Stain with the Alcian blue-PAS procedure (Appendix 10, p. 673) or with the high iron diamine-Alcian blue method (Appendix 10, p. 664).

(5) Dehydrate, clear, and mount in a synthetic medium.

Result

Comparison of control and test sections indicates removal of sialomucins. (Change from blue to red with AB/PAS or from purple to brown with HID/AB).

COLLAGENASE EXTRACTIONS

(*Ethanol, Carnoy, paraffin sections*)

Method

(1) Bring sections to distilled water (3 changes).

(2) Cover sections with collagenase (1 mg/ml.) dissolved in 0·9 per cent NaCl.

(3) Incubate in a moist chamber at 50° for 45–60 minutes.

(4) Incubate serial control sections, in saline only, under the same conditions.

(5) Wash in distilled water.

(6) Stain with Gordon and Sweet's reticulin stain (Appendix 8, p. 641) or with Masson's trichrome.

Result

Collagen and reticulin should be totally absent from collagenase-treated sections.

APPENDIX 26

MASSON-FONTANA METHOD FOR MELANIN (WITH HEXAMINE-SILVER VARIANT)

(*various, not chromate; paraffin sections*)

Silver Solutions

(1) Fontana

To 20 ml. of 10 per cent $AgNO_3$ add strong ammonia drop by drop until only a few granules of the first formed precipitate remain. Add 20 ml. of distilled water and allow to settle for 24 hours. Decant into a dark bottle. Filter each jar before use and do not use for more than a few sections. The solution keeps for about a month.

(2) Hexamine-silver (Gomori)

To 100 ml. of 3 per cent hexamine add 5 ml. of 5 per cent $AgNO_3$. A precipitate forms and redissolves. Add 5 ml. of borate buffer of approximately pH 8. (Add a drop of phenolphthalein to 3 per cent boric acid and titrate with N NaOH until just pink.) Make up to 200 ml. with distilled water.

This M/5 borate buffer tends to crystallize. Burtner and Lillie (1949) recommend the substitution of boric acid-borate buffer (pH 7·8), 25 ml. per 100 ml. hexamine silver solution (Appendix 1).

Method

(1) Bring paraffin sections to water.
(2) Treat with Gram's iodine solution for 10 minutes and wash in running distilled water.
(3) Leave sections for 18–24 hours in one of the two silver solutions given above, in the dark in a covered jar.
(4) Rinse in distilled water.
(5) Fix in 5 per cent sodium thiosulphate, 1–2 minutes.
(6) Counterstain nuclei with 1 per cent carbol safranin,* 10 minutes.
(7) Differentiate in 70 per cent alcohol.
(8) Dehydrate rapidly in alcohol, clear in xylene and mount in Canada balsam.

Result (Fig. 219)

Melanin granules appear black, nuclei red, keratin orange.

DIAMMINE SILVER STAIN

(after Lillie, 1965)

(*formalin-fixed, paraffin sections*)

Diammine Silver

Place 2 ml. 28 per cent ammonia in a 100 ml. beaker. Add about 35 ml. 5 per cent silver nitrate, quickly at first, shaking between each addition to dissolve the dark brown silver oxide and, cautiously, the last 5 ml. Add just enough silver nitrate to produce a faint permanent turbidity.

* Melt 0·5 g. phenol in a dry flask, add 0·2 g. safranin and mix well. Dissolve the sludge in 50 ml. starch-dextrin solution. (Grind 0·25 g. starch with 0·25 g. dextrin and add 50 ml. of distilled water while continuing to grind. Heat to 80°, cool and filter.)

Method

(1) Bring paraffin sections to water (distilled).

(2) Immerse sections for 15 minutes to 2 hours at 22° in the dark. (For bile pigments, at 60° for 30 minutes).

(3) Wash in distilled water and immerse for 5 minutes in 0·2 per cent gold chloride ($HAuCl_4$).

(4) Rinse in distilled water and place in 5 per cent sodium thiosulphate for 2 minutes.

(5) Wash in running water for 10 minutes.

(6) Counterstain, if required. Mount in Apathy or glycerine jelly.

Result

Cutaneous and ocular melanins stain strongly in 2 minutes. Other melanins require longer treatment.

SILVER METHOD FOR NORADRENALIN

(after Tramezzani *et al.*, 1964)

(*Glutaraldehyde, frozen sections*)

Silver Solution

To 100 ml. of 10 per cent silver nitrate add 300 ml. of 5 per cent sodium carbonate. A precipitate forms. Add conc. ammonia solution drop by drop until only slight turbidity remains. Add 100 ml. distilled water. Alternatively, use the Fontana solution described above.

Method

(1) Fix thin tissue slices in 3 per cent glutaraldehyde in 0·1 M phosphate buffer (pH 7·6) for 12–18 hours.

(2) Wash in distilled water for 10 minutes.

(3) Cut frozen (cryostat) sections directly into distilled water.

(4) Treat with alkaline silver solution for 20 seconds.

(5) Rinse in distilled water.

(6) Treat with 1 per cent sodium thiosulphate for 2 minutes.

(7) Wash in tap water.

(8) Dehydrate, clear and mount in Canada balsam.

Result

Noradrenalin-containing granules stain black.

THE FERROUS IRON TECHNIQUE FOR MELANIN

(Lillie, 1957)

(*Formalin, Carnoy, etc., not chromate; paraffin section*)

Method

(1) Bring paraffin sections to water.

(2) Immerse for 1 hour in 2·5 per cent ferrous sulphate ($FeSO_4.7H_2O$)

(3) Wash for 20 minutes in 4 changes of distilled water.

(4) Immerse for 30 minutes in 1 per cent potassium ferricyanide in 1 per cent acetic acid.

(5) Wash in 1 per cent acetic acid.
(6) Counterstain, if desired, in Van Giesons picric-acid fuchsin.
(7) Dehydrate, clear and mount in synthetic resin.

Result

Melanins, dark green; background faint green or colourless (unless counterstained).

BLEACHING METHODS FOR MELANIN

(1) Permanganate

Treat sections with 0·1 per cent $KMnO_4$ for 12–24 hours, rinse in water and treat with 1 per cent oxalic acid for 1 minute.

(2) Chlorate

Add a small quantity of $KCl.O_4$ to a Coplin jar containing 40–50 ml. of 50 per cent alcohol. Add a few drops of conc. HCl. Immerse sections for 24–48 hours. Wash.

(3) Bromide

Treat sections with 1 per cent bromine water for 12–24 hours.

(4) Chromic Acid

Treat sections with a mixture of equal parts of 1 per cent chromic acid and 5 per cent $CaCl_2$ for 8–12 hours.

(5) Peroxide

Treat sections with 30 vol. (10 per cent) hydrogen peroxide for 24–48 hours.

(6) Peracetic acid

Treat sections with 40 per cent peracetic acid for 2–16 hours.

Of all these methods the last is probably the best. Most melanins in paraffin sections are bleached within 16 hours and if the sections are affixed with egg

TABLE 92

Bleaching Times for Different Melanins

	Skin Melanin	Tricho-xanthin	Ocular Melanins	Neuro-melanin
0·25% $KMnO_4$ (25°)	20 min.	0–5 min.	2–4 hr.	1–5 min.
70% HNO_3 (25°)	6–8 hr.	30 sec.	3–4 hr.	4–6 min.
5% CrO_3 (25°)	2 hr.	15 min.		1 hr. 40°
1% H_5IO_6 (60°)	4 hr.	Unchanged 4 hr.		
Bromine Water	8 hr.	3 hr.		
Peracetic acid (25°)	2–12 hr.	1 hr. 3°		

albumin to the slides no harm is done to the tissues. Other pigments (e.g. lipofuscins) tend to resist for longer than 16 hours. The permanganate and chromic acid routines bleach all pigments and the rate of bleaching does not allow any differentiation to be made between one pigment and another. Table 92, derived from Lillie (1963) gives further details of bleaching routines, including those for trichoxanthin.

CHROMAFFIN REACTION FOR ADRENALIN AND NORADRENALIN

(Hillarp and Hökfelt)

(*Fresh tissues*)

Method

(1) Treat thin slices of tissue with a solution containing 10 vols. 5 per cent $K_2Cr_2O_7$ and 1 vol. 5 per cent K_2CrO_4 (pH 5·6). Leave for 16 hours at room temperature.

(2) Wash in distilled water for 30–60 minutes; 3 changes.

(3) Mount in glycerine jelly after transferring sections to clean slides.

(4) Alternatively sections can be dehydrated, cleared and mounted in Canada balsam.

Result

Adrenalin-containing cells stain dark brown; noradrenalin usually stains more lightly (yellow or yellowish-brown).

IODATE METHOD FOR NORADRENALIN

(after Hillarp and Hökfelt, 1955)

(*Fresh tissues*)

Method

(1) Transfer thin (0·5 mm.) slices of adrenal gland to 10 per cent aqueous potassium iodate. Leave for 16 hours at room temperature.

(2) Transfer to 10 per cent formalin for 2 hours.

(3) Cut frozen sections 10–20μ thick.

(4) Wash well in distilled water.

(5) Counterstain nuclei if required.

(7) Mount sections on slides; dehydrate, clear and mount finally in Canada balsam.

Result (Fig. 221)

Cells which contained noradrenalin show a diffuse brown staining which, in counterstained preparations, affects nuclei as well as cytoplasm.

MODIFIED GIEMSA METHOD FOR ADRENOCHROME (CHROMAFFIN)

(after Sevki)

(*Müller, Orth, Zenker etc., paraffin sections*)

The original author fixed his material in formalin and his results differed considerably from those which are obtained with chromate-fixed material. Whereas after formalin fixation the granules of the adrenal medullary cells are acidophilic and stain rose-red with Giemsa, after chromate fixation they are moderately basophilic and take up some of the azure component.

Method

(1) Bring sections to distilled water and leave for 30–60 minutes.
(2) Stain in dilute Giemsa (1 drop to 1 ml. distilled water) for 24 hours.
(3) Rinse in distilled water.
(4) Differentiate quickly in 90 per cent alcohol.
(5) Pass through absolute alcohol to xylene and mount in Canada balsam.

Result

Adrenochrome-containing (chromaffin) granules stain greenish-yellow, nuclei dark blue, cytoplasm of other cells, pink to red. (Similar results can be obtained by brief staining with methylene blue except that the cytoplasm of other cells is unstained.)

Chromaffin granules have moderately strong reducing properties and they will progressively darken if kept in ammoniacal silver solutions. Ferricyanide is quickly reduced and Schmorl's test (see below) is therefore positive.

THE SCHMORL METHOD FOR LIPOFUSCIN

(*Formalin; frozen or paraffin sections*)

Preparation of Ferricyanide Solution

Mix three parts of 1 per cent ferric chloride or ferric sulphate and one part of freshly prepared 1 per cent potassium ferricyanide. Use within 30 minutes.

Method

(1) Bring sections to water.
(2) Immerse in ferricyanide solution 5 minutes.
(3) Wash in running water.

For strictly histochemical investigations the section should be examined at this stage in water.

(4) Counterstain nuclei in 1 per cent neutral red, 3 minutes.
(5) Dehydrate rapidly in alcohol, clear in xylene and mount in Canada balsam, or in a suitable synthetic resin.

Result (Fig. 225 and Plate XXIVd)

Substances which reduce ferricyanide to ferrocyanide appear dark blue. These include melanin, argentaffin granules and lipofuscin as well as tissue components containing active sulphydryl groups.

METHOD FOR DISTINGUISHING MELANINS FROM LIPOFUSCINS

(Hueck, 1921

(*Various fixatives; Paraffin sections*)

Method

(1) Bring sections to water.
(2) Stain in a freshly prepared saturated aqueous solution of Nile Blue sulphate, 30 minutes.

(3) Rinse in distilled water.
(4) Transfer to 3 per cent (10 per cent is better) H_2O_2 for 24 hours.
(5) Wash in running water.
(6) Mount in glycerine jelly or corn syrup.

Result

Lipofuscins, blue. Melanins, colourless.

ALTERNATIVE NILE BLUE METHOD

(Lillie, 1956)

(*Various fixatives; paraffin sections*)

Method

(1) Bring paraffin sections to water.

(2) Stain for 20 minutes in 0·05 per cent Nile Blue A in 1 per cent sulphuric acid.

(3) Wash in running water for 10–20 minutes.

(4) Alternatively rinse quickly in 1 per cent sulphuric acid and dehydrate in 4 changes of absolute acetone.

(5) Mount from stage 3 into glycerine jelly and from stage 4, through xylene, in a synthetic resin.

Result

If stage 3–5 is employed lipofuscins stain dark blue or green, melanins dark green.

If stage 4–5 is employed melanins stain dark green and lipofuscins appear in their original colours.

THE CHROME ALUM HAEMATOXYLIN METHOD FOR LIPOFUSCINS

(after Gomori, 1941)

(*Formalin, etc.; paraffin sections*)

Preparation of the Staining Solution

Mix equal parts of a fresh 1 per cent aqueous haematoxylin and 3 per cent chrome alum. Add to each 100 ml. of the mixture 2 ml. 5 per cent potassium dichromate and 1 ml. 5 per cent sulphuric acid. Allow the mixture to ripen for 2–3 days before use. (The solution retains its staining properties for 4–8 weeks.)

Method

(1) Bring sections to water.

(2) Oxidize sections for 1–2 minutes in acid permanganate solution (47·5 ml. 0·5 per cent $KMnO_4$ and 2·5 ml. 3 per cent sulphuric acid).

(3) Bleach for 1 minute in 1 per cent oxalic acid and wash in water.

(4) Stain in chrome alum haematoxylin for 10 minutes.

(5) Differentiate in 1 per cent acid alcohol.

(6) Counterstain in 1 per cent aqueous eosin, 5 minutes.

(7) Wash in running water.

(8) Dehydrate rapidly in alcohol, clear in xylene and mount in DPX.

Result

Lipofuscins, blue-black. Nuclei, purple. Cytoplasmic structures, pink or red.

LONG ZIEHL-NEELSEN METHOD FOR ACID FAST LIPOFUSCINS

(*Formalin, etc.; paraffin sections*)

Method

(1) Bring sections to water.

(2) Stain in carbol fuchsin solution for 3 hours at 60°.

Basic fuchsin	10 g.
Phenol	50 g.
Alcohol	100 m.
Distilled water	1,000 ml.

(3) Wash in running water.

(4) Differentiate in 1 per cent acid alcohol until the red cells are just faint pink.

(5) Counterstain lightly with Mayer's haemalum or 0·5 per cent toluidine blue.

(6) Wash in running water.

(7) Dehydrate in alcohol, clear in xylene, mount in DPX.

Result

Acid fast lipofuscins, bright red. Lipoproteins, pink. Nuclei, dark blue or blue.

If counterstaining is too strong the basophilia of the lipofuscin will lead to development of a purple colour in positive lipofuscins. If a cytoplasmic counterstain is desired, the best is Lendrum's (1947) saturated tartrazine in cellosolve.

INDOPHENOL METHOD FOR LIPOFUSCINS

(after Alpert *et al.*, 1960)

(*Formalin-fixed, or fresh, frozen sections: paraffin*)

Preparation of the Stain

Immediately before use, prepare a 0·1 per cent solution of sodium 2,6-dichlorophenolindophenol in 50 per cent ethanol. Add 1 per cent HCl until the solution becomes red (approximately pH 2).

Method

(1) Sections should be mounted on slides; deparaffinize if necessary.

(2) Treat with acid indophenol stain for 5 minutes.

(3) Wash by dipping in tap water until background is clear.

(4) Mount in glycerine jelly.

Result

Lipofuscin red. Erythrocytes blue. Other tissues colourless, blue or faintly pink. (Fades in 24 hours.)

DAM METHOD FOR FAT PEROXIDES

(Glavind *et al.*, 1949)

(*Fresh frozen, formalin-fixed frozen sections*)

Preparation of Solutions

Solution A. 40 mg. haemin.
10 ml. pyridine.
20 ml. glacial acetic acid.

(This solution is stable.)

Solution B. Just before use, dissolve 25 mg. leuco-3 : 5-dichloro-4 : 4′-dihydroxy-phenylamine in 3·5 ml. absolute alcohol and add 5 ml. distilled water.

(The leuco compound is prepared by reducing 2 : 6-dichlorophenolindophenol with the equivalent of ascorbic acid in 50 per cent alcohol. The sodium salt is precipitated with NaCl and purified by repeated solution in alcohol and precipitation by the addition of water.)

Method

(1) Mount fresh frozen sections on slides.
(2) Fix, if desired, for 5 minutes in 40 per cent formalin, 1 part, 40 per cent alcohol, 3 parts.
(3) Wash in water, remove and blot dry.
(4) Mix 0·74 ml. Solution A and 8·5 ml. Solution B and apply for 3–5 minutes.
(5) Wash in distilled water.
(6) Mount in glycerine jelly.

Result

Fat peroxides appear deep red.

DIAMINOFLUORENE METHOD FOR FAT PEROXIDES

(Feigl, 1954)

(*Fresh sections*)

Method

(1) Mount 15μ fresh frozen sections on slides or coverslips.
(2) Cover section with 2 or 3 drops of the reagent* and apply coverslip.
(3) Examine under the microscope.

Result

A violet, blue or greenish-blue colour constitutes a positive reaction.

DIPHENYLCARBAZIDE REACTION FOR HYDROXY FATTY ACIDS

(Feigl, 1954)

(*Fresh sections*)

Method

(1) Treat fresh frozen sections with 0·5 per cent diphenylcarbazide in 50 : 50 alcohol-ether.
(2) Rinse in alcohol.
(3) Bring to water and mount in glycerine jelly.

Result

Hydroxy fatty acids give a red colour.

* Dissolve 100 mg. 2, 7-diaminofluorene and 5 mg. haemin in 5 ml. acetic acid. This reagent must be freshly prepared.

PHLOROGLUCINOL METHOD FOR EPIHYDRINALDEHYDES

(Feigl, 1954)

(*Fresh sections*)

Method

(1) Mount fresh sections on slides or coverslips. Store in air for 7–28 days.

(2) Immerse in 0·1 per cent phloroglucinol in 96 per cent alcohol containing 10 per cent HCl. Incubate at 40° for 15 minutes.

(3) Rinse in alcohol and bring to water

(4) Mount in glycerine jelly.

Result

Epihydrinaldehydes give a deep red colour.

BENZIDINE METHOD FOR HAEMOGLOBIN

(after Pickworth, 1934–35)

(*Formalin etc.; paraffin sections*)

Preparation of Solutions

Benzidine. Dissolve 0·2 g. benzidine and a small (5 mm.^{3}) crystal of sodium nitroprusside in 15 ml. methanol. Add 4 drops of glacial acetic acid and shake.

Ozonic Ether.	3 per cent H_2O_2	50 ml.
	Methanol	100 ml.
	Ether	50 ml.

Method

(1) Bring sections to water.

(2) Rinse in methyl alcohol.

(3) Invert the slide over benzidine solution on a staining tile. Leave for 5–10 minutes.

(4) Wash off benzidine with ozonic ether.

(5) Leave slide for 10 minutes in fresh ozonic ether.

(6) Wash in running water 10–15 minutes.

(7) Counterstain nuclei in 1 per cent aqueous neutral red, 3 minutes.

(8) Wash, dehydrate in alcohol, clear in xylene and mount in DPX.

Result

Haemoglobin and certain "oxidase" granules in leucocytes, dark blue. Nuclei, red.

FLUOROGENIC METHOD FOR HAEMOPROTEINS

(after Granick and Levere, 1965)

(*Air-dried smears; monolayer cultures*)

Preparation of the Reagent

Add to 87·2 ml. H_2O, 4·5 g. mercaptoethylamine hydrochloride and 12·8 ml. 70 per cent $HClO_4$.

Method

(1) Flood slides with 0·2 ml. reagent solution and cover with a coverslip. Blot off excess solution. Rim coverslip with paraffin wax to prevent drying.

(2) Locate required cell or cells, if necessary with phase contrast.
(3) Switch to UV (365–440 nm) and irradiate for 10–200 seconds.
(4) Photograph when maximum porphyrin fluorescence appears.

Result

Red fluorescence, not present before irradiation, indicates the presence of haemoproteins.

OXIDIZED AMINE METHOD FOR HEMES

(after Hirai, 1968)

(*Aldehyde-fixed, cryostat sections*)

Preparation of Sections

Fix small blocks for 24 hours at 0·4° in 4 per cent formaldehyde in 0·1 M cacodylate buffer (pH 7·2), containing 1 per cent $CaCl_2$. Dialyse for 2 hours against 0·1 M Tris-HCl containing 0·25 M sucrose. Cut 8μ cryostat sections.

Preparation of the Reagent

Dissolve 3 g. of 3,3′-diaminobenzidine-tetrahydrochloride (DAB) and 430 mg. sucrose in 5 ml. of 0·1 M Tris-HCl buffer (pH 8·4) and add 0·1 ml. N-HCl to bring pH to 7·2. Keep in the light at 22°. Oxidation occurs and a precipitate may form. Filter.

Preparation of the Medium

4·5 ml. Oxidized DAB solution
0·5 ml. 0·1 M Tris-HCl (pH 7·2), or
0·5 ml. 0·1 M Tris-HCl (pH 7·2) containing
0·1 M 3-amino-1, 2, 4-triazole (catalase inhibitor).

Method

(1) Incubate sections for 3–6 hours at 30°.
(2) Wash in 0·25 M sucrose.
either (3) Mount in glycerine jelly
or (4) Post-fix for 2 hours in cold 1 per cent OsO_4 in veronal buffer (pH 7·4) containing 0·25 M sucrose.
(5) Dehydrate in graded ethanol and propylene oxide.
(6) Embed in Epon. Cut 0·5 to 1 micron sections.

Result

Erythrocytes, dark brown; mitochondria, pale brown; microbodies dark brown. In aminotriazole-treated material microbodies are unstained.

THE GMELIN REACTION FOR BILIRUBIN AND HAEMATOIDIN

(*Formalin, alcohol; paraffin sections*)

Method

(1) Bring sections to water.
(2) Apply a few drops of a mixture of equal parts of conc. nitric acid and absolute alcohol.
(3) Apply a coverslip and wipe off excess reagent with blotting-paper.

(4) Ring with paraffin wax to seal edges (optional).
(5) Examine under microscope for development of colour.

Result

A spectrum of colours from red through purple to green is observed to spread outwards from masses of bilirubin and from haematoidin crystals.

MODIFIED GMELIN REACTION FOR BILE PIGMENTS

(after Lillie and Pizzolato, 1967)

(Formalin-fixed, paraffin sections)

Method

(1) Deparaffinize sections in xylene.
(2) Wash in CCl_4.
(3) Brominate for 10 minutes in 0·5 per cent Bromine in CCl_4 (v/v).
(4) Wash in CCl_4, then in xylene.
(5) Mount in xylene-cellulose caprate.

Result

Small bile pigment granules are converted to rose-red up to dark reddish purple. Larger bile casts are pale rose, through gray-violet to violet-black.

(Stable for some weeks, with minimal change of colour.)

THE IODINE METHOD FOR BILE PIGMENTS

(Stein, 1935)

(Formalin, alcohol; paraffin sections)

Solutions

(1) **Iodine Reagent.**	Lugol's iodine	2 parts
	Tinct. iod.	1 part

(9) **Five per cent Sodium Thiosulphate.**

Method

(1) Bring sections to water.
(2) Treat with the iodine reagent for 12–18 hours.
(3) Wash in running water for 5 minutes.
(4) Decolorize with thiosulphate solution, 30 seconds.
(5) Wash and counterstain with Mayer's carmalum, 3–18 hours.
(6) Wash in water, dehydrate in absolute acetone, clear in xylene, mount in balsam.

Result

Bile pigments appear dark green. Nuclei, red.

MODIFIED STEIN METHOD FOR BILE PIGMENTS

(after Leibnitz, 1964)

(96 per cent ethanol; paraffin sections)

Method

(1) Bring sections to methanol and treat with 1 per cent iodine in methanol (absolute) for 12 hours.

(2) Wash in water and counterstain in haematoxylin.
(3) Dehydrate, clear and mount in synthetic resin.

Result

Unconjugated bilirubin stains green.

FERRIC IRON METHODS FOR BILIRUBIN

(Kutlik, 1957)

(*Formalin, Carnoy etc.; paraffin sections*)

Method

(1) Bring paraffin sections to water.

(2) Immerse for 15 minutes in 5 per cent aqueous iron alum or for 10 minutes in 5 per cent $FeCl_3$ in 2 per cent acetic acid.

(3) Wash briefly in distilled water.

(4) Rinse in 96 per cent alcohol.

(5) Counterstain with 1 per cent Brilliant Yellow (Geigy) in 96 per cent alcohol, 10 minutes.

(6) Rinse in 96 per cent alcohol.

(7) Blot dry, transfer to xylene and mount in Canada balsam.

Result

Bilirubin is converted into green biliverdin. Background yellow.

TRICHROME METHOD FOR BILIRUBIN, HAEMOSIDERIN AND LIPOFUSCIN

(Glenner, 1957)

(*Fresh sections*)

Method

(1) Cut thin (10μ) cold microtome sections and mount on slides.

(2) Immerse for 5 minutes in 2 per cent potassium ferricyanide.

(3) Place in acetic-ferricyanide for 20 minutes. (Freshly mixed, equal parts of 2 per cent ferricyanide and 5 per cent acetic acid.)

(4) Rinse in running water and immerse in 3 per cent potassium dichromate at pH 2·2 for 15 minutes.

(5) Wash in water.

(6) Place sections in freshly filtered Oil red O in isopropanol (see p. 854). Stain for 20 minutes.

(7) Wash in running water for 5 minutes.

(8) Mount in Apáthy's syrup.

Result

Lipofuscins dark orange red, haemosiderin blue, bilirubin dark emerald green.

DIRECT AND INDIRECT VAN DEN BERGH REACTIONS

(after Raia, 1967)

(*Cold formalin-fixed, cryostat sections*)

Preparation of Stock Solutions

A. Stock diazo reagent

Dissolve 200 mg. 2,4-dichloroaniline in 100 ml. water containing 2 ml. conc. HCl and filter.

B. 1 per cent aqueous sodium nitrite

These two solutions must be freshly prepared every 2 weeks and kept in the dark at 0–4°.

C. Accelerator solution

Dissolve 6 g. caffeine, 10 g. sodium benzoate and 10 g. urea in 100 ml. water. To 35 ml. of this solution add 25 ml. 40 per cent formaldehyde and make up to 85 ml. with water.

Working Solutions

These should be freshly prepared.

Direct diazo reagent: Add 1 volume of solution B to 50 volumes solution A and leave for 20 minutes in an ice-bath before use.

Indirect diazo reagent: Add 2 volumes solution C to 1 volume of the direct diazo reagent. If a precipitate forms, dissolve by heating to 38–40°.

Method (Direct Reaction)

(1) Immerse sections for 30 minutes in direct diazo reagent.
(2) Wash for 3 minutes in running water.
(3) Counterstain with haematoxylin.
(4) Dehydrate rapidly in alcohol, clear in xylene and mount in D.P.X.

Result (Plate XXIVb, p. 1075)

Conjugated bilirubin, deep rose-red to almost black.

Method (Indirect Reaction)

Use indirect diazo reagent and proceed as above.

Result

All bilirubin is stained in various shades of red to reddish-black.

DIAZO METHOD FOR CONJUGATED BILIRUBIN

(after Desmet *et al.*, 1968)

(*Fresh or cold-formalin, cryostat sections*)

Preparation of Diazo Reagent

Mix 10 ml. 0·3 N-HCl and 0·2 ml. ethylanthranilate and shake continuously for 3 minutes. Add 10 ml. water and 0·6 ml. 0·5 per cent sodium nitrite. Shake for 2 minutes. Add 0·2 ml. 18 per cent ammonium sulphamate to destroy excess HNO_2. Place in a 37° incubator for 5 minutes, with occasional shaking.

Warm 15 ml. Glycine-HCl buffer (0·4 N-HCl, pH 2·5) for 10 minutes at 37°. Add an equal volume of the freshly prepared diazotate solution. Keep at 37° and use within 40 minutes.

Preparation of Zinc (II)-Ammonia Solution

Make a 2 M solution of zinc acetate in 0·1 M ammonia. Adjust pH to 6·0 with 25 per cent ammonium hydroxide.

Method

(1) Immerse in diazo reagent at 37° for 8 minutes.
(2) Transfer to zinc-ammonia solution at 22° for 2–3 minutes.
(3) Wash in water.
(4) Counterstain briefly in 2 per cent celestin blue, 1 minute and in haemalum for 10 seconds.
(5) Mount in glycerine jelly

Result

Conjugated bilirubin stains purplish-red.
(Stable, in the dark, for several days.)

NORMAL DIAZOTIZATION PROCEDURE

(Lillie and Pizzolato, 1969)

Dissolve 1 mmole of amine in 3 ml. N-HCl and 6 ml. distilled water. Cool to 3°, add 1 ml. N-$NaNO_2$ and keep at 3° for 20 minutes with occasional agitation. With some amines 3–5 ml. ethanol can be substituted for an equivalent amount of water to facilitate solution.

CLAUS DIAZOTIZATION PROCEDURE

(Lillie and Pizzolato, 1969)

TABLE 93

Na_3PO_4 1M (ml)	pH	Na_3PO_4 1M (ml)	pH
0	3·1	20	6·2
1	3·8	21	6·4
2	4·3	22	6·5
3	4·4	23	6·6
4	4·5	24	6·7
5	4·6	25	6·8
6	4·7	26	6·85
7	4·75	27	6·9
8	4·8	28	7·3
9	4·9	29	7·6
10	4·95	30	7·7
11	5·0	31	7·8
12	5·2	32	7·85
13	5·3	33	7·9
14	5·4	34	8·0
15	5·5	35	8·6
16	5·6	36	9·1
17	5·7	37	9·5
18	5·9	38	9·9
19	6·0		

The 1.0M Na_3PO_4 must be warmed to 30° to dissolve it completely. The amount of water to be added is 39 ml. less the number of ml. of Na_3PO_4.

Dissolve 1 mmole of amine in glacial acetic acid to make 9 ml. Cool to 10–15°. Add 0·5 ml. 20 per cent aqueous $NaNO_2$. A little brown gas may be evolved and a distinct colour change may occur. After 20 minutes, add 50 mg. urea in 0·5 ml. water to destroy excess HNO_2. The mixture, now at 100 mM concentration, should be diluted directly with N-acetic acid 1:39 for acid coupling (pH 2·0 to 2·3), or with special buffers for other pH levels. Table 93 gives the quantities of water and 1.0M-Na_3PO_4 which must be added to 1 ml. acetic acid diazotate to provide a 10 mM concentration at the pH indicated.

APPENDIX 27

FIF METHOD FOR NORADRENALIN

(Eränkö, 1955)

(*Fresh tissues*)

Method

(1) Cut fresh frozen sections at 50μ by means of the cold knife procedure or with a cold microtome.

(2) Allow sections to thaw on the surface of a solution containing 1 vol. 40 per cent formaldehyde, 5 vols. 2 per cent $CaCl_2$ and 4 vols. distilled water.

(3) After 2–6 hours rinse the sections in distilled water, mount on slides and cover with glycerol and a coverslip.

Result (Fig. 222)

Cells which contain noradrenalin show a strong yellowish-green fluorescence. The nuclei do not fluoresce.

HOT FORMALDEHYDE VAPOUR TECHNIQUE (FIF)

(after Falck and Owman, 1965)

PREPARATION OF TISSUES

Use a standard freeze-drying technique such as the one described in Vol. 1, p. 593.

PREPARATION OF WATER-EQUILIBRATED PARAFORMALDEHYDE

Place 5–6 g. aliquots of paraformaldehyde in small open containers on the shelf of a desiccator over sulphuric acid of the strength required to achieve the relative humidity selected. Table 94, below, derived from Hamberger (1967) should be used. Leave for 10 or more days. Use each sample once only.

TABLE 94

Relative humidity, density and g. per litre conc. H_2SO_4 for Equilibration Procedure

Relative Humidity %	Density	Conc. H_2SO_4 grams/litre
10	1·58	1064
20	1·49	887
30	1·44	779
40	1·39	678
50	1·34	586
60	1·29	503
70	1·25	428
80	1·20	341
90	1·14	231

CONDENSATION PROCEDURE

Transfer the freeze-dried blocks to a closed vessel or box (approximately one litre capacity) containing 5–6 g. water-equilibrated paraformaldehyde (or paraformaldehyde from normal supply in countries with correct humidity). Close the vessel tightly and maintain at 60, 70 or 80°, as required, for 1, 2 or 3 hours, as required.

FURTHER PROCESSING

Remove tissues from formaldehyde vapour chamber and vacuum-embed in 52° paraffin wax. This should take 10–15 minutes. Prepare blocks. Cut sections, preferably on a sliding microtome, and mount on clean slides. Transfer to 60° hotplate to melt wax and allow sections to spread. Remove as soon as the process is complete. Allow wax to harden again. Cover sections with liquid paraffin and apply a coverslip. Warm on the hotplate for 20–30 minutes to dissolve the wax in the mounting medium.

As an alternative, the sections can be covered with a mixture of 3 ml. xylene and 30 ml. Entellan (Merck), enclosed with a coverslip.

PERFUSION FIXATION FIF TECHNIQUE FOR CATECHOLAMINES

(after Laties *et al.*, 1967)

PERFUSION PROCEDURE

Use freshly prepared 4 per cent buffered paraformaldehyde (pH 7·2, 4°), made by adding 4 g. paraformaldehyde to 83 ml. 2·26 per cent NaH_2PO_4 and 17 ml. 2·52 per cent NaOH and heating to 60°. Perfuse selected organ with the above medium for 2–3 minutes (volume approximately 50 ml. for rat heart or 20 ml. for mouse heart; similar quantities for other organs).

PREPARATION OF SECTIONS

Remove organ and prepare small blocks on cryostat tissue holders. Cut 10μ sections at −10° and mount on glass slides. Place slides in a Coplin jar containing 1 g. paraformaldehyde. Seal and heat to 70–80° for 1–2 hours. Remove and mount in xylene. Ring coverslip if necessary.

TISSUE SLICE FIF TECHNIQUE FOR CATECHOLAMINES

(after Ehinger *et al.*, 1969)

PREPARATION OF SLICES

Use a standard tissue slicer, Sorvall TC–2 or Mickle Tissue Chopper. Trim tissue to a suitable size (3–10 × 2 × 2 mm.) and fix to the supporting base of the slicer with Eastman 910 glue, or similar preparation. Alternatively mount the tissue in 7 per cent agar. To prevent slices from sticking place a drop of saline on the tissue before cutting. Cut slices 50–100μ thick and stretch them on glass slides in saline (or buffer). Dry over P_2O_5 for one hour and treat with formaldehyde vapour, as in preceding methods. The author recommended paraformaldehyde equilibrated at 50 per cent relative humidity.

METHOD FOR DISTINCTION OF DA AND NA

(after Björklund *et al.*, 1968)

PROCEDURE

Prepare sections by the standard method of Falck and Owman (1965). Expose to vapour of concentrated HCl (fresh), in a closed vessel of 22° for 15 seconds to 2 hours. For some tissues better results are obtained with treatment at 80° for 30 minutes. Remove sections and mount in liquid paraffin, heating at 60° to dissolve the wax. Entellan (Merck) is not suitable for this procedure.

Result. The fluorescence peak of DA (370 nm) remains unaltered. That of NA changes gradually from 370 to 330 nm.

THE BOROHYDRIDE SPECIFICITY TEST

(after Corrodi *et al.*, 1964)

PROCEDURE

Select control and test pairs of sections prepared by the standard FIF procedure (p. 1394). Transfer control sections to 80 per cent ethanol or isopropanol. Transfer test sections similarly to 80 per cent ethanol or isopropanol containing 0·5 per cent sodium borohydride. Leave for 15–30 minutes. Wash in fresh ethanol or isopropanol, dry in air, and mount in liquid paraffin for examination.

THE BOROHYDRIDE-PERIODIC ACID TEST

(after Mukherji *et al.*, 1966)

PROCEDURE

Prepare sections by the standard FIF procedure (p. 1394). Note and photograph fluorescence. Remove liquid paraffin with isopropanol and reduce with 0·5 per cent sodium borohydride in 80 per cent isopropanol (15–30 minutes). Wash in isopropanol, dry, mount in liquid paraffin and rephotograph. Remove liquid paraffin with isopropanol, bring to water, treat with 0·69 per cent acid sodium periodate (0·69 per cent sodium periodate in 0·3 per cent HNO_3 for 15 minutes. Wash in water, dehydrate in isopropanol and dry in air. Re-expose sections to hot formaldehyde vapour at 80° for 1 hour. Mount in liquid paraffin and rephotograph.

Result. Borohydride reduction quenches all specific catecholamine fluorescence. After periodic acid treatment only DA and A fluorescence is restored.

DIAZO METHOD FOR ARGENTAFFIN CELL GRANULES

(*Formalin; paraffin sections. Avoid mercury*)

Method

(1) Bring sections to water.
(2) Treat for 30 seconds with a dilute (1 mg/ml.) solution of the stable diazotate of 5-nitroanisidine (Fast Red salt B, I.C.I. Ltd.), also known as Echtrotsalz B, in 0·1 M veronal acetate buffer at pH 9·2.
(3) Wash thoroughly in running water.
(4) Stain nuclei with Mayer's haemalum, 6–10 minutes.
(5) Wash in running water for 30 minutes.
(6) Dehydrate in alcohol, clear in xylene and mount in DPX.

Result (Plate XXVIb)

Argentaffin (E.C) cell granules, fiery orange red. Nuclei, dark blue. Cytoplasmic structures, yellow.

(Any one of a large variety of stable diazotates can be used; most satisfactory results are obtained with those giving reddish azo dyes. The colours given with granules in carcinoid tumours are more usually brownish-red.)

ACID DIAZONIUM REACTION FOR AROMATIC AMINES AND PHENOLS

(*Formalin; paraffin sections*)

Method

(1) Bring sections to water.

(2) Immerse in 0·5 M acetate buffer (pH 5·2) containing 1 mg/ml. of diazotized *p*-nitroaniline. Leave for 20–40 minutes.

(3) Wash in running water.

(4) Counterstain nuclei with Mayer's haemalum, 3 minutes.

(5) Wash in running water for 30 minutes.

(6) Dehydrate in alcohol, clear in xylene and mount in DPX.

Result

Argentaffin granules in the enterochromaffin cells or in carcinoid tumours stain orange-red. Other tissue components which possibly contain aromatic amines (e.g. pseudomelanin in the intestine) are coloured to a lesser extent.

THE ALKALINE THIOINDOXYL METHOD FOR QUINONEIMINES

(*Formalin; paraffin sections*)

Method.

(1) Bring sections to water.

(2) Oxidize for 10 minutes in either 5 per cent iodine in 95 per cent ethanol, or in 2·5 per cent aqueous sodium iodate at 56°.

(3) After iodine, bring to water and treat for 2 minutes with 5 per cent aqueous sodium thiosulphate.

(4) Rinse briefly in water.

(5) Incubate for 30 minutes at 56° (or for 2–4 hours at 37°) in the alkaline thioindoxyl solution.*

(6) Wash in water and counterstain nuclei, if required, with haemalum.

(7) Dehydrate, clear and mount in a suitable synthetic resin.

Result. E.C. granules appear orange-red to reddish-brown.

STABLE DIAZOTATES

Stable diazotates are produced in four different ways to give products with very different stabilities.

(1) The diazonium salt is warmed with alkali to form the anti-diazotate. This is very stable but easily convertible by the addition of mineral acid to the free diazoic acid. The latter breaks down rapidly to a nitrosamine.

$$C_6H_5.N{=}NOK \longrightarrow C_6H_5N{=}NOH \longrightarrow C_6H_5NHNO$$

Diazoic acid Phenyl nitrosamine

The nitrosamines are yellow compounds and they are partly responsible for the staining of the background structures in tissue sections exposed to diazotates in solution.

* Dissolve 10–15 mg. purified 4-methyl-6-chlorothioindoxyl in 2–5 ml. acetone and add to 35 ml. 0·2 M-Tris buffer at pH 8·5. Filter before use. Use only freshly made solutions.

(2) The diazonium salt is coupled with sarcosine or methyl taurine, to give a very stable compound which releases the free diazonium salt on acidification.

(3) By the addition of naphthalene disulphonic acids to diazonium salts stable derivatives are formed. In this way Manheimer and Seligman (1949) produced a stable diazotate of α-naphthylamine for use in their modification of the alkaline phosphatase technique of Menten *et al.* (1944).

(4) By the addition of zinc salts diazo double salts are produced. These are relatively stable and they do not require acidification for activation. Aluminium, or other metallic salts, may be present as stabilizers.

The majority of stable diazotates are produced by this last method and they contain some 5–6 per cent of zinc chloride and often some aluminium salt in addition. A number of stable diazotates are available commercially and a list of these is given in the Table below. Additions are constantly being made as new demands occur, so that the list is inevitably incomplete.

Some Common Stable Diazotates

(see also Appendix 15, p. 711)

Seller*	Trade name	Diazotate of	Country†
Ciba . . .	Fast Red Salt V	5-nitroanisidine	Sw.
G. T. Gurr (I.C.I.) .	Fast Red Salt B	5-nitroanisidine	U.K.
I.G. Farb. . .	Kernechtrotsalz B	5-nitroanisidine	G.
Bayer . . .	Azorotsalz B	5-nitroanisidine	G.
DuPont . . .	Naphthanil Diazo Red B	5-nitroanisidine	U.S.
G.D. Corp. . .	Diazo Fast Blue RR	4-benzoylamino-2:5-dimethoxyaniline	U.S.
Dajac (DuPont) .	Naphthanil Diazo Blue B	*o*-dianisidine	U.S.
G. T. Gurr (I.C.I.) .	Fast Blue B Salt	*o*-dianisidine	U.K.
G.D. Corp. . .	Fast Blue Salt BN	*o*-dianisidine	U.S.
Dajac (DuPont) .	Naphthanil Diazo Red RC	4-chloro-2-anisidine	U.S.
Dajac . . .	(None)	α-naphthylamine	U.S.
G. T. Gurr (I.C.I.) .	Fast Black Salt K	*p*-nitrobenzene-azo-dimethylaniline	U.K.
G. T. Gurr (I.C.I.) .	Fast Garnet	*o*-aminoazotoluene	U.K.
G.D. Corp. . .	Diazo Garnet GBC Salt	*o*-aminoazotoluene	U.S.

* Ciba — Ciba-Geigy, Basel.
Bayer — Bayer, Leverkusen.
Dajac — Dajac Laboratories.
DuPont — DuPont de Nemours.
G.D. Corp. — General Dyestuff Corporation.
G. T. Gurr — George T. Gurr Ltd., Searle Scientific Ltd.
I.C.I. — Imperial Chemical Industries Ltd., Manchester 9, England.
I.G. Farb. — I.G. Farbenindustrie.

† G. Germany. U.K. United Kingdom.
Sw. Switzerland. U.S. United States of America.

GIBBS' METHOD FOR ARGENTAFFIN CELL GRANULES

(after Gomori)

(*Formalin; paraffin sections. Avoid mercury*)

Preparation of Staining Solution

Dissolve 20 mg. 2:6-dichloroquinone-chloroimide (Gibbs' reagent) in 20 ml. veronal acetate buffer at pH 9·2 by warming to 70° and shaking. Cool before use and filter.

Method.

(1) Bring paraffin sections to water.
(2) Treat with the above solution for 10–15 minutes.
(3) Wash well in running water.
(4) Counterstain nuclei with 1 per cent aqueous neutral red, 3 minutes.
(5) Dehydrate and differentiate the nuclear stain by rapid passage through 70 per cent and absolute alcohol to xylene.
(6) Mount in DPX.

Result (Fig. 229). Argentaffin granules, black. Nuclei, red. (Granules in carcinoid tumours are often brownish.)

THE ALKALINE THIOINDOXYL METHOD FOR BETA-CARBOLINES

(Pearse, 1956)

(*FDFV; formalin or glutaraldehyde, paraffin sections*)

PREPARATION OF THIOINDOXYL SOLUTION

Add to 2 ml. absolute acetone, 20 mg. of unpurified 4-methyl-6-chlorothioindoxyl (the crude product from the Dyestuff Industry invariably contains some of the red thioindigo). Add this, drop by drop, with shaking, to 40 ml. 0·2M tris-HCl buffer (pH 8·5). If the final solution is cloudy add acetone, drop by drop, until it clears. Other substituted thioindoxyls may be used.

Method.

(1) Bring sections to water.
(2) Oxidize for 10–30 minutes in either (a) 5 per cent iodine in 95 per cent ethanol or (b) in 2·5 per cent aqueous sodium acetate, at 56–60°,
(3) After (a), bring to water and treat for 2 minutes with 5 per cent aqueous sodium thiosulphate.
(4) Rinse (a and b) in water.
(5) Incubate for 30 minutes at 56–60°, or for 2–4 hours at 37° in the alkaline thioindoxyl solution.
(6) Wash in water.
(7) Counterstain nuclei in Mayer's haemalum.
(8) Wash, differentiate quickly in 1 per cent acid alcohol, dehydrate, clear, and mount in synthetic resin.

Result. Reddish-brown thioindigoid dyes indicate the presence of β-carbolines (formaldehyde-fixed preparations). The colours produced with glutaraldehyde-fixed material are somewhat darker. The positive result in such cases indicates that ring closure to the carboline is not necessary for condensation.

FIF-OZONE METHOD FOR TRYPTAMINES

(After Björklund *et al.*, 1968)

Method.

(1) Place tissue blocks, freeze-dried in the usual manner, freeze-dried cryostat sections or air-dried cryostat sections, together with 5–6 g. water-equilibrated paraformaldehyde (see p. 1394) in a 1 litre chamber or box.
(2) Provide an electric discharge, from a high tension coil, within the chamber or box, for 1–45 minutes.
(3) Close the chamber or box and heat for 1–2 hours at 70–80°.

Alternatively.

(1) Expose tissue blocks or sections to formaldehyde vapour in the standard manner (p. 1395), for 1–2 hours.
(2) Transfer to another chamber, in which ozone has been generated.
(3) Close the chamber and heat to 80° for one hour.
(4) Proceed in the usual manner to prepare sections for viewing.

Result. Tryptamine and some related amines can be expected to give fluorescence (370/495 nm) which is absent from controls not exposed to ozone.

OPT METHOD FOR HISTAMINE

(after Ehinger and Thunberg, 1967; Ehinger *et al.*, 1968; Cross *et al.*, 1971)

PREPARATION OF SECTIONS

Quench tissues as for normal freeze-drying sequence. Mount small blocks on tissue holders (cryostat) without allowing temperature to rise above −30°. Cut cryostat sections (20μ) at this temperature. Freeze-dry for 2–8 hours at −40° (thermoelectric dryer) or at −30° over P_2O_5 in a desiccator *in vacuo*. Mount sections on albuminized slides and maintain at 37° for 18 hours to ensure adhesion. As an alternative, freeze-dried paraffin-embedded sections can be employed, after removal of wax with light petroleum.

Method.

(1) Place 25–50 mg. of crystalline *O*-phthaldehyde (recrystallized from hot heptane) in a closed glass vessel (500 ml. capacity) preheated to 100°, and heat to 100° for 10 minutes.
(2) Remove OPT vessel from oven and place slides in it for 10 seconds.
(3) Remove slides to a moist chamber at 22°, for 2 minutes.
(4) Dry sections at 80° for 5–10 minutes; cool.
(5) Dip briefly in xylene and mount in D.P.X.
(6) Examine in a fluorescence microscope, using the mercury 366 nm line for excitation (UG1 filter) and a Zeiss 41 or Leitz K430 barrier filter.

Result. Histamine gives a yellow fluorescence if present at high concentration. Blue fluorescence indicates lower levels of the amine.

BISCHLER–NAPIERALSKI TECHNIQUES

(Rost and Ewen, 1971)

PREPARATION OF TISSUES

Use a standard freeze-drying technique such as the one described in Vol. **1**, p. 593. The condensation procedure can be carried out on either freeze-dried blocks, or alternatively the blocks can be embedded and individual paraffin sections exposed to acid or aldehyde vapour.

CONDENSATION PROCEDURE

Blocks are exposed for 2–4 hours, or sections for 2 hours, in hot vapour (60°–80°) generated in a chamber similar to that described in Vol. **1**, p. 51. Sections exposed to aldehyde vapours need not be dewaxed before exposure.

Acetaldehyde vapour is generated in a manner similar to formaldehyde vapour, using metaldehyde ("Meta") instead of paraformaldehyde.

Glutaraldehyde vapour can be obtained by heating 50% aqueous glutaraldehyde, water vapour in the chamber being reduced as necessary by phosphorous pentoxide or other dessicant.

Acetic acid vapour is obtained by heating glacial acetic acid. It may be necessary to use phosphorous pentoxide to reduce water vapour in the chamber.

APPENDIX 28

PERLS' METHOD FOR FERRIC IRON

(after Lison and Bunting)

(*Formalin-fixed, paraffin sections preferable*)

ACID fixatives must be avoided, chromates also interfere with the preservation of iron.

Method. (1) Bring sections to distilled water.

(2) Expose sections to a fresh mixture of equal parts of 2 per cent potassium or sodium ferrocyanide and 2 per cent HCl for 30–60 minutes. (If the longer period is to be used, change the solution after 30 minutes.)

(3) Wash in distilled water.

(4) Counterstain nuclei with 1 per cent aqueous neutral red, 3 minutes.

(5) Wash, dehydrate in alcohol, clear in xylene and mount in DPX.

Result. Ferric iron, deep Prussian blue. Nuclei, red.

TURNBULL BLUE METHOD FOR FERRIC IRON

(after Tirmann and Schmeltzer)

(*Formalin-fixed, paraffin sections*)

Method. (1) Bring sections to distilled water.

(2) Treat the section for 1–3 hours with a dilute solution of yellow ammonium sulphide.

(3) Rinse in distilled water.

(4) Treat with a freshly-prepared mixture of equal parts of 20 per cent potassium ferricyanide and 1 per cent HCl for 10–20 minutes.

(5) Rinse in distilled water.

(6) Counterstain nuclei red, if desired.

(7) Dehydrate, clear and mount.

Result. Ferrous iron, and ferric iron converted to this state, deep blue.

THE DINITROSORESORCINOL METHOD FOR FERRIC IRON

(after Humphrey, 1935)

(*Formalin-fixed, paraffin sections*)

Method. (1) Bring sections to distilled water.

(2) Treat with 10 per cent yellow ammonium sulphide for 1 hour.

(3) Rinse in distilled water.

(4) Treat with a 3 per cent solution of dinitroresorcinol in 50 per cent alcohol for 6–18 hours.

(5) Wash in 50 per cent alcohol.

(6) Dehydrate in absolute alcohol, clear in xylene, mount in DPX.

Result. Ferrous iron, and ferric iron converted by treatment with sulphide, appear greenish-blue. Background structures brown.

This method is insensitive by comparison with Perls' technique and the deep colour of the background obscures some of the finer deposits. Diffuse iron is not demonstrated by this method, which is only occasionally useful when avoidance of iron containing reagents is thought necessary.

HAEMATOXYLIN-LAKE METHOD FOR HAEMOSIDERIN IRON

(after Macallum, Mallory and Bunting)

Method. (1) Proceed through stages 1–3 of the dinitroresorcinol method.

(2) Stain for 2–6 hours in a freshly-prepared 0·5 per cent haematoxylin in 30 per cent ethanol. (Prepare by dissolving 10 mg. haematoxylin in 2 ml. absolute ethanol and adding 4 ml. distilled water.)

(3) Wash in absolute alcohol, then in 50 per cent alcohol and in distilled water.

(4) Counterstain (optional) in Mayer's carmalum 1–6 hours.

(5) Wash in water, dehydrate, clear and mount in DPX.

Result. Haemosiderin appears blue-black. Nuclei, red.

BATHOPHENANTHROLINE METHOD FOR IRON

(after Hukill and Putt, 1962)

(*Formalin; paraffin sections*)

PREPARATION OF THE REAGENTS

Add 100 mg. bathophenanthroline (4,7-diphenyl-1,10-phenanthroline) to 100 ml. 3 per cent acetic acid. Agitate vigorously and store at 60° overnight (8–12 hours). Cool to 22°, add 0·5 ml. thioglycollic acid, filter. (Stable for 4 weeks; replenish thioglycollic acid with each use.)

Methylene blue. Dissolve 0·5 g. of the dye in 100 ml. distilled water and filter.

Method. (1) Bring sections to water.

(2) Stain in bathophenanthroline for 2 hours.

(3) Rinse in distilled water.

(4) Counterstain in methylene blue, 3 minutes.

(5) Rinse in distilled water, 3 changes, 1 minute each.

(6) Dry at 60° for 1 hour.

(7) Mount in synthetic resin.

Result. Iron, red; nuclei, blue.

CADMIUM SULPHATE METHOD FOR FERRITIN AND APOFERRITIN

(Granick, 1946)

(*Fresh frozen, cold knife or cold microtome, teased*)

Method. (1) Teased portions of the fresh tissues, weighing about 0·2 g., or fresh frozen sections should be kept overnight at room temperature in 10 per cent aqueous cadmium sulphate. This may be done on the slide, under a ringed coverslip, or in a small amount of the cadmium solution in a closed vial.

Result. If ferritin is present the typical yellow octahedral crystals of the cadmium-ferritin complex will be visible under the microscope.

TECHNIQUES FOR REMOVAL OF IRON

(1) 5 per cent oxalic acid, 2–6 hours.

(2) 10 per cent H_2SO_4, 24 hours.

(3) 1 per cent Sodium dithionite in 0·2 M acetate buffer (pH 4·5) for 5 minutes (Lillie *et al.*, 1963).

ALIZARIN RED S METHOD FOR CALCIUM

(Dahl, 1952)

(*Alcohol-fixed, paraffin sections*)

Mount sections without adhesive.

PREPARATION OF SOLUTIONS

(1) *Alizarin Red S*. Stir 0·5 g. of the dye in 45 ml. distilled water and add 5 ml. of a 1 : 100 dilution of 28 per cent NH_4OH, with constant stirring. The pH should be between 6·3 and 6·5 and the solution keeps for at least a month.

(2) *Differentiation*. 10^{-3} M-HCl in 95 per cent ethanol (conc. HCl 1 part; 95 per cent ethanol 10,000 parts).

Method. (1) Bring sections to water.

(2) Stain with Alizarin solution for 2 minutes.

(3) Wash with distilled water for 5–10 seconds.

(4) Rinse in acid ethanol for 15 seconds.

(5) Dehydrate, clear in xylene and mount in cedarwood oil.

Result. Calcium deposits appear orange-red. Inorganic iron (not haemosiderin), purple.

With the older alizarin techniques stress was laid on the avoidance of acids for differentiation because of the danger of removal of calcium. The acid wash employed here is of brief duration and it is improbable that any calcium is dissolved.

GYPSUM METHOD FOR CALCIUM

(*Formalin-fixed, paraffin sections*)

Method. (1) Bring sections to acetone and allow to dry in air.

(2) Place a drop of 5 per cent sulphuric acid on the area of the section where the deposits are located.

(3) Observe under low power objective for the formation of the characteristic long thin needles of hydrated calcium sulphate ($CaSO_4.2H_2O$). The needles usually change into thin prisms and finally into masses of plates piled on top of one another.

CALCIUM RED METHOD FOR CALCIUM

(McGee-Russell, 1955)

(*Formalin or alcohol, paraffin sections*)

PREPARATION FOR STAINING SOLUTION

Wash 2 g. Nuclear Fast red (G. T. Gurr; Batch 3569 recommended), twice, with about 100 ml. distilled water. Take up the residue (about 0·25 g.) with 100 ml. distilled water.

Method. (1) Bring sections to water.

(2) Stain in calcium red solution 1–10 minutes.

(3) Wash in distilled water.

(4) Dehydrate in alcohols, clear in xylene and mount (synthetic media).

Result. Calcium deposits, red. Tissues, various shades of pink.

PHTHALOCYANIN METHOD FOR CALCIUM

(*Formalin or alcohol, paraffin sections*)

Method (1) Bring two sections to water.

(2) Treat one section with 1 per cent aqueous HCl for 20 minutes, wash briefly in water.

(3) Stain for 30–60 minutes in a 0·1 per cent aqueous solution of Durazol Fast blue 8 G with or without 0·1 per cent Alcian blue 8 GS (I.C.I. Ltd.), or of any suitable sulphonated copper phthalocyanin dyes.

(4) Wash in running water for 1 minute.

(5) Differentiate in 0·05 per cent lithium carbonate until the tissues in general are pale blue.

(6) Dehydrate, clear and mount in a suitable synthetic medium.

Result (Fig. 227). Calcium deposits appear bright blue in the untreated section only; tissues stain blue in various degrees, unaffected by acid treatment.

Crystalline, birefringent, calcium phosphate deposits, and some amorphous ones, may stain more readily with dilute solutions of spirit-soluble copper phthalocyanins in 85–95 per cent ethanol. Methanol Fast blue 2 G (I.C.I. Ltd.) and Luxol Fast blue MBS (Dupont) are examples of dyes of this type.

ERIOCHROME BLACK T METHOD FOR CALCIUM
(Invertebrates only)

(Formalin, etc.; paraffin sections)

Method. (1) Bring paraffin sections to water.

(2) Stain for 3–5 minutes in 0·5 per cent aqueous Eriochrome black T buffered to pH 9·5 with N-NaOH.

(3) Wash in distilled water.

(4) Mount in glycerine jelly.

Result. Calcium deposits stain bright red. Tissues generally stain in various shades of blue.

THE VON KÓSSA METHOD FOR "CALCIUM" DEPOSITS

(Formalin-fixed, paraffin sections, etc.)

Method. (1) Bring sections to distilled water and rinse thoroughly.

(2) Immerse in 0·5–1 per cent aqueous $AgNO_3$ for 10–15 minutes. (This procedure should be carried out in sunlight or ultraviolet light.)

(3) Rinse in distilled water.

(4) Immerse in 5 per cent aqueous sodium thiosulphate, 30 seconds.

(5) Wash in tap water.

(6) Counterstain nuclei with 1 per cent neutral red, 30 seconds.

(7) Dehydrate in alcohol, clear in xylene and mount in Canada balsam or DPX. (The latter causes slow fading.)

Result. Phosphates and carbonates appear black. Nuclei, red.

SILVER-H_2O_2 METHOD FOR CALCIUM OXALATE
(after Pizzolato, 1964)

(Formalin; paraffin sections)

Method. (1) Bring sections to water and treat for 15 minutes with 2 N-acetic acid to remove phosphate and carbonate.

(2) Treat with 1 per cent silver nitrate in 15 per cent H_2O_2 (equal parts of 30 per cent hydrogen peroxide and 2 per cent $AgNO_3$) for 15 minutes at 22°. (Extra illumination is not usually necessary.)

(3) Wash in distilled water.

(4) Counterstain with 2 per cent safranin for 3 minutes.
(5) Dehydrate, clear and mount in synthetic resin.

Result. Calcium oxalate deposits black, background red.

SILVER-RUBEANATE METHOD FOR CALCIUM OXALATE

(after Yasue, 1969)

(*Formalin; paraffin sections*)

Method. (1) Bring sections to water and treat with 5 per cent acetic acid for 30 minutes.
(2) Wash in distilled water.
(3) Treat with 5 per cent aqueous $AgNO_3$ for 10–20 minutes.
(4) Rinse in running distilled water.
(5) Immerse in saturated rubeanic acid in 70 per cent ethanol, containing 2 drops conc. ammonia per 100 ml., for 1 minute.
(6) Rinse in 50 per cent ethanol.
(7) Counterstain in 2 per cent methyl green, 2 minutes.
(8) Dehydrate rapidly, clear in xylene, mount in a suitable synthetic resin.

Result. Calcium oxalate deposits black, nuclei green.

GBHA METHOD FOR CALCIUM

(after Kashiwa and Atkinson, 1963)

(*F.D or F.S in acetone; paraffin sections*)

Preparation of Reagents

(1) Glyoxal bis (2-hydroxyanil), 0·4 g. in 100 ml. absolute ethanol.
(2) 5 per cent NaOH.
(3) Saturated Na_2CO_3 and KCN in 90 per cent ethanol.
(4) Fast green FCF; 0·08 g. in 100 ml. 95 per cent ethanol. All solutions are stable.

Method. (1) Cover the undeparaffinized section with a freshly made solution containing 2 ml. alcoholic GBHA and 0·3 ml. of 5 per cent NaOH. Allow to stand for 3 minutes.
(2) Rinse in 70 per cent ethanol.
(3) Immerse in Na_2CO_3/KCN solution for 15 minutes.
(4) Rinse in 95 per cent ethanol.
(5) Counterstain in alcoholic Fast green FCF, 3 minutes.
(6) Rinse in 3 changes of 95 per cent ethanol.
(7) Dehydrate in ethanol, deparaffinize and clear in 4 changes of xylene.
(8) Mount in a neutral medium.

Result. (Examine using a didymium green filter.) Red calcium deposits appear black.

SOLOCHROME AZURINE METHOD FOR ALUMINIUM

(*Formalin-fixed paraffin, etc.*)

Method. (1) Bring sections to water.
(2) Stain for 15–20 minutes in 0·2 per cent aqueous Solochrome azurine (I.C.I. Ltd.).
(3) Rinse in distilled water.

(4) Dehydrate in alcohols, clear in xylene and mount in a suitable synthetic medium.

Result (Plate XXVIIb). Aluminium deposits appear deep steel blue; beryllium stains similarly. If distinction is required this can be made by means of alkali pretreatment which removes aluminium hydroxide.

FLUORESCENT (MORIN) METHOD FOR CALCIUM AND ALUMINIUM

(Cold knife, cold microtome; formalin-fixed frozen sections, formalin-fixed paraffin sections)

Cold-knife or cold-microtome sections may be delivered straight into the staining solution. Other sections are brought to 85 per cent ethanol. A control section pretreated with 1 per cent aqueous HCl for 10 minutes should always be included.

Method. (1) Stain for 2–5 minutes in 0·2 per cent morin in 85 per cent ethanol containing 0·5 per cent acetic acid.

(2) Wash briefly with 70 per cent ethanol, or, if overstained, with 1 per cent acid alcohol.

(3) Wash in distilled water.

(4) Mount in water and ring coverslip with paraffin wax. If a more permanent mount is required the section should rapidly be dehydrated in the alcohols, cleared in xylene and mounted in fluormount (E. Gurr Ltd.).

Result (Fig. 230). In the section untreated with HCl a bright greenish-white fluorescence, absent from the control section, indicates the presence of calcium. Aluminium hydroxide, which is acid-stable, gives a more green fluorescence than calcium. On occasion the edges of the section may show an even fluorescence of greenish colour regardless of structure. In such cases only the inner part of the section, where the background fluorescence varies from dull to bright yellow, should be used for critical assessments.

AURINE METHODS FOR ALUMINIUM HYDROXIDE

(Irwin, 1955)

(Formalin-fixed paraffin sections, etc.)

PREPARATION OF SOLUTIONS

(1) *Aurine buffer Solution.* Mix 60 ml. 5 M-ammonium chloride, 60 ml. 5 M-ammonium acetate, 10 ml. 6 N-HCl.

This should give ±pH 5·2.

Dissolve 2 g. aurine tricarboxylic acid (Aluminon) in a few ml. of the buffer and make up to 100 ml. Heat to 75° for 15 minutes and filter while hot.

(2) *Differentiating Solution.* Mix 50 ml. of the buffer solution (above), 22 ml. of 1·6 M-ammonium carbonate.

The pH of this solution should be about 7·2.

Method. (1) Bring sections to water.

(2) Stain for 5–10 minutes in buffered aurine at 75°.*

(3) Rinse in distilled water.

(4) Differentiate in ammonium carbonate solution for 3–5 seconds.

* At this temperature there is a tendency for the sections to float off the slides. Longer incubation at 60° may be tried.

(5) Wash in distilled water.

(6) Counterstain in 1 per cent aqueous methylene blue.

(7) Rinse in distilled water.

(8) Dehydrate in alcohols, clear in xylene and mount in a suitable permanent medium.

Result (Fig. 231). Aluminium hydroxides appear as the bright red aurine lake. Nuclei, blue.

FLUORESCENT (SOLOCHROME) METHOD FOR ALUMINIUM SALTS

(*Formalin, frozen or paraffin sections, etc.*)

Method. (1) Bring paraffin sections to water.

(2) Treat with a saturated solution of Solochrome black B (I.C.I. Ltd.) in absolute ethanol containing 2 per cent HCl, for 20–60 minutes.

(3) Rinse in absolute ethanol (paraffin sections) or in 1 per cent acid alcohol (frozen sections).

(4) Bring frozen sections to water and mount in glycerine. Clear paraffin sections in xylene and mount in fluormount.

Result (Plate XXVIIa). Aluminium salts fluoresce in colours which vary from yellowish-orange to bright brick-red on a light blue background.

Calcium salts are first dissolved and then reprecipitated diffusely as the red Solochrome lake in approximately their original location. These Calcium-Solochrome lakes do not fluoresce.

ACID SOLOCHROME CYANINE METHOD FOR ALUMINIUM

(*Formalin-fixed paraffin, etc.*)

Method. (1) Bring sections to water.

(2) Stain in 0·2 per cent Solochrome cyanine RS (I.C.I. Ltd.) in 1 per cent aqueous HCl for 15–30 minutes.

(3) Rinse in running hot tap water until the section has changed from yellow, through red to a pale grey-blue colour.

(4) Dehydrate in alcohols, clear in xylene and mount in a suitable synthetic medium.

Result (Plate XXVIIc). Aluminium deposits appear deep rose red to pink, depending on the amount of aluminium present. Nuclei, bright steel blue. Red blood cells, bright red. Basic proteins, pinkish-red or red.

Differentiation in 1 per cent acid alcohol before this stage removes the nuclear and red cell counterstain, leaving only aluminium deposits stained.

NAPHTHOCHROME GREEN B METHOD FOR ALUMINIUM, CALCIUM AND BERYLLIUM

(after Denz, 1949)

(*Formalin-fixed frozen sections, formalin- or alcohol-fixed paraffin sections*)

Method. (1) Bring sections to water.

(2) Stain in 0·1 per cent Naphthochrome green B in 0·1 M-phosphate buffer (pH 5·0) for 30 minutes at 37° (beryllium).

(2A) Stain in 0·1–0·2 per cent aqueous Naphthochrome green B at room temperature for 5–10 minutes (calcium, aluminium).

(3) Wash in distilled water.

(4) Differentiate for up to 30 minutes (if necessary) in absolute alcohol (beryllium).

(4A) Dehydrate, clear and mount in synthetic medium (calcium, aluminium).

(5) Wash in distilled water (beryllium).

(6) Counterstain if desired (Denz recommended 1 per cent acridine red).

(7) Dehydrate, clear, and mount, etc.

Result (Fig. 229). Beryllium (hydroxide) or beryllium salt/protein complexes stain apple green.

Calcium salts appear brown to brownish-red and aluminium hydroxide a rich deep green. (If other metals (Fe^{3+}) are present calcium deposits stain green. After treatment with mercuric salts most calcium deposits appear green but in the case of bony trabeculae the older parts stain red and the newer parts green. Osteoid does not stain.

A method for the purification of this dye is given by Aldridge and Liddell (1948). I have not found purification of presently available dye samples at all necessary.

COPPER METHOD FOR ARSENITES AND ARSENATES

(Castel, 1936)

(*Formalin-fixed paraffin sections*)

Method. (1) Fix tissues (thin blocks) in 10 per cent formalin containing 2·5 per cent $CuSO_4 5H_2O$ for 5 days.

(2) Wash in running water for 24 hours.

(3) Dehydrate, clear and embed in paraffin wax.

(4) Cut sections (preferably thick ones), remove wax and mount without counterstaining.

Result. A green granular deposit indicates the presence of salts of arsenic.

ALKALINE SOLOCHROME AZURINE METHOD FOR BERYLLIUM

(*Formalin-fixed paraffin sections, etc.*)

Method. (1) Bring sections to water.

(2) Stain for 30 minutes in a 0·2 per cent solution of Solochrome azurine (I.C.I. Ltd.) in 2 per cent NaOH.

(3) Rinse in distilled water.

(4) Dehydrate in alcohols, clear in xylene and mount in a suitable synthetic medium.

Result (Plate XXVIId). Beryllium deposits appear almost black; treatment with acid converts them to a brownish-orange. Aluminium deposits ($AlOH_2$) are dissolved in the staining baths, calcium deposits by treatment with dilute acids.

ALKALINE QUINALIZARIN METHOD FOR BERYLLIUM AND CALCIUM

(*Formalin-fixed frozen sections, formalin- or alcohol-fixed paraffin sections, etc.*)

Method. (1) Bring sections to water.

(2) Stain in 0·2 per cent quinalizarin (1,2,5,8-tetrahydroxy anthraquinone) in 4 per cent NaOH for 3–5 minutes.

(3) Wash carefully in distilled water.

(4) Mount frozen sections in glycerine jelly, preferably without counterstain.

(4A) Dehydrate paraffin sections in alcohols, clear in xylene and mount in a suitable synthetic medium.

Result (Fig. 232). Beryllium and calcium deposits stain deep purple. Nuclei may begin to stain, especially if time of staining is prolonged or if the concentration of NaOH is reduced.

RHODIZONATE METHOD FOR BARIUM SALTS

(*Formalin, frozen or paraffin sections, etc.*)

Method. (1) Bring sections to water.

(2) Transfer to a freshly-prepared 0·2 per cent aqueous solution of sodium rhodizonate. Incubate at 37° (frozen sections) or at 60° (paraffin sections) for 1–2 hours.

(3) Wash briefly in water.

(4) If desired convert the black rhodizonate to the red acid salt by immersion in 20 per cent HCl for a few minutes.

(5) Wash again in water.

(6) Dehydrate, clear and mount in a synthetic medium.

Result. Barium deposits black, converted to scarlet-red by acid treatment. Strontium salts, which react similarly, would be dissolved by this treatment.

BRUCINE-IODIDE METHOD FOR BISMUTH

(after Wachstein and Zak)

(*Formalin-fixed, frozen or paraffin sections*)

PREPARATION OF SOLUTIONS

(1) **Modified Castel Reagent.** Dissolve 0·25 g. brucine sulphate in 100 ml. distilled water with 3 drops conc. sulphuric acid. Stir until the salt dissolves and add 2 g. potassium iodide. Store in a brown bottle and filter before use.

(2) **Dilute Castel Reagent.** Dilute the stock reagent 1 : 3 with distilled water.

(3) **Thirty per cent. (100 vol.) H_2O_2.**

(4) **Laevulose Solution.** Dissolve 30 g. laevulose in 20 ml. water by warming at 37° for 24 hours and add a drop of the dilute Castel reagent.

(5) **Counterstain.** Add 1 ml. 1 per cent aqueous light green SF to 100 ml. undiluted Castel reagent. Filter before use.

Method. (1) Bring sections to water and blot dry.

(2) Apply 30 per cent H_2O_2 for 10–15 seconds.

(3) Wash in running water.

(4) Place in modified Castel reagent for 1 hour.

(5) Transfer to dilute Castel reagent and shake gently to remove precipitates.

(6) Counterstain for 4 minutes and wash in water.

(7) Mount in laevulose solution.

Result. Bismuth appears as an orange-red deposit whose colour may darken on standing.

DIMETHYLGLYOXIME METHOD FOR NICKEL SALTS

(Feigl, 1954)

This method has been applied by Wells (1956) to formalin-fixed or fresh frozen skin sections treated for short periods with 1 per cent aqueous nickel sulphate.

Method. (1) Treat sections with a saturated solution of dimethylglyoxime in alcohol at 37° for 10–20 minutes.

(2) Rinse in absolute alcohol.

(3) Clear in xylene and mount in a suitable synthetic medium.

Result. Sites of nickel adsorption appear in shades of pink.

RUBEANIC ACID METHOD FOR COPPER

(Okamoto, 1938; Uzman, 1956)

(*Cold knife and cold microtome, fresh slices, formalin-fixed, frozen or paraffin sections, etc.*)

The method as described is for fresh slices 1–2 mm. thick. It may easily be adapted for application to fresh frozen or formalin-fixed frozen sections, or to paraffin sections. The latter may first be treated as described below in order to make bound copper available for reaction.

Method. (1) Immerse thin slices of fresh tissues in 0·1 per cent rubeanic acid in 70 per cent ethanol for 10 minutes.

(2) Add sodium acetate (analytical reagent grade) at 200 mg./100 ml. to the above solution and leave for 24–48 hours.

(3) Wash with 70 per cent ethanol and stand in this for 60–90 minutes.

(4) Replace with absolute ethanol and stand for 24 hours.

(5) Clear in chloroform and embed in paraffin wax.

(6) Cut sections and mount on slides (egg-albuminized).

(7) Remove wax with xylene and mount in Canada balsam.

Result (Fig. 233). Copper salts or protein complexes appear as black precipitates.

Application to Paraffin Sections

In order to release protein-bound copper the sections are dewaxed with xylene and then placed face downwards over a beaker of conc. HCl for 15 minutes. They are then treated with absolute ethanol for 15 minutes and then transferred to the rubeanic acid bath.

DIETHYLDITHIOCARBAMATE METHOD FOR COPPER

(after Waterhouse, 1945)

(*Cold microtome, cold knife, teased, or freshly-dissected whole specimens*)

Method. (1) Stain in aqueous 0·1–0·2 per cent sodium diethyldithiocarbamate, with or without 2 per cent HCl, for 1–3 hours.

(2) Wash in distilled water and fix in 10 per cent formalin. (In the case of fresh frozen sections 1 ml. of 40 per cent analytical reagent formaldehyde may be added to 10 ml. of the staining medium after staining is complete. The tissues are allowed to fix before washing and mounting.)

(3) Transfer to clean slides and mount in glycerine jelly.

Result (Fig. 235). Copper salts appear yellowish-brown.

O-TOLIDINE-THIOCYANATE METHOD FOR COPPER

(*Cold knife, cold microtome, formalin- or formol-calcium-fixed frozen sections, formol-calcium-fixed paraffin sections*)

Method. (1) Bring paraffin sections to water.

(2) Flood the section with a mixture containing acetone, 5 ml.; water, 5 ml.; *o*-tolidine, 50 mg.; ammonium thiocyanate, 500 mg. Frozen sections should be

immersed in a small dish of this medium, and they should be mounted on slides and blotted dry before the next stage.

(3) Rinse in 70 per cent alcohol.

(4) Dehydrate in absolute alcohol and clear in xylene.

(5) Mount in Canada balsam or other suitable neutral medium.

Result. Copper salts are indicated by the presence of a bright blue colour, diffuse or more often crystalline. This is unstable and it disappears rapidly in mounting media which contain oxidizing or reducing agents.

BENZIDINE METHOD FOR COPPER

(Formalin, paraffin sections)

Method. (1) Bring sections to water.

(2) Cover with a solution containing 10 mg. benzidine hydrochloride and 30 mg. ammonium thiocyanate in 5 ml. distilled water. Leave for 5–10 minutes.

(3) Rinse in distilled water.

(4) Counterstain in 1 per cent aqueous neutral red, 15 seconds.

(5) Rinse in water.

(6) Dehydrate in the alcohols, clear in xylene and mount in Canada balsam or Gurr's neutral mounting medium.

Result. Copper, if present as Cu^{2+}, shows as an amorphous or sometimes crystalline deposit of benzidine blue.

This method has not been applied successfully to mammalian tissues except when copper salts have been introduced artificially.

REDUCTION METHOD FOR INORGANIC GOLD SALTS

(Christeller, 1927)

(Formalin- or alcohol-fixed paraffin sections)

Method. (1) Bring sections to water.

(2) Incubate for 36 hours at 56° in a solution containing 40 ml. freshly-prepared 5 per cent $SnCl_2$ and 5 ml. of concentrated HCl.

(3) Wash carefully in water.

(4) Counterstain the background if desired (carmalum).

(5) Dehydrate, clear and mount in the usual way.

Result. Gold appears in the form of black or brownish granules.

SULPHIDE-SILVER METHOD FOR HEAVY METALS

(after Timm, 1958)

PREPARATION OF DEVELOPER

(1) 20 per cent gum arabic (acacia). Dissolve and allow to stand for 7–14 days (100 ml.).

(2) 10 per cent aqueous $AgNO_3$, freshly prepared (1 ml.).

(3) 5 g. citric acid and 2 g. hydroquinone per 100 ml. (10 ml.).

Mix the three solutions, in the amounts stated in brackets, immediately before use.

Method. (1) Fix small blocks of tissue in 70 per cent ethanol saturated with H_2S and made alkaline by addition of 2 drops per 100 ml. of conc. ammonia. Leave for 8–12 hours at room temperature.

(2) Dehydrate blocks in alcohols, clear in xylene and embed in paraffin. Cut 6–8 μ sections.

(3) Bring sections to water.

(4) Treat with developer solution, in the dark, at 22°, for 20–60 minutes (until sections are slightly brown. Up to 6 hours' development has been recorded as necessary.)

(5) Wash in distilled water.

(6) Counterstain in 2 per cent safranin, or with Mayer's haemalum, 2 minutes.

(7) Dehydrate, clear, and mount in Canada balsam.

Result (Plate XXVIIIa). Heavy metals (Pb, Pt, Au, Ag, Fe, Cd, Cu, Co, Ni, Zn, Hg) brownish-black. Background red (safranin). Nuclei blue (haemalum).

CRYOSTAT SULPHIDE-SILVER METHOD

(after Brunk and Sköld, 1967)

Method. (1) Cut small slices of fresh tissues (max. 2 mm. thick).

(2) Place in moist chamber and expose to H_2S vapour for 20 minutes at 22°.

(3) Cut 6–8 μ cryostat sections and mount on slides.

(4) Proceed as for standard Timm method (above) with developer solution.

H_2S VAPOUR MODIFICATION FOR SULPHIDE-SILVER

Method. (1) Freeze-dry tissues according to normal procedure (Vol. 1, p. 27 and Appendix 3).

(2) Enclose dried blocks in a sealed chamber, at 22 or 37°, with relative humidity at least 40 per cent, in an atmosphere of H_2S, for 30–90 minutes.

(3) Allow humidity to fall to below 20 per cent by heating to 50°.

(4) Vacuum embed blocks. Cut paraffin sections and proceed with development, as in standard Timm (1958) method.

SULPHIDE-SELENIUM-SILVER METHOD FOR THALLIUM

(after Giusti and Fiori, 1969)
(*Formalin or ethanol; paraffin sections*)

Method. (1) Bring sections to water.

(2) Treat in moist chamber with H_2S gas for 15 minutes.

(3) Immerse in 20 per cent yellow ammonium sulphide containing excess black powdered selenium. Leave for 10 minutes or until sections are coloured greenish.

(4) Wash in distilled water.

(5) Treat with 20 per cent H_2O_2 until colourless.

(6) Proceed with development by standard procedure (above).

CHROMATE METHOD FOR LEAD SALTS

(Crétin, 1929; modified by Fairhall, 1940)
(*Formalin-fixed paraffin sections or fresh tissues*)

Method. (1) Fix fresh tissues in Regaud's (80 ml. 3 per cent potassium dichromate with 20 ml. 40 per cent formaldehyde freshly added), 24–48 hours.

(1A) Treat paraffin sections, after bringing to water, with a 2 per cent potassium chromate solution in 1 per cent aqueous acetic acid. A time of several days is recommended.

(2) Wash tissues in running water for 6 hours.

(3) Dehydrate, clear and embed in paraffin wax.

(4) Cut paraffin sections at 5–10 μ and bring to water.

(5) Counterstain sections from either source with 0·5 per cent toluidine blue (1 minute).

(6) Dehydrate, clear and mount in a suitable synthetic medium.

Result. Lead salts appear as yellow crystals of the chromate, on a pale blue background.

RHODIZONATE METHOD FOR LEAD SALTS

(*Cold microtome or cold knife, fresh sections; formalin-fixed paraffin sections*)

Bones containing lead salts might well be decalcified in 5–10 per cent H_2SO_4 containing 5–10 per cent Na_2SO_4 (Gomori, 1952). This procedure should transform Pb deposits into insoluble $PbSO_4$.

Method. (1) Transfer frozen sections (15–20 μ) directly into a freshly-prepared 0·2 per cent solution of sodium rhodizonate in 1 per cent acetic acid. Leave for 30–60 minutes.

(2) In the case of unfixed sections add 40 per cent formaldehyde to the rhodizonate medium and leave to fix for 30–60 minutes.

(3) Transfer sections to slides and counterstain if desired with 0·1 per cent light green S.F. in 1 per cent acetic acid (30 seconds).

(4) Mount in glycerine jelly.

Result. Lead salts appear scarlet-red, background light green.

THE TITAN YELLOW METHOD FOR MAGNESIUM

(*Mounted fresh frozen sections, formalin-fixed frozen and paraffin sections, etc.*)

Method. (1) Cover the section with a few drops of 0·2 per cent aqueous Titan yellow.

(2) Add a similar quantity of 2 N-NaOH and leave for 45–60 minutes.

(3) Wash off the reaction mixture.

(4) Dehydrate in alcohols.

(5) Clear in xylene and mount in DPX or other suitable synthetic medium.

Result (Fig. 236). Magnesium salts appear in a rich flame-red colour which is unstable and fades in about 24 hours.

THE MAGNESON METHOD FOR MAGNESIUM

(*Fresh frozen, formalin-fixed frozen or paraffin sections, Carnoy, etc.*)

Method. (1) Incubate sections at 56–60° for 30 minutes in a 1 per cent solution of magneson II* in 5 per cent NaOH (pH 12·2).

(2) Wash in dilute NaOH.

(3) Dehydrate in alcohols.

(4) Clear in xylene and mount in DPX or other suitable synthetic medium.

Result (Fig. 237). Magnesium salts are coloured bright blue.

* *p*-Nitrobenzene-azo-l-naphthol.

p-DIMETHYLAMINOBENZYLIDENE RHODANINE METHOD FOR SILVER

(Okamoto *et al.*, 1939)

(*Cold knife or cold microtome sections, fresh slices, formalin- or alcohol-fixed paraffin sections*)

PREPARATION OF SOLUTIONS

Rhodanine Solution. Dissolve 0·2 g. *p*-dimethylaminobenzylidene rhodanine in 100 ml. of 10 per cent ethanol containing 0·1 per cent nitric acid.

Method. (1) If possible, cut fresh tissue sections (20 µ) into the staining solution. Alternatively, fix and stain thin slices of tissue in it, in both cases for 1–3 hours at 37°. If no other is available paraffin sections may be treated for 24 hours at 37° in Okamoto's original mixture (saturated alcoholic solution of the reagent, 3·5 ml.; N-HNO_3, 3 ml.; water, 93·5 ml.).

(2) Wash the free floating sections in 70 per cent ethanol.

(3) Bring to water and mount on slides in glycerine jelly.

Result. Silver deposits appear reddish-violet.

METHOD OF GERSH FOR POTASSIUM

(*Freeze-dried material only*)

All operations, the first excepted, must be carried out in a cold room at 0°.

Method. (1) Cut paraffin sections at 15 µ and mount dry on coverslips by the application of finger pressure.

(2) Remove wax with light petroleum and dry section by burning off excess petroleum (condensation produced by evaporation is thus avoided).

(3) Cover the section with a drop of 12 per cent aqueous sodium cobaltinitrite.

(4) Allow to dry and mount directly in glycerine.

(5) Examine microscopically at 0°.

Result. Yellowish crystals and microcrystals of potassium cobaltinitrite indicate the presence of potassium salts.

The sensitivity of this reaction is not high (Feigl, 1954, gave 4 γ potassium as the limit in spot tests), and unless there is a high concentration of potassium in a given situation conditions necessary for the formation of a precipitate will not be present. Under these circumstances caution is necessary in interpreting results.

MODIFIED GERSH METHOD FOR POTASSIUM

(Crout and Jennings, 1957)

Method. (1) Cut freeze-dried sections at 10 µ and transfer directly into light petroleum at room temperature. Pick up sections and transfer to a fresh bath of light petroleum.

(2) Mount sections on a clean dry slide. Remove from light petroleum and allow to dry in the air. Press section firmly on to the slide.

(3) Immerse the slide in ice-cold cobaltinitrite* stain for 15 minutes.

(4) Withdraw the slide and blot off excess stain. Rinse in two changes of ice water until no more yellow material flows from the section (about 15 seconds).

* Dissolve 25 g. of cobaltous nitrite in 50 ml. distilled water; add 12·5 ml. acetic acid. Dissolve 120 g. sodium nitrite in 180 ml. distilled water. Add to the solution of cobalt nitrate 210 ml. of the nitrite solution and draw air through the mixture until all the nitric oxide passes off (1–2 hours). Store in refrigerator and filter before use.

(5) Dehydrate in absolute alcohol.
(6) Clear in xylene and mount in a suitable synthetic resin.

Result. Fine yellowish-brown granules indicate potassium, but the degree of accuracy of localization is uncertain.

TETRAPHENYLBORON METHOD FOR POTASSIUM

(after Collewijn, 1963)

(*FD or FS in ethanol; paraffin sections*)

Method. (1) Remove wax with light petroleum and allow to dry.
(2) Treat sections with 2 per cent aqueous sodium tetraphenylboron, at 22° for 5 minutes.
(3) Wash in 3 changes of distilled water.
(4) Stain in 1 per cent $A\mu Cl_3$ for 10 minutes.
(5) Rinse in 3 changes of distilled water.
(6) Dehydrate, clear, and mount in a suitable neutral medium.

Result. Blue-black deposits indicate location of potassium ions. Heavy metals may interfere.

MOLYBDATE METHOD I FOR INORGANIC PHOSPHATES

(after Serra and Feigl)

(*Formalin-fixed paraffin sections*)

Method. (1) Bring sections to water, and shake off excess.
Cover with a mixture of equal parts of 5 per cent ammonium molybdate and 1 per cent nitric acid, for 3–5 minutes, not longer.
(3) Wash in running water.
(4) Cover with benzidine solution (0·05 g. benzidine and 10 ml. of glacial acetic acid, made up to 100 ml. with distilled water). Leave for 1 minute.
(5) Pour off benzidine solution and flood section with 30 per cent sodium acetate.
(6) Mount in sodium acetate and ring the coverslip with paraffin wax.
(Permanent sections cannot be obtained.)

Result. A blue colour indicates the presence of phosphate. Localization is not strictly accurate.

MOLYBDATE METHOD II FOR INORGANIC PHOSPHATES

(Cheng, 1956)

(*Formalin-fixed frozen sections*)

Method. (1) Fix tissue slices 1 mm. thick in 50 per cent formalin neutralized by passage through a column of activated ion-exchange resin (De-Acidite; Permutit Co. Ltd.) before use.

(2) Cut frozen sections 20–50 μ thick and rinse these in acetate buffer at pH 4·0 (0·1 N-acetic acid and 0·025 N-sodium acetate).

(3) Transfer to acid molybdate solution (1·5 per cent ammonium molybdate in 0·005 N-sulphuric acid) adjusted to pH 4·0 with an equal volume of acetate buffer before use. Incubate at 37° for 10 minutes.

(4) Mount sections in slides, drain, and cover with 1 or 2 drops of ascorbic acid solution (2 per cent ascorbic acid) adjusted to pH 4·0 with an equal volume of acetate buffer. Apply a coverslip and observe at once under the microscope.

Result. Inorganic phosphates appear blue.

METHOD FOR CARBONATES

(after Bunting, 1951)

(*Formalin-fixed paraffin sections*)

The presence of gas bubbles arising from any deposit treated with dilute acid and examined under a coverslip is usually taken to indicate carbonate. Bunting described the following method for the positive identification of gas as CO_2.

Method. (1) Bring sections to acetone and allow to dry.

(2) Cover the selected area with clear mineral oil.

(3) Inject a microdrop of dilute acid (5 per cent sulphuric) under the oil.

(4) If a gas bubble forms the acid drop is pushed aside and replaced by a drop of saturated barium hydroxide.

Result. The slow formation of a white precipitate at the interface between the barium solution and the gas bubble indicates CO_2 and therefore carbonate.

DITHIZONE METHOD FOR ZINC

(after Mager *et al.*, 1953)

(*Freeze-dried, paraffin sections; cold knife or cold microtome, mounted sections*)

STOCK SOLUTIONS

A. 0·01 per cent dithizone in absolute acetone.

B. Complex-forming buffer. Dissolve 550 g. sodium thiosulphate (10 H_2O), 90 g. sodium acetate (3 H_2O) and 10 g. KCN in 1 litre distilled water. Adjust pH to 5·5 with acetic acid and make up to 2 litres. Extract with dithizone in CCl_4 to remove traces of zinc.

WORKING SOLUTIONS

A. Dilute stock solution A 1·5 : 1 with distilled water.

B. Mix 24 ml. stock solution A with 18 ml. distilled water. Adjust to pH 3·7 with N-acetic acid. Add 5·8 ml. stock solution B and 0·2 ml. 20 per cent sodium potassium tartrate.

Method. (1) Stain pairs of sections in A and B for 10 minutes.

(2) Wash in chloroform and rinse quickly in water.

(3) Mount in Karo syrup (Vol. 1, p. 578).

Alternatively, for cold microtome sections, stain for 3 minutes in equal parts of stock tris buffer (pH 10) and 0·05 per cent dithizone in absolute ethanol.

Result. (Figs. 238 and 239.) A red or reddish-purple granular precipitate indicates the presence of zinc.

HYDROXYQUINOLINE METHOD FOR ZINC

(after Smith *et al.*, 1969)

(*Air-dried smears*)

PREPARATION OF THE REAGENT

Add 1 ml. 3 per cent 8-hydroxyquinoline in absolute ethanol to 25 ml. Michaelis buffer, pH 8·0 (19·428 g. sodium acetate, 29·428 g. sodium phenobarbital in 1 litre. To 100 ml. add 40 ml. 0·85 per cent NaCl and adjust to pH 8·0 with 0·1 N-HCl).

Method. (1) Stain for 15 minutes in the above reagent.

(2) Drain and wash in distilled water.

(3) Dry in air.
(4) Mount in non-fluorescent immersion oil.

Result. Granules of leucocytes fluoresce pale greenish-yellow. Controls treated with 1 per cent acetic acid fail to fluoresce.

PAS METHOD FOR DEXTRAN

(after Mowry and Millican, 1953)
(*Cold ethanol; paraffin sections*)

The original authors recommend that after 48 hours fixation in cold ethanol tissues should be transferred to fresh alcohol for 18 hours before clearing in light petroleum and embedding in paraffin.

Method. (1) Cut sections at 5 μ and float sections on to clean slides moistened with 95 per cent alcohol.
(2) Dry at 37° for a few hours and then at 56–60° for 20 minutes.
(3) Remove paraffin with xylene and transfer to absolute alcohol.
(4) Incubate with 1 per cent periodic acid in 90 per cent ethanol for 2 hours.
(5) Wash in 95 per cent alcohol and then in 50 per cent alcohol.
(6) Immerse in Schiff's reagent for 10 minutes.
(7) Wash in acid metabisulphite for 5 minutes.
(8) Dehydrate in alcohol, clear in xylene and mount in a suitable synthetic resin.

Result. Dextran and other polysaccharides stain magenta-red.
Control slides stained by the aqueous PAS method (Appendix 10, p. 659) should show loss of staining where this is due to dextran.

METHODS FOR DEXTRAN SULPHATE

(Mowry, 1954)
(*Cold ethanol; paraffin sections*)

Method I. (1) Prepare sections as in the preceding method.
(2) Bring sections to 95 per cent alcohol, containing 0·5 per cent toluidine blue. Leave for 20 minutes.
(3) Transfer to 70 per cent alcohol containing toluidine blue and then to 50 per cent alcohol containing toluidine blue.
(4) Stain finally in 0·5 per cent aqueous toluidine blue for 1–2 minutes.
(5) Rinse in 70 per cent alcohol, dehydrate in 95 per cent and absolute alcohol; clear and mount in a synthetic resin.

Method II. (1) Prepare sections as above.
(2) Stain in 0·005 per cent toluidine blue in 70 per cent alcohol for 5–10 minutes.
(3) Rinse in tertiary butyl alcohol.
(4) Clear in xylene and mount in a synthetic resin.

Result. Dextran sulphates stain (metachromatically) deep red.

IODINE METHOD FOR PVP

(after Jancsó, 1954)

Method I. (1) Fix small pieces of tissue in a mixture containing 20 ml. saturated ammonium sulphate, 20 ml. saturated KIO_4 and 0·3 g. KI. Leave for 24 hours at room temperature, changing the fixative after 5 hours.
(2) Cut frozen sections, mount on slides in glycerine jelly containing enough Lugol's iodine to impart a brownish-yellow colour.

Method II. (1) Transfer tissues from stage 1 of Method I to 10 per cent aqueous uranyl nitrate containing 1 drop per 10 ml. of Lugol's iodine. (The tissues should rest on cotton wool.) Leave for 24–36 hours.

(2) Transfer to a mixture of 3 parts alcohol and 7 parts xylene containing a small concentration of iodine.

(3) Transfer to xylene-alcohol 9 : 1 containing iodine as before.

(4) Embed in paraffin wax *in vacuo*.

(5) Cut sections 5–7 μ thick, remove wax with xylene and mount in Canada balsam.

Result. PVP stains brown, polyvinyl alcohol black.

CONGO RED METHOD FOR PVP

(Freiman and Gall, 1955)

(*Formalin; paraffin sections*)

Method. (1) Bring sections to water.

(2) Stain in 2 per cent Congo red in 10 per cent ethanol for 30 minutes.

(3) Wash in water.

(4) Counterstain in 1 per cent aqueous light green SF for 2 minutes.

(5) Rinse quickly.

(6) Blot and mount in Apathys gum syrup.

Result. PVP stains orange to cherry-red. Elastic tissue brown. Nuclei brown to green, other structures green.

CHLORAZOL FAST PINK METHOD FOR PVP

(*Formalin; paraffin sections*)

Method. (1) Bring sections to water.

(2) Stain in 1 per cent Chlorazol fast pink in 50 per cent ethanol at 60° for 10–30 minutes.

(3) Rinse in water.

(4) Counterstain nuclei with Mayer's haemalum, 3 minutes.

(5) Wash, differentiate in 1 per cent acid alcohol.

(6) Wash in water until nuclei are blue.

(7) Dehydrate, clear and mount in a synthetic resin.

Result. PVP stains pink to cherry-red; nuclei blue.

XANTHYDROL METHOD FOR UREA

Method. (1) Fix thin blocks (1–2 mm.) of tissue for 3–6 hours in 5 per cent xanthydrol in acetic-ethanol (Acetic acid 60 ml.; ethanol 40 ml.) at 56–60°.

(2) Wash in water for 1 hour.

(3) Fix in 10 per cent formalin, 6 hours.

(4) Wash briefly in water.

(5) Cut frozen sections 5–10 μ thick and mount on slides.

(6) Blot dry and transfer to 95 per cent alcohol.

(7) Cover with a thin film of 1 per cent celloidin in alcohol-ether.

(8) Stain briefly in 1 per cent eosin and wash.

(9) Dehydrate, clear and mount in balsam.

(10) Examine under ordinary and polarized light.

Result. Yellowish-brown birefringent, crystals of urea-xanthydrol indicate sites where urea was present.

HEXAMINE SILVER METHOD FOR URIC ACID

(*Alcohol; paraffin sections*)

Method. (1) Fix fresh tissues in absolute alcohol for 24 hours, changing the alcohol at least once during this time.

(2) Embed in paraffin. Cut sections 5–7 μ thick and float on to slides with 95 per cent alcohol.

(3) Dry at 37° for 2 hours.

(4) Remove wax with light petroleum and wash in absolute alcohol.

(5) Transfer directly to hexamine-silver solution (Appendix 26, p. 1379) at 37°. The slides should remain in this solution until uric acid (urate) deposits are black (15–60 minutes).

(6) Rinse in 5 per cent sodium thiosulphate, 3 minutes.

(7) Counterstain with 1 per cent eosin, 1 minute.

(8) Wash. Dehydrate, clear and mount in a synthetic resin.

Result. Urates black. Calcium phosphate and carbonate deposits will also stain black if large enough to withstand solution in the silver bath.

APPENDIX 29

STOWARD'S METHYLATION-PERIODIC ACID-SO_2-DYE METHOD

(Stoward, 1967)

(1) (Methylation) Take deparaffinized sections through two changes of methanol and methylate with a 2 per cent solution of thionyl chloride in methanol (1–7 days old) for 4 hours at room temperature. Rinse in methanol and take down to water.

(2) Oxidize in 1 per cent periodic acid (pH previously adjusted to be within range 3–5) for 10 minutes. Rinse briefly in water.

(3) Immerse in sulphur dioxide solution for 20 minutes (0·25 per cent v/v thionyl chloride in distilled water, pH adjusted with sodium hydroxide to pH 3). Rinse briefly in water.

(4) Stain in solution of a fluorescent basic dye at appropriate pH, usually 3, for 20 minutes.

(5) Blot dry, dehydrate in isopropyl alcohol, rinse in 3 changes of xylene, and mount in DPX.

Basic fluorochrome solutions:

Acridine orange, 1/10000	pH 3
Coriphosphine O, 1/10000	pH 3
Acridine yellow, 1/1000	pH 3
Benzoflavine, 1/1000	pH 2·0–2·3
Auramine O, 1/1000	pH 3
Basic Fuchsin	
Pararosaniline	

METHODS FOR MASKED METACHROMASIA

(Solcia *et al.*, 1968; Bussolati *et al.*, 1969; Rost and Maunder, 1971)

FIXATION

Small blocks of fresh tissue are fixed in glutaraldehyde-picric acid fixative for 24 hours, rinsed, and processed to conventional paraffin sections.

GLUTARALDEHYDE-PICRIC ACID FIXATIVE

Picric acid, saturated aqueous solution	3 parts
Glutaraldehyde, 25 per cent aqueous solution . . .	1 part
Sodium acetate	1 per cent

HYDROLYSIS

The amount of hydrolysis required depends upon the tissue. We have found 5 N-hydrochloric acid at 60° for 10 minutes suitable for most tissues; muscle requires much longer.

STAINING

Metachromasia is demonstrated by staining in Toluidine blue (C.I.52040) or Coriphosphine O (C.I. 46020)—the latter is a fluorochrome. Alcian blue has been used to give a permanent preparation, but does not show metachromasia.

The method for staining in Coriphosphine O is as follows:

Stain in 0·01 per cent Coriphosphine O in 0·1 M-acetate buffer (pH 5) for 10–30 minutes. Wash in buffer for not more than 5 minutes, blot gently and dry thoroughly in air. Dip in xylene and mount in DPX. Alternatively, mount in buffer

TABLE 95

Name	C.I.	Chemical Type	Fluorescence Colour	Notes†			
Acridine orange	46005	Acridine	G, R	Bs	M	V	S
Acridine yellow	46025	Acridine		Bs			S
Acriflavine	46000	Acridine	Y	Bs			S
Atebrin, see Quinacrine							
Auramine O	41000	Arylmethane	Y	Bs			
BAO*		Oxdiazole	B				S
Berberine sulphate	75160		Y	Bs			
Benzoflavine	46065	Acridine			M		S
Congo red	22120	Disazo	R	Ac	D		
Coriphosphine O	46020	Acridine	G, R	Bs	M		S
Eosin (Y)	45380	Fluorone	G	Ac			
Evans blue	23860	Disazo	R			V	
Flavophosphine N	46065		Y–O	Ac			S
Fluorescein	45350	Fluorone	G	Ac			
Fluorescein isothiocyanate		Fluorone	G				
Fuchsin, acid	42685	Arylmethane	R	Ac	D		
Fuchsin, basic	42510	Arylmethane	R	Bs	D		S
Geranine B	15080	Monoazo, thiazole					
Lissamine Rhodamine B200	45100	Rosamine	R				
Mercurochrome		Fluorone			D		
Methyl green	42585	Arylmethane	R				
Morin	75660		G				
Neutral red	50040	Quinone-imine	R	weak BS DV			S
Pararosaniline	42500	Arylmethane	R	Bs	D		S
Phenosafranin	50200	Quinone-imine	R				S
Phloxine (B)	45405	Fluorone			D		
Phosphine, Phosphine 3R	46045	Acridine	B	Bs			S
Phosphine GN	46056	Acridine	O–brown				S
Primulin	49000	Thiazole		Ac		V	
Pseudoisocyanin		Quinoline	Y		D		
Pyronin Y	45005	Aminoxanthene	R	Bs	D		S
Quinacrine		Acridine	Y				
Quinacrine mustard		Acridine	Y				
Rheonin A	46075	Acridine	R			V	S
Rhodamine B	45170	Rhodamine	R	Bs			
Rhodamine S	45050	Xanthene	Y	Bs	D		
Rhodamine 3G	45210	Rhodamine	O	Bs	D		S
Safranin	50240	Quinone-imine	R	Bs			S
Thioflavine T	49005	Thiazole	B–white	Bs			
Thioflavine S	49010	Thiazole	B–Y	Ac		V	
Uranin = sodium fluoresceinate							

† Ac = acid dye; Bs = basic dye; V = vital dye; S = usable in pseudo-Schiff; D = diachrome; R = red; Y = yellow; G = green; B = blue; O = orange; M = metachromatic.

* 2,5 bis[4′-aminophenyl-(1′)] = 1,3,4-oxdiazole.

solution, but this preparation fades more quickly under irradiation and the dye soon dissolves out. View in a fluorescence microscope with blue-light excitation and a barrier filter with a cut-off not higher than 530 nm.

TABLE 96

Fluorochrome excitation and emission maxima

Fluorophore	Excit. max. nm	Reference	Emiss. max. nm	Reference
FITC conjugates	495	Nairn, 1969	520	Nairn, 1969
DANS conjugates	340	Nairn, 1969	525 (450–620)	Nairn, 1969
Lissamine Rhodamine B (RB200) conjugates	575	Nairn, 1969	595, 710	Nairn, 1969
Acridine orange (C.I. 46005)	430–500	Rigler, 1966	530, 650	Rigler, 1966
Coriphosphine O (C.I. 46020)				
Green fluorescence	500	Bussolati, Rost, and Pearse, 1969	535	Bussolati, Rost and Pearse, 1969
Red fluorescence	450 (440–460)	Bussolati, Rost and Pearse, 1969	650	Bussolati, Rost and Pearse, 1969
Thioflavine T or amyloid (C.I. 49005)			c. 450	Rost (unpublished)
Pararosaniline Feulgen	565–590	(absorption) Hardonk and van Duijn, 1964		

QUINACRINE STAINING OF CHROMOSOMES

(modified, after Pearson *et al.*, 1971)

STANDARD CHROMOSOME PREPARATIONS ARE USED.

Stain for 5 minutes in quinacrine hydrochloride, 0·5 per cent in distilled water or ethanol. Rinse *briefly* in tap water to differentiate. Shake off excess fluid and mount in distilled water or buffer pH 7·2–7·4. Examine in a fluorescence microscope, using the 436 mercury line for excitation and a barrier filter transmitting wavelengths above 490 nm (e.g. HBO200 mercury lamp, Schott AL436 excitation filter, Leitz "Ploem" epi-illuminator in position 2, barrier filter Leitz K490).

Neither the concentration of the staining solution nor the time of staining is critical, the end result being easily judged by inspection. First-class optical equipment is essential if the banding pattern is to be observed.

RECIPROCITY CHARACTERISTICS OF PHOTOGRAPHIC FILMS

The corrections required for long exposures, such as are usually required for photographing fluorescence, are given below in respect of some black-and-white and colour reversal films.

BLACK-AND-WHITE

Ilford HP4. The exposure indicated by an exposure meter should be increased as follows:

Indicated exposure sec.	Corrected exposure sec.
1/5000 to 1	indicated exposure
2	3
4	7
8	18
16	48

Ilford FP4. The graph below can be used to calculate the increase in exposure time to compensate for reciprocity failure. No correction is required for exposure times in the range 1/1000 to ½ sec. The times on the horizontal axis represent the exposure times, the vertical axis gives the corrected exposure times which should be used in practice. By entering the graph the other way round, it is possible to calculate a correction to the speed value (ASA or DIN) to be set on a fully automatic camera (such as the Leitz Orthomat). For example, if experience suggests that the exposure should take about 30 seconds, the graph gives that the "indicated exposure" should be 12 seconds; the "ASA" rating should therefore be set, not to 125 but to 125 × 12/30.

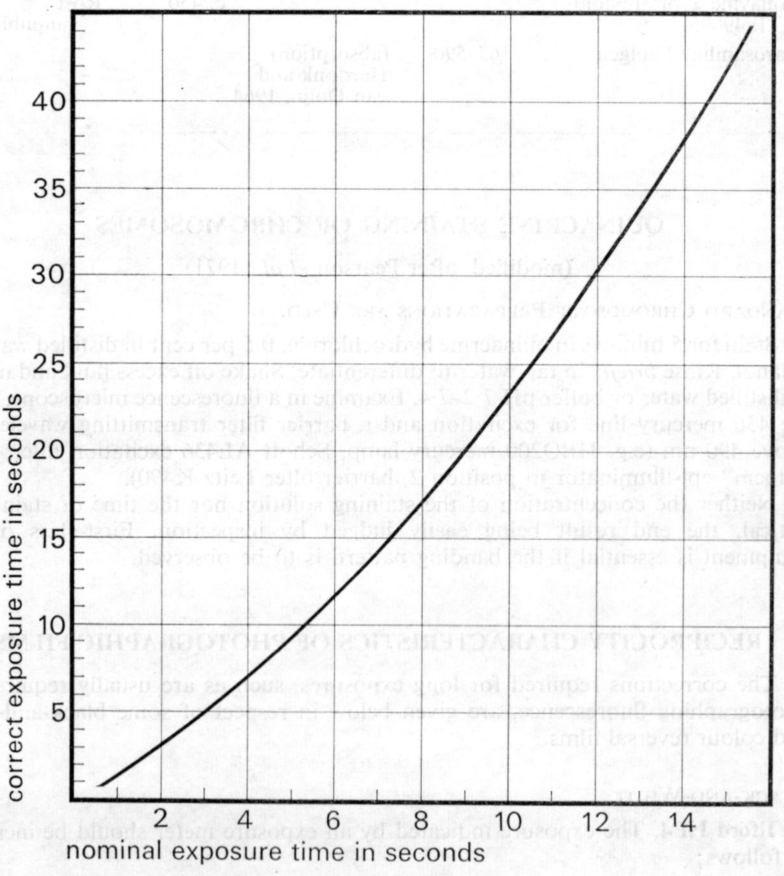

Colour Reversal Films

The graph below indicates the approximate "ASA" value to be set on a fully automatic camera for various exposure times. It should be noted that deviations from correct colour balance will occur at long exposure times, because of differing reciprocity failures in each of the colour layers. For fluorescence photomicrography, however, these changes in colour balance are unlikely to be noticeable.

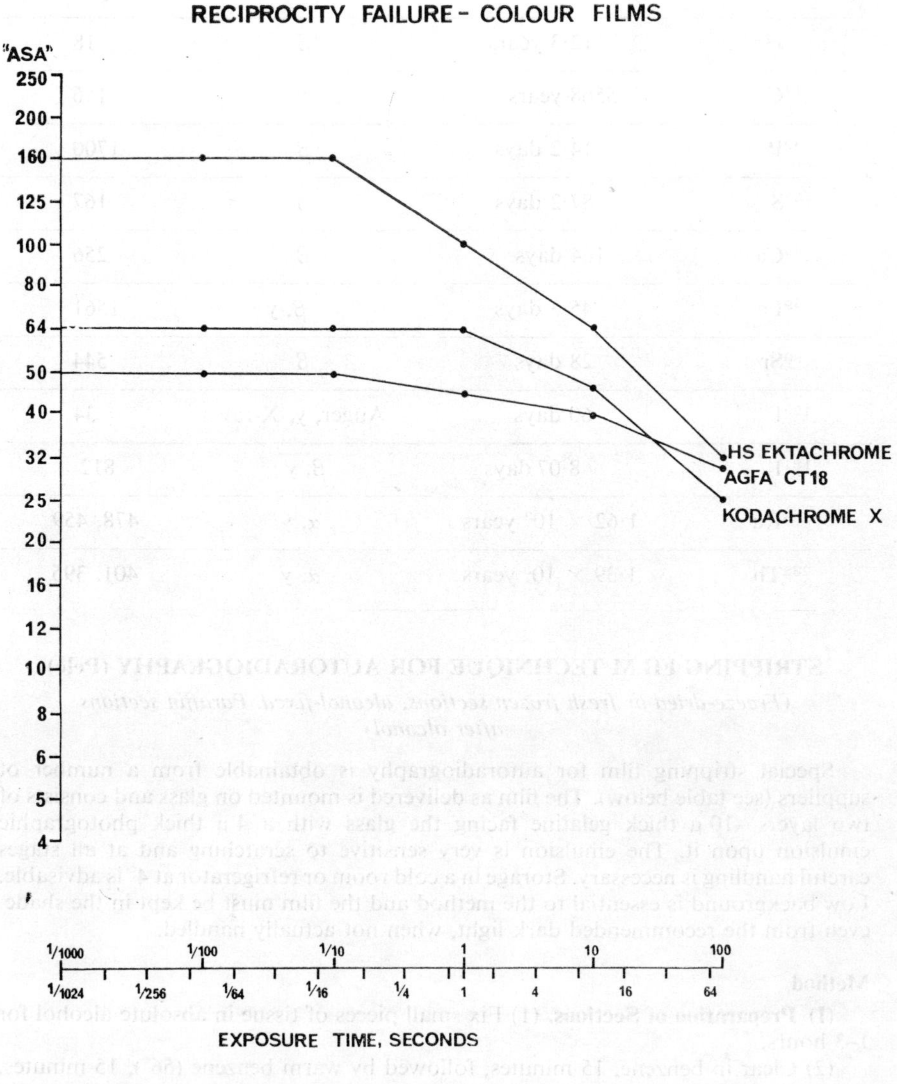

APPENDIX 30

TABLE 97

Isotopes in common use

Atom	Half-life	Emitted Radiation(s)	Max. Energy keV
^{3}H	12·3 years	β	18
^{14}C	5568 years	β	156
^{32}P	14·2 days	β	1700
^{35}S	87·2 days	β	167
^{45}Ca	164 days	β	256
^{59}Fe	45·3 days	β, y	1561
^{90}Sr	28 days	β	544
^{125}I	60 days	Auger, y, X-ray	34
^{131}I	8·07 days	β, y	812
^{226}Ra	$1{\cdot}62 \times 10^{3}$ years	α, y	478, 459
^{232}Th	$1{\cdot}39 \times 10^{3}$ years	α, y	401, 395

STRIPPING FILM TECHNIQUE FOR AUTORADIOGRAPHY (Pelc)

(*Freeze-dried or fresh frozen sections, alcohol-fixed. Paraffin sections after alcohol*)

Special stripping film for autoradiography is obtainable from a number of suppliers (see table below). The film as delivered is mounted on glass and consists of two layers—10 μ thick gelatine facing the glass with a 4 μ thick photographic emulsion upon it. The emulsion is very sensitive to scratching and at all stages careful handling is necessary. Storage in a cold room or refrigerator at 4° is advisable. Low background is essential to the method and the film must be kept in the shade, even from the recommended dark light, when not actually handled.

Method

(I) Preparation of Sections. (1) Fix small pieces of tissue in absolute alcohol for 1–3 hours.

(2) Clear in benzene, 15 minutes, followed by warm benzene (56°), 15 minutes.

(3) Transfer to four changes of molten paraffin wax (56°), for 30 minutes in each case.

TABLE 98

Glass Plate Emulsions

Type		Grain size (nm)	Indications
Eastman-Kodak	NTB	290	α and β (low energy)
	NTB2	260	β up to 200 keV
	NTB3	340	High sensitivity
	No-screen	1250	Low resolution
Ilford	G5	270	High resolution
	K2	200	β up to 200 keV
Kodak (UK)	AR10	200	α and β-particles

(4) Embed, block and cut sections at 5 μ or less and float out on warm water.

(5) Pick up sections on to slides which have previously been dipped in an aqueous solution of 1 per cent gelatin and 0·1 per cent chrome alum, and then allowed to dry.

(II) Application of Film (Fig. 279). (1) Cut film in squares with scissors, razor blade or scalpel, allowing for a generous overlap over the edge of the slide; ¼ to ½ inch from the edge of the slide should be discarded. Leave in the dark for 5–10 minutes.

(2) Strip one square slowly, starting at one corner and float on dust-free, fresh distilled water at 23°–25°, emulsion-side *downwards* for 2–3 minutes. (The water should only be used for one run of coatings.)

(3) Place a deparaffinized slide in the water, section or squash facing upwards, and lift out of the water so that the film covers the specimen.

(4) Dry in front of a fan or hair-dryer at room temperature.

(5) Store in the dark for exposure. (Some workers expose at room temperature, others at 4°. Warm rooms should be avoided.)

(6) Photographic processing should be carried out according to the maker's instructions. Use chemicals of analytical purity dissolved in distilled water and filtered. All processing should be done at 17·5±0·5°.

(7) Rinse in distilled water, and in running tap-water if this is sufficiently clear and cool, for 1 hour. Otherwise use 4–6 changes in distilled water.

(8a) For phase contrast microscopy mount in glycerine jelly; glychrogel (Appendix 1) is suitable for this purpose.

(8b) Staining with haemalum: place moist preparation in celestin blue (see PAS technique, Appendix 10) for 5 minutes, wash briefly and immerse in Mayer's haemalum, 5 minutes.

(9) Differentiate in 1 per cent aqueous HCl for 16–60 seconds, or until there is minimal staining of the gelatine compatible with retention of nuclear staining.

As an alternative, deparaffinized preparations can be stained before stage 3, by the Feulgen reaction (Appendix 9) or with Neutral red-Carbol fuchsin (30 ml. 2 per cent Neutral red with 10 ml. Ziehl-Neelsen's carbol fuchsin). Haematoxylin may also be used if hardener and fixer without acetic acid are used in processing. Objections have been made to the use of iron haematoxylins for use at stage 8b since silver is removed by iron salts, especially in acid solution. Removal of silver does not apparently occur with the iron alum-mordanted celestin blue solution recommended above.

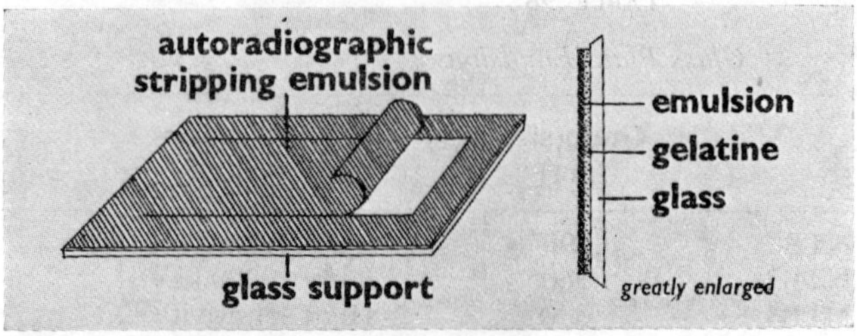

(1) Cut with knife and strip film as shown.

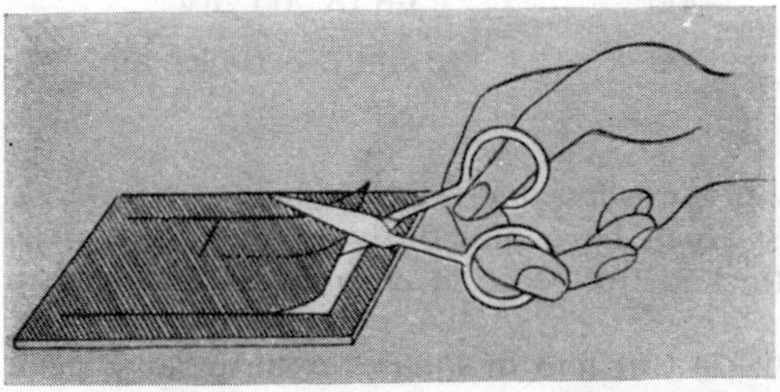

(2) Cut off the area of the emulsion required (usually 1 in. square).

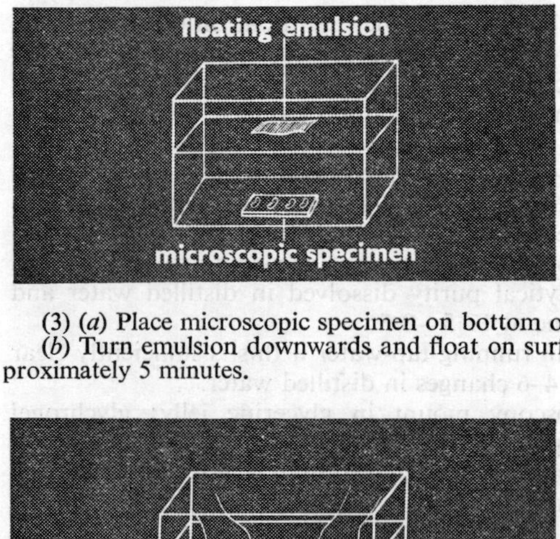

(3) (*a*) Place microscopic specimen on bottom of dish filled with distilled water.
(*b*) Turn emulsion downwards and float on surface of distilled water (23°–25°) for approximately 5 minutes.

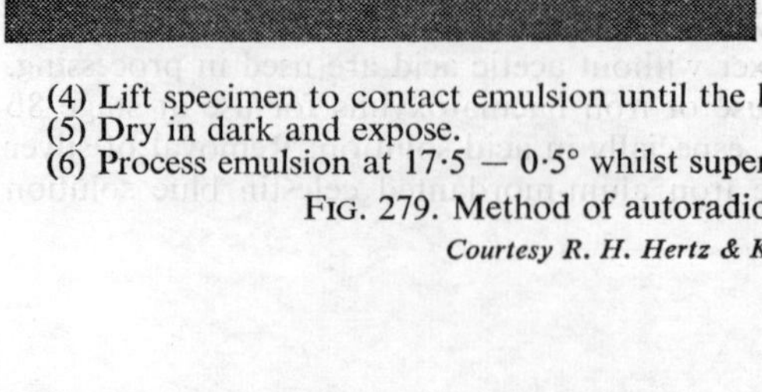

(4) Lift specimen to contact emulsion until the latter covers the specimen.
(5) Dry in dark and expose.
(6) Process emulsion at 17·5 — 0·5° whilst superimposed on specimen.

FIG. 279. Method of autoradiography after Pelc.

Courtesy R. H. Hertz & Kodak Ltd.

CONTACT METHOD FOR AUTORADIOGRAPHY (Pelc)

(*For discussion of the technique of contact autoradiography see Lotz* et al., *1951*)

Method

(1) Prepare specimen as in Stages 1–5 of the previous method.

(2) Remove wax by immersion in light petroleum.

(3) Bring the section into contact with X-ray film (for highest speed) or fine-grained film (for better resolving power).*

(4) Apply even pressure by putting film and specimen into a printing frame, X-ray cassette, or simply by putting a weight on it.

(5) Store in the dark for exposure.

(6) Process the film according to the maker's instructions, paying particularly strict attention to cleanliness throughout.

DIP-COATING TECHNIQUE FOR AUTORADIOGRAPHY

(Kopriwa and Leblond, 1962)

Preparation of Tissue Sections

Slides carrying sections, stained or unstained, are dipped twice in 1 per cent celloidin in alcohol/ether (equal parts). After drying overnight at room temperature the coated slides are smeared with a small quantity of egg albumin: glycerine (1:1), except over the section.

Emulsion Coating

This is carried out in a completely light-proof dark room, maintained at 16°. The emulsion is kept at 40–45°, either in a Coplin jar on a water-bath, or in an incubator. Before dipping the slides may be warmed to 40° on a hot-plate. Alternatively they may be fixed in plastic holders and kept with the liquid emulsion in the incubator (as shown in Fig. 280). A Wratten safelight No. 2 may be used, preferably intermittently.

The slides, held at the label end, are dipped for 1–2 seconds in the melted emulsion. Excess emulsion is allowed to drain back into the container, and the slide is allowed to dry in the vertical position (about 1½ hours) with the safelight off.

Exposure

The slides are stored in black plastic boxes containing a metal capsule filled with Drierite, or other drying agent. The box is sealed with black adhesive tape and kept at 4° for the period of exposure, in such a position that the slides are in the horizontal position, emulsion side down.

Processing

The exposed preparations are developed for 1½ minutes in Kodak Dektol (D-72) developer. Fixation is carried out for 10 minutes in acid fixer with hardener. The slides are then washed in running water at about 18° for 15 minutes and dehydrated in graded alcohols. Unless counterstaining is required they are then rinsed in acetone, placed in a 1:4 acetone dilution of "vinylite cement" and dried at 60° in the vertical position. When dry, a drop of Canada balsam and a coverslip are added.

* Film is preferable to plate because of the unevenness of the latter. If necessary, a thin film of cellophane (less than 0·2 mm. thick) can be inserted between the film and the specimen.

DRY-MOUNTING TECHNIQUE FOR CRYOSTAT SECTIONS

(Appleton, 1964)

Preparation of Coverslips

Coverslips are used in preference to slides because frozen sections can be picked up more easily from the microtome knife and because the relative position of emulsion and tissue of the completed preparation becomes the same as in the standard stripping film autoradiograph.

Number 1 coverslips are smeared on both sides with glycerine-albumen and covered with Kodak AR 10 stripping film, emulsion side *upwards*, as shown in the diagram. The filmed coverslips are dried in a stream of air at 22° for 2–3 hours.

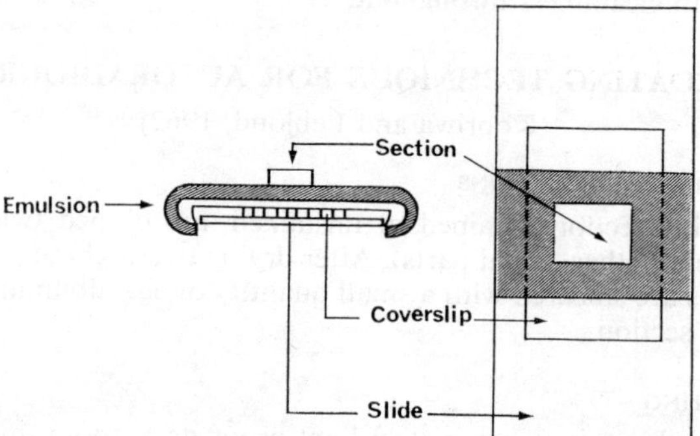

They can be stored in light-tight boxes at 4°, and cooled to −5 to −10° when required for use. The undersides are marked with a felt pen.

Preparation of Sections

Cryostat sections, 5 μ thick, are cut at −20° in the darkroom (Wratten series 1 safelight). Sections are transferred to the cooled filmed coverslips by the usual type of procedure (see Vol. 1, p. 591).

Exposure

The completed preparations are exposed in light-tight plastic boxes at −20°.

Processing

After suitable exposure the attached tissue sections are fixed in acetic-ethanol for 1 minute at room temperature. They are then washed very thoroughly and developed in Kodak 19b at 17° for 5 minutes. Fixation is carried out in Johnson's Fixsol (1:10) for 12 minutes at 17°.

Staining

Since the section is on top of the film staining presents no problem. Toluidine blue (0·1 per cent aqueous) for 30 seconds is a useful routine procedure. After staining the preparations are air-dried, cleared with xylene and mounted on a drop of Euparal on a glass slide.

DRY-MOUNTING TECHNIQUE FOR FREEZE-DRIED BLOCKS

(Nagata, Nawa and Yokota, 1969)

Quenching Procedure

Small blocks of tissue are placed on a cold plate (thermoelectric platen excellent for this purpose). With two razor blades they are cut down to 1 × 0·5 × 0·5 mm. The blocks are attached to small (5 × 5 mm.) pieces of aluminium foil. Cultivated cells are centrifuged at 500 r.p.m. for 10 minutes and the pellet is then mounted on aluminium foil, as above.

The foil-mounted tissues or cells are plunged into melting isopentane (−160°), ladled out with a small aluminium foil cup, into liquid nitrogen and, finally, transferred to the freeze-drying apparatus at −50° in the same foil cup. This cup should be formed on a mould with a flat bottom, to ensure perfect apposition to the platen surface. The authors described their own thermoelectric freeze-drying apparatus and the cold (−65°) platen of the Edwards thermoelectric dryer (Vol. 1, p. 50) is ideally suited to the purpose.

Drying Procedure

The authors recommended 24–48 hours at −50° followed by several hours gradually raising the temperature of the platen (by reversing the current) to 22°.

Embedding Procedure

An Epon mixture is prepared consisting of 9·1 ml. Epon 812, 6·2 ml. 2-dodecenylsuccinic anhydride (DDSA), 4·7 ml. methyl nadic anhydride (MNA) and 0·4 ml. tridimethylaminomethylphenol (DMP30). This is placed in the reservoir of the resin embedding accessory (Vol. 1, p. 596) and degassed (before commencing the drying run).

On completion of drying the embedding medium is allowed to drip into the chamber to fill the cup containing the dried tissues. The chamber is then shut off from the pump and the infiltrated tissues sink to the bottom of the cup. The chamber is then let down to air, the tissues removed and infiltrated with a new batch of embedding medium overnight at 22°. Polymerization is carried out for 3 hours at 37° and at 50° for 48 hours.

Sectioning

Ultrathin sections are cut with ethylene glycol instead of water in the knife trough (Pease, 1966). They are picked up therefrom on to collodion-coated grids.

Dry-Mounting Procedure

The grids, carrying sections, are placed on a grid holder made of glass rods (mounted vertically upwards on a glass slide). Diluted Kodak NTE or Ilford L4 emulsion (1:10, 40°) is used. Before use, to 10 ml. of emulsion is added 0·2 ml. 2 per cent aqueous dioctyl sodium sulphosuccinate and the mixture is maintained at 40° for several minutes. A thin film of emulsion is taken up in a platinum wire loop and allowed to dry for 1–2 minutes in the horizontal position. While the peripheral zone of the film is still wet (opaque) and the central zone dry (translucent), it is applied to the grids (mounted on top of their holders). The grids are then transferred to Petri dishes and warmed to 37° for 1 hour to promote adherence of the film.

EXPOSURE AND DEVELOPMENT

Expose at 4° in a light-tight box containing desiccant. The development procedure recommended by the authors is that introduced by Salpeter and Bachmann (1964) under the name of gold "latensification" and Elon-ascorbic acid.

GOLD "LATENSIFICATION"

(This process deposits metallic gold on to the latent images before actual development.)

Dilute 0·5 ml. of 2 per cent stock gold chloride ($AuCl_3.HCl.3H_2O$) to 10 ml. and dissolve therein 0·125 g. potassium thiocyanate. Add 0·15 g. potassium bromide and dilute to 250 ml. This solution should be used for one day only.

Dip the exposed grids in water. Transfer to a 1:20 dilution of the gold thiocyanate mixture for 30 seconds. Wash in distilled water. (Sensitivity is increased as a function of strength and time of application.)

ELON-ASCORBIC ACID DEVELOPER

This consists of 0·045 per cent Elon (Metol), 0·3 per cent ascorbic acid, 0·5 per cent borax and 0·1 per cent potassium bromide. It is used at 24° and is unstable. Development time varies with age of the developer. Between 5 and 48 hours-old, the grain size is approximately 50 nm at 8 minutes.

DRY-MOUNTING TECHNIQUE FOR FREEZE-DRIED SECTIONS

(Stumpf and Roth, 1966)

PREPARATION OF SECTIONS

Small pieces of tissue are quenched, in the usual manner for freeze-drying, mounted on cryostat tissue holders. Sections (1–2 μ thick) are cut at −30 to −70° in a low temperature cryostat (see Vol. 1, p. 592). They are transferred, with a fine brush, either to a tissue carrier suitable for the cooled specimen chamber of the Stumpf and Roth cryosorbtion-pumping freeze-drier, or to a desiccator over P_2O_5 contained in the cryostat chamber. In either case, after evacuation of the container, drying is allowed to occur at −40 to −70°. The freeze-dried sections can be sorted at room temperature in a desiccator.

APPLICATION TO THE EMULSION

Emulsion coated slides are prepared and stored in a refrigerator at 0–4°. A convenient modification (Fig. 280) is the provision of a sliding door. The sections are transferred to a polytetrafluoroethylene (Teflon) plate. It is convenient to have both radioactive and control sections together. The plate is positioned under a dissecting microscope (Fig. 281) and, with fine needles, any folds or ridges are smoothed out. An emulsion-coated glass slide is then placed on top of the Teflon plate and the two are pressed together. When pressure is released, the Teflon plate falls away.

EXPOSURE

The sections are placed in a light-tight plastic box (the authors recommend a modified Clay-Adams box with a Drierite compartment and a humidity plug, so that the internal conditions can be monitored. Exposure is carried out at −15°. Following exposure the box is allowed to come to room temperature before opening. Emulsion and tissue are briefly moistened (with the breath) before development.

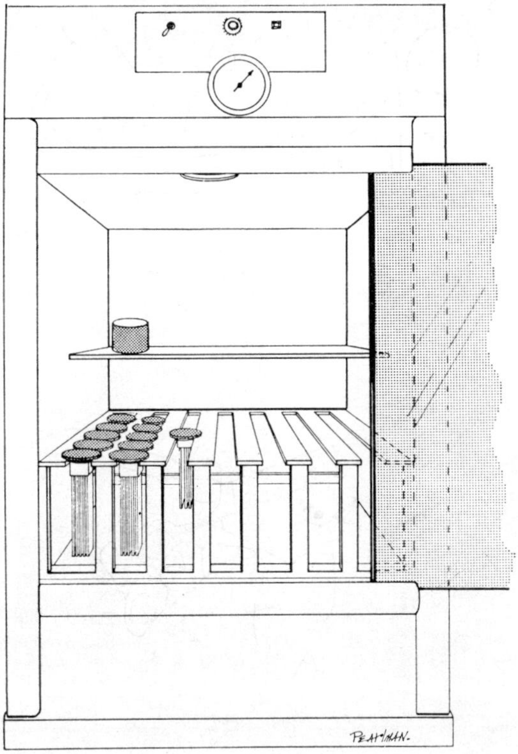

FIG. 280. Incubator (40°) with sliding door for liquid emulsion and slide racks.

DEVELOPMENT

Kodak D19 for 1 to 1½ minutes at 22° is followed by a brief rinse in water, and by Kodak fixer for 5–7 minutes at 22°. Staining is then carried out. Finally the preparation is air-dried and mounted in Permount.

HIGH RESOLUTION, EM AUTORADIOGRAPHY

(Caro and van Tubergen, 1962; Van Kleef *et al.*, 1969)

PREPARATION OF EMULSION

Colloidal silver (about 1 g.) is peptisized in 100 ml. of warm (70°) 1 per cent photographic gelatin. The solution is centrifuged at $30^{1}000$ g. and 40° for 30 minutes to remove coarse silver particles. The supernatant is then centrifuged under the same conditions for 6 hours. The sediment, consisting of 60 Å silver particles, is taken up in 30 ml. glass-distilled water at 70°. Stored at 4° this emulsion is stable for 6 months.

PREPARATION OF SECTIONS

Sections are fixed in buffered O_sO_4, dehydrated in alcohols and embedded in epoxy resin. They are picked up on a grid, coated with collodion and carbon, and

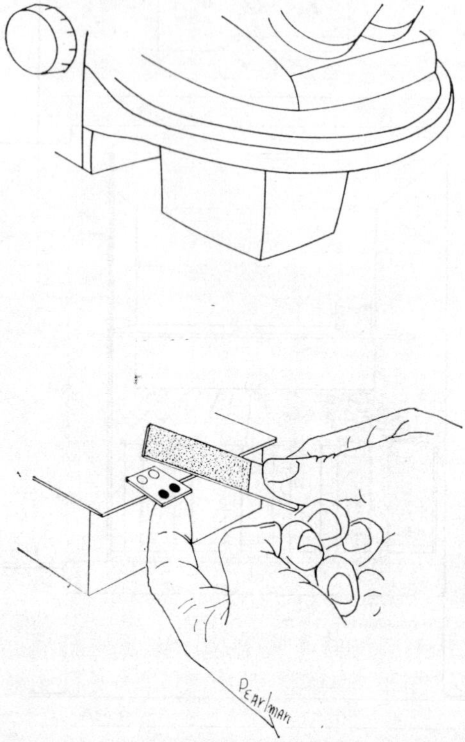

FIG. 281. Dry mounting of freeze-dried sections on Teflon support under dissecting microscope. Radioactive ● and non-radioactive control O sections are mounted simultaneously.

attached to a glass slide with masking tape. Several grids can be placed side by side on a single glass slide.

APPLICATION OF EMULSION

The emulsion (prepared as above or alternatively Ilford L4, 10 g. in 20 ml. of water) is melted at 45° and then cooled until viscous. A thin film of emulsion is taken up on a wire loop (4 cm. diameter) and this is applied to the slide carrying the grids.

EXPOSURE

Slides are treated as for autoradiography with the optical microscope. The time of exposure is much longer, however, for EM autoradiography 8 to 16 times as much exposure may be necessary.

DEVELOPMENT

See Table 98.

PROCESSING SCHEDULE FOR VARIOUS PREPARATIONS

(Caro and van Tubergen, 1962)

TABLE 98

Preparation	Developer	Stop	Fixer	Rinse
Optical microscope K5	D19 2 min	1% acetic 10 sec.	Kodak (rapid) 5 min.	Running water, 5 min. Distilled water 1 min.
Optical microscope L4	D19 4 min.	1% acetic 10 sec.	Kodak (rapid) 5 min.	Running water, 5 min. Distilled water, 1 min.
Electron microscope L4	Microdol 5 min.	1% acetic 10 sec.	Kodak (rapid) 5 min.	Running water, 5 min. Distilled water, 1 min.
Electron microscope L4 (fine grain)	Physical Developer* 1 min. 20°	1% acetic 10 sec.	Kodak (rapid) 5 min.	Running water, 5 min. Distilled water, 1 min.

* Freshly prepared 0·1 M sodium sulphite with 10 mM paraphenylenediamine.

APPENDIX 31

STANDARD FLUORESCEIN SOLUTION FOR MICROFLUORIMETRY

(Jongsma, Hijmans and Ploem, 1971)

SOLUTION A

Add 20·80 mg. fluorescein diacetate to 10 ml. of freshly prepared 5 per cent sodium hydroxide in 96 per cent ethanol in a round-bottomed flask of 50 ml. capacity. Warm the flask under reflux in a water-bath and swirl gently until the fluorescein diacetate is completely hydrolyzed and dissolved (about 10 minutes). Allow the mixture to cool to room temperature. Add 20 ml. distilled water. Adjust to pH 8·5 with 1 N HCl added dropwise (about 12 ml. will be required). Transfer the contents of the flask completely to a 50 ml. volumetric flask and make up the volume to 50 ml. with distilled water.

FLUORESCEIN STANDARD

In a 150 ml. beaker, place:

1·21 g. trishydroxymethyl aminomethane (Tris)
82·0 g. glycerol
2 ml. 1 NHCl
10 ml. distilled water

Stir to dissolve the Tris (about 30 minutes).

Adjust to pH 8·5 (using electrode suitable for Tris buffers) with 1 N HCl added dropwise. Transfer to a 100 ml. volumetric flask. Add 5 ml. of Solution A, mix, and make up the volume to 100 ml. with distilled water.

Store in a glass stoppered dark bottle at 4°C.

APPENDIX 32

SILVER-METHENAMINE METHOD FOR CYSTINE
(after Swift, 1968)

PREPARATION OF THE REAGENT

Stock Solution A. 5ml of 5 per cent silver nitrate + 100ml of 3 per cent hexamethylene-tetramine (hexamine).

Stock Solution B. 10ml of 1·44 per cent boric acid + 100ml of 1·9 per cent borax. For use, mix 25ml of solution A and 5ml of solution B. Add 25ml distilled water. The pH of this reagent should be 9·2. Can be stored in the dark at 0–4° for up to 7 days.

Method. (1) Cut sections (from Araldite or other embedding medium) at 50–100 nm thickness. Collect on collodion-coated gold grids.

(2) Place section-covered grids in a Petri dish and cover with the reagent. Leave for 30 seconds upwards. (It is possible to continue for up to 24 hours).

(3) Wash in distilled water and dry.

GLUTARALDEHYDE PROTEIN COUPLING TECHNIQUE
(after Avrameas, 1969)

Method. To 1ml 0·1M phosphate buffer (pH 6·8), containing 5mg antibody, add 12mg peroxidase. This should be of the highest degree of purity available. Stir gently until dissolved and continue to stir while adding, dropwise, 0·05ml of 1 per cent aqueous glutaraldehyde. Allow to stand for 2 hours at 22–25° and then dialyze against two changes of 5 litres of buffered physiological saline at 4°, overnight. Remove the precipitate which forms by centrifugation (30 minutes at 4° and 20′000 rev/min).

The product can be kept at 4° until required for use. A similar technique is applicable to the conjugation of other enzymes to antibody (e.g. glucose oxidase, cytochrome c, alkaline phosphatase, acid phosphatase, tyrosinase).

TABLE 99

Reactant List for Protein-Enzyme Complexes

Protein	Quantity mg	Enzyme	Quantity mg	Vol. Enzyme Protein Solution ml	1%(w/v) Glutaraldehyde ml
Human IgG	5	Peroxidase	12	1	0·05
" "	5	Phosphatase	10	1·3	0·10
" "	12	Phosphatase	50	2	0·40
" "	12	Glucose oxidase	50	2	0·40
Human serum albumin	5	Glucose oxidase	50	1	0·80
Sheep anti-rabbit γ-globulin	5	Phosphatase	10	2	0·15
"	5	Phosphatase	10	1	0·15
"	5	Glucose oxidase	25	1	0·15
"	5	Tyrosinase	12	1	0·05
Rabbit anti-serum γ-globulin	5	Peroxidase	12	1	0·05

FNPS ENZYME CONJUGATION METHOD

(after Nakane and Pierce, 1967)

Method. To 50mg horseradish peroxidase (Sigma Type VI) and 50mg γ-globulin, dissolved in 2ml of 0·5M cold carbonate buffer (pH 10), add 0·25ml of 0·5 per cent difluoro-dinitrophenyl sulphone in acetone. Gently agitate the mixture for 6 hours at 4° and then dialyze against a large excess of phosphate-buffered saline, overnight. Remove the precipitate by centrifugation. Separate the enzyme-labelled antibody, and unreacted antibody, from unreacted peroxidase by precipitation with an equal amount of saturated ammonium sulphate in distilled water. Collect the precipitate by centrifugation, and wash twice with 50 per cent saturated ammonium sulphate in phosphate-buffered saline (PBS). Dissolve in PBS and dialyze against PBS to remove ammonium sulphate.

PEROXIDASE LABEL TECHNIQUES (OPTICAL AND E.M.)

(after Nakane and Pierce, 1966 and 1967)

Method (Optical Microscopy). (1) Prepare tissue sections in an appropriate manner. (This varies with the antigen it is proposed to demonstrate). For many antigens cryostat sections post-fixed in cold acetone or in phosphate-buffered 10 per cent formalin are optimal, in other cases carbodiimide fixation alone may allow successful demonstration (Kendall, Polak and Pearse, 1971).

(2) Wash sections in phosphate buffered saline (PBS).

(3) React for 30 minutes with antibody globulin or (control) with antibody globulin absorbed with its own antigen.

(4) Wash three times in PBS.

(5) React with peroxidase-labelled anti-rabbit (or appropriate species) γ-globulin for 30 minutes.

(6) Wash three times in PBS.

For the optical microscopical demonstration of the peroxidase label the usual technique is as follows:

(7) Incubate sections for 10–30 minutes at 22° in a solution containing 75mg 3,3′-diaminobenzidine (DAB) and 0·001 per cent hydrogen peroxide in 100ml 0·05M Tris buffer (pH 7·6).

(8) Wash in Tris buffer.

(9) Osmicate in 2 per cent OsO_4 in distilled water.

(10) Wash in distilled water, dehydrate, clear and mount.

Method (Electron Microscopy). (1) Prepare 50 micron sections (unfixed or suitably fixed) with a Tissue Chopper or Vibratome.

(2) Treat with peroxidase-labelled antibody (direct method) or with antibody (indirect method), 3–24 hours.

(3) If the indirect method is being used, wash blocks for 2–3 hours in PBS and

(4) Treat with peroxidase-labelled anti-globulin antibody for 3–24 hours.

(5) In both cases wash for a further extended period in PBS.

(6) Fix blocks in 5 per cent glutaraldehyde for 1 hour.

(7) Stain by the DAB procedure (above).

(8) Wash three times in distilled water.

(9) Treat with 2 per cent OsO_4 in distilled water.

(10) Dehydrate and embed in Epon.

MULTIPLE TISSUE ANTIGEN METHOD. PEROXIDASE LABEL
(after Nakane, 1968)

This method is based on an indirect procedure using sheep anti-rabbit (or guinea-pig) γ-globulin, labelled with peroxidase by one of the given techniques, and specific rabbit (or guinea-pig) antiserum to the antigens it is desired to demonstrate.

Alternative staining techniques to the DAB method (above) are used, together with successive elutions of peroxidase-labelled antibody from each specific antigenic site. Alternative procedure A must be used before procedure B, if three different coloured products are required.

Solutions for Elution of Specific Antibody

(1) 5M sodium iodide in 0·05M Tris buffer, pH 9·0.
(2) 5M potassium iodide in 0·05M Tris buffer, pH 9·0.
(3) 0·1M HCl.
(4) Glycine-HCl buffer, pH 2·0.

Alternative Staining Procedures

(A) Treat sections for 1–3 hours with a freshly prepared solution containing 1g of 1-naphthol, 10ml of 40 per cent ethanol and 0·2ml of 30 vol H_2O_2. Rinse and stain for 3 hours in alcoholic pyronin (pyronin G, 100mg; 40 per cent ethanol, 96ml; aniline oil, 4ml). If the whole routine is to be completed, proceed, after washing in water, to the application of the next specific antibody. If not, rinse, dehydrate, clear and mount (final colour, pinkish red).

(B) Treat sections with a freshly prepared solution of 4-chloro-1-naphthol, for 30–90 minutes. (Dissolve 40mg 4-chloro-1-naphthol in 0·5ml ethanol and then mix with 100ml 0·05M Tris buffer, at pH 7·6, containing 0·001 per cent H_2O_2. A precipitate forms which is removed by filtration and the filtrate is used at once). After staining, rinse in water, dehydrate, clear and mount. (The final colour produced by this stage is grayish-blue.)

PEROXIDASE LABEL (ULTRATHIN SECTION) TECHNIQUE
(after Kawarai and Nakane, 1970)

This technique is applicable only in the case of antigens which are known, or can be shown, to survive fixation and resin embedding procedures. The method, as given, refers specifically to the anterior pituitary protein hormones.

Method. (1) Fix small pieces of tissue for 4–8 hours in 4 per cent paraformaldehyde solution, buffered with phosphate to pH 7·4. Alternatively, use 4 per cent paraformaldehyde-picric acid (Zamboni and De Martino, 1967).

(2) Wash tissues in PBS for 30–60 minutes.
(3) Dehydrate in graded alcohols.
(4) Impregnate in a mixture of methyl and butyl methacrylate.
(5) Embed in prepolymerized methacrylate.
(6) Polymerize at 4° with ultraviolet light.
(7) Cut ultrathin sections and pick up on grids coated with collodion and carbon.
(8) Remove some of the embedding medium with water-saturated xylene or benzene.
(9) Wash, for 5–10 minutes, in several changes of PBS.
(10) Place, face downwards on a droplet of antiserum. Leave for 10–30 minutes.

(11) Wash in several changes of PBS.

(12) Place on a drop of peroxidase-labelled anti-globulin antibody (e.g. sheep anti-rabbit γ-globulin) for 10–30 minutes.

(13) Wash in PBS for 10 minutes.

(14) Stick the reacted grids to the edge of an 11 × 22 mm coverslip with double-sided adhesive tape and place in a 5ml syringe.

(15) Pump DAB reaction mixture at a constant rate (determined by experiment) through the syringe, for 10–30 minutes.

(16) Wash, in the syringe, with distilled water and dry.

(17) Examine, with or without counterstaining with lead hydroxide or lead citrate.

PREPARATION OF MICROPEROXIDASE
(after Feder, 1970)

Pepsin Digestion Procedure

Prepare a solution of 27g horse heart cytochrome c. in 2 litres of water with 160ml N-HCl. To this add 500mg 3X-crystallized pepsin in 30ml 0·1N-HCl. Leave at 22° for 24 hours. Bring pH to 8·5 with N-NaOH and add 610g ammonium sulphate per litre (85 per cent saturation). Collect the red precipitate by filtration and wash with 85 per cent saturated ammonium sulphate. Dissolve the precipitate in 1 litre of 0·01N-HCl. Filter and bring to pH 1·6 with 0·5N-HCl. Add 250mg pepsin in 30ml 0·1N-HCl. Leave for 24 hours at room temperature. Bring pH to 8·5 with N-NaOH and make 85 per cent saturated with ammonium sulphate. Collect the precipitate by filtration, wash in 85 per cent ammonium sulphate and dissolve in sufficient 0·01M ammonia to give 600ml of solution.

To this solution add 1200g ammonium sulphate in 1970ml water and then 81·3ml 20 per cent (w/w) trichloroacetic acid. Collect the precipitate by centrifugation, and dissolve in 0·02M ammonia to give a total volume of 800ml. Dissolve 374g ammonium sulphate in this solution and add 25·4ml of 20 per cent trichloroacetic acid. Dissolve the resulting precipitate in enough 0·04M ammonia to give 350ml of solution.

Place the solution in benzoylated dialysis tubing (directions given below) and dialyze in the cold against 0·1M sodium phosphate buffer (pH 5·0, three changes), and then against 0·01M phosphate buffer (pH 5·0, three changes) and then against distilled water (6 changes). Total dialysis time, 5 days.

The product, at this stage consists of 860g of a red-brown slurry, pH 5·7, which is adjusted to pH 5·0 with 0·1N-HCl.

Collect the precipitate by centrifugation, wash twice with water and suspend in 300ml water. Bring pH to 7·3 with N-NaOH. Stir at 4° overnight, warm to 22° and adjust to pH 7·3 with N-NaOH. Filter by suction through Millipore filters (0·45μ pore size) replacing clogged filters as necessary.

Freeze-dry the solution (350ml) to give a fluffy material. Redissolve in a smaller volume of water (90ml) and freeze-dry once more to yield a more compact product.

Benzoylation of Dialysis Tubing

Soak dialysis tubing (size 27, flat width 1·31 inches, thickness 0·001 inch) for 17 hours at 22° in water, then in four changes of methanol for 6 hours, then in triethylamine (3 changes, 12 hours total).

Treat with 10 per cent (w/v) benzoic anhydride in triethylamine (22°, 24 hours).

Wash successively in triethylamine (3 changes, 6 hours), methanol (3 changes, 3 hours) and water (24 hours).

Carry out a partial hydrolysis by soaking the tubing for 24 hours in 0·1M phosphate buffer at pH 10·0. Rinse in phosphate buffer at pH 7·0, in water, in methanol and, finally, store in water.

PERIODIC ACID-METHENAMINE SILVER METHOD

(after Rambourg, 1967)

Preparation of Reagents (Stock Solutions)

Methenamine: A freshly prepared 3 per cent aqueous solution.
Silver nitrate: A freshly prepared 5 per cent solution in distilled water.
Periodic acid: A 1 per cent aqueous solution (Stable).
Sodium thiosulphate: A 5 per cent aqueous solution (Stable).
Sodium borate: A 2 per cent solution in distilled water (Keeps one week).

Preparation of Reagents (Working Solutions)

Silver Methenamine: All glassware must be new, freshly washed in soap and water and rinsed in distilled water. The preparation, and staining, are carried out in the darkroom.

Add 5ml silver nitrate solution to 45ml methenamine solution in a 100ml Pyrex cylinder. A white precipitate appears which may be dissolved by shaking. Add 5ml sodium borate solution from a clean 10ml measuring cylinder. The final solution must be completely clear. After thorough mixing the solution is filtered through two sheets of Whatman No. 42 paper in a funnel inserted into an Erlenmayer flask. (The funnel is cleaned with sulphochromic mixture and rinsed with water).

Preparation of Sections

Fix in 2·5 per cent glutaraldehyde and embed in Epon. Cut ribbons (pale yellow) and store by floating on distilled water.

Staining Procedure. (1) Transfer some of the ribbons to periodic acid solution for 29 minutes. (Unoxidized sections are controls.)

(2) Rinse briefly 3 times in distilled water.

(3) Leave for 18–24 hours in a bath of distilled water.

(4) Heat a Petri dish and cover to 60° and pour in 25ml of the silver methenamine filtrate.

(5) Transfer the ribbons from distilled water with a platinum loop and allow them to float on the silver solution, at 60°, for 30 minutes, with the cover in place. A precipitate may form in the dish and the oxidized sections will be visible against a white background. Their interference colour changes from yellow to gold.

(6) Remove ribbons, rinse quickly in distilled water and transfer to a second bath of silver methenamine. Allow reaction to proceed for 30 minutes, at 60°. Sections now appear brownish (controls now yellow).

(7) Float all sections for a few minutes on distilled water.

(8) Transfer to thiosulphate solution for 5–10 minutes.

(9) Rinse in 2 or 3 baths of distilled water.

(10) Mount ribbons on Formvar-coated grids (0·3 per cent Formvar solution in ethylene dichloride) by simply touching the ribbon with the coated surface. Drain off excess water with filter paper.

Result. Only those reactions which appear in oxidized sections, and are absent from controls, can be attributed to *vic*-glycol groups of carbohydrates.

THE PATO REACTION

(after Hanker *et al.*, 1964)

Method. (1) Fix tissues in formaldehyde or glutaraldehyde and embed in Epon or methacrylate.

(2) Cut ultrathin sections and store on distilled water.

(3) Treat with 1 per cent aqueous periodic acid at 22° for 1 hour.

(4) Wash in 3 changes of distilled water, 10 minutes each.

(5) Immerse in 1 per cent thiocarbohydrazide in 5 per cent acetic acid for 90 minutes at 60°.

(6) Wash in 5 per cent acetic acid at 40° for 2–5 minutes.

(7) Rinse again in distilled water.

(8) Immerse in 1 per cent OsO_4 at 22° for 1 hour.

(9) Rinse in three changes of distilled water.

(10) Mount on grids as in previous method.

THE PATCSP METHOD

(after Thiery, 1967)

Method. (1) Carry out procedures 1 to 7 above.

(2) Immerse in silver protein reagents (a, b, or c) for 46 minutes, in the dark.

(a) 1 per cent aqueous solution of silver proteinate (pH 6·4).

(b) 1 per cent aqueous solution brought to pH 4·4 with acetic acid.

(c) 1 per cent solution in borate buffer at pH 9·2.

(3) Carry out procedures 9 and 10 above.

THE FLUOROPHENYLHYDRAZINE METHOD

(After Bradbury and Stoward, 1967)

PREPARATION OF THE REAGENT

Dissolve 0·25g *p*-fluorophenylhydrazine hydrochloride in 10ml 2N-HCl. Add 35ml mixed phosphate buffer (2·5g $NaH_2PO_4.2H_2O$ and 2·5g Na_2HPO_4 in 100ml water, and filter). Adjust pH to 5·0 to 6·0 with 2N-HCl. Use within 2 hours.

Method. (1) Fix small blocks in cold buffered formalin, 18 hours. (83ml 2·26 per cent sodium dihydrogen phosphate dihydrate and 17ml 2·5 per cent NaOH, with 4g paraformaldehyde. Dissolved by gentle heating and pH adjusted to 7·3).

(2) Store blocks, if necessary, in the above phosphate buffer with 7·5 per cent added sucrose.

(3) Wash blocks in buffer and transfer to 1 per cent periodic acid (pH adjusted to 4·0 with NaOH) for one hour.

(4) Wash in distilled water, 5 minutes.

(5) Treat with fluorophenylhydrazine reagent for 2 hours.

(6) Wash, twice, in distilled water for a total of 10 minutes.

(7) Treat with 1 per cent OsO_4 in distilled water, 1 hour.

(8) Rinse in 70 per cent alcohol and dehydrate *via* graded alcohols to epoxypropane.

(9) Leave blocks overnight in 50:50 Araldite-epoxypropane.

(10) Embed in fresh Araldite.

(11) Cut thin sections (80 nm) and mount, unsupported, on 200 mesh grids.

RUTHENIUM RED TECHNIQUE

(after Pihl *et al.*, 1968)

Method. (1) Fix for 1 hour to 14 days in cold 2·5 per cent cacodylate buffered methanol-free formaldehyde (Control).

(2) Fix for 1 hour to 14 days in the above medium in which 1mM ruthenium red has been dissolved.

(3) Dehydrate and embed in glycol methacrylate according to Leduc and Bernhard (1967), doubling each stage of dehydration and embedding.

(4) For optical microscopy cut 1μ sections into 1mM ruthenium red in 0·1M phosphate buffer at pH 7·4 and stain for 30 minutes.

(5) For electron microscopy cut sections into the trough containing buffered ruthenium red with 10 per cent ethanol added. Stain for 15 minutes.

(6) Mount on Formvar-coated grids and rinse briefly in distilled water.

LEAD CITRATE METHOD FOR ALKALINE PHOSPHATASE

(after Mayahara *et al.*, 1967)

Preparation of Incubating Medium

This consists of the following: 1·4ml 0·2M Tris-HCl buffer, pH 8·5 (final concentration 28mM), 2·0ml 0·1M sodium β-glycerophosphate (20mM), 2·6ml 15mM magnesium sulphate (3·9mM) and 4·0ml of saturated (about 0·5 per cent) alkaline lead citrate solution (K and K Laboratories Inc) pH 10·0 (2·0mM). Sucrose may be added up to 8 per cent. Final pH is adjusted to pH 9·2–9·4 with 0·1N NaOH. Alternatively, use 4·0ml Reynold's lead citrate reagent and adjust pH to 9·2–9·4 with 0·1N HCl. For tissues with high enzyme activity the level of substrate can be reduced to 3–5mM.

Method. (1) Fix small blocks in either 2 per cent glutaraldehyde in 0·1M cacodylate buffer (pH 7·4) for 1 hour at 0–4° or in 4 per cent formaldehyde in 0·1M cacodylate buffer (pH 7·4) for 1 hour, with 8 per cent sucrose in either case.

(2) Rinse briefly in buffer and cut thick frozen (cryostat) sections at 40–60μ. Alternatively use tissue chopper or vibratome.

(3) Incubate in substrate medium for 5–20 minutes at 22°.

(4) Refix tissues in 1 per cent OsO_4 for 1 hour at 4°.

(5) Dehydrate through graded alcohols and propylene oxide.

(6) Embed in Epon.

(7) Mount sections directly on 400-mesh grids.

TECHNIQUE FOR MITOCHONDRIAL ATP-ASE

(after Ogawa and Mayahara, 1969)

Preparation of Incubating Medium

Prepare medium as for alkaline phosphatase technique but with the sodium salt of adenosine triphosphate (3mM) in place of glycerophosphate.

Preparation of Control Media

(1) Substrate-free medium.

(2) Medium containing DNP (0·1mM) and oligomycin (0·1mM).

(3) Medium containing DNP (0·1mM) and *p*-chloromercuribenzoate (10mM).

Oligomycin is dissolved in methanol and 0·05ml methanolic solution is added to 5ml medium.

Method. As for alkaline phosphatase. Incubate 20–30 minutes at 37°.

LEAD METHOD FOR ACID PHOSPHATASE

(after Holt and Hicks, 1961; Etherton and Botham, 1970)

Since the standard Gomori (1952) technique, as modified by Holt and Hicks (1961) gives variable results at the E.M. level it appears preferable to follow the example of Etherton and Botham (1970) and give several alternative procedures. Table 100, below, provides these.

TABLE 100

Alternative Procedures for Acid Phosphates

Method,
(1)

Stage	Time	Materials	Variations
Prefixation (4°)	24 hrs	4% depolymerized paraformaldehyde in 67mM phosphate, pH 7·2, with 7·5% sucrose	(1) + 10% DMSO (v/v) (2) 50μ frozen sections cut at −30° after fixation
	30 mins	3% glutaraldehyde in 0·1M cacodylate, pH 7·2	
Buffer Wash (4°)	30 mins	50mM acetate + 7·5% sucrose	pH as in incubating medium
Incubation (37°)	30 mins	50mM acetate + 7·5% sucrose + 1mg/ml lead acetate + substrate	(1) pH 4·0, 4·5, 5·0, 5·5, 6·0 or 6·5. (2) Different substrates: (a) 1mgEml *α*-glycerophosphate + 1mg/ml *β*-glycerophosphate (b) 0·2mg/ml Naphthol AS-BI phosphate (c) 0·4mg/ml *α*-naphthyl phosphate
Buffer Wash (4°)	30 mins	50mM acetate + 7·5% sucrose	pH as in incubating medium
Post-fixation (4°)	60 mins	1% OsO_4 in 0·21M cacodylate, pH 7·4	

Which ever variation is employed, from those suggested above, continue as follows.
(2) Dehydrate in graded alcohols and embed in Araldite.
(3) Section at 50 nm and mount directly on uncoated grids.
(4) Examine unstained.

OSMIUM BRIDGING TECHNIQUE FOR ACID PHOSPHATASE

(after Hanker *et al.*, 1971)

PREPARATION OF INCUBATING MEDIUM

Prepare the following medium *immediately* before use. Dissolve 5mg di-dicyclohexylammonium 2-naphthylthiolphosphate (DDNTP) in 0·1ml dimethylformamide, on a vortex mixer. To the magnetically stirred solution add, in order, the following reagents:

0·06N (0·008%) sodium acetate	.. .. ..	7·9ml
0·1N (2·7%) acetic acid	.. .. ..	0·25ml
0·1M (2·9%) sodium citrate	.. .. ..	0·6ml
and dropwise,		
30mM (0·75%) copper sulphate	.. .. ..	1·25ml
5mM (0·17%) potassium ferricyanide	.. ..	1·25ml

Adjust the pH, if necessary, to 5·5 to 5·6. Filter through a fast, pre-folded, paper directly on to the rinsed tissues.

Method. (1) Fix small blocks at 0–4° for 8–96 hours in 8 per cent formaldehyde containing 2 per cent calcium chloride dihydrate mixed with an equal portion of 0·2M acetate buffer (pH 5·6).

(2) Rinse in several changes of 0·44M sucrose containing 1 per cent gum acacia, for 8–96 hours at 0–4°.

(3) Incubate the blocks, or air-dried 4μ cryostat sections for optical microscopy, for 10 minutes, at room temperature.

(4) Rinse in 3 changes of distilled water.

(5) Immerse tissues in a freshly prepared 0·5 per cent solution of thiocarbohydrazide, for 5 minutes at room temperature.

(6) Rinse 6 times, 5 minutes each, in distilled water.

(7) Treat the wet tissues for 15 minutes at 50–55° with OsO_4 vapour or by immersion in 2 per cent unbuffered OsO_4.

(8) Dehydrate blocks and embed in Epon. (Rinse cryostat sections in distilled water, dehydrate and mount in a suitable synthetic resin).

OXIDATIVE POLYMERIZATION TECHNIQUE FOR ACID PHOSPHATASE

(after Hanker, Anderson, and Bloom, 1971)

Method. (1) Proceed up to stage 4 as in the preceding method.

(2) Treat tissues for 1 hour with a freshly prepared solution of 0·05 per cent DAB in 0·05M phosphate buffer, pH 6·8.

(3) Rinse 3 times, 5 minutes each, in distilled water.

(4) Osmicate as in the above method.

(5) Proceed as stage 8 above.

The freezing artifact, induced by frozen sectioning, is substantially reduced by the addition of DMSO to the fixative. Frozen sections allow much superior penetration of reagents. For most purposes, therefore, fixation with added DMSO; cryostat sections at 50μ (− 30°), and incubation with lead acetate and α-naphthyl phosphate are recommended.

THIOCHOLINE-COPPER METHOD FOR CHOLINESTERASES

(after Lewis and Shute, 1969)

Preparation of Incubating Medium

Dissolve 100mg acetylthiocholine iodide (or butyrylthiocholine iodide) in 4·0ml distilled water and add, drop by drop, 7·0ml 0·1M copper sulphate. Centrifuge and add to 4ml of the supernatant 25mg glycine, 4ml isotonic sodium sulphate, 1ml water *or* 1ml ethopropazine hydrochloride (3·5mg in 5ml distilled water), and 0·5ml 0·5M succinic acid. Bring the pH to 5·3 by adding 1·0N-NaOH, slowly. About

0·5ml is required. Final concentrations are: substrate 13mM, copper, 18mM glycine, 33mM, succinate buffer, 25mM (ethopropazine, if added, 200 μM).

Two media are used routinely:

(a) Acetylthiocholine + ethopropazine for AChE.
(b) Butyrylthiocholine for ChE.

Method. (1) Fix, preferably by perfusion with 2 per cent glutaraldehyde in 50mM cacodylate with 2mM calcium acetate.

(2) Remove tissues, slice, and fix in the above fixative at 4° for 2–3 hours.

(3) Transfer to an equivalent medium containing 4 per cent formaldehyde instead of glutaraldehyde for 2–3 hours at 4°.

(4) Wash for at least 18 hours in the following medium:

25ml 0·2M sodium cacodylate, 10ml 0·2M cacodylic acid, 1ml 0·2M calcium acetate, made up to 100ml with isotonic sodium sulphate.

(5) Take thin slices from the surface of the tissue block (50–250μ).

(6) Incubate in medium (a) for 2–4 hours at 4° or in medium (b) for 4–6 hours at 4°.

(7) Wash for 60 minutes (2 changes) in the following medium:

10ml 0·5M succinic acid, 1ml 0·2M calcium acetate, 60ml isotonic sodium sulphate, adjusted to pH 5·3 with 1·0N NaOH (approximately 7ml) and made up to 100ml.

(8) Transfer to a fresh filtered solution made as above but brought to pH 5·3 by the addition of freshly prepared 4 per cent of reagent grade sodium sulphide (approximately 25ml required). Change solution twice, total time 60 minutes.

(9) Return slices to washing solution and dissect out areas of interest.

(10) Post-fix in Dalton's fluid at 22° for 3 hours.

(11) Dehydrate in alcohols and embed in Araldite.

COPPER-LEAD-THIOCHOLINE TECHNIQUE FOR AChE and ChE

(after Kása and Csillik, 1966)

Preparation of Incubating Medium

The medium has the following composition:

Acetyl or butyrylthiocholine iodide	7 mg
0·1N CH_3COONa	2·5ml
0·1N CH_3COOH	0·3ml
3·8% glycine	0·1ml
0·1M $CuSO_4$	0·1ml
0·5% Pb $(NO_3)_2$	0·1ml

Adjust to pH 5·4

Method. (1) Fix small blocks in cold (4°) Ringer (20ml) formalin (2ml), 0·1M Na_2HPO_4 (4ml) 0·1M KH_2PO_4 (1ml) with sucrose (0·5g) for 1–2 hours.

(2) Cut frozen sections (40–60μ) and rinse in 10 per cent sucrose for 30 mins.

(3) Incubate in substrate medium at 22° or at 4° for 10–30 minutes.

(4) Rinse in 10 per cent sucrose and develop (3 mins) in Ringer saturated with H_2S.

(5) Post-fix in 1 per cent OsO_4.

(6) Dehydrate in alcohols; embed in Araldite.

AuThCh METHOD FOR SPECIFIC CHOLINESTERASE

(after Koelle and Gromadzki, 1966)

PREPARATION OF STOCK SOLUTIONS

(1) $AuNa_3(S_2O_3)_2.2H_2O$ – 0·1M. Dissolve 100mg (supplied in 100mg ampoules as Sanocrysin, or sodium aurothiomalate) in 1·9ml H_2O.

(2) Acetyl or butyrylthiocholine iodide—50mM. Dissolve 46mg AThCh (or 50mg BuThCh) in 2·05ml H_2O. Add 1·15ml 0·1M $AgNO_3$ and shake. This causes aggregation and precipitation of AgI.

(3) Eserine salicylate .. ..mM solution
BW 284 .. ..mM solution
Nu 683 .. ..mM solution

This last is incorporated in the preincubation and incubation solutions, at 0·03 μM, for selective inhibition of ChE.

(4) $NaH_2PO_4.H_2O$ and K_2HPO_4 .. 6M

(5) DFP 0·3M solution in anhydrous propylene glycol, stored in a desiccator. For the irreversible inhibition of ChE (0·1μM) or AChE (0·01mM) DFP is added freshly to 0·85 per cent saline and sections are preincubated for 30 minutes.

(6) Acid-alcoholic $(NH_4)_2S$. Mix 3·6ml yellow ammonium sulphide with 20ml ethanol and 6ml glacial acetic acid. Add 1 drop 0·1M $AuNa_3(S_2O_3)_2$.

(7) Alcohol-formalin. Add 2·0ml 40 per cent formaldehyde and 4·0ml distilled water to 14ml ethanol.

PREINCUBATION, INCUBATION AND RINSING SOLUTIONS

These are prepared by adding the numbered reagents in the sequence and amounts listed in Table 101, below:

TABLE 101

Solutions for AuThCh Method

Solution	Stock Solutions					
	1.	H_2O(ml)	2.	3.	4. NaH_2PO_4/K_2HPO_4	
Final Concentration	4mM	—	4mM	0·01mM	4M	
Preincubation	0·4ml	2·9(or 2·8) a.			4·6ml	2·1ml
Incubation b.	0·4ml	2·1(or 2·0) a.	0·8ml	0·1ml	4·6ml	2·1ml
Rinse		3·3	c.		4·6ml	2·1ml

a. With inhibitor (eserine or BW 284)

b. Stand until visibly colloidal. Filter immediately before use (Whatman No. 3).

c. Saturated with gold thiocholine phosphate obtained by allowing incubating medium to stand for 3 days at 37° and collecting the precipitate by centrifugation.

Method. (1) Fix small pieces of tissue for 2–24 hours at 4° in 4 per cent formaldehyde-sucrose (7·5 per cent), buffered with sodium hydrogen maleate (28mM), pH 7·4.

(2) Rinse briefly in buffered saline.
(3) Incubate for 15–60 minutes at 22°.
(4) Rinse briefly and treat for 5 minutes with alcoholic $(NH_4)_2S$.
(5) Post-fix in 1 per cent OsO_4 for 2–4 hours.
(6) Dehydrate in alcohols and embed in Epon.

INDOXYL-BAXD METHOD FOR CHOLINESTERASE

(after Kawashima and Murata, 1969)

PREPARATION OF INCUBATING MEDIUM

Prepare freshly, immediately before use, as follows:

Indoxyl acetate (in 0·1ml acetone)	4mg
0·2M phosphate buffer (pH 7·3)	2·5ml
Distilled water	2·5ml
0·85% saline	5ml
50mM tetrazotized BAXD	0·2ml

PREPARATION OF TETRAZOTIZED BAXD

Dissolve 0·32g of *N,N'*-bis(*p*-aminophenyl)-1,3-xylylenediamine in 10ml 20 per cent HCl. Add 10ml freshly prepared 4 per cent sodium nitrite. Shake and adjust pH to 7·3 with 1·0N NaOH.

Method. (1) Fix blocks in cold buffered formalin-sucrose.
(2) Rinse in 0·85 per cent saline.
(3) Incubate for 90 minutes at 4°.
(4) Wash for 60 minutes in cold (4°) saline (3 changes).
(5) Select suitable small blocks.
(6) Post-fix in 2 per cent OsO_4 at 37° for 90–120 mins.
(7) Wash in distilled water.
(8) Dehydrate *via* ethanols to propylene oxide.
(9) Embed in Epon.

BARIUM-NCS METHOD FOR ARYLSULPHATASE

(after Hopsu-Havu *et al.*, 1967)

PREPARATION OF INCUBATING MEDIUM

Dissolve 160mg 2-hydroxy-5-nitrophenyl sulphate (*p*-nitrocatechol sulphate, NCS) in 4ml distilled water.

Add 12ml 0·1M acetate buffer (pH 5·5) and 4ml 5 per cent $BaCl_2$. Adjust pH to 5·5 with 0·2M acetic acid.

Method. (1) Fix small blocks in 5 per cent glutaraldehyde, buffered to pH 7·4 with 0·1M cacodylate, at 4° for 1 hour.

(2) Wash in 0·1M cacodylate (pH 7·4) containing 7·5 per cent sucrose for 12 hours or longer.

(3) Incubate for 30–60 minutes at 37°.

(4) Rinse blocks in several changes of 0·1M cacodylate buffer at pH 7·4 containing 7·5 per cent sucrose for 2–3 hours, or 0·1M acetate for 1 minute.

(5) Post-fix in 2 per cent cacodylate buffered OsO_4 for 1–2 hours.

(6) Dehydrate in graded ethanols and embed in Epon 812.

(7) Cut sections 70–90 nm.

DAB METHOD FOR CYTOCHROME OXIDASE

(after Seligman *et al.*, 1968)

PREPARATION OF INCUBATING MEDIUM

Dissolve 5mg 3,3′-diaminobenzidine tetrahydrochloride in 9ml phosphate buffer (0·05M, pH 7·4) and add 1ml catalase (20 µg/ml), 10mg cytochrome c, and 750mg sucrose.

Method. (1) Fix small blocks in cold 4 per cent formaldehyde (from paraformaldehyde) neutralized with alkali, for 60 minutes.

(2) Wash for 15 minutes in 0·22M sucrose with 0·05M phosphate buffer at pH 7·4. (Alternatively use fresh, unfixed, tissues).

(3) Incubate for 1 hour at 37° (Fresh blocks, 30 minutes).

(4) Wash, 3 times, (5 mins each), in buffered sucrose phosphate.

(5) Post-fix in 1 per cent buffered OsO_4 (Millonig*).

(6) Dehydrate and embed in Araldite.

DAB METHOD FOR PEROXIDASE

(after Beard and Novikoff, 1969)

INCUBATION MEDIUM (pH 9·0)

10·0ml 0·05M 2-amino-2-methyl-1,3-propanediol buffer, pH 9·4. 0·2ml 1 per cent freshly prepared H_2O_2 (from 30 per cent solution). 20mg DAB.

Filter and adjust pH to 9·0.

INCUBATION MEDIUM pH 6·0

10.0ml 0·05M sodium acetate—acetic acid buffer, pH 5·0.
0·1ml 0·1 per cent freshly prepared H_2O_2.
20mg DAB.
1·0ml 0·05M manganous chloride.

Method. (1) Fix blocks in 5 per cent buffered glutaraldehyde 1–3 hours.

(2) Cut 25µ sections with tissue chopper or vibratome, after a brief rinse in buffered sucrose.

(3) Incubate at 37° for 60 minutes in either of the two incubation media, with added sucrose (0·22M).

(4) Rinse in 0·1M cacodylate buffer (pH 7·4) containing 0·22M sucrose.

(5) Post-fix in Millonig's OsO_4 medium, for 1 hour at room temperature.

(6) Dehydrate *via* graded alcohols and propylene oxide.

(7) Embed in Epon 812; mount on uncoated 300-mesh copper grids, with or without counterstaining.

For the demonstration of peroxisomes the incubation medium at pH 9·0 is preferred.

NADH DEHYDROGENASE BY BRIDGING OSMIUM TO HATCHETT'S BROWN

(Hanker, Neff, Bloom and Pearse, 1971)

TISSUE PREPARATION

Blocks of tissue, 1 mm³, or air-dried fresh-frozen cryostat sections are incubated for NADH dehydrogenase demonstration in the following medium.

*Mix 415ml 2·26 per cent $NaH_2PO_4.H_2O$ and 8·5ml 2·52 per cent NaOH. to 45ml of this add 5ml 5·4 per cent glucose and 0·5 g OsO_4.

Preparation of Incubation Medium

This medium is prepared immediately before use.

To 5 mg NADH, add with constant stirring on a magnetic stirrer:

0·06N (0·008%) sodium acetate	6·4ml
0·1M (2·9%) sodium citrate	0·6ml
30mM (0·75%) copper sulfate	1·25ml
dimethylsulfoxide	1·5ml
5mM (0·17%) potassium ferricyanide, dropwise ..	1·25ml

The pH is titrated to pH 6·6–6·8 using 1N NaOH and 0·1N acetic acid if necessary. The medium is then filtered through a fast pre-folded filter paper on to the rinsed tissues.

Incubation and Post-fixation

After the tissues are incubated for 45 minutes to one hour at 37°, they are then rinsed for a short period in 4 per cent formaldehyde which is 0·1M in phosphate buffer, pH 7·4. They are then post-fixed overnight in the buffered formaldehyde. The fixative is removed by washing in several changes of distilled water for one hour.

Bridging with Thiocarbohydrazide (TCH)

The tissues are immersed in a freshly prepared 0·5 per cent solution of TCH for 5 minutes at room temperature. The TCH is removed by six rinses of 5 minutes duration each with distilled water.

Osmication

After rinsing, the wet tissues are osmicated by treating for 15 minutes with OsO_4 vapour at 50–55° or by immersion in 2 per cent unbuffered OsO_4 solution at 50–55° for 15 minutes.

Embedding or Mounting

Sections for light microscopy are then rinsed, dehydrated and mounted in Permount. For electron microscopy, blocks 1 mm^3, are dehydrated and embedded in a suitable resin such as Epon.

OSMIUM BRIDGING TECHNIQUE FOR LACTATE DEHYDROGENASE

(after Hanker *et al.*, 1971)

Preparation of Incubation Medium

Prepare this medium *immediately* before use. Add 6·0ml 0·5M sodium potassium tartrate to 1·6ml 0·1M phosphate buffer, pH 7·4. Stir with a magnetic stirrer and add, in order, the following reagents:

(dropwise) 0·3M (7·5%)	copper sulphate	0·7ml
10M (85%)	lactic acid	0·025ml
	NAD	5mg
	Dimethylsulphoxide	1·5ml

Adjust pH to, 6·6 to 6·8.

Add (dropwise) 0·05M potassium ferricyanide 0·3ml.

Filter directly on to the rinsed tissues.

Method. (1) Incubate for 45 minutes to 1 hour at 37°.
(2) Rinse in 4 per cent formaldehyde in 0·1M phosphate buffer (pH 7·4).
(3) Post-fix for 8 hours in buffered formaldehyde.
(4) Wash in several changes of distilled water.
(5) Bridge with TCH and OsO_4 as described above.

COPPER-FERROCYANIDE METHOD FOR SUCCINOXIDASE

(after Kerpel-Fronius and Hajós, 1968)

INCUBATION MEDIUM

Mix, with vigorous shaking, in the following order:
3ml 0·5M sodium-potassium tartrate in 0·1M phosphate buffer (pH 7·6).
0·35ml 0·3M $CuSO_4$.
0·8ml 0·1M phosphate buffer, pH 7·6.
0·7ml 1·0M sodium succinate.
0·15ml 0·05M potassium ferricyanide.

The solution is green, and completely clear. (Stable for several hours). The final pH, before addition of ferricyanide is 6·6 to 6·7, due to the formation of copper tartrate chelate.

Method. (1) Briefly wash thin slices of tissue in 0·1M phosphate buffer (pH 7·6).
(2) Incubate for 45 minutes at 37°.
(3) Wash briefly in phosphate buffer.
(4) Transfer to Millonig's 1 per cent OsO_4 for 1 hour.
(5) Wash in buffer.
(6) Dehydrate and embed in Durcupan.
(7) Orientate blocks with care, as only the surface 50–100μ contains reaction product.

Controls. Tissues are pretreated, before incubation, with either 5mM *p*-chloromercuribenzoic acid or with 0·1M sodium malonate. In the second case, malonate is also added to the incubation medium at the same concentration as the substrate. Alternatively, substrate-free incubation medium can be used.

THE SULPHIDE-SILVER METHOD FOR HEAVY METALS

(after Timm, 1958; Pihl and Falkmer, 1967; Ibata and Otsuka, 1969)

Method. (1) Fix in 3 per cent glutaraldehyde, buffered to pH 7·3 with phosphate, and saturated with H_2S, for 2–3 hours. Perfusion-fixation is desirable, followed by dissection and immersion-fixation.
(2) Wash in 0·2M phosphate buffer, pH 7·3, for 24 hours.
(3) Prepare small blocks of tissue and wash these in distilled water.
(4) Develop in Timm's developer at 22°, in daylight, for 30–90 minutes.
Timm's solution has the following composition:

20% gum arabic solution,	100 parts
1·0M $AgNO_3$	1 part
2% Hydroquinone/5% citric acid,	20 parts

(5) Post-fix in 1 per cent phosphate-buffered OsO_4.
(6) Dehydrate in graded alcohols and embed in Epon.

Controls. The usual control is H_2S-free fixative.

Result. Small, intensely electron dense, silver deposits indicate sites of heavy metal sulphides superimposed with silver.

TRIPLE FIXATION METHOD FOR MONOAMINES

(after Tranzer *et al.*, 1969)

The fixation procedure consists of a 1 hour treatment in glutaraldehyde-paraformaldehyde mixture (A), followed by dichromate treatment for 4–18 hours (B), then a routine post-fixation in osmium tetroxide solution for 1–2 hours (C).

DETAILS:

A. Aldehyde fixation.

Conditions: 1 hour, 4°, under slow agitation.

1. Glutaraldehyde 1% (final concentration) = 4ml commercial solution of 25% strength
2. Paraformaldehyde 0·4% (final concentration) = 5ml
   (2g/25ml H_2O)
   (at 60°C + 1 drop)
   (1N.NaOH → 8% solution)
3. Krebs-Ringer solution = 91ml

(a) NaCl 9g/1000ml	100ml
(b) KCl 1·25g/100ml	4ml
(c) KH_2PO_4 2·1g/100ml	1ml
(d) $MgSO_4.7H_2O$ 3·8g/100ml	1ml
(e) Na_2HPO_4, $2H_2O$ $Na_2HPO_4.2H_2O$ (4·45g/100ml) (+ 1N.HCl 5ml) (+ H_2O 145ml)	21ml

(f) 130g glucose/127ml of Krebs Ringer solution.

Check the pH and if necessary adjust to pH 7·4, with a few drops of saturated $NaHCO_3$ solution.

Comments. (a) Krebs-Ringer solution yields more dense core vesicle than phosphate or cacodylate when used as buffer. This is true even if no dichromate treatment follows.

(b) A fixation in solutions containing higher aldehyde concentrations than noted above, or fixation tissues exceeding 1 hour lead to a poorer preservation of the dense cores in the vesicles.

B. Dichromate treatment (Wood and Barrnett, 1964)

After aldehyde fixation, rinse the tissues quickly with Krebs-Ringer solution (without aldehydes) and incubate at 4° for 4–18 hours in the following solution:

0·2M Acetate buffer (pH 4·1–4·2), 100ml.
2·5g Potassium dichromate.
1·0g Sodium sulphate.
Acetate Buffer Sol. A: 27·2g/1000ml $CH_3COONa.3H_2O$
Sol. B: 12ml/1000ml acetic acid 100%
Final Buffer pH 4·1–4·2. 13·2ml Sol. A.
36·8ml Sol. B.
50ml H_2O.

Remove the solution and post-fix in OsO_4 (do not rinse between the two treatments).

Comments. The pH (4·0–4·2) of the dichromate solution seems to be absolutely necessary for an optimal preservation of the dense cores. Unfortunately a treatment with solutions having higher pH yields less satisfactory results. The treatment should be of at least 4 hours duration.

C. Post-fixation (OsO_4).

Two per cent in 0·1M phosphate, 1–2h, 4° as used routinely. Dehydration in alcohols, propylene oxide and embedding in Epon. Contrasting of ultrathin sections with lead citrate alone looks mostly better than uranium and lead.

General Comments. This triple fixation technique yields by far the best preservation of dense core vesicles in adrenergic nerves compared to all other techniques used to date: Aldehyde(s)—OsO_4 or $KMnO_4$. There are two disadvantages:

(1) The endoplasmic reticulum especially in smooth muscle cells is often swollen.

(2) At higher magnifications (×100′000) the membranes appear digested: probably, due to its strong oxidative and acidic nature, dichromate is the prime cause of this.

LIST OF SUPPLIERS

Key	*Supplier*	*Agent or Agents*
AL	Aldrich Chemical Co. Inc., 2371 North 30th Street, Milwaukee, Wis. 53210, U.S.A.	Ralph N. Emanuel Ltd., 264 Water Road, Wembley, HA0 1PY, Middlesex, England.
A	Alfa, P.O. Box 159, Beverly, Mass. 09195, U.S.A.	Ralph N. Emanuel Ltd., 264 Water Road, Wembley, HA0 1PY, Middlesex, England.
B	The Boehringer Corporation (London) Ltd., Bilton House, 54/58 Uxbridge Road, Ealing, London, W.5, England.	
BC	The Borden Chemical Company, 5000 Langdon Street, Philadelphia, Pa., 19124, U.S.A.	
BDH	The British Drug Houses Ltd., B.D.H. Laboratory Chemicals Division, Poole, Dorset, England.	
CB	Calbiochem, P.O. Box 331, 5 Remsen Avenue, Monsey, New York 10952, U.S.A.	Calbiochem Ltd., 10 Wyndham Place, London, W1H 1AS, England.
CL	CIBA Ltd., Basel, Switzerland.	
CO	Columbia Organic Chemicals Co. Ltd., P.O. Box 5273, Columbia, S.C. 29209, U.S.A.	Kodak Limited, Kirkby, Liverpool, England.
CC	Cyclo Chemical Corporation, 1922 East 64th Street, Los Angeles, Calif., U.S.A.	
DP	E.I. Du Pont de Nemours & Co. Inc., Wilmington, Delaware, U.S.A.	
EK	Eastman Organic Chemicals, Eastman Kodak Company, Rochester, N.Y. 14650, U.S.A.	Kodak Limited, Kirkby, Liverpool, England.
FS	Fisher Scientific Co., 3514 Delaware Avenue, Buffalo, N.Y., U.S.A.	
FL	Fluka AG., Chemische Fabrik, CH-9470 Buchs, Switzerland.	Fluorochem Limited, Dinting Vale Trading Estate, Dinting Lane, Glossop, Derbyshire, SK13 9NU, England.

Key	*Supplier*	*Agent or Agents*
GL	Glaxo Laboratories Ltd., Greenford, Middlesex, England.	
G	Edward Gurr Ltd., 42 Upper Richmond Road West, London, S.W.14, England.	
GG	George T. Gurr Ltd., Searle Scientific Ltd., Coronation Road, Cresswick Industrial Estate, High Wycombe, Bucks, England.	
HC	Halewood Chemicals Ltd., Stanwell Moor, Staines, Middlesex, England.	
HB	Harrington Bros., Reeve Angel Scientific Ltd., 14 New Bridge Street, London, E.C.4, England.	
HW	Hopkin and Williams Ltd., Freshwater Road, Chadwell Heath, Essex, England.	
HY	Hyland, P.O. Box 2214, 3300 Hyland Avenue, Costa Mesa, California 92626, U.S.A.	Hyland Division, Baxter Laboratories Ltd., Thetford, Norfolk, England.
ICI	Imperial Chemical Industries, Dyestuff Division, P.O. Box 42, Blackley, Manchester 9, England.	
IC	International Chemical and Nuclear Corp., 2727 Campus Drive, Irvine, California 92664, U.S.A.	
Ka	Kallestad Laboratories, Inc., 4005 Vernon Avenue, Minneapolis, Minnesota 55412, U.S.A.	Charles Druce Ltd., 94 York Street, London, W.1, England.
KK	K. & K. Laboratories, Inc. 121 Express Street, Engineers Hill, Plainview, New York 11803, U.S.A.	Kodak Ltd., Kirkby, Liverpool, England.
KL	Koch-Light Laboratories Ltd., Colnbrook, Buckinghamshire, England.	
LP	Laboratories Plan S.A., 1211 Aïre, Geneva, Switzerland.	Ralph N. Emanuel Ltd., 264 Water Road, Wembley, HA0 1PY, Middlesex, England.

Key	*Supplier*	*Agent or Agents*
RL	Raymond A. Lamb, 12 The Viaduct, Ealing Road, Alperton, Wembley, HA0 4LN, Middlesex, England.	
MB	May and Baker, Dagenham, Essex, England.	
M	E. Merck Aktiengesellschaft, 61 Darmstadt, Hügelstrasse 8–16, Germany.	Aderman & Co., 87 Tooley Street, London, S.E.1, England.
MA	Microbiological Associates Inc., Bethesda, Maryland, U.S.A.	Amhangio, 6380 Ober-Eschbach, Frankfurt, Germany. Micro-Bio Laboratories Ltd., 46 Pembridge Road, London, W.11, England.
MS	Miles-Seravec Laboratories (Pty) Ltd., Moneyrow Green, Holyport, Maidenhead, Berkshire, England.	Miles Laboratories, Inc., Research Products Division, Elkhart, Indiana 46514, U.S.A.
NB	Nutritional Biochemicals Corporation, Cleveland, Ohio 44128, U.S.A.	Micro-Bio Laboratories Ltd., 46 Pembridge Road, London, W.11, England.
Ph	Pharmacia (Great Britain) Ltd., Paramount House, 75 Uxbridge Road, London, W.5 5SS, England.	
PL	P.L. Biochemicals Inc., 1037 West McKinley Avenue, Milwaukee, Wis. 53205, U.S.A.	International Enzymes Limited, 11a St. John's Hill, London, S.W.11, England.
PI	Polysciences Inc., Paul Valley Industrial Park, Warrington, Pa. 18976, U.S.A.	Polaron Equipment Ltd., 4 Shakespeare Road, Finchley, London, N3 1XH, England.
Ra	The Radiochemical Centre, Amersham, Bucks.	
RC	Regis Chemical Company, 1101 N. Franklin Street, Chicago, Ill. 60610, U.S.A.	
R	Roche Products Ltd., 15 Manchester Square, London, W.1, England.	
S	Schuchardt, 8 Munchen 13, Ainmillerstrasse 25, Germany.	Cambrian Chemicals, 73 Cherry Orchard Road, Croydon, CR9 6AG, England.

Key	*Supplier*	*Agent or Agents*
SM	Schwarz/Mann, Orangeburg, New York 10962, U.S.A.	Serlabo Pure Chemical Products, 25 Rue Saint-Gilles, 75 Paris 3ème, France. Becton, Dickinson UK Limited, York House, Empire Way, Wembley, HA9 0PS, Middlesex, England.
SF	Serva Feinbiochemica GmbH & Co., 6900 Heidelberg 1, P.O.B. 1505, Römerstr., 118, Germany.	Micro-Bio Laboratories Ltd., 46 Pembridge Road, London, W.11, England.
SIG	Sigma Chemical Co., P.O. 14508, St. Louis 63178, Missouri, U.S.A.	Sigma London Chemical Co. Ltd., 12 Lettice Street, London, S.W.6, England.
WR	Wellcome Reagents Limited, Wellcome Research Laboratories, Beckenham, Kent, BR3 3BS.	
WB	Wessex Biochemicals Limited, Castle Road, Bournemouth, BH9 1PH, England.	

The Mol. Wt. and Source List which follows is not intended to be comprehensive. For common compounds no source may be indicated; likewise no source is quoted for those which are commercially unobtainable. In all other cases sources are quoted up to a maximum of four, selected usually to give a proportion of *cis* and *trans*-Atlantic suppliers. It follows that there could be cases where all the firms quoted in the suppliers list could offer the compound in question. Omission of any supplier thus implies no inferiority in any respect, the method of selection, subject to the above-mentioned proviso, being largely arbitrary.

MOLECULAR WEIGHT AND SOURCE LIST

Compound	*Mol. Wt.*	*Source**
Acetazolamide (Diamox Sodium Salt)	265·25	EK; KK
Acetic anhydride	102·09	—
Acetobromoglucose	411·22	FL; KL
Acetonitrile	41·05	—
Acetophenone	120·15	FL; KL
2-Acetoxy-3-naphthoic anilide (Naphthol AS acetate)	321	BC; PI
2-Acetoxy-3-naphthoic-o-anisidide (Naphthol AS-OL acetate)	352	—
2-Acetoxy-3-naphthoic-o-toluidide (Naphthol AS-D acetate)	336	PI
O-Acetyl-5-bromoindoxyl	254	KL; PI
Acetyl chloride	78·50	BDH; HW; KL
O-Acetyl-4-chloro-5-bromoindoxyl	289	—
Acetylcholine chloride	181·7	BDH; HW; KL; PI
Acetyl-D-glucosamine	237	KL; SIG; SM
N-Acetylhomocysteine thiolactone (AHTL)	—	CB; EK; KK; SM
N-Acetylindoxyl	175·2	PI
Acetylthiocholine iodide	289·18	BDH; KL; PI
Acid haematein	—	BDH; GG; HW; RL
cis-Aconitic acid anhydride	174	CB; NB; SIG; SM
Acridine orange (C.I.46005)	301·83	EK; G; RL; PI
Acridine yellow (C.I.46025)	—	EK; G; GG; KK
Acriflavine (C.I.46000)	259·74	HW; KK; NB; RL
Acrolein	56·06	BDH; EK; FL; KK
^{14}C-Adenine	135·13	Ra; SM
Adenosine-5-diphosphate	427·21	B; FL; KL; SIG
Adenosine-5-phosphate (Adenylic acid)	347·2	FL; HW; NB; SIG
Adenosine-5-triphosphate	507·2	B; KK; NB; SIG
Adenosine-5′-triphosphate (dipotassium salt)	583·38	FL; SIG
Adenosine-5-triphosphate (disodium salt)	623·21	FL; KL
L-Adrenalin (A)	183·21	BDH; FL
Adrenalin hydrochloride	217·5	CB; SIG
Adrenochrome	179·17	KL; NB
DL-Alanyl-*β*-naphthylamide	—	CB; KK; KL; SIG
L-Alanyl-*β*-naphthylamide	—	KL; SIG
Albumin (egg) Cryst	—	CB; SIG
Alcian blue-8GX (C.I.74240)	—	FL; HW; RL; PI
Alginic acid	(176)n	FL; KK; KL
Alizarin red S (C.I.58005)	342·26	FL; HW; KK; PI
Alloxan (monohydrate)	160·09	BDH; KK; KL; PI
Aluminium ammonium sulphate	453·33	—
Aluminium isopropoxide	204·25	BDH; FL; KK; KL
Aluminium oxide	101·96	—
Aluminium potassium sulphate (Potash alum) $(12H_2O)$	474·39	—
Aluminon (triammonium aurine tricarboxylic acid)	473·44	BDH; KK; MB
Aminoacetic acid (Glycine)	75·07	BDH; HW; KK; SIG
5-Aminoacridine hydrochloride	230·70	BDH; HC; KK
2-Aminoanthraquinone	223·2	BDH; FL; KK
3-Amino-carbazole	181·21	KK
4-Amino-1-*N*,*N*-dimethylnaphthylamine (ADN)	—	EK
p-Aminodiphenylamine	184·24	FL; KK; KL
3-Amino-9-ethylcarbazole	210·28	AL; SIG
3(*β*-Aminoethyl)indole (Tryptamine)	160·22	KL; RC
Aminofluorescein I	—	KK; KL
Aminofluorescein II	—	KK; KL

* See page 1454

Compound	Mol. Wt.	Source*
2-Amino-2(hydroxymethyl)1,3-propanediol (Tris)	121·2	BDH; EK; KL
2-Amino-2-methyl-1,3-propanediol (buffer)	105·14	AL; FL
8-Amino-1-naphthol-3,6-disulphonic acid (H-acid)	319·3	BDH; KK
8-Amino-1-naphthol-5-sulphonic acid (S-acid)	239·25	EK; G
M-Aminophenol	109·13	AL; BDH; KK; KL
p-Aminophenyl arsonic acid	216·9	BDH; KK
p-Aminophenylmercuric acetate	351·76	AL; NB
6-Aminopurine sulphate	404·4	FL; KK; SIG
8-Amino-1,2,3,4-tetrahydroquinoline	148·21	EK; KK
3-Amino-1,2,4-tetrazole	84	KL
2-Aminothiazole	100	KK; KL
3-Amino-1,2,4-triazole	84·08	AL; FL
Ammeline (2,4-diamino-6-hydroxy-*s*-triazine)	127·11	FL; KK
Ammonium molybdate	1235·9	BDH; FL; HW; SIG
Ammonium thiocyanate	76·12	BDH; EK; FL; HW
α-Amylase	—	FL; KL; NB; SIG
β-Amylase	—	FL; KL; NB; SIG
iso-Amyl nitrite	117·15	BDH
p-Anisidine	123·16	AL; EK; KL; SIG
L-Arginine	174·20	KK; KL; NB; SIG
L-Arginine β-naphthylamide hydrochloride (ANA)	299·2	NB; PI
L-Asparagine	132·12	CB; FL; SIG
Auramine O (C.I.41000)	321·86	FL; G; PI; RL
Aurine tricarboxylic acid, ammonium salt (Aluminon)	473·44	BDH; FL; HW
L-Ascorbic acid	176·12	AL; BDH; HW; SM
Azoene fast red salt (Ponceau fast L salt) (C.I. 37151)	—	BDH; PI; RL
Azoene fast violet B (C.I.3765)	—	PI; RL
Azure A (C.I.52005)	291·80	G; GG; HW; RL
BAO (bis-Aminophenyl-oxidiazole)	252·28	CL; FL
Barium hydroxide	315·48	—
Bathophenanthroline (4,7-diphenyl-1,10-phenanthroline)	332·41	FL; KK; KL
Benzaldehyde phenylhydrazone	196·2	BDH
Benzene sulphohydroxamic acid	152	HW
Benzidine	184·24	FL; KK; KL; SIG
Benzidine dihydrochloride	257·16	FL; KK; KL
Benzoin	212	BDH
Benzoquinone	108·1	BDH
N_α-Benzoyl-L-Arginine	278·31	FL; KL
N_α-Benzoyl-L-Arginine amide hydrochloride (BAA)	331·80	FL; KK; KL; SIG
N_α-Benzoyl-L-Arginine ethyl ester hydrochloride (BAEE)	342·83	FL; KK; KL; SIG
N_α-Benzoyl-DL-Arginine-β-naphthylamide hydrochloride (BANA)	439·95	FL; KL
Benzoyl chloride	140·57	—
6-Benzoyl-2-naphthol	249·29	KL; PI
6-Benzoyl-2-naphthol sulphate (K salt)	—	KL; PI
6-Benzoyl-2-naphthyl phosphate (Disodium)	—	KK; KL; PI; SIG
6-Benzoyl-2-naphthyl sulphate (potassium)	—	KL; PI
N-Benzoyl-*p*-phenylene diamine (BPDA)	—	PI
Berberine sulphate	822·85	FL; KK; SIG
Biebrich Scarlet (C.I.27195)	760·58	G; GG; KK; RL
Bismuth trichloride	315·34	A; BDH; HC; KK
Blue tetrazolium (BT)	727·67	AL; KL; PI; RL
Bordeaux GP (2-Nitro-*p*-anisidine)	168·15	G; GG; KK; RL
Boric acid	61·84	—
Brilliant Cresyl blue (C.I.51010)	332·84	BDH; HW; RL; SIG
Brilliant Crystal Scarlet 6R (Acid red 44) (C.I.16250)	—	G; GG; HW; RL
Bromoaniline	172·03	AL; EK; KK; KL
5-Bromo-4-chloroindoxyl acetate	289·54	PI; SIG
5-Bromo-4-chloroindoxyl-3-β-D-galactopyranoside	—	SIG

* See page 1454

Compound	*Mol. Wt.*	*Source**
5-Bromo-4-chloroindoxyl phosphate	—	PI
5-Bromoindoxyl acetate	254·09	FL; KL; PI; SIG
5-Bromoindoxyl phosphate	—	PI
6-Bromo-2-naphthol	223·07	KL; PI; SIG
6-Bromo-2-naphthyl acetate	—	PI; SIG
6-Bromo-2-naphthyl-α-D-galactopyranoside	—	KL; PI
6-Bromo-2-naphthyl-β-D-galactopyranoside	—	KL; PI; SIG
6-Bromo-2-naphthyl-α-D-glucopyranoside	—	KL; PI; SIG
6-Bromo-2-naphthyl-β-D-glucopyranoside	385·22	BDH; KL; SIG
6-Bromo-2-naphthyl-β-D-glucuronide	—	KL; SIG
6-Bromo-2-naphthyl-α-D-mannopyranoside	—	KL
6-Bromo-2-naphthyl sulphate (K salt)	—	KL; PI
6-Bromo-2-naphthyl-β-D-xylopyranoside		KL
N-Bromosuccinimide	178·0	HW; KL; PI
Brucine sulphate	1013·13	FL; KK
Butyrylthiocholine chloride	193·72	KL; PI
Butyrylthiocholine iodide	317·23	KK; KL; PI
Cacodylic Acid (Hydroxydimethylarsine oxide)	137·99	FL; HW; KK; SIG
Cacotheline	427·42	AL; BDH; FL; KK
Calcichrome (Cyclo-tris-7-(1-azo-8-hydroxy-naphthalene-3:6-disulphonic acid)	990·89	BDH; KK; PI
Carbazole	167·2	KL; PI
Catalase	225·000	B; KL; SIG
Celestin blue B (C.I.51050)	363·80	G; GG; KK; RL
Celloidin	—	BDH; G; HW; KK
N-Cetylpyridinium chloride	358	BDH; HW; SIG
Cetyltrimethyl ammonium bromide	364·46	FL; KK; KL
Cetyltrimethyl ammonium chloride	320·03	BDH; HW; KK
3-Chloroacetoxy-2-naphthoic acid anilide (Naphthol AS Chloroacetate)	339·78	FL
3-Chloroacetoxy-2-naphthoic acid *o*-toluidide (Naphthol AS-D-chloroacetate)	—	NB; PI
Chloramine-T	281·68	BDH; EK; FL; KK
p-Chloroanilidophosphonic acid	206·5	KK; KL; PI
Chlorantine fast red 5B (C.I.22310)	—	G; GG; HW; RL
1-Chloro-2,4-dinitrobenzene	202·6	BDH
p-Chloromercuribenzoic acid (PCMB)	357·16	FL; PI
p-Chloromercuribenzoic acid-Na Salt	379·14	FL; KK; KL; SIG
o-Chloromercuriphenol	329·17	AL; KK; PI
1-(4-Chloromercuriphenylazo-2-naphthol) (Mercury orange)	482	KL; SIG; SM
p-Chloromercuriphenylsulphonic acid	394	KL; SIG
6-Chloro-4-methyl-thionaphthene		DP; CL
DL-*p*-Chlorophenylalanine	199·64	PI; SF
Chlorosulphonic acid	116·52	BDH; EK; FL; HW
Cholesterol (5-Cholesten-3β-ol)	386·6	HW; KL
Cholesteryl acetate	428·7	BDH; KL
Cholesteryl oleate	651·12	BDH; KL
Choline chloride	139·63	BDH; KL; PI
Chromyl chloride	154·92	AL; HW; KK
Cocarboxylase (tetrahydrate)	496·37	FL; KL; SM
S-Collidine(2,4,6-trimethyl pyridine)	121·18	BDH; FL; KK
Coriphosphine O	—	G; GG; KK; RL
Crotonaldehyde	70·09	BDH; EK; FL; HW
p-Cyanoaniline	118·13	KK
Cyanogen bromide	105·93	AL; EK; FL; KK
Cyclic 3′-5′-AMP	365·24	SIG
Cyclohexane	84·16	AL; BDH; FL; EK
1-Cyclohexyl-3-(2 morpholinyl-(4)-ethyl)-carbodiimide-*p*-toluol-methyl sulphonate (CME-CDI)	423·58	AL; FL
Cysteic acid	169·2	BDH; KL

* See page 1454

Compound	*Mol. Wt.*	*Source**
Cysteine hydrochloride	157·62	BDH; HW; KK; SM
Cytidine-5′-monophosphate	323·20	FL; KK; SIG
Cytochrome c	13′000 to 16′000	BDH; FL; KL; SIG
Dehydroepiandrosterone (DHA)	288·43	FL; KK; SIG
Deoxycholic acid	392·58	FL; KK; KL
Deoxyribonuclease	—	B; SIG; SM
Diacetin	176·2	BDH
m-Diacetoxybenzene	135	BDH
Dialysed iron	—	BDH; HW
3,3′-Diaminobenzidine	214·27	PI
3,3′-Diaminobenzidine tetrahydrochloride	360·1	KK; SIG
1,2-Di(2-Aminoethoxy)ethane-N,N,N′,N′-tetra-acetic acid (EGTA)	380·35	BDH
2,7-Diaminofluorene	196·25	FL; KK; KL
Diamox (Acetazolamide)	222·25	KK
Diastase	—	BDH; KL
Diazo fast blue RR (4-Benzoylamine-2:5-dimethoxyaniline, Zinc double salt)	639·50+ 136·28	KK
1-Diazo-2-naphthol-4-sulphonic acid	251	KL
5,6-Dibenzyloxyindole	—	RC
2,4-Dibromo-1-naphthol	302	BDH
Dichloro-difluoromethane (Arcton 12, Freon 12)	120·92	DP; ICI; KK
D7-Dichlorofluorescein diacetate	—	SF
2,ichlorohydrin(1,3-dichloropropanol)	129	BDH
2,4-Dichloro-α-naphthol	213·06	BDH; EK; HW; KK
2,6-Dichlorophenolindophenol	290·1	BDH; FL; KK; KL
2,6-Dichloroquinonechloroimine (Gibbs' reagent)	210	BC; BDH; KK
Dicoumarol (Bis-hydroxycoumarin)	336·29	KK; KL; PI
Di-(dicyclohexyl)ammonium-2-naphthyl thiophosphate (DD-NTP)	—	PI
3-(Diethoxyphosphinyloxy)-N-Methylquinolinium methyl sulphate (3-0422)	—	R
m-Diethylaminophenol	165·24	AL; EK; FL; KK
N,N′-Diethyl-6,6′-dichloropseudoisocyanin chloride	431·80	—
Diethyldithiocarbamic acid (Diethylammonium salt)	—	KK
Diethylene dioxide (Dioxan)	88·1	BDH
Diethylene glycol	106·12	FL
Diethylene glycol dimethacrylate	242·27	PI
Diethyl-*p*-nitrophenyl phosphate (E600)	275·21	KK; KL
N,N′-Diethyl pseudoisocyanin chloride	362·90	FL
Digitonin	1229·3	BDH; CB; KK; SM
2,4-Dihydroxybenzaldehyde	138·12	BDH
2,2′-Dihydroxy-6,6′dinaphthyl disulphide (DDD)	350·46	BDH; FL; KL; PI
4,5-Dihydroxy-2,7-naphthalene disulphonic acid	356·33	EK
DL-3,4-Dihydroxyphenylalanine (DOPA)	197·19	BDH; KL; SIG
3,5-Dihydroxytoluene	124·14	HC; HW; KK; KL
Diisopropyl fluorophosphate (DFP)	184·15	FL; KL
Diisopropyl *p*-nitrophenyl phosphate (DINP)	303·26	—
N,N′-Diisopropyl phosphodiamidic fluoride (Mipafox)	182·20	KL
Di-Lithium carbamoyl phosphate	152·88	FL
Dimedone	140·18	AL; HW; KL; NB
2,3-Dimercaptopropanol (BAL)	124·21	BDH; KK; KL
2,5-Dimethoxyaniline	153·18	AL; FL; KL; PI
3,3′-Dimethoxybenzidine (*o*-Dianisidine)	244·29	HW; IC; FL; KL
4-Dimethylaminoazobenzene	225·30	FL; KK; KL
p-Dimethylaminobenzaldehyde	149·2	AL; HC; KK
p-Dimethylaminobenzylidene	264·37	BDH; HW; KK
p-Dimethylaminocinnamaldehyde	175·23	BDH
1-Dimethylaminonaphthalene-5-sulphonic acid	251·31	BDH; FL; KK

* See page 1454

Compound	*Mol. Wt.*	*Source**
1-Dimethylaminonaphthalene-5-sulphonyl chloride (Dansyl chloride)	269·75	KL; PI; SIG; SM
Dimethyl formamide	73	AL; BDH; EK; SIG
Dimethylglyoxime	116·12	BDH; HW; KK; KL
3,3′-Dimethylnaphthidine	312	BDH; KL
N,N-Dimethyl-m-phenylenediamine dihydrochloride	209·12	EK; KK
Dimethyl sulphate	126·13	AL; BDH; EK; HW
3-(4,5-Dimethyl-thiazolyl-2)-2,5-diphenyl tetrazolium bromide (MTT)	414·34	PI; RL; SIG
2,4-Dinitroaniline	183·12	EK; FL; HC; KL
2,4-Dinitrofluorobenzene	186	AL; FL; KK; SM
2,4-Dinitrophenol	184·11	BDH
2,4-Dinitrophenylhydrazine	198·14	FL; HW; KL; SM
Diphenyl carbazide	242·3	FL; HW; KK; KL
Diphenyl carbazone	240·27	AL; FL; HW; KL
1,5-Diphenylcarbohydrazide	242·28	HW; PI
Diphenylthiocarbazone (Dithizone)	256·33	FL; HW; KL
Dipicrylamine	439·21	FL; KK; KL
α,α′-Dipyridyl	156·19	FL; KK; KL;
2,2-Diquinolyl	395·5	KL
Disodium ethylenediamine tetraacetate	390·04	BDH; FL
Disodium glucose-1-phosphate	376·17	CB; NB; SM; WB
Dithiol (Toluene-3,4-dithiol)	156·27	BDH; FL; KK; KL
Eriochrome Black T (C.I.14645)	461·397	FL; FS; RL; SIG
Eserine	275·36	CB; FL; KK; SIG
Eserine salicylate	413·45	FL; KK; SIG
Eserine sulphate	648·76	CB; KK; SIG
Estradiol-17*β*	272·37	CB; KK; KL; SIG
Estrone	270·36	CB; KK; KL; SIG
Ethanol-perchloric acid	46·07	BDH; HW
Ethidium bromide	394·32	BDH; KK; KL
Ethopropazine hydrochloride	348·93	MB
Ethopropazine methosulphate (Lysivane methosulphate)	359·53	MB
Ethylene chlorohydrin (2-Chloroethanol)	80·51	FL
Ethylenediamine	60·10	BDH; FL; KL
Ethylenediamine tetra-acetic acid (Disodium) (EDTA)	373·2	FL; KK; KL
Ethylene dichloride (1,2-Dichloroethane)	98·96	AL; EK; HW
Ethylene glycol monomethyl ether (2-Methoxyethanol)	76·10	FL; FS; KL
1-Ethyl-3-(3-diemthylaminopropyl)-Carbodiimide hydrochloride (EDAP-CDI)	191·70	KK; PI; SIG
N-Ethyl maleimide	125·13	AL; FL; KL; SM
Ethyl phenyl carbamate	165·2	BDH
Evan's blue	960·81	G; GG; KK; RL
Fast black K salt (Zinc double salt) (C.I.37190)	699·48+ 136·28	FL; G; KL; RL
Fast blue B salt (Tetrazotized diorthoanisidine, Zinc double salt)	339·18+ 136·28	G; KL; RL; PI
Fast blue BB salt (Zinc double salt) (C.I.37175)	695·60+ 136·28	FL; GG; RL; SIG
Fast blue BBN (Zinc double salt)	695·60+ 136·28	FL; PI
Fast blue RR salt (Zinc double salt) (C.I.37155)	639·50+ 136·28	BDH; FL; G; PI
Fast garnet GBC (Acid sulphate) (C.I.37210)	237·29+ 97·07	BC; G; GG; SIG
Fast red B salt (Acid 1,5-Naphthalene disulphonate) (C.I.37125)	180·14+ 287·29	G; GG; RL; SIG

* See page 1454

Compound	*Mol. Wt.*	*Source**
Fast red 3GL (Zinc double salt) (C.I.37040)	440·46+ 136·28	FL; G; GG; RL
Fast red ITR (Zinc double salt)	611·58+ 136·28	G; GG; ICI; RL
Fast red RC salt (Zinc double salt) (C.I.37120)	410·08+ 136·28	BC; G; GG; RL
Fast red TR (Acid 1,5-napthalene disulphonate) (C.I.37085)	153·60+ 287·30	G; GG; RL; SIG
Fast red-violet LB (Diazonium chloride)	308·18	RL; SIG
Fast violet B (Zinc double salt) (C.I.37165)	579·48+ 136·28	FL; G; NB; RL
Ferrocene (Dicyclopentadienyl-Iron)	186·04	FL; KK; KL; SIG
Fettrot 7B (C.I.26050)	—	FL
Fluorescein	332·31	AL; EK; FL; PI
Fluorescein amine (I or II)	347·33	BC; KK; KL; NB
Fluorescein diacetate	465·43	PI
Fluorescein dibutyrate	501·54	SF
Fluorescein isothiocyanate	389·39	KL; NB; PI; SIG
Fluorescein phosphate	—	PI
Fluoroboric acid	—	BDH; FL; HW
1-Fluoro-2,4-dinitrobenzene	182	KL; PI
Fructose-1,6-diphosphate (Disodium)	340·13	KL; SIG
D-Fructose-1,6-diphosphate (Trisodium salt)	406·06	CB; FL
D-Fructose-6-phosphate (Barium salt)	395·46	CB; KK; SIG
Fumaric acid	116·08	BDH; KL
D-Galactose-1-phosphate (barium)	274·16	B; SIG
D-Galacturonic acid	194·14	KL; SIG
Gallocyanin(e)	300·27	FL; G; GG; RL
Globulin (Goat anti-guinea pig) FITC-labelled	—	HY; WR
Globulin (Goat anti-rabbit) FITC-labelled	—	HY; WR
1,5-Gluconolactone	178·15	BDH
D-Glucosamine hydrochloride	215·64	BDH; PI
Glucose pentaacetate	390·34	KL
Glucose-1-phosphate (Dipotassium)	372·36	B; BDH; SIG
D-Glucose-1-phosphate (Disodium salt)	376·17	FL; KK; SIG
D-Glucose-6-phosphate (calcium salt)	298·20	FL; KK
D-Glucose-6-phosphate (Disodium)	304·11	FL; HC; SM
D-Glucose-6-phosphate (diNa,3H_2O)	358·2	PI
Glucose-6-phosphate (G-6-P dehydrogenase)	—	B; KL; NB; SIG
β-Glucosidase	—	KL; SIG
D-Glucuronolactone	176·13	FL; KK; KL
L-Glutamic acid (Na salt) monohydrate	187·13	KL; PI
Glutamic dehydrogenase	—	B; SIG
L-Glutamine	146·15	BDH; PI; SIG
N-(γ-L-Glutamyl)-4-methoxy-2-naphthylamide	—	PI
α-L-Glutamyl-β-naphthylamide	—	PI
γ-L-Glutamyl-α-naphthylamide	—	KL; PI; SIG; SM
γ-L-Glutamyl-β-naphthylamide	—	KL; PI
Glutaraldehyde (EM grade)	—	PI
Glutathione	307·33	GG; KK; PI; SIG
DL-Glyceraldehyde	90·1	KL; PI
Glyceraldehyde-3-phosphate dehydrogenase	—	CB; NB; SIG; SM
Glyceraldehyde-3-phosphate diethylacetal (monobarium salt)	379·51	CB; FL; SIG
Glycerol-3-phosphate (Disodium salt)	315·12	FL
Glyceryl monostearate	358·57	BDH; G; HC; KK
Glycine	75·07	AL; FL; KL; SM
Glyoxal-bis-(2-hydroxyanil)	240·26	FL; SIG
Glyoxal sodium bisulphite	145	BDH
Glyoxylic acid	74·04	BDH
Guaiacol	214·14	BDH; HW; PI
Guanidinium chloride	95·53	WR

* See page 1454

Compound	*Mol. Wt.*	*Source**
Guanine	151·13	PI; SIG
Haemin	651·9	FL; KK; KL; SIG
Harmaline	214·27	FL; KK
Heparin	6′800 to 20′000	HW; NB; SIG; SM
Hexacyanoferrate (Potassium salt)	329·26	FL
Hexamethylenetetramine (Hexamine)	140·19	BDH; KK; KL; PI
Hexokinase		B; KL; SIG
Histamine dihydrochloride	184·07	FL; KK; SIG
L-Histidine	155·16	KK; KL; SIG
L-Histidine dihydrochloride	288·08	FL; KK
DL-Homocysteine thiolactone hydrochloride (HTL)	153·63	FL; KK; KL
Hyaluronic acid	200′000 to 500′000	FL; KK; SIG
Hyaluronidase	—	BDH; KL; SIG; SM
Hydrazine dihydrochloride	104·98	BDH
Hydrazine hydrate	50·06	BDH
Hydroquinone	110·11	
2-Hydroxy-adipaldehyde	130·14	FL; KL; PI
β-Hydroxybutyrate (Sodium salt)	126·09	FL; KK; KL
DL-*α*-Hydroxybutyric acid	104·1	KL
DL-*β*-Hydroxybutyric acid	104·1	BDH; HW; KL
6-Hydroxydopa	213	RC
N-2-Hydroxyethylpiperzine-*N*-2-ethanesulphonic acid (HEPES)	238·31	PI; SIG; WR
Hydroxylamine hydrochloride	69·5	HC; HW; KK; SM
3-Hydroxy-2-methyl-1,4-naphthoquinone (Phthiocol)	187·2	KL
2-Hydroxy-1-naphthaldehyde	172·18	KL; PI
2-Hydroxy-3-naphthaldehyde	172·18	PI; SIG
1-Hydroxy-2-naphthoic acid	188·17	EK; KK; KL; SIG
2-Hydroxy-3-naphthoic acid anilide (Naphthol AS)	263·30	FL; KL; SIG
2-Hydroxy-3-naphthoic acid hydrazide	202·21	KL; PI
8-Hydroxy-1,4-naphthoquinone	174·15	KK
8-Hydroxyquinoline	145·16	AL; HC; HW; KK
8-Hydroxyquinoline-*β*-D-glucuronide	338·3	KK; KL
8-Hydroxyquinoline sulphate	388·4	BDH; KK
5-Hydroxytryptamine (Creatinine sulphate)	405·43	KK; KL; R; SIG
L-5-Hydroxytryptophan	220·23	FL; KL; SIG
Hypoxanthine	136·11	FL; KK; KL; PI
Imidazole	68·1	AL; FL; KL; SIG
Indole	117·14	KL; PI
3-Indolyl aldehyde	131	KL; PI
Indoxyl acetate (*o*-acetylindoxyl)	175·2	KK; PI; SIG
Indoxyl butyrate (*o*-Butyrylindoxyl)	203·2	FL; KK; KL; PI
Inosine	286·25	FL; KK; KL
Inosine-5′-diphosphate (Disodium salt)	490·17	FL; KK; KL; SM
Iodoacetamide	184·96	FL; HW; KL; PI
4-Iodoacetamido-1-naphthol	—	PI
Iodoacetic acid	185·96	BDH; KL; PI
2(*p*-Iodophenyl)3-(*p*-nitrophenyl)-5-phenyltetrazolium chloride (INT)	505·71	BC; FL; KL; PI
DL-Isocitrate (Trisodium salt)	258·07	FL; KK; SIG; SM
Isocitric acid (Trisodium) $2H_2O$	276·12	KL; SIG
Isonicotinic acid	123·11	KL; PI
Isonicotinic acid hydrazide	—	KL; PI
α-Ketoglutaric acid	146·1	PI; SIG
β-Ketoglutaric acid	146·1	KL; PI
DL-Lactate (Sodium salt)	112·07	FL; KK; KL
Lanthanum nitrate, $7H_2O$	324·93	KL; PI

* See page 1454

Compound	*Mol. Wt.*	*Source**
Laurent's acid (1-Aminonaphthalene-5-sulphonic acid)	223·24	G
Lead Tetraacetate	443·37	A; G; KK; KL
L-α-Lecithin	734·06	FL; KL
D-Leucine	131·2	KL
L-Leucine	131·2	KL; PI
L-Leucine-2-naphthylamide hydrochloride	292·81	FL; KK; KL; SIG
L-Leucyl-4-methoxy-2-naphthylamide-hydrochloride	355·81	CB; KK; KL; SIG
Lissamine fast yellow 2G (Acid Yellow 17) (C.I.18965)	—	ICI; RL
Lissamine Rhodamine B200 (C.I.45100)	—	GG; KK; RL
2,6-Lutidine	107·15	BDH
Luxol fast blue ARN	—	PI
Luxol fast blue MBS	—	G; HW; RL; SF
Magneson II (*p*-Nitrobenzene-azo-1-naphthol)	293·28	FL; KK
L-Malate (Sodium salt)	187·07	HW; KK
Maleic acid	116·08	BDH; KL
Maleic anhydride	98·06	FL; KL; SIG
L-Malic acid	134·09	EK; FL; KK; KL
Menadione (2-Methyl-1,4-naphthoquinone)	172·19	KK; SIG
Menadione sodium hydrogen sulphite	276·25	FL; KK; SIG
2-Mercaptoethylamine (Cysteamine)	77·15	FL; SIG
2-Mercaptoethylamine hydrochloride	114·6	KK; KL
Mercury orange (1(4-Chloromercuriphenylazo)-2-naphthol)	482	G; KL; NB; SIG
DL-Methionine	149·2	KL; SIG
o-Methoxyphenol	124·14	BDH; HW
DL-Methoxyphenylacetic acid	166	EK
Methyl benzoate	136·15	AL; EK; FL; HC
Methyl cellosolve (ethylene glycol monomethyl ether acetate)	118·13	EK; HW; KL; NB
20-Methylcholanthrene	268·36	FL; KL; SIG
4-Methyl-6-chlorothionaphthene	182	ICI
Methylene blue (zinc free)	374	BDH; RL
Methyl hydrazine	46·07	AL; EK; FL; KK
Methyl hydroquinone	124·14	AL; EK; FL; KL
β-Methylindole	131·2	KK
Methyl iodide (iodomethane)	141·94	BDH; HC; HW; KK
N-Methylphenazonium methyl sulphate (Phenazine methosulphate)	306·34	HW; KL; PI; SM
9-Methyl-2,6,7-trihydroxyfluorone	228·2	KL
4-Methyl umbelliferone	177·15	BDH; HW
4-Methylumbelliferyl acetate	235·22	PI
4-Methylumbelliferyl nonanoate	333·41	PI
4-Methylumbelliferyl phosphate	254·25	PI
Mipafox (bis-monoisopropylamine fluorophosphine oxide)	182	KL
Morin (2′,3′,4′,5,7-Pentahydroxy-flavone)	338·27	BDH; FL; KK; KL
MTT (Thiazolyl blue tetrazolium bromide)	414·34	KL; PI; RL; SM
Myristic acid	228·4	BDH
Myristoyl choline chloride	349·99	FL; KK; KL; SIG
1-Naphthol	144·17	—
2-Naphthol	144·17	—
Naphthol AS-chloroacetate	339·78	FL; NB; PI
Naphthol AS nonanoate	320·54	KL; SIG
Naphthol ASBI *N*-acetyl-β-glucosaminide	481·40	SIG
Naphthol AS-BI β-D-glucuronide	454·33	CB; CC; KL; SIG
Naphthol AS-BI phosphate	341·19	KK; PI
Naphthol yellow S	—	BDH; G; GG; HW
1,2-Naphthoquinone	158·16	EK; FL; KK; KL
1,2-Naphthoquinone-4-sulphonic acid (sodium salt)	260·20	AL; HC; KK; KL

* See page 1454

Compound	*Mol. Wt.*	*Source**
1-Naphthoxyacetic acid	202·21	HW; PI
1-Naphthyl acetate	186·21	BDH; KK; NB; RL
2-Naphthyl acetate	186·21	FL; KK; KL; NB
1-Naphthylamine (1-Aminonaphthalene)	143·19	—
2-Naphthylamine	143·19	—
1-Naphthyl-*N*-acetyl-*β*-D-glucosaminide	363·39	SIG
2-Naphthyl butyrate	230·28	PI; SIG
N-1-(Naphthyl) ethylenediamine dihydrochloride	259·18	EK; HW; KL; SM
1-Naphthyl-*β*-D-glucopyranoside	322·34	PI
2-Naphthyl-*β*-D-glucuronide	336·32	PI
1-Naphthyl isocyanate	169·2	BDH
2-Naphthyl laurate	342·50	FL; NB; PI; SIG
2-Naphthyl myristate	354·52	KK; KL; PI
1-Naphthyl phosphate (DiNa salt)	224·15	HW; PI
2-Naphthyl phosphate (DiNa salt)	224·15	FL; KK; KL; PI
2-Naphthyl stearate	410·63	KK; KL; PI
1-Naphthyl sulphate	350·39	PI
α-Naphthyl thymidine-5′-phosphate	464·46	SIG
Neocuproine (2,9-Dimethyl-1,10-phenanthroline hemihydrate)	208·25	FL; KK; KL; SIG
Neuraminidase (*Clostridium perfringens*)	—	NB; SIG
Neuraminidase (*Vibrio cholerae*)	—	KL; SIG
Neotetrazolium chloride (NT)	667·60	KK; NB; PI; SIG
Nicotinamide	122·12	KL; SIG
Nicotinamide adenine dinucleotide (NAD)	663·44	FL; HW; KL; SIG
Nicotinamide adenine dinucleotide phosphate (NADP)	743·4	CC; FL; KL; SIG
Ninhydrin	178·14	—
p-Nitrobenzaldehyde	151·12	KK; KL
p-Nitrobenzoyl chloride	185·57	—
Nitro blue tetrazolium chloride (Nitro-BT) (2,2′-di-*p*-Nitrophenyl-5,5′-diphenyl-3,3′-3,3′-dimethoxy-4-4′-biphenylene) ditetrazolium chloride	819·67	AL; KL; NB; SIG
p-Nitrocatechol sulphate	374·4	NB
4-Nitro-1-naphthylamine	188·19	FL; KK
p-Nitrophenyl phosphate (Di-sodium salt)	263·06	KK; KL; PI; SIG
Nitrophenyl-*β*-D-galactoside (2- and 4-)	301·26	FL; KK; KL; SIG
p-Nitrophenyl sulphate (K salt)	—	KL; PI; SIG
Nitrophenyl-*β*-xylopyranoside (2- and 4-)	271·23	KK; KL
4-Nitrophthalic acid	211·14	AL; KK; KL; SIG
4-Nitrophthalic acid anhydride	193·12	FL; KK
1-Nitroso-2-naphthol	173·17	—
1-Nitroso resorcinol	139·11	BDH
3-Nitro-*o*-toluidine	152·15	FL; KL
Noradrenalin bitartrate	317	KL
Oxalacetic acid	132·18	KL; SIG
Oxalyl chloride	126·93	BDH
Octyl alcohol (1-Octanol)	130·23	BDH; FL; KK; KL
Oil blue N (C.I.61555)	—	GG; KK; RL
Oil red O (C.I.26125)	—	G; GG; RL; SIG
iso-OMPA (Tetraisopropylpyrophosphoramide)	342	KK; KL
Orcinol (5-methylresorcinol)	142·15	—
Ouabain (*γ*-Strophanthin)	584·67	CB; FL; NB; SIG
2-Oxoglutaric acid	146·10	BDH; FL; SIG
Paraldehyde	132·16	AL; BDH; EK; HW
Pectinase		CB; NB; SM
Pentacyano-ammine ferroate (Disodium)	266·99	AL; FL; HW
3,5,7,2′,4′-Pentahydroxyflavanol (Morin)	338·27	BDH; FL; KK; KL
Pepsin	—	CB; SIG; SM
Peracetic acid (40 per cent)	76·05	—
Perchloric acid	100·46	BDH; EK; FS; HW

* See page 1454

Compound	*Mol. Wt.*	*Source**
Periodic acid ($2H_2O$)	227·96	
o-Phenanthroline	198·23	AL; BDH; KL; HW
Phenazine methosulphate (PMS)	306·34	BDH; CB; NB; SIG
p-Phenetidine hydrochloride	127·7	HW
Phenol	94·11	—
Phenolphthalein	318·33	—
DL-Phenylalanine	165·2	—
p-Phenylenediamine	108·14	—
Phenylhydrazine	108·14	—
Phenylhydrazine hydrochloride	144·61	—
α-Phenyl indole	193·14	KL
Phenyl isocyanate	119·12	BDH
Phenylmercuric acetate	336·74	—
Phenyl mercuric chloride	313·15	—
1-Phenyl-3-(M-nitrobenzamidopyrazolone)	—	EK
N-Phenyl-*p*-phenylenediamine (4-Amino diphenylamine)	184·24	FL; KK; KL
β-Phenyl-propionic acid (Hydrocinnamic acid)	150·8	FL; HW; KK; KL
1-Phenyl semicarbazide	151·17	KK
9-Phenyl-2,3,7-trihydroxyfluorone	320·30	FL; HW
N-Phenylurethane (Ethyl-*N*-phenylcarbamate)	165·19	BDH; FL; KK
Phloroglucinol (1,3,5-Trihydrobenzene)	162·14	BDH; FL; KL
6-Phosphogluconate (Ba salt)	430	SIG
6-Phosphogluconate (Tri-Na salt)	342·08	FL; KK; KL; SIG
Phosphomolybdic acid	3939·45	—
Phosphorus oxychloride	153·33	—
Phosphorus pentachloride	208·24	—
Phosphorus pentoxide	141·95	—
Phosphotungstic acid	6498·93	—
o-Phthaldehyde (Benzene 1,2-dialdehyde)	134·14	KL; PI; SF
Phthalic anhydrine	148·11	BDH; FL
Piperazine	86·14	—
Polyethylene glycol (Carbowax 20M)	—	BDH; FL; HC; RL
Polyvinyl pyrrolidone	(varies)	BDH; FL; KL; SM
Potassium hydrogen saccharate	248	BDH
Proflavine hemisulphate	552·62	BDH
1,2-Propane-1,2-diol (Propylene glycol)	76·10	BDH; KL
β-Propiolactone	72·06	FL; KL
Propionyl choline chloride	212	KK; KL
Propionyl choline iodide	287·14	FL; KK; KL
Propionyl thiocholine iodide	303·21	KL; PI
Pteroylglutamic acid	441·41	FL; KK; KL; SIG
Purpurin	256·22	AL; GG; HW; KK
Pyridine-2-aldoxime methiodide (2-PAM)	264·07	CB; FL; SIG
Pyridoxal-5-phosphate (Codecarboxylase)	265·16	FL; KK; KL; SIG
Pyrocatechol (1,2-Benzenediol)	110·11	BDH; KL
Quinolyl-8-glucuronide	338·3	KL; SIG
Quinacrine hydrochloride	508·91	GG; SIG
Quinacrine mustard	577·81	PI
Quinalizarin (1,2,5,8-Tetrahydroxy anthraquinone)	272·2	BDH; FL; KK
Raney nickel	—	BDH; FL; KK
Resorcinol	110·11	
Rhodamine B isothiocyanate		BDH; PI; RL; SIG
Rhodizonic acid (Na salt)	214·05	BDH; FL; HW; KL
Ribonuclease	—	B; KL; SIG; SM
D-Ribose-5-phosphate (Barium salt)	401·47	CB; KK; KL; SIG
Rotenone	394·41	KK; SIG
Rubeanic acid (Dithiooxamide)	120	EK; HC; HW; KL
Ruthenium red	293·14	BDH; FL; HW; KK
Salicylaldehyde	122·12	—
Salicyloyl (syn: salicyl) hydrazide	152·15	EK; FL; KK; SIG

* See page 1454

Compound	*Mol. Wt.*	*Source**
Semicarbazide hydrochloride	111·53	FL; HW; KL; PI
Semicarbazide nitrate	138·09	BDH
Sephadex G50	—	SIG
Serotonin creatine sulphate	405·43	FL; KL
Silver proteinate	—	RL
Sinigrin (Potassium myronate)	415·48	KL
Snake venom (*Crotalus atrox*)	—	KK; KL; SIG
Sodium arsenate	185·91	BDH; HB; HW; SIG
Sodium aurothiomalate (Myocrisin)	—	MB
Sodium azide	65·02	—
Sodium barbitone	206·18	—
Sodium bismuthate	279·97	BDH; HB; HW
Sodium borohydride	37·85	FL; HW; KK; KL
Sodium diethyl dithiocarbamate ($3H_2O$)	225·31	BDH; HW; KL
Sodium dimethyldithiocarbamate	179·24	FL; KK
Sodium L-glutamate	169	
Sodium *β*-glucerophosphate ($5\frac{1}{2}H_2O$)	315·15	BDH; KK; KL; SIG
Sodium hydrogen citrate	263·12	
Sodium hydrogen malate	155·08	
Di-Sodium hydrogen orthophosphate	141·97	
Di-Sodium hydrogen orthophosphate ($12H_2O$)	358·16	—
Sodium DL-isocitrate	258	KK
Sodium phenyl phosphate	254·10	
Sodium rhodizonate	214·05	HB; KK; KL
Sodium D-tartrate	230·09	
Sodium taurocholate	537·70	BDH; HB; KK; KL
Sodium tetraphenylboron	342·24	BDH; HB; FL; KK
Solochrome black 6B (Eriochrome black B) (C.I.14640)	—	BDH; G; RL
Spermidine trihydrochloride	254·63	FL; SIG
Starch (hydrolyzed)	—	FL; KL; SIG
Streptodornase	—	NB
Sudan black B (C.I.26150)	—	G; GG; PI; RL
L-Tartaric acid	150·09	
Terephthalaldehyde (Benzene,4-dialdehyde)	134·13	FL; KK; KL
Tetraethyl ammonium hydroxide	147	—
Tetraethyl pyrophosphate (TEPP)	290·20	—
Tetrahydrofuran (Tetramethylene oxide)	72·11	BDH; HW; KK; KL
Tetraisopropylpyrophosphoramide (iso-OMPA)	342	KK; KL
Tetramethyldiaminotriphenyl methane	330	HW
Tetramethyl-*p*-phenylene diamine dihydrochloride	237·18	FL; KK
Tetramethylrhodamine isothiocyanate (isomer R)	—	SM
Tetranitro-BT (TNBT) (2,2′,5,5′-tetra-*p*-nitrophenyl-3,3′-3,3′-dimethoxy-4,4-biphenylene)-ditetrazolium chloride	907·65	KK; PI; RL; SIG
Thiamine pyrophosphoric acid	496·37	FL; KK; KL
1-(2-Thiazolylazo)-2-naphthol (TAN)	255·30	BDH; KK
Thiazolyl blue (MTT)	414·33	BDH; NB; PI; SIG
Thioacetamide	75·13	FL; KK; KL; SIG
Thioacetic acid (Thiolacetic acid)	76·12	—
2-Thiobarbituric acid	144·16	FL; KK; KL; SIG
Thiocarbamyl nitroblue tetrazolium (TCNBT)	935·83	NB; PI; SIG
Thiocarbohydrazide	106·15	KK; PI
DL-Thioctic acid (*α*-Lipoic acid)	206·33	—
M-Thioglycolic acid (Mercapto acetic acid)	92·12	BDH; FL; KL; SIG
Thiolacetic acid (Thioglycolic acid)	92·12	—
6-Thio-2-naphthol	160·24	FL; KK
Thiophosgene	114·98	FL; KK; KL
Thiosemicarbazide	91·14	BDH; KL; PI; SIG
Thiosemicarbazone	236·29	—
DL-*α*-Tocopherol acetate	472·73	KK; KL
o-Tolidine	212·3	BDH; FL; KK

* See page 1454

Compound	*Mol. Wt.*	*Source**
o-Tolidine dihydrochloride	285·21	PI
Toluene-2,4-diisocyanate (TC)	174·16	FL; KK; KL
Toluene-3,4-dithiol (Dithiol)	156·27	—
Tolyl sulphonyl methyl nitrosamide	214·24	FL; KK; KL
1,2,4-Triazole	69·07	FL; KK; KL
Tri-*N*-Butylamine	185·33	FL; KK; KL
Tributyrin (Glyceryl tributyrate)	302·37	FL; KK
2,4,5-Trichlorophenol	197·45	FL; KL
2,6,8-Trichloropurine	223·45	FL; KK
2,4,6-Tridi(methylaminomethyl)phenol (DMP30)	—	KK; KL
Triketohydrindene hydrate	178·15	—
Triphosphopyridine nucleotide (reduced) (NADPH)	869·39	B; KL; SIG
2,4,6-Tri-(2′pyridyl)-1,3,5-triazine	312·34	FL; KL
Triolein (Glyceryl trioleate)	885·46	FL; KK; SIG
Tri-orthocresyl phosphate	368·37	FL
Triphenyl tetrazolium chloride (TTC)	334	BDH; KL; NB; PI
Triphosphopyridine nucleotide (TPN)	734·4	B; KL; NB; SIG
Triphosphopyridine nucleotide (Sodium salt)	801·43	FL; KK; SIG
Tris(Hydroxymethyl)aminomethane	121·14	BDH; FL; KK; KL
N-Tris-(Hydroxymethyl)-Methyl-2-Amino-Ethanesulphonic acid (TES)	229·25	BDH; FL; KK; KL
Trypsin	—	B; CB; SIG
Tryptamine hydrochloride	196·69	—
Tween 20, 40, 60 and 80	—	FL; KL; SIG
Tyramine hydrochloride	173·65	—
Ubiquinone-30	—	FL; KK; KL; SIG
Uranyl nitrate ($6H_2O$)	502·18	—
Urease	—	KL; SIG
Uridine-5-diphosphate (Trisodium salt)	506·14	B; FL; SIG
Uridine-5-diphosphate glucose (UDPG)	—	B; SIG
Uridine-5-triphosphate (Trisodium salt)	586·12	B; FL; SIG
Vinyl alcohol (Polymerized)	—	BDH
Xanthydrol	198·22	FL; KK; KL
M-Xylylene diisocyanate (XC)	156·19	FL

* See page 1454

AUTHORS INDEX

Aarseth, P., 793
Abdulla, Y. H., 950
Abe, M., 786, 787, 1310
Abello, J., 1178
Aberg, B., 1161
Abood, L. G., 881
Abraham, R., 992, 1363
Abrahams, V. C., 800
Abrams, R., 850
Ackerman, G. A., 1012
Ackermann, J., 1080
Adams, B. J., 1273
Adams, C. W. M., 794, 800, 964, 972, 1010, 1042, 1081, 1273, 1369
Adams, D. H., 767, 768, 799
Adams, E., 998
Adams, E C., 1135, 1136
Adams, L. R., 1232
Adams, M. H., 859
Adler, E., 904
Adler, I., 1233
Adlersberg, D., 1086
Adrian, E. D., 799
Affonso, O. R., 874
Aghajanian, G. K., 1262
Agner, K., 851
Ahsmann, W. B. A. M., 1249
Aikat, B., 1043
Aitken, W. M., 1213
Akagi, T., 1152
Albersheim, P., 1273
Albert, E. N., 1263
Albert, S., 1209
Albert, Z., 973, 1356
Alberti, P., 34
Albrecht, M., 888
Aldridge, W. G., 1273
Aldridge, W. N., 762, 763, 764, 768, 987, 1409
Alexander, J., 766
Ali, S. Y., 963
Alin, K., 1178
Al-Khalidi, U. A. S., 874
Allen, J. M., 800, 941
Allen, R. D., 1246
Allison, A. C., 1178
Alm, P., 1112
Alpen, E. L., 1231
Alpert, M., 1082, 1084, 1385
Altmann, F. P., 885, 889, 890, 891
Altschul, A. M., 850
Altschuler, C. H., 1031
Amano, H., 1012
Amano, M., 1022, 1025
Amarasingham, C. R., 922, 923, 1346
Ammon, R., 767
Amprino, R., 1211
Andersen, H., 936, 1348
Anderson, F. B., 1325
Anderson, P. J., 1074
Anderson, R. J., 1082
Anderson, T. F., 1262
Anderson, W. A., 1274
Anderson, W. E., 1277, 1281, 1283, 1445
Andresen, C. C., 1217
Andresen, N., 1217
Andrews, M., 1354
Angelakos, E. T., 1109, 1121, 1122
Angevine, D. M., 1031
Änggård, E., 935
Anlyan, A. J., 809
Antopol, W., 881
Anwar, R. A., 1042
Aoki, T., 1284
Aoshima, Y., 942
Aoyama, T., 1155
Appleton, T. C., 1214, 1430
Arase, M., 974
Archer, H. E., 1138
Archer, R. K., 851
Argyris, T. S., 800, 922
Arioka, I., 867
Armitage, F. L., 853
Armstrong, D., 989
Armstrong, J. A., 1186
Armstrong, J. M., 995
Arnold, J. S., 1212
Arnold, M., 1149, 1275, 1291, 1293
Aronsohn, R. B., 1084
Arstila, A. U., 1293
Arvy, L., 799, 1067
Asboe-Hanson, G., 1032
Asero, B., 1118
Ashley, J. N., 894
Ashwell, G., 931
Ashworth, C. T., 1279
Astrup, T., 1011
Atherton, G. W., 793
Atkin, N. B., 1228, 1229
Atkinson, E., 881
Atkinson, W. B., 1137, 1406
Attramadal, A., 1215
Augustinsson, K.-B., 764, 766, 767
Auricchio, F., 818
Auricchio, S., 821
Austin, J., 989
Austin, L., 770
Austoni, M. E., 1211
Avers, C. J., 853
Avrameas, S., 1267, 1437
Axelrod, D. J., 1207
Axelrod, D. T., 1211
Baar, H. S., 921
Babson, A. L., 984, 985
Bachmann, L., 1217
Baggenstoss, A. H., 1068
Bahn, R. C., 893
Bahr, G. F., 1226, 1227, 1261, 1262
Bailar, J. C., 1129
Bailar, J. C., jr., 1269
Baillie, A. H., 941, 943
Baker, F. A., 948
Baker, J. R., 810, 814, 817
Baldwin, E., 1018
Balfour, B. M., 1027, 1031
Ball, E. G., 841, 893
Ball, S., 1088
Ballantyne, B., 799
Baló, J., 1041, 1042
Balogh, K., 926, 931, 933, 942, 943, 954, 971, 1345, 1348
Bals, M. G., 1171, 1195
Baltscheffsky, M., 913, 914
Banga, I., 1041, 1042
Baranowski, T., 929
Barbaglia, V., 1156
Barbosa, P., 1180
Barden, H., 1279
Barenboim, G. M., 1173, 1177, 1238
Barer, R., 1228, 1246, 1247
Barka, T., 1074
Barnard, E. A., 1019, 1218
Barnett, J. E. G., 820
Barnicot, N. A., 1055
Baron, C., 868
Baron, D. N., 940, 943
Barone, P., 1089
Baroni, B., 1052
Barrett, A. J., 963, 966, 967, 970
Barrnett, R. J., 777, 778, 779, 884, 1062, 1063, 1263, 1276, 1278, 1279, 1280, 1281, 1282, 1284, 1286, 1289, 1291, 1452
Barrowman, J. A., 763, 786, 796
Barry, J., 800
Barter, R., 1102, 1103, 1104, 1116, 1118, 1246
Baserga, R., 1209, 1211, 1215, 1216, 1217
Baskys, S. B., 787
Bastos, A. L., 1180
Batelli, F., 872, 948
Baudhuin, P., 812, 854
Bauer, K., 1033
Baum, H., 988
Baumann, A., 1207
Baust, P., 801
Baxandall, J., 1264, 1267
Baxter, A. O., 943
Baxter-Grillo, D. L., 954
Bayer, E., 1137
Bayley, S. T., 1208
Bayliss, B. J., 770
Bayliss, O. B., 1010, 1042
Beard, M. E., 1286, 1449
Beatty, C. H., 835
Beaufay, H., 812, 872
Beck, C., 819
Beck, S., 1178
Becker, B., 811, 812, 813
Becker, S. W., 860, 1286, 1338
Beckett, E. B., 922
Bednář, B., 1085
Beer, C. S., 857
Beer, M., 1194, 1273
Beer, R. J. S., 1051
Behal, F. J., 964, 970
Behrens, B., 1207
Bélanger, L. F., 794, 1010, 1208, 1210, 1211, 1212, 1215, 1319
Beletskaya, L. V., 1043
Bell, C., 793, 1282
Beller, F. K., 1012
Bellman, S., 1247
Belt, W. D., 1194
Benditt, E. P., 971, 974, 1040, 1116, 1357
Beneke, G., 1233, 1246
Beneš, K., 801
Benesch, R., 1270
Benesch, R. E., 1270
Benitez, L., 923, 1346
Bennett, H. S., 1337
Bensch, K. G., 1276
Bensley, R. R., 1076, 1077
Berg, J. W., 1081
Berger, J., 962, 963
Berger, R., 1138
Berggren, H., 1208
Bergman, R. A., 1281, 1283
Bergman, S., 941
Bergmann, F., 763, 764, 766, 767, 775, 976
Bergmann, M., 963
Bergner, A. D., 800
Bergstrand, H., 1085
Berlin, M., 1207
Bernfeld, P., 811
Bernhard, W., 1290
Bernsohn, J., 800
Bernstein, M. H., 1262
Berriman, R. W., 1214
Berry, W. K., 770
Bertalanffy, F. D., 1186
Berthelot, M., 821
Bertolini, R., 946
Berwick, L., 1275
Bettendorf, A., 1141
Betts, A., 1186
Beyer, H., 849, 897
Bianchi, U. A., 1241, 1251
Bidwell, O., 1043
Bierther, M., 1291
Bigelow, L. B., 1121

Bigelow, R., 809
Billett, F., 809, 814
Billing, B. H., 810, 1071, 1072
Bilski, R., 933
Binkley, F., 973
Birk, Y., 1040
Birkbeck, M. S. C., 1055
Birks, R. I., 793, 1282
Birnbaum, S. M., 967
Biro, V., 898
Bischler, A., 1107
Bizzozero, E., 1057
Bjersing, L., 942
Björklund, A., 1052, 1109, 1110, 1112, 1120, 1395, 1399
Björklund, U., 844, 1235, 1237 1238, 1240
Black, M. M., 881
Blackberg, S. N., 860
Blackburn, S., 1041
Blacklock, J. W. S., 1060
Blackwood, C., 963
Blake, A., 1188
Blakley, R. L., 930
Blaschko, H., 863, 864, 867, 871, 1061, 1063
Blaschko, K., 772
Blasius, W., 921
Blazicek, G., 1187
Bleyl, U., 996
Bloch, B., 858, 860, 861, 862
Bloem, J. H., 997, 999, 1365, 1367
Blondé, C., 855
Blonk, D. I., 954
Bloom, F. E., 1262, 1277, 1282, 1286, 1288, 1445, 1449
Bloom, W., 1207
Blumberg, B. S., 1041
Blumberg, J. M., 795
Blumenthal, W. B., 1129
Bo, W. J., 835
Boadle, M. C., 1286
Boas, N. F., 1032
Bobrow, M., 1184
Bocek, R. M., 835
Bock, R., 1241
Bodansky, O., 768
Bodian, D., 1020
Boddy, R. G. H. B., 1228
Bog, R., 1144
Boggart, W. A., 1064
Bogoroch, R., 1208, 1214
Böhm, N., 1183, 1189, 1234, 1235 1239, 1241, 1249, 1251, 1252
Bois, I., 970
Bois-Svensson, I., 970, 971, 972
Bokdawala, F. D., 785, 1310
Bollier, M. E., 1148
Bonati, F., 853
Bone, A. D., 1121
Bonnichsen, R., 917
Bontke, E., 801, 1080
Booth, H., 851
Boothroyd, B., 1293
Borchardt, H., 1085
Borei, H. G., 844
Borgers, M., 1278
Borgström, R., 763, 821
Borst, M., 1067
Boseck, S., 1294
Bosshard, U., 1237, 1238, 1241
Boström, H., 1210
Botham, C. M., 1281, 1444
Bots, G. Th. A. M., 815, 834, 1137
Botte, V., 943
Bourland, A., 800
Bourne, G. H., 800, 801, 874, 921 922, 932, 954, 1134
Bowen, V. T., 1148
Bowie, D. J., 1217
Bowman, R. L., 1104
Boxer, G. E., 929
Boyd, G. A., 1207, 1208, 1209, 1211
Brachet, J., 1019, 1020, 1023, 1024
Bradbury, S., 1274, 1275, 1280, 1442
Bradley, H. C., 1144, 1154, 1187
Brady, R. O., 820
Brahn, B., 1085
Brand, I., 1147
Brandau, H., 943
Brandes, D., 801
Brandino, G., 1151
Brandt, E. A., 1042
Bratt, H., 943
Brattgård, S. O., 1247
Brault, J. W., 1246
Braun-Falco, O., 834, 951, 954, 971, 995
Braunstein, A. A., 983
Brauss, E., 832
Bray, R. C., 873
Brecher, A. S., 962, 965
Breckenridge, B. M., 830
Brenner, S., 1186
Breuer, H., 943
Breslow, E., 932
Brière, N., 1023
Brinkley, W. R., 1273
Broda, B., 1150
Brody, I., 1262
Brody, I. A., 935
Brooke, M. H., 887, 924
Broome, J., 851
Brosowski, K. H., 991
Brown, D. M., 1018
Brown, G. W. jr., 867
Brown, L. M., 793, 1282
Broyer, M., 1138
Brun, A., 824, 1326
Brunet, P. C. J., 822
Brunk, U., 1149, 1413
Bruttin, H., 903
Bryan, J. H. D., 1273
Bryson, M. J., 941
Bubis, J. J., 950
Bublitz, C., 932
Bucher, N. L. R., 1214
Bücher, Th., 929
Buell, M. V., 829
Bukatsch, F., 1186
Bulger, R. E., 1158
Bull, B., 821
Bull, G., 793
Bull, J. P., 1161
Bulliard, H., 1208
Bulmer, D., 815, 816
Bu'lock, J. D., 1051, 1052, 1054, 1061
Bu'lock, J. H., 1336
Bumstead, J. H., 1070
Buño, W., 784, 922
Bunting, H., 1027, 1031, 1032, 1130, 1134, 1158, 1402, 1403, 1417
Burgen, A. S. V., 767, 770
Burkhalter, A., 800
Burkl, W., 942
Burnett, F. M., 1033
Burstone, M. S., 774, 775, 801, 848, 849, 850, 857, 916, 967, 968, 970, 971, 1285, 1304, 1305, 1331, 1332, 1357
Burt, A. M., 982, 1360
Burt, R. C., 1084
Burtner, H. J., 855, 867, 893, 1118, 1379
Burton, J. F., 813, 814, 816, 817
Burton, K., 917, 1025
Burton, V., 1183
Bussolati, G., 1189, 1190, 1191, 1241, 1251, 1421, 1423

Cabrini, R. L., 834
Cahn, R. D., 934
Caley, E. R., 1151
Calman, K. C., 943
Cambel, P., 1036
Cameron, G. R., 1134
Campani, M., 1027
Campbell, H., 883, 885
Campbell, J. G., 809, 813
Campbell, L. B., 1119
Campo-Aasen, I., 787, 800
Cannata, M. A., 1063, 1291
Capella, C., 1119, 1189, 1190
Caputto, R., 932
Carbonell, L. M., 791, 834
Cardini, C. E., 826, 827, 832
Carere-Comes, O., 1157
Carlson, L., 1246
Carlsson, A., 1111
Carmichael, G. G., 1003, 1004
Carneiro, J., 1044
Caro, L. G., 1218, 1219, 1433, 1435
Carranza, F. A., jr., 834
Carr Barnes, C., 910
Carroll, B. R., 942
Carruthers, C., 842
Carstensen, H., 942
Carter, S. D., 888
Carvalheira, A. F., 800, 1111, 1119, 1189
Cascarano, J., 909, 922
Cason, J. E., 1036, 1377
Caspersson, T., 1184, 1185, 1226, 1228, 1229, 1230, 1231, 1232, 1235, 1237, 1238, 1249, 1251
Casselman, W. G. B., 1084
Castaing, R., 1233
Castel, P., 1141, 1143, 1409
Castellani, P., 830, 1327
Castellano, M. A., 951, 1351
Castoldi, G. L., 948, 1351
Catchpole, H. R., 1030, 1031, 1037, 1043
Catchside, D. G., 1023
Cauna, N., 800
Cavallero, C., 941, 1189
Cavallini, O., 871
Cavanagh, J. B., 798
Cavasso, N., 1290
Cazal, P., 853
Čech, S., 870
Celani, R. B., 853
Chaffee, E., 1026
Chaglassian, T. H., 874
Chain, E., 1025, 1026
Chalkley, H. W., 1248
Chambers, T. C., 1037
Champy, C., 1063
Chance, B., 913, 1177, 1237, 1238, 1242
Chan-Curtis, V., 1194, 1273
Chang, J. P., 946, 1279
Chapman, J. A., 1267
Chapman, J. H., 1240, 1241
Chapman, N. B., 854
Charèst, J., 1013, 1373
Chargaff, E., 1023
Chatterjee, I. B., 932
Chauball, K. A., 1248
Chauncey, H. H., 1156
Cheng, K-K., 1142, 1158, 1416
Cheronis, N. O., 881
Chessick, R. D., 796, 800
Chèvremont, M., 1024
Chiappino, G., 941
Chibnall, A. C., 1190
Chieffi, G., 943
Chinoy, N. J., 1067
Chiocchio, S. A., 1291
Chisholm, G. D., 1138
Chittayasothorn, K., 1085
Chou, L. Y., 1138
Chou, T. C., 772
Chouchkov, H., 954
Christeller, E., 1143, 1148, 1412
Christian, W., 904, 925
Christie, A. C., 1103, 1115
Christie, G. A., 834, 943, 972
Chu, E. J.-H., 1067
Chu, T. C., 1067
Churg, J., 1274
Cichocki, T., 1080
Cimasoni, G., 771, 799
Ciotti, M. M., 902
Clapp, D. H., 1281, 1283
Clara, M., 1057, 1113
Clark, A. M., 994
Clark, D. E., 1209
Clarke, A. J., 865

Clarke, B. L., 1146, 1148
Clarke, D. E., 1111
Claude, A., 1026
Cloetens, R., 1154
Close, R. A., 1136
Clowes, F. A. L., 1209
Cobb, J., 828
Cochran, J., 954
Coers, C., 800
Coey, W. E., 1084
Cogan, D. G., 888
Cohen, B. B., 1042
Cohen, L. A., 974, 1357
Cohen, P. P., 983
Cohen, R. B., 817, 821, 822, 835, 930, 932, 933, 937, 948, 954, 990, 1321, 1350, 1362
Cole, P. G., 810, 1071, 1072
Coleman, J. E., 994
Coleman, M. H., 771
Coleman, R. F., 927
Colfer, H. F., 1208
Collewijn, H., 1157, 1416
Collins, P. F., 1132
Collins, W. P., 941, 942
Colowick, S. P., 910
Colvin, J. R., 1263
Comar, C. L., 1210
Commoner, B., 1054, 1228
Comstock, E. G., 862
Conchie, J., 819
Conn, E. C., 953
Conover, T. E., 905
Contier, L., 1186
Cook, R., jr., 862
Coons, A. H., 1171
Cooper, O., 841
Cooper, W. G., 922
Cope, G. H., 1276
Copenhaver, W. M., 922
Copp, F. C., 770
Cordier, R., 1060, 1113, 1117, 1148
Cori, C. F., 826
Cori, G. T., 826, 827, 830
Cornelius, C. E., 1089
Corran, H. S., 904
Corrodi, H., 1103, 1104, 1105, 1107, 1110, 1249, 1396
Cosslett, V. E., 1233
Cotson, S., 780, 781, 782
Couceiro, S., 799
Coujard, R., 1063
Coujard-Champy, C., 1063
Coupland, R. E., 1057, 1063, 1291
Couteaux, R., 772, 792, 799
Coutinho, H. B., 765
Craig, C. P., 1279
Cramer, W., 1178
Crane, F. L., 1000
Crane, R. D., 1161
Crane, R. K., 818, 821
Crawford, D. T., 847, 967, 968, 971, 1353
Crawford, J. D., 937
Creasey, N. H., 864, 865
Crétin, A., 1133, 1134, 1149, 1154, 1413
Crevier, M., 794, 1319
Cripps, D. J., 1177, 1200
Cronkite, E. P., 1212
Cross, S. A. M., 997, 1288, 1365, 1400
Crout, J. R., 1157, 1415
Crowley, N., 1026
Cruickshank, C. N. D., 922
Csillik, B., 793, 794, 795, 800, 1283, 1316, 1446
Cunningham, L., 1010, 1011, 1369
Curd, M. R., 1056
Curran, R. C., 1275
Curri, S. B., 800
Curzen, P., 972
Czitober, H., 1177

Daddi, C., 1074
Dahl, L. K., 1135, 1404
Dahlqvist, A., 821, 824, 1013, 1326
Dahlström, A., 1111, 1112
Dales, S., 1267
Dalgaard, O. Z., 801
Dalgleish, C. E., 1119
Dalldorf, G. J. G., 1085
d'Almeida, D. F., 799
Dalton, A. J., 1083, 1262
Dalziel, K., 951
Dam, H., 1078, 1079
Dance, N., 819, 821
Dancheva, K. I., 829, 835
Daniel, O., 1211
Danielli, J. F., 1016, 1246
Daoust, R., 1009, 1012, 1371, 1372
Darden, E. B., 1262
Daria Haust, M., 1267
Darnton, S. J., 786, 796
Datta, N., 792
Davidson, F. M., 1044
Davies, D. R., 800
Davies, D. V., 1030, 1031, 1210
Davies, H. G., 1228, 1246
Davis, B. J., 781, 782
Davis, J. C., 922
Davis, R., 795, 1283
Davison, A. N., 768, 798
Dawson, A. P., 869, 944, 945
Dawson, E. R., 762
Dawson, I., 1111
Dawson, I. M. D., 941
Dawson, M. H., 1026
Dawson, R. M. C., 787
Day, T. D., 1040
Dayan, A. D., 1245
Deamer, D. W., 946
Deane, H. W., 941, 1020
Debenedetti, S., 1151
Debruyn, P. P. H., 1186
Decker, L. E., 985
de Duve, C., 812, 854, 872, 966
Deeley, E. M., 1229
Defendi, V., 782, 921
De Galantha, A., 1165
de Giamcomo, P., 800
Deguchi, Y., 894
Deimling, O., 776, 1305
Dejardin, M., 800
De La Haba, G., 963
de la Torre, L., 1251
Delellis, R., 783, 1307
De Lellis, R. A., 812, 834
de Lerma, B., 1067, 1171
Dellovo, M. C., 1033
Delory, E. E., 1154
De Martino, C., 1274, 1439
Dempsey, E. W., 851, 1020
Denham, S. W., 922
Denker, H. W., 1013
Dennis, P. M., 943
Denny, J. J., 1140
Den Tonkelaar, B. M., 1232
Denz, F. A., 789, 790, 1136, 1137, 1142, 1408
De Pascale, A., 1003
De Paoli, A. M., 1183
De Robertis, E., 851, 853, 854, 1334
Desmet, V. J., 1075, 1391
Desnuelle, P., 785, 787
Dessau, F. I., 1024
Dettmer, N. I., 1274
Devlin, M., 795
Devlin, T. M., 929
Dewan, J. G., 904
Diamantstein, T., 996
Diamond, L., 1178
Dible, J. H., 1043
Dick, D. A. T., 1247
Diculescu, I., 926, 951, 952, 954, 998, 999, 1366
Diegenbach, P. C., 798
Diehl, H., 1132
Diezel, P. B., 1132
Dilley, R. A., 1002
Di Pasqua, A., 1074
Ditmars, W. E., 1187
Dixon, A. D., 1282
Dixon, F. J., 1209
Dixon, M., 842, 899, 900, 901, 910, 948, 949, 1000
Dobb, M. G., 1263
Dobriner, K., 1067
Dobyns, B. M., 1208, 1213
Dodgson, K. S., 810, 988, 989, 1032
Doell, R. G., 821
Dohlman, C.-H., 988
Dols, M. J. L., 1207
Domanskii, A. N., 1173, 1177, 1238
Donati, E. J., 1270
Doniach, I., 1178, 1208, 1211, 1212, 1214
Donn, J. D., 1158
Donné, A., 1138
Donohue, W. L., 1138
Dorfman, A., 1030
Dormer, P., 1217, 1243
Dorn, A., 1279
Doty, P., 1188
Douglas, L. T., 800
Douglas, W. H. J., 1290
Doxey, D. L., 855
Drabikowska, A. K., 944
Dresner, E., 1043
Drews, G. A., 1067
Drouin, M., 800
Dubach, U. C., 1226
Dubin, I. N., 1088
Dubos, R. J., 1019, 1026
Du Buy, H. G., 1055
Duckett, S., 869, 954
Dudley, R. A., 1207, 1208, 1217
Duff, T. A., 994
Dufrenoy, J., 881
Dumont, L., 800
Duncomb, P., 1293
Dunlap, C. E., 922
Dunn, R. C., 856, 1065, 1335
Durand, S. C., 1041
Duran-Reynals, F., 1025, 1026
Durante, M., 800
Durocher, J. G., 1012
Dutcher, R. A., 881
Duthie, E. S., 1025, 1026
Duthie, R. B., 1212
Dutton, G. J., 932
Dutton, R. W., 1119
Dyson, J., 1246
Dziewiatkowski, D. D., 1210

Eadie, M. J., 888, 894
Eapen, J., 784, 785
East, M. E., 820
Ebel, A., 1042
Ebschner, K. J., 1189
Echave, D., 1144
Eckfors, T. D., 973
Eder, M., 867, 1155, 1339
Edmonson, V. R., 1217
Edsall, J. T., 994
Edson, N. L., 930
Edwards, B. B., 1151
Edwards, J. E., 1083
Edwards, R. G., 892, 954
Eegriwe, E., 1152
Egami, F., 989
Egli, K., 835
Ehinger, B., 800, 1052, 1109, 1110, 1111, 1123, 1235, 1237, 1238, 1240, 1395, 1400
Eidinger, D., 1036
Eijkman, C., 1041
Einarson, L., 1085
Elftman, A. G., 1148
Elftman, H., 1148
Ellinger, P., 1067
Elliot, K. A. C., 842
Ellis, G. W., 1217
Ellis, R. A., 829, 834
Ellis, S., 965
Ely, R. V., 1233

Emmart, E. W., 947
Endahl, G. L., 940
Endicott, K. M., 1083, 1207, 1212, 1213
Enerbäck, L., 776, 1123, 1353
Engel, P. C., 951
Engel, W. K., 887, 924, 935, 1067
Engfeldt, B., 1210
Engström, A., 1228, 1233, 1246, 1247
Eränkö, O., 793, 800, 830, 870, 1062, 1101, 1102, 1107, 1108, 1209, 1226, 1228, 1247, 1282, 1318, 1327, 1394
Ericson, L. E., 1220
Ericsson, J. L. E., 1280
Ernster, L., 903, 905, 913, 937
Erös, G., 1102
Errera, M., 1210
Erspamer, V., 1101, 1118, 1119, 1122
Erwin, H. L., 996
Esilä, R., 800
Essex, H. E., 1208
Essner, E., 1080, 1280
Estabrook, R. W., 913
Esterley, J. R., 818, 819
Etcheverry, G. J., 1291
Etherton, J. E., 1281, 1444
Ettori, J., 853
Evans, H. M., 1209
Evans, R. A., 786
Evans, T. C., 1212, 1213
Evans, W. C., 859
Ewen, S. W. B., 1107, 1250, 1400

Fabergé, A. C., 1037
Fagundes, L. A., 835
Fahimi, H. D., 854, 922, 923, 1286, 1337, 1346
Fahmy, N. I., 915
Fairbanks, V. F., 938
Fairhall, L. T., 1413
Fajberg, V., 1210
Falck, B., 1052, 1101, 1107, 1108, 1109, 1110, 1111, 1112, 1235, 1237, 1238, 1240, 1251, 1394, 1395
Falk, G. J., 1215
Falk, H., 1279
Falk, S., 1268
Falkmer, S., 1154, 1290, 1451
Fand, S. B., 801, 996
Fantin, A. M. B., 1110
Farber, E., 872, 907, 915, 922, 925, 946
Farr, R. S., 1186
Farrant, J. L., 1069
Fasske, E., 947
Fattorusso, E., 900
Fautrez, J., 856
Fauvel, P., 1058
Favard, P., 1290
Favilli, G., 1026
Fawns, H. T., 1031
Fearon, W. R., 1064
Feder, N., 1268, 1440
Feeley, J., 952
Feess, E., 991
Feigl, F., 777, 1079, 1130, 1137, 1141, 1142, 1143, 1144, 1145, 1146, 1149, 1151, 1152, 1157, 1165, 1386, 1387, 1410, 1415, 1416
Feldberg, E., 799
Feldberg, W., 1063
Feldman, D. G., 1262
Felgenhauer, K., 970
Felt, V., 954
Fennell, R. A., 801
Fenton, R. H., 1131
Ferguson, M. M., 927, 943
Fernand, V. S. V., 1010
Figq, A., 1210
Figge, F. H. J., 1056, 1068
Finck, H., 1270, 1272
Fine, A., 843, 846
Fiori, A., 1156, 1413
Firket, H., 1142
Fischer, E. H., 827
Fischer, F. G., 1023
Fischer, H., 888, 1070
Fischer, J., 954
Fischer, M. I., 1153, 1154
Fischer, R., 776, 923, 1085, 1346
Fisher, E. R., 1040, 1086
Fisher, H. F., 904
Fishman, W. H., 783, 808, 809, 810, 811, 812, 814, 815, 817, 834, 1272, 1281, 1284, 1307, 1322
Fite, G. L., 1082
Fitzgerald, P. J., 1210, 1211, 1214, 1228
Fitzpatrick, T. B., 862, 1051
Fleischer, E., 1263
Fleischhauer, K., 890
Flexner, L. B., 1132
Flitney, F. W., 923, 1217
Florey, H. W., 1210
Fodor, P. J., 762, 766
Folk, J. E., 967, 968, 973, 974, 1356
Fontaine, R., 1042
Foraker, A. G., 922
Formanek, H., 1272
Formanek, S., 1272
Formisano, V. R., 921
Foroglou-Kerameos, C., 1283
Foster, M., 862
Fourman, J., 799
Fowler, B. A., 1275
Fowler, C. B., 1074
Fox, B., 1086
Fox, M. D., 790
Franchi, F., 853
Francis, C. M., 800, 868
Francois, C. J., 1191
Frankenberger, Z., 1149
Franz, H., 1269, 1270
Fratello, B., 1011, 1370
Fredricsson, B., 800
Freeman, J. A., 1262
Freeman, L. W., 1055
Freiman, D. G., 1162, 1419
French, J. E., 1040
Freudenberg, E., 1133
Friberg, U., 1161, 1210
Frič, P., 934, 954
Fridovich, I., 873
Fried, G. H., 922
Friede, R. L., 954
Frieden, C., 951
Friedenwald, J. S., 789, 791, 811, 812, 813, 988
Friedmann, H. C., 948
Fromme, H. G., 1218
Frommer, J., 1163
Frost, J. L., 801, 1179
Fruton, J. S., 962, 963, 966, 976, 1154
Fuchs, U., 887
Fujii, S., 966
Fujita, H., 981
Fujita, S., 851
Fullerton, P. M., 1086
Fullmer, H. M., 1042
Furth, A. J., 994
Futterman, S., 952
Fuxe, K., 1108, 1111, 1112

Gabay, S., 864
Gabler, W., 893, 946, 1349
Gabriel, L. O., 999
Gaddum, J. H., 1118
Gahan, P. B., 922, 923, 950
Gahrton, G., 1241
Gaillard, J. L. J., 1232
Gal, E. M., 1119
Galjaard, H., 1246
Gall, E. A., 1162, 1419
Gallico, E., 842
Gallyas, F., 1276
Ganote, C. E., 1280
Garcia, A. M., 1228, 1231, 1249
Garcia Blanco, J., 852
Garcia-Hermandez, M., 983
Gareau, R., 971
Garrett, J. R., 799
Garrod, A. E., 1064
Gasic, G., 1275
Gautier, A., 1273
Gawron, O., 771
Gay, H., 1016, 1021
Gedick, P., 801, 1069, 1074, 1080, 1085, 1152
Gehauf, B., 1148
Geiger, J., 1156
Gentshev, T., 835
George, J. C., 785, 835, 1310
George, L. A., 1218
Gérard, P., 1060, 1148
Gerebtzoff, M. A., 792, 794, 799, 800
Germino, N. I., 922
Gersch, N. F., 1187
Gersh, I., 1020, 1030, 1031, 1037, 1038, 1040, 1043, 1157, 1159, 1415
Gershon, M. D., 1111
Gertler, A., 1042
Geyer, G., 776, 1033, 1260, 1263, 1273, 1290
Ghiringhelli, F., 1118
Ghislandi, E., 1062
Ghoos, Y., 1013, 1374
Ghosh, A., 1036
Giacobini, E., 800
Gianetto, R., 988
Giarman, N. J., 1118
Gibbs, H. D., 1114, 1117
Gibson, J. G., 1155
Gillespie, R. E., 818
Gillman, J., 1067, 1070, 1082
Gillman, T., 1067, 1070, 1082
Giroud, A., 1057
Giusti, G. V., 1156, 1413
Glassman, A., 1237, 1238, 1239
Glaubach, S., 881
Glauert, A. M., 1276
Glavind, J., 1078, 1385
Glenner, G. G., 776, 832, 867, 868, 869, 942, 964, 970, 972, 973, 974, 975, 977, 1072, 1073, 1328, 1340, 1354, 1356, 1357, 1390
Glick, D., 871, 886, 888, 953, 1128, 1131, 1225, 1228, 1275
Glimcher, M. J., 1191
Glimstedt, G., 1247
Glock, G. E., 851
Glogner, P., 946, 1349
Glover, P., 1183
Glücksmann, A., 1210
Gmelin, L., 1072, 1073
Goddard, D. R., 953
Goddard, J. W., 921
Godlewski, H. G., 829, 830, 834, 933, 1134
Goebel, A., 995
Goessner, W. *See* Gössner, W.
Gofstein, R., 881
Goland, P. G., 1178, 1186
Goldberg, B., 828, 941, 942
Goldberg, J. A., 822
Goldenberg, M., 1161
Goldenberg, S. H., 827
Goldenson, J., 1148
Goldfischer, S., 855, 992, 1080, 1280, 1285, 1363
Goldman, L., 881
Goldman, M., 1171, 1237, 1238, 1241, 1251
Goldman, R. D., 954
Goldman, S. S., 815
Goldring, D., 1138
Goldstein, A., 767
Goldstein, D., 1137
Goldstein, D. J., 1228, 1230, 1244, 1248
Goldstein, T. P., 847

Gomori, G., 762, 773, 774, 775, 783, 784, 785, 786, 788, 789, 790, 796, 800, 828, 855, 866, 967, 974, 988, 1057, 1066, 1072, 1080, 1102, 1113, 1114, 1117, 1118, 1130, 1134, 1138, 1142, 1147, 1164, 1165, 1274, 1278, 1280, 1284, 1303, 1304, 1309, 1312, 1384, 1398, 1444
Goodpasture, F. W., 853
Goodwin, T. W., 1067
Gorbman, A., 1209
Gordon, E., 809
Gordon, M. P., 1187
Gore, M. B. R., 943
Gorski, J., 930, 931, 933, 1345, 1346
Gortner, R. A., 1051
Gössner, W., 774, 946, 991, 1069, 1349
Gosztonyi, T., 893
Gottschalk, A., 1013
Gouinlock, E. V., 1191
Gracheva, N. D., 1210
Gracie, M., 1122
Graf, W., 1161
Gräff, S., 844, 845
Graffi, A., 1178
Grafflin, A. L., 1067
Graham, G. S., 853, 855
Graham, R. C., 1267, 1286, 1338
Graham, R. C., jr., 854, 857, 869, 870, 872, 873, 1337, 1341
Granados, H., 1078
Grandis, E., 1134
Granick, S., 1066, 1132, 1387, 1403
Grassmann, W., 1043
Grasso, R., 851, 853, 854, 1334
Graves, J. L., 948
Gray, C. H., 1071
Grazi, E. 999
Greco, J., 1033, 1034
Green, A. A. 827
Green, C. L., 1147
Green, D. E., 904, 944, 1061
Green, D. S., 1156
Green, J. A., 1043
Green, M. N., 967, 968
Green, R. C., 873
Green, S., 810, 815
Greenbaum, A. L., 812
Greenbaum, L. M., 963
Greenberg, D. M., 952
Greenstein, J. P., 967, 1018, 1051
Gregory, D. W., 1265
Greig, M. E., 842
Greiling, H., 1029
Gresham, G. A., 1085, 1086
Greville, G. D., 946
Gries, G., 1074
Griese, A., 904
Griffith, K., 941
Grillo, T. A. I., 830, 835
Grimm, R. B., 853
Grisolia, S., 852
Grollman, A. P., 931
Gromadzki, C. G., 793, 1283, 1316, 1319, 1447
Grose, F., 942
Gross, J., 1208, 1211, 1215
Grossman, I. W., 825, 1279
Gruber, M., 963
Grunfeld, A., 834
Grzycki, S., 972
Guarino, A. J., 811
Gudbjarnason, S., 876
Guha, S., 829, 830, 946
Guinier, A., 1233
Guinnenbault, M., 1211
Gullberg, J. E., 1217
Gunter, J. M., 799
Gunther, R. E., 871
Gürtner, Th., 800
Gustafson, B. E., 821
Gustafson, G. T., 1275
Güthert, H., 1156
Gutmann, H. R., 963
Gutowska, M. S., 996
Gutzeit, G., 1152

Hack, M. H., 1056
Hadjiolov, A. A., 829, 835
Hagen, P., 800, 1063
Hager, G., 1275, 1291
Hahn, L., 820, 1028, 1029
Haitinger, M., 1171, 1177, 1179, 1180, 1186
Hajós, F., 1277, 1285, 1288, 1451
Håkanson, R., 1111, 1123
Håkansson, A., 1247
Hakim, A. A., 1021
Hale, A. J., 1014, 1233, 1246, 1293, 1374
Hale, C. W., 1029
Hall, B. D., 1025
Hall, C. E., 1261
Hall, M. J., 1072
Hallén, O., 1247
Halton, S. W., 801
Hamberger, B., 1108, 1109, 1394
Hamilton, J. G., 1207, 1211
Hamlisch, R. E., 767
Hamm, D. I., 942
Hammarströn, V., 1213
Hammes, E. M., jr., 1068
Hammett, L. P., 1140
Hamperl, H., 1113, 1119, 1178
Hand, W. C., 1151
Hanes, C. S., 973
Hanker, J. S., 1263, 1269, 1274, 1277, 1281, 1282, 1283, 1288, 1442, 1444, 1445, 1449, 1450
Hansen, R. E., 829
Hanson, H., 970
Hanson, H. H., 965
Hanson, H. T., 1154
Hansson, C. G., 776, 971, 1353
Hansson, H. P.-J., 997, 1288, 1365
Harbinson, R. J., 1182
Hard, W. L., 790
Hardonk, M. J., 941, 1137, 1249, 1423
Hardrodt, W., 1362
Hardwick, D. C., 800
Harley-Mason, J., 1336
Harms, H., 1180
Harris, C., 799
Harris, W. H., 1179
Harrison, C. V., 1140
Harrower, J. R., 1178
Hart, D. McK., 943
Hartley, B. S., 962
Hartlieb, J., 1022
Hartmann, G., 1187
Hartree, E. F., 841, 842, 843, 872, 912
Hartrodt, W., 991, 1364
Hartroft, W. S., 1084
Hartsough, G. R., 1084
Hasebroek, K. A., 858
Hashimoto, K., 870
Hashimoto, P. H., 869
Hashimoto, T., 916, 917
Hassid, W. Z., 829
Hassler, O., 1042
Hastings, A. B., 1225
Hatefi, Y., 924
Hatem, S., 1063
Haugaard, N., 1282
Häusler, G., 995, 996, 997, 1288 1365
Hausser, I., 896
Haustein, U. F., 1012
Hawgood, R. S., 1177, 1178, 1200
Hawkins, R. D., 768, 799
Hay, A. J., 810
Hayashi, M., 814, 816, 824, 834, 991, 1284, 1322, 1325
Haydon, G. B., 1288
Heard, B. E., 1138
Heath, D. F., 762
Heath, I. D., 1057
Hebb, C., 800
Hebb, C. O., 982
Heene, R., 1109
Heidenreich, O., 1077
Heilbron, E., 771
Heirwegh, K. P. M., 1075
Heitkamp, D. H., 1279
Helander, S., 1178
Hellauer, H. F., 772
Heller, D., 1211
Heller, J. H., 1121
Hellmann, K., 799, 864, 867
Hellström, H., 904
Hellströmm, B., 972
Helmreich, M. L., 941
Helmy, F. M., 1056
Henderson, H. J., 884
Henderson, J. R., 793, 1316
Henle, J., 1059
Henschen, F., 1085
Henson, J. P. G., 1118
Heppel, L. A., 1019
Herbertson, S., 1042
Herbst, F., 794
Hercules, D. M., 1173
Hermance, H. W., 1146, 1148
Hernandez, F., 1004
Hernandez, W., 1274
Hershberger, L. R., 1075
Hershey, F. B., 922
Herveg, J. P., 812
Hess, R., 782, 783, 797, 800, 835 887, 888, 902, 906, 907, 913, 922, 924, 926, 933, 937, 938, 945, 954, 973, 976
Hess, S. M., 1120
Heyworth, R., 819
Hickman, E. M., 1085
Hicks, R. M., 1280, 1281, 1284 1444
Hieronymi, G., 1086
Higashi, H., 994, 1364
Higashi, K., 828
Higgins, J. A., 1289
Hijmans, W., 1240
Hill, C. R., 801
Hill, K. J., 800
Hill, R. L., 1190
Hillarp, N.-Å., 1062, 1063, 1104, 1105, 1107, 1251, 1382
Hillary, B. B., 1250
Himmelhoch, S. R., 947, 993 1350
Hines, M., 800
Hirai, K., 845, 854, 855, 858, 1004, 1066, 1286, 1337, 1388
Hirano, G., 1288
Hirano, H., 800, 1003, 1367
Hird, F. J. R., 973
Hirs, C. H. W., 1021
Hirvonen, J., 1112
Hitzeman, J. W., 887
Hoare, R., 1154
Hobbiger, E. E., 763, 771, 781
Hobbiger, F., 767, 771
Hobby, G. L., 1026
Hoch, F. L., 929, 1154
Hoch-Ligeti, C., 842, 921
Hodge, A. J., 1262
Hoffman, P., 1029, 1032
Hofmann, T., 1042
Hogeboom, G. H., 846, 859, 937
Hogness, T. R., 850
Hökfelt, B., 1062, 1382
Hökfelt, T., 1292
Holborow, E. J., 1172, 1192, 1226, 1241
Holcenberg, J., 1116
Holden, M., 1035, 1037, 1038
Holimann, S., 932
Holland, P., 798
Hollande, A. C., 845
Hollinger, D. M., 809
Hollinshead, W. H., 772
Hollmann, S., 930
Holmes, B., 1023
Holmes, R. S., 765
Holmstedt, B., 762, 767, 771, 792, 798, 1314

Holt, M. W., 1213
Holt, S. J., 777, 778, 779, 780, 781, 782, 800, 812, 1280, 1281, 1284, 1306, 1307, 1444
Hopkins, F. G., 910
Hoppe-Seyler, F., 841
Hopsu. V. K., 776, 870, 964, 970, 972, 974, 975, 976, 977, 991, 1285, 1354
Hopsu-Havu, V. K., 966, 972, 973, 992, 1285, 1363, 1448
Hopwood, D., 954, 1063, 1291, 1292
Hori, S., 986
Hori, S. H., 829, 830, 831, 946, 1279
Horn, R. S., 795, 1319
Horning, E. S., 1128
Hornsey, S., 1210
Horobin, R. W., 1180
Horwitz, C., 950
Hosoda, S., 801
Houck, C. E., 1262
Houck, J. C., 1030
Howard, A., 1208, 1210
Howell, J. S., 1147
Howling, D. H., 1228
Hsia, D. Y. Y., 820
Hsu, K. C., 1266
Hsu, Y. T., 921
Hueck, W., 1057, 1082, 1085, 1383
Huennekens, F. M., 952
Huff, J. W., 1078
Huggins, C., 773, 988, 990
Hughes, A., 922
Hughes, D., 851
Hughes, W. L., jr., 1154
Hugon, J., 1278
Huijing, F., 833
Hukill, P. B., 1086, 1130, 1132, 1403
Hull, S., 800
Hülser, D. F., 1218
Humphrey, A. A., 1130, 1131, 1402
Hunter, L., 896
Hunter, R. L., 765
Hurley, H. J., 799, 800
Hurlock, B., 940, 942
Hussein, K. A., 941
Husselmann, H., 1162
Huxley, H., 1273
Huxley, J. S., 809
Hydén, H., 1023, 1024
Hyyppä, M., 997

Ibaraki, H., 1155
Ibata, Y., 1156, 1290, 1451
Ibrahim, M. Z. M., 830, 1327
Idahl, L.-Å., 972
Idelman, S., 1276
Ikeda, S., 853
Ikonen, M., 971
Inagaki, S., 851, 863
Ingelman, B., 1161
Iorio, R., 1249
Irving, J. T., 1041
Irwin, D. A., 1140, 1407
Isherwood, F. A., 973
Ishihara, M., 972
Ishii, S., 1282
Ishikawa, Y., 1280
Ishizaki, M., 997
Ishizaki, Y., 1085
Itoiz, M. E., 801
Iverson, F., 771, 799
Iype, P. T., 785, 1310
Izumiya, N., 963

Jackson, B., 1024
Jackson, R. H., 1179
Jacobsen, N. O., 935, 1347
Jacobson, W., 1102, 1104
Jacoby, W. B., 946
Jacyszyn, K., 973
Jailer, J. W., 1178
Jalling, O., 913
Jambor, B., 883
James, N. T., 1067
Jancsó, N., 1162, 1163, 1178, 1418
Jancsó-Gabor, A., 1163
Janigan, D. T., 814, 815, 816
Jännes, L., 922
Jansen, E. F., 1035
Jansen, M. T., 1230
Jarrett, A., 998
Jasmin, G., 971
Jatzkewitz, H., 989, 990
Jeckeln, E., 1162
Jee, W. S. S., 1212
Jeffree, G. M., 1242
Jennings, M. A., 1210
Jennings, R. B., 1157, 1415
Jensen, C. O., 881
Jensen, J. M., 1209
Jensen, W. A., 855
Jepson, J. B., 1103, 1116
Jerchel, D., 880, 883, 888, 896
Jeuniaux, C., 1038
Jevons, F. R., 820
Jewell, G. G., 1274
Joftes, D. L., 1211, 1215, 1216
Johnson, A. B., 908, 930, 931, 1345
Johnson, F. B., 1034, 1075, 1088, 1138
Johnson, M. J., 962, 963, 1154
Johnson, P. M., 1210
Johnson, R. M., 1085
Johnston, M. H., 922
Jones, G. R. N., 882, 888, 891, 892, 893
Jones, H. W., jr., 828
Jones, M. E., 976
Jones, O. T. G., 914
Jones, P. A., 1063
Jones, W., 1019
Jongsma, A. P. M., 1240, 1436
Jonnard, R., 1115
Jonsson, G., 1103, 1104, 1105, 1106, 1107, 1108, 1110, 1111, 1120, 1249, 1251, 1292
Joó, F., 793, 1282, 1316
Jordan, D. O., 1187
Jos, J., 825, 826
Jowsey, J., 1179
Jucker, C., 1086
Juhlin, L., 1123
Jurand, A., 1273

Kabat, E. A., 1264
Kahani, S. E., 999
Kalimo, H. O., 1277
Kalina, M., 908, 914, 922, 923, 950, 1288
Kaluza, J. S., 916
Kambli, E., 780, 781, 782, 1307
Kanczak, N. M., 1260
Kane, P. O., 883, 885
Kannas, O., 1248
Kantonen, U.-M., 972
Kaplan, H. S., 1019
Kaplan, N. O., 902, 908, 910
Kaplow, L. S., 851
Karinkantha, H., 870
Karnovsky, M. J., 793, 854, 869, 870, 872, 873, 947, 993, 1262, 1267, 1268, 1282, 1286, 1317, 1337, 1338, 1341, 1350
Karrer, P., 1086, 1087
Karunairatnam, M. C., 809, 810, 811
Kàsa, P., 982, 1283, 1446
Kashiwa, H. K., 1137, 1406
Kasten, F. H., 1024, 1180, 1183, 1184, 1186, 1250
Katchburian, A. V., 991
Katchburian, E., 991, 1085
Katz, R., 1178
Kaufmann, B. P., 1016, 1021, 1022, 1039
Kaufmann, H. E., 1135, 1136
Kawarai, Y., 1268, 1439
Kawashima, T., 1284, 1448
Kawiak, J., 992
Kaye, G. I., 1158
Kazimierczak, J., 1281
Kearney, E. B., 841, 910, 916, 949
Keil, J., 771
Keilin, D., 841, 842, 843, 872, 995
Kelemen, M. V., 818
Keller, M., 1327
Keller, R. N., 898
Kellogg, D. A., 942
Kelly, J. W., 1231
Kelsey, F. E., 762
Kember, N. F., 1212
Kemplay, J. R., 1238
Kendall, P. A., 1265, 1270, 1271, 1272, 1438
Kent, P. W., 822
Kern-Bontke, E., 954
Kernohan, J. C., 995
Kerpel-Fronius, S., 1277, 1285, 1288, 1451
Kerr, D. N. S., 1265
Kerr, L. M. H., 809, 810
Kertész, D., 859, 860, 862
Khau van Kien, L., 1153
Khera, K. S., 800
Kickhöfen, B., 1042
Kiefer, G., 1228, 1250
Kilby, B. A., 799
Killias, U., 1273
Kim, I. C., 1041
Kimball, R. F., 1246
Kimura, A. T., 946
King, E. J., 1140, 1350
King, M. P., 1121, 1122
King, R. C., 1215
King, R. J., 1228
King, T. E., 930
Kinoshita, K., 834
Kinter, W. B., 1214
Kishi, K., 833, 1329
Kishino, Y., 983, 985, 986, 1360
Kivalo, P., 883, 884
Kitz, R., 771
Klein, H. J., 937
Klein, P. D., 1085
Kleiner, I. S., 881
Kleiner, N., 972
Kleinwachtler, V., 1187
Klemperer, P., 1032
Klingenberg, M., 929
Klinger, J., 1178
Klüver, H., 1068
Knapp, S. E., 800
Knight, B. C. J. G., 1044
Kobayashi, I., 857, 1333
Kochakian, C. D., 910, 941, 942
Kodousek, R., 1149
Koelle, G. B., 789, 791, 792, 793, 795, 799, 800, 866, 867, 1283, 1313, 1316, 1318, 1319, 1339, 1447
Koen, A. L., 911
Kohen, C., 1242
Kohen, E., 1242
Köhl, W., 1080
Koide, S. S., 942
Koike, M., 1269
Kokko, A., 800, 1279, 1283
Koller, T., 1194
Kolthoff, I. M., 1157
Komatsu, K., 1155
Komaya, G., 1143
Komnick, H., 1157, 1290, 1291, 1293
Komnick, U., 1157
Konev, S. V., 1177, 1245
Königsdorfer, H., jr., 1067
Kopp, R., 1266
Kopple, K. D., 1192
Kopriwa, B., 1212, 1215, 1216, 1429
Kopriwa, B. M., 1218
Korhonen, E., 996, 997
Korhonen, L. K., 971, 996, 997
Korsgaard, B., 831
Kotzareff, A., 1207
Koudelka, J., 1187
Kozlowska, K., 834

Kramer, S. P., 786, 787, 1310
Kraml, I., 824
Krebs, E. G., 827
Krebs, H. A., 983
Kress, L. F., 966
Krestinskaia, T. V., 954
Kretchmer, K., 821
Kritzmann, M. G., 983
Krogh, R. E., 1189
Krug, W., 1246
Kühn, K., 1262
Kuhn, R., 880, 896
Kukita, A., 862
Kun, E., 881, 983
Kunde, D., 882, 887, 890
Kunitz, M., 1019, 1023, 1038
Kupfer, C., 800
Kurata, J., 987, 991
Kurata, Y., 995, 996
Kuriaki, H., 828, 830, 1327
Kurnick, N. B., 1024
Kurono, K., 988
Kusnetz, J., 1241
Kussäther, E., 800
Kusyk, C. J., 1288
Kutlík, I. E. 1072, 1073, 1390
Kuusisto, A. N., 921
Kuwabara, T., 888
Kuznetsov, V. J., 1152
Kuzuya, H., 870
Kwaan, H. C., 1011, 1012

Laas, E., 1138
Labedsky, L., 954
Lacassagne, A., 1207
Ladoulis, C. T., 1194
La Du, B. N., 1064
Lagerstedt, S., 1022
Lagunoff, D., 974, 1122, 1357
Laham, Q. N., 800
Laidlaw, G. F., 860
Lajtha, A., 966
Lake, B. D., 766
Lakon, G., 881
Lallemant, C., 868
Lambert, P. P., 856
Lammanna, C., 1082
Lampe, L. T., 938
Lamprecht, W., 935
Lamy, F., 1042
Landells, J. W., 1031
Lane, B. P., 1293
Lang, W., 822, 1216
Lange, N. A., 1081
Lange, P. W., 1247
Lange, R., 1294
Langemann, H., 863
Langer, E., 1178
Langeron, M., 1134
Langley, T. J., 820
Lapresle, C., 967
Larner, J., 818, 826
Larson, B., 1111
Lartique, D. J., 1082
Laskey, A., 1034
Laskowski, M., 1025
Lasnitsky, I., 1210
Lathe, G. H., 810, 1071, 1072
Laties, A. M., 1109, 1395
Lattes, J., 1207
Laude, P. P., 1212
Laufberger, V., 1132
Laursen, T., 973
La Valle, M., 1158
Lavalloy, J., 1130, 1131
Lawn, A. M., 1262
Lazarus, S. S., 1279
Lea, A. J., 1056
Lea, L., 1210
Leaback, D. H., 820, 1325, 1326
Leblond, C. P., 1044, 1208, 1210, 1212, 1215, 1216, 1429
Leder, L. D., 775
Ledoux, L., 1019
Leduc, E., 1290
Leduc, E. H., 921, 1268
Lee, C. S., 784, 1083, 1084
Lee, J. B., 936
Lee, S. H., 983, 1360
Leene, W., 884
Legallais V., 1237, 1238, 1242
Legg, P. G., 1286
Legge, J. W., 1067, 1072
Lehninger, A. I., 935
Lehninger, A. L., 931
Lehrer, G. M., 923, 1281
Leibnitz, H. L., 1072, 1073, 1248, 1389
Leiner, G., 1154
Leiner, M., 1154
Leley, D. D., 771
Leloir, L. F., 826, 827, 832, 904 998
Lembeck, F., 1118
Lemberg, M. R., 842
Lemberg, R., 1067, 1072
Lendrum, A. C., 1065, 1385
Lennartz, K.-J., 1216
Lennon, B., 1213
Lennox, B., 1057
Leonardelli, J., 800
Leong, J. L., 945
Lepehne, G., 1065
Leplat, G., 800
Lepoutre, C., 1138
Leppi, T. J., 1033
Lerman, L. S., 1186, 1187, 1188
Lerner, A. B., 859, 860, 1051
Lettré, H., 888
Leuchtenberger, C., 1024, 1039
Leuthardt, F., 903
Lever, J. D., 1262
Levere, R. D., 1066, 1387
Levi, H., 1208, 1217
Levine, H. J., 996
Levine, M. D., 1026
Levine, R. J., 1119
Levine, W., 905
Levonen, E., 800
Levvy, G. A., 809, 810, 811, 820
Levy, H., 941
Levy, L., 841
Lewis, C., 922
Lewis, M. R., 1178, 1186
Lewis, P. R., 800, 1282, 1445
Lewis, U. J., 1042
Lhotka, J. F., 1156
Liddell, H. F., 1409
Liersch, M., 1187
Lierse, W., 954
Lignac, G. O. E., 1054, 1058, 1059, 1070
Lilienfeld, L., 1158
Lillie, R. D., 855, 862, 1033, 1034, 1056, 1057, 1058, 1060, 1061, 1069, 1072, 1073, 1074, 1075, 1081, 1082, 1083, 1085, 1118, 1130, 1131, 1133, 1134, 1149, 1164, 1180, 1379, 1380, 1384, 1389, 1392
Lindberg, L.-A., 831
Lindberg, O., 913
Lindenbaum, A., 1207
Linderstrøm-Lang, K., 1213
Lindquist, T., 1087
Lindskog, S., 994, 995
Lindström, B., 1246
Lineweaver, H., 1035, 1036
Ling, K. H., 944
Ling, V., 1042
Linker, A., 820, 1029, 1030
Linnane, A. W., 903
Linnartz-Niklas, A., 1217
Linstrom, F., 1137
Lionetti, F., 1155, 1156
Lipkin, D., 1228
Lippi, U., 972
Liquier-Milward, J., 1218
Lishajko, F., 1121
Lison, L., 844, 845, 852, 855, 856, 1060, 1062, 1063, 1065, 1066, 1067, 1088, 1113, 1114, 1115, 1117, 1130, 1131, 1133, 1134, 1140, 1144, 1151, 1154, 1158, 1159, 1402
Livingston, D. C., 1281
Lloyd, A. G., 1032
Lloyd, J. B., 991
Lloyd, K. O., 1265
Lobel, B. L., 941
Lodin, Z., 1248
Loele, W., 845, 853
Loeser, C. N., 1238
Loevenhart, A. S., 761
Löffler, H., 801
Logan, M. A., 1043
Logothetopoulos, J., 1155
Lojda, Z., 818, 822, 824, 825, 851, 923, 924, 934, 954, 1249, 1324, 1326
Lomakka, G., 1229, 1230, 1235, 1237, 1238, 1246
Lombardo, C., 1150
Long, C., 1044
Longuet-Higgins, H. C., 1054
Longley, J. B., 893, 1162
Lopez, J. A., 1158
Lorbacher, P., 834
Lothian, G. F., 1227
Lotz, W. E., 1429
Loudon, E. S., 1207
Louviere, C. D., 915, 916
Love, R., 1022, 1025, 1294
Lovenberg, W., 1119
Low, F. N., 1262
Low, H., 903
Löw, H., 913
Lowry, J. R., 1211
Lowry, O. H., 1158, 1225, 1226
Lubarsch, O., 1085
Lubínska, L., 800
Luciano, L., 1277, 1278, 1279
Lucy, J. A., 1276
Ludowieg, J., 1029
Ludwig, A. W., 1032
Luft, J. H., 1262, 1275, 1337
Lukáš, Z., 870
Lundegårdh, H., 911
Lundholm, U., 1338
Lundin, S. J., 800
Lusky, L. M., 1162
Luzzati, V., 1187
Luzzatto, L., 921

MacAllum, A. B., 1140, 1156, 1157, 1403
MacDonald, H. M., 1213, 1215
MacDonald, R. A., 1070
MacDonnell, L. R., 1035
MacFarlane, R. G., 1042, 1044
Machida, H., 800
Mack, W. S., 877
MacKenzie, A., 834
MacLennan, J. D., 1042
Macleod, C. M., 1019
MacMann, C. A., 841
MacPherson, C. R., 834, 921
Maddox, G. E., 1207
Maddy, A. H., 1180
Madinaveitia, J., 820, 1028, 1029
Maeda, R., 1085
Maender, O. W., 894
Magasanik, L. G., 954
Mager, M., 1155, 1417
Magnus, I. A., 800, 1177, 1178, 1200
Mahler, H. R., 872, 904
Main, A. R., 771, 799
Mainini, A., 1134
Mäkinen, K. K., 972
Mäkinen, P. L., 971
Malamud, D., 1211, 1217
Malaty, H. A., 800, 921, 922
Malendowicz, L., 763, 800
Malhorta, O. M., 820
Mallory, F. B., 1082, 1130, 1131, 1144, 1147, 1149, 1403
Mallucci, L., 1178
Malmgren, H., 792, 953, 971
Malmström, B. G., 1228
Man, J. C. H., 1279
Mancini, R. E., 853
Mander, S. T. K., 1003
Mandl, I., 963, 1042, 1043

Manheimer, L. H., 1398
Mann, P. J. G., 865, 945
Mann, T., 930, 995
Mannozzi Torini, M., 1031
Manocchio, I., 1189, 1191
Manocha, S. L., 954
Mansberg, H. P., 1241
Mapson, L. W., 932, 953
Marcus, A., 952
Marcus, P. I., 902, 940, 942
Marcuse, P. M., 922, 954
Maren, T. H., 995
Margoliash, E., 854, 1067
Marino, R. G., 784
Marinozzi, V., 1273, 1274
Mark, D. D., 784, 786, 1309
Markert, C. L., 765, 934
Märki, F., 905
Marks, N., 966
Marmur, J., 1188
Marples, E. A., 787
Marsh, C. A., 810, 811
Marshak, A., 1208
Martin, B. F., 784, 1309
Martin, E., 1293
Martin, G. M., 1116
Martinez de Morentin, J., 1004
Martinez Rodriguez, R., 793
Martius, C., 905
Masek, B., 923
Masin, F., 1186
Masin, M., 1186
Mason, H. S., 1051, 1054, 1056
Mason, K. E., 1084
Massey, V., 841, 905, 916, 917, 925, 953
Masson, F., 1187
Masson, M. P., 1057
Masters, C. J., 765
Mattson, A. M., 881
Maunder, C., 1189, 1241, 1421
Maurer, W., 1216
Mautner, H. G., 1284
Mawson, C. A., 1153, 1154
Maxwell, E. S., 998, 999
Mayahara, H., 1278, 1279, 1443
Mayall, B. H., 1228
Maynard, D. M., 800
Maynard, E. A., 800
Mayner, D. A., 1012
Mazia, D., 1020, 1039, 1210, 1214, 1217
Mazur, A., 768, 1132
McAllan, A., 810, 811, 820
McAllister, R. A., 1159
McAlpine, J. C., 924, 1274
McCabe, M., 908
McCarthy, W. J., 1173, 1234
McCarty, M., 1023, 1024
McClean, D., 1025, 1026, 1028, 1029
McClean, F. C., 1179
McCombs, H. L., 1013, 1373
McConnell, D. G., 1002
McCorkindale, J., 930
McCrory, P., 1138
McCurdy, H. M., 1055
McDonald, J. K., 966
McDonald, M. R., 1016, 1021, 1038
McElroy, W. D., 1173, 1226
McGadey, J., 999
McGavran, M. H., 863, 1338, 1339
McGee-Russell, S. M., 814, 1135, 1404
McGill, D. B., 821
McGregor, M. M., 1112
McIntosh, J. E. A., 994
McIntyre, N., 800
McIsaac, R. J., 1211
McJunkin, F. A., 853
McMahon, P., 819
McManus, J. F. A., 1034, 1035, 1036, 1058, 1083, 1377
McMillan, P. J., 934, 935, 973
McMunn, C. A., 1059
McNary, W. F., jr., 1155
McQuade, H. A., 1210
McShan, W. H., 842
Meany, J. E., 995
Meek, G. A., 1280
Mehl, E., 989, 990
Meijer, A. E. F. H., 830, 831, 832, 986, 997, 999, 1361, 1365, 1367
Meirowsky, E., 862, 1055
Meisel, E., 853, 854, 921, 922, 1335
Meister, H., 776
Meldrum, N. U., 994
Melford, D. A., 1293
Mellon, M. G., 1227
Mellors, R. C., 1237, 1238, 1239
Mendel, B., 766, 768, 772, 789, 799
Mendel, L. B., 1144, 1154
Mendelsohn, M. L., 1227, 1228, 1230, 1231
Menten, M., 1156
Menten, M. L., 899, 1398
Mercer, E. H., 1055
Merenmies, E., 850, 954
Merker, H.-J., 981, 1359
Mescon, H., 799, 800
Messier, B., 1215
Metrione, R. M., 963
Meulon, A., 1059
Meyer, K., 820, 1025, 1026, 1027, 1028, 1029, 1030, 1032
Meyer, R. K., 842
Meyer, W., 954
Michael, A. I., 1076
Michael, J. C. Im., 954
Michaelis, L., 899, 1187
Middlebrook, W. R., 1041
Midorikawa, O., 1155
Miescher, G., 862
Mietkiewski, K., 763, 800
Migicovsky, B. B., 1010
Milch, R. A., 1179
Millar, M. J., 1211
Miller, D., 818, 821
Miller, R. W., 948
Miller, S. A., 1269
Millican, R. C., 1162, 1418
Millichap, J. G., 994
Milligan, C. W., 1137
Millonig, G., 1262
Millott, N., 1059
Mills, G. T., 809, 810, 811
Milsom, D. W., 988
Miranda, M., 799
Mirsky, A. E., 1039, 1250
Mitchell, C. A., 996
Mitsudo, S. M., 942
Mitsui, T., 853, 1286
Mittermayer, Ch., 943
Mizushima, T., 997
Mizutani, A., 981, 1278, 1359
Mladenov, S., 794
Modest, E. J., 1185
Mogensen, K. R., 1213
Mohanty, G. P., 1140
Möhle, W., 883
Mohr, E., 792, 799
Mölbert, E. R. G., 1278
Møller, F., 934
Møller, K. M., 871
Molnar, J., 1149
Moloney, W. C., 775, 801
Moncorps, C., 862
Monis, B., 801, 823, 834, 969, 971, 1353
Montagna, W., 829, 834, 921
Monti, A., 1158
Moog, F., 846, 850, 1330
Moore, A. T., 1164
Morales, D. R., 952
Morel, F., 1211
Morell, D. B., 873
Morgan, C. 1264, 1267
Morgenstern, S., 985
Mori, M., 817, 954
Mori, S., 1282
Morihara, K., 1040
Morita, S., 1067
Morrison, M., 1044
Morrow, P. F. W., 989
Morton, R. A., 915, 1000, 1088
Moses, H. L., 1280
Moss, D. W., 1242
Mott, M., 1266
Mottet, K. N., 971
Mottram, J. C., 1178
Moulton, S. H., 773
Mounter, L. A., 762, 763
Movat, H. Z., 1274
Mowry, R. W., 1162, 1418
Moyer, E. S., 1234
Moyer, F. H., 1055
Mudd, S., 884, 922, 1262
Muir, A. R., 1265
Muir, R., 1068
Mukherjee, M., 1061
Mukherji, M., 1110, 1396
Müller, A., 1290
Müller, D., 1156, 1246
Müller, E., 870
Müller, P., 1071, 1149
Müller, W., 1085
Mundell, D. B., 766
Mundkur, B., 1263, 1272
Murata, F., 1284, 1448
Murphy, N. H., 1072
Mustakallio, A., 870
Mustakallio, K. K., 864, 883, 884, 921, 922, 954, 996, 1163, 1178
Mycek, M. J., 963
Myers, D. K., 762, 763, 765, 766, 793, 976

Nachlas, M. M., 762, 772, 773, 789, 800, 847, 850, 888, 907, 908, 913, 916, 922, 926, 967, 968, 969, 971, 974, 990, 1330, 1353
Nachmansohn, D., 766, 789
Nachmias, V. T., 922
Nagata, T., 1214, 1431
Nagatsu, I., 970
Nagatsu, T., 870
Nägel, L. C. A., 1242
Nagel, W., 971
Nagelschmidt, G., 1140
Nagy, K., 1186
Nairn, R. C., 1171, 1423
Najjar, V. A., 828
Nakajima, Y., 932, 1346
Nakamura, N., 973
Nakamura, T., 794
Nakane, P. K., 1267, 1268, 1438, 1439
Nakano, J., 935
Naora, H., 1228
Napieralski, B., 1107
Narayanan, A. S., 1042
Narumi, M., 1155
Natochin, J. V., 954
Naughton, M. A., 1042
Navazio, F., 937
Nawa, T., 1431
Nayyar, S. M., 886, 888
Negelein, E., 841
Negri, L., 1031
Nelson, A. A., 1162
Nemeroff, K., 1215, 1216
Nene, R. V., 835
Neth, R., 1149
Neubauer, J., 932
Neuberg, C., 988
Neugebauer, F. A., 894
Neumann, E., 1068
Neumann, K., 1233
Neurath, H., 976, 1154
Newcomer, A. D., 821
Nicander, L. A., 972
Nichol, C. A., 952
Nickerson, W. J., 1041
Nicolaus, R. A., 1051
Nielsen, T. W., 771
Niemi, M., 800, 850, 954, 971, 972, 1231
Nienhuis, A. W., 927

Nilson, B., 1063
Nirenberg, M., 946
Nisch, G., 1147
Nishiyama, Y., 857, 1333
Nissen, H. M., 936, 1348
Nitowsky, H. M., 834
Niven, J. S. F., 1068, 1186
Nixon, W. C., 1233
Nomarski, G., 1248
Nomoto, M., 1040
Norberg, K.-A., 1109
Nordén, G., 1178, 1193, 1234, 1241, 1252
Norman, J. H., 830
Northrop, J. H., 1038
Norton, I. L., 998
Novikoff, A. B., 921, 923, 924, 950, 1080, 1280, 1286, 1449
Novogrodsky, A., 854
Nunn, A., 1370
Nyberg-Hansen, R., 1283
Nye, S. W., 1085
Nygaard, A. P., 1044
Nylander, G., 1132

O'Brien, P. J., 873
O'Brien, R. T., 1218
O'Carra, P., 1074
Ochi, J., 1290
Oda, T., 891, 913
Odeblad, E., 1210
Ofuji, S., 975, 1358
Ogata, A., 1059
Ogata, T., 954, 1059
Ogawa, K., 864, 911, 1003, 1279, 1280, 1288, 1367, 1443
Ogawa, Y., 921, 1136
Ogle, J. D., 1043
Ogston, A. G., 1041
O'Hara, M., 987, 991
Öhman, S., 1123
Okada, M., 834
Okamoto, K., 1072, 1144, 1145, 1151, 1152, 1155, 1411, 1415
Okano, K., 954
Okazaki, H., 913
Őkrös, I., 1276
Okui, S., 893
Okumura, H., 1059
Okun, M. R., 851, 860, 862, 863, 1055, 1057, 1286, 1339
Oledzka-Slotwińska, H., 1279
Oleson, J. J., 855, 1335
Olivecrona, H., 1052
Olson, C. K., 973
Olson, R. A., 1235, 1237, 1238
Onicescu, D., 952, 998, 999, 1351, 1366
Ono, T., 970
Opler, S. R., 881
Orchardson, R., 999
Ord, M. G., 768
Ornstein, L., 781, 782, 1230, 1231, 1261
Ortmann, R., 921
Osborn, M. J., 952
Oshima, G., 808, 810
Oster, G., 1245
Oster, K. A., 866
Osterberg, H., 1245, 1248
Ostrow, J. D., 1072
Ostrowski, K., 1218
Otsuka, N., 1156, 1290, 1451
Otto, K., 970
Otto, L., 816, 1321
Ottolenghi, A., 788
Ottolenghi, P., 871, 1311
Owman, C., 1107, 1108, 1251, 1394, 1395
Ozone, K., 835

Packer, D. M., 1128
Packer, L., 946
Padykula, H. A., 921, 922
Paget, G. E., 1178
Painter, T. S., 1020
Palade, G., 1218
Palade, G. E., 846, 884, 1276, 1282
Palkama, A., 792, 830, 831, 1279, 1281, 1327
Palm, E., 1208
Palmer, J. M., 914
Palmer, J. W., 1025
Pani, K., 1138
Panitz, E., 800
Papadimitriou, J. M., 1289
Papanicolaou, G. N., 1237, 1238, 1239
Papermaster, B. W., 1192, 1242
Pappenheimer, A. M., 1083, 1086
Parker, C. A., 1235, 1237, 1238, 1239
Parker, F., jr., 1144
Parry, R. W., 898
Parsons, D. F., 1262
Partridge, S. M., 1041
Pascoe, E., 1249
Passwater, R. A., 1173
Pastor, E. P., 801
Patau, K., 1230
Patrick, A. D., 766
Pattee, H. H., 1233
Patterson, E. K., 964, 970
Paul, J., 809, 810
Paul, K. G., 851
Pauling, L., 1264
Pavlic, M., 771
Peacock, P. R., 1178
Peacocke, A. R., 1188
Pearce, N. A. M., 872
Pearce, R. H., 1030, 1032
Pearse, A. G. E., 774, 782, 783, 786, 787, 795, 797, 798, 799, 800, 801, 812, 813, 814, 815, 816, 817, 834, 835, 856, 869, 870, 882, 887, 888, 897, 902, 903, 906, 907, 910, 913, 916, 921, 922, 923, 924, 926, 930, 933, 937, 938, 945, 950, 954, 971, 976, 991, 996, 1001, 1003, 1004, 1013, 1020, 1052, 1061, 1074, 1079, 1080, 1084, 1085, 1102, 1103, 1104, 1107, 1110, 1111, 1112, 1115, 1116, 1118, 1119, 1131, 1136, 1156, 1157, 1189, 1190, 1191, 1192, 1225, 1235, 1237, 1238, 1239, 1240, 1241, 1242, 1251, 1252, 1260, 1265, 1276, 1288, 1362, 1367, 1399, 1423, 1438, 1449
Pearson, B., 782, 812, 817, 818, 822, 823, 824, 921, 941, 942, 943, 950, 969, 1323, 1354
Pearson, C. K., 792
Pearson, P. L., 1184
Pease, D. C., 1275, 1290
Peche, K., 987
Pecher, C., 1207
Peck, S. M., 862
Pedersen, K. J., 800
Pelc, S. R., 1208, 1210, 1211, 1212, 1213, 1217, 1428, 1429
Pellegrino, C., 809
Penar, B., 834
Penner, D. W., 1154
Penney, J. R., 1027, 1031
Pennington, R. J., 873, 916
Penny, I. F., 1044
Penttilä, A., 870, 1112, 1119
Pepe, F. A., 1270, 1272
Pepler, W. J., 782, 797, 798, 800, 976, 1042
Perceval, N. L., 1215
Percival, W. L., 1208
Perkins, E. M., 831
Perkins, E. M., jr., 1327
Perls, M., 1130
Perner, E. S., 846
Pernow, B., 1118
Perry, M., 965
Perry, M. M., 1275
Persijn, J.-P., 1232, 1249, 1279
Person, P., 843, 846
Personne, P., 1274
Persson, B., 1161
Pescetto, G., 800, 921
Peters, H. A., 1177
Peters, T. M., 1180
Peterson, A. C., 790
Peterson, R. E., 1148
Petzoldt, D., 951, 954
Phillipart, M., 820
Phillips, H., 1041
Philpot, F. J., 864
Piantelli, A., 785
Piattelli, M., 1051
Pick, L., 1085, 1091
Pickett-Heaps, J. D., 1274
Pickworth, F. A., 1065, 1387
Pictet, A., 1107
Pierce, G. B., 1264, 1267, 1438
Pihl, E., 1275, 1290, 1443, 1451
Pilgrim, C., 1216
Pillat, G., 1228
Pilny, J., 1248
Pioch, W., 1152
Piringer-Kuchinka, L., 1085
Pittman, B., 1241
Pizzolato, J. M., 1140
Pizzolato, P., 1072, 1073, 1074, 1075, 1138, 1140, 1389, 1392, 1405
Plagemann, P. G. W., 934
Plane, R. A., 929
Planta, R. J., 963
Planteydt, H. T., 971
Plapinger, R. E., 850, 974, 1286
Plaut, W., 1210
Plaut, W. S., 1217
Ploem, J. S., 1109, 1175, 1196, 1199, 1200, 1240
Pocker, Y., 995
Polak, J. M., 1052, 1192, 1250, 1265, 1438
Policard, A., 1128
Pollister, A. W., 1021, 1024, 1039, 1226, 1231
Pontinen, J., 776
Poole, A. R., 970
Poppen, K. J., 1156, 1157
Popper, H., 1088, 1161
Porter, M. T., 989
Porter, V. S., 907
Portugalov, V., 922
Potter, R. L., 1159
Potter, V. R., 842, 913
Potts, A. M., 993
Poulson, D. F., 1148
Pratt, R., 881
Praver, L. L., 860
Prenant, M., 846, 853
Prenna, G., 1183, 1241, 1251, 1252
Prescott, D. M., 1215
Presnov, M. A., 954
Press, W. M., 966
Pressman, D., 1209, 1264
Price, G. R., 1171
Price, H. L., 1121
Price, J. L., 1140
Price, M. L., 1121
Price, R. G., 819, 821
Prttchard, E. T., 1078
Pritchard, N. J., 1188
Pritschow, A., 1156
Przibram, H., 858
Puchtler, H., 995, 1135
Puera, B., 1004
Pugh, C. E. M., 863
Pugh, D., 812, 816, 820, 823, **824**
Pulido, E., 1003
Pullman, A., 1054
Pullman, B., 1054
Pullman, M. E., 828
Pupkin, M., 943
Putt, F. A., 1130, 1132, 1403
Pyl, T., 897
Pyper, A. S., 1063
Pyrah, L. N., 1138

Quaroni, E., 1116
Quastel, J. H., 863, 945
Queiroz Lopes, A., 1158
Quibell, T. H. H., 1028
Quinke, H. I., 1130, 1131
Quinlan-Watson, F., 1154
Quintarelli, G., 1033

Raaflaub, J., 902, 903
Rabl, C. H. R., 1133
Rabotti, G., 1022, 1025
Rachlin, L., 1178
Racker, E., 899, 904, 907, 910, 946
Radaeli, G., 862
Raekallio, J., 800, 971
Rafelson, M. E., jr., 1033
Raia, S., 1074, 1075, 1390
Raikhlin, N. T., 954
Räisänen, L., 1318
Rajewsky, M. F., 1218
Rall, D. P., 1179
Ramadan, M. A., 1076
Rambourg, A., 1274, 1441
Rand, R., 1074
Randall, L.-O., 851
Rao, K. R., 967
Rao, S., 986
Raper, H. S., 859
Rappaport, B. Z., 862, 1338
Rapport, M., 1026, 1027
Rasch, E. M., 1231
Rasch, R. W., 1231
Rath, F. W., 816, 1321
Rath, H., 991
Rathjens, B., 995
Ratzenhofer, M., 1118, 1119
Rau, E. M., 985
Raunio, V., 821
Rautschke, R., 1155
Reale, E., 800, 921, 1277, 1278, 1279
Rebollo, M. A., 785
Rechardt, L., 1279, 1281
Redaelli, G., 1055
Reed, L. J., 905, 917
Reedy, M. K., 1262
Reeves, R. E., 1082
Reggianini, O., 1027
Regoli, D., 954
Reif, A. E., 913
Reindel, F., 1070
Reinhold, J. C., 1074
Reiss, J., 849, 855, 951, 953, 954
Rendina, G. R., 945
Renwick, J. H., 1014
Reynolds, E. S., 1262
Rhoads, C. P., 1067
Rice, D. F., 1268
Rich, A. R., 1068, 1070, 1071
Richards, O. W., 1246
Richards, W. C. D., 1086
Richardson, K. C., 1292
Richter, D., 1061
Richter, G. W., 1069, 1133
Richterich, R., 784
Rickli, E. E., 994
Rieche, A., 858, 1333
Riecken, E. O., 1149, 1156
Ried, W., 883
Riedel, A., 974
Riekkinen, P. J., 977
Rienitz, J., 1246
Rifkind, R. A., 1266, 1267
Rigler, R., jr., 1186, 1187, 1191, 1192, 1234, 1235, 1236, 1237, 1238, 1239, 1241, 1423
Riley, P. A., 862
Rimon, S., 763, 764, 976
Ringertz, N. R., 1210
Ringler, D., 944, 945, 1155
Rintola, P., 966
Ripley, R. C., 1290
Rippon, J. W., 1042
Ris, H., 1250
Ritter, H. B., 855, 1335
Ritzén, M., 1111, 1120, 1175, 1213, 1217, 1237, 1240, 1249, 1251, 1252
Robb-Smith, A. H. T., 1042, 1043
Roberts, C. B., 896
Roberts, L. W., 881, 916
Robertson, J. D., 1262
Robertson, R. C., 1186
Robinson, D., 819, 821
Robinson, J. P., 828
Robinson, K. L., 1084
Robinson, N., 869
Robson, W. D., 1140
Rodkey, F. L., 893
Rodriguez, H. A., 863, 1338 1339
Rodriguez, R. M., 800
Roembeld, R., 972
Rogers, A. W., 1211, 1218, 1243
Rogers, G. E., 921
Rogers, H. J., 1026, 1028, 1029
Rogister, G., 800
Rohdenburg, G. L., 1156
Röhlich, P., 799
Romney, S. L., 941
Roncoroni, G., 1143
Roots, L., 793, 1282, 1317
Roppel, R. M., 888
Rorsman, H., 1052
Rosa, C., 882
Rosa, C. G., 921, 922, 1286
Rosa, F., 1034
Roschlau, G., 1189
Rose, F. A., 988, 989
Rosenblatt, D. H., 968, 969
Rosenfeld, M. G., 976
Rosenthal, A. S., 1280
Ross, K. F. A., 1245, 1246
Ross, L. L., 1111
Rosselet, A., 1241
Rossi, F., 800, 921
Rossiter, R. J., 809
Rost, F. W. D., 1052, 1107, 1110 1189, 1190, 1191, 1192, 1235, 1237, 1238, 1239, 1240, 1241, 1242, 1250, 1251, 1252, 1254, 1400, 1421, 1423
Roth, H., 1179
Roth, L. J., 1211, 1215, 1432
Rothenberg, M. A., 766
Rothman, A. H., 1275
Rotman, B., 1192, 1242
Roughton, F. J. W., 994
Roulet, F., 1060
Roure, A., 853
Roy, A. B., 988, 989
Rubin, B. L., 941, 942
Ruch, F., 1183, 1184, 1234, 1235, 1237, 1238, 1239, 1241, 1244, 1245, 1251, 1252
Ruddell, C. L., 1057
Rudney, H., 766, 772, 789
Rudolph, G., 921, 937
Runge, W. J., 880, 1235
Ruponen, S., 971
Russell, G. A., 894
Rutenberg, A. M., 822, 823, 881, 888, 921, 973, 989, 990, 1289, 1321, 1323, 1324, 1355, 1362
Rutenburg, S. H., 817, 822
Ruthmann, A., 1227, 1228, 1245, 1246
Rutland, J. P., 800
Ryan, T. J., 1012, 1370
Ryder, M. L., 1211
Rzeszowska, G., 972

Sabatini, D. D., 1276
Sacchi, S., 1058, 1183
Sachs, C., 1111
Sackler, M. L., 874, 1341
Sacktor, B., 825
Sadler, P. W., 778, 779, 781, 1307
Saikkonen, J. I., 921, 1163, 1178
Saito, T., 1280
Sakae, Y., 800
Salfeld, K., 971
Salisbury, G. W., 1251
Salpeter, M. M., 1217
Sampietro, R., 1117, 1119, 1189, 1190
Samuels, L. T., 941
Samuelsson, B., 935
Sanadi, D. R., 905
Sander, G., 1175
Sanders, F., 1020
Sandler, M., 1106, 1251
Sandritter, W., 800, 1189, 1227 1241, 1246
Sanger, F., 1042
Sani, G., 1186
San Pietro, A., 908
Santini, D., 1055
Santti, R., 975, 977
Saqueton, A. C., 1042
Sarimo, S. R., 973
Sasai, Y., 998
Sasaki, M., 1284, 1285
Sasse, D., 832, 833, 1293, 1329
Sato, A., 853
Saunders, B. C., 851, 854
Saunders, J. C., 1034, 1035, 1036
Saunders, L., 1311
Savage, N., 904
Savant, P. L., 988
Sávay, G., 793, 794, 795, 800, 1316
Sawada, H., 1292
Sawada, T., 1076
Sawicki, W., 1216
Sawyer, C. H., 772
Sawyer, D., 831, 1327
Saxton, C. A., 1274
Scarpelli, D. G., 888, 902, 903, 907, 910, 913, 924, 926, 1260, 1279
Schaaf, F., 861
Schaberg, A., 1232
Schaefer, H. E., 776
Schäfer, G., 935
Schaffer, N. K., 764
Schajowicz, F., 834
Schauer, A., 1122
Schechter, I., 1264
Scheidegger, J. P., 870
Schien, J., 1086
Schick, A., 1265
Schillinger, H., 1089
Schittenhelm, A., 872
Schlossman, N. C., 866
Schlüns, J., 996
Schlüter, G., 1241
Schmeltzer, W., 1130, 1402
Schmid, R., 1177
Schmidt, R., 1084, 1155
Schmidt, U., 1226
Schmidt, W., 893
Schmidt, W. J., 1245
Schmit, R. W., 800
Schmitt, F. O., 1262
Schmitz-Moormann, P., 1033
Schnabel, R., 1147, 1162
Schnack, H., 1177
Schneider, W. C., 842, 846, 921, 937, 1020
Schnieden, H., 1010
Schoenberger, M. D., 1238
Scholl, O., 921
Schor, N. A., 953
Schornagel, H. E., 1084
Schorr, E., 1132
Schrieb, D., 812
Schubert, M., 1043
Schuit, Henrica R. E., 1200
Schuler, D., 1042
Schultz, G., 1246
Schultze, B., 1211, 1217
Schultze, W., 1279
Schultze, W. H., 844
Schumacher, H. H., 921
Schümmelfeder, N., 1186, 1189
Schütte, E., 1029
Schwartz, M. A., 864
Schwartz, S., 1171, 1177
Schwarz, W., 1274

Schwert, G. W., 976
Scott, G. H., 1128
Scott, J. E., 1275
Scott, J. F., 1182
Searls, R. L., 905
Sears, M., 1279
Sebruyns, M., 1154
Sedar, A. W., 1286
Seelig, H.-P., 972
Segal, R., 763, 764, 976
Seggel, K. A., 1068
Seidler, E., 882, 887, 890, 1003
Seiji, M., 1055
Sekiyama, S., 1274
Sela, M., 1264
Seliger, H. H., 1173, 1226
Seligman, A. M., 762, 772, 773, 775, 777, 778, 779, 780, 786, 789, 800, 812, 815, 817, 821, 822, 823, 847, 849, 881, 907, 908, 921, 923, 926, 967, 968, 969, 971, 974, 989, 990, 1264, 1269, 1273, 1274, 1285, 1286, 1287, 1289, 1310, 1321, 1332, 1353, 1362, 1398, 1449
Sellinger, O. Z., 821
Semenoff, W. E., 881
Semenza, G., 821
Semm, K., 972
Sernetz, M., 1240, 1242
Serra, J. A., 1020, 1158, 1416
Sevki, K., 1060, 1382
Seyhan, M., 899
Shantha, T. R., 932
Sharma, D. R., 954
Shapiro, H. S., 1023
Sharlet, H., 860
Sharples, W., 1113
Shaver, J. R., 1020, 1023, 1024
Shaw, C. R., 911
Shear, M., 1013
Shearer, L., 989
Shelley, W. B., 1123
Shelton, E., 921
Shibata, Y., 1155
Shibko, S., 812
Shimizu, N., 834, 869, 1282
Shinonaga, V., 911
Shiose, Y., 1279
Shnitka, T. K., 775, 780, 932
Shoden, A., 1070
Shore, P. A., 1122
Shoshan, S., 1083
Shugar, D., 1012
Shuster, L., 910
Shute, C. C. D., 800, 1282, 1445
Siebert, G., 1077
Siegel, B. V., 1055
Siegel, G. J., 799
Sierakowska, H., 1012
Sikorski, J., 1263
Silver, R., 1239
Silver, V. L., 1140
Silverman, L., 1275
Simpson, W. L., 1178
Singer, M., 1020
Singer, S. J., 1265
Singer, T. P., 841, 910, 916, 944, 945, 949
Singh, H., 1078
Siurala, M., 785
Sivaram, S., 954
Siverama Sastry, K., 1187
Sizer, I. W., 951, 1051
Sjoerdsma, A., 864
Sjöqvist, F., 771, 798, 800
Sjöstrand, F., 1067
Sjövall, K., 986
Sköld, G., 1149, 1413
Slack, J., 842
Slater, E. C., 949
Slater, T. F., 888, 913, 914
Slayter, E. M., 1226, 1244
Slezak-Klemencic, E., 942
Slorach, S. A., 1119
Smith, A. A., 831, 1327
Smith, A. D., 1191
Smith, A. N., 1119
Smith, D. R., 988, 990
Smith, D. S., 1279
Smith, E. E., 964
Smith, E. E. B., 809, 810, 811
Smith, E. L., 962, 963, 964, 1154
Smith, F. H., 1246
Smith, G. L., 1156, 1417
Smith, L., 841
Smith, L. H., jr., 948
Smith, M., 835
Smith, R. E., 1272, 1281, 1284
Smith, S. O., 1288
Smoler, M. H., 1207
Smyth, E. M., 1026
Sneath, P. H. A., 784
Snell, R. S., 799, 800
Snellman, O., 963, 970
Sognnaes, R. F., 1212
Solari, A. J., 1294
Solcia, E., 1117, 1119, 1189, 1190, 1421
Solomon, A. K., 1211
Sommer, J. R., 1279
Somogyi, E., 1279
Sonenberg, M., 1209
Sørensen, M., 831, 1328
Sörensen, S. P. L., 821
Sotonyi, P., 1279
Sottery, C. T., 1140
Søvik, Ø., 832
Spach, M. S., 1279
Spackman, D. H., 962, 964
Speare, G. S., 1085
Spector, S., 864
Speer, F. D., 881
Speigelman, S., 1025
Spencer B., 811, 988, 989
Spencer, J. M., 940, 942
Spengler, T., 1107
Spicer, S., 1263
Spicer, S. S., 1033
Spiro, H. M., 1086
Spors, S., 981, 1359
Sprenger, E., 1183, 1234, 1235, 1239, 1241, 1249, 1251, 1252
Springell, P. H., 973
Sriram, J., 1267
Stammler, A., 1085
Standen, A. C., 824
Stansfield, D. A., 786
Stark-Mayer, C., 1137
Staübli, W., 1290
Steenbock, H., 1087
Steendijk, R., 1179
Steigleder, G. K., 801
Stein, J., 1072, 1073, 1389
Stein, W. D., 1051, 1055
Steinberg, D., 1211
Steinberger, E., 943
Steinmann, C., 1003
Steins, I., 947
Stenevi, U., 1120
Stenram, U., 1217
Stern, K. G., 857
Stern, L., 872, 948
Sternberg, W. H., 872, 922
Sternberger, L. A., 1264, 1265, 1267, 1268, 1269, 1270
Stevens, B. J., 1103, 1116
Stevens, C. E., 1208
Stevenson, J., 1186
Stewart, M. J., 1085
Stieglitz, E. J., 1159
Stiller, D., 930, 931, 933, 1345 1346
Stirling, C. E., 1214
Stirling, J. L., 821
Stollar, B. D., 1264
Stone, A. B., 1025
Stone, J. D., 1033
Stone, J. T., 995
Storey, I. D. E., 932
Story, M. N., 970
Stotz, E., 841
Stoward, P. J., 1183, 1184, 1187, 1194, 1274, 1275, 1421, 1442
Stowell, R. E., 784, 1016, 1021, 1022
Strassmann, G., 1068
Straub, F. B., 904, 906, 908, 925
Straus, E., 881
Straus, F. H., 881
Straus, W., 853. 857
Strauss, G., 1069
Strecker, H. J., 941
Striker, G. E., 1266
Strömblad, B. C. R., 799
Strominger, J. L., 932
Struck, J., jr., 951
Strufe, R., 905
Strugalska-Cynowska, M., 835
Strugger, S., 1186, 1263
Stubblefield, E., 1217
Studer, A., 870
Studzinski, G. P., 1294
Stumpf, W. E., 1211, 1215, 1432
Sturgeon, P., 1070
Su, H. C. F., 780, 781, 893
Sugino, M., 834
Sun, C. N., 828
Sundberg, R. D., 1177
Suntzeff, V., 842
Suszkin, J. B., 962, 965
Sutherland, E. W., 829
Sutton, J. S., 1262
Suzuki, S., 989
Suzuki, T., 1274
Svalander, C., 971
Svanborg, A., 800
Swan, G. A., 1051, 1052, 1054
Sweat, M. L., 941, 942
Sweet, T. R., 1136
Swift, J. A., 1263, 1273, 1274, 1437
Sylvén, B., 792, 953, 970, 971, 972
Szarkowska, L., 944
Szemplinska, H., 1012, 1013
Szentágothai, J., 800

Taft, E. B., 1056
Takagi, Y., 894
Takahashi, N., 989
Takamatsu, H., 845, 975, 1004, 1358
Takamori, T., 843
Takase, S., 801
Takats, I., 1147
Takaya, K., 1123
Takeuchi, T., 828, 829, 830, 832, 834, 1284, 1285, 1327, 1328
Takeya, K., 884
Taki, I., 954
Talalay, P., 902, 940, 942, 943
Täljedal, I.-B., 972, 1242
Tallan, H. H., 963, 976
Tanaka, H., 1264, 1266
Tandler, C. J., 1158, 1294
Tanimukai, H., 867
Tanka, D., 1327
Tapia Freses, A., 941
Tappel, A. L., 812, 819, 1084
Tashian, R. E., 994, 995
Tawde, S. S., 1265
Taxi, J., 792
Taylor, A. N., 1020
Taylor, G., 1267
Taylor, J. H., 1210, 1211
Telenius, U., 821
Telford, I. A., 1085
Telkkä, A., 921
Teräväinen, H., 800, 1282
Ternynck, T., 1267
Tewari, H. B., 800
Thaer, A., 1235, 1238, 1239, 1240, 1251, 1252
Thaï-Tuong, 1153
Thatcher, H., 860
Theorell, H., 917, 929
Thiele, E. H., 1078
Thiele, H.-J., 849, 1332
Thieme, G. A., 1235, 1238
Thiéry, J. P., 1274, 1442
Thomas, E., 954
Thomas, G. H., 943
Thomas, I. L., 1311
Thomas, J., 988, 989, 1041

Thomas, J. A., 1130, 1131
Thomas, M., 987
Thomas, R. S., 1293
Thompson, E., 1263
Thompson, E. C., 856, 1065
Thompson, J. H., 1119
Thompson, R. H. S., 767, 768, 787, 799, 800, 1019
Thomson, D. L., 821
Thomson, R. H., 1059
Thorell, B., 1177, 1241, 1242
Thorne, C. J. R., 944, 945
Thorsén, G., 1161
Thunberg, T., 948, 1123, 1400
Thurston, J. M., 1216
Thybusch, D., 775, 868, 870
Tice, L. W., 1263, 1276, 1279, 1280
Tiedemann, F., 1072
Tikkala, A. O., 1153
Timm, F., 1149, 1150, 1290, 1412, 1451
Tipton, K. F., 864, 869
Tirmann, J., 1402
Titus, E., 864
Tjeder, A., 965
Tobie, J. E., 1179
Todd, A. R., 820, 1018
Todd, A. S., 1011, 1012, 1370
Todrick, A., 770
Toffler, A. H., 1086
Tolles, W. E., 1217
Tonew, E., 1024
Tonna, E. A., 1212
Torack, R. M., 983, 1158, 1276
Tóth, G., 1038
Tousimis, A. J., 1233
Towe, K. M., 1265
Towers, R. P., 1163
Tracey, M. V., 1037, 1038
Tramezzani, J. H., 1063, 1291, 1380
Tranzer, J.-P., 950, 1001, 1003, 1004, 1278, 1292, 1367, 1452
Traverso, G., 1118
Treadwell, A. de G., 1207
Tremblay, G., 954, 971, 1013, 1373
Trop, M., 1040
Tsou, K.-C., 780, 781, 817, 823 882, 893, 1286, 1321
Tsukamoto, S., 954
Tsuzuki, H., 1040
Tu, A. T., 852
Tubbs, R. K., 1187
Tudball, N., 989
Tung, T. C., 944
Tupper, R., 1154
Tuqan, N. A., 1010, 1369
Turbow, M. M., 800
Turchini, J., 1134, 1159
Turner, H.-B., 1052
Turner, R. H., 1012, 1370
Turoverov, K. K., 1173, 1177 1238
Tuzson, P., 1087
Tverdy, G., 1086
Tyrer, J. H., 870, 892

Udenfriend, S., 864, 1104, 1118 1120, 1121, 1173, 1236
Ugolev, A. M., 821
Ullberg, S., 1213
Umbrath, G., 772
Underhay, E., 781, 800
Uotila, U., 1248
Urist, M. R., 1179
Urnäs, B., 1119
Usuku, G., 1262
Utamura, M., 1144, 1152
Uusitalo, R., 1279
Uzman, L. L., 1147, 1411

Valcourt, A. J., 864
Valentine, R. C., 954, 1260, 1261, 1276
Valk, A. de T., jr., 866, 867, 1339
Vallee, B. L., 929, 1154, 1155, 1156
Vallyathan, N. V., 835
Van Dalen, J. P. R., 1249
Van Den Bergh, A. A. H., 1071, 1074
Van Der Ploeg, M., 856, 861, 1249, 1253, 1336
Van Duijn, P., 853, 854, 856, 861, 862, 1059, 1231, 1232, 1249, 1253, 1289, 1334, 1336, 1423
Van Eys, J., 908
Vanha-Perttula, T. P. J., 966, 970, 976
Van Harreveld, A., 1159
Van Herwerden, M. A., 1019
Van Iterson, W., 884
Van Kleef, B. H. A., 1219, 1433
Van Lancker, J. L., 812
Van Orden, L. S., 1235, 1292
Van Roy, F., 1075
Vantrappen, G., 1013, 1374
Van Tubergen, R. P., 1218, 1433 1435
Van Wijhe, M., 935, 954
Van Winkel, Q., 1187
Van Woert, M. H., 1057
Varadi, D. P., 1042
Vasquez, J., 823
Vassallo, G., 1119, 1189, 1190
Vaughan, J., 1210
Velardo, J. T., 921, 922
Velculescu, V. G., 1195
Velican, C., 1044
Vennesland, B., 948, 953
Vercauteren, R., 855, 862, 922 1024
Verhoeff, F. H., 1040
Verkataraman, K., 1129, 1164
Verne, J., 800, 1059
Verpoorte, J. A., 995
Vethamany, V. G., 1279
Vevers H. G., 1059
Viala, R., 988
Viale, G. L., 933
Vialli, M., 1101, 1110, 1116, 1118 1122, 1252
Vickerstaff, T., 1164
Vickery, A. L., 1162
Victor, J., 1083, 1086
Villamil, M. F., 853
Villani, G., 809
Villanueva, A. R., 1179
Villar-Palasi, C., 826
Villee, C. A., 940, 942
Villela, G. G., 874
Virchow, R., 1064, 1070
Vischer, W., 1138
Vitry, G., 1292
Vogt, A., 1266
Voigt, G. E., 1144, 1156
Vollrath, L., 972
Von Bertalanffy, L., 1186
Von Euler, H., 904
Von Euler, U. S., 1121
Von Kóssa, J., 1138
Von Pechmann, H., 880
Von Recklinghausen, R., 1082
Von Sallman, L., 1208
Vorbrodt, A., 1280
Vosa, C. G., 1184
Vosburgh, G. I., 1132
Vulpian, A., 1115, 1117

Wachstein, M., 784, 795, 801, 853, 854, 884, 921, 922, 972, 1143, 1335, 1410
Wächtler, K., 991, 1362
Wade, D. R., 828
Wagner, J., 988
Waidl, E., 972
Wajda, I., 772
Wald, G., 929, 1087
Waldenström, J., 1118
Waldrop, F. S., 1075
Walford, R. L., 1042
Walker, B. E., 1210
Walker, D., 1074
Walker, D. G., 923, 926, 986
Walker, P. G., 812, 816, 819, 820, 823, 824, 1326
Walker, W. H., 907, 908
Wallenfels, K., 820
Wallerstein, J. S., 1121
Walsh, R. R., 771
Warburg, O., 841, 904, 925, 1154
Warren, L., 1033
Warren, S., 1209, 1215
Washburn, A. H., 853
Wasserkrug, H., 834, 1353
Watanabe, K., 820, 834
Watanabe, Y., 858, 1333
Watanuki, S., 828, 834
Waterhouse, D. F., 1142, 1145, 1148, 1411
Watson, E. M., 1032
Watson, M. L., 1262, 1263, 1273, 1275
Watson, W. A., 914
Wattenberg, L. W., 887, 941, 945, 1348
Wattiaux, R., 812
Watts, R. W. E., 1154
Webb, E. C., 842, 948, 949, 989, 1000
Webb, T., 967
Weber, G., 800, 801, 1031, 1084
Weber, R., 812
Webb, E. C., 762, 899, 900, 901
Wegmann, R., 829, 830, 946
Weigert, F., 1178
Weil-Malherbe, H., 1121
Weinberg, T., 801
Weiner, N., 864
Weisburger, E. K., 1357
Weissbach, H., 869
Weissmann, B., 820, 1029
Weitsen, H. A., 1282
Welch, A. D., 1063
Weller, R. O., 1276
Wellings, S. R., 1055
Wells, C. L., 1069
Wells, G. C., 800, 1144, 1410
Welsch, U., 1189
Wendler, D., 893
Werkheiser, W. C. 952
Wenk, H., 954
Werle, E., 1122
West, S. S., 1235, 1238, 1245 1251
West, T. S., 1136
Wetterdal, B., 1211
Wetzel, M., 1275
Wewalka, F., 1177
Whelan, W. J., 818
White, H. J., 1013, 1373
White, J., 1083
White, J. C., 1020, 1022
White, J. G., 1267
White, R. F., 1027, 1031
Whitney, P. L., 995
Whittaker, V. P., 762, 767
Whur, P., 1122
Wiechowski, D., 872
Wied, G., 1226, 1227
Wielenga, G., 972
Wieme, R. J., 934
Wiener, H., 1135
Wiest, W. G., 942
Wiggans, D. S., 963
Wilk, M., 883
Wilkin, P., 972
Wilkins, M. F., 1228
Wilkinson, G., 1269
Williams, B. W., 1026
Williams, C. H., 953
Williams, G. R., 913
Williams, M. A., 1265, 1276
Williams, P. S., 1128
Williams, R. T., 811
Williams, W. L., 1084, 1085
Williams-Ashman, H. G., 930
Willig, F., 971
Willighagen, R. G. J., 971, 972
Wills, E. D., 811
Willstaedt, H., 1087

Wilson, D. W., 948
Wilson, I. B., 764, 766, 767, 771, 794
Wilson, T. H., 917
Wilson, V. W., jr., 1033
Winckler, J., 1052
Winkelmann, R. K., 800
Winkleman, J., 1263
Winkler, H., 1191
Winnail, D. S., 1085
Wise, J. B., 935
Wislocki, G. B., 921, 1020
Withers, R. F. J., 779, 780, 781, 782, 800, 1306
Wizinger, R., 898
Wöckel, W., 1150
Woessner, J. F., 966
Woessner, J. F., jr., 810
Wohlfarth-Bottermann, K. E., 1262
Wohlrab, F., 868, 874, 887, 916, 932, 947, 951, 952, 1349
Wolf, A., 1083
Wolf, G., 801
Wolf, I., 1144
Wolf, P., 823, 1354
Wolfe, H. J., 954
Wolff, H., 1155
Wolff, H. H., 954
Wolken, J. J., 1069
Wollenberger, A., 1279
Wolman, M., 921, 950, 1083, 1135, 1138
Wood, E. M., 1085
Wood, J., 1291, 1292, 1452
Wood, J. G., 1062, 1063
Wood, R. L., 1286
Woodward, G. E., 973
Woohsmann, H., 868, 991, 1362, 1364
Wormall, A., 811, 1154
Wrenn, H. T., 1156
Wu, R., 948
Wulff, H. R., 831, 1328
Wünsch, E., 974
Wyllie, J. C., 1267
Wynn, C. H., 988
Wynne, E, S., 862

Yagoda, H., 1207, 1212, 1213
Yamada, H., 864
Yamada, K., 1275
Yamada, M., 975, 1358
Yamada, Y., 997, 1251
Yamamoto, Y., 851
Yano, Y., 987, 1361
Yasue, T., 1140, 1406
Yasunobu, K. T., 864
Yasutake, S., 1262
Yasutaki, W. T., 1085
Yasuzumi, G., 1292
Yataganas, X., 1241
Yates, P. E., 1281, 1283, 1289
Yin, H. C., 828
Yokota, S., 1288, 1431
York, J. L., 931
Yoshioka, W., 801, 834
Young, A., 1136
Young, B. A., 1218
Young, L., 1210
Young, M. R., 892
Yudkin, W. H., 1154

Zacks, S. I., 795, 800
Zagury, D., 1277
Zaidel, L., 1083
Zajicek, J., 792, 909
Zak, F. G., 1143, 1410
Zakrzewski, S. F., 952
Zamboni, L., 1274, 1439
Zamecnik, P. C., 1044
Zannoni, V. G., 1064
Zanotti, L., 1241
Zebe, E., 929, 1279
Zech, L., 1185
Zechmeister, L., 1038, 1087
Zeigel, R. F., 1262
Zeitler, E., 1261
Zelena, J., 800
Zelenin, A. V., 1172, 1186
Zeller, E. A., 863, 864, 870, 872
Zenker, W., 794
Zermatten, H. L., 1143
Zhdanov, U. M., 1270
Ziegler, D. M., 903, 925
Zieher, L. M., 1291
Zimmermann, H., 910, 921, 930
Zobel, R. C., 1273
Zorzoli, A., 1016, 1021, 1022
Zubay, G., 1273
Zugibe, F. T., 1032, 1033
Zweifach, B. W., 909, 922

SUBJECT INDEX

Absorptiometry, 1226, 1227 et seq.
Absorption cytophotometry, errors in, 1228
"Accelerator", in bilirubin techniques, 1072, 1074
Acetaldehyde, in production of fluorophores, 1107
removal of, by trapping reagents, 901, 902
Acetamide, 820
4-Acetamido-4′-isothiocyanostilbene-2,2′-disulphonic acid, 1180
Acetic acid, as alternative to formaldehyde, 1107
Acetic-haematoxylin, stain for bile pigments, 1074
2-Acetylamino-1,3,4-thiadiazol-5-sulphonamide, 995
***O*-Acetyl-5-bromoindoxyl,** 780, 783, 1306
Acetylcholine, as substrate for cholinesterases, 766
Acetylcholinesterase, 764, 766, 772, 798
quantitation by DFP-^{3}H method, 1218
thiocholine methods for, 1312 et seq.
Acetylcholinesterase, inhibitors for, 798
Acetyl CoA carboxylase, 1289
Acetyl-CoA: choline *O*-acetyltransferase, 982
Acetyl disulphide, active contaminant of thiolacetic acid, 795
Acetylesterases, 761, 763
Acetylglucosaminases, 1038
***N*-Acetyl-glucosamine,** 820
Acetyl-*β*-glucosaminidase, methods for, 1325
***N*-Acetyl-*β*-glucosaminidase,** 820
methods for, 823, 824
Acetyl-*β*-glucosaminide, of Naphthol AS–BI, 824
***N*-Acetylhomocysteinthiolactone,** (AHTL), 1270
***O*-Acetylindoxyls,** 778 et seq., 798
Acetyl-*β*-methylcholine, 766
as specific AChE substrate, 768
Acetyl-*β*-methyl thiocholine, EM method for cholinesterases, 1283
Acetylselenocholine, EM method for cholinesterases, 1283
Acetylthiolcholine methods, 791 et seq.
Acid diazo reaction, 1397
Acid fast method, for lipofuscins, 1385
Acid fastness, 1082
Acid fast pigment, 1078
Acid formalin pigment, 1075
Acid haematein reaction, of EC granules, 1115
Acid hydrolysis, effect of, on carboxamido groups, 1190
Acid lipase, 766
and indoxylesterase, 977
Acid peptidases, biochemistry of, 962
Acid phosphatase, labelling of antibody, 1267
lysosomal, 824
microdensitometry of, 1230
Acid polysaccharides, acridine orange and, 1186
Acid proteinase, lysosomal, 966
Aconitase, method for, 997, 998
Acridine orange, 1024, 1421, 1422
as fluorochrome, 1186 et seq.
as vital fluorochrome, 1186
stoichiometry for DNA, 1187
structure of, 1185
Acridine yellow, 1183, 1251, 1421, 1422
Acridines, demonstration of, in tissues, 1161
Acriflavine, 1241, 1251, 1422
Acriflavine-phosphotungstate, 1194
Acriflavine-phosphotungstic acid complex, for DNA and RNA, 1273
Acrylamide gel medium, for dehydrogenases, 923
ACTH cells, 1112
Acylase I, 973
Acyl naphthylamidase method, 1354
Adenyl cyclase system, 830
ADN method, 1330
for cytochrome oxidase, 847
Adrenal, cholinesterases of, 800
Adrenal cortex, glucuronidase in, 834
Adrenal medulla, amine oxidase in, 863
chromaffin and, 1059
rat, catecholamine-FIF and, 1252
rat, non-specific Pb absorption by storage granules of, 1281
Adrenalin, chromaffin reaction for, 1382
demonstration of, by trihydroxyindole method, 1121
demonstration of, by FIF, 1108
distinction from noradrenalin, 1062
EM demonstration of, 1291, 1292
fluorescence maxima, 1106
Adrenalin granules, EM recognition of, 1063
Adrenergic fibres, simultaneous demonstration of ChE and, 1109
Adrenergic nerves, results of crushing, 1111
Adrenochrome, Giemsa method for, 1382
Adrenochromes, 1059
Adrenolutin, 1121

A-Esterase, of rabbit serum, 764
A-Esterases, heavy-metal method for, 1305
A and B Esterases, 763
***Agkistrodon* venom,** 1311
Alanyl-alanine naphthylamide, 966
DL-Alanyl β-naphthylamide, 967
Alcian blue, 1013, 1032
  for EM demonstration of carbohydrates, 1275
  and masked metachromasia, 1189
  after neuraminidase, 1033
Alcian blue 8GS, 1136
Alcohol, as transglycosylase inhibitor, 832
Alcohol dehydrogenase, 899, 901, 904
  inhibitors of, 929
  method for, 927, 929
  as "nothing dehydrogenase", 911
  standard method for, 1344
Alcoholic haematoxylin method, for calcium, 1134
Alcoholic PAS, for water-soluble dextrans, 1162
Alcoholic toluidine blue, for dextran sulphate, 1162
Aldehyde dehydrogenase, method for, 946, 947, 1349
Aldehyde fixatives, and EM enzyme cytochemistry, 1277
Aldehyde groups, stoichiometry of, and Schiff reaction, 1182, 1183
Aldehyde oxidase, 865
Aldehydes, polymerization of, and lipofuscins, 1077
  from tyramine oxidation, 866
Alginate, 1161
  and pectinase, 1036
Ali-esterase, 763
Alizarin, metal complexes of, 1135
Alizarin-purpurin methods, for calcium, 1134
Alizarin red S, 1135, 1310
  for demonstration of calcium soaps, 785
Alizarin staining, mechanism of, 1135
Alkaline diazo reaction, 1112, 1396
Alkaline pH levels, effect of, on NAD, 910
Alkaline quinalizarin, method for beryllium, 1143, 1409
Alkaline tetrazolium reaction, 1093
Alkaline thioindoxyl,
  method, for β-carbolines, 1399
  method for quinoneimines, 1397
  reaction, 1115, 1116
Alkaptonuria, 1064
Allantoin, 872
Aluminium, acid Solochrome cyanine method for, 1408
  aurine method for, 1407
  fluorescent Solochrome method for, 1408
  methods for, 1140, 1141, 1406, 1407, 1408
Aluminium, Naphthochrome B method for, 1408
  Naphthochrome green staining of, 1136
  Solochrome azurine method for, 1406
Aluminium-Morin complex, 1137
Aluminon, 1140
Ambenonium, as AChE inhibitor, 798
Amethopterin, 952
Amidase, liver, similarity of to aliesterase, 766
Amine-amine, method for cytochrome oxidase, 848, 1331
Amine-free radical coupling methods, for cytochrome oxidase, 847
Amine precursor uptake, 1111
Amine-storage granules, EM of, 1291
Amines, biogenic, 1101 et seq.
  microfluorimetry of, 1241
  primary, oxidation of, 863
Aminoacid naphthylamidases, methods for, 1353
D-Amino-acid oxidase, 841
  method for, 872
L-Aminoacid dehydrogenase, method for, 951
L-Aminoacid tetrazolium reductase, method for, 1351
***O*-Aminobenzaldehyde,** 871
Aminocaproic acid, naphthol AS esters of, 776
3-Amino-9-ethylcarbazole, 857, 869
3(β-Aminoethyl) indole, 1102
4-Amino-*N,N'*-dimethylnaphthylamine, 847
Aminopeptidase, indigogenic method for, 1354
  microsomal, 966
  oxidation method for, 1357
Aminopeptidase B, 972
Aminophenols, 1113
***N,N'*-bis(p-Aminophenyl)-1,3-xylylenediamine,** 850
Aminophthalhydrazide (diazotized), 1289, 1290
Aminopolysaccharase activity, in hyaluronidase, 1029
8-Amino-1,2,3,4-tetrahydroquinoline, 848
Ammeline, 873
Ammonium oxalate-formalin, 1133
Ammonium sulphide, stage in thiocholine reactions, 792
Ammonium thiocyanate, 1143
Amoebocytes, echinochromes in, 1059
Amphibian tissues, arylsulphatase in, 991
***Amphiuma*,** pigment in liver of, 1056
Amylamine, and rabbit MAO, 864
Amylase, starch film method for, 1373
α-Amylase, 808 et seq., 832
β-Amylase, 808 et seq., 832
Amylases, as histochemical reagents, 1033

Amylases, substrate film methods for, 1012
Amyloid, dichroism of, 1243, 1244
mucosubstances in, 1274
Amylopectin, 828
Amylophosphorylase, 826
Amylose, 828
Amylo-1,4-1,6-transglycosylase, 826, 829
Amytal, and mid-chain block, 913
and tetrazolium reductase, 906
Analytical electron microscopy, 1128
Androstane, 939
Androsterone, 940
Angiotensinase A, 973
Anionic site, of esterases, 767
Anodonta cygnea, gill plates, calcium in, 1136
Anterior pituitary gland, hormone localization in, 1268
Anthracenes, and pseudomelanosis, 1085
Anthranilic acid, for antibody coupling, 1269
Anthraquinone, 1129
Anthraquinone dyes, as calcium stains, 1134
Anthraquinone-1-azo-4-dimethylaniline hydrochloride, 1152
Antibiotics, localization of, by fluorescence microscopy, 1178
Antibody, electron opaque, 1265 et seq.
Antibody-combining site, 1264
Antibody conformation, immuno-EM and, 1264
Antibody labelling, with enzyme proteins, 1267
Anti-cholinesterases, 767, 798
Anti-esterases, 796
Antigen-antibody, short range forces and, 1264
Antimycin, and electron chain block, 913
Antimycin A, 912, 914
Ansco 500 film, 1203
Apoferritin, 1069, 1132
Aposiderin, 1076
APUD cells, 1113, 1190
argyrophilia of, 1119
APUD series, of endocrine polypeptide cells, 1111, 1112
Apudomas, 1193
Arabinosidase, 819
Arabitol, 930
Araldite, ^{3}H-labelled, as internal standard for quantitative autoradiography, 1218
Arbacia, pigments in, 1059
Area, measurement of, 1248
Arenichrome, 1058
Argentaffin, EM techniques and, 1291
Argentaffin cells, 1057, 1090
Argentaffin cell tumours, 1113
Argentaffin (EC) granules, diazo method for, 1396
Argentaffin reaction, of EC granules, 1112, 1113
Argentaffinomas, esterases in, 976
5-HT in, 1118
Arginine, activation of esterase and, 762
quantitation of, by microfluorimetry, 1241
Arginyl amides, 974
Argyrophil, non-argentaffin, cells, 1113
Argyrophil cells, as EC cell precursors, 1113
Argyrophilia, 1113
Argyrosis, 1151
Arom-esterase, 763
Arsenates, methods for, 1141, 1142, 1409
Arsenites, methods for, 1141, 1142, 1409
Arthropoda, copper in, 1144
Arylamidases, characteristics of, 964, 965
Arylesterases, 761, 763
Arylethylamines, FIF and, 1107
Arylhydrazines, fluorinated, for EM demonstration of mucosubstances, 1274
Arylsulphatase, EM method for, 1448
post-coupling method for, 1362
Arylsulphatase A, 988, 989
Arylsulphatase B, 988, 989
Arylsulphatase C, 988, 989, 992
Arylsulphatase A and B, Naphthol AS method for, 1362
NCS method for, 1363
Arylsulphatases, biochemistry of, 987 et seq.
EM method for, 1285
function of, 989
hydroxyquinoline method for, 1364
methods for, 990, 991, 992
substrates for, 988
Ascites tumour cells, phosphorylase in, 834
Ascorbic acid, 881
as glucuronidase inhibitor, 811
and silver reduction, 1057
Asparaginase, 992
Aspartate aminotransferase, 1289
diazo method for, 1360
lead method for, 1360
methods for, 983, 984
Aspartate carbamoyl transferase, 981
method for, 1359
Aspartic acid, 1112
β-Aspartyl cross-links, 1191
Aspergillus, emulsins and, 819
Aspergillus oryzae **glucosidases,** 1037
Atebrin, 1161, 1163
fluorescence of, 1178
Atheroma, ceroid in, 1084
Atmungsferment, 841, 843
Atoxyl, inhibition of esterases and lipases, 762
ATP, mitochondrial, 903
ATPase, membrane, EM method for, 1279, 1280

ATPase, mitochondrial, EM methods for, 1279, 1443
Auramine O, 1183, 1251, 1421, 1422
Aurine tricarboxylic acid, 1140
Autofluorescence, 1176 et seq.
of ceroid, 1084
in pituitary gland, 1112
Autolysis, postmortem, effects of on lipases, 785
Autolytic vacuoles, demonstration by LNA methods, 972
Auto-oxidation, in cytochrome-rich tissues, 846
Autophagic vacuoles, 1080
Autoradiographs, coated, 1213, 1215
mounted, 1213
stripping film, 1213, 1214
Autoradiography, 1207 et seq.
EM, 1218 et seq.
high resolution, 1215
methods for, 1211 et seq., 1426 et seq.
quantitative, 1217
stripping film technique, 1426
Autoxidation, unsaturated fats and, 888
of reduced menadione, 906
Avogadro's number, 1261
Axial organ, of *Diadema*, *Arbacia* and *Paracentrotus*, 1059
Azobilirubins, 1075
Azo-coupling, reaction for indoxyls, 780
Azoene Fast red salt, 985
Azohaematoidins, 1075

Bacteria, dehydrogenases in, 884
BAEE, 963
BANA, 963, 974
BAO, 1241, 1251, 1422
as fluorochrome, 1179, 1183
Baranowski enzyme, 929
Barium, methods for, 1142
rhodizonate method for, 1410
Barium-anthraquinone lakes, 1135
Barium-GBHA complex, 1137
Basic dyes, chart of applications, 1181
Basic Fuchsin, 1421, 1422
as fluorochrome, 1179, 1183
Basophils (blood), phenol oxidase in, 863
Bathophenanthroline, 1130, 1132
method, 1403
Bauer-Feulgen procedure, 1161
BAXD, 850, 1284, 1286
in oxidase reactions, 850
Beer-Lambert law, 1228, 1253
Benzedrine, 863
Benzenediazonium chloride, 1071
Benzene-saturated water, for etching, 1268
Benzidine, 1139
method for copper, 1412
Benzidine, procedure for EM peroxidase demonstration, 1286
reaction of copper salts and, 1146
reactions for peroxidases, 852
substitutes for, 854
Benzidine blue, 852
Benzidine-nitroprusside, 1065
Benzidine-peroxide, methods, 853
Benzoflavine, 1421, 1422
1,4-Benzoquinone, 948
Benzoquinones, as electron acceptors, 906
4-Benzoylamino-2:5-dimethoxyaniline, diazotate of, 816
α-*N*-Benzoyl-L-arginine ethyl ester, 963
α-*N*-Benzoyl-L-arginine β-napthylamide, 963
Benzoylcholine, 766
Benzpyrene, identification of, in tissues, 1241
fluorescence of, 1178
***N*-Benzyl-*p*-phenylenediamine,** 849, 1285
Berberine sulphate, 1422
Beryllium, haematoxylin lakes of, 1143
methods for, 1142, 1143, 1409
naphthochrome green staining of, 1136
Beryllium lake, of aurine tricarboxylic acid, 1140
Beryllium-Morin complex, 1137
Best's carmine, 828
Betaine aldehyde, 945
dehydrogenase method for, 947, 1349
Bicarbonate, ^{14}C-labelled, in bone, 1207
Bidentate ligands, 1269
Biladienediols, 1073
Bilatriene, 1072
Bile canaliculi, CAH in, 1288
Bile pigments, 1070 et seq.
bleaching of, 1074
chemistry of, 1071
histochemistry of, 1072
iodine method for, 1389
Stein method for, 1389
Bile salts, action of, on esterases, 762
Bilharzial pigment, 1075
Bilirubin, 1070 et seq.
conjugated, 1071
diazo methods, 1391, 1392
direct reaction for, 1072
Ferric iron methods for, 1390
glucuronide of, 810
indirect reaction for, 1072
methods for, 1388 et seq.
oxidizing agents for, 1072
unconjugated, 1071
Bilipurpurine, 1073
Biliverdin, 1071
Biogenic amines, 1101 et seq.
EM methods for, 1291, 1292
fluorescence characteristics of, 1106
microfluorimetry of, 1241

Biopsy, muscle, 1248
Biorhythms, and liver phosphorylase, 835
Birefringence, 1243, 1244
Bischler-Napieralski condensations, 1107
techniques, 1400
Bis-cyclohexanone oxalyldihydrazone, 1148
Bis(cyclopentadienyl) iron, 1269
Bis-diazotized benzidine, 1265
Bismuth, brucine-iodide method for, 1410
as EM stain for chromatin, 1273
method for, 1143
Bisulphite, on bilirubin, 1074
as carbonyl trapping reagent, 902
Blatta, tanning of ootheca in, 822
Bleaching methods, for melanins, 1381
Blood brain barrier, 5-HT and, 1118
Blood films, phosphorylase technique for, 1328
Bodian silver technique, 1113, 1119
Bone, radioactive Pb in, 1207
radiophosphorus in, 1207
tetracyclines and, 1178
Bone marrow, cholinesterases of, 800
glucuronidases in, 834
Bone matrix, esterases of, 775
Bones, lead in, 1149
Borohydride reduction, 1110
specificity tests, 1396
Borohydride-periodic acid sequence, 1110
test, 1396
Bound enzymes (oxido-reductases), standard method for, 1342, 1343
BPDA, 849
Brain, monoamine oxidase in, 869
porphyrins in, 1068
"pseudo calcium" in, 1137
radiopotassium in, 1208
rodent, glycosidases in, 834
Brain homogenates, mitochondrial function in, 906
Branching factor, 826
Brilliant Cresyl blue, and induced fluorescence, in tumour cells, 1180
Bromelain, 962
5-Bromo-4-chloro-indoxyl β-D-glucopyruroniside, 817
DL-**N-(5-Bromoindol-3-yl) leucinamide hydrobromide,** 969
3-(5-Bromoindoxyl)-β-D-glucopyranoside, 822
6-Bromo-2-naphthol-β glucoside, 821
6-Bromo-2-naphthol glucuronide, 817
6-Bromo-2-naphthyl-α-D-galactopyranoside, 823
6-Bromo-2-naphthyl-β-D-galactoside, 822
6-Bromo-2-naphthyl-α-D glucopyranoside, 822
6-Bromo-2-naphthyl-β-D-glucopyruronoside, 1321
Brucine, method for bismuth, 1143
Brunner's glands, 823, 1210
BT, 881, 882, 883 et seq., 893, 907, 914
Butyrylthiocholine, 766
methods, 791 et seq.
methods for cholinesterases, 1313 et seq.

62C47, as AChE inhibitor, 770, 798
284C51, as AChE inhibitor, 770, 798
^{45}Ca, 1211
Cabbage, Savoy, lecithinase in, 1045
Cacodylate-buffered glutaraldehyde, 1285
Cacotheline method, for tin salts, 1152
Cadmium sulphate, and ferritin, 1132
CAH, erythrocyte, 994
reactions catalysed by, 995
Calcichrome, 1136
Calcitonin M, human, molecular model of, 1190
Calcium, alizarin red method for, 1404
calcium red method for, 1404
EM demonstration of, 1290
Eriochrome black T method for, 1405
fluorescence methods for, 1137, 1138
GBHA method for, 1406
gypsum method for, 1404
insoluble forms of, 1134
ionized, 1133
methods for, 1133 et seq.
phthalocyanin method for, 1404
soluble forms of, 1133
substitution methods for, 1138
Calcium-cobalt method, for EM demonstration of alkaline phosphatase, 1278, 1279
Calcium ions, functions of, in indoxyl reactions, 782
Calcium red, 1135
Calliactus parasitica, chromaffin cells of, 1110
Canaliculi, bile, CAH in, 1288
Capture reactions, for indoxyls, 779
Carbamoyl transferases, 981
Carbazole reagents, 857
Carbodiimide, and EM demonstration of carboxyls, 1263
as fixative for immuno-EM, 1265
Carbohydrases, in snail stomach fluid, 1037
Carbohydrates, EM methods for, 1273 et seq.
β-Carboline, in EC cells, 1103
β-Carbolines, fading of fluorescence of, 1107
Carbon-Sulphur lyases, 998
Carbon tetrachloride, effect of on liver, 1083
Carbonate, demonstration of, by von Kóssa method, 1138
methods for, 1158
Carbonates, methods for, 1417
Carbonic anhydrase, biochemistry of, 994
EM method for, 1288

Carbonic anhydrase, methods for, 995, 1365
in rat prostate, 1153
reaction, mechanism of, 997
Carboxy acids, in production of fluorophores, 1107
***o*-Carboxy formazyl-2-quinoline,** copper chelation by, 899
Carboxylases, 994
Carboxylic esterases, common properties with proteases, 765
Carboxylic ester hydrolases, 761 et seq.
Carboxy methyl cellulose, 1161
β-Carboxy-propionyl-L-arginyl-2-naphthylamide, 974
Carboxyl groups, EM methods for, 1263
free, in polyamide fibres, 1164
Carboxyl side chains, and amine uptake, 1112
Carcinoids, esterases in, 976
5-HT in, 1118
Carcinoid tumours, 1113
autofluorescence of, 1102
Cardiac muscle, MAO in, 870
Carnitine, acetyltransferase, 1289
Carnosinase, Zin activation of, 1154
Carnoy-sodium sulphide fixative, 1149
Carotid body, MAO in, 870
Carotenoids, 1086, 1087
histochemistry of, 1087, 1088
in lipids, 1077
Carr-Price test, 1088
Cartilage, arylsulphatase C in, 992
cathepsin in, 963
ChS-A in, 1027
phosphorylase in, 828
UDPG dehydrogenase in, 933
Cascara sagrada, and pseudomelanosis, 1085
Cassius, purple of, 1148
Catalase, 841, 850, 857, 865
in hydroquinone-MTT reaction, 1001
method for, 1333
microbody, 854
staining of, by DAB, 1066
substrate film method for, 1014, 1374
Catalysts, oxidation, for indoxyls, 779
Catecholamine fluorescence, distinction from 5-HT, 1109
Catecholamines, breakdown pathways, 865
FIF, and fading, 1107, 1251
methods for, 1121
perfusion fixation, and FIF technique, 1395
primary, 1104
secondary, 1105
Catechol (DOPA) oxidase methods, 1338
Catechol-*O*-methyltransferase, 865
Catechol oxidase, 841
block method for, 1339
methods for, 860 et seq.
Catechols, Vulpian reaction and, 1115
Cathepsin A, in bovine spleen, 963
Cathepsin B, demonstration of, by LNA hydrolysis, 970
and indoxyl esterase, 976
thiol dependency of, 963
Cathepsin C, 976
chloride requirement of, 966
pig kidney, 963
Cathepsin D, 966
demonstration of, by substrate film method, 1011
Cathepsin E, 967
Cathepsins, 962, 963 et seq.
C cells, 1111
Cell membranes, kinetic studies of, 1242
Cellobiose, 819
Cells, specific radioactivity of, 1217
Cerebrosides, sulphated, 1273
Cerebroside sulphate esters, 989
Ceroid, 1083 et seq.
acid fastness of, 1081
experimental production of, 1084
Cervix uteri, phosphorylase in, 834
C-esterase, 763
Chalkley ratios, 1248
Chelates, 1129
Chick embryo, aminopeptidases of, 971
cytochrome oxidase in, 850
lipases in, 784
Chitinases, 1038
Chloramphenicol, fluorescence of, 1178
Chlorazol fast pink, for PVP, 1162
for PVA, 1163
Chloride, and phosphate-carbonate, combined technique for, 1159
Chloride ion, methods for, 1158
Chloride ions, EM demonstration of, 1291
Chloroacetyl-2-hydroxy-3-naphthoic acid, 971
Chloroacetyl naphthylamines, 967
4-Chloro-5-bromo acetylindoxyl, synthesis of, 781, 1307
2-Chloroethanol, 1056
***p*-Chloromercuri-benzoate,** as C-esterase inhibitor, 797
6-Chloro-4-methyl thionaphthene, 1116
4-Chloro-1-naphthol, in peroxidase-label technique, 1267
Chloroplasts, 1055
difluorescence and, 1245
UQ in, 1000
Chlorpromazine, 914
Cholane, 939
Cholestane, 939
Cholesterol, 940
Cholesterol esterase, 762, 766
Choline acetyltransferase, 982
method for, 1360

Choline dehydrogenase, method for, 945, 946, 1349
Choline oxidase, 916
Cholinergic fibres, simultaneous demonstration of adrenalin and, 1109
Cholinesterase, brain, and inverse relation to MAO, 863
methods for distinguishing AChE from ChE, 797
thiolacetic acid method for, 1319
Cholinesterases, acetylthiocholine methods for, 1312 et seq.
applied histochemistry of, 799
azo-dye EM techniques for, 1281
biochemistry of, 766
competitive inhibition of, by Tris, 771
distribution of, 772
EM methods for, 1281 et seq., 1445 et seq.
esterase sites of, 795
inhibitors of, 766, 768, 770, 771, 790, 795 et seq., 798
methods for, 788 et seq.
myristoyl choline method for, 1312
pH optima of, 766
Chondroitin sulphates, 1026, 1027
Chondroitinase, as histochemical reagent, 1032
Chondrosulphatase, 987
Chromaffin, EM techniques for, 1291
and the PAS reaction, 1060
Chromaffin granules, relationship of, to mitochondria, 1063
Chromaffin reaction, 1113, 1114, 1382
historical views on, 1059
Chromaffinity, four criteria of, 1060
Chromate method, for Pb, 1149
Chromatophores, bacterial, 1000
Chrome haematoxylin method, 1384
Chromic acid, effect of, on RNases, 1022
Chromium, in chromaffin granules, 1059
hexavalent, and diphenylcarbazide, 1145
Chromium-anthraquinone lakes, 1135
Chromium compounds, as electron stains for nucleic acids, 1262
Chromolipoids, 1077
Chromosome, Y, fluorochromy and, 1184
Chromosomes, banding in, fluorochromy of, 1184
bean-root, 1023
DNA measurements in, 1232
quinacrine staining of, 1423
Chymotrypsin, 765, 962, 976
Chymotrypsin-like enzymes, 974
Chymotrypsin-like esterase, method for, 1357
Cirrhosis, hepatic, 1074
Citrate, and inhibition of glucuronidases, 811
Claus diazotization, 1075, 1392
Clostridiopeptidase A, 1043
Clostridium **hyaluronidases,** 1028
Clostridium perfringens, **neuraminidase,** 1033
C-N Hydrolases, 992
C.N.S., esterases and cholinesterases in, 800
^{58}Co, 1211
Coatings, anti-reflection, for objectives, 1202
Cobalt, methods for, 1143, 1144
Cobalt hydroxide, 997
Cobaltinitrite method, for potassium ions, 1156, 1157
Cobalt sulphate, degree of hydration of, and CAH reactions, 1288
Cockroach, colleterial glands, 822
Coenzyme A, 982
Coenzyme levels, high, in diaphorase reactions, 907
Coenzyme Q, 1000
Coenzymes, binding of, by dehydrogenases, 908
histochemistry of, 1000
pyridine nucleotide, pH optima and, 902
role of, 903 et seq.
Coenzymes I & II, 881
Collagen, absence of masked metachromasia and, 1191
autofluorescence and, 1177
Collagenase, 1030
effects of fixation and, 1043
extraction, 1378
as histochemical reagent, 1042, 1043
***s*-Collidine,** 1337
Colloid, thyroid, ^{131}I in, 1208
Colloidal iron, 1069
as EM stain for carbohydrates, 1275
Colour, appreciation, and human eye, 1175
Colour filters, drawbacks of, for fluorophores, 1109
Coloured film methods, for proteases, 1010, 1011
Computers, digital and analogue, 1237
in microdensitometry, 1230
Condensation reactions, and binding of fluorochromes, 1181
Conformation, tertiary, antibodies and, 1264
Congo red, 1422
and amyloid, 1244
as fluorochrome, 1179
method, for PVP, 1419
stain for PVP, 1162
Conjugated bilirubin, 1075
diazo method for, 1391
Connective tissues, peribronchial, anaphylactic shock and, 1209
radioactive oestradiols in, 1209
Contact autoradiography, 1212, 1213
Contact microradiography, 1233
Contrast, EM image and, 1260, 1261 et seq.
Coordination compounds, 1129

Copper, applied histochemistry of, 1147
benzidine method for, 1412
methods for, 1144 et seq., 1411 et seq.
unreactive, 1147
Copper ferricyanide technique, for ChE's, 793
Copper ions, binding of, by Ranvier's nodes, 794
as oxidation catalysts for indoxyls, 779
Copper methods, sensitivities of, 1147
Copper-Protein oxidases, 858
Copper thiocholine, conversion of, to sulphide, 792
CoQ-lecithin, and SD reactions, 950
Coriphosphine O, 1189, 1251, 1421, 1422
structure of, 1185
Coronary arteries, ceroid in, 1084
Corpus striatum, MAO in, 869
Corriedale sheep, hepatic pigment in, 1089
Cortisol, 940
Cortisone, 940
Cottus pancreatic islets, zinc in, 1154
Coupled oxidation, reaction sequence, 825
Coupled peroxidation, for D-Amino-acid oxidase, 873
for uricase, 872
Coupled tetrazonium reaction, EM method, 1263
and melanin, 1056
Cozymase, 903
Creatine kinase, method for, 986
Crétin's method, for calcium, 1133, 1134
Cristae mitochondriales, staining of, by oxidized DAB, 854
Cryostat procedures, damage to mitochondria and, 907
Cryostat sections, catechol oxidase method for, 1338
for diazo reactions on bilirubin, 1074
freeze-drying of, 1109
Crushing condenser, 1228
Cupric ions, oxidation of 5-hydroxyindoles by, 1116
Cuprizone, 864
Cyanide, as carbonyl trapping reagent, 902
effects of, on cobalt chelation by formazans, 898
as succinic oxidase inhibitor, 846
as trap for carbonyls, 898
Cycle, glucuronic acid, 931
Krebs', 949
Cyclo-tris-7-(1-azo-8-hydroxynaphthalene-3 : 6-disulphonic acid), 1136
Cylinder lens method, for grain counting, 1217
Cylindrite, in pneumokoniosis, 1152
Cystamine, 871
Cysteine, 881
desulphurase, method 998
Cystine, EM method for, 1437
Cytidine-^{3}H, 1212
Cytidine monophosphate, as alkaline phosphatase substrate, 1279
Cytochrome C, as electron acceptor, 905
O-R potentials and, 944
peroxidase, 850
staining of, by DAB, 1066
as ultrastructural tracer, 1268
Cytochrome oxidase, 841 et seq.
ADN method for, 1330
amine-amine method for, 1331
DAB method for, 1332
EM methods for, 1285, 1286, 1449
G-Nadi reaction for, 1330
inhibitors of, 842, 843
methods for, 847
naphthol-amine method for, 1331
nitronaphthylamine-amine method for, 1332
O-R potentials, 842
quinoline-amine method for, 1331
thiazolyl thiosemicarbazide method for, 1332
Cytochromes, 841 et seq.
Cytolipochrome, 1082
β-Cytomembranes, 1280
Cytophotometry, 1226 et seq.

DA and NA, distribution technique, 1395, 1396
DAB, for heme proteins, 1066
and "Nadi" reaction, 849
for osmiophilic polymer generation, 1277
oxidized, affinity of heme enzymes for, 854
reagent, for myoglobin, 1067
DAB method, for EM demonstration of cytochrome oxidase, 1286
and peroxidase-labelling technique, 1267
Dam reaction, for lipid peroxides, 1078, 1079
Dark current, 1238
DDNTP, 1281
Debranching enzyme, 1034
Decarboxylation, 1111
by pituitary cells, 1112
Dehydroepiandrosterone, 940
Dehydrogenase, alcohol, 899, 901
endogenous, 907
reactions, kinetics of, 899
Dehydrogenases, 921 et seq.
activation and inhibition of, 902
applied histochemistry of, 954
copper-ferrocyanide methods for EM demonstration of, 1288
EM methods for, 1286
histochemical table of, 928
histochemical significance table for, 953

Dehydrogenases, histochemistry of, historical aspects, 921
methods for, 927 et seq.
soluble and insoluble, 923
soluble, gel medium method for, 1345
steroid, pH optima of, 902
Dehydroindigo, formation from indigo during esterase reactions, 779
production of in indoxyl reactions, 782
1,2-Dehydro thiamorpholine, 871
Densitometry, photometric, in autoradiography, 1217
Deoxyribonuclease, extractions, 1375
as histochemical reagent, 1023
Deoxyribonucleases, substrate film methods for, 1012, 1371
Descemet's membrane, 1147
copper in, 1233
Deuterium-labelled substrates, for dehydrogenases, 904
Development, in autoradiography, 1216
Dextran, method for phosphorylase, 1328
PAS methods for, 1418
Dextrans, as glucosyl acceptors, 830
methods for, 1160, 1161
staining of, with Niagara sky blue, 1162
water-soluble, alcoholic PAS for, 1162
Dextran sulphate, 1161
method for, 1418
γ-Dextrins, as glucosyl acceptors, 830
DFP, 763, 767, 771, 791, 789, 799
DFP-ase, 764
DFP-^{3}H, for AChE quantitation, 1218
Dia-, dark-ground, and epi-illumination, comparisons, 1197
Diachromes, 1179
Diadema antillarum, echinochromes in, 1059
Diamine oxidase, 841
inhibitors of, 870
methods for, 870, 871
3,3′-Diaminobenzidine, 849
microbodies and, 854
2,7 Diaminofluorene, 1079
Diaminoguanidine, 871
2,4-Diamino-6-hydroxy-*S*-triazine, 873
Diammine silver, and bilirubin, 1074
stain, 1379
Diamox, 995
Diaphorases, 904 et seq., 921 et seq.
histochemistry of, 907 et seq.
Diastase, malt, 1033
Diastase digestion, of glycogen, 1275
Diazine green S, method, for tin salts, 1152
Diazonium chlorides, Pb-containing, for EM enzyme cytochemistry, 1281
Diazonium coupling methods, for histamine, 1122
Diazonium reaction, of EC granules, 1114
Diazonium salt, choice of, in amidase methods, 968
Diazonium salts, inactivation of naphthylamidases by, 974
Diazo-phthalocyanins, 1263
Diazo reactions, for bilirubin, 1074
Diazo-safranin method, 1118
Diazotates, stable, 1397, 1398
Diazotized sulphanilic acid, 1071
Diazyme, 1034
2,4-Dichloro aniline diazotate, 1074
Dichlorophenol indophenol, 906
2,6-Dichlorophenol indophenol, 925
2,6-Dichloroquinone chloroimine, 1114
4,5-Dichloro-2-trifluoromethyl benzimidazole, 914
Dichroism, 1243, 1244
Dicoumarol, 905
on α-GPD activity, 945
Di-dicyclohexyl ammonium 2-naphthylthiol phosphate, 1281
***N,N′-bis*(2-Diethylaminoethyl) oxamide-*bis*-2-chlorobenzyl chloride,** as AChE inhibitor, 799
Diethyldithiocarbamate, 1139
method, for copper, 1145, 1411
Diethyl *p*-nitrophenyl phosphate (see E600) 763 et seq.
Difluorescence, 1243, 1245
***p,p′*-Difluoro-*m,m′*-dinitro phenyl sulphone,** 1265, 1267
Diffraction analysis, for electron micrographs, 1294
Diglucuronide, of bilirubin, 1072
N^{α}-cbz-Diglycyl-L-arginine-2-naphthylamide, 974
Dihalogen-*O*-acetylindoxyls, synthesis of, 1307
Dihydric phenols, 1113
Dihydrofolate dehydrogenase, method for, 952
Dihydrofolate reductase, method for, 1351
3,4-Dihydroisoquinolines, 1104, 1105
reduction of, by borohydride, 1110
6,7-Dihydroxyisoquinoline, 1110
Dihydrolipoate, as diaphorase substrate, 926
Dihydroxy-α-naphthoquinone, 1059
3,4-Dihydroxy-phenylalanine, as melanin precursor, 1051
oxidation of, 856, 858
3,4-Dihydroxy-phenylethylamine, 1051
Dihydro-orotate dehydrogenase, 1350
method for, 947, 948
in smears, method for, 1351
Dihydro-orotate ubiquinone reductase, 948
Dihydrostreptomycin, fluorescence and, 1178
Dihydrotachysterol, 971
Dihydrothioctyl dehydrogenase, 925

5,6-Dihydroxyindole, 856, 861, 862
as melanin precursor, 1051
synthesis of, 1336
11,15-Dihydroxy-9-oxaprosta-5,13-dienoic acid, 936
3,4-Dihydroxy-phenylserine, 1106
Dihydroxyacetone phosphate, 944
6,7-Dihydroxy-1,2,3,4-tetrahydroisoquinoline, 1104
5,6-Dihydroxy tryptamine, 1106
Diisopropyl fluorophosphate (DFP), 763, 767, 771, 791, 798
Dimedone, as trapping reagent for carbonyls, 902
Dimethylaminoazo-benzene, and hepatoma, 809
Dimethylaminobenzylidine rhodanine, 1139
method, for copper, 1145
2,6-Dimethyl benzoquinone-4-(2′,4′,6′-trimethylanil), 854
Dimethylbutyl-thioacetate, EM method for cholinesterases, 1283
Dimethyl formamide, 942
4,5-Dimethylthiazolyl tetrazoles, 892
Dimethylthioacetate, method for, ChE's, 793, 794
Dinitro-monotetrazoles, 890
Dinitrosoresorcinol, 1130, 1131
Dip-coating, 1215
Dipeptidases, 962
Dipeptide naphthylamidases, 972
Diphenol oxidases, biochemistry of, 858
Diphenylcarbazide, method, for copper, 1145
method, for mercury, 1151
***sym*-Diphenyl carbazide,** 1079
4,7-Diphenyl-1,10-phenanthroline, 1132
Dipicrylamine, 1157
α,α′-Dipyridyl, 929
Direct colouring method, for cholinesterases, 793, 1282, 1317
Direct van den Bergh reaction, 1390
Disaccharidases, 821
coupled oxidation methods for, 824, 825, 1326
Disodium phenyl phosphate, 1278
Dispersed dyes, as fibre stains, 1164
Disse, space of, 1288
Dissociation constants, in enzyme reactions, 900
Distributional error, in absorptiometry, 1228, 1229
Disulphides, EM methods for, 1263
Ditetrazoles, contamination of, by monotetrazoles, 893
half reduction and, 890
Dithiol, method for tin salts, 1153
Dithionite, 1056
Dithiooxamide, 1145
Dithiothreitol, 965, 966
Dithizone, 1139
and copper ions, 1146
as intravital stain for zinc, 1155
DMAB reaction, for EC granules, 1117
DNA, acridine orange and, 1186
Feulgen-stained, absorptiometry of, 1229
^{3}H-adenine in, 1210
sedimentation coefficient of, 1187
DNA films, preparation of, 1371
DNA/RNA hybridization, 1025
DNA synthesis, ^{14}C-adenine and, 1210
DNase I, crystalline, 1024
DOPA, as cofactor for "tyrosinase", 863
fluorescence maxima of, 1106
as peroxidase substrate, 856
Dopachrome, 1051
DOPA factor, 861
Dopamine, EM demonstration of, 1291, 1292
fluorescence maxima of, 1106
Dopamine granules, silver reactions and, 1063
DOPA-oxidase, as catechol oxidase, 860
EM method for, 1286
nature of reactions for, 861
peroxidase-dependent, in melanogenesis, 1055
reaction, applications of, 862
DOPA quinone, 859, 1064
Double helix, of DNA, 1188
Dowex-50 resin, 947
Drosophila, larvae, copper in, 1148
Dry mass, by interferometry, 1246
Dry mounting techniques, in autoradiography, 1214
DT diaphorase, 905
Dubin-Johnson pigments, 1088
Duodenum, alkaline phosphatase in, 1278
xanthine oxidase in, 874
Dupanol C, 1214
Durazol fast blue, 89, 1136
Dye conversion and elution methods, 1232
Dye-substrate binding, 1241
Dyes, fluorescent, list of, 1422

E 600, 763 et seq., 797, 798
EC cells, applied histochemistry of, 1119
granules of, 1063
EC granules, glutaraldehyde fixation of, 1117
historical considerations on, 1117
Echinochromes, 1059
EDTA, 829
as dehydrogenase inhibitor, 929
on phosphorylase activity, 829
Ektachrome, high-speed, 1203
Elastase, 962
as histochemical reagent, 1041

Elastic fibres, electron stains and, 1262, 1263
Elastin, autofluorescence of, 1177
Electrodes, cadmium, rotating disc, 892
Electron acceptors, competitive, 914
 intermediate, 915
Electron autoradiography, quantitative, 1217
Electron diffraction techniques, 1293
Electron histochemistry, 1260 et seq.
Electron microprobe analysis, 1293
Electron microscopy, analytical, 1128
Electron probe analysis, 1128
 microanalyser, 1233
Electron staining, 1260, 1261 et seq.
Electron track autoradiographs, 1210
Electron transfer, in epimerization, 998
 sites of, 911
 sites of, in plant and animal mitochondria, 914
Electron transport, diaphorases and, 904
Electron transport chain, hydrogen acceptance from, by tetrazoles, 890
Electrophoresis, disk, LNAses and, 972
Electrophorus, ChE's in, 799
Eledone moschata, fluorophores in, 1110
EM autoradiography, 1219 et seq.
EM reactions for monoamines, specificity of, 1292
Embryo, chick, lipases in, 784
 tissues, UDPG transglycosylase in, 835
Embryos, esterases in, 800
Emulsin, 969
Emulsion, silver bromide, 1218
Emulsions, photographic, in autoradiography, 1213
 photographic, Hurter-Driffield curves and, 1231
 stripping film, 1427
Endocrine cells, decarboxylation by, 1112
Endocrine polypeptide cells, 1112
Endometrium, human, steroid metabolism in, 941
Endopeptidases, 962
 silver proteinate method for, 1358
End-point reactions, fluorescent, 1176 et seq.
Entellan (Merck), 1109
Enterochromaffin, 1058
 cells, 1101, 1102
 granules, 1112 et seq.
Enterocyte, invertase in, 821
Enzyme activity, relationship of formazan production to, 888
Enzyme electron cytochemistry, 1260, 1276 et seq.
Enzyme proteins, as EM labels, 1267 et seq.
Enzymes, EM histochemical techniques for, 1276
 quantitative autoradiography of, 1218
Enzymal analysis, 1016
Eosin (Y), 1422
Eosin, as fluorochrome, 1179
Eosinophil leucocytes, horse, 855
Epididymis, rabbit, aminopeptidases in, 972
 rat, galactosidases in, 821, 823
 β-glucosidase in, 819
Epihydrinaldehydes, 1079
 phloroglucinol method for, 1387
Epi-illumination, 1176, 1195, 1196, 1197 et seq., 1200, 1202, 1235, 1238, 1239, 1243, 1245
Epithelial mucins, ^{35}S and, 1210
Eriochrome Black T, 1136
Erythritol, as glycosidase inhibitor, 818
Erythrocyte AChE, 769, 772
Erythrocytes, ceroid production from, 1084
 DOPA reaction of, 862
Erythrose-4-phosphate, 999
"Escape" phenomenon and formazan crystallization, 897
Eserine, 765, 767, 982
Esterase, heavy-metal method for, 1305
 non-specific, in lipofuscins, 1080
 non-specific methods for, 1303 et seq.
 red cell, and indigogenic methods, 781
 thiolacetic acid method for, 1319
Esterase, A, of rabbit serum, 764
Esterase, C, 763
Esterase inhibitors, species differences and, 795
Esterase-Peptidase relationships, 976
Esterases, applied histochemistry of, 800
 biochemistry, 762
 as cathepsins, 775
 distribution of, by naphthyl acetate method, 773
 indigogenic methods for, 1306
 indoxylazo methods for, 1307
 inhibitors of, 764, 795 et seq.
 isoenzymes of, 765
 naphthol AS-HPR methods for, 774, 775
 non-specific, 761 et seq.
 non-specific, distinction of types, 796
 non-specific, EM methods for, 1281 et seq.
 non-specific, methods for, 772
 substrate specificity of, 762
Esterases, A, heavy-metal method for, 1305
 A & B, 763
Esteratic site, of esterases, 767
Ester glucuronides, of bilirubin, 1071
Ester sulphates, in melanin, 1057
Estradiol-17β, 940
Estrane, 939
1,3,5(10),6,8-Estrapentene, 939
1,3,5(10) Estratiene, 939
Estrone, 940
Ethyl anthranilate, 1075
1-Ethyl-3-(3-dimethylamino propyl)-carbodiimide, 1267
Ethylene glycol, and oxalosis, 1138

Ethylene chlorohydrin, 1056
Ethylenediamine condensation, 1121
Ethyl hydrogen peroxide, synthesis of, 1333
***n*-Ethyl-*o*-hydroquinoline,** 1142
***N*-Ethyl maleimide,** 911
Euchrysine, as vital fluorochrome, 1180
Evans blue, 1422
as vital fluorochrome, 1180
Excitation filters, 1174
Exocrine pancreas, uptake of L-DOPA by, 1112
Exogenous substances, localization of, 1160 et seq.
Exopeptidases, 962
Exposure, in autoradiography, 1216
Extraction techniques, and EM demonstration of nucleic acids, 1290
Exudates, purulent nucleases in, 1019

FAD, 873, 925
Fading, of fluorescence, 1175
and quantitation errors in microfluorimetry and photometry, 1250
Fading characteristics, fluorescence, 1107
Fading errors, minimization of, in microfluorimetry, 1252
Fast blue B salt, 823, 1303
Fast blue BBN salt, 1283
Fast Garnet GBC, 824
Fat peroxides, Dam method for, 1385
diaminofluorene method for, 1386
Fats, unsaturated, autoxidation of, 888
Fatty acid, short chained, esters of, 761
Fatty peroxides, in leucocytes, 855
^{59}Fe, 1211
Ferric hydroxyquinoline, method for β-glucuronidase, 813
method, specificity of, 814, 815
Ferric iron, methods for, 1402 et seq.
Ferric salts, as electron stains, 1262
Ferricyanide, as electron acceptor, 906, 925
Ferritin, 1132, 1133
antibodies to, 1132
methods for, 1403
tissue content of, 1132
Ferritin-labelling technique, 1265, 1266
Ferrocene, protein coupling and, 1269
Ferrocyanide, and Perls' reaction, 1130
Ferro-ferricyanide, as oxidation catalyst for indoxyls, 780
Feulgen-stained material, microdensitometry of, 1230
Feulgen-stained nuclei, dye conversion methods and, 1232
fading and, 1250
Fibres, textile, demonstration of, 1163
Fibroblast cultures, in metachromatic leucodystrophy, 990
Fibrinolysin, method for, 1370
and substrate film method, 1011
Ficin, 962
FIF, 1101 et seq.
chemistry of, 1104
in endocrine cells, 1112
fading and, 1251
in melanomas, 1052
methods, 1394
FIF maxima, comparison with THI maxima, 1122
FIF-ozone method, for tryptamines, 1399
Film ratings (ASA, DIN), 1200
Films, choice of, for fluorescence microscopy, 1203
Filters, barrier, 1199
for fluorescence microscopy, 1198, 1199
interference, 1198
Fixation, for autoradiography, 1212
for cholinesterases, 792
minimal, for dehydrogenases, 923
optimum, for inorganic iron, 1130
Fixatives and RNase, 1022
significance of in enzymal analysis, 1017
Fixed tissues, diaphorases in, 924
Flavobacterium, chondroitinase, 1032
Flavophosphine N, 1422
Flavoprotein, diaphorase, 904, 905
diaphorases, 924 et seq.
Fluorescein, and immunofluorescence fading, 1251
standard solution of, 1436
Fluorescence, in adrenal medulla, 1062
and electron microscopy, 1193
fading, 1175, 1176
formaldehyde-induced, 1101 et seq.
induced, 1122, 1176, 1179
of melanomas, 1052
metachromasia, 1185
Fluorescence microscope, optical diagram of, 1174
Fluorescence microscopy, 1171 et seq.
applications of, 1176
illumination systems for, 1195 et seq.
instrumentation for, 1194 et seq.
text books of, 1171
Fluorescent antibody technique, 1171
Fluorescent end-point reactions, 1176
Fluoride, inhibition of phosphoglucomutase, 828
as peroxidase inhibitor, 853
as phosphorylase phosphatase inhibitor, 829
reversible inactivation of ChE by, 799
Fluorides, as ChE inhibitors, 771
Fluorimetry, 1226, 1227
Fluorochromes, basic dye, 1180 et seq.
excitation and emission maxima of, 1423
and fading, 1175

Fluorochromy, direct, 1176, 1179
indirect, 1176, 1179
vital, 1180
Fluorophenylhydrazine method, 1442
***p*-Fluorophenylhydrazine hydrochloride,** 1275
Fluorophore, definition of, 1173
Fluorophore-fluorescence relationships, 1235
Fluorophores, augmentation of, 1110
methods of distinction for, 1109, 1110
production of, by aldehydes, 1107
solubility of, 1109
Flying-spot scanning, in quantitative autoradiography, 1217
FMN, 925
FNPS coupling method, 1438
Foreign substances, 1160 et seq.
fluorescence of, 1178, 1179
Formaldehyde, alternatives to, for induced fluorescence, 1107
gas, 1107
reaction of with tryptamine, 1120
Formaldehyde-HCl vapour, for tryptamines, 1120
Formaldehyde-induced fluorescence, 1101 et seq.
technique, 1394
Formalin pigment, acid, 1075
Formazan, NT, reflex colours of, 890
Formazan, production from tetrazoles by light, 888
Formazans, absorption spectra of, 891
crystal size of, 894
light sensitivity of, 896
lipid solubility and, 896
metal chelated, 894
metal chelation and, 896, 897
molecular extinctions of, 892
physical and chemical characteristics of, 890
red and blue, from ditetrazoles, 893
"single pattern" theory and, 922
substantivity for protein and, 896
Form birefringence, 1244
Formic acid, fluorophore production, 1107
***N*-Formyl-L-kynurenine,** 975
Fraunhofer diffraction principle, 1294
Free radical properties, of melanins, 1054
Free radical stage, in tetrazole reduction, 894
Freeze-dried duodenum, rat, ninhydrin reaction of, 1116
Freeze-dried sections, carotenoids in, 1087
dry-mounting ARG technique and, 1432
and potassium ion localization, 1157
Freeze-dried tissues, acetone-dithizone reagent and, 1155
Bischler-Napieralski reaction and, 1107
demonstration of histamine in, 1122
dry-mounting ARG method for, 1431
Freeze-dried tissues, exogenous drugs in, 1163
formaldehyde vapour and, 1102
galactose-^{3}H in, 1214
guinea-pig duodenum, autofluorescence of, 1104
inactivation of SD in, 1001
Lowry techniques and, 1226
naphthylamidases in, 967
ninhydrin reaction and, 1103
porphyrins in, 1067
SD in, 950
silicone plastic embedded, 1214
sodium ions in, 1157
water-soluble isotopes and, 1213
Freeze-drying, for FIF, 1107
FRP, quantitation of, 1226 et seq.
β-Fructofuranosidase, 821
Fuchsin, acid, 1422
basic, 1422
Fuchsin-sulphurous acid, 1182, 1183
Fucose, 819
Fucosidase, 818
method for, 824
β-Fucosidase, 820

Galactitol, 930
β-Galactosidase, methods for, 822
post-coupling method for, 1323
Galactosidases, 808 et seq., 819 et seq.
β-Galactosidases, intestinal, 820
two varieties of, 820
Gallic acid-metaldehyde method, for calcium, 1134
Gallium, radioactive, in bone, 1207
Gallocyanin-chromalum, in autoradiography, 1217
Ganglia, sympathetic, cholinesterases in, 772, 790
Gastric mucosa, enterochromaffin-like cells in, 1111
lipases in, 785
rat, histamine in, 1123
Gastric tumours, LNAses in, 971
tetracycline uptake and, 1178
Gastrin, 1190
Gastrocnemius, rat, LDH isoenzymes in, 935
Gastrointestinal tract, APUD cells in, 1113, 1119
endocrine polypeptide cells in, 1111
ferritin in, 1132
radiosulphur uptake by, 1210
Gastropods, copper in, 1144
Gaucher's disease, 820
GBHA method, for calcium, 1137
G cells, masked metachromasia of, 1189
Gelatin, and masked metachromasia, 1191

Gelfoam, 1161
Gel medium method, for soluble dehydrogenases, 1346
Geranine B, 1422
Germanin, 1161
Gevaert Scientia NUC, 307, 1218
G.H., ultrastructural localization of, 1268
Gibb's method, 1398
Gibb's reaction, for EC granules, 1114
Giemsa method, for chromaffin tissues, 1060
Glass, temperature-coloured, for filters, 1199
γ-Globulin, rabbit, mercury-labelling of, 1271
Globus pallidus, ferritin in, 1133
Glomeruli, dextran accumulation in, 1161
α-Glucan-branching glycosyl transferase, method for, 831, 832
Glucan phosphorylase, biochemistry, 827
α-Glucan phosphorylase, 808 et seq., 826
methods for, 828 et seq., 1327
α-Glucan-UDP glucosyltransferase, method for, 832
Gluco-cerebroside, 820
Glucokinase, method for, 986, 1361
Glucosaminidase, lysosomal, 821
β-Glucosaminidase, naphthol AS-BI methods for, 1325
Glucose oxidase, in disaccharidase methods, 824
Glucose phosphate isomerase, methods for, 999, 1367
Glucose-6-phosphatase, acid, 1281
EM method for, 1280
Glucose-6-phosphate, 832
Glucose-6-phosphate dehydrogenase, 923, 924
method for, 937
standard method for, 1344
Glucose phosphate isomerase, methods for, 999, 1367
Glucose phosphomutase, method for, 987
α-Glucosidase, methods for, 822
post-coupling method for, 1324
β-Glucosidase, indigogenic method for, 822
methods for, 821, 822
Glucosidases, biochemistry of, 818
α and β Glucosidases, 808 et seq.
β-Glucosidases, as histochemical reagents, 1037
Glucosulphatase, 987
Glucosyl acceptors, in phosphorylase reaction, 830
Glucuronate dehydrogenase, method for, 1345
Glucuronic acid cycle, 931
β-Glucuronidase, 808 *et seq*
dual localization of, 812
indigogenic method for, 1323
lysosomal, 824
β-Glucuronidase, methods for, 812 et seq., 1321 et seq.
post-coupling azo dye methods for, 816
simultaneous coupling azo dye methods for, 815
Glucuronidases, activators and inhibitors of, 810 et seq.
biochemistry of, 808 et seq.
intracellular localization of, 812
β-Glucuronidases, EM methods for, 1284
in regressing tissues, 812
Glucuronide synthesis, cycle, 933
Glucuronyl transferase, 1071
Glutamate decarboxylase, method for, 994, 1364
Glutamate dehydrogenases, 923, 950
methods for, 950
standard method for, 1342
Glutamic acid, 1112
enzyme reactions of, 992, 993
Glutamic-oxaloacetic transaminase, 983
Glutaminase, methods for, 993
γ-Glutamyl cross-links, 1191
γ-L-Glutamyl-L-cysteinyl glycine, 973
γ-Glutamyl-4-methoxy-2-naphthylamide, 973
***N*-(γ-L-Glutamyl)-4-methoxy-2-naphthylamide,** 1289
γ-L-Glutamyl naphthylamides, 973
α-L-Glutamyl peptidase, method for, 973
γ-Glutamyl transpeptidase, 973
method I, 1355
method II, 1355
method III, 1356
Glutaraldehyde, antibody conjugation and, 1267
as fixative for dehydrogenases, 923, 924
and immuno-EM, 1265
and noradrenalin, 1062
Glutaraldehyde-dichromate, 1063
Glutaraldehyde-silver reaction, 1292
Glutathione, 881
Glutathione reductase, method for, 953
Glyceraldehyde phosphate dehydrogenase, 947, 1350
α-Glycerophosphate dehydrogenase, standard method for, 1342
α-Glycerophosphate dehydrogenase (mitochondrial), 932, 924, 944, 945
method for, 944
α-Glycerophosphate dehydrogenase (NAD-linked), inhibitors and activators, 929
methods for, 929
Glyceryl trinitrate, as MAO inhibitor, 864
Glycine, ^{14}C-labelled, 1209
Glycogen, branched and unbranched, 829
digestion of, by amylases, 1034
intrinsic, in phosphorylase reactions, 831, 832
lead staining of, for EM, 1275

Glycogen, localization of, by PA-TSC-SP method, 1274
newly formed and native, 828
periodate-oxidized, diastase digestion of, 1275
as phosphorylase reaction primer, 830
Glycogen cycle, 827
Glycogen synthetase, 826
Glycolipids, in Gaucher's disease, 820
Glycol methacrylate, 1290
Glycosidases, competitive inhibition of, 818
β-Glycosidases, indigogenic method for, 1324
Glycoside hydrolase techniques, applications of, 833, 834
Glycoside hydrolases, 808 et seq.
Glycoside transferase techniques, application of, 833
Glycosyltransferases, 808 et seq., 826 et seq.
Glycylglycine, 973
Glycyl-DL-prolyl-β-naphthylamide, 972
Glyoxal-bis-(2-hydroxyanil), 1137
Gmelin reaction, 1073, 1388
modified, 1389
G-Nadi oxidase, 844, 845, 846
G-Nadi reaction, 1330
Goblet cells, radiosulphur uptake by, 1210
Gold, methods for, 1148, 1412
Goldblatt clamp, 937
Gold-Thiocholine, EM method for cholinesterases, 1283
method, 793, 1316
Gold-Thiolacetic acid method, 1319
EM method for cholinesterases, 1283
Golgi region, and melanogenesis, 1055
metal accumulation in, 1149
Gonane, 939
G-6-P dehydrogenase, 922
solubility of, 922
Grain counting, automatic, 1217
in autoradiography, 1217
Grain distribution curves, in EM autoradiography, 1219
L-Gulonate dehydrogenase, method for, 931
L-Gulonolactone oxidase, method for, 1346
L-Gulono-γ-lactone oxidase, method for, 932
Gum sucrose medium, 924
Gypsum crystals, and ionized calcium, 1133

^{3}H, as radioactive label, 1210 et seq.
Haematins, 1075, 1076
Haematoidin, 1065, 1068, 1069, 1070 et seq.
crystalline, and Gmelin's reaction, 1073
methods for, 1388 et seq.
Haematoxylin, iron lake, 1131
Haem compounds, peroxidase activity of, 850
Haemochromatosis, 1070
ferritin and, 1133
Haemofuscin, 1082
Haemoglobin, 1065
benzidine method for, 1387
content, of erythrocytes, 1231
demonstration of, by peroxidase activity, 851
Haemolysis, iron and, 1132
Haemoprotein oxidases, 841 et seq.
Haemoproteins, fluorogenic method for, 1387
Haemosiderin, negative reaction of, with *o*-tolidine, 1146
solubilities of, 1069
staining by haematoxylin, 1131
Haemosiderin granule, carrier substance of, 1069
electron microscopy of, 1069
Haemosiderin-iron, haematoxylin-lake method for, 1403
Haemosiderins, 1065, 1068 et seq.
differential diagnosis of, 1070
Half Nitro-BT, 883
substantivity of, 887
Half reduction, Nitro-BT and, 893
Half-TNBT, 883
Half wave potential, of β waves, 883
Half wave potentials, of tetrazolium salts, significance of, 886
Halogen substituted indoxyl esters, as substrates for esterases, 780
Harderian glands, ChE in, 772
porphyrins in, 1067
Harding-Passey melanoma, 859
Harmaline, 1103
Hatchett's brown, 1277, 1283
as product of ChE reactions, 793
Heart, conducting system of, and ChE's, 791
conducting system of, phosphorylase in, 834
Heart muscle, diaphorase, 904
fluorescent pteridines and, 1178
lipoic acid dehydrogenase in, 917
potassium ions in, 1157
succinic oxidase system and, 842
HEBES buffer, 982
α-Helix conformation, of proteins, 1191
Helix aspersa, β-galactosidase in, 820
Helix pomatia, albumen gland of, 819
glucosidases, 1037
Heme-copper-protein, 842
Heme proteins, oxidized DAB method for, 1337
Hemes, conversion of, to porphyrins, 1066
oxidized amine method for, 1388
Hepatitis, bile pigments in, 1071
Hepatocytes, CAH in, 1288

Hepato-lenticular degeneration, 1147
Hepatoma, rat, glucuronidases and, 809
Hepatopancreas, molluscan, copper in, 1146
Hexacyanoferroate, 1130
Hexamine-silver method, 1113, 1379
Hexazotized new Fuchsin (HNF), 783
Hexazotized pararosanilin (HPR), 783, 824, 1281, 1303, 1311, 1322
Hexokinase, method for, 986, 1361
Hexose monophosphate shunt, 949
Hippocampus, rat, zinc in, 1156
Histaminase, 870 et seq.
Histamine, methods for, 1122
OPT method for, 1400
Histamine fluorophore, fluorescence maxima of, 1123
Histidine, activation of esterases, 762
EM methods for, 1263
Histiocytes, LNAses in, 971
Histiocytoma, esterase reaction of histiocytes in, 776
Histochemistry, electron, 1260 et seq.
quantitative, 1225 et seq.
Histone, 843
Histones, detection of, 1023
Histo-photochemical techniques, 1066
Histospectrography, emission, 1128
Homocysteine thiolactone (HTL), 1271
Homogentisic acid oxidase, 1064
Horseradish, peroxidase, 851
HPR, 783, 824
in esterase methods, 1303
in β-glucosaminidase method, 1325
in glucuronidase techniques, 816
and β-glucuronidase method, 1322
modification of esterase methods, 774, 775
in phospholipase method, 1311
Hq-MTT, reaction mechanisms, 1002, 1003
5-HT, biological importance of, 1118, 1119
EM demonstration of, 1291, 1292
fluorescence maximum of, 1104, 1106
formaldehyde condensation and, 1103
5-HT-FIF, fading of, 1252
^{3}H-5-HTP, 1111
5-HTP, 1106, 1111, 1118
Hueck method, for lipofuscin/melanin, 1383
Humidity, and paraformaldehyde, 1107
Humphrey method, for inorganic iron, 1131
Hurter-Driffield curves, 1231
Hyaluronic acid and PAS reaction, 1031
preparation of substrate film, 1373
structure of, 1026
Hyaluronidase, distinction from acetyl-β-glucosaminidase, 820
extractions, 1376
sources of, 1026
substrate film method for, 1013
two types of activity of, 1028
Hyaluronidase-labile substances, 1031, 1032
Hyaluronidases, action of, on various substrates, 1027
biochemistry of, 1026 et seq.
choice of, for histochemical use, 1030
history of, 1025, 1026, 1029 et seq.
method for, 1373
pH optima, 1028
Hyamine, 970, 1355
Hydrazine, as cytochrome oxidase inhibitor, 843
Hydrobromic acid, reaction of, with copper salts, 1144
Hydrocarbons, tetracyclic, 939
Hydrochloric acid, conversion of isoquinolines and, 1110
Hydrogen, pick-up, by tetrazoles, 890
Hydrogen peroxide, concentration of, in peroxidase reactions, 853
Hydrolases, C-N, 992
Hydroperoxidase, method for, 1333
Hydroquinone, 844
Hydroquinone-MTT reaction, 1001 et seq., 1327
Hydrosulphite, effect of on ferritin, 1132
Hydroxyamphetamine, 1106
3-Hydroxybutyrate dehydrogenase, method for, 935
β-Hydroxybutyrate dehydrogenase, standard method for, 1342
β-Hydroxybutyrate dehydrogenase, 923, 935
5-Hydroxydopamine, 1110
Hydroxy fatty acids, diphenyl carbazide method for, 1386
5-Hydroxyindole, 1118
4-Hydroxyindole acetaldehyde, 864
5-Hydroxyindole acetic acid, 864
Hydroxylamine, 871
as cytochrome oxidase inhibitor, 843
as trapping agent, for retinene, 929
1-Hydroxy-2-naphthoic acid, 848
2-Hydroxy-3-naphthoic acid hydrazide, 858, 866
8-Hydroxy-1,4-naphthoquinone, 848
1-*o*-Hydroxyphenylazo-2-naphthol, 813
3-Hydroxyphenylethylamines, 1120
Hydroxyproline-2-epimerase, method for, 998, 1366
Hydroxypropyl methacrylate, 1290
15-Hydroxyprostanoate dehydrogenase, method for, 935
Hydroxyquinoline, as MAO inhibitor, 864
8-Hydroxyquinoline, 813, 814, 929
azo-dye, methods for β-glucuronidase, 1321
method, for iron, 1130, 1131
method, for zinc, 1156
8-Hydroxyquinoline glucuronide, 816

8-Hydroxyquinoline sulphate, method for sulphatases, 991
3α-Hydroxysteroid dehydrogenase, 940
method for, 942, 1348, 1349
3-β-Hydroxysteroid dehydrogenase, method for, 1348
20α-Hydroxysteroid dehydrogenase, method for, 942
Hydroxythionaphthene, 1116
3-Hydroxythionaphthenes, 856
5-Hydroxytryptamine, 1102
fluorescence maxima, 1106
6-Hydroxytryptamine, 1106
5-Hydroxytryptophan, 1118
fluorescence maxima, 1106
Hypophysis, esterases of, 801
Hypothalamus, MAO in, 869
Hypoxanthine, 833

Iditol, 930
IDP, non-enzymic dephosphorylation of, 1280
IgG, mercury-labelling and, 1271
Ilford G5 and K5, emulsions, 1215
Ilford L4, nuclear track emulsion, 1218
Illumination, Köhler, 1196
Imidazole, as active centre of A-esterase, 764
Imidazole alkylamines, 1122
Imidazole buffer, 987
Immunodiazo-thioether-OsO_4, 1269
Immuno-electron microscopy, 1260, 1264 et seq.
Immunoglobulin, labelled, 1266
Immunoglobulins, EM localization of, 1268
Immunofluorescence, 1192
fixation for, 1264, 1265
identification of GAP dehydrogenase, 947
quantitation of, 1240, 1241
quantitative, 1226
Immunospecific protection, 1270
Indigogenic method, for galactosidase, 823
for β-glucosidase, 822
for glucuronidase, 817, 818, 1323
for β-glucuronidase, 1323
for β-glycosidases, 1324
Indigogenic methods, for esterases, 777 et seq., 1306
principles of, 777, 778
Indigoids, in urine, 1064
Indirect van den Bergh reaction, 1390
Indoanilines, osmiophilia of, 849
Indogenic method, for aminopeptidase, 969
Indole-5,6-quinone, 861, 1051
Indole reaction, pigments and, 1064
Indole reactions, of EC granules, 1117
Indole reagents, 856
Indolylethylamines, 1120
Indolylethylamine-FIF, fading of, 1251
Indolyl-3-pyruvic acid, 966
Indonaphthol purple, 847
Indophenol blue, 843, 845
Indophenol method, for lipofuscins, 1082
Indophenol oxidase, 843
Indophenol reaction, of EC granules, 1114
Indoxyl acetate-BADX, EM method for esterases, 1284, 1448
Indoxylazo methods, applications of, 783
for esterases, 1307
principles of, 782 et seq.
Indoxyl butyrate, 780
Indoxyl esterase, organophosphorus-resistant, and cathepsin C, 976
Indoxyl glucuronides, 817
Indoxyl-HPR method, 1284
Indoxyl propionate, 780
Indoxyl substrates, synthesis of, 780, 1307
Indoxyls, formation of indigo from, 778
Influenza virus, neuraminidase, 1033
Infra-red, photomultipliers for, 1232
Inhibitor method, labelled, for quantitative enzymology, 1218
methods for cholinesterases, 1314, 1315
Inhibitor safety factor, 768
Inhibitors, cholinesterase, concentration table of, 1316
in enzymal analysis, 1017
organophosphorus, 767
Inhibitor Ratios, for ChE's, 768
INT, 869, 873, 881, 892, 893, 894, 913, 914 916
two formazans from, 893
Inorganic anions, histochemistry of, 1158
Inorganic cations, 1129 et seq.
Inorganic constituents, 1128 et seq.
list of methods for, 1160
Inorganic ions, EM techniques for, 1290 et seq.
Inorganic iron, methods for, 1130 et seq.
Inorganic phosphates, molybdate methods for, 1416
Inosine, as substrate for purine nucleotide phosphorylase, 833
Inositol, 930
Instrumentation, for microfluorimetry, 1234
Insulin, as phosphorylase activator, 828
zinc in, 1153
Integrating microdensitometry, 1229
Interference filters, 1198, 1199
for distinguishing fluorophores, 1109
Interference microscopy, 1245, 1246
Interferometry, 1226, 1227
Intestine, brush border, disaccharidases in, 825
β-galactosidases in, 820
α-glucosidase in, 822
invertases in, 821
MAO in, 870

Intrinsic birefringence, 1244
Iodides, methods for, 1159
Iodine, as glycogen stain, 831
Iodoacetate, 911
inhibition of SH groups, in esterases, 764
Iodo-adrenochrome, 1061
Iproniazid, 864
Iron, inorganic, methods for, 1130 et seq.
removal of, methods for, 1403
storage of, as ferritin, 1132
Iron-anthraquinone lakes, 1135
Iron sulphide, 1130
Isatins, from arylamines, by Sandmeyer synthesis, 781
dihalogen substituted, 781
intermediate synthesis of, 1308
and pink colour in indoxyl reactions, 782
Isocarboxazide, 864
Isocitrate dehydrogenase, standard method for, 1344
Isocitrate dehydrogenase (NAD-linked), 937
Isocitrate dehydrogenase (NADP-linked), 937
Isocitrate dehydrogenases, 923, 937, 949
Isoenzymes, of esterases, 765
Isomerases, 998, 999
Isopropanol, as substrate for secondary alcohol dehydrogenase, 941
Isoquinolines, synthesis, from phenylethylamines, 1107
Isotopes, list, 1426
protein-bound, 1212
water-soluble, 1213

Jaundice, acholuric, 1071
haemolytic, 1071
obstructive, 1071
Jejunal biopsies, LDH isoenzymes in, 934
Jejunal glycosidases, inhibition of, 819
Jejunal mucosa, ceroid in, 1085
Jejunum, disaccharidases in, 821
fucosidase in, 824

^{42}K, 1211
Kairine method for arsenates, 1142
Kayser-Fleischer rings, 1147
Keilin-Hartree, heart muscle preparation, 846
Keratin, electron microscopy of, 1263
Keratinase, on wool fibres, 1041
Keratins, affinity of for nickel, 1144
Keratohyalin, phenoloxidase reaction of, 862
Kernechtrot, 1136
Kernicterus, 1073
α-Ketoglutarate, 949
Kidney, esterases in, 800
mouse, localization of A-esterases in, 776
Kinetic studies in microfluorimetry, 1242
Kinnier-Wilson's disease, 1147
Kiton red-almond green, 1066
Km Values, in dehydrogenase histochemistry, 901
Kodachrome X, 1203
Kodak AR 10, stripping film, 1214
Kodak NTB emulsions, 1215
Kodak NTE, 1218
Köhler, fluorescence and, 1171
Komaya-Christeller method, 1143
Krebs' cycle, 949
Kultschitsky cells, 1101
Kupfer cells, peroxidases in, 857
Kurloff bodies, 1021
Kynurenine formamidase, 975

"Labelled-Unit" counting, in autoradiography, 1217
Lactate dehydrogenase, EM technique for, 1450
osmium bridging technique for, 1288
standard method for, 1344
Lactate dehydrogenase (NAD-linked), 923
L-Lactate dehydrogenase, methods for, 933 et seq.
Lakes, 1129
calcium, 1134
Lamps, cadmium arc, 1195
high pressure mercury arc, 1195
iodine-quartz, 1195
tungsten filament, 1235
tungsten-halogen, 1195
xenon arc, 1195, 1235
Langerhans cell, lipase in, 787
LAP, characteristics of, 964, 965
identity of histochemically demonstrable enzyme, 969, 970
pH optimum of the reaction, 968
Lauroyl choline, as ChE substrate, 789
Lauryl sulphate, 1214
LDH, multiple molecular forms of, 934
LDH isoenzymes, method for, 1347
Lead, methods for, 1149, 1150
non-specific binding of, 794
Lead citrate, effect of, on enzyme reaction products, 1277
as EM glycogen stain, 1275
Lead ions, and EM contrast, 1262
hydrolysis of ATP by, 1280
Lead-reactive material, in synapses, 793
Lead salts, chromate method for, 1413
coloured chelate reactions for, 1149, 1150
rhodizonate method for, 1414
Lecithin, hydrolysis of, by phospholipase B, 788

Lecithinase, as histochemical reagent, 1044
Leech extract, 1026
Leech hyaluronidase, 1028
Leitz K460 and K470 filters, 1200
Leptodactyline, method for, 1122
Leucine aminopeptidase, biochemistry of, 962
indigogenic methods for, 969
Leucine naphthylamidase, 968
method, 1353
L-Leucine β-naphthylamide, 963, 967
Leucoadrenochrome, 1061
Leuco-Cyanol, 1065
method for peroxidase, 856
Leuco-dye methods, for peroxidases, 855, 856
Leucocyte arylsulphatase, in metachromatic leucodystrophy, 989
Leucocyte granules, peroxidases in, 851
Leucocytes, lysozyme in, 1013
zinc-protein complexes in, 1154
Leuco-methylene blue, 881
Leuco-Patent Blue, 1065
for peroxidases, 856
preparation of, 1335
Leucylglycylglycine, activation of lipases, 762
L-Leucyl-4-methoxy-β-naphthylamide, 968
Leukaemic cells, DHO dehydrogenases in, 948
Lewisite, arsenic-labelled localization of, 1207
L.H., ultrastructural localization of, 1268
Lieberkühn, crypts of, 1210
Ligands, bidentate, 1269
Ligases, EM methods for, 1289
Light pen, use of in quantitative histochemistry, 1254
Light sources, for fluorescence microscopy, 1195
for microspectrofluorimetry, 1235
Limnaea, magnesium in mantle of, 1150
Linear aromatic substituents, tetrazoles and, 887
Linearity, of tetrazole reduction, 888
Line sampling, 1248
Linkages, covalent, 1129
Lipase, 1303
acid, 766
methods for, 1309, 1310
mitochondrial localization of, 785
naphthol AS nonanoate method for, 1310
pancreatic, method for, 786
Tween method for, 1309
Lipase-esterase, modified Tween method for, 1309
Lipases, activators of, 787, 796
biochemistry, 762
distinction from esterases, 761 762
Lipases, effects of fixation on, 784
EM method for, 1284
as histochemical reagents, 1044
inhibitors of, 796, 797
methods for, 783 et seq.
Lipid droplets, formazan production in, 887
Lipid peroxides, 1078
Lipid pigments, 1057, 1076 et seq.
Lipid rims, acid fastness and, 1084
Lipids, EM methods for, 1276
extraction of, by EM preparative techniques, 1276
staining of, by indophenol blue, 845
Lipoamide dehydrogenase, 905, 926
Lipoate dehydrogenase, 925
Lipochrome, cardiac, 1077
Lipochromes, 1086, 1087
"Lipochrome jaune", 1058
Lipofuscin, autofluorescence of, 1177
Schmorl method for, 1383
Lipofuscinosis, intestinal, 1086
Lipofuscins, 1056, 1076 et seq.
chrome haematoxylin method for, 1384
distinction from melanins, 1082
enzymes in, 1080
esterases in, 801
histogenesis of, 1078
indophenol method for, 1385
long Z-N method for, 1385
significance of tests for, 1081
α-Lipoic acid, 905
"Lipomelanin", 1088
Lipoprotein envelope, of red cells, 1066
Lissamine rhodamine B200, 1422
Lithium carbonate, for removal of urates, 1138
Liver, mouse, glucuronidase in, 809
ox, arylsulphatases in, 988
ox, β-galactosidase in, 820
rat, arylsulphatases in, sex differences, 988
UDPG dehydrogenase in, 933
Liver cell pigment, in *Rana,* 1080
Livers, cirrhotic, pigment in, 1083
foetal, copper in, 1144
LNA, 963, 1353
LNAses, histochemistry of, 970 et seq.
lysosomal, 970
Locus coeruleus, MAO in, 869
Lowry techniques, for microanalysis, 1225, 1226
Lucilia cuprina, copper in, 1145
Luminescence, 1239
Lung, prostaglandin dehydrogenase in, 935
rat, aluminium oxides in, 1140
Lupus erythematosus, 1032
Lycopene, 1087
Lymphatic tissues, β-glucuronidase and, 809
Lysine, activation of esterases, 762

Lysolecithin, action of, on chromaffin granules, 1063
preparation of, 1311
Lysosomal esterase, 780
Lysosomal membranes, and sulphatase reactions, 991
Lysosomal sulphatase, method for, 1363
Lysosomal sulphatases, 988
Lysosomes, and β-glucuronidase, 812
hydrocarbon carcinogens in, 1178
Lysozyme, 843
substrate film method for, 1013, 1374

Macrophages, pigment in, 1085
Macula densa, of rat kidney, G-6-PD in, 937
Magnesium, methods for, 1150, 1414
Magnesium ions, and mitochondrial stability, 903
Magneson, 1139
method, for magnesium, 1150
Malarial pigment, 1075
Malate dehydrogenase, 923, 936, 949
standard method for, 1344
L-Malate dehydrogenase (NAD-linked), method for, 936
Maleylacetoacetic acid, 1064
Malic enzyme, method for, 936
Mallory's method, for inorganic iron, 1131
Malonate, 914
Maltase, intestinal, 818
Maltases, 821
Malt diastase, 1033
Mammary gland, involution of, glucuronidases and, 812
MAO, applied histochemistry of, 869
substrate specificity and, 864
Marsalid, 864, 867, 868
Masked iron, in haemoglobin, 1130
Masked metachromasia, 1119, 1189 et seq.
method for, 1421
Masson-Fontana technique, 1113
Masson-Hamperl technique, 1113
Mast cell, amidase, 974
histamine, Reineke salt and, 1122
Mast cells, affinity of, for diazonium salts, 776
esterases in, 776
hydroxylase and, 1119
peroxidase in, 862
trypsin-like enzymes in, 974
May-Gruenwald Giemsa, in autoradiography, 1216
"M-band" enzyme, 1282
Media, mounting, non-fluorescent, 1194
Medulla, brain, MAO in, 870
Megacolon, congenital, cholinesterases in gut, 800
Megakaryocytes, glycosidases in, 834
Melanoblasts, 860
Melanin, association with protein of, 1051
bleaching methods for, 1381
from 5,6-dihydroxyindole, 856, 861
from DOPA, 858
ferrous iron technique for, 1380
formation of, in lysosomes, 1080
granule, structure of, 1054, 1055
Masson-Fontana method for, 1379
retinal, 1056, 1057
synthesis of, by mast cells, 862
Melanins, basophilia of, 1057
biochemistry of, 1050 et seq.
bleaching of, 1056
ferrous iron uptake and, 1058
free radical properties of, 1054
histochemistry of, 1055
PAS reaction and, 1058
pial, 1057
and silver reduction, 1056
solubility of, 1055, 1056
Melanocyte, EM studies on, 1055
Melanocytes, FIF excitation and emission spectra of, 1052
peroxidase in, 862
Melanoma, FIF fluorescence spectra of, 1052, 1053
Harding-Passey, 1286
human, phenol oxidases of, 862
Melanomas, diagnostic pathology of, 1052
malignant, masked metachromasia and, 1192
Melanophages, PAS reaction in, 1058
Melanosomes, 1055
Melibiose, 819
Membrane, plasma, fluorochromy of, 1180
Membrane technique, for soluble enzymes, 934
Membranous cisternae, G-6-Pase in, 1280
Menadione, 905, 906, 917
effect of on α-GPD, 945
in G-6-PD reaction, 938
Mepacrine, 1161, 1163
2-Mercaptoethanol, 965
2-Mercaptoethanol-amine, 965, 966
Mercuric chloride, as transglycosylase inhibitor, 832
Mercuric ions, as peptidase activators, 975
Mercurochrome 220, 1193, 1422
Mercury, methods for, 1150
protein-labelling and, 1270
reduction methods for, 1150
sublimation problems and, in EM, 1272
Mercury electrode, in polarography, 883
Mercury ions, effect of, on C-esterases, 764
Mercury orange, 1263, 1272
Merthiolate, as lipase inhibitor, 785
Mesidine, 854

Metachromasia, with acridine orange, 1188
of basic dyes, 1181
fluorescence, 1185, 1241
masked, 1182
Metachromatic leucodystrophy, arylsulphatases in, 989
***Meta*-diphenols,** and diazo reaction, 1118
Metal chelation, for antibody coupling, 1269
of formazans, mechanism of, 897
of formazans, rate of reaction, 898
method for A-esterases, 776
in naphthylamidase methods, 968
Metal complexes, of GBHA, 1137
Metal-labelling, for antibodies, 1268 et seq.
Metallic ions, autoradiography of, 1210, 1211
Metalloprotein labelling, for antibodies, 1265
Metalloproteins, zinc, 1154
Metal peptidases, biochemistry of, 962
Metal precipitation methods, for arylsulphatases, 992
Metals, demonstration of, by silver-sulphide method, 1149, 1150
heavy, sulphide silver method for, 1412
reagents for, 1139 et seq.
Metal shadowing, 1260
Metaraminol, 1106
Methanamine silver, as EM technique, 1274
and SS groups of keratin, 1263
Methotrexate, 952
3-Methoxyphenylethylamines, 1120
5-Methoxytryptamine, 1106
3-Methoxytyramine, 1106
2-Methylamino-4-amino-6,7-diphenylpteridine, 1178
Methylation, to block acidic groups, 1182
Methylation-PAS-SO_2-Dye method, 1421
Methylcholanthrene, 1178
α-Methyldopa, 1106
α-Methyldopamine, 1106
Methylene blue, as redox acceptor, 915
staining of bilirubin, 1074
Methyl green, 1422
Methyl green-pyronin method, and RNases 1020
α-Methyl-5-hydroxytryptamine, 1106
1-Methyl-2-mercaptoimidazole, 1159
Methylmercuric chloride, 1272
Methyl mercury mercaptide, 1269, 1270, 1272
***S*-Methyl-β-*N*-[4-methylthiazolyl-(2)] isothiosemicarbazide,** 849
α-Methyl noradrenaline, 1106
***N*-Methyltryptamine,** 1106
Methyl umbelliferone, 1192
α-Methyne groups, in bilirubins, 1071
Micelles, iron hydroxide, 1133
Michaelis constant, dehydrogenases and, 899, 900
Microbodies, 1337
catalase inhibitors and, 1286
DAB and, 1066
uricase in, 872
Microdensitometry, integrating, 1229
Microdissection, and microanalysis, 1225
Microdroplets, fluorochrome, as standards for microfluorimetry, 1240
Microfluorimeter, Chance-Legallais, 1242
Microfluorimetry, 1233 et seq.
quantitative, 1241
standards for, 1239, 1240
Micro-incineration, 1128
and alizarin red, for calcium, 1135
for calcium, 1134
and electron microscopy, 1292
oxalate crystals and, 1138
and Perls' reaction, 1131
Microperoxidase, as EM Label, 1268
preparation of, 1440
Microradiography, 1233
Microscopy, interference, 1245, 1246
phase-contrast, 1245
polarization, 1243 et seq.
Microspectrofluorimeters, 1234 et seq.
Microspectrofluorimetry, analytical, 1240
and fluorescence fading, 1107
after HCl, for NA and DA distinction, 1109
Mid-chain block, by respiratory inhibitors, 912, 913
Mikrokator, for section thickness, 1247
Millon's reaction, 1056
Mipafox, 768, 769, 771, 797, 798
use of to distinguish A and C esterases, 797
Mitochondria, dihydrofolate dehydrogenase in, 952
protection of, by PVP, 903
relationship of to chromaffin granules, 1063
swollen, dehydrogenase reactions and, 903
ubiquinones in, 950
Mitochondrial cristae, ATPase in, 1279
Mitochondrial α-GPD, mechanism of electron transfer and, 945
Mitochondrial matrix, Hq-MTT reaction in, 1004
Mitochondrion, differentiation from melanin granule, 1055
membrane penetration by tetrazoles, 886, 887
pH effects on, 902
Mitoquinone, 1000
^{52}Mn, 1211
M-Nadi oxidase, 844, 845, 846

MNA method, 1353
Model systems, for calibration, 1249
Molecular energy levels, and fluorescence, 1172
Molecules, excited and fluorescence, 1173
Mollusca, copper in, 1144
Molluscan hepatopancreas, zinc in, 1154
Molybdate, as catalyser in peroxidase reactions, 853
Molybdate methods, for phosphate ions, 1158
Molybdenum, reaction of with diphenylcarbazide, 1145
Moniliformis dubius, surface mucopolysaccharides of, 1275
Monoamine oxidase, 841
coupled peroxidatic method for, 1341
EM method for, 1286
inhibitors of, 864
methods for, 866 et seq.
naphthoic hydrazide method for, 1339
tetrazolium method for, 1340
Monoamine oxidases, biochemistry of, 863 et seq.
Monoamines, triple fixation (EM) method for, 1452
Monochromation, 1198, 1199
Monochromators, filter, 1234
grating, 1198, 1234
prism, 1198
Monoethyl hydrogen peroxide, 857
Monoglucuronide, of bilirubin, 1072
Monophenol oxidases, biochemistry of, 858
Morin, 1422
Morin method, for aluminium salts, 1141
for beryllium, 1143
for calcium, 1137
tin salts and, 1152
Morin-calcium complex, distinction from other metals, 1138
Moths, pigmentation in, 858, 859
Motor end plates, AChE in, 772, 1281
and ChE in, 790, 799
non-specific metal binding and, 794
trypsin-like enzymes in, 776
Mounting media, general considerations, 1250
MSH-cells, 1112
MTT, 885, 891, 892, 898, 906, 910, 913, 914, 916, 924, 933, 938, 950
and electron acceptance from menadione, 906
linearity of reduction and, 888
MTT reduction, polarogram of, 885
Mucic acid, 819
Mucosubstances, EM histochemistry of, 1274, 1275
Mullerian ducts, regression of, and glucuronidases, 812
Multistep method, for glutaminase, 993
Muramidase, 808
Murexide, 1135
Murexide test, 1102
Muscle, myopathic, phosphorylases in, 835
skeletal, cholinesterases in, 800
skeletal, lipases of, 785
Muscle fibres, track autoradiography and, 1218
Mushroom tyrosinase, 1051
Mustard gas, sulphur-labelled, 1207
Mutagenesis, mechanism of, 1188
Myeloid leucocytes, Nadi reaction and, 845
naphthoquinones in, 845
Myeloid series, blood cells, esterases in, 775, 776
Myoglobin, 1066, 1067
in skeletal muscle, 855
Myometrium, plasminogen activator in, 1012
Myosin ATPase, EM methods for, 1279
Myristoyl choline, as ChE substrate, 789, 790
Myristoyl choline method, applications of, 790
for ChE, and localization of enzyme, 789
for cholinesterases, 1312
Myrosulphatase, 987
Myxoedema, 1032

^{22}Na, 1211
NAD, 903 et seq.
factors causing destruction of, 910
NAD concentration, substrate affinity and, 909
NADH, autofluorescence and, 1177
NADH cytochrome c reductase, 925
amytal and, 913
NADH dehydrogenase, EM technique for, 1449
NADH diaphorase, 925
localization of, 907
methods for, 925 et seq.
standard method for, 1342
Nadi oxidase, M, 844 et seq.
Nadi reactions, history of, 844, 845
inhibitors and, 846
specificity of, 845, 846
NAD (NADP) tetrazolium reductase, 908
NADP, 903 et seq.
factors causing destruction of, 910
NADPH diaphorase, 925
localization of, 907
methods for, 925
standard method for, 1342
Naphthochrome green B, 1139
for beryllium, 1142
for calcium, 1136

Naphthochrome green B, method, for aluminium salts, 1141
Naphthoic hydrazide methods, for MAO, 866, 867
Naphthol-amine method, for cytochrome oxidase, 848, 1331
Naphthol AS acetate, method for esterase, 1303
methods, 774 et seq.
synthesis of, 1303, 1304
Naphthol AS ε-aminocaproate, 974
Naphthol AS-BI *N*-acetyl-β-glucosaminide, synthesis of, 1325
Naphthol AS-BI β-D-glucosiduronic acid, synthesis of, 1322
Naphthol AS-BI glucuronide, 816
method, 1322
Naphthol AS-D acetate, method for esterase, 1305
Naphthol AS glucuronides, 816 et seq.
Naphthol AS-LC acetate, method for esterase, 1304
synthesis of, 1304
Naphthol AS-LC acetyl β-glucosamide, 824
Naphthol AS-LC glucuronide, 816
Naphthol AS nonanoate, as lipase substrate, 787
method for lipase, 1310
Naphthol AS phenylpropionoxy ester, 974
Naphthol AS sulphate, method for arylsulphatase, 991
α-Naphthol, methods for peroxidases, 855 et seq.
and non-enzymatic oxidative coupling, 846
α-Naphthol-pyronin sequence, for peroxidase-label technique, 1267
Naphthoquinone, and Nadi reaction of leucocytes, 845
Naphthoquinones, and Nadi reaction, 1004
α-Naphthyl acetate, methods for esterase, 1303
Naphthyl acetates, as esterase substrates, 772 et seq.
Naphthylamidases, biochemistry of, 962
as indicators of cell injury, 972
methods for, 967
α-Naphthyl-5-glutamine, 993
β-Naphthyl laurate, 763
β-Naphthyl oleate, 763
Naphthyl sulphates, as substrates, 990
2-Naphthylthiol acetate (NTA), 1283
Negative contrast objectives (phase-contrast), 1245
Neocuprine, 864
Neoplasia, fluorescence microscopy and, 1193
β-glucuronidase and, 809
Neoplastic cells, phosphorylase in, 835
Neoplastic tissues, succinic oxidase system and, 842
Neotetrazolium, 874
Nerves, cholinergic, 772
Newt, California, neural crest in, 1055
Neural crest, melanocytes and, 1055
Neuraminidase, 808
Neuraminidase extractions, 1378
Neuraminidases, as histochemical reagents, 1033
Neurohypophysis, ^{131}I-labelled thyroxine in, 1209
Neuromelanins, PAS reaction and, 1058
Neutral red, 1422
as vital fluorochrome, 1180
NH_2 groups, and Hq-MTT reaction, 1003
Niagara sky blue, for dextrans, 1162
Nicotinamide, as nucleotidase inhibitor, 910
Nickel, methods for, 1143, 1144, 1410
Nickel ammonium sulphate, 853
Nickel salts, rubeanic acid reaction of, 1145
Nile blue methods, for lipofuscins, 1383, 1384
Nile blue sulphate, 1082
Nine-position test, 1249
Ninhydrin, fluorescence and, 1179
Ninhydrin reaction, 1103
for tryptamines, 1116
Nissl's granules, basophilia of and nucleases, 1020
***p*-Nitroanilide,** 965
***p*-Nitrobenzene-azo-1-naphthol,** 1150
Nitro-BT, 882, 883 et seq., 892, 896, 907, 910, 913, 914, 916, 924, 926, 932, 946, 950, 1286
red and blue formazans from, 893
reduction products of, 894
Nitro-BT reduction, polarogram of, 886
Nitrocatechol sulphate, 988
Nitro groups, in tetrazoles, substantivity and, 887
5-Nitro-1-naphthylamine, 848
Nitronaphthylamine-amine method, for cytochrome oxidase, 1332
Nitro-NT, 896
substantivity for protein and, 896
Nitroprusside, as reagent for zinc, 1154
1, Nitroso-2-naphthol, 1129, 1143
α-Nitroso-β-naphthol, 1129, 1143
Noradrenalin, chromaffin reaction for, 1382
distinction from adrenalin, 1062
EM demonstration of, 1291, 1292
and FIF, 1102
FIF method for, 1394
fluorescence maxima, 1106
iodate method for, 1382
PAS reaction of, after oxidation, 1061
silver method for, 1380

Noradrenochrome, ^{131}I and, 1209
reaction of, with Schiff's reagent, 1061
Noradrenochromes, 1059
Norharman, 1120
Nothing dehydrogenase, 910, 930, 931, 933, 936
effect of PMS on, 916, 917
identity of, 911
NT, 881, 882, 883 et seq., 893, 894, 907, 908, 914
red formazan from, 893
NT formazan, reflex colours of, 890
Nuclear fast red, 1135
Nucleases, substrate film methods for, 1012
Nuclei, melanin in, 1055
Nucleic acids, breakdown of, 1018
electron stains for, 1262
EM methods for, 1272, 1273
extraction techniques and EM demonstration of, 1290
fluorescence cytochemistry of, 1172
^{32}P and, 1208
staining of, by aminoacridines, 1187
Nucleoli, plant cell, orthophosphate in, 1294
Nucleotidases, destruction of coenzymes by, 910
Nucleotide phosphatases, EM cytochemistry of, 1280
Nucleotides, as phosphorylase activators, 829
Nucleus ambiguus, MAO in, 869
Nylon fibres, dichroism of, 1164

Objectives, choice of, for fluorescence photomicrography, 1201
Ochronosis pigment, 1064
Octopus vulgaris, fluorophores in, 1110
salivary glands of, 1102
Octyl alcohol, as MAO inhibitor, 868
Ocular tissues, pigmented, zinc in, 1154
Okamoto method, for copper, 1147
for zinc, 1155
Old yellow enzyme, 925
Oligosaccharides, degradation of, 820
iso-OMPA, 768, 770
sensitivity of ChEs to, 769, 798
One-wavelength methods, in absorptiometry, 1228
Onion skin, cells of, and cytochrome oxidase, 846
Onion root tips, chromosomes in, 1021
Opsin, 929
OPT, vapour phase method for histamine, 1123
Optical quantitation, 1226 et seq.
Optical systems, calibration of, 1249
OPT method, for histamine, 1122, 1400
OPT reaction, of pancreatic islet α_2 granules, 1123
Orcein, for squash preparations, 1215
Organophosphorus inhibitors, 762
Orlon, birefringence of fibres of, 1163
Ornithine, carbamoyl transferase, 981
method for, 1359
Ornithine-keto acid aminotransferase, method for, 985
***Ortho*-diphenols,** Vulpian reaction and, 1117
***Ortho*-phthalaldehyde,** 1122
Os-DMEDA, 1264
Osmiophilic diazoether, EM techniques for esterases, 1283
Osmiophilic polymer generation, 1277
Osmium-black, 850, 1269, 1273, 1274, 1278, 1283, 1285, 1287, 1289
Osmium bridging technique, for acid phosphatase, 1281
for cholinesterases, 1282
Osmium tetroxide-potassium iodide, for adrenalin, 1063
Osteocytes, proteases in, 1010
tetracyclines and, 1179
Os-TNST, 1287
Ouabain, inhibition of membrane ATPase, 1280
Ovary, ChE in, 772
Oxalate, conversion of, to carbonate, 1138
methods for, 1405, 1406
Oxalate crystals, birefringence of, 1138
Oxalate method, for calcium, 1133
Oxalates, in plant tissues, 1139
Oxalic acid, for removal of iron, 1069
Oxalosis, 1138, 1139
Oxazolines, and competitive inhibition of AChE, 771
Oxidases, 841 et seq.
Oxidation catalyst reactions, for copper, 1146
Oxidation catalysts, use of, in indigogenic methods, 779
Oxidation methods, for bilirubin, 1073
Oxidative deamination, 872
Oxidative enzymes, microfluorimetry of, 1242
Oxidized DAB method, for hemes, 1337
Oxidoreductases, 841 et seq., 880 et seq., 921 et seq.
Oxidoreductases (bound), standard method, 1342
Oxidoreductases (soluble), standard method, 1343, 1344
2-Oxoglutarate, oxidation of, and lipoate, 905
Oxoglutarate dehydrogenase, 926, 927
Ox spleen, glucuronidases of, 810
Oxytocinase, placental, 972
Oyster, iodides in, 1159

Ozone-formaldehyde reaction, for tryptamine, 1120

Palmitoyl choline, as ChE substrate, 789
2-PAM, 1218
PA-Methenamine silver, EM method, 1441
Pancreas, ChE in, 772
ox, carboxypeptidase of, 1154
Pancreatic elastase, 1042
Pancreatic islet cells, masked metachromasia of, 1189
Pancreatic lipase, method for, 786
Paneth cells, zinc in, 1155
Papain, 962, 973
as histochemical reagent, 1040, 1041
Paraformaldehyde, 1107
water equilibration procedure and, 1108
Pararosaniline, 1422
Parasites, cholinesterases in, 800
Parathyroid gland, 823
naphthylamidase in, 969
Parathyroid hormone, LNAses and, 971
Parietal cells, alcohol dehydrogenase in, 929
gastric, 823
rat stomach, CAH in, 1288
Parotid gland, protease inhibitors, 966
Pars intermedia, APUD qualities of, 1112
Particles, α, β and γ, 1211
submitochondrial, formazan production in, 888
β-Particle tracks, ^{14}C-glycine, 1208
PAS reaction, 831
of ceroid, 1083
of dextrans, 1161, 1162
PATCO method, 1273, 1274, 1277
PATCSP method, 1442
Patella vulgaris, β-fucosidase in, 820
β-glucuronidase and, 810
PATO method, 1273, 1274
PATO reaction, 1442
PA-TSC-SP method, 1274
PCMB, 911
as C-esterase inhibitor, 797
inhibition of arylesterases by, 765
inhibition of MAO by, 864
Pectinase, digestions, 1377
histochemistry of, 1035
Pectinases, biochemistry of, 1034, 1035
Pectinesterase, 1035
Penetration, of antisera, problems of, 1268
Pentadecanoyl choline, as ChE substrate, 791
Pentafluorophenylhydrazine, 1274
3,5,7,2′,4′-pentahydroxy-flavonol, 1137
Pentalysine, 974
Pentitol, 930
Pepsin, 962, 973
digestion of tissues by, 1039
Pepsin, extractions, 1377
Peptidase-Esterase relationships, 976
Peptidases, 962 et seq.
EM methods for, 1289, 1290
γ-Peptide linkage, hydrolysis of, 973
Peracetic acid, 1056
Performic acid, 1056
Perfusion fixation, 1109
Periodic acid-silver, 1272, 1273
Periodic acid-sulphonation-basophilia method, 1184
Periplaneta, tanning reactions and, 822
Perls' reaction, 1065, 1130, 1131, 1402
and haemosiderins, 1069
Permanganate method, for EM demonstration of monoamines, 1292
Peroxidase, 841
aminoethyl carbazole method for, 1338
benzidine methods for, 1334, 1335
DAB method for, 1337
DAB (EM) technique for, 1449
dihydroxyindole method for, 1336
EM methods for, 1286
(Haemoglobin) leuco-Patent blue method for, 1335
red cell, and indigo production from indoxyls, 781
Peroxidase-labelling technique, 1267 et seq., 1438, 1439
Peroxidases, 850 et seq.
methods for, 852 et seq.
Peroxides, fatty, 855
and Nadi reactions, 845
PFAS reaction, lipids and, 1091
PGE_2, 936
pH, effect of on indigo formation, 782
effect of on isoquinoline fluorescence, 1105
Phaeochromocytoma, 1060, 1062
Phase-contrast microscopy, 1245
***O*-Phenanthroline,** 864, 929
Phenazine methosulphate, 824, 872, 873, 916, 935, 1326, 1344
in MAO reactions, 868
Phenol oxidase, nature of, 862
Phenoloxidases, 844
Phenosafranin, 1422
stain for suramin, 1163
Phenylalanine-2-^{14}C, and nucleic acids, 1210
Phenyl butyrate, as esterase substrate, 762
Phenylhydrazine, as carbonyl trapping reagent, 902
as cytochrome oxidase inhibitor, 843
test for urea, 1165
β-Phenylisopropyl hydrazine, 864
Phenylmercuric chloride, as alternative to PCMB, 798
***N*-Phenyl-*p*-phenylamine,** 848
β-Phenyl propionate, 1355
3-Phenylpyruvic acid, 966

Phenylurethane, 846
Phloroglucinol, 1079
Phloxine B, 1422
Phosphatase, acid, EM techniques for, 1444, 1445
 microdensitometry of, 1230
 in rat pancreas, 1153
Phosphatase, alkaline, EM methods for, 1278, 1279, 1443
 leucocyte, determination of, 1253
 measurement of, in single cells, 1242
 quantitation of, by interferometry, 1246
 in rat prostate, 1153
 as ultrastructural tracer, 1268
Phosphatases, acid, EM methods for, 1280, 1281
Phosphate, demonstration of, by von Kóssa method, 1138
Phosphate-binding, non-specific and ATPase, 1280
Phosphate deposits, inorganic, 1158
Phosphate ion, methods for, 1158
Phosphates, insoluble, in animal tissues, 1138
3′-Phosphoadenosine diphosphate-pantoyl-*β*-alanyl-cysteamine, 982
Phosphocozymase, 904
Phosphocreatine, 986
Phosphine GN, 1422
Phosphine 3R, 1422
Phosphoglucomutase, biochemistry of, 828
 method for, 987, 1361
Phosphogluconate dehydrogenase, method for, 937
6-Phosphogluconate dehydrogenase, 923, 937
 standard method for, 1344
Phospholine iodide, 982
Phospholipase B, 761
 method for, 787, 1311
Phospholipases, A, B, C and D, 787
Phospholipid, bound naphthoquinones and, 845
 in cytochrome oxidase, 842
 in EC granules, 1115
 and Hq-MTT reaction, 1003
Phospholipids, radiophosphorus exchange and, 1208
Phosphorylase, activators of, 829
 biochemistry, 827
 blocking reactions and, 829
 dextran technique for, 1328
 distinction from transglycosylase, 832
 EM methods for, 1284
 inactive B form of, 827
 liver, nucleotide activation of, 829
 methods for, 828 et seq.
 permanent iodine stains for, 1327
 phosphatase, 827
 primer requirements of, 826
Phosphorylase, purine nucleoside, method for, 1329
 standard technique for, 1327
Phosphorylase kinase, 827, 830
Phosphorylases, 808, 827 et seq.
 action of EDTA on, 829
Phosphotungstic acid, as constant/contrast stain for carbohydrates, 1275
 as electron stain, 1262
Photographic colorimetry, 1231
Photographic emulsions, Hurter-Driffield curves of, 1231
Photographic plates, for protease methods, 1010
Photometric densitometry, in autoradiography, 1217
Photometry, for fluorescence photomicrography, 1202
Photomicrography, fluorescence, 1200 et seq.
 equipment for, 1201, 1202
Photomultipliers, 1232, 1238
 EMI 9558 and 9558QA, 1237
Photon counting, 1238
Photoreceptor cells, "histamine" in, 1123
***o*-Phthalaldehyde,** fluorescence and, 1179
Phthalate, as glucuronidase activator, 811
Phthalocyanin dyes, for calcium, 1136
Phylloerythrins, 1068
Physostigmine, 765, 767
Pictet-Spengler reaction, of DA and NA, 1104
Pigment, acid fast, 1078
 bilharzial, 1075
 Dubin-Johnson, 1088
 Hopkins-Cole, 1064
 malarial, 1075
 pseudomelanosis, 1085
Pigment formation, in techniques for MAO, 867, 868
Pigments, 1050 et seq.
 carotenoid, 1086
 diagnostic schedule for, 1092
 differential diagnosis of, 1089, 1090
 haem, 1065 et seq.
 lipid, 1076 et seq., iron-containing, 1086
 tryptophan, 1064
Pineal gland, LNAse in, 971
 tryptophan hydroxylase and, 1119
Pink-tooth, disease of cattle, 1067
Pituitary gland, histamine in, 1123
 mammalian, amine precursor uptake by, 1112
Placenta, allantoic, ferritin in, 1132
 aminopeptidases in, 972
 ceroid in, 1085
 cholinesterases of, 800
 glucuronidase in, 834
 histaminase in, 870
 iron transport in, 1132

Placenta, steroid dehydrogenases in, 942
Planimetry, 1248
Plant cells, orthophosphate in nuclei of, 1294
Plant cell walls, EM demonstration of disulphides in, 1263
Plant tissues, amine oxidase in, 865
catalase in, 850
G-Nadi reaction in, 845
oxalates in, 1140
tetrazolium reduction in, 881
and zinc deficiency, 1154
Plasmal reaction, 1078
Plasma membrane, fluorochromes and, 1180
Plasmin, 962
action of, on fibrin, 1011
Plasminogen, activation of, 1011
Plasminogen activator, method for, 1370
Plutonium, 1207
PMS, 824, 872, 873, 916, 935, 1326, 1344
reaction with succinate dehydrogenase, 950
Pneumokoniosis, tin salts in, 1152
^{210}Po, 1218
Point counting, 1226, 1227
Point sampling, 1248
automated, 1248
Polar groups, of tetrazoles, effects on penetration, 887
Polarization microscopy, 1243 et seq.
Polarized fluorescence, 1234, 1243
Polarographic studies, on tetrazoles, 883
Polaroid, as polarizer and analyser, 1243, 1244
Polonium, 1207
Polyacrilamide films, for model systems, 1249
Polyacrylonitriles, 1163
Polyamide fibres, 1164
Polyamines, 1113
Polyepoxides, water-soluble, 1290
Polygalacturonase, 1035
Polygalacturonic acid, 819
Polyglucose sulphate, 843
Polyhydric phenols, 1113
Polyisoprenes, 1086
Polylysine, 843
Polymerizing diamines, and cytochrome oxidase, 849
Polyols, reduction of, 930
Polyoxyethylene sorbitan trioleate, 1310
Polypeptides, tyrosine-rich, in melanogenesis, 1055
Polyphenol dyes, and keratohyalin, 862
Polyvinyl alcohol, 1161
method for, 1163
Ponceau 2R, 1023
Porphyrin, fluorescence of, 1177
Porphyrinuria, congenital, 1067
Porphyrins, 1065, 1067, 1068
product of from hemes, 1066
Positive contrast objectives (phase contrast), 1245
Post-coupling method, for β-glucuronidase, 1321
Potassium, methods for, 1415, 1416
Potassium *p*-acetyl phenyl sulphate, 988
Potassium 6-benzoyl-2-naphthyl sulphate, 990
Potassium 6-bromo-2-naphthyl sulphate, 990
Potassium indoxyl sulphate, 988
Potassium *p*-nitrophenyl sulphate, 988
Potassium-2-hydroxy-5-nitrophenyl sulphate, 988
Potassium iodate, and chromaffin reaction, 1060
Potassium ions, methods for, 1156, 1157
Potassium permanganate, as EM fixative, 1262
Precursor uptake, 1111
Pregnane, 939
Premelanosomes, 1055
Preputial gland, of female rat, as source of glucuronidase, 810
Primary conformations, of proteins, 1192
Primary fluorescence, 1177
Primulin, 1422
Proflavine, 1187
Progesterone, 940
Projection microradiography, 1233
Prolactin, EM localization of, 1268
radio-labelled, 1209
Proline dehydrogenase, 952
Pronase, 1040
8-Propionoxy-5-nitroquinoline, synthesis of, 1305
Propionylcholine, as substrate for avian ChE's, 793
Propylene glycol, as substrate for secondary alcohol dehydrogenase, 941
Prostaglandin dehydrogenase, 1348
method for, 935
Prostate, dog, steroid dehydrogenases in, 941
rat, zinc levels in, 1153
Prostate gland, human, acid phosphatase in, 1154
zinc in, 1154
Protamine sulphate, 843
Protease inhibitors, parotid gland, 966
Proteases, colour film method for, 1370
dyed gelatin method for, 1369
gelatin method for, 1369
substrate film methods for, 1010
Protection, immunospecific, 1270
mitochondrial, by PVP, 903
Protective media, for mitochondria, 847
Proteinases, as histochemical reagents, 1038
Protein, dried, and amine fluorescence, 1106

Protein, end-groups, EM histochemistry of, 1263
Protein coupling, glutaraldehyde technique for, 1437
Proteins, acid and basic, EM methods for, 1264
autofluorescence of, 1177
mercury-labelling of, 1270
respiratory, in Mollusca, 1144
Protocatechuic acid, 822
Protoporphyrins, 1067
Proteus vulgaris, chondroitinase, 1032
Prussian blue, 1057, 1130
Pseudoisocyanin, as fluorochrome, 1189, 1422
Pseudomelanin, 1091
Pseudomelanosis pigment, 1085
Pseudoperoxidase, 1066
versus peroxidase, 852
Pseudo-Schiff reagents, fluorescent, 1241
Pteridines, 1102
Ptyalin, 1033
Purine bases, reaction with aminoacridines, 1187
Purine nucleoside phosphorylase, method for, 833, 1329
Puromycin, 970
PVP, 824
Chlorazol fast pink method for, 1419
Congo red method for, 1419
iodine method for, 1418
methods for, 1162, 1163
and mitochondrial protection, 903
in preparation of model systems, 1250
as stabilizer in phosphorylase reactions, 830
Pyloric glands, radio sulphur uptake by, 1210
Pyocyanin, as redox acceptor, 871, 915
Pyridine-2-aldoxime methiodide, 1218
Pyridine nucleotides, fluorescence of, 1192, 1193
Pyridoxal-5-phosphate, 985
Pyrimidine bases, reaction with amino-acridines, 1187
Pyroantimonate, 1157
Pyroantimonate method, electron probe analysis and, 1293
Pyrogallol, as hydrogen donor, 855
Pyronin Y, 1422
Pyrophosphate, 826
as cofactor, 927
Pyruvate dehydrogenase, 926, 927
Pyruvate-lactate conversion, by H & M enzymes, 934

Q30, 948
Q-enzyme, 826
Quantitation, in autoradiography, 1217
tetrazoles and, 889
Quantitative histochemistry, 1225 et seq.
fixation and, 1253
Quantum converter, (Rhodamine B), 1237
Quantum efficiency, recovery after fading, 1252
Quartz slides, for fluorescence microscopy, 1194
Quaternary ammonium, substrates for ChE, 1283
Quaternary ammonium bases, as ChE inhibitors, 767
Quinacrine, 1422
Quinacrine hydrochloride, as fluorochrome for chromosomes, 1184
Quinacrine mustard, 1184, 1422
Quinacrines, demonstration of, in tissues, 1161, 1163
fluorescence of, 1178
Quinalizarin method, alkaline, 1143
Quinhydrones, in pigments, 1058, 1060
Quinine, inhibition of esterases and lipases, 762
Quinke method, 1130, 1131
Quinoline-amine, method for cytochrome oxidase, 848, 1311
Quinone-amine, method for cytochrome oxidase, 848
Quinoneimines, conversion to thioindigoid dyes, 1116
Quinone reductases, 905, 906

Rabl's method, for soluble calcium, 1133
Radioautography, *syn* Autoradiography, 1207 et seq.
Radio-iodine, and thyroid gland, 1208
Radiopotassium, in brain, 1208
Rana esculenta, gall-bladder, naphthylamidases in, 972
liver pigment in, 1080
Random coil structure, of polypeptides, 1191
Ranvier's nodes, metal binding and, 794
Rare earth metals, effects of on A-esterases, 764
Rat adrenal medulla, storage granules, and Pb^{2+} uptake, 1281
Rat kidney, galactosidase in, 822, 823
Rat stomach, parietal cells in, 1112
Reagents, blocking, in phosphorylase techniques, 829
carbonyl trapping, in dehydrogenase reactions, 902
Reciprocity characteristics, (photographic films), 1423 et seq.
Reciprocity failure, 1200
Red cells, demonstration of, 1065
staining of by indigogenic methods, 781

Redox compounds, lipid soluble, 1004
soluble, as electron acceptors, 915
Redox potentials, of tetrazoles, 883 et seq.
of tetrazoles, and coenzymes, 917
Reduced coenzymes, as diaphorase substrates, 926
Reductases, brain, soluble and particulate, 906
Reference channels, in microspectrofluorimeters, 1237
Reflecting optics, 1232
Reflectiometry, 1226, 1227
Reflection microradiography, 1233
Reflex microscopy, 1243
Refractive index, 1228
determination of, 1246
Reineke salt, 1122
Rennin, 962
Resolution, in autoradiography, 1211
Resorcinols, 1116
Respiratory chain, 911 et seq.
Retardation microscopy, 1176
Reticulo-endothelial system, dextran storage in, 1161
Retina, alcohol dehydrogenase in, 929
ATPase in, 1279
mammalian and avian, redox compounds in, 1001
Retinene, 929, 1087
Rheonin A, 1422
Rheumatoid factor, localization of, 1267
Rhodamine B, 1422
Rhodamine 3G, 1422
Rhodamine S, 1422
Rhodanine methods, for silver, 1152
Rhodizonate, 1139
method for Ba and Sr, 1142
method for lead salts, 1149
Rhodopsin, 929, 1087
Ribonuclease, as cytochrome oxidase inhibitor, 843
extractions, 1376
fixatives and, 1022
as histochemical reagent, 1018 et seq.
uses of, in histochemistry, 1019, 1020
optimum conditions for, 1022
specificity of, for RNA, 1021
Ribonuclease A, 1021
Ribonucleases, actions of, 1019
substrate film methods for, 1012, 1372
Ribosomes, EM staining and, 1275
Ring and Blob, in thyroid autoradiographs, 1208
RNA, acridine orange and, 1186
RNA films, preparation of, 1372
Roehl and Leutert method, 1134
Rose bengal, 1193
Rotenone, and electron chain blockage, 913
Rubeanic acid, 1139, 1143
method for copper, 1145, 1411
method for silver, 1152
"Rule of 110", for FIF, 1108
Russell bodies, 1021
Ruthenium red, EM method, 1443
for EM demonstration of carbohydrates, 1275

Saccharic acid, 811, 819
as glucuronidase inhibitor, 809
Saccharo-1,4-lactone, as glucuronidase inhibitor, 810, 811
Safranin, 1422
Salivary gland, ducts, lipases in, 784
ChE in, 772
cholinesterases of, 799
(***Octopus***), 5-HT in, 1102
Salmine, 843
Salt linkages, and binding of fluorochromes, 1181
electrovalent, 1129
Sandmeyer synthesis, 1307
for isatins, 781
Sarcoma, rat, arylsulphatase in, 988
Scanning methods, in absorptiometry, 1229
Schiff reaction, mechanism of, 1182, 1183
Schiff reagent, fluorescent, 1241
Schiff-type reagents, 1183
Schistosoma, cholinesterases of, 800
Schistosomiasis, 1076
Schmorl's reaction, 1057, 1058
of EC granules, 1115
of ochronosis pigment, 1064
Schott AL 406 interference filter, 1200
Scrapie, glucuronidases and, 834
Schwarzchild-Villiger effect, 1228
Scintillators, and optical system calibrations, 1249
Secondary alcohol dehydrogenase, 941, 947, 1349
Secretion granules, hormones in, 1268
Semen, zinc in, 1154
Semicarbazide, as carbonyl trapping reagent, 902
Sepia officinalis, melanin of, 1051
Serine, in active centre of β-esterase, 764
Serine peptidases, 962
Serotonin, 864
Serum GOT, 983, 984
SH groups, esterase, sensitivity of to heavy metals, 764
SH inhibitors, effect of on xanthine oxidase, 873
Shock, anaphylactic, 1209
Shuttle, α-glycerophosphate, 929, 944
Sialic acid, 821
Sialidases, as histochemical reagents, 1033

Side-chain carboxyls, and masked metachromasia, 1190
Siderosis, 1068
Siderosomes, 1070
Silver, methods for, 1151, 1152
DMABR method for, 1415
reduction of, by melanins, 1056
Silver-H_2O_2, method for calcium oxalate, 1138, 1405
Silver intensification procedure, in thiocholine method, 1316
Silver methenamine, as EM technique, 1274
Silver methods, von Kóssa, 1138
Silver nitrate-hydrogen peroxide, for oxalates, 1138, 1405
Silver nitrate methods, for urates, 1165
Silver proteinate, 1274
method for proteases, 975
Silver reaction, for chloride ion, 1158
Silver-rubeanate, method for calcium oxalate, 1406
Silver solutions, preparation of, 1379
Singlet electronic states, 1172
Sinigrin, 987
Skeletal muscle, aminoacid dehydrogenase in, 951
ATPase in, 1248
birefringence and, 1244
calcium in T system of, 1290
creatine kinase in, 986
electron stains for, 1262
EM localization of phosphorylase in, 1284
ferritin in, 1132
hydroxyproline epimerase in, 998
LDH in, solubility of, 922
LDH subunits in, 934
lipases of, 785
locust, α-GPD in, 945
mitochondrial ATPase in, 1279
myoglobin, 1066, 1067
PMS and dehydrogenases in, 916
quantitation of AChE in, 1218
quantitative histochemistry of, 1254
selective binding of Nitro-BT to, 887
Skeletal muscles, ceroid and, 1085
Skin, cholinesterases of, 800
esterases in, 800
freeze-dried, DOPA reaction in, 862
phosphorylases in, 834
^{35}S-methionine, in acid polysaccharides, 1210
Smooth muscle, gastrointestinal, ceroid in, 1085
Snail, Roman, β-glucuronidase and, 814
stomach fluid, 1035
Snake venom, lecithinases, 1044
Snake venoms, 1026
Soaps, calcium, in lipase techniques, 784, 785
Sodium alizarin sulphonate, 1134
Sodium ions, EM demonstration of, 1290
methods for, 1157, 1158
Sodium tetraphenylboron, 1157
Sodium tungstate, for EM staining of nucleic acids, 1273
Solid (Gel) media, for dehydrogenases, 923
Solid medium method, for CAH, 1365
Solochrome-aluminium chelates, 1141
Solochrome azurine, 1139
method for beryllium, 1143
Solochrome blue-black R method, 1141
Solochrome cyanine, acid, 1139
Soluble enzymes, technique for, 934
(oxido-reductases) standard method for, 1343, 1349
Sorbitol dehydrogenase, 910
method for, 1345
D and L-Sorbitol dehydrogenases, 930
Specific radioactivity, of cells, 1217
Spectra, emission, plotting of, 1241
excitation, plotting of, 1240
fluorescence, of DOPA models, 1053
fluorescence, excitation and emission, 1173
Spectral curves, fading and, 1251
Sperm, mid-piece, succinate dehydrogenase in, 892
Spermatozoa, bovine, Feulgen-fading and, 1251
glycogen in, 1274
zinc in, 1154
Spleen, horse, ferritin in, 1132
Stabilizers, in phosphorylase reactions, 830
Stable Sudanophilia, method for, 1335
Staining procedures, in autoradiography, 1216
Standard fluorescein unit, 1240
Stannic salts, lakes of, 1152
Stannite, in pneumokoniosis, 1152
Starch, soluble, as glucosyl acceptor, 830
Starch films, for amylase, 1013
Stearoyl choline, as ChE substrate, 789
Stein's test, for bilirubin, 1073
Stereomicroradiography, 1247
Steroid dehydrogenase, methods for, 938 et seq.
3-β-ol steroid dehydrogenase, method for, 941, 942
17-β-ol steroid dehydrogenase, method for, 942
Steroid dehydrogenases, applied histochemistry of, 943
various, 943
Steroids, metabolism of, and glucuronidases, 810
nomenclature, 938 et seq.
stereochemistry of, 939

Steroids, trivial names of, 940
Sterol sulphatase, 987, 989
Stoichiometry, of aldehydes, and Schiff reaction, 1182, 1183
Stokes shift, 1173
Streptodornase, 1024
Streptokinase, 1024
***Streptomyces* chitinases,** 1038
Strontium, methods for, 1142
Strontium-anthraquinone lakes, 1135
Strontium-GBHA complex, 1137
Strychnine, inhibition of esterases and lipases, 762
Substantia nigra, 1057
Substantivity, of tetrazoles, 887
Substrate concentration curves, 901
Substrate film methods, 1009 et seq., 1369 et seq.
Succinate, oxidation of by swollen mitochondria, 902
Succinate dehydrogenase, 842, 923, 924, 945, 948
  method for, 948 et seq.
  standard method for, 1342
Succinic oxidase system, 841 et seq.
  distribution of, 842
Succinoxidase, copper-ferrocyanide method for, 1451
Succinyl-lecithin, 950
Sudanophilia, of chromolipoids, 1077
  stable, 855
Sulphatases, biochemistry of, 987
  methods for, 1362, 1363, 1364
Sulphide-silver methods, 1148, 1150, 1412, 1413
  (EM), 1415
  for EM demonstration of heavy metals, 1290
Sulphide-silver reaction, for zinc, 1156
Sulphonic acid groups, in melanin, 1057
Sulphosalicylic acid, 1161
Sulphydryl, EM methods for, 1263
Sulphydryl groups, nothing dehydrogenase and, 910
Suramin, 1161, 1163
  as glucuronidase inhibitor, 811
Surface antigens, localization of, 1267
Sweat glands, EM demonstration of sodium ions in, 1290
Sycotopus canaliculatus, copper in, 1144
Synapses, cholinesterases in, 800
  lead-reactive material in, 793
Syntheses, Bischler-Napieralski, 1107
Systems, model, for calibration, 1249

TAB, as EM esterase substrate, 1283
Tadpole, tail, resorption phase and glucuronidases of, 812
Tadpole, thyroid glands, autoradiography of, 1209
Taka-diastase, 1037
Talc, 1161
***Taricha torosa*,** neural crest in, 1055
Taurocholate, activation of lipase, 762
  inhibition of esterase, 762
  as lipase activator, 796
Teeth, developing, glucuronidases in, 834
  ^{24}Na in, 1208
Television Image Analyser (MRC), 1248
Tellurite, 884
Terminal chain block, by azide and cyanide, 912
Testicular extracts, hyaluronidase in, 1029
Testicular hyaluronidase, specificities of, 1029
Testosterone, 940
Tetra-(acetoxy mercuri) arsanilic acid, 1270
Tetracyclines, fluorescence of, 1178
Tetrahydrofolate, 953
Tetrahydronorharman, 1120
1,2,5,8-Tetrahydroxy-anthraquinone, 1134
Tetranitro-BT, 882, 883 et seq., 892, 893, 896, 914, 1286
Tetraphenylboron method, for potassium, 1416
  for potassium ions, 1157
Tetra phenyl-porphine sulphonate, 1263
Tetrapyrroles, 1067, 1068
  structure of, 1071
Tetra thiocyano-diammonochromic acid, 1122
Tetrazolium methods, for EM demonstration of dehydrogenases, 1286
  for MAO, 868 et seq.
Tetrazolium reductase, electron pathways and, 922
Tetrazolium salts, concentration effects and, 888
  choice of, 924
  early development of, 880
  as electron acceptors in enzyme reactions, 881
  half reduction and, 892 et seq.
  hydrogen pick-up and, 889, 890
  isomers of, 893, 894
  light sensitivity and, 888
  lipid solubility of, 887
  redox potentials of, 883 et seq.
  reducibility of, 890
  reduction products of, 894
  R_F values of, 892
  solubility of, 886
  substantivity of, 887
  toxicity of, 888
Textile fibres, demonstration of, 1163
  solubility of, 1164
Thalamus, MAO in, 870
Thallium, methods for, 1156

Thallium nitrate, 1279
Thiamine pyrophosphatase, EM method for, 1280
Thiazol blue, method for cytochrome oxidase, 849
Thiazolines, and competitive inhibition of AChE, 771
1-(2-Thiazolylazo)-2-naphthol, 1148
Thiazolyl tetrazoles, metal chelation and, 897
stability of to light, 888
Thickness, section, measurement of, 1246 et seq.
Thin layer chromatography, tetrazoles and, 892
Thiocarbamyl nitro-BT (TC-NBT), 1287
Thiocarbohydrazide, 1263, 1269, 1273, 1274, 1277, 1288
Thiocholine method, for ChE's, modifications of, 792
direct colouring, 1317
mechanisms of, 792
Pb modification of, 1316
silver intensification procedure for, 1316
Thiocholine-copper, EM method for cholinesterases, 1282
Thiocholine-copper ferrocyanide, EM method for cholinesterases, 1282
Thiocholine-copper-lead, EM method for cholinesterases, 1282
Thiocholine and inhibitor method, for cholinesterases, 1314, 1315
Thiocholine-lead ferrocyanide method, 1318
Thiocyanatochromate, 777
Thioflavine S, 1422
Thioflavine T, 1176, 1179, 1422
Thioindigoid dyes, 1116
from quinoneimines, 1116
Thioindoxyls, as hydrogen donors for oxidases, 856
Thiolacetic acid, acetyl disulphide in, 795
EM method, 1281
Thiolacetic acid method, non-specificity of, 795
(Pb modification), 1319
Thiolacetic acid methods, for esterases, 794 et seq.
2-Thiolacetoxy-benzanilide (TAB), 1283
Thiolating reagents, 1270
Thiol peptidases, biochemistry of, 962
2-Thiolpropionoxy-benzanilide (TPB), 1283
Thionin, as redox acceptor, 915
Thionyl chloride, 1110
Thiourea, as peroxidase inhibitor, 851
Thiosemicarbazide, 1273, 1274
Thorium-Morin complex, 1137
Thrombin, 765, 962
Thymidine-^{3}H, 1214
Thyroid, follicular cells, 823
Thyroid, gland, cathepsin D in, 966
C cells in, 1111
MAO in, 870
parafollicular (C) cells, 1190
peroxidases in, 851
radio-iodine in, 1208
Thyroxine, ^{131}I-labelled, 1209
Tin, methods for, 1152, 1153
Tirmann-Schmeltzer method, 1130
Tissue-bound tetrazoles, false localization due to, 923
Titan Yellow, 1139
for magnesium, 1150
TNTTC, 883, 890
α-Tocopherol, in retina, 1002
***o*-Tolidine,** 1139
as "benzidine" reagent, 854
reaction for copper, 1146
Toluene-2-4-diisocyanate, 1265
Toluene-3,4-dithiol, 1153
Toluidine blue, masked metachromasia and, 1189
***Torpedo* electric organ,** AChE in, 789
Track counting, in autoradiography, 1217
Transferases, 981 et seq.
EM method for, 1289
Transhydrogenases, 940
Transiodination, and thyroid gland, 851
Translocation, isotopes and, 1212
Transmittance, in absorptiometry, 1227
Trapping reagents, in dehydrogenase histochemistry, 902
Trasylol, 966
Trehalase, 825
Trehalose, 1326
1,2,4-Triazole, 854
Nα-cbz-L-Triarginyl-2-naphthylamide, 974
Tributyrin, as lipase substrate, 762
Tricarboxylic acid cycle, 949
2,6,8-Trichloropurine, 872
Trichohyalin, phenol oxidase in, 862
Trichoxanthin, 1058
Trifluoroacetyl-α-naphthylamide, 977
1,2,7-Trihydroxy-anthraquinone, 1152
4,6,7-Trihydroxy-3,4-dihydro isoquinoline, 1110
Trihydroxyindole method, for catecholamines, 1121
Trinitro-monotetrazoles, 890
2,4,5-Tri(*p*-nitrophenyl) tetrazolium bromide, 890
Triolein, 763
Triphenyl-*p*-aminophenylethyl-lead, 1281
Triplet electronic states, 1172
2,4,6-Tri-(2′-pyridyl)-1,3,5-triazine, 1132
Tritium, as radioactive label, 1210 et seq.
***Triton nodiferus*,** sulphatases of, 989
***Triton rubicundus*,** copper in, 1145
hepatopancreas of, 1146

Triturus viridescens, leucocytes peroxidase in, 1286
Trophoblast, aminopeptidases in, 972
Trypsin, 765, 962, 976
digestion of tissues by, 1039
extractions, 1377
Trypsin-like enzymes, 974
demonstration of, 776
method for, 1357
Tryptamine, 1102, 1106
fluorescence maxima of, 1106
in MAO reaction, 886
as MAO substrate, 868
methods for, 1119, 1120
storage of, in pituitary gland, 1112
Tryptamine-FIF, fading of, 1251
Tryptamines, FIF-ozone method for, 1399
unsubstituted ninhydrin reaction and, 1103
Tryptochrome, 1064
Tryptophan, and autofluorescence, 1177
EM methods for, 1263
fluorescence maxima of, 1106
Tryptophan hydroxylase, in various tissues, 1119
TTC, 880, 881, 883 et seq., 893, 914
half-wave potential of, 884
Tubercle bacillus, lipids of, 1082
Tumour cells, vital fluorochromy and, 1180
Tungsten filament lamps, 1235
Tungsten-halogen lamps, for fluorescence microscopy, 1192
Turnbull blue, 1130
method, 1402
reaction, 1058
TV, 892
Tween method, for lipases, 1309
modified for lipase-esterase, 1309
Tween 85 method, for lipase, 1310
Tween methods, application of, 786
specificity of, 785, 786
Tweens, as lipase substrates, 784, 785
Two-wavelength methods, in absorptiometry, 1230
Tyramine, 863, 864
fluorescence maxima of, 1106
oxidation pathways for, 865
Tyrosinase, 859, 860
EM demonstration of, 1286
in melanogenesis, 1055
"Tyrosinase", tyrosine-DOPA reaction for, 1339
Tyrosine, and autofluorescence, 1177
EM methods for, 1263
iodination of, 1209
melanin from, 852
as substrate for "tyrosinase", 862
Tyrosine-DOPA reaction, 863
for "tyrosinase", 1339
Tyrosine group pigments, 1050
Tyrosine-*o*-sulphate, 989

Ubiquinone, 915
biochemistry of, 1000
Ubiquinone-30, 948
Ubiquinones, Hq-ferricyanide method for, 1367
method for, 1001
MTT method for, 1367
UDGP dehydrogenase, method for, 1346
UDP glucose-α-1,4-glucan α-4-glucosyl transferase, 826
UDP glucosyl transferase, 832
UDPG dehydrogenase, method for, 932, 933
UDP Glucose-4-epimerase, method for, 998
UDP Glucuronate-4-epimerase, method for, 999
UDPG glycogen transglycosylase, method for, 1328, 1329
UDPG transglycosylase, 808, 832
Ultimobranchial bodies, 1111, 1209
Unconjugated bilirubin, 1075
Uranin, 1422
Uranium, labelling for antibodies, 1269
Uranium glass, as standard for microfluorimetry, 1240
Uranyl acetate, EM staining for nucleic acids and, 1273
Uranyl cation, and EM contrast, 1263
Uranyl glass, nine-position test, 1249
Urate oxidase, 841
method for, 872
Urea, method for, 1164, 1419
Urease, 992
Ureidosuccinate, 948
Urethans, as ChE inhibitors, 767
Uric acid, 833
confusion with calcium, by silver methods, 1138
method for, 1420
and urates, test for, 1165
Uricase, 869, 872
coupled peroxidatic method for, 1341
Uridine diphosphogalactose epimerase, method for, 1366
Uridine diphosphoglucose, 826
Uterus, cervix, phosphorylase in, 834
glucuronidase and, 810
β-glucuronidases in, 834
U.V.-Schiff method, 1182

Vacuoles, autophagic, 1080
Valence bonds, 1129
Vanadium, catalysed oxidations and, 1116
Van den Bergh reactions, 1390
Vapour fixation, glutaraldehyde and, 1107

Variamine blue B base, 1331
Varidase, 1024
Verdoperoxidases, 850
Versene, as zinc chelator, 1155
Vibrio cholerae **neuraminidase,** 1033
Visking dialysis tubing, 934
Vital fluorescence, induced, in tumour cells, 1180
Vitamin A, 929
fluorescence of, 1088
in liver, 1087
Vitamin E deficiency, 1078
ceroid and, 1083
Vitamin K, 1001
von Kóssa method, 1138, 1405
Vulpian reaction, 1117
of EC granules, 1115

Warburg's, old yellow enzyme, 925
Wavelength, in absorptiometry, 1232
Wavenumber, 1240
Whole body autoradiography, 1213

Xanthine, 833
Xanthine oxidase, 841
methods for, 873, 874
tetrazolium method for, 1341
Xanthinuria, 874
Xanthophyll, 1087
Xanthopterin, 1102
Xanthydrol, technique for 5-HT, 1112
test for urea, 1164
Xenon arc lamps, 1235
Xenopus laevis, embryo, β-glucuronidases in, 809
X-ray film, for contact autoradiography, 1213
X-ray microscopy, 1233
Xylitol, 930
dehydrogenase, method for, 930, 1345
Xylose, 819
Xylosidase, 819
method for, 824
***m*-Xylylene diisocyanate,** 1265

Y-Chromosome, fluorochromy and, 1184
Yeast cell, disaccharidase in, 821
Yolk sac membrane, ferritin in, 1132
YT, 882, 891

Zea mays, nucleic acid metabolism of, 1210
Zeiss 41 filter, 1200
Ziehl-Neelsen method, 1081
long, 1385
Zinc, intravital staining for, 1155
methods for, 1153 et seq., 1417
and positive staining with CAH methods, 996
radioactive, 1211
Zinc-bilirubin complex, 1074
Zinc deficiency, aldolase levels in, 1154
Zinc-leuco methods, 856, 1066
Zirconium-Morin complex, 1137
^{65}Zn, 1211
uptake of, by avian eggs, 1154
Zolon red, for copper, 1148
Zoom systems, 1202
Zymogen granules, 1112
Zymograms, esterases and, 765